Encyclopedia of Applied Physics

SPONSORS

AMERICAN INSTITUTE OF PHYSICS
DEUTSCHE PHYSIKALISCHE GESELLSCHAFT
JAPAN SOCIETY OF APPLIED PHYSICS
PHYSICAL SOCIETY OF JAPAN

DEUTSCHE
PHYSIKALISCHE
GESELLSCHAFT

JAPAN SOCIETY
OF APPLIED PHYSICS

PHYSICAL SOCIETY
OF JAPAN

ENCYCLOPEDIA OF APPLIED PHYSICS

VOLUME 12
Nuclear Waste Management to Optics, Underwater

Edited by
GEORGE L. TRIGG

Associate Editors
EDUARDO S. VERA
WALTER GREULICH

Managing Editor
EDMUND H. IMMERGUT

Assistant Managing Editor
CHRISTOPHER THOMAS MORAN

Editorial Assistant
HILARY GRANSON

George L. Trigg
275 Beaver Dam Road
Brookhaven, New York 11719

Walter Greulich
Ziegeleiweg 30
D-69488 Birkenau
Federal Republic of Germany

Eduardo S. Vera
Science and Technology Information Center (ICT)
University of Chile
Beaucheff 850
Santiago, Chile

Edmund H. Immergut
Christopher Thomas Moran
Hilary Granson
2 Sidney Place
Brooklyn, New York 11201

Library of Congress Cataloging-in-Publication Data

Encyclopedia of applied physics.
(Revised for Vol. 12)

"Sponsors, American Institute of Physics . . . [et al]."
Includes bibliographical references and index.
1. Physics—Encyclopedias. 2. Engineering—Encyclopedias. I. Trigg, George L. II. Vera, Eduardo S. III. Greulich, Walter. IV. American Institute of Physics.
QC5.E543 1991 530'.05 91-8738
ISBN 1-56081-071-8 (VCH Publishers)

British Library Cataloguing in Publication Data

Encyclopedia of applied physics.
Vol. 12
I. Trigg, George L. (George Lockwood) II. Vera, Eduardo S. III. Greulich, Walter
621
ISBN 1-56081-058-0 (set)
ISBN 1-56081-071-8 (vol. 12)

Printed in the United States of America.

ISBN 3-527-28134-7 (Volume 12) VCH Verlagsgesellschaft mbH
ISBN 3-527-26841-3 (set) VCH Verlagsgesellschaft mbH

Printing History:
10 9 8 7 6 5 4 3 2 1

Published jointly by:

VCH Publishers, Inc.
220 East 23rd Street
New York, NY 10010

VCH Verlagsgesellschaft mbH
P.O. Box 10 11 61
69451 Weinheim
Federal Republic of Germany

VCH Publishers (UK) Ltd.
8 Wellington Court
Cambridge CB1 1HZ
United Kingdom

ADVISORY BOARD

EDITORIAL CONSULTANTS

MAIN ENTRIES

The subject matter in the *Encyclopedia of Applied Physics* is presented in approximately 500 individual articles, arranged alphabetically. The topics can be classified into 20 sections, similar to the AIP Physics and Astronomy Classification Scheme (PACS):

01	General Aspects: Mathematical, Computational, and Information Techniques	11	Condensed Matter B: Thermal, Acoustic, and Quantum Properties
02	Measurement Science, General Devices and/or Methods	12	Condensed Matter C: Electronic Properties
03	Nuclear and Elementary Particle Physics	13	Condensed Matter D: Magnetic Properties
04	Atomic and Molecular Physics	14	Condensed Matter E: Dielectrical and Optical Properties
05	Electricity and Magnetism	15	Condensed Matter F: Surfaces and Interfaces
06	Optics (classical and quantum)	16	Materials Science
07	Acoustics	17	Physical Chemistry
08	Thermodynamics and Properties of Gases	18	Energy Research and Environmental Physics
09	Fluids and Plasma Physics	19	Biophysics and Medical Physics
10	Condensed Matter A: Structure and Mechanical Properties	20	Geophysics, Meteorology, Space Physics, and Aeronautics

Each article has been assigned a code number consisting of two digits which denotes the section, and a letter which gives the type of article. There are six types: A = Devices, Equipment; B = Materials; C = Methods, Processes; D = Phenomena, Effects; E = Scientific or Technological Fields; F = Institutions, Companies, Societies and other organizations.

CONTRIBUTORS

Leo Beiser, Leo Beiser Inc., Flushing, NY 11354, U.S.A.
Optical Scanners

Craig F. Bohren, Department of Meteorology, Pennsylvania State University, University Park, PA 16802, U.S.A.
Optics, Atmospheric

James J. Burke, Optical Data Storage Center, University of Arizona, Tucson, AZ 85721, U.S.A.
Optical Storage

Peter C. Chu, Naval Postgraduate School, Monterey, CA 93943, U.S.A.
Oceanography

Curtis A. Collins, Naval Postgraduate School, Monterey, CA 93943, U.S.A.
Oceanography

Robert W. Collins, Materials Research Laboratory and Department of Physics, The Pennsylvania State University, University Park, PA 16802, U.S.A.
Optical Properties of Solids

J. A. Dobrowolski, Institute for Microstructural Sciences, National Research Council of Canada, Ottawa, Ontario, Canada K1A 0R6
Optical Filters

Lev S. Dolin, Institute of Applied Physics, Nizhnij Novgorod, Russia
Optics, Underwater

A. Dorsel, 140 Santa Margarita Avenue, Menlo Park, CA 94025, U.S.A.
Optics, Linear

Christos Flytzanis, Laboratoire d'Optique Quantique, Ecole Polytechnique, F-91128 Palaiseau, Cédex, France
Optics, Nonlinear

Maxim D. Frank-Kamenetskii, Department of Biomedical Engineering and Center for Advanced Biotechnology, Boston University, Boston, MA 02215, U.S.A.
Nucleic Acids

Thomas Hellmuth, 30 Tyson Court, Danville, CA 94526, U.S.A.
Optical Microscopy

Gabor T. Herman, Department of Radiology, University of Pennsylvania, Philadelphia, PA 19104, U.S.A.
Optical Tomography

Dietmar Hoeschen, Physikalisch-Technische Bundesanstalt, D-38116 Braunschweig, Germany
Optical Instrumentation

Yoshiki Ichioka, Department of Applied Physics, Osaka University, Osaka 565, Japan
Optical Computing

J. A. Klein, Chemical Technology, Oak Ridge National Laboratory, Oak Ridge, TN 37861, U.S.A.
Nuclear Waste Management

Herwig Kogelnik, AT&T Bell Laboratories, Holmdel, NJ 07733, U.S.A.
Optical Communications

Iosif M. Levin, P. P. Shirshov Institute of Oceanology, 199053 St. Petersburg, Russia
Optics, Underwater

Gerd Litfin, Spindler and Hoyer GmbH & Co., D-33070 Göttingen, Germany
Optical Components and Systems

Takao Matsumoto, NTT Transmission Systems Laboratories, Kanagawa 238-03, Japan
Optical Interconnections

Werner Mirandé, Physikalisch-Technische Bundesanstalt, D-38116 Braunschweig, Germany
Optical Instrumentation

Hugh J. Miser, 199 South Road, Farmington, CT 06032, U.S.A.
Operations Research

Paul A. Petzrick, 1116 Gumbottom Road, Crownsville, MD 21032, U.S.A.
Oil Shales and Tar Sands

Rainer Röhler, Waldschmidtstrasse 12, D-82327 Tützing, Germany
Optics, Physiological

Rainer Schuhmann, Spindler and Hoyer GmbH & Co., D-33070 Göttingen, Germany
Optical Components and Systems

Roland Shack, Optical Sciences Center, University of Arizona, Tucson, AZ 85721, U.S.A.
Optics, Geometrical

Jun Tanida, Department of Applied Physics, Osaka University, Osaka 565, Japan
Optical Computing

K. Vedam, Materials Research Laboratory and Department of Physics, The Pennsylvania State University, University Park, PA 16802, U.S.A.
Optical Properties of Solids

Conway Yee, Department of Radiology, University of Pennsylvania, Philadelphia, PA 19104, U.S.A.
Optical Tomography

NUCLEAR WASTE MANAGEMENT

J. A. KLEIN, *Chemical Technology, Oak Ridge National Laboratory,* Oak Ridge, Tennessee, U.S.A*

INTRODUCTION

Because nuclear (radioactive) waste decays to background levels of radioactivity with time, the ultimate goal of radioactive waste management is quite obvious—store the material until it is no longer harmful. The difficulty with this approach is in the implementation of the theory. What does "no longer harmful" mean in both technical and public perception terms? What are the "best" ways to store material safely and in a cost-effective

*Managed by Martin Marietta Energy Systems, Inc. for the U.S. Department of Energy under Contract DE-AC05-84OR21400.

3-527-28134-7/95/$5.00 + .50

manner? Only recently has public insistence been directed toward efforts to find environmentally acceptable methods of managing radioactive waste. This insistence has led to a more nearly complete understanding, by both scientists and policymakers, of the importance of developing cost-effective technologies to dispose of radioactive waste.

1. CLASSIFICATION OF RADIOACTIVE WASTE

A number of different schemes exist for the classification of radioactive waste. Some of these are based on the origin of the waste (defense/nondefense wastes, medical wastes, utility wastes, nuclear fuel cycle wastes, etc.); the type of material present in the waste (transuranic, mixed wastes, etc.); the level of radioactivity (high, low, intermediate, etc.); the form of the waste (dry solid, wet solid, liquid, etc.); and, in some cases, legally mandated definitions that may or may not be identical to a straightforward technical or intuitive definition. An example of the latter is the presence in the United States of waste that is defined as low-level but which may have a relatively high level of radioactivity. A general classification of the various radioactive waste categories currently used in various radioactive-waste–producing countries follows.

1.1 Spent Fuel

Spent fuel consists of irradiated fuel assemblies discharged from a nuclear reactor. Spent fuel may be stored at the reactor for reuse in the future, scheduled for reprocessing at some later date, or considered to be eligible for either permanent or temporary storage at a geological repository. In some definitions, especially if long-term disposal is being considered, the spent fuel may also be considered high-level waste. Spent fuel may derive from either commercial (utility) or noncommercial (research or defense) reactors. Spent fuel is usually highly irradiated (high activity), but it may be irradiated to a lesser degree.

1.2 High-Level Waste

High-level waste is usually defined as highly radioactive material resulting from the reprocessing of spent nuclear fuel. This material includes primarily the liquid waste remaining from the recovery of uranium and plutonium in a fuel reprocessing plant. High-level waste may also be in the form of sludge, calcine, vitrified glass, or other products into which such liquid waste is converted to facilitate its handling, storage, or disposal. High-level waste contains fission products that result in the release of considerable decay energy. For this reason, heavy shielding is required to control penetrating radiation, and provisions (e.g., cooling systems) are often needed to dissipate decay heat from the high-level waste.

1.3 Transuranic Waste

Only the U.S. Department of Energy (DOE) manages material specifically classified as "transuranic waste." In the United States, this term refers to radioactive waste that contains more than 100 nCi/g of alpha-emitting isotopes with atomic numbers greater than 92 and half-lives greater than 20 years. Such waste results primarily from fuel reprocessing and from the fabrication of plutonium weapons and plutonium-bearing reactor fuel. Generally, little or no shielding is required ("contact-handled" transuranic waste), but energetic gamma and neutron emissions from certain transuranic nuclides and fission-product contaminants may require shielding or remote handling ("remote-handled" transuranic waste). Although the United States is one of the few countries that recognize transuranic waste as a separate waste classification, any country with a plutonium nuclear weapons program or extensive spent-fuel reprocessing activities will generate similar material. Depending on the country and the specific radioactivity characteristics, this waste may be managed as either low- or high-level waste. Outside the United States, waste of this type is usually categorized as intermediate- or medium-level waste, but it also may be called transuranic-bearing waste or any of a number of similar terms.

1.4 Intermediate- or Medium-Level Waste

Intermediate- or medium-level waste is generally a non-U.S. category for waste that contains higher levels of radioactivity than low-level wastes and so requires some shielding. The boundary between this category and low-level waste varies depending on the country. Treatment and disposal of intermediate-level waste vary depending on the waste form and whether it contains short- or long-lived radioactive isotopes. In general, short-lived intermediate-level waste can be disposed of as low-level waste, but long-lived intermediate-level waste will need to be disposed of similarly to high-level waste.

1.5 Low-Level Waste

Low-level waste is generally considered radioactive waste not classified as spent fuel, high-level waste, transuranic waste, or any other specific category of radioactive waste. The radiation levels from this type of waste are generally quite low, although occasionally the radiation levels may be high enough to require shielding for handling and transport.

1.6 Mixed Waste

Mixed waste contains concentrations of both radioactive materials and nonradioactive hazardous chemicals. All radioactive waste forms may have the potential for being classified as mixed waste, but usually only mixed low-level radioactive waste requires the consideration of management practices separate from low-level radioactive waste. Management practices for spent fuel, high-level waste, and transuranic waste are generally stringent enough to minimize any problems arising from the hazardous components. The United States considers both the radioactive and the hazardous components as separate materials requiring independent regulation. However, most European communities treat low-level mixed waste as if it were strictly low-level waste. For them, the disposal risks for the hazardous components are less than the risks associated with the radioactive components.

1.7 Uranium Mill Tailings

Commercial uranium mill tailings are the earthen residues that remain after the extraction of uranium from ores. Tailings are generated in very large volumes and contain low concentrations of naturally occurring radioactive materials. Because they provide a potential health hazard, the isotopes that concern us most are ^{226}Ra and its daughter, ^{222}Rn.

2. SPENT-FUEL MANAGEMENT

Historically, storage facilities for both commercial and noncommercial spent fuel were limited in capacity. It was generally understood that the spent fuel would be stored only on-site for a limited time and then shipped to a reprocessing facility. However, only a few countries have reprocessed any spent fuel; and although a limited number of countries have decided to reprocess all their spent fuel, many countries have not committed to either the reprocessing or the direct disposal of spent fuel. Currently, direct disposal appears to be gaining favor. Because the fabrication, irradiation, handling, storage, possible fuel reprocessing, and long-term management of spent fuel are all integral parts of the nuclear fuel cycle, these topics are discussed in more detail in NUCLEAR FUELS AND ISOTOPES.

2.1 Spent-Fuel Storage

Storage sites for spent fuel usually consist of deep-water pools; however, many of these pools are rapidly reaching capacity. Utilities in the United States are currently dealing with this problem by reracking spent-fuel storage modules in existing pools, by shipping the spent fuel to other reactor sites that have surplus storage capacity; and by studying alternative storage options, including the use of dry storage casks for the on-site storage of spent fuel. It may also prove feasible to reuse these casks for the future shipment of spent fuel to a centralized storage facility or repository if one becomes available. The pros and cons of on-site storage vs a centralized storage facility or repository are actively being debated in the United States and in many other international communities.

2.2 Spent-Fuel Packaging and Transportation

Because spent fuel might not be reprocessed and because, even if it is to be reprocessed, it is unlikely to be reprocessed at the generation site, the transportation of spent fuel is inevitable. Because of the great deal of attention directed to the shipping of spent fuel, the shipping casks or packages used to transport it must satisfy a variety of societal, political, and technical requirements. Foremost among the latter are requirements that will prevent the release of any radioactive material into the environment and that will provide for radiation and criticality protection. Because the construction of nuclear facilities, fuel fabrication, and shipment of spent fuel are international businesses, the rules and regulations affecting the shipment of it throughout the world are quite similar.

As spent-fuel packages must withstand serious accidents, the criteria of no loss of shielding and no loss of containment have been taken by the international community to mean that certain test conditions must be met by any package designed to transport spent fuel. These mandated requirements are remarkably similar throughout the world. For the United States, the U.S. Nuclear Regulatory Commission (NRC) has determined that spent-fuel packages must experience no loss of integrity upon

1. a 30-ft free-fall drop onto an unyielding surface;
2. a puncture test of at least a 40-in. free-fall drop onto a 6-in.-diameter steel pin;
3. exposure to a thermal environment of 1475 °F for 30 min; and
4. immersion in water for 8 h.

Because of the various regulations and design requirements, spent-fuel shipping casks are invariably made from steel or steel composite (with lead or uranium) thick enough to meet structural and gamma-shielding requirements. Additional elastomer or resin material may also be included for neutron shielding. An appropriately designed liner, specific to the spent fuel being transported, holds the assemblies in place and prevents damage during transport. These casks typically weigh between 25 and 100 metric tons with rail, barge, or ship transportation being required for the larger casks.

In deference to the current uncertainties of whether reprocessing will occur and to the uncertainties of on- or off-site storage of the waste, a number of spent-fuel casks are being designed to meet both storage and transport requirements. These casks are basically similar to the transportation casks described previously, but they have been modified as necessary to provide longer storage life.

2.3 Spent-Fuel Disposal

If the direct disposal of spent fuel is eventually determined to be the disposal method of choice, the fuel will be sent either immediately or after some period of interim storage to a geological repository. For those countries that must dispose of both spent fuel and high-level waste, these materials will be handled in a common repository with possible differences only in specific handling methods and the final repository cells. The differences are usually trivial. The following discussion pertains to both.

Inherently, all proposals for the safe disposal of spent fuel require the safe storage of radioactive isotopes until they decay to nonradioactive elements or to nonhazardous levels of radioactivity (as defined by the host country). Depending on the specific materials and radionuclides present, this requirement mandates storage times of thousands to millions of years. Because these times are much greater than the design lifetimes of conventional manufactured structures, alternative methods of disposal are often required. The obvious method of choice is a deep geological repository, especially in underground rock formations that have not changed for millions of years and thus are considered inherently stable.

Although candidate geological formations have isolated material for sufficiently long periods of time, representative repository designs generally include additional engineered barriers that add greater long-term assurances that material will not migrate out of the formation. These barriers may include long-lived container compositions, incorporation of the spent fuel into inert waste forms, and special backfill material in completed waste cells.

Similar geological repositories have already been used for low-level and hazardous waste. However, the disposal of spent fuel differs considerably in that the heat content of the waste packages (due to radioactive decay) may damage both the package and the structural integrity of the geological formation. Thus, spent-fuel repository designs call for the spreading out of the packages over large areas of the repository. In contrast, low-level or transuranic repository designs call for relatively large rooms filled with waste packages.

Once a suitable geological formation has been selected, underground space needs to be excavated along with suitable access shafts or tunnels. The excavation of the final waste storage areas may occur before storage or may occur simultaneously with storage, with new excavation occurring in areas remote from ongoing waste emplacement. Throughout the world, a number of different rock types are being considered for individual geological repositories; examples include salt, clay, crystalline granites, and tuff. None of these rock types can be called a perfect choice, because each type has its own inherent advantages and disadvantages. There are difficulties in mining, in isolating the waste from groundwater, and in the long-term stability of the formation factor in the choice of geologic formation. Salt is generally considered the easiest and fastest material to excavate, and granite, the slowest. Granite, however, is more stable and allows also for deeper repositories. None of the rock types is overwhelmingly superior for waste isolation. The total effectiveness of a repository is determined by the combination of a favorable geology and an engineering design that mitigates any disadvantages.

Before the emplacement of spent fuel within the repository, the waste will have to be packaged in final canisters or packages. Various options are being considered for the required packages. One option consists of having the spent fuel shipped to the repository in typical spent-fuel shipping casks and transferred to the final disposal package at the repository. Other designs incorporate the use of the actual shipping container as a self-shielded waste package for direct emplacement within the underground repository. The packaging of the spent fuel can be done at the repository, at the reactor before shipment, or at a centralized spent-fuel storage facility where the economics of scale may make packaging more cost effective.

The time interval from spent-fuel generation to actual emplacement within a geological repository may be short or may encompass several decades. No specific requirements determine the time frame involved, although interim storage costs may dictate shorter time intervals, while the desirability of taking advantage of the drop-off in decay heat with time favors longer times.

2.4 Representative Spent-Fuel Repository Designs

Although, currently, no country has an operating geological repository for spent fuel, a number of countries have repositories either under construction or far enough along in their design to permit a discussion of the proposed repositories. Several, though not all, of the various international designs are discussed to present some indication of the similarities and differences among the various designs. The information presented here is as reported in the open literature and is therefore not definitive, but rather gives a working understanding of the proposed designs (Forsberg, 1991). Several countries with major nuclear programs (e.g., France, the United Kingdom, and Japan) have not yet published any detailed plans on their respective repository programs. In a majority of countries, the final decision whether to reprocess spent fuel has not been made. Thus, most of the repository designs are flexible enough to allow for the disposal of spent fuel or high-level waste.

2.4.1 United States The United States is considering building its first repository for both spent fuel and high-level waste in a tuff (fused volcanic deposits) formation at the Yucca Mountain Repository Site in Nevada. The current design includes provisions for disposing of commercial and defense-related spent fuel, high-level waste from defense activities, and a small amount of high-level waste from the historical commercial reprocessing of spent fuel. This repository will have a capacity of some 70 000 metric tons initial heavy-metal equivalent (MTIHM). Typical depths for placement of packages into this repository will be 300 m. The spent-fuel or high-level waste packages (identical in outer dimensions) are to be placed in widely separated bore holes placed in the floor of the disposal tunnels. These tunnels are to be arranged in a series of panels (compartments) with one panel receiving waste, a second ready to receive waste, and a third under construction. Because of heat dissipation requirements, the bore holes are to be widely spaced. The U.S. design calls for the potential retrieval of the waste if the repository does not prove suitable. This requirement has resulted in engineered designs for bore-hole liners and other handling and emplacement components to allow for future retrieval.

2.4.2 Sweden The Swedish design for the final disposal of spent fuel is unique in that it relies heavily on the waste package to provide the primary barrier to environmental releases. The reference waste package design consists of a solid copper package with the spent-fuel assemblies embedded within it. These packages are to be placed 500 m underground into crystalline rock (granite, gneiss, or gabbro). This unique design is based on the knowledge that native copper in granitic formations is quite stable for long geological periods of time. Copper can deform over time; therefore, the package fabrication procedure involves embedding the spent-fuel assemblies within a solid copper cylinder, filling the remaining void spaces with copper powder, and hot-pressing the entire package to produce a single solid copper cylinder. Because copper in nature has been shown to be stable at relatively low temperatures, current plans call for a minimum of 30 to 40 years of interim storage before final packaging to allow for the drop-off in decay heat. Operation is expected around the year 2020.

2.4.3 Germany The German design calls for the joint disposal of high-level waste and spent fuel from reactors of western design at a depth of some 870 m in the Gorleben salt dome. The high-level waste will result from the reprocessing of German spent fuel in France and Great Britain. Direct emplacement of spent fuel will result from unique situations that would make reprocessing difficult or very expensive. Although current German law calls for the reprocessing of spent fuel, direct spent-fuel emplacement will also occur if the reprocessing of reactor fuel is terminated. For spent fuel generated at Soviet-designed reactors within Germany, the preferred option is to return the spent fuel to Russia for reprocessing. The German design calls for the use of the Pollux cask concept, which includes the option for use as a transport cask, for long-term interim storage, or as a final disposal package. The Pollux design provides excellent radiation shielding, thus permitting the direct emplacement of packages at the end of tunnels, backfilling around the package with salt, and repeating the process until the tunnel is full. Salt slowly creeps (deforms) with time and, thus, over tens of years, the tunnels will be self-sealing. Current plans call for initial emplacement of material in the year 2008.

2.4.4 Canada One Canadian design for a spent-fuel repository is based on the direct disposal of spent fuel 500 to 1000 m underground in plutonic formations sometime after the year 2025. Because the Canadian commercial reactor program is based on CANDU reactors that use natural uranium fuel, several unique factors are involved in the Canadian repository design. The low burnup of CANDU fuel implies very large volumes of spent fuel with low decay-heat releases from the spent-fuel packages. The combination of these factors is a reference Canadian repository design of over 190 000 MTIHM capacity, but with larger spent-fuel waste packages and closer spacing of waste packages in the repository than occur with other countries' repository designs.

3. HIGH-LEVEL WASTE MANAGEMENT

High-level waste, which is generated by the reprocessing of spent nuclear fuel and irradiated targets, generally contains more than 99% of the nonvolatile fission products produced in the fuel or targets during reactor operation. The high-level waste from a facility that recovers uranium and plutonium contains about 0.5% of these elements, while the high-level waste from a facility that recovers only uranium contains about 0.5% of the uranium and essentially all of the plutonium. Most of the current U.S. inventory of high-level waste is the result of defense-related programs and is stored at DOE sites. A small amount of commercial high-level waste was generated during the period 1966–1972. Other countries that have or plan to have a significant commercial fuel reprocessing capability (i.e., France and Great Britain with the world's only commercial reprocessing facilities) might well have a different mix of commercial and defense-program-related high-level waste. The production of high-level waste as a waste product is an inherent part of the nuclear fuel cycle, and, as such, the various spent-fuel reprocessing technologies that generate high-level waste are discussed in detail in NUCLEAR FUELS AND ISOTOPES.

3.1 High-Level Waste Treatment

A number of highly publicized recent events in the United States concerning the existence of leaking liquid high-level waste storage tanks highlight the potential limitations and inherent difficulties in the long-term storage of liquid high-level waste. Although the tanks in question were of the old single-shell design, a number of countries (e.g., France, Australia, and the United States) have embarked on vigorous research and development efforts to stabilize high-level waste safely (American Nuclear Society, Inc., and American Society of Civil Engineers, 1992). These efforts are directed toward the conversion of liquid high-level waste to a form that simplifies the long-range waste management problems and often includes a major reduction in waste volumes. In the United States, high-level wastes are currently stored as neutralized sludge, salt, acid liquid, alkaline liquid, and calcine. Only the calcine is a form that can be stored safely in stainless steel bins in concrete vaults. Current plans call for all of the U.S. high-level waste eventually to be incorporated into a stable matrix using borosilicate glass in several different melters and processes. The French use a closely coupled calcination and borosilicate glass vitrification process that appears to be relatively flexible and inexpensive. Several other processes have also gathered attention on an international basis, including the Australian SYNROC process. SYNROC is basically a synthetic rock that consists of a number of titanate mineral phases. The main benefit of this process is that it offers exceptionally high resistance to leaching by groundwater. Although a number of countries have shown interest in alternative waste forms, only the borosilicate matrix has been, and is being, developed commercially.

3.2 High-Level Waste Disposal

To date, it appears that none of the major producers of high-level waste have directly disposed of any of this material. In numerous cases, this liquid waste has been stabilized and solidified by a number of processes, including evaporation, calcination, and vitrification of the waste into a glass matrix. Ultimately, high-level waste probably will be disposed of similarly to spent fuel in a geologic repository. Although a number of countries, including France and the United States, have some of their high-level waste in a form ready to be disposed of, none has apparently done so. The final repository designs generally call for the emplacement of high-level waste and spent fuel depending on each country's decision whether to reprocess spent fuel. These designs are discussed in Sec. 2.4.

4. TRANSURANIC WASTE MANAGEMENT

4.1 Processing and Handling of Transuranic Waste

Transuranic waste is generated primarily by activities associated with research and development, spent-fuel reprocessing, plutonium recovery, weapons manufacturing, environmental restoration, and decontamination and decommissioning projects. About half of the countries having nuclear reactors produce transuranic waste from at least one of these sources. Most transuranic waste exists in solid form (e.g., protective clothing, paper trash, rags, glass, miscellaneous tools, and equipment). However, some transuranic waste is in liquid form (sludges), the result of the chemical processing for recovery of plutonium or other transuranic elements. Before 1970, all U.S. DOE–generated transuranic waste was disposed of on-site in shallow landfill-type configurations. In 1970, DOE concluded that waste containing long-lived alpha-emitting radionuclides should have greater confinement from the environment. Thus, all transuranic waste generated in the United States since the early 1970s has been segregated from other waste types and placed in retrievable storage pending shipment and final disposal in a permanent geologic repository. This retrievable stored transuranic waste is contained in a variety of packages (metal drums and wooden and metal boxes) in earth-mounded berms, concrete culverts, or other types of facilities.

The majority (>90%) of transuranic waste contains primarily plutonium, which emits alpha particles and low-energy photons. Therefore, the packaging is designed to provide sufficient containment and shielding to minimize personnel exposure problems. This waste form is referred to as "contact handled." Less than 3% of transuranic waste contains activation materials and fission products that decay by beta emission and produce penetrating gamma and x-ray radiation. This waste is referred to as "remote handled" if the radiation level at the surface of the packaging exceeds 200 mrem/h.

It is estimated that as much as 50 to 60% of the transuranic waste is also mixed waste in that it contains hazardous constituents defined and regulated by the U.S. Environmental Protection Agency (EPA). Examples of transuranic mixed waste are radionuclide-contaminated spent solvents, discarded materials contaminated with both solvents and radioactive materials, scintillation fluids, and discarded contaminated lead shielding. However, the proposed disposal plans for transuranic waste are such that any concerns about the management of the hazard constituents should be minimal.

Current plans call for the retrievable stored transuranic waste and newly generated transuranic waste from U.S. defense-related activities to be shipped to the Waste Isolation Pilot Plant (WIPP), located near Carlsbad, New Mexico. The disposition of buried transuranic wastes and transuranic waste generated from nondefense activities is uncertain at this time.

4.2 Waste Isolation Pilot Plant (WIPP)

Except for those countries that manage transuranic or transuranic-like waste as low-level waste, most transuranic-waste–generating countries have adopted an interim storage concept with the final disposal as yet to be decided upon. In the past, Germany disposed of transuranic waste in the Asse salt mine, but it has discontinued that operation. Other countries are considering repository projects; for example, the United States is proceeding with the WIPP. The WIPP has been designed as a facility for the permanent isolation of U.S. defense-generated transuranic wastes, but it will also provide an underground test facility in which the safe disposal of high-level waste in salt can be demonstrated. Initial construction of the underground facility was begun in 1981. Currently, the WIPP facility is fully prepared to receive some transuranic waste to conduct various experiments related to regulatory compliance of the repository. The final decisions permitting the initial emplacement of transuranic waste have not yet been made.

Because of the public's and regulators' strong interest in the initial disposal of transuranic waste, a detailed set of waste acceptance criteria has been drafted. These rules are concerned with a wide variety of issues, including gas generation, combustibility, toxic and corrosive compounds, surface dose rate, surface contamination, and criticality. Generator, interim storage, and shipping sites must maintain detailed documentation to meet these criteria and be allowed to ship material to the WIPP.

Contact-handled transuranic waste will be transferred to the WIPP in specially designed and constructed packaging containers referred to as TRUPACT-II containers. Although designed for easily shielded contact-handled transuranic waste, the container consists of several sealed containment vessels and a sandwichlike foam and stainless steel shell that increases the package strength and safety to withstand postulated accidents associated with transport. TRUPACT-II has a capacity of 14 standard 55-gal drums and is about 3 m in diameter and 3 m tall. Three TRUPACT-IIs can be carried on a flatbed truck trailer. Shipment of the remote-handled transuranic waste will take place in a special shipping container that is somewhat similar to the various designs for a spent-fuel cask.

The WIPP site, which is located in the Permian Basin deposits of the southwestern United States, contains salt formations that are about 900 m thick, some of the thickest formations in the United States. Present mining activities take place 650 m below the surface near the middle of the salt bed. This formation was chosen because the area appears to be geologically stable with little earthquake activity, salt deposits demonstrate the absence of any circulating groundwater, salt is relatively easy to mine, and the salt has the ability to heal fractures and creep to fill voids and seal a waste repository. Aboveground facilities house the site personnel and equipment needed for WIPP operations and research activities. Included are separate areas for the contact- and remote-handled transuranic wastes. A series of four shafts connect the surface facilities to the underground area, making it possible to lower transuranic waste to its destination and to transport workers, equipment, mined salt, and fresh air.

Underground WIPP facilities consist of transuranic waste storage rooms 4 m high, 10 m wide, and 90 m long; these rooms are separated from each other by 30-m pillars. As one room is filled with waste, the next room is being excavated, and some of the salt rubble is used to backfill the voids in the room that is being filled. Contact-handled transuranic waste containers will be removed from the TRUPACT-II transporter and transferred to the underground storage area. There they will be moved to and stacked in the appropriate storage room. Remote-handled transuranic waste, however, will need to be handled by specially designed and shielded handling equipment. Actual waste containers will be remotely emplaced within predrilled holes in the storage room wall and immediately plugged. Other nonstorage rooms have been excavated for experimental high-level waste heat-dissipation tests and as shop and maintenance areas.

Near the end of the five-year demonstration and evaluation period, a decision will be made regarding permanent disposal of transuranic wastes at the WIPP. If a decision for permanent disposal is made, the facility will operate as a repository for an additional 20 years. The total anticipated volume of transuranic waste to be stored at the WIPP is about 200 000 m^3. Remote-handled transuranic waste is expected to require only 5000 m^3 of the total available volume.

Plans are to use both a traditional cement-based grout and natural salt plugs to cap the waste-filled underground rooms and to seal the vertical shafts. The cement grout provides a well-characterized short-term seal, and the salt plugs provide a deformable seal that will fuse with the surrounding formation. Plans are in place for the final decommissioning of the WIPP, including decontaminating the surface areas and dismantling of the surface facilities. Permanent surface markers will be placed on the site, and a long-term monitoring program will be conducted.

5. LOW-LEVEL WASTE MANAGEMENT

5.1 Low-Level Waste Generation

As indicated in the classification section, the usual way of determining what is included in low-level waste is to describe what it is not. In most countries, low-level waste is that radioactively contaminated material that is not covered by the definitions for spent fuel, high-level waste, transuranic waste, uranium mill tailings, or any other form of waste. Generally, low-level waste contains relatively small amounts of radioactivity, but occasionally it does have a level of radioactivity high enough to warrant special treatment and disposal. Principal generators of low-level waste include academic and medical institutions, nuclear power plants, and other industries. Of obvious importance is the generation of low-level waste in defense-related programs for those countries having such programs. A majority of low-level waste contains very low levels of radioactivity and generally requires no shielding to protect workers or the general public and decays in fewer than 100 years to levels that are generally considered acceptable. The remainder (3–5%) requires some degree of shielding and takes longer to decay to acceptable levels of radioactivity. This material may not be acceptable for near-surface disposal without including additional engineering barriers. The highest-activity low-level waste is often slated for disposal in a geologic repository along with spent fuel and high-level waste.

The types of low-level waste being generated include the widest range of materials, including contaminated trash, paper, glass, clothing, plastics, equipment, tools, sludges and slurries, and organic and aqueous liquids. Solids, liquids, or gases can all be included as low-level waste. Normally liquid wastes are stabilized or solidified before disposal, and gaseous wastes need to be filtered or scrubbed to remove contaminants before discharge to the environment.

5.2 Low-Level Waste Treatment

The best way to manage low-level waste is not to generate it. The use of substitute materials or good housekeeping practices can go a long way toward eliminating, or at least minimizing, the generation of low-level waste. Sometimes, however, it is not possible to eliminate a particular low-level waste; then various treatment methods have to be relied upon to reduce the volume or simplify the handling of the low-level waste. Numerous such technologies are available, and it would be impossible to describe all of them in detail. Generally, they can be grouped according to their principal objectives: dewatering, thermal transformation, sorting and segregation, decontamination, mechanical volume reduction, and stabilization and solidification.

5.2.1 Dewatering Low-Level Waste Liquid wastes can be concentrated and, thus, their volumes reduced by evaporation, incineration, ion exchange, filtration, reverse osmosis, and precipitation. Other technologies (e.g., distillation, ultrafiltration, centrifugation, and crystallization) can be and probably have been used to treat low-level waste, but the former are more commonly called upon.

5.2.2 Thermal Transformation Incineration of low-level waste is one of the two most widely used technologies for the volume reduction of solid low-level waste; the other technology is compaction. Incineration is also heavily used for a number of liquid low-level waste streams, especially those involving combustible organic materials. Some incineration of noncombustible liquid streams can be handled by mixing these streams with combustible low-level waste or by using a supplemental fuel. Although highly efficient (with reduction ratios of 100 to 1 being achieved for liquid low-level wastes), the capital costs are significant, and secondary waste streams are produced that have to be disposed of safely. These wastes include waste gases that need to be monitored and, if necessary, scrubbed. Fly ash and bottom ash are produced. The radioactivity present in the liquid low-level waste tends to concentrate in the bottom ash, and this material will need to be stabilized and disposed of safely.

5.2.3 Sorting and Segregation Sorting and segregation processes can be highly cost effective when the disposal costs for low-level waste are relatively high. The sorting process can take place either before the discarded material is labeled "low-level" waste or before actual disposal of the material. The effective use of sorting of radioactively contaminated items (and parts of items) from the bulk of the discarded material has been shown to cut down drastically on waste volumes. Alternatively, known contaminated material can be sorted into combustible material, waste suitable for decontamination, compactable waste, and noncompactable waste.

5.2.4 Decontamination Decontamination is the process whereby the surface contamination of solid low-level waste is either removed or reduced. The ultimate goal is to reduce the contamination to a level whereby reuse or nonradioactive waste disposal is feasible. Decontamination methods include mechanical (steam, water, or sand blasting), chemical (acids, alkalies), and ultrasonic (vibrating liquid baths). In some cases, the reuse or recycle of the decontaminated material is possible. Surface-contaminated scrap metal is often reclaimed by smelting. In some cases, minimally contaminated metal ingots that can be used in the production of shielding for spent fuel and high-level waste storage, shipping, and disposal casks are produced.

5.2.5 Mechanical Volume Reduction Numerous mechanical volume-reduction techniques [baling, shredding, and compaction (both conventional and supercompaction)] are available for use on dry low-level waste. Volume reduction values of 3 to 10 are not unusual. Mechanical volume reduction and other forms of volume reduction are heavily used when the disposal costs are high. Often, the final products are high-density disks that can be repacked in standard barrels to facilitate handling and disposal.

5.2.6 Stabilization and Solidification Solidification is generally used to convert low-level waste to a form that, upon its disposal, resists degradation and any release of contaminants to the environment. The waste to be treated can be liquid, slurry, sludge, wet solid, or dry solid. In most cases, the material to be stabilized is simply mixed with an appropriate solidification agent and allowed to solidify in the final disposal container. Some of the most widely used solidification agents are cements, asphalt, polymers, and molten glass. The form of the waste (liquid or solid) and the chemical and thermal stability of the waste determine the optimal solidification agent.

5.3 Low-Level Waste Disposal Methods

The various methods for the final disposal of low-level waste can be grouped into a number of generic categories: above-ground disposal, near-surface underground disposal, intermediate-depth disposal, deep disposal in geologic repositories, and ocean disposal. All specific methods for the disposal of low-level waste have their own advantages and disadvantages. All the low-level waste-producing countries have assumed the responsibility of disposing of their own low-level waste (i.e., very little, if any, transferring of low-level waste from one country to another exists) and, thus, need to optimize the cost-benefit trade-offs (including safety) of the various options under their particular set of circumstances.

5.3.1 Above-Ground Disposal For above-ground disposal, reinforced concrete vaults or buildings are constructed directly on the surface. Any waste stored in the vault will be isolated from the environment as long as the integrity of the building is maintained. Above-ground disposal vaults are relatively inexpensive and can be easily located and monitored. However, they do lack the environmental protection and the barrier to radionuclide migration that any surrounding soil would offer. This type of disposal is often used for both short- and long-term interim storage of low-level waste when permanent disposal is to occur at a later time.

5.3.2 Near-Surface Underground Disposal Technologies fitting this category have historically been used for the disposal of the vast majority of low-level waste throughout the world. Various modifications of this category are generally the current method of choice for most countries that generate low-level waste because of the low cost and adequate isolation of the low-level waste. For most of these technologies, the waste is simply emplaced within about the first 10 m of the surface and then capped with a layer of soil. The most common near-surface disposal methods include trenches and below-grade or soil-mounded vaults or bunkers. In the past, water infiltration into near-surface disposal units has been linked to the loss of integrity in the overlying cap caused by the compaction and settling of the waste. Regulations requiring the compaction of the low-level waste before its emplacement, along with the use of high-integrity containers and concrete overpacks, help prevent many of these problems. In addition, a number of designs call for the stabilization of the emplaced waste with either a loose soil, sand, or gravel fill, or an injected concrete grout around, and in some cases into, the waste packages. After a disposal unit is filled, a cap composed of single or multiple layers of soil or impermeable synthetic membranes is used to seal the unit from precipitation and ground-water runoff. Although most low-level waste disposal units have been the below-grade type, where the elevation of the surface drainage is above the buried waste, a number of countries have used above-grade facilities, where the surface drainage elevation is below the lowest waste level. Earth-covered bunkers have been used successfully in France for over 20 years, and above-grade tumuli are currently being evaluated for some of the new generation of low-level waste disposal units being considered for use in the United States (Fig. 1).

5.3.3 Intermediate-Depth Disposal Augered holes in the earth's surface have been successfully used for the emplacement of low-level waste at a number of DOE sites (Fig. 2). This method involves the construction of concrete- or metal-lined holes, up to 3 m in diameter and up to 20 m or more in depth, in suitable soil. After the emplacement of the low-level waste, the holes are typically filled with a cement or concrete grout, and a concrete cap is added. Although this is a possibly acceptable method for the disposal of low-level waste, the difficulty in locating and monitoring suitable sites will generally favor other disposal methods. Also included in this category are the use of deep trenches and old pit mines. Unless readily available, these methods are costlier than shallow trenches and yet do not add any protection against releases to the environment.

5.3.4 Deep Disposal in Geologic Repositories Geologic repositories located up to 1500 m below the surface are generally being considered as sites for the disposal of most of the world's high-level waste. These deep repositories, although expensive, generally provide excellent barriers to the migration of radionuclides into the environment and to human intrusion into the disposal unit. A number of European countries are also planning to use such repositories for the disposal of low-level waste. Germany disposed of low-level waste in the Asse salt mine from 1967 to 1978. Sweden has built, and the United Kingdom plans to develop, geologic repositories for low-level waste. The United States is considering

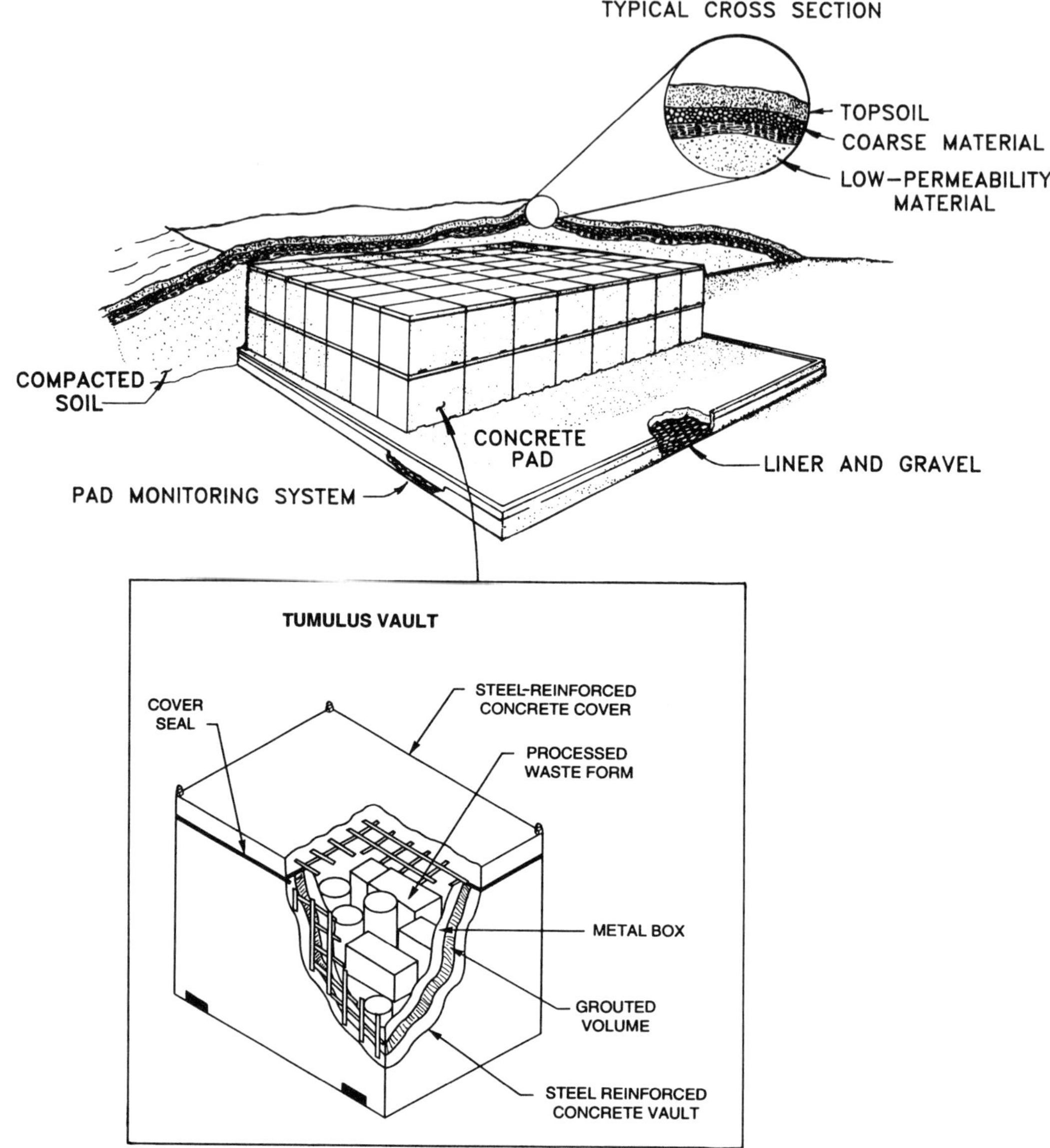

FIG. 1. Above-grade tumuli for the disposal of low-level radioactive waste (as being considered in the United States).

such repositories only for spent fuel, high-level waste, and transuranic waste. In most cases, deep geologic repositories are not considered cost-effective for disposing of low-level waste. The additional protection gained by disposing of material at such depths is often not necessary, and locating such repositories can be difficult and time-consuming.

5.3.5 Ocean Disposal In the 1950s and 1960s, the United States disposed of about 90 000 containers of low-level waste at various depths and distances from shore at several Atlantic and Pacific Ocean sites. Since 1967, a number of European countries have cooperated in a joint disposal of some 36 000 containers off the English coast. Although this is not currently practiced in the United States, several other countries do practice limited ocean disposal of radioactive waste, including low-level waste, without a great deal of publicity.

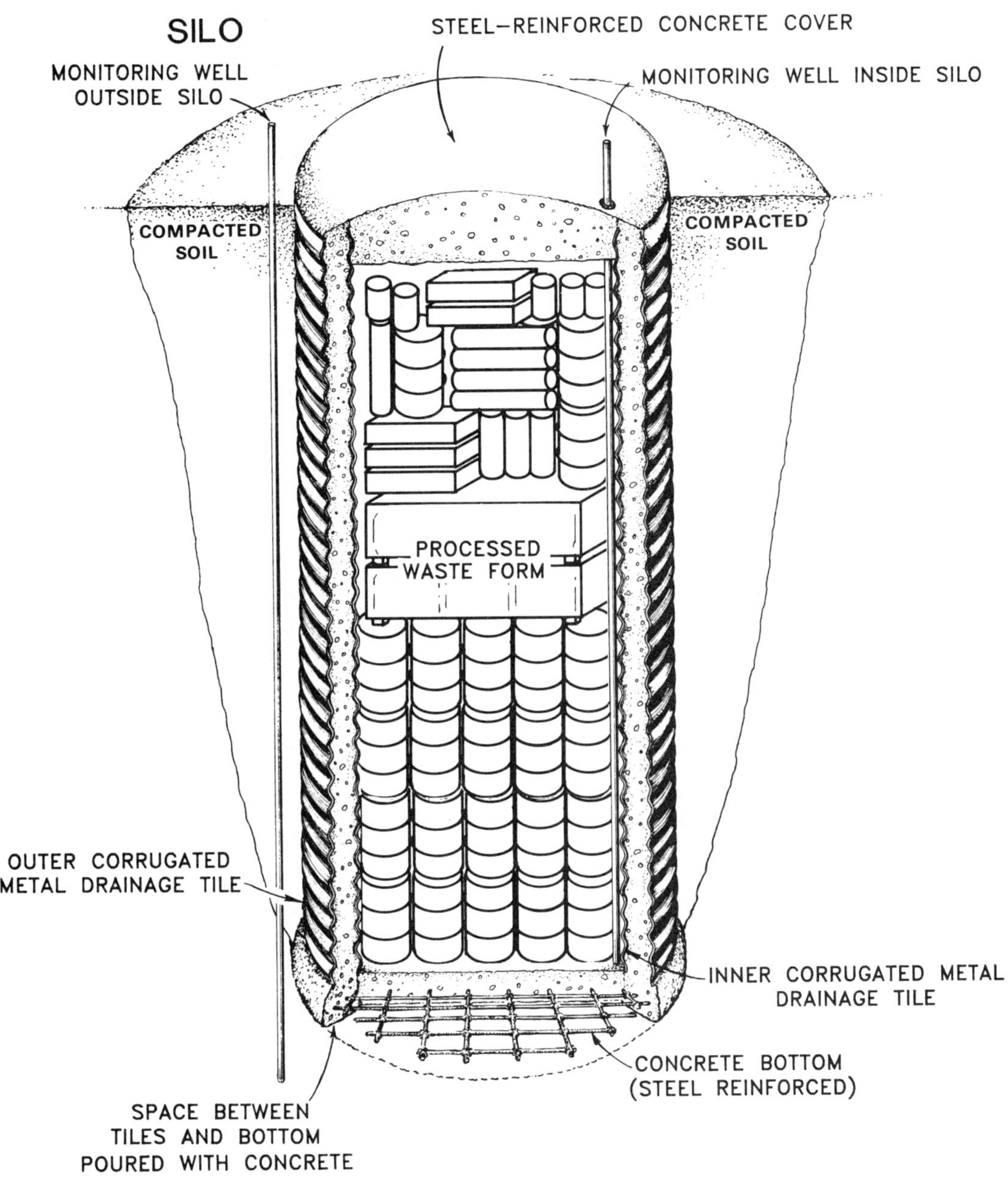

FIG. 2. Augured holes (silos) for the disposal of low-level radioactive waste.

5.4 Low-Level Waste Disposal in the United States

The disposal of commercial low-level waste in the United States dates back to 1962 with the opening of the Beatty, Nevada, disposal site. Since then, a number of other sites have been opened and subsequently closed; only two (Hanford, Washington, and Barnwell, South Carolina) are still accepting low-level waste. The shipping of the country's waste to the two remaining sites developed into a general concern that the responsibility for accepting low-level waste for disposal should not rest with only a few states and that a coordinated national plan was needed. In response, in 1982, the U.S. Congress passed an act (the Low-Level Radioactive Waste Policy

Act) making each state responsible for its own low-level waste and encouraging the formation of regional interstate compacts to deal with the disposal problems. Although the law laid out the milestones leading to the operation of new regional facilities by 1 January 1993, none have been built, and it remains to be seen what the full impact of the law will be on those industries producing low-level waste.

6. MIXED-WASTE MANAGEMENT

Because mixed waste contains both radioactive materials and nonradioactive hazardous chemicals, it raises a number of legal, regulatory, institutional, and management issues that are unique. All radioactive waste types may, in part, contain some mixed wastes. However, the dominating radioactive characteristics of spent fuel, high-level waste, and transuranic waste generally dictate the methods by which these materials need to be treated, handled, stored, and disposed of. Thus, only mixed low-level waste is generally considered for mixed-waste management.

While low-level waste is somewhat ill-defined throughout the world, the definition of what constitutes a hazardous material is even less exact. Thus, the combination of the two definitions can be either extremely specific or quite confusing, or there may not be a definition at all. A majority of countries consider the low-level components to be of dominant concern and treat their low-level mixed waste as if it were only a radioactive waste. Other countries highlight the hazardous chemicals and manage the waste primarily on the basis of its hazardous components.

In the United States, the radioactive components of mixed waste are subject to the Atomic Energy Act, which, for government sources, is administered by DOE. Commercial sources are administered by the Nuclear Regulatory Commission (NRC) (unless a state has obtained agreement-state status). The hazardous components of all mixed wastes in the United States are subject to two federal statutes that are administered by the Environmental Protection Agency (EPA) (unless a state has obtained an authorization status): the Resource Conservation and Recovery Act [RCRA (1976, 1984)] and the Toxic Substances Control Act [TSCA (1976)]. Thus, the treatment, handling, and disposal of mixed wastes are subject to the regulations of the EPA and NRC (or the authorized and agreement states) or DOE.

Typically, mixed waste includes a variety of contaminated materials—for example, liquid scintillation fluids, air filters, cleaning materials, engine oils and grease, chlorinated or fluorinated organics, pesticides, lead wastes, mercury wastes, other toxic metals, paint residues, photographic materials, soils, building materials, and decommissioned plant equipment. It has been estimated that about 5 to 10% of U.S. low-level waste may, in fact, be properly classified as mixed waste. The characterization as mixed wastes is often quite difficult because such wastes may have different blends of hazardous (chemical or physical) and radioactive components that preclude accurate characterization. In addition, several processes may be involved in generating the wastes, including accidents that may not be amenable to characterization. Further complicating the issue is the fact that regulatory procedures for determining the hazardous components and their toxicities are rapidly changing.

6.1 Availability of Mixed-Waste Treatment Facilities

If, in a particular country, mixed waste is to be treated as either low-level waste or hazardous waste, then the treatment facilities or methods used for disposal are somewhat well defined. If, however, the mixed waste is jointly regulated—both as a radioactive material and as a hazardous waste (as in the United States)—then the nature and availability of mixed waste treatment and disposal facilities become much more difficult to assess.

In most cases, the usual procedure is to treat or remove the hazardous component by the most expeditious and effective treatment technology. The resulting treated waste, residue, or immobilized material can then be managed as if it were a low-level waste. In some cases, such as waste being held for the decay of radioactive isotopes or the presence of low concentrations of certain isotopes such as 0.05 μCi/g or less of ^{3}H or ^{14}C in the United States, the waste may be managed as if it were a hazardous material. The meager number of waste management facilities that meet all the technical requirements to handle both hazardous and radioactive materials and that have all the necessary permits and licenses to do so limits the availability of mixed-waste management facilities to the generators of this waste. In the United States, the nonavailability of licensed or permitted facilities to treat certain categories of mixed waste and the general prohibition on the long-term storage of hazardous wastes have placed numerous mixed-waste generators in a difficult position.

7. URANIUM MILL TAILINGS MANAGEMENT

Uranium mill tailings are the residual wastes of milled ore that remain after the uranium values have been recovered. Depending on the ores' chemical characteristics, uranium mills use either an acid or alkaline leach process to recover uranium. Currently, mills having more than 96% of the U.S. milling capacity use the acid leach process; the rest use the alkaline process or a number of nonconventional methods. Acid leaching is preferred for ores with low lime content. Those with high lime content require excessive quantities of acid for neutralization and, for economic reasons, are best treated by alkaline leaching. In either leach process, most of the uranium is dissolved together with other materials present in the ore (e.g., iron, aluminum, and other impurities). After the ore is leached, the uranium-laden leach liquor is removed from the tailings solid by decantation. Mill tailings from both processes consist of slurries of sands and clay-like particles called slimes.

Because the amount of uranium (by weight) extracted from the ore during milling is relatively small, the dry weight of the tailings produced is nearly equal to the dry weight of ore processed. Typically, dry tailings are composed of 70 to 80 wt % sand-sized particles and 20 to 30 wt % finer-sized particles. The tailings contain other naturally radioactive substances (e.g., radium) that are responsible for more than three-fourths of the radioactivity that was originally present in the ore. After thorough washing, the tailings are pumped as a slurry to a tailings pond where the emplaced waste is drained and the water evaporated.

Because huge amounts of tailings are produced from even a modest uranium mill industry, large tailings have historically been produced. In the United States, over 100×10^6 m^3 of commercial uranium mill tailings are distributed over about 24 tailings sites. An equal number of DOE sites possess an estimated additional 30×10^6 m^3 of mill tailings. Several of the largest commercial sites contain over 10×10^6 m^3 of tailings, in piles covering over 100 ha each. Only two of these are currently active, adding some 360 000 m^3 of tailings in 1991. Of the inactive sites, their respective status ranges from unstabilized, to covered with wood chips, to partially stabilized, to fully stabilized with soil and surface plantings. In the future, most of the U.S. production of uranium is expected to come from nonconventional extraction operations (*in situ*, by-product, etc.) with imports and U.S. inventory drawdown projected to make up 80% of U.S. requirements through the year 2005. On a worldwide basis, at least 200×10^6 m^3 of tailings exist, with an additional $(10–20) \times 10^6$ m^3 of tailings being added each year.

In the past, failures of the holding dams at mill tailings impoundments have occurred. In addition, there was the historical use of the easy-to-handle, readily available, and expensive mill tailings as a source of landfill. Individuals and institutions used the mill tailings in the construction of homes, commercial buildings, and even schools. Because of such events and situations, the creation of large tailings piles is now widely regarded by the public and various regulators as unacceptable. Alternative disposal practices are being actively pursued. The obvious solution is to require the burial of tailings below grade in specially excavated pits or as backfill in deep pit and underground mines so as to reduce radioactive exposure levels to natural background levels. Use of cost-benefit, as low as reasonably achievable (ALARA), and sound engineering analyses leads to the conclusion that the various uncertainties in the predicted values of collective doses to the surrounding population and the costs associated with various waste management options often make sound choices for the future management of mill tailings quite difficult.

In the United States in 1986, the EPA issued rules on ^{222}Rn emissions from tailings piles. Mill owners will have to phase out the use of large existing tailings piles. New tailing piles must be contained in small (i.e., less

Table 1. Spent-fuel production through 1991.

Country	MTIHM
Argentina	1 800
Belgium	1 400
Brazil	150
Canada	15 000
Finland	450
France	16 000
Germany	4 000
India	1 800
Italy	2 000
Japan	9 000
Netherlands	250
Slovenia	140
South Africa	400
South Korea	1 200
Spain	3 300
Sweden	3 000
Switzerland	1 200
Taiwan	1 200
United Kingdom	34 000
United States	24 000

than 16 ha) impoundments or disposed of by continuous dewatering and burial with no more than 4 ha uncovered at any one time.

8. QUANTITIES OF RADIOACTIVE WASTE

8.1 International Radioactive Waste Generation

Information on the quantities of radioactive waste being managed by the various international waste producers is often difficult or impossible to obtain. Accurate numbers for only spent fuel are available for most countries. Table 1 presents data on the amount of spent fuel generated through 1991 (The Uranium Institute, 1992). A comment needs to be made that some of this quantity has been reprocessed in France, Great Britain, and the United States and that additional material may be reprocessed in the future. Thus, these numbers represent total generation, not material in storage awaiting disposal. Additional quantities of spent fuel have been produced in Soviet-supplied reactors, but these numbers are not included.

8.2 U.S. Radioactive Waste Generation

The DOE annually publishes a reference document that quantifies the inventories of all radioactive material being generated, stored, or disposed of in the United States (Martin Marietta Energy Systems, Inc., 1994). Table 2 provides a summary of the quantities of radioactive waste being stored in the United States as of 31 December 1993. It must be noted that some differences in definitions for the various waste categories exist between the commercial and the DOE waste-generating industries. In the United States, low-level waste accounts for over 85% of the stored radioactive waste but less than 1% of the radioactivity.

Table 2. Radioactive waste inventories as of 31 December 1993.

Waste category	Total amount in storage	Comment
Spent fuel	28 000 MTIHM[a]	Does not include approximately 2500 MTHM of DOE-owned spent fuel
High-level waste	403 500 m^3	Material predominantly unsolidified waste
Transuranic waste	232 600 m^3	Includes both buried material and waste to be sent to WIPP
Low-level waste	4 561 400 m^3	Includes both commercial and defense material
Uranium mill tailings	118 600 000 m^3	Includes only commercial tailings.
Mixed waste	190 700 m^3	Does not include 2,100 m^3 of commercial mixed waste as of 31 December 1990

[a]Metric tons initial heavy-metal equivalent.

GLOSSARY

Activation Product: A radioactive material produced by bombardment with neutrons, protons, or other nuclear particles.

Burnup (Spent Fuel): The total energy released per initial unit mass of reactor fuel as a result of fission.

Calcine: A form of high-level waste produced from spent-fuel reprocessing waste by heating to a temperature below the melting point to bring about a loss of moisture and oxidation to a chemically stable form.

Canister: A metal container used for the storage or disposal of heat-producing solid radioactive waste.

Contact-Handled Waste: Radioactive waste with a low surface dose rate and minimal heat generation that permits handling by contact methods.

Decay, Radioactive: The transition of a nucleus from one energy state to a lower one, usually involving the emission of a photon, an electron, or a neutron.

Enrichment (Reactor Fuel): A nuclear fuel cycle process in which the concentration of fissionable uranium (i.e., ^{235}U) is increased above its natural level of 0.71%.

Fertile Nuclide: A nuclide capable of being transformed into a fissile nuclide by neutron capture.

Fissile Nuclide: A nuclide capable of undergoing nuclear fission with neutrons.

Fission Products: Nuclides produced either by fission of heavier nuclides or by the subsequent decay of the nuclides thus formed.

Fuel Assembly: A grouping of nuclear fuel rods that remains integral during the charging and discharging of a reactor core.

Geologic Repository: A facility that has an excavated subsurface system for the permanent disposal of spent fuel, high-level waste, and possibly other waste forms.

Half-Life, Radioactive: For a single radioactive decay process, the time required for the activity to decrease to half its initial value.

Heavy Metal: Mass of the fissile and fertile (heavy metal) content of nuclear fuel.

High-Level Waste: Generally, the highly radioactive material resulting from the reprocessing of spent nuclear fuel, including the liquid waste produced directly in reprocessing and the solid material derived from such liquid waste.

Intermediate-Level Waste: Radioactive waste that contains higher levels of radioactive than most low-level waste.

Low-Level Waste: For most countries, radioactive waste not classified as spent fuel, high-level waste, transuranic waste, or any other specific category of radioactive waste. Radiation levels are usually quite low, but occasionally the radiation levels may be high enough to require shielding.

Mixed Waste: Waste that includes concentrations of both radionuclides and hazardous chemicals.

Nuclear Fuel Cycle: The complete series of steps involved in supplying fuel for nuclear reactors. It includes mining, refining, enrichment, fabrication of fuel elements, use in a reactor, chemical processing to recover the fissionable material in the spent fuel, reenrichment of the fuel material, refabrication of new fuel elements, and management of radioactive waste.

Remote-Handled Waste: Radioactive waste with a high surface dose rate and/or heat generation that requires remote handling and/or shielding.

Reprocessing, Fuel: The chemical/mechanical processing of irradiated nuclear reactor fuel to remove fission products and recover fissile and fertile material.

Spent Fuel: Nuclear fuel that has been permanently discharged from a reactor after it has been irradiated.

Transuranic Waste: Radioactive waste that contains elevated concentrations of alpha-emitting isotopes with atomic numbers greater than 92. Some countries (e.g., the United States) may have a more specific definition.

Uranium Mill Tailings: Earthen residues that remain after the extraction of uranium from its ores.

Vitrification: The conversion of radioactive waste materials into a glassy or noncrystalline solid for subsequent disposal.

Works Cited

American Nuclear Society, Inc., and American Society of Civil Engineers (1992), *High-Level Radioactive Waste Management, Proceedings of the Third International Conference, Las Vegas, Nevada, April 12–16, 1992*, New York: ASCE. Third in a series of international conferences devoted solely to the subject of high-level waste management.

Forsberg, C. W. (1991), *Representative Geological*

Repository Designs for Spent Fuel and High-Level Waste, K/ITP-393, Oak Ridge, TN: Martin Marietta Energy Systems, Inc.

Martin Marietta Energy Systems, Inc. (1994), *Integrated Data Base Report—1994: U.S. Spent Fuel and Radioactive Waste Inventories, Projections, and Characteristics,* DOE/RW-0006, Rev. 10, Oak Ridge, TN: Oak Ridge National Laboratory.

The Uranium Institute (1992), *Radioactive Waste and the Nuclear Fuel Cycle,* London: The International Industrial Association for Energy from Nuclear Fuel.

Further Reading

League of Women Voters Education Fund (1985), *The Nuclear Waste Primer—A Handbook for Citizens,* New York: Nick Lyons Books. Although somewhat out of date, a comprehensive view of the political implications in managing radioactive waste. A newer version of this publication may be available.

Moghissi, A. A., Godbee, H. W., Hobart, S. A. (Eds.) (1986), *Radioactive Waste Technology,* New York: American Society of Mechanical Engineers. Contains a great deal of information on radioactive waste management for spent fuel, high-level waste, and low-level waste.

Post, R. G., Wacks, M. E. (Eds.) (1992), *Waste Management '92, Proceedings of the Symposium on Waste Management at Tucson, Arizona, March 1–5, 1992,* Tucson: Arizona Board of Regents. The latest in a series of annual waste management symposia that cover all aspects of radioactive waste management.

Yang, Y. S., Saling, J. H. (1990), *Radioactive Waste Management,* New York: Hemisphere Publishing Corporation. A detailed look at the entire subject of radioactive waste management.

NUCLEAR WEAPONS

See WEAPONS, NUCLEAR

NUCLEIC ACIDS

Maxim D. Frank-Kamenetskii, *Department of Biomedical Engineering and Center for Advanced Biotechnology, Boston University, Boston, Massachusetts, U.S.A.*

3-527-28134-7/95/$5.00 + .50

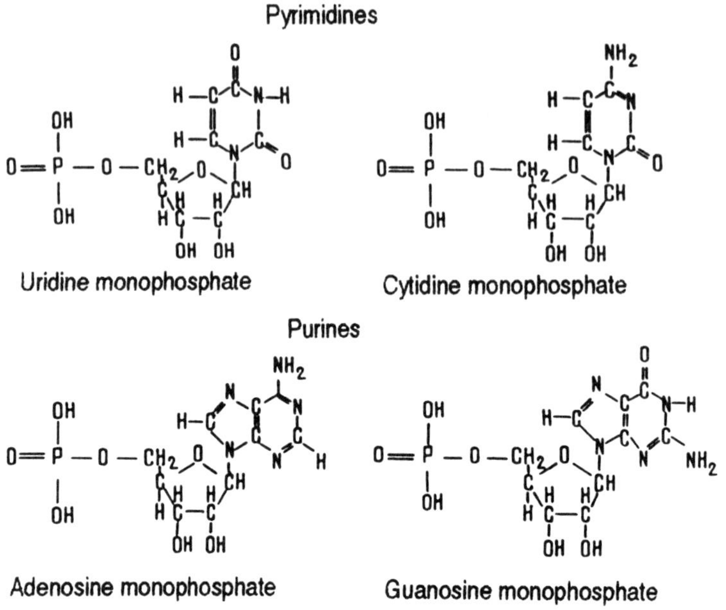

Pyrimidines

Uridine monophosphate

Cytidine monophosphate

Purines

Adenosine monophosphate

Guanosine monophosphate

FIG. 1. Chemical formulas of residues of RNA, i.e., of nucleotides. At the top are the pyrimidine nucleotides (U and C), and below, the purine nucleotides (A and G). Nucleotides within DNA differ in that, instead of the right-hand lower OH group, they simply have H. In addition, DNA, instead of the uridine nucleotide, includes the thymidine nucleotide whose top C-H group in the ring is replaced by the C-CH_3 group.

INTRODUCTION

Nucleic acids play a crucial role in all living organisms because they are the key molecules responsible for storage, duplication, and realization of genetic information. There are two types of nucleic acids; ribonucleic acid, or RNA, and deoxyribonucleic acid, or DNA. They are both heteropolymeric molecules consisting of residues (nucleotides) of four types. Four RNA nucleotides are shown in Fig. 1. Each consists of a phosphate group, sugar (these two elements are identical for all four nucleotides), and a nitrous base, which determines the type of the nucleotide. DNA nucleotides are very similar to RNA nucleotides. The main difference is that the right-hand OH group in the sugar is replaced with just H (hence, "deoxy"). Three bases (A,G, and C) are the same for DNA and RNA. Only the fourth base is different: Instead of uracil (U), DNA carries thymine, T (see Fig. 2). The DNA chain is shown in Fig. 3.

In living nature, as we observe it today, DNA plays the most important role because genetic information is stored and duplicated in the form of DNA in all living cells and organisms, without any exceptions. Only in some viruses, which are not living creatures because they cannot reproduce themselves outside cells, is genetic information carried in the form of RNA molecules. The most notable class of such viruses are retroviruses, which include many cancer-inducing viruses as well as HIV, the AIDS virus.

The genetic message is "written down" in the form of continuous text consisting of four letters (DNA nucleotides A, G, T, and C). This continuous text, however, is subdivided, in its biological meaning, into sections. The most significant sections are genes, parts of DNA, which carry information about the amino-acid sequences of proteins. Whereas DNA stores

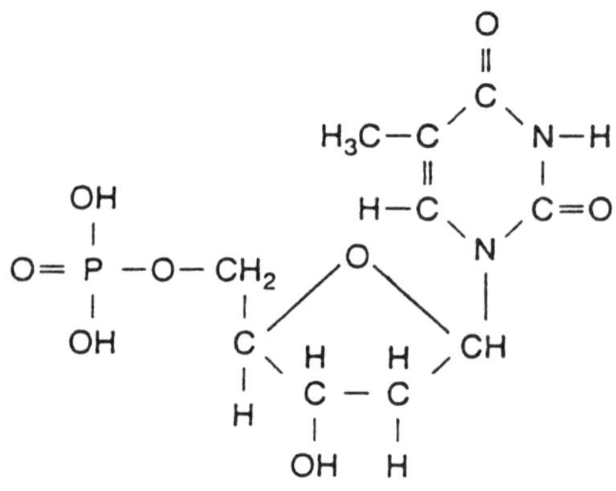

FIG. 2. Thymidine monophosphate is a thymine nucleotide that is part of DNA. The remaining three DNA nucleotides have similar structure, but each has a nitrous base of its own (the top group). These three bases (adenine, guanine, and cytosine) are similar in DNA and RNA (see Fig. 1).

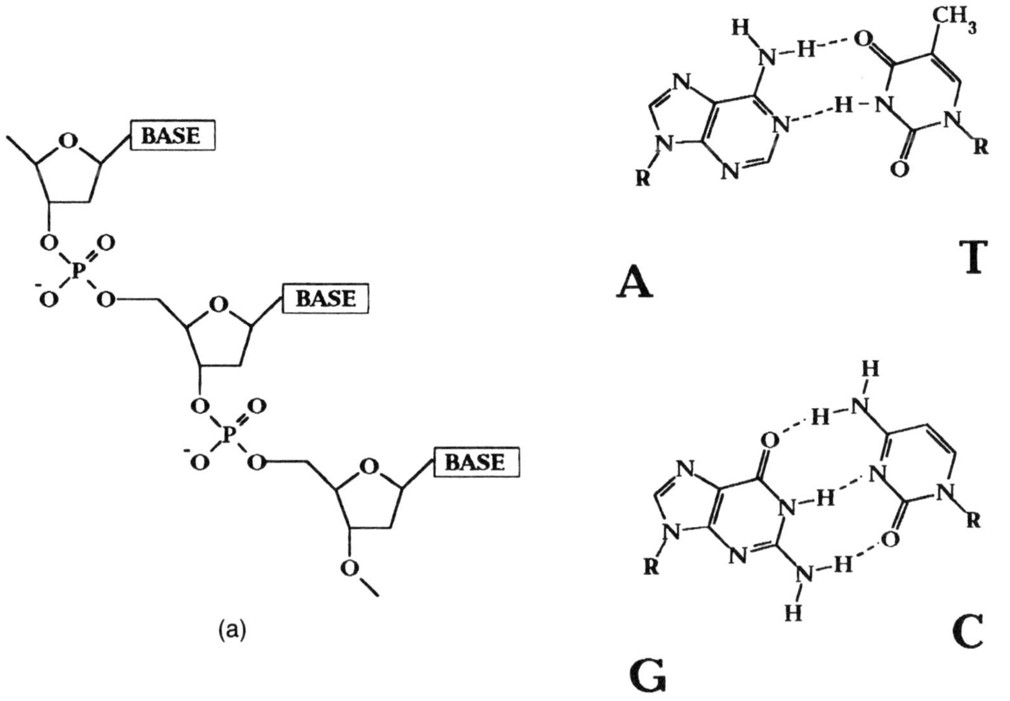

(a)

A T

G C

(b)

FIG. 3. (a) The DNA single strand and (b) the Watson–Crick complementary base pairs.

and duplicates genetic information, RNA plays a key role in its realization, i.e., in the process of synthesis of a protein molecule in accordance with the nucleotide sequence of the corresponding gene.

This process consists of two major steps. First, a RNA copy of the gene is made with the help of a special enzyme, RNA polymerase. The process is called transcription, and the RNA copy of the gene is called messenger RNA (mRNA). Messenger RNA is an exact copy of the gene, in which all thymines (T) are replaced with uracils (U). In the second step, called translation, mRNA is used as a template for protein synthesis in a special cell machinery, the ribosome. The ribosome is a complex aggregate of special ribosomal RNA (rRNA) and a number of proteins. It translates the nucleic-acid text into amino-acid language. In so doing, it uses a special dictionary, the genetic code. In the process of translation, an additional very important class of small RNA molecules plays a crucial role. These RNA molecules consist of about a hundred nucleotides each and are called transfer RNA (tRNA).

The above scheme, known as the central dogma of molecular biology, gives a very general description of how the nucleotide sequence of a gene is realized in the form of the amino-acid sequence of a protein molecule. The scheme is really very general because, for instance, it does not show any difference between prokaryotes, unicellular organisms carrying no cell nucleus, and eukaryotes, complex organisms that have cell nuclei. In practice, however, these two major classes of living organisms significantly differ in the organization of their genomes and its functioning. This difference manifested itself most vividly during the advent of the applied branch of molecular biology, genetic engineering. When pieces of DNA from eukaryotes were incorporated into prokaryotic DNA, they failed to produce any proteins. Only after the fundamental difference was realized in the genome organization of simple microorganisms on the one hand and of complex higher organisms on the other were methods developed that permitted genes of eukaryotes to be expressed in prokaryotes. This opened the way for innumerable applications of molecular biology and marked the beginning of a new era of technological revolution—the era of biotechnology.

The following are the objectives of this article:

Provide an overview of our present knowledge of physical structures of nucleic acids. We describe the most common structures, as well as some new structures discovered recently. The most important structural transitions, such as DNA melting and the B–Z transition, are explained. Special attention is paid to topological properties of DNA, which play a significant role in its functioning within the cell.

Describe some of the major experimental methods used to study nucleic acids, including x-ray diffraction, nuclear magnetic resonance (NMR), electron microscopy, spectroscopic methods, gel electrophoresis, and chemical, photochemical, and enzymatic probing. The potencies of the methods are compared and their advantages and disadvantages discussed.

Describe major biological phenomena where nucleic acids play a crucial role, DNA replication, transcription, and translation. We describe these processes not only in bacteria, where they obey the central dogma of molecular biology, but also in higher organisms and their viruses, for which significant deviations from the central dogma are common, such as RNA splicing, synthesis of DNA on the RNA template, etc.

Provide the background of biotechnology and its prospective. We give the basic ideas for breakthrough discoveries that led to the advent of biotechnology, including restriction enzymes and DNA cloning, DNA sequencing methods, and the polymerase chain reaction (PCR).

1. NUCLEIC ACID STRUCTURES

Traditionally, nucleic acids' structural versatility was considered much narrower than the structural versatility of proteins. This viewpoint well fitted "the division of power" between nucleic acids and proteins in the present-day living creatures. The major function of nucleic acids is to store and reproduce genetic information, while proteins perform an innumerable variety of reactions within the cell and also are its main construction blocks. Major discoveries of the last two decades, however, have significantly changed this attitude. The structural and functional versatility of nucleic acids have proved to be much greater than anybody anticipated before. This is especially true for RNA, which not only adopts extremely complex spatial structures, rivaling the complexity of proteins' spatial structures, but can perform catalytic functions. Such catalytically active RNA molecules have received the special name of ribozymes. The discovery of ribozymes and of the remarkable versatility of RNA structure completely changed the attitude toward the pivotal question of which type of major biopolymers, proteins or nucleic acids, had been the prebiotic molecule, the forefather of the living cell. Before, everybody believed that only protein could fulfill this function. Now, the common view is that this should be the RNA molecule.

However, in the present-day cell, RNA plays just an auxiliary role. The major nucleic acid now is DNA. As a result, DNA structures attract the most attention.

1.1. Major Helical Structures of DNA

In spite of the enormous versatility of living creatures and, accordingly, variability of genetic texts that DNA molecules in different organisms carry, they all have virtually identical physical, spatial structure: the double-helical B form discovered by Watson and Crick (1953). Sequences of the two strands of the double helix obey the complementary principle. This principle is the most important law in the field of nucleic acids, and, probably, is the most important law of the living nature. It declares that, in the double helix, A always opposes T and vice versa, whereas G always opposes C and vice versa.

1.1.1 B-DNA B-DNA (see Fig. 4) consists of two helically twisted sugar-phosphate backbones stuffed with base pairs of two types, AT and GC. The helix is right-handed with ten base pairs per turn. The base pairs are isomorphous: The distances between glycosidic bonds, which attach bases to sugar, are virtually identical for AT and GC pairs. Because of this isomorphism, the regular double helix is formed for an arbitrary sequence of nucleotides, and the fact that DNA should form a double helix imposes no lim-

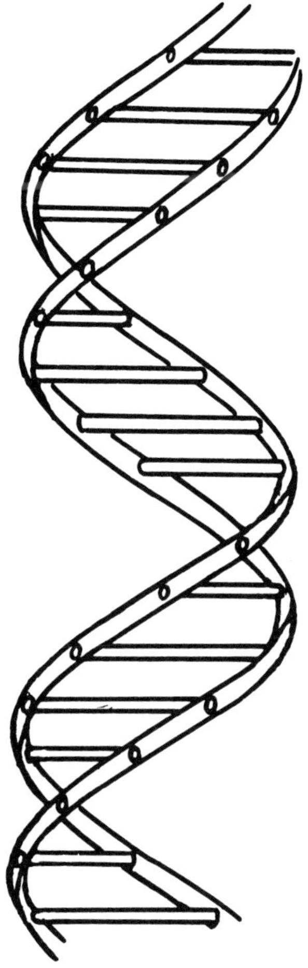

FIG. 4. The double helix.

itations on DNA texts. The surface of the double helix is by no means cylindrical. It has two very distinct grooves: the major groove and the minor groove. These grooves are extremely important for DNA functioning because, in the cell, numerous proteins recognize specific sites on DNA via binding with the grooves.

Each nucleotide has direction: If we treat the sugar as a ring lying in the plane, then the CH_2 group will be above the plane. This carbon is designated as 5′, whereas the carbon atom connecting the sugar to the next nucleotide in the polynucleotide chain is designated as 3′. Thus, the chemical direction is inherent in a single-stranded nucleic acid. In the DNA double helix, the two strands have opposite directions. In B-DNA, base pairs are planar and perpendicular to the axis of the double helix.

Under normal conditions in solution, often referred to as "physiological" (neutral pH, room temperature, about 200 mM NaCl), DNA adopts the B form. All available data indicate that the same is true for the totality of DNA within the cell. It does not exclude, however, the possibility that separate stretches of DNA carrying special nucleotide sequences would adopt other conformations.

1.1.2 B′-DNA Up to now, only one such conformation is demonstrated, beyond any doubts, to exist under physiological conditions. When several A residues in one strand (and, accordingly, several T in the other DNA strand) occur, they adopt the B′ form. In many respects, the B′ form is similar to the classical B form, but there are also significant differences. The main difference consists in the fact that base pairs in B′-DNA are not planar: They form a kind of propeller with the propeller twist of 20°.

Stretches of A residues produce bends in the double helix. Such bends play a very important role in DNA functioning. Although the structural basis of these bends is not fully understood the involvement of the B′ form in the DNA bending is very probable. In spite of its importance, the B′ form does not differ dramatically from the B conformation. Other helical conformations have been found in solution that are significantly different from B-DNA.

1.1.3 A-DNA Similarly to B-DNA, the A form can be adopted by an arbitrary sequence of nucleotides. As in B-DNA, in A-DNA the two complementary strands are antiparallel and form right-handed helices. Normal DNA undergoes a transition from the B to A form under drying. In A-DNA, the base pairs are planar, but their planes form a considerable angle with the axis of the double helix. In so doing, the base pairs shift from the center of the duplex, forming a channel in the center.

If any, A-DNA plays a rather modest role in DNA functioning. There are data indicating that some proteins induce the transition from B to A form. The reason for this remains to be elucidated.

1.1.4 Z-DNA Z-DNA presents the most striking example of how different from the B form the DNA double helix can be. Although in Z-DNA the complementary strands are antiparallel as in B-DNA, unlike in B-DNA, they form left-handed, rather than right-handed, helices. There are many other dramatic differences between Z- and B-DNA.

Not any sequence can adopt the Z form. To adopt the Z form, the regular alternation of purines (A or G) and pyrimidines (T or C) along one strand is strongly preferred. However, even this is not enough for Z-DNA to be formed under physiological conditions. Nevertheless, Z-DNA can be adopted by DNA stretches in a cell as a result of DNA supercoiling (see below). The biological significance of Z-DNA, however, remains to be elucidated.

1.1.5 psDNA The complementary strands in a DNA duplex can be parallel. Such parallel-stranded (ps) DNA is formed most readily if both strands carry only adenines and thymines, and their sequence excludes formation of the ordinary antiparallel duplex. If these requirements are met, the parallel duplex is formed under quite normal conditions. It is right-handed, but the AT pairs are not regular, Watson–Crick ones, but rather so-called reverse Watson–Crick.

1.2 RNA Structure

Quite naturally, similarity between DNA and RNA in their chemical nature entails significant similarity in structures they adopt. RNA can also form the double helix with complementary base pairs AU and GC. However, the major RNA duplex conformation resembles A-DNA rather than B-DNA. In general, because of the bulkier OH group in the sugar ring, the versatility of RNA duplexes is rather narrower than those of DNA.

In the cell, different types of RNA molecules (mRNA, rRNA, tRNA, etc.) exist as single strands. Of course, these strands are not unfolded but fold into complex spatial structures. Mutually complementary stretches of the same molecules form numerous short duplexes, which are the major structural motif of RNA molecules. Single-stranded regions form loops of different types. Sometimes, very complex folding patterns occur, called pseudoknots.

Spatial structure plays an important role in RNA functioning. For instance, rRNA organized in a complex spatial structure forms a scaffolding to which numerous ribosomal proteins attach to form the functional ribosome. The spatial structure of tRNA molecules is specially designed to permit them to fulfill the central role in realization of the genetic code. One part of tRNA carries an anticodon, a trinucleotide complementary to a codon (a trinucleotide corresponding to one amino acid in the genetic code), whereas the amino acid that corresponds to the codon in the genetic code is attached to one terminus of the molecule.

1.3 DNA Topology

DNA very often operates within the cell in the form of a closed circular molecule in which both strands form closed circles (Fig. 5). The closed circular state of DNA is extremely important because, even if DNA is linear as a whole, like the chromosomal DNA of eukaryotes, its parts are organized in closed domains, the closure being fulfilled by proteins. The physical properties and physiological behavior of closed circular (cc) DNA are different in many respects from those of linear DNA molecules. These differences stem from the fact that ccDNA acquires, as compared to linear DNA, a new fundamental feature, topology. There are two levels of DNA topology.

1.3.1 Knots of DNA The double helix as a whole may form a simple circle (i.e., a trivial knot) or be knotted (see Fig. 6). The knotting of DNA may be accomplished by ran-

FIG. 5. In a circular closed DNA, two complementary strands form linkage of a high order.

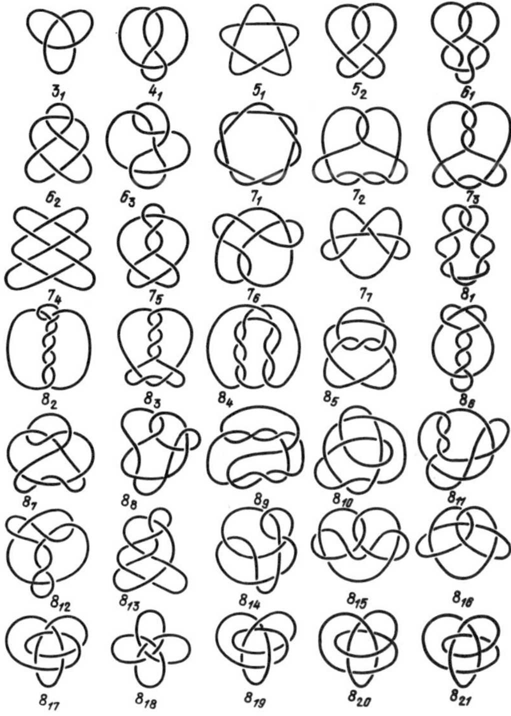

FIG. 6. Knots.

dom closure of linear molecules with "cohesive" ends (mutually complementary single-stranded overhangs). There are also various enzymes that do the job. The most important class of such enzymes is known as DNA topoisomerases II. The type of knot is a topological invariant; i.e., it cannot be changed (converted to another type of knot) without breakage of both DNA strands. As a result, identical ccDNA molecules belonging to different knot types may be separated in gel (Shaw and Wang, 1993).

1.3.2 DNA Supercoiling The two complementary strands in ccDNA are topologically linked with one another (see Fig. 5). The degree of linkage, which is designated as *Lk*, the DNA linking number, is a topological invariant, which cannot be changed by any deformation of the DNA strands without strand breaks. The *Lk* value is easily calculated as the number of times one strand pierces the surface bounded by the other strand. If N is the number of base pairs in DNA and Γ_0 is the number of base pairs per turn of the double helix under given ambient conditions, then the difference

$$\tau = Lk - N/\Gamma_0 \tag{1}$$

is the number of superhelical turns in ccDNA under given conditions. If $\tau \neq 0$, then ccDNA cannot form a planar ring with the helical twist corresponding to the Γ_0 value. Either the helical twist should change, or the DNA molecule as a whole should writhe in space. Actually, with $\tau \neq 0$, both things happen: DNA adopts writhe, and its twist changes. The distribution of energy associated with DNA supercoiling (i.e., with $\tau \neq 0$) between twist and writhe is determined by the torsional and bending rigidities, which have been measured with good accuracy (Frank-Kamenetskii, 1990a). As a result, theoretical knowledge of DNA supercoiling is fairly accurate (Vologodskii *et al.*, 1992).

Closed circular DNA molecules extracted from cells of eukaryotes are always negatively supercoiled ($\tau < 0$; the double helix is "underwound"). Usually, superhelicity is characterized by superhelical density:

$$\sigma = \tau\Gamma_0/N. \tag{2}$$

Special enzymes, topoisomerases, regulate DNA superhelicity in the cell. Supercoiling is also significantly affected by replication and transcription. DNA topology plays an extremely important role in DNA functioning (Cozzarelli and Wang, 1990).

1.4 Unusual Structures and Structural Transitions

If negative supercoiling in ccDNA becomes too high, the bending and torsional degrees of freedom may become insufficient to accommodate the superhelical energy. Under these circumstances, the integrity of the DNA double helix breaks down in its most "weak" sites. These sites, which have the potential to adopt structures significantly different from B-DNA, can undergo structural transitions into unusual (non-B-DNA) structures.

1.4.1 Melting of DNA On heating or under superhelical stress, the DNA complementary strands separate, forming single-stranded loops. In the case of linear duplex molecules, the process ends up, at sufficiently high temperature, with complete separation of the two strands. The process has been comprehensively studied both experimentally and theoretically (Wada and Suyama, 1986) and is known as DNA melting (the terms denaturation or helix-coil transition can be also found in the literature). AT pairs have significantly lower stability than GC pairs.

With increasing temperature, long DNA sections melt out. The melting of each such section is a cooperative, all-or-none phenomenon. However, the melting of a long enough molecule becomes a series of such cooperative processes, each occurring at its own temperature (Fig. 7). Roughly speaking, sections more enriched with CG pairs melt out at higher temperatures. Thus, DNA melting is not a classical phase transition because two different "phases," helical and melted, coexist with each other in one and the same molecule.

When two mutually complementary single-stranded DNA molecules are mixed, the process reverse to melting occurs. It is called renaturation or annealing.

DNA melting is an exceptionally important phenomenon. It occurs in the cell when DNA replicates and when DNA polymerase "reads" the mRNA copy of the gene. Cycles of melting and annealing are utilized in PCR machines, the most remarkable tools of the current biotechnological revolution (see Sec. 4.1.5).

Melting of AT-rich regions in ccDNA may occur with increasing negative supercoiling. However, this does not happen often because other unusual structures defeat melting in competition for the energy of supercoiling.

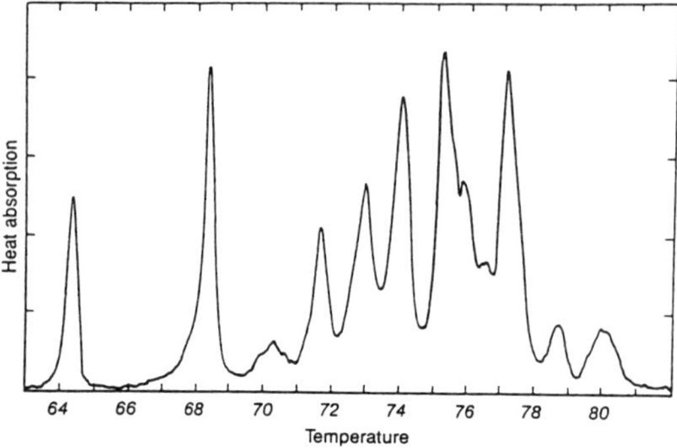

FIG. 7. Melting of DNA. This curve is also often called the differential melting curve. The curve was obtained for DNA that has the code name of ColEI and contains about 6500 nucleotide pairs.

1.4.2 B–Z Transition Negative supercoiling most favors formation of Z-DNA because, in this case, the maximal release of superhelical stress per base pair adopting a non-B-DNA structure is achieved. As a result, although under physiological conditions the Z form is energetically very unfavorable as compared with B-DNA, it is easily adopted in negatively supercoiled DNA by appropriate DNA sequences (with alternating purines and pyrimidines).

Linear DNA with the appropriate sequence adopts the Z conformation at a very high salt concentration (about 3 M NaCl).

1.4.3 Cruciforms Another structure readily formed under negative supercoiling is cruciform, which requires palindromic regions (see Fig. 8). To form a cruciform, a palindromic region must be longer than a certain minimum. For example, six-base-pair–long palindromes recognized by restriction en-

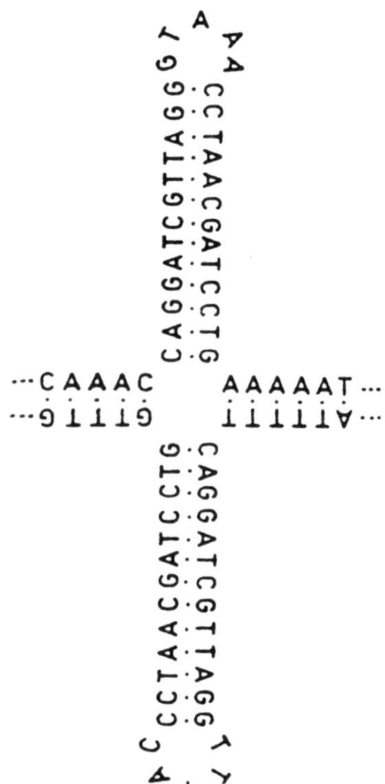

FIG. 8. A cruciform formed in ColEI DNA when the molecule is in a superhelical state.

zymes do not form cruciforms under any conditions.

1.4.4 H-DNA H-DNA forms a special class of unusual structures, which are adopted under superhelical stress by sequences carrying purines (A and G) in one strand and pyrimidines (T and C) in the other, i.e., homopurine-homopyrimidine sequences (Frank-Kamenetskii, 1990b). The major element of H-DNA is a triplex formed by one half of the insert adopting the H form and by one of two strands of the second half of the insert (Fig. 9). Two major classes of triplexes are known—pyrimidine-purine-pyrimidine (PyPuPy) and pyrimidine-purine-purine (PyPuPu). Figure 10 shows the canonical base triads entering these triplexes.

Always two isomeric forms of H-DNA are possible, which are designated as H-y3, H-y5, H-r3, and H-r5, depending on which kind of triplex is formed and which half of the insert forms the triplex (see Fig. 9). H-DNA may be considered as an intramolecular triplex (it is often referred to in this way). Its formation under physiological conditions occurs only under superhelical stress.

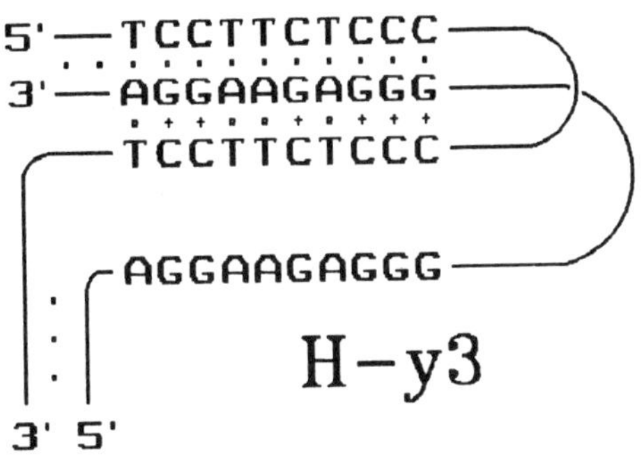

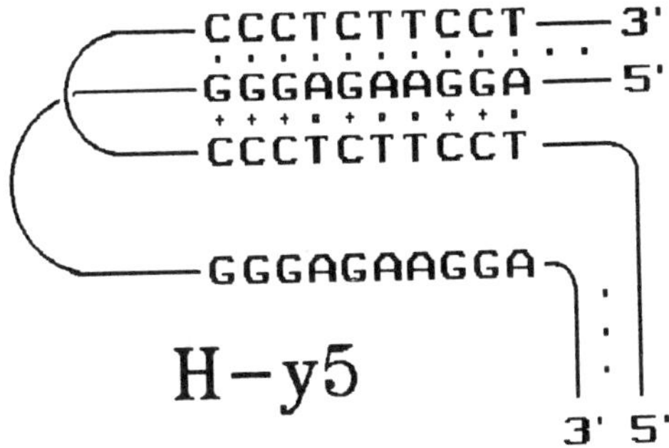

FIG. 9. DNA's H-form structure. Two possible "isomeric" variants of the structure are shown. The Watson–Crick pairing is designated by filled circles, while the GC Hoogsteen pairing, involving the presence of an extra proton, is designated by plus symbols.

The discovery of H-DNA stimulated studies of intermolecular triplexes, which may be formed between homopurine-homopyrimidine regions of duplex DNA and corresponding pyrimidine or purine oligonucleotides. These studies have led to the development of very promising approaches to sequence-specific targeting of duplex DNA (Cherny *et al.*, 1993).

1.4.5 Quadruplexes Of all nucleotides, guanines are the most versatile in forming different structures. They may form GG pairs, but the most stable structure, which is formed in the presence of monovalent cations (especially potassium), is the G4 quadruplex (see Fig. 11). G quadruplexes may exist in a variety of modifications: all-parallel, all-antiparallel, and others. As a result, G quadruplexes are easily formed both intermolecularly and intramolecularly, again with a variety of modifications.

A totally unusual quadruplex was discovered by Gehring *et al.* (1993). It contains two hemiprotonated parallel-stranded duplexes consisting of unusual CC^+ pairs. The two parallel-stranded duplexes are associated in a mutually antiparallel manner so that CC^+ base pairs from one duplex are "layered" by CC^+ pairs from the other duplex, thus alternating along the structure.

1.4.6 RNA Unusual Structures Most DNA unusual structures have their analogs in the RNA world. Specifically, the formation of triplexes was first demonstrated by Felsenfeld *et al.* (1957) for artificial RNA chains. Solitary RNA base triads are often met in tRNA structures. Triplexes are formed by mixed DNA-RNA hybrids, and some of them are rather stable (Roberts and Crothers, 1992).

RNA molecules form G quadruplexes, and there are exciting data indicating that RNA G quadruplexes could be an important element in the structure of HIV RNA. However, in general, studies of RNA unusual structures lag behind the corresponding DNA structures because methods to study unusual structures are much better developed in the case of DNA than in the case of RNA.

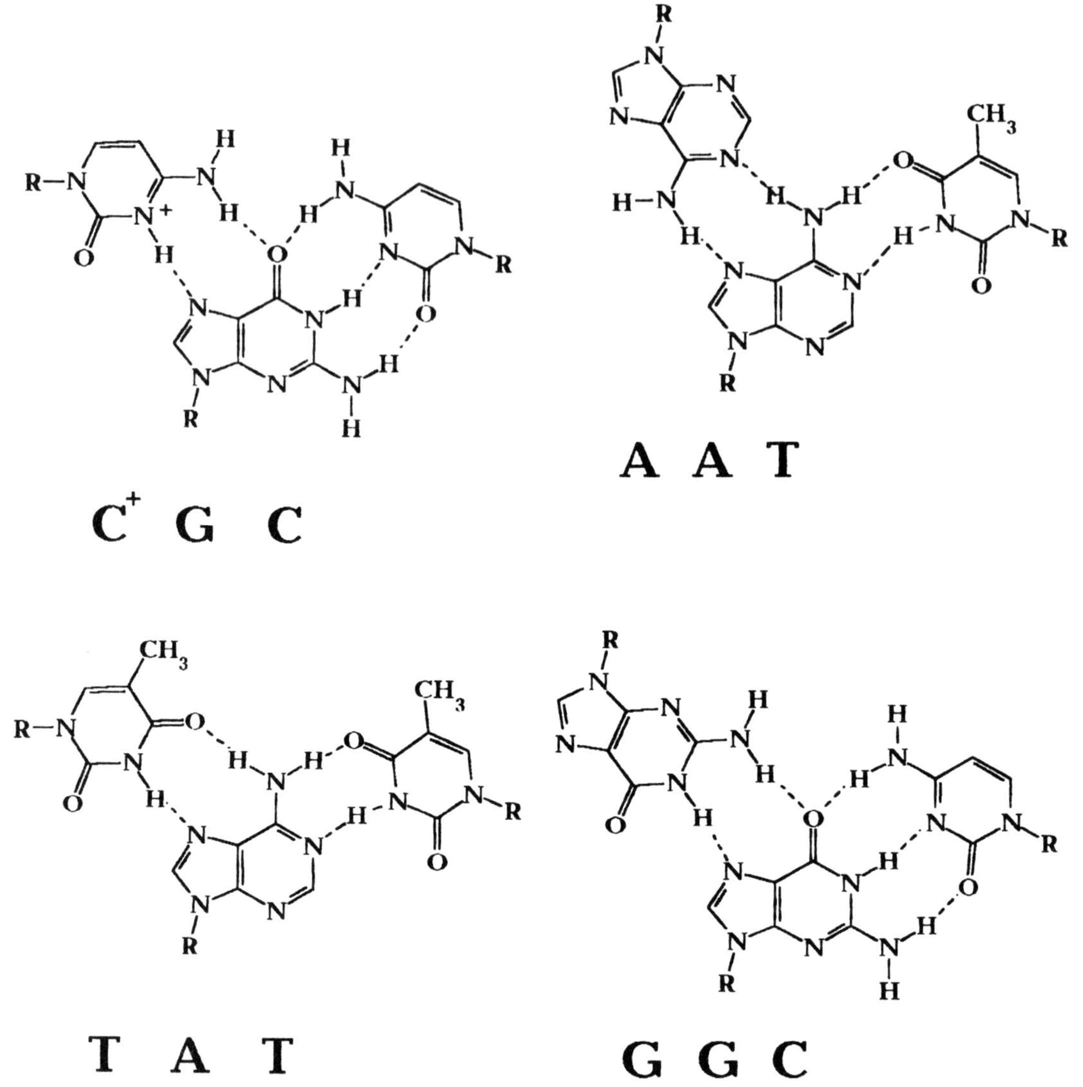

C⁺ G C

A A T

T A T

G G C

(a)

(b)

FIG. 10. The structure of (a) pyrimidine (TAT and C^+GC) and (b) purine (AAT and GGC) base triads of which the DNA triple helix is made.

2. METHODS TO STUDY NUCLEIC ACID STRUCTURES

The number of methods that are used to study the structure of nucleic acids is truly enormous. Methods to study DNA alone occupy two volumes of the Methods in Enzymology series (Lilley and Dahlberg, 1992a,b). We survey only those of them that are most informative and are widely used. Along with general methods to study molecular structures, a number of special methods have been elaborated to study nucleic acids. But even the regular methods, such as x-ray analysis or electron microscopy, are often significantly modified to fit the needs of studying such complex molecules as nucleic acids.

2.1 General Methods

In this section, we discuss methods that are generally used to study molecular structures. We do not describe the background of

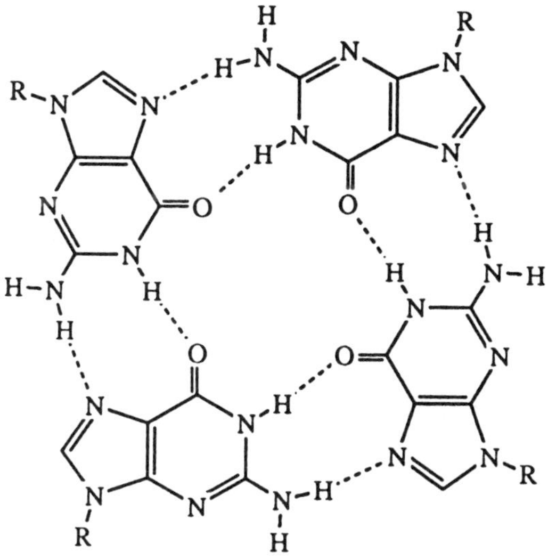

FIG. 11. G quadruplex.

the methods, because they are widely known, but rather concentrate on specific questions of their application to studing nucleic acids.

2.1.1 X-Ray Analysis As in other fields where molecular structure is essential, x-ray analysis occupies a unique position among methods of studying nucleic acid structure as the only direct method that permits elucidation of the structure in all details. X-ray crystallography is absolutely indispensable in the study of the detailed structure of complexes of nucleic acids with proteins, which is most essential for understanding how nucleic acids function in the cell.

However, for a long period of time, the whole edifice of molecular biology relied on one of the indirect versions of x-ray analysis, fiber diffraction, rather than on classical x-ray crystallography.

2.1.1.1 Fiber Diffraction. Long molecules of nucleic acids cannot be crystallized. As a result, from the early 1950s till the mid-1970s, only x-ray diffraction from nucleic acid fibers was used to elucidate the structure of nucleic acids. Such data were used by Watson and Crick (1953) to propose structures for B- and A-DNA. Fiber diffraction is an essentially indirect method, and, to elucidate structure from the data on fibers, one must heavily rely on theoretical approaches, such as conformational analysis.

2.1.1.2 X-Ray Crystallography. The first good crystals of nucleic acids were obtained for the case of relatively short nucleic acids, the tRNA molecules, and those were the first nucleic acids for which structure was solved by classical x-ray crystallography.

The first DNA crystals became available only in the late 1970s, after remarkable progress in chemical synthesis of short DNA pieces had been achieved. This led to many discoveries, first of all of left-handed Z-DNA by Rich and co-workers (Wang *et al.*, 1979).

In recent years, a lot of very detailed data on the structure of DNA have been obtained by x-ray crystallography, including detailed study of the B, B', A, and Z forms (Dickerson, 1992). Among the most recent achievements of the method is solution of the structure of the G quadruplex (Kang *et al.*, 1992).

In spite of remarkable accomplishments, serious limitations are inherent in the method. It is extremely hard to obtain good crystals of a nucleic acid even if it adopts only a unique structure. For instance, for DNA or RNA triplexes, sufficiently good crystals are still unavailable. Even if this difficulty is overcome, it sometimes appears that the structure in a crystal is significantly perturbed by interaction with neighboring molecules. It is especially true of such subtle, but very important from a biological viewpoint, deformations as bending of the double helix. Thus, the data obtained by the methods of x-ray crystallography should always be correlated with the results of indirect methods, which permit the study of nucleic acid in solution. X-ray crystallography is totally useless for studying such biologically significant problems as DNA supercoiling or the structure of large RNA molecules.

2.1.2 Nuclear Magnetic Resonance (NMR) The role of NMR (see MAGNETIC RESONANCE AND QUADRUPOLE RESONANCE, NUCLEAR) constantly increases, and, in recent years, it has even started to compete successfully with x-ray crystallography. This has become possible as a result of the development of two-dimensional proton NMR techniques, especially nuclear Overhauser-effect spectroscopy (NOESY). The great advantage of NMR, as compared with x-ray crystallography, consists in the fact that it does not require crystals. As a result, structures of some nucleic acids that resist crystallization, like triplexes, have become subjects of very fruitful study by NMR (Rajagopal and Feigon, 1989).

A great advantage of NMR consists in the possibility of studying structural fluctuations, or "breathing," of nucleic acids by following proton exchange in nucleic acid bases. Such studies made it possible to find out important characteristics of base-pair fluctuational openings in DNA (Gueron *et al.*, 1987).

Turning to the limitations of the method, it should be emphasized that resolution of even the most powerful NMR spectrometers permits the study of only short nucleic acids containing about a dozen distinguishable nucleotides. Although crystals are not needed, only very concentrated solutions can be studied. Therefore, like x-ray crystallography, NMR is useless for studying many biologically relevant problems of nucleic acid structure. In spite of its limitations, proton NMR has firmly occupied the second position, after x-ray crystallography, among methods to study nucleic acid structure.

2.1.3 Microscopy Although microscopy looks like the most direct way to visualize structure, in application to nucleic acids it has too numerous limitations to occupy a position ahead of x-ray crystallography or NMR. Nevertheless, in recent years, the role of microscopy in the field of nucleic acids has significantly increased as a result of progress in regular electron microscopy of nucleic acids and their complexes and the development of new techniques, cryo-electron microscopy and scanning tunneling microscopy.

2.1.3.1 Regular Electron Microscopy. In regular electron microscopy, nucleic acids are placed on the grid, dried, and contrasted by one or another method. In recent years, the most popular technique of contrasting consists in staining with uranyl acetate. As a result, for instance, duplex DNA molecules are clearly seen (see Fig. 12).

Regular electron microscopy permits one to see complexes of nucleic acids with proteins. Poor resolution usually does not permit one to observe the internal structure of the complex but permits mapping of the location of the protein on DNA (see Fig. 12).

A limitation of regular electron microscopy stems from poor resolution and the possibility of significant perturbation of nucleic acid structure in the process of sample preparation. And still, in many cases, the method provides the most convincing evidence.

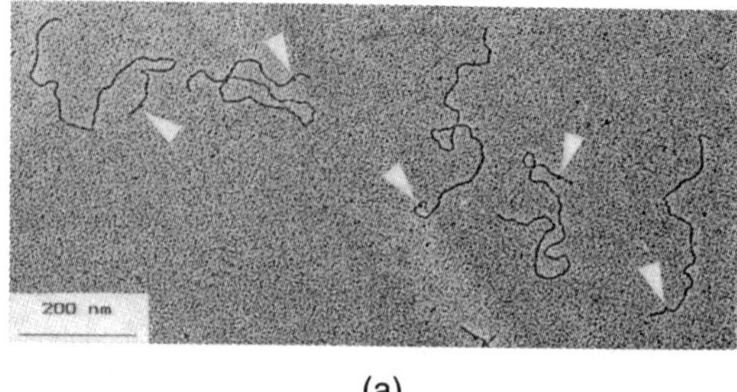

(a)

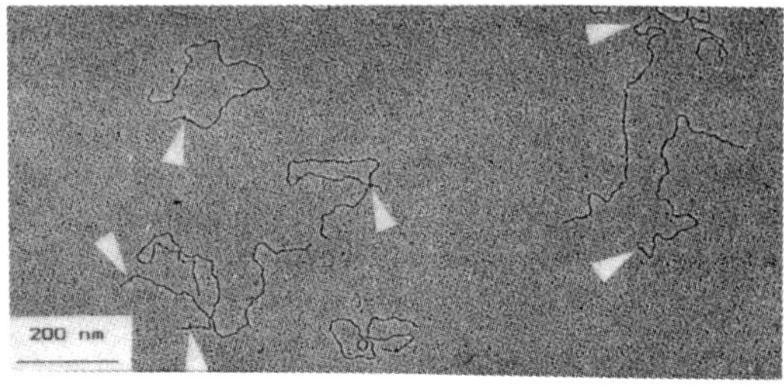

(b)

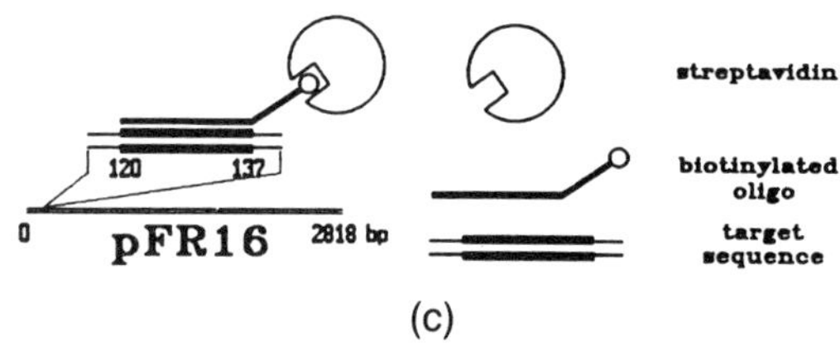

(c)

FIG. 12. Electron-microscopy visualization of triplexes formed between duplex DNA and corresponding oligonucleotides [(a) pyrimidine or (b) purine]. The oligonucleotides were chemically tagged by biotin molecules, and then the protein streptavidin was added, which strongly binds to biotin [see the chart, (c)]. The sites of triplex formation are seen as "beads" due to streptavidin molecules and are indicated by the white arrowheads. The data are from Cherny *et al.* (1993).

2.1.3.2 Cryo-Electron Microscopy. The method is based on obtaining vitrified water solutions via very quick cooling of an extremely thin (in the micron range) sample. As a result, the molecule is "frozen" in the state it adopted in solution before cooling. In recent years, the method has received numerous applications in molecular biology in general and in the field of nucleic acids and their complexes with proteins in particular (Dubochet *et al.*, 1992).

The great advantage over regular electron microscopy consists in avoiding the harsh procedures of sample preparation, which strongly limit the value of the data obtained by regular electron microscopy.

The major problem of cryo-electron microscopy stems from the low contrast of biological molecules. Without staining or other contrasting, they are barely visible in an electron microscope. Nevertheless, DNA molecules and their complexes with proteins are extensively studied by the method (Dubochet *et al.*, 1992).

2.1.3.3 Scanning Tunneling and Force Microscopy. These are very new methods in general and in the field of nucleic acids in particular. Spectacular images of DNA helices and even bases obtained by scanning tunneling microscopy (STM) have been widely publicized and created great excitement. However, the nature of these images remains obscure because DNA molecules are not conductors and are not expected to be seen in STM. The recent data of Dunlap *et al.* (1993) strongly indicate that the STM images of DNA could be artifacts.

By contrast, scanning force microscopy (SFM) provides reliable images of DNA molecules and their complexes with proteins (Bustamante *et al.*, 1993). At present, the resolution of this method is not much higher than that of regular electron microscopy. However, there are reasonable hopes that the resolution will be significantly improved in the course of further development of the method.

2.1.4 Optical Methods All optical methods that are traditionally applied to study molecular structures are widely used to study nucleic acids. They are indispensable in routine investigations because they are cheap and quick. The great advantage of those that use light in the visible and UV region consists in the possibility of study of very dilute solutions of nucleic acids, for which intermolecular interaction can be completely neglected. On the other hand, all these methods are essentially indirect and provide any structural information only after careful assignment of particular spectra or special changes with the help of more direct methods described above.

2.1.4.1 UV Spectroscopy. Nucleic acid bases absorb UV radiation around 260 nm. The intensity of this absorption, which is easy to measure with regular spectrophotometers, changes when, for instance, the DNA double helix melts. Hence, UV spectroscopy has been extensively applied to study DNA melting. In application to RNA, the method is not very informative because, in single-stranded RNA molecules, the variety of structures is too large to be elucidated by this method. Changes of UV absorption are too small for B-to-A and B-to-Z transitions.

2.1.4.2 Circular Dichroism (CD). CD spectra in the vicinity of 260 nm are much more sensitive to DNA helical structure than UV absorption. B-DNA, A-DNA, and Z-DNA have characteristic and very different CD spectra (Johnson, 1990), and this fact is extensively used in the study of structural transitions in DNA between different helical structures (Ivanov and Krylov, 1992). As with UV spectra, CD spectra are much less informative in the case of RNA than in the case of DNA.

2.1.4.3 Infrared and Raman Spectroscopy. Infrared and Raman spectra are sensitive to nucleic acid structure. Correspondingly, IR and Raman spectroscopies are used to study nucleic acids. However, the main limitation of these methods stems from the fact that they require high concentrations of nucleic acids. As a result, these methods are much less popular than UV and CD spectroscopies.

2.1.4.4 Fluorescence. Nucleic acids practically do not emit absorbed radiation. However, some RNA molecules, along with the four canonical bases, contain unusual bases, and some of them are fluorescent. This fact is extensively used to study the RNA molecules that carry such bases. Regularly, fluorescence methods are used via binding to nucleic acids of strongly fluorescent dye molecules, such as ethydium.

Fluorescence sensitization and quenching due to excitation energy transfer between the donor molecule of electronic excitation and the acceptor molecule of the excitation are also used. For instance, to check the polarity of strands the donor molecule is tagged to, say, 3'-end of one strand, whereas the acceptor molecule is tagged to the 3' end of the complementary strand. Then the two strands anneal, forming a duplex. Strong excitation energy transfer and acceptor indicates that a parallel-stranded duplex is formed.

2.1.5 Theoretical Methods The paper that signified the beginning of extensive studies of nucleic acids and their biological role was purely theoretical (Watson and Crick, 1953). Since then, theory has been playing a very important role in study of nucleic acids.

2.1.5.1 Conformational Analysis. Conformational analysis was especially important during the era of fiber diffraction. In fact, what Watson and Crick did in their classical paper (Watson and Crick, 1953) was a very simple, but exceptionally efficient, variety of conformational analysis. Since then, the method has been extensively used to refine structures solved by x-ray crystallography (Dickerson, 1992). The method is also often used to predict new structures. For instance, parallel-stranded DNA duplexes were first predicted theoretically and then found experimentally.

2.1.5.2 Theoretical Models. As in the study of any important physical object, a number of simplified theoretical models of nucleic acids exist, different models being used to analyze different properties. Figure 13 presents schematics of some of these models.

The DNA double helix may be treated as an isotropic elastic rod [Fig. 13(a)]. In the framework of this model, the DNA molecule is described by only three parameters: bending and torsional rigidities and the diameter. The model has proved to be extremely useful to analyze hydrodynamics and other properties of linear DNA, when it behaves as a usual polymeric molecule. It also permitted provision of the comprehensive theoretical treatment of DNA topology of both levels, knotting (Klenin *et al.*, 1988; Shaw and Wang, 1993) and supercoiling (Vologodskii *et al.*, 1992).

A quite different, but also very successful, model treats the DNA double helix as consisting of base pairs of two types: closed and open [Fig. 13(b)]. This is the helix-coil model, which has permitted quantitative explanation of all major features of DNA melting (Wada and Suyama, 1986).

The polyelectrolyte model [Fig. 13(c)] treats DNA itself just as a charged cylinder but allows for mobile counterions surrounding the double helix (Frank-Kamenetskii *et al.*, 1987).

To predict RNA structure, a simplified model of RNA folding is widely used [Fig. 13(d)].

There are many other theoretical models of DNA and RNA. Usually, the theoretical models are treated within the framework of statistical mechanics. However, in many cases, such consideration has proved to be insufficient because very slow relaxation is rather typical for nucleic acids. Such slow relaxational processes were very thoroughly studied both theoretically and experimentally in the case of DNA melting (Anshelevich *et al.*, 1984). One cannot correctly describe RNA folding without allowance for slow relaxational processes. Such processes are very essential in many cases of formation of DNA

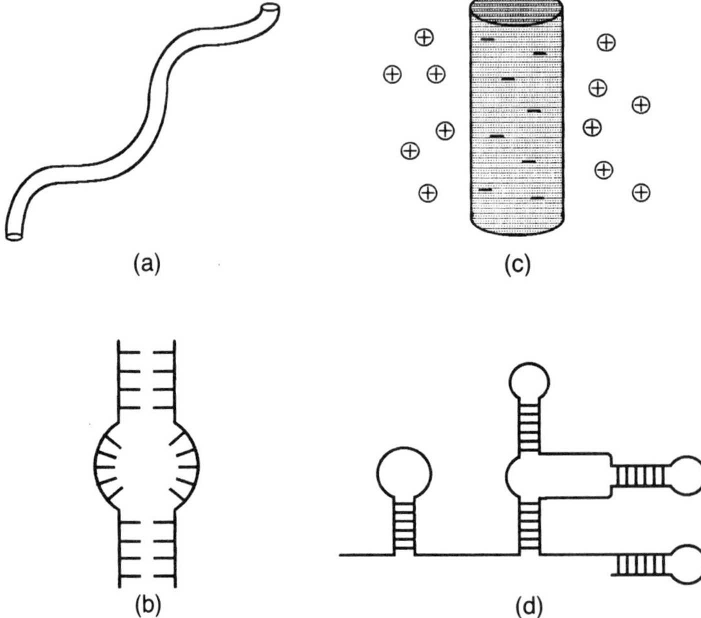

FIG. 13. Theoretical models of nucleic acids: (a) elastic-rod model, (b) helix-coil model, (c) polyelectrolyte model, (d) model of RNA folding.

unusual structures under negative superhelical stress.

2.2 Special Methods

In this section, we consider the most important methods specially developed to study nucleic acids. These methods have been introduced relatively recently (in the past 20 years), but, in many cases, they have pushed aside the traditional methods.

2.2.1 Gel Electrophoresis Gel electrophoresis is a simple technique, introduced in the early 1970s, which truly revolutionized the study of nucleic acids and, subsequently, the whole field of molecular biology. All developments in this field in the past 20 years are connected, directly or indirectly, with the gel electrophoresis method. Gel electrophoresis has pushed aside ultracentrifugation as the method used to separate nucleic acids.

2.2.1.1 Background. Gel electrophoresis differs from electrophoresis in solution only in the nature of the medium in which molecules are separated in the electric field. In the case of gel electrophoresis, the medium is a gel, a polymer network. The most popular in the field of nucleic acids are gels made of polyacrylamide or agarose. Originally, the great advantage of gels in the separation of nucleic acids was discovered purely empirically. The understanding came later after some ideas of P.-G. De Gennes were borrowed from polymer physics, namely the notion of reptation of polymer molecules.

As in regular electrophoresis, the electrophoretic mobility μ is defined as the proportionality coefficient between velocity of movement, v, and the electric field E:

$$v = \mu E. \qquad (3)$$

The electric force applied to a nucleic acid of length L is proportional to L (because each residue carries a negative charge). Hence

$$\mu \sim LD, \qquad (4)$$

where D is the diffusion coefficient:

$$D = \langle x^2 \rangle / \tau, \qquad (5)$$

where τ is the characteristic time for a nucleic acid molecule to go out of its original "tube" and $\langle x^2 \rangle$ is the mean-square shift of the molecule after it leaves the tube.

All movements other than within the "tube" are forbidden in the gel. As a result, the molecule experiences Brownian motion only along its own axis (i.e., within the "tube"). Because the friction in the course of such movement is proportional to L,

$$\tau \sim LL^2 = L^3. \qquad (6)$$

For the ideal polymer coil,

$$\langle x^2 \rangle \sim L,$$

and we finally obtain

$$\mu \sim 1/L, \qquad (7)$$

whereas, without the gel, similar considerations would lead to the lack of dependence of μ on L. Equation (7) explains why the gel is so efficient a medium to separate molecules of nucleic acids according to their lengths during electrophoresis.

The consequences of Eq. (7) are really far-reaching. The entire idea of genetic engineering, i.e., reshuffling of DNA pieces extracted from different organisms, became feasible only after two major breakthroughs: the discovery of restriction enzymes, which cut long DNA molecules into shorter pieces recognizing special short nucleotide sequences, and the implementation of gel electrophoresis to separate the pieces obtained after cutting. Each piece forms its own band in the gel after the electric field is switched off. Then the gel is cut by an ordinary razor to obtain one unique piece of DNA.

2.2.1.2 Regular Gel Electrophoresis and DNA Sequencing. Gel electrophoresis is indispensable in all methods of DNA sequencing, which have revolutionized the entire field of biology. We explain the basic idea of application of regular gel electrophoresis to DNA sequencing using the chemical method of Maxam and Gilbert (1977) as an example. We explain the essence of the techniques using the example of single-stranded DNA fragments. The technique also happens to be applicable to double-stranded DNA fragments.

So, suppose we have a sample consisting of identical single-stranded DNA fragments of an unknown sequence. First, using a specific enzyme, radioactive phosphorus ^{32}P is attached to one (particular) end of the fragment. We shall refer to the molecule's labeled end as the beginning, the other end being unlabeled. The sample is then divided into four parts.

Then a substance which breaks the DNA strand after an adenine nucleotide (A) is added to the first fraction. In so doing, the reaction conditions are chosen in such a way that, on the average, about one A per fragment will be attacked during the reaction. The reagent will transform the original mixture of uniform-length fragments into one of fragments of varying length. In the process, we are concerned solely with the labeled fragments. If, say, adenines occupy positions 1, 3, 7, 13, 21, 25, and 26 on the original fragment, we may assume that the reagent is added to result in the appearance of labeled fragments with lengths of 1, 3, 7, 13, 21, 25, and 26 nucleotides. Not a single fragment of a different length is expected to appear.

Similarly, the three other parts of the original solution are treated with substances breaking the chain after T, G, and C. After this, all four samples are separated in a parallel manner in the same apparatus for gel electrophoresis. After the electric field is cut off, a photographic plate is placed on the gel's surface to imprint on it the labeled bands in the gel.

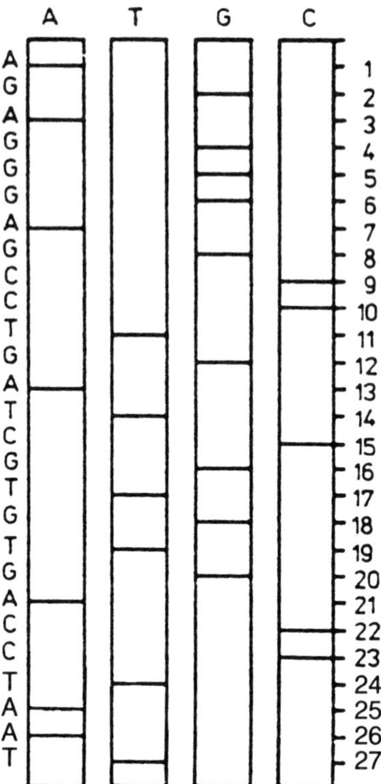

FIG. 14. The Maxam–Gilbert method (chart).

The experiment's result is diagrammed in Fig. 14. The sequence that can be read directly on the electrophoregram is shown to the left. (Actually, three, rather than four, different reagents are sufficient.)

The DNA sequencing techniques have been developed to perfection, based on different conceptual approaches. The most fertile approach proved to be that of Sanger *et al.* (1977), who developed a method of DNA sequencing that uses enzymes working on DNA. Instead of radioactive labels, the approach uses nucleotide analogs linked to different dyes, which makes sequencing an ecologically much cleaner technique.

2.2.1.3 Pulse Electrophoresis. Even gel electrophoresis has its limitations. According to Eq. (7), with increasing length of molecules, their electrophoretic mobility decreases. Therefore, to separate very long molecules in a practically acceptable time scale, one needs to increase the electric field. However, the treatment in Sec. 2.2.1.1 is valid for the case of very small electric fields, which do not deform the molecules [otherwise Eq. (4) would not be valid]. At high fields, the DNA molecule (we talk about DNA because only DNA molecules could be so long that the problems arise) straightens along the electric field. As a result, it moves not like a polymer coil but like a rodlike, straight object. The electrophoretic mobility of such a molecule does not depend on its length independently of whether it moves in pure solvent or in gel, because, in this case, the friction and the driving force are both proportional to L. Thus, in a strong field, separation with respect to length occurs only for a short time before the DNA molecules are straightened. It is totally senseless to conduct electrophoresis longer than this time because all molecules, independently of their length, will just shift by the same distance. Does this mean that long DNA molecules cannot be separated in gel?

Schwartz and Cantor (1984) found a simple way out of the deadlock. If, soon after the molecules are straightened, the direction of the electric field is significantly changed, say by 90°, then, before the molecules will be straightened in the new direction, they assume again the shape of a polymer coil and will be separated for the same short time as while moving in the first direction. After straightening in the second direction, the field again is switched to the first direction, etc. As a result of such cycles or pulses, the molecules effectively move in the diagonal direction, and the separation takes place throughout the duration of the experiment.

Pulse electrophoresis dramatically increased the range of lengths of DNA molecules that can be separated in gels. The method permits separation of entire chromosomal DNA molecules. Implementation of this method opened the way to such ambitious projects as the Human Genome Project, which is designed to sequence the entire human genome (see Sec. 4.2.6).

2.2.1.4. Separation of DNA Topological Conformers. In the field of DNA topology and supercoiling, gel electrophoresis created a kind of revolution similar to those in other branches of nucleic acid research and applications. Although for a different reason than linear DNA, closed circular DNA molecules belonging to different topological classes move in gel with different velocities. As a result, knots of different types can be separated in gel (Shaw and Wang, 1993). The same is true for different topoisomers, ccDNA molecules differing in the linking number *Lk* (see Fig. 15). There it is not the *Lk* value according to which DNA molecules are actually separated, but the writhing, i.e., the spatial shape of the double helix (see Sec. 1.3.2). More precisely, the mobility depends on the absolute value of writhing.

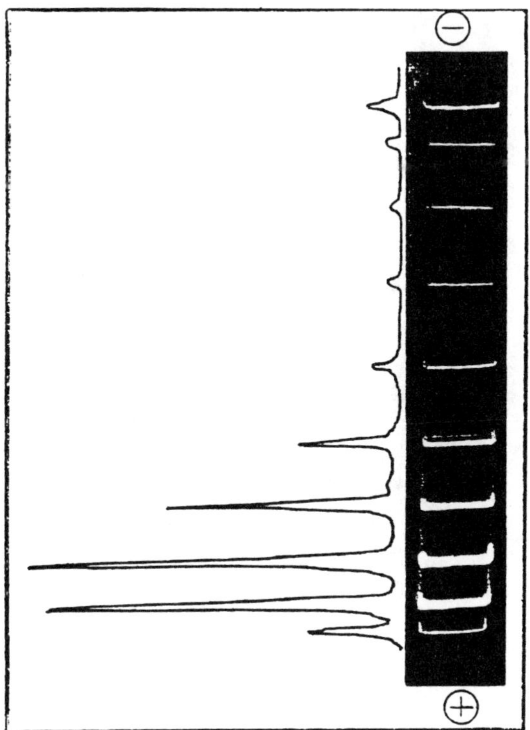

FIG. 15. Separation of DNA molecules differing by the number of superhelical turns, done by the gel electrophoresis technique. The experiment was conducted with DNA of a small pA03 plasmide, containing 1683 nucleotide pairs. Originally, the molecules were inserted at the top, near the negative electrode (the place is not shown in the figure).

2.2.1.5. Two-Dimensional Gel Electrophoresis. The regular, one-dimensional gel electrophoresis does not separate supercoiled molecules that have the same absolute value of the number of superhelical turns τ but different sign because those molecules have the same absolute value of writhing. When a structural transition into an unusual structure occurs, although the *Lk* value does not change, both twisting and writhing change (their sum, which is the linking number, remaining unchanged). As a result, a topoisomer carrying an unusual structure may move in gel with the same speed as another topoisomer without an unusual structure. To avoid such confusion, two-dimensional gels are used.

A specially prepared mixture of different topoisomers of one and the same DNA, carrying an insert capable of changing into an alternative structure, is placed in the left top angle of a quadrangular gel plate (see Fig. 16). Then an electric field is applied to force DNA molecules to move from top to bottom along the left edge of the plate. Following the separation of topoisomers in the first direction, the gel is saturated with chloroquine molecules, which lessen superhelical stress. The chloroquine concentration is chosen in such a way as to make the superhelical stress insufficient for the formation of an unusual structure. After that, the direction of the electric field is changed to force molecules to move from left to right. As a result, the sequence of spots in the second direction corresponds to the topoisomers' sequence.

The uppermost spot in Fig. 16 corresponds to zero topoisomer, i.e., to a relaxed and nonsuperhelical DNA. The spots coming from that spot clockwise correspond to positive topo-

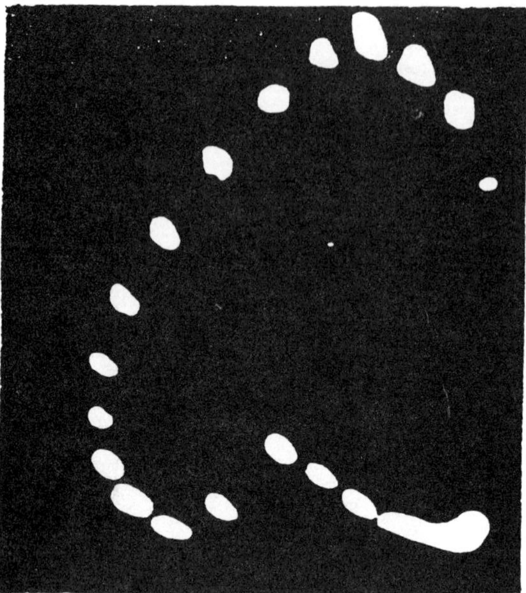

FIG. 16. A typical pattern of two-dimensional gel electrophoresis, observed during formation in DNA of an unusual structure.

isomers; those going anticlockwise, to the negative. One can clearly see the mobility drop, observed in this case between −10 and −12 topoisomers. This means that in topoisomers −12, −13, . . ., an unusual structure is present, while in topoisomers . . ., −9, −10, it is absent. Topoisomer −11 occupies an intermediate position: It carries the unusual structure during, roughly, half the time of its movement in the gel.

2.2.2 Chemical, Photochemical, and Enzymatic Probing A great variety of special approaches have been elaborated to study nucleic acid structures (Lilley and Dahlberg, 1992a,b). They are grounded on the different reactivities of nucleic acids adopting different structures with respect to chemical, photochemical, and enzymatic reactions. In many cases, these methods make it possible to arrive at very specific conclusions about the structure of a particular region of a nucleic acid under conditions that totally exclude application of not only x-ray crystallography or NMR but even spectroscopy and other indirect physical methods. Sometimes, the methods under consideration are applicable even *in vivo*.

To explain the general ideology underlying these methods, let us consider a specific example. In Sec. 1.4.4, we mentioned intermolecular triplexes, which are formed between homopurine-homopyrimidine regions of DNA and corresponding homopurine or homopyrimidine oligonucleotides. If such a complex is actually formed, the reactivity of the N7 position of guanine should dramatically decrease because this nitrogen is sheltered in the triplex by the Hoogsteen pairing (see Fig. 10).

The chemical agent used is dimethyl sulfate (DMS), which reacts with the N7 position of guanine, alkylating it. This alkylation occurs in single-stranded as well as in duplex nucleic acids. However, it cannot take place in triplexes. As a result, in the complex of duplex DNA with an oligonucleotide, which forms a triplex, all guanines in the duplex outside the triplex zone will be alkylated by DMS, whereas guanines within the triplex zone will remain unmodified.

Then the DNA piece under study is end-labeled and subjected to hot piperidine treatment, just as in the Maxam–Gilbert sequencing technique (actually, it is DMS that is used in the Maxam–Gilbert method as a reagent on guanines). Piperidine will convert the sites of alkylated guanines into chain breaks. Such breaks will never occur in the triplex zone. Figure 17 shows the pattern that is obtained after separation of the fragments in gel and radioautography.

The above example is a specific case of the footprinting assay. Such assays can be applied, for instance, to complexes of DNA with proteins to find out which sequences are recognized by the proteins. Instead of DMS, DNAase I, which cuts the uncovered DNA duplex, is often used. The yield of some photoproducts that can also be converted to strand breaks dramatically decreases when a duplex region is covered by a protein or an oligonucleotide. Hence the photo-footprinting assay is useful.

Some chemical reagents, like diethyl pyrocarbonate, potassium permanganate, and osmium tetroxide, do not react with bases in the double helix but react with open bases. The products can be converted into chain breaks. These reagents are widely used to detect open regions. Single-strand–specific nucleases, which digest single strands but do not digest duplex, are used in a similar way.

Chemical, photochemical, and enzymatic

probing are extremely powerful methods of detecting unusual structures, such as Z-DNA, cruciforms, H-DNA, and G quadruplexes (Lilley and Dahlberg, 1992a,b).

3. BIOLOGICAL FUNCTIONS OF NUCLEIC ACIDS

Nucleic acids occupy the most basic position in all living organisms. In this section, we briefly consider how nucleic acids fulfill the most significant of their numerous physiological functions.

3.1 Replication

The most basic function of nucleic acids consists in storage and reproduction of genetic information, which is encoded in the sequence of nucleotides. In principle, one strand of a nucleic acid carries the complete information, and many viruses actually carry only single-stranded DNA or RNA molecules. However, in all living cells, information is stored and reproduced in the form of the DNA double helix.

3.1.1 DNA Replication The ability of nucleic acids to replicate is inherent in the complementary principle. In autonomous living organisms, from unicellular up to the human, genetic information is stored in the form of duplex DNA. Schematically, replication consists in separation of the complementary DNA strands and synthesis on each of the strands of complementary ones (see Fig. 18). In practice, the process is conducted by a special replicating machinery consisting of a number of proteins. The central protein of the machinery is the DNA polymerase enzyme. Other important proteins are single-strand–binding (SSB) and DNA-helicase, which unwind the double helix. Topological problems, which are inevitable because of the necessity of unwinding of the two parent strands, are solved by topoisomerases.

Replication is initiated by binding to the specific sites on DNA of special short RNA oligonucleotides, the primers. On one of the two DNA strands, the synthesis of the complementary strand is carried out continuously. On the opposite strand, the synthesis is often interrupted because the polarity of the second strand is opposite to the direction

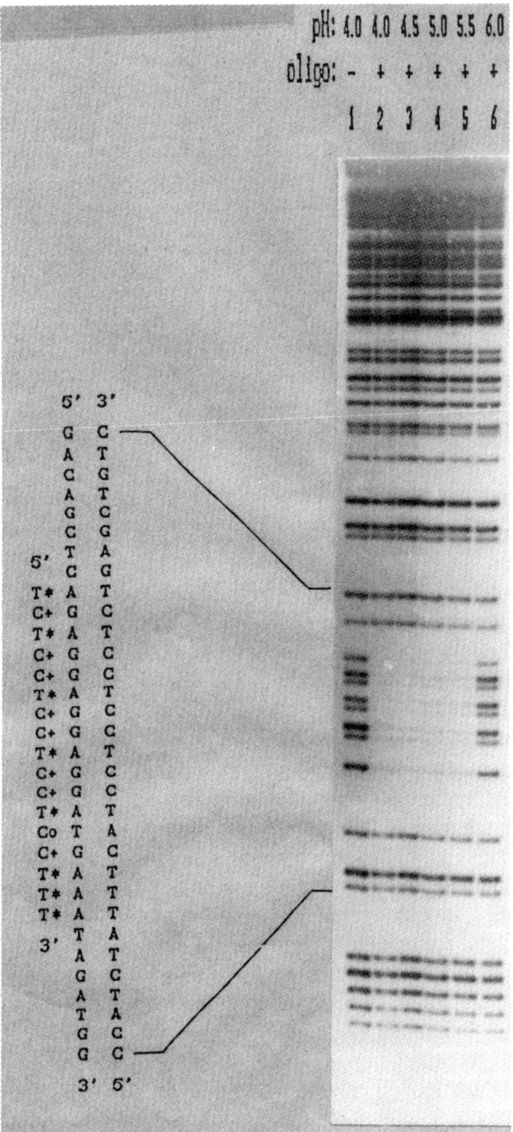

FIG. 17. The result of a footprinting experiment with dimethyl sulfate of a complex of duplex DNA carrying a homopurine-homopyrimidine insert with corresponding pyrimidine oligonucleotide. The data are from Cherny *et al.* (1993).

of movement of the replication machinery along DNA. The gaps between fragments formed on the second strand are filled by DNA ligase.

3.1.2 Replication of Viruses The above scheme is universal for cells. Viruses, the cellular parasites, cannot reproduce themselves autonomously. They penetrate the cell and

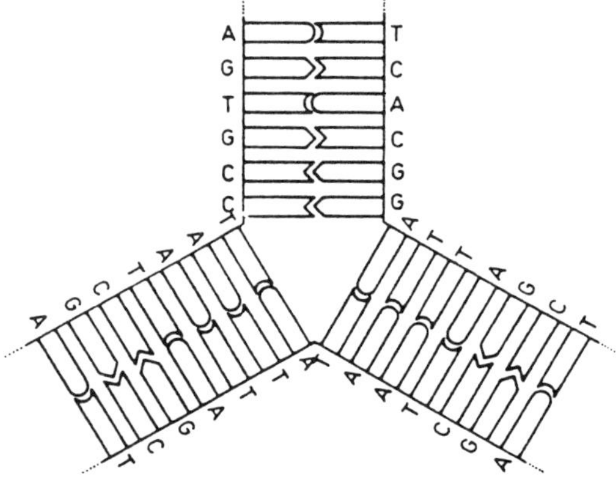

FIG. 18. Replication of duplex DNA.

replicate using the cellular resources. Viruses exhibit extreme versatility in the forms of storage of genetic information and, accordingly, in the fashion of replication. The genetic material of viruses can be in the form of linear duplex DNA, closed circular duplex DNA, linear single-stranded DNA, closed single-stranded DNA, or, last, linear single-stranded RNA.

Some RNA-containing viruses replicate with the help of RNA-dependent RNA polymerase. In many cases, however, first DNA is synthesized on the RNA template by the special enzyme reverse transcriptase, or RNA-dependent DNA polymerase. Then the closed circular double-stranded, so-called replicative, form of DNA is obtained. Then DNA replicates, and, in the final stage, virus RNA is synthesized on the DNA template. This complicated process is typical for retroviruses, the most important class of viruses from the medical viewpoint, which comprises many oncogenic (cancer inducing) viruses and HIV, the AIDS virus.

3.2 Transcription

Genetic information is expressed via the process of protein synthesis. The first step of this process is synthesis of the RNA copy of a gene, i.e., of messenger RNA (mRNA). This process is significantly different for prokaryotes and eukaryotes.

3.2.1 Transcription in Prokaryotes Prokaryotes generally obey the central dogma of molecular biology mentioned in the beginning of this chapter. The special enzyme RNA polymerase recognizes its binding sites on duplex DNA (promoters), binds to them, and synthesizes the full-length copy of the gene. The process is regulated by various mechanisms. Positive regulation occurs either via synthesis of entirely new RNA polymerase, which recognizes other promoters, or via protein factors, which bind with RNA polymerase and change its specificity. Negative regulation occurs via special sites on DNA (operators) to which proteins, called repressors, can bind. This binding prevents RNA polymerase from synthesizing RNA.

Messenger RNA serves as a template for protein synthesis on ribosomes. In prokaryotes, these two processes are coupled (see Fig. 19). The transcription process is often associated with formation of positive supercoils ahead of RNA polymerase and negative supercoils behind it. These topological problems are resolved by topoisomerases (Cozzarelli and Wang, 1990).

3.2.2 Transcription in Eukaryotes In eukaryotes, the process of synthesis of messenger RNA is much more complicated than in prokaryotes. First of all, in eukaryotes, DNA sequences corresponding to protein molecules are not continuous. They are interrupted by other sequences. The pieces corresponding to proteins have been named exons. The interrupting sequences were named introns. The role of introns still remains obscure, although the intron-exon or-

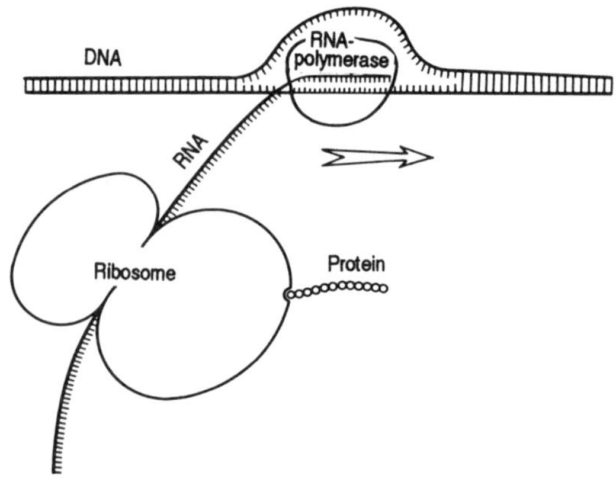

FIG. 19. RNA polymerase moves along DNA, synthesizing RNA. The ribosome copies information from RNA by synthesizing protein, in accordance with the genetic code.

ganization of eukaryotic genes has been known since the late 1970s.

The eukaryotic cell has several different RNA polymerases and numerous proteins (transcription factors) that modulate their activity and specificity. As a result, the mRNA synthesis is carried out by a complicated machinery consisting of many protein molecules.

The full-length copy of the entire sequence of the gene, including all its exons and introns, is synthesized by the transcription machinery. This full-length copy of the gene, which is called pre-mRNA, is converted into mRNA through a special maturation process, or splicing. In the course of splicing, all introns are deleted, and exons are "spliced" to form the maturated molecule of mRNA with the sequence corresponding to the amino-acid sequence of the protein molecule.

This ready-made mRNA is transported, in the form of a complex with special proteins, from the nucleus to the cytoplasm, where it strips the transporting proteins and serves as a template in the process of protein synthesis on ribosomes, i.e., translation. Thus, in contrast to prokaryotes, in eukaryotes transcription and translation are completely separated from each other.

Protein synthesis is extensively regulated on the mRNA level. Special protein factors, which bind with mRNA recognizing its various structural elements, modulate the rate of degradation of mRNA and its affinity to ribosomes. Thus, the yield of a particular protein is regulated the entire way from DNA to the spot where the protein molecule is synthesized, and even beyond this (see the next section).

3.3 Translation Process and Genetic Code

The translation process is the final stage of the protein synthesis. In this process, the genetic code plays the key role. The message brought by mRNA in the form of a nucleotide text is read by a ribosome, and the corresponding protein molecule is synthesized. Because a nucleotide text consists of four different letters, whereas sequences of proteins consist of 20 different amino-acid residues, the genetic code consists of triplets; i.e., one amino acid is encoded by a sequence of three adjacent nucleotides, which are called codons. The entire number of codons is $4^3 = 64$. Therefore, the genetic code is degenerate, several codons corresponding to one amino acid. There are also termination codons. There are no initiation codons, their role under appropriate conditions being played by the codons AUG and GUG, which normally correspond to methionine and valine. Figure 20 shows the genetic code.

In the realization of the translation process, the tRNA molecules play the key role. Each tRNA molecule carries an anticodon, the triplet complementary to the codon. A special enzyme recognizes the particular tRNA molecule and "charges" it with the corresponding amino acid. The ribosome creates conditions for binding (via codon-anticodon complementary recognition) of mRNA codons with the charged tRNA molecules and then adds to the growing protein chain the amino-acid residue brought by the tRNA molecule.

After the protein chain is synthesized, special large protein molecules, called chaperones, create conditions for folding of the chain into a biologically active spatial structure. After this, the protein molecule is ready to fulfill its function. However, even after the protein molecule is ready, its rate of degradation continues to be controlled by the cell.

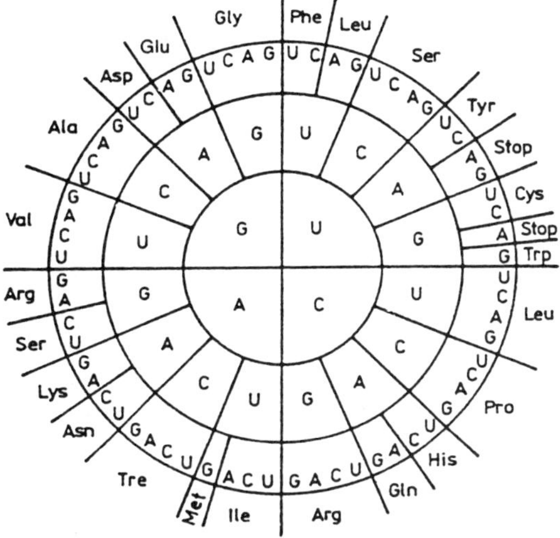

FIG. 20. The genetic code. The first letter of the codon is located in the central ring, the second in the first ring, and the third in the second. Written in the outer ring are the abbreviated designations of amino acids.

4. BIOTECHNOLOGY

The advent of the biotechnology era is the result of the entire accumulated knowledge about nucleic acids and the enzymes working with them. However, among countless achievements that have led to biotechnology and genetic engineering, there are a very few that have most contributed to the elaboration of techniques permitting interference into the most basic processes of life. In this section, we briefly survey these landmarks and then outline the major achievements and prospective of biotechnology.

4.1 Major Tools of Genetic Engineering

The biotechnology revolution is grounded on genetic engineering, a set of relatively simple chemical, biochemical, and biophysical techniques that have been developed during the past two decades. Some of them, like gel electrophoresis, are explained in other sections of this article. In this section, we briefly survey other very significant methods of genetic engineering.

4.1.1 Chemical Synthesis of DNA Achievements in the chemical synthesis of DNA had led, by the early 1980s, to the creation of robots that quickly and efficiently synthesize DNA oligonucleotides containing up to several dozen residues. This opened the way for obtaining synthetic genes. However, the most widespread application of DNA synthesis consists in the creation of so-called DNA probes, relatively short pieces that serve as primers for polymerases in numerous projects, most notably in PCR machines (see below).

4.1.2 Restriction Enzymes These enzymes recognize short sequences of duplex DNA (usually containing six base pairs) and specifically cut the duplex at these sites. Restriction enzymes are, along with gel electrophoresis (see Sec. 2.2.1), the major tools of genetic engineering.

4.1.3 Reverse Transcriptase The intron-exon organization of the genes of eukaryotes prevents their direct utilization in genetic engineering. When transferred to prokaryotes, these genes just do not express because, in prokaryotes, the mRNA maturation process cannot occur. Therefore, eukaryotic genes suitable for transfer to prokaryotes should not contain introns. This is often achieved by extracting maturated mRNA and synthesizing on it the corresponding DNA with the help of reverse transcriptase (RNA-dependent DNA-polymerase).

4.1.4 Plasmids Usually, alien genes are introduced not into the chromosome of a prokaryotic cell but into special small circular molecules, plasmids, which multiply together with the cell. Genetic engineers have constructed many artificial plasmids, which carry multiple recognition sites for restriction enzymes. When DNA fragments obtained as the result of cutting by a restriction enzyme are mixed with a plasmid cut with the same enzyme, the result is new plasmids carrying the fragments inserted into them. This is most common way of obtaining recombinant DNA molecules.

4.1.5 Polymerase Chain Reaction (PCR) This technique has revolutionized genetic engineering in recent years and had an impact comparable to that of the discovery of restriction enzymes and the implementation DNA sequencing methods. It makes it possible to multiply pieces of DNA starting from literally a single DNA copy.

The amplification is performed in a special PCR machine. The machine is loaded with the DNA molecule containing the piece to be amplified (it could be really only *one* DNA molecule to begin with), thermally stable *Taq* DNA-polymerase, and two DNA primers specially synthesized for the particular piece. One primer is complementary to the left-hand flanking sequence of the piece, the second one to the right-hand flanking sequence, but to the opposite strand. The PCR machine works as a cycler, each cycle consisting in heating the sample up to subboiling temperature (94 °C) with subsequent cooling down to 72 °C (the optimal temperature for *Taq* polymerase isolated from thermophilic bacteria). As a result of the first heating, the original DNA molecule melts, and the two strands separate (see Sec. 1.4.1). The subsequent cooling entails the binding of the primers, which are present in macroscopic quantity, to the complementary sites on the two single-stranded DNA molecules. The polymerase immediately starts to extend the primers from their 3′ ends using the complementary strands as the templates. Thus, the first cycle ends up with two partially double-stranded DNA molecules. The common duplex part of these two molecules encompasses only the piece to be amplified plus the two flanking primers. In the subsequent heating-cooling cycles, only this common part is amplified. After n cycles, one obtains 2^n duplex molecules corresponding to the piece flanked by the two primers.

4.2 The Coming of the Biotechnology Age

Biotechnology applications of the methods of genetic engineering are truly breathtaking and hardly embraceable. We provide a very brief survey of only some of them.

4.2.1 Protein Superproduction By incorporating genes of virtually any organism into bacteria, genetic engineers have made it possible to produce any protein in considerable quantities. On the one hand, this has enormously stimulated basic research on proteins and their functions, and, on the other hand, it has opened fantastic opportunities for producing proteins for medicine.

4.2.2 Protein Engineering A special technique has been elaborated to substitute nucleotides in cloned genes. The method consists in synthesis of a DNA probe that embraces the site to be changed and carries the desired substitution. This oligonucleotide is annealed with the single-stranded copy of the gene cloned in a special vector DNA. Then DNA polymerase is used to synthesize the double-stranded DNA, which will carry a mismatch in the site of substitution. After replication of the vector within the cell, two versions of the gene are obtained: the original, wild-type copy and the changed, mutant copy. The mutant copy is selected by special technique, again using the DNA probe.

As a result of the advent of protein engineering, unprecedented opportunities have appeared to change purposefully the amino-acid sequences of proteins. This has completely changed the way to study protein structure and function.

4.2.3 New Pharmacology Genetic engineering has made it possible to obtain in large quantities any proteins, even those that are present in the entire human body in very few copies. This opportunity is widely used by pharmaceutical companies to produce a rapidly increasing variety of biologically active proteins unavailable before, such as human growth hormone, human insulin, human interferon, etc.

4.2.4 Forensic Applications The implementation of PCR and other routine techniques of manipulation with DNA promises to revolutionize forensic science. Indeed, all people, with the exception of identical twins, differ from each other in a significant part of their DNA texts. So if a criminal left a tiny portion of his cells in any form (blood, sperm, etc), his DNA can be multiplied by PCR and then analyzed and compared with the DNA analysis of suspects. Although the new DNA forensic methods still need development in order to be widely applied, there is no doubt that they will become very useful in the nearest future.

4.2.5 Transgenic Animals Using microinjections of alien DNA into the cell nucleus and other methods, researchers routinely modify cells' genetic material. If cells are modified at the early stage of their differentiation, entire animals with directionally changed genomes can be produced. This has opened breathtaking opportunities for studying the role of separate genes on the level of the entire organism. Special laboratory animals with significant modifications of heredity, which are not encountered in natural animals, have been engineered. This can lead to a revolution in methods of creation of new species of domestic animals.

4.2.6 Human Genome Project The most ambitious idea that has emerged from achievements in nucleic acid research was sequencing the entire human genome, i.e., the Human Genome Project. The Human Genome Project is expected to be completed in the very early years of the next century. This landmark achievement will undoubtedly lead to a new revolution in medicine.

4.2.7 Diagnostics and Correction of Genetic Defects As a result of the Human Genome Project and related studies on human genetics, remarkable progress is underway in the tracing and sequencing of genes responsible for various hereditary diseases. In parallel, more data have been accumulating about a substantial genetic component in such very common diseases as cancer and atherosclerosis. As a result, the coming revolution in medicine is foreseen as the development and implementation in humans of methods of diagnosis and correction of wrong genes. Very efficient diagnostic methods have been developed and implemented for some genetic defects—for instance, in the case of sickle-cell anemia.

However, efficient methods to correct the established genetic defects are still lacking. Their development is the most challenging unsolved problem in the field of nucleic acids.

GLOSSARY

Adenine: A chemical group that is part of DNA and RNA. It is one of the four bases of nucleic acids. The abbreviated designation is A.

Amino Acid: A chemical compound with the structure H_2N–CHR–COOH, where R is any radical. Amino acid is the starting product for protein synthesis.

Amino-Acid Residue: A chemical group with the structure –HN–CHR–CO–, which is a residue of protein chain. It is what remains of an amino acid when it is built into the protein chain.

Base (Nucleic or Nitrous): A class of chemical compounds that includes adenine, guanine, thymine, cytosine, and uracil.

Chaperones: Specialized proteins that carry out inside the cell the folding of a polyamino-acid chain, newly synthesized on a ribosome, into a native protein molecule.

Chimerical DNA: An artificial molecule, put together by genetic engineering techniques with sections of different natural DNA. The terms "hybrid" and "recombinant" DNA have the same meaning.

Codon: A term connected with the genetic code. Denotes a trinucleotide corresponding to one amino-acid residue. There are several meaningless (nonsense) codons not corresponding to any amino acid. They play the role of stop signals during protein synthesis by mRNA on ribosomes. They are called terminating codons. Initiating codons serve as signals for starting protein synthesis.

Coil (Polymer): A notion of polymer physics. Serves to denote the form a polymer molecule assumes in space. In consequence of thermal motion, the form of a polymer coil constantly changes.

Cruciform: A DNA structure that can be formed in palindrome sequences.

Cytosine: A chemical group that is part of DNA and RNA. One of the four bases of nucleic acids. The abbreviated designation is C.

Degeneracy of the Code: One of the properties of the genetic code, which is that several codons may correspond to the same amino acid.

Deoxyribonucleic Acid: The full name of the DNA molecule.

DNA-Polymerase: An enzyme responsible for DNA synthesis on a DNA template. The process is called DNA replication.

Eukaryotes: Organisms provided with a cell nucleus.

Exon: A DNA segment that stores information about a part of the amino-acid sequence of protein.

Gel: A polymer network saturated with a solvent.

Gene: The principal notion of classical genetics which for a long time understood the term to mean an indivisible particle of heredity. During the 1950s and 1960s, the word "gene" was taken to mean a continuous DNA section, recorded on which, in the form of a nucleotide sequence, is information about the amino-acid sequence of one protein. The word is still used as the name for a DNA segment. But in some cases, it denotes a continuous segment that corresponds to only a part of the protein chain, while in others, it means a set of segments corresponding to a complete protein molecule. And it may well be that the same DNA segment simultaneously belongs to two and even three genes.

Genome: The entirety of genetic information of the organism.

Guanine: A chemical group that is part of DNA and RNA. It is one of the four nucleic acid bases. The abbreviated designation is G.

Hybrid DNA: An artificial molecule put together through genetic engineering techniques with segments of different natural DNA. The terms "recombinant" and "chimerical" DNA have the same meaning.

Intron: A DNA section dividing exons.

Ligase: An enzyme curing ruptures in DNA.

Linking Number: A quantitative characteristic of the degree to which two contours are linked. The linking number equals the number of times that one contour pierces the surface spread tightly across the other contour. It is denoted by *Lk*.

Nuclease: An enzyme that splits DNA or RNA.

Nucleotide: A residue of DNA and RNA.

Oligonucleotide: A short single-stranded piece of nucleic acid.

Oncogene: A gene causing cancer.

Palindrome: A phrase that reads the same from left to right or from right to left. In the context of DNA texts, such palindromes are called "mirror repeats." A real DNA palindrome is a segment of double helix that has the same sequence when reading either strand in the same direction, dictated by the chemical structure of DNA strands—for example,

$$\begin{array}{c} \rightarrow \\ \text{ATGCGCAT} \\ \cdots\cdots \\ \text{TACGCGTA} \\ \leftarrow \end{array}$$

The notion is crucial for the theoretical understanding of DNA separation in gel.

Plasmid: A circular DNA molecule multiplying together with bacteria and capable of passing from cell to cell.

Primer: A single-stranded oligonucleotide that binds, via complementary pairing, to DNA or RNA and serves for the priming of polymerases working on both DNA and RNA.

Prokaryotes: Single-cell organisms with no cell nuclei.

Promoter: A DNA segment with which RNA polymerase binds to set in motion mRNA synthesis.

Purine: A class of chemical compounds that includes adenine and guanine.

Pyrimidine: A class of chemical compounds that includes thymine, uracil, and cytosine.

Recombinant DNA: An artificial molecule put together by genetic engineering techniques from segments of different natural DNA. The terms "hybrid" and "chimeric" DNA have the same meaning.

Replication: Doubling of genetic material. DNA synthesis on DNA.

Repressor: A protein binding very strongly with an appropriate DNA segment between the promoter and the gene itself. By associating with DNA, repressor prevents the progress of RNA polymerase from promoter to gene and thus blocks mRNA synthesis. Serves to regulate transcription.

Reptation: Snakelike movement of a polymer molecule through the polymer network.

Restriction Endonuclease: An enzyme cutting the double helix in places with a definite nucleotide sequence. It is the chief instrument of genetic engineering, and often is called just restriction enzyme.

Restriction Fragment: A piece of DNA cut out of the molecule with the help of restriction endonucleases.

Reverse Transcriptase: An enzyme responsible for DNA synthesis on an RNA template. The process is called reverse transcription.

Ribosome: An involved complex of RNA and proteins, which is responsible for the translation process in the cell.

RNA Polymerase: An enzyme synthesizing mRNA on a DNA template. Carries out the transcription process.

Splicing: The process of mRNA maturing in eukaryotes, which results in the discharge of introns, while exons combine into one RNA chain.

Superhelicity: A characteristic of a circular closed DNA. Superhelicity arises when the linking number Lk in the DNA differs from the value N/Γ_0, where N is the number of base pairs in the DNA and Γ_0 is the number of base pairs per turn of double helix in linear DNA placed in the same conditions.

Template: A polymer molecule whose sequence is used for charting a sequence for another polymer molecule. DNA serves as a template for DNA synthesis during replication and for RNA during transcription. RNA serves as a template for protein synthesis during translation and for DNA during reverse transcription.

Topoisomerases: A class of enzymes that change the topology of a circular closed DNA.

Topoisomers: DNA molecules identical in chemical terms but different in topology (by the type of knot, or by the linking number).

Transcription: RNA synthesis on a DNA template. Reverse transcription is DNA synthesis on a RNA template.

Translation: Protein synthesis by an mRNA template on ribosome.

Thymine: A chemical group that is part of DNA. Is one of four DNA bases. The abbreviated name is T.

Uracil: A chemical group that is part of RNA. It is one of four RNA bases. The abbreviated designation is U.

Vector: A genetic engineering term. This is the name of a carrier DNA molecule (plasmid, virus, and so on), within which the desired gene is cloned.

Virus: A cellular parasite, one of the simplest objects of living nature. Outside the cell, a virus is a molecular complex consisting of nucleic acid (DNA, sometimes RNA) and several proteins forming the virus membrane. The cell's penetration by a virus (or its nucleic acid) switches the resources of the cell over to synthesizing the virus nucleic acid and proteins. When the cell's resources are exhausted, its membrane bursts, spilling out "ready-made" virus particles. Animals' viruses have a much simpler structure than bacteria viruses (bacteriophages). Animal viruses are incapable of injecting their nucleic acid into the cell, and they get inside the cell together with food. Viruses cause many infectious diseases, such as the grippe, smallpox, poliomyelitis, hepatitis, and AIDS. In some cases, finding itself inside the cell, the virus does not undo it but builds its DNA into that of the cell, after which the virus DNA begins to multiply together with the cell DNA. The behavior of the cell itself may, however, undergo a sharp change in the process.

Works Cited

Anshelevich, V. V., Vologodskii, A. V., Lukashin, A. V., Frank-Kamenetskii, M. D. (1984), *Biopolymers* **23**, 39–58.

Bustamante, C., Keller, D., Yang, G. (1993), *Curr. Opinion Struct. Biol.* **3**, 363–372.

Cherny, D. I., Malkov, V. A., Volodin, A. A., Frank-Kamenetskii, M. D. (1993), *J. Mol. Biol.* **230**, 379–383.

Cozzarelli, N. R., Wang, J. C. (Eds.) (1990), *DNA Topology and Its Biological Effects*, Cold Spring Harbor, NY: Cold Spring Harbor Laboratory.

Dickerson, R. E. (1992), in: D. M. J. Lilley, J. E. Dahlberg (Eds.), *DNA Structures. Part A. Synthesis and Physical Analysis of DNA*, Methods in Enzymology Vol. 211, San Diego: Academic, pp. 67–111.

Dubochet J., Adrian, M., Dustin, I., Furrer, P., Stasiak, A. (1992), in: D. M. J. Lilley, J. E. Dahlberg (Eds.), *DNA Structures. Part A. Synthesis and Physical Analysis of DNA*, Methods in Enzymology Vol. 211, San Diego: Academic, pp. 507–518.

Dunlap, D. D., Garcia, R., Schabtach, E., Bustamante, C. (1993), *Proc. Natl. Acad. Sci. USA* **90**, 7652–7655.

Felsenfeld, G., Davies, D. R., Rich, A. (1957), *J. Am. Chem. Soc.* **79**, 2023–2025.

Frank-Kamenetskii, M. D., Anshelevich, V. V., Lukashin, A. V. (1987), *Usp. Fiz. Nauk* **151**, 595–618. [*Sov. Phys. Usp.* **30**, 317–330].

Frank-Kamenetskii, M. D. (1990a), in: W. Saenger (Ed.), *Landolt-Börnstein Series, Group VII: Biophysics, Nucleic Acids, Subvol. 1C*, Heidelberg: Springer, pp. 228–240.

Frank-Kamenetskii, M. D. (1990b), in: N. R. Cozzarelli, J. C. Wang (Eds.), *DNA Topology and Its Biological Effects*, Cold Spring Harbor, NY: Cold Spring Harbor Laboratory, pp. 185–216.

Gueron, M., Kochoyan, M., Leroy, J.-L. (1987), *Nature* **328**, 89–92.

Gehring, K., Leroy, Y.-L., Gueron, M. (1993), *Nature* **363**, 561–565.

Ivanov, V. I., Krylov, D. Yu. (1992), in: D. M. J. Lilley, J. E. Dahlberg (Eds.), *DNA Structures. Part A. Synthesis and Physical Analysis of DNA*, Methods in Enzymology Vol. 211, San Diego: Academic, pp. 111–127.

Johnson, W. C. (1990), in: W. Saenger (Ed.), *Landolt-Börnstein Series, Group VII: Biophysics, Nucleic Acids, Subvol. 1C*, Heidelberg: Springer, pp. 1–59.

Kang, C. H., Zhang, X., Ratliff, R., Moyzis, R., Rich, A. (1992), *Nature* **356**, 126–131.

Klenin, K. V., Vologodskii, A. V., Anshelevich, V. V., Dykhne, A. M., Frank-Kamenetskii, M. D. (1988), *J. Biomol. Struct. Dyn.* **5**, 1173–1185.

Lilley, D. M. J., Dahlberg, J. E. (Eds.) (1992a), *DNA Structures. Part A. Synthesis and Physical Analysis of DNA*, Methods in Enzymology Vol. 211, San Diego: Academic.

Lilley, D. M. J., Dahlberg, J. E. (Eds.) (1992b), *DNA Structures. Part B. Chemical and Electrophoretic Analysis of DNA*, Methods in Enzymology Vol. 212, San Diego: Academic.

Maxam, A. M., Gilbert, W. (1977), *Proc. Natl. Acad. Sci. USA* **74**, 560–564.

Rajagopal, P., Feigon, J. (1989), *Nature* **339**, 637–640.

Roberts, R. W., Crothers, D. M. (1992), *Science* **258**, 1463–1466.

Sanger, F., Nicklen, S., Coulson, A. R. (1977), *Proc. Natl. Acad. Sci. USA* **74**, 5463–5466.

Schwartz, D. C., Cantor, C. R. (1984), *Cell* **37**, 67–75.

Shaw, S. Y., Wang, J. C. (1993), *Science* **260**, 533–536.

Vologodskii, A. V., Levene, S. D., Klenin, K. V., Frank-Kamenetskii, M. D., Cozzarelli, N. R. (1992), *J. Mol. Biol.* **227**, 1224–1243.

Wada, A., Suyama, A. (1986), *Prog. Biophys. Mol. Biol.* **47**, 113–157.

Wang, A. H.-J., Quigley, G. J., Kolpak, F. K, Crawford, J. L., van Boom, J. H., van der Marel, G., Rich, A. (1979), *Nature* **282**, 680–685.

Watson, J. D., Crick, F. H. C. (1953), *Nature* **171**, 737–738.

Further Reading

Alberts, B., Bray, D., Lewis, J., Raff, M., Roberts, K., Watson, J. D. (1989), *Molecular Biology of the Cell*, 2nd ed., New York: Garland.

Cantor, C. R., Schimmel, P. R. (1980), *Biophysical Chemistry*, Parts 1–3, San Francisco: Freeman.

Cozzarelli, N. R., Wang, J. C. (Eds.) (1990), *DNA Topology and Its Biological Effects*, Cold Spring Harbor, NY: Cold Spring Harbor Laboratory.

Frank-Kamenetskii, M. D. (1993), *Unraveling DNA*, New York: VCH.

Lewin, B. (1990), *Genes IV*, Oxford: Oxford University.

Lilley, D. M. J., Dahlberg, J. E. (Eds.) (1992), *DNA Structures. Part A. Synthesis and Physical Analysis of DNA*, Methods in Enzymology Vol. 211, San Diego: Academic.

Lilley, D. M. J., Dahlberg, J. E. (Eds.) (1992), *DNA Structures. Part B. Chemical and Electrophoretic Analysis of DNA*, Methods in Enzymology Vol. 212, San Diego: Academic.

Ptashne, M. (1992), *A Genetic Switch*, 2nd ed., Palo Alto, CA: Cell Press & Blackwell.

Watson, J. D., Gilman, M., Witkowski, J., Zoller, M. (1992), *Recombinant DNA*, 2nd ed., Scientific American Books, New York: Freeman.

Watson, J. D., Hopkins, N. H., Roberts, J. W., Steitz, J. A., Weiner, A. M. (1987), *Molecular Biology of the Gene*, 4th ed., Menlo Park, CA: Benjamin/Cummings.

Weaver, R. F., Hedrick, P. W. (1992), *Genetics*, 2nd ed., Dubuque, IA: Wm. C. Brown.

NUMERICAL METHODS

See SCIENTIFIC COMPUTING BY NUMERICAL METHODS

OCCUPATIONAL HEALTH AND SAFETY

See SAFETY AND HEALTH IN THE WORKPLACE

OCEANOGRAPHY

Curtis A. Collins and Peter C. Chu, *Naval Postgraduate School, Monterey, California, U.S.A.*

INTRODUCTION

Oceanography deals with the application of fundamental scientific principles to the ocean environment. This results in a variety of subfields in oceanography (physical, chemical, and biological oceanography and marine geology), which must often be integrated to understand ocean processes. This report will focus on physical oceanography. Physical oceanography is the study of the physical structure, motion, and mixing of ocean waters.

The two traditional approaches used in

3-527-28134-7/95/$5.00 + .50

physical oceanography are the descriptive approach and the theoretical approach. The descriptive approach uses observations to determine the structure of the ocean while the theoretical approach applies Newton's laws of motion and thermodynamical laws to understand how forces affect the structure of the ocean. These two approaches complement one another.

As a recognizable scientific discipline, oceanography is young, beginning with the voyage of *HMS Challenger* from 1872 to 1876. The first systematic large-scale expedition to study primarily the physical aspects of the ocean circulation was the German *Meteor* expedition to study the South Atlantic from 1925 to 1927. Worldwide surveys of the ocean were conducted during the International Geophysical Year in 1957–1958, and the first satellite mission that was dedicated to studies of the ocean was *Seasat*, which observed the oceans for 100 days in 1978.

This report is organized into nine sections. First, we describe the physical characteristics of seawater, then the character of the ocean's waters and currents, and then the instruments used in the practice of physical oceanography. Next, we introduce marine hydrodynamics and then develop the physics of the mean ocean circulation. Ocean waves are then described. The final sections deal with mesoscale motions, small-scale processes, and concluding remarks. A list of references and additional reading is provided for those who need more detailed information.

1. PHYSICAL CHARACTERISTICS OF SEAWATER

The density of water is determined by the temperature T, salinity S, and pressure p. Temperature is measured in degrees Celsius. Salinity represents the weight of dissolved solids (in grams) in a kilogram of seawater. In 1978, oceanographers adopted the Practical Salinity Scale, which defined a practical salinity unit (psu) in terms of the conductivity of seawater at 15 °C referenced to the conductivity of a potassium chloride solution at 15 °C (see UNESCO, 1981). This provided an accurate, reproducible primary standard and a better measure of the effect of the dissolved solids on density, and supported *in situ* measurement methods. The unit of pressure used by oceanographers is the decibar, db, where 1 db = 10 000 Pa.

In addition to modifying the conductivity of water, salt modifies the physical characteristics of water in another important way. The temperature of maximum density of seawater decreases with increasing salinity at a rate greater than the freezing point, so that for salinity greater than 24.7 psu, seawater freezes before the maximum density is reached. Other physical properties of water, e.g., viscosity, surface tension, and thermal conductivity, are also modified.

Typical ranges of salinity and temperature found at the sea surface in the open ocean are 32–37 psu and −1 to 29 °C, respectively. Salinities lower than these values are associated with melting ice, rivers, or intense rainstorms. Larger salinities are encountered in semienclosed evaporative basins such as the Red Sea or Mediterranean. Since water is slightly compressible, *in situ* temperature increases below about 3500 db because of adiabatic effects. These adiabatic changes can be taken into account by using a potential temperature, θ, which nominally uses the sea surface as a reference pressure. The mean temperature and salinity for the world ocean are 3.5 °C and 34.7 psu.

1.1 Equation of State

The equation relating density to temperature, salinity, and pressure is called the *equation of state*. The equation of state in current use, EOS80, was established by the Joint Panel on Oceanographic Tables and Standards (UNESCO, 1981) and uses the functional form

$$\rho(S,T,p) = \frac{\rho(S,T,0)}{1 - p/K(S,T,p)}, \qquad (1)$$

where $\rho(S,T,0)$ is the density of seawater at one standard atmosphere and $K(S,T,p)$ is the bulk modulus, and both are nonlinear functions of S, T, and p. Typical density values for the ocean range from 1020 to 1060 kg/m^3. The density is usually represented by a density anomaly γ:

$$\gamma(S,T,p) = \rho(S,T,p) - 1000 \text{ kg/m}^3. \qquad (2)$$

The nonlinearity of the equation of state

has several important properties. These can be deduced from the curved shape of constant-density lines on the *T-S* diagram (Fig. 1). At temperatures close to zero, the constant-density lines become nearly vertical, indicating that salinity changes have a much greater effect on density than they do at warmer temperatures. At warmer temperatures, the slope of these constant-density lines decreases, indicating that temperature changes contribute more to density variability. When two different water masses of the same density but differing temperature and salinity are mixed, temperature and salinity are conserved so that the resulting water mass will lie on a line connecting these two points on the *T-S* diagram; but, as indicated by the curve of the constant-density lines, the resulting mixture will be heavier (more dense) than the original waters. This provides a mechanism of generating internal buoyancy changes that is called either cabelling or thermobaric instability.

1.2 Hydrostatic Equation

In most oceanographic applications, vertical accelerations can be neglected so that pressure and depth z are related by the hydrostatic equation

$$\frac{\partial p}{\partial z} = -\rho g, \tag{3}$$

where g represents gravitational acceleration. Zonal variations of g are usually considered, and g is estimated as follows:

$$g(\phi) = 9.790\,318(1.0 + 5.2788 \times 10^{-3} \sin^2\phi + 2.36 \times 10^{-5} \sin^4\phi) \text{ m}^2\text{s}, \tag{4}$$

where ϕ is latitude (in degrees). Note that 1 db corresponds to about 1 m and that 1010 db is about 1000 m.

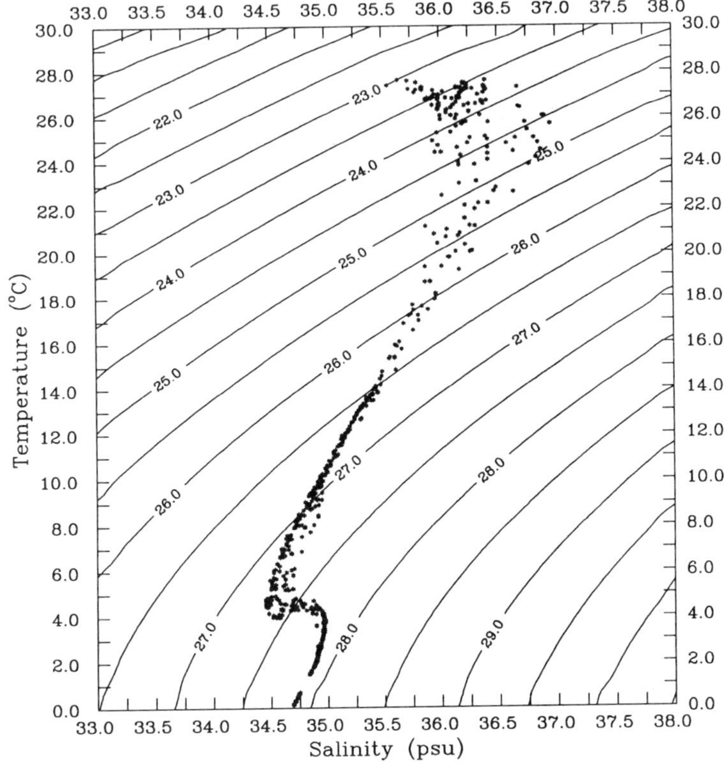

FIG. 1. *T-S* diagram for waters of the Tropical Atlantic along 23° W from 6° N to 9° S. Readily identifiable water masses include subtropical subsurface water at 26 °C, 36.5 psu, Antarctic intermediate water at 4 °C, 34.5 psu, North Atlantic deep water at 4 °C, 35 psu, and Antarctic bottom water at 2 °C, 35 psu. The curved lines are contours of a density anomaly, γ_t which is referenced to atmospheric pressure and has units of kg/m^3.

1.3 Sound in the Ocean

The speed of sound in seawater, U_a, in units of m/s, is given by

$$U_a^2 = \left(\frac{\partial p}{\partial \rho}\right)_a, \tag{5}$$

where $(\partial p/\partial \rho)_a$ is the reverse adiabatic density gradient. In practice, the speed of sound is usually not computed in this manner; instead, empirical formulas are used that express sound velocity as a function of S, T, and p and that are derived from laboratory measurements. Typical sound-speed values range from 1400 to 1700 m/s. To a first approximation, sound is nondispersive in the ocean. Refraction of sound is important, creating both convergence and shadow zones. A simple model for the attenuation of sound (which does not take into account geometric spreading, refractive loss, and volume and boundary scattering loss) is

$$I(x) = I_0 e^{-\kappa x}, \tag{6}$$

where $I(x)$ is the intensity of sound at distance x from the source and I_0 is the intensity of sound at the same frequency at the location of the source. The attenuation coefficient κ, in units of dB/m, is given as

$$\kappa = 4.343\left[\frac{1.7 \times 10^8(\frac{4}{3}\mu_F + \mu_F')f^2}{\rho_F c_F^3} + \left(\frac{SA' f_{rm} f^2}{f^2 + f_{rm}^2}\right)(1 - 1.25 \times 10^{-2} p) + \frac{A'' f_{ra} f^2}{f^2 + f_{ra}^2}\right],$$

where μ_F is the dynamic coefficient of shear viscosity for fresh water, μ_F' is the dynamic coefficient of bulk viscosity for fresh water, f_{rm} is the relaxation frequency (kHz) for magnesium sulfate, f_{ra} is the relaxation frequency (kHz) for boric acid, A' is 2.03×10^{-5} dB/(kHz psu m), A'' is 1.2×10^{-4} dB/(kHz m), S is salinity (psu), and f is the frequency in kHz. At intermediate frequencies (100–1000 kHz), absorption is mainly due to boric acid and magnesium sulfate chemical relaxation processes. At low frequencies (below 100 Hz), attenuation is almost independent of frequency, and the attenuation coefficient is about 0.003 dB/km. This means that low frequencies can easily insonify the ocean, i.e., 15 W of transmitted power can be detected at a distance of 2000 km. In part, this is because sound speed increases with increasing pressure, creating a duct that focuses acoustic energy (called the SOund Fixing And Ranging or SOFAR channel) at about 1-km depth.

Current applications include not only finding submarines and fish, and mapping the seafloor, but inverse problems including acoustic tomography and ocean acoustic thermometry. The latter application may lead to the first quantitative measure of "greenhouse" warming of the ocean.

1.4 Light in the Ocean

The behavior of light in the ocean affects not only the heating of the upper layer but also the growth of plants (and hence the oceanic food chain). Light also has marine applications for charting, and finding fish, submarines, and mines, or measuring bioluminescence.

Seawater has both inherent and apparent optical properties (see OPTICS, UNDERWATER). Inherent properties can be measured on a seawater sample in the laboratory (if, of course, suspended particles, etc., were not affected by removal from the sea). Apparent properties must be measured *in situ*. The inherent properties include the beam attenuation coefficient c, the volume absorption coefficient a, and the volume scattering coefficient b, where $c = a + b$. A beam transmissometer is used to measure c (m^{-1}) over a fixed distance, so that for 1 m

$$T = I_1/I_0 = e^{-c}; \tag{7}$$

and c may be written

$$c = c_\lambda = a_{\lambda w} + a_{\lambda d} + a_{\lambda p} + b_{\lambda w} + b_{\lambda d} + b_{\lambda p}, \tag{8}$$

where the subscript λ indicates dependence on wavelength, w indicates pure water, d is dissolved matter, and p is suspended particles.

Apparent properties include the radiance (W m^{-2}) and the diffuse attenuation coefficient for downwelling (indicated by a superscript consisting of an arrow pointing downward) irradiance, $k_\lambda^\downarrow$ (W m^{-2}), which are both strongly λ dependent. Downwelling irradiance attenuation $k_\lambda^\downarrow(z)$, where z is depth, better represents the depth to which a lidar can detect an object than does $c_\lambda(z)$. It is always true that $a_\lambda < k_\lambda < c_\lambda = a_\lambda + b_\lambda$. A good estimate of the upwelling solar irradiance $k_\lambda^\uparrow$ can be obtained from satellite observations of the sea surface (see Sec. 3.4.3). The apparent properties are connected to the inherent ones via the radiative transfer equation [see Eq. (13)].

Within the ocean, light is mostly absorbed but is also scattered, by both seawater and particles within the water including plankton. The absorption results in heating of the water. The euphotic zone refers to the upper layer of the ocean in which the intensity of light is greater than 1% of its value at the sea surface. The attenuation of light with depth depends not only on the wavelength of light but also the clarity of the water. The clarity is mostly controlled by biological activity although at the coast, sediments transported by rivers or resuspended by wave action can strongly attenuate light.

In terms of clarity, waters may be classified according to Jerlov (1976) as coastal (types 1–9) or oceanic (types I–III). For oceanic waters, blue light (400–500 nm) is transmitted to the greatest depth, while coastal waters favor transmission of green-yellow light (650–580 nm) because plankton and organic compounds absorb blue light. Diffuse attenuation coefficients may vary from 0.01 to 1 m^{-1} for the oceanic and coastal regimes, respectively, in the latter case limiting light to the upper few meters.

2. OCEAN STRUCTURE

The physical structure of the ocean is largely governed by the shape of the ocean basins and atmospheric forcing and to a lesser extent by mixing and internal dynamic processes. Continents prevent ocean waters from freely circulating around the Earth except in the Southern Ocean between 56° S and 63° S. A system of midocean ridges splits the major oceans (Pacific, Atlantic, Southern, and Indian) into a series of smaller basins. In addition to these ridges, major topographic features include abyssal plains, trenches, submarine canyons, seamounts, and islands. The mean depth of the ocean is 3.8 km; the greatest depth, 12 km, is found in Marianas Trench. Note that these dimensions, e.g., 4 km in the vertical and 4000 km or more in the horizontal, imply aspect ratios of 1/1000 for vertical sections presented in an atlas or report. Semienclosed seas such as the Mediterranean, Red, Black, Baltic, Arctic, Japan, and Bering Seas or gulfs such as the Persian, Mexico, or California exchange water with adjacent oceans to varying degrees, often introducing waters of distinct characteristics.

Next to the coast, shallower waters overlie the continental shelf. The continental shelf is found inshore of a sharp break in topography that occurs at an average depth of 130 m. Continental shelves are regions of high fishery productivity, oil production, and recreational boating. Because of the shallow water, circulation on the continental shelf can have a different character from that found offshore. Oceanographers refer to the region that is found further inshore, where ocean swell begins to feel the bottom in about 10 m of water, as the "nearshore." Here surf forms, and waves run up onto the beach. Other distinct geographical features along the coast include bays, estuaries, fjords, and river mouths. Note that continental shelves are separated from deep ocean basins by a region where depth increases rapidly, which is called the continental slope.

The atmosphere drives the ocean both through the stress that the winds exert upon the water and through the buoyancy effects of heat, evaporation, and rainfall. Over much of the globe, the atmosphere has a distinct meridional character consisting of trade winds in the tropics and westerlies in the midlatitudes. In the Indian Ocean, these are modified by the Southwest Monsoon, which begins in late May, and the Northeast Monsoon, which begins in October. Continents serve to modify air masses further in such a way that cold, dry air may be found over the ocean off east coasts during the winter.

2.1 Vertical Structure

A major feature of the ocean is its layered structure. Radiation impinging on the ocean surface heats both the ocean surface and the lower atmosphere. As noted above, seawater is a good absorber of radiation, so that typically 60% of the penetrated radiation is absorbed within the first meter of the ocean, around 80% is absorbed within 10 m, and only about 1% of the entering radiation remains at 140-m depth in the clearest tropical ocean waters. The turbulent fluxes of heat, water, and momentum across the ocean surface result in vigorous mixing in the uppermost layer of the ocean. The turbulent kinetic energy generation by the wind stress and the upward buoyancy flux will mix the seawater vertically, eventually forming a vertically uniform layer called the upper mixed layer.

Below the upper mixed layer, the water can be divided into two zones in terms of the rate at which temperature decreases with depth: a region of rapid change of temperature with depth, called the thermocline, and a region where temperature changes slowly with depth, called the deep zone (Fig. 2). In mid- and high latitudes, the thermocline can be subdivided into an upper or seasonal thermocline and a lower or permanent thermocline.

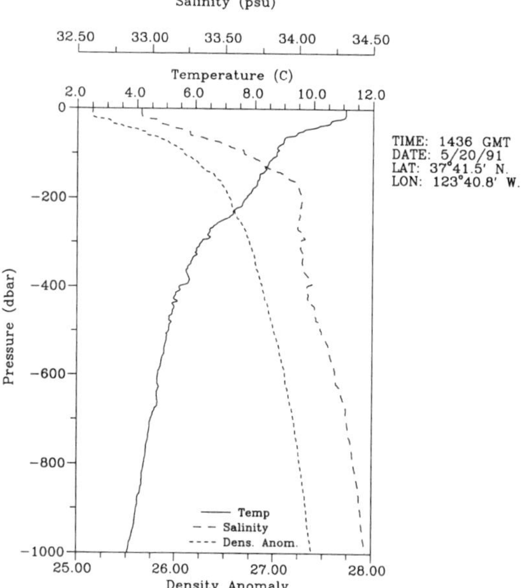

FIG. 2. Vertical profiles of temperature, salinity, and density for an oceanographic station off San Francisco, California. An oceanographic station consists of observations of temperature and salinity at a fixed location. Data were obtained by lowering an instrument that measures conductivity, temperature, and pressure (or depth), called a CTD, from the *R/V Pt. Sur* on May 20, 1991. The mixed layer is confined to the upper 15 m, and the seasonal thermocline/halocline/pycnocline extends to a depth of 80–100 m at this location.

At low and mid latitudes, despite a steady input of heat and downward mixing by eddy diffusion, the depth of the thermocline does not appear to change at any given location. It has been postulated that this is due to upward advection of water of the order of cm/day due to sinking of water at high latitudes.

Density increases with depth and has the same kind of vertical zonation as temperature. The zone of rapid density increase is called the pycnocline. The stability E is defined as

$$E = -\frac{1}{\rho}\frac{\partial \rho}{\partial z} - \frac{g}{U_a^2}. \tag{9}$$

The first term is the *in situ* vertical density gradient and the second term the adiabatic gradient ($\approx 400 \times 10^{-8}$ m^{-1}). The requirement of static stability means that E must be positive. Situations where E is negative and the water column is not stable can exist, e.g., convection due to surface cooling or subsurface hot springs; when using a surface reference pressure, deep densities (below 4 or 5 km) can appear to be less than those at shallower depths, but this is due to inaccuracy of EOS80 and the situation is corrected by using a reference pressure that is near the observed density inversion. Equation (9) can be expanded to include vertical gradients of temperature and salinity:

$$E = \alpha\left(\frac{\partial T}{\partial z} + \Gamma\right) - \tilde{\beta}\frac{\partial S}{\partial z}, \tag{10}$$

where $\alpha = -\rho^{-1}\partial\rho/\partial T$ is the thermal expansion coefficient of seawater at the surface, $\tilde{\beta} = \rho^{-1}\partial\rho/\partial S$ is the haline contraction coefficient, and Γ is the adiabatic temperature gradient. This indicates that subsurface warming (called a temperature inversion) does not violate the requirement for stability if it is less than the adiabatic gradient or the salinity increases sufficiently.

The vertical distribution of salinity cannot be summarized as easily as that for density and temperature. Upper-layer salinity tends to be high in low to mid latitudes, decreasing to subsurface minimum at 600–1000 m. In the tropics, a shallow salinity maximum is usually found at about 100-m depth.

At high latitudes, upper ocean waters are fresher than those at depth. The region of rapid salinity change with depth is called the halocline. Since salinity usually does not dominate density changes, it can be used to trace the horizontal flow pattern.

2.2 T-S Diagram

The temperature-salinity (*T-S* or *θ-S*) diagram is used as a characteristic diagram by oceanographers. This is because temperature and salinity are conservative properties when a water mass is not in the mixed layer. This means that water masses must acquire temperature and salinity properties at the ocean surface and that these properties are determined by the local climate; when the water later sinks along density surfaces, it carries these properties with it. As a consequence, only a limited number of temperature and salinity combinations occur. These can be recognized by plotting observations for a series of oceanographic stations as a function of temperature and salinity.

A water type appears as a point on a *T-S* diagram, and a water mass is represented by a line and is formed from the mixing of water types. Using this representation, the global ocean has a distinct geography with specific regions readily identifiable by their *T-S* characteristics (Fig. 1).

2.3 Ocean Currents

In the mean, a steady ocean circulation exists. This circulation is the result of both thermohaline forcing and forcing by the wind. Currents are classified by their depth (surface, intermediate, deep, and bottom) or by their forcing mechanism. This is not unambiguous because some currents extend from the surface to the bottom of the ocean. The pattern of surface currents was first studied because of their effect on navigation. Today, the principal focus for such studies is to understand the poleward transport of heat by the ocean.

The steady-state circulation is characterized by subpolar and subtropical gyres with zonal tropical currents and an eastward flow around Antarctica called the Circumpolar Current. The strongest currents occur as poleward flow at the western boundaries of subtropical gyres and include the Gulf Stream in the North Atlantic, the Kuroshio in the North Pacific, and the Agulhas in the South Indian Ocean. The eastern-boundary currents of subtropical gyres are weaker, flow equatorward, and are characterized by upwelling at the coast. These include the California Current in the North Pacific, the Canary Current in the North Atlantic, the Benguela Current in the South Atlantic, and the Humboldt or Peru Current in the South Pacific. In the South Indian Ocean, the eastern boundary current, called the Leeuwin Current, flows poleward.

The system of equatorial currents includes the westward-flowing North and South Equatorial Currents. These currents are separated by the eastward-flowing North Equatorial Counter Current, which, in the Pacific, is found between 4° N and 10° N. At the equator, a strong subsurface eastward flow is usually found between 75 and 200 m depth and is called the Equatorial Undercurrent. It is usually about 400 km wide, and speeds in excess of 1 m/s are not uncommon.

The Indian Ocean circulation is dominated by the monsoons. A strong western-boundary current, the Somali Current, develops in response to the Southwest Monsoon. The monsoon currents in the North Indian Ocean reverse and flow in the direction of the monsoon. When eastward winds occur at the equator, a strong eastward current develops at the surface as well.

Intermediate, deep, and bottom currents are the result of thermohaline processes. These currents are also intensified at western boundaries.

3. OCEAN MEASUREMENTS

The equations of motion and state specify the variables that must be measured experimentally. These include temperature, salinity, and pressure for the equation of state and the velocity of the water. Related variables include the height of sea level and wind stress. Although the ocean environment is hostile, e.g., cold, great pressures, and often large accelerations at the sea surface, recent technical developments have greatly improved our ability to obtain data of the requisite accuracy and precision for study of most physical processes. Our major challenge is to obtain sufficient synoptic data to be able to identify unambiguously the processes that control ocean circulation.

3.1 Shipboard Profiling

Instruments that are routinely used from shipboard include conductivity-temperature-depth (CTD) profilers, expendable bathythermographs (XBT), and acoustic-Doppler current profilers (ADCP). The ship itself is a key element of these systems. Research vessels range in length from 30 to 100 m and have special capability for maneuvering at low speed, maintaining station, and handling equipment over the side. Protection from wind and seas is provided for scientists working on deck, and special electronic, biological, and chemical laboratories are needed. Larger ships must be capable of extended cruises of more than 40 days at sea and 20 000-km ranges.

CTDs are lowered by a winch and conducting cable when the ship is hove to (CTDs can also be towed, cycling up and down through the water column, but depth is limited to about 200 m by cable drag) (Fig. 3). Temperature measurement is by means of a platinum resistance thermometer (PRT) and a faster-response thermistor. Conductivity is measured by means of paired electrodes mounted inside a 4-mm-diam channel that is 30 mm long. Pressure is measured by a strain gauge. Data are sampled at 30 Hz and transmitted up the cable as a FM signal. The fast-response thermistor compensates for the mismatch in time constants of the PRT and the conductivity electrodes, allowing the computation of salinity without severe spikes. The accuracy of the system is 0.005 mS, 0.005 °C, and 0.1 of full scale for pressure. Corresponding precisions are 0.001 mS, 0.005 °C, and 0.1. The CTD is usually attached below a rosette that allows water samples to be collected at discrete depths by Niskin bottles for shipboard analysis. The Niskin bottles are open cylinders with endcaps, which are held together with an elastic band. During descent, the rosette holds the endcaps back, allowing water to flow freely through the bottles. During ascent, the rosette is electrically triggered, releasing the endcaps and closing the bottles. The Niskin bottles shown in Fig. 3 hold 5 L. Salts from these samples are necessary to maintain the accuracy of the CTD conductivity measurements. Final data processing produces a table of temperature and salinity values averaged over successive 2-db intervals. A typical CTD cast takes about 45 minutes per 1000 meters.

FIG. 3. Ocean profiling system. Vertical profiles of conductivity and temperature are obtained by a conductivity-temperature-depth (CTD) instrument and water sampling bottles. The CTD is the stainless steel cylinder that is mounted in the lower part of the cage. Water sampling (Niskin) bottles are held by a rosette at the top of the cage. The oceanographers standing on deck are filling seawater sample bottles from the Niskin bottles while the oceanographer standing on the cage is resetting the rosette.

XBTs are designed to be launched while a ship is under way. They were originally designed for use in antisubmarine warfare by naval vessels, but the simplicity and reliability have led to widespread use by merchant vessels in climate-related ocean studies. Temperature measurement is by means of a thermistor mounted on the nose of a 1-kg lead weight. Depth is determined by using an empirically determined rate of fall. Data are transmitted back to the ship by means of two-strand copper wire. XBTs are available for 450-m, 750-m, and 1500-m depths. Systematic errors of 0.2–0.4 °C occur in these measurements, and random errors are about half this amount; depth accuracy ranges from 2% to 10%, depending on the type (depth) of probe used (the increase with depth is due to errors in fall rate and microscopic leaks in wire insulation). Expendable probes that measure sound velocity and velocity shear are also available.

ADCPs are mounted in the hull of a ship, run continuously, look downward, and in good weather allow sampling to 400-m depth. Paired transducers look fore–aft and port–starboard, and the Doppler shift in gated returns (caused by plankton drifting in the water) is transformed to horizontal velocity. Typically, a pulse repetition rate of 1 Hz is used, and data are resolved in 8-m depth bins. Averaging over 3-min intervals reduces random errors to 1 cm/s. Current velocities are obtained by removing the ship's motion from ADCP output. ADCPs can also be lowered on a wire with a CTD, yielding velocity profiles that extend throughout the entire water column. The Global Positioning System (GPS) provides accurate navigational data for ADCP operation, including ship's heading.

3.2 Moored Instruments

Current measurements that are made at a fixed point in space are called "Eulerian" measurements. Eulerian measurements are made by mooring current meters. Currents are measured either

1. mechanically by means of orthogonal rotors or a rotor and vane,
2. acoustically using changes in travel time or an ADCP, or
3. electromagnetically in higher flow regimes associated with waves and surf.

In deep-sea applications, the mooring may be 4–6 km long, and is held to an anchor by means of a release that is interrogated acoustically from shipboard. A deep-sea mooring is typically left in place for two years.

Temperature and pressure measurements are easily made from moorings, but conductivity measurements of the precision and accuracy required for most deep-sea applications are not, because of problems of biological growth and electrode oxidation.

3.3 Lagrangian Measurements

Measurements that are of quantities that are traveling with the fluid are called "Lagrangian" measurements. Lagrangian measurement systems include surface (called drifters) and subsurface (called floats) drifting buoys. Surface drifting buoys must be properly coupled to the surface layer of the ocean, requiring a subsurface drogue that has a drag that is at least 40 times that exerted on the surface float. Drifter positions are currently determined by satellite at least six times per day with position accuracy (2σ) of 600 m. Future applications will use GPS with continuous position keeping of 100-m accuracy. Observations of near-surface temperature and vertical temperature gradients can also be made from drifters.

Subsurface floats are positioned acoustically. Floats that signal are called SOFAR floats while those that listen are called RAFOS floats. These floats may be isobaric or isopycnal or profile a short distance above and below these levels to measure potential vorticity. Moored autonomous listening stations (sound sources) or fixed Navy listening stations provide the requisite positioning network for SOFAR (RAFOS) floats. RAFOS floats surface when their mission is completed, relaying data via satellite.

A new system of Autonomous Lagrangian Circulation Explorer (ALACE) floats is currently being developed. These floats drift at depth but surface at regular intervals so that their position can be determined by satellite and subsurface ocean observations relayed to shore. Current floats are capable of 50–100 depth cycles. Further developments of this type of float, adding CTD, electromagnetic measurements, and perhaps "wings" to maneuver the float, promise to make ALACE a key component of a global ocean-observing system.

Anthropogenic chemical tracers are also an effective Lagrangian technique for determining ocean circulation. Tritium, produced by atmospheric atomic bomb testing, and cesium-137 and strontium-90, associated with nuclear power plant operations, have been used to trace horizontal and vertical circulation. Since tritium decays to helium-3 at a known rate, the ratio of tritium to helium-3 can be used to determine the length of time since the water parcel was exposed to the ocean surface; the utility of this technique is limited by the introduction of helium-3 into the ocean by submarine vents. Freon compounds (10, 11, 12) can also be used to trace and date ocean waters.

3.4 Satellite Measurements

Satellite techniques promise continuous global coverage. The principal problem with these techniques is that the atmosphere gets in the way and the ocean is nearly opaque to electromagnetic radiation. Satellite applications that we will discuss include satellite altimetry, synthetic-aperture radar, and ocean-surface temperature measurements. Other satellite techniques effectively map ice cover, and new satellite-positioning systems [the Global Positioning System (GPS)] and data relay capabilities are essential for modern experimental programs. Two kinds of sensors (active and passive) are available for measurements. The active sensors transmit radio signals, comparing the transmitted with the reflected or scattered signals that are received by the sensors from an area on the earth's surface. Passive sensors do not transmit a signal; instead, the emitted radiance from the earth's surface is compared to that received from space or a blackbody. Satellite altimetry and the synthetic-aperture radar are active sensors. Satellite radiometers are passive sensors.

3.4.1 Satellite Altimetry If the ocean were motionless, the shape of its surface would be determined entirely by the gravitational acceleration of the Earth. The spatial variation of the gravity field generates a sea-surface topography with a global range in height of nearly 160 m relative to the center of the earth. But forces associated with ocean currents produce departures from this static topography called *dynamic ocean topography*. The dynamic ocean topography has a global range in height of about 2 m, about 1% of the height range of the static or gravitational topography (Plate 1).

Satellite altimetry maps the dynamic ocean topography in three steps. First, the radar altimeter sends short pulses of microwave energy toward the ocean; the round-trip travel times of the reflected pulses yield the distance between the spacecraft and the sea surface (additionally, the shape of the reflected pulse can be used to determine wave height and sea-surface wind speed). Second, precise determination of satellite orbit permits these distances to be translated into a global map of sea level relative to the center of the Earth. Third, a map of the gravitational sea-surface topography (which must be obtained independently) must be subtracted from the satellite-derived topography, yielding the dynamic ocean topography. Methods for determining ocean current speed and direction from the dynamic ocean topography are described below.

Errors are associated with absorption of microwave energy by atmospheric water vapor and pulse-arrival delays due to electron release by ionization of gases in the ionosphere caused by sunlight. The largest source of error is associated with the determination of satellite orbit.

Satellite altimeter measurements have been made on Skylab (1973), Goes-3 (1975–1978), Seasat, and Geosat (1985–1989), and are currently being made by *Topex/Poseidon*. Topex has an orbital altitude of 1136 km and observes the Earth from 66° N to 66° S. Measurement tracks across the Earth are reproduced within 1 km over every 10-day repeat cycle. Widest spacing between tracks, at the equator, is 315 km. The altimeter operates at dual frequencies, 5.3 and 13.6 GHz, permitting correction for ionospheric electron delays. Topex also carries a microwave radiometer that operates at frequencies of 18, 21, and 37 Ghz to estimate total atmospheric water vapor content. Topex is tracked by three systems: ground-based lasers, measurement of Doppler shifts of microwave frequencies transmitted by a global network of ground-based beacons, and GPS. Sea-level topography is being measured with 2-cm accuracy.

3.4.2 Synthetic-Aperture Radar A synthetic-aperture radar (SAR) is a pulse-Doppler radar that transmits short radio pulses and maps the radar reflectivity of the sea over areas 50–100 km on a side, with a resolution of 10–40 m. From SAR images of ocean surface, we can "see" and study meanders of currents, eddies, storm waves, ice floes, leads, ice ridges, and polynyas. The perturbed sea surface ζ modulates the reflectivity of the radio pulse. The radar cross section becomes

$$\sigma = \langle\sigma\rangle + \delta\sigma, \tag{11}$$

where $\langle\sigma\rangle$ is the spatially averaged cross section per unit area of the sea, and $\delta\sigma$ is the deviation induced by surface long waves. A complex modulation transfer function R is used to link $\delta\sigma$ to ζ:

$$\delta\sigma/\langle\sigma\rangle = R\zeta. \tag{12}$$

When R is determined, the surface fluctuations can be measured by $\delta\sigma$.

Three processes cause variability in radar cross section: surface tilting, hydrodynamic interaction, and velocity bunching. The modulation caused by surface tilting and hydrodynamic interactions depends on the wave direction. Consider a SAR viewing the sea at an incidence angle of 23°: With an amplitude of 3 m and a wavelength of 200 m, the modulation caused by tilting is roughly ±3 dB for waves traveling toward the range direction (detectable by SAR), and nearly ±0.3 dB for waves traveling in the azimuthal direction (not detectable by SAR).

3.4.3 Satellite Radiometer Sea-surface temperature (SST) has a 150-year history of global measurement and is intimately linked with climate and global change. Although historical measurements were made by naval and merchant vessels, space-based measurements of SST have improved remarkably in the past decade and now represent one of our most important data sets for monitoring the global environment. Satellite measurement of SST is made possible by satellite radiometers (Plate 2).

The physical principle for the satellite radiometer is radiative transfer through an emitting and absorbing atmosphere with the absorption coefficient κ_λ, which is depicted by the radiative transfer equation

$$L^\lambda = L_S^\lambda \exp[-\tau(0,H)] + \int_0^H L_B^\lambda(z)\, \kappa_\lambda(z) \exp[-\tau(z,H)]\, dz, \qquad (13)$$

where

$$\tau(z,H) = \int_z^H \kappa_\lambda(z')\, dz'$$

is the optical thickness (or opacity) of the atmospheric layer with thickness H. The superscript λ indicates the wavelength of the radio waves. The terms L_S^λ, L^λ, and L_B^λ are the source radiance, the radiance seen by an observer viewing the source, and the radiance of the blackbody at the same temperature as the absorbing air, respectively. The first term on the right-hand side of Eq. (13) gives the total attenuation due to absorption; the second term gives the emission $L_B^\lambda\kappa_\lambda$ from any point z along the path attenuated by the air along the remainder of the path (Plate 2). The radiative transfer equation is complicated, and one needs to know L_B^λ (or temperature) and κ_λ (or atmospheric constituents) to solve it. At the atmospheric windows, however, the absorption coefficient κ_λ is small such that the second term of the right-hand side of Eq. (13) is negligible. Therefore, the source radiance (or the surface temperature) can be easily determined by the satellite radiometer.

The primary instrument for production of operational satellite-derived SST is the Advanced Very High Resolution Radiometer (AVHRR) carried on two polar-orbiting meteorological satellites. These satellites are sun-synchronous with local equator crossing times of 0730 and 1330 for descending orbits. The AVHRR scans the Earth perpendicular to the satellite direction of travel, providing twice-daily global coverage at 4-km ground resolution. The AVHRR has four or five channels, which use the 3–4-μm or 10–12-μm atmospheric windows, and carries an on-board blackbody viewed by the instrument, and this, along with a view of space, is used to perform a linear calibration. Multichannel observations are used to correct for atmospheric effects to produce a global SST data set. This is done after navigation of the images, land masking, eliminating observations of clouds, and averaging into 8-km areas. The bias and scatter of these data, when compared to drifting buoy measurements, are less than 0.3° and 0.7° C, respectively. Some of this difference is due to the fact that the radiometers measure the skin temperature of the ocean while buoy measurements are made in the upper meter of the ocean.

AVHRR images with 1-km resolution are broadcast directly to ground stations. These high-resolution images are useful for delineating fronts and eddies in the ocean that are associated with biological productivity and that also have tactical significance for naval warfare.

4. MARINE HYDRODYNAMICS

The equations of motion for the ocean are the momentum equations (application of Newton's law to a unit mass of seawater) and the continuity equation indicating conservation of mass per unit volume:

$$\frac{d\mathbf{V}}{dt} = \mathbf{F}_C + \mathbf{F}_p + \mathbf{F}_t + \mathbf{F}_g + \mathbf{F}_v, \tag{14}$$

$$\frac{d\rho}{dt} + \rho\nabla \cdot \mathbf{V} = 0, \tag{15}$$

where $\mathbf{V}$ is the velocity vector relative to the Earth, ρ is water density, and $\mathbf{F}_C$, $\mathbf{F}_p$, $\mathbf{F}_g$, $\mathbf{F}_t$, and $\mathbf{F}_v$ are various forces (per unit mass) exerted on the water parcels: Coriolis force, pressure-gradient force, gravity, tidal forces, and viscous forces.

4.1 Coriolis Force

The Earth is a rotating planet with the angular velocity of 2π/day. Any water parcel at rest relative to the Earth in fact moves with a velocity of as much as 460 m/s relative to a "fixed" star. Rather than include these velocities and accelerations, a coordinate system is generally chosen that is fixed with respect to the Earth. In this coordinate system, any moving water parcel "feels" an apparent force called the Coriolis force due to the rotation of the Earth:

$$\mathbf{F}_C = -2\mathbf{\Omega} \times \mathbf{V}, \tag{16}$$

where Ω is the angular velocity of the Earth.

4.2 Pressure-Gradient Force

If the pressure (p) on a level surface is inhomogeneous, the pressure gradient leads to a force exerted on the water parcel:

$$\mathbf{F}_p = -(1/\rho)\,\nabla p, \tag{17}$$

which is called pressure-gradient force.

4.3 Tidal Forces

In principle, the ocean responds to all gravitational fields exerted on it, but in practice, except for the Earth's own gravity, only the gravitational forces exerted by sun and moon are strong enough to induce detectable tidal excursions. If the ocean is attracted by the sun or moon, water parcels at different locations experience different forces that lead to a tidal force, which is decomposed into components normal and parallel to the ocean surface

$$\mathbf{F}_t = \frac{GMR}{D^3}\left(3\cos^2\chi - 1,\ -\tfrac{3}{2}\sin 2\chi\right). \tag{18}$$

Here G is the Newtonian gravitational constant, χ is the angle between the local vertical axis and the line between the center of the Earth and the external terrestrial body, M is the mass of the external terrestrial body, D is the distance between the Earth and the external terrestrial body, and R is the radius of the Earth. The 11 most important tides with commonly accepted nomenclature are listed in Table 1.

Table 1. Major tidal modes.

	Tidal mode	Period, T (h)
Semidiurnal Tides		
M_2	principal lunar	12.421
S_2	principal solar	12.000
N_2	elliptical lunar	12.658
K_2	declination luni-solar	11.967
Diurnal Tides		
K_1	declination luni-solar	23.935
O_1	principal lunar	25.819
P_1	principal solar	24.066
Q_1	elliptical lunar	26.868
Long-Period Tides		
Mf	fortnightly lunar	327.86
Mm	monthly lunar	661.31
Ssa	semiannual solar	4383.04

If the Earth were uniformly covered with a fluid envelope, these forces would give rise to an equilibrium tide having a maximum elevation of 0.35 m for the lunar (M2) tide and 0.16 m for the solar (K1) tide. In the ocean, this behavior is considerably modified. The phase speed of the tidal wave is approximated by $(gH)^{0.5}$ (where H is the water depth), ocean basins have natural periods of oscillation, tides are affected by the Coriolis force, and land masses can reflect the tides. A given tidal constituent may have amphidromic points where no vertical displacements occur.

Tides can behave as either standing or progressive waves. Tides are generally amplified in shallow water, and here tides are classified as either diurnal, semidiurnal, or a combination of diurnal and semidiurnal called "mixed." Tidal currents associated with rising sea level are called *flood*, while those associated with falling sea level are called *ebb*. Tides cause considerable mixing in shallow water, can generate internal tidal waves when waters are stratified, and, with the rate of river discharge, serve to control the character of estuaries and bays. Bores occur on some rivers, and, at a few locations, tides are used for power generation.

4.4 Gravity

Gravity is the strongest force acting on the sea. It is a body force exerted on the water column, and is the sum of the Earth's gravitational force and the centrifugal force. It is taken in the vertical direction downward:

$$\mathbf{F}_g = -g\mathbf{k} = \frac{d\Phi}{dz}, \tag{19}$$

where $\mathbf{k}$ is the vertical unit vector (directed positive upward from a level surface), g is the gravitational acceleration given by Eq. (4), and Φ is called the *geopotential*, which has units of $m^2\,s^{-2}$. The geopotential defines a level surface to which a motionless ocean would conform (see discussion in Sec. 3.4.1). Gravity provides the restoring force for sea and swell and the energy for thermohaline circulation. It also acts to compress water at depth.

4.5 Viscous Forces

The molecular viscosity of seawater is the mechanism by which mechanical energy is removed. The viscous force is always a resistance force and is expressed by

$$\mathbf{F}_\nu = \nu\nabla^2\mathbf{V}; \tag{20}$$

here ν is the kinetic molecular viscosity, which is a function of temperature, salinity, and pressure and has a value of about $10^{-6}\ \mathrm{m^2\,s^{-1}}$ for seawater.

Although damping due to molecular viscosity is small, it is important for small-scale processes, e.g., acoustic and capillary waves. It is neglected for larger-scale processes for which the damping due to eddy mixing and diffusivity is represented by means of eddy viscosity.

4.6 Geostrophic Balance

For the large-scale motions in the ocean, the Coriolis force and the pressure-gradient force are the two dominating forces, and the momentum equation becomes geostrophically balanced. The water moves along isobars.

The pressure-gradient force is estimated through the use of the geopotential. From Eqs. (19) and (3),

$$d\Phi = g dz = -(1/\rho)\,dp = -\alpha dp, \tag{21}$$

and here α is specific volume. Integrated over a given pressure interval, this yields *dynamic height*. In practice, an anomaly δD is used; results are divided by ten, and the resulting unit is called *dynamic meters*. Since currents at depth in the ocean are typically slow, an assumption of a *level of no motion* is made, i.e., the lower-pressure surface is assumed to be parallel with a geopotential surface, and a geostrophic current velocity v_g, normal to two CTD stations, a and b, can be calculated:

$$v_g = (\delta D_a - \delta D_b)/fL, \tag{22}$$

where L is the distance between the stations.

4.7 β Effect

The Coriolis parameter ($f = 2\Omega\,\sin\phi$) depends on latitude. If the latitudinal range of motion exceeds 20°, the Coriolis parameter cannot be treated as a constant. To keep this dependence as simple as possible, a linear function is obtained by expanding f into a Taylor series about $y_0 = a\phi_0$ ($a \sim$ Earth's radius),

$$f = f_0 + \beta(y - y_0) + \cdots$$
$$= f_0\left[1 + \frac{\beta}{f_0}(y - y_0) + \cdots\right], \tag{23}$$

where

$$\beta = \left.\frac{df}{dy}\right|_{\phi_0} = \frac{2\Omega \cos\phi_0}{a}. \tag{24}$$

This linear relationship between f and y is called the β effect. In the equatorial region, $f_0 \simeq 0$, the second term on the left-hand side of Eq. (23) becomes dominant and the β effect is evident. In the polar regions, $\beta \simeq 0$, the β effect is negligible. If the zonal variation is much smaller than the Earth's radius, $|y - y_0 \ll a$, the latitudinal variation of the Coriolis parameter is negligible:

$$|(\beta/f_0)(y - y_0)| \ll 1.$$

Therefore, it is reasonable to replace f by the constant f_0 except where it is differentiated by y and to change df/dy to the constant β. This is called the β-plane approximation.

4.8 Conservation Principles

With neither a heat source nor sink, the second law of thermodynamics requires entropy to be conserved, and therefore the potential temperature θ should also be conserved:

$$\frac{d\theta}{dt} = 0. \tag{25}$$

Similarly, with no source or sink of salt, salt should be conserved, i.e.,

$$\frac{dS}{dt} = 0. \tag{26}$$

Further, if water is incompressible, mass conservation leads to

$$\frac{\partial u}{\partial x} + \frac{\partial v}{\partial y} + \frac{\partial w}{\partial z} = 0, \tag{27}$$

which indicates that horizontal divergence will induce vertical convergence, and vice versa.

If there are neither heat nor salt sources or sinks, nor friction nor diffusion, a mathematical manipulation of momentum, heat, and continuity equations leads to conservation of potential vorticity (Π),

$$\frac{d\Pi}{dt} = 0. \tag{28}$$

4.9 Topographic Forcing

The ocean has sides (e.g., coasts) as well as a bottom. These boundaries are generally assumed to be rigid,

$$\mathbf{V} \cdot \mathbf{n} = 0, \tag{29}$$

indicating that no water particle can penetrate the boundaries. Here $\mathbf{n}$ is the unit vector in the outward normal direction to the rigid boundary.

In the open ocean, bathymetry is expressed by a two-dimensional function $z = -H(x,y)$. Therefore, the rigid-bottom boundary condition becomes

$$w = -u\frac{\partial H}{\partial x} - v\frac{\partial H}{\partial y}, \quad \text{at } z = -H(x,y). \tag{30}$$

Vertical motion is then generated by horizontal currents flowing over the bottom. This forced vertical motion is important in studies of deep-water processes.

For coastal regions, the water is generally considered as barotropic. Its potential vorticity is $\Pi = (f + \zeta)/H$, where $\zeta = \partial v/\partial x - \partial u/\partial x$ is the vertical component of the relative vorticity. Since the magnitude of ζ is much smaller than f, $|\zeta| \ll |f|$, potential vorticity conservation becomes

$$\frac{d\Pi}{dt} = u\frac{\partial}{\partial x}\left(\frac{f}{H}\right) + v\frac{\partial}{\partial y}\left(\frac{f}{H}\right) = 0, \tag{31}$$

which means that the current vectors are parallel to the isolines of the f/H field. This forcing is very important for topographic waves in coastal regions.

5. OCEAN CIRCULATION THEORY

Motions are generated both by surface wind stress (the wind-driven circulation) and by buoyancy flux (the thermohaline circulation). While it is difficult to say that one type of forcing is more important than the other, aspects of wind forcing have been easier to treat analytically.

5.1 Wind-Driven Circulation

The atmospheric forcing on the ocean takes two forms. One, called the *inverted barometer effect*, is the force exerted on the irregular ocean surface that is associated with atmospheric-pressure differences. The second is the wind stress. Except for severe weather systems (e.g., typhoons) in which the horizontal pressure gradient is large, wind stress is generally considered the major mechanical forcing of the ocean.

Winds blowing over the ocean surface drive different kinds of motion (gravity and capillary waves, turbulence, circulation, etc.) through frictional effects near the ocean surface. The surface wind stress, defined as the turbulent momentum flux at the ocean surface, is on the order of 1 dyn/cm^2 and is related to the surface wind speed $\mathbf{V}_a$ by the drag law

$$\boldsymbol{\tau} = C_D \rho_a |\mathbf{V}_a| \mathbf{V}_a, \tag{32}$$

where C_D ($\sim 10^{-3}$) is the drag coefficient, whose magnitude is affected by the roughness, the buoyancy flux, and the waves and turbulence near the ocean surface.

5.1.1 Ekman Transport When the wind blows over the ocean surface, the friction between the air and the ocean will slow the wind and cause ripples, waves, and currents in the ocean. The stress of the wind on the water involves a transfer of turbulent fluxes of momentum and kinetic energy from the atmosphere to the ocean. In the ocean, the turbulent kinetic energy transferred from the atmosphere decreases from the surface downward and will vanish at some depths. A depth of 100 m is usually considered the practical lower limit for wind-driven currents. The layer that contains the turbulent flux is called the surface layer, or the Ekman layer.

Imagine that the ocean surface layer is divided into thin horizontal layers. The frictional force (per unit area) exerted between the two neighboring layers is represented by a horizontal stress (X,Y), which is the retarding force for the upper layer and the driving force for the lower layer. At the ocean surface, (X,Y) is the wind stress. If these stresses are balanced by the Coriolis force, the Ekman velocity (u_E,v_E) satisfies

$$\frac{\partial u_E}{\partial t} - f v_E = \frac{1}{\rho}\frac{\partial X}{\partial z}, \tag{33a}$$

$$\frac{\partial v_E}{\partial t} + f u_E = \frac{1}{\rho}\frac{\partial Y}{\partial z} \tag{33b}$$

[this solution was derived by Ekman (1905) to explain why Arctic icebergs drifted 40° to the right of the wind]. Notice that currents driven by the pressure-gradient force are not included in the Ekman flow. For steady state ($\partial/\partial t = 0$), the vertical variation of the Ekman velocity, called the Ekman spiral, shows that the surface current is directed 45° to the right of the wind stress in the Northern Hemisphere and 45° to the left in the Southern Hemisphere. As depth increases, the direction of water transport deviates to the right in the Northern Hemisphere relative to the overlying water in the same fashion as the surface water is transported by the wind. Friction causes each succeeding layer to decrease in speed.

The total water transport in the Ekman layer is calculated by the vertical integration of the Ekman velocity:

$$(U_E,V_E) = \int (u_E,v_E)dz,$$

which is also called the Ekman transport. Vertical integration of Eq. (33) leads to

$$\rho\left(\frac{\partial U_E}{\partial t} - fV_E\right) = \tau_x, \quad \rho\left(\frac{\partial V_E}{\partial t} + fU_E\right) = \tau_y, \tag{34}$$

where (τ_x,τ_y) is the surface wind stress. In steady conditions, the Ekman transport is directed 90° to the right (or left) of the surface wind stress in the Northern (or Southern) Hemisphere. The discontinuity of the Ekman transport at the equator results in equatorial upwelling (downwelling) for westward (eastward) winds.

The wind stress varies from place to place, and hence so does the Ekman transport. This leads to convergence or divergence of water mass in the Ekman layer, i.e., causes the water mass to flow horizontally into some regions and out of others. This will result in vertical motion. For example, if the Ekman transport is horizontally convergent in a particular region, vertical downward motion away from the Ekman layer is induced in order to conserve mass. If we assume no significant distortion of the ocean free surface, a vertical integration of the continuity equation across the Ekman layer yields the vertical velocity (called Ekman pumping velocity) at the base of the Ekman layer:

$$w_E = \nabla \cdot \mathbf{U}_E = \mathbf{k} \cdot \mathrm{curl}(\boldsymbol{\tau}/\rho f). \tag{35}$$

This vertical velocity, driven by the surface wind stress, exchanges water between the surface layer (under the direct effect by the atmosphere) and the deeper water. In subtropical gyres (equatorward of 45 deg) w_E is usually negative: The surface water is pumped downward into the deeper layer. In subpolar gyres, w_E is generally positive: The water mass is pumped upward from the interior into the surface layer. Therefore, Ekman pumping is a crucial link between the surface wind forcing and deep-ocean dynamics.

5.1.2 Sverdrup Transport Ekman flow excludes the currents forced by the pressure gradient. The Sverdrup transport includes both the Ekman transport and the geostrophic transport that results from the pressure gradients set up by the convergent or divergent Ekman-layer flow. Let the mass transport function with respect to the ocean bottom (H) and the integrated pressure (P) be defined as

$$\frac{\partial M}{\partial x} = \int_{-H}^{0} v dz, \quad \frac{\partial M}{\partial y} = -\int_{-H}^{0} u dz,$$
$$P = \int_{-H}^{0} \frac{p}{\rho} dz. \tag{36}$$

Vertical integration (from $-H$ to 0) of the momentum equation (14) after neglecting the advection term and the tidal forcing leads to

$$-\frac{\partial^2 M}{\partial t \partial y} - f\frac{\partial M}{\partial x} = -\frac{\partial P}{\partial x} + \tau_x, \tag{37a}$$

$$\frac{\partial^2 M}{\partial t \partial x} + f\frac{\partial M}{\partial y} = -\frac{\partial P}{\partial y} + \tau_y. \tag{37b}$$

Sverdrup (1947) first obtained the steady-state solution of Eq. (37) for the mass transport M under the β-plane approximation:

$$\frac{\partial M}{\partial x} = \frac{1}{\beta}\left(\frac{\partial \tau_y}{\partial x} - \frac{\partial \tau_x}{\partial y}\right), \tag{38}$$

which shows the congruity between the north–south ocean flow and the wind-curl patterns. Note that the Sverdrup transport was obtained only when the zonal dependency of the Coriolis parameter is considered. If the β effect is negligible (f = const), there is no Sverdrup transport. In the low to middle latitudes of the Northern Hemisphere (15°–50° N), the surface wind-stress curl is usually anticyclonic, and drives a southward Sverdrup transport. Sverdrup further showed that if Eq. (38) is integrated for M from an east coast, where the mass transport diminishes, a current counter to the prevailing winds would occur in the Eastern Pacific Ocean under the Inter-Tropical Convergence Zone (ITCZ). Such a current is observed, the North Equatorial Counter Current. (Oceanographers have subsequently defined a unit of transport as a Sverdrup, i.e., 10^6 m^3/s = 1 Sv.)

5.1.3 Westward Intensification If you look at a pilot chart of ocean currents, you immediately notice that the strongest currents (e.g., the Gulf Stream and Kuroshio) appear at the western boundaries of the ocean and that currents elsewhere are noticeably weaker. This east–west asymmetry is called *westward intensification*.

As noted in Sec. 4.7, the zonal change of the Coriolis parameter (β effect) must be taken into account. In low to middle latitudes, the anticyclonic surface wind-stress curl drives equatorward Sverdrup transport. This equatorward flow must return poleward somewhere in order to satisfy continuity. Will this occur at the eastern or western boundary? Because the Coriolis effect causes equatorward flow to turn toward the west, the return flow must take place at the western boundary. Stommel (1948) and Munk (1950) offered theoretical explanations for the westward intensification. Both models are based on potential vorticity conservation,

$$\frac{d\Pi}{dt} = \frac{\partial \Pi}{\partial t} + (\mathbf{V}\cdot\nabla)\Pi$$
$$= \frac{1}{\rho h^2}\mathbf{k}\cdot(\text{curl }\boldsymbol{\tau}|_0 - \text{curl }\boldsymbol{\tau}|_{-h}) + A\nabla^2\Pi, \qquad (39)$$

which indicates that the time-rate change of the potential vorticity is caused by the surface wind stress ($\boldsymbol{\tau}|_0$), bottom friction ($\boldsymbol{\tau}|_{-h}$), and the horizontal diffusion ($\nabla^2\Pi$). Here, h is the ocean depth and is assumed to be a constant. The magnitude of the relative vorticity (ζ) for large-scale motions in the ocean is much smaller than the magnitude of the planetary vorticity f, so that the advection of the potential vorticity can be linearized:

$$\mathbf{V}\cdot\Pi = (\beta/h)\,v. \qquad (40)$$

For the steady state, potential vorticity advection is balanced by friction and diffusion. The difference between the two models is that there is no horizontal diffusion in Stommel's model, and no bottom friction in Munk's theory.

5.1.3.1 Stommel's Model. By neglecting horizontal diffusion and parametrizing the bottom friction by the Rayleigh friction, $\tau_{-h}/\rho = R(u,v)$, the barotropic potential vorticity equation (39) becomes

$$\nabla^2\psi + \frac{h}{R}\beta\frac{\partial\psi}{\partial x} = \frac{\mathbf{k}}{\rho R}\text{curl }\boldsymbol{\tau}_0,$$
$$u = -\frac{\partial\psi}{\partial y},\quad v = \frac{\partial\psi}{\partial x}. \qquad (41)$$

Stommel (1948) applied this model to a rectangular ocean. The shores of the ocean were at $x = 0, L_x$, $y = 0, L_y$. The surface wind stress was chosen as a sinusoidal function $\mathbf{k}\cdot\boldsymbol{\tau}_0/\rho = F\sin(\pi y/L_y)$. The model was integrated for three cases: $f = 0$, $f = \text{const}$, and $f = f_0 + \beta y$. The nonrotating, wind-driven system features a broad circulation exhibiting no tendency toward western boundary intensification. The uniformly rotating, wind-driven system has a similar pattern (i.e., no westward intensification). The final case, which used a β plane, has intense crowding of streamlines toward the western boundary of the ocean.

5.1.3.2 Munk's Model. By neglecting bottom friction, Munk (1950) simplified the barotropic potential vorticity equation (39):

$$\left[-A\nabla^4 + \beta\frac{\partial}{\partial x}\right] = \mathbf{k}\cdot\text{curl }\boldsymbol{\tau}_0. \qquad (42)$$

His solutions show westward intensification similar to Stommel's model.

5.2 Thermohaline Circulation

Thermohaline circulation problems have proven more difficult, both theoretically and experimentally. On the experimental side, this has been due to the relatively weak flows and the depth at which they occur, as well as the fact that the polar regions, where much of the convective forcing occurs, are not easily accessible. The two major deep-water masses are the North Atlantic Deep Water and Antarctic Bottom Water. North Atlantic Deep Water is formed at high latitudes in the North Atlantic from overflow waters from the Nordic seas (a production rate of about 6 Sv) and in the Labrador Sea (about 8 Sv). Most Antarctic Bottom Water is formed in the Weddell Sea at a rate of 2–4 Sv.

Stommel (1958) proposed a circulation model for the deep circulation given these two sources of deep water. Upwelling in the interior is fed by slow eastward flow from equatorward-flowing deep western-boundary currents, which carry water from the deep-water formation regions. Measurements subsequently confirmed the deep western-boundary currents.

5.2.1 Thermocline Theory Broecker (1991), Munk (1966), Schmitz and McCartney (1993), and Luyten *et al.* (1983) argue that the difficulty in modeling the thermocline is due to the tendency of the ocean to preserve both density and potential vorticity over great distances from source areas of both fields. The problem of the vertical structure of thermocline must be coupled to the horizontal structure of the ocean circulation.

A thermocline equation can be derived from the geostrophic and hydrostatic balance of the large-scale ocean circulation, and from the advective–diffusive balance of the density field:

$$J\left(\frac{\partial Q}{\partial z}, \frac{\partial^2 Q}{\partial z^2}\right) + \cot\phi \frac{\partial Q}{\partial \lambda}\frac{\partial^3 Q}{\partial z^3} = \mu \frac{\partial^4 Q}{\partial z^4} \cos\phi \sin\phi, \tag{43}$$

where ϕ and λ are latitude and longitude, respectively, $J(F,G)$ is the Jacobian for any two functions F and G; $\mu \equiv \delta_D/\delta_A$ is the ratio of the diffusive to advective length scales; Q is defined as

$$Q(\phi,\lambda,z) = -\int_z^0 p dz' + \sin^2\phi \int_0^\lambda w_E d\lambda'. \tag{44}$$

The thermocline equation (43) is very difficult to solve, although solutions have been obtained for special cases.

5.2.2 Thermohaline Forcing The combined effect of fluxes of heat and salt on the ocean causes buoyancy contrasts, and induces motion called the thermohaline circulation. The most important thermodynamic forcing is the buoyancy flux at the air–ocean interface:

$$B = \alpha(LE + Q)/c_{pw} + \beta(E - Pr)\, S_0 + \beta(F - M)(S_0 - S_f), \tag{45}$$

where Q is the sum of the radiative and turbulent heat fluxes at the air–ocean interface (upward positive), E the evaporation rate, Pr the precipitation rate, F the ice freezing rate, L the latent heat of vaporization, M the ice melting rate, C_{pw} the specific heat of seawater, S_0 the upper-ocean salinity, and S_f the ice salinity ($S_f \sim 0.7$ g/kg). Since $\alpha L/c_{pw} \simeq 0.12$ and $S_0 \simeq 0.03$, the contribution of evaporation rate to the buoyancy flux from the first term on the right-hand side of Eq. (45) is 4 times greater than that from the second term. The values of B are on the order of 10 g/cm^2 yr, which corresponds to a turbulent kinetic energy generation rate of the order of 3×10^{-4} cm^2 s^{-3}. Upward surface buoyancy flux B (heat loss from the ocean surface, or excess evaporation over precipitation, or salt injection due to ice formation) increases the surface density and makes the water parcel descend. The circulation that is driven by the buoyancy flux is called the thermohaline circulation.

5.2.3 Ventilation Ventilation is any process that brings subsurface water to the surface so that heat, momentum, and gases can be exchanged with the atmosphere. Ventilation of deeper water occurs primarily through two processes. The first is the Ekman pumping of water onto density surfaces. The region on an isopycnal surface affected by this process is called the ventilation region. The second process is deep convection. Although a number of estimates of the mean ventilation time for the deep ocean have been proposed, that of about 1400 years, based on radiocarbon, is the most widely quoted. The ventilation time T of the deep ocean is usually measured by the carbon isotope ratios:

$$T = \frac{V_D}{R} = \left[\frac{(^{14}C/C)_S}{(^{14}C/C)_D} - 1\right] T_{^{14}C}, \tag{46}$$

where V_D and R are the volume and the ventilation rate of the deep sea, respectively. Here $(^{14}C/C)_S$ and $(^{14}C/C)_D$ are the carbon isotope ratios for the mean surface and mean deep ocean. The term T_{14_C} is the mean half-life of ^{14}C.

Ventilation is fundamental to understanding the large-scale motion in the deep sea. To understand deep-sea circulation, one needs to answer the dual questions of how a fluid below the surface layer (which is directly driven by the wind) is set into motion, and what determines the depth to which that motion penetrates. Since the ocean has no significant frictional stresses between layers, only the surface water that is immediately acted upon by the wind stress can brought into motion. The existence of subsurface motion therefore requires either ventilation or convective mixing. Ventilation requires the water parcel trajectories to thread back and intersect near-surface regions directly affected by Ekman pumping. Convective mixing leads to momentum transfer between different layers.

This scheme creates at least two regions in the thermocline of the North Atlantic. One region is connected directly to northern latitudes where water is ventilated to the atmosphere and forced downward by Ekman pumping, subsequently being subducted and conserving potential vorticity. Another region is not ventilated to the atmosphere. The unventilated region sweeps across the entire low-latitude North Atlantic from the eastern boundary at the depth of 27.4 kg/m^3 isopycnal. Observed distributions of dissolved oxygen and tritium on this surface are indicative of poor ventilation in this region.

Subduction is a companion concept to ventilation. In the subtropical gyre, surface waters are pumped down into the main thermocline (ventilation) by the action of the prevailing winds. Under mechanical and buoyancy forcing, the mixed layer cools and burrows into the thermocline, entraining thermocline water. Stratification existing in the thermocline determines how the mixed layer responds to the forcing. Alternatively, the mixed layer may warm and retreat, leaving behind newly stratified fluid, which then subducts into the deeper layers.

5.2.4 Convection Generally the ocean is stable, i.e., the lighter water lies over denser water. In polar regions and the Mediterranean Sea, the stratification sometimes becomes quite weak. Slight negative buoyancy in the surface water caused by atmospheric cooling or salt injection will make the water column statically unstable and trigger deep convection. Water columns with vertically homogeneous hydrographic structure (called "chimneys") have been observed in the Arctic, Antarctic, and Mediterranean; they have been observed near ocean boundaries as well as in the open ocean.

Near-boundary convection requires three ingredients:

1. a reservoir,
2. a source of dense water within the reservoir,
3. a reason for the dense water to leave the reservoir and to descend the slope to the deep layer (Killworth, 1983).

Open-ocean convection usually requires a cyclonic gyre and "preconditioning." Upward "doming" of isopycnals in the center of the cyclonic gyre reduces static stability of water column within the gyre. The preconditioning creates a region of very weak static stability within the dome of the gyre where a sufficient surface forcing can drive deep convection. Many processes can be thought of as preconditioning, e.g., topographic forcing, baroclinic instability, etc.

An example of open-ocean convection is a column of water (14-km radius) that was observed within the central region of the Weddell Gyre west of Maud Rise (Gordon, 1978). The normal Antarctic stratification was absent, and the column appeared as a cold, low-salinity, high-oxygen, cyclonic flow (surface velocity about 50 cm/s), which extended to at least 4000 m.

5.2.5 The Conveyor Belt Recent observations and numerical modeling suggest the existence of a "conveyor belt" in the ocean that may serve as a mechanism for long-term climate variability. Deep waters in the northern North Atlantic are convected, sinking to depth and flowing southward along the western boundary of the North and South Atlantic Oceans as the North Atlantic Deep Water (NADW) with high salinity content, hence forming the conveyor belt's lower limb. After passing 30° S, this lower limb of the conveyor joins the Circumpolar Current along the perimeter of the Antarctic continent, along with older deep waters recirculated into the Antarctic from the deep Pacific and Indian Oceans. The lower limb water returns to the surface in the northern Indian and Pacific Oceans, feeding the upper limb of the conveyor. The imported water that feeds the upper limb of the conveyor belt in the Atlantic can have three components: Antarctic surface waters passing through the Drake Passage, Indian surface waters passing around the tip of Africa via the Agulhas Current, and intermediate waters formed at the northern perimeter of the Atlantic segment of the Antarctic. Different conveyors (or climatic regimes) can be envisioned in which only one or two of these three components are active.

6. OCEAN WAVES

An ocean basin filled with slightly compressible and electrically conducting seawater on a weakly magnetized rotating Earth will support a variety of waves. An ocean wave is generally depicted by a periodic function. A one-dimensional wave propagating in the x direction is usually described by a sinusoidal function

$$\zeta = A\cos(kx - \omega t), \tag{47}$$

where ζ is any wave variable, A the wave amplitude, k the wave number, and ω the wave frequency. The quantity $L = 2\pi/k$ is called the wavelength; $T = 2\pi/\omega$ is called the period. The nondimensional quantity $kx - \theta t$ is called the phase angle. The wave propagates with a phase speed

$$C = \omega/k. \tag{48}$$

When the phase speed (C) varies with the wave number (k), the waves with different wavelengths travel with different phase speeds, which is called the wave dispersion. The wave packet, a group of waves with different wavelengths, travels with the group velocity

$$C_g = \frac{\partial \omega}{\partial k}. \tag{49}$$

Hydroelectromagnetic waves (Alfvén waves) are possible in the ocean, but the Earth's magnetic field is so weak that the electromagnetic restoring forces are insignificant compared to other restoring forces; so Alfvén waves are not important in the ocean. Acoustic waves exist as a result of compressibility. Gravity waves arise through the restoring force of gravity. Short, high-frequency capillary waves occur at the interface between air and water through the restoring force of surface tension. Inertial waves appear as a result of the Coriolis force, acting at right angles to the velocity vector. Finally, very slow, planetary waves develop from the variation of quasigeostrophic potential vorticity.

These five basic types of oceanic waves (acoustic, capillary, gravity, inertial, and planetary waves) generally occur together, with five basic restoring forces all acting simultaneously to produce more complicated and mixed waves, such as inertial–gravity waves, mixed Rossby–gravity waves, etc. The effect of the lateral boundary on the inertial–gravity waves leads to a special type of wave named after Lord Kelvin, who first found a solution describing the wave.

6.1 Acoustic Waves

The propagation of sound in the ocean was discussed in Sec. 1.3. The propagation of acoustic waves in the ocean is governed by

$$\nabla^2 p - \frac{1}{U_a^2}\frac{\partial^2 p}{\partial t^2} + \frac{\mu}{\rho U_a^2}\nabla^2\frac{\partial p}{\partial t} = 0, \tag{50}$$

where μ is molecular viscosity of seawater. The ocean surface is a very good reflector/scatterer of sound energy. The ocean bottom, composed of hard rock to soft muds, has large variability in reflection, scattering, and absorption. We usually solve the wave-propagation equation subject to various surface and bottom boundary conditions and obtain information about the propagation and attenuation.

6.2 Gravity and Capillary Waves

The wind generates surface waves. At very low wind speeds, small capillary waves develop on the surface. Surface waves have frequencies ω much higher than the inertial frequency (Coriolis parameter) f and buoyancy frequency N. Therefore, the Coriolis, buoyancy, and viscous terms are usually neglected in the momentum equations. Under these conditions, the fluid motion will be *irrotational.* A scalar velocity potential ϕ can be defined to represent the vector velocity:

$$\mathbf{V} = \nabla\phi. \tag{51}$$

The incompressibility condition, $\nabla \cdot \mathbf{V} = 0$, implies that the velocity potential (ϕ) satisfies Laplace's equation,

$$\nabla^2\phi = 0. \tag{52}$$

To solve this equation, we need surface and bottom boundary conditions. Consider a uniform, frictionless ocean of depth H, whose surface is subject to a small wavelike perturbation

$$\xi(x,y,t) = \xi_0 \cos(kx + ly - \omega t). \tag{53}$$

The pressure across the air–ocean interface should be continuous, i.e.,

$$p = p_a - \tau_s\left(\frac{\partial^2 \xi}{\partial x^2} + \frac{\partial^2 \xi}{\partial y^2}\right). \tag{54}$$

Here p and p_a are oceanic and atmospheric pressures at the ocean surface, and τ_s is surface tension. If $\tau_s = 0$, we have the pure gravity waves; if $\tau_s \neq 0$, we have the gravity-capillary waves. Combining these two conditions, the surface boundary ($z = 0$) condition for ϕ can be derived:

$$\frac{\partial^2\phi}{\partial t^2} + \frac{1}{\rho}\left[-\tau_s\left(\frac{\partial^2}{\partial x^2} + \frac{\partial^2}{\partial y^2}\right) + \rho g\right]\frac{\partial\phi}{\partial z} = 0, \quad z = 0. \tag{55}$$

The ocean bottom is generally considered a rigid boundary. Since this is a linear system, ϕ should have the same harmonics as ξ, i.e.,

$$\phi = \phi_0(z)\sin(kx + ly - \omega t). \tag{56}$$

Substitution of Eq. (56) into Eq. (55) leads to an ordinary differential equation for ϕ_0 with z. By using the surface and bottom boundary conditions, we will have the dispersion relation

$$\omega^2 = (gk_h + k_h^3\tau_s/\rho)\tanh k_h H, \tag{57}$$

where $k_h = \sqrt{k^2 + l^2}$ is the horizontal wave number. The dispersion relation shows that the gravity-capillary waves are generally dispersive ($c_g = \partial\omega/\partial k \neq 0$).

For the pure gravity waves, $\tau_s = 0$, and therefore, the dispersion relation becomes

$$\omega^2 = gk_h \tanh k_h H. \tag{58}$$

If the horizontal wavelength $L_h = 2\pi/k_h$ is much shorter than the ocean depth (deep water), $k_h H = 2\pi H/L_h \gg 1$ and $\tanh k_h H \simeq 1$, and so the deep-water gravity-wave phase speed becomes

$$c = (g/k_h)^{1/2}, \quad k_h H \gg 1, \tag{59}$$

while the group velocity is

$$c_g = c/2. \tag{60}$$

The deep-water gravity waves are dispersive with the phase speeds depending inversely on the square root of the wave number; therefore, longer waves travel faster than short ones. In addition, the group velocity is $\frac{1}{2}$ of the phase speed, so that within a traveling wave packet, wave crests first appear toward the rear of the group. An initially localized packet consisting of many wave components spreads out in space and time as it propagates.

If the horizontal wavelength is much longer than the ocean depth (shallow water), $k_h H = 2\pi H/L_h \ll 1$ and $\tanh k_h H \simeq k_h H$, and so the shallow-water gravity-wave phase speed becomes

$$c = c_g = (gH)^{1/2}, \quad k_h H \ll 1. \tag{61}$$

Thus, the shallow-water gravity waves are nondispersive. Earthquakes generate a special kind of gravity wave called a tsunami, which behaves as a shallow-water wave even in deep water.

6.3 Rossby Waves

For large-scale motion, departure from geostrophic balance (sometimes called geostrophy) is small. Such flow is called quasigeostrophic flow. The quasigeostrophic potential vorticity (QGPV) Π_{QG} is

$$\Pi_{QG} = f_0 + \beta y + \zeta_g + \Lambda\frac{\partial}{\partial z}\left(\frac{f_0^2}{N^2}\frac{\partial\Phi}{\partial z}\right), \tag{62}$$

where $\zeta_g = (\partial^2/\partial x^2 + \partial^2/\partial y^2)\Phi$ is the geostrophic vorticity, Φ is the geopotential computed by Eq. (19), $N^2 = E$ is the square of the buoyancy frequency, $\Lambda = 0$ refers to the barotropic mode, and $\Lambda = 1$ to the baroclinic mode.

Rossby waves are associated with the β effect. The conservation of quasigeostrophic potential vorticity,

$$\left(\frac{\partial}{\partial t} + u\frac{\partial}{\partial x} + v\frac{\partial}{\partial y}\right)\Pi_{QG} = 0, \tag{63}$$

leads to a special form of oscillation on the planetary scale. An individual water parcel experiences no change in its QGPV. If a water parcel moves poleward, then as f increases, ζ_g has to decrease for the barotropic mode ($\Lambda = 0$) in order to keep Π_{QG} constant. If, furthermore, the geostrophic vorticity is mostly curvature, this means that a poleward-moving water parcel loses cyclonic curvature and ultimately must curve anticyclonically and move equatorward. The converse is true for a water parcel moving equatorward. This mechanism leads to the generation of Rossby waves (Rossby, 1936).

If the disturbances are sufficiently small, the advection terms in the QGPV conservation equation (63) can be neglected,

$$\frac{\partial}{\partial t}\left[f_0 + \beta y + \nabla_H^2 \Phi' + \Lambda \frac{\partial}{\partial z}\left(\frac{f_0^2}{N^2}\frac{\partial \Phi'}{\partial z}\right)\right] = 0, \tag{64}$$

where Φ' is the geopotential perturbation and $\nabla_H^2 = \partial^2/\partial x^2 + \partial^2/\partial y^2$ is the horizontal Laplacian. If we assume no vertical velocity at either the surface or bottom, the sinusoidal wave solution has the form (when N = const)

$$\Phi' = \mathrm{Re}\left[\cos\frac{n\pi z}{H}\, e^{i(kx+ly-\omega t)}\right], \tag{65}$$

where H is the ocean depth. Substitution of Eq. (65) into Eq. (64) leads to the Rossby-wave dispersion relation

$$\omega = -\frac{\beta k}{k^2 + l^2 + \Lambda m_n},$$
$$m_n = \frac{n\pi f_0}{NH}, \quad n = 1, 2, \ldots . \tag{66}$$

For baroclinic Rossby waves, $\Lambda = 1$, formula (66) shows the dispersion relation of the nth baroclinic mode. For barotropic Rossby waves, $\Lambda = 0$, the dispersion relation becomes

$$\omega = -\beta k/(k^2 + l^2). \tag{67}$$

Rossby waves are highly dispersive and propagate westward. Rossby-wave phase speed and group velocity can be easily computed from the dispersion relation:

$$\mathbf{C} = \left(\frac{\omega}{k}, \frac{\omega}{l}\right), \quad \mathbf{C}_g = \left(\frac{\partial\omega}{\partial k}, \frac{\partial\omega}{\partial l}\right). \tag{68}$$

If the barotropic Rossby waves propagate only in the x direction ($l = 0$), the phase speed is

$$C_x = \omega/k - \beta/k^2 \tag{69}$$

and group velocity is

$$C_{gx} = \frac{\partial\omega}{\partial k} = \frac{\beta}{k^2}. \tag{70}$$

The highly dispersive nature of these waves is indicated by the fact that the phase speed and group velocity have the same magnitude but opposite directions.

6.4 Solitary Rossby Waves

Rossby waves are usually nonlinear and dispersive. The degree of nonlinearity depends on the amplitude and length of the waves, while the phase dispersion is determined by their length relative to the vertical length scales related to stratification and water depth. In general, these two dynamic effects tend to balance in such a way that nonperiodic, solitary waves emerge from initial disturbances.

Consider barotropic Rossby waves ($\Lambda = 0$) being excited from a zonal basic flow ($\bar{U}$) and propagating in a zonal channel bounded by southern and northern boundaries $y = y_S$, $y = y_N$. Introducing a scaled coordinate $\xi = \epsilon^{1/2}(x - Ct)$ moving with the wave at speed C, we seek a steady solution

$$\Phi = -\frac{1}{f_0}\int_{y_S}^{y} \bar{U}dy + \epsilon\Phi_1(\xi,y) + \epsilon^2\Phi_2(\xi,y) + \cdots, \tag{71a}$$

$$C = C_0 + \epsilon C_1 + \epsilon^2 C_2 + \cdots, \tag{71b}$$

where ϵ is a measure of the wave amplitude. Notice that the disturbances are no longer sinusoidal functions. Substitution of Eq. (71) into Eq. (63) leads to nonlinear equations for the components Φ_n, $n = 1, 2, \ldots$. If the first component is separated into

$$\Phi_1 = A(\xi)\psi(y), \tag{72}$$

the function $A(\xi)$ satisfies the steady Korteweg–de Vries (KdV) equation

$$\frac{dA}{d\xi} + rA\frac{dA}{d\xi} + s\frac{d^3A}{d\xi^3} = 0, \tag{73}$$

and solvability to $O(\epsilon^2)$ is insured if r and s satisfy certain conditions.

The soliton solutions are a linear sum of a barotropic soliton and a baroclinic soliton (first baroclinic mode). If a rigid lid is imposed, the barotropic solitary Rossby wave has an eastward phase speed ($c > 0$), and the baroclinic solitary Rossby wave has a phase speed that is either westward with a faster speed than βa_1^2 ($c < -\beta a_1^2$), or eastward ($c > 0$). Here a_1 is the Rossby radius of deformation for the first baroclinic mode. When the barotropic and baroclinic solitons are combined, only the baroclinic part moves west.

6.5 Kelvin Waves

When planetary waves approach the coastal zone, their behavior changes. The lateral boundary supports a different type of wave called a Kelvin wave. Consider a straight coastline, placing the x axis along the coast and the y axis perpendicular to the coast so that offshore is positive. Assume that the offshore variation of wave height and alongshore velocity exponentially decays. Since the coast is a rigid boundary, v vanishes at $y = 0$; therefore, v should vanish for the other values of y as well. Kelvin-wave solutions can be obtained by letting $v = 0$ in the momentum and continuity equations,

$$\frac{\partial u}{\partial t} = -g\frac{\partial \eta}{\partial x}, \tag{74a}$$

$$\frac{\partial \eta}{\partial t} + H\frac{\partial u}{\partial x} = 0, \tag{74b}$$

$$fu = -g\frac{\partial \eta}{\partial y}, \tag{74c}$$

where η is the surface elevation. Solutions of Eq. (74) are

$$\eta = A_\eta e^{-y/a}\cos(kx - \omega t),$$
$$u = \eta(g/H)^{1/2}, \quad v = 0, \tag{75}$$

where A_η is the amplitude for wave variable η, and a is the barotropic Rossby radius of deformation [see Eq. (76)]. The solution shows two important features of coastal trapped Kelvin waves:

1. The amplitude is greatest at the boundary and decreases exponentially away from it, and
2. propagation is with the coast at the right (left) in the Northern (Southern) Hemisphere.

Kelvin waves also occur along the equator where f changes sign, propagating from west to east. Abrupt changes in topography can also support Kelvin waves; in this case, they are called double Kelvin waves because there is motion on both sides of the depth change.

7. INSTABILITIES, EDDIES, AND FRONTS

The kinetic energy of the ocean is dominated by synoptic or mesoscale eddies, rotating about nearly vertical axes and extending throughout the water column. The scale of these eddies is controlled by the baroclinic radius of deformation. Typical scales are 100 km and 100 d with a rotational velocity of about 10 cm/s. Mesoscale eddies are generated by baroclinic and barotropic instabilities.

7.1 Rossby Radius of Deformation

Two fundamentally important frequencies in ocean dynamics are f and N, representing buoyancy and rotational effects. For small-scale motion, buoyancy dominates, while for large-scale motion, the rotational effect is dominant. There must exist a horizontal length scale at which rotation becomes important. This length scale, a, is called the Rossby radius of deformation, and the barotropic mode is given by

$$a = \sqrt{gH}/|f| \tag{76}$$

while the nth baroclinic mode is

$$a_n = \frac{NH/n\pi}{|f|}. \tag{77}$$

If $f = 10^{-4}$ s^{-1} (about 45° N), and $H = 4$ km, then the barotropic Rossby radius $a \simeq 2000$ km. For shallower water on the continental shelf, $H \simeq 40$ m, then $a \simeq 200$ km. The baroclinic Rossby radius is much shorter. The typical value for the first baroclinic mode is 10–30 km, which is a natural scale for the ocean and is often associated with eddies, fronts, and coastal processes.

7.2 Baroclinic Instability

Baroclinic instability originates from the horizontal inhomogeneity of the density field and allows eddies to grow with energy coming from the mean potential energy field. The horizontal density gradient is related to the vertical shear of the geostrophic flow (U_g, V_g) by the *thermal wind* equation:

$$\frac{\partial U_g}{\partial z} = \frac{g}{f\rho}\frac{\partial \rho}{\partial y}, \quad \frac{\partial V_g}{\partial z} = -\frac{g}{f\rho}\frac{\partial \rho}{\partial x}. \tag{78}$$

The existence of vertical shear leads to baroclinic instability and eddy generation. In this instability process, energy is taken from the large-scale potential energy distribution and put into disturbance-scale kinetic and potential energies. Stommel (1965) suggested that the enormous amount of potential energy present in the stratification of the ocean and associated with the mean circulation is a possible source for baroclinic instability.

7.3 Barotropic Instability

Barotropic instability originates from the lateral shear of the flow and is an important mechanism to generate eddies in the tropics. For simplicity, consider a zonal current U that changes with latitude, i.e., $U = U(y)$. If the lateral shear satisfies the condition (Kuo, 1949)

$$\beta - \frac{d^2U}{dy^2} = 0 \tag{79}$$

somewhere in the flow, then the flow U is barotropically unstable. Eddies generated by the shear in the mean flow gain energy from the mean flow kinetic energy.

Within the ocean, eddy generation is the result of a mixture of baroclinic and barotropic instability.

7.4 Mesoscale Eddies

Mesoscale eddies are energetic disturbances of currents and the associated fields of salinity, temperature, and density (Robinson, 1983). More kinetic energy is contained in these features than in any other form of oceanic motion. These eddies are partly organized and yet highly irregular, with spatial scales of 10–100 km and temporal scales of weeks to months. The energy is distributed unevenly in space, with the greatest energy associated with western-boundary currents. Mesoscale eddies provide an important mechanism for the horizontal stirring of the ocean, and for transfers of energy, momentum, and water properties. Numerical model results (Schmitz and Holland, 1986) suggest that these eddies may also drive the recirculation region associated with subtropical gyres. The mesoscale features can also be found from satellite altimetry data. Geostrophic currents are maintained by horizontal pressure gradients and are expressed as sea-surface slopes relative to the level of constant geopotential (the geoid). All mesoscale current systems exhibit high levels of sea-height variability due to meandering, changes in speed or current reversal, vortex shedding, and migration of eddies. For example, western-boundary currents such as Gulf Stream and Kuroshio are characterized by high rms sea-height variability (Plate 3).

An important category of mesoscale eddy is called a *ring*. Rings associated with the Gulf Stream have been well documented (see Plate 2) (Richardson, 1983). These rings appear to form from cut-off Gulf Stream meanders. Two types of rings have been observed, warm-core rings to the north of the Stream and cold-core rings to the south. Cold-core rings are cyclonic, are carried to the west by the Gulf Stream recirculation, and are reabsorbed by the Gulf Stream to the south of Cape Hatteras. A young cold-core ring has rotational velocities of 1.5 m/s and is remarkably similar to a Rankin vortex: Speed increases with distance from the center (a nearly solid rotating core), reaching a maximum near a radius of 30–60 km, and then drops off toward the outer limits of the ring. Ten cold-core rings have been observed at a given time to the south of the Stream.

Warm-core rings are anticyclonic and are formed at a rate of about five per year. They may be reabsorbed by the Gulf Stream, but a few impact the continental shelf/slope where they are the most energetic source of variability on the Slope Water (Joyce, 1991). Rings are also observed in other strong western-boundary currents, most notably the Kuroshio and Agulhas.

7.5 Quasigeostrophic Turbulence

It is possible to extend the concepts of eddies and planetary waves to two-dimensional quasigeostrophic turbulence. Beginning first with an f plane in the absence of friction or forcing, conservation equations for energy E,

$$\frac{\partial E}{\partial t} = 0, \quad E = \int E(k)dk, \tag{80}$$

and for enstrophy Z,

$$\frac{\partial Z}{\partial t} = 0, \quad Z = \int k^2E(k)dk, \tag{81}$$

require an energy transfer to smaller wave numbers and an enstrophy transfer to larger wave numbers. This means that vorticity and eddy fields change with time, with eddies increasing in size and vorticity contours increasing in length. The rate of cascade controls the spectral slope so that for the energy cascading range,

$$E(k < k_0) \sim k^{-3/5}, \tag{82}$$

and for the enstrophy cascading range,

$$E(k > k_0) \sim k^{-3}. \tag{83}$$

There is some suggestion that the ocean behaves in this manner, but in numerical models, spectral slope is related to the friction parameterization.

With β, linear Rossby waves will propagate to the west and disperse energy in space, but can also concentrate energy at a specific λ. The transition between turbulence and waves occurs for wave steepness

$$k_\beta = (\beta/2U)^{1/2}, \tag{84}$$

and energy and enstrophy cascades may be impeded. For a typical midlatitude synoptic motion, $k_\beta \sim 70$ km. This means that eddies in the energy cascade region increase scale to k_β, then propagate as Rossby waves. The waves undergo transition to zonal anisotropy, eventually creating a field of alternating zonal jets with a width scale of $2U_{RMS}/\beta$.

Stratification and topography further modify the behavior of quasigeostrophic turbulence. With stratification, vertical transfers of energy occur through interface vortex stretching and nonlinear modal coupling. Transfer of energy to both smaller and larger scales is possible. The latter occurs through interaction of barotropic and baroclinic wave triads. Barotropic triads involve energy transfer to larger scales only, while for baroclinic triads layers act independently for $k > k_{Rd}$, and energy is transferred to smaller scales for $k < k_{Rd}$, where k_{Rd} is the wave number for the Rossby radius of deformation for the first baroclinic mode.

In general, the effect of topography on two-dimensional turbulence is dependent on the ratio

$$r = \left(\frac{V}{fL}\right)\left(\frac{f}{H}\frac{\partial H}{\partial x}\right)^{-1} \tag{85}$$

and the amount of energy in each layer. With strong lower-layer flow, cascade occurs until they reach the wavelength of the topography and become locked to f/H.

7.6 Submesoscale Coherent Vortices

Submesoscale coherent vortices are also observed in the ocean (McWilliams, 1985). Lenses of Mediterranean water are observed throughout the Eastern North Atlantic and are readily identified by their water properties and are called *Meddies*. Meddies have diameters of 30–70 km and thicknesses of 0.5–1 km; they are anticyclonic, 15–20 exist at any one time, and they cover about 5% of the area of the Canary Basin. Their core appears to be in solid-body rotation with a period of five days. They appear to erode from their sides and have 2–4-year life-time (Armi *et al.*, 1989). Stratification is minimum in their center, and their motion is primarily through advection by the mean flow. Similar features have been observed in other locations where strong water mass contrasts exist, e.g., the Gulf of California (Bray, 1988).

7.7 Fronts

Fronts are regions with sharp horizontal gradients in temperature, salinity, or density. They occur on a variety of scales: The Antarctic Polar Front circumscribes the Earth at 60° S while in coastal waters fronts associated with internal waves or river plumes may be less than 1 km in length. Fronts may occur in the temperature or salinity field as well as the density field. More generally, fronts are formed by horizontal convergence, by a water-mass boundary, or by coastal upwelling. The Antarctic Polar Front and the Kuroshio/Oyashio Front are regions where intermediate water masses are formed. Surface jets are associated with density fronts.

8. SMALL-SCALE TURBULENCE AND MIXING

Although eddies and quasigeostrophic turbulence serve to stir the ocean horizontally, smaller-scale motions are necessary to connect these scales to scales on which molecular diffusion can act. At the boundaries of the ocean, wind stress and shear provide mechanisms for mixing, while at fronts intrusive features serve to enhance mixing, and within the interior of the ocean, processes associated with internal waves provide energy for mixing. Since turbulence is discussed elsewhere in this Encyclopedia (see TURBULENT FLOWS), we will concentrate on those aspects that are unique to the ocean.

8.1 Instability and Turbulence

There are two major types of instabilities in the ocean relating to the onset of turbulence: static (gravitational) and dynamic. The static stability is given by Eq. (9), so that instability occurs when $E = N^2 < 0$. Dynamic instability usually appears at the interface of two layers moving initially at different velocities. Such an instability is called Kelvin–Helmholtz instability. Two nondimensional numbers, Reynolds number (*Re*) and Richardson number (*Ri*), are used to identify the onset of turbulence in the ocean:

$$Re = LU/\nu, \quad Ri = N^2/U_z, \tag{86}$$

where L, U, and U_z are scales for length, velocity, and vertical gradient of velocity, ν is the kinematic coefficient of molecular viscosity, and N is the buoyancy frequency. If the Reynolds number is greater than some criterion

$$Re > Re_{cr}\,(\sim 10^6), \tag{87a}$$

or if the Richardson number is smaller than some criterion

$$Ri < Ri_{cr}\,(\sim 0.25), \tag{87b}$$

turbulence will occur. These criteria provide useful guidelines for the ocean.

8.2 Generation and Maintenance of Turbulence

The onset of turbulence may occur suddenly through a breakdown of originally nonturbulent flow in certain localized regions. The cause of the breakdown is an instability caused by processes naturally occurring within the ocean, e.g., breaking internal waves, convective overturning, etc. Once turbulence is produced and becomes fully developed in the sense that its statistical properties achieve a steady state, the instability mechanism is no longer required to sustain turbulence. The turbulent kinetic energy (TKE) is converted from the mean flow kinetic energy by the shear production (*S*), and from the potential energy by the buoyancy production (*B*), damped by viscous dissipation (*D*), and transported from or to other regions of flow (*Tr*). This leads to the TKE equation

$$\frac{d(\mathrm{TKE})}{dt} = S + B - D + Tr. \tag{88}$$

Thus, in order to maintain turbulence, one or more generating mechanisms must be active continuously.

8.3 Statistical Characteristics of Turbulence

To describe a turbulent field, it is necessary to know the probability density functions (pdf), which require measurements with sensitive and stable instruments for a sufficiently long period of time and at a fine spatial grid. The problem should become much simpler if the turbulent field could be assumed homogeneous and stationary. In such a case, the turbulent field can be depicted by its spectral density function, $E(k)$, which is generally in power-law form:

$$E(k) \simeq k^{-n}, \tag{89}$$

where k is the wave number. Laboratory results indicate that the exponent n equals 5/3 for the inertial–convective range, and equals 1 in the convective–diffusive range. Oceanic spectra of velocity and temperature appear to behave in this manner.

8.4 Double Diffusion

The ocean generally has a multilayer structure, caused by the stratification of two different components: heat and salt. The rate of molecular diffusion of heat is about 100 times that for salt. Consider a layer of warmer, saltier water above cooler, fresher water, such that the upper layer has the same density as or less density than the lower layer. Then, the saltier water at the interface will lose heat much faster than it will lose salt. If the density difference between the layers is small, the saltier water above may become heavier than the cooler, fresher water below, sink into the lower layer, and form tall thin convective cells, called "fingers." In these cells, descending branches of water diffuse heat horizontally more rapidly than salinity and, therefore, become heavier than surrounding water and continue to sink. A similar mechanism occurs in the rising branch. This process is called double-diffusive convection. The density ratio,

$$R_\rho = \alpha \Delta T / \tilde{\beta} \Delta S, \tag{90}$$

is a controlling parameter in the growth rates of salt fingers and the resulting fluxes of heat and salt. The terms ΔT and ΔS are the temperature and salinity jumps between the two layers. There exists a criterion $R_{\rho cr}$ such that if

$$R_\rho \leq R_{\rho cr},$$

double-diffusion takes place. Here $R_{\rho cr}$ ranges from 1.2 to 2.2. In the laboratory, salt fingers have lengths of 20–30 cm with a spacing between fingers of about 1 cm and a temperature difference of 0.1 °C. In the ocean, this produces a "staircase" structure in temperature, salinity, and density where the steps are meters thick.

Although the occurrence of double diffusion is related to $R_{\rho cr}$, the rate of double diffusion has not been parametrized in terms of large-scale oceanic fields. The mass flux associated with double diffusion is downward so that it does not help to maintain the thermocline.

8.5 Microstructure

Bounds for mixing processes can be inferred from large-scale distributions. For example, a vertical advective–diffusive model of the ocean interior,

$$K_v \frac{d^2T}{dz^2} - w \frac{dT}{dz} = 0, \tag{91}$$

governs the vertical distribution of temperature so that for the vertical velocity $w \simeq 4$ m/yr, the vertical mixing coefficient $K_v = 10^{-4}$ m^2/s provides a bound for other estimates of vertical mixing. The term K_v can be estimated from sufficient measurements of temperature microstructure from free-fall probes. In regions with no lateral gradients, we find

$$K_v = \mu k C, \tag{92}$$

where the Cox number $C = \overline{(\partial T'/\partial z)^2} (\partial \bar{T}/\partial z)^2$, and the multiplier μ takes a value between 1 and 3 depending upon layering or isotropy at small scales (Garrett, 1979). Measurements yield $K_v \simeq 10^{-6}$ m^2/s, implying that mixing takes place either at boundaries or water-mass formation regions but at rates that do not seem sustainable. Internal waves or internal tides or biological activity may provide sufficient energy for the missing mixing energy. In general, vertical mixing appears to be inversely proportional to N.

9. CONCLUDING REMARKS

In this review, there were a number of important areas that are normally considered within the purview of physical oceanography that we were unable to describe. These include coastal (Huyer, 1990) and estuarine (Officer, 1976) circulation, sea ice (*q.v.*) and related processes, internal waves, and electromagnetics and associated measurements (Sanford, 1971). The latter may be increasingly important as abandoned submarine cables are provided for research use.

Oceanographers are currently engaged in a number of important programs to understand ocean circulation better. Global surveys using carefully calibrated CTDs and water samplers for collection of nutrient, tritium, radiocarbon, and Freon are underway. These data, with *Topex/Poseidon* altimetric measurements, will provide a better description of thermohaline circulation and poleward transport of heat. They will also serve to constrain numerical ocean circulation models, which in turn can be used to study physical processes and design future experiments. In the Tropical Pacific, moored buoys are providing real-time oceanographic data via satellite, better to understand and predict *El Niño*. Modern supercomputers allow the coupling of global ocean and atmosphere models and hence a better understanding of climate. These computers also allow ocean models to be used for forecasts, which can be substantially improved through better ocean data assimilation techniques.

The ocean dominates the surface of the Earth. It has tremendous capacity for heat, carbon dioxide, and organic waste as well as biological productivity. We need to be able to use this resource in a balanced, ecologically sound manner without inducing anthropogenic change. Toward this end, we need a global ocean-observing system that will allow a continuous quantitative assessment of the state of the ocean.

ACKNOWLEDGMENTS

Ching-Sang Chiu provided valuable comments on Sec. 1.3, and Stevens Tucker assisted with Sec. 1.4. Albert Semtner and Carl Wunsch graciously provided Plates 1 and 3, respectively, and Paul Jessen constructed Figures 1 and 2.

GLOSSARY

Baroclinic: The state of the ocean in which surfaces of constant pressure (isobars) intersect surfaces of constant density (isopycnals).

Baroclinic Instability: A hydrodynamic instability arising from a baroclinic flow and vertical shear, allowing eddies to grow with energy coming from the mean potential energy field.

Barotropic: The state of the ocean in which surfaces of constant pressure (isobars) coincide with surfaces of constant density (isopycnals).

Barotropic Instability: A hydrodynamic instability arising from lateral shear in a current, which allows eddies to grow with energy coming from the mean flow kinetic energy.

Ekman Layer, Ekman Transport, and Ekman Pumping: The Ekman layer is the oceanic planetary boundary layer (typically 100 m thick) through which the stresses exerted on the water surface are transmitted into the ocean. For steady wind stress in the open ocean, the surface wind stress is balanced by the Coriolis force. The Ekman transport is the vertically integrated Ekman-layer currents, and is to the right (left) of the wind in the Northern (Southern) Hemisphere and depends only on the magnitude of the wind stress and the latitude. Ekman pumping is the vertical motion in the Ekman layer forced by convergence and divergence of the surface flow, associated with spatial variability of wind stress.

Convection: Vertical motion caused by dense water lying over lighter water.

CTD: An instrument to measure vertical profiles of conductivity and temperature vs depth (pressure).

Double Diffusion: Convection caused by different diffusion coefficients for heat and salt.

Geostrophic Flow: Flow due to the balance between Coriolis and pressure-gradient forces.

Mesoscale Eddies: Horizontal vortices whose size is controlled by the Rossby radius of deformation, e.g., 10–100 km.

Nutrients: Nonconservative chemicals (phosphate, silicate, nitrate) that identify water properties and are used to trace flow patterns.

Ocean Acoustic Tomography: A technique for determining ocean density and velocity structure by measuring the arrivals of specific acoustic rays.

Potential Density, and Potential Temperature: The density and temperature a water parcel would attain if raised (or lowered) adiabatically to the surface or other specified reference pressure.

Quasigeostrophic Flow: For large-scale motion, departure from geostrophic balance is small. Such flow is called quasigeostrophic flow.

Stratification: The vertical structure of the ocean.

Subduction: Newly stratified water leaving the ocean-surface mixed layer to enter deeper layers.

Subtropical Gyre: A basin-scale subtropical anticyclone.

Sverdrup Transport: The wind-driven meridional transport in the ocean interior. It includes both the Ekman transport and the geostrophic transport caused by Ekman pumping.

Thermohaline Circulation: The flow field associated with the sinking of dense water at high latitudes, its subsequent equatorward flow in deep western-boundary currents, and finally deep flow into the ocean interior and upwelling.

T-S Diagram: A characteristic diagram used to study water properties and to identify water masses and water types.

Ventilation: Any process that brings deep water to the surface so that it exchanges heat, momentum, and gases with the atmosphere.

Western-Boundary Intensification: The appearance of strong currents (e.g., the Gulf Stream and the Kuroshio) along the western boundaries of the oceans, as contrasted to weak currents in the interior and at the eastern boundary. This large-scale east–west asymmetry of the ocean circulation is called western-boundary intensification.

Works Cited

Armi, L., Herbert, D., Oakey, N., Price, J. F., Richardson, P. L., Rossby, H. T., Ruddick, B. (1989), "Two Years in the Life of a Mediterrean Salt Lens," *J. Phys. Oceanogr.* **19**, 354–370.

Bray, N. A. (1988), "Water Mass Formation in the Gulf of California," *J. Geophys. Res.* **93**, 9223–9240.

Broecker, W. S. (1991), "The Great Ocean Conveyor," *Oceanography* **4**, 79–89.

Ekman, V. M. (1905), "On the Influence of the Earth's Rotation on Ocean Currents," *Arch. Math. Astron. Phys.* **2**, (11).

Garrett, C. (1979), "Mixing in the Ocean Interior," *Dyn. Atmos. Oceans,* **3**, 239–265.

Gordon, A. L. (1978), "Deep Antarctic Convection West of Maud Rise," *J. Phys. Oceanogr.* **8**, 600–612.

Huyer, A. (1990), "Shelf Circulation," in: B. Le Mehaute and D. M. Hanes (Eds.), *The Sea*, Vol. 9A, New York: John Wiley and Sons, pp. 423–466.

Jerlov, N. G. (1976), *Marine Optics*, Amsterdam: Elsevier Scientific Publishing Co.

Joyce, T. M. (1991), "Review of U.S. Contributions to Warm-Core Rings," *Rev. Geophys.* Suppl. 610–616.

Killworth, P. D. (1983), "Deep Convection in the World Ocean," *Rev. Geophys. Space Phys.* **21**, 1–26.

Kuo, H. L. (1949), "Dynamical Instability of Two-Dimensional Nondivergent Flow in a Barotropic Atmosphere," *J. Meteorol.* **6**, 105–122.

Luyten, J. R., Pedlosky, J., Stommel, H. (1983), "The Ventilated Thermocline," *J. Phys. Oceanogr.* **13**, 292–309.

McWilliams, J. C. (1985), "Submesoscale Coherent Vortices in the Ocean," *Rev. Geophys.* **23**, 165–182.

Munk, W. H. (1950), "On the Wind-Driven Ocean Circulation," *J. Meteorl.* **7**, 79–93.

Munk, W. H. (1966), "Abyssal Receipes," *Deep-Sea Res.* **13**, 707–730.

Officer, C. B., (1976), *Physical Oceanography of Estuaries and Associated Coastal Waters*, New York: Wiley.

Richardson, P. L. (1983), "Gulf Stream Rings," in: A. R. Robinson (Ed.), *Eddies in Marine Science*, New York: Springer-Verlag, pp. 16–65.

Robinson, A. R. (1983), *Eddies in Marine Science*, New York: Springer-Verlag.

Sanford, T. B. (1971), "Motionally Induced Electric and Magnetic Fields in the Sea," *J. Geophys. Res.* **76**, 3476–3492.

Schmitz, W. J., Jr., Holland, W. R. (1986), "Observed and Modeled Mesoscale Variability Near the Gulf Stream and Kuroshio Extension," *J. Geophys. Res.* **91**, 9624–9638.

Schmitz, W. J., Jr., McCartney, M. S. (1993), "On the North Atlantic Circulation," *Rev. Geophys.* **31**, 29–49.

Stommel, H. (1948), "The Westward Intensification of Wind-Driven Ocean Currents," *Trans. Am. Geophys. Union* **99**, 202–206.

Stommel, H. (1958), "The Abyssal Circulation," *Deep-Sea Res.* **5**, 80–82.

Stommel, H. (1965), *The Gulf Stream: A Physical and Dynamical Description*, 2nd ed., Berkeley: Univ. of California Press.

Sverdrup, H. U. (1947), "Wind-Driven Currents in a Baroclinic Ocean, with Application to the Equatorial Currents of the Eastern Pacific," *Proc. Natl. Acad. Sci. U.S.A.* **33**, 318–326.

UNESCO (1981), *Tenth Report of the Joint Panel on Oceanographic Tables and Standards*, UNESCO Technical Papers in Marine Sci., Vol. 36, Paris: UNESCO.

Further Reading

Apel, J. R. (1987), *Principles of Ocean Physics*, New York: Academic Press.

Chu, P. C., Gascard, J. C. (1991), *Deep Convection and Deep Water Formation in the Oceans*, Elsevier Oceanography Series, Vol. 57, Amsterdam: Elsevier.

Clay, C. S., Medwin, H. (1977), *Acoustical Oceanography*, New York: John Wiley and Sons.

Gill, A. E. (1982), *Atmosphere-Ocean Dynamics*, New York: Academic Press.

Godin, G. (1972), *The Analysis of Tides*, Toronto: Univ. of Toronto Press.

Kinsman, B. (1965), *Wind Waves*, Englewood Cliffs, NJ: Prentice Hall, Inc.

Pedlosky, J. (1987), *Geophysical Fluid Dynamics*, 2nd ed., New York: Springer-Verlag.

Pickard, G. L., Emery, W. J. (1990), *Descriptive Oceanography: An Introduction*, 5th ed., Oxford, England: Pergamon Press.

Pond, S., Pickard, G. L. (1983), *Introductory Dynamical Oceanography*, 2nd ed., Oxford, England: Pergamon Press.

Preisendorfer, R. W. (1976), *Hydrologic Optics*,

Vols. I–IV, Honolulu, HI: National Oceanic and Atmospheric Administration.

Robinson, A. R. (1983), *Eddies in Marine Science*, New York: Springer-Verlag.

Semtner, A. J., Chervin, R. M. (1992), "Ocean General Circulation from a Global Eddy-Resolving Model," *J. Geophys. Res.* **97**, 5493–5550.

Stommel, H. (1965), *The Gulf Stream: A Physical and Dynamical Description*, 2nd ed., Berkeley: Univ. of California Press.

Stommel, H. (1987), *A View of the Sea*, Princeton: Princeton University Press.

Stommel, H., Moore, D. W. (1989), *An Introduction to the Coriolis Force*, New York: Columbia University Press.

Warren, B. A., Wunsch, C. (1981), *Evolution of Physical Oceanography*, Boston: MIT Press.

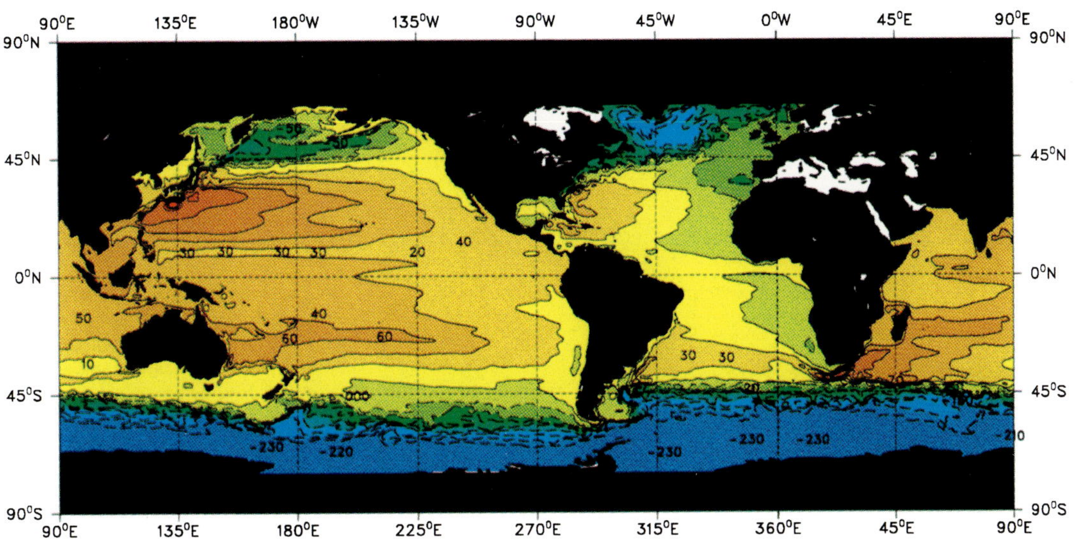

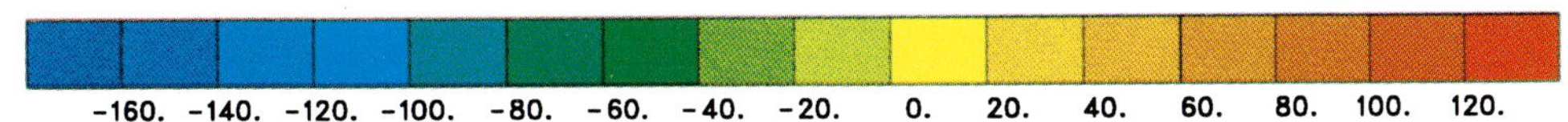

PLATE 1. Dynamic topography of an ocean model sea surface. Results were obtained from a global eddy-resolving ocean model. Units are centimeters. (Courtesy A. Semtner.)

PLATE 2. Composite Advanced Very High Resolution Radiometer (AVHRR) image of the Western North Atlantic during April 1984. The image was produced by compositing data from 35 satellite passes obtained during the first week in April; this was necessary to remove clouds. The image is printed with an artificial color scale; Reds and oranges are warm, 24°–28° C; yellows and greens 17°–23° C; blues 10°–16° C; and purples are cold, 2°–9° C. The Gulf Stream appears as a band of warm water, originating in the Florida Straits, separating from North America at Cape Hatteras, and thence flowing west in the North Atlantic, developing meanders. Two cold-core rings appear south of the Stream while a warm-core ring appears to be almost pinched off from the Stream to the south of the southern tip of Nova Scotia. (Courtesy NASA.)

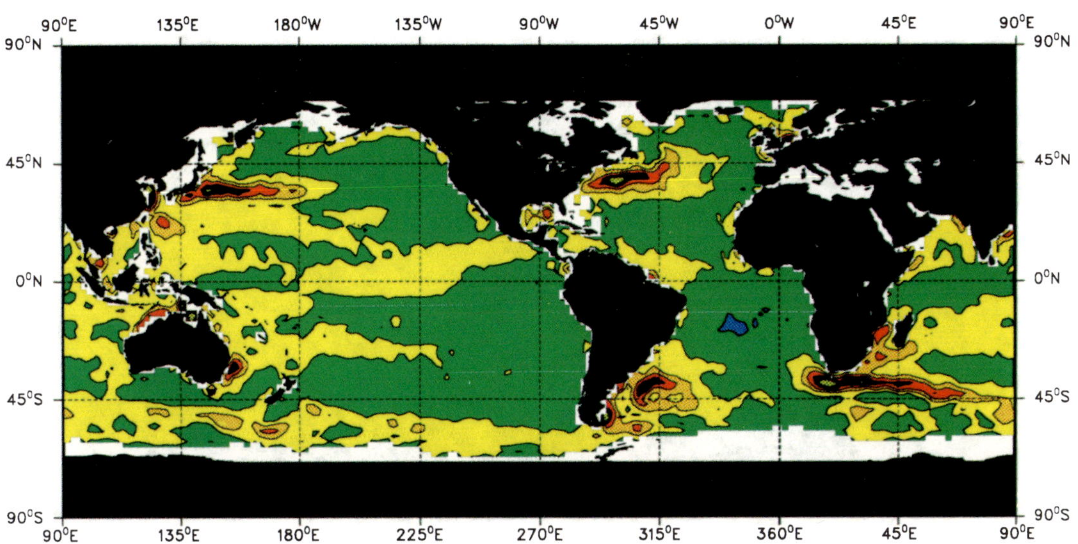

PLATE 3. Root mean square variability of sea surface height determined by *Geosat*. Contours are centimeters. (Courtesy C. Wunsch.)

OIL SHALES AND TAR SANDS

PAUL A. PETZRICK, *1116 Gumbottom Road, Crownsville, Maryland, U.S.A.*

INTRODUCTION

Oil shale and tar sand are relatively recent terms for rocks containing hydrocarbon materials that are of marginal economic interest to present society but have been used for centuries. Interest in oil shales and tar sands (OS&TS) has declined over the last decade, repeating historic cycles related to oil prices and the prognosis for petroleum supplies. The profile of OS&TS use since prehistoric time reflects widespread availability and relative ease of extraction of useful products from these resources by primitive technologies. Large-scale production of liquid fuel from OS&TS using modern technologies, however, has not generally been competitive with modern petroleum operations. The history of

3-527-28134-7/95/$5.00 + .50

small-scale operations exploiting these ubiquitous solid hydrocarbons, including asphaltites and lake asphalt, to produce a variety of products has been more stable. The components of the hydrocarbon spectrum, whether progressing from oil shale to petroleum to tar sands or from peat to coals, are formed and transformed under a great variety of conditions from diverse materials. As a result, the hydrocarbon resources found today have an enormous variety of physical properties and compositions. The terminology that has evolved for these resources is inconsistent. Labels, from ancient use of the word "asphalt" to the modern misnomers, oil shale and tar sand, are usually related to first use, discoverer, or location of first discovery rather than to composition. Historically, the efficiency of OS&TS extraction technologies has continually improved, whereas use of these substances has reflected cycles in markets and world economics. Just as OS&TS became important resources when wood and whale oil became scarce, they may become crucial resources as petroleum becomes harder to find and recover. OS&TS are comparable to other natural resources in that the occupational health, environmental, and legal issues associated with their extraction and use are receiving more attention as the world's growing population becomes more sensitive to global health and environmental issues and organizes to deal with them.

1. ORIGIN, PROPERTIES, AND CHARACTERISTICS

OS&TS are important components of the spectrum of hydrocarbon resources found at or near the Earth's surface. Modern geology has established with certainty that these ubiquitous resources are part of a continuum of organic materials and the conversion products thereof present in the Earth's crust. This spectrum of sometimes valuable materials graduates from just-fallen plant, animal, and marine life, to rocks containing the petroleum precursor kerogen, to petroleum (natural bitumens, oil, and natural gas), and on to petroleum degradation products resulting from the varying degrees of geologic processing to which the original buried organic materials were subjected. The resultant resources grade from one to another with great variety depending upon how the important parameters of depositional condition, pressure, temperature, and geochemistry changed over time. This continuum, as currently understood by geologists, is depicted in Fig. 1 with annotations of theoretical estimates of the order of magnitude of kerogen or equivalent energy value present in each of the important hydrocarbon forms (Durand, 1980, p. 29).

1.1 Origin and General Characteristics

Most deposited organic material is initially converted to the solid hydrocarbon called kerogen by diagenesis as the combination of descending mineral and organic materials is transformed into sedimentary rock. A minor portion is converted directly to natural bitumen. Rocks containing significant quantities of kerogen are called oil shales. In addition to being insoluble in most solvents, kerogen does not melt upon heating but is converted to vapors, some of which can be condensed to oil.

Kerogen is widely recognized as the precursor of petroleum. If, however, the host rock is not subsequently subjected to the geologic pressures and temperatures required to convert the kerogen to oil and gas, the oil shale (and coal) remains for man to find and exploit for its hydrocarbon content. All sedimentary rocks laid down since life began on Earth contain some kerogen. In most of the great depositional basins, the processes of deposition, burial, and geologic change continued to the point that some kerogen underwent catagenesis, and great quantities of petroleum were produced. In many cases, the petroleum migrated to the traps that are today's oil and natural gas reservoirs. As petroleum migrated, some of the liquids saturated porous or unconsolidated formations, or filled joints or voids; there the petroleum underwent thermal or biological degradation. The lighter fractions evaporated, leaving the heavy fractions to cool and with their new host material to form a reservoir with zero mobility. These reservoirs that do not produce petroleum by conventional primary methods are now called tar sands.

Most hydrocarbon materials that fit the definition of tar sands are natural asphalts,

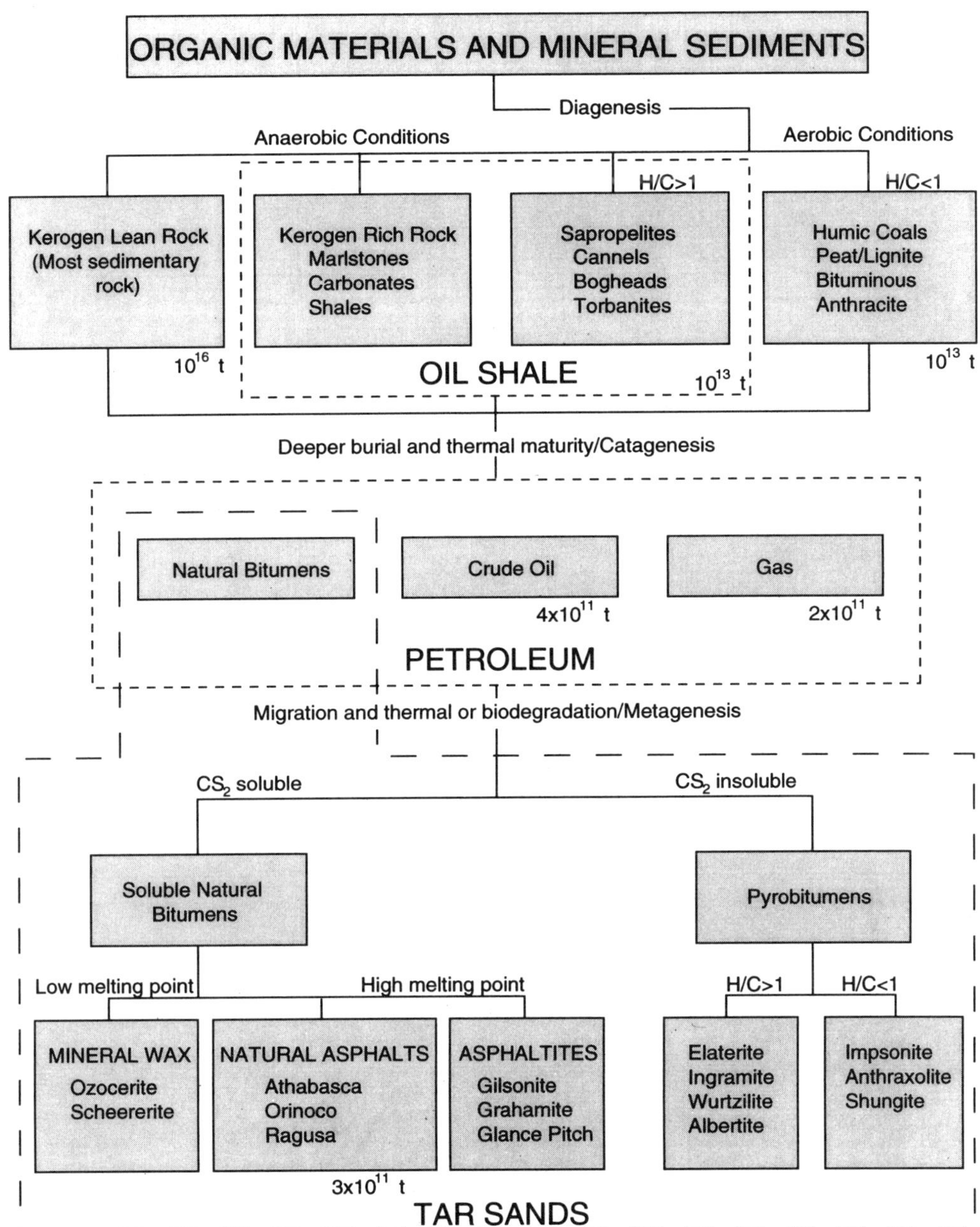

FIG. 1. The spectrum of hydrocarbons in the Earth's crust.

which are soluble, melt on heating, vaporize at higher temperatures, and can be cooked out of their host materials in boiling water. In a few cases, nearly pure petroleum moved into joints in adjacent rock and degraded into highly paraffinic material called mineral wax. In other isolated cases, petroleum carrying fine sediment escaped into joints in adjacent rock beds and was converted into solid hydrocarbons called asphaltites. At a few famous locations, such as Pitch Lake on Trinidad Island, a substantial amount of petroleum carrying fine sediment found its way to the surface, where it ponded and cooled to form asphalt lakes. At a few locations, such as New Brunswick, Canada, geologic conditions were such that *insoluble* natural bitumens formed, usually in veins or in voids in rock. These solid hydrocarbons have characteristics much like the kerogen in oil shale or coal and are called pyrobitumens.

As geologists, geophysicists, and geochemists better recognized the differences between kerogen as a precursor of petroleum and the products of petroleum degradation, the latter became known as "asphalts and allied substances." As the term asphalt came to be used for products manufactured from petroleum, the natural products became known as "natural asphalts and related substances." The modern term "tar sands" initially referred to deposits such as those in Canada and Venezuela that are mixtures of natural asphalt and sand, but the definition of tar sands evolved to recognize all host mineral matrices and to rely on viscosity to differentiate tar sands from heavy oil. Whether intended or not, the resultant definition includes all of the solid products of natural petroleum degradation, as well as the natural bitumens left behind in petroleum source rock, petroleum migration paths, and petroleum reservoirs. Tar sands are so delineated in Fig. 1.

In most cases, petroleum lost its mobility by degradation or metamorphosis after moving up along faults and layers of porous materials from where it was formed. Both oil shales and tar sand, therefore, are generally found at shallower depths than petroleum.

Resources cannot be classified as oil shale solely on the basis of composition because oil shale is an economic term. Although frequently misused as a lithologic term, oil shale means only that the rock is of interest for the production of oil. No minimum yield is specified, nor should one be. For instance, in China lean shale of 20 L/t has been processed for oil for 70 years because it had to be removed to reach the coal below it. At the rich end of the spectrum, there has been a 150-year controversy about what is coal and what is shale or pyrobitumen. To avoid this controversy, Fig. 1 includes only kerogenous rocks that are good sources of oil as oil shale. Humic coals have hydrogen-to-carbon atomic ratios of less than unity and were not considered good sources of oil using the simple pyrolysis practiced when the term oil shale came into use. The sapropelites, which can be lithologically shale or coal depending on the amount of mineral matter they contain, have hydrogen-to-carbon atomic ratios greater than unity and are excellent sources of oil using simple pyrolysis. The Australian and European oil industries were initiated on sapropelites and considered them the feedstock of choice. The early North American oil industry preferred the famous pyrobitumens from New Brunswick, albertite and sterralite, as feedstock but quickly turned to using sapropelites as pyrobitumen resources were exhausted (Ramsay, 1910).

1.2 Relevant Theories Regarding the Origin of Hydrocarbons

Theories about the origin of hydrocarbons are important for predicting their future availability and finding additional resources. Historically, petroleum and asphalt were believed to be primordial materials of the Earth that rose from its bowels through fissures in the crust for man to find and use. However poorly founded on knowledge, this was an abiogenic theory for the origin of hydrocarbons. As it became necessary to find and develop oil reservoirs systematically, petroleum geologists associated petroleum with organic materials deposited in sedimentary basins, and the biogenic theory of the origin of petroleum became dominant. In recent decades, as space exploration has provided new insights into the existence of hydrocarbons in the universe, the possibility of at least an abiogenic contribution to the availability of hydrocarbons in the Earth's crust is gaining credibility. A deep biogenic gas theory suggests that organic material falling to the ocean floor is pushed under the continental plates at subduction zones and is ultimately converted to methane or hydrogen-rich thermogenic gas.

The combination of these theories suggests that descending sedimentary materials *may be* enriched with hydrogen by ambient gas rising in the crust of the Earth. The nature of the sediment falling with the organic material and the geochemistry of the sedimentary basin will determine the type of rock formed with the kerogen. The role of hydrogen enrichment in the conversion of organic material to kerogen and the subsequent conversion of kerogen to petroleum is uncertain. Many petroleum geochemists now believe that water plays an important role in petroleum formation. When kerogen is converted to petroleum at greater depths, escaping gases may enrich younger layers of sediment; therefore hydrogen enrichment does not depend on deep gas or an abiogenic source of hydrogen or hydrocarbons.

1.3 Specific Characteristics and Properties

The specific characteristics and properties of OS&TS are important for planning extraction techniques and predicting economics. Table 1 is a summary of mineral content of the three geologic types of oil shale present in the world's largest known deposits. Some physical characteristics and properties of these shales are presented in Table 2. The properties of tar sands vary with those of the host rock and the impregnating hydrocarbon. Table 3 presents a summary of the known important properties of the tar sands in three of the world's largest deposits. The properties presented in these tables are only representative of the world's OS&TS; each deposit is unique, and the characteristics and properties of each must be determined to plan extraction. Characteristics and properties also vary by location and horizon within the deposit.

2. RESOURCES

The kerogen in sedimentary rocks is converted to petroleum as depositional basins mature, and petroleum degrades into tar sands as basins become overmature. OS&TS deposits are therefore best understood when the depositional basin in which they were formed is viewed as a whole.

2.1 Major Deposits

The geography and geology of six of the great depositional basins of the world are discussed below with a brief description of the OS&TS deposits found in them. These descriptions should be viewed as important data points in the Earth's crust and its history that illustrate the great variety and broad distribution of OS&TS. The quantity of kerogen in oil shales and asphalt in tar sands in just these six basins is sufficient to serve man for many centuries.

2.1.1 Western Canada Basin The Western Canada Basin (Meyer, 1987, p. 6) covers 770 000 km^2 in Alberta; tar sands deposits cover 31 000 km^2. Hydrocarbon deposits are found in this basin in the entire range of sedimentary rocks from Devonian to late Cretaceous. The depth and thermal maturity of

Table 1. Typical composition of oil shales.

Component	Location: Green River Formation	Appalachian Basin	Baltic Basin
	Composition (% by weight)		
Kerogen	13.4	13.9	25–30
Quartz	17	30	6.5
Analcine	12		
Feldspars			4.9
Orthoclase	3	9	
Albite	7		
Buddingtonite	3.5		
Clays			10.1
Illite	4	22	
Kaolinite	0.4	8	
Chlorite		2	
Mixed Clays		4.8	
Carbonates			
Dolomite	26	2	
Calcite	13		40.7
Siderite		0.7	
Pyrites	0.7	7.6	3
Al, Ca, Mg, Fe, and Ti oxides (not accounted for above)			5.3

Table 2. Typical characteristics and properties of oil shales.

Characteristic or Property	Location: Green River Formation	Appalachian Basin	Baltic Basin
Depositional condition	Lake Basin	Epicontinental Sea	Marine Platform
Geological era	Early and Middle Tertiary	Silurian and Devonian	Ordovician
Typical oil yield (L/t)	100	50	233
Porosity (% of bulk vol.)	0.14	0.57	0.1
Gross heating value (GJ/t)	5.38	4.85	12.45
H/C ratio of kerogen	0.13	0.09	0.12
Water yield (L/t)	58.4	23–44	94
Compressive strength (kg/cm^2)			
Perpendicular to bedding planes	700	1900	914
Parallel to bedding planes	750	1650	907

the basin are greatest along its southwest edge, next to the Rocky Mountain uplift. Numerous oil and gas reservoirs have been delineated throughout the basin. The world's greatest deposits of tar sands are found in the shallow northeast portion of the basin. One of the geologic mysteries of this basin is that the combined volume of oil (1.35×10^{11} barrels) in the four largest tar sands deposits (Athabasca, Cold Lake, Wabasca, and Peace River) and the numerous oil reservoirs in the basin cannot be explained by the volume of source rock present. The asphalt of these tar sands deposits is biodegraded and is most logically explained as the residue of once-mobile petroleum that saturated the extensive deposits of unconsolidated sand on the northeast boundary of the basin. Current reservoir temperatures in the tar sands deposits are 13–21 °C. At these temperatures the hydrocarbon residue is not mobile. Considerable tar sands in this area are exposed or have limited overburden; therefore, they are suitable for large scale surface mining.

2.1.2 Eastern Venezuela Basin The Eastern Venezuela Basin (Meyer, 1987, p. 6) covers 155 000 km^2 and contains the largest combined accumulation of oil, heavy oil, and tar sands in the world. The tar sands deposits cover 41 000 km^2. The basin extends north from the Orinoco River to off-shore beyond the north coast of South America and off-shore to the east to include Trinidad. As in the Western Canadian Basin, most of the tar sands have accumulated in the shallow part of the

Table 3. Typical characteristics and properties of tar sands. Source: principally Meyer (1987).

Property	Location: Western Canada Basin	Eastern Venezuela Basin	Aldan Arch
Host material	Unconsolidated sands	Sandstone and surface lakes	Voids and joints
Bitumen condition	Biodegraded	Biodegraded	Thermal-degraded
Reservoir age	Cretaceous	Tertiary, Cretaceous	Riphean, Vendian, Cambrian
API gravity	3°–8°	<6°–10°	<10°
Composition		% by weight	
Sand	83	65–70	[a]
Bitumen	13	24–27	0.8–50
Water	4	6–8	[a]
Sulfur content of bitumen	4–5	3.2–3.8	[a]

[a]The Aldan Arch Region in Russia has only recently been recognized as surpassing in magnitude the tar sands resources of the Melekess Basin, thereby becoming the third largest source of tar sands in the world. The solid hydrocarbons of the Aldan Arch Region are distributed in nine zones and have highly variable properties.

basin, in this case the southern boundary area, formerly called the Orinoco Tar Belt. Some of the material, normally defined as tar sands with API gravities less than 10° at normal temperatures, can be exploited by primary methods because reservoir temperatures in the basin range from 38 to 85 °C. The name of the area has been officially changed to Orinoco Heavy Oil Belt to reflect the presence of some oil that can be produced by primary methods.

Two famous lake asphalt deposits, Trinidad's Pitch Lake and Bermudez Pitch Lake near Guanoco, Venezuela, punctuate the northeast corner of the Eastern Venezuela Basin. These unusual occurrences result from very sediment-laden, heavy, warm oil flowing up along faults in this area to the surface.

2.1.3 Aldan Arch (Russia) The Aldan Arch (Meyer, 1987, p. 77) is a large uplift in Russia's Siberian Platform. This single structure with 1 300 000 km^2 underlain by asphalt rocks and asphaltite deposits will probably prove to be at least the third largest tar sands deposit in the world. The origin of the collection of tar sands in this relatively undeveloped area is more easily explained than is the origin of the Alberta tar sands. The Siberian Platform accumulated 9000 m of sedimentary rock from Riphean time (925×10^6 years ago) to Ordovician time. Considerable oil had formed in this great mass of sediment when magma moved under the area, creating the Aldan uplift. Heat from that intrusion drove the volatiles from large oil deposits that had moved into porous formations or the voids and joints created as the area was disturbed, leaving behind extensive asphaltites and other tar sands.

2.1.4 Green River Formation (U.S.A.) The Green River Formation (Russell, 1990, p. 85) of the Rocky Mountain Region of the United States is the largest and richest resource known on Earth. The term resource is used to describe this complex geologic formation instead of OS&TS deposit because of the unique collection of minerals that are commingled with the kerogen. The formation contains enough kerogen to produce 2×10^{12} barrels of oil, but the formation is also a remarkable collection of massive quantities of unique saline minerals that exist only in minute quantities elsewhere in the world. The formation covers 41 000 km^2 and is up to 2200 m deep.

The genesis of this anomaly was a large, inland stratified body of water that remained remarkably stable for 10^7 years from 60 to 50 Ma ago during the Tertiary Period. With no outlet, the water became very salty at depth but supported verdant algae blooms. These blooms produced the large volume of organic matter that became kerogen as 10 Ma of sediment piled up. The water existed for most of the period in two lakes: Lake Uinta in what is now northwest Colorado and northeast Utah, and Lake Gosuite in what is now southwest Wyoming. These lakes gradually diminished in size and became localized into six basins; clockwise around the Uinta Mountains the basins are the Green River, Great Divide, Washakie, Sand Wash, Piceance Creek, and Uinta. Each basin received slightly different runoff and developed different geochemistry.

The Piceance Creek Basin supported the most consistent growth of algae and was warmer and deeper than the other basins; consequently, surface effects did little to disturb normal stratification of the water. The deep center of the lake became a giant chemical system producing massive quantities of the saline minerals nahcolite, dawsonite, and sodium chloride in what is known as the salt cap area. The production and deposition of algae produced an amazingly uniform 640 m of oil shale. Natural gas has been found in and under the oil shale horizons of the Piceance Basin.

The Uinta Basin was not as deep, and algae growth was not as great as in the Piceance Basin. As a result, oil shale in the Uinta Basin is not as rich, except for the Mahogany Zone of rich shale that pervades the entire Green River Formation. An earlier sediment basin near the Uinta Mountains was buried and heated sufficiently to produce the Cedar Rim and Blue Bell oil fields of the Altamont Trend. This earlier basin was also the likely source of the petroleum that migrated upward and formed the tar sands of Asphalt Ridge; the famous vein asphaltite, gilsonite; the pyrobitumen, wurtzilite; and the mineral wax, ozocerite, in and near the Uinta Basin.

Algae growth in the more northern basins was not as great nor as persistent; thus, these basins have only layers of rich oil shale. In the Green River Basin, varying periods of the unique geochemistry laid down layers of trona (a hydrous acid sodium carbonate) of sufficient thickness to be mined adjacent to the richest oil shale horizons. The entire Green River Formation was subsequently lifted in the mountain-building process, and the great quantities of oil shale remain as originally formed.

2.1.5 Great Appalachian Basin (U.S.A.) When the North American continent was still connected to a great land mass to the east during middle Devonian time, great rivers fed the giant inland Chattanooga Sea, which stretched from the present Appalachian Mountains to the center of the Great Plains. Organic growth in the inland sea was sufficient to produce some rich oil shale and enormous quantities of lean oil shale over a wide area. Variations in river flows produced clastic intrusions from the east, and subsequent glaciers and erosion buried or carved away great parts of the oil shale. Earlier depositions produced oil and gas under parts of the shale, but the Devonian/Mississippian/Pennsylvanian shales were never buried deep enough or otherwise matured to convert the kerogen in them to petroleum. Collectively, what remains of these shales has been estimated to contain sufficient kerogen to produce 2.6×10^{12} barrels of oil. Scattered sapropelic and tar sands deposits, including some asphaltite veins, dot the large basin. The shales range over 650 000 km^2 and vary in depth from a few meters to 250 m. Overburden varies from zero at outcroppings to hundreds of meters of glacial till. Much of the area has a limestone cap over the oil shale. Even the richest shales in this complex basin have a serious deficiency of hydrogen and leave considerable carbon when pyrolyzed to produce oil (Russell, 1990, p. 85).

2.1.6 Baltic Basin (Europe) The Baltic Basin (Russell, 1990, p. 487) contains the most-used oil shale deposit of this century. The basin contains two oil shale seams. The rich middle Ordovician carbonate or kukersite provides the feedstock for the Estonian oil shale industry. The lower Ordovician Dictyonema shale contains phosphorite but has not been exploited. The kukersite extends over just 5200 km^2 in the area between the Gulf of Finland and Lake Peipus in Estonia and extends east in Russia. Overburden varies from zero at the outcrop near the Baltic Sea to 250 m at the southern limits of the shale of commercial interest. Mining is beset by problems with groundwater and weakened areas that must be left as pillars during underground mining. Numerous conflicting estimates of the amount of kerogen in the basin suggest the potential to produce 0.6×10^{12} barrels of oil.

2.2 Historically Significant Resources

The above discussion of six of the world's giant OS&TS deposits confirms the enormous magnitude of these resources in the Earth's crust. Several smaller deposits have had unique roles in the history of use of OS&TS and are noted to illustrate the broad distribution and historical importance of even small resources. The known history of native asphalt begins with hydrocarbon-saturated soil that became known as the slime pits of Hetite in ancient Babylon, and nearly pure asphalt fished out of the Dead Sea (Abraham, 1960, p. 9). Native asphalt oozed into the Dead Sea, solidified, and broke off to rise to the surface of the water in chunks the size of horses. This source was economically valuable in its time; authorities are reported to have issued permits to fish for asphalt in the tepid water of the Dead Sea. These sources and numerous other seeps in the Middle East provided the material for the applications identified in ancient Egypt and Mesopotamia dating from before 5000 B.C. (Abraham, 1960, p. 23).

Oil shale has an even longer history of use. Bracelets of French Autun oil shale found on skeletons of Neolithic man of the late stone age may have been used to repel insects (Russell, 1990, p. 545). The first confirmed production of shale oil has been traced by its name, "Giant's Blood," to a legend that dates to 800 A.D. and the Seefeld oil shale of Austria (Russell, 1990, p. 532). Oil from this shale high in the Austrian Alps near Tyrol gained fame for the treatment of skin diseases and ultimately became the only authentic source of Ichthyol (Blade, 1938). A modern sulfonated version of this medication is still produced at outcroppings of the Seefeld oil shale in Austria and Italy.

The first recorded European use of hydrocarbons in the New World was the caulking of ships at Lake Trinidad during Columbus' third voyage (Abraham, 1960, p. 41). Although discovered first by Europeans, this sticky lake asphalt proved too difficult to haul back to Europe, and so a harder hydrocarbon deposit near Havana, Cuba, became the primary source of exports to Europe for the next century. Europe had numerous small tar sand deposits, and many, such as Seyssel, were worked to exhaustion as the demand for asphalt grew. The Seyssel tar sand deposit in France's Rhone Valley was discovered in 1735. Production there led to the discovery of modern road paving, as pieces of asphalt that fell off the wagons were compacted by subsequent traffic. Seyssel became the most important source of asphalt in Europe until it was exhausted. The asphalt rocks of Ragusa, Sicily, are reported to have been mined sporadically since Roman time. The Wietzer tar sand deposit in Germany was the scene of J. Taube's 1769 development of the hot-water process of extraction and distillation of asphalt that is the basis of modern commercial production of oil from tar sands (Abraham, 1960, p. 45).

The Scottish oil shale deposits near the Firth of Forth are the most famous as sources of oil. Although devoid of any advantage for mass production, these resources supported the Scottish industry for its entire 103 years of competition with the American petroleum industry. The deposits are actually nine thin seams of shale buried in 900 m of calciferous sandstone at the base of a carboniferous system of rocks (Sneddon, 1938, p. 54). The entire system is highly folded and steeply dipping. The shales are lenticular deposits cut by severe faults and subject to flooding by groundwater. The yield varied from 450 L/t initially to as low as 60 L/t—not likely candidates for a role in history.

The best-known oil shale of the 19th century was the rich torbanite of New South Wales, Australia. At the peak of the gas-lighting era, it was discovered that the addition of just 5% of this ultrarich (up to 900 L/t) shale to coal increased the luminosity of the resulting gas by a factor of 6 (Carne, 1903, p. 97). Demand for this unique shale pushed the industry for its production into the wilderness of New South Wales. Pieces of shale left in river valleys by floods were traced upstream to outcroppings in the Sydney Permian Basin. Individual deposits were thin seams of ox-bow shape exposed by the intense erosion of the area. Exposures usually occurred at the top of long talus slopes topped by sheer sandstone cliffs. The shale seams were under intense pressure from extensive overburden but could be mined easily by hand methods because the pressure caused the shale to splinter when struck with a pick. The industry moved to ever more remote areas, solving onerous materials handling problems, as the small deposits were exhausted and new ones discovered. Ultimately, the industry spawned its own railroad to the port at Sydney, where daily boat loads of raw shale were shipped to the gas light systems of the world. The seconds, or leaner shale (less than 300 L/t), were discarded at the mine mouth or processed in local plants for oil, wax, and other products.

The fame of the Australian torbanites led to testing of North American sapropelic rocks of similar appearance as a substitute for the imported product. The spread of agriculture led to discovery of numerous small deposits of cannel and boghead materials in the eastern United States and Canada that were initially identified as coal. Even albertite and stellarite were mistakenly identified by the courts as coal (Gesner, 1865, p. 49). North American sapropelic rocks, whether subsequently proven to be coal or shale, and pyrobitumens became the feedstock of choice for the American pyrolysis industry that blossomed in 1855 and that ended quickly with large-scale production of petroleum in 1859.

2.3 Summary of World Resources

Any tabulation of world OS&TS must recognize that these resources have been neglected for the last hundred or so years during which petroleum has dominated the supply of hydrocarbons worldwide. Except in the most recent two decades, when various oil crises generated interest in OS&TS, these solid hydrocarbons received little attention by the world's major hydrocarbon producers, who were in the position to define and explore them. Even recent interest has been limited to the giant deposits that could produce enough oil to replace petroleum. Most discoveries of deposits and extension of data about known deposits have been anecdotal to the search for petroleum. Generally, the more thoroughly a depositional basin is explored the more OS&TS are found. Thus in the 20 years between 1964, when Ball (1965) did a comprehensive survey of tar sand resources in the United States, and 1983, when Lewin and Associates, Inc. (1983) completed a similar survey, U.S. tar sand resources grew almost tenfold from 5.5×10^9 barrels to 54×10^9 barrels. This suggests that largely unexplored developing countries and undeveloped frontier areas such as arctic Alaska, Canada, and Russia, where some oil has been found, may have enormous undiscovered OS&TS resources. Table 4 presents estimates of the order of magnitude of oil in total resources of OS&TS.

3. PRODUCTION

As depicted in Fig. 1, oil shales, petroleum, and tar sands form a spectrum of hydrocarbon resources. Man has used various forms of these natural resources for centuries as fuels, lubricants, paving, building and manufacturing materials, cosmetics, and medicine. Archaeologists and historians have traced the use of petroleum, natural asphalt, and asphalt rocks to the earliest civilizations in the Middle East prior to 5000 B.C. Archaeologists have also found evidence of centuries of use of natural hydrocarbons by Chinese, Japanese, and aboriginal Americans of both continents. As the demand for hydrocarbons reached massive quantities during the industrial age, petroleum, which is the easiest form to extract and transport, became the dominant resource used and the use of OS&TS declined.

Table 4. Ultimate recoverable OS&TS resources.

	10^9 barrels of oil	
Region	Oil shale	Tar sands
Africa	4,000	1
Asia	5,500	492
Europe	1,400	21
North America		
Canada	10	2,516
United States	2,990	56
Oceania	1,000	
South America		
Brazil	2,000	
Venezuela		1,182
Total	17,000	4,000

Sources: Preparation of this summary of OS&TS resources revealed numerous conflicts in data regarding these resources. The oil shale amounts are from Duncan (1965) for just the richest (over 104 L/t) shale, and the tar sands amounts are from Meyer (1987). The totals exceed the amounts derived by Durand (1980) of 3.7×10^{12} barrels for oil shale and 2.2×10^{12} barrels for tar sands. Duncan's total for oil shales yielding over 21 L/t is 2.092×10^{15} barrels.

3.1 Historical Production

Cycles of use of OS&TS predate their competition with petroleum. The ancient uses of asphalt in the Middle East faded as Roman building technology using cement became dominant. Subsequently, the growth of cities fostered the broad use of charcoal as fuel, and societies remote from petroleum sources learned about liquid hydrocarbons as a byproduct of charcoal production. A demand for liquid products, initially easily met with byproducts of charcoal, had been created, and when the forests were used up in populated areas, the availability of fuel of any type became a daily problem. Artificial tar sands, prepared by saturating earth with petroleum or other liquid hydrocarbons, were burned as a substitute for charcoal. Various patented or manufactured solid fuels using hydrocarbon binders became items of commerce. In 1838, builders, especially of railroads, rediscovered the advantages of pickling wood with liquid hydrocarbons. Coupled with the growing demand for liquid fuel for lighting, the resultant demand for liquid hydrocarbons became overwhelming. Prices for whale oil in the United States rose to $16/L by 1846.

The oil shortage of 1838 triggered the first modern effort to perfect mass production of oil from solid hydrocarbons. Much was already known from the gas-lighting industry, which had started to produce gas from coal, other solid hydrocarbons including oil shales, and various organic materials by pyrolysis during the late 18th century. Petroleum was not unknown at the time, but its transportation problems had not been solved. As early man used more hydrocarbons, he found a more widespread and certain source in mining and cooking oil from rocks than in traveling great distances to skim oil from water wells, ponds, springs, and streams or primitive oil wells. Considering the primitive nature of transportation at the time, this was probably the correct economic choice. Even so, where liquid hydrocarbons could be found close to the Earth's surface, this relatively convenient resource was exploited. For instance, in 1838 there were at least 82 crude wells near Baku and 3500 pits for the production of oil on an island in the Caspian Sea. Marco Polo had described exporting oil from this area by animal train in 1298. Muspratt (1860, Vol. 1) also reported 520 oil wells operating in Burma exporting 400 000 hogsheads [(0.6–1.4)$\times 10^6$ barrels) of oil per year prior to 1859.

It is a staggering paradox, therefore, that the petroleum industry waited until 1859 to explode onto the world from North America. This paradox did give rise to a short but flourishing and relatively modern oil shale industry. After a jump start from this short market pull and subsequent government protection from the predatory marketing practices of the U.S. petroleum industry, oil shale plants operated in Scotland, France, Spain, Estonia, Sweden, Germany, Brazil, and Australia. Some plants survived for many years in competition with the petroleum industry. In North America, there would be no such longevity. The initial oil shale plants of 1855 in Boston and New York, which operated on imported torbanites and pyrobitumens, and the Breckenridge, Kentucky, plant, which used cannel coal as feedstock, quickly grew to over 55 plants in the eastern United States by 1859. The largest was a 25 000-L/d plant near Pittsburgh. All were shut down or converted to oil refineries as large-scale petroleum production started in Pennsylvania in 1859.

Tar sands were well known in the 1830s, but this sticky material that caked so easily was hard to work with for the production of oil or fuel. Concerning this period, Muspratt (1860) reported what he considered many impractical attempts to use native asphalt in applications for which it was not suited. Natural asphalts, which appeared to have found their broadest use in early civilizations, made a dramatic comeback as road-paving material during the late 19th century. As the demand for paved roads spread across North America, dozens of local tar sands deposits were mined to supply paving material. The Kyrock Mine in Kentucky was a typical operation that shipped asphalt rock by river barge and rail to pave hundreds of miles of roads in midwest America. Two such mines still operate in Uvalde, Texas, supplying paving material to a distance of 175 km from the mines. Paving was the dominant use of tar sands for 50 years from 1870 to 1920. Production of paving material reached 0.25 Mt per year in both Europe and North America. The production of asphalt from petroleum was patented in 1894 and was gradually perfected. After 1920, the use of synthetic asphalt prepared from petroleum and local aggregates became more economical for most paving. In Russia, where distribution of petroleum products is still a problem, the local production and use of natural asphalts are reaching the levels reported during the early part of this century in Europe and the United States.

Among the tar sands, asphaltites in particular have been the basis of stable industries producing highly varied products for years. Grahamite, initially mined in West Virginia during the 1850s and burned as coal, has since been used to produce tar, lubricants, illuminating gas, and varnish. The gilsonite industry of Utah, in operation since 1888, has produced ink, insulation, paving material, roofing, varnish, asphalt additives, sealers, gasoline, and carbon control rods for nuclear reactors.

3.2 Current Production

OS&TS were worked in at least eight countries in 1992.

3.2.1 Brazil Brazil has nine significant oil shale formations. The Brazilian oil shale industry started in 1881 with 20 Scottish retorts operating on Tertiary shale at Taubate in San Paulo State. This plant operated until 1920, sat idle until 1941, and was then renovated and expanded for wartime production of oil, wax, and gas. A second company operating on Cretaceous shale in Bahia erected a plant of 52 Scottish retorts in 1884 and produced illuminating gas, oils, paraffin, candles, soap, and sulfuric acid. Production was equivalent to about a thousand barrels of oil per day. Several other plants produced shale oil products at intermittent times throughout World War II. Oil shale was also used as railroad locomotive fuel (Russell, 1990, p. 172). The high water content of the Cretaceous oil shale was a problem. In an attempt to improve the economics of Brazil's oil shale operations, the government investigated the dryer Irati shale. The Taubate plant was close to the Irati Formation; it was purchased by the national government in 1951 and became the start of the Petrobras research and development program that is currently demonstrating large-scale production of oil from the Irati Formation using the Petrosix technology (see Sec. 4.4.1).

3.2.2 Canada In 1967, after a half century of diverse trials, Great Canadian Oil Sands Ltd. (now Suncor) put the first modern commercial plant for the production of oil from tar sands into operation in Alberta. The Suncor plant is producing at design capacity of 65 000 barrels per day. A more recent effort, the joint venture Syncrude Canada Ltd., attained sustained production of 109 000 barrels per day during 1983 and produced at design capacity of 129 000 barrels per day during 1985. These operations use open-pit mining and hot-water separation of the tar and sand in surface plants. The development of horizontal drilling technology may result in larger-scale oil production from these massive resources by *in situ* methods.

3.2.3 China Oil shale is widely distributed in China; deposits are known to exist in more than a dozen provinces. Two deposits are currently worked. Fushun, in Manchuria, is the most famous and is the world's oldest continuous oil shale operation currently in production. The shale is lean, 20 L/t, but generally has to be removed to get to coal beds. Started as a local activity shortly after discovery of the shale in 1909, the operation has been expanded and renewed several times. In 1924, the Japanese added facilities at Fushun based on Estonian technology. The Japanese plant produced 5500 barrels per day during World War II. The Chinese took over the plant after the war, rebuilt it during 1954, and expanded it during the 1970s. Peak production had reached 7.5×10^6 barrels per year but has dropped to less than 0.5×10^6 barrels per year. A second oil shale operation was started in Maoming, southern China, in the 1950s using richer shale of 75 L/t.

3.2.4 Estonia Kukersite of the Baltic Basin has long been recognized for its fuel value and has been used extensively because of the unusual richness of so large a deposit. Farmers also burned it on their fields to use the ash for fertilizer and as a soil conditioner. Estonia became an independent country after World War I; three private companies and the Estonian government built plants for oil production. By 1938, the Estonian portion of the basin produced a million barrels of oil per year and employed 6000 people, meeting the fuel needs of two-thirds of Estonia's heavy industry. The industry was devastated during World War II, but produced at the rate of 1.75×10^6 barrels of oil per year for a period during German occupation. In 1945, Estonia became part of the Soviet Union. As part of implementing a succession of Five Year Plans, the Soviet central government built major gas and electric generation plants fueled by kukersite in the Baltic Basin. These plants produced gas and over 3000 MW of electricity for Leningrad and the northwest Soviet Union. The industry is now in decline because natural gas pipelines have been extended to the area, and the power plants have major environmental problems.

3.2.5 Russia Since the 17th century, many projects in Russia have produced natural bitumen and oil shale products (Russell, 1990, p. 470). None can be classified as large-scale oil producers, but the diversity of resources, lack of transportation, and great local demand for a wide range of products have prompted the greatest diversity in use of oil shale and tar sand products in the world. There are numerous local tar sands operations in the former republics of the Soviet Union. For example, a surface mining operation started in the 1950s in Kazakhstan is now producing more than 4 Mt per year. A large portion of the material is used directly for road paving. A 1991 report (Vakhitov, 1991) cited uses for fuels (in liquid, gas, and solid forms), petrochemical feedstock, varnishes, paints, insulating material, metalliferous products, biostimulants, building materials, and paving material.

3.2.6 Trinidad Asphalt from Trinidad's Pitch Lake continues to be produced by the government-owned Trinidad Lake Asphalt Company, Ltd. at the rate of 820 barrels per day, down from peak production of 3100 barrels per day during 1964.

3.2.7 United States Local paving still supports continuous operation of the two tar sand mines near Uvalde, Texas. Limited local operations of this type, once widespread, also continue intermittently, as needed for local paving, at Asphalt Ridge near Vernal,

Utah. In the eastern Uinta Basin, commercial production of gilsonite began during 1888 and continues. Also, in the Green River Formation, the important saline minerals associated with oil shale are being exploited; trona has been mined in Wyoming since 1947, and solution mining of nahcolite in Colorado was initiated during 1990.

3.3 Prospects for Future Production

In addition to continuation of the operations listed in Sec. 3.2 at about their present levels, the following projects probably will go into production during the next decade.

3.3.1 Australia A large multinational oil company and an Australian oil shale resource company have persisted for several years in preparations to produce oil from shale in Queensland. Assuming reasonable cooperation from the Australian government, this group is expected to operate an open-pit mine and to retort shale using a modern surface pyrolysis technology within the next decade.

3.3.2 United States With enormous, rich oil shale deposits and widespread shallow tar sands and sapropelic deposits, the United States is a logical location for increased use of these resources. Small projects producing premium hydrocarbon products, mineral coproducts, and chemical by-products are more likely to be feasible earlier than large-scale projects competing with the petroleum industry.

3.3.3 Venezuela The great Orinoco tar sands have not yet been put into production. The high viscosity (100 000 mPa s at ambient temperature) of the bitumen is a problem, but Venezuela has learned a great deal about handling heavy oils from its experience throughout the Eastern Venezuelan Basin since the 1950s. Neither heating nor dilution with lighter crudes is considered to be a cost-effective method for extracting the viscous tar. The decision has been made to try to produce, transport, and burn the bitumen directly as an emulsion using a brine present in the deposit. Venezuela is testing world markets for the new power plant fuel called Orimulsion.

4. PROCESSING AND APPLICATIONS

The method of extracting useful products from OS&TS depends on the product desired, the formation geometry, and the physical properties and characteristics of the pay zone. Attempts to classify extraction methods tend to obscure the range of choices and combinations available to operators. The general categories of mining and direct combustion, mining and surface processing, *in situ* processing, and combinations of these methods have all been used or demonstrated at large scale.

4.1 Mining

Mining methods for OS&TS are very sensitive to the properties of the particular deposit under consideration and the overburden covering the pay zone. The two general categories of mining are surface and underground.

4.1.1 Surface Mining Open-pit mining is the method of choice for mining both OS&TS when depths of overburden permit. Most OS&TS in consolidated formations require blasting and mucking to remove the ore. For unconsolidated tar sand formations and weak shales, direct loading with large shovels or ripping and bulldozing is possible. Suncor's bucket-wheel loader and conveyor system in its Alberta mine is the leading example of large-shovel mining of tar sands. The rich shale of the Baltic Basin can be ripped directly from the mine face and loaded to rail cars or trucks with large shovels. The tar-impregnated Anacacho Limestone of Uvalde, Texas, is blasted and mucked with large shovels and load-haul-dumps, but the weak sandstone of Asphalt Ridge near Vernal, Utah, is ripped with bulldozers and pushed directly to the mixing table for further processing. The unique properties of gilsonite permit very efficient mining of this asphaltite even from very narrow vertical veins using high-pressure water jets. In each case, the physical properties of the ore body, volume of operation, and economics determine the exact choice of method and equipment.

4.1.2 Underground Mining Where depth of overburden requires underground mining, the operator is faced with a far broader set of problems. Access, whether by vertical, horizontal, or sloping adit, must be planned and mined out. A strong competent horizon of rock must be selected for the mine roof. Ventilation, dewatering (if needed), and emergency egress for every stage of mine development must be carefully planned. Room-and-pillar or chamber-and-wall mining have been the methods of choice for most large underground oil shale mines. The heights of rooms or chambers in the Green River Formation have dictated special drilling and blast hole loading equipment to reach the entire face of the mine. Similarly, specialized roof bolting equipment and techniques have evolved for large oil shale mines. The U.S. Bureau of Mines did considerable development work to perfect equipment and techniques for large production mines in the Green River Formation. Cleveland-Cliffs, Inc., Mobil Oil, and Unocal have all operated room-and-pillar mines safely in the Green River Formation. Numerous mechanical miners to replace blasting and mucking have been proposed for OS&TS operations, but although several have been carefully tested, none have been used in large-scale production. The Scottish and Australian oil shale mines worked only narrow seams of a meter or two compared to the 27-m cuts used in the hundreds of meters deep ore body of the Green River Formation. During 100 years of operation, the Scottish mines evolved from crude hand-pick operations to very sophisticated electrical operations (Sneddon, 1938, p. 53).

4.2 Crushing

An extensive choice of crushing equipment designed primarily for crushing limestone is available and very satisfactory for most OS&TS. Crushing tar sands from unconsolidated formations can be as simple as driving over the run-of-mine material with a bulldozer, as is done to produce road paving material at Asphalt Ridge. For harder ores, gyratory, jaw, cone, and roll crushers have been used successfully. Particle size is usually dictated by the choice of subsequent oil extraction process or firing method.

4.3 Direct Use

Historical evidence suggests that oil shales have been used directly as fuel for hundreds of years. The use of kukersite as power plant fuel in Estonia is the leading contemporary example of this practice.

The outstanding direct use of tar sands has been for road paving. Processing for this application may range from simple compaction of run-of-mine material to some crushing and softening with hydrocarbon solvent, as is done at Uvalde, Texas, or blending with water, as is done at Asphalt Ridge in Utah.

4.4 Methods of Oil and Gas Production

Converting of kerogen in oil shale to oil and separating bitumen from the host tar sand material have many similarities, but the properties of bitumen offer more choices for extracting useful products. The bitumen in tar sands melts, can be decomposed or cracked at high temperatures, can be washed off the host material with hot water in some cases, and is soluble. Kerogen in oil shale does not melt and is insoluble; therefore, it is generally separated from the host rock by application of heat to decompose the solid kerogen into gases and residual carbon. This process is called pyrolysis and is completed in a reaction chamber called a retort for both surface and *in situ* operations. The atmosphere in which pyrolysis occurs is critical because the newly formed products of pyrolysis are very reactive with any other elements present in the pyrolysis zone, particularly oxygen. Retorts are therefore designed to have inert gases sweep the product gases out of the pyrolysis zone to a condenser or condensing zone in the retort where the oil and uncondensable gas fractions are separated. The residual carbon and uncondensable gases may be burned to provide heat for the process. Maximum oil production requires pyrolysis at the lowest possible temperature (about 480 °C) to avoid unnecessary cracking of the hydrocarbon molecules, which reduces oil yield. Higher temperatures can be used to produce lighter oil and a greater volume of gas at the expense of oil yield. Supercritical extraction techniques have been tested on oil shale to improve oil yield. This involves washing intermediate products of kerogen conversion out of the host rock at lower temperatures than used for pyrolysis.

Where permeability of the host rock is sufficient or can be created, several methods of *in situ* heating are technically feasible for both oil shales and tar sands. The basic objective of both surface and *in situ* operations is to complete the minimum decomposition of the hydrocarbon component necessary to get it free of the mineral matrix and do as little heating of the mineral matter as possible. Minimum heating of the mineral matter is particularly important for carbonate shales because the carbonate decomposition reaction if triggered is highly endothermic and generates unwanted carbon dioxide but no useful products.

4.4.1 Surface Pyrolysis The most recently developed surface pyrolysis technology is the Alberta Oil Sands Technology and Research Authority (AOSTRA) Taciuk Processor (ATP). Developed for tar sands, the ATP has also been tested on oil shale and has been used to remove hydrocarbons from contaminated soil. This rotary retort, developed for thermal treatment of Alberta tar sands, is the method that has been chosen for the Australian oil shale project previously mentioned (Kuss, 1994). The ATP is illustrated in Fig. 2.

Another surface pyrolysis technology suitable for both oil shales and tar sands is the Lurgi–Ruhrgas technology developed in Germany to process coal fines. This technique, which features a lift pipe to recycle hot solids, has been successfully tested for processing the diatomaceous earth tar sands of the McKittrick Formation in California and for Green River oil shale. It has been found useful on a wide range of feedstocks. Figure 3

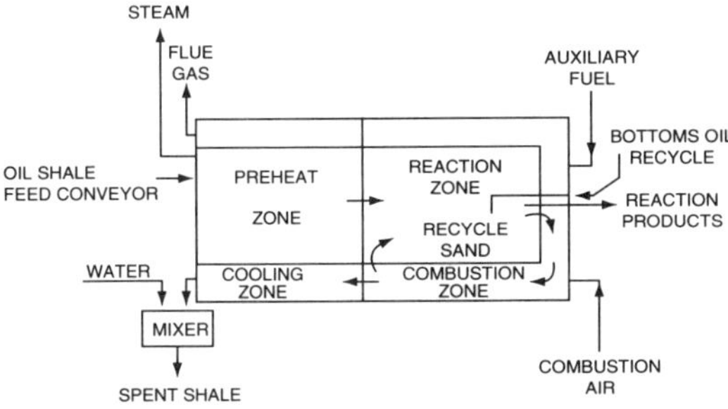

FIG. 2. AOSTRA Taciuk Processor (ATP). The ATP consists of a single, horizontal, rotating vessel in which separate compartments or zones are maintained by the geometry of the equipment and flow of gases. The reaction zone is heated by a combination of conduction from the combustion zone, where residual carbon is burned off the spent feedstock, and a portion of the hot spent feedstock, which is recycled to the reaction zone.

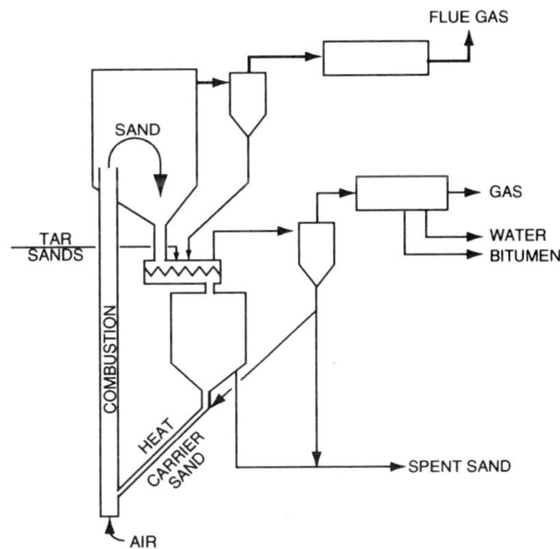

FIG. 3. Lurgi–Ruhrgas process. The unique feature of this system is the lift pipe in which residual carbon is burned off spent feedstock to provide heat. Burned feedstock is carried to the retort for solid-to-solid heat transfer to raw feedstock.

shows a configuration of the Lurgi–Ruhrgas technology suitable for processing tar sands.

The largest surface oil shale pyrolysis reactor currently operating is the Petrosix 11-m vertical-shaft retort used in Brazil's development program. The modern vertical-shaft retort is the product of years of evolution that can be traced to the original change from horizontal- to vertical-shaft retorts in the Scottish oil shale industry. James Young and the Scottish oil shale companies shifted from the fixed horizontal retorts that had been adopted from the coal gasification industry to higher-throughput, vertical-shaft retorts to produce oil. A succession of outstanding engineers following Young improved the Scottish retorts and set the standard of efficiency for the world. The Scottish designs were exported to France, Australia, Spain, Brazil, and the United States. These retorts were indirectly heated by coal. When U.S. Bureau of Mines engineers set out to develop a high-efficiency, high-throughput oil shale retort specifically for the Green River Formation shale, they studied the Scottish design and elected to develop a vertical-shaft retort, called the Gas Combustion Retort (GCR), that would burn the uncondensable gases of the retorting process as fuel. Engineers who worked on that program went on to develop and build the Petrosix GCR for Brazil and the Paraho version of the GCR, which has been used extensively at pilot scale for various research and testing programs in the United States. The Paraho version is illustrated in Fig. 4 as an example of a modern, high-throughput, vertical-shaft retort.

Upon completion of the Bureau of Mines GCR demonstration program, a consortium of major oil companies leased the pilot and demonstration plants at Anvil Points, Colorado, and conducted proprietary testing. Subsequently, three of the major oil companies developed alternative concepts. Union Oil Company (later Unocal) perfected its rock pump retort to the demonstration stage. During 1949, Standard Oil Company (later Exxon) tested and patented a fluid-bed retort. Chevron demonstrated an ebullient or boiling-bed retort during the 1970s. The latter two designs are extensions of oil company experience in operating sand crackers and fast catalytic crackers in modern refineries.

During the most recent period of high interest in oil shale, the Kiviter retort replaced older equipment for the production of liquids from kukersite in Estonia. China continued to use the Pintsch retorts. The Institute for Gas Technology, with support from the U.S. government, developed hydroretorting in a vertical-shaft retort for the hydrogen-deficient Devonian shales.

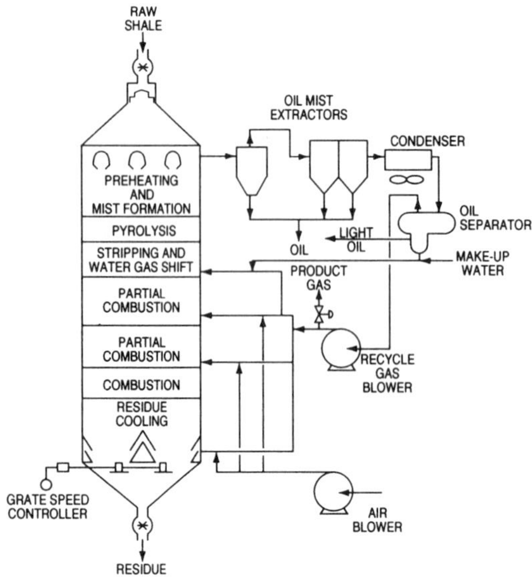

FIG. 4. Paraho retorting process. The Paraho process is typical of vertical-shaft retorts in which crushed shale with the fines removed descends through the retort under the influence of gravity. Zones for each step in processing the shale are maintained by managing gas flow in the retort. The direct combustion mode is shown here. When operated in the indirect mode, process gas is burned in a separate furnace and hot gases carry heat to the retort.

4.4.2 *In situ* Extraction *In situ* pyrolysis or heating of both oil shales and tar sands in place to produce oil has been tested extensively. One particularly successful operation involved using cheap hydroelectric power to heat shallow beds of shale in Sweden. The U.S. government provided extensive support of *in situ* OS&TS research and development during the 1970s. The cooperative program with industry resulted in the Geokinetics horizontal *in situ* process for shallow beds of oil shale and the Occidental vertical modified *in situ* process for deep, thick oil shale beds being tested to demonstration scale. These concepts are illustrated in Figs. 5 and 6.

4.5 Upgrading and Refining

Extensive testing and practice in the former oil shale industries of the world and Canada's tar sand industry have adequately proven that shale oil and bitumen from tar sands can be refined into the full range of petroleum products. Some shale oils contain high concentrations of nitrogen or arsenic and must be hydrotreated to be suitable as feedstock for conventional refineries. Trace metals, sulfur, and chlorides can also be a problem, just as they are with some crude oils. Historically, most oil shale plants were so isolated that small refineries of custom design were collocated with them to process their output. This was also the case for the Utah gilsonite industry. When it was decided to make gasoline from this unique resource, a small refinery of special design was constructed at Fruita, Colorado. The larger oil shale plants contemplated during the 1970s would have included upgrading facilities, as did Unocal's demonstration plant. Minimal treatment in custom-designed upgrading facilities is required to make feedstock suitable for modern refineries.

5. ECONOMIC ASPECTS

The monumental works of Abraham (1960) and Russell (1990) chronicled the rise and fall of OS&TS industries around the world. The

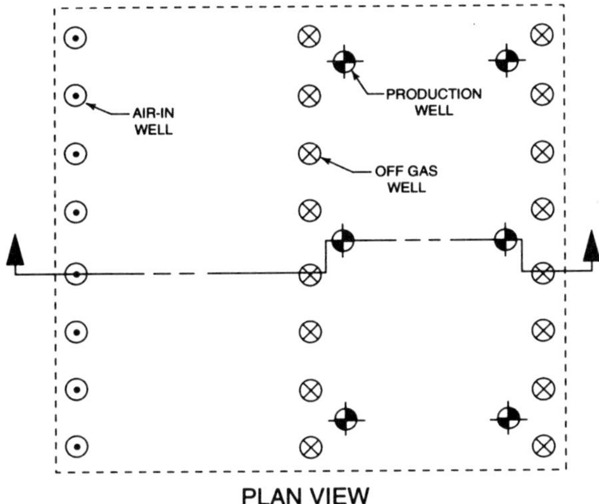

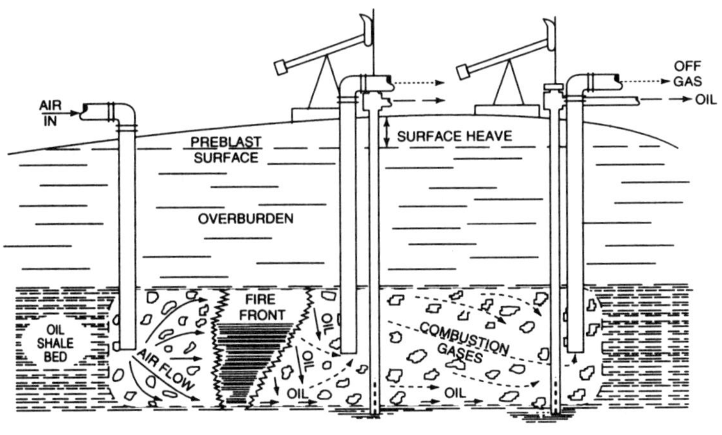

FIG. 5. Plan view and cross section of typical Geokinetics horizontal *in situ* process. This scheme for retorting shallow beds of shale in place depends on uniform blasting of the shale and surface heave to create void space for gas to flow through the shale. The shale is set afire at the air inlet wells, and air flow is managed to move hot inert combustion products through the balance of the shale ruble.

existence of an industry cannot be assumed to prove independent economic viability because special tax structure and tariff policy are well known to have sustained industries that otherwise would not have been profitable. The simplicity of just heating rock to get oil accounts for the long history of use of OS&TS. The economic competition between petroleum and even this simple method of extraction accounts for the enormous variety of methods tried, as inventors attempted to improve the economics of operating with solid feedstock. No "cotton gin" of pyrolysis technology, however, was developed to fix the cost of producing shale or tar sands oil at or near that of petroleum extraction. At the time of the birth of the modern petroleum industry (1859), Muspratt (1860, Vol. 2, p. 99) reported a dozen methods for heating both hydrocarbons that melted (resins and asphalt) and others that vaporized or decomposed (kerogen and organic waste). By the time the U.S. Bureau of Mines surveyed oil shale patents a century later (Klosky, 1949, 1958), more than 1500 patents had been registered, and there are now about 2000. No commercial-scale production of oil from shale or tar sands since the emergency fuel production required during World War II, other than projects that received government subsidies, has survived more than a few years in the era of cheap Middle East oil. Neither have the opportu-

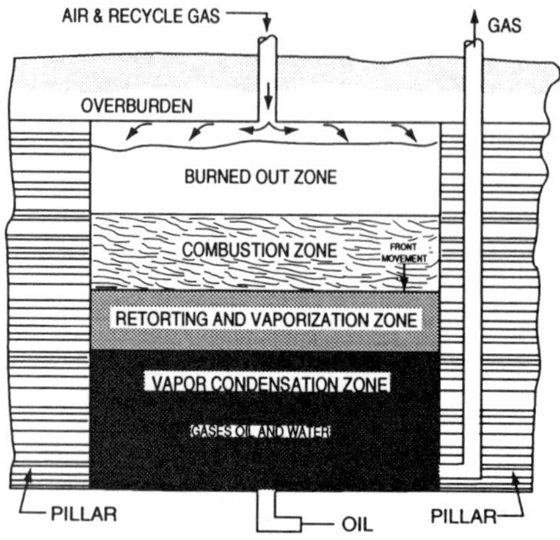

FIG. 6. Occidental vertical modified *in situ* process. This concept was developed specifically for the deep, thick shale beds of the Green River Formation. About 20% of the shale in the retort area is mined; the balance of the shale is then carefully blasted using the mined-out volume to permit expansion and uniform distribution of void space throughout the retort. A combustion zone is started at the top of the retort and moved down through the shale rubble by management of combustion air and recycled gases. Full-scale retorts would contain 350 000 m^3 of shale rubble.

nities offered by the fact that the bitumen in tar sands is soluble resulted in success. Most solvent projects failed because of the loss of expensive solvent adhering to the host rock or sand. The economics of all OS&TS projects were seriously impaired by the extra cost of mining and handling solid feedstock and disposal of spent feedstock. The materials handling problem has forced the development of ever more efficient surface plants and provided the impetus to experiment with *in situ* methods.

5.1 Historical Profitability

The Australian oil shale industry of New South Wales is a concise example of an industry's response to competition, outside influences, government policy, and technological change. As shown in Fig. 7, the small industry started experimentally in 1855, reached continuous production status in 1865, and flourished in relative isolation until the financial panic of 1873. After that, the struggling industry needed tariff protection from the predatory marketing practices of Standard Oil Company. The Provincial Government of New South Wales provided that protection, and the fledgling industry recovered and grew with the local economy. During 1902, central government came to Australia and canceled the protective tariffs in response to perceived greater national interests, and the oil shale industry went into decline. The gas mantle, invented in 1911, eliminated the export market for Australia's raw shale, and the tenacious little industry died shortly thereafter. It sprang to life again in response to the emergency needs in World War II and disappeared when the emergency was over (Sell, 1951).

Other OS&TS industries cannot be so neatly capsulized, but all that had reasonably sustained operations began and grew under special circumstances and declined when those conditions ended.

5.2 Current Profitability

The current large-scale operations in Brazil and Canada are not economical in competition with the petroleum industry. These projects barely meet operating costs and are currently paying little return on capital investment. Newell (1993) most recently reported operating costs of \$15/bbl (Canadian) in a facility that would now cost (\$8–\$10)$\times 10^9$ to build. World oil prices at the time of this announcement were in the range of \$17/bbl. The industries in Russia, Estonia, and China are not yet operating in true free-market economies. The proposed project for Queensland, Australia, has requested special tax treatment to compete with the petroleum industry. No current large-scale OS&TS operation can compete with the petroleum industry as a source of refinery feedstock as long as petroleum prices remain low. However, small tar sands operations, producing local paving material, have a long history of profitability. The gilsonite industry of Utah has a checkered history but overall has been profitable.

5.3 Prospects for Future Financial Feasibility

Both upstream and downstream costs are increasing for the petroleum industry and can be expected to increase at even greater rates as convenient reserves are used up and more production must come from sensitive and remote areas. Nevertheless, world petroleum

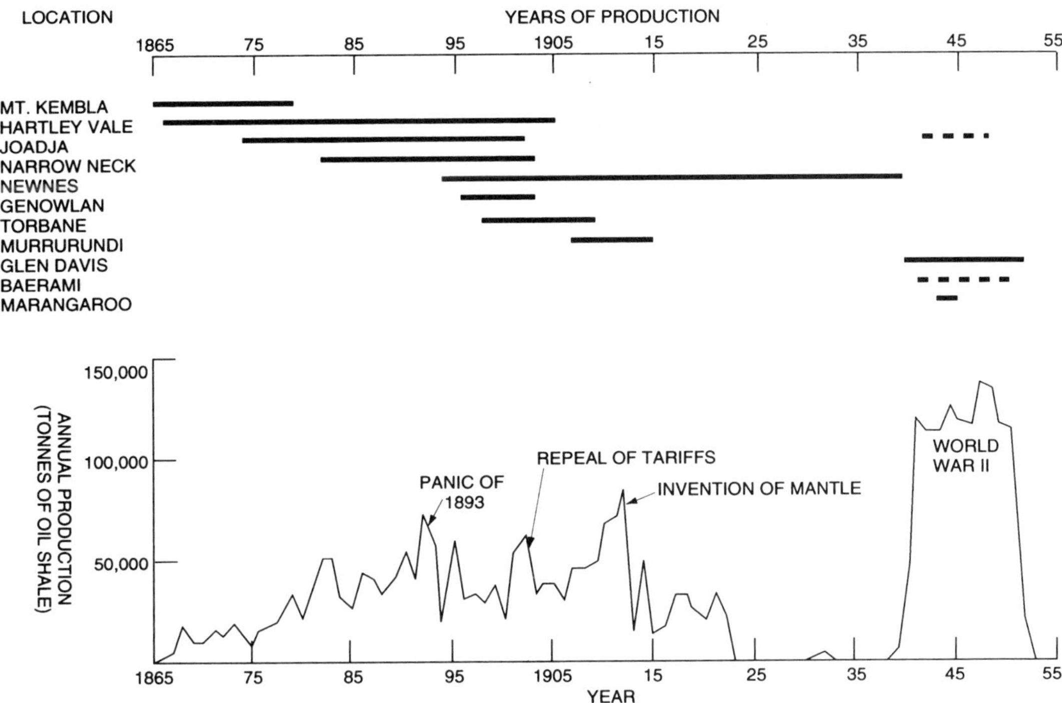

FIG. 7. Australian oil shale industry production.

producers can be expected to continue to operate as an imperfect Von Stackelburg monopoly (Greene and Leiby, 1993) and manage petroleum prices to a degree that makes large-scale production of refinery feedstock from OS&TS a financially unacceptable venture. Petroleum producers will keep prices between the long-term and short-term cost of producing petroleum, and both of these boundaries will increase with time. Unfortunately, justification of large OS&TS operations to produce oil in competition with petroleum requires a sharp increase in world petroleum prices at a predictable time. Sharp increases in petroleum prices, such as occurred in the 1970s, are certain to occur again, but their timing is uncertain. The most likely scenario will be more projects that expand the envelope of primary, secondary, and tertiary recovery of crude oil into the most convenient and most easily worked tar sands, including depleted shallow petroleum reservoirs. This movement forward along the geologic spectrum from conventional petroleum to heavy oil, extra heavy oil, and then tar sands is more likely to happen than any giant step to oil shale operations in competition with petroleum. Canadian *in situ* tar sands operations using horizontal drilling are the most likely candidates for success in this scenario. The production cost for this extraction technology is expected to be \$6.00/bbl (Canadian). In any country where modern transportation is available, oil shale can compete with petroleum as a source of liquid fuel only if given special tax treatment, import protection, royalty forgiveness, or even direct government support. In some countries, these incentives may be provided to develop domestic resources, improve balance of trade, and create domestic employment.

5.4 Role of Byproducts and Specialty Products

The wide range of products and characteristics of shale oils and bitumen from tar sands makes them excellent sources of petrochemical feedstocks. The current liquid production in Estonia is used almost entirely for petrochemical production. The historically successful oil shale industries usually produced a range of products in addition to oil and gas. Oil shale development is most likely to occur as small projects justified by the high dollar value of specialty products, special petrochemical feedstocks, coproducts, and byproducts. The oil shale resources of the United States are most amenable to this type of development. The Green River Formation, as mentioned in Sec. 2.1.4, is a veritable warehouse of saline minerals and hydrocarbons. The New Paraho Corporation has found the high nitrogen content of Green River Formation shale oil, which is usually a problem for refining, to be an advantage in producing modern asphalt paving modifiers that greatly improve the performance of conventional asphalt produced from petroleum derivatives. Several test strips of highway paving using their Shale Oil Modified Asphalt Technology (SOMAT) under a variety of climatic condition show that the shale oil–derived modifier produces asphalt pavement of such superior characteristics compared to conventional asphalt that its production at commercial scale is economically justified (Lukens, 1993).

In addition to the current production of trona and nahcolite from the Green River Formation, dawsonite in the Piceance Basin is an enormous resource for potential domestic production of alumina. The potential for multimineral operations improves the prospects for overcoming the barriers to successful oil shale operations. The University of Kentucky is pursuing a similar approach in investigating integrated production of several products from Devonian shale.

5.5 Relationship to Petroleum Recovery Technology

Eventual large-scale production of shale oil in competition with petroleum will hinge on a better understanding of how nature converts kerogen to petroleum, including the role of water and ambient gases that may be present in the natural process. This requires fundamental research in aqueous organic chemistry and how the atmosphere in which pyrolysis occurs influences reaction rates and pyrolysis products. Similarly, large-scale production of refinery feedstock from tar sands in competition with petroleum will depend on better understanding of the thermal and biological degradation processes that generate tar sands. Some large oil companies and governments are sponsoring research in these areas.

The terms "oil shale" and "tar sands" may be misnomers, but they identify the largest and most widely distributed hydrocarbon resources in the world, with a close technological and geological relationship to petroleum. These resources represent an assured future source of petrochemical feedstock, even if hydrocarbons are no longer used as fuels for environmental or economic reasons. The technology for extraction of useful products from these ubiquitous resources, whether by mining and surface processing or by *in situ* methods, is particularly crucial when related to present tertiary petroleum extraction technology. Reservoir engineers concur that current primary petroleum extraction methods leave behind as much as two-thirds of the oil in the ground. Although the world has extracted between (435 and 440)$\times 10^9$ barrels of oil from petroleum reservoirs and will probably extract that amount again, twice as much oil will remain in petroleum reservoirs and have to be extracted by methods akin to the extraction technologies for OS&TS. This fact alone establishes an enormous economic reward for developing the most efficient OS&TS extraction technologies. The real value of such technologies will only become apparent in the future, as they are used to clean out petroleum reservoirs. Economics will most likely dictate that exposed deposits and shallow reservoirs will be first worked by these methods, followed by ever deeper operations.

Ultimately, mankind may have to resort to use of the less economic portions of the spectrum of hydrocarbon materials available to us, such as fuels manufactured from renewable biomass, unconventional gas resources, and dirtier coals. Large-scale OS&TS operations have favorable economics in such scenarios.

6. OCCUPATIONAL, ENVIRONMENTAL, AND LEGAL ISSUES

The coincidence of renewed interest in OS&TS during the 1970s, and implementation of the environmental ethic of the 1960s, created a clash of interests that delayed many OS&TS projects. The long history of OS&TS operations, frequently at isolated locations, provides many opportunities for case studies of the effects of long-term exposure to raw OS&TS and their products on flora, fauna, and human populations. The common source and close relationship between raw OS&TS and their products and petroleum and petroleum products provide little reason to expect surprise environmental impacts from future OS&TS operations.

6.1 Historical Health Concerns

Health problems in the early oil shale industries depended on the nature of the feedstock and had much to do with the primitive state of technology and industrial hygiene. Antisell (1859) reported on a Rhenish feedstock that "on removing the cap of the retort, a strong smell of arsenic is perceived, and the workmen suffer from colic, ulcers at the root of the nose, of the joints, and an irritable condition of the skin." As early as 1920, Scottish shale oils, particularly the high-temperature fractions, were shown to have higher levels of carcinogenicity than other mineral oils tested (Key, 1974). As a consequence, the health histories of Scottish oil shale workers have been the subject of extensive studies. These studies confirmed that there was a higher incidence of skin cancer, especially scrotal cancer, among workers with prolonged exposure to Scottish shale oil and shale oil products. Classic high rates of scrotal cancer reported among British textile workers who oiled spinning jennies with shale-derived lubricants decreased as alternative lubricants were used. In general, the incidence of disease decreased as special precautions were instituted and improved hygiene became standard practice. Studies of the health of oil shale workers and residents in the area of the modern oil shale industry in Estonia reveal that an excess of skin cancer occurred only in women who worked in the industry for 10 or 20 years (Coffin, 1981). Investigations of health problems at isolated oil shale operations and villages in South Africa and Australia show that the workers and their families were generally very healthy but suffered from lack of immunity to communicable diseases brought in by visitors (Knapman, 1988). No unusual incidence of disease has been reported from current oil shale operations in Brazil or tar sand operations in Canada.

6.2 Environmental Issues

OS&TS share a common environmental problem with coal and petroleum: their use as fuel releases carbon dioxide and sulfur compounds to the environment. The carrying capacity of the environment, therefore, may limit future use of these fuels. In addition, OS&TS have the unique problem that the mining and extraction of their hydrocarbon content at commercial scale leaves enormous quantities of solid residue. Some shale oils, like some crude oils, contain enough heavy metals to be of concern. Many OS&TS contain connate water or water of hydration that results in an overall excess of water from their processing, which must be treated as wastewater. Air emissions are a potential problem for processing or burning OS&TS just as they are for any other hydrocarbon. Pyrolysis products such as shale oil are generally viewed as containing more carcinogens and toxic substances than other hydrocarbons. Special precautions must be taken regarding pyrolysis products from any source until their exact nature is known. The emissions from direct combustion of kukersite in Estonia are reported to be a severe environmental problem. The Estonian power plants do not have modern emission controls. Regardless of location, large-scale OS&TS operations will produce major socioeconomic impacts, population displacements, and transportation problems.

All of the above problems have been studied extensively (Bates, 1981). For the most part, there are technical solutions to the environmental issues of OS&TS development. Ultimately, each nation and the world collectively will have to make decisions regarding the tradeoffs involved in development of natural resources as population growth and world development demand greater quantities and varieties of products. The risks, dangers, and environmental damage of producing petroleum in sensitive frontier areas and transporting petroleum and petroleum products over great distances must be weighed against the environmental costs of meeting society's demands from more distributed alternatives.

6.3 Legal Issues

The marginal economics of OS&TS can make ordinary legal issues into insurmountable barriers to development. Among the legal issues are ownership of resources, mining and environmental laws, use of public lands and resources, and tax structure. The same laws and tax structure that are a nuisance to petroleum development can be an absolute barrier to OS&TS development because of the longer lead time and higher capital cost of the latter. Governments that seek to promote development of these alternatives to petroleum must be sensitive to the expectations of entrepreneurs evaluating higher-risk projects. Government ownership of the oil shale plant in Brazil, near-partnership operation of the tar sands plants in Canada, and support of clean fuels production in the United States are examples of government actions needed to promote large-scale production of alternatives to petroleum.

GLOSSARY

API Gravity: An arbitrary scale expressing the gravity or density of petroleum and petroleum products. The measuring scale is calibrated in terms of degrees API and is calculated by the following formula:

$$\text{Degrees API} = 141.5 \times [\text{specific gravity at } 15.56\ ^\circ\text{C}]^{-1} - 131.5.$$

Catagenesis: Regressive stage in organic metamorphism in which chemical bonds in large molecules of kerogen are broken by thermodynamic factors to form smaller petroleum (oil and gas) molecules.

Coal: Sedimentary rock containing sufficient carbonaceous material, usually of vegetable origin, to be of interest as fuel.

Dawsonite: A mineral consisting of basic sodium carbonate occurring in white, bladed crystals.

Diagenesis: Reconstructive process by which changes are produced in sedimentary rock during or immediately after their deposition and which is caused by such forces as weight of overburden or hot water. Stage in organic metamorphism in which disseminated organic material is converted to large molecules of kerogen.

Kerogen: The insoluble hydrocarbon material in sedimentary rock that upon destructive distillation forms oil and gas.

Lithologic: Relating to the character of rock in terms of its structure, mineral composition, color, and texture.

Metagenesis: Stage of organic metamorphism in which petroleum is altered by thermal or biological degradation under conditions that change the carbon structure and produce solid or nearly solid hydrocarbons.

Nahcolite: A mineral consisting of natural sodium bicarbonate.

Natural Asphalt or Bitumen: Soluble natural hydrocarbon material having less than 10° API gravity at ordinary temperatures.

Oil shale: Rock containing sufficient kerogen to be of economic interest as a source of oil.

Pyrobitumens: Natural bitumens that as a result of metamorphosis are insoluble in carbon disulfide and have characteristics similar to kerogen but are degradation products of petroleum.

Retort: A closed vessel or chamber with an outlet, used for distillation, sublimation, or decomposition by heat.

Sapropelites: Coal or oil shale formed from organic-rich sediment of marine, estuarine, or lacustrine deposition consisting of largely organic debris from aquatic plants and animals.

Tar Sands: Hydrocarbon-bearing materials formed by petroleum impregnation of mineral matter or voids along its migratory path in which the petroleum is subsequently degraded or metamorphosed leaving hydrocarbon material having an *in situ* viscosity greater than 10 000 cP at reservoir conditions. (*In situ* viscosity data are frequently not available, and so a gravity cutoff value of less than 10° API gravity or zero mobility at ordinary temperatures is used as a default value. This definition has evolved to rely on *in situ* viscosity as the key physical property and omits reference to mineral matter present with the hydrocarbon. Thus, the definition includes all of the solid products of petroleum degradation as well as natural bitumen left behind as petroleum is expelled or removed from reservoirs or migratory paths.)

Trona: Hydrous acid sodium carbonate in crystals or fibrous or columnar masses as a deposit from various soda-brine springs and lakes.

Works Cited

Abraham, H. (1960), *Asphalts and Allied Substances, Their Occurrence, Modes of Production, Uses in the Arts, and Methods of Testing*, 6th ed., vol. I, New York: Van Nostrand.

Antisell, T. (1859), *The Manufacture of Photogenic or Hydro-Carbon Oils from Coal and Other Bituminous Substances, Capable of Supplying Burning Fluids*, New York: D. Appleton and Co., p. 128.

Ball Associates, Ltd. (1965), *Surface and Shallow Oil-Impregnated Rocks and Shallow Oil Fields in the United States*, Oklahoma City: Interstate Oil Compact Commission.

Bates, E. R., Thoem, T. L. (Eds.) (1981), *Environmental Perspective on the Emerging Oil Shale Industry*, Washington, DC: U.S. Government Printing Office, EPA-600/2-80-205a.

Blade, O. C. (1938), *Ichthyol—Its Source and Properties*, Washington, DC: U.S. Bureau of Mines (USBM Information Circular 7042).

Carne, J. E. (1903), *The Kerosene Shale Deposits of New South Wales*, Sydney, Australia: Geological Survey of New South Wales, p. 97.

Coffin, D. (1981), in: Bates E. R., Thoem, T. L. (Eds.), *Environmental Perspective on the Emerging Oil Shale Industry*, Washington, DC: U.S. Government Printing Office, EPA-600/2-80-205a, p. 100.

Duncan, D. C., Swanson, V. E. (1965), *Organic-Rich Shale of the United States and World Land Areas*, Washington, DC: U.S. Geological Survey (USGS Circular 523).

Durand, B. (1980), *Kerogen Insoluble Organic Matter from Sedimentary Rocks*, Bayeux: Institut Francais du Petrole.

Gesner, A. (1865), *A Practical Treatise on Coal, Petroleum, and Other Distilled Oils*, New York: Bailliere Brothers, p. 49.

Greene, D. L., Leiby, P. N. (1993), *The Social Costs to the U.S. of Monopolization of World Oil Market, 1972–1991*, Oak Ridge, TN: Oak Ridge National Laboratory (ORNL-6744).

Key, M. M. (1974), in: Committee on Science and Technology, *Oil Shale Technology*, Washington, DC: U.S. Government Printing Office, p. 458.

Klosky, S. (1949), *An Index of Oil-Shale Patents*, Washington, DC: U.S. Bureau of Mines (USBM Bulletins 467 and 468).

Klosky, S. (1958), *Index of Oil-Shale and Shale-Oil Patents: 1946–56, Parts I, II, and III*, Washington, DC: U.S. Bureau of Mines (USBM Supplements to USBM Bulletins 467 and 468).

Knapman, L. (1988), *Joadja Creek, The Shale Oil Town & Its People 1870–1911*, Sydney, Australia: Hale & Iremonger Pty Ltd., p. 60.

Kuss, V. H. (1994), *1993 Annual Report—Central Pacific Minerals, N.L.*, Canberra City, Australia: Central Pacific Minerals, N.L., p. 4.

Lewin and Associates, Inc. (1983), *Major Tar Sand and Heavy Oil Deposits of the United States*, Oklahoma City: The Interstate Oil Compact Commission.

Lukens, L. A. (1993), *SOMAT Field Evaluation Program, Fourth Year Annual Report, U.S. Highway 191, Moab, Utah*, Lakewood, CO: The New Paraho Corporation.

Meyer, R. F. (Ed.) (1987), *Exploration for Heavy Crude Oil and Natural Bitumen, AAPG Studies in Geology #25*, Tulsa: The American Association of Petroleum Geologists.

Muspratt, S. (1860), *Chemistry, Theoretical, Practical, and Analytical, as Applied and Relating to the Arts and Manufactures*, London: William Mackenzie.

Newell, E. (1993), "Oilsands Play Larger Role in Canadian Oil Production," *Oil Gas J.* **91** (31), 25–30.

Ramsay, W. (Ed.) (1910), *Seventh International Congress of Applied Chemistry, Section IVa 1. Organic Chemistry and Allied Industries*, London: Partridge and Cooper, Ltd., p. 23.

Russell, P. L. (1990), *Oil Shales of the World, Their Origin, Occurrence & Exploitation*, New York: Pergamon Press.

Sell G. (Ed.) (1951), *Oil Shale and Cannel Coal*, Vol. 2 London: The Institute of Petroleum, p. 186.

Sneddon, J. B. *et al.* (1938), in: A. E. Dunstan (Ed.), *Oil Shale and Cannel Coal*, London: The Institute of Petroleum.

Vakhitov, G. G. (1991), "Soviet Production of Natural Bitumen," in: *Pace Synthetic Fuels Report*, December 1991, Houston: The Pace Consultants, Inc., pp. 3–25.

Further Reading

Descriptions of continuing developments regarding tar sands can be found in the publications of The United Nations Institute for Training and Research/United Nations Development Programme Information Centre for Heavy Crude and Tar Sands and the publications of the Alberta Oil Sands Technology Authority. The *Pace Synthetic Fuels Report* and *Oil and Gas Journal* provide continuous coverage of developments in both oil shales and tar sands. The Oil Shale Association supports the continued collection of oil shale literature in the Tell Ertl Oil Shale Archives at the Colorado School Of Mines. A list of relatively recent textbooks covering kerogen, conversion of kerogen to petroleum, and degradation of petroleum to tar sands follows.

Brooks, J., Welte, D. (1984), *Advances in Petroleum Geochemistry*, Vol. 1, Orlando, FL: Harcourt Brace Jovanovich.

Lee, S. (1991), *Oil Shale Technology*, Boca Raton, FL: CRC Press, Inc.

Schumacher, M. M. (1982), *Heavy Oil and Tar Sands Recovery and Upgrading International Technology*, Park Ridge, CA: Noyes Data Corporation.

Shell International Petroleum Company Ltd. (1983), *The Petroleum Handbook*, Amsterdam: Elsevier.

OPERATIONS RESEARCH

HUGH J. MISER, *Professor Emeritus of Industrial Engineering and Operations Research, University of Massachusetts at Amherst, Amherst, Massachusetts, U.S.A.*

INTRODUCTION

Operations research is unusual among the sciences in that its beginnings—and even its name—can be traced unambiguously to an event in history: military operational trials being conducted on England's east coast in the summer of 1938.

In 1935, as part of Great Britain's effort to respond to the growing threat of German air power, scientists began an urgent series of developments and experiments aimed at detecting flying aircraft at a distance; by 1937, devices that came to be known as radar were in place and experiments were being conducted. Their technical success led in 1938 to large-scale operational trials involving both the radar sets and the fighter-plane defenses; they were aimed at testing the combined effectiveness in actual operations against an attacking force. In connection with this work, A. P. Rowe referred to its analysis as "operational research"—a term he is thought to have originated.

When the term crossed the Atlantic, it became "operations research," and the two synonymous usages remain today: operational research in the United Kingdom and parts of Europe, and operations research elsewhere; both are commonly abbreviated as OR. In the early 1950s, the term *management science* (MS) was introduced in the U.S. with a virtually synonymous meaning; thus, by the middle 1980s, the common subject was often referred to as OR/MS.

The 1938 partnership between the Royal

3-527-28134-7/95/$5.00 + .50

Air Force operational people and the scientists, who not only worked with the radar but also analyzed the results of the operational trials, was successful both in characterizing the results of the trials and in showing how the operations could be improved. The next section sketches the growth of this seminal experience into the activities of an important science and an influential profession.

As operations research developed, its field of inquiry came to be recognized as the operations of systems consisting of nature, men, and machines, this last term including not only man's technical artifacts but also his laws, common practices, human behavior, and social structures and customs.

Attempts to generalize this description based on the concept of a system have not proved to be fruitful. Rather, the concept of an *action program* (Boothroyd, 1978; summarized in Miser and Quade, 1988, pp. 489–507) is more useful: a function, operation, activity, or response that is related to and given coherence by a human objective, need, or problem, together with the system of people, equipment, portion of nature, organizational elements, and management or social structure involved. OR can then be defined as the science of action programs, its work including both developing new knowledge and applying it to the problems of real action programs.

Two U.S. pioneer operations analysts (as OR workers engaged in practice are sometimes called), the physicist Philip L. Morse and the chemist George E. Kimball, observed early that "large bodies of men and equipment carrying out complex operations behave in an astonishingly regular manner, so that one can predict the outcome of such operations to a degree not foreseen by most natural scientists." Much of the success of the wartime work arose from the ability of the analysts to identify aspects of complex operations that exhibited such regularities.

1. DEVELOPMENT OF THE PROFESSION

The success of the small British team that was conducting operations research in 1938 set a pattern that was soon widely imitated; well before the end of World War II, most of the military services and their major commands in Britain, Canada, and the U.S. had established multidisciplinary teams to do such work (McCloskey, 1987). G. Majone has observed that the success of these teams did not arise from deep new theories of operational phenomena but rather the craft skills of the scientists applied to analyzing complex operations and assembling theories from readily available scientific tools.

However, one new theoretical development did emerge: search theory (Stone, 1989). It deals with an activity common in warfare: searching from a moving platform (such as a ship or an aircraft) for a fixed or moving target (such as a hostile submarine, ship, or aircraft). The goal is to detect, locate, and identify the target so that appropriate action can be taken. In a peacetime setting, this theory has been applied successfully to finding a sunken treasure ship off the U.S. Atlantic coast (Stone, 1992).

The wartime analysts emerged with the strong feeling that, if the man/machine/nature systems of warfare could be subjected to successful analysis, there should be many operations in peacetime society that could also benefit from such work. This feeling was soon converted to reality, as the community of analysts grew and the context of their inquiries spread widely, not only into varied operating situations, but also around the globe. Soon valuable new theories and applications were emerging from the work.

1.1 Professional Societies

Within the dozen years after the war, professional OR societies had been established in the United Kingdom (1948), the U.S. (1952), France (1956), India (1957), and Japan (1957). An International Federation of Operational Research Societies (IFORS) came into being in 1959; by 1973, 26 national OR societies were members, and by 1993 there were 41, with five other professional groups also adhering to the federation (including The Institute of Management Sciences, founded in the U.S. in 1954). The world-wide OR/MS community in 1993 was estimated to consist of over 35 000 persons.

The principal IFORS activities are these:

1. Conducting triennial international meetings for the exchange of information and discussion of international OR concerns.

The first meeting, at which IFORS was designed and agreed upon, occurred in 1957; the 13th meeting took place in 1993. For the first 12 of these meetings there has been a published volume of proceedings.
2. Publishing, since 1961, *International Abstracts in Operations Research (IAOR)*, a journal devoted to abstracts of the world OR literature. Three independently published volumes (Batchelor, 1959, 1962, 1963) contain abstracts of the literature for the years before 1961.
3. Publishing, beginning in 1994, *International Transactions in Operational Research*, a journal that will publish refereed papers from the 13th and later triennial meetings, as well as other essays of international interest.

1.2 Journals

The first two journals in the field were the *Operational Research Quarterly* (now the *Journal of the Operational Research Society*), founded in the United Kingdom in 1950, and *Operations Research*, founded in the U.S. in 1952. By 1973, there were 15 primary OR journals and 22 others of significant interest to OR workers. In 1993, *IAOR* covered the literature in 35 primary journals, 76 supplemental journals, and 67 specialist journals.

This literature and the professional activities that have prompted its growth since 1938 represent a tremendous development of interest in OR, both the validity of its theories and the effectiveness of their applications to real-world problems.

1.3 Books

Three books were written at the end of the war about the wartime work. Only one, however, was published promptly (Morse and Kimball, 1951); the other two (Johnson and Katcher, 1973; Waddington, 1973) were held up for many years by military classification concerns. Thus, the Morse and Kimball book was the first to offer a coherent account of a major body of OR work.

By the early 1950s, several anthologies of articles had appeared, but the first textbook emerging from an educational program did not appear until 1957 (Churchman *et al.*, 1957). However, the decade of the 1950s saw some 20 more specialized treatises appear. In the ensuing decades, this stream became a flood, with both textbooks and specialized treatises emerging in numbers almost too great to count.

Of the textbooks, Harvey M. Wagner's volume (Wagner, 1969) was widely regarded as setting a new standard and was awarded the Operations Research Society of America's Lanchester Prize. Recently, Hillier's (1990) has been widely used.

1.4 Education

In the early postwar years, a person interested in working in OR could learn about it only from the scattered literature and/or by joining a group engaged in such work, where the knowledge and craft skills of experienced analysts could be passed along. By the early 1950s, however, short courses of varying durations appeared in several countries, and by the end of the decade a few more formal university curricula had been established. By 1973, there were 53 such curricula in the U.S., and by 1993 the number in the U.S. and Canada had grown to over 100. Some included undergraduate as well as graduate training. Growth in other countries had also occurred on roughly the same scale.

Further, well before the 1990s, graduate training in these OR curricula had become recognized as an important qualification for entry into professional OR work, whether teaching, research, or practice.

2. OPERATIONS RESEARCH AS A SCIENCE

Following Ravetz (1971; for a summary focused on OR, see Miser and Quade, 1988, pp. 469–489), science may be described in general as "craft work operating on intellectually constructed objects," each object defining a class of phenomena. Scientific work is thus aimed at establishing new properties of the objects and verifying that they reflect the reality of the classes of phenomena that they represent.

This description has four implications:

1. The intellectual objects—which most scientists call theories but which OR workers usually call models—are created by the imagination, informed by earlier knowl-

edge of the phenomena and objects that have described aspects of them successfully, as well as innovative ideas or new evidence from reality. For example, there is a queuing theory that describes a wide variety of situations in which there are arrivals that demand service, which may be obtained immediately, or after a wait until earlier arrivals are served.

2. There is a continuing reference to the phenomena of reality, in OR's case those of action programs, or nature/man/machine systems. For queuing theory to be representative of some reality, such as the operations of a bank's customer services, the behavior of both the arrivals and the servers would have to be observed and characterized so that the theory could be based on these characterizations.
3. Scientific inquiry then becomes the search for new properties of the classes of phenomena both by manipulating the objects (that is, the theories or models), and by seeking new evidence from reality as a basis for revising them. In the banking example, after the queuing theory has been adapted to the observed behavior of the arrivals and servers, it can be manipulated to explore what might occur if the arrangements in the bank were changed, such as adding more servers, or improving their skills in various ways.
4. The new properties deduced from the objects—theories or models—must then be compared with the appropriate aspects of the phenomena of reality. In the banking example, the theory's predictions for changed conditions would be compared with an actual realization of these conditions in order to confirm the theory as a representation of reality, or to suggest how it might have to be revised.

Since all of science shares this generic description, it is important to note how the different sciences—such as physics, chemistry, biology, or operations research—are distinguished: not by their theories, models, or methods, many of which are widely shared, but by the portions of reality that they are undertaking to understand, explain, and solve problems in (Kemeny, 1959).

Within the framework established by this conception, it is possible to distinguish three classes of problems, depending on their goals: to paraphrase Ravetz, scientific problems, where the goal of the work is to establish new properties of the objects of inquiry, and the ultimate function is to achieve new knowledge; technical problems, where the function to be performed specifies the problem; and practical problems, where the goal of the task is to serve or achieve some human purpose and the problem is brought into being by recognizing a problem situation in which some aspect of human welfare could be improved.

In the queuing situation being used as an example, the scientific problem is to develop a model describing a queuing process, the technical problem is to achieve practical calculations from the model (usually in this case a rather complicated and difficult process), and the practical problem is to use the results to solve practical problems, such as those of improving a bank's service, or—to mention the purpose for which the theory was originally developed—to design an efficient and effective telephone exchange.

In some sciences, the professional groups that solve scientific problems are distinguished from those that work on technical and practical problems (although there is, of necessity, considerable overlap in these activities); for example, chemists solve scientific problems, while chemical engineers work primarily on technical and practical problems.

It is important to note that, in the case of operations research, there has not yet been such a split: operations research comprehends the work in all three classes of problems, with an attendant richness of interaction among its workers.

Thus, we can now describe operations research as the science of action programs, encompassing work on scientific, technical, and practical problems. It should be noted that, when the basic phenomena underlying a problem situation are not adequately understood, to solve a practical problem may involve solving scientific problems to obtain the needed models and technical problems to fashion the technical means needed to support the work toward ameliorating the practical problem.

3. THE CONTEXT

Within the broad contextual framework suggested by the term action program, it is possible to distinguish two major categories in which OR work takes place: processes common to a variety of action programs, and action programs distinguished by their practical context. Table 1 lists common processes to which OR work has contributed, and Table 2 shows areas of application in which significant OR work has been done. Since these tables list the standard terms of classification used by *International Abstracts in Operations Research*, work in any of these areas can be traced through this journal.

4. THEORIES AND TECHNIQUES

In contrast with the wartime years of operations research, when only one essentially new theory was devised, the postwar years have been a time of extraordinary fecundity for the development of models and solution techniques (often called algorithms in OR). While many have been borrowed from other fields and adapted for use in OR contexts, others have grown up largely in OR. Also, the uses of many have spread to other scientific fields.

Table 3 offers a list of the model and technique areas being actively pursued in OR in 1993. The terms shown are the standard classifications used in *International Abstracts in Operations Research*, so that, again, work in any of these areas can be traced through this journal.

As this table suggests, the most extensive borrowing from other fields is from probability theory (*q.v.*) [including especially stochastic processes (*q.v.*)] and statistics; indeed, surveys of OR practitioners show that, of the various techniques they employ, most are from probability and statistics. This is a natural consequence of the activities of exploring the behavior of action programs that not only yield numerical data—often in great volume—but also significant variations in the observations.

The reader will also note, however, bor-

Table 1. Common processes studied by operations research.

Accidents	Maintenance, repair, and replacement
Allocation: resources	Management
Behavior	Marketing
Bidding	Materials
Communication	Measurement
Computers	Organization
Calculation	Performance
Data structure	Personnel and manpower planning
Information	Planning
Control	Production
Cybernetics	Flexible mfg. systems
Decision	Just-in-time
Applications	Material requirements planning
Rules	Project mgmt.
Studies	Quality and reliability
Demand	Queues: applications
Design	Research
Distribution	Risk
Equipment	Scheduling
Experiment	Search
Facilities	Service
Financial	Storage
Forecasting: applications	Supply
Gaming	Systems
Growth	Testing
Information	Timetabling
Innovation	Values
Inspection	Vehicle routing and scheduling
Inventory	Work
Order policies	
Storage	
Investment	
Learning	
Location	

Source: *International Abstracts in Operations Research* **44** (1993), p. 13.

Table 2. Areas of application studied by operations research.

Advertising	Marketing
Agriculture and food	Medicine
Biology	Meteorology
Commerce	Military and defense
Communications	Mineral industries
Community OR	Petroleum
Construction and architecture	Politics
Cutting stock	Public service
Ecology	Recreation and tourism
Economics	Science
Education	Social
Energy	Space
Engineering	Sports
Finance and banking	Transportation
Forestry	Air
Geography and environment	General
Government	Rail
Health services	Road
Law and law enforcement	Water
Libraries	Urban affairs
Manufacturing industries	Water
	World affairs

Source: *International Abstracts in Operations Research* **44** (1993), p. 14.

Table 3. Model and technique areas OR is actively exploring.

Adaptive processes	Critical path
Artificial intelligence	Dynamic
Decision support	Fractional
Expert systems	Geometric
Calculus of variations	Goal
Combinatorial analysis	Integer
Computational	Markov decision
analysis	Mathematical
Parallel computers	Multiple criteria
Personal computers	Network
Supercomputers	Nonlinear
Control processes	Parametric
Cost-benefit analysis	Probabilistic
Decision theory	Quadratic
Multiple criteria	Transportation
Differential equations	Traveling salesman
Fuzzy sets	Queues: theory
Game theory	Sets
Gradient methods	Simulation
Graphs	Analysis
Heuristics	Applications
Information theory	Languages and
Lagrange multipliers	programs
Markov processes	Statistics
Matrices	Data envelopment
Networks	analysis
Flow	Decision
Path	Distributions
Scheduling	Empirical
Neural networks	Experiment
Numerical analysis	General
Optimization	Inference
Simulated annealing	Multivariate
Probability	regression
Programming	Sampling
Assignment	Stochastic processes
Branch and bound	Time series and
Convex	forecasting methods

Source: *International Abstracts in Operations Research* **44** (1993), p. 14.

rowings from classical mathematics, such as the calculus of variations, combinatorial analysis (which has been spurred significantly by OR applications), differential equations, Lagrange multipliers, matrices (an essential ingredient in programming models), numerical analysis, optimization (a field whose development has been greatly spurred by OR work), probability (much used in view of the variabilities often encountered in the phenomena of action programs), and set theory.

In what follows, there is not space for even a cursory survey of all of the models and techniques commonly used in operations research. Thus, the ensuing subsections take up only a few of the most important that are regarded as especially associated with OR.

4.1 Linear Programming

The structure of the linear programming model, together with a procedure for obtaining solutions from it, was devised by George B. Dantzig in 1947 (Dantzig, 1963). While the original setting and motivation for it were largely economic, it soon became clear that it had many applications elsewhere, including business and industry, as well as in other sciences.

The simplest form of the linear programming model is this:

$$\text{Maximize} \quad c_1x_1 + \cdots + c_nx_n,$$

where the variables $x_1, \cdots, x_n$ satisfy the conditions

$$x_j \geq 0, \quad j = 1, \cdots, n,$$

$$a_{11}x_1 + \cdots + a_{1n}x_n \leq b_1,$$

$$\cdot \quad \cdot \quad \cdot$$

$$a_{m1}x_1 + \cdots + a_{mn}x_n \leq b_m,$$

and the c_j, b_i, and a_{ij} are constants.

One interpretation of this model that encompasses a wide variety of applications is this: There are n activities X_j, each producing x_j units of product, where each unit yields a value c_j. Each unit of the jth product consumes an amount a_{ij} of the ith resource, of which there is a maximum amount b_i that may not be exceeded by the sum of all of the activities. The problem then is to program a mix of activities that maximizes the total value (spoken of as the objective function) while remaining within the constraints on the resources.

The linearity implied by the statement of the model implies two basic assumptions underlying it: proportionality (that is, the resources used by an activity and their contribution to the objective function are proportional to the level of the activity) and additivity (that is, the combined use of resources by all of the activities is the sum of the individual uses). The third assumption is non-negativity (that is, negative quantities of activities are not possible).

The linear inequalities the variables x_j must

satisfy correspond to the geometrical fact that a point $x = (x_1, \ldots, x_n)$ satisfying the model's conditions lies within or on the surface of a convex polyhedron (that is, one in which a line connecting any two points in or on the polyhedron also lies in or on it). If the objective function does not become arbitrarily large for points x on the surface of the polyhedron, then the maximum occurs at a vertex. Thus, the solution to a linear programming problem is one of finding the vertex, and therefore the point x, at which the objective function attains its maximum.

In addition to formulating the problem in this way, Dantzig devised a systematic and relatively economical procedure, called the *simplex method*, for discovering the point at which the objective function attains its maximum (Dantzig, 1963).

The model stated above has a dual:

$$\text{Minimize} \quad b_1y_1 + \cdots + b_my_m,$$

where the variables $y_1, \ldots, y_m$ satisfy the conditions

$$y_i \geq 0, \quad i = 1, \ldots, m,$$

$$a_{1j}y_1 + \cdots + a_{mj}y_m \geq c_j, \quad j = 1, \ldots, n.$$

If there is a maximum for the first problem there is a minimum for its dual and these two numbers are equal. This duality theorem is essentially equivalent to the minimax theorem of game theory (*q.v.*).

With the programming problem now expressed as minimizing an objective function, other applications come to mind. For example, there is a classical diet problem for mixing livestock and poultry feed that can be formulated in this way: If a_{ij} denotes the number of units of the jth nutrient in one unit of the ith food, c_j is the minimum amount of the jth nutrient needed for satisfactory health, b_i is the unit price of the ith food, and y_i is the number of units of this food to be purchased, then find the point $(y_1, \ldots, y_m)$ that minimizes the objective function and produces a satisfactorily mixed diet at minimum cost.

Since manual calculations by the simplex method become very laborious even for problems of modest size, computer programs for this method were introduced as soon as computers became powerful enough to handle them. By now, both the methods of solution and the computer programs for them have evolved to the point where problems of fairly large size can be handled on personal computers and problems with thousands of variables and constraints can be dealt with successfully at modest cost on large machines.

The resulting tools have greatly expanded the possible applications in many fields, including agriculture, business, government, manufacturing and service industries, transportation, and economics, and OR workers have achieved thousands of such applications. The theoretical basis for such work has also expanded similarly. By 1993, hundreds of papers were appearing yearly on both theory extensions and their important applications.

4.2 Descendant Programming Models

The value of a theory or model and its associated solution technique can be judged by the vigor with which they are extended either by theoretical developments or a widened coverage of the phenomena of reality. Thus, the ensuing value can be judged in part, first, by the scope of the phenomena that the model helps to understand and, second, by the importance of the problems the technique can deal with; but this judgment must also include, third, their fecundity in prompting the development of descendant models and techniques, that is, ones suggested by the basic model and its central technique but demanding important new approaches. The previous subsection suggests that linear programming has high value on the basis of the first two of these criteria; this subsection describes some descendant theories that support its value based on the third. In each case, there are also descendant solution techniques.

4.2.1 Convex Programming Let x be a set $(x_1, \ldots, x_n)$ of points in n-dimensional space. Then, a function $f(x)$ is a convex function if it is defined over a set of points x in a convex set and if the set of points $z \geq f(x)$ is convex in the $(n + 1)$-dimensional space (x,z).

Early investigators were able to amend the simplex method to apply to programs where the constraints and the objective function are convex.

4.2.2 Quadratic Programming A quadratic form in n variables is the sum of terms $a_{ij}x_ix_j$, where both i and j take on values from 1 to n. Such a form is positive semidefinite if its value is greater than or equal to zero for all values of the x's. By suitable transformations of the variables, it is possible to set such a form equal to the sum of squares of the new transform variables.

The transformed quadratic form can now be recognized as a convex function, which allows programs with positive semidefinite objective functions to be solved for the minimum value of this function by adjusting the methods for convex functions. A similar statement can be made for programs seeking the maximum of the negative semidefinite quadratic objective function. These special cases allow some improvements in the efficiency of the solution procedure.

4.2.3 Flow through Networks Consider a network connecting two nodes, a source and an destination, by means of a series of intermediate arcs that meet at intermediate nodes. Flows can take place in the arcs and through the intermediate nodes. Each arc has two flow capacities associated with it, one for each direction. The problem is to find the maximal flow that the network permits from the source to the destination. Efficient procedures for solving this problem are available.

It has also been generalized—for example, for a network with many sources and destinations and stipulated amounts at both, to minimize the costs of the flows that will be required for the flows from the sources to achieve the stipulated amounts at the destinations at minimum transportation cost—and suitable algorithms have been developed. This model and its associated solution algorithm have been applied, for example, to the problem of assigning children to schools.

4.2.4 Integer Programming It was recognized early that there were many attractive problems for which the variables in a program are constrained to be integers. Unfortunately, simply approaching these problems with continuous variables and their solution techniques does not work satisfactorily, so that special procedures must be devised for them.

The pioneer work in this regard was done by R. E. Gomory in 1958 (Gomory, 1991). Since then, tremendous advances have been made in developing solution algorithms, and they have been applied to numerous important problems, some quite large.

The well-known traveling salesman problem can also be formulated as an integer program. This problem arises when a salesman desires to keep his total travel distance or time to a minimum while making one visit to each of a number of cities and returning to his starting point at the end of his trip. Similar problems can occur in a wide variety of situations.

Further, this line of research has generated its own descendant models and their applications, including those for so-called mixed-integer problems (where some of the variables are continuous but others are restricted to integral values).

In cases where it is possible *a priori* to state upper and lower bounds for the maximum value, there is a branch-and-bound procedure that has had many applications. It has also been much developed since it was first set forth.

4.2.5 Other Developments As Table 3 indicates, there are a number of other descendant programming models and solution techniques. While there is not space here to describe them all, a few may be mentioned.

It can happen that a programming problem is stated such that there is no vector x that satisfies all of the constraints. If one can establish penalty costs for violating one or more of the constraints, then goal programming finds a solution that violates the constraints in such a way as to minimize the total penalty cost.

The programming structure has also been used to formulate models in which the functions are nonlinear (as, for example, in convex and quadratic programming, as mentioned earlier), and various special cases of this situation have yielded reasonably economical solution techniques.

So far, the programming formulations have treated the coefficients in the constraints as completely known. In many practical situations, however, these coefficients may not be known with precision, or may in fact be subject to random variations. Chance-constrained programming considers such cases under a few simplifying assumptions. Stochastic programming deals with more general cases.

For a modern overview of linear programming and its descendant theories, see Nemhauser *et al.* (1989).

4.3 Dynamic Programming

The word programming in the name of this subject might suggest that it is another theory that descended from linear programming, but this is not the case. Rather, dynamic programming is concerned with a sequence of decisions that must be made over time.

Unlike linear programming, which has specific models that can be solved by various techniques, dynamic programming deals with a structure and an analytic approach that can be applied to a variety of models.

The dynamic programming approach deals with problems that can be put into this structure:

1. The decision variables and their associated constraints are grouped according to stages to be considered sequentially.
2. For a given stage, the only information about previous stages relevant to selecting optimal values for the current decision variables is summarized by a quantity called the state variable.
3. For a given stage in its current state, the current decision has a forecastable influence on the state at the next stage.
4. The optimality of the current decision is judged in terms of its forecasted impact on both the present stage and all later stages.

There is a canonical form for the central recursion relation that can be developed in this way: Consider an N-stage process that is n stages from completion. Let s stand for a state of the system, S_n for the set of all possible states at the nth stage, d_n for the decision made at the nth stage, and $D_n(s)$ for the set of all the feasible values of d_n for the system in the nth state. Also for the nth state, let $R_n(s,d_n)$ be the immediate value of decision d_n and $T_n(s,d_n)$ the transformed state of the system at stage $n - 1$. If $f_n(s)$ is the value of the optimal decision at the nth stage, then we can say that it depends not only on the state the process is in at the nth stage, but also on the number of stages left to be completed. Then the dynamic programming recursion states that $f_n(s)$ is the optimum (that is, either the maximum or minimum) over all $d_n(s)$ in $D_n(s)$ of $R_n(s,d_n) + f_{n-1}(T_n(s,d_n))$ for every s in S_n.

Perhaps the most striking aspect of this procedure is that it begins with the last stage and works toward the beginning. Its advantage over the complete enumeration of all possibilities at all of the stages is that it optimizes at each stage over a limited number of decision variables.

Here are some problems that can be formulated with a dynamic programming structure: inventory rules indicating when to replenish an item and by what amount; capital budgeting for allocating scarce resources to new ventures; scheduling for routine and major overhauls of complex machinery; and formulating a long-range strategy for replacing depreciating assets. For further examples, see the classic book by Bellman and Dreyfus (1962).

4.4 Queuing Theory

Queuing is one of the commonest processes in developed societies: telephone calls arrive at an exchange, customers arrive at lunch counter, vehicles arrive at a car wash, aircraft arrive at an airport, and so on, each demanding service; when others are ahead of them, they wait in a queue until served. The theory of this process began early in this century in connection with telephony, but did not move significantly out of this context until it was adopted in the early postwar years by operations research, after which it experienced an explosive growth.

The simple examples just mentioned suggest the three aspects of a queue that are important for developing a theory of it: the arrival process, the service mechanism, and the queue discipline.

Consider the simplest case, in which the arrivals are random over time, and the service times are exponentially distributed. If λ and μ are the average arrival and service rates, respectively, then the average traffic density ρ is λ/μ. For the number N of customers in the system, the queue length Q, the waiting time W, and the total time D the customer is in the system, if we denote the probability that $Q = j$ ($j = 0, 1, \ldots$) by $Pr[Q = j]$, the average or expected number of customers in the system by $E[N]$, the expected length of the queue by $E[Q]$, the expected time a customer waits until service begins by $E[W]$, the expected total delay for a customer (waiting time plus service time) by $E[D]$, and the variance of the queue length by $Var[Q]$, then the following relations can be derived from the theory:

$$Pr[Q = j] = (1 - \rho)\rho^j, \quad E[N] = \rho/(1 - \rho),$$
$$E[Q] = \rho^2/(1 - \rho), \quad E[W] = \rho/(1 - \rho)\mu,$$
$$E[D] = 1/(1 - \rho)\mu, \quad Var[Q] = \rho/(1 - \rho)^2.$$

There are a tremendous number of other queuing processes with other assumptions about the arrival processes, the service-time distributions, and the queue disciplines, but, unfortunately, only a few yield simple formulas like those just set forth. Rather, more complicated expressions must be dealt with, approximations employed, or the queuing process simulated on a computer (Gross and Harris, 1985).

4.5 Simulation

Any theory or model is, of course, a representation—that is, a simulation of a sort—of some aspect of the phenomenon that it purports to describe. However, the term simulation here means something different: instead of representing a process by a model, it is a somewhat simplified ideal process that imitates selected aspects of the one from reality.

Consider an operational process where it is possible to subdivide it into a series of stages or steps, such as the queuing process just discussed: An arriving customer either takes a place at the end of the queue if there is one, or, if there is no queue, proceeds to obtain the desired service; whenever a service is completed, the queue, if there is one, is reduced by one waiting customer. This example can easily be simulated by hand, provided the analyst has been able first to establish a series of random arrivals and exponential service times to draw from as the simulation proceeds. However, it can also be programmed on a computer.

Similarly, many processes, such as for highway toll-collection operations, air-traffic arrivals at airports, banking service operations, or telephone ordering operations, can be simulated, either to solve problems or to see how revised procedures might work.

It is usual to represent the chosen aspects of the operations by a computer program that imitates them; OR workers usually call such a representation a simulation, but in other fields it is also called a Monte Carlo process. This procedure has been used quite often by OR workers when no simplifying theories have existed to make the analysis more compact and efficient. They do this in spite of the procedure's shortcomings: that specific assumptions about the situation have to be made instead of more general ones that a theory could incorporate, that a large number of trials (or "runs") may be needed to achieve a stable set of results, and that complicated situations may call for very large computer programs whose results may be difficult to interpret and apply.

Indeed, there is a standing—if informal—rule of action among OR workers that simulation is to be shunned if any analytic models can be made to represent the situation adequately, and that therefore simulation should be a last resort. Nevertheless, OR workers have found many situations in which simulation has been helpful to decision makers, and there has been substantial development in this field. Whereas in the early days of OR it was necessary to construct computer simulations from the basic elements and principles, by 1993 it was far more common for an analyst to adapt a commercially available simulation package. The journal *OR/MS Today* (December 1993, pp. 62–78) describes the essential properties of 55 such packages (see also Law and Kelton, 1991).

Simulation has an advantage in that it can portray to a user or client the steps in the process, thus conveying an important sense of reality. And it is a technique that can be used when analytic models cannot cover the problem situation adequately. Historically, successful simulations in OR have run all the way from relatively simple ones conducted by hand on a large sheet of paper to very complicated ones managed on large computers.

4.6. Other Theories and Techniques

As Table 3 indicates, there are many other theories and techniques that occupy a place in modern OR work; STOCHASTIC PROCESSES (for a modern overview, see also Heyman and Sobel, 1990), GAME THEORY, ARTIFICIAL INTEL-

LIGENCE, COMPUTER DATA BASES, and COMPUTER GRAPHICS are discussed elsewhere in this Encyclopedia. Space in this introductory overview does not permit others to be described, but they can be traced easily in the literature through the references and *International Abstracts in Operations Research*. For a survey of computing from an OR perspective, see Coffman *et al.* (1992).

4.7 A Perspective on Standard OR Models and Techniques

Over the last four decades, the community of OR workers primarily interested in scientific problems has not only developed many new models, but also algorithms (that is, solution techniques) for them where they were appropriate. This growing community of OR scientists has exhibited great skill and inventiveness in this process. Thus, such tools exist in many of the categories listed in Table 3 (as hinted at in the brief introductory descriptions above).

Many of these models and algorithms can be thought of as standard in the sense that they are reference points for OR workers looking at the phenomena of action programs with a view to constructing models to represent them and algorithms to obtain specific results. Thus, they can be considered to be an OR analyst's tool kit. And, of course, they form the substance of advanced treatises and courses in OR theory.

On the other hand, OR workers facing technical or practical problems can relatively seldom simply apply one of these standard tools; rather, the problem structure and its elements usually demand that the tools be adapted to the new situation. Indeed, how to be effective in adapting the tools in the kit of standard models and algorithms can be considered one of the essential craft skills of the OR practitioner.

It has been pointed out that OR comprehends work on all three classes of problems: scientific, technical, and practical. The close association of work on all three classes of problems has an important benefit for operations research: Not only do the model and algorithmic tools help practitioners quite directly, but also the process of looking at new situations and their phenomena has a feedback benefit to OR's theorists, as it often brings new modeling and algorithmic possibilities to them that might otherwise be overlooked.

5. PRACTICE

An OR analyst working to solve a practical problem is said to be involved in practice. Such work may entail, as has frequently been the case, work on technical problems when instrumentalities need to be designed to achieve desired goals, or on scientific problems when the underlying phenomena are not yet understood well enough. In any case, regardless of what subproblems the work may require solving, the overall goal of practicality qualifies the whole as being practice.

An OR analyst is often consulted when someone with a suitable responsibility or interest in an action program discerns a problem situation that needs improvement. While this person—and perhaps others—may have diagnosed the problem and may even have a belief about the course to take toward its possible solution, it is commonly the case that the forces actually at the core of the problem situation lie buried deeply enough to make such an early diagnosis questionable, and the preconceived fix inappropriate. Thus, typically it is best for the analyst—or the team of analysts if the problem situation is large and complex—to approach it with an open mind and explore it thoroughly before deducing its essential properties and using them to devise a scheme for ameliorating its undesirable aspects.

5.1 The Practice Processes

While each situation in practice may properly be regarded as unique, it is nevertheless possible to describe a central analysis process that contains elements central to most—if not all—of practice, as follows.

Figure 1 offers a synoptic view of these elements. It proceeds from the general recognition that a problem situation exists to the implementation of some policy or course of action and the evaluation of its effects. Since each situation has its own unique properties, few OR practice engagements follow such a procedure exactly—and sometimes not even approximately. However, it is common for at least some of these elements to occur during the work. Thus, Fig. 1 is not intended to be

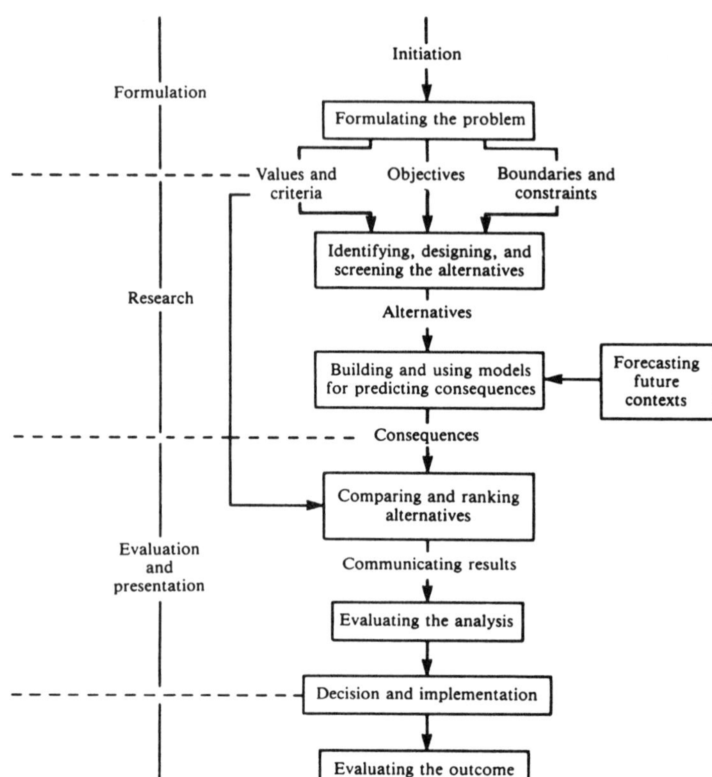

FIG. 1. Important elements in an operations research practice engagement that runs from problem formulation through research and implementation to evaluating the outcome. [From Miser and Quade (1988), p. 23; reproduced by permission.]

prescriptive; rather, it offers a suggestive structure as the basis for discussion.

5.1.1 Formulation The work begins with a thorough exploration of the problem situation. It is vital for the client and people in the action program to cooperate in this work. The purpose is to formulate the problem to address. Once this is done and the client has agreed with the analysis team on the diagnosis, it is possible to plan the work to be done.

This early work also identifies the values and criteria that should inform the choice of what eventually will be done to ameliorate the client's concerns, sets up the objectives to be sought by the solution, and agrees with the client on the boundaries and constraints that must be observed in devising it.

5.1.2 Research This stage extends the information- and data-gathering that began in the formulation stage. The findings that emerge from processing these results allow the analysis team to identify, design, and screen possible alternative courses of action that may help with the problem. Against this background, the analysis team can build models capable of deducing the consequences of adopting each of the alternatives chosen for further investigation within the framework of possible future conditions.

5.1.3 Evaluation and Presentation With estimates of the consequences in hand, the analysts may compare—and possibly rank—the alternatives against the criteria chosen earlier in the analysis, plus any new ones that may have emerged during the work. These findings must then be presented to the client and other parties at interest in a way that enables them not only to appreciate the results but also have at least a broad overview of the logic and its factual basis that produced them. These understandings, combined with the client's wider and more general knowledge of the situation, may then enable him or her to adopt a suitable policy or course of action—and, when necessary, persuade the various parties at interest to accept it.

5.1.4 Implementation and Evaluation The client, and not the analysts, must decide on what to do and how to carry it out effectively, and the client must see to the implementation and take full responsibility for it. Nevertheless, experience shows that it is very important for the analysis team, or at least some analysts who understand and appreciate what was done, to work cooperatively throughout the implementation stage.

Later, if an evaluation of the efficacy of the results is undertaken, it is also desirable for someone familiar with the input analysis to be involved.

5.1.5 Variations While it is possible to specify a core diagram of the principal elements of OR practice, it must be admitted immediately that few, if any, such engagements follow this outline exactly. Rather, since each problem situation is different, the analysis must be adapted to it. Thus, in studying a series of cases, one finds variations such as these:

1. Instead of proceeding linearly from the top to the bottom of the diagram, the work cycles from intermediate stages back to earlier ones as the research program brings new insights and fresh intermediate results that may prompt reconsidering the beginning foundations of the work.
2. Some work may be aimed more at fleshing out the client's understanding of the situation than prompting a significant change, so it may stop at one of the intermediate stages.
3. The relative effort expended in the various stages may vary tremendously from case to case: one case may have to expend its major effort in just the information- and data-gathering stage, after which what needs to be done may be fairly apparent without much further analysis. Another case may proceed fairly expeditiously through the outline of Fig. 1 and then have a very long and complicated period of work to achieve what may appear to an outsider to be the implementation of a fairly simple set of proposals.
4. In some cases, an intermediate stage may dominate the work, as a result of such factors as technical difficulty in devising proper models, major uncertainties in forecasting future conditions, complexities of the underlying situation, and so on.

In any case, the procedure outlined here as the basis for discussion must be regarded as one that has stitched together the key elements that may enter OR practice to varying extents depending on the peculiarities of the situation being analyzed.

5.2 Systems Analysis

There is a branch of operations research devoted to problems of such large scale that they almost always involve major efforts by teams of analysts drawn from a variety of disciplines in addition to operations research. In the mid-1950s, it came to be called systems analysis (in this meaning not to be confused with the same term used in the computer community). While there is a separate literature about such work (Miser and Quade, 1985, 1988; Miser, 1995), its essential properties are sufficiently close to those of OR as it is usually understood that inferences back and forth between the two kinds of work are quite appropriate.

5.3 The Relation between Analyst and Client

Based on close study in a variety of fields of the ideal relation that should exist between professional practitioner and client, Schön (1983; for a relevant summary, see Miser and Quade, 1988, pp. 508–520) advocates a "reflective contract" that works in this way: ". . . in a reflective contract between a practitioner and client, the client does not agree to accept the practitioner's authority but to suspend disbelief in it. He agrees to join the practitioner in inquiring into the situation for which the client seeks help; to try to understand what he is experiencing and to make that understanding accessible to the practitioner; to confront the practitioner when he does not understand or agree; to test the practitioner's competence by observing his effectiveness and to make public his questions over what should be counted as effectiveness; to pay for services rendered and to appreciate competence demonstrated. The practitioner agrees to deliver competent performance to the limits of his capacity; to help the client understand the meaning of the professional's advice and the rationale for his actions, while at the same time he tries to learn the meanings his actions have for the client; and to reflect on his own tacit understanding when he needs to do so in order to play his part in fulfilling the contract."

Not all OR practice engagements realize this ideal, of course, but successful practitioners achieve a suitable form of it in their work. In any case, the practical arrangements aimed at achieving the ideal of the reflective contract must, of necessity, be evolved in the light of the circumstances and people peculiar to each engagement. It is common, as one way of supporting this concept in practice, for a member of the client's staff to work with the analysis team throughout the OR engagement. And, of course, it is wise for the client and the analysis team to have frequent interactions as the work proceeds.

5.4 Examples of Good Practice

Fortunately for someone interested in seeking out cases of good practice, there is a convenient central source: Since 1973 the journal *Interfaces* has specialized in publishing papers on such work. Assad *et al.* (1992) offer a selection of 19 cases from this source, accompanied by discriminating discussions. Miser (1995) centers his attention on cases, and Miser and Quade (1985, 1988) accompany their general discussions of analysis principles with illustrative cases.

Two areas often associated with operations research are discussed elsewhere in this Encyclopedia: MANUFACTURING ENGINEERING, and COMPUTER-AIDED MANUFACTURING. Three areas of application had by 1994 been given overviews in handbook volumes: logistics of inventory and production (Graves *et al.*, 1993), marketing (Eliashberg and Lilien, 1993), and problems in the public sector (Pollock *et al.*, 1994).

Each year there is a muscular and carefully administered competition in the U.S. for the best papers on practice. Since 1975, *Interfaces* has published the finalist papers in this competition; since there are five or more finalists in each competition, this competition alone offers a large number of papers and a wide variety of work. For example, in recent years there have been papers on computer-aided train dispatching, planning the management of the freshwater resources of the Netherlands, how OR is used in a large petroleum company, the size and deployment of a sales force, scheduling police patrols in a major city, assigning operators in a large urban transportation system, analyzing the basis for U.S. Medicare payments to hospitals, improving passenger traffic management at a major international airline, and considering the effectiveness of a needle-exchange system as a device for reducing the incidence of AIDS among drug addicts in American cities. Two contrasting cases, the first of which won this prize, are described briefly below.

5.4.1 Improving the Distribution of Blood to Hospitals Human blood, a living tissue of unique medical value, is a perishable product. In the U.S. when this study was being done, it had a fixed lifetime during which it could legally be used for transfusion to a patient of the proper type and after which it had to be discarded; this lifetime, since extended, was 21 days.

The blood is collected in units of one pint (approximately a half liter) from volunteer donors at various collection sites such as a regional blood center (RBC). After a series of typing and screening tests, it is shipped to a hospital blood bank (HBB) in the region of the RBC. The units are then stored at the HBB, where they are available to satisfy the random daily demands for transfusions to patients. Since in general not all units demanded and assigned to a patient are used, a unit can be assigned several times during its lifetime before it is transfused or outdated and discarded.

To manage the blood resources of a region is a difficult task. The blood distribution problem is complex, because of the blood's perishability, the uncertainties of its availability to the RBC, and the random nature of the demands and usages at the HBB's. Superimposed on these complexities are the large variations in the sizes of the hospitals—and therefore their HBB's—to be supplied, in the relative occurrences of the eight different blood groups, and in the mixes of whole blood and red blood cells called for when this study was done (since then the development of other blood products has complicated matters further). Finally, on the one hand, there is the imperative of having blood available when and where needed (although elective surgery can be postponed because of a blood shortage, this act incurs additional costs for all concerned), while on the other hand there is the desire to operate efficiently and economically, and thus to avoid wasting the precious individual gifts of blood.

The two most common performance measures for an HBB are the shortage rate (that is, the proportion of days when supplementary unscheduled deliveries from the RBC have to be made to satisfy the hospital's demand), and the outdate rate (that is, the proportion of the hospital's blood supply that is discarded because of its becoming outdated). Suitable calculations convert these measures for individual hospitals to similar ones for a region.

In 1973, two analysts in the U.S., one a blood expert and the other an OR analyst, reported a study that went most of the way toward solving this problem of managing blood supplies efficiently for the Greater New York Blood Program, which supplies over two hundred hospitals. After studying the patterns of demand and supply, they were able to characterize the situation with relatively simple models that were variants of standard ones adapted to suit this new purpose. On the basis of these models they were able to devise a decision support system for an RBC that addressed these questions: What are the minimum achievable outdate and shortage rates that can be set for the region? What distribution policy will achieve these targets? What levels of supply are needed to achieve alternative targets?

The blood management system they devised was characterized by centralized management at the RBC, prescheduled deliveries to the HBB's supplemented by emergency deliveries when needed, and a distribution policy according to which some blood was rotated among the hospitals. The operation of this system is based on a programming model whose objective was to optimize the allocation of the regional blood resources while observing policy constraints.

The final step was to implement this programmed blood distribution system (PBDS) in a trial region on Long Island near New York City. The success of this trial led to the system being spread throughout the region served by the Greater New York Blood Program.

Before PBDS came into effect on Long Island, the RBC made an average of 7.8 blood deliveries per week to each hospital in the region, all unscheduled, and the outdate rate was 0.20. After PBDS was implemented, the average number of deliveries dropped to 4.2, of which only 1.4 were unscheduled, and the average outdate rate fell to 0.04, which appears to be about the lowest possible for this situation (some wastage being an inevitable consequence of the random demand and limited supply). These management improvements represented substantial cost savings.

The analysis designed the PBDS so that it could easily be adapted to other regions, and some adaptations were made.

For the prize-winning account, see Brodheim and Prastacos (1979); for an extended summary accompanied by discussions of its lessons for practice, see Miser and Quade (1985, pp. 68–79 and *passim*).

While this analysis was conducted by a team of two analysts, one with an expertise in blood properties and detailed management and the other with a background in OR, the next example involved a large interdisciplinary team led by an OR expert and including representatives of many scientific specialities.

5.4.2 Protecting an Estuary from Flooding In 1953 a severe storm in the North Sea flooded much of the delta region of the Netherlands, killing several thousand people. Determined not to allow this to happen again, the government started a program to increase the protection from floods by constructing a new system of dams and dikes. By the mid 1970s this system was complete except for the protection of the largest estuary, the Oosterschelde. Three alternatives for this last segment were under consideration: building an impermeable dam to close off the estuary from the sea, building a flow-through dam with gates that could be closed during a storm, and building large new dikes around the estuary.

In 1975, the Netherlands Rijkswaterstaat (the government agency responsible for water control and public works) and The Rand Corporation of Santa Monica, California (a research institution with substantial experience in systems analysis), began a joint project with a view to helping decide what should be done. It set out to determine the major consequences that would follow from implementing each of the three alternatives for protecting the estuary.

These consequences were grouped into eight categories: financial costs; security from flooding; effects on jobs and profits in the fishing industry; changes in recreational opportunities; savings to carriers and customers of the inland shipping industry; changes in production, jobs, and imports for the 35 industrial sectors of the national economy; changes of the populations of the species that make up the ecology of the region; and, finally, the social impacts—displacement of households and activities and significant effects on the regional economies. A major uncertainty was the severity and frequency of the super-storms that make the provisions for protection necessary.

By intention, the study did not conclude by recommending a particular alternative. Rather, it clarified the issues by comparing, in a common framework, the many different impacts of the three alternatives. The choice among them was left to the political process, where the responsibility properly resides. (To go further by trying to rank the alternatives was inappropriate—as it usually is in a complex study of policy.)

There was no dominant alternative; rather, each was found to have a major disadvantage that might be considered serious enough to render it politically unacceptable. The storm-surge barrier alternative (that is, the flow-through dam) was by far the most costly; the impermeable dam was the worst for the ecology; and the open case, with new dikes around the estuary, lacked security.

The Rijkswaterstaat supplemented this work with several special studies of its own and submitted a report to the Cabinet recommending the more costly storm-surge barrier plan to the Parliament. It was adopted in 1976, and construction was completed about a decade later.

This study was not only a major effort, involving dozens of analysts and specialties, it also developed some novel ideas about how to present multifaceted results to policy makers with divergent backgrounds and concerns. Thus, in addition to helping the Dutch government reach a decision about an overwhelmingly important problem, it broke new ground on how to carry through and present the results of such a complicated analysis. A summary volume (Goeller *et al.*, 1977) is supported by a variety of volumes on specialized contributing analyses; for an overview and comments on the professional implications of the work, see Miser and Quade (1985, pp. 89–109 and *passim*).

6. PRESENT STATE AND FUTURE PROSPECTS

There is a tendency in mature knowledge-based professions for the academic theorists to have acquired dominance over the professional view of quality, relevance, indices of prestige, and rewards (Abbott, 1988). While this force is visible in operations research, it has yet to achieve the dominance shown elsewhere. Thus, OR workers devoted to successful practice continue to be honored along with their colleagues who center their attention on solving scientific problems. The resulting mix of interests and the communication among them still can be seen as one of the OR profession's great strengths, on which much of its current success has been built.

On the other hand, the creation of a profession with an identified body of knowledge and clearly understood foci of practice, while it has produced the strengths leading to the profession's success, has also tended to inhibit its willingness to look outward to entirely new contexts in which many of society's new and pressing problems are found. Major problems seldom occur in unidisciplinary form; rather, their successful analysis demands multidisciplinary inquiry. While OR has traditionally either led or played an important part in such teams, this form of effort has not grown commensurate with the needs that exist for this form of work.

In spite of the tremendous success of the profession, both in developing theoretical tools and applying them to important problems, it has yet to develop a widely shared underlying philosophy for its work, or a well-based epistemology of practice.

Thus, the vitality of the profession in the future depends to an important extent on maintaining the balance between theory and practice, reaching out to more important and diverse problems as part of the team efforts that will be needed to address them, and developing widely shared underlying concepts as the basis for the profession's work. Some efforts to meet these challenges are being made, but it is not yet clear what their outcomes will be.

ACKNOWLEDGMENTS

The discussion of practice processes in Sec. 5.1 is drawn from an article written for Gass and Harris (1995), and is reproduced by permission of Kluwer Academic Publishers. Figure 1 is copyright by Elsevier Science Publishing Co. of New York and is reproduced by permission. Tables 1, 2, and 3 are drawn from *International Abstracts in Operations Research* **44** (1993), pp. 13–14; they are reproduced by permission of the International Federation of Operational Research Societies. The case summaries in Secs. 5.4.1 and 5.4.2 are drawn by permission from Miser and Quade (1985).

GLOSSARY

Action Program: A function, operation, activity, or response that is related to and given coherence by a human objective, need, or problem, together with the system of people, equipment, portion of nature, organizational elements, and management or social structure involved.

Algorithm: A set of rules or a procedure for solving a problem; in operations research the term usually refers to a method for obtaining numerical solutions from a model or theory.

Convex Polyhedron or Set: A set (or polyhedron) is convex if any two points in it or on its boundary can be connected by a straight line that lies in it or on its boundary.

Descendant Model or Theory: A variant or extension of an earlier model (or theory) that extends it to a new class of phenomena.

IFORS: The acronym for the International Federation of Operational Research Societies.

Linear Programming: A model, together with its solution algorithm, that seeks the maximum (or minimum) of a linear function of n variables where these variables are positive and subject to linear constraints.

Model: The term used by operations research workers for a theory.

Operations Research: The science devoted to understanding the behavior of operating systems, including nature, men, and machines, this last term including not only the artifacts of modern technology but also laws, common practices, human behavior, and social structures and customs. More briefly, operations research is the science of action programs. Its activities include work on scientific, technical, and practical problems.

OR: The acronym for operations research.

Practical Problem: A problem brought into being by recognizing a problem situation in which some aspect of human welfare could be improved and where the goal is to serve or achieve some human purpose.

Scientific Problem: A problem whose solution has the goal of establishing the properties of the object of inquiry, the ultimate function being to achieve new knowledge.

Systems Analysis: The portion of operations research devoted to addressing large-scale problems, usually with interdisciplinary teams.

Technical Problem: A problem specified by the function to be performed.

Theory: An intellectual construct representing a class of phenomena.

Works Cited

Abbott, A. (1988), *The System of Professions: An Essay on the Division of Labor*, Chicago: University of Chicago Press.

Assad, A. A., Wasil, E. A., Lilien, G. L. (Eds.) (1992), *Excellence in Management Science Practice: A Readings Book*, Englewood Cliffs, NJ: Prentice-Hall.

Batchelor, J. H. (1959), *Operations Research: An Annotated Bibliography*, St. Louis: St. Louis University Press. Covers the literature through 1957.

Batchelor, J. H. (1962), *Operations Research: An Annotated Bibliography*, Vol. 2, St. Louis: St. Louis University Press. Covers the literature for 1958 and 1959.

Batchelor, J. H. (1963), *Operations Research: An Annotated Bibliography*, Vol. 3, St. Louis: St. Louis University Press. Covers the 1960 literature.

Bellman, R. E., Dreyfus, S. E. (1962), *Applied Dynamic Programming*, Princeton: Princeton University Press.

Brodheim, E., Prastacos, G. (1979), *Interfaces* **9** (1), 3–20.

Boothroyd, H. (1978), *Articulate Intervention*, London: Taylor and Francis.

Churchman, C. W., Ackoff, R. L., Arnoff, E. L. (1957), *Introduction to Operations Research*, New York: Wiley.

Coffman, E. G., Jr., Lenstra, J. K., Rinooy Kan, A. H. G. (Eds.) (1992), *Computing*, Amsterdam: North-Holland.

Dantzig, G. B. (1963), *Linear Programming and Extensions*, Princeton: Princeton University Press.

Eliashberg, J., Lilien, G. L. (Eds.) (1993), *Marketing*, Amsterdam: North-Holland.

Goeller, B. F., Abrahamse, A. F., Bigelow, J. H., Bolten, J. G., de Ferranti, D. M., DeHaven, J. C., Kirkwood, T. F., Petruschell, R. L. (1977), *Protecting an Estuary from Floods—A Policy Analysis of the Oosterschelde: Vol. I*, Summary Report, R-2121/1-NETH, Santa Monica, CA: The Rand Corporation.

Gomory, R. E. (1991), in: J. K. Lenstra, A. H. G. Rinooy Kan, A. Schrijver (Eds.), *History of Mathematical Programming*, Amsterdam: North-Holland, pp. 55–61.

Graves, S., Rinooy Kan, A. H. G., Zipkin, P. (Eds.) (1993), *Logistics of Production and Inventory*, Amsterdam: North-Holland.

Gross, D., Harris, C. M. (1985), *Fundamentals of Queuing Theory*, 2nd ed., New York: Wiley.

Heyman, D. F., Sobel, M. J. (Eds.) (1990), *Stochastic Models*, Amsterdam: North-Holland.

Hillier, F. S. (1990), *Introduction to Operations Research*, New York: McGraw-Hill.

Johnson, E. A., Katcher, D. A. (1973), *Mines against Japan*, Silver Spring, MD: Naval Ordnance Laboratory.

Kemeny, J. G. (1959), *A Philosopher Looks at Science*, New York: Van Nostrand Reinhold.

Law, A. M., Kelton, W. D. (1991), *Simulation Modeling and Analysis*, 2nd ed., New York: McGraw-Hill.

McCloskey, J. F. (1987), *Oper. Res.* **35**, 143–152, 453–470, 910–925.

Miser, H. J. (Ed.) (1995), *Handbook of Systems Analysis: Cases*, Chichester, England: Wiley.

Miser, H. J., Quade, E. S. (Eds.) (1985), *Handbook of Systems Analysis: Overview of Uses, Procedures, Applications, and Practice*, Chichester, England: Wiley.

Miser, H. J., Quade, E. S. (Eds.) (1988), *Handbook of Systems Analysis: Craft Issues and Procedural Choices*, Chichester, England: Wiley.

Morse, P. M., Kimball, G. E. (1951), *Methods of Operations Research*, New York: Wiley.

Nemhauser, G. L., Rinooy Kan, A. H. G., Todd, M. J. (Eds.) (1989), *Optimization*, Amsterdam: North-Holland.

Pollock, S. M., Barnett, A., Rothkopf, M. H. (Eds.) (1994), *Operations Research and the Public Sector*, Amsterdam: North-Holland.

Ravetz, J. R. (1971), *Scientific Knowledge and Its Social Problems*, Oxford: Oxford University Press.

Schön, D. H. (1983), *The Reflective Practitioner: How Professionals Think in Action*, New York: Basic Books.

Stone, L. D. (1989), *Oper. Res.* **37**, 501–506.

Stone, L. D. (1992), *Interfaces* **22**(1), 32–54.

Waddington, C. H. (1973), *OR in World War 2: Operations Research against the U-Boat*, London: Elek Science.

Wagner, H. M. (1969), *Principles of Operations Research with Applications to Management Decisions*, Englewood Cliffs, NJ: Prentice-Hall.

Further Reading

Gass, S. I., Harris, C. M. (Eds.) (1995), *Encyclopedia of Operations Research and Management Science*, Boston: Kluwer Academic Publishers.

Miser, H. J. (1980), *Science* **209**, 139–146. Also in: P. H. Abelson, R. Kulstad (Eds.) (1980), *Science Centennial Review*, Washington, DC: American Association for the Advancement of Science, pp. 121–128.

Mitchell, G. (1993), *The Practice of Operational Research*, New York: Wiley.

Moder, J. J., Elmaghraby, S. E. (Eds.) (1978a), *Handbook of Operations Research: Foundations and Fundamentals*, New York: Van Nostrand Reinhold.

Moder, J. J., Elmaghraby, S. E. (Eds.) (1978b), *Handbook of Operations Research: Models and Applications*, New York: Van Nostrand Reinhold.

Quade, E. S. (1989), *Analysis for Public Decisions*, 2nd ed., New York: North-Holland.

Rivett, P. (1994), *The Craft of Model Building for Decision Analysis*, New York: Wiley.

OPTICAL AMPLIFIERS

See AMPLIFIERS, OPTICAL

OPTICAL COMMUNICATIONS

HERWIG KOGELNIK, *AT&T Bell Laboratories, Holmdel, New Jersey, U.S.A.*

INTRODUCTION

Optical fiber communications, also known as lightwave technology, is now the preferred technology for the transmission of information over long distances. This technology is based on semiconductor lasers and optical fibers. It is the technology of the information superhighways used for the transmission of voice, data, and video information in the networks of local telephone exchange loops, for terrestrial long-distance trunking, for undersea transmission between continents, for computer interconnections, for local-area networks, and for cable television. Proposed lightwave applications range from distances of 1 cm to intercontinental distances of 10 000 km. Details of the envisioned distance scales are shown in Table 1.

An extensive selection of textbooks is now available on this subject. Among these are the books by Miller and Chynoweth (1979), Midwinter (1979), Personick (1981), Miller and

3-527-28134-7/95/$5.00 + .50

Kaminow (1988), and Agrawal (1992). The reader is referred to these for details on the history and fundamentals of lightwave technology. Selected historical highlights are the following:

- 1854, Tyndall's optical fiber
- 1880, Bell's photophone
- 1960, the first laser
- 1962, the semiconductor junction laser
- 1970, the low-loss fiber
- 1983, the U.S. Northeast Corridor system
- 1988, the TAT-8 transatlantic system.

Compared to the earlier transmission media such as coaxial cables or copper pairs, the optical fiber medium offers several advantages. The principal ones are large transmission capacity (high bandwidth), large repeater spacings, small cable size and weight, and immunity from electromagnetic interference (EMI).

Rapid technological progress has advanced fiber systems through four successive technological generations. The first generation started with multimode fibers and gallium arsenide sources operating at 0.8-μm wavelength. The second generation shifted to single-mode fibers and indium phosphide lasers operating at 1.3 μm. The third generation employs single-frequency lasers operating in the low-loss fiber window near 1.55 μm, and the fourth generation uses erbium-doped fiber amplifiers. More details on this evolution are given in Sec. 7 for undersea systems and Sec. 8 for terrestrial long-haul systems. Figure 1 depicts the exponential pace of this technological progress in terms of the bit-rate $\times$ distance product for lightwave systems demonstrated in the research laboratory. This is an economical figure of merit that will be discussed below in more detail. The figure shows that this product has doubled every year during the past two decades.

Table 1. Distance ranges of proposed applications of lightwave technology.

Application	Distance
Undersea transmission	1 000–10 000 km
Terrestrial long-haul	20–1 000 km
CATV links	10–20 km
Fiber in the loop	10–20 km
Optical data links	0.1–10 km
Optical interconnects	0.1–100 m
Optical crossconnects	1–100 cm

The elements of a lightwave transmission system are shown in Fig. 2. For a single link one needs an optoelectronic transmitter (T) that converts the electrical data into an optical signal and launches sufficient power P_T for transmission through the fiber. The receiver (R) detects the received signal arriving with power P_R and converts it back to electrical data. More detail on transmitters is given in Sec. 3 and on receivers in Sec. 4. A minimum power P_R is required at the receiver to allow detection with an acceptable bit-error rate. This power level is called the receiver sensitivity. The transmission system has to meet the power budget

$$10\log(P_T/P_R) = \alpha L + \text{coupling loss} + \text{penalties} + \text{margin}, \quad (1)$$

where all quantities are measured in decibels. The term αL is the fiber loss, coupling losses are due to splices, connectors, and the like, and power penalties arise from signal distortions.

An illustrative example for the power budget of a 2.5-Gb/s (Gb stands for gigabits) transmission system using an avalanche photodetector receiver and spanning a length of L = 100 km might be the following:

Launched power P_T	0 dBm (1 mW)
Fiber loss	22 dB
Coupling loss	3 dB
Power penalty	2 dB
System margin	9 dB
Receiver sensitivity P_R	−36 dBm

where the dBm measure indicates power levels in decibels relative to 1 mW.

Section 1 summarizes the properties of optical fibers. This includes fiber dispersion, which is one cause of signal distortion. Other sources of distortion are fiber nonlinearities, which are discussed in Sec. 6. This section also gives a brief summary of soliton propagation in fibers.

Figure 2 also shows a repeatered transmission link, which consists of a series of single links. The repeaters used to regenerate the signal are essentially transmitter and receiver pairs.

The sketch of Fig. 3 gives the background for an explanation of the economic value of the bit-rate $\times$ distance product BL of a single

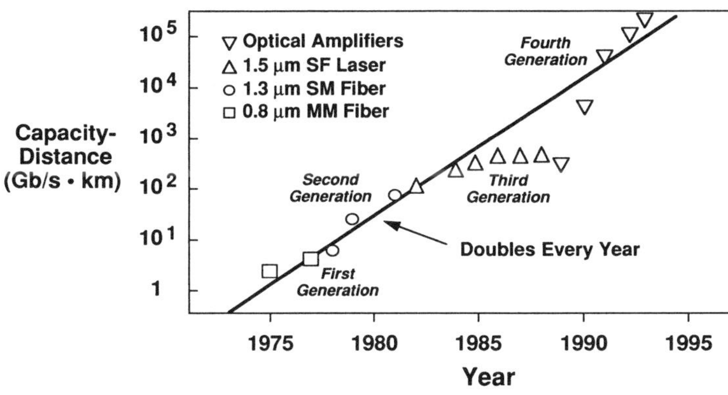

FIG. 1. Exponential growth of the capacity-distance product for lightwave systems over time as demonstrated in research laboratories. MM = multimode, SM = single-mode, SF = single-frequency.

link. It shows a transmission system with the task of transmitting information with a total bit rate B_{TOT} over a total distance L_{TOT}. The task is partitioned between single links that can each transmit a bit rate of B over a distance L. The number of transmitter-receiver pairs N required for the whole task is

$$N = B_{TOT}L_{TOT}/BL. \quad (2)$$

The largest BL product will lead to the smallest number of required T/R pairs, and hence its value as a figure of merit.

The use of optical amplifiers can help to compensate for the loss terms in the power budget and increase the distance between optoelectronic repeaters. This is discussed in Sec. 2.

Three more sections not yet mentioned round out our description of lightwave technology. Section 9 describes fiber applications in the local loop of the telephone exchange, and Sec. 10 discusses CATV transmission via optical fibers. Section 5 summarizes the modulation and multiplexing technologies used in the various lightwave applications. This includes a discussion of increasing capacity through wavelength-division multiplexing and subcarrier modulation.

1. OPTICAL FIBERS

The subject of optical fibers is treated by an extensive literature. Among the textbooks are the ones by Kapany (1967), Unger (1977), Adams (1981), Okoshi (1982), Snyder and Love (1983), Li (1985), Izawa and Subdo (1987), Neumann (1988), and Marcuse (1991a). See also FIBER OPTICS.

The fibers for lightwave transmission systems are made of fused silica (SiO_2) glass. Dopants such as germania (GeO_2) or fluorine (F) are used to produce small changes in the refractive index of the glass. Figure 4 shows

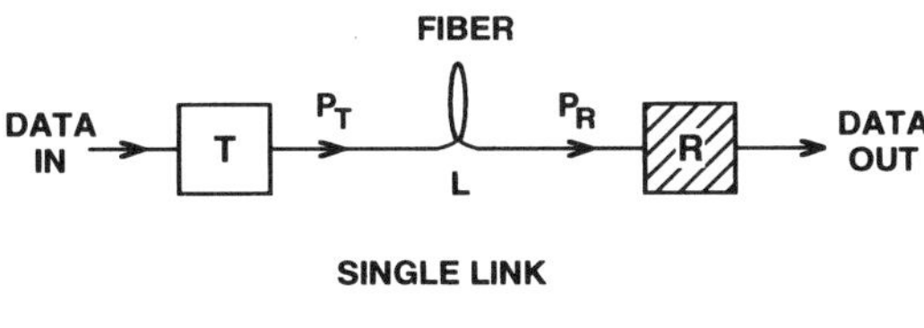

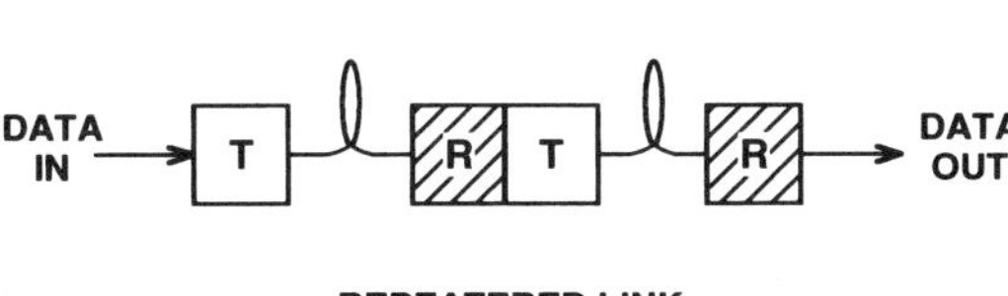

FIG. 2. Elements of a transmission system, for a single (unrepeatered) link and a repeatered link. The repeater contains the elements of a receiver and a transmitter. T = transmitter, R = receiver.

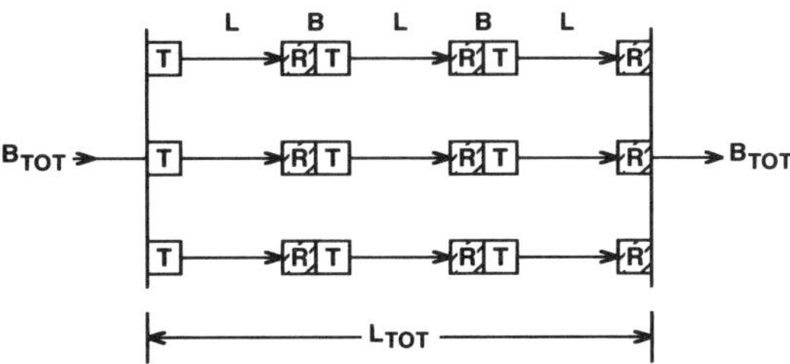

FIG. 3. Transmission of B_{TOT} over the distance L_{TOT}. The task is partitioned between single-link systems, each capable of transmitting the bit rate B over a distance L.

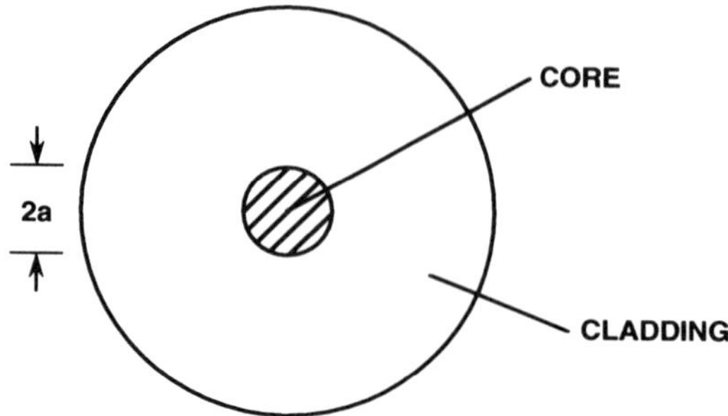

FIG. 4. Typical fiber cross section. The core region is doped to a higher refractive index than the cladding.

a typical cross section of a fiber with a central core and a surrounding cladding, which has a diameter of about 125 μm. The core is doped to a higher index than the cladding to confine the light in the core region by total internal refraction. Single-mode fibers have a core diameter of about 8 μm and index differences between core and cladding of a few tenths of a percent. Various index profiles are used in the core to achieve desired fiber characteristics such as a spectral shift in the fiber dispersion. A sampling of index profiles is sketched in Fig. 5.

The fundamental mode of the light guided by the step-index fiber of Fig. 5(a) has a radial electric field distribution E_x described by Bessel functions:

$$\begin{aligned} E_x(r) &= J_0(\kappa r) \quad \text{for } r \leq a, \\ &= K_0(\gamma r) \quad \text{for } r > a, \end{aligned} \tag{3}$$

where x, y are the Cartesian coordinates in the cross section, and r is the radial coordinate. Linear polarization ($E_y = 0$) is assumed. The core radius is a, and

$$\begin{aligned} \kappa^2 &= n_1^2k^2 - \beta^2, \\ \gamma^2 &= \beta^2 - n_2^2k^2, \end{aligned} \tag{4}$$

where n_1 and n_2 are the refractive indices of core and cladding, respectively, and $k = 2\pi/\lambda$ is the propagation constant of free space. The propagation constant β of the fiber mode is related to the phase velocity v_p by

$$\beta = \omega/v_p, \tag{5}$$

where ω is the radian frequency ($=2\pi\nu$). The term v_p is often expressed in terms of the effective index n of the mode:

$$v_p = c/n, \tag{6}$$

which measures the reduction of speed relative to the velocity of light in free space $c = 3 \times 10^8$ m/s. For a guided mode n is bounded by

$$n_1 > n > n_2. \tag{7}$$

Fibers with larger core diameters can support several guided modes, which all travel at different velocities. This reduces the bandwidth of transmission systems. Therefore, for high-bandwidth systems, one uses single-mode fibers, which can support only one guided

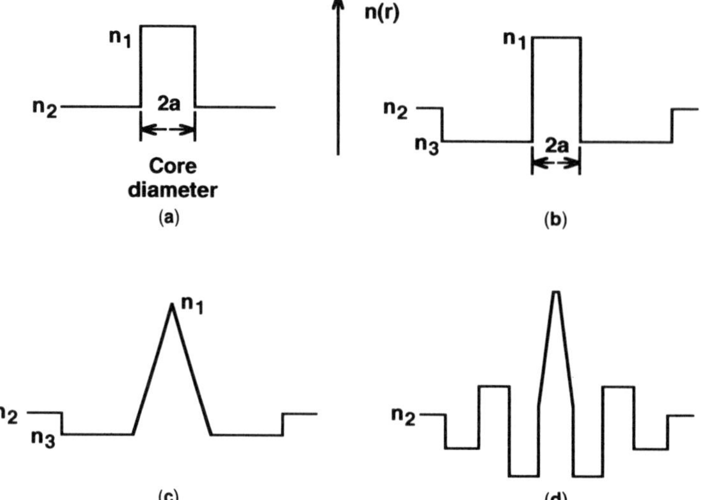

FIG. 5. Schematic of index profiles of several fiber types: (a) step-index, (b) depressed-cladding, (c) depressed-cladding triangular-core, (d) quadruply clad triangular-core.

mode (in each polarization). The cutoff condition for single-mode operation is

$$ka\sqrt{n_1^2 - n_2^2} < 2.405 \tag{8}$$

for the step-index fiber design of Fig. 5(a).

1.1 Fiber Loss

Remarkable advances in the technology of fiber making have reduced fiber losses to values that approach the theoretical limits for SiO_2. The technology employs careful control of transition-metal and water impurities, which must be reduced to a level of few parts per billion. Figure 6 shows the loss spectrum of a single-mode fiber with a loss minimum near the 1.55-μm wavelength low as 0.157 dB/km (Csencsits *et al.*, 1984). The figure also indicates the contribution to loss resulting from Rayleigh scattering due to submicroscopic density and composition fluctuations in the glass, and the infrared absorption due to molecular vibrations in the glass lattice.

1.2 Chromatic Dispersion

High-capacity fiber systems require the transmission of very short pulses over long fiber distances. The pulse envelopes carried by the light travel at the *group velocity* v_g of the fiber mode, which is

$$v_g = \frac{d\omega}{d\beta} = \frac{c}{n_g}, \tag{9}$$

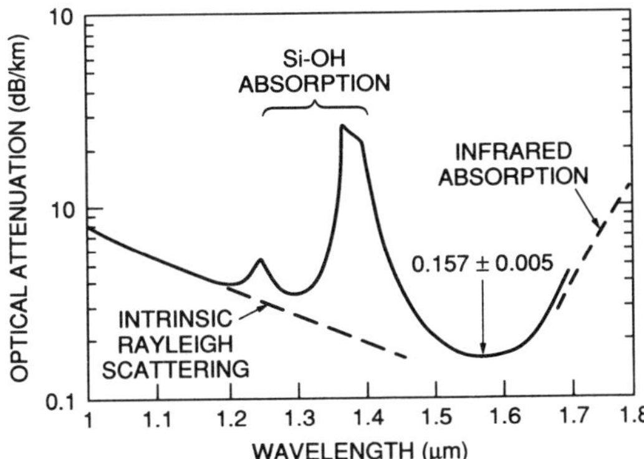

FIG. 6. Loss spectrum of a low-GeO_2 core single-mode fiber.

where n_g is the group index given by

$$n_g = n - \lambda \frac{dn}{d\lambda}. \tag{10}$$

For silica fibers, n and n_g have values of approximately 1.46. The difference between the two is due to dispersion. The propagation delay t_g of an optical pulse transmitted through a fiber of length L is

$$t_g = L/v_g, \tag{11}$$

which is about 30 ms across the Atlantic Ocean. Because of group dispersion or "chromatic dispersion" in the fiber, this delay is different for different wavelengths. This has important implications for high bit-rate transmission as each signal pulse contains different spectral components. Each of the components arrives at the fiber output at a different time. This leads to a spreading Δt_g of the pulse given by

$$\Delta t_g = LD\Delta\lambda, \tag{12}$$

where $\Delta\lambda$ is the spectral width of the transmitted signal, and D is the chromatic dispersion parameter defined as

$$D = \frac{1}{L}\frac{dt_g}{d\lambda} = \frac{1}{c}\frac{dn_g}{d\lambda} = -\frac{\lambda}{c}\frac{d^2n}{d\lambda^2}. \tag{13}$$

The spreading must be kept to values much smaller than the signal pulse width to avoid signal impairments. Figure 7 shows the spectral dependence of D for a conventional single-mode fiber and a dispersion-shifted (DS) fiber. The value of D is measured in ps/nm-km. The conventional single-mode fiber has zero dispersion near 1.3 μm, and a value of about 15 ps/nm-km in the low-loss window near 1.5 μm. The location of the dispersion zero of the conventional fiber is mostly due to the dispersion of the glass material. The DS fiber has been designed such that dispersion due to the waveguide geometry shifts the dispersion zero to the low-loss window in order to allow for higher bit-rate transmission.

It is interesting to note that the zero-dispersion wavelength λ_0 of the material can be simply linked (Wemple, 1979) to the average energy gap E_0 of the glass, its electronic oscillator strength E_d, and its lattice oscillator

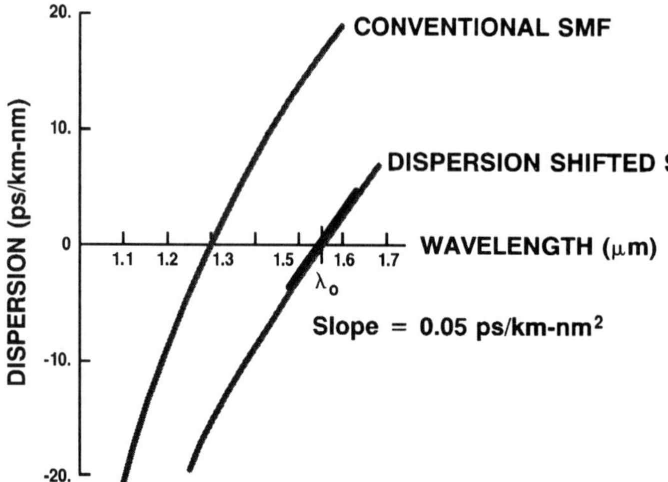

FIG. 7. Spectrum of the group dispersion D for a single-mode fiber (SMF) and a dispersion-shifted fiber.

strength E_e (all measured in electronvolts):

$$\lambda_0 = 1.63(E_d/E_0^3E_e^2)^{1/4}\ \mu\text{m}. \tag{14}$$

For silica one gets $\lambda_0 = 1.28\ \mu$m, which is slightly smaller than the value shown in the figure because of the effect of dopants and guide dispersion.

1.3 Dispersion Limits

The dispersive broadening of optical pulses in fibers leads to limitations of the bit-rate × distance product BL in transmission systems. This is because energy spreads from a given bit interval to adjacent intervals and interferes with adjacent signal bits. The power penalty caused by this interference depends somewhat on the modulation format. A good rule of thumb (Henry *et al.*, 1988) is that the penalty reaches 1 dB when the full width Δt_g of the pulse spreading becomes equal to a half of the bit interval $1/B$, implying a limit

$$\Delta t_g B \leq \frac{1}{2}. \tag{15}$$

This can be expressed as a limit for the BL product

$$BL \leq 1/2D\Delta\lambda, \tag{16}$$

which depends on the spectral width $\Delta\lambda$ of the signal source. The BL product can be maximized by minimizing $\Delta\lambda$, i.e., by careful control of the signal spectrum using single-frequency lasers and related techniques. The minimum bandwidth $\Delta\nu$ of the signal is, approximately, equal to the bit rate B:

$$B \cong \Delta\nu = \Delta\lambda c/\lambda^2. \tag{17}$$

With this, one obtains a single-frequency dispersion limit for the B^2L product of the system. For the special case of the non-return-to-zero (NRZ) format (see Sec. 5) this limit is

$$B^2L \leq 0.79\,(c/D\lambda^2) \tag{18}$$

as determined by a detailed computer evaluation (Elrefaie *et al.*, 1988). It is somewhat higher than that predicted by the general rule of thumb above. For systems operating at $\lambda = 1.55\ \mu$m with conventional fibers of about $D = 15$ ps/nm-km, the dispersion limit is

$$B^2L \leq 6000\ (\text{Gb/s})^2\ \text{km}. \tag{19}$$

Figure 8 shows this dispersion limit for conventional fibers. The figure also indicates the limits of dispersion-shifted fibers for two cases corresponding to operating wavelength shifts of 10 nm and 2 nm away from the zero-dispersion point. In addition, the figure shows the loss limit for a relatively ambitious transmission system with 10 mW launched power, a fiber loss of 0.25 dB/km, and a receiver sensitivity of 100 photons per bit (such as that of an excellent optical preamplifier). Note that the lower bit-rate systems are loss

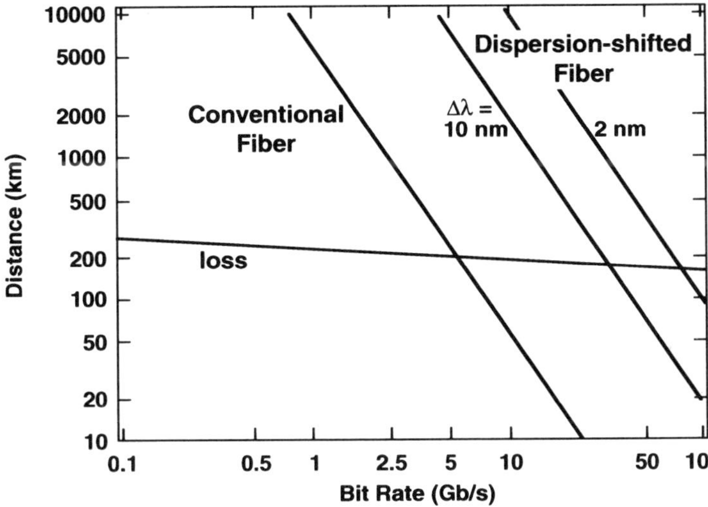

FIG. 8. Chromatic dispersion limits for chirp-free sources at λ = 155 nm.

limited, while the higher bit-rate systems are dispersion limited. Note also that these dispersion limits apply to ideal spectrum-limited sources. Spectral broadening such as that due to laser chirp will lower the dispersion limits.

2. OPTICAL AMPLIFIERS

Optical amplifiers enhance the power levels of optical signals. In transmission systems, amplifiers have found four applications, namely as

1. power or booster amplifiers to provide high levels of launch power for transmission over very long spans of fiber,
2. in-line or repeater amplifiers to periodically compensate for fiber loss in ultralong fiber systems,
3. preamplifiers to enhance the sensitivity of optoelectronic receivers, particularly at high (>1 Gb/s) bit rates, and
4. compensators for the loss in optical power splitters.

The first three cases are sketched in Fig. 9.

Because of their large optical bandwidth (tens of nanometers), amplifiers provide transparency for lightwave systems: they amplify signals without regard to their bit rate or modulation format, and they amplify all channels of a wavelength-division-multiplexed (WDM) signal stream (as illustrated in the lower half of Fig. 9).

Several types of amplifiers are under exploration:

1. Doped-fiber amplifiers where the glass host of the fiber core is doped with active ions that are optically pumped. Dopants include the elements erbium, neodymium, and praesodymium.
2. Semiconductor laser amplifiers, which use the active waveguide of a junction laser. Their gain is provided by electrical pumping via electron-hole recombination.
3. Raman-fiber amplifiers, which use optical pumping of conventional glass fibers where gain is obtained by coupling through molecular vibrations in glass.

Key characteristics of amplifiers are their gain, gain saturation, and noise. To illustrate these features we use the example of the erbium-doped fiber amplifier (EDFA). This device has seen a rapid move from explorations in the research laboratory (Mears *et al.*, 1987; Desurvire *et al.*, 1987) to telecommunications applications today. EDFA amplifiers offer

1. a broad gain spectrum (30–40 nm) in the 1550-nm spectral range,
2. high gain (30–50 dB) that is nearly polarization independent,
3. high saturated output powers of several hundred milliwatts,
4. nearly ideal noise performance with noise figures approaching the theoretical limit of 3 dB,
5. slow gain dynamics causing minimal

SINGLE CHANNEL

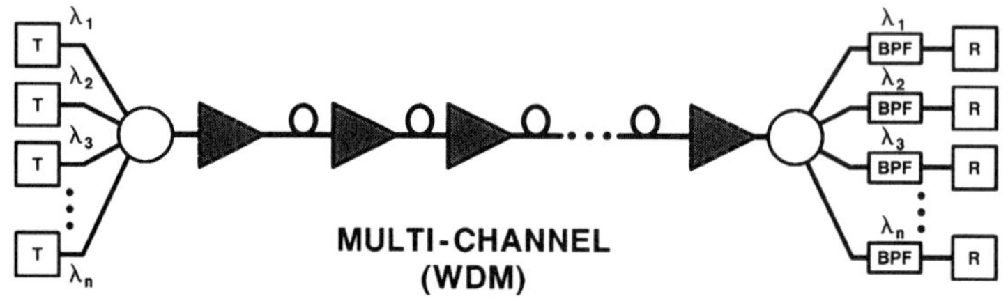

FIG. 9. Sketch of applications of optical amplifiers as power amplifiers, in-line amplifiers (repeaters), and preamplifiers. BPF = band-pass filter.

crosstalk between WDM channels because of gain saturation and recovery times that are in the 0.1- to 1-ms range,

6. pump bands at 980 and 1480 nm, which are compatible with semiconductor laser pumps,
7. good coupling efficiency to transmission fibers.

Figure 10 shows a sketch of an EDFA subsystem. The key element is a specially prepared optical fiber whose core is doped with erbium (Er). This fiber is typically tens of meters in length and coupled to the transmission fiber. The optical pump power is supplied through a wavelength-division multiplexer that allows the optical signal to pass undisturbed. Optical isolators are used to prevent oscillations and excess noise due to unwanted reflections in the fiber system.

Figure 11 shows the relevant energy diagram for the three-level system of the active Er^{3+} ions in glass. The 980- and 1480-nm pump bands are indicated together with the laser transitions near 1550 nm. The corresponding absorption and emission (gain) spectra are also shown. Note that the type of core codopant used in the fiber (such as germanium or aluminum) can change the erbium gain spectra somewhat.

2.1 Amplifier Gain

The gain G of the amplifier is defined as the ratio of the output and input powers of the signal:

$$G \equiv \frac{P_{\mathrm{OUT}}}{P_{\mathrm{IN}}} = \exp\left(\int_0^L dz g(z)\right). \tag{20}$$

It is related to the gain factor $g(z)$, which changes along the EDFA length z as a result of factors such as pump depletion and gain saturation. The gain factor measures the local growth of optical power P:

$$g \equiv \frac{1}{P}\frac{dP}{dz} = \Gamma_\rho(\sigma_e N_2 - \sigma_a N_1) = \frac{g_0}{1 + P/P_{\mathrm{SAT}}}, \tag{21}$$

where N_2 and N_1 are the populations of the upper and lower laser levels of the active Er^{3+} ions, ρ is the active ion density, and Γ is a confinement factor measuring the overlap of the signal field with the doped core. The emission and absorption cross sections of the laser transition at the signal frequency ν are designated σ_e and σ_a. Their frequency dependence is shown in Fig. 11. The right-hand side of Eq. (21) indicates the homogeneous satu-

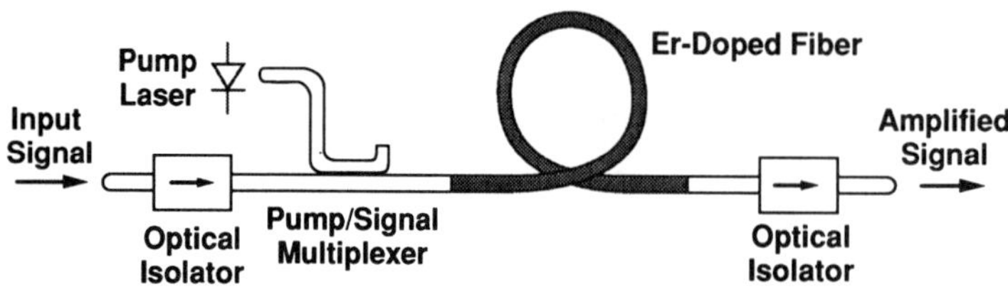

FIG. 10. Sketch of an EDFA subsystem with a diode laser pump and two isolators.

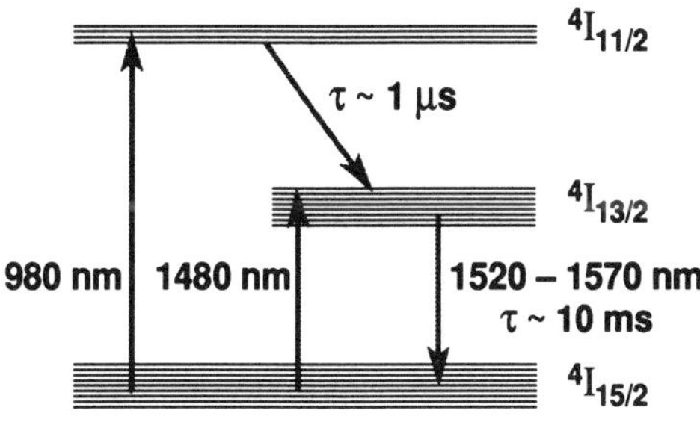

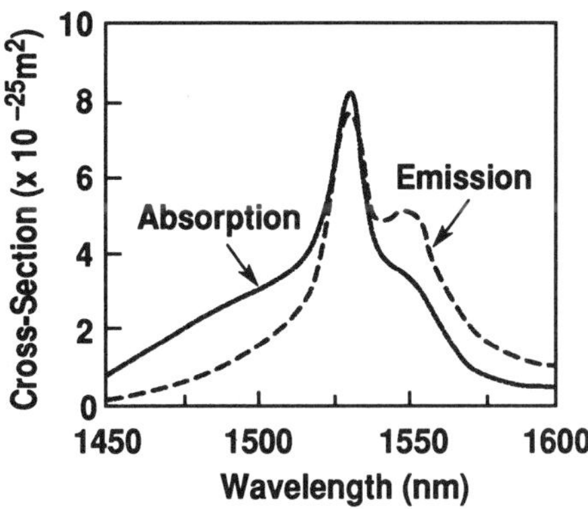

FIG. 11. Energy-level diagram of an EDFA. The associated absorption and emission spectra are shown for the case of Al and Ge core co-dopants.

ration of the gain in the usual form (Siegman, 1986).

The small-signal gain factor g_0 is conveniently expressed in the form

$$g_0 = \frac{\rho\Gamma\sigma_e(P_p - P_{\rm TH})}{P_p + P_{\rm TH}\sigma_e/\sigma_a}, \tag{22}$$

where the subscript p indicates the parameters of the pump. The gain is negative for pump powers P_p lower than the threshold power $P_{\rm TH}$, and reaches a maximum of $\Gamma\rho\sigma_e$ for very high pump levels. The threshold power is given by

$$P_{\rm TH} = (\sigma_a/\sigma_e)(h\nu_p A/\Gamma_p\tau\sigma_p), \tag{23}$$

where τ is the spontaneous life time of the upper laser level, A is the core area, and h is Planck's constant. A simple three-level laser model is assumed for our discussion, where $\sigma_{ap} = \sigma_p$ and $\sigma_{ep} = 0$ is used for simplicity. For further detail on gain modeling of EDFAs we refer the reader to Saleh *et al.* (1990); Zyskind (1991); Giles and Desurvire (1991).

2.2 Gain Saturation

The above expression for g also indicates the homogeneous gain saturation of the amplifier: g is reduced to half its value when the signal power P reaches the saturation power $P_{\rm SAT}$, where

$$P_{\rm SAT} = \frac{h\nu A}{(\sigma_e + \sigma_a)_s\Gamma\tau}\left[1 + \frac{\sigma_a}{\sigma_e}\frac{P_p}{P_{\rm TH}}\right]. \tag{24}$$

Note that $P_{\rm SAT}$ can be increased by driving the laser with higher pump powers P_p, which is a unique feature of a three-level laser. An illustration of the saturation characteristics of an EDFA is given in Fig. 12. Recently, quantum conversion efficiencies greater than 90% have been achieved in highly saturated EDFAs.

2.3 Amplifier Noise

Spontaneous emission adds noise to the amplified signal. This noise increases with G and with the optical bandwidth B. The total amplified spontaneous emission noise power $N_{\rm ASE}$ at the amplifier output is

$$N_{\rm ASE} = 2n_{\rm sp}(G - 1)h\nu B = h\nu BFG, \tag{25}$$

where the fiber guide is assumed to support guided modes of two polarizations. Here, $n_{\rm sp}$ is the spontaneous emission factor given by

$$n_{\rm sp} = \sigma_e N_2/(\sigma_e N_2 - \sigma_a N_1), \tag{26}$$

where σ_e and σ_a are the emission and absorption cross sections. For complete inversion ($N_1 \approx 0$), $n_{\rm sp}$ reaches its minimum value of unity.

An important measure for an amplifier, or a chain of amplifiers, is the optical signal-to-noise ratio at the system output,

$$\mathrm{SNR_{OUT}} = P_{\rm OUT}/N_{\rm ASE}. \tag{27}$$

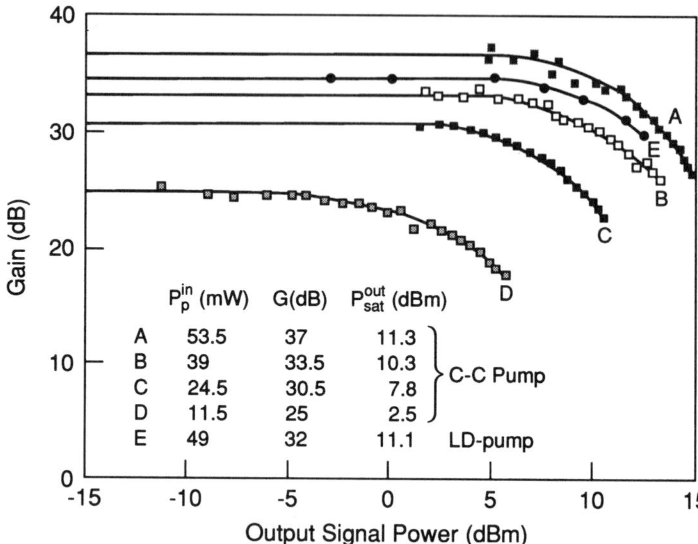

FIG. 12. EDFA saturation characteristics for various pump power levels. After Desurvire *et al.* (1989).

This is often compared to the input optical SNR,

$$\mathrm{SNR_{IN}} = P_{\mathrm{IN}}/h\nu B, \tag{28}$$

where the equivalent input shot noise $h\nu B$ is used as a reference. This comparison is expressed by the noise figure of the system defined as

$$\begin{aligned} F &\equiv \mathrm{SNR_{IN}}/\mathrm{SNR_{OUT}} = N_{\mathrm{ASE}}/h\nu BG \\ &= 2n_{\mathrm{sp}}(G-1)/G, \end{aligned} \tag{29}$$

where the expression on the outer right applies to a single amplifier stage (note that several slightly differing definitions occur in the literature). For high gain, the minimum possible noise figure is 2, or 3 dB. Systems applications tend to use two or more amplifier stages, or amplifier chains such as illustrated in Fig. 13. A two-stage amplifier with loss L_1 between the stages has a net gain of $G = G_1L_1G_2$ and a noise figure of

$$F = F_1 + F_2/L_1G_1, \tag{30}$$

where F_1 and F_2 are the noise figures of the two stages. For a chain of n stages, we have, similarly, a net gain of $G = G_1L_1G_2L_2 \cdots L_{n-1}G_n$ and a noise figure of

$$F = F_1 + \frac{1}{G_1L_1}\left\{F_2 + \frac{1}{G_2L_2} \times \left[F_3 + \cdots (F_{n-1} + F_n/G_{n-1}L_{n-1})\right]\right\}. \tag{31}$$

The noise figure is independent of the optical bandwidth B of the system and, thus, a convenient measure of ASE noise. This noise affects the sensitivity of the optical receiver and its bit-error rate. Refer to Sec. 4 for more detail.

For the special case where $G_1L_1 = G_2L_2 = \cdots = G_{L-1}L_{n-1} = 1$, Eq. (31) assumes the simple form

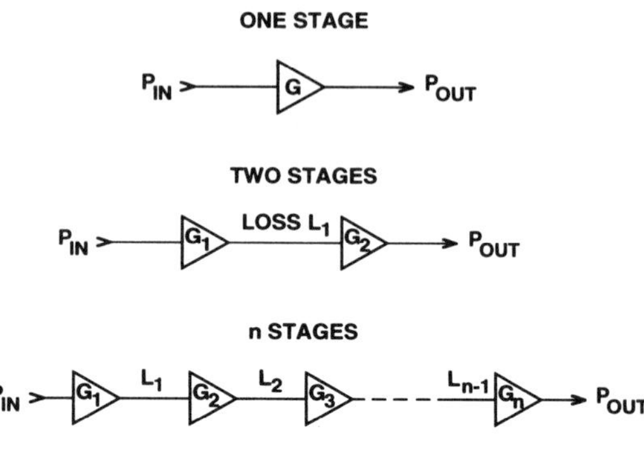

FIG. 13. Sketch of systems with one, two, and n amplifier stages of gain G_i. The term L_i is the transmission loss between the stages.

$$F = \sum_{1}^{n} F_i. \tag{32}$$

3. OPTICAL TRANSMITTERS

In the following, we will sketch the principal laser characteristics of importance for lightwave systems. Further background and detail can be found in texts such as Kressel and Butler (1977), Casey and Panish (1978), Thompson (1980), Kressel (1982), Yariv (1985), Tsang (1985), Agrawal and Dutta (1986), and Kaminow and Tucker (1988).

Transmitters prepare the optical signal for transmission through the fiber. They contain a laser source, a modulator function, and electronic circuitry. The laser needs to generate light of sufficient power (mW), at the desired wavelength (e.g., 1.3 or 1.55 μm), and of the required spectral quality. The modulation of the light is accomplished by direct modulation or by external modulation. In direct modulation, the drive current of the laser is modulated to change the intensity of the optical signals. Typical external modulators are made on electro-optic $LiNbO_3$ crystals or in semiconductors. The latter can be integrated on the laser chip into a simple form of a photonic integrated circuit. One of the obvious functions of the electronic circuitry is to supply the drive powers for the laser and for the modulation function.

3.1 *L-I* Laser Characteristics

Figure 14 shows the output characteristics of a semiconductor junction laser. It is called the *L-I* curve and depicts the power of the output light (mW) as a function of the drive current (mA). Note that there is a threshold pump current I_{th} (~20 mA) below which there is no laser output. Note also the nearly linear dependence of the laser output as a function of the drive current. This aspect is used for direct modulation of the laser output by modulating the drive current. The figure also shows the temperature dependence of the *L-I* characteristics.

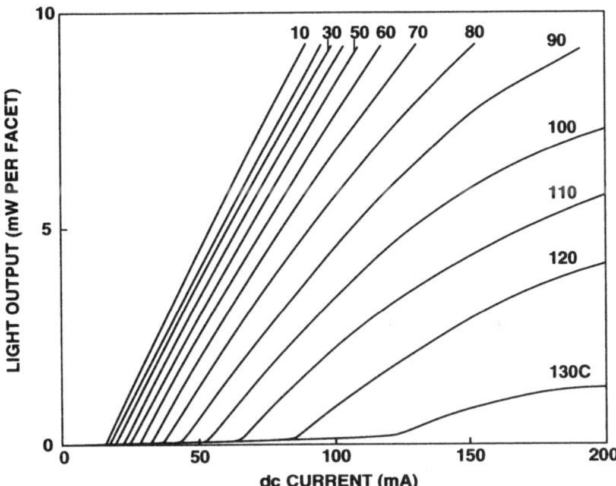

FIG. 14. *L-I* curves of a semiconductor junction laser at $\lambda = 1.3\ \mu$m for a selection of temperatures. The slope of the response curves indicates the quantum efficiency of the device.

3.2 The InGaAsP Material

Semiconductor materials are required for constructing reliable lasers and detectors for the low-loss fiber windows near the 1.3- and 1.55-μm wavelengths. The quaternary material indium gallium arsenide phosphide (InGaAsP), which can be grown lattice matched on the substrate crystal indium phosphide (InP), has emerged as the primary material for these devices. Figure 15 shows the band-gap energy E_g of this compound, and

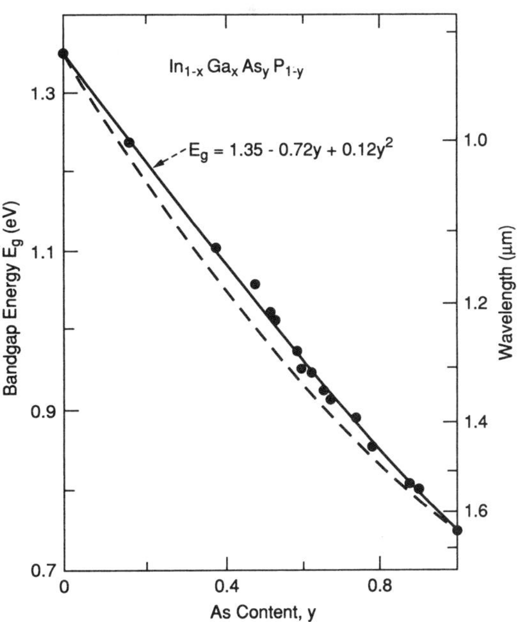

FIG. 15. Band-gap energy E_g of InGaAsP lattice matched to InP as a function of As content (y). The corresponding laser wavelengths are indicated on the right.

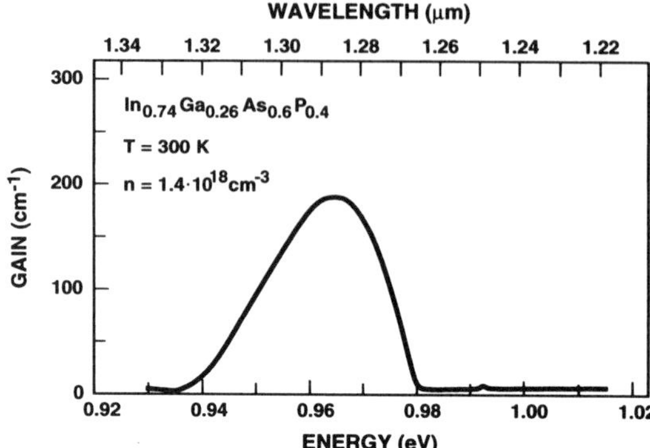

FIG. 16. Spectral dependence of the gain of a junction laser.

how it can be tailored to the desired wavelength by a change in composition. For this purpose, indium content is exchanged for gallium content (x) and phosphorus for arsenic content (y). For a good lattice match to the substrate one requires, approximately,

$$x = 0.47y. \tag{33}$$

By varying the composition, heterostructure junction lasers have been made for wavelengths between 1.1 and 1.65 μm, which covers both low-loss fiber windows. The relationship between the band gap and the corresponding wavelength λ_g is

$$V_g\lambda_g = E_g\lambda_g/e = 1.24 \text{ V } \mu\text{m}. \tag{34}$$

To achieve laser operation a voltage $V_a > V_g$ is applied in forward bias. Its value is, approximately,

$$V_a = V_g + IR_s, \tag{35}$$

where R_s is the resistance of the diode.

In semiconductor junction lasers, the gain is produced by carrier-induced stimulated emission, i.e., by driving a current through the junction. The gain covers a spectral band of about 30 nm. This is illustrated in Fig. 16 for a quaternary laser of composition $y = 0.6$. The gain factor g_{PEAK} at the peak of the gain spectrum increases with the injected carrier density N as

$$g_{\text{PEAK}} = \sigma(N - N_{\text{th}}), \tag{36}$$

where σ is the gain cross section and N_{th} the threshold carrier density. For InGaAsP, N is of the order of 10^{18} cm^{-3} and $\sigma \cong 10^{-16}$ cm^2.

3.3 Laser Output Spectrum

The sketch of Fig. 17 shows a cut of a typical junction laser used for telecommunications. The semiconductor crystal structure contains a waveguide with a higher-index core that confines the light similarly as in a fiber guide. In advanced laser structures, the active layer providing optical gain is made of quantum wells, each about 100 Å thick. In a simple laser design (called Fabry–Pérot laser), oscillation is achieved because the amplified

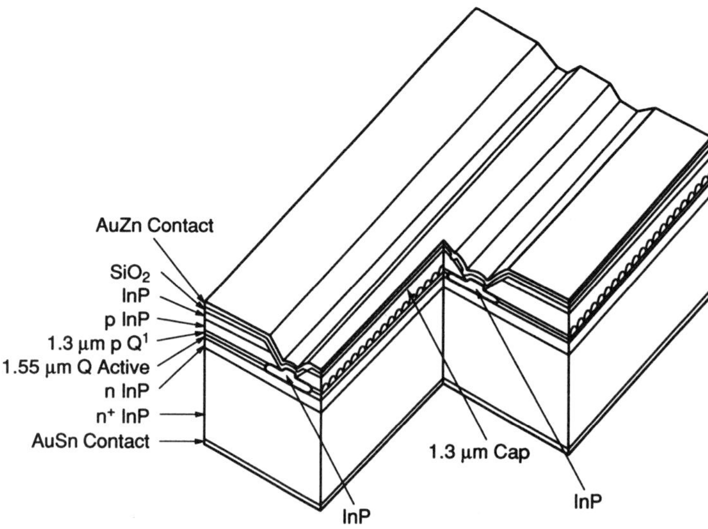

FIG. 17. Sketch of the cross section of a semiconductor distributed-feedback laser. The composition of the epitaxial layers is designed to confine both the charge carriers and the optical wave. The active layer provides the gain. The corrugation of the composite waveguide provides single-frequency operation.

light is reflected back and forth between the end facets of the crystal. The facets form a laser resonant cavity with a typical cavity length L of 250 μm. Figure 18 depicts the multifrequency output spectrum of such a Fabry–Pérot laser. It contains several spectral lines spread over a band of about 10 nm. The lines correspond to those resonances of the laser cavity that are supplied with sufficient gain to oscillate. For these resonances the laser gain equals the cavity loss:

$$g = \alpha_{\text{guide}} - \ln(R_1 R_2)/2L, \tag{37}$$

where R_1 and R_2 are the reflectivities of the facet mirrors and α_{guide} is the attenuation constant of the guide. Major contributors to this guide loss are free-carrier absorption and scattering.

The frequency spacing $\Delta\nu$ between the lines of the spectrum is the inverse of the round-trip delay time t_g of the light pulse shuttling back and forth between the laser facets:

$$\Delta\nu = 1/t_g = v_g/2L. \tag{38}$$

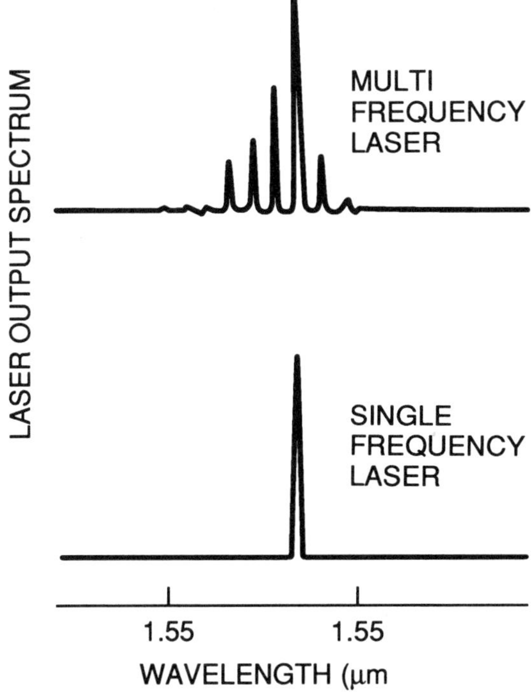

FIG. 18. Output spectrum of a typical multifrequency Fabry–Pérot laser, and the spectrum of a single-frequency laser.

This corresponds to a wavelength spacing of

$$\Delta\lambda = \lambda^2/2n_g L, \tag{39}$$

where n_g is the group index and v_g the group velocity in the laser guide. For InGaAsP laser material, n_g is 3.9 and 3.7 at 1.3 and 1.55 μm, respectively.

To minimize the effect of fiber dispersion, one needs to suppress oscillations at all laser resonances except one. This is the object of single-frequency lasers. An example of a single-frequency laser spectrum is shown in the lower part of Fig. 18.

Most single-frequency lasers in use today are distributed-feedback lasers. In these devices, the thickness of the laser waveguide is corrugated to form a high-resolution grating. This is indicated in the sketch of Fig. 17. For operation at 1.55 μm, the corrugations have a period of about 240 nm. In this technique, Bragg reflection from the grating is used as a feedback mechanism, taking the place of the mirror reflections in Fabry–Pérot lasers. The frequency selectivity of the Bragg effect helps to insure single-frequency operation.

3.4 Laser Chirp

For high-speed transmission, one must pay attention to yet another spectral artifact that occurs when a single-frequency laser is directly modulated. This artifact is called chirping. This happens because the modulation of the drive current changes carrier injection and thereby the gain by Δg. Because of the Kramers–Kronig relations, an increase in gain will lead to a decrease $\Delta n'$ of the refractive index of the active laser layer. This is described as a change of the complex index,

$$\Delta n = \Delta n' - j\Delta n'', \tag{40}$$

where

$$\Delta g = -2(\omega/c)\Delta n''. \tag{41}$$

As a result of the above index change, the laser frequency is chirped by an amount of approximately (Koch and Bowers, 1984; Koch and Linke, 1986)

$$\Delta\nu(t) = -(\alpha/4\pi)\frac{d}{dt}\ln P(t), \tag{42}$$

where $P(t)$ is the modulated laser output power. This chirp must be kept to a minimum to avoid dispersive pulse broadening. The alternative is, of course, the use of a laser with constant output power (and no chirping) combined with a chirp-free external modulator. The factor α appearing in Eq. (42) was introduced by Henry (1982). It is defined by the ratio

$$\alpha = \Delta n' / \Delta n'', \tag{43}$$

which, in InGaAsP, has a magnitude of about 3–5 depending on the laser structure.

3.5 Laser Linewidth

The spectral linewidth of a cw, unchirped, single-frequency laser oscillation is not infinitely small. It is broadened by spontaneous emission, which causes amplitude and phase fluctuations in the optical field. The amplitude fluctuations cause, in turn, fluctuations in the carrier density N, the refractive index, and the cavity resonance. The end result (Henry, 1982) is a Lorentzian line with a full width at half-maximum of

$$\Delta \nu = R_{\text{spont}}(1 + \alpha^2)/4\pi I, \tag{44}$$

where α is Henry's α-factor of Eq. (43), I is the number of photons in the cavity, and R_{spont} is the rate of spontaneous emission into the lasing mode. The latter is given by

$$R_{\text{spont}} = n_{\text{spont}}\, v_g \Gamma g, \tag{45}$$

where g is the gain factor, Γ is the confinement factor ($\lesssim 1$) that measures the confinement of the light to the active laser layer, and

$$n_{\text{spont}} = N/(N - N_{\text{th}}) \tag{46}$$

is the spontaneous emission factor (of order 1–2). For equal facet reflectivities $R = R_1 = R_2$, the photon number I is related to the laser output power P per facet by

$$P = h\nu I(v_g/2L)\ln(1/R). \tag{47}$$

The laser linewidth thus scales with the inverse of the laser power. For conventional InGaAsP lasers the linewidth is about 100 MHz for a power of 1 mW.

4. LIGHTWAVE RECEIVERS

The purpose of digital lightwave receivers is to detect the incoming optical signal and to convert it to an electrical signal for amplification and processing. This processing includes the error-free recovery of the information carried by the light. Speed of response and sensitivity are two important characteristics of receivers. The optical signal arrives as a stream of light pulses signifying binary "ones" and "zeros." These pulses have been attenuated and distorted along the transmission path. Receiver noise interferes with the error-free recovery of the weakened pulses.

The sensitivity of the receiver measures its ability for "error-free" detection, or, more precisely, for detection with a bit-error rate (BER) of 10^{-9}, i.e., one error per billion pulses. It is quoted in terms of the input light level required for this BER and expressed in watts, dBm, or photons per bit. The dBm measure describes the incident power level P in terms of decibels compared to 1 mW,

$$\text{dBm} = 10 \log_{10}[P/(1\ \text{mW})]. \tag{48}$$

The photon flow per second is $P/h\nu$, and the number of photons N received per bit interval is

$$N = \frac{P}{h\nu B} = 5.03 \times 10^6 \times \left(\frac{P}{1\ \text{mW}}\right)\left(\frac{\lambda}{1\ \mu\text{m}}\right)\left(\frac{1\ \text{Gb/s}}{B}\right), \tag{49}$$

where B is the bit rate and $1/B$ the bit interval. At $\lambda = 1.3\ \mu$m, $P = 1$ mW, and $B = 1$ Gb/s, there are 6.5 million photons per bit. Other samples are shown in Table 2.

Figure 19 shows a block diagram of a typical digital lightwave receiver. The light impinges on the photodiode, which converts the

Table 2. Photons per bit (N) and dBm values for a set of typical power levels at $\lambda = 1.3\ \mu$m. A bit rate of 1 Gb/s is assumed.

$\lambda = 1.3\ \mu$m, $B = 1$ Gb/s		
P	dBm	N
1 mW	0	6.5×10^6
1 μW	−30	6500
1 nW	−60	6.5

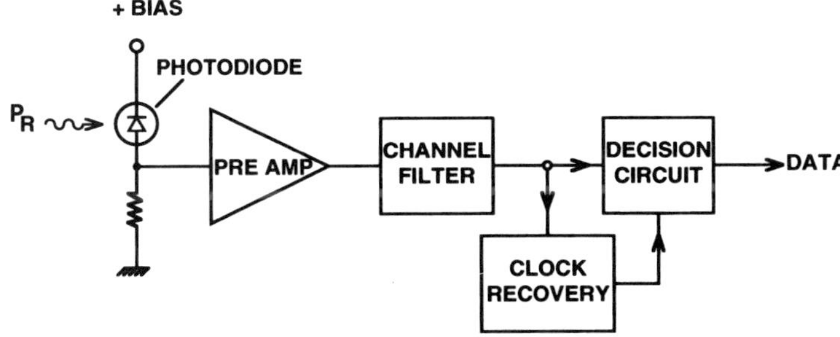

FIG. 19. Block diagram of a digital lightwave receiver.

signal to a photocurrent that reproduces the power envelope of the received optical signal. The electrical signal is first amplified in the low-noise preamplifier and then processed in the channel filter (or "linear channel"). This contains equalizer circuitry, automatic gain control, and a low-pass filter to limit the noise. The clock recovery circuit extracts timing information to synchronize the process of making decisions between "ones" and "zeros." These decisions are made in the decision circuit, which uses a threshold level to distinguish the two. More background and detail on receivers are available in review texts including Smith and Personick (1982), Kasper (1988), Forrest (1988), and Williams (1991). The following will provide a brief sketch of the subject.

4.1 Quantum Limit, Poisson Distribution

The corpuscular nature of light sets a lower limit on receiver sensitivity. To determine this limit, assume that we signal a "one" by N photons (per bit interval), and a "zero" by sending no light at all. The optical particles are random and have to obey Poisson statistics. This tells us that we will observe N photons per bit as an average over many bit samples. An individual observation, however, will see a number of n photons per bit, with a probability $P(n)$ given by the Poisson distribution

$$P(n) = N^n \exp(-N)/n!. \tag{50}$$

An average signal stream has the same number of "zeros" and "ones." The Poisson distribution predicts no errors on "zero" bits where $N = 0$; they are always seen as "zeros." However, there is a finite probability

$$P(0) = \exp(-N) \tag{51}$$

of observing a "one" as a "zero." With no errors on "zeros," we can get an average BER of 10^{-9} by allowing $P(0) = 2 \times 10^{-9}$ errors on "ones." This sets the quantum limit on sensitivity as $N = 20$ photons per bit interval of the "ones." The corresponding average number of photons per bit interval of the entire signal pulse stream is $\bar{N} = N/2 = 10$. The comparison of Sec. 4.5 shows that current *p-i-n* and avalanche photodetector (APD) receivers are more than two orders of magnitude less sensitive than this quantum limit.

4.2 Photodiodes

The majority of lightwave systems operating in the long-wavelength (1.3–1.6 μm) region employ either *p-i-n* photodetectors or APDs. *p-i-n* detectors can be very fast and simple, while APDs offer more sensitivity at Gb/s rates. Both devices are grown on InP substrates. Figure 20 shows sketches of the two devices.

The *p-i-n* detectors use *n*-type InGaAs material as a light-absorbing layer. A *p*-type region is created by doping with Cd or Zn. The resulting *p-n* junction is reverse biased. This results in a region near the junction that is free of carriers and, consequently, low in current in the absence of light. An absorbed photon creates an electron and a hole, which are rapidly drawn to the electrodes. This leads to a photocurrent I_R in response to the received light power P_R given by

$$I_R = \eta(e/h\nu)P_R, \tag{52}$$

where η is the quantum efficiency of the detector. The quantity $e/h\nu$ can be expressed as

$$e/h\nu = e\lambda/hc = (0.804 \text{ A/W})\, \lambda/(1\ \mu\text{m}). \tag{53}$$

For $\lambda = 1.3\ \mu$ this is 1.05 A/W, which means

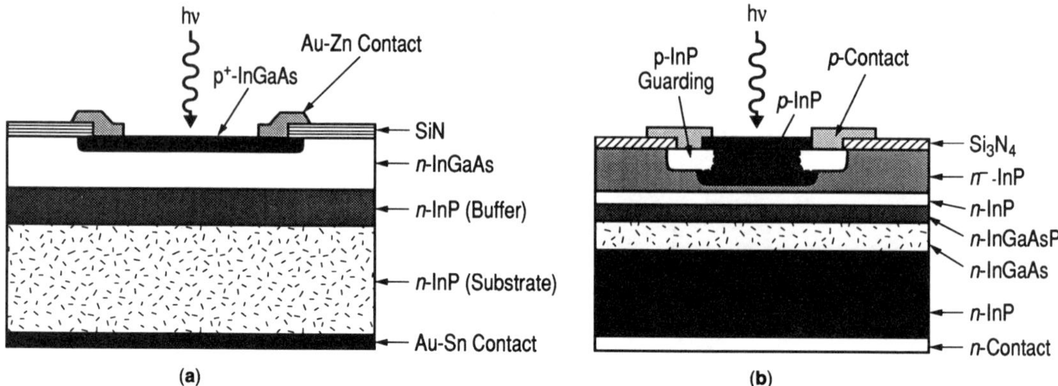

FIG. 20. Sketch of the structure of (a) a *p-i-n* photodiode and (b) an avalanche photodiode. Both use InP substrates. The light ($h\nu$) is incident from above.

that about 1 μA of current is generated for each microwatt of received light.

In an APD one provides for avalanche multiplication of the received photocurrent to minimize the influence of noise from the receiver amplifier. APDs are also operated under reverse bias. However, in contrast to *p-i-n* detectors, the applied voltages are made sufficiently large that photocurrent gain due to impact ionization of carriers occurs.

The APD device shown in the Fig. 20(b) is a so-called SAM structure, deriving its name from the Separate Absorption and Multiplication layers. The structure is optimized to provide simultaneously low noise, high speed, and avalanche multiplication at sufficient gain. The narrow-gap InGaAs layer serves to absorb the incident light. The wide-gap *n*-type InP region provides the carrier multiplication. Experimental sensitivities of such devices are shown in Sec. 4.5.

4.3 Receiver Noise and Sensitivity

We have already mentioned that the sensitivity levels of practical receivers exceed the quantum limit by 20 dB or more. The principal reason for this is noise from the electronic amplifiers following the photodiode. This noise is usually described by the variance or mean square value of the equivalent input noise current $\langle i^2 \rangle$. An example is the noise of a low-impedance front end that is given by

$$\langle i^2 \rangle_{\mathrm{AMP}} = (4kTF/R_L)\Delta f, \tag{54}$$

where Δf is the circuit bandwidth, R_L the load resistor (see Fig. 19), and F the noise figure of the amplifier. The contributions of other noise sources depend on the details of the receiver architecture. The reader is referred to the literature cited above for detailed analyses. In the following, we assume that the sum of all noise contributions is known and represented in $\langle i^2 \rangle$.

Our next task is to express the analog signal-to-noise ratio ($I^2/\langle i^2 \rangle$) as a digital BER. For this we refer to Fig. 21, where we have indicated the current levels I_0 and I_1 for the "zeros" and "ones" of the signal. The presence of noise causes the actual current to

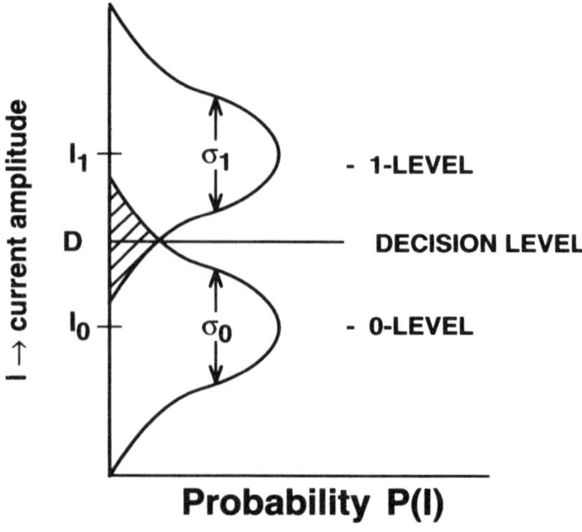

FIG. 21. Probability distributions of the current amplitudes for the "zero" and "one" signals. The decision level is *D*.

fluctuate around these mean values. This is described by Gaussian probability distributions

$$P_i(I) = \frac{1}{\sqrt{2\pi}\sigma_i} \exp\left[-\frac{(I-I_i)^2}{2\sigma_i^2} \right], \tag{55}$$

where the index $i = 0$ or 1 refers to the "zero" and "one" levels, and

$$\sigma_i^2 = \langle i^2 \rangle_i \tag{56}$$

are the variances of the corresponding noise currents. The decision level D of the receiver is also shown. Currents below D are judged a "zero" signal, and currents above it a "one." The shaded areas in the probability distributions lead to errors. If D is set for equal errors on the "zeros" and the "ones," one calculates a BER of

$$\text{BER} = \tfrac{1}{2} \operatorname{erfc}(Q/\sqrt{2}), \tag{57}$$

where erfc is the complementary error function, and

$$Q = (I_1 - I_0)/(\sigma_1 + \sigma_0). \tag{58}$$

Figure 22 shows the values of BER as a function of Q. A BER of 10^{-9} requires a value of $Q = 6$. For equal variances $\sigma_1^2 = \sigma_0^2$, one gets a decision level half-way between I_1 and I_0. If, furthermore, $I_0 = 0$, we have $Q = I_1/2\sigma_1$. This relates Q to the analog signal-to-noise ratio:

$$\text{SNR} = \frac{I_1^2}{\sigma_1^2} = \frac{I_1^2}{\langle i^2 \rangle_1} = 4Q^2. \tag{59}$$

For BER $= 10^{-9}$, therefore, we need a SNR of $4 \times 6^2 = (12)^2$ or 21.6 dB.

4.4 Optical Preamplifiers

The section on optical amplifiers discusses the characteristics of these devices. When preamplifiers or optical amplifier chains are used in the transmission system, their ASE noise will tend to dominate other noise sources in the optoelectronic receiver. There is a received signal current I_R due to the received optical signal power P_R, but there

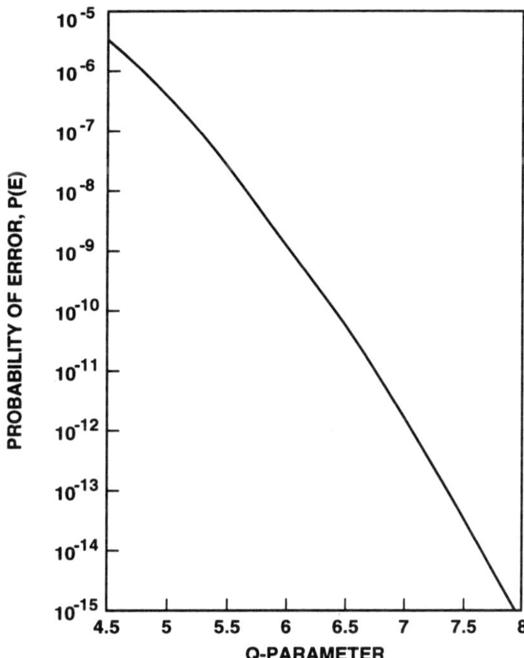

FIG. 22. Bit-error rate (BER) as a function of the parameter Q.

is also a received current

$$I_{\text{ASE}} = \eta(e/h\nu)N_{\text{ASE}} \tag{60}$$

due to the received spontaneous noise N_{ASE}. The square-law nature of the detection process leads to signal–spontaneous beat noise between the ASE and signal. The mean square of this current is

$$\langle i^2 \rangle_{\text{S-SPONT}} = 2I_{\text{ASE}} I_R B_E / B_0, \tag{61}$$

where B_E is the electrical bandwidth of the receiver and B_0 is the optical bandwidth of the system. In most practical cases, this noise is much larger than other noise components. Under these conditions, the electrical signal-to-noise ratio is given by

$$\text{SNR}_E = \frac{I_R B_0}{2I_{\text{ASE}} B_E} = \frac{B_0}{2B_E} \text{SNR}_0 = \frac{P_{\text{IN}}/h\nu B_E}{2F}, \tag{62}$$

where SNR_0 is the optical signal-to-noise ratio discussed in Sec. 2.3. The term P_{IN} is the input power to the optical preamplifier or amplifier chain and F is the corresponding noise figure.

In the ideal limit, we have $I_1 = I_R$, $I_0 = 0$, $\sigma_0 = 0$, $F = 2$, and $B_E = B/2$, where B is the bit rate. This leads to

$$\mathrm{SNR}_E = Q^2 = (\bar{P}_{\mathrm{IN}}/h\nu B), \tag{63}$$

where $\bar{P}_{\mathrm{IN}} = P_{\mathrm{IN}}/2$ is the average input signal power. From this one calculates the quantum limit for optical preamplifier receivers as 36 photons per bit for a BER of 10^{-9} ($Q = 6$). For more background on optical preamplifiers see Olsson (1989), Ruhl and Ayre (1993), Laming *et al.* (1992), and Li (1993).

4.5 Comparison of Receiver Sensitivities

Figure 23 provides a comparison of the sensitivity of various receivers as a function of bit rate. Shown are values for the average received power in dBm required for a BER of 10^{-9}. These values represent state-of-the-art results obtained in research laboratories for *p-i-n* receivers, APD receivers, and receivers using optical preamplifiers (EDFAs). As a reference, the chart indicates also the optical amplifier limit of 36 photons per bit.

Note that there is a decrease in sensitivity (i.e., an increase in the required optical power) as the bit rate increases. The quantum limit ($\bar{N} = 10$) and the optical amplifier limit ($\bar{N} = 36$) suggest that the number of photons per pulse should stay constant as the bit rate goes up. Then, the power should scale linearly with bit rate. The actual data show a somewhat faster rise than that.

The sensitivity of APD receivers tends to be about 5 dB better than that of the *p-i-n*'s, while EDFA amplifiers appear to offer more than 10 dB improvement over APDs in the multi-Gb/s range.

5. MODULATION FORMATS

The light transmitted through the fiber acts as a carrier for the signals that represent information such as voice, video, or data. These signals are modulated onto the light by the transmitters. Various modulation formats are in use. The digital format predominates in practice. However, analog modulation is used for CATV transmission (see Sec. 10). Among the digital formats are ASK (amplitude-shift keying), FSK (frequency-shift keying), and PSK (phase-shift keying), where either the amplitude, the frequency, or the phase of the light is changed to signal "ones" or "zeros."

5.1 RZ and NRZ

Two ASK formats are in use and illustrated in Fig. 24. One is the return-to-zero (RZ) format, and the other the non-return-to-zero (NRZ) format. RZ signals return their amplitude to zero after each "one," NRZ signals return their amplitude to zero only if a "zero"

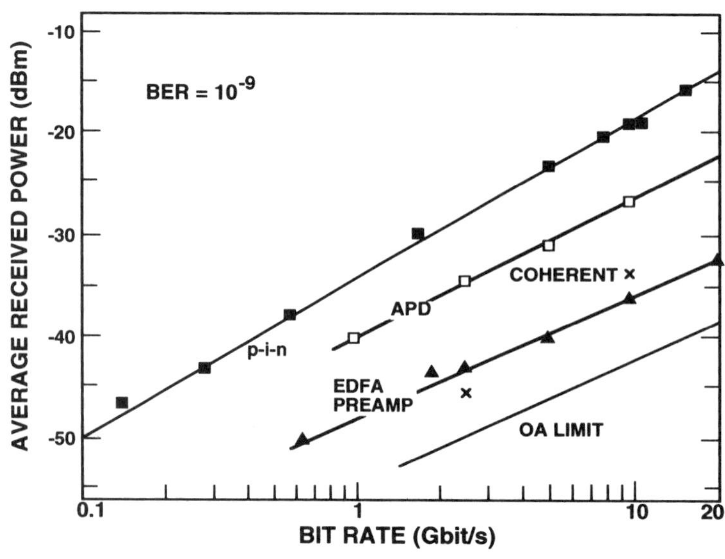

FIG. 23. Sensitivity vs bit rate for *p-i-n* detectors, APDs, EDFA preamps, and coherent detection. The data shown are the best achieved in research laboratories. The quantum limit for optical preamps is also shown.

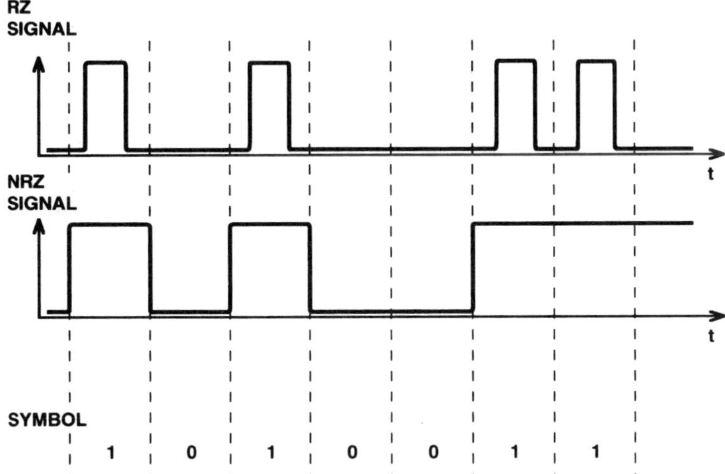

FIG. 24. RZ and NRZ signal amplitudes for a typical sequence of "one" and "zero" symbols.

follows a "one." RZ signals require higher bandwidth at the transmitter.

5.2 TDM and WDM

Higher capacities can be transmitted over the fiber by combining two or more digital signals of a given bit rate. Three methods can be used: TDM (time-division multiplexing), WDM (wavelength-division multiplexing), or SCM (subcarrier multiplexing). TDM and WDM are illustrated in Fig. 25. In TDM, the two incoming pulses of the two signals are shortened and interleaved in time, resulting in a bit stream of higher rate. In WDM, several wavelength channels are used, each carrying a different signal. A combination of TDM and WDM can be used to increase capacity further. This is shown in Fig. 26, which also indicates the approximate trend in the increase of fiber capacity in long-haul systems.

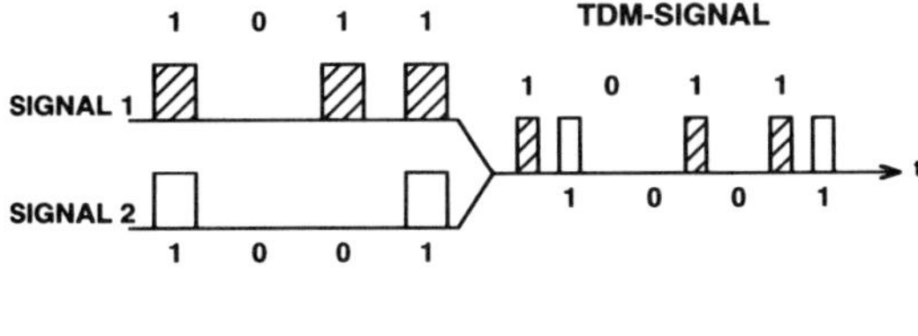

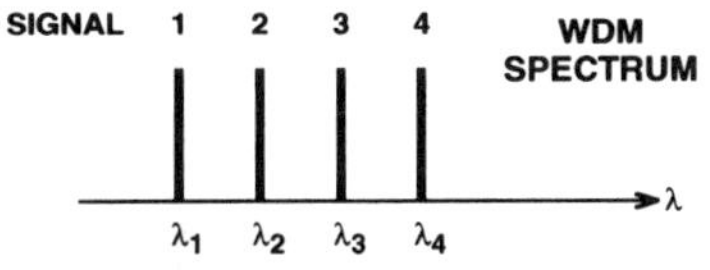

FIG. 25. TDM for two illustrative symbol sequences. The two signals are combined into one. Also shown is the spectrum of a WDM signal where four signal streams are carried on four different wavelengths.

5.3 Subcarrier Modulation

This format is depicted in Fig. 27. The digital (or analog) signal is first modulated onto an RF carrier (e.g., 500 MHz). The resulting signal is then modulated onto the optical carrier. Different RF carriers can be used to combine several channels into a SCM signal as shown.

6. OPTICAL NONLINEARITIES IN FIBERS

Nonlinearities in optical fibers can provide useful effects such as Brillouin amplification, Raman amplification, and soliton propagation, but they can also lead to significant degradations and performance limitations in long-haul transmission systems. System impairments that need to be understood and minimized include spectral distortions due to self-phase modulation, cross-phase modulation, and four-photon mixing. The nonlinearities of the glass material of fibers are small by the standards of nonlinear optics. However, nonlinear effects can accumulate to great strength in the fiber because of the large interaction lengths, which measure hundreds or even thousands of kilometers. Review texts discussing the impact of

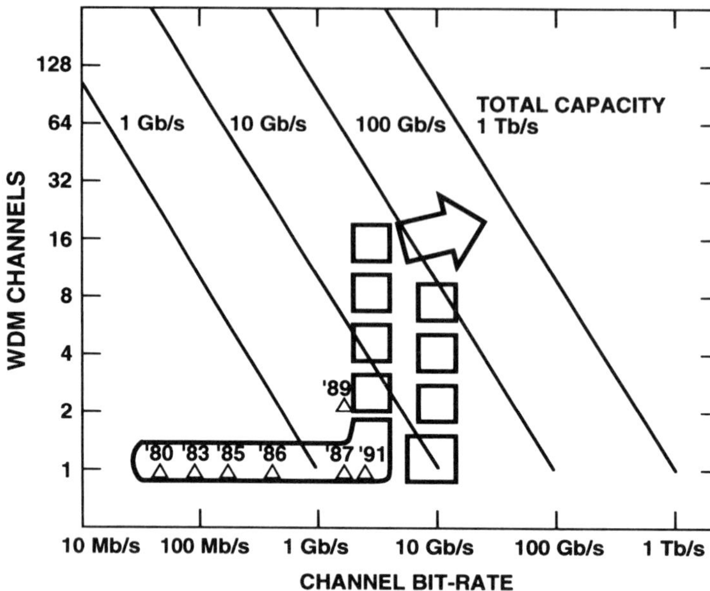

FIG. 26. Approximate trend in the capacity increase of long-haul lightwave systems. Total capacity per fiber is increased by a combination of higher bit-rate TDM and WDM. The triangles indicate the years of initial commercial installation.

nonlinearities on lightwave systems include those by Smith (1972), Stolen (1979, 1980), Chraplyvy (1990), Marcuse (1991b), Tkach (1992), and Chraplyvy and Tkach (1993).

The power of a typical nonlinear effect grows exponentially with length and pump intensity P_p/A_{eff}:

$$P_{\text{OUT}} = P_{\text{IN}} \exp(\gamma L_{\text{eff}} P_p/A_{\text{eff}}), \tag{64}$$

where γ is the gain coefficient of the nonlinearity and L_{eff} is the effective length of the fiber. This length is related to the actual length L of the fiber and the fiber loss α (which attenuates the effectiveness of the pump) by

$$L_{\text{eff}} = [1 - \exp(-\alpha L)]/\alpha. \tag{65}$$

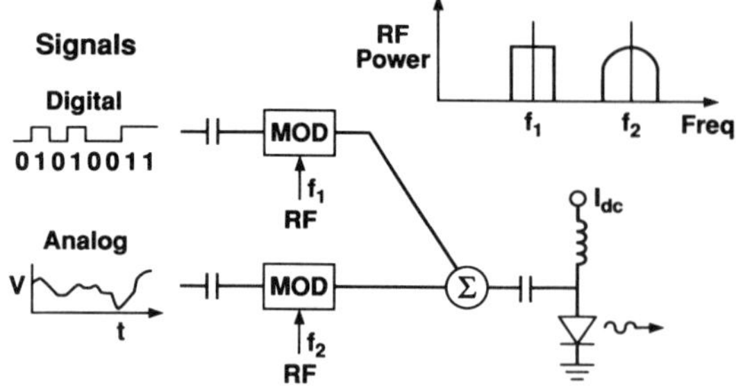

FIG. 27. Schematic of subcarrier modulation using rf subcarriers f_1 and f_2. The resulting spectra are indicated. Both digital and analog signals can be SC modulated onto the optical carrier.

For large L, we have $L_{\text{eff}} \approx 1/\alpha$, which is about 20 km for a single fiber span with $\alpha = 0.22$ dB/km.

The distribution of the actual pump intensity $I(x,y)$ across the fiber cross section is determined by the corresponding fiber mode. Important for the nonlinear interaction is the effective area of this distribution,

$$A_{\text{eff}} = \left[\int I dA\right]^2 \left[\int I^2 dA\right]^{-1}, \tag{66}$$

which relates the effective intensity to the power P_p of the pump. For a Gaussian distribution $\exp(-r^2/w^2)$ of the modal field, the effective area is $A_{\text{eff}} = \pi w^2$. A typical step-index single-mode fiber has an A_{eff} of about 80 μm^2, whereas dispersion-shifted fibers tend to have smaller areas (50 μm^2).

6.1 Stimulated Brillouin Scattering (SBS)

This nonlinearity involves backward scattering of light from acoustic phonons in the glass with the phonon frequency

$$\nu_B = 2nV/\lambda. \tag{67}$$

Here, V is the velocity of the longitudinal sound waves, n is the refractive index of the glass, and λ is the optical wavelength. At $\lambda = 1.55$ μm, we have $\nu_B = 11$ GHz. The SBS interaction is sketched in Fig. 28(a). Pump light of frequency ν_p enters the fiber from the left. The pump provides gain for a counterpropagating signal of frequency ν in a spectral range near

$$\nu = \nu_P - \nu_B. \quad (68)$$

Near 1.55 μm, the bandwidth of this Brillouin gain is about

$$\Delta\nu_B \approx 20 \text{ MHz}. \quad (69)$$

This bandwidth scales with the square of the phonon frequency. At the gain peak, the Brillouin gain coefficient in fused silica is

$$\gamma_B = 4 \times 10^{-9} \text{ cm/W}, \quad (70)$$

This applies to well-aligned polarizations of the pump and signal light. If the polarizations are scrambled, then the SBS gain coefficient is reduced to half of this value.

SBS imposes a limit P_B on the input signal power that can be transmitted through a fiber. This is the Brillouin limit

$$P_B = 42A_{eff}/\gamma L_{eff}. \quad (71)$$

Here, the signal intended for transmission acts as a pump for the SBS amplification of spontaneously Brillouin scattered light. When the signal power reaches P_B, the amplification of this spontaneous emission will deplete the signal (= pump) by about 50% and cause severe signal degradation.

The above formula applies to scrambled polarizations and signal bandwidths $\Delta\nu$ that are smaller than $\Delta\nu_B$. For a typical L_{eff} = 20 km and A_{eff} = 80 μm^2 one, then, gets P_B = 4.2 mW. When $\Delta\nu_s \gg \Delta\nu_B$, and the signal spectrum is uniformly distributed over $\Delta\nu_s$, the effective gain coefficient γ is reduced to

$$\gamma = \gamma_B\Delta\nu_B/\Delta\nu_s, \quad (72)$$

and the Brillouin limit is increased accordingly.

6.2 Stimulated Raman Scattering (SRS)

SRS involves light scattering by molecular vibrations in the glass or high-frequency optical phonons. This process provides broadband gain for signals in either the forward or backward direction relative to the pump wave [see Figs. 28(a) and 28(b)]. The spectrum for the Raman gain coefficient is shown in Fig. 29. This applies to λ = 1.55 μm and aligned single polarizations of pump and signal. For scrambled polarizations, the coefficient is reduced to half the values shown. The peak of the gain curve is reached at a frequency

$$\nu_R = 400 \text{ cm}^{-1} = 12 \text{ THz} \quad (73)$$

lower than the pump frequency. At the peak, the gain coefficient for SRS is about

$$\gamma_R = 7 \times 10^{-12} \text{ cm/W}. \quad (74)$$

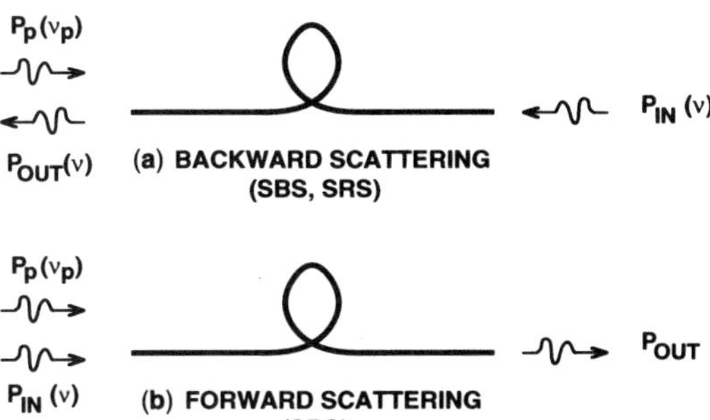

FIG. 28. Schematic for (a) backward and (b) forward scattering of light in fibers. The pump light (P_p) propagates to the right in both cases.

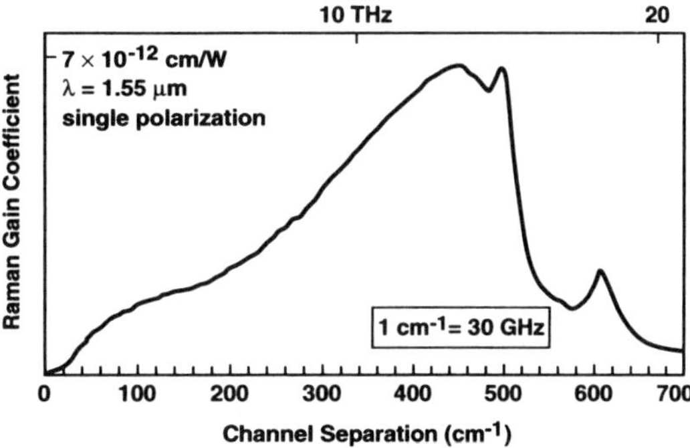

FIG. 29. Spectrum of the Raman gain coefficient for fused silica fibers. The channel separation between pump and signal is shown in THz as well as in the inverse centimeter (cm^{-1}) measure used by spectroscopists.

This value scales inversely with the wavelength λ, and depends somewhat on the doping.

SRS also imposes limits on the power of signals that can be transmitted before significant signal depletion (~50%) and degradation occurs as a result of the Raman amplification of spontaneously scattered light. For scrambled polarizations and a single-frequency signal the Raman limit is

$$P_R = 32A_{eff}/\gamma_R L_{eff}. \tag{75}$$

For L_{eff} = 20 km and A_{eff} = 80 μm^2 the Raman limit is 1.8 W.

When a WDM format is used to transmit n signals at n different frequencies spaced $\Delta\nu$ apart, the Raman limit P_{WDM} is much lower. In this case, the higher-frequency signals act as pumps for all lower-frequency signals. Signal depletion is greatest at the highest frequency. To insure that the power penalty due to this depletion is less than 0.5 dB, the total power $P_{WDM} = nP$ must obey the inequality

$$P_{WDM}B_{WDM}L_{eff} < 9 \text{ W THz km}. \tag{76}$$

Here, $B_{WDM} = (n - 1)\Delta\nu$ is the total bandwidth occupied by the WDM signal, P is the power in each channel, and an A_{eff} = 80 μm^2 is assumed. Note that L_{eff} can be much larger than 20 km when amplifier chains are used in the transmission system.

6.3 Self-Phase Modulation (SPM)

Third-order nonlinearities in glass give rise to such effects as self-phase modulation, cross-phase modulation, four-wave mixing, and soliton propagation. A complete description of these effects requires the use of a susceptibility tensor of rank 4. However, for the simple case of linear polarization, we can use a scalar description

$$n = n_0 + n_2P/A_{eff}, \tag{77}$$

which indicates the change of the refractive index n of the glass as a function of the light intensity P/A_{eff}. For fused silica, the nonlinear index is approximately

$$n_2 = 3 \times 10^{-20} \text{ m}^2/\text{W}. \tag{78}$$

Because of the power-dependent refractive index, intensity modulation of the signal will lead to modulation of the phase of the light:

$$\Delta\phi = (2\pi n_2 L_{eff}/\lambda A_{eff})P. \tag{79}$$

For A_{eff} = 50 μm^2 and L_{eff} = 1000 km (i.e., 50 spans with L_{eff} = 20 km each) a signal power of 1 mW will lead to a phase shift of $\Delta\phi = \pi$, approximately. This phase modulation leads to a frequency modulation

$$\Delta\nu_{SPM} = \frac{1}{2\pi}\frac{d(\Delta\phi)}{dt} = \frac{n_2 L_{eff}}{\lambda A_{eff}}\frac{dP}{dt}, \tag{80}$$

and, consequently, to a broadening of the signal spectrum. Figure 30 shows this spectral broadening for a Gaussian signal pulse for various values of the phase shift $\Delta\phi_{MAX}$ at the peak of the pulse. The case of $\Delta\phi_{MAX} = \pi$ indicates, approximately, the intensity level for which the spectral width is doubled due to SPM. Note that the details of SPM spectral broadening depend on the pulse shape of the signal and the dispersion of the fiber. In the presence of fiber dispersion, the SPM broadening of the signal spectrum will lead to a more rapid degradation of the transmitted signal.

6.4 Cross-Phase Modulation (XPM)

The phenomenon of XPM occurs when two or more signals are present in the fiber, such as in a WDM transmission system. Then, a signal of frequency ν_i will experience not only SPM, but also XPM due to the presence of

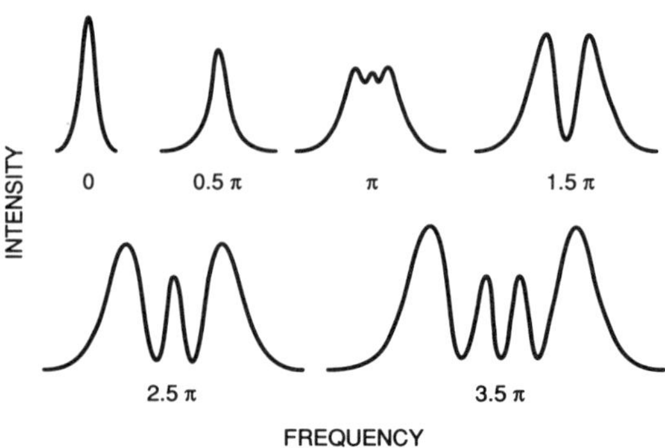

FIG. 30. Spectra of a Gaussian pulse broadened by SPM (calculated). The labels refer to the maximum phase shift $\Delta\phi_{MAX}$ (that occurs at the peak of the pulse).

another signal of power P_j. The resulting frequency modulation is

$$\Delta\nu_i = 2\frac{n_2 L_{\text{eff}}}{\lambda A_{\text{eff}}}\frac{dP_i}{dt}. \tag{81}$$

Note that XPM is twice as strong as SPM. However, fiber dispersion will tend to reduce its effect as "j" pulses overtaking "i" pulses produce both upshifts in frequency (due to the leading edge of the pulses) and downshifts (due to the trailing edge).

6.5 Four-Photon Mixing (4PM)

When two or more frequency channels are present in the fiber, nonlinear mixing creates sidebands at new frequencies

$$\nu = \nu_i + \nu_j - \nu_k, \tag{82}$$

where i, j, k designate the original channels. This process is called four-photon mixing and is illustrated in Fig. 31. In WDM systems with n channels all combinations occur, leading to a number

$$n^2(n-1)/2 \tag{83}$$

of mixing products. Examples for this are given in Table 3. The appearance of the 4PM products causes signal depletion and crosstalk in the transmission system.

The effect of 4PM is strongest when signals and 4PM products travel in synchronism (phase-matched). Strong fiber dispersion will, therefore, weaken the 4PM efficiency, as will a larger frequency separation between the channels. Figure 32 shows the 4PM mixing efficiency (i.e., the maximum 4PM product power relative to the signal power) as a function of the channel separation for selected values of the fiber dispersion D. Note that dispersion-shifted fibers ($D = 0$) have the highest 4PM efficiencies and, thus, suffer the largest signal degradation due to 4PM.

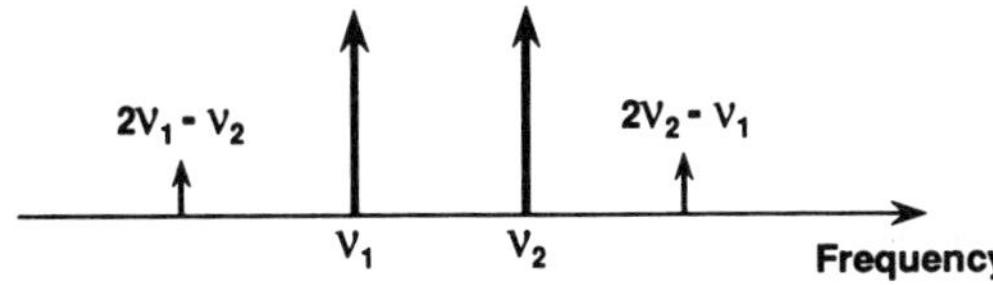

(a) Two Signals

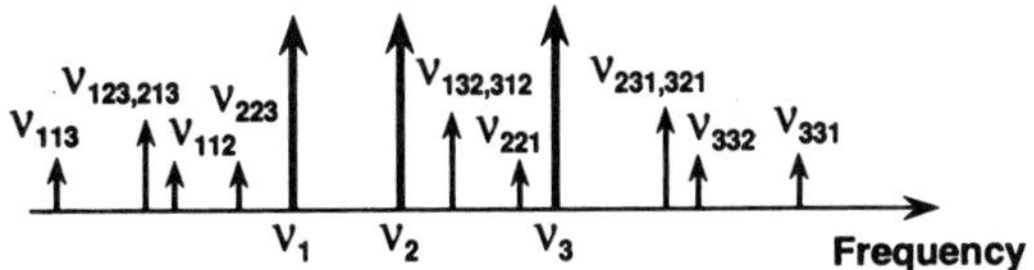

(b) Three Signals

FIG. 31. Spectra generated by 4PM for two incident signals (ν_1,ν_2), and for three incident signals (ν_1,ν_2,ν_3).

Table 3. Number of four-wave mixing products for typical numbers of WDM channels.

WDM channels	4PM products
3	9
4	24
8	224
16	1920

6.6 Solitons in Fibers

Texts discussing the applications of solitons to optical fiber communications include Hasegawa and Tappert (1973), Mollenauer *et al.* (1980), Gordon and Mollenauer (1991), and Mollenauer *et al.* (1991). Solitons are optical pulses that can propagate over very long distances in the fiber without pulse spreading and virtually free of nonlinear distortions. In the soliton mode, fiber dispersion and nonlinear SPM cancel each other, and two solitons of different frequencies can pass each other without generating four-photon mixing products. Necessary requirements for the formation of solitons are a positive fiber dispersion D, and a specific peak power level for the pulse, the so-called soliton power

$$P_{\text{SOL}} = \lambda A_{\text{eff}}/4n_2 Z_0. \tag{84}$$

Here, Z_0 is the soliton period defined by

$$Z_0 = \pi^2\tau^2 c/\lambda^2 D, \tag{85}$$

where c is the velocity of light. The soliton pulse width τ is related to the full width at half of the maximum intensity by

$$\tau_{\text{FWHM}} = 1.76\tau. \tag{86}$$

Table 4 lists some sample values for these

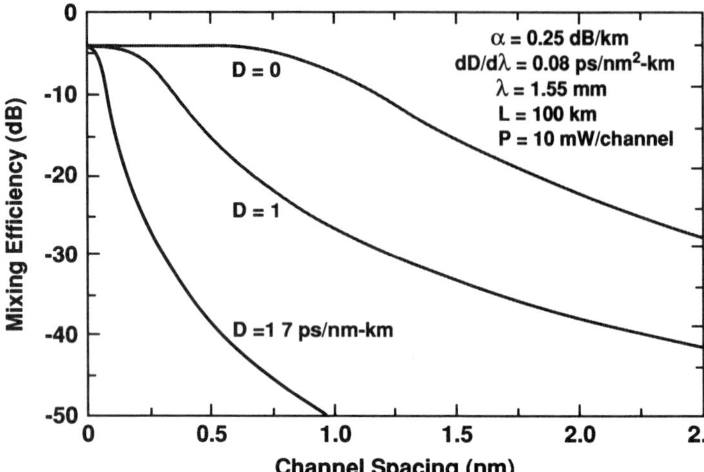

FIG. 32. Four-wave mixing efficiency as a function of channel spacing for three typical values of the fiber dispersion *D*.

parameters typical for long-distance transmission systems. The soliton period Z_0 is a characteristic distance for the soliton process. It is, e.g., the distance at which higher-order soliton pulses reassume their original pulse width and pulse shape after undergoing a pulse compression and expansion cycle.

Solitons are governed by the nonlinear Schrödinger equation for the pulse envelope $u(z,t)$:

$$j\frac{\partial u}{\partial z} = -j\beta'\frac{\partial u}{\partial t} - \frac{1}{2}\beta''\frac{\partial^2 u}{\partial t^2} + \kappa|u^2|\,u, \tag{87}$$

where $v_g = 1/\beta'$ is the group velocity in the fiber, $D = -2\pi c\beta''/\lambda^2$ is the fiber dispersion, and $\kappa = \pi c n_0 n_2/4\lambda$. A simple solution of this gives the pulse envelope of the fundamental soliton:

$$u = \text{sech}\,[(t - z/v_g)/\tau] \tag{88}$$

for a power level P_{SOL}. Higher power levels give rise to higher-order soliton solutions. Figure 33 shows experimental results for the pulse envelopes of some low-order solitons at $z = Z_0/2$.

Table 4. Sample values for soliton pulse width τ_{FWHM}, soliton period Z_0, and soliton power P_{SOL}. $D = 1$ ps/nm km, $\lambda = 1.56\ \mu$m, $A_{eff} = 50\ \mu m^2$.

τ_{FWHM}/ps	Z_0/km	P_{SOL}/mW
25	240	2.6
50	980	0.65

Solitons have shown remarkable resistance to perturbations of all kinds such as effects of higher-order group dispersion or fiber loss. However, the loss must be compensated periodically by amplifiers to maintain the average power level at P_{SOL}. In this case, the amplifier spacings have to be smaller than Z_0. Recent soliton transmission experiments indicate that 20-Gb/s capacities over transoceanic distances are achievable using the technique of sliding-frequency guiding filters (Mollenauer *et al.*, 1992, 1993).

7. UNDERSEA FIBER SYSTEMS

Undersea fiber systems are information links that span distances of the order of 1000 to 10000 km. They have become the major information highways between continents. The first large system of this kind was TAT-8, which was placed into service in 1988 linking the United States and Europe. This transmission technology is the successor to the transatlantic telegraph cable, first installed in 1858, and the transatlantic telephone (TAT) cable whose first system TAT-1 began operation in 1956. TAT-1 used coaxial cables and carried the information in analog form. It had an information capacity of about 40 voice circuits across the ocean. The new fiber systems use digital technology, which provides a considerable improvement in quality. At the same time, they offer an information capacity that is at least 1000 times larger than the early TAT-1. The rapid introduction of fiber

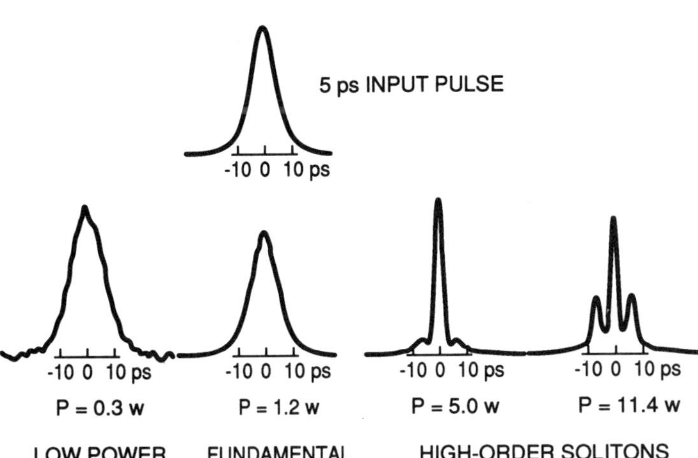

FIG. 33. Output pulse shapes at a fiber distance $z = Z_0/2$ as a function of pulse power level. Fiber length is 700 m. The pulse shapes of the fundamental and two higher-order solitons are shown. After Mollenauer *et al.* (1980).

in global networks is illustrated by the fact that, already in 1991, fiber carried more information traffic over international digital links than satellites. Overviews of the subject of undersea fiber systems are provided by texts such as Runge and Trischitta (1986), Runge (1992), and Thiennot *et al.* (1993).

7.1 Undersea Fiber Cables

The typical construction of an undersea fiber cable is shown in Fig. 34. The diameter of the cable is approximately 25 mm. At the center of the cable's core is a steel wire, which is surrounded by an elastomer. Up to six single-mode fibers are embedded in the elastomer. The core is surrounded by interlocked steel wires to protect the fibers from ocean pressure and give the cable more tensile strength. The copper jacket provides a hermetic seal and is also used to conduct electric power to the undersea repeaters.

The power is supplied from terminal stations on land, which use high-voltage supplies of about 7500 V that carry constant currents of about 1.6 A. The water of the ocean is used as a return conductor.

Near the coast, the cable is buried to a depth of about 1 m for protection. The cable is laid down to ocean depths of 8000 m.

The polyethylene layer surrounding the copper jacket also has two functions: it provides electric insulation and mechanical protection for the copper.

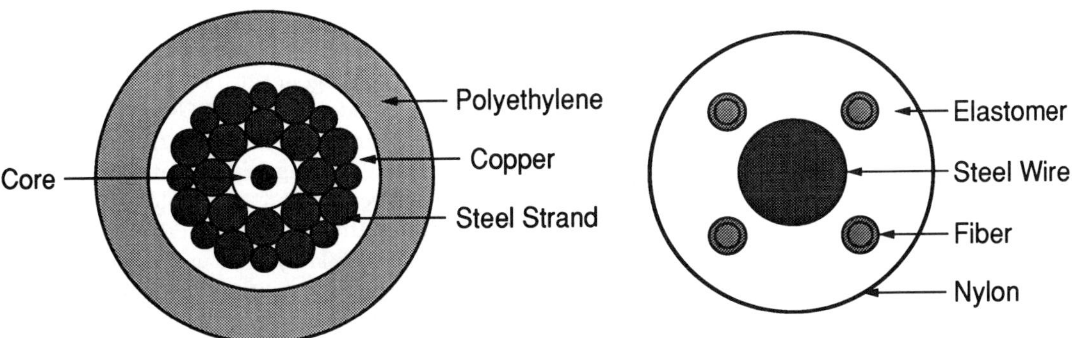

FIG. 34. Left: cross section of a typical undersea cable. Right: detail of the core. The cable diameter is about 25 mm.

Another layer of steel tape is added to the structure shown in the figure to protect the cable against shark bites. This provision traces back to an experiment with the first undersea system in 1985 between Tenerife and Grand Canaria in the Canary Islands. This cable was attacked by sharks for reasons that are still not completely understood.

7.2 Undersea Repeaters

The purpose of undersea repeaters is to boost the strength of the optical signal that has been attenuated by the fiber and to minimize the effect of distortions and noise. The first two generations of undersea fiber systems used optoelectronic repeaters as shown in Fig. 35. They consist of a detector and associated electronic receiver circuitry, followed by automatic gain control (AGC), decision and retiming circuitry used to regenerate the signal, transmitter circuitry controlling a laser, and finally a semiconductor laser emitting a strong and refreshed stream of optical signal pulses.

The third-generation technology will use an all-optical repeater to amplify the signal. It will use an erbium-doped fiber, a semiconductor pump laser to supply optical power to the amplifier, a wavelength-division multiplexer (WDM), an isolator, and a filter, as shown in Fig. 36. The purpose of the WDM is to feed the pump light into the erbium fiber without attenuating the signal (which is carried at a different wavelength). The purpose of the isolator is to suppress light reflections from the output port.

7.3 Undersea Systems Technologies

Figure 37 shows a plan for the installation of major undersea lightwave links by about 1996. Several of these systems are already in

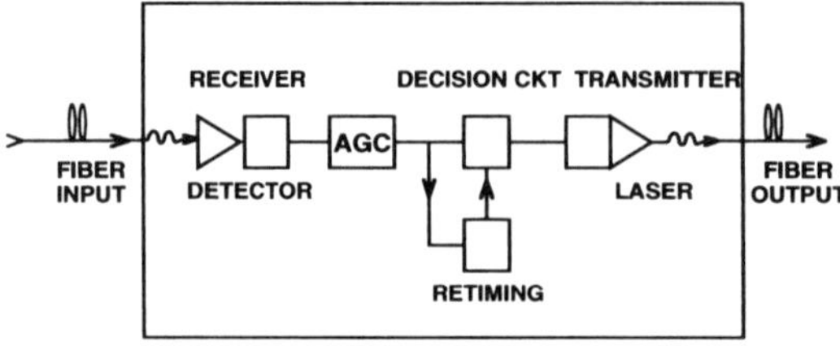

FIG. 35. Principal building blocks of an optoelectronic regenerator.

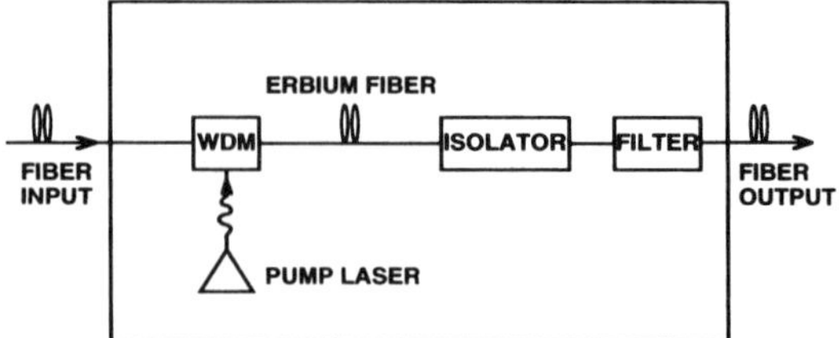

FIG. 36. Principal building blocks of an optical repeater using an optical amplifier. The optical filter shown is often not required.

place. In fact, over 45 000 km of undersea lightwave cable were installed by the end of 1992. Typical system lengths are about 6000 km in the Atlantic and 9000 km in the Pacific. The rapid pace of progress in this technology is reflected by the fact that we can, already, distinguish three different technological generations.

The first generation is used in TAT-8 (1988), TPC-3 (1989), and HAW-4. Here, TPC stands for "Trans-Pacific Cable." The year of first service is shown in parentheses. This undersea technology uses conventional single-mode (SM) fiber operating at a wavelength of 1.3 (μm.) The system uses the optoelectronic regenerators of Fig. 35 with InGaAsP semiconductor lasers and *p-i-n* photodetectors of the same material. The typical laser output is 1 mW, the receiver sensitivity is -37 dBm, and the typical regenerator spacing is 70 km. The bit rate of the optical signal stream in each fiber is 280 Mb/s.

The second generation undersea technology is used in the TAT-9 (1991), TAT-10, TAT-11, and TPC-4 installations. Here, SM fibers are used as before, but the operating wavelength is moved to 1.55 μm to benefit from lower fiber losses. Single-frequency distributed-feedback (DFB) lasers are used in the optoelectronic regenerators together with avalanche photodetectors (APDs). The bit rate per fiber is doubled to 560 Mb/s and the typical repeater spacing is 150 km.

A third-generation technology is being prepared for the TAT-12, TAT-13, TPC-5, and TPC-6 systems. They will use optical repeaters with optical amplifiers, such as shown in Fig. 36. These systems will operate at 1.55-μm wavelengths and use dispersion-shifted fiber (DSF) that will provide near-zero dispersion in the operating range. The system

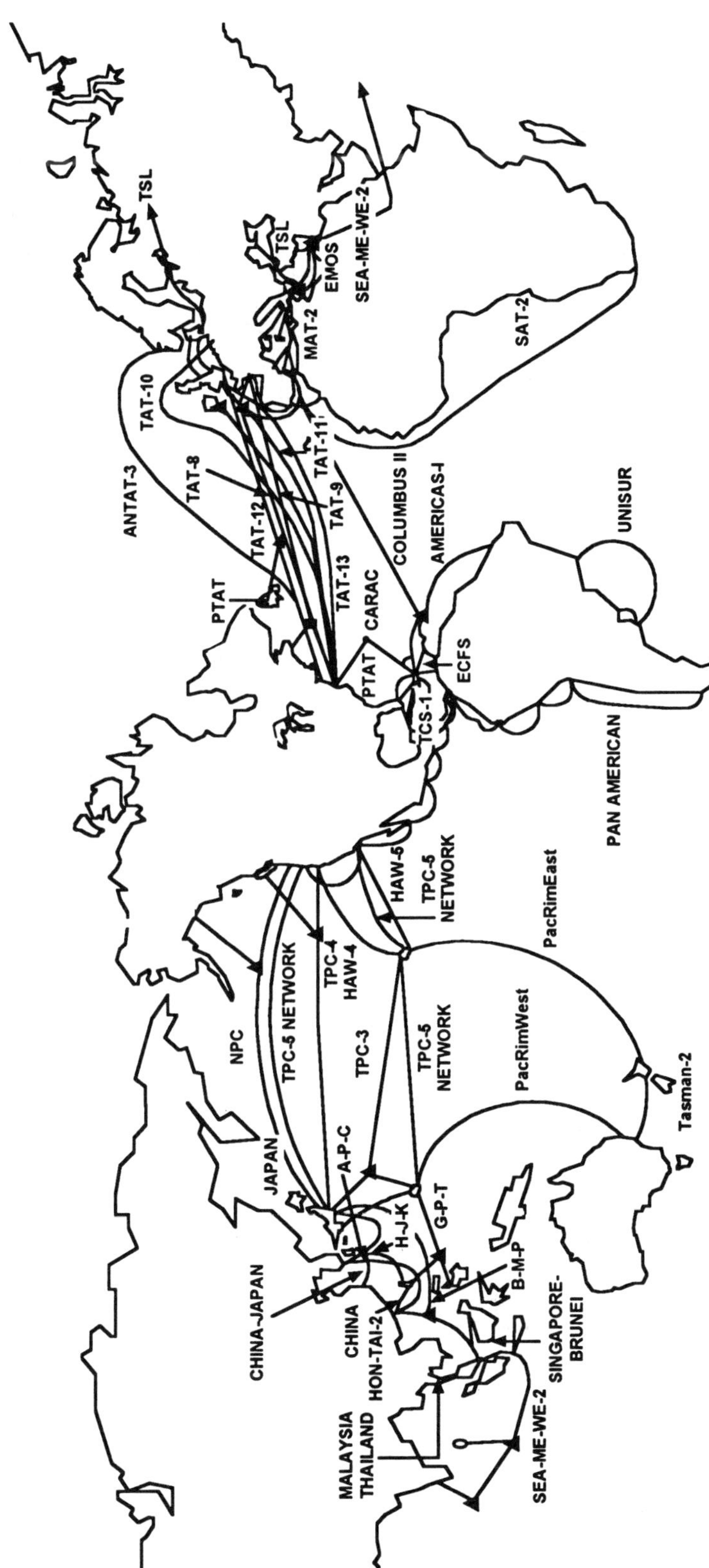

FIG. 37. Map showing the planned installations of major undersea lightwave systems by about 1996.

will be designed to operate at a bit rate of 5 Gb/s per fiber.

Table 5 lists the major undersea technologies and provides a compact overview of the three generations. It shows the total number of two-way voice circuits (channels) per cable for four fibers (two each way). Also shown is the total voice traffic capacity of the cable when the TASI circuit multiplication technique is used. TASI stands for "time-assignment speech interpolation" and is a technique that enhances the cable capacity by a factor of 5 by taking advantage of the natural pauses in conversations and by using more efficient coding of speech. Note that the TAT-12 cable will carry more than half a million voice channels, an astounding improvement over the 40 channels of the coaxial TAT-1 technology of 1956.

8. TERRESTRIAL LONG-HAUL SYSTEMS

Optical fiber systems used for long-distance transmission on land (also called "trunking") span distances ranging from about 50 km to several hundred kilometers. They provide the links between metropolitan telephone offices, between cities, and across the continent. For this application, the optical technology offers several advantages over the earlier transmission technologies that sent electrical analog signals over copper pairs or coaxial cable. These include the small size and weight of the cables, large repeater spacings, high transmission capacity, immunity from electromagnetic interference, and digital transmission quality. Early metropolitan applications, such as those in the San Francisco Bay area, have highlighted these advantages: The FT-3 system, introduced in 1980, provided digital transmission at a rate of 45 Mb/s and a repeater spacing of 7 km (Jacobs and Stauffer, 1980). This spacing was already sufficient to allow direct links between telephone offices with no need for outside electronics. This system also helped save space in metropolitan cable ducts where *one fiber* replaced 28 copper pairs of the earlier T1 carrier, which has a 1.5-Mb/s capacity per pair. The list of reviews of terrestrial systems technologies includes Kogelnik (1985), Sanferrare (1987), Nakagawa and Nosa (1987), DeDuck and Johnson (1992), Heidemann *et al.* (1993), and Li (1993).

The first major lightwave installation in the United States was the Northeast Corridor system linking Washington and New York in 1983 and New York and Boston in 1984, a total length of 747 miles. It employed a technology that had a still higher capacity, the FT3C system, which carried 90 Mb/s per fiber or the equivalent of 1344 digital voice channels. This system confirmed (again) the higher quality inherent in digital technology, but it brought out another point: an excellent technical and economic synergy between the digital lightwave system and the digital telephone switches along the route. No analog-to-digital conversion was needed to interconnect the fiber system to the 23 existing digital electronic switching systems along the Northeast route, resulting in considerable cost savings.

8.1 Terrestrial Fiber Installations

The Northeast Corridor was soon followed by more large-scale fiber installations using rapidly advancing technology. An example is the 1993 status of the AT&T fiber network, which is shown in Fig. 38. Note the dense web of connections in the large metropolitan areas and the existence of several links between the east coast and the west coast.

In addition to this, there have been a considerable number of other major fiber installations in the world, a summary of which is given in Table 6. This shows that more than 50 million km of fiber were installed in the world at the end of 1993.

Table 5. Generations of undersea lightwave technologies.

Tech. generation	First system	Year	Wavelength	Fiber type	Bit rate/fiber	Voice ckts/cable	Voice ckts with TASI
1	TAT-8	1988	1.3 μm	SM	280 Mb/s	8 000	40 000
2	TAT-9	1991	1.55 μm	SM	560 Mb/s	16 000	80 000
3	TAT-12	1996	1.55 μm	DSF	5 Gb/s	123 000	614 000

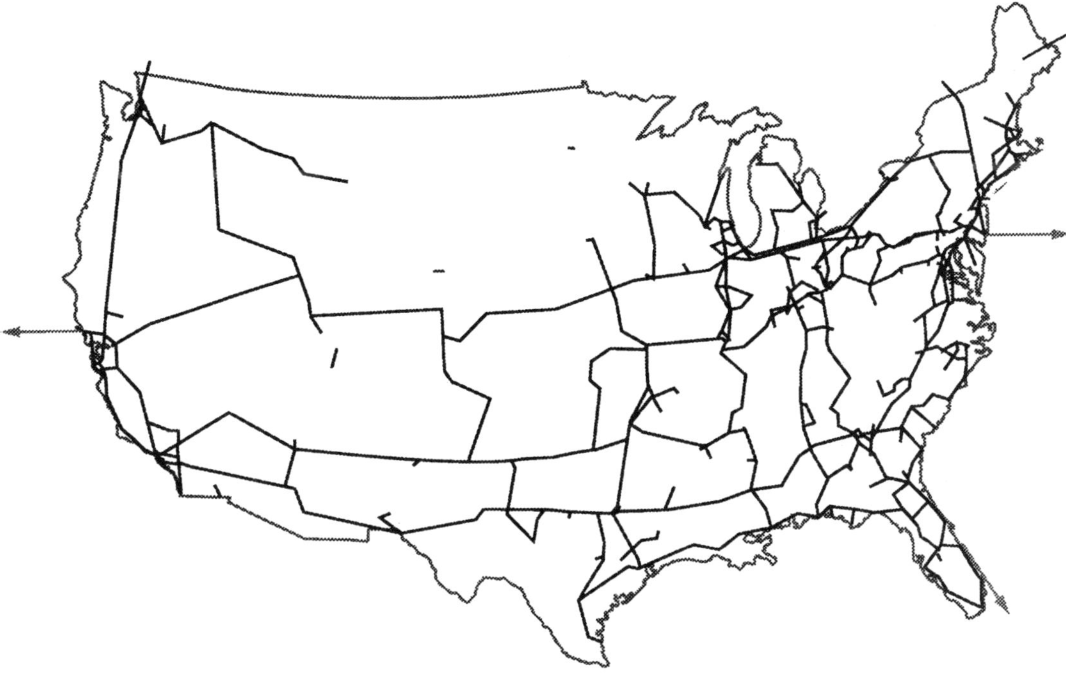

FIG. 38. The status of the AT&T fiber network in the United States at the end of 1993.

8.2 Terrestrial Lightwave Technologies

As a result of intense worldwide research and development (R&D) there have been rapid advances in the transmission capacity and the regenerator spacings of optical fiber systems. We distinguish several generations of terrestrial technology (which are somewhat different from the generations of undersea systems).

The first generation is represented by the FT3 system and the FT3C system of the Northeast Corridor. It operates with multimode (MM) fibers at 0.82-μm wavelength. This technology employs silicon detectors and gallium-arsenide–based semiconductor lasers. Typical transmission capacities are 45 and 90 Mb/s with repeater spacings of about 7 km.

The second-generation technology is represented by the FT Series G and the FT-2000 systems. It utilizes the higher-capacity single-mode (SM) fibers, and operates at wavelengths near 1.3 μm where the fiber loss is lower. Single-mode lasers based on InP-based substrates are used for this technology together with InP-based avalanche photodetectors. Typical transmission capacities are 417 Mb/s, 1.7 Gb/s, and 2.5 Gb/s, the last corresponding to over 30 000 one-way voice channels per fiber. Repeater spacings are 40–50 km.

The third-generation systems employ the wavelength band near 1.55 μm where fused-silica fiber losses are lowest. To combat fiber dispersion at this wavelength one has to control the laser spectrum. This is done by using single-frequency lasers based on InP, such as the distributed-feedback (DFB) laser. These systems typically run at similar bit rates as the 1.3-μm systems, e.g., at 2.5 Gb/s, but they can be used together with the 1.3-μm system in a wavelength-division multiplexing (WDM) configuration. This doubles the transmission capacity per fiber, e.g., to 5 Gb/s, which corresponds to over 64 000 digital one-way voice

Table 6. Worldwide fiber deployment in terrestrial systems as of January 1994 (source: KMI Corp., Newport, RI, U.S.A.).

Area	Total length of fiber (10^6 km)
North America	24.1
Europe	14.2
Pacific Rim	14.8
Rest of world	1.6
World total	54.7

Table 7. Generations of Terrestrial Lightwave Technologies (AT&T example).

Tech. Generation	System	Year	Fiber	Wavelength	Bit rate/fiber	Voice channels per fiber
1	FT3	1980	MM	0.82 μm	45 Mb/s	672
	FT3C	1983	MM	0.82 μm	90 Mb/s	1 344
2	FTG-417	1985	SM	1.3 μm	417 Mb/s	6 048
	FTG-1.7	1987	SM	1.3 μm	1.7 Gb/s	24 192
	FT-2000	1992	SM	1.3 μm	2.5 Gb/s	32 256
3	FTG-1.7 WDM	1989	SM	1.3 +1.55 μm	3.4 Gb/s	48 384
	FT-2000 WDM	—	SM	1.3 +1.55 μm	5 Gb/s	64 512

channels per fiber. Examples for this technology are the WDM upgrade systems of FT Series G and FT-2000.

Table 7 summarizes these terrestrial technology generations using the example of AT&T systems. It also lists the year of introduction and the number of one-way voice channels per fiber carried by those systems. Note that, in contrast to undersea systems, terrestrial systems use considerably more fiber per cable. Depending on the installation, from about 20 to over 100 fibers are contained in each cable.

The fourth generation of lightwave systems is currently in the R&D stage. Several new elements are under exploration. Foremost is the use of optical amplifiers (see Sec. 2). This will allow longer spaces between electronic regenerators. It will also favor the use of dense WDM techniques, employing, e.g., four or eight different wavelengths to multiply the fiber capacity four- or eightfold. Other options under consideration are the use of dispersion-shifted fiber in combination with data rates of 10-Gb/s speed. In the distant future we will, most likely, see a combination of WDM and high-speed techniques used to increase transmission capacity further (see Fig. 26).

9. FIBER IN THE LOOP

Fiber in the loop, or FITL, is a term used for fiber applications in the access portion of the telephone network, i.e., between the central telephone office and the home or customer. Since the early 1980s, fiber systems, called subscriber loop carrier or SLC systems, have begun to serve the feeder portion of this access network (Bohn *et al.*, 1984). This reaches from the central office to a distribution point from which a given geographical area of customers is served. The lightwave technology used for this application is similar to the second-generation long-haul technology, and based on single-mode fibers operated at 1.3 μm. Bit rates are typically 90 Mb/s, and span lengths range over 10 km. These systems are now in extensive use.

The distribution technology from the feeder to the customer is based on copper wires at present, but considerable R&D efforts worldwide are attempting to provide a broadband fiber distribution technology as a replacement and upgrade. A statistic for the distribution distances in the United States is shown in Fig. 39. It indicates that a range of 5–10 km is of interest for this application.

The 1980s have seen a large number of field trials and explorations of various options to bring the fiber to the home (FTTH) or at least to the curb (FTTC). However, since 1990, the press has reported plans for major FITL installations in East Germany (1.2 million homes), New Jersey (50 000 homes), and Japan (a comprehensive installation lasting

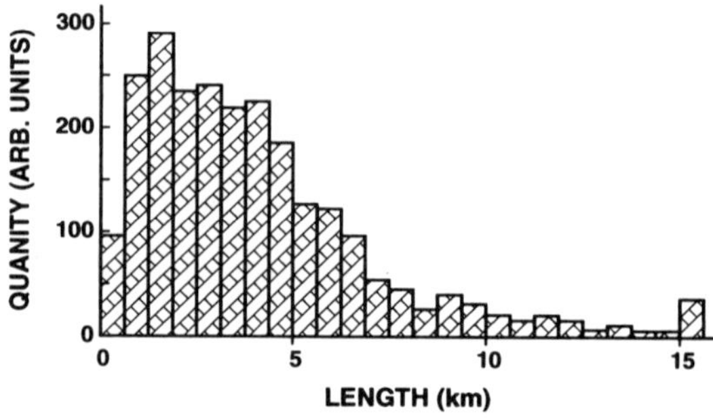

FIG. 39. Distribution of the lengths of loop transmission links in the United States. From the 1983 Loop Survey (Lemberg, 1993).

about 20 yr). This would indicate that FITL technology has already become cost effective in some situations. For overviews of this technology, see, for example, Faulkner *et al.* (1989), Wagner and Lemberg (1989), Bohn *et al.* (1992), Kashima (1993), and Smith *et al.* (1993).

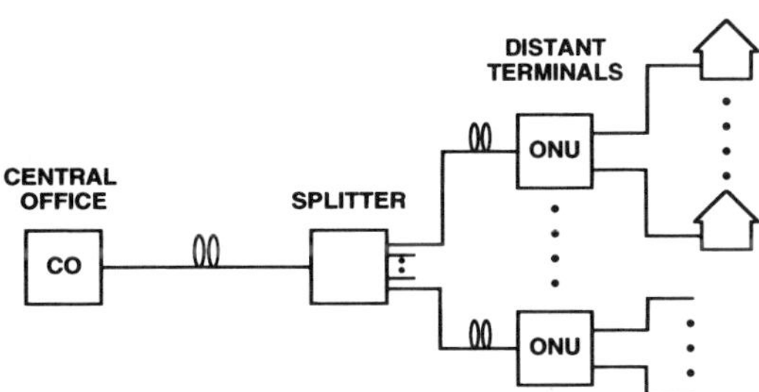

FIG. 40. Sketch of a passive optical network (PON). There is no outside plant electronics between the central office (CO) and the optical network unit (ONU). The ONU terminates the optical fiber path and serves about 1 to 16 homes.

9.1 FITL Issues and Challenges

FTTH or FTTC face both a cost and a services challenge. Many potential users require that FITL must be cost-effective for the delivery of traditional telephone service, i.e., they require cost parity with copper. However, they also demand that FITL architectures must also support later upgrades for future broadband services.

The need for new broadband services and their technical feasibility is still under active evaluation. The list of proposed services includes video distribution (CATV), two-way video, multimedia services, high-resolution images, video on demand, high-resolution video (HDTV), digital audio, high-volume data, and various interactive services.

Note also that there have been considerable advances in the technology of digital video compression. This is projected to result in a reduction of the bit rate required for digital video transmission by about a factor of 30. This promises a considerable simplification and cost reduction of video-related FITL technology.

Another FITL issue is related to the fact that optical fiber *per se* cannot carry electrical power to the home or the distant terminal. The FITL network will either have to provide power separately, or use local power with adequate provision for emergencies. Yet another important issue is the cost of maintenance over the life cycle of the system.

9.2 FITL Architectures

The search for cost-effective FITL systems is exploring a considerable number of architectural options. First, there is the option between the active star system and passive optical networks (PONs). PONs avoid the need for outside plant electronics and power supplies, but they appear to be more difficult for broadband upgrades. Figure 40 shows a sketch of a PON network. Note the passive optical splitter, which distributes the optical signals to a number of optical network units (ONUs). In an active star architecture, this splitter is replaced by active optoelectronic circuitry.

Two splitting ratios influence the technical feasibility and cost effectiveness of FITL systems: the number of ONUs per splitter, and the number of homes per ONU. Large splitting ratios are technically more difficult, but they allow cost sharing of costly components. For FTTH the home/ONU ratio is one, while for FTTC systems the typical home/ONU ratios under consideration are 4, 8, and 16.

The options for the links between the ONU and the home include copper wires, coaxial cable, or wireless transmission. Systems under consideration also differ in the number of fibers from the CO to each ONU. Our sketch shows one fiber/ONU that transmits information in both directions. Other architectures use two fibers/ONU.

In PON systems, each ONU receives the same signal. This is a time-division multiplexed (TDM) composite of all signals to be transmitted. It has a typical bit rate of 20 Mb/s. The electronics of the ONU selects only those signals that it requires. Special care has to be taken to avoid collisions between the upstream signals from the ONU to the central office. For TDM, this requires careful timing of all signal components.

Options for broadband upgrades include the use of high speed TDM, WDM, and subcarrier techniques, or combinations of the above. WDM proposals can be distinguished between those that provide a different wavelength for each service and those that provide a different wavelength for each customer.

Many of the above options require the development of a low-cost WDM technology. For optimum compactness, this demands integration of components on semiconductor chips such as the technology of photonic integrated circuits, which is subject of worldwide R&D.

10. CATV LIGHTWAVE SYSTEMS

The acronym CATV is derived from Community Antenna Television; however, today it is commonly understood to mean Cable Television. The application of optical-fiber technology to CATV transmission has had a relatively late start. However, following the first U.S. field experiments in 1988, the acceptance of CATV lightwave systems has grown rapidly, and more than 2×10^5 km of fiber have been installed by the end of 1992. The market for this application is widespread, as there are now more than 100 million TV sets in the United States alone, and more than 55% of them are reached by cable television. Lightwave technology offers an improvement of the video quality of current services as well as extended bandwidth for the introduction of new services.

The old CATV technology uses coaxial cable to transmit the signals electrically from the head-end office to the users. In these systems, the signal is repeatedly boosted by long cascades of rf amplifiers (with 30, 40, or even 50 amplifiers). These add noise to the signal (and thereby reduce image quality) and also reduce the bandwidth of the transmission system. The upper limit for the bandwidth of coaxial systems is about 500 MHz. The use of optical fiber extends the system bandwidth to 1 GHz and reduces the addition of noise.

A review of CATV lightwave technology is provided by the articles of Olshansky *et al.* (1988), Chiddix *et al.* (1990), McGrath (1992), and Darcie (1993). Figure 41 shows a typical architecture for CATV lightwave transmission. Fiber is used to link the head-end to a distribution node, a distance which is typically 10 km. At the node, the optical signal is converted to an electrical signal. This is distributed to between 500 and 2000 users via coaxial cable over distances of about 1 km. The number of cascaded rf amplifiers in this architecture is limited to a maximum of three, which enhances both video quality and bandwidth. Like fully coaxial distribution systems, this is a one-way broadcast architecture. However, possibilities for upstream transmission to the head-end are under exploration.

10.1 AM-VSB Transmission

While other lightwave applications use the digital transmission format, the first generation of optical CATV technology needs to use analog modulation (AM). Its purpose is to avoid any modification of current TV receivers.

The spectrum of analog (NTSC) CATV signals is 6 MHz wide. Each analog signal is modulated onto an rf carrier, one for each channel. The typical channel spacing is 6 MHz in the United States, and 8 MHz in Europe.

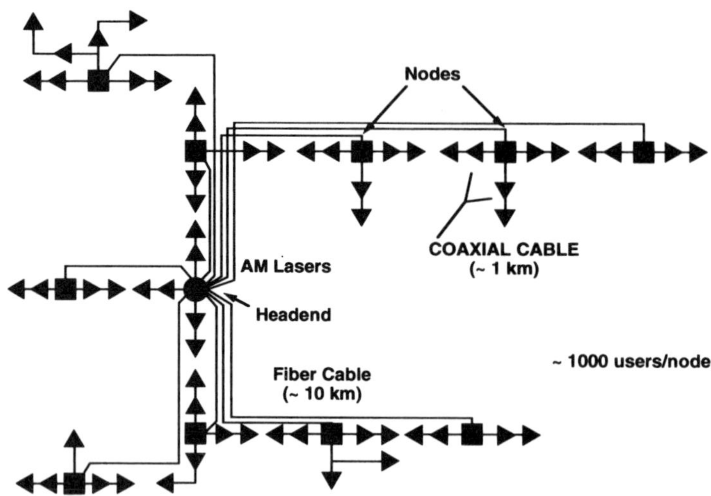

FIG. 41. Sketch of a CATV fiber-optic backbone trunk. The triangles designate rf amplifiers and the squares are the local nodes terminating the fiber.

Filters are used to clip almost all of one sideband of each channel to limit the bandwidth for transmission. What is left is called a vestigial sideband or VSB signal. The composite of these modulated rf carriers (or subcarriers) is amplitude modulated onto the optical carrier for transmission through the fiber. A high degree of linearity is required to preserve signal quality.

The sketch of Fig. 42 indicates the AM of the output light P of an ideal linear semiconductor laser in response to the modulating drive current I. For an average light output P_0 one writes for the case of one carrier of frequency Ω

$$P = P_0(1 + m\cos\Omega t). \tag{89}$$

This introduces an important parameter called the modulation index m. Practical values of m are 3%–4% for each TV channel.

High-quality AM-VSB video transmission requires carrier-to-noise ratios (CNRs) of at least 50 dB. This places critical limits on all sources of additive noise and distortions. An important fundamental limit is due to shot noise, which has its origin in the corpuscular nature of light. An average optical power P_R incident upon the photoreceiver generates a photocurrent $I_R = \eta(e/h\nu)P_R$ (see Sec. 4). The shot noise created in this event is described by the variance of the noise current,

$$\langle i^2\rangle_{\mathrm{SHOT}} = 2eBI_R, \tag{90}$$

where e is the electron charge. The shot noise

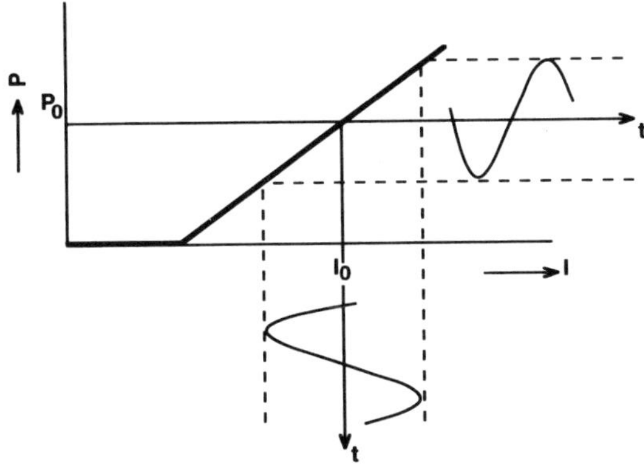

FIG. 42. Amplitude modulation (AM) of the output of an ideal semiconductor laser. The drive current is I.

is proportional to the bandwidth B of the received TV channel. The mean square current of the received carrier signal is $\frac{1}{2}m^2I_R$, which yields a shot-noise limit for the CNR of

$$\mathrm{CNR}_{\mathrm{SHOT}} = \frac{\frac{1}{2}m^2I_R^2}{\langle i^2\rangle_{\mathrm{SHOT}}} = \frac{m^2I_R}{4eB} = \frac{\eta m^2P_R}{4h\nu B}. \tag{91}$$

For typical values of $m = 3\%$, $B = 6$ MHz, $\lambda = 1.3$ μm, and CNR $= 10^5$ (50 dB), one gets a shot-noise limit of about 1 mW. Compared to digital systems, this is a surprisingly high limit for the required received optical power.

Other important sources of noise are the noise $\langle i^2\rangle_{\mathrm{AMP}}$ of the receiver amplifier (front end noise; see Sec. 4) and the relative intensity noise (RIN) which is proportional to I_R^2,

$$\langle i^2\rangle_{\mathrm{RIN}} = \mathrm{RIN}\cdot I_R^2B. \tag{92}$$

There are several sources of RIN: noise from the laser source, from reflections along the transmission path, and from Rayleigh backscattering in the fiber. RIN is usually quoted in dB/Hz. CATV systems require RIN of −150 dB/Hz or better.

The overall CNR of the system,

$$\mathrm{CNR} = \frac{m^2I_R^2}{2\langle i^2\rangle_{\mathrm{NOISE}}}, \tag{93}$$

is determined by the sum of all noise contributions:

$$\langle i^2\rangle_{\mathrm{NOISE}} = \langle i^2\rangle_{\mathrm{AMP}} + \langle i^2\rangle_{\mathrm{SHOT}} + \langle i^2\rangle_{\mathrm{RIN}}. \tag{94}$$

Other sources of interference that have to be kept to a minimum include various nonlinearities and clipping. A high degree of linearity is required from the modulation technique at the source, where either directly modulated lasers or external modulators are used. Another source of nonlinear distortion is the dispersion of the fiber.

Clipping presents a fundamental limit rooted in the fact that optical power can only be positive. For a single carrier, this means that m cannot exceed 100% (refer to Fig. 42). When there is a large number N of channels (typically $N = 60$), each modulated with an index m (typically 3%–4%), the peak modulation at any point in time can exceed 100%, which results in clipping. This leads to interference described by a carrier-to-interfer-

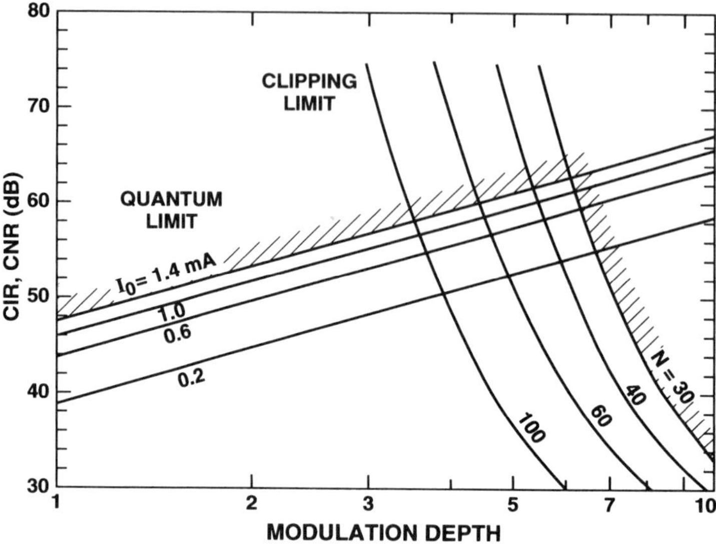

FIG. 43. Noise and nonlinearity limits of a typical CATV lightwave system as a function of the modulation index m (Darcie, 1992). The shading emphasizes the carrier-to-noise limits of a 30-channel system with an average received current of I_0 = 1.4 mA.

ence ratio (CIR) given by

$$\mathrm{CIR} = \sqrt{2\pi}\,\frac{(1 + 6\mu^2)}{\mu^3}\exp\left(\frac{1}{2\mu^2}\right), \tag{95}$$

where

$$\mu = m\sqrt{N/2}, \tag{96}$$

is the rms modulation index. This imposes a limit, the so-called Saleh limit, on the number of channels and on m (Saleh, 1989). This limit is shown in the sketch of Fig. 43 together with the quantum limit from shot noise for the case of a typical system.

10.2 Compressed Digital Video and QAM

A future generation of lightwave CATV transmission is likely to use digital techniques, which are less sensitive to noise and distortion. Digital coding is known to consume more bandwidth. NTSC-quality video, for example, requires currently a bit rate of about 100 Mb/s. Compressed digital video (CDV) techniques are thought to be able to reduce the required bit rate to less than 5 Mb/s without the loss of image quality. Relatively complex compression algorithms need to be implemented on inexpensive integrated circuit chips to make CDV feasible.

An additional level of bandwidth compression is offered by rf techniques such as quadrature-amplitude modulation (QAM) (see, e.g., Shi, 1993). This employs multilevel modulation formats. 64-QAM, for example, uses eight amplitude levels each for the in-phase and out-of-phase quadrature components. It is thought to require only 1 Hz of bandwidth for 5 b/s of information. With increasing number of levels, QAM becomes more sensitive to noise and distortion than standard digital systems. 64-QAM requires a CNR of about 30 dB.

Note that the combination of CDV and QAM may offer as many as 500 quality TV channels in the band from 500 to 1000 MHz opened up by optical fibers.

In closing, we should remark on a fascinating trend in the business and technology of telephony and CATV. These two have been very distinct in the past. However, there are now strong signs that both the business aspects and the related FITL and CATV lightwave technologies discussed here will tend to get much closer to each other.

ACKNOWLEDGMENT

It is a pleasure to acknowledge with thanks the commentary and contributions to this manuscript made by Andy Chraplyvy, Randy Giles, Marty Pollack, and John Zyskind. They have resulted in valuable improvements.

GLOSSARY

Bit-Error Rate (BER): Measure of the performance of a digital transmission system. Count of error bits as a fraction of total bits transmitted. Usually expressed in powers of 10, e.g., BER = 10^{-9}.

Bit-Rate × Distance Product (*BL*): Figure of merit for the performance of a lightwave transmission system, defined for one link from transmitter to receiver.

CATV: Cable television (originally "community antenna television").

Digital Voice Channel: Stream of signal pulses with a bit rate of 64 Kb/s carrying pulse-code modulated speech signals.

Dispersion-Shifted Fiber (DSF): Fiber design with a group dispersion zero near a wavelength of 1.55 μm.

Erbium-Doped Fiber Amplifier (EDFA): Optical amplifier employing as the gain element a fiber having a core doped with erbium (Er^{3+}).

Fiber Dispersion: Also called chromatic dispersion (D) of a fiber, or its group dispersion. Measure of the difference of pulse delay in the fiber due to a difference in the optical frequency (in ps/nm · km).

Fiber in the Loop (FITL): Term used for fiber applications in the telephone network between the local central office and residences or businesses.

Laser Chirping: Change of the output frequency during a pulse emitted by a laser. Leads to a broadening of the output spectrum.

Passive Optical Networks (PON): Fiber transmission networks with no active electronics or optoelectronics between the central distribution point and the terminals.

Receiver Sensitivity: Minimum optical power of a digital signal required at the receiver for error-free detection of the signal.

Single-Mode Fiber (SMF): Fiber with a refractive index profile and cross-sectional dimensions designed such that only one confined mode of propagation is allowed (at a specified wavelength). Usually, two polarizations exist for this guided mode.

Single-Frequency Laser: Semiconductor diode laser emitting a controlled output spectrum containing only one longitudinal mode of the laser cavity.

Wavelength-Division Multiplexing (WDM): A technique to enhance the information carrying capacity of a fiber. Several information channels are transmitted simultaneously, each carried by light of a different wavelength.

Works Cited

Adams, M. J. (1981), *An Introduction to Optical Waveguides*, New York: Wiley.

Agrawal, G. P. (1992), *Fiber-Optic Communication Systems*, New York: Wiley.

Agrawal, G. P., Dutta, N. K. (1986), *Long-Wavelength Semiconductor Lasers*, New York: Van Nostrand Reinhold.

Bohn, P. P., Brackett, C. A., Buckler, M. J., Rao, T. N., Saul R. H. (1984), *AT&T Bell Laboratories Tech. J.* **63**, 2389–2416.

Bohn, P. P., Kania, M. J., Nemchick, J. M., Purkey, R. C. (1992), *AT&T Bell Laboratories Tech. J.* **71**, 31–43.

Casey, H. C., Jr., Panish, M. B. (1978), *Heterostructure Lasers, Parts A and B*, New York: Academic Press.

Chiddix, J. A., Laor, H., Pangrac, D. M., Williamson, L. D., Wolfe, R. W. (1990), *IEEE J. Selected Areas Commun.* **8**, 1240.

Chraplyvy, A. R. (1990), *J. Lightwave Technol.* **8**, 1548–1557.

Chraplyvy, A. R., Tkach, R. W. (1993), *IEEE Photonics Tech. Lett.* **5**, 666–668.

Csencsits, R., Lemaire, P. J., Reed, W. A., Shenk, D. S., Walker, K. L. (1984), in: *Conference on Optical Fiber Communications. Digest of Technical Papers*, Washington, DC: Optical Society of America, p. 54.

Darcie, T. E. (1993), *Electrical Engineering Handbook 65*, Boca Raton, FL: CRC Press, pp. 1417–1427.

Darcie, T. E. (1992), *Opt. Photonics News* **3**, 18–23.

DeDuck, P. F., Johnson, S. R. (1992), *AT&T Bell Laboratories Tech. J.* **71**, 14–22.

Desurvire, E., Giles, C. R., Simpson, J. R., Zyskind, J. L. (1989), *Opt. Lett.* **14**, 1266–1268.

Desurvire, E., Simpson, J. R., Becker, P. C. (1987), *Opt. Lett.* **12**, 888–890.

Elrefaie, A. F., Wagner, R. E., Athas, D. A., Daut, D. G. (1988), *J. Lightwave Technol.* **6**, 704–709.

Faulkner, D. W., Payne, D. B., Stern, J. R., Ballance, J. W. (1989), *J. Lightwave Technol.* **7**, 1741–1750.

Forrest, S. R. (1988), in: S. E. Miller, I. P. Kaminow (Eds.), *Optical Fiber Telecommunications II*, New York: Academic Press, 569–599.

Giles, C. R., Desurvire, E. (1991), *J. Lightwave Technol.* **9**, 271–283.

Gordon, J. P., Mollenauer, L. F. (1991), *J. Lightwave Technol.* **9**, 170–173.

Hasegawa, A., Tappert, F. D. (1973), *Appl. Phys. Lett.* **23**, 142–144.

Heidemann, R., Wedding, B., Veith, G. (1993), *Proc. IEEE* **81**, 1558–1567.

Henry, C. (1982), *IEEE J. Quantum Electron* **18**, 259–264.

Henry, P. S., Linke, R. A., Gnauck, A. H. (1988), in: S. E. Miller, I. P. Kaminow (Eds.), *Optical Fiber Communications*, New York: Academic Press, p. 781.

Izawa, T., Subdo, S. (1987), *Optical Fibers, Materials and Fabrication*, Boston: Reidel.

Jacobs, I., Stauffer, J. R. (1980), *Proc. IEEE* **68**, 1286–1290.

Kaminow, I. P., Tucker, R. S. (1988), in: T. Tamir (Ed.), *Guided-Wave Optoelectronics*, Berlin: Springer, 211–315.

Kapany, N. S. (1967), *Fiber Optics: Principles and Applications*, New York: Academic Press.

Kashima, N. (1993), *Optical Transmission for the Subscriber Loop*, Boston: Artech House.

Kasper, B. L. (1988), in: S. E. Miller, I. P. Kaminow (Eds.), *Optical Fiber Telecommunications II*, New York: Academic Press, pp. 689–722.

Koch, T. L., Bowers, J. E. (1984), *Electron. Lett.* **20**, 1038–1040.

Koch, T. L., Linke, R. A. (1986), *Appl. Phys. Lett.* **48**, 613–615.

Kogelnik, H. (1985), *Science* **228**, 1043–1048.

Kressel, H. (Ed.) (1982), *Semiconductor Devices for Optical Communications, Topics in Applied Physics*, Berlin: Springer.

Kressel, H., Butler, J. K. (1977), *Semiconductor Lasers and Heterojunction LEDS*, New York: Academic Press.

Laming, R. I., Gnauck, A. H., Giles, R. C., Zervas, M. N., Payne, D. N. (1992), *IEEE Photonics Tech. Lett.* **4**, 1348–1350.

Lemberg, H. L. (1993), *Passive Optical Networks, Shortcourse Notes*, Optical Fiber Conference, San Jose, CA.

Li, T. (Ed.) (1985), *Optical Fiber Communications*, Vol. 1, New York: Academic Press.

Li, T. (1993), *Proc. IEEE* **81**, 1568–1579.

Marcuse, D. (1991a), *Theory of Dielectric Optical Waveguides*, 2nd ed., New York: Academic Press.

Marcuse, D. (1991b), *J. Lightwave Technol.* **9**, 121–128.

McGrath, C. J. (1992), *AT&T Bell Laboratories Tech. J.* **71**, 23–30.

Mears, R. J., Reekie, L., Jauncey, I. M., Payne, D. N. (1987), *Electron. Lett.* **23**, 1026–1028.

Midwinter, J. E. (1979), *Optical Fibers for Transmission*, New York: Wiley.

Miller, S. E., Chynoweth, A. G. (Eds.) (1979), *Optical Fiber Telecommunications*, New York: Academic Press.

Miller, S. E., Kaminow, I. P. (Eds.) (1988), *Optical Fiber Communications*, New York: Academic Press.

Mollenauer, L. F., Evangelides, S. G., Haus, H. A. (1991), *J. Lightwave Technol.* **9**, 194–197.

Mollenauer, L. F., Gordon, J. P., Evangelides, S. G. (1992), *Opt. Lett.* **17**, 1575–1577.

Mollenauer, L. F., Lichtman, E., Neubelt, M. J., Harvey, G. T. (1993), *Electron. Lett.* **29**, 910–911.

Mollenauer, L. F., Stolen, R. H., Gordon, J. P. (1980), *Phys. Rev. Lett.* **45**, 1095–1098.

Nakagawa, K., Nosa, K. (1987), *IEEE J. Lightwave Technol.* **5**, 1498–1504.

Neumann, E. G. (1988), *Single-Mode Fibers*, Berlin: Springer.

Okoshi, T. (1982), *Optical Fibers*, New York: Academic Press.

Olshansky, R., Laurisera, V., Hill, P. (1988), in: *Fourteenth European Conference on Optical Communications (ECOC 88)*, Stevenage, U.K.: Institute of Electrical Engineers, pp. 143–146.

Olsson, N. A. (1989), *J. Lightwave Technol.*, 1071–1082.

Personick, S. D. (1981), *Optical Fiber Transmission Systems*, New York: Plenum Press.

Rühl, F. F., Ayre, R. W. (1993), *IEEE Photonics Tech. Lett.* **5**, 358–361.

Runge, P. K. (1992), *AT&T Bell Laboratories Tech. J.* **71**, 5–13.

Runge, P. K., Trischitta, P. R. (Eds.) (1986), *Undersea Lightwave Communications*, New York: IEEE Press.

Saleh, A. A. M. (1989), *Electron. Lett.* **25**, 776–777.

Saleh, A. A. M., Jopson, R. M., Evankow, J. D., Aspell, J. (1990), *IEEE Photonics Tech. Lett.* **2**, 714–717.

Sanferrare, R. J. (1987), *AT&T Bell Laboratories Tech. J.* **66**, 95–107.

Shi, Q. (1993), *IEEE Photonics Tech. Lett.* **5**, 1452–1455.

Siegman, A. E. (1986), *Lasers*, Mill Valley, CA: University Science Books.

Smith, P. J., Faulkner, D. W., Hill, G. R. (1993), *Proc. IEEE* **81**, 1580–1587.

Smith, R. G. (1972), *Appl. Opt.* **11**, 2489–2494.

Smith, R. G., Personick, S. D. (1982), in: H. Kressel (Ed.), *Semiconductor Devices for Optical Communication*, Berlin: Springer, pp. 89–160.

Snyder, A. W., Love, J. D. (1983), *Optical Waveguide Theory*, London: Chapman and Hall.

Stolen, R. H. (1979), in: S. E. Miller, A. G. Chynoweth (Eds.), *Optical Fiber Telecommunications*, New York: Academic Press.

Stolen, R. (1980), *Proc. IEEE* **68**, 1232–1236.

Thiennot, J., Pirio, F., Thomine, J.-B. (1993), *Proc. IEEE* **81**, 1610–1623.

Thompson, G. H. B. (1980), *Physics of Semiconductor Laser Devices*, New York: Wiley.

Tkach, R. W. (1992), "System Implications of Optical Fiber Nonlinearity," in: *Technical Digest of OSA Annual Meeting*, Paper WXI, Washington, DC: Optical Society of America, p. 116.

Tsang, W. T. (Ed.) (1985), *Lightwave Communications Technology*, Vol. 22, Parts B and C, New York: Academic Press.

Unger, H. G. (1977), *Planar Optical Waveguides and Fibers*, Oxford: Clarendon Press.

Wagner, S. S., Lemberg, H. L. (1989), *J. Lightwave Technol.* **7**, 1759–1768.

Wemple, S. H. (1979), *Appl. Opt.* **18**, 31.

Williams, G. F. (1991), in: T. Li (Ed.), *Topics in Lightwave Transmission Systems*, New York: Academic Press, pp. 79–199.

Yariv, A. (1985), *Optical Electronics*, 3rd ed., New York: Holt, Rinehart and Wilson.

Zyskind, J. L. (1991), in: M. J. Digonnet (Ed.), *Fiber Laser Sources and Amplifiers II*, SPIE Proceedings No. 1373, Bellingham, WA: SPIE, pp. 80–92.

Further Reading

Agrawal, G. P. (1989), *Nonlinear Fiber Optics*, New York: Academic Press.

Desurvire, E. (1994), *Erbium-Doped Fiber Amplifiers*, New York: Wiley.

France, P. W. (Ed.) (1991), *Optical Fiber Lasers and Amplifiers*, Boca Raton, FL: CRC Press.

Green, P. E. (1993), *Fiber Optics Networks*, Englewood Cliffs, NJ: Prentice Hall.

Hasegawa, A. (1989), *Optical Solitons in Fibers*, Berlin: Springer.

Iga, K. (1994), *Fundamentals of Laser Optics*, New York: Plenum Press.

Jeunhomme, I. B. (1990), *Single-Mode Fiber Optics*, 2nd ed., New York: Marcel Dekker.

Kaiser, P. (Ed.) (1993), *Proc. IEEE* **81** (11). Special issue on optical communication network trends.

Kogelnik, H. (Ed.) (1985), *Phys. Today* **38** (5). Special issue on optoelectronics.

Plevyak, Thomas J. (Ed.) (1994), *IEEE Commun. Mag.* **32** (2). Special issue on fiber-optic subscriber loops.

Runge, Peter K., Trischetta, Patrick R. (Eds.) (1984), *J. Lightwave Technol.* **2**, 741–1015. Joint special issue on undersea lightwave communications.

Tech. Staff of CSELT (1980), *Optical Fiber Communications*, Torino: CSELT.

OPTICAL COMPONENTS AND SYSTEMS

GERD LITFIN AND RAINER SCHUHMANN, *Spindler and Hoyer GmbH & Co., Göttingen, Germany*

INTRODUCTION

Geometrical optics is one of the classical areas of technical sciences. This article gives an overview for the practicing engineer who requires technical information on optical design philosophies. The general tendency in this paper is the isolation of principal background information with a focus on a few systems of general interest.

The influence of optical components such as lenses, mirrors, fibers, and thin films on light beams occurs from reflection, refraction, or diffraction. In all these cases, the direction of light propagation, described by rays, can be changed. Transmissive components like prisms, plane plates, and lenses consist of different types of glasses, crystals, plastics, or amorphous materials. The variation in transmission and refraction properties gives rise to a broad diversity of physical features of the optical components. The combination of single components in order to form achromats, triplets, or even more complicated optical systems enables the design of optical properties to meet the needs of the special imaging application. In addition, transmissive components can be coated with antireflection or beam-splitter coatings. Reflecting components like mirrors consist of metal or dielectric films normally evaporated on glass or crystal substrates. By use of a proper coating, the optical properties of mirrors and lenses can be optimized with respect to wavelength selectivity, transmission, or reflection.

In the following, some typical optical components are described. A short section on optical bench systems summarizes the necessary support of optics by mechanical components. Some well-known optical systems are outlined in the final part. Camera lenses, microscopes, and more sophisticated optical systems are described in their general features. A short side glance at the requirements of the optical components of interferometers considers instruments that cannot be de-

3-527-28134-7/95/$5.00 + .50

scribed by the geometrical optics approach. More details on the physical background of geometric optics as well as on special components like diffractive and integrated optics are given in the article OPTICS, LINEAR.

1. OPTICAL COMPONENTS

In the following considerations, the light will propagate from left to right. This direction means that all distances from left to right are positive, and distances from right to left negative. Optical image formation starts from an object (in object space left from the optical system). The optical system forms the image (in image space). In general, primed symbols apply to the image space.

1.1. Transmissive Imaging Components

A spherical interface between two different media has the refractive power

$$D = \frac{n' - 1}{r} = -\frac{n}{f} = \frac{n'}{f'}, \tag{1}$$

where n is the refractive index in front of the surface, f the focal length in object space, n' the refractive index behind the surface, f' the focal length in image space, and r the radius of surface curvature.

A single lens in air ($n_1 = n_2' = 1$) consisting of two surfaces of refractive powers D_1 and D_2 has the total refractive power

$$D = D_1 + D_2 - \frac{d}{n} D_1 D_2 = \frac{1}{f'}, \tag{2}$$

where n is the refractive index of the lens, $f' = -f$ is the focal length in image space, and d is the lens thickness.

A lens or a lens system as shown schematically in Fig. 1 is characterized by its focal length f', measured from the principal plane P', and in addition to this by its numerical aperture

$$NA = n \sin\sigma \quad \text{(object space)} \tag{3a}$$

$$NA = n' \sin\sigma' \quad \text{(image space)} \tag{3b}$$

or by its f number

$$k = f'/2\, r_{EP}, \tag{4}$$

where n is the refractive index in object space (normally 1), σ the aperture angle in object space, n' the refractive index in image space (normally 1), σ' the aperture angle in image space, and 2 r_{EP} the entrance pupil diameter (r_{EP} = EP radius). The entrance pupil characterizes the free diameter of a system through which light can be transferred. The heights and the positions of objects and images can be calculated by first-order geometrical optics.

Especially for a lens or a lens system in air the following formulas are relevant:

An object of height u will be imaged by a lens into the image height u' with the magnification

$$\beta' = u'/u = NA/NA'. \tag{5}$$

The positions of object and image are described either by the Newtonian formula as

$$\beta' = -z'/f' = f/z \tag{6}$$

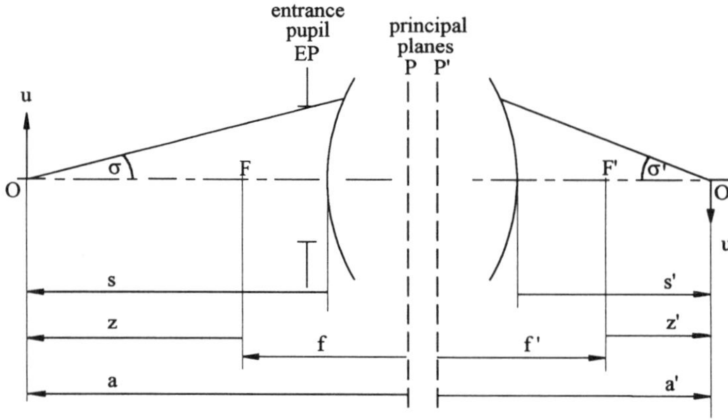

FIG. 1. First-order optic of a lens system.

or by the lens formula as

$$\frac{1}{a'} - \frac{1}{a} = \frac{1}{f'}, \tag{7}$$

where z is the object distance from focal point F, z' the image distance from focal point F', a the object distance from principal point P, and z' the image distance from principal point P'.

The exact transfer of any points of the object field into points of the image field can be calculated by ray tracing through a lens or a lens system. Therefore, rays are characterized by vectors that are determined from surface to surface by considering the refractive law (Snell's law). The calculations start from the coordinates of the object points and result in the coordinates of the corresponding points in the image field. Because of the geometry of the spherical surfaces, normally a real point is not transferred into an exact point but into a blurred image. The deviations of the image coordinates from the target values calculated by first-order optics are called aberrations.

The basic aberrations are spherical aberration (longitudinal and lateral), coma (sagittal and meridional), astigmatism, field curvature, and distortion. The aberrations depend on the aperture of the lens and the image height. Furthermore, they depend on the considered wavelengths, so that the differences of all these aberrations for different wavelengths are called chromatic aberrations.

In complex optical systems, a combination of several lenses is optimized in such a way that the aberrations that are of importance for a special imaging task are minimized. The specification of a system depends on external parameters like image size and image position, aperture size and pupil position, wavelength range, and mechanical boundaries. The quality requirements depend on the special system usage and the resolution characteristics of the image detectors [film, charge-coupled device (CCD), eye, etc.).

1.1.1 Single Lenses For solving simple imaging tasks or as components of complex optical systems, single lenses are used. The different lens types are shown in Fig. 2.

Positive refractive power of a lens is given when the edge thickness is less than the center thickness; otherwise negative power results, when the edge thickness is higher than the center thickness. Lenses with one planar surface can be plano-convex or plano-concave. Two refractive surfaces can be arranged as biconvex or biconcave (symmetrical or nonsymmetrical) or as meniscus lenses. Lenses for which the focal points of the two surfaces coincide have no refractive power.

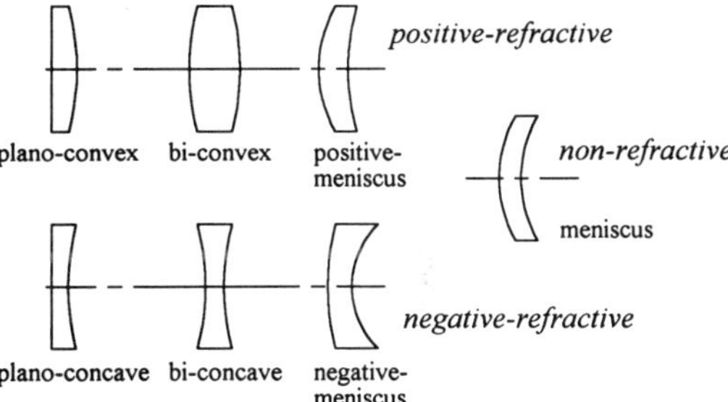

FIG. 2. Single lenses.

1.1.1.1 Best-Form Lenses. A single lens that will be used to image an axial point can be optimized for minimum spherical aberration by choosing the best-fitted radii. For this aim, the refraction angles should be minimized or, in other words, the change of the ray directions should be distributed equally onto both surfaces. For the following two special magnifications, best-form lenses can be characterized easily.

For an image magnification of 1:1, that means $\beta' = -1$, a uniform distribution of refractive power minimizes the spherical aberration. So in this case, a symmetrical biconvex or biconcave lens with two equal radii has the best form [Fig. 3(a)]. To image an axial point at infinity, so that light is incident parallel to the optical axis and the magnification is $\beta' = 0$, the optimum refractive-power distribution between the two surfaces depends on the refractive index of the glass material. For normal glasses, the surface curvature ratio is approximately 6:1, which means that the first radius is 6 times shorter than the second [Fig. 3(b)].

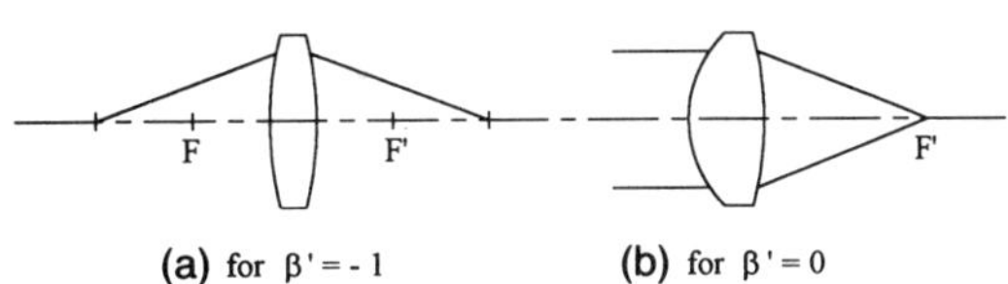

FIG. 3. Best-form lenses.

1.1.2 Doublets To use more than one lens gives additional degrees of freedom for the improvement of image quality. Two uncemented lenses have four optical surfaces, three if cemented, and two glass materials are characterized by the refractive indices n_1 and n_2 and the dispersion constants $\nu_{d,1}$ and $\nu_{d,2}$, also called Abbe numbers:

$$\nu_d = \frac{n_d - 1}{n_F - n_C}, \tag{8}$$

with the wavelengths (spectral lines) $d = 587.6$ nm, $F = 486.1$ nm, $C = 656.3$ nm.

Approximately, the total refractive power can be calculated under the supposition of thin lenses (lens thickness = 0) with Eq. (2), where D_1 and D_2 are the refractive powers of the lenses and d is the lens distance:

$$D = D_1 + D_2 - d\,D_1D_2. \tag{9}$$

The longitudinal chromatic aberration of one lens is the difference between the image distances s' from the last surface for the two wavelengths λ_1 and λ_2. This can be calculated approximately by

$$\frac{1}{s'_{\lambda_1}} - \frac{1}{s'_{\lambda_2}} = \frac{D}{\nu}. \tag{10}$$

For two cemented thin lenses with thickness $d = 0$, the difference of the total refractive powers for two wavelengths will be

$$D_{\lambda_1} - D_{\lambda_2} = \frac{D_1}{\nu_1} + \frac{D_2}{\nu_2}. \tag{11}$$

For achromatic imaging Eq. (10) must be zero (achromatism condition). If this is done for the wavelengths corresponding to the C and F lines the residual chromatic aberration (secondary spectrum) of such an achromat for the wavelength $d = 587.6$ nm is approximately $f'/2000$.

An achromat [Fig. 4(a)] also can be optimized for a minimum spherical aberration, not only for one magnification but for a wide range. This can be done to a higher grade of image quality than for a single lens. For a noncemented achromat [Fraunhofer achromat, Fig. 4(b)] with a small air space, additional optimization of image quality is possible because one more surface radius is available. Therefore, the air space between the two lenses acts like a third (air) lens. Optical systems with more than one element are described in Sec. 3.

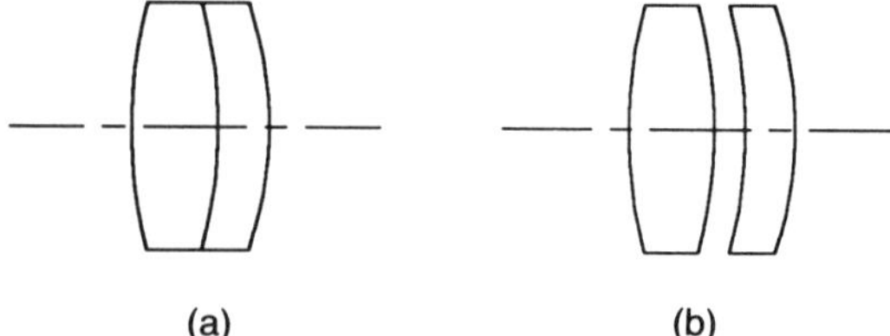

FIG. 4. Achromats: (a) cemented, (b) air-spaced (Fraunhofer type).

1.2 Transmissive Plane Optics

Optical elements with only planar surfaces have no refractive power if the refractive index is uniform over the material volume. Light can be influenced by filtering, polarizing, or beam deflection with different elements like plane-parallel plates, wedges, or prisms.

1.2.1 Filters The incident light intensity Φ_0 will be divided by an optical filter into the reflected fraction Φ_r, the absorbed fraction Φ_a, and the transmitted fraction Φ_t (scattering neglected), so that

$$\Phi_0 = \Phi_r + \Phi_a + \Phi_t, \tag{12}$$

where Φ_r consists of the contributions from both filter boundaries—normally interfaces between air and glass, plastics, or other transparent materials.

Reflection, transmission, and absorption normally depend on the wavelength. The transmission factor $\tau(\lambda)$ is the ratio of the transmitted and the incident intensities:

$$\tau(\lambda) = \Phi_t/\Phi_0. \tag{13}$$

Filters are defined by their spectral transmission characteristics as

1. edge filters (long-pass or short-pass), where wavelengths shorter or longer than a given wavelength are blocked,
2. band-pass filters, where a special small or large wavelength range is transmitted, and
3. line filters, where very small wavelength

ranges are transmitted, reflected, or absorbed.

The different spectral transmittances of these filters are shown in Fig. 5. See also OPTICAL FILTERS.

For band-pass filtering absorbing filters made of colored filter glass are in use. To reduce the intensity uniformly over a large range of wavelengths, neutral-density filters (with τ = const approximately) can be used. The spectral transmission characteristics of absorbing filters are—in contrast to interference filters—insensitive to variations of the angle of incidence.

For the selection of small bandwidths, interference filters must be used. These filters consist of highly reflective thin films separated by dielectric "spacer" layers. The reflective films can either be metallic films or dielectric multilayers.

Edge filters based on light interference mostly are also realized by dielectric multilayers. In connection with these films, absorbing glass can be used to cut off short and/or long wavelengths.

1.2.2 Prisms Plane surfaces that are not parallel to each other form wedges and prisms. Depending on the geometrical arrangement of the surfaces, these elements can be used for beam deviation by refraction only or by reflection only or by a combination of both.

Prisms will deviate light beams by refraction on two surfaces. Because of the dispersion of the glass material, i.e., the dependence of the refractive index on the wavelength, the deviation angle is also wavelength dependent as shown in Fig. 6. The refractive index n_λ, the prism angle ω, and the

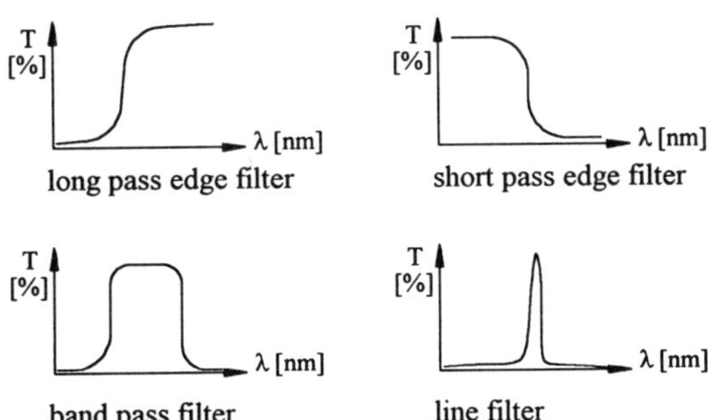

FIG. 5. Spectral filter characteristics.

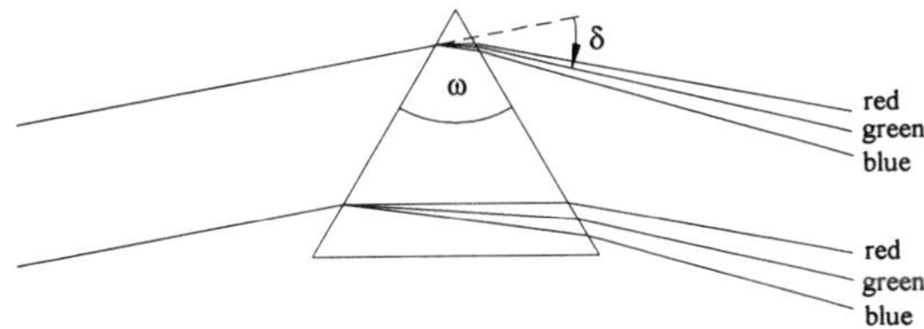

FIG. 6. Light deviation by a prism.

incident angle of the light beam determine the deviation angle δ_λ. The minimum deviation is reached by a symmetrical ray tracing through the prism (as shown in Fig. 6), so that the beam deviation of a prism in air is

$$\delta_{\lambda,\min} = (n_\lambda - 1)\omega. \tag{14}$$

A dispersion prism splits the light into its spectral components. Such prisms can be used for monochromators or for spectrographs.

Reflection prisms are optical components where internal reflections are used for changing the position or orientation of an image. This can also be done by plane surface mirrors; however, the manufacture of a stable mirror arrangement is in most cases more expensive. Especially when long path lengths are used, the disadvantage of a prism is the dispersion of the glass, which results in chromatic aberrations if the prism is used in imaging systems. Because a prism acts like a thick plane-parallel plate, astigmatism will influence the image quality, too.

The reflecting surfaces of a prism can be mirror coated or only the total internal reflection at these surfaces will be used.

Fig. 7 shows different kinds of reflection prisms and shows how the position and the orientation of an image is changed.

For the purpose of splitting beams, two 90° prisms can be combined by cementing where one of the interface surfaces is coated with a partially transmitting mirror. In such a beam-splitter cube (Fig. 8), one beam is deflected by 90° and one beam is transmitted without changing its direction. These cubes are used, for example, in interferometers (see Sec. 3.10).

1.3 Mirrors

Besides refracting elements, also reflecting surfaces can be used for changing beam directions and for imaging. High reflection is

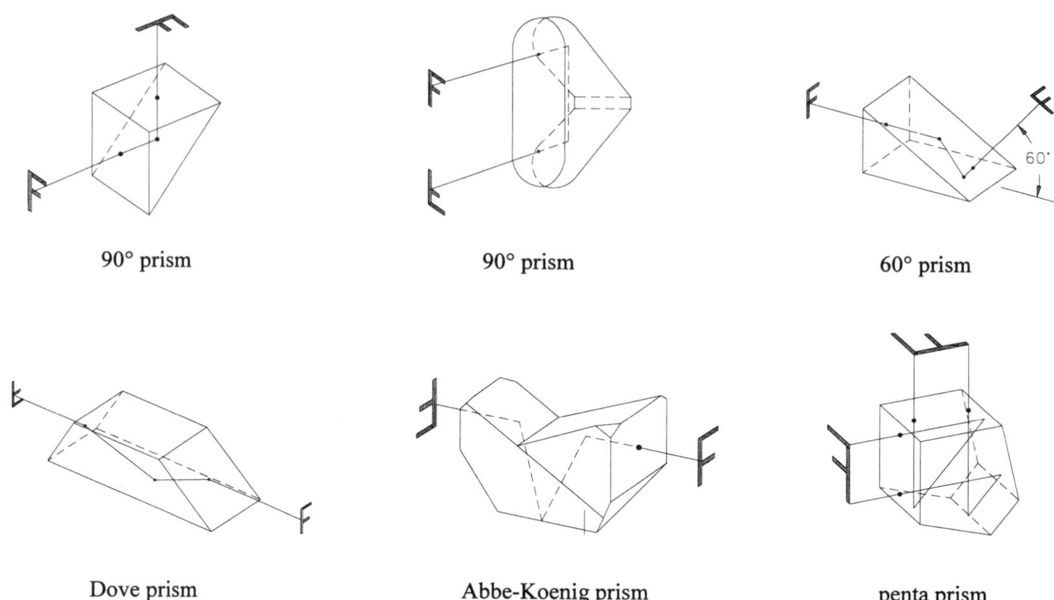

FIG. 7. Reflection prisms.

achieved by special metallic or dielectric coatings of thin films. Dielectric mirrors can be produced to have an extremely low absorption and an extremely high reflectivity. The main advantage of mirrors compared to lenses and prisms is the absence of chromatic aberrations.

The reflection at a surface (Fig. 9) is described by the reflection law

$$\epsilon' = -\epsilon, \tag{15}$$

where ϵ and ϵ' are the ray angles to the surface normal of the incident and reflected rays.

1.3.1 Plane Mirrors For ray deflection with high reflection efficiency, plane mirrors

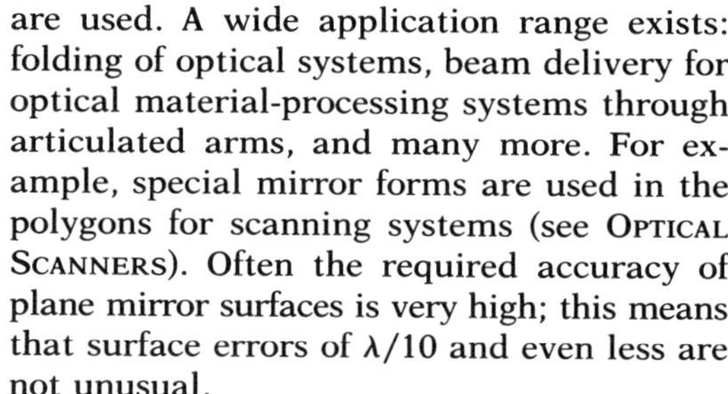
are used. A wide application range exists: folding of optical systems, beam delivery for optical material-processing systems through articulated arms, and many more. For example, special mirror forms are used in the polygons for scanning systems (see OPTICAL SCANNERS). Often the required accuracy of plane mirror surfaces is very high; this means that surface errors of $\lambda/10$ and even less are not unusual.

1.3.2 Curved Mirrors Imaging by mirrors requires curved surfaces. The advantage compared to lenses is mainly that the deflection does not depend on the wavelength, which means that complex imaging systems

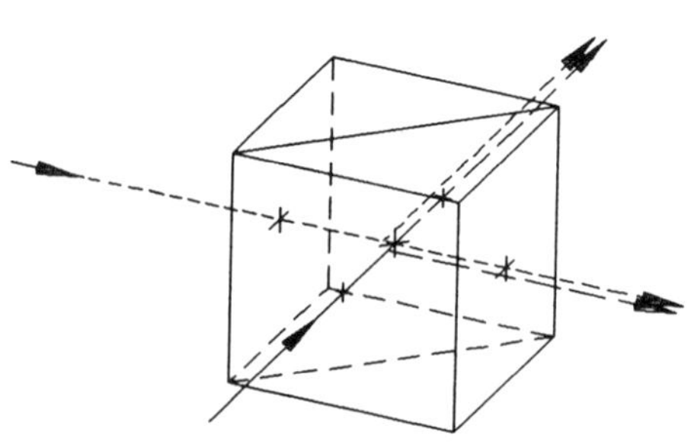

FIG. 8. Beam-splitter cube.

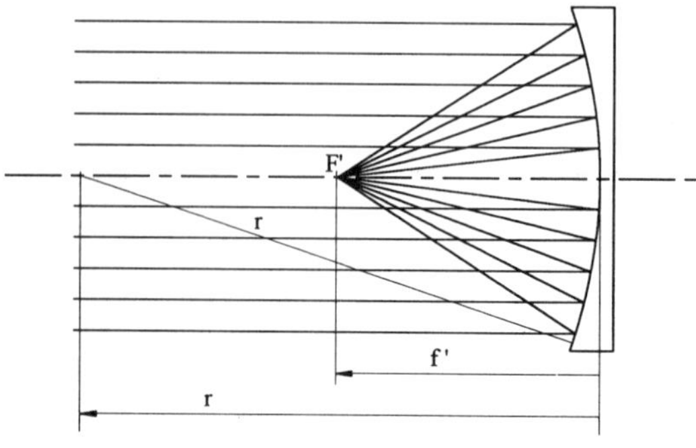

FIG. 9. Parabolic mirror.

with long focal lengths, free of chromatic aberrations, can be built.

The focal length of a mirror with a radius of curvature r is given by

$$f' = r/2. \quad (16)$$

The object and image positions s and s' from the mirror surface are described by

$$\frac{1}{s'} + \frac{1}{s} = \frac{1}{f'}. \quad (17)$$

For spherical mirrors, the aperture and the image field are limited by spherical aberration. For high-aperture systems, the spherical aberration can be corrected by giving the mirror an aspherical shape. Especially a parabolic surface shape of a mirror eliminates its spherical aberration completely (see Fig. 9). Such mirrors are mostly used in astronomical systems, where long focal lengths with small fields of view and high apertures are required.

1.4 Crystal Optical Components

There exist many crystals whose properties differ completely from those of glass and other amorphous materials as far as the propagation of light is concerned. In glass, the atoms are arranged randomly, i.e., the orientations of their axes are uniformly distributed over all directions in space. In crystals, on the other hand, the atoms are ordered, forming a periodic lattice with certain privileged directions. This makes the lattice optically anisotropic. There is a large number of different types of lattices to be found in nature. The type of lattice determines the crystal's properties, which exhibit a more or less marked directional dependence.

In the following, the properties of crystals as they pertain to light polarization will be discussed. Because of different refractive indices for different polarization directions (= orientations of the electric field vector), the velocities of light propagation in a given direction are different for these polarization directions. This phenomenon is called birefringence (see Fig. 10).

In birefringent crystals there exist two directions of propagation for which the propagation velocity does not depend on the polarization orientation. These directions are the

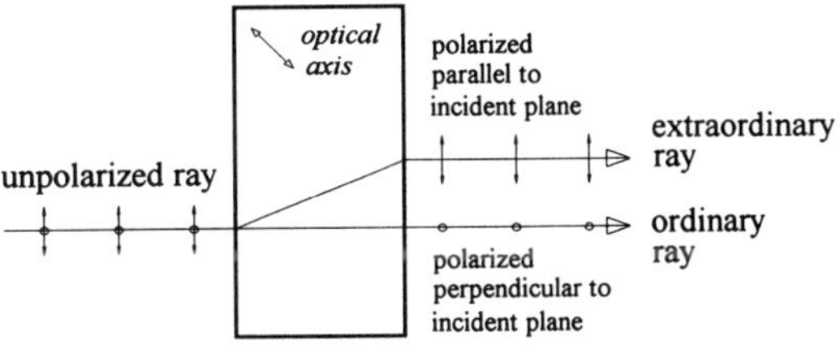

FIG. 10. Birefringence in a uniaxial crystal.

optical axes. In uniaxial crystals, like quartz or calcite, the two axes coincide. Light propagation in uniaxial crystals is described by two refractive indices: the ordinary index n_o corresponding to a polarization direction perpendicular to the direction of the optical axis, and the extraordinary index n_e corresponding to the polarization direction perpendicular to the ordinary polarization direction. The term n_e depends on the direction of propagation. Therefore, when light propagates in an arbitrary direction with arbitary polarization in a uniaxial crystal, it is split into an ordinary ray with the propagation speed c/n_o (with c light velocity in vacuum) and an extraordinary ray with the propagation speed c/n_e. Both are polarized linearly and perpendicularly to each other and, when incident on a crystal surface, are refracted through different angles because of the different indices of refraction. The refraction of the extraordinary ray does not obey Snell's law. Only rays that enter the medium parallel or perpendicular to the direction of the crystal axis will not be split.

1.4.1 Prism Polarizers Polarized light can be produced from unpolarized light with the aid of one or two prisms made of a birefringent material. In Figs 11 and 12, the orientation of the optical axis a is represented

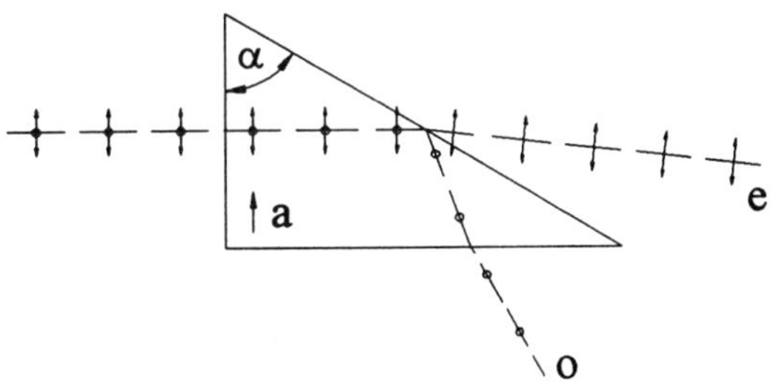

FIG. 11. Single-prism polarizer.

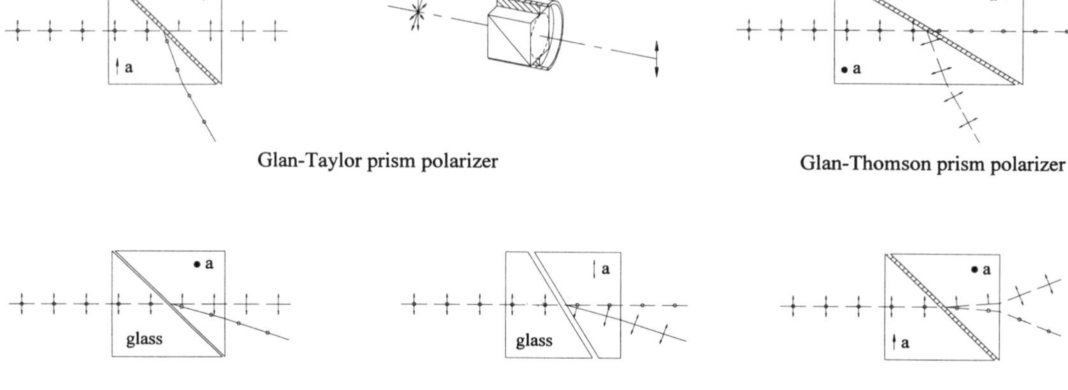

Rochon prism polarizer

Sénarmont prism polarizer

Wollaston prism polarizer

FIG. 12. Prism polarizers.

by an arrow (axis in drawing plane) or a point (axis perpendicular to drawing plane). The orientation of the polarization is represented in the same way, where the combination of arrow and point means unpolarized light. The principle of a single-prism polarizer (Glan-Taylor prism) is shown in Fig. 11. The unpolarized light beam strikes one of the entrance faces of the prism and is split in the crystal into the ordinary (o) and extraordinary (e) beams, which are perpendicularly polarized. Angle α is selected such that the ordinary beam is totally reflected on the hypotenuse and the extraordinary beam, on the other hand, is refracted and can emerge.

By different combinations of two prisms, it is possible to separate one direction of polarization without beam deflection or to produce beam separations between 0.5° and 30° typically. Fig. 12 shows different types of prism polarizers. Except for the Rochon prism, where a crystal prism is cemented on a glass prism, all other types contain two crystal prisms, where the optical axis of the two prisms are parallel or perpendicular to each other.

1.4.2 Retardation Plates When polarized light strikes a birefringent crystal, it is separated into ordinary and extraordinary beams, which are polarized perpendicular to each other. After passage through the crystal, they exhibit a phase difference that is proportional to the thickness of the crystal and the difference in the indices of refraction. A special selection of the material and the thickness of the crystal allow conversion of linearly polarized light into any polarization state desired.

The retardation can be described by the optical path difference $\Delta n \times d$ for the ordinary and extraordinary beams as

$$k\lambda = \Delta n d. \tag{18}$$

Half-wave plates ($\lambda/2$ plates, with $k = 1/2$) and quarter-wave plates ($\lambda/4$ plates, with $k = 1/4$) are the retardation plates commonly used. These plates can be produced as first-order plates with the exact thickness according to Eq. (17), or as higher order plates where the thickness is an odd multiple of $k\lambda$. The most common materials used are quartz, mica, and magnesium fluoride. Normally the crystal plates are cemented between planar glass plates.

As an example, the following polarization states will be described: A light beam incident normally onto a plane-parallel quartz plate is linearly polarized at an angle of 45° with respect to the optical axis of the crystal which lies in the plane of the crystal plate. The beam is split into the ordinary and the extraordinary components and, as a result of the 45° angle between the direction of polarization and the optical axis, both have the same intensity. As the index of refraction n_e of the extraordinary beam is greater than that of the ordinary beam n_o, the extraordinary beam propagates more slowly through the plate. This creates a phase difference and therefore an optical path difference Δnd between the two beams. If the difference is $\frac{1}{4}\lambda$, then the beam resulting from the superposition of the two beams is circularly polarized [see Fig. 13(a)]. This means that the direction of polarization is not constant but is rotating with constant length of the electric field vector. A path difference of $\lambda/2$ [see Fig. 13(b)] results in linearly polarized light beam where the direction of polarization is rotated by exactly 90° with respect to the incident beam.

1.5 Optical Fibers

An optical waveguide is a transparent material that allows conduction of light from one point to another via internal multireflection at the surfaces. The simplest type of waveguide is a dielectric cylinder surrounded by air. Light striking the walls will be totally internally reflected. The angle of incidence at each reflection has to be larger than the critical angle for total reflection. As total reflection always happens in the medium with higher refractive index, cylinders made of glass or plastic are suited as optical waveguides (see FIBER OPTICS). As long as the diameter of these rods is large compared to the wavelength, the process obeys the laws of geometrical optics. If the diameter is getting close to the wavelength of radiation, the transport of radiation is better explained in terms of wave propagation.

As the propagation of light in optical fibers depends on total internal reflection, the fiber core itself cannot be treated as optically isolated. In contact with other dielectric media, the total reflection will be disturbed so that a loss of light energy will result. Therefore, the fiber core is generally enclosed in a transparent cover of lower index called cladding.

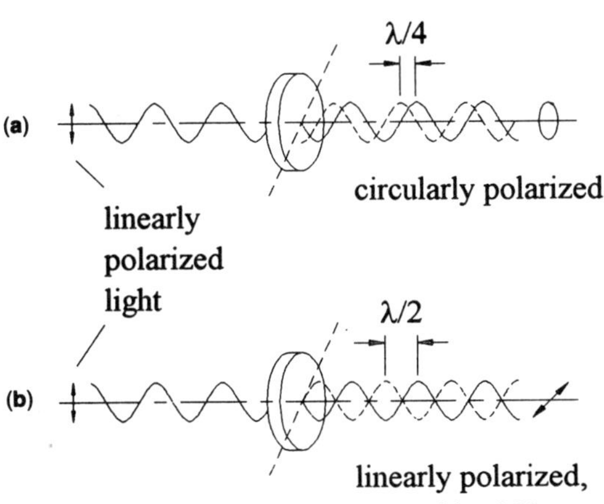

FIG. 13. Principle of a $\frac{1}{4}$-wave and a $\frac{1}{2}$-wave plate.

The principal layout of an optical fiber is shown in Fig. 14. For a fiber core with a refractive index n_f and a cladding with an index n_c, total internal reflection appears with angles of incidence of greater than the critical angle for total internal reflection θ_c. The numerical aperture NA of a fiber is defined by the maximum angle of incidence at which light is propagated through the fiber core:

$$NA = \sqrt{(n_f^2 - n_c^2)} = n_i \sin \theta_{\max}. \tag{19}$$

Thus, for air surrounding the fiber ($n = 1$), the numerical aperture cannot exceed the value of unity. $NA = 1$ means that the half angle $\theta_{\max} = 90°$ and all light entering the face of the fiber is totally internally reflected. For low-loss light transmission, it is required that core and cladding materials are as free of losses as possible.

Besides fibers where the change of the index of refraction between core and cladding is a step function (step-index), graded-index fibers are widely used. This type is characterized by its radial gradient of the index of refraction, so that core and cladding are no longer separate partitions of the fiber. The advantage of a graded-index fiber is that it has much less dispersion than a step-index multimode fiber.

For a more detailed description of the physics of optical fibers see FIBER OPTICS.

With regard to applications fibers are classified as suited for

1. information transmission (low dispersion, low loss), or
2. power transmission, or beam delivery for

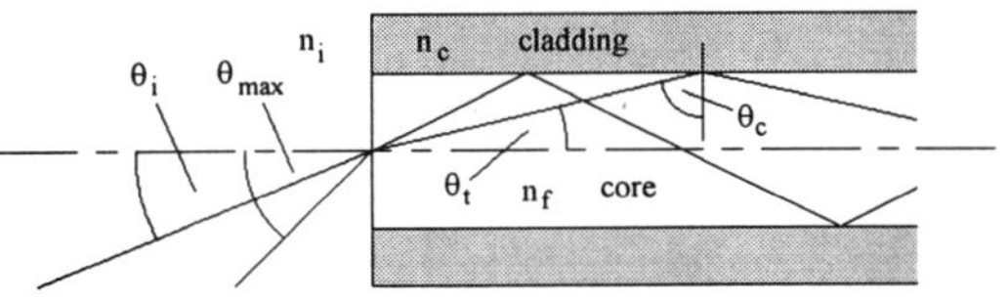

FIG. 14. Principal fiber layout.

material processing and medical therapy (high damage threshold, high aperture).

From the material point of view, fibers can be classified as follows:

1. all plastic, consisting of a plastic core with different plastic material as cladding;
2. all glass, consisting of a glass core cladded with a glass with slightly lower refractive index;
3. plastic-cladded silica fibers;
4. all plastic, consisting of a plastic core and cladding;
5. crystal fibers; or
6. fibers made from infrared transmitting material.

In correspondence to their major physical properties, the following classification is of interest: multimode fibers or single-mode fibers.

Thin waveguides and optical fibers as transmission media were suggested and developed even before the development of lasers. The once severe problem of energy loss in the fiber has been solved in the meantime. Today's single-mode fibers have a transmission of more than 90% per kilometer. For physical reasons, in multimode fibers the transmission is somewhat lower, around 70% per kilometer. The loss of modern fibers is close to that limited by Rayleigh scattering. The diameter of the fiber core ranges between a few microns and approximately 1 mm.

Most of the fibers are manufactured from glass. The fibers made of plastic still have low transmission of about 1% per kilometer.

Fibers offer a high degree of security and privacy. The tapping of a fiber is difficult. It can be detected by the losses in transmission that it introduces. Cross-talk problems as known from copper cables are excluded by design. In addition, fibers are immune to electromagnetic noise generated by radio stations, lightning, and other signals caused by electronic or electric equipment. Fibers are flexible and light weight, and they can be easily used in difficult environments. In communications, the advantages come from the high carrier frequency so that, for instance, thousands of telephone conversations can be transmitted via a single optical fiber. Because SiO_2 is fairly inexpensive and the production process is comparatively simple, one of the major advantages of optical fibers is their low cost.

1.5.1 Multimode Fibers If we consider the propagation of light in a waveguide as an electromagnetic field representing mathematically a solution of Maxwell's wave equations, we find that under certain boundary conditions a set of eigenfunctions results as solutions of the equation. Each of these modes has its correspondence to a zig-zag path in the ray-optics approach. It can be easily seen that the different modes are reflected differently in the core. Some waves follow longer paths than the others. The higher the order of a mode, the closer the angle of incidence at the core cladding interface is to the critical angle. For these modes, the travel path is significantly longer than for the lowest order mode, which travels without reflection in the center of the core. The z components of their phase velocity are different. This results in modal dispersion. For example, when a short pulse is coupled into the multimode fiber, various fiber modes will be excited, with different propagation velocities. At the end of the fiber, the shape of the short pulse is broadened.

1.5.2 Single-mode Fibers The best solution of the problem of intermodal dispersion, however, is to reduce the core diameter so that it will provide only one mode wherein the rays travel parallel to the central axis. Typically such single-mode fibers have core diameters of only a few microns (both step-index and the newer graded-index fibers). In a single-mode fiber, the light is only affected by the dispersion of the glass material used. As the light transmitted in such a fiber does not have zero bandwidth, a small dispersion results due to the dispersion. For special applications like optical communications, single-mode fibers are operated at 1.55 mm, where the material dispersion is close to zero and the attenuation is in the order of 0.2dB/km. Single-mode fibers are also available as polarization preserving fibers. This can be achieved by introducing mechanical stress in one axis, so that a small variation of the refractive index results, causing the polarization plane to be preserved. Fibers have a broad variety of applications in communications and fiber optic sensors with increasing potential.

1.5.3 Image-Transmitting Ordered Fiber Bundles An interesting application for optical fibers is in image transmission through ordered fiber bundles. Ordered fiber bundles whose ends are bound together with epoxy, ground, and polished give the opportunity to be used as flexible image carriers. This feature gives rise to a number of applications, especially in the medical area. The examination of internal cavities is done with these endoscopes. This category includes bronchoscopes, gastroscopes, and colonoscopes, which are named considering their individual application. Between 10^3 and 10^5 fibers of ~10 mm diameter and a length of between 0.5 and 2 m are combined in these instruments. A resolution of up to 100 lines per millimeter is achieved.

2. OPTICAL BENCH SYSTEMS

For the development of optical systems, prototyping of optical devices, the assembly of experimental equipment, the construction of special precision measurement and control devices, and physical optics research, optical components, light sources, detectors, and mechanical subassemblies have to be connected and mounted so that stability and adjustment requirements are fulfilled. This is achieved by the use of opto-mechanical systems. A classification of these systems in three groups gives a general overview:

1. One-dimensional systems (optical benches);
2. Two-dimensional systems (optical tables);
3. Three-dimensional systems (combinations of benches and tables, structural systems).

2.1 Optical Benches

There is a big variety of optical bench systems available. All these systems consist of the bench itself acting as a one-dimensional guide for the alignment of opto-mechanical components and the carrier mating with the structure of the bench. Benches and carriers can be classified by material and profile. Depending on the application, beginning with the light-weight aluminum profile up to cast iron profiles and solid granite benches, different physical properties are obtained. The highest surface flatness is achievable with solid granite structures. Granite can be perfectly ground to a flatness of less than 1 mm per meter length. Cast iron structures in principle can be ground with nearly the same accuracy; however, the temperature behavior and the homogeneity of the material are disadvantageous. Metal surfaces also show the disadvantage of nonreversible deformation of the surface due to forces. Aluminum profiles can be made very light weight. The stability of such benches depends on the profile used. A very well designed profile, the X-95 structure, is shown in Fig. 15.

The design of the carrier is mainly determined by the bench profile. For easy handling, it is important that the carriers can be removed from the profile and also that they slide smoothly over longer distances. This is required in order to perform z-axis alignment along the optical axes without influencing the x and y positions. On top of the carrier all kinds of opto-mechanical equipment can be used to mount the optical component.

2.2 Optical Tables

For two-dimensional setups, especially with larger extensions, optical tables are suited. Again, granite material results in the highest flatness available for optical tables. The disadvantage is the high weight of solid granite. Also, the damping characteristics for acoustical waves is not sufficient for many optical experiments. For practical reasons, stainless steel honeycomb tables have gained wide acceptance among users of optical equipment. Different combinations of skin, core, sidewalls, and bonding materials result in significant variations of the key performance

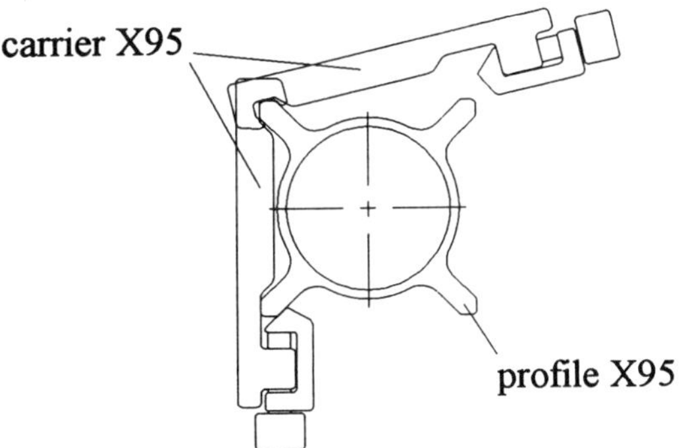

FIG. 15. Optical bench profile.

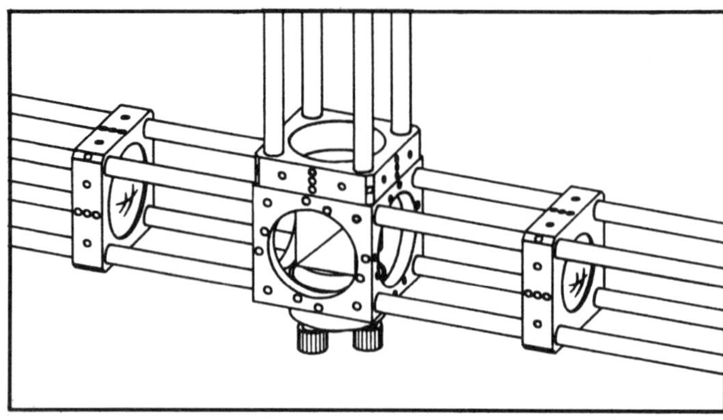

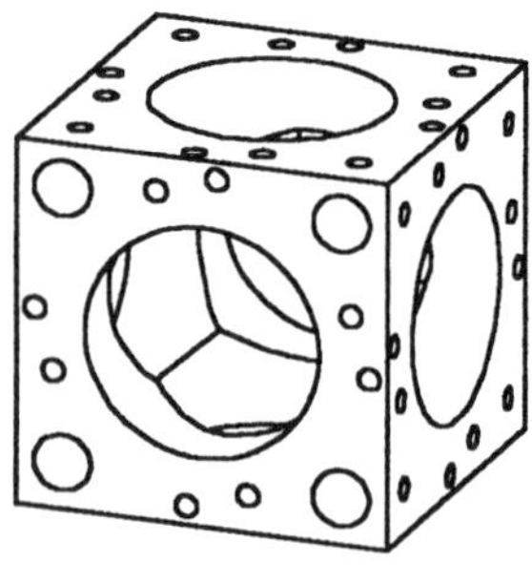

FIG. 16. Three-dimensional setup using a structural system.

properties. These are stiffness, flatness, structural damping, and durability. Structural damping is required to minimize sensitivity and response to dynamic inputs, and to dissipate quickly any vibration energy that might be transmitted directly to the table.

The mounting of optical components is done either by magnetic posts or by use of clamps. The latter require a hole pattern in the table top. Another way of fixing optical components to optical tables, which also gives a small degree of tridimensionality, is by columns that are directly fixed to the hole pattern. Optical tables are often used in research as they provide a high versatility for the experimental setup. Also, heavy instruments like lasers or experimental apparatus are carried by an optical table. In order to prevent vibrations of the building from influencing the experiment, vibration damping systems are often necessary. They are in general mounted on top of the table stand. The combination of tables and bench systems also allows for limited three-dimensional setups.

2.3 Structural Systems

The most versatile systems for optical setups are the structural systems. Structural systems can be used in one-, two-, or three-dimensional applications. They have the advantage that optical axes are defined by the system. They also can be made small enough for micro-optics and other applications that need mounting of many optical components close to each other. Because of their flexibility, such systems can be easily used for prototyping. Figure 16 shows a three-dimensional structural system based on four rods, cubes, and mounting plates.

As an example of a real three-dimensional opto-mechanical setup, Fig. 17 shows a test device for quality control of prisms, which is realized with a Microbench system. These structural systems have the advantage of high mechanical stability. Therefore, they are even suited for interferometric applications.

3. SELECTIONS OF OPTICAL SYSTEMS

In the following, a special selection of the most well known optical systems is described.

3.1 Objectives

For the imaging of finite-sized object fields, normally optical systems are necessary that have several optical surfaces. The complexity of these systems depends on the required image quality in connection with different parameters of the imaging task such as focal length, relative aperture (or numerical aperture or f number), object or image size or field angle, magnification, back focal length, wavelength range, or object or image characteristics. The image quality can be specified by resolution and contrast (modulation transfer or spot diameters), distortion, relative illumination (or vignetting), and others.

Most of the well-known lens types have been developed for photographic applications. Here, the quality requirement for a relatively large field angle cannot be satisfied with only two elements like achromats. The

FIG. 17. Prism test system realized with Microbench.

next more complicated lens type is a triplet consisting of three single-lens elements [see Fig. 18(a)]. For a standard field angle, the spherical, chromatic, and astigmatic aberration can be reduced so that a good image quality results.

Further enhancements in image quality also for larger field angles can be obtained by varying these single-lens elements into cemented or noncemented doublets. One variation is the Tessar-type lens consisting of four lenses in three elements [see Fig. 18(b)].

For reprographic tasks where the magnification is nearly $\beta' = -1$, lens arrangements that are symmetrical to a stop [Dialyt type; see Fig. 18(c)] are used. The advantage of these symmetrical lenses is that the aberrations coma and distortion are eliminated by compensation of the two halves.

High-quality camera lenses (taking lenses) also have nearly symmetrical lens arrangements for the minimization of coma and distortion. Two typical refractive-power distributions of the elements (+ - - + and - + + -) are shown in Figs. 18(d) and 18(e) as examples of double-Gauss lenses of first and second type.

To image objects under small or large angles, special telephoto lenses must be used. With a tele-type lens, long focal lengths can be realized where the focal length is longer than the system length from first surface to image plane [see Fig. 18(f)]. A retrofocus type, where the focal length is shorter than the back focal length, is used for wide-angle lenses [see Fig. 18(g)].

3.2 Microscopes

To magnify small objects for visual viewing, simple eyepieces can be used or, for higher magnification, a compound microscope. This is a two-component system (see Fig. 19) where the components are more or less complex depending on the required image quality. The objective produces an image with a relatively high lateral magnification β'_{ob}. This intermediate image is imaged to infinity for visual viewing by an eyepiece (ocular) where the viewing angle w' (intermediate object angle) is magnified into the image angle w'' by the angular magnification

$$\Gamma_{ey} = \frac{\tan w''}{\tan w'}. \tag{20}$$

The total angular microscopic magnification is

$$\Gamma_M = \beta'_{ob}\,\Gamma_{ey} = \frac{\tan w''}{\tan w'}. \tag{21}$$

More details of microscopes are described in the articles OPTICS, LINEAR, and OPTICAL MICROSCOPY.

3.3 Telescopes

Two lens elements can be positioned in such a way that the back focal point of the first and the front focal point of the second element coincide. Then, the residual total re-

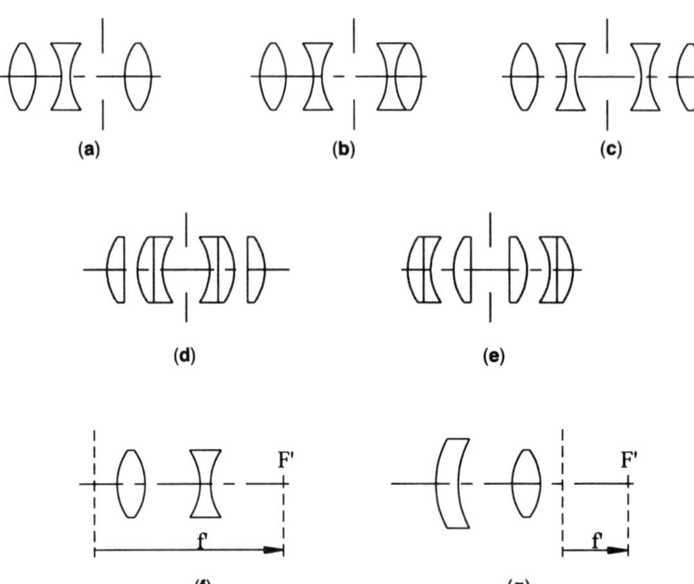

FIG. 18. Photographic lenses: (a) triplet, (b) Tessar, (c) Dialyt, (d) double-Gauss (first type), (e) double-Gauss (second type), (f) telephoto, (g) retrofocus.

fracting power is zero, which means that the total focal length is infinite. Such a system is nonfocal and it can be used as a telescope where an incoming parallel beam with the diameter of $2r_{EP}$, according to the entrance pupil of the system, is transferred into a parallel beam with the exit pupil diameter $2r_{EX}$. The entrance lens (objective) with a focal length f'_{ob} produces an intermediate image that is visually viewed by an eyepiece (ocular) with a focal length f'_{ey} like in a microscope. The object angle w is magnified into the image angle w'' by the angular telescopic magnification

$$\Gamma_T = \frac{2r_{EP}}{2r_{EX}} = \frac{f'_{ob}}{f'_{ey}} = \frac{\tan w''}{\tan w}. \tag{22}$$

Two principal configurations are possible, which are shown in Fig. 20. With the Kepler type, with two positive lens elements, a real intermediate image is produced where a field stop can be positioned so that the viewed image has sharp edges. The image is viewed upside down; the angular magnification is negative. The Galilean type produces a virtual intermediate image that is viewed right way up but has no sharp edges. Here, the magnification is positive. This telescope has a short system construction length. Objectives and eyepieces can be made up by simple elements like single lenses or achromats or by more complex systems to reach higher image qualities. More details of telescopes are described in the articles OPTICS, LINEAR, and ASTRONOMICAL TELESCOPES.

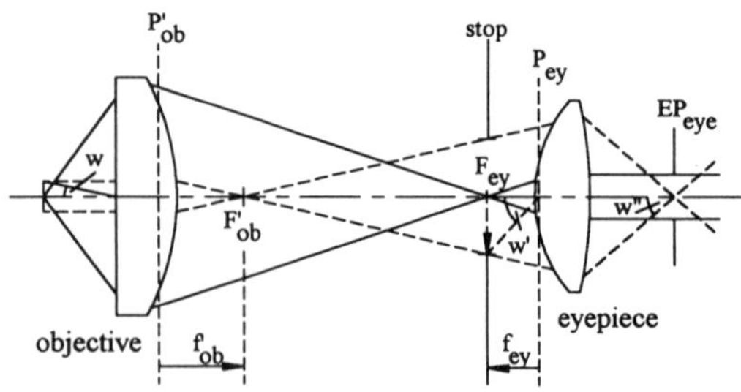

FIG. 19. Microscope.

3.4 Beam Expanders

Expanding of laser-beam diameters is mainly necessary for illuminating extended objects with plane waves (directed light) or for decreasing the beam divergence. Small beam divergence is required for laser light transfer to retain high power density over long distances, and for achieving small laser spots in combination with a focusing lens. The beam diameter W is defined as the diameter where the light intensity drops to $1/e^2$ of the center intensity. The beam waist is the smallest cross section of a laser beam. The wavelength λ and

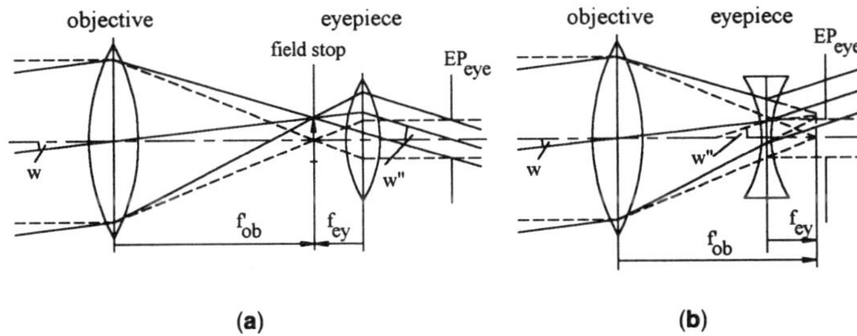

FIG. 20. Telescopes: (a) Kepler type, (b) Galilean type.

the beam waist diameter W determine the divergence angle:

$$\sigma = 2\,\lambda/\pi W. \tag{23}$$

The beam waist diameter W' in the image space of a lens (Fig. 21) is approximately proportional to the focal length f' of the lens and the incoming beam divergence angle σ:

$$W' \sim 2f'\sigma. \tag{24}$$

The bigger the beam diameter W in the object space, the smaller the beam divergence and the smaller the spot size of the focus of a lens.

Expanding a laser beam is possible with telescope systems like those shown in Fig. 20, used in reversed position (see Fig. 22). The angular magnification given in Eq. (22) is modified to the expansion factor

$$\Gamma_e = \frac{2r_{EX}}{2r_{EP}} = \frac{f'_1}{f'_2} = \frac{\tan\sigma}{\tan\sigma'}. \tag{25}$$

The divergence angle σ of the laser beam is decreased by this expansion factor Γ_e to σ'. The Kepler type produces an intermediate real image—this means a laser focus—between the entrance and the exit lens, where a spatial filter can be positioned. The very high intensity in the focus of high-power lasers can cause a breakdown in air. In this case, the Galilean type should be used. This type is mechanically shorter but offers no possibility of inserting spatial filters. It should be mentioned that for high-power lasers, the entrance lens of beam expanders must not consist of cemented elements. Otherwise, the high energy density of the incoming small laser beam can produce irreversible damage on the optical surfaces caused by absorption in the cement layers.

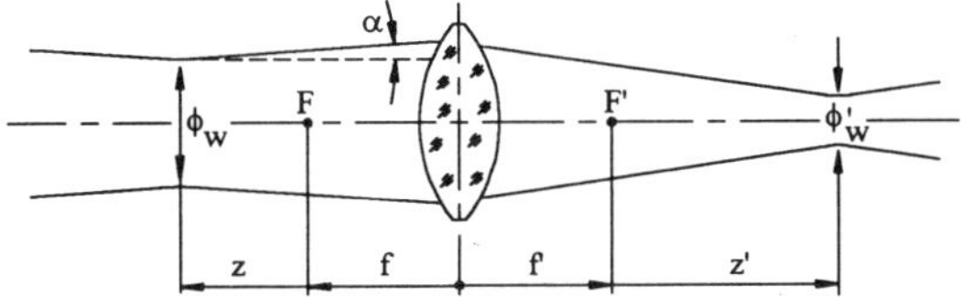

FIG. 21. Focusing a laser beam by a lens.

3.5 Scanning Systems

For writing or for reading pixel information, scanning systems are used. In such systems, a focused light beam, normally an expanded laser beam, is guided over the image plane. The beam guiding can be realized by rotating deflector elements like galvo mirrors, polygon mirrors, or diffractive elements (see OPTICAL SCANNERS). These deflectors can be positioned behind or in front of a scanning lens. In the first case, the image is not plane but curved. Systems for flat-field scanning have the deflecting elements in front of the scanning lens. Such a system is shown in Fig. 23 for scanning in one dimension. The entrance pupil diameter is identical to the beam diameter. The mirror is positioned in the entrance pupil of the lens.

To generate a flat scan line, the lens must be optimized for a minimum field curvature. Furthermore, for many applications, a linear dependence between the object angle θ, which is produced by the mirror rotation, and the image height u' is required. Normally the image height is a function of the focal length and the tangent of the object angle:

$$u' = f'\tan\theta. \tag{26}$$

The required scan linearity is expressed by the f-theta condition

$$u' = f'\theta. \tag{27}$$

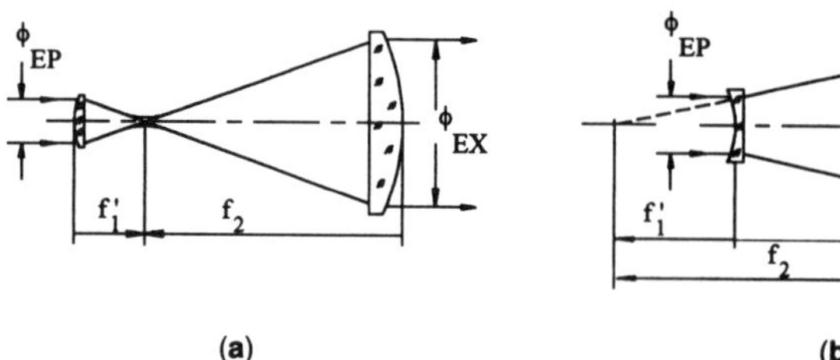

FIG. 22. Beam expanders: (a) reversed Kepler type; (b) reversed Galilean type.

The linearization can be achieved by introduction of a finite negative distortion (barrel distortion). Finally, such a lens design system is like the right half part of a symmetric, more complex system such as a reprographic lens [compare Figs. 18(d) and 18(e)], where field curvature and astigmatism are corrected in each half and the distortions of the two parts compensate each other. Besides on an accurate deflector rotation, the resolution of writing or reading by a scanning system depends on the image spot diameter. Therefore, spherical aberration and coma must be minimized.

3.6 Telecentric Systems

Applications in metrology require special optical systems, for which defocusing will not result in measuring errors. To achieve this, the aperture stop can be positioned in one focal point so that the stop is imaged into infinity. Figure 24 shows the principle of a profile projection lens where the aperture stop is positioned at the image-side focal point. In the object space, the principal (or chief) rays that pass the stop and the pupil centers are parallel to the optical axis. The image plane will be a screen with a measuring scale or a sensor. For a constant image distance, the systems will be focused by varying the object distance. Defocusing, which means wrong object distance, results in a blurred image of course, but the measured image height always corresponds to the exact object height.

3.7 Projection Systems

To image transparent or nontransparent objects onto a screen, systems are used that consist of a projection lens in combination with an illuminating system. For maximum and uniform image brightness, a special arrangement of the combination of illumination and projection must be realized as shown in Fig. 25 for a slide projector.

The transparent object (slide) is illuminated by a condenser system. For uniform illumination, the slide must be positioned near the exit pupil of the illuminating system EX_{light}, where the light beams are not too divergent. To obtain maximum image brightness, the lamp filament is imaged into the entrance pupil of the projecting lens EP_{lens} while the slide is imaged by the lens onto the screen.

3.8 Autocollimator

For measuring small angles, the principle of autocollimation can be used. As shown in Fig. 26, an illuminated pinhole or scale is

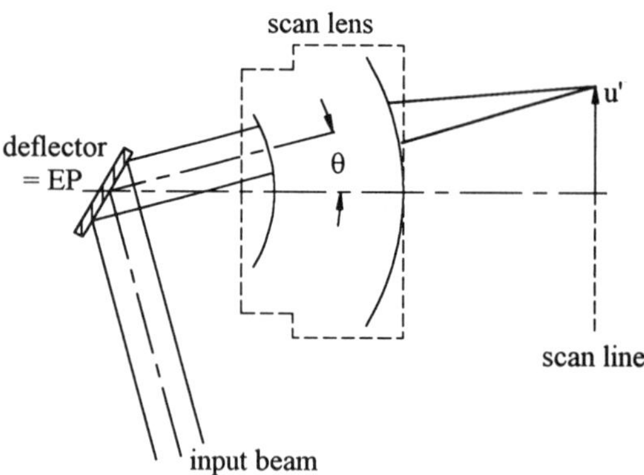

FIG. 23. Scanning system.

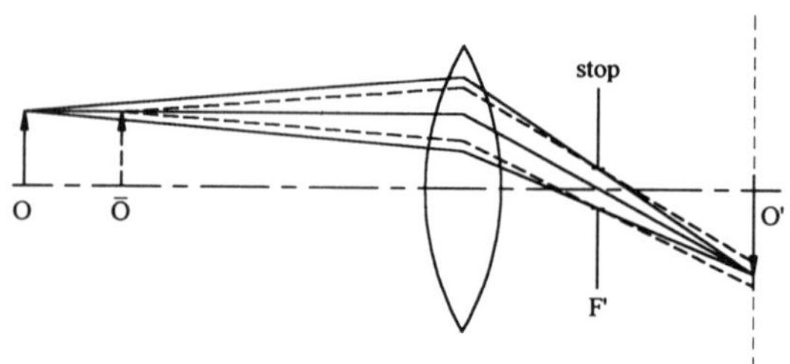

FIG. 24. Telecentric lens for profile projection.

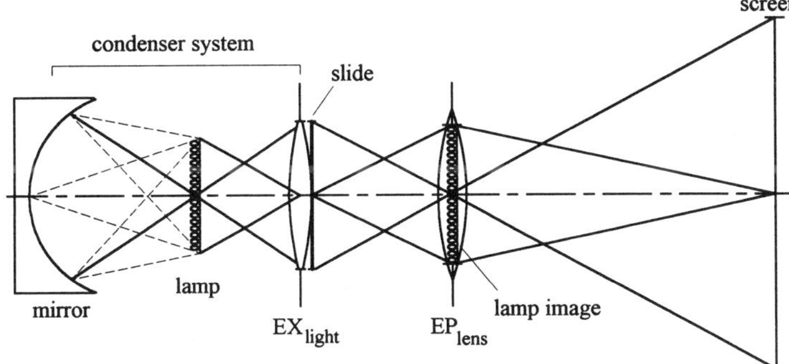

FIG. 25. Projection system.

collimated by a lens, normally an achromat. This means that the object is imaged to infinity. The imaging rays are then reflected by a plane mirror back through the lens. After reflection at a beam-splitter surface, the rays form the image of the pinhole or scale in the focal plane.

If the reflecting plane is exactly perpendicular to the collimator axis, object and image coincide. A tilt of the reflector plane results in a lateral shift of the pinhole or scale image. The image shift $\Delta u'$ is directly proportional to the tilt angle variation $\Delta\alpha$ and the collimator focal length f'_{coll} so that the angle variation is

$$\Delta\alpha = f'_{\text{coll}}\,\Delta u'/2. \tag{28}$$

The image shift can be measured visually by an eyepiece with scale or electronically by a CCD line sensor. Typical applications of an autocollimator are testing prism angles, alignments, vibrations, scanning angles, accuracy of spindles, and surface flatness.

3.9 Optical-Beam Delivery System for Material Processing

Lasers with high power can be used for material processing like engraving, marking, welding, etc. Also, for different medical applications such systems are used. Besides the light source, the optical head of all these systems normally consists of three main elements: collimator lens, beam expander, and focusing element. As an example, Fig. 27 shows a typical system.

The laser light is transferred to the system through an optical fiber. The light leaving the fiber is collimated by a lens system. On the way further on, the light can be deflected by prisms or mirrors. The following beam expander reduces the beam divergence for better focusing, as described in Sec. 3.4. Finally, the light is focused by a lens system to a small high-energy spot that burns into the material target.

3.10 Interferometers

Most of the instruments described so far do not need highly monochromatic light. Interferometers do. Lasers are high-intensity monochromatic sources. Therefore, today, interferometers for metrology purposes are supplied with lasers, the classical geometries included. However, lasers have also brought about novel interferometric techniques such as speckle interferometry and holographic interferometry.

In an interferometer, two, or in some cases

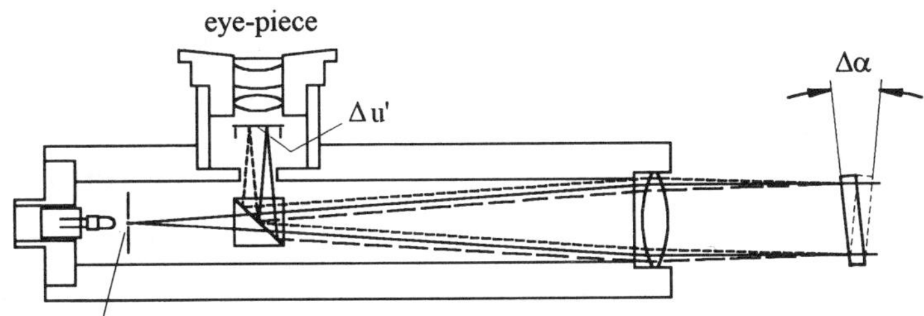

FIG. 26. Autocollimator.

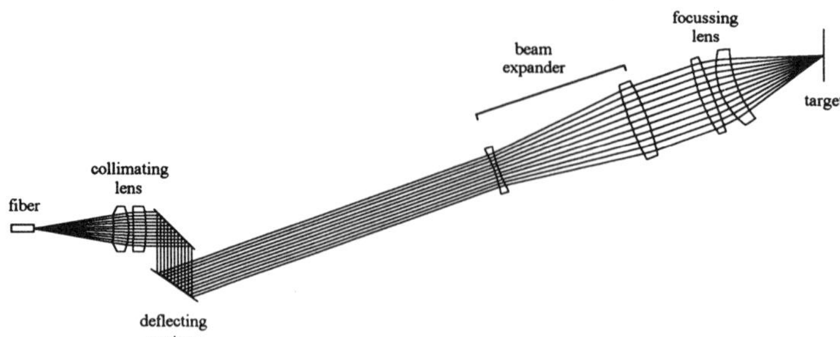

FIG. 27. Optical system for material processing.

several, light beams from the same light source that have been traveling through different optical paths are recombined, so that the phase differences between corresponding points in the the two beams result in interference patterns. From the fringe patterns, the changes of the optical path length (= geometrical path length multiplied by the refractive index) in the different arms can be determined. Thus, it is possible to measure extremely small changes in distance and refractive indices.

The Michelson interferometer shown in Fig. 28 consists of a beam splitter and two mirrors as its main parts. The laser light is divided into two portions by the beam splitter, one traveling a distance $2l_1$ to mirror M_1 and back to the beam splitter and the other one traveling over a distance of $2l_2$ to mirror M_2 and back. From the beam splitter a portion of light is reflected to the observation screen. As long as l_1 is equal to l_2 over the beam cross sections, the two light beams are in phase and constructive interference appears, showing a bright spot. With increasing l_2, interference becomes destructive, whenever the difference between l_2 and l_1 is $m\lambda/4$ (m = odd integers). Such instruments similar to the Michelson interferometer are used to measure exactly the travel of translation stages in the machine tool industry. The measurement is possible with very high accuracy because the wavelength of a frequency-stabilized laser is the reference. The resolution is a small fraction of the wavelength ($\lambda \sim 600$ nm).

A different setup is shown in Fig. 29. The Mach–Zehnder interferometer is constructed from two beam splitters and two mirrors. The source beam is divided by the first beam splitter. Two different light paths via mirrors M_1 and M_2 are combined with a second beam splitter so that the interference is observed on a screen. In this case, the path difference between the two paths of $m\lambda/2$ produces destructive interference. For a more detailed description of interferometry, see INTERFEROMETERS AND INTERFEROMETRY.

Whenever the wave fronts of the two beams are not perfectly planar, structures of bright and dark areas appear on the screen. Aberrations from nonperfect optical components influence the wave fronts and cause distortions of the interference patterns. Therefore the quality of optical components can be measured by use of interferometers. How-

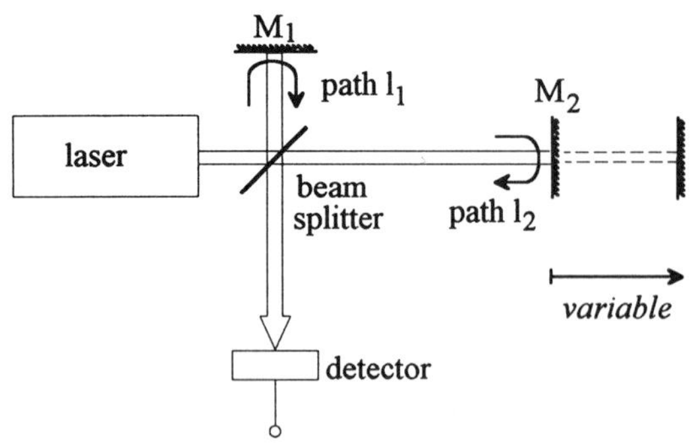

FIG. 28. Michelson interferometer.

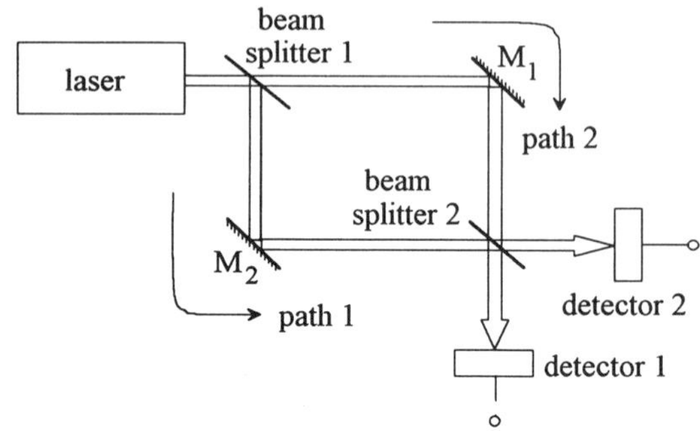

FIG. 29. Mach–Zehnder interferometer.

ever, because the optical components of the interferometer itself, e.g., beam splitters and mirrors, act as reference surfaces, the requirements on the optical quality of interferometer components are extreme. Wave front deformations of less than $\lambda/50$ are often necessary. Especially for reference lenses to be used in Twyman–Green interferometers, such extreme requirements are sometimes beyond the physical and technical limits for design and production.

GLOSSARY

Aberrations: Ray aberrations are all deviations of the coordinates of the real image and the perfect image. Wave aberrations are all deviations of the real and the perfect wave form.

Damage Threshold: Power or energy limit where physical damage of coatings or bulk materials occurs.

Dispersion: Wavelength dependence of the refractive index of a medium. Therefore the refraction angle of a light beam at an interface of two different media is different for different wavelenghts.

Endoscopes: Optical instruments for the examination of internal cavities, characterized by a small outer diameter (e.g. < 10 mm) and a length starting from approx. 10 mm.

Frequency-Stabilized Laser: Laser with stabilized carrier frequency. With constant refractive index wavelength stabilization results. Different stabilisation schemes are available. As frequency standard narrow atomic transitions are well suited.

Principal Plane: Theoretical subsidiary plane of a lens or a lens system, where a paraxial ray coming in from infinity crosses the corresponding image ray that crosses the optical axis in the focal point. The focal length is the distance from the principal plane to the focal point.

Resolution: Minimum distance of two image points which can be identified as separate points.

Thin-Film Coating: Thin dielectric or metallic films coated on a substrate by thermal or chemical evaporation in a vacuum chamber.

Further Reading

Born, Max, Wolf, Emil, (1990), *Principles of Optics*, London: Pergamon Press.

Hecht, E., (1987), *Optics*, Reading, MA: Addison-Wesley.

Naumann, H., Schröder, G., (1992), *Bauelemente der Optik*, München: Carl Hanser Verlag.

Smith, Warren J., (1966), *Modern Optical Engineering*, New York: McGraw-Hill.

Young, M., (1984), *Optics and Lasers*, Berlin: Springer-Verlag.

Optical Society of America (1964), *Proceedings of the International Lens Design Conference*, Washington, DC: Optical Society of America. These conferences have been held approximately every five years since 1966.

OPTICAL COMPUTING

JUN TANIDA AND YOSHIKI ICHIOKA, *Department of Applied Physics, Osaka University, Osaka, Japan*

INTRODUCTION

Optical computing is a new area of technology for information processing using desirable features of light. The goals and methodology in optical computing are different from those in conventional electronic computing because the former has much more flexibility in data representation and processing. At the present stage, it is difficult to point out concrete differences in goals and methodology of optical computing from those of electronic computing. However, it can be said that optical computing aims to realize a kind of information processing for extremely large amounts of data. The methodology of optical computing is characterized as the efficient use of the multiple dimensionality appearing in the image plane and in free space. Although it is a young area, optical computing has progressed tremendously and many results have already been reported.

Historically, the root of optical computing is optical information processing. After the development of the laser in the 1960s, various techniques for information processing were proposed with optical implementations (Horner, 1987). A spatial filtering technique

3-527-28134-7/95/$5.00 + .50

using the Fourier transform property of a lens, and holography as a phase recording technique, are typical examples of earlier optical information processing, which are also used in current optical computing.

Attempts at fabricating a novel computing system based on optical techniques are found in several references, and they are called optical computers. Among them, the Tse computer shown in Fig. 1 (Schaefer and Strong, 1977) is considered the origin of today's optical computing scheme. "Tse" is the English transliteration of the Chinese word for a pictograph character, and was adopted as the name of the proposed computer in recognition of the fact that the pictograph is processed as an image rather than as a composition of bit information. In the system, array devices executing logical operations for two-dimensional data are connected by parallel signal lines implemented by fiber bundles or fiber plates. Using parallelism and digital data representation, the Tse computer was expected to be a powerful tool for massively parallel processing.

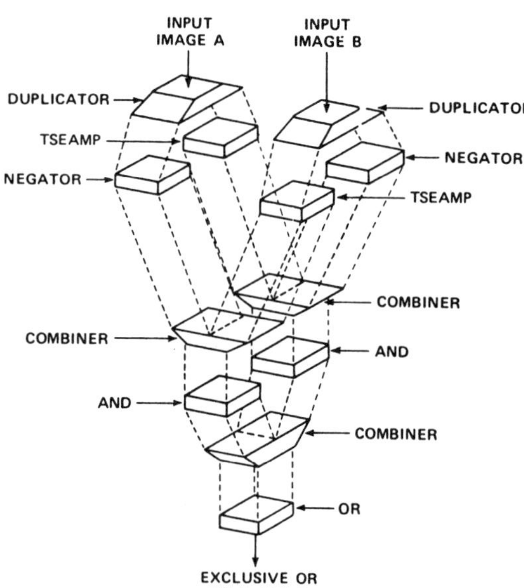

FIG. 1. Conceptual diagram of the Tse computer. An exclusive OR circuit is constructed with Tse components connected by parallel optical signal lines. AND, OR, NEGATOR, and tseAMP are called Tse devices, and execute logical operations for all pixels on the image plane in parallel. DUPLICATOR and COMBINER are interconnection elements composed of multiple sheets of image fibers (Schaefer and Strong, 1977).

Stimulated by the Tse computer, many research efforts on optical devices for parallel logic operations were started (McAulay, 1991). Studies on optical bistability and optical nonlinear phenomena were also accelerated by the same motivation. On the other hand, other ideas for optical parallel logic operations were proposed. These are methods in which information is converted into spatial code patterns and processed with optical operations. Optical shadow casting shown in Fig. 2 (Tanida and Ichioka, 1983) and spatial filtering are examples of the optical parallel logic. Related to optical logic, logical theories directed to two-dimensional data processing have been proposed, e.g., symbolic substitution and optical array logic (McAulay, 1991).

As a computational model other than the digital scheme, the neural network has been proposed. From the early stage of the neural network, it has been recognized that this scheme is quite suitable for optical computing because of the interconnection capability of optics, the low functionality required in the neural network, and the robustness of the neural computing scheme. Consequently, optical neural computing has become one of the major trends of optical computing.

Recently, as fabrication technologies on optoelectronic devices have developed, a practical approach for optical computing has arisen. That is an approach to achieve high performance by efficient use of current electronic device technology. Optics is suitable for communication; therefore, interface elements converting optical signals to electronic (or electronic to optical) are prepared on the device. On the other hand, electronics is suitable for processing, so that processing itself is executed by electronic circuits. Combining optical communication with electronic processing, effective computing architectures can be implemented (see OPTICAL INTERCONNECTIONS).

1. CONCEPTS OF OPTICAL COMPUTING

1.1 Definition

Optical computing has several definitions, resulting from broad and narrow interpretation of the words. According to the definition by Sawchuk and Strand (1984), optical

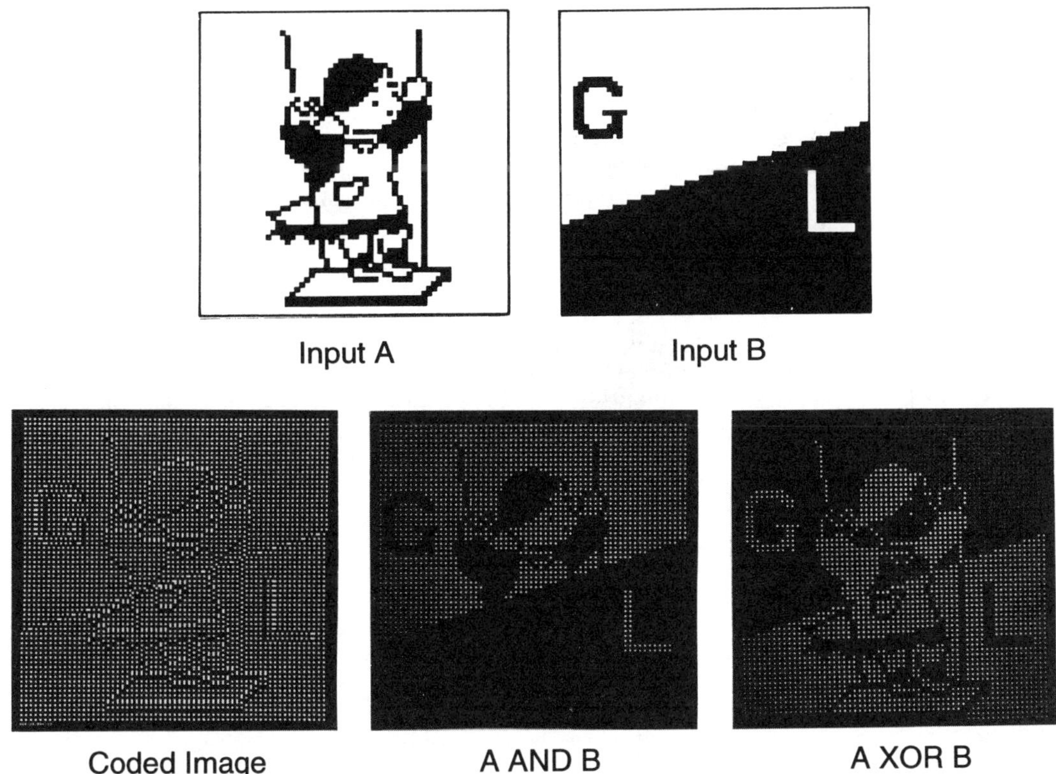

FIG. 2. Parallel logic operations implemented by optical shadow casting. Two binary images consisting of 64 × 64 pixels are converted into a coded image composed of 64 × 64 spatial code patterns. The code pattern consists of 2 × 2 subpixels in which one of the subpixels is transparent and the others are opaque according to the combination of the corresponding two pixels to be coded. The coded image is illuminated by point light sources arranged in a 2 × 2 square form. As a result, multiple shadowgrams are projected so as to overlap on the screen. By sampling of the shadowgram at every two structuring cells, the result of parallel logic operation is obtained as bright and dark signals. In the arrangement, 16 individual combinations of switching of the sources produce all 16 logic functions for two variables. As examples, results of logical AND and exclusive OR are shown.

computing is the use of optical systems to perform numerical computations on one-dimensional or multidimensional data that are generally not images. This definition is a rather narrow one.

In contrast, Goodman (1989) gives a very broad definition that optical computing consists of all methods for intentionally manipulating optically represented data for useful purposes. As is well known, many optical techniques have been investigated for a variety of information processing. Optical Fourier transforms, optical spatial filtering, and optical correlation are typical examples of optical information processing. Although these methods have been categorized into optical information processing, they are collectively included in optical computing.

Another definition of optical computing is based on methods or techniques aiming at the development of an optical computer. The optical computer is an information processing system of the future, suitable for massively parallel processing using the properties of light mentioned in Sec. 1.3. In the definition, research on all-optical or all-optoelectronic devices as basic elements of the optical computer is also included. In this article, we mainly adopt this definition of optical computing.

1.2 Motivation

The motivation of optical computing is rather clear compared with its confusing definitions.

It can be said that today is the era of electronics. Advances in electronics have brought us many benefits in life, science, society, and so on. Although many researchers are dedicated to the improvement of performance of electronic systems and to breaking the limitations of current electronics, it begins to be recognized that fundamental limitations of electrons prevent us from extending the current electronic technology. For example, the limiting factor in an electronic system is not the delay in processing but in communication. Since signal lines must be mapped onto a planar structure in an electronic integrated system, tremendous efforts are required in the design. Electrons are easily affected by electromagnetic fields and other electrons, so that noise and crosstalk problems become serious, especially in high-speed operation. Therefore, the importance of technologies based on other than electronics increases.

Among many candidates for the future technologies, an optical system is a promising one because of the desirable features of light. The advantages provided by light are parallelism, high speed, large time- and space-bandwidth products, noninteraction in free-space propagation, high-impedance signal lines, resistance against electromagnetic noise, high-speed interaction via material, and so on.

For optical computing, the maturity of related technologies, including optical communication, optical devices, and optoelectronics, is also an important factor in advancing the studies. Optical communication (*q.v.*) is one of the most successful technologies based on optics. Optical fiber and high-speed optical-electronic signal converters are investigated actively, and high-performance products are commercially available. Related to the advances in optical communication, optoelectronic devices and optical devices have advanced tremendously. All of these technologies can be used in optical computing.

The final motivation of optical computing is the challenge of a new technology. Especially, the development of a computer based on optical technology can be considered as the symbol or dream of researchers engaged in optics.

1.3 Advantages

The advantages of optical computing are mainly provided by the physical characteristics of light.

1.3.1 Parallelism Light has inherent parallelism when it propagates in free space or materials. Parallelism is expected to be used in massively parallel processing, which is a principal goal of computer science. An effective parallel computing system is expected to be constructed by use of parallelism in optics.

1.3.2 High Speed The high speed of light propagation is an attractive feature. According to Einstein's theory of relativity, no object can move faster than light propagating in vacuum. This suggests that the highest speed is attained by using light as the medium of information in the material of interest.

1.3.3 Large Time- and Space-Bandwidth Product Light has a large information capacity. Visible light is a kind of electromagnetic wave whose frequency is in the range $(3.84–7.69) \times 10^{14}$ Hz. From the point of view of information theory, the information capacity of a wave medium is given by the product of frequency and duration time. This measure is called the time-bandwidth product. Compared with radio frequency or microwaves used in conventional communication, light has from thousands to millions of times as much information capacity, which provides an essential advantage to optical communication. In addition to the large time-bandwidth product, the parallel nature of light offers two-dimensionality in data representation. In this case, a space-bandwidth product is defined. In a typical imaging system for a photographic camera, the space-bandwidth product is of the order of several million. As a result, a large amount of information can be treated in an optical computing scheme.

1.3.4 Noninteraction in Free-Space Propagation Light propagates in free space without interaction between wave fronts of the light. This is true even if two light waves cross each other. This characteristic is a result of the boson nature of the photon. As a result, flexible information transfer can be implemented by free-space optical interconnections.

1.3.5 High-Impedance and Noise-Resistant Interconnections In a typical implementation of optical interconnections, a light source emits an optical signal converted from an electronic signal, the optical signal runs in an optical signal line, and the signal is received by a detector. In this configuration, the signal lines can be considered as having high impedance. Therefore, the power for transmission can be reduced as compared with electronic transmission. Moreover, optical interconnections are resistant to electromagnetic noise from both internal and external sources.

1.3.6 High-Speed Interaction via Material Although optical signals cannot interact with each other in free space, they do interact via an intermediary medium. Subpicosecond interaction can be achieved by this arrangement.

1.4 Goals

Optical computing has several goals. These are numerical processing, parallel image processing, massively parallel computation, construction of a communication processor, and development of a general-purpose optical computer.

Numerical processing is the initial motivation for optical computing. With the use of the parallelism offered by optics, large processing capabilities are expected. Image processing is one of the most suitable applications of optical computing. Until now, many methods and techniques have been studied for image processing by optical methods. Massively parallel computation is an important application in optical computing. Database management (see COMPUTER DATA BASES), artificial intelligence (*q.v.*), and intelligent memory systems have been mentioned as typical examples.

Besides processing-level goals, those of system level are considered important in optical computing. One of the practical goals of optical computing is to develop a communication processor, such as a telecommunications switching network system. Since information is transferred as optical signals in optical communication, direct processing on the information medium provides high processing efficiency. In addition, compared with a general-purpose computer, such a special-purpose computer is simple and easy to fabricate. Thus, as a near-future goal, communication processors are being actively studied.

The ultimate goal of optical computing is to develop a general-purpose massively parallel computer. The most serious problem in current von Neumann computers is known as the von Neumann bottleneck. As shown in Fig. 3, the processing speed of the von Neumann computer is limited by the transfer capacity of the signal line between the central processing unit and the memory. As long as the same architecture is adopted, the bottleneck cannot be removed. To overcome this problem, various architectures and techniques have been considered (Hwang and Briggs, 1984). Since optical computing has inherent parallelism, it is expected to be used to implement the architectures and techniques. Especially, massively parallel architectures are promising candidates for optical implementation.

1.5 Classifications

The implementation methods of optical computing have a different basis compared with conventional electronic information-processing schemes. This is a reason why many interesting methods and techniques have been studied and developed in the field of optical computing. To grasp the whole image of this field, classification in several aspects is helpful.

1.5.1 Data Representation Light has various attributes to be used in data representation—for example, intensity, phase, wavelength, polarization state, spatial posi-

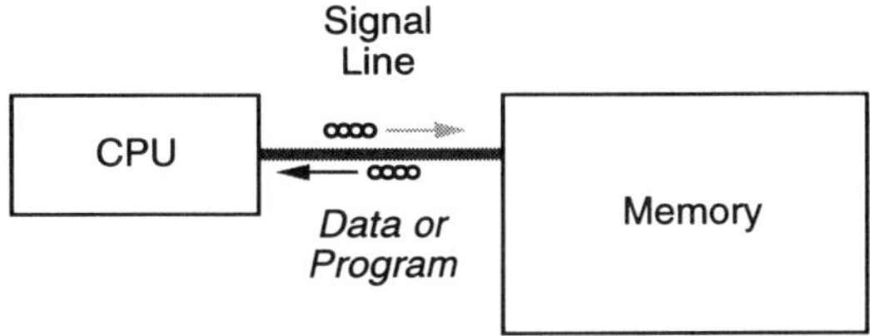

FIG. 3. Von Neumann bottleneck. A von Neumann computer, a typical architecture of current computers, is composed of a central processing unit (CPU) and a memory storing data and programs. In this architecture, program codes and data stored in the memory must be loaded into the CPU at every clock cycle. Thus, the signal line between the CPU and the memory is a bottleneck for data flow and processing itself. This problem is called the von Neumann bottleneck.

tion, and spatial pattern. For these attributes, both digital and analog values can be used. In addition, the format of identifying position is also classified into two classes, discrete and continuous. These concepts are shown in Fig. 4. A photographic picture or image written on an optically addressable spatial light modulator can be treated as continuous data. Note that digital representation is inherently discrete.

1.5.2 Computing Schemes According to the method of data representation, computing schemes are categorized into two classes: digital and analog. Because of the nature of the data format, the digital scheme treats discrete information on images with digital values, whereas the analog scheme handles both discrete and continuous images with analog values. In the digital scheme, individual discrete data are treated as logical variables and processed by logical operations. In the analog scheme, transformations or nonlinear functions are applied to the data.

1.5.3 SIMD and MIMD Another categorization in optical computing is that following the classification used in parallel processing. The major categories are SIMD (single-instruction stream, multiple-data stream) and MIMD (multiple-instruction stream, multiple-data stream). The former is an operation style of parallel processing in which the identical operation is applied to multiple data at one time, and the latter is that in which different operations are executed for individual data concurrently. Generally, SIMD operation is suitable for optical implementation because of the nature of light. To implement MIMD operation, some mechanism for selecting the operation is required on each processing element. Then multiple sets of both instructions and data must be transferred to the individual processors to execute MIMD operation. Since each processing element becomes complicated, the degree of parallelism is expected to decrease.

2. TRENDS IN OPTICAL COMPUTING

While many trials have been carried out on optical computing, five trends exist today, namely, Fourier image processing, acousto-optic signal processing, digital optics, neural computing, and the smart pixel.

2.1 Fourier Image Processing

According to the definition of optical computing offered by Goodman (1989) (see Sec. 1.1), traditional optical information-processing techniques are included in the category of optical computing. The most useful technique in optical information processing is

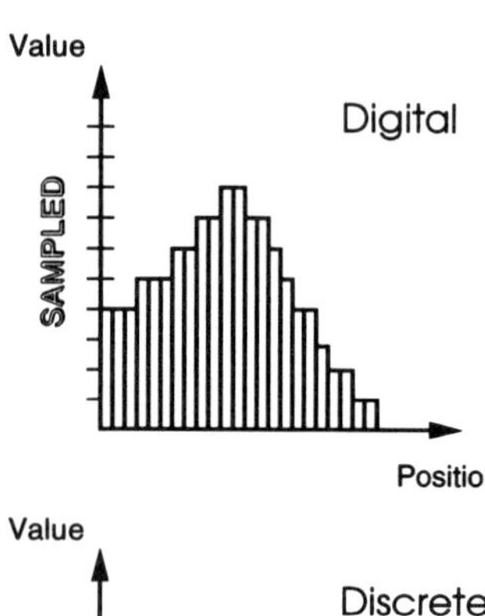

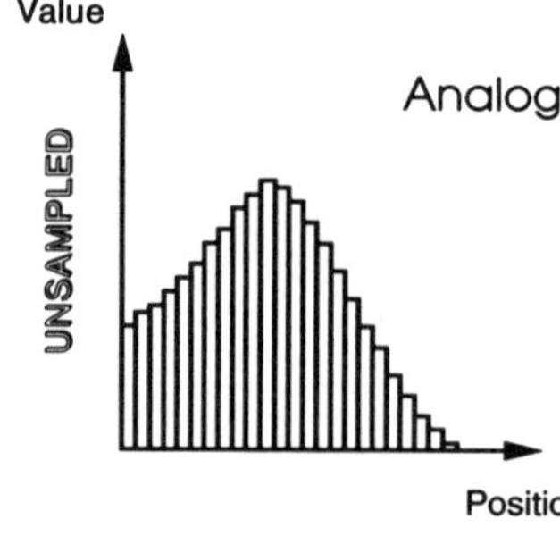

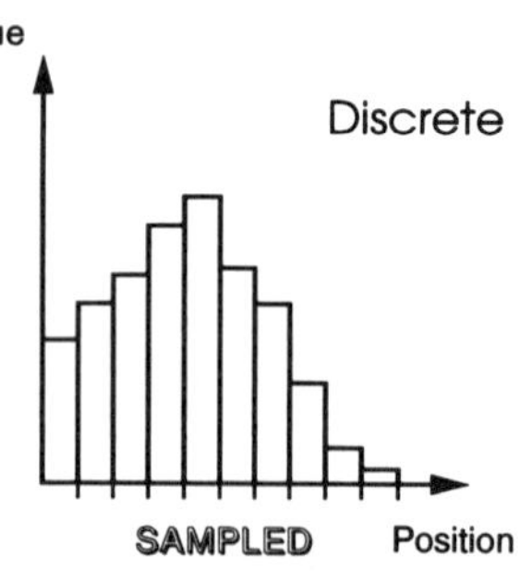

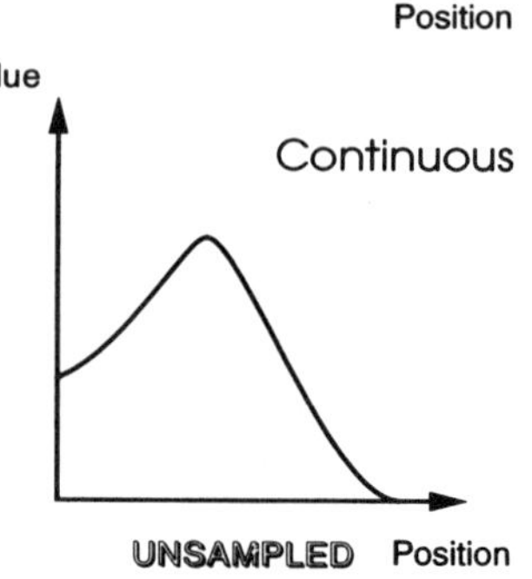

FIG. 4. Classifications of data representation. On magnitude and spatial position of individual data, two sets of classes can be considered. Digital and analog are used for discriminating the representation method on value, whereas discrete and continuous are on spatial position. In the figure, "Sampled" and "Unsampled" indicate continuity and discontinuity, respectively, of data along the axis. Note that digital representation must be discrete.

optical Fourier transformation by means of a lens. Based on this technique, various kinds of Fourier image processing have been investigated (see OPTICAL FILTERS).

Figure 5 shows a typical setup of a frequency-plane optical correlator that is the foundation of Fourier image processing. Although this optical system executes correlation only between the input and the filter, various kinds of functionality such as image enhancement, image restoration, and pattern recognition can be achieved by changing the filter.

Among these, pattern recognition is the most important application because of its tremendous processing capability, which cannot be achieved by conventional electronics. In the basic setup, the complex conjugate of the Fourier transform of the pattern to be searched, called a matched filter, is used as a filter. When the target image is set at the input plane, bright spots are produced at points corresponding to positions where the search pattern exists. Since all processes can be executed quite fast (in the propagation time of light from the input to the output planes) and in parallel, a large processing capability is expected.

For pattern recognition based on Fourier image processing, filter design is an important issue. To improve recognition capability, sophisticated filter design is required. In general, a matched filter produces a broad and dark peak distribution, which may cause errors in recognition. Ideally, the distribution should be narrow and bright. In addition, flexibility in the shape of search patterns and alignment of the target image will increase the availability of the technique in practical use. To meet demand, various kinds of filters have been investigated—for example, phase-only, binary phase-only, composite, and synthetic discriminant function filters. Another important issue in pattern recognition is real-time implementation. For this purpose, spatial light modulators are used. For more detail on Fourier image processing, see Flannery and Horner (1989).

2.2 Acousto-optic Signal Processing

For fast signal processing, acousto-optic techniques have been studied. These techniques use the interaction of light and acoustic waves propagating in material media (see MODULATORS AND DEMODULATORS, OPTICAL). Figure 6 shows an acousto-optical cell spatial light modulator in which water, glass, or a crystal is used as a photoelastic interaction material. The transducer set at the bottom converts electrical signals into acoustic waves that propagate in the medium. The acoustic waves in the material work as a phase diffracting element, so that the light waves are modulated behind the cell. In this configuration, acoustic signals propagating in the acousto-optic cell are multiplied by the light signals incident on the cell. On the basis of this operation, various functionalities can be realized. Spectrum analysis, spatial or temporal integrating correlation, and synthetic aperture radar are typical applications of acousto-optic processing.

Attractive features of acousto-optic processing are low power consumption and high processing throughput in multiplication. For multiplication with precision similar to digital 8-bit operation, the acousto-optic multiplier requires orders of magnitude less power than a digital multiplier. In terms of processing throughput, of the order of 10^{15} multiplications per second is achievable with

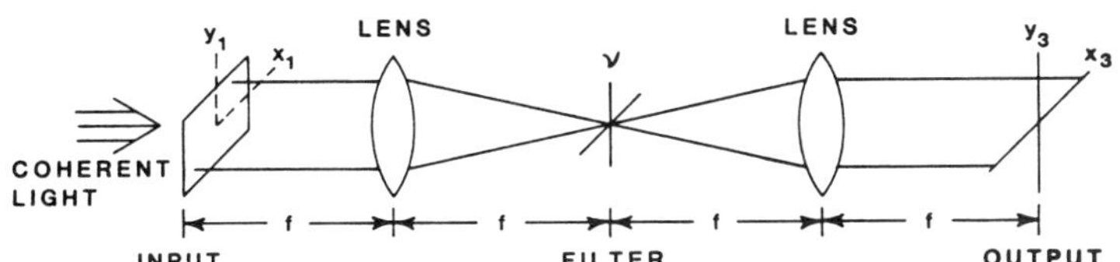

FIG. 5. Optical correlator based on frequency plane multiplication. A target image is set at the input plane and illuminated by coherent light. On the filter plane, the Fourier transform of the target image is obtained by the Fourier transform property of the lens and multiplied with a filter located on the plane. If the filter is the complex conjugate of the Fourier transform of a filter function, correlation of the target image and the filter function is obtained on the output plane (Flannery and Horner, 1989).

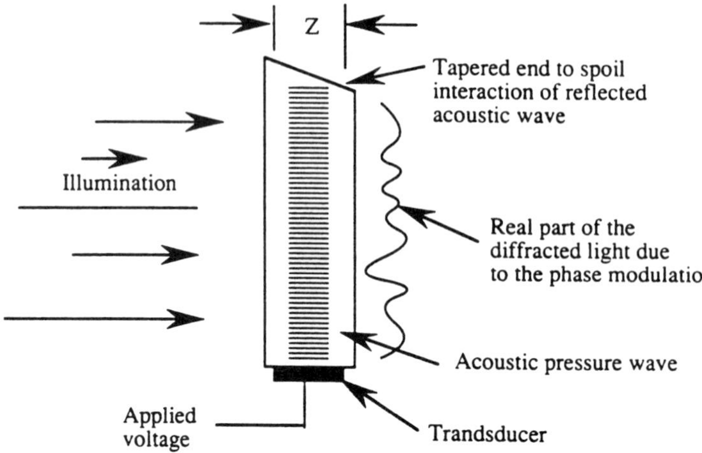

FIG. 6. Acousto-optical cell spatial light modulator. For mode of operation, see text.

$LiNbO_3$. For a detailed discussion, see Lee and VanderLugt (1989).

2.3 Digital Optics

Digital optics is a class of optical computing in which a digital scheme and logical theory such as Boolean algebra are employed. As a benefit of the use of logical theory, any complicated function can be constructed by combination of elemental devices, or logic gates. Techniques and resources accumulated in current computer science can be used. Moreover, the parallelism in optics provides large computational capabilities.

To implement a logic gate, a device executing a nonlinear function is required. Thus, many optical devices and spatial light modulators have been developed using various types of nonlinear optical phenomena. However, an optical device usually requires more power than an electronic device, so that it is difficult to increase density because of the heat problem. Therefore, low power consumption is an important requirement for optical devices. To solve this problem, the use of a modulator type of device and concurrent drive of electronic and optical signals are considered effective and have proceeded. Self–electro-optic effect devices (SEEDs) (Miller *et al.*, 1984), semiconductor etalons based on dispersive optical bistability, and liquid-crystal spatial light modulators based on birefringence controlled by an electric field are typical examples of modulator-type devices. A vertical to surface transmission electrophotonic device (VSTEP) (Kasahara *et al.*, 1988) is mentioned as an instance of a device based on the concurrent drive of electronic and optical signals. For more detail, see QUANTUM OPTOELECTRONIC DEVICES.

For construction of logical circuits by logical gates, there are two strategies. One is small fan-in and fan-out construction, and the other is large fan-in and fan-out construction. Generally, the modulation ratio of a modulator or a transducer decreases as the operational speed increases. The modulation ratio determines the number of fan-in and fan-out lines. Thus, small fan-in and fan-out construction implies high-speed operation of the system. On the other hand, large fan-in and fan-out construction has an advantage in the connectivity of optical interconnection.

To design a logical circuit for a desired function, special ideas are required in digital optical computing. The multistage interconnection network (MIN) is a typical design method for small fan-in and fan-out construction. In MIN, multiple sets of arrays of logical devices are connected by regular interconnection. For this purpose, a perfect shuffle, banyan, or crossover network is used. The required number of fan-in and fan-out lines for an individual gate device is two. For large fan-in and fan-out construction, various logical theories for two-dimensional data can be applied such as symbolic substitution, optical-array logic, cellular logic, binary image algebra, or image logic algebra.

The goal of digital optical computing is to develop an optical digital computer. Several architectures of optical digital computers have been presented. Details are discussed in Sec. 3.

2.4 Optical Neural Computing

The neural network is a novel computing paradigm imitating the information system of the brain (Rumelhart *et al.*, 1986) (see also NEURAL NETWORKS). As shown in Fig. 7, a neural network consists of a large number of simple processing elements, called neurons, and interconnections among them. An individual neuron sums input signals and generates an output signal with a kind of threshold function of the input. The result is broadcast to all neurons connected to the outputting one. A neural network is configured by the individual connecting weights of the interconnection network. Dynamic control of the interconnection weights provides learning capability.

A neural network is quite suitable for optical implementation for the following reasons. First, optical interconnection is promising for implementing the complicated interconnection pattern required in a neural network. Second, simple functionality, sum and threshold, is good enough for each neuron. Third, the distributed nature of a neural network compensates for the low uniformity of massively parallel devices. The last two reasons are useful to allow for the immaturity of device technology in optical computing.

Applications of optical neural networks include associative memory, computation of optical problems, and pattern classification. By means of learning capability, computation without programming can be realized. Although the theoretical background of neural computing is not mature, optical methods are expected to be useful in future implementations of such systems.

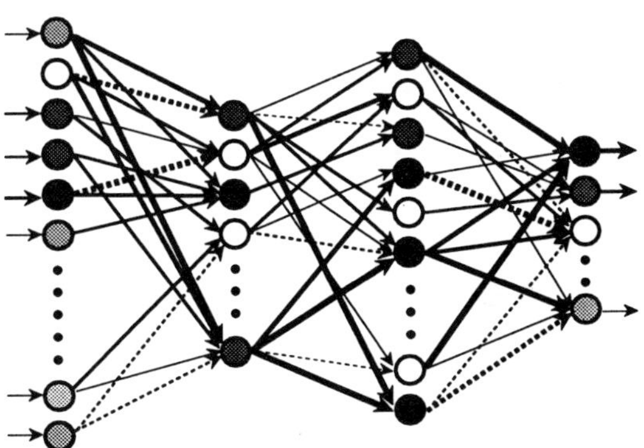

FIG. 7. Conceptual diagram of neural network. Circles represent neurons, and the density indicates the activity of the individual neuron. The neurons are connected through interconnection lines with positive and negative connecting weight. Solid and dashed lines show positive and negative connections, respectively, and the width of each line indicates the magnitude of the connection.

2.5 The Smart Pixel

The smart pixel is an optoelectronic array device consisting of light detectors, electronic amplifiers, processing circuits, and light emitters or modulators. Figure 8 shows the general concept. The smart pixel combines advantageous features of optics and electronics. Optics is advantageous in the long and complicated interconnections, whereas electronics is advantageous in nonlinear functions including signal amplification and logical processing. Thus, both technologies can be used effectively on a smart pixel.

As fabrication technologies for smart pixels, multiple quantum well SEEDs, optoelectronic integrated circuits (OEICs), vertically integrated array devices, and liquid-crystal modulators integrated with very large-scale integration (VLSI) electronics are under investigation. In terms of function complexity, smart pixels involve from simple fixed functions to complicated numerical or logical computations. Thus, some of the processing devices for digital optics are classified as smart pixels.

Applications of smart pixels are as routing or switching nodes in optical interconnection networks, processing elements for optoelec-

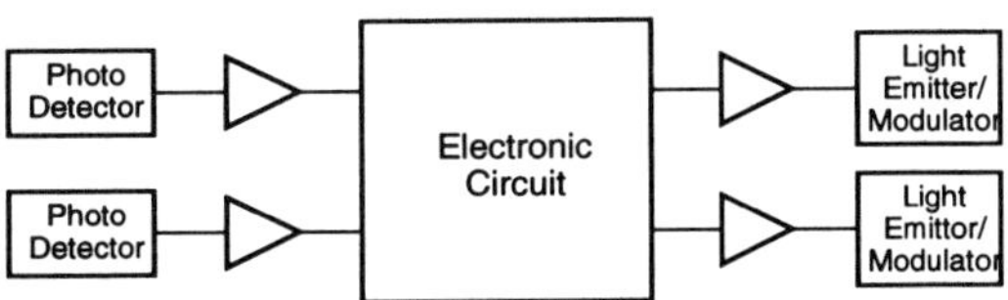

FIG. 8. General concept of smart pixels. Data represented by optical signals are fed into the device through photodetectors and converted into electronic signals. After amplification, the data are processed by an electronic circuit. Various kinds of processing are considered for the electronic circuit such as logical operation, thresholding, or packet processing. Although some of them can be carried out optically, processing efficiency and fabrication flexibility are the main reasons for electronic implementation. The result is converted back into optical form with light emitters or modulators.

tronic multiple processor systems, optical competitive learning neural computing systems, and so on. Examples of concrete smart pixels are FET-SEED (field-effect transistor–SEED) (Miller *et al.*, 1989), VSTEP, and competitive learning chip with liquid-crystal modulator (Wagner and Slagle, 1993).

3. ARCHITECTURE OF OPTICAL COMPUTERS

Since an optical computer is considered the final goal of optical computing research, several interesting architectures have been presented. These architectures are based on different principles and strategies. Figure 9 shows present trends of investigation on optical computers.

3.1 Free-Space Multistage Interconnection Network Architecture

Free-space MIN architecture is based on optical MIN (see Sec. 2.3) and a feedback loop as shown in Fig. 10 (Murdocca, 1990). Using the dimensionality of the image plane, multiple MIN's are integrated and operate together. As a result, the capability of functional specification is increased, and the required volume of the system can be reduced. This architecture is considered a pipeline processor composed of fine-grain stages. Demonstration systems have been constructed with S-SEED (symmetric SEED) and a reflective optical system (Prise *et al.*, 1991).

3.2 OPALS (Optical Parallel Array Logic System)

OPALS is an architecture composed of a multifunctional processor based on optical array logic, an encoder, input-output ports, and a feedback loop (Tanida and Ichioka, 1986). Compared with the free-space MIN architecture, the OPALS processes image data in parallel while keeping their structure. Parallel logical operations, including neighborhood operations for image data, are executed by the OPALS. As instances of the OPALS, several systems have been considered—e.g., H-OPALS (hybrid OPALS), M-OPALS (modular OPALS), and P-OPALS (pure OPALS).

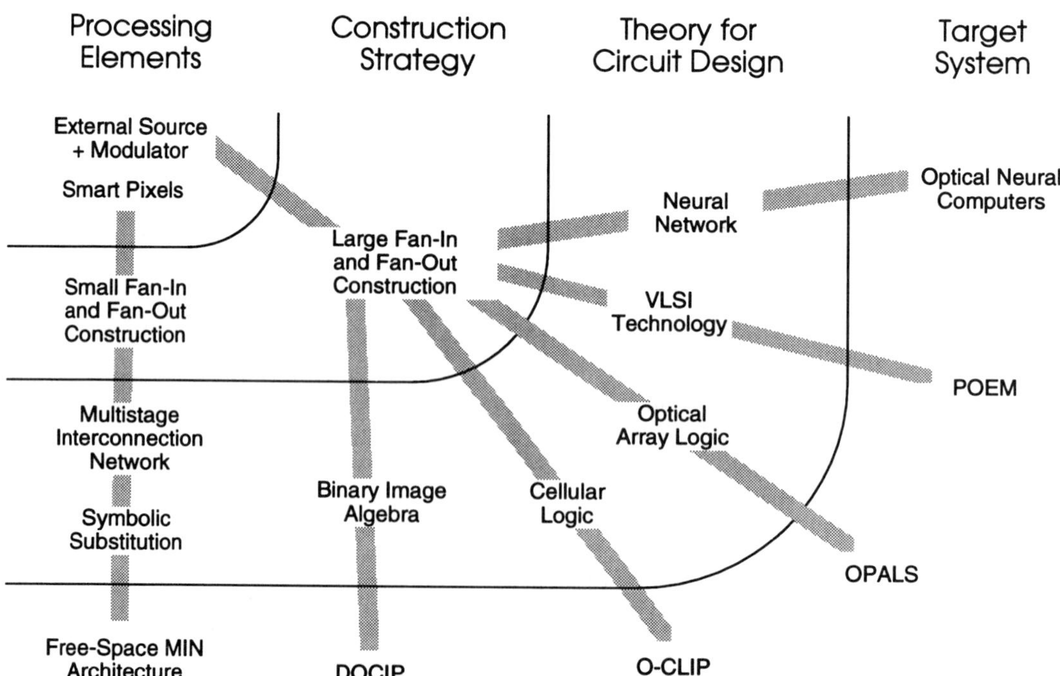

FIG. 9. Current trend of investigations on optical computers. Since a computer is a highly developed system, various kinds of technologies must be applied. Details of each item are described in the text.

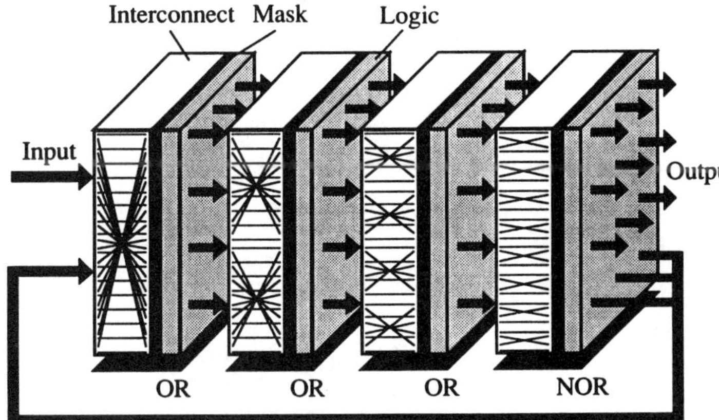

FIG. 10. Free-space MIN architecture. Several stages are connected by different sizes of regular interconnection networks, e.g., crossover networks. Just behind each network, a mask is inserted to specify the logical circuit. Logic elements execute a simple logical operation, OR or NOR, for their two inputs. Some signals are fed back to increase the processing capability of the architecture (Murdocca, 1990).

Preliminary experimental H-OPALS's have been constructed combining an electronic circuit and an optical correlator.

3.3 O-CLIP (Optical Cellular Logic Image Processor)

O-CLIP (optical cellular logic image processor) (Wherrett *et al.*, 1990) is designed to execute cellular logic, a kind of logical system for multidimensional processing described in Sec. 4.1. The O-CLIP is composed of sequentially connected optical devices and holographic optical elements for space-invariant interconnection. As the optical devices, ZnSe BEAT (bistable etalon with absorbed transmission) devices and S-SEEDs are used.

3.4 DOCIP (Digital Optical Cellular Image Processor)

DOCIP is also designed to execute cellular logic (Huang, 1990). While the O-CLIP uses space-invariant interconnection, the DOCIP adopts space-variant interconnection using an array of hologram facets to establish optical signal lines. The principle used in the DOCIP is to map a logical circuit composed of NOR gates onto an array of optical NOR gates and optical signal lines connecting individual gates as shown in Fig. 11. Design of a processor executing operations for one pixel has been presented and experimentally demonstrated with a liquid-crystal SLM (spatial light modulator).

3.5 POEM (Programmable Optoelectronic Multiprocessor)

POEM is an architecture based on multiple VLSI microprocessors and an optical dynamic interconnection system (Kiamilev *et al.*, 1989). In the POEM, processing is mainly executed by microprocessors and communication between the microprocessors is achieved by electronic or optical interconnection. Optical input/output ports on each microprocessor are prepared for optical interconnection.

4. PROGRAMMING IN OPTICAL COMPUTING

Programming is an important functionality required for any computing system. Effective programming theories have been presented for two-dimensional data or image processing. Such programming theories also take important roles in optical computing research.

4.1 Cellular Logic

Cellular logic is a logical system to describe logical processing on multidimensional data. Originally, von Neumann proposed the cellular automaton as a self-reproducing cellular array machine. Cellular automata have been extended to treat general processing on multidimensional data, as reviewed by Preston and Duff (1984). In a cellular automaton, an array of data is transformed into another array of data on inspection of neighboring

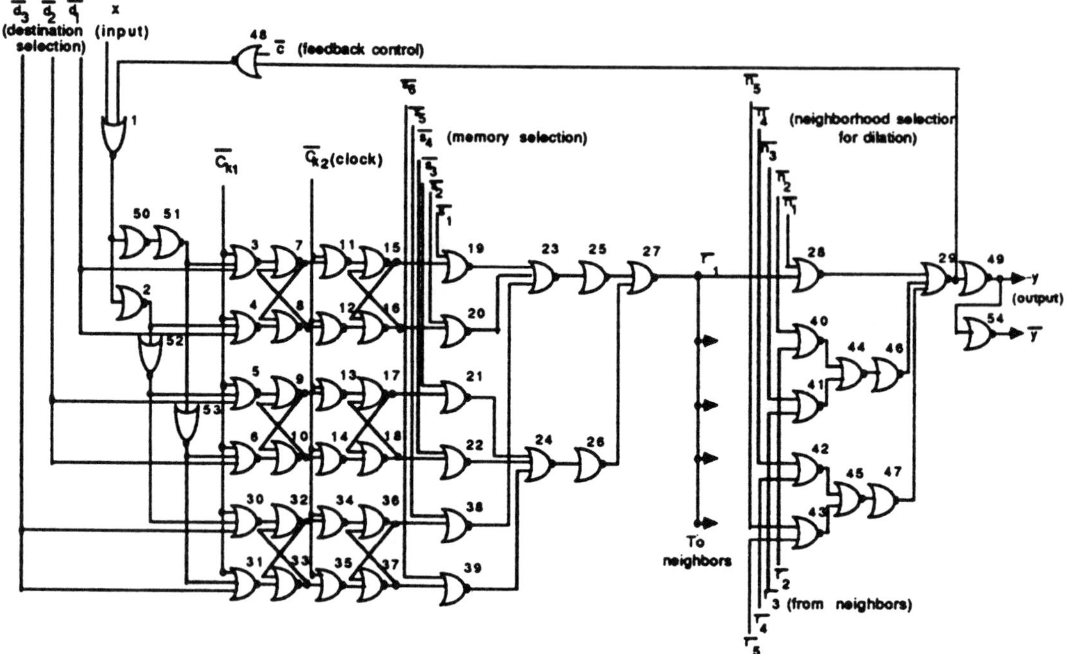

FIG. 11. Circuit design of a processor corresponding to one pixel in the DOCIP. Only NOR gates are used to achieve the required functions for the processor. The number on each NOR gate indicates the address of the gate in the array device. All gates are located on the same device and connected by space-variant interconnection with optical feedback. Note that to process image data, the same circuits must be arranged on a uniform plane (Huang, 1990).

data. Since a cellular automaton is a kind of hardware, the programming method for the hardware should be distinguished from it. The programming method for a cellular automaton is called cellular logic. Although greater than two-dimensionality can be considered in cellular logic, the two-dimensional case is the most popular in optical computing. In particular, image processing is an important application of cellular logic and much research is found in the optical computing field.

4.2 Symbolic Substitution

Symbolic substitution is a technique for executing logical operations with pattern transformation, presented by Huang (1983). Figure 12 shows the processing procedure of symbolic substitution. This technique is proposed as an effective computing paradigm for optical implementation. In the paradigm, information is represented by binary spatial patterns on the image and processed by space-invariant transformation for the patterns. The transformation is achieved in two stages, recognition and substitution, both of which are implemented by optical correlators. Any logical and higher level processing can be programmed by configuration of the substitution rules. As applications of symbolic substitution logic, the following subjects have been reported: numerical processing, image processing, emulation of a Turing machine, modified-signed-digit computing, residue arithmetic, systolic array computing, and others.

4.3 Optical Array Logic

Optical array logic is based on parallel logic gates by optical shadow casting and analogy between the scheme and array logic in electronics (Tanida and Ichioka, 1985). Figure 13 illustrates the processing procedures used in array logic. In this scheme, a pair of images is converted into a coded image composed of spatial patterns corresponding to the values of individual data in the im-

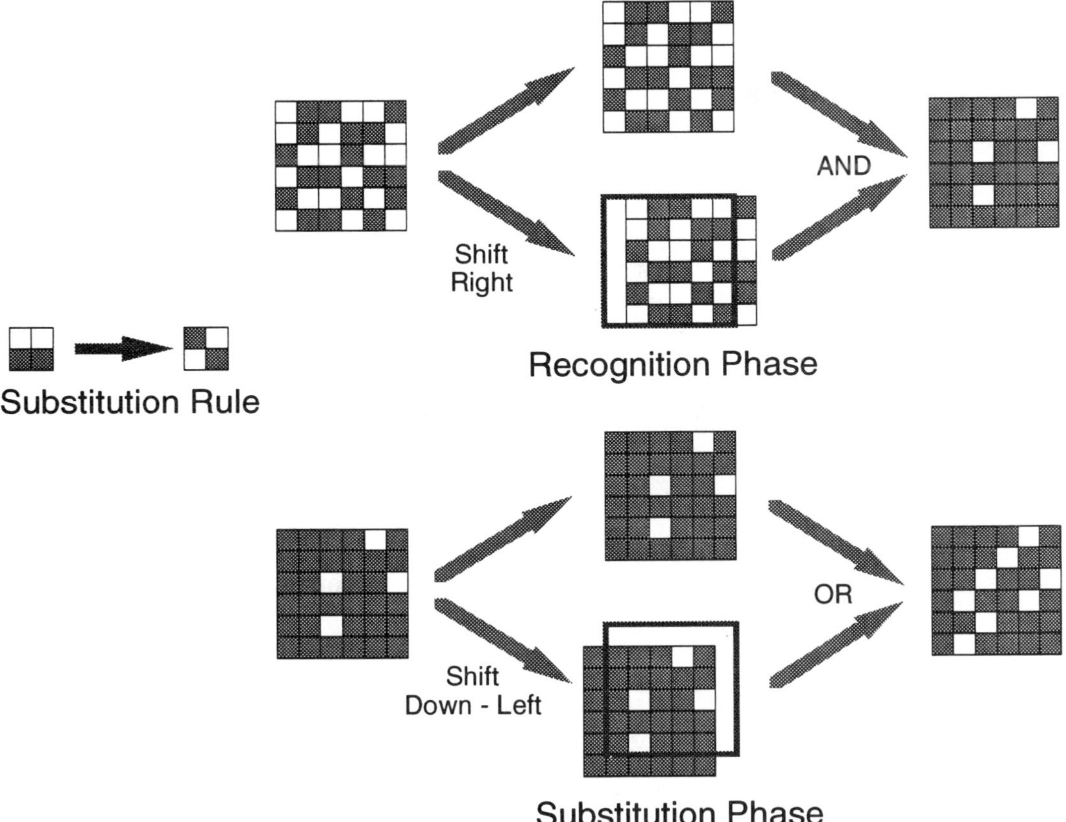

FIG. 12. Processing procedure of symbolic substitution. The substitution rule is shown on the left, the left-hand and right-hand parts being the search and substitute patterns, respectively. In the recognition phase, the locations of the search pattern in the target image are extracted. To execute the process, the target image is duplicated and one of the two duplicates is shifted to the right. An AND operation for the two images provides the top right corner of the occurrence of the search pattern as a white pixel. In the substitution phase, the result of the recognition phase is processed by the same procedure as used in the recognition phase except for the direction of the shift (down and left) and the logical operation (OR). As a result, all occurrences of the search pattern in the target image are converted into the substitution pattern. For different substitution rules, the number of duplications and the directions of shifts are changed.

ages. Several sets of discrete correlations between the coded image and discrete kernels, called operation kernels, provide any logical operations for neighboring data on the images. In optical array logic, a combination of the operation kernels is used to specify the processing to be executed. To describe the operation in optical array logic, a native programming language has been developed. Applications of optical array logic cover a wide variety of fields, including numerical processing, image processing, structured data computing by systolic array, inference, database management, and so on. Figure 14 depicts an example of parallel processing by optical array logic.

4.4 Binary Image Algebra

Binary image algebra is an algebraic system for describing operations on binary image data compactly (Huang, 1990). Binary image algebra is composed of an image space and a family of operations, consisting of three fundamental operations and five elementary images. The fundamental operations are complement, union, and dilation. Each elementary image is an image consisting of a pixel point located at either the origin or one of its neighboring points. Binary image algebra is applied not only to description of image and numerical processing but also to design of effective logical circuits for the DOCIP.

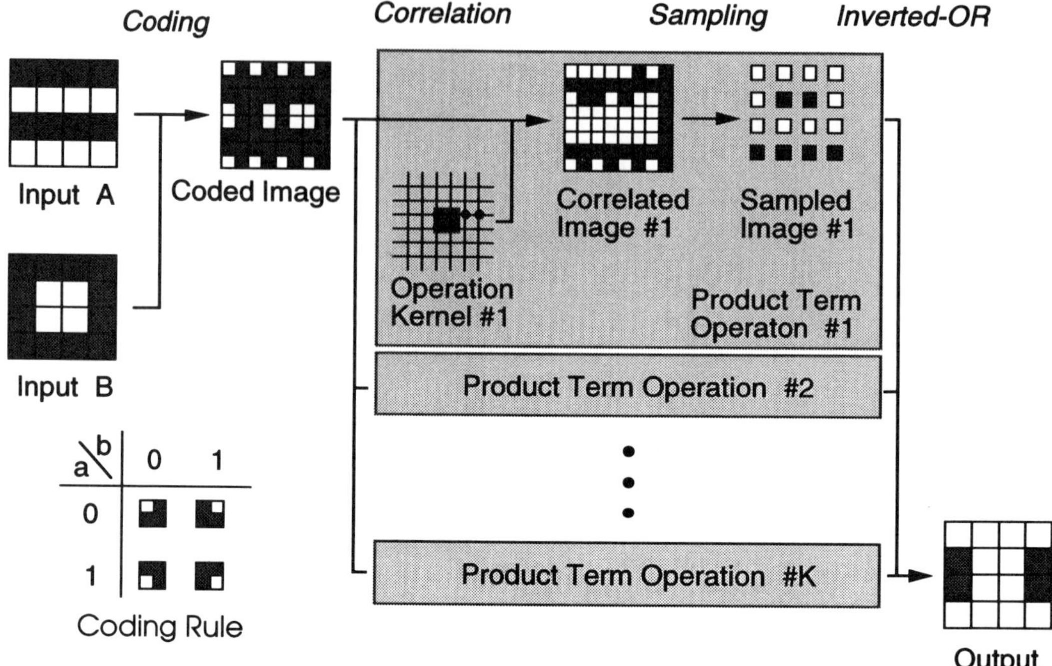

FIG. 13. Processing procedure of optical array logic. The input images are converted into a coded image by application of the coding rule in individual pixel pairs at the same position in the inputs. The coded image is correlated with an operation kernel specifying a logical operation for neighboring pixels. The operation kernel is a kernel of correlation whose elements are composed of several delta functions located at the grid points. As the effect of correlation with the operation kernel, the coded image is duplicated and each of the duplicates is shifted and overlapped. By the functional meaning of the operation kernel, the shaded portion corresponds to the center pixel of the neighborhood area and the dot pattern specifies the operation. After correlation, sampling of the correlated image provides an intermediate result. To complete the general logic operation, several sets of correlations with different operation kernels are executed and all intermediate results are collected by an inverted-OR operation.

4.5 Image Logic Algebra

Image logic algebra is designed as a generic language for optical parallel processing (Fukui and Kitayama, 1992). Image logic algebra describes transformation of an image by neighborhood configuration patterns. The transformation is executed by correlation between an object image and a neighborhood configuration pattern. As applications of image logic algebra, image processing, numerical processing, and wire routing problems have been demonstrated.

5. FUTURE APPLICATIONS OF OPTICAL COMPUTING

Since optical computing is not a mature research field, it is difficult to predict its effective applications. However, considering the motivation and features of optical computing, massively parallel processing is a promising application. Of course, massively parallel electronic computers will be developed as extensions of current computers, which accomplish the same tasks as those that optical computing aims at. However, to clarify the position of optical computing methods, it is valuable to mention concrete applications for optical computing.

At present, three promising application fields are considered. These are high functional computing, real-world computing, and high-reliability computing. High functional computing is linked with current optical communication technology. An optical computing system is used as a controller or a processor for information carried by light signals. Thus, without changing the medium of information, quite high throughput is expected. Real-world computing is a theme of

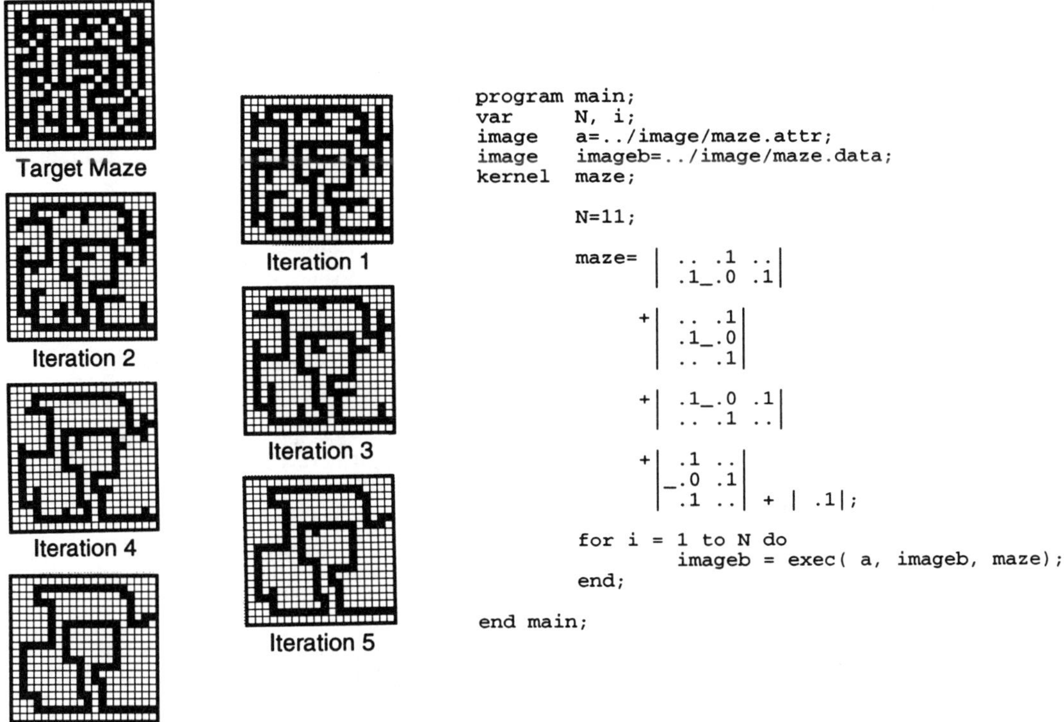

FIG. 14. Maze solution by optical array logic. For a given maze, in which paths are represented by dark pixels, processing for eliminating dead-end paths is executed in parallel. Dead ends are searched by logical template matching and converted into white pixels. After iterative application of the same processing, the correct path is extracted. The processing trace and the program written in optical array logic language are shown.

a national program promoted by the Japanese government. In real-world computing, multidimensional data appearing in the real world should be processed in real time. To treat such massive information, the processing capabilities offered by optical computing are promising. High-reliability computing is expected from the features of optical computing of high resistance against electromagnetic noise and of redundancy obtained by massive parallelism.

6. CONCLUSIONS

Optical computing is a promising area for massively parallel computing. Although it is young, many fruitful results have been obtained and active studies are going on. Progress in fabrication technologies on optical and optoelectronic devices is tremendous and makes it possible to clarify the advantages of optical computing over conventional electronics. As a benefit of device progress, some architectures of optical computers are in the demonstration stage. Results of the demonstrators will be used for further investigation on optical devices. Future applications of optical computing are expected to be more effective as a result of cooperation with electronic systems. Compensating for the weak points of electronics, optical computing is expected to play an important role in the future of the information-processing area.

GLOSSARY

Birefringence: An optical phenomenon in which a material exhibits different refractive indices for the two orthogonal components of light.

Boolean Algebra: A unified mathematical system for propositional logic, logic circuits, and power sets, which is the basis of digital computing schemes.

Correlation: Linear operation between two functions indicating the degree of similarity.

Fan-In (-Out): The maximum number of input (output) signals allowable for a device.

Fourier Transform: The transformation between the spatial (or temporal) and the frequency domains.

Fourier Transform Property of Lens: The specific function of a lens of implementing a two-dimensional Fourier transform on image data.

Holography: An optical technique to record light waves, including interference phenomena between the object and reference light beams.

Optical Bistability: Existence of two stable states, one transparent and the other absorbing (or scattering), of light signals in material.

Optical Nonlinear Phenomena: Phenomena appearing in the case of strong light power. Although weak light signals can be treated with linear functions in regard to propagation and interaction with materials, nonlinear effects become dominant in the high-power regime.

Photoelasticity: The production of optical birefringence in an elastic body as a result of deformation of the body.

Spatial Code Pattern: Patternlike representation of information suitable for optical computing.

Spatial Filtering: A method of optical information processing based on spatial frequency, the number of lines in unit length, of image data. Specific components of spatial frequency are cut off or modulated by optical Fourier transformation by a lens to provide various processing of the image such as image enhancement, edge detection, and so on.

Spatial Light Modulator: A device producing a position-dependent modulation of optical signals passing through it or reflected from it.

Time- (Space-) Bandwidth Product: Measure of the amount of information treated by a system or apparatus, defined as the product of the maximum values in the temporal (or spatial) and frequency domains.

Works Cited

Flannery, D. L., Horner, J. L. (1989), *Proc. IEEE* **77**, 1511–1527.

Fukui, M., Kitayama, K. (1992), *Appl. Opt.* **31**, 4645–4656.

Goodman, J. W. (1989), in: B. S. Wherrett, F. A. P. Tooley (Ed.), *Optical Computing*, Edinburgh: Scottish Universities Summer School Press. p. 7.

Horner, J. L. (1987), *Optical Signal Processing*, New York: Academic.

Huang, A. (1983), in: *Technical Digest, IEEE Tenth International Optical Computing Conference*, pp. 13–17.

Huang, K.-S. (1990), *A Digital Optical Cellular Image Processor*, Singapore: World Scientific.

Hwang, K., Briggs, F. A. (1984), *Computer Architecture and Parallel Processing*, New York: McGraw-Hill.

Kasahara, K., Tashiro, Y., Hamao, N., Sugimoto, M., Yanase, T. (1988), *Appl. Phys. Lett.* **52**, 679–681.

Kiamilev, F., Esener, S. C., Paturi, R., Fainman, Y., Mercier, P., Guest, C. C., Lee, S. H. (1989), *Opt. Eng.* **28**, 396–409.

Lee, J. N., VanderLugt, A. (1989), *Proc. IEEE* **77**, 1528–1557.

McAulay, A. D. (1991), *Optical Computer Architectures*, New York: Wiley.

McClelland, J. L., Rumelhart, D. E., the PDP Research Group (1986), *Parallel Distributed Processing*, Vol. 2, Cambridge, MA: MIT Press.

Miller, D. A. B., Chemia, D. S., Damen, T. C., Gossard, A. C., Wiegmann, W., Wood, T. H., Burrus, C. A. (1984), *Appl. Phys. Lett.* **45**, 13–15.

Miller, D. A. B., Feuer, M. D., Chang, T. Y., Shunk, S. C., Henry, J. E., Burrows, D. J., Chemia, D. S. (1989), *IEEE Photo. Tech. Lett.* **1**, 62–64.

Murdocca, M. (1990), *A Digital Design Methodology for Optical Computing*, Cambridge, MA: MIT Press.

Prise, M. E., Craft, N. C., Downs, M. M., LaMarche, R. E., D'Asaro, L. A., Chirovsky, L. M. F., Murdocca, M. J. (1991), *Appl. Opt.* **30**, 2287–2296.

Preston, Jr., K., Duff, M. J. B. (1984), *Modern Cellular Automata*, New York: Plenum.

Rumelhart, D. E., McClelland, J. L., the PDP Research Group (1986), *Parallel Distributed Processing*, Vol. 1, Cambridge, MA: MIT Press.

Sawchuk, A. A., Strand, T. C. (1984), *Proc. IEEE* **72**, 758–779.

Schaefer, D. H., Strong, III, J. P. (1977), *Proc. IEEE* **65**, 129–138.

Tanida, J., Ichioka, Y. (1983), *J. Opt. Soc. Am.* **73**, 800–809.

Tanida, J., Ichioka, Y. (1985), *J. Opt. Soc. Am. A* **2**, 1245–1253.

Tanida, J., Ichioka, J. (1986), *Appl. Opt.* **25**, 1565–1570.

Wagner, K., Slagle, T. M. (1993), *Appl. Opt.* **32**, 1408–1435.

Wherrett, B. S., Craig, R. G. A., Snowdon, J. F., Buller, G. S., Tooley, F. A. P., Bowman, S., Pawley, G. S., Redmond, I. R., McKnight, D. J., Taghizadeh, M. R., Walker, A. C., Smith, S. D. (1990), in: R. Arrathoon (Ed.), *Digital Optical Computing II*,

Proceedings Vol. 1215, Bellingham, WA: SPIE, pp. 264–279.

Further Reading

Arrathoon, R. (Ed.) (1989), *Optical Computing*, New York: Marcel Dekker.

Arsenault, H. H., Szoplik, T., Macukow, B. (Eds.) (1989), *Optical Processing and Computing*, Boston: Academic.

Caulfield, H. J., Gheen, G. (Eds.) (1989), *Selected Papers on Optical Computing*, Washington: SPIE.

Feitelson, D. G. (1988), *Optical Computing*, Cambridge, MA: MIT Press.

Karim, M. A., Awwal, A. A. S. (1992), *Optical Computing: An Introduction*, New York: Wiley.

Lee, S. H. (Ed.) (1981), *Optical Information Processing*, Berlin: Springer-Verlag.

Wherrett, B. S., Tooley, F. A. P. (Eds.) (1989), *Optical Computing*, Edinburgh: Scottish Universities Summer School Press.

OPTICAL FILTERS

J. A. Dobrowolski, *Institute for Microstructural Sciences, National Research Council of Canada, Ottawa, Ontario, Canada*

INTRODUCTION

In the broadest sense of the term, an optical filter is any device or material that is deliberately used to change the spectral intensity distribution, the phase, or the state of polarization of the electromagnetic radiation incident upon it. The change in the spectral intensity distribution may or may not depend on the wavelength. The filter may act in transmission, reflection, or both.

Currently, the spectral region for which optical filters are constructed extends from about 0.005 to 1000 μm, although in this article the emphasis will be on filters for the visible and adjacent spectral regions (0.2–10 μm). For simplicity, in the following, all electromagnetic radiation, regardless of wavelength, will be referred to as "light."

Polarizers and phase retarders, although they fall within the broad definition of a filter given above, are not discussed here. Prism, diffraction, and holographic grating monochromators and Fabry-Pérot and Fourier-transform interferometers can all be used to filter incident light. So can certain fiber-optic and waveguide devices. However, these are instruments rather than filters, and thus lie outside the scope of the topic.

In this article, the basic theory of one or more transmission or reflection filters placed in series or in parallel is first reviewed. Next, the most important physical principles used in the construction of optical filters are outlined and their principal advantages and disadvantages discussed. Typical performances of various specific filter types based on the above physical principles are then presented. This is followed by a discussion of the parameters that need to be specified when designing or ordering filters. Finally, attention is drawn to some related topics that, for lack of space, are not discussed in this article.

1. GENERAL THEORY OF FILTERS

There are three different wavelength-dependent elements that have to be considered when dealing with filters. The first of these

3-527-28134-7/95/$5.00 + .50

is the spectral energy distribution $S(\lambda)$ of the incident light. The light source may be, for example, the sun, an incandescent lamp, a spectral line source, or a laser. The next element, $F(\lambda)$, represents the combined response of the filter and of the associated optical system. It may involve a simple quantity like the spectral transmittance or reflectance of one or more filters placed in series or in parallel. It may be a more complex colorimetric or photometric function. The third element is the spectral responsivity $D(\lambda)$ of the device that is used to detect the filtered light. Typical detectors include the human eye, a photocell, or a photographic emulsion. The quantity that is detected is

$$\int_{\lambda} S(\lambda)F(\lambda)D(\lambda)d\lambda. \tag{1}$$

This integral is defined over a wavelength range that extends from zero to infinity. However, $F(\lambda)$ need only be defined over the spectral region in which the product $S(\lambda)D(\lambda)$ is significant. If the spectral characteristics of the primary filter are not controlled over this entire wavelength range, additional blocking filters might have to be used.

1.1 Transmittance, Reflectance, Absorptance, and Scatter

When a light of intensity I_0 is incident on an optical filter, the light may be partially transmitted (I_T), reflected (I_R), absorbed (I_A), or scattered (I_S) by the filter. If the filter is a parallel plate, the transmitted beam will propagate in the same direction as the incident beam. Reflection from a smooth surface is specular and the angles of incidence and reflection are the same. The intensity of scattered light varies with angle and takes place in both the forward and backward directions.

In general, the most complete specification of a filter is in terms of the spectral and angular variation of the transmittance T, reflectance R, absorptance A, and scatter S, defined as follows:

$$T = I_T/I_0; \quad R = I_R/I_0;$$
$$A = I_A/I_0; \quad \text{and} \quad S = I_S/I_0. \tag{2}$$

These expressions are valid for normal incidence of light and when the entrance and exit media are the same. Because energy is conserved in optical filters, the following relationship holds for any wavelength λ of the incident light:

$$T + R + A + S = 1. \tag{3}$$

This expression assumes that S is the total scatter integrated over all angles in the forward and backward directions. T, R, A, and S are dimensionless quantities that are sometimes expressed as percentages of the incident light I_0. A related quantity, the density of a filter, sometimes also called the absorbance, is given by

$$D = \log(1/T). \tag{4}$$

Filtering may occur at the surface of the filter, or may be distributed throughout it, depending on the physical mechanism on which the filter is based.

1.2 Surface Reflection Effects

In addition to the primary filter mechanism, there are secondary effects that affect the performance of a filter. One such effect is the reflection at the interfaces between the filter and the surrounding media. For example, consider the overall transmittance T_{tot} of the filter of Fig. 1, whose surfaces may be coated. Let the transmittances and reflectances of the two interfaces be T_1, R_1 and T_2, R_2, respectively. The internal transmittance

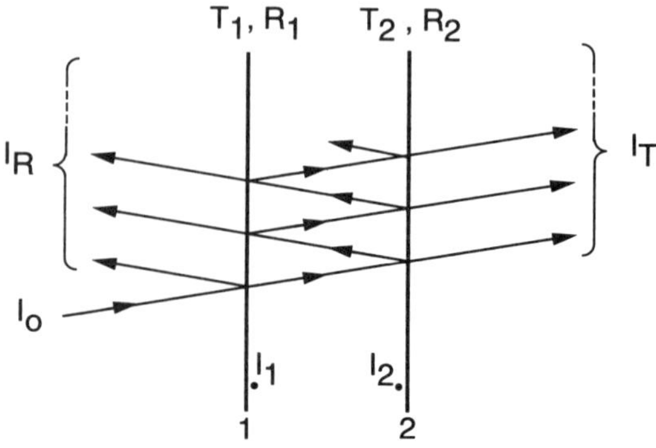

FIG. 1. Multiple reflections in a filter with two interfaces (see text).

τ of the plate is defined by

$$\tau = I_2/I_1. \tag{5}$$

Here, I_1 and I_2 are the intensities of the light just beyond the first and just before the second interface, respectively. If the reflected beams are not coherent with one another, the overall transmittance is given by

$$T_{\text{tot}} = \frac{T_1 T_2 \tau}{1 - R_1 R_2 \tau^2}. \tag{6}$$

For an uncoated interface between air and a nonabsorbing or weakly absorbing filter of refractive index n the normal incidence reflectance R is given by the expression

$$R = \left(\frac{n-1}{n+1}\right)^2. \tag{7}$$

For glass $R_1 = R_2 \approx 0.04$, $T_1 = T_2 \approx 0.96$, and therefore Eq. (6) simplifies to $T_{\text{tot}} \approx 0.92\tau$. However, for coated interfaces and for materials with a higher refractive index, expression (6) must be used in calculations if significant errors are to be avoided.

1.3 Transmission Filters in Series and in Parallel

To obtain a desired spectral transmittance, it is often necessary to combine several filters. One common approach is to place several filters in series. An approximate expression for the resultant transmission T_{series} of k filters placed in series is

$$T_{\text{series}} \approx T_1 T_2 T_3 \cdots T_i \cdots T_k. \tag{8}$$

This expression assumes

1. that T_i is the total transmission of the ith filter evaluated according to Eq. (6);
2. that the filters are slightly inclined to each other so that multiple reflections between the various interfaces are avoided; and
3. that only the directly transmitted beam reaches the detector or the observer.

Reflection losses can be reduced with antireflection coatings or by cementing the various filter components.

Some spectral transmittance curves cannot be easily achieved by placing filters in series alone. For certain applications it is quite acceptable to place filters in parallel. This introduces areas as additional design parameters. The expression for the resultant transmission of filters placed in parallel, T_{parallel}, is

$$T_{\text{parallel}} = \left(\frac{a}{A} T_a + \frac{b}{A} T_b \cdots\right), \tag{9}$$

where A is the overall area of the filter, a, b, $\cdots$ are the areas of the individual zones, and T_a, T_b, $\cdots$ are the transmittances of the zones. Great care must be exercised when using such filters. This arrangement assumes that the intensity of the incident beam is uniform. Errors due to a lack of beam uniformity will be reduced if the individual filters are cut into a large number of small, regular elements and reassembled in the form of a mosaic. Care must be also be used when employing the filtered light. To ensure that there is no spatial variation in the spectral composition of the beam it is best to pass the light through an integrating sphere.

For solution of problems with very complex spectral transmittance requirements a series/parallel filter arrangement may be necessary.

1.4 Reflection Filters in Series

If light is reflected from k different filters, the resultant reflectance R_{series} will be given by

$$R_{\text{series}} = R_1 R_2 R_3 \cdots R_k, \tag{10}$$

which is analogous to Eq. (8) for the resultant transmittance of filters placed in series. It is clear that R_{series} will be significant only at those wavelengths at which every one of the filters has a significant reflectance. Schematic representations of some multiple-reflection arrangements are given in Fig. 2. Other arrangements are possible. For the sake of convenience, the number of different reflecting surfaces used is normally restricted. Reflection filters placed in series require more space and accurate alignment and hence are more complicated to use than transmission filters. However, if in a given application these shortcomings can be accepted, multiple-reflection filters offer important advantages,

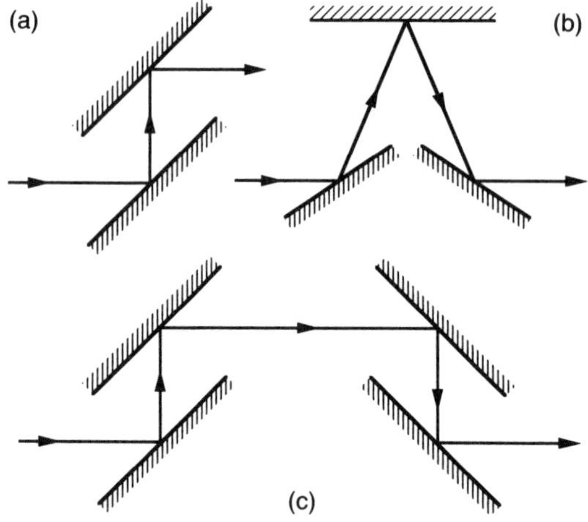

FIG. 2. (a) Two, (b) three, and (c) four reflection filters placed in series.

which stem mainly from the nature of the reflectors available for their construction.

2. PHYSICAL PRINCIPLES USED FOR THE CONSTRUCTION OF FILTERS

Filters can be based on many different physical phenomena, including absorption, refraction, interference in thin films and holographic emulsions, diffraction, scattering, and polarization—principles that are briefly outlined in this section.

2.1 Absorption

Absorption filters are constructed from plane-parallel plates of one or more materials with refractive indices n and extinction coefficients k that vary with wavelength in such a way that, through the selection of appropriate thicknesses of the individual materials, the desired transmittance is obtained.

An enormous number of colored glasses, crystals, sintered materials, gelatin films with dyes, inorganic and organic compounds, semiconductors, liquids, gases, and thin films exist with interesting absorption features. The main advantages of absorption filters are that they are simple to use, insensitive to angle of incidence, and cheap or moderate in cost. Disadvantages are that their spectral features depend on the existence of suitable materials. It follows that elaborate spectral transmittance curves cannot be achieved through the use of absorption filters only.

When light passes through a homogeneous, isotropic, but absorbing material, its intensity is reduced. Bouguer's law relates the internal transmittance [Eq. (5)] of an absorbing material to its thickness d:

$$\tau_\lambda = e^{-\alpha_\lambda d}. \tag{11}$$

The spectral absorption coefficient a_λ depends on the nature of the material and is related to the extinction coefficient k as follows:

$$\alpha_\lambda = 4\pi k/\lambda. \tag{12}$$

The most important absorption filters are colored glasses. They consist of a base glass to which colorants have been added. Typical base glasses are potassium, sodium, phosphate, borate, borosilicate, and silicate glasses. The colorants may take the form of metal ions, metal atoms, or some nonmetallic elements or their compounds. The spectral properties of the filter glasses depend not only on the combination of the colorant and base glass used and the furnace atmosphere, but also on the temperature and duration of the subsequent annealing. The optical properties of most colored glass filters are controlled in the 0.3- to 2.4-μm spectral region. However, variations in colorant concentration, and hence in the transmission of a certain thickness of glass, exist from one batch to another. The transmission of some glass filters depends on the ambient temperature [Fig. (3a)].

A large number of solid and liquid organic dyes also have spectral-absorption characteristics that can be utilized for filter construction. Although such filters are generally less stable than glass filters, their spectral features are often much sharper. Gelatin filters are the most common type of solid organic filter. The gelatin film is either lacquered or cemented between glass for protection. Some gelatin filters can be quite fragile and their transmission characteristics can change on exposure to light. Many are destroyed by heating beyond 50–55 °C or immersion in water. However, gelatin filters are cheap. In the lacquered version they can be cut to any size and are thin, flexible, and convenient to

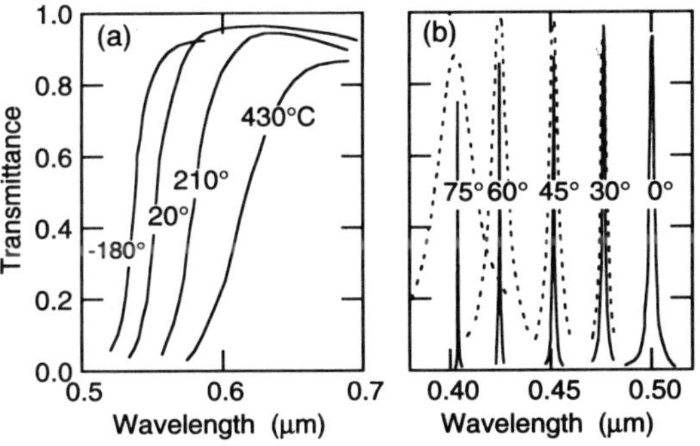

FIG. 3. Shortcomings of optical filters. (a) Effect of temperature on the spectral transmittance of an orange selenium glass filter. (b) Shift in transmission peak and polarization splitting in a thin-film interference filter with oblique incidence of light. The broken and solid curves correspond to light polarized parallel (p) and perpendicular (s) to the plane of incidence.

use. Most organic filters are designed for use in the 0.3- to 0.8-μm spectral region.

For a liquid filter, α_λ is replaced by $\epsilon_\lambda c$ (Beer's law), where c is the concentration of the solution and ϵ_λ is the spectral absorption coefficient per unit concentration. A number of liquids and gases are suitable for the construction of filters. However, many of these lack adequate stability with time and temperature or are decomposed by the light to be filtered. Liquid and gaseous filters are less convenient to use than solid filters. However, they are useful for the isolation of the ultraviolet spectral region. Liquid-filter absorption characteristics are sometimes very sensitive to temperature and to the amount of oxygen dissolved in the liquid. Some solutions show a marked fluorescence, while others become turbid as the material dissolved in the liquid precipitates.

A number of metals in thin-film form exhibit strong absorption effects that can be used to construct optical filters. Although the primary process is absorption, interference of light reflected at the interfaces of the layer also affects the spectral transmission characteristics. The films can be produced by different processes and they may be self-supporting. They are especially useful as reflectors, attenuators, beam splitters, and cutoff filters for the ultraviolet, visible and infrared spectral regions.

Yoshinaga described another form of absorption filter. It consists of powders of materials with strong infrared absorption bands and reflection peaks embedded in polyethylene. This host material does not absorb above 14 μm and can be pressed into thin uniform sheets. Although scattering also plays a part, the main filtering mechanism in Yoshinaga filters is absorption. By using mixtures of powders of one or more suitable materials, one can construct short-wavelength cutoff filters for the 20–500-μm spectral region. These filters work in transmission and therefore are more convenient to use than the *Reststrahlen* reflection filters described below. In addition, the number of potential materials for their construction is much larger because size is not important and only small amounts of powder are needed.

2.2 Reflection

At normal incidence the Fresnel reflection coefficient of an interface between air and a material with a complex refractive index $n - ik$ is given by

$$R = \frac{(n-1)^2 + k^2}{(n+1)^2 + k^2}. \tag{13}$$

If the material is opaque, the light that is not reflected is absorbed within the material. Metals and *Reststrahlen* materials are the two most frequently used materials for the construction of reflection filters.

The refractive index and/or the extinction coefficient of metals are frequently much larger than unity over an extended spectral region. It follows from Eq. (13) that this results in a high reflectance. Metals are therefore used whenever a high reflectance over a wide wavelength region is required. However, different metals have to be used in different parts of the spectrum for the highest possible reflectance. Reflectors are most commonly made by vacuum deposition of the metal onto a suitable glass, quartz, ceramic, or metal substrate. For good adhesion of the layers, the substrate surface must be specially prepared or precoated.

Metals are less stable than ceramic materials and some are quite soft. The thin oxide layers that form on some of the metals do not offer sufficient protection against abrasion and chemical attack. Metal reflectors are frequently coated with additional thin oxide layers to provide a more durable surface to protect them from this "aging." However, single-layer protective coatings usually reduce the reflectance of the metal layers. Multiple layers can enhance the reflectance, but usually only over a relatively narrow spectral region.

Most inorganic compounds have a small extinction coefficient and a refractive index that is constant over extended spectral regions. However, in the vicinity of wavelengths that excite atomic or lattice vibrations, the extinction coefficient rises steeply and the refractive index undergoes a rapid variation. It follows from Eq. (13) that high reflectance peaks may be observed at these wavelengths. This effect is called the *Reststrahlen* (residual-rays) effect and it occurs mostly in the 6- to 300-μm region. Since few satisfactory filters exist at these wavelengths, multiple reflections from *Reststrahlen* materials placed in arrangements of the type shown in Fig. 2 can be used to extract a more or less narrow spectral region from a light continuum. The number of reflections needed to achieve a certain spectral purity can be reduced by auxiliary filtering techniques.

The peak reflectance of *Reststrahlen* materials increases markedly at low temperatures. *Reststrahlen* materials in thin-film form exhibit interference maxima and minima on the short-wavelength side of the reflectance peak with their number and location depending on the thickness of the film.

2.3 Interference in Thin Films

When light falls onto a multilayer structure consisting of thin plane-parallel homogeneous films of different absorbing and nonabsorbing materials deposited onto a suitable substrate, multiple reflections take place between the various interfaces. Depending on the thicknesses of the individual layers and on the light source used, the reflected beams may be coherent and interfere with one another. Similar effects are observed in thin inhomogeneous layers in which the optical constants vary continuously with thickness. The theory of thin-film interference coatings is based on the electromagnetic theory of light and can be found in a number of standard texts. Elaborate computer programs based on this theory exist for the design of such coatings.

The spectral performance of a multilayer depends on the number of layers in the system, their thicknesses d, and their optical constants $n - ik$, as well as on the adjoining media. The refractive index profile of a typical system consisting of homogeneous thin films made of many different materials is shown in Fig. 4(a). In this figure the thicknesses are expressed in terms of some convenient central wavelength λ_0 and the interface with the substrate is at 0. Optical filters with elaborate performance characteristics can be obtained by the proper selection of the above parameters. However, the more demanding the filter specifications, the more complex the resulting multilayer system.

Interference is by far the most versatile filtering method. It has been used to construct filters for the widest range of wavelengths (0.005–500 μm) and for the largest variety of performance characteristics. Interference filters are one of the few filter types that can be used with high-energy laser sys-

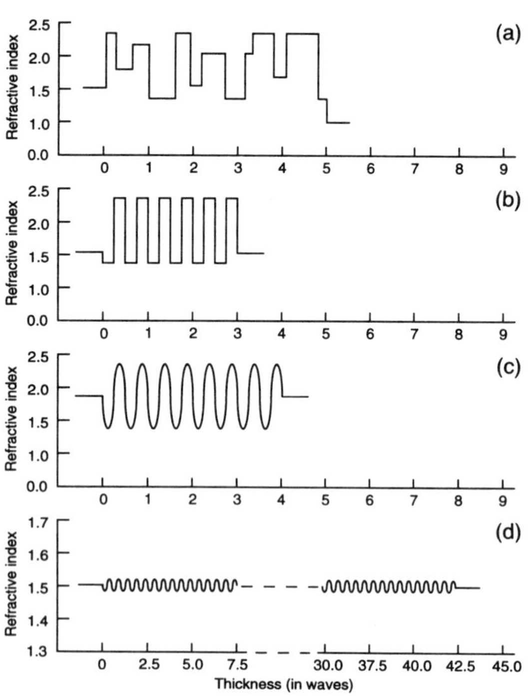

FIG. 4. Refractive-index profiles of some typical thin-film and holographic filters: (a) homogeneous multilayer system made of many materials; (b) quarter-wave stack composed of two materials; (c) periodic inhomogeneous layer (rugate filter); and (d) holographic rejection filter. Note the change in the ordinate and abscissa scales in (d).

tems. They have a number of interesting properties. For example, for nonabsorbing systems composed entirely of layers for which the optical thickness nd is a multiple of $\lambda_0/4$, the transmittance curves are symmetrical about λ_0 when plotted on a relative wave-number scale λ_0/λ. A proportional change of all the thicknesses in a nonabsorbing multilayer results merely in a linear displacement of the transmittance curve on a wave-number scale. This can be used to construct linear or circular wedge filters in which the transmission spectrum varies in a line or around the circumference of a circle, permitting easy tuning of the spectral features of interest. In general, for an interference filter containing absorbing materials, the reflectance and absorptance, but not the transmittance, will depend on which side of the filter the light is incident upon.

Multilayer interference filters also have some disadvantages. The most serious of these is that the spectral properties vary with the angle of incidence of the light. As the angle of incidence is increased, the features move toward shorter wavelengths [Fig. 3(b)]. This makes possible a limited tuning, but only at the expense of a reduced angular field.

Further, at non-normal incidence the properties will, in general, depend on the plane of polarization of the incident light [Fig. 3(b)]. For example, the average transmittance and reflectance of a multilayer for obliquely incident randomly polarized light is given by

$$T = \tfrac{1}{2}(T_p + T_s), \quad R = \tfrac{1}{2}(R_p + R_s). \tag{14}$$

In the above expressions, the subscripts p and s denote light linearly polarized parallel and perpendicular to the plane of incidence, respectively.

Finally, unless the layers are deposited by an energetic process to ensure that they are not porous, the spectral features can shift in time, or with change in temperature or humidity. Although these shifts are small, they can render useless any filter with a sharp transition between a transmission and rejection region.

2.4 Holography

Holographic filters are produced by depositing a uniform photosensitive gelatin film onto a suitable substrate, forming a standing-wave pattern with blue or blue-green light within it, and developing and fixing the emulsion. Under suitable conditions, a sinusoidal refractive-index modulation results. Because the gelatin can swell on exposure to high humidity, it is protected with a glass cover plate. Filters packaged in this way are said to withstand temperature cycling between −50 and 80 °C.

If the standing-wave pattern was parallel to the substrate surface during the exposure, a volume hologram is formed with a refractive-index profile resembling that of a thin-film rugate rejection filter [see Figs. 4(c) and 4(d)]. However, because the refractive-index fluctuation in the hologram is usually only of the order of 0.02 to 0.1, many more periods and a much higher overall thickness are required to achieve a rejection comparable to that of a rugate filter. The transmittance is low only in the immediate neighborhood of λ_0 and the resulting rejection band is very narrow. No ripples are observed in the transmission regions of holographic rejection filters because the refractive-index modulation is small and the average index of a volume hologram is close to those of the substrate and cover glass. Holographic rejection filters with transmissions down to 10^{-4} to 10^{-6} having high laser damage thresholds are now available commercially for a spectral range extending from the ultraviolet to the near infrared.

Holographic beam splitters and narrow-bandpass filters have also been described.

2.5 Scattering

Light is lost through scattering whenever it passes through a suspension of particles in a transparent medium of different refractive index. If there is no absorption, the transmission is given by

$$T = \exp(-dS), \tag{15}$$

where d is the thickness of the suspension and S is the scattering coefficient. Here, T depends on the refractive index of the medium n_0, the refractive index n_p, and the size of the particle α, as well as on the wavelength λ of the incident light. Various empirical and theoretical expressions for S have been given in the past. Two expressions given by Rayleigh are

$$S = 9\left(\frac{n^2 - 1}{n^2 + 2}\right)^2 \frac{\alpha^3}{(\lambda/n_0)^4}, \quad \alpha/\lambda \ll 1, \qquad (16)$$

$$S = 9\left(\frac{n^2 - 1}{n^2 + 2}\right)^2 \frac{\alpha}{(\lambda/n_0)^2}, \quad n \approx 1, \qquad (17)$$

where $n = n_p/n_0$. This dependence on wavelength can be exploited to construct filters.

Christiansen filters consist of small particles of a transparent optical material that are suspended in a liquid of similar refractive index but widely different dispersion. This suspension is homogeneous only at the wavelength of light for which the two dispersion curves intersect. Light of other wavelengths is scattered according to Eq. (17). A filter with a narrow-band transmission peak results.

Christiansen filters are very simple, and they have been constructed for wavelengths ranging from 0.2 to 9.0 μm. The narrowest band reported for the visible part of the spectrum was 0.0002 μm, although 10 or 100 times wider half-widths are more typical. Christiansen filters have a number of drawbacks. An aperture must be used to remove the scattered light. As the particles precipitate, the filter deteriorates. Temperature control is necessary to maintain a constant transmission wavelength because the refractive indices of liquids change more rapidly than those of solids.

The refractive index of crystals with absorption bands in the infrared part of the spectrum frequently assumes a value of 1.0 at a wavelength below that corresponding to the absorption maximum. Powders of such materials form a Christiansen filter with air at that wavelength. Christiansen powder filters are much less temperature sensitive than the solid-liquid filters described above.

Surface scattering can be used to remove shorter wavelengths from a beam of light. Transmission devices have been described that consist of powders of inorganic compounds uniformly spread over the surfaces of glass and rock salt plates. Alternatively, ground surfaces of various materials with a carefully controlled roughness can be used. For reflection devices, powders on speculum metal as well as aluminized ground glass surfaces have been used. These devices operate in the 1.0- to 20.0-μm spectral region, the cutoff wavelength being determined by the size of the powder or the roughness of the ground surfaces.

2.6 Diffraction

Light of a wavelength λ incident at an angle θ onto a diffraction grating with a grating constant g will be diffracted into different orders, unless

$$\lambda > g(1 + \sin\theta). \qquad (18)$$

In this case, the light will propagate only in the direction of the zero transmitted or reflected orders. This property can be used to construct short-wavelength cutoff filters if the blaze angle is chosen to minimize the zero-order content of light not satisfying the above equation.

The statement made above about diffraction gratings is valid also if the rulings are replaced by metallic grids consisting of wires or strips. Grids of this type are used as infrared polarizers because, while they are practically transparent to light linearly polarized parallel to the grid elements, they reflect light linearly polarized perpendicular to the grid elements at wavelengths greater than the grid constant g. If the grid is replaced by wire or electro-formed meshes [Fig. 5(a)], the properties no longer depend on the polarization of the incident light because of the rectangular symmetry. Furthermore, meshes are more readily available and easier to handle, and so they are more attractive than grids for the construction of filters for middle- and far-infrared. Meshes of this type can be used in transmission to isolate fairly broad spectral regions or in reflection as short-wavelength cutoff filters.

The meshes described above are called inductive meshes to distinguish them from capacitative meshes. These latter consist of metallic squares deposited in rectangular symmetry onto a suitable thin transparent substrate [Fig. 5(b)]. Capacitative meshes have a transmittance that is equal to the reflectance of an inductive mesh, and vice versa.

By using cross-shaped rather than square-shaped holes [Fig. 5(c)] or platelets [Fig. 5(d)] it is possible to construct meshes that generally reflect or transmit in the nondiffraction region, except in the neighborhood of one particular wavelength. Since rectangular

FIG. 5. Schematic representation of (a) inductive and (b) capacitative metallic meshes. (c) Inductive and (d) capacitative resonant meshes.

symmetry has been maintained, such resonant inductive and capacitative meshes are nonpolarizing. Their properties can be varied widely by adjusting the dimensions of the crosses.

Metallic meshes of the type described above in a certain sense behave like thin films and can be combined to construct a variety of interference filters for the infrared part of the spectrum. The more meshes used, the greater the control over the spectral characteristics of the resultant filter. In practice, the performance and the wavelengths for which filters can be constructed are limited by the flatness and parallelism of the meshes, the Ohmic losses in the meshes, and the dielectric losses in the substrates.

2.7 Interference of Polarized Light

Interference of polarized light is mainly used to construct bandpass filters with halfwidths that are measured in fractions of angstroms. Such filters are used by astrophysicists. Polarization interference filters are composed of birefringent crystal plates and of polarizers. Both uniaxial and biaxial crystals can be used, provided that they are cut in such a way that both the fast and slow axes of the crystal lie in the plane of the plate. When light passes through such a plate, the retardation

$$\xi = \mu d/\lambda \tag{19}$$

between the components of the beams polarized parallel to the fast and slow axes is maximized. In Eq. (19), μ is the birefringence of the crystal and d is its thickness. The most commonly used materials for the construction of polarization interference filters are quartz, calcite, and ammonium dihydrogen phosphate (ADP) with μ values of 0.009, 0.172, and 0.015, respectively.

Two basic types exist. In the Lyot-Öhman filter, a series of N parallel crystal plates are interspersed with polarizers and the retardation of each crystal is twice that of the preceding one. The crystals are aligned so that their axes are parallel to one another. The planes of polarization of the polarizers are inclined at 45° to the axes. It can be shown that the spectral transmittance of such a filter is given by the expression

$$T = \tfrac{1}{2}T_0\left(\frac{\sin 2^N \pi\xi_1}{2^N \sin\pi\xi_1}\right)^2 . \tag{20}$$

Here ξ_1 is the retardation of the thinnest component. T_0 is a factor that allows for the light losses (mostly in the polarizers) and is approximately equal to 0.85^{N+1}. Equation (20) represents a series of transmission peaks of half-width $\lambda/2^N\xi_1$ that are separated from one another by a wavelength interval λ/ξ_1. Light transmitted between any two adjacent peaks amounts to about 11% of the transmission within one peak. There are ways of reducing this parasitic transmission.

The Šolc filter is the second type of polarization interference filter. It is composed of N identical birefringent crystal plates placed between two polarizers. Two variants of this filter exist. In the fan configuration the two polarizers are parallel to one another and the ith plate is inclined at an angle $(i - 0.5)\pi/2N$ to them. The polarizers in the folded configuration are crossed, and the ith plate makes an angle $(-1)^i\pi/4N$ with the first polarizer. The spectral transmittances of both configurations resemble the transmittances of Lyot-Öhman filters of similar overall thickness and with the thinnest crystal plate of the same thickness as those of the components of the Šolc filter. The main departures are that T_0 for the Šolc filters is about 0.72, which is much higher than that of Lyot-Öhman filters, and that the parasitic transmission is about 2.5 times as high. This latter is a serious disadvantage for solar photography, but can be significantly reduced through a fine-tuning of the orientations of the axes of the individual components.

Like any other bandpass filter with a very narrow half-width, polarization filters must be placed in a constant-temperature enclosure if the wavelength of the transmission band is not to drift by more than a fraction of its half-width. Thin-film interference filters with solid spacers have a similar performance to the basic polarization filters described above. Furthermore, they can be constructed at much lower cost. The angular fields of both devices are similar and, unfortunately, are too small for many applications. However, if the thickest elements of Lyot-Öhman filters are replaced by more complicated structures, the angular field can be increased by as much as a factor of 20. No other type of filter can match this combination of small half-width and wide angular field.

3. PERFORMANCE OF SPECIFIC FILTER TYPES

In this section, examples are given of the spectral performances of various generic filter types based on the physical principles outlined above. Ways of specifying these basic filter types are also described.

3.1 Antireflection Coatings

Multiple reflections occurring between untreated surfaces of optical lens systems can result in significant amounts of stray light arriving in the image plane. This can completely obscure the image under unfavorable conditions. Even in nonimaging systems, losses of light due to surface reflections can be prohibitive. Surface reflection losses in laser systems can outweigh the gain of the laser medium. These and other problems are overcome by the application of suitable antireflection coatings to the surfaces. Such coatings are based on thin films, or on some suitable approximation of thin films, such as a structured surface with lateral and transverse dimensions that are of the order of the wavelength of light.

A single homogeneous layer is the most widely used antireflection coating. A zero reflectance is obtained at a wavelength λ_0 for which the optical thickness nd of the layer is equal to $\lambda_0/4$, provided that the refractive index n satisfies the relationship $n = \sqrt{n_m n_s}$. Here, n_m and n_s are the refractive indices of the medium and the substrate, respectively. Because of a lack of suitable coating materials, this relationship cannot be satisfied with dense films for substrates having a refractive index that is less than 1.9. Nevertheless, even with the available materials, a useful reduction in reflection is possible for all common glass types. An even lower reflection can be achieved with porous films. Low–refractive-index coating materials can also be simulated by the deposition or etching of subwavelength structures. However, the gain in performance is at the expense of mechanical strength and long-term stability.

If more than one layer is used, the additional parameters can be used to reduce the reflection at a single wavelength or to increase the width of the low-reflection region. As the number of layers and the overall thickness of the antireflection coating increases, it becomes easier to find solutions that meet the optical requirements and are at the same time based on the use of mechanically satisfactory coating materials. The widest antireflection region and the least sensitivity to angle of incidence is obtained with thick inhomogeneous antireflection coatings in which the refractive index varies continuously from n_s to n_m.

The performances of some typical antireflection coatings are shown in Fig. 6.

When specifying antireflection coatings, it is necessary to state the wavelength range over which the coatings are to be effective and the maximum reflection in that range that can be tolerated.

3.2 Neutral Attenuators

Neutral attenuators can be based on mechanical devices and on absorption in solid materials or thin films.

Densities of up to 2.0 can be achieved with self-supporting opaque grids and screens. This is one of the most uniform means of attenuating light, provided that the ratio of the filter aperture to the wavelength is large. Higher densities can be obtained with evaporated or etched metallic-film grids on transparent substrates. However, the neutrality and spectral range of the device will now depend on the substrate material. The uncertainty arising from the fact that the incident beam is intercepted by a finite number of grid openings can be reduced by vibrating the grids. Similar results can be obtained with rotating choppers, provided that the intermittent nature of the resulting signal can be tolerated.

A number of absorbing glasses, gelatin filters and photographic emulsions are suitable for making neutral-density filters, with densities of up to 5.0. However, their spectral transmission curves are not very uniform.

Evaporated thin films of metals such as aluminum, chromium, palladium, platinum, rhodium, and tungsten and alloys such as Chromel, Nichrome, and Inconel can be used to produce filters with densities of up to 6.0. A disadvantage of such filters is their high specular reflection. Inconel films are commonly used for high-precision neutral-density filters. Chromium is favored when tough, unprotected coatings are required. The operating range and neutrality of the attenuators depend on the substrate materials. Linear and circular metal-film wedges with a variable neutral density as well as step filters are available commercially.

The performances of typical neutral-density filters are shown in Fig. 7.

When specifying a neutral attenuator, it is important to state the range of wavelengths

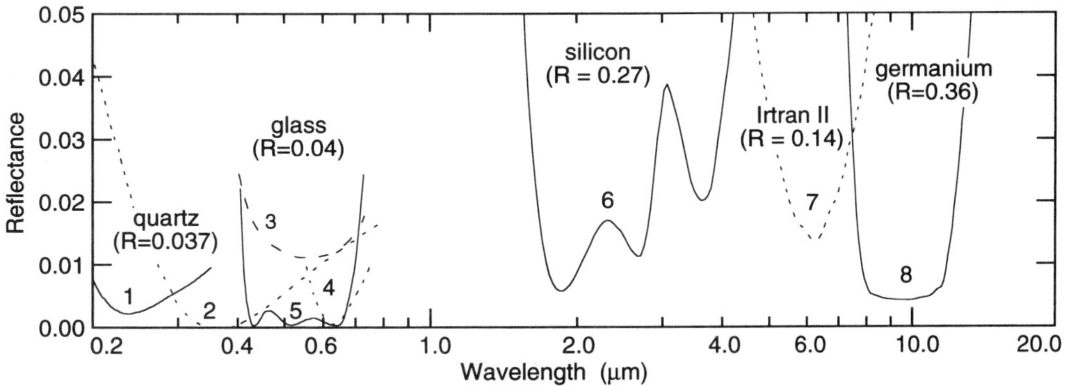

FIG. 6. Antireflection coatings for different substrates. Curves 1, 2: single homogeneous porous layers on quartz produced by two different processes; curves 3, 4, and 5: homogeneous single-layer, two-layer, and multilayer coatings on glass; curves 6, 7, and 8: antireflection coated silicon, Irtran II, and germanium substrates. The numbers in parentheses represent the reflectance of one surface of the bare substrates.

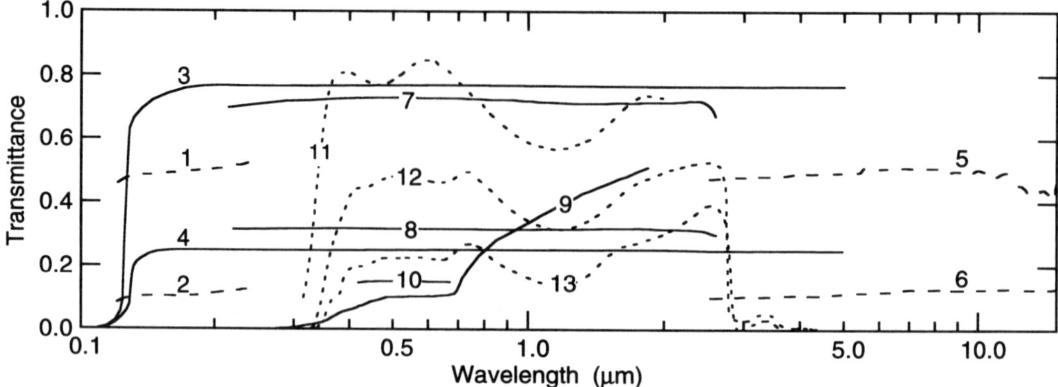

FIG. 7. Neutral attenuators. Alloy films on magnesium fluoride (curves 1, 2); on calcium fluoride (curves 3, 4); on germanium (curves 5, 6); and on quartz (curves 7, 8). Curves 9 and 10: gelatin filter with organic dye and photographic emulsion. Curves 11, 12, and 13: neutral density glasses.

over which the density is to be constant and the required density value.

3.3 Narrow- and Intermediate-Bandwidth Reflectors

Reflectors of narrow and intermediate bandwidth can be based on various periodic or aperiodic thin-film multilayers and on *Reststrahlen* materials.

Thin-film multilayers are used in various scientific instruments where a high reflection is required at a particular wavelength, and as components of other, more elaborate thin-film filters. For the ultraviolet, visible, and infrared spectral regions, periodic nonabsorbing two-material multilayers of the type

$$\underset{1}{AB} \cdot \underset{2}{AB} \cdot \underset{3}{AB} \cdots \underset{N}{AB} = [AB]^N \qquad (21)$$

have a high reflection at and around the wavelength λ_0 for which

$$n_A d_A + n_B d_B = \lambda_0/2. \qquad (22)$$

With a given set of materials, the highest reflectance is achieved when $n_A d_A = n_B d_B = \lambda_0/4$. The refractive index profile of such a multilayer is shown in Fig. 4(b). Such layer systems are called quarter-wave stacks. They have higher-order reflection peaks at wavelengths $\lambda_0/3$, $\lambda_0/5$, . . ., provided both materials are transparent at these wavelengths. The reflectance increases with N and with the ratio n_A/n_B. The width of the high-reflection region also increases with this ratio. The highest reflectance recorded for a quarter-wave stack is of the order of 0.999998. This is much higher than what can be achieved with metal mirrors and is important for some laser applications.

There are no completely absorption-free materials in the x-ray and extreme ultraviolet region. Furthermore, the refractive indices of materials at those wavelengths are close to unity. It is still possible to achieve a relatively high reflectance, provided that two materials are selected for which the Fresnel reflection coefficient is as high as possible at the wavelength of interest. However, a very large number of layers may be required. In general, the systems are not quarter-wave stacks, but Eq. (22) is approximately satisfied.

The measured performances of a number of reflectors based on the above principles are shown in Fig. 8.

With asymmetric metal/dielectric thin-film systems, it is possible to produce reflection filters with quite narrow reflectance bands. The performance of such reflectors is shown in Fig. 9.

Reststrahlen reflectors are used for isolating spectral regions through multiple reflection. The measured spectral reflectance of a number of *Reststrahlen* materials is also shown in Fig. 9. The spectral purity can be enhanced through multiple reflections and/or with auxiliary filters, such as the reflection from roughened surfaces. The peak reflectance can be enhanced with thin films at

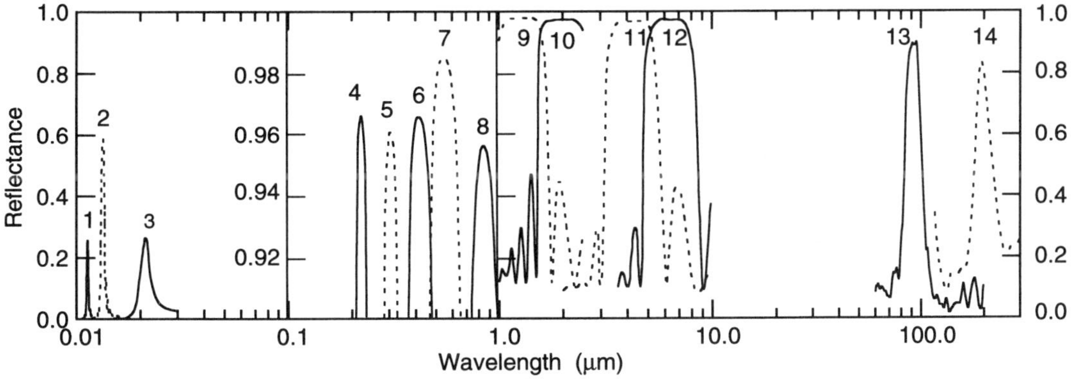

FIG. 8. Performance of periodic multilayer interference reflectors. Curves 1–3: x-ray mirrors composed of Pb-B_4C, Mo-Si, and Mo-Si metal layer pairs, respectively. Curves 4–12: all-dielectric quarter-wave stacks. Curves 13, 14: periodic reflectors composed of coated polyethylene sheets.

wavelengths for which adequate coating materials exist.

When specifying narrow and medium-width reflectors, it is necessary to state the range of wavelengths over which the reflectance is to be high and the minimum acceptable reflectance within this band.

3.4 Broadband Reflectors

Broadband reflectors can be based on thin-film interference and on the reflection from metal surfaces.

If the high-reflection region of the quarter-wave stacks described above is not wide enough, it is possible to create a wider high-reflection region by superposing two or more quarter-wave stacks centered at such wavelengths that the high-reflection zones are contiguous. Alternatively, it is possible to design reflectors in which the thicknesses of adjacent layers gradually increase in a geometric or arithmetic progression.

When even broader high-reflection regions are required, metal reflectors must be used. These are most commonly produced by deposition of thin metal films onto a suitable glass or quartz substrate. The actual metal used depends on the spectral region for which the reflectance is to be high. The most important metal reflectors are made of aluminum, silver, and gold.

The measured performances of a number of wide-band reflectors are shown in Fig. 10.

Once again, the specifications for such reflectors should state the range of wave-

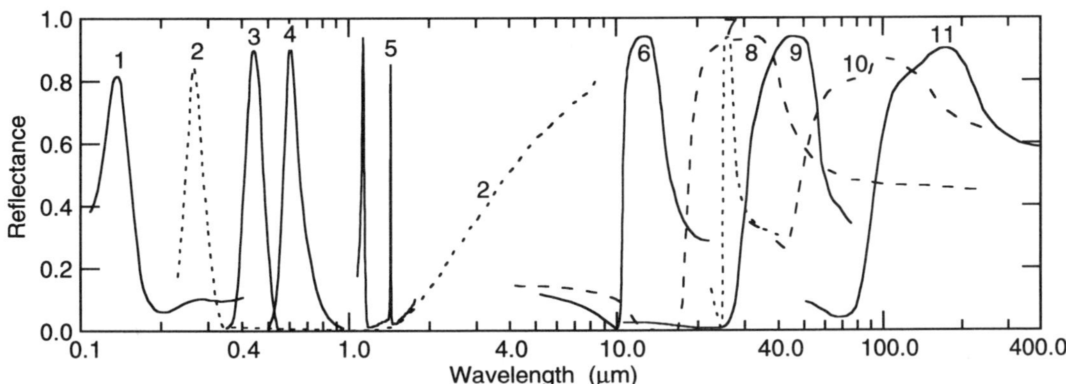

FIG. 9. Narrow-band interference and *Reststrahlen* reflectors. Curves 1–5: metal/dielectric thin-film interference reflectors. Curves 6–11: *Reststrahlen* reflectance peaks of AlN, GaP, UO_2, BaF_2, PbS, and TlBr, respectively.

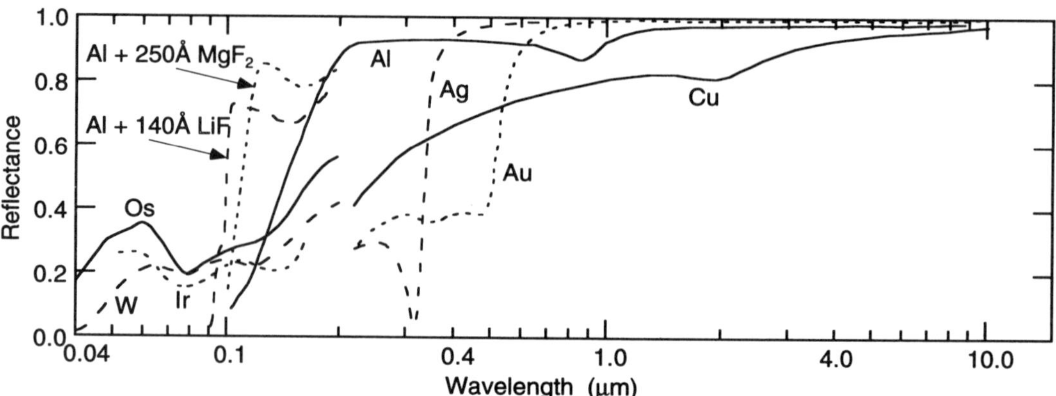

FIG. 10. Metallic reflectors. LiF and MgF layers deposited on Al films prevent the formation of an oxide layer, which absorbs at wavelengths shorter than 0.2 μm.

lengths over which the reflectance is to be high and the minimum acceptable reflectance within this band.

3.5 Short-Wavelength Cutoff Filters

Short-wavelength cutoff filters can be based on absorption, reflection, scattering, diffraction, or interference in thin films.

Transmission short-wavelength cutoff filters based on thin-film interference are essentially quarter-wave stacks (Sec. 3.3) with the thicknesses of some or all of the layers slightly adjusted to maximize the transmission at longer wavelengths. Reflection short-wavelength cutoff filters absorb or transmit light of shorter wavelengths while maintaining a high reflection for longer wavelengths.

The measured performances of short-wavelength cutoff filters based on a number of different physical principles are shown in Fig. 11.

An ideal cutoff filter would reject all the light below, and transmit all that above, a certain wavelength, or vice versa. Real cutoff filters are not perfect, and the transmission and rejection regions are limited in extent. In addition to the cutoff wavelength, it is therefore also necessary to specify the slope of the transition region. One definition used at times is $(\lambda_{0.8} - \lambda_{0.05})/\lambda_{0.5}$, where $\lambda_{0.05}$, $\lambda_{0.5}$, and $\lambda_{0.8}$ correspond to the wavelengths at which the

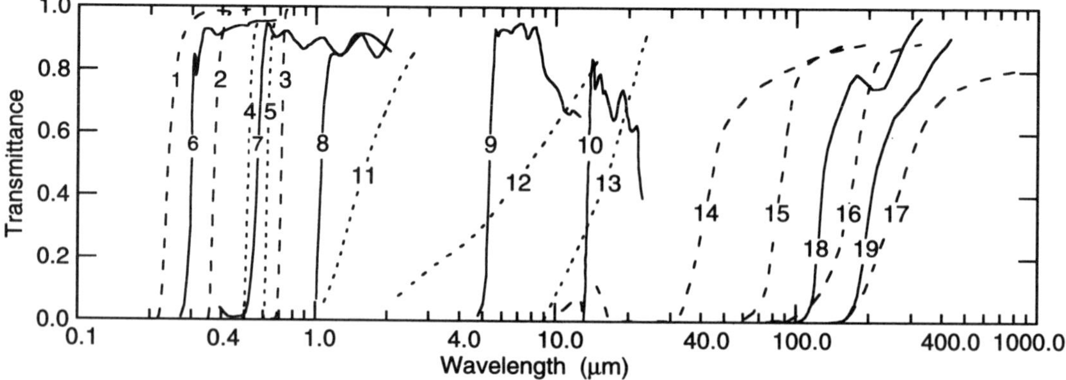

FIG. 11. Short-wavelength cutoff filters. Curves 1–3: colored glass; curves 4, 5: gelatin with organic dyes; curves 6–10: thin-film interference filters; curves 11–13: scatter filters; curves 14–17: Yoshinaga filters; curves 18, 19: metallic meshes.

transmittance amounts to 0.05, 0.5, and 0.8 of the maximum, respectively. In addition, it is necessary to specify the minimum widths and the desired transmittances of the transmission and rejection regions.

3.6 Long-Wavelength Cutoff Filters

Long-wavelength cutoff filters based on colored glasses are not nearly as effective as the corresponding short-wavelength cutoff filters. Those based on reflection and diffraction are also not very effective. Only the performance of long-wavelength cutoff filters based on interference in thin films compares to that of the corresponding short-wavelength cutoff filters.

The measured performances of different long-wavelength cutoff filters are shown in Fig. 12.

The specification of long-wavelength cutoff filters is analogous to that of short-wavelength filters.

3.7 Narrow-Band Transmission Filters

Narrow-band transmission filters can be based on absorption, interference in thin films, diffraction, and interference of polarized light. Half-widths ranging anywhere from 0.002 to 25% of the peak wavelength are possible.

Bandpass filters based on absorption usually consist of several materials placed in series [Eq. (7)] and have a transmission that is much smaller than unity.

Once again, narrow-bandpass filters based on thin films offer the most control over the performance. Different approaches to the design of thin-film narrow band transmission filters exist. It is possible not only to construct bandpass filters of high transmission and a specified width for a given wavelength, but also to control the shape of the transmission band. In the far infrared, multimesh filters offer a similar control over the transmission characteristics.

An ideal bandpass filter would transmit all the incident light in one spectral region and reject all the other light. Such a filter would be completely described by the width of the transmission region and the wavelength at which it is centered. Many more parameters are needed to describe adequately the performance of a practical filter. Unfortunately, a standardized terminology does not yet exist. First, the position of the transmission band is variously specified by the wavelength λ_{max} at which the maximum transmission T_{max} occurs, the wavelength λ_0 about which the filter passband is symmetrical, or the spectral center of gravity λ_c of the band. Differences between these different specifications become significant with an increase in the width and asymmetry of the band. The half-width, sometimes also called full width at half maximum (FWHM), of a narrow-band filter is the difference $\Delta\lambda_{0.5}$ in wavelengths at which the transmittance is $0.5T_{max}$. There are several ways of describing the shape of a transmission band. One makes use of a ratio called the shape factor $\Delta\lambda_{0.01}/\Delta\lambda_{0.5}$, in which the term $\Delta\lambda_{0.01}$ is sometimes

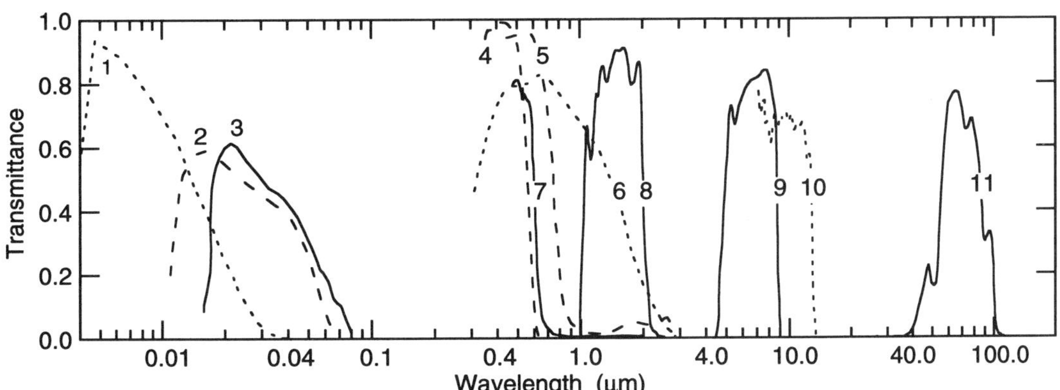

FIG. 12. Long-wavelength cutoff filters. Curves 1, 2, and 3: C, Si and Al thin films; curve 4: $CuSO_4$ solution; curve 5: colored glass filter; curve 6: indium tin oxide thin film; curves 7–10: multilayer thin-film interference filters; curve 11: metallic mesh filter.

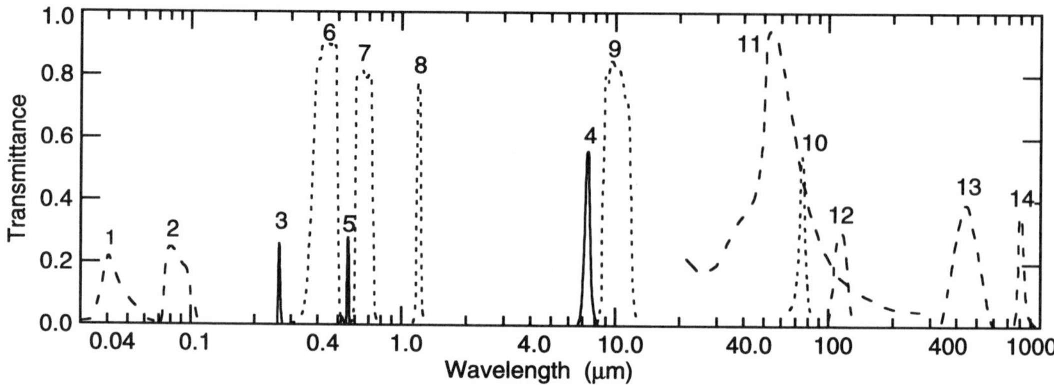

FIG. 13. Narrow-band transmission filters. Curves 1, 2: thin Ti and In metal films; curves 3, 4: Christiansen scatter filters; curve 5: several colored glass filters placed in series; curves 6–10: thin-film interference filters; curve 11: inductive metal mesh; curves 12–14: interference filters formed from inductive meshes.

called the base width of the filter. For the simplest type of thin-film interference filter, the shape factor is about 10, whereas that for an ideal bandpass filter would be 1.0.

Other quantities that need to be specified are the minimum acceptable transmittance in the passband and the maximum transmittance outside it. In many narrow-band filters, the transmittance rises again at some distance on either side of the peak. The range of wavelengths at which the transmission is low is called the rejection region. Sometimes there are several transmission peaks in the rejection region. The distance between the closest transmission bands on either side of the desired transmission peak is called the free spectral range. Auxiliary blocking will be required if either the rejection region or the free spectral range is not wide enough.

The performance of a number of narrow-band transmission filters is shown in Fig. 13. The spectral transmittances of four thin-film interference filters of very narrow half-width are shown in Fig. 14.

3.8 Rejection Filters

In principle, a rejection filter transmits or reflects all the light incident upon it, except that corresponding to a specified wavelength band that is to be rejected. Rejection filters are based on thin-film interference, holography, or diffraction.

Excellent narrow-band rejection can be achieved if light is reflected at a small angle of incidence from two or more identical narrow-band transmission interference filters. However, the range of wavelengths reflected

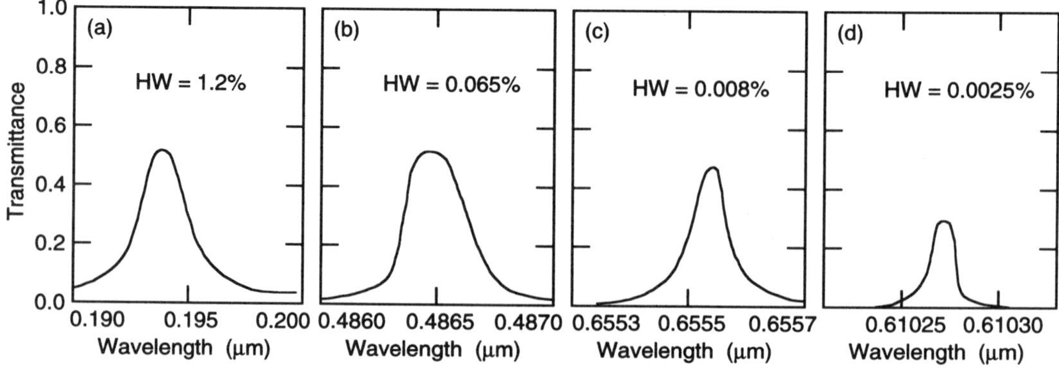

FIG. 14. Very narrow-band interference filters. Filters in (a), (b) are produced entirely by evaporation. Filters in (c) and (d) consist of very thin mica and quartz plates that have been coated on both sides with thin films.

will depend on whether all-dielectric or metal reflectors were used in the construction of the interference filter.

Transmission rejection filters based on nonabsorbing layer systems can be designed with properties that resemble those of quarter-wave stacks, except that the ripples in the transmission bands on either side of the rejection region have been removed. It follows from Sec. 3.3 that additional higher-order rejection peaks will occur at wavelengths $\lambda_0/3$, $\lambda_0/5, \ldots$. It is possible to eliminate these peaks through the use of more complicated multilayer structures composed of more than two materials, or based on the use of inhomogeneous layers in which the refractive index varies in a periodic manner. Such inhomogeneous layers are sometimes called rugate filters [see Fig. 4(c)].

Examples of rejection filters are shown in Fig. 15.

When specifying a rejection filter operating in transmission (reflection), it is necessary to state, in addition to the location, width, and maximum acceptable transmittance (reflectance) of the rejection region, the extent and minimum acceptable transmittance (reflectance) in the region surrounding it.

3.9 Neutral and Color-Selective Beam Splitters

Neutral beam splitters are used whenever the incident light has to be divided into two beams of approximately the same spectral composition propagating in different directions. In a color-selective (dichroic) beam splitter the spectral compositions of the two beams are different. Both types are based on thin films that may be deposited onto a very thin membrane, a thin plate, or the interface between two cemented prisms. Neutral beam splitters frequently employ thin metal layers.

The angle of incidence in neutral and color-selective beam splitters is usually 45°, but not always. Unless special precautions are taken, the transmittances and reflectances of light polarized parallel and perpendicular to the plane of incidence will be different. Polarization effects can be reduced at the expense of a considerable increase in complexity of the coatings.

The measured performance of four neutral beam splitters is shown in Fig. 16. Color-selective beam splitters are shown in Fig. 17.

When specifying beam splitters, the angle of incidence at which they are to operate must be specified. It is important to remember that the smaller this angle, the smaller the polarization effect. For neutral beam splitters, it is necessary to specify the relative intensities of the transmitted and reflected beams for both polarizations and the width of the spectral region over which these properties should remain constant. For color-selective beam splitters, the spectral composition of the transmitted and reflected beams must be specified for each polarization.

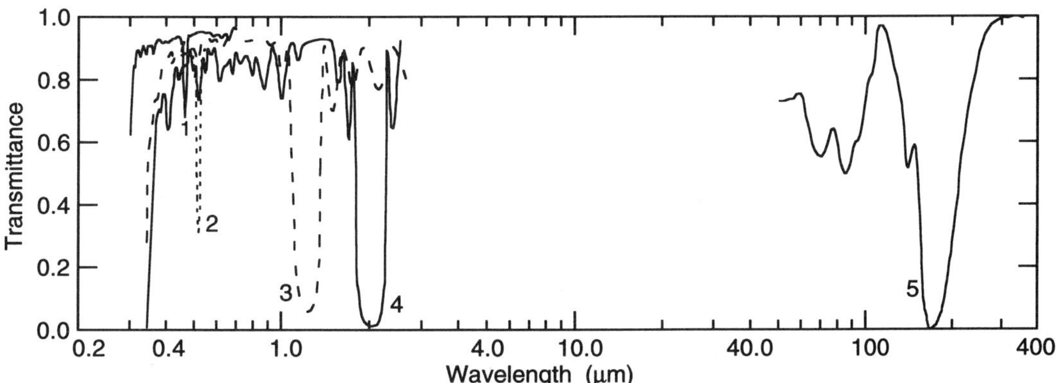

FIG. 15. Rejection filters. Thin-film interference filters composed of homogeneous layers (curves 1, 2) and of periodic inhomogeneous coatings (curves 3, 4). Curve 5 represents an interference filter composed of two capacitative resonant metallic meshes.

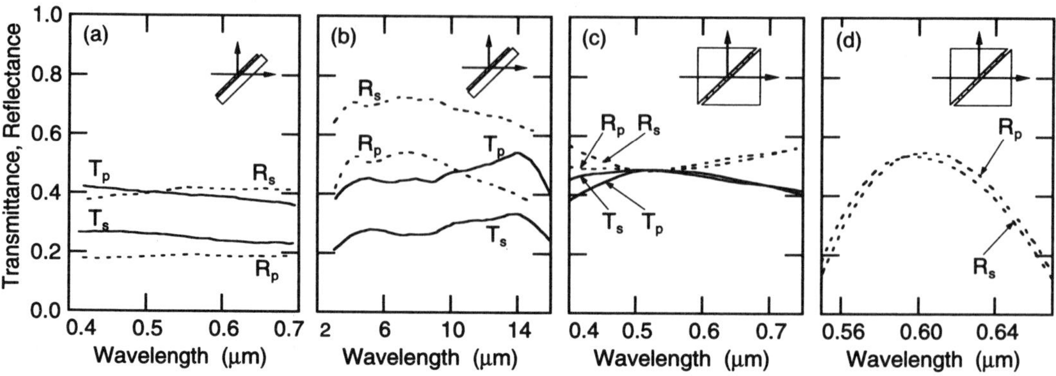

FIG. 16. Neutral beam splitters. (a), (b) Metal and all-dielectric thin-film beam splitting plates. (c), (d) Metal/dielectric and all-dielectric thin-film polarization-insensitive beam splitting cubes.

4. SPECIFYING FILTERS

It should be emphasized that Figs. 6 through 17 show only a few representative examples of available filters. Filters with similar properties at other wavelengths, or with intermediate properties, are possible. In particular, it is possible to construct various generic filters with much higher attenuations (Fig. 18).

It is always a good idea to consult a filter expert at the conceptual stage of the design of an apparatus, and to present the problem in the most basic way. At that stage, given the $S(\lambda)$, $F(\lambda)$, and $D(\lambda)$ of Eq. (1), the filter designer may be able to offer suggestions that would result in less obvious, but better or cheaper solutions to the problem. For example, if multiple-reflection filters can be accommodated, efficient blocking over very wide spectral regions is possible.

The parameters that need to be specified for specific filter types have been introduced in Sec. 3. Here, it is necessary to emphasize that upper and lower tolerances are essential for each parameter (wavelength, transmittance, reflectance, rejection, half-width, gradient, etc.). They should result in a just acceptable performance. If the tolerances are too tight, solutions may be found that are far more elaborate and expensive than necessary, or they may not be found at all.

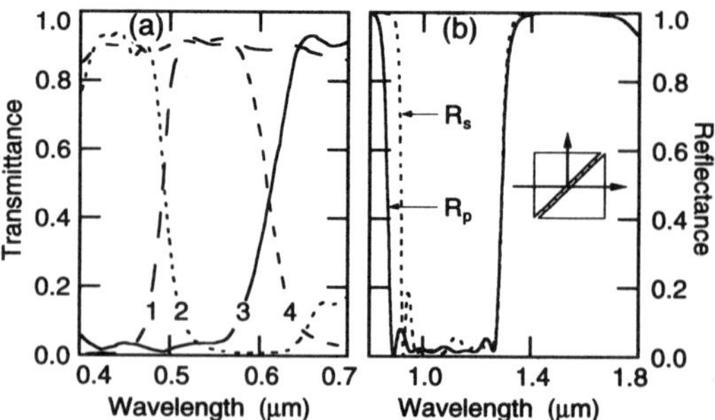

FIG. 17. Color-selective beam splitters. The multilayer coating in (b) is polarization independent.

The following questions must be answered in the process of specifying a filter: What is the spectral range of the incident light? What is the angle or range of angles of incidence of the light? Is the incident beam parallel or convergent? What is the convergence angle? For non-normal incidence, is the incident light randomly polarized or polarized parallel or perpendicular to the plane of incidence? Is the light coherent or not? Are solutions based on filters placed in series and/or in parallel acceptable? Is there a minimum laser damage threshold that the component must withstand? Is light scattering a concern? It is also necessary to specify the environment in which the filters are to be used: the lowest and highest temperature and humidity, exposure to vacuum and/or space, to corrosive chemicals, solvents, etc. What mechanical stability and, in the case of coatings, what adhesion and scratch resistance are required? Is protection through cementing of a cover glass an option? Are there any military or commercial standards that must be satisfied?

Other questions to be considered are as follows: What are the dimensions of the filter? In the case of coatings, is the substrate flat or curved? Is a single component or a large

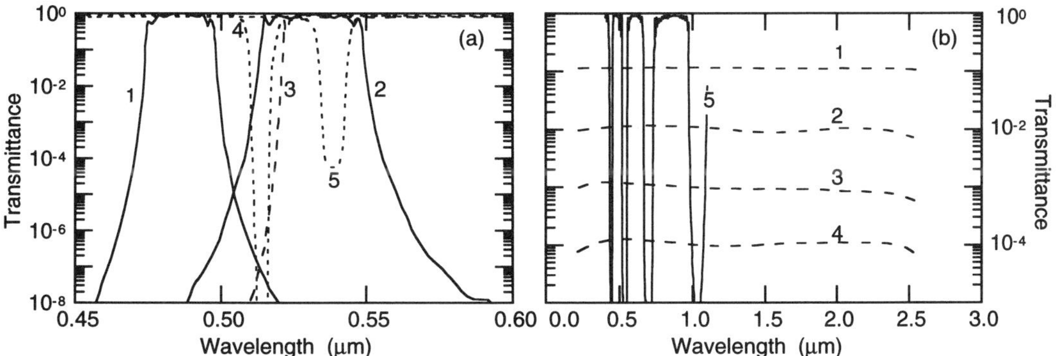

FIG. 18. Filters with high attenuation. (a) Curves 1, 2: bandpass filters; curve 3: short-wavelength cutoff filter; curves 4, 5: rejection filters. (b) Curves 1–4: neutral attenuators; curve 5: rejection filter for four laser wavelengths. Curve 5 in (a) corresponds to a holographic filter. All other filters in this diagram are based on thin films.

number of filters required? Are there any materials that must be either used or avoided when seeking a solution to the problem? Are there any cost constraints?

5. CONCLUDING REMARKS

The filters described in this article are generic filters with standard functions and fixed optical properties. More details about many of the filters presented in the diagrams will be found in Cook and Stokowski (1986) and Dobrowolski (1978).

There are many complex filters for specific applications, such as colorimetric measurement, energy conservation and conversion, architectural and automotive applications, or use in specialized scientific or technical instruments. There exist also many materials whose optical properties can be changed by external electric or magnetic fields, temperature, or illumination level. The optical characteristics of filters constructed from such materials can be tuned continuously. Because of space considerations, these two topics could not be discussed in this article.

GLOSSARY

Absorptance: The ratio of the energy absorbed by an optical filter to that incident upon it.

Internal Transmittance: The ratio of the intensity of a beam of light just before reaching the second surface of an absorption filter to the intensity just after passing through its first surface.

Optical Filter: Any device or material that is deliberately used to change the spectral intensity distribution, the phase, or the state of polarization of electromagnetic radiation incident upon it.

Optical Density: The logarithm of the inverse of the transmittance of a filter.

Reflectance: The ratio of the reflected intensity to that incident upon the optical filter.

Scatter: The ratio of the intensity of the radiation that is not absorbed by an optical filter or transmitted, or reflected by it in the specular directions, to the intensity of the radiation incident upon it.

Spectral Energy Distribution: The variation with wavelength of the energy emitted by a given radiation source.

Spectral Responsivity: The variation with wavelength of the ratio of the output of a detector to the energy of the incident radiation.

Surface Reflection: The reflection from a single surface of an optical component, not taking into account the contributions of reflections from other surfaces of the device.

Transmittance: The ratio of the transmitted intensity to the intensity incident upon the optical filter.

Further Reading

Cook, L. M., Stokowski, S. E. (1986), "Filter Materials," in: M. J. Weber (Ed.), *Handbook of Laser Science and Technology. Optical Materials: Part 2*, Boca Raton, FL: CRC Press, pp. 185–219.

Dobrowolski, J. A. (1978), "Coatings and Filters," in: W. G. Driscoll, W. Vaughan (Eds.), *Handbook of Optics*, 1st ed., New York: McGraw-Hill, pp. 8.1–8.124.

Dobrowolski, J. A. (1994), "Optical Properties of Films and Filters," in: M. Bass, (Ed.), *Handbook of Optics*, 2nd ed., New York: McGraw-Hill, Vol. 1, pp. 42.1–42.130.

Geffcken, W. (1957), "Lichtfilter," in: E. Schmidt (Ed.), *Landolt-Börnstein: Zahlenwerte und Funktionen aus Physik, Chemie, Astronomie, Geophysik und Technik*, Vol. 4, Part 3, Berlin: Springer, p. 925.

Macleod, H. A. (1986), *Thin Film Optical Filters*, New York: McGraw Hill, 519 pp.

Yeh, P. (1988), *Optical Waves in Layered Media*, New York: Wiley, 406 pp.

OPTICAL HOLOGRAPHY

See HOLOGRAPHY, OPTICAL

OPTICAL INSTRUMENTATION

DIETMAR HOESCHEN AND WERNER MIRANDÉ, *Physikalisch-Technische Bundesanstalt, Braunschweig, Germany*

INTRODUCTION

"Optical instrumentation" means that branch of technology where more or less sophisticated instruments are constructed that make use of optical phenomena. These instruments are designed to investigate materials, make measurements, transmit energy or information, image the real world. The common basis of all these instruments is optical phenomena, which are treated by specific scientific disciplines.

The term "optical" is derived from the ancient Greek words "optike techne," meaning "the science and skill concerning seeing." Looking at the history of optical instrumentation, we are therefore dealing with phenomena where the detector of the radiation is the human eye. Nowadays, the term "optical" is not limited to that part of the electromagnetic spectrum in which the human eye is sensitive, that is the wavelength range between 380 and 780 nm, but one is talking about ultraviolet and infrared optics and even about x-ray and even electron and neutron optics.

In this article we will not limit the term "optics" to the visible part of the electromagnetic spectrum; we will extend the spectral range of "optics" to boundaries where significant changes of optical phenomena occur. Such phenomena may be found in the interaction of radiation with matter. The energy of the light particles, the photons, is equal to $h\nu = hc/\lambda$, where h is the Planck constant ($h = 6.626\,075\,5 \times 10^{-34}$ J s), c the velocity of light ($c = 299\,792\,458$ m s^{-1}) in vacuum, and ν the frequency. If the wavelength increases—equivalent to a decrease in frequency—the energy of the photons will approximate the thermal energy kT, where k is the Boltzmann constant ($k = 1.380\,608 \times 10^{-23}$ J K^{-1}). There, a distinction between processes caused by photons and those caused by temperature is no longer possible. A limiting value to longer wavelengths will therefore be for optics at room temperatures a wavelength λ of about $\lambda = hc/kT \approx 50$ μm. That is rather arbitrary, since the boundary may be moved by lowering the temperature. But most optical instrumentation is used at room temperatures and under "natural conditions," meaning in a natural atmosphere. The absorption of radiation by oxygen in the atmosphere limits the range of optics to wavelengths above approximately 200 nm. Beyond that limit, optics can be performed only under special environmental conditions,

3-527-28134-7/95/$5.00 + .50

for instance in vacuum. Therefore, optical instrumentation is limited to the wavelength range between 200 nm and 50 μm in this article, assuming room temperatures and natural atmosphere.

The history of optical instrumentation is very old (Rosenberger, 1965). Refraction was already used in the ancient world to concentrate energy by refracting elements, for instance in Nineveh (640 B.C.) or in Greece. The Romans used magnifying lenses. Archimedes (287–212 B.C.) investigated the reflection of light by mirrors in theory and built practical devices. The highlight of optical instrumentation in the Middle Ages was the work of the Arabian scientist Alhazen (965–1039 A.D.), who made a camera obscura, lenses, and spectacles. Starting around 1600 A.D., more and more optical phenomena were systematically investigated and used to construct new instruments: Kepler (1611) described the construction of a telescope with lenses, Newton built one based on mirrors. Newton also explained the nature of colors, using the refraction and dispersion of light by a prism. The analysis of the spectra of optical radiation became one of the most powerful tools in chemistry in the 19th century. Diffraction was discovered in 1665 by Grimaldi, polarization in 1669 by Bartholinus, interference in 1801 by Young. Despite their early discovery, these three phenomena were not put to use in the manufacture of optical instrumentation before the second half of the 19th century. The phenomenon of coherence of optical radiation, which is the basis for many modern applications in optics, became more common in the 20th century. The particle character of light, assumed already by Newton but in disfavor in the 19th century, was revived by Einstein in 1905 with his interpretation of the photoeffect. This phenomenon is used for the wide field of light detectors and for cooling by radiation.

1. CHARACTER OF OPTICAL RADIATION

The question of the nature of light is a very old one. It arose and was answered particularly in the context of visual perception. Ancient Greek philosophers thought that light consists of rays emerging from the eye and acting like feelers on objects. It was the middle of the 17th century before ideas arose that were in broad agreement with our understanding of the nature of light today. Men such as Francesco Maria Grimaldi, Robert Hooke, and Ignace-Gaston Pardies expounded the idea that light is a wavelike motion, only some years before Christiaan Huygens in 1678 wrote down his principle of elementary waves (published in 1690). At the beginning of the 19th century, brilliant scientists such as Thomas Young, Augustin Fresnel, Joseph Fraunhofer, and others discovered that light may be considered as a transverse wave, able to be polarized, reflected, refracted, and diffracted. As James Clark Maxwell had formulated the theory of electromagnetics in 1862, and as it was recognized that light was an electromagnetic wave, the "wave theory of light was certain," as Heinrich Hertz declared at the end of the 19th century.

But in spite of the fact that he used some aspects of the wave theory to explain his experiments, Isaac Newton had always favored the idea that light consists of small particles. For a century, the scientific authority of Newton guaranteed that light was mainly considered to consist of particles. But this idea was abandoned in the light of the overwhelming impression gained from experiments done by Young and others, which supported the wave theory. Another century was to pass before Albert Einstein posed a hypothesis that light might have particle character. Arthur H. Compton indeed showed by experiment in 1922 that "light" of very high frequency (x rays) behaves as particles when scattered by matter. Niels Bohr was the first to attempt, in 1927, to unify the apparently contradictory aspects by his "principle of complementarity," where he stated that wave and particle theory complement each other. The development of quantum electrodynamics by Bethe, Schwinger, Feynman, and Dyson finally solved the problem, showing that the phenomenon of "light" may be completely described by this theory, and that light behaves as a particle whose behavior is described by a probability function, which is proportional to the squared wave function of electromagnetics (Feynman, 1961).

Nevertheless, in the application of optical phenomena for optical instrumentation light is used either as a wave or as a particle. A transverse wave (see Fig. 1) is characterized by the wavelength λ, the frequency ν, the

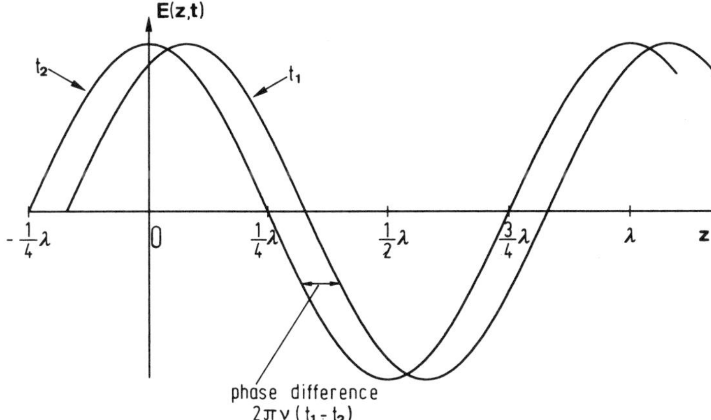

FIG. 1. Part of an infinitely extended wave train at times t_1 and t_2. The wavelength λ is the distance between points where the amplitudes E of the wave again have the same value after one cycle.

speed c, and the polarization. The amplitude E, which is the amount of the electric field vector—which is perpendicular to the propagation of the wave, taken in the z direction—changes according to

$$E(z,t) = E_0 \cos 2\pi\nu(t - z/c) \tag{1}$$

as a function of time and space. Besides the movement in the z direction with speed c, the vector E generally rotates around the z axis, changing its length during rotation. Depending on its rotation, the light is said to be elliptically (when E changes its length), circularly (when E rotates, but does not change length), or linearly (when E does not rotate) polarized light. The description of the wave by Eq. (1) assumes that the wave is extended in both directions of the z axis to infinity. The argument of the cosine function is called the phase and describes, for instance, the displacement of the wave along the z axis for different times t_1 and t_2. The speed c is therefore called the phase velocity v_p in vacuum. In vacuum it is constant for all electromagnetic waves. The phase velocity generally changes when light travels through material and becomes a function of wavelength.

In practice, the train of waves is finite and may be described by a superposition of wave trains with different wavelengths. It is called a wave group. The group as a whole moves with a speed v_g, which is equal to or less than the phase velocity v_p with

$$v_g = v_p - \lambda \frac{dv_p}{d\lambda}. \tag{2}$$

Since v_p is constant in vacuum, phase and group velocity are equal in vacuum.

The other aspect of light is the fact that in certain experiments, it has the character of particles. The release of electrons from matter under exposure to light is the best known example. These particles, called photons, may be described by their mass, energy, momentum, and angular momentum (Table 1). While the angular momentum is complementary to the polarization of the light, the other parameters are given by λ, h, and c, the relationships being derived from quantum theory and the theory of relativity. It should be

Table 1.

Light as a wave		Light as a particle		
Wavelength λ λ (μm)	Frequency ν $\nu = \frac{c}{\lambda}$ (s^{-1})	Energy E $E = \frac{hc}{\lambda}$ (J)	Momentum p $p = \frac{h}{\lambda}$ (kg m s^{-1})	Mass m $m = \frac{h\nu}{c^2}$ (kg)
50	5.996×10^{12}	3.793×10^{-21}	1.325×10^{-29}	4.42×10^{-38}
1	2.998×10^{14}	1.986×10^{-19}	6.626×10^{-28}	2.21×10^{-36}
0.2	1.499×10^{15}	9.932×10^{-19}	3.313×10^{-26}	1.11×10^{-35}

noted that the mass of the photon differs from zero because it moves with the speed of light; the fictive mass of the photon at rest is zero.

2. PHENOMENA AND APPLICATIONS

As in many other physico-technical disciplines, mankind has a twofold interest in optical radiation: to understand the principles and to exploit the phenomena, whereby use without thorough understanding is common, but better knowledge often results in a more sophisticated application. In the following, many phenomena resulting from the wave and particle character of optical radiation are described and applications are discussed.

2.1 Wavelength

As pointed out in Sec. 1, light has a physical character similar to that of radio waves. On an electromagnetic wave that has a single frequency and that propagates in vacuum or a homogeneous material, points of equal separation can be found along the direction of propagation where the values of the electric or magnetic field strength and the direction of change of these parameters are the same (see Fig. 1). The distance between two of these points is called the wavelength.

In the case of common radio waves, it is no great problem to measure, for example, the variation of the electric or magnetic fields directly and it is rather easy to determine the frequency by counting the periods in a certain interval of time. Optical radiation, however, has frequencies 10^4 to 10^7 times higher than that of common radio broadcast bands (see Table 1), and detectors for such radiation can only detect the intensity, which is defined as the time average of the amount of energy that crosses in a unit of time a unit area perpendicular to the direction of energy flow. Moreover, it is still impossible to determine the extremely high frequencies by counting. But interference effects caused by diffraction or produced by special instruments such as interferometers give rise to variation of the intensity in combination with coherent radiation. Thus Fraunhofer, for example, calculated the wavelength of the unresolved sodium yellow lines at 589.2 nm from the period of the grating he used to produce the spectrum. The first comparison of the wavelength of the red line (643.8 nm) of cadmium with the International Prototype Meter was performed in 1892 by Michelson and Benoit, who used a special type of two-beam interferometer. Later, in 1913, Benoit, Fabry, and Pérot repeated the measurement using a set of multiple-beam interferometers.

Since the wavelengths of atomic spectral lines are invariant and very well reproducible if they are excited under prescribed, suitable conditions, and as interferometric measurements can be made with very high precision, it was natural to consider replacing the prototype meter with a definition of length in terms of the wavelength of some selected spectral line. This is why in 1960, the meter was defined as equal in length to 1 650 763.73 wavelengths in vacuum of the radiation corresponding to the transition between the levels $2p_{10}$ and $5d_5$ of the ^{86}Kr atom. The wavelength of this line was chosen because of its spectroscopic advantages. This definition of the unit of length was reproducible with an uncertainty of 4×10^{-9}, and the unit could be transferred directly by interferometric methods to physical or technical measurements of length or distance.

When precision length measurements using interferometric techniques or comparisons of wavelengths are performed in air instead of vacuum, the refraction index and the dispersion of the air must be taken into account. For a relative precision of about 10^{-8} the correction can be calculated by means of equations given by Edlén (1966), which give correction factors for the dependence of the refraction of air on its pressure, temperature, and humidity.

A totally new definition of the meter was given in 1983. Since then the meter is defined as equal to the length that is traversed by light during a time of 1/299 792 458 s in vacuum. This new definition, which relates the meter to the unit of time represented by the cesium frequency standard (see Sec. 2.2.1), can be realized with an uncertainty several orders of magnitude lower than that of the former ^{86}Kr wavelength standard. For practical purposes, secondary wavelength standards are still in use, especially as nowadays certain stabilized laser systems provide wavelengths with an uncertainty similar to that of the cesium frequency standard.

2.1.1 The Laser "LASER" is an acronym and stands for "light amplification by stimulated emission of radiation." The device commonly referred to as a laser, however, is not an amplifier of light but an oscillator generating electromagnetic radiation in a limited frequency range. If an amplifying process is available, it is possible to construct an oscillator by returning a small portion of the output of the system to the input in the correct phase. An essential part of an oscillator is a resonant system that defines the frequency of the oscillator and stores energy that produces the output power and the feedback power. In the range of optical frequencies, the resonator consists of a cavity formed by highly reflecting mirrors. Such a cavity resembles that of a Fabry–Pérot interferometer (see INTERFEROMETERS AND INTERFEROMETRY) except that its length is generally large in comparison with the linear size of its mirrors.

The key to successful laser operation is, of course, the active medium that amplifies the wave. Qualitatively, a material that fluoresces or exhibits luminescence is an obvious candidate. In either process, stimulated emission of radiation can occur only if a population inversion of the energy states is produced. This means that, for instance, the number of electrons that have been stimulated to the upper-energy state for the radiating transition by a suitable excitation process must be higher than the number of electrons in the lower state, as in this case, an incident photon will stimulate the further transition of electrons and amplification will result. The amplifying medium may be a suitable mixture of gases, a fluorescing liquid or solid material, or the junction region of a special semiconductor diode (see LASER PHYSICS and other special articles concerning this topic).

Because of the special process of its generation, laser radiation has some unique properties in comparison with most of the other sources of optical radiation. With all precautions taken to select only one mode of resonance, the bandwidth of laser radiation can be made extremely small. For example, light from the best single-line low-pressure gas-discharge lamp has a spectral width of about 1 MHz. The radiation of a single-mode stabilized laser, on the other hand, may have a bandwidth of some tens of hertz. In fact, the mechanical vibrations of the mirrors of the resonator are one of the limitations to the spectral purity of laser radiation. In spite of the extreme frequency purity of a few parts in 10^{14}, which is possible, the absolute frequency and thus the absolute wavelength are, however, known only to the accuracy to which the reference frequency is known to which the laser resonator is locked.

Another significant property of laser radiation is its high spatial coherence, which is why laser radiation can be collimated and focused to a degree limited only by diffraction. The high spectral purity and high spatial coherence make laser radiation attractive for special applications in spectroscopy and interferometry. The development of the holographic techniques in particular has been considerably influenced by lasers. On the one hand, these can provide very long coherent wave trains. On the other hand, by means of Q-switching and mode-locking the radiation, they can also produce pulses of extremely short duration (10^{-15} s) for investigations of short time processes.

Finally, a special example of laser application in a measuring system will be mentioned, the laser gyro. This system is based on the Sagnac effect (Sagnac, 1913; see also Post, 1967). The laser gyro consists basically of a ring resonator combined with a He-Ne plasma as the amplifying medium. A rotational movement of the resonator gives rise to a virtual change Δl of the cavity length for photons traveling in opposite directions, leading to frequency shifts and hence to a difference frequency proportional to the angular velocity. The multiple circulation of the light beams in the cavity makes the device highly sensitive to the detection of low angular velocities (smaller than 0.1 deg/hr).

2.1.2 Interferometry If electromagnetic radiation emerging from some source is divided by a suitable device into two or more beams that are superposed after having traveled different paths, the resultant amplitude and intensity may be very different from the sum of those contributed by the beams added separately. This phenomenon is called interference. It has had a considerable influence on the development of physics. The observations and explanations of the interference of light that Thomas Young (1801) made in context with his famous experiments provided the basis for Fresnel's wave theory of light, and the same experiment has formed the foundation of the modern theory of coherence (Marathay, 1982). Interferometry techniques today are important methods used in experimental physics with applications extending into other branches of science and technology.

Interferometry is chiefly applied either as a tool for obtaining knowledge of the nature of radiation itself (spectroscopy, measurement of the degree of spatial coherence or correlation effects) or for carrying out measurements on material, physical bodies, optical components, and lens systems (refractive index measurements, length or surface profile measurements, and optical testing).

In conventional optical interferometers, basically two methods are used to divide into several beams a beam emerging from a light source. These methods often provide the basis for classifying the systems used to produce interference.

In his fundamental investigations on the interference of light, Thomas Young divided the beam by diffraction at two small apertures S_1 and S_2 placed side by side in an opaque diaphragm (see Fig. 2). This technique is one of the methods that are labeled as division of wave front. The visibility of the interference fringes obtained with this arrangement largely depends on the degree of coherence of the field of radiation that strikes the diaphragm with the apertures. Consequently the source must have rather limited angular dimensions, and a small band of frequencies may be admitted for good visibility of the fringes if an incoherent light source is used for illuminating the apertures. Young's experiment or similar modified arrangements can also be used particularly for determining the degree of coherence or for measuring the angular dimensions of distant incoherent sources such as double stars or cosmic nebulae.

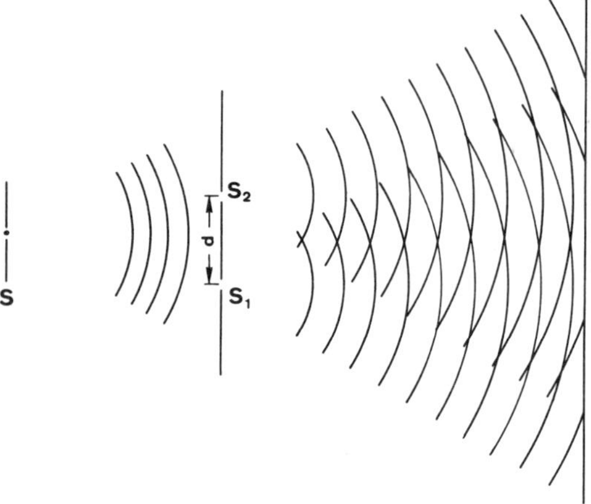

FIG. 2. Principle of Young's interference experiment.

Lloyd's mirror, Fresnel's double mirror or biprism, Billet's split lens (see Fig. 3) or—as a modification of the last mentioned—Meslin's lens are all examples of devices used for dividing wave fronts. While in Young's experiment the two apertures that are illuminated by a more or less spatially coherent wave front act as secondary sources, virtual or real images S_1, S_2 of the primary source S_p are generated by mirrors, prisms, or split lenses in the other systems mentioned above.

Alternatively, two or more images of a primary source or two or more beams can be produced by means of partially reflecting surfaces, at each of which a part of the light is reflected and some is transmitted. This method is called division of amplitude. In either case, different distributions of intensity for two-beam or multiple-beam interference patterns result. This provides yet another basis for the classification of interferometric systems. While two-beam interferometers produce interference patterns with a sinusoidal variation of intensity, in a multiple-beam interferometer (see Fig. 4), each pair of interfering beams contributes a Fourier component to the fringe pattern. This is why multiple-beam interferometers produce interference fringes whose intensity profiles approximate those of narrow spikes. The most well-known multiple-beam interferometer is the Fabry–Pérot interferometer, which is mainly used for high-resolution spectrometry. Multiple-beam interference also takes place in Fizeau interferometers (see INTERFEROMETERS AND INTERFEROMETRY), which in particular are used for the determination of profiles of smooth surfaces. In microscopy, Fizeau and Tolansky interference patterns (Tolansky, 1973) are applied to provide measurements of the microtopography of surfaces.

Interference filters eventually are a particular variety of the Fabry–Pérot interferometer. The main difference of these devices, in comparison with common Fabry–Pérot in-

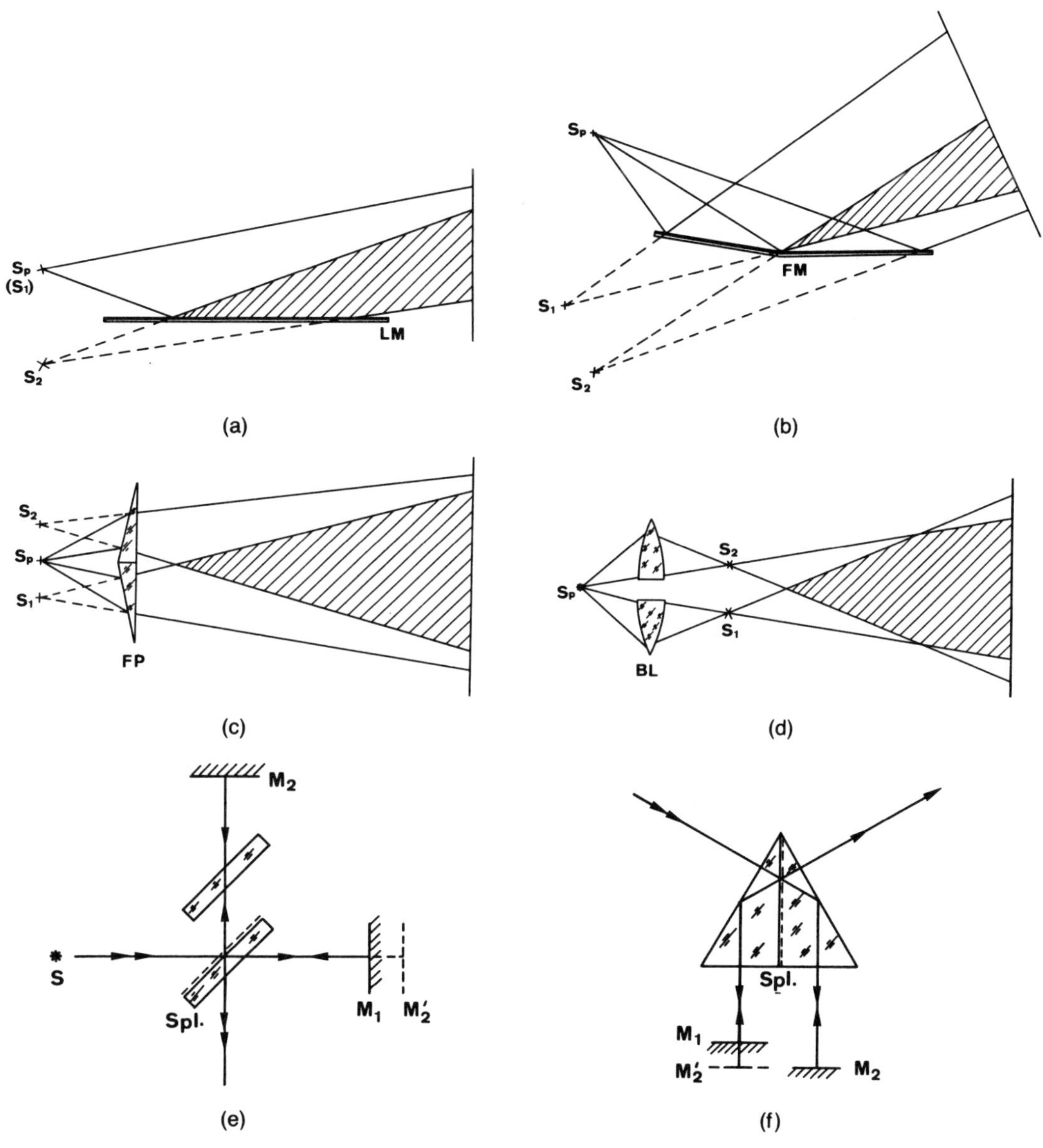

FIG. 3. Schematic diagram of some interferometers: (a) Lloyd's mirror (LM), (b) Fresnel's double mirror (FM), (c) Fresnel's biprism (FP), (d) Billet's split lens (BL), (e) Michelson interferometer, (f) Köster's interferometer.

terferometers, consists in the fact that the spacing between the two partially reflecting mirrors is reduced to a distance comparable to the wavelengths of light. Thus they are operated in a much lower order of interference, and therefore they have a larger free spectral range and a larger bandwidth than Fabry–Pérot interferometers that are used in spectroscopy.

The best known and most versatile two-beam interferometers with amplitude division are the Michelson interferometer (see Fig. 3) and modifications of the original arrangement, such as for example Köster's interferometer or the Twyman–Green interferometer. These instruments are suitable for measurements of length, using the wavelength as a standard. They may also be used for Fourier spectroscopy by recording the visibility of the interference fringes in depen-

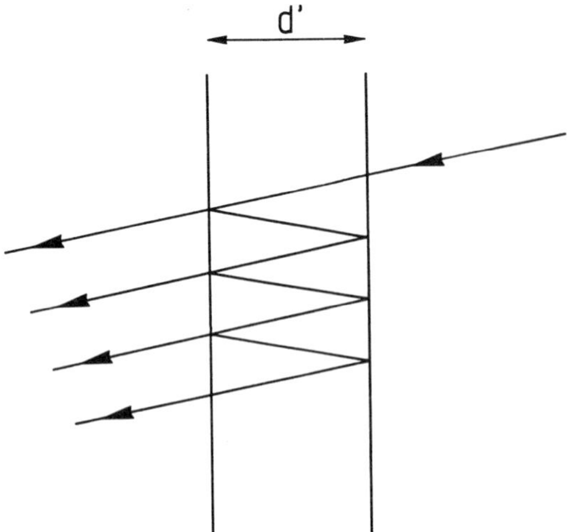

FIG. 4. Principle of the Fabry–Pérot interferometer.

dence on the displacement of one of the mirrors. The Twyman–Green interferometer in particular is applied to test certain optical components such as prisms or even complete imaging systems.

Modified Michelson interferometer arrangements are also applied in microscopy to measure the microtopography of surfaces by means of two-beam interferometry. These systems are in general more versatile in use than those based on multiple-beam interference. If only low magnification and moderate lateral resolution are needed, systems with a beam divider in front of the microscope objective, which must have a large working distance, are used. Microinterferometers according to the Linnik type (Linnik, 1933), however, may be operated with objectives that have numerical apertures up to 0.9. In these microscopes, two carefully selected objectives with almost identical correction have to be used, one in each path of this interferometer.

Wave front division and amplitude division may take place if a diffraction grating is used as a beam splitter. When a diffraction grating acts as a dispersing element in a spectrometer, multiple-beam interference results from the division of the wave front at the grating structures, while a grating used as a beam splitter in an achromatic fringe interferometer (Strutt, 1889) may produce a two-beam interference pattern by dividing the amplitude. Instead of periodic gratings, diffracting elements with a random distribution of the transmittance or reflectance may also serve as beam splitters.

New interferometric analysis techniques become possible when holography is used, and new interferometer systems may be designed by introducing waveguide structures such as optical fibers and integrated optics.

Holographic techniques allow observation of interference effects resulting from the superposition of different wave fronts that emerge from an object at different times. Thus, for instance, actual variations of a wave front can be analyzed in a suitable interferometer arrangement in real time by interference with a wave front that has been recorded on a hologram at a previous time (for holography see Holography, Optical). Moreover, well-defined wave fronts that are difficult to generate by real optical systems can be provided by computer-generated holograms and can be compared to actual wave fronts—for example, for testing aspherical surfaces. Other methods of hologram interferometry make use of the fact that two or more hologram patterns can be exposed on the same photographic plate. So alterations of the wave fronts emerging from an object at different moments are made visible by interference fringes in the reconstructed image without use of an additional interferometer. If these alterations are produced by varying the index of refraction of the medium surrounding the object, or by use of slightly different wavelength for the different hologram recordings, contour lines of the surface of objects with rough, diffusely scattering surfaces can be generated (Varner 1971; Hildebrand and Haines, 1967).

A hologram taken from an object with a vibrating surface provides time-average fringes in the reconstructed image, which in principle show the stationary nodal regions. More details on holographic interferometry are to be found in a rather extensive article on this topic by Ennos (1970).

Other methods that can be used to detect and to measure changes of the profiles or translations of objects with rough diffusely scattering surfaces are speckle interferometry and speckle photography. By use of photographic, video, or electronic techniques, different single speckle patterns can be correlated to give an image that makes visible any alterations of the surface profile.

A very simple technique consists in recording speckle images on a photographic plate without a reference beam (speckle photography) and evaluating them by diffraction experiments. If, for example, two speckle images are superposed on the same photographic plate, a lateral movement of the object that has taken place between the two exposures gives Young's fringes in the Fraunhofer diffraction image of the speckle photograph. For further details on this topic see Yamaguchi (1985). Another interesting application of speckle photography in astronomy has been proposed by Labeyrie (1976).

By using waveguiding structures like optical fibers (for details see FIBER OPTICS) or structures that are employed in integrated optics, very compact interferometer systems can be built (for integrated optics see Tamir, 1975). Waveguides in integrated optics have the form of thin films—so-called slab or planar guides when they are effectively limited only in one dimension—or they consist of laterally ridged thin film stripes with an index of refraction differing from that of the substrate material—so-called strip or ridged guides—which are deposited above or embedded in a substrate.

In these interferometers, all or part of the beam paths may be passed through the waveguiding material. If this material is adequately chosen, the optical path can be influenced by temperature, mechanical stress, or electric and magnetic fields, for example. Thus sensors or modulators that make use of the high efficiency of interference effects can be constructed.

2.1.3 Spectrometry Spectrometry or spectroscopy is used to characterize and measure the spectral composition of radiation in order to obtain information on the emission, absorption, or reflection properties of various materials. Spectrometric methods are used in many areas of physics, chemistry, and biology, and in industrial applications. The most widespread application of spectrometry is spectrochemical analysis. In 1826 H. Fox-Talbot recognized the advantages of spectroscopy for chemical analysis, and later Bunsen and Kirchhoff laid the foundations of this by thorough investigations showing that certain spectral lines could always be associated with the same element. Quantitative chemical analysis can be performed by measuring the power of selected spectral lines in comparison with the corresponding lines of standard samples. Organic materials can be detected in particular by their infrared spectra, and the structure of molecules can be determined by their infrared bands.

Astrophysics is an area in which spectrometry is put to important use, the best known early application here being the recognition of elements present in the atmosphere of the sun by so-called Fraunhofer lines. The determination of the velocity of galaxies by measuring the Doppler shift of their spectra, or the measurement of the temperature of stars from the spectral distribution of the light they emit, are other examples. The first spectra were observed by Sir Isaac Newton, who dispersed sunlight using a prism and introduced the name "spectrum" for the spread of colors. Later, Fraunhofer used gratings as dispersing elements for investigating the spectrum of sunlight and discovered the dark lines that have been named after him.

Prisms or gratings are still the most frequently used devices in spectrometers. A conventional prism or grating spectrometer consists mainly of an entrance slit, the dispersing element, and certain optics that collimate the radiation emerging from the entrance slit and focus the dispersed wave fronts to the exit plane, where the spectrum can be recorded by means of an image receiver such as a photographic layer or an array of photodiodes, or where the radiation flux can be detected with some other suitable detector behind an exit slit (see Fig. 5).

Depending on the particular application, different instrument parameters such as spectral resolution, radiation collecting and transmitting capacity, or free spectral range are the deciding factors. The spectral resolution and light-collecting capacity of the spectrometers mentioned above are determined by the angular dispersion, the widths of the entrance and exit slits, and the relative aperture of the spectrometer. When such an instrument is used for the measurement of the spectra of extended sources, the light-collecting capacity is rather limited because of the mismatch of the image of the source and the narrow entrance slit, which is needed to give an acceptable resolving power. In order to overcome this problem, spectrometers with multiple slits in combination with special modulation and decoding techniques have been developed (Decker, 1971).

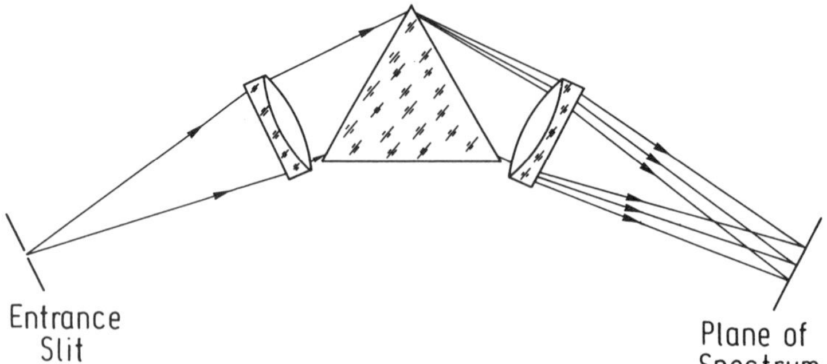

FIG. 5. Principle of a spectrometer.

For the study of the Zeeman effect or of the hyperfine isotope structure extremely high resolving power is needed, which is practically impossible to obtain using prism or grating spectrometers, as these are ultimately limited by diffraction. Fabry–Pérot interferometers, which make use of multiple-beam interference effects between highly reflecting mirrors, provide such a high resolution.

Another spectrometry method based on the use of a two-beam interferometer is Fourier spectrometry (Vanasse and Sakai, 1967), which involves the measurement of fringe visibility as a function of path difference. The advantage of this method regarding the signal-to-noise ratio has been pointed out by Fellgett (1958).

The use of lasers in spectrometry (Demtröder, 1981) has introduced totally new techniques and possibilities for measurements on spectral lines. Laser spectrometry methods have resulted in significant progress in the resolution, the accuracy of wavelength measurements, and the sensitivity in detecting small quantities of material.

2.1.4 Colorimetry As color is in fact not a physical quantity but a sensory perception, it actually cannot be measured by means of a physical detector. In principle only the intensity and spectral energy distribution of optical radiation can be measured. Apart from this, there does not exist an unambiguous relation between a certain color perception and the spectral composition of light that enters the eye. Whereas light with a certain spectral energy distribution produces a color stimulus that gives rise to a certain sensation of color, the opposite is not true. Colored light of a certain subjective quality may namely be composed of waves of different frequency distributions.

To define the color stimulus consistently in terms of some color system, the color specification, a set of parameters, has been introduced. Consequently colorimetry is the science of measurement of color specification. Because of the trivariance of human color vision only three of such parameters are needed. The Commission Internationale de l'Eclairage (C.I.E.), for example, has proposed and established an international system of objective specifications. The CIE system specifies a series of standard illuminants, a standard observer, and three imaginary primary reference stimuli into which, for convenience of calculation, color matching data obtained with real red, green, and blue "lights" can be converted.

On this foundation physical colorimetric measurements can be performed, for instance, by using three photoelectric receivers that have spectral response functions according to the CIE equal-energy distribution functions $\bar{x}(\lambda)$, $\bar{y}(\lambda)$, and $\bar{z}(\lambda)$ (see Fig. 6). If additionally the responses of the receivers were linearly related to the light input, the reading would be proportional to the tristimulus values X, Y, and Z. The precision and accuracy of a colorimeter working on this principle depends on the quality of approximations to the desired responses.

Alternatively, if it is possible to use spectrometric methods for determining the spectral energy distribution $S(\lambda)$ of a self-luminous object, the spectral luminance factor $\beta(\lambda)$ of an opaque object, or the spectral transmission factor $\tau(\lambda)$ of a transparent sample then the CIE color specification can be evaluated by computation. Colorimetry based on

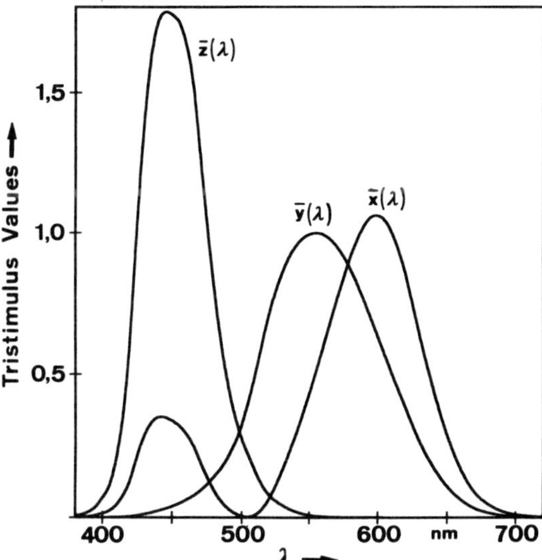

FIG. 6. Tristimulus curves.

this procedure is capable of providing measurements of highest accuracy.

To complete the picture it shall be mentioned that the terms colorimetry and colorimeter are widely applied to methods of chemical analysis or instruments used for performing quantitative analysis in chemistry or biochemistry. In most cases, these methods are concerned with the measurement of the spectral absorption, which is proportional to the concentration of a certain absorbing substance, rather than with the measurement of color specification.

2.2 Frequency

Optical radiation frequencies are very high (see Table 1) compared with those experienced in daily life, such as acoustic and radio frequencies or even frequencies used for television. But there are the atoms and molecules that possess characteristic frequencies, when oscillating in a crystal or rotating, which are just of the same order. Thus the interaction of optical radiation with matter results—according to the same frequencies—in resonance, giving rise to numerous phenomena such as absorption, reflection, color, etc. These resonance phenomena are used either to study the properties of matter with the help of radiation or to form and stabilize radiation with the help of matter. One of the latter applications is the use of frequency-stabilized radiation to construct highly accurate clocks.

2.2.1 Optical Frequency Standards The unit of time is the second, which is a base unit of the International System of units. It is defined as the duration of 9 192 631 770 periods of the radiation corresponding to the transition between the two hyperfine levels of the ground state of the ^{133}Cs atom. The unit of time is realized by national laboratories, running "atomic clocks" as primary standards (De Marchi, 1989). Because the meter is defined since 1983 as the distance that light travels in vacuum in 1/299 792 458 s, and precise measurements of length up to several meters are done by interferometry in the visible part of the spectrum, frequency standards in the visible are very important. Moreover, the high frequency of light—more than four magnitudes greater than that of the ^{133}Cs clock—promises a lower relative uncertainty $\Delta\nu/\nu$.

The transfer of the Cs frequency to optical frequencies is done by a "frequency chain" (see Fig. 7). The principle of the chain is (Jennings *et al.*, 1986), starting with the Cs frequency, to mix the lower well-known fre-

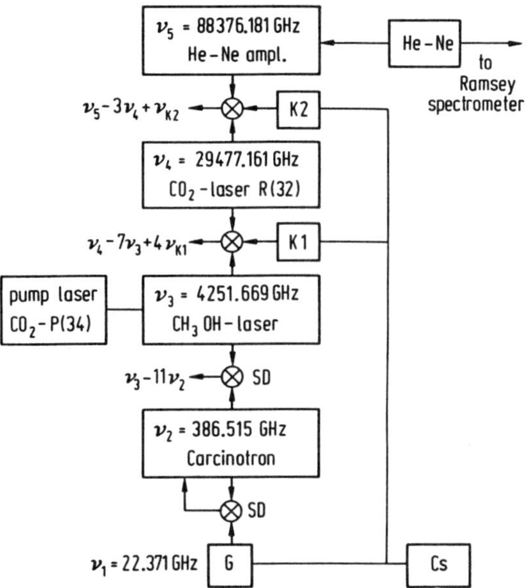

FIG. 7. Principle of frequency chain (from Weiss and Ni, 1992). G: Gunn oscillator, 22.371 GHz; K1: Klystron, 71.1 GHz; K2: Klystron, 55.3 GHz; x SD: Schottky diode; x: W-Ni diode.

quency with the unknown frequency in a nonlinear element (diode, nonlinear crystal, etc.), where all linear combinations of the two frequencies are produced. The beat frequency of a higher harmonic of the low frequency and the unknown frequency is measured with high accuracy, and the unknown frequency is calculated. Highest accuracy is only possible if the coupling of frequencies to the Cs frequency is phase coherent. These chains have been established and are in operation in some national metrological institutes up to the methane-stabilized H-Ne laser, which has a frequency of 88 376.181 599 5 × (1 ± 10^{-11}) GHz. A breakthrough to extend a phase-coherent frequency chain to the visible part of the spectrum has not yet taken place, but is expected in the near future.

Another approach to couple frequencies of the optical spectral domain to the rather low frequency of the primary Cs standard is a frequency interval division. The principle is as follows (Telle *et al.*, 1990): The frequency ν_1 of a laser in the visible or near infrared and its second harmonic $2\nu_1$ are mixed to yield $3\nu_1$. The frequency ν_2 of a second laser is also doubled and $2\nu_2$ is made equal to $3\nu_1$. The frequency interval $\nu_2 - \nu_1 = \frac{1}{2}\nu_1$ is half the value of ν_1. With the second laser and other ones, the division is repeated until the frequency interval ν^* can be determined by direct comparison with one of the frequencies of the frequency chain, say the CO_2 laser. After n steps, the frequency of the first laser can be calculated from the formula

$$\nu_1 = 2^n \nu^*. \tag{3}$$

Semiconductor lasers are suitable candidates for such a "division chain." A feasibility study supported by experiments has shown that this principle can be realized.

2.3 Diffraction

If an opaque obstacle is placed in the path of a beam of optical radiation emerging from a pointlike source, some part of the radiation is found in the area of the geometrical shadow. And at some distance from the obstacle, the distribution of intensity shows a system of dark and light fringes in the immediate vicinity of the shadow image of the edge of the obstacle (see Fig. 8). This phenomenon is known as diffraction. The term diffraction has been appropriately defined by Sommerfeld as "any deviation of light rays from rectilinear paths which cannot be interpreted as reflection or refraction."

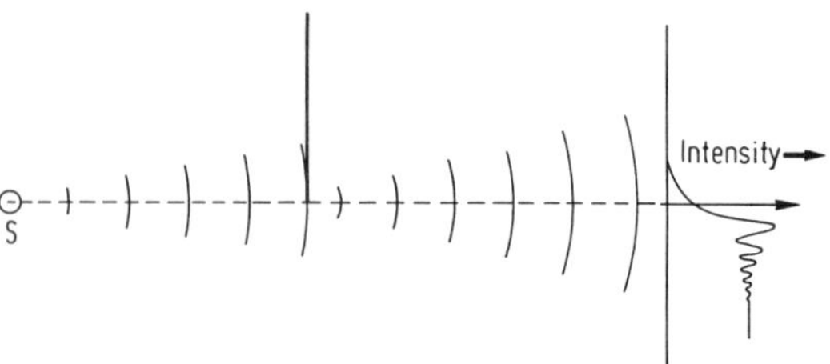

FIG. 8. Diffraction pattern caused by an edge.

Diffraction phenomena play a role of outstanding importance in all branches of physics and technology where the propagation of waves is involved. In the field of optics, diffraction effects can be observed whenever a beam of optical radiation encounters an obstacle or aperture of finite linear dimensions. This is why they are always present when light passes through an optical instrument, and so it is essential that diffraction and the limitations it imposes on the system's performance must be taken into account for a full understanding of the properties of optical systems.

Until now only a fairly small number of exact solutions of special diffraction problems have been found. All these solutions require solving sets of integral or differential equations with known boundary values. In most cases of practical interest, approximate methods must be used because of the mathematical difficulties involved in rigorous diffraction theories (Maystre, 1984). The simplified approaches ignore polarization, and the electromagnetic properties of matter are no longer described by the constants of solid-state physics but are replaced by phenomenological optical constants such as transmittance, reflectance, and the refraction index.

An early approximate theory of diffraction is Kirchhoff's formulation of the diffraction problem. It is based on a certain integral theorem, which had been derived for scalar waves in acoustics by Helmholtz and leads after some approximations and simplifications to the Fresnel–Kirchhoff diffraction

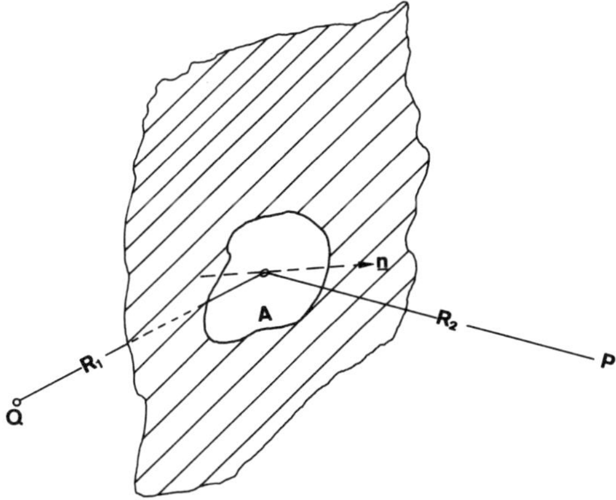

FIG. 9. Scheme for the derivation of the Kirchhoff integral. Q, source of radiation; A, aperture in a diaphragm; P, point of observation.

formula,

$$U_P = C \int_A \frac{\exp\{(2\pi i/\lambda)(R_1 + R_2)\}}{R_1 R_2} \times [\cos(\eta,R_1) - \cos(\eta,R_2)]dA, \qquad (4)$$

which describes the amplitude of the diffracted wave at P (see Fig. 9). Further simplifications, which are discussed in more detail in many textbooks (see, e.g., Klein, 1986), yield the equation

$$U(x',y') = C_1 \iint_A \exp\left\{\frac{2\pi i}{\lambda z}(x'\xi + y'\eta)\right\} \times d\xi d\eta, \qquad (5)$$

which gives the amplitude $U(x',y')$ in the far-field or Fraunhofer diffraction pattern of the diffracting aperture. Here x' and y' are the coordinates of a point P in the plane of observation and ξ and η are the coordinates of a point in the object plane (see Fig. 10). An essential assumption in the derivation of Eq. (5) is that the distances z_0, z of source and point of observation P from the diaphragm are large in comparison to the linear dimensions a of the diffracting aperture. That means that z satisfies

$$z \gg a^2/\lambda. \qquad (6)$$

If Eq. (6) is not valid, a term of the form $(\xi^2 + \eta^2)/2\lambda z$ must be introduced in the exponent of the integral. This characterizes the areas of Fresnel diffraction. On the other hand, it is not necessary to go to the distance given by Eq. (6) for observing a Fraunhofer diffraction pattern, as it can easily be shown that at least the intensity distribution of a far-field diffraction pattern is produced in the image

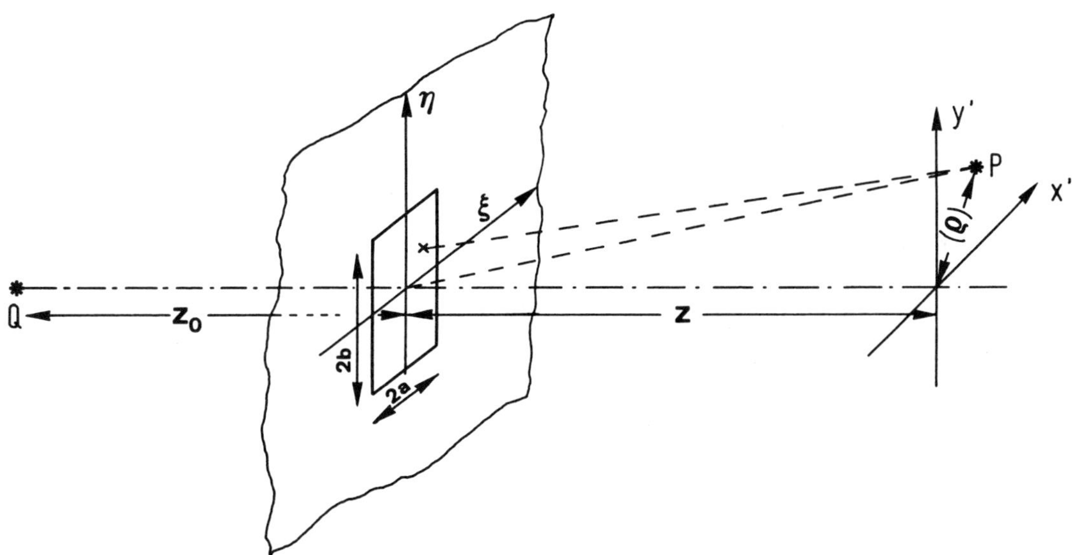

FIG. 10. Fraunhofer diffraction.

of a point source if an aperture is installed in the ray path of the imaging system (see, e.g., Goodman, 1968). When the aperture is illuminated by a plane wave and a lens is put up at a distance behind the aperture equal to its focal length, even the amplitude distribution in the focal plane of the lens may be described by Eq. (5).

When applying Eq. (5) to the simple case of a rectangular aperture with the side lengths $2a$ and $2b$ (see Fig. 10), the double integration can be separated and split into two factors. The complex amplitude of the diffraction pattern now may be written as

$$U(x',y') = C_2 \int_{-\infty}^{+\infty} \exp\left\{\frac{2\pi i}{\lambda}\frac{x'\xi}{z}\right\} d\xi \times \int_{-b}^{+b} \exp\left\{\frac{2\pi i}{\lambda}\frac{y'\eta}{z}\right\} d\eta. \quad (7)$$

Since the intensity is the squared modulus of the complex amplitude, the relative intensity in the diffraction pattern is

$$I(x',y') = C_2^2 \left\{\frac{\sin\left(\dfrac{\pi a x'}{\lambda z}\right)}{\pi a x'/\lambda z}\right\}\left\{\frac{\sin\left(\dfrac{\pi b y'}{\lambda z}\right)}{\pi b y'/\lambda z}\right\}. \quad (8)$$

This pattern exhibits a central maximum of intensity surrounded by subsidiary maxima and minima, the distances between the maxima and minima being proportional to the wavelength and inversely proportional to the linear dimension of the diffracting aperture. A fairly similar result is obtained if a circular opening with the diameter $2r$ acts as diffracting aperture (see Fig. 10). In this case, the amplitude and intensity patterns are radially symmetric around the optical axis, and the intensity pattern has the form

$$I(\rho) = C_K \frac{J_1(2\pi 2r\rho/\lambda z)}{2\pi 2r\rho/\lambda z}, \quad (9)$$

where J_1 denotes the Bessel function of order one and ρ is the distance to the origin of the coordinate system. This intensity distribution is generally referred to as the Airy pattern after the astronomer Sir George Airy (1801–1892), who calculated it as the theoretical distribution of intensity of a star image.

In the derivation of the Fresnel–Kirchhoff integral, the amplitude over the opening has been assumed to be constant. If Eq. (5) is generalized by introducing the function $f(\xi,\eta)$ describing a complex distribution of transparency in the aperture, and if the boundaries of integration are extended to infinity, the complex amplitude in the Fraunhofer diffraction pattern may be written in the form

$$U(x',y') = C_F \iint_{-\infty}^{+\infty} f(\xi,\eta) \times \exp\left\{\frac{2\pi i}{\lambda z}(x'\xi + y'\eta)\right\} d\xi d\eta. \quad (10)$$

This equation shows that the amplitude of the Fraunhofer diffraction pattern is expressed mathematically by a Fourier transformation (see FOURIER AND OTHER MATHEMATICAL TRANSFORMS). This relation and similar mathematical structures play an essential part in context with other theoretical approaches in many branches of optics, such as for instance spectroscopy, imaging, or theory of partial coherence (Hopkins, 1951).

2.3.1 Gratings Any arrangement that is equivalent in its effect to a number of parallel structures is called a diffraction grating. Apart from some special cases that will be discussed later, the slits or equivalent periodic structures are equidistant, are of the same width, and have the same distribution of transmittance or reflectance. Since diffraction-grating-like structures are used in a large number of applications, it is worth treating them in some detail. Gratings operate by mutual interference of the radiation beams diffracted by the individual grating structures. According to this, the distribution of intensity in the Fraunhofer diffraction image is the result of multiple-beam interference.

The first gratings Fraunhofer used for his early spectroscopic experiments were made by winding a thin wire around two parallel screws. Later, he made gratings by ruling grooves through gold films deposited on a glass plate. Both types of gratings were transmission gratings, which modulated the incident radiation field only by absorption in the areas of the bars or wires. Such gratings basically have a rather low diffraction efficiency. From some simple calculations based on Fourier analysis, it can be deduced that in the most favorable case only about 10% of the radiation flux that hits the grating is diffracted into each of the first diffraction orders. Nevertheless, transmission gratings of this type have still been used for certain specific applications.

A considerably large part of the radiation, however, can be directed into desired diffraction orders if the grating structure imposes a periodic phase modulation on the radiation field. Gratings with a surface profile as shown in Fig. 11 have this property. They are called blazed or echelette gratings. If such a structure is used in reflection and the radiation is incident at angle α relative to the normal of the surface with a period $2d$ of the grooves, the path difference for the radiation incident on any two adjacent grooves is $d \sin\alpha$ (Fig. 11). When this radiation is diffracted at some other angle by the amount β, the path difference of the range is further increased by the amount $d \sin\beta$. Thus the total phase difference is the algebraic sum $d(\sin\alpha \pm \sin\beta)$. The positive sign is valid when both rays are on the same side of the normal to the surface, while the negative sign applies when they are on opposite sides. The diffracted radiation is in phase over the entire wave front when the path difference for rays incident on adjacent grooves is an integral multiple of the wavelength. For a given radiation of wavelength λ, this is fullfilled for a particular angle of incidence when the radiation reflected from all grooves is in phase only at certain angles. The number of wavelengths that is equivalent to the path differences for radiation from adjacent grooves is called the order of interference, m. The preceding formulation leads to the grating equation:

$$m\lambda = d(\sin\alpha \pm \sin\beta). \quad (11)$$

With the help of this equation, it is possible to calculate the angles at which radiation of a given wavelength will appear in various orders when the radiation is incident at angle α to the normal of the grating surface. The angle of reflection from the face of the groove and the angle of diffraction are the same at the blaze wavelength, and for this wavelength (λ_B) the grating has its highest diffraction efficiency. The efficiency at the blaze wavelength can be higher than 80% but will approach a very low level in the first-order spectrum at half the blaze wavelength.

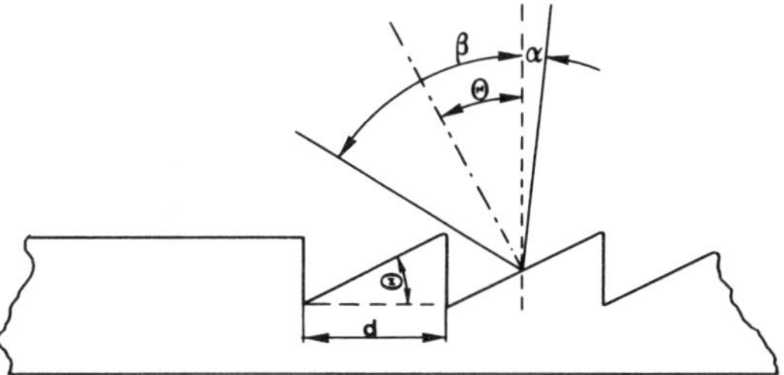

FIG. 11. Schematic diagram of an echelette grating.

Groove profiles shown in Fig. 11 can be obtained by ruling them with an adequately shaped diamond. Another technique for making gratings with surface profiles that exhibit a blaze effect consists in exposing a suitable interference pattern onto a photoresist layer. Such gratings are usually called holographic gratings. By using specially formed wave fronts to generate the interference pattern, gratings with imaging properties can be produced. In those gratings the distances of the grooves are not equal.

Although gratings are mostly used as dispersing elements in spectrometers or monochromators, it is worth mentioning some other interesting applications of their properties. Grating structures can for example be used for measuring and for recording linear and angular displacements of moving parts and automatic positioning in machine tools. In this application, either the so-called moiré effect is used to detect the relative displacement of gratings with identical period (Burch, 1963), or an interferometer is built up, which produces achromatic fringes by using gratings as beam splitters and for beam combination. These interferometers can be used with broadband sources, as the dispersion of the beam-splitter grating is compensated for by the same dispersion of the gratings used for recombination. If in such an interferometer the beam-splitting grating is displaced, and the first diffraction orders are recombined, the intensity in some point in the area of interference will vary according to half the grating period. By recombining higher diffraction orders and suppressing the lower ones, the resolution of the interferometer, which acts as a length transducer, can be multiplied.

If ultrasonic waves are applied to some solid-state material, standing waves can be generated in the material. These waves give rise to a periodic variation of the refraction index, thus forming a gratinglike phase structure. A system using this effect can act as high speed modulator for the intensity of a laser beam diffracted at the phase structure, when the amplitude of the ultrasonic wave is varied (Willard, 1949). In a similar way, a moving periodic phase pattern can be generated that can be used for shifting the frequency of a diffracted laser beam.

2.3.2 Imaging As a principal function most optical systems have the formation of an areal image either for detection and recording by an image sensor such as photographic film, or for visual evaluation. In many cases, the performance of the complete system depends essentially on the image quality supplied by the image-forming components. The prospective user of such an optical system will, of course, be interested in having the image quality specified in terms related to its application. This is why it is useful to take at least a brief look at the principal way in which these components work and the fundamental and physical limits of their performance.

The terms "object" and "image" in optical terminology have formalized meanings, which differ slightly from common usage. Both refer to distributions of radiation fields in space or particularly at surfaces, rather than to real objects. In this context, an image is the distribution of electric or magnetic excitation or intensity produced by bringing rays or wave fronts of electromagnetic radiation from an object point to a focus in an image point.

In an ideal optical imaging system, every point P of a three-dimensional object space should produce a corresponding stigmatic image P' in the image space. If P describes a curve C in the object space and P' runs through a corresponding curve C', and if each image curve is geometrically similar to the corresponding object curve, the imaging may be called perfect. According to the theorems of Maxwell and Caratheodory, perfect imaging can only be of a trivial kind, producing a plane mirror image of the object with unit magnification, when, for example, the object space and image space are homogeneous and when they have equal refraction indices. This implies that for nontrivial imaging between homogeneous object and image spaces, the requirements of perfect stigmatism or of strict similarity between object and image must be abandoned. Although most imaging systems are not perfect in the sense discussed above or even stigmatic, a considerable variety of systems exists, which provide more or less acceptable "sharp" images over a considerable volume of object and image spaces. In some systems, moreover, only a very restricted object field or a single, simple surface has to be imaged onto another surface in the image space. Thus the demands made on the imaging system are to a large degree relaxed. On the other hand, it may be pointless to have perfect imaging in some cases, such as, for instance in photographic cameras when a three-dimensional object has to be condensed to a "sharp" two-dimensional image on the film plane. This could not be achieved by means of a perfect imaging system.

The most common components suitable for focusing beams or wave fronts are lenses with curved refracting surfaces. As there are practical difficulties encountered in the production of surfaces of complicated shapes with the high degree of precision required in optics, the refracting or reflecting components of optical systems are generally plane, spherical, elliptical, or paraboloidal in form. In most cases only spherical surfaces are used.

A single lens made of glass or another transparent material that has two spherical surfaces, or a spherical mirror, produces acceptable stigmatic images only for those rays that are confined close to the optical axis of the lens or mirror, the region where the paraxial approximation is valid. In this region, the equations for ray tracing are fairly simple, and by using the paraxial approximation, many useful theorems for lens performance can be worked out. But in many practical applications, image forming systems that work well for an extended field of view and that have a large relative aperture are required. The field of view, when specified for an optical system, refers to the maximum size of the lateral object dimensions that may be imaged with acceptable image quality. It is proportional to the angle subtended by the object as seen from the optical system. The relative aperture refers to the half-angle of the cone of rays converging to an axial object or image point. The sine of this angle is usually called the *numerical aperture* NA of the system. If the relative aperture is large, the system is capable of collecting and transmitting a large amount of radiation, and, assuming the geometrical aberrations are compensated for, it has a high capacity for transmitting high spatial frequencies.

When single optical elements, with only plane or spherical surfaces, are used with high relative apertures for imaging large object fields, all image points suffer from some aberration. As a result of spherical aberration, object points situated on the optical axis of the system produce only more or less blurred, rotationally symmetric patches in the image. Object points at some distance from the axis, however, are focused to nonrotationally spread-out spots, and these image spots become larger with increasing distance from the axis, added to which the geometry relations are increasingly distorted. These off-axis aberrations are called astigmatism, coma, and distortion, according to the classification by Seidel who published the earliest systematic treatment of aberrations (see, e.g., Born and Wolf, 1975; OPTICS, GEOMETRICAL). In refractive systems, the images may also be degraded by chromatic aberrations due to the dependence of the refractive index on the wavelength. In principle, it is possible to suppress the spherical aberration by replacing a spherical surface with an asphere, but this usually increases the off-axis aberrations, thus restricting the usable field of view. This unwanted effect can, however, be suppressed by using a suitable combination of a spherical surface with an asphere. A well-known example of an efficient compensation for spherical aberration without impairing the off-axis imaging properties is the Schmidt camera. This type of telescope essentially consists of a spherical mirror combined with an aspherical corrector plate that is placed at the center of the mirror's curvature. Thus, approximately the same correction applies to all field angles, and the rather large field of view of a spherical mirror is conserved.

If only lenses with flat or spherical surfaces are available, a large number of lenses of different radii of curvature, consisting of different materials with different refraction indices and different dispersions, have to be arranged in appropriate sequences and at suitable relative distances in order to build up an imaging objective that has a high relative aperture and a large field of view. In addition to this, the best correction of aberrations can be performed only for fixed object and image distances and thus for fixed magnification. For this reason and as virtually the total compensation for all aberrations cannot be achieved simultaneously, the corrections have to be optimized depending on the particular application.

In the far-ultraviolet and far-infrared spectral regions, it is rather difficult or even impossible to make well-corrected objectives with a high numerical aperture and a large field of view because of the lack of a transparent material with suitable optical properties. In this and in other cases where the compensation for the off-axis aberrations is problematic, image recording by scanning is an efficient method. When the object is scanned by moving it in a suitably controlled manner while imaging it point by point to a detector, the imaging system must be corrected only for spherical aberration. Most electrophotographic or xerographic systems as well as image processing equipment used in the printing industry exploit the advantages of this method, as in principle arbitrarily large object fields can be recorded in this way. Moreover, the absence of image distortion makes the method attractive for use in measurement microscopes.

It should be remembered that there are still other ways of obtaining images or preparing image information for its storage or evaluation besides using refractive or reflective optics. The pinhole camera and the Fresnel zone plate or Soret plate are historical examples. The image-forming element of the pinhole camera is only a small, sharp-edged hole in a diaphragm, which confines the radiation emerging from each object point to a small bundle. The best possible resolution is achieved if the diameter of the pinhole is approximately $d = (\lambda D)^{1/2}$, where D is the distance of the image surface from the pinhole. The Soret plate is a kind of annular grating with opaque and transparent zones according to the Fresnel zones. Its focusing or imaging properties, discovered by Soret (1875), are based on diffraction and interference. The so-called holographic lenses can be seen as modern equivalents of that optical element. Because of the strong dispersion effects of diffracting structures, this element has a rather large chromatic aberration and should be used with nearly monochromatic radiation only.

Finally, image information can be recorded by interferometric techniques without use of any focusing device. In each surface at some distance from an object, the information on the amplitudes and phases of the wave fronts emerging from the object can be recorded by registering the intensity distribution that is obtained when a coherent reference wave is superimposed with the object wave fronts. If a photographic film or some other material capable of recording the interference pattern with sufficiently high resolution is used as the storage medium, the original wave front can be reconstructed by illuminating the record of the interferogram with the reference wave. This process is known by the name of holography (Leith and Upatnieks, 1967; see also HOLOGRAPHY, OPTICAL). Instead of reconstructing the image information by the process described above, the information contained in the intensity pattern can also be converted to digital information in a computer by means of a suitable receiver, and the computer can be used for the further processing or evaluation of the information.

The concept of geometric optics has for a long period of time provided the foundation for the development of efficient imaging devices. At the same time, techniques for testing of optical systems on this concept have successfully been used. The basis of modern image quality analysis, however, is Fourier optics. This approach has its roots in the works of Ernst Abbe (1873) and Lord Rayleigh (1896). After an essential stimulus supplied by Duffieux (1946), Fourier optics (Stößel, 1993) became the most powerful tool for describing the merit parameters such as the point spread function and the optical transfer function, which are related by Fourier reciprocity.

2.4 Refraction

There is the simple observation that a bundle of light when entering a transparent medium—say water or glass—changes its direction (see Fig. 12). This phenomenon, first quantitatively described by W. Snell in 1629, is called *refraction*. Snell's law states that

$$n_1 \sin\alpha_1 = n_2 \sin\alpha_2, \quad (12)$$

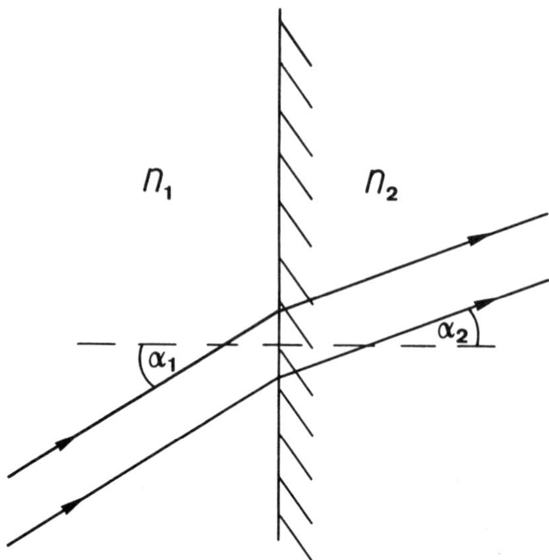

FIG. 12. Principle of refraction at a surface that separates two media 1 and 2, the refractive indices of which are n_1 and n_2, respectively.

where the n are constants related to materials 1 and 2. For wavelengths of optical radiation the constants—called refractive indices—usually have values >1, where *per definitionem* the refractive index of vacuum is 1. Sometimes the term "refractive number" is used for the ratio n_2/n_1, where n_1 is the refractive index of air under defined conditions; this is valuable for calculations in geometric optics, where air is generally on one or both sides of the optical element.

Theoretical considerations on the interaction of an electromagnetic wave with matter, especially with electrons, show that the phase velocity v_p of light is changed by a factor of $1/n$, which means that in most cases it is diminished. But there are also cases—depending on wavelength and material—where the refractive index n is lower than 1, and therefore the phase velocity is greater than c. But it can also be shown that the group velocity v_g, which transmits an optical signal, can never exceed the value c.

The dependence of the refractive index n on the wavelength is called *dispersion*. The value of n usually decreases with increasing wavelength, and this behavior is called *normal dispersion*. Because an increase of n with increasing wavelength is more rare, it is referred to as *anomalous dispersion*. Dramatic changes in the refractive index always occur where absorption bands of the material are observed. Kundt's rule states that in the vicinity of absorption bands the refractive index is decreased on the side of shorter wavelengths, but is increased sharply on the side of longer wavelengths adjacent to the absorption band. This behavior can be explained by the rigorous theory of the interaction of electromagnetic waves with matter. This theory also requires that the refractive index be written as a complex number, when absorption cannot be neglected:

$$\tilde{n} = n(1 - i\kappa), \qquad (13)$$

where n is the refractive index, i.e., the ratio of the phase velocities in vacuum and material, while κ is the index of absorption, describing the damping of the amplitude of the electromagnetic wave in matter. This complex presentation of the refractive index is therefore necessary for absorbing materials, for instance for metals.

The possibility of studying the internal properties of matter by measuring the refractive index and, conversely, the use of matter with known refractive index to shape, analyze, and guide optical radiation has led to numerous applications. Only a few of them can be discussed here.

2.4.1 Refractometry The term "refractometry" describes the methods for measuring the refractive index of materials. All the methods are based on the measurement of angles or differences of angles that occur when a bundle of light is refracted or reflected according to total reflection.

In Fig. 13, the scheme of a device called a *goniometer* is shown. This allows refractive numbers to be measured with high accuracy. The most important element—as in many other methods—is a prism. The light enters the prism, which has a so-called "refracting" angle α, at an angle β_1 and leaves it at an angle β_2, changing the original direction by an angle of δ. It can be shown that, as one changes the entrance angle β_1, δ goes through a minimum δ_{min}. If δ_{min} is known, the refractive number of the material of which the prism is made can be calculated by the formula

$$n = \frac{\sin[\frac{1}{2}(\alpha + \delta_{min})]}{\sin(\frac{1}{2}\alpha)}. \qquad (14)$$

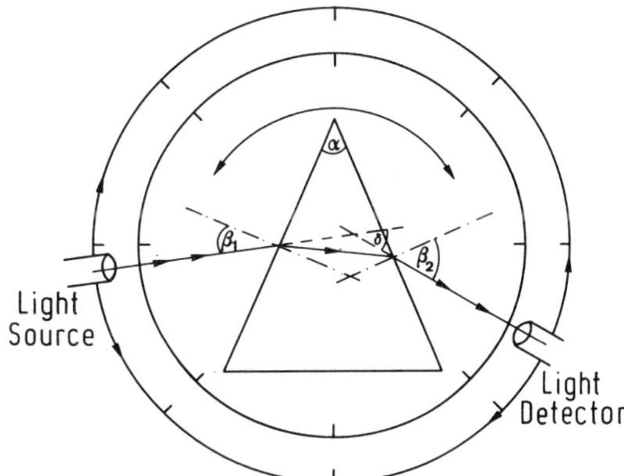

FIG. 13. Scheme of a goniometer. α: angle of prism; β_1 and β_2: entrance and exit angles; δ: angle of deviation.

The angle of refraction of the prism is measured by an autocollimation telescope on the same precisely circular table as the minimum deviation angle δ_{min} is determined by a movable lamp and detector device. In Table 2, the most important parameters and their possible uncertainties are listed, together with their influence on the measured refractive number of a glass material.

The uncertainties result in a total uncertainty less than 10^{-6} for the refractive number n. This goniometer method, the principle of which was first mentioned by Fraunhofer, is a tedious but the most accurate one, and is used to calibrate reference materials accurately. The method may also be used for the measurement of refractive indices of liquids, which are filled in a hollow glass prism.

Table 2. Uncertainties of parameters and their influence on the accuracy of refractive number measurement.

Parameter	Uncertainty	Uncertainty in n
Refracting angle α	$\Delta\alpha = \pm 0.03''$	1×10^{-7}
Minimum deviation angle δ_{min}	$\Delta\delta_{min} = \pm 0.1''$	3×10^{-7}
Wavelength λ	$\Delta\lambda = \pm 0.001$ nm	1×10^{-7}
Temperature of the prism T_p	$\Delta T_p = 0.01$ K	1×10^{-7}
Temperature of the air T_a	$\Delta T_a = 0.1$ K	1×10^{-7}
Refractive index of air n_a	$\Delta n_a = 2 \times 10^{-7}$	2×10^{-7}

2.4.2 Refractive Optics The ability of matter to bend light by refraction is widely exploited in optics. When a certain material of appropriate design is used, a wave front of an arbitrary shape can be formed. For the sake of simplicity, it is assumed that the dimensions of optical elements such as lenses, prisms, apertures, etc. are large compared with the wavelength of the radiation. In such cases, we refer to *geometric optics* (see OPTICS, GEOMETRICAL), where optical radiation emanating from a source can be considered as propagating rays. If all the rays are considered as originating from a "point source" at a fixed time after the light has started, all the tips of the rays form a surface (see Fig. 14), which is called the *geometric wave front* or the *eikonal S*. It can be shown that this wave front obeys the equation

$$(\text{grad } S)^2 = n^2. \tag{15}$$

The directions of the rays in an optically isotropic medium are always parallel to the normals on the surface, while the electric and magnetic field vectors are tangential to S. The "eikonal equation" demonstrates that according to the dispersion of n the surfaces depend on the wavelength. It is the art of the optician to select the appropriate material and shape of the optical elements to form a desired geometric wave front.

Since the materials have quite different refraction and absorption properties, special materials have to be chosen to build optical elements for the different wavelength regions. Table 3 contains some of the materials most used for optical elements.

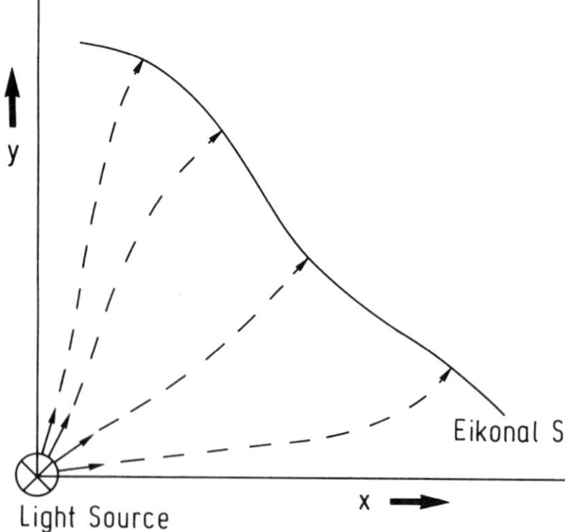

FIG. 14. The eikonal *S*.

There are mainly two methods to check to what an extent a desired geometric wave front has been verified. One method is to analyze the wave front directly by interferometry. The other one is to measure the optical transfer function $T(r_1,r_2)$. $T(r_1,r_2)$ is defined as the two-dimensional Fourier transform of the point spread function, which is the normalized light distribution in the image of a point source; r_1 and r_2 are the spatial frequencies in the two-dimensional frequency domain. Generally, $T(r_1,r_2)$ is a complex function. The modulus of $T(r_1,r_2)$ describes how the modulation of a sinusoidally shaped signal is changed with the spatial frequencies when the signal is transferred by optical elements. Very sophisticated optical systems such as high-resolution objectives for lithographic purposes are so well designed that they transfer optical signals of spatial frequencies higher than 2000 mm^{-1}, which is near the theoretical resolution given by the so-called Rayleigh criterion.

2.5 Polarization

Optical radiation is called "polarized" (see ELECTROMAGNETIC WAVE PROPAGATION; POLARIMETERS AND POLARIZATION SPECTROMETERS) if a fixed relation of the phase exists between the components of the electrical vector E of the wave propagating in the z direction (see Figs. 1 and 15). The tip of the vector E may describe a straight line, a circle, or an ellipse; the radiation is then called linearly, circularly, or elliptically polarized radiation. If one is looking on a piece of paper, and the wave rises upward in the direction of the reader, and E turns around clock-

Table 3. Materials used for the fabrication of optical elements.

Material	Wavelength (μm)	Material	Wavelength (μm)
Glass	0.35–4	KBr	0.28–37
Quartz	0.15–4	KJ	0.30–30
Al_2O_3	0.17–5.5	CsJ	0.26–60
LiF	0.11–8	Si	1.2–15
CaF_2	0.12–12	Ge	1.8–23
NaCl	0.21–20		

wise, then it is right-hand polarized radiation, otherwise it is left-hand polarized radiation.

If optical radiation interacts with matter, the charged particles, especially the electrons, react with this electric field. These interactions give rise to numerous phenomena, particularly with regard to polarization.

In the simplest case, the material is an isotropic, nonabsorbing material such as glass. The directions of the incoming and the reflected radiation define the plane of incidence. The Fresnel formulas (see ELECTROMAGNETIC RADIATION) describe the ratios of the reflected (E_r) or transmitted (E_t) components of the electric field vector to those components of the incoming (E_i) wave:

$$\frac{E_{r\perp}}{E_{i\perp}} = -\frac{[(n^2 - \sin^2\alpha_1)^{1/2} - \cos\alpha_2]^2}{n^2 - 1} = -\frac{\sin(\alpha_1 - \alpha_2)}{\sin(\alpha_1 + \alpha_2)}, \qquad (16)$$

$$\frac{E_{r\parallel}}{E_{i\parallel}} = \frac{n^2 \cos\alpha_1 - (n^2 - \sin^2\alpha_1)^{1/2}}{n^2 \cos\alpha_1 + (n^2 - \sin^2\alpha_1)^{1/2}} = -\frac{\tan(\alpha_1 - \alpha_2)}{\tan(\alpha_1 + \alpha_2)}, \qquad (17)$$

$$\frac{E_{t\perp}}{E_{i\perp}} = \frac{2 \cos\alpha_1(n^2 - \sin^2\alpha_1)^{1/2} - 2 \cos^2\alpha_1}{n^2 - 1} = \frac{2 \cos\alpha_1 \sin\alpha_2}{\sin(\alpha_1 + \alpha_2)}, \qquad (18)$$

$$\frac{E_{t\parallel}}{E_{i\parallel}} = \frac{2n \cos\alpha_1}{n^2 \cos\alpha_1 + (n^2 - \sin^2\alpha_1)^{1/2}} = \frac{2 \cos\alpha_1 \sin\alpha_2}{\sin(\alpha_1 + \alpha_2) \cos(\alpha_1 - \alpha_2)}, \qquad (19)$$

Here n denotes the ratio of the refractive indices of the media 1 and 2 (see Fig. 12) and the symbols $\perp$ and $\parallel$ are used when the electric field vector oscillates perpendicularly or parallel to the plane of incidence. If $\alpha_1 + \alpha_2 = 90°$, which means that the reflected and the refracted waves propagate perpendicularly to each other, the formulas show that no radiation oscillating in the plane of incidence is reflected. This characteristic angle of incidence is called the *Brewster angle* α_P, and it can be shown that

$$\tan\alpha_P = n; \qquad (20)$$

n can be determined by measuring α_P.

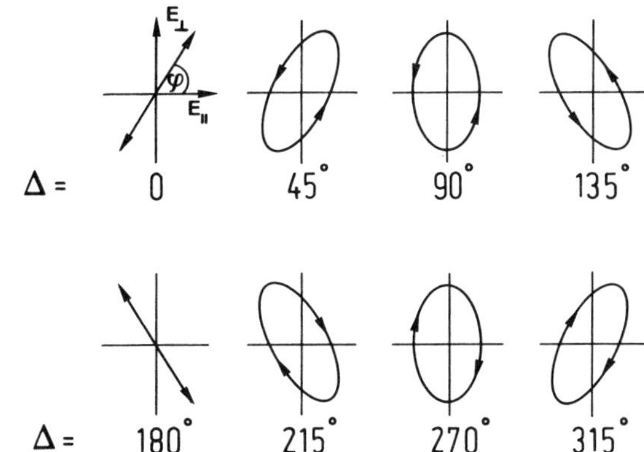

FIG. 15. Movement of the tip of the electric vector if radiation is polarized. The ratio of the mutually perpendicular components $E_\perp$ and $E_\parallel$ of the vector E is $\tan\varphi$, the phase difference between $E_\perp$ and $E_\parallel$ is Δ.

The phenomenon that one component of the reflected field vector may be eliminated by an appropriate choice of the angle of incidence is used to produce linearly polarized light, especially where no other methods are suitable.

If the material is an absorbing one, the Fresnel formulas are also valid; but the real refractive index must be replaced by the complex one, which includes the absorption. Therefore, the ratio of the components of the vector E becomes complex, and the Fresnel Eqs. (16) to (19) only hold for the equality of the two left-hand terms.

If one drops the assumption that the material is isotropic, some interesting phenomena may occur. If, for instance, long molecules are lined up in a material, the components of the electric field may move with different velocities along or perpendicular to the molecules, and therefore the refractive index is different for those directions. The same holds for crystals, when the symmetry of the unit cell is twofold or manyfold. With such materials, which are called birefringent, phase differences between the mutually perpendicular components of the electric field vector may be introduced. If, by choosing the right thickness and orientation of the material, a phase difference of 90° is introduced, this material is called a quarter-wave plate. Linearly polarized light can be transformed into circularly polarized and vice versa; two quarter-wave plates may turn the polarization of linearly polarized light by 180°.

Birefringence may be introduced when molecules are forced to change their order under the influence of stress, temperature, and electric fields. Polarized light can therefore be used to investigate whether there are regions of different order in a transmitting material.

A special case of birefringence occurs when the unit cells of crystals with twofold or manyfold symmetry are slopingly orientated in relation to the surface; spar ($CaCO_3$) is a famous example. Such crystals are said to have an anomalous refraction, because a light ray hitting the surface along the normal is split into two rays: one that goes straight through the crystal and is called the "ordinary" ray, and one that is deflected by the crystal and is called the "extraordinary" ray. The polarization of the outgoing light is perpendicular. Depending on the symmetry of the crystal, with lower degree of symmetry a second extraordinary ray may evolve.

These phenomena lead to instruments that create polarized light. Some examples of these are the Nicol and the Glan–Thomson prisms (see POLARIMETERS AND POLARIZATION SPECTROMETERS; OPTICAL SYSTEMS), which are carefully cleaved and sawed $CaCO_3$ crystals cemented together in such a way as to eliminate the ordinary ray and to transmit only the extraordinary one or, as in the case of the Wollaston prism, to separate the two rays by an angle. Separation of rays with different polarization may also be achieved by dichroism, where, because of different dispersion and absorption of the two polarizations, one ray is eliminated by absorption. Foils made of small, oriented dichroic crystals embedded in a thin carrier material or colored foils where birefringence is introduced by stress are widespread as a means of creating or eliminating polarized radiation.

Electric fields may cause anisotropy in isotropic materials and thus cause birefringence. We call this the Kerr effect. The effect is used to build very fast switches (down to 10^{-9} s) for laser physics.

An important class of matter for polarization is the *optically active* materials. These materials—sugar solutions and quartz are well known examples—either contain molecules (sugar) that are not invariant if viewed in a mirror, or are anisotropic crystals (quartz), where the optical axis is parallel to the light beam. The orientation of linearly polarized light is rotated around either to the left- or to the right-hand side, depending on the substance. This effect, which depends on the wavelength, is widely used in polarimetry. It can also be introduced into originally isotropic materials such as glass, when an external magnetic field is put on the glass parallel to the beam (Faraday effect). The glass becomes optically active; the amount of rotation can be controlled up to a resolution of 10^{-3} degrees by choosing the appropriate field strength.

2.5.1 Polarimetry The term *polarimetry* describes that field of metrology in which the phenomena of polarization are used either to measure properties of matter itself, such as material constants, or to analyze the concentration of chemical substances (see POLARIMETERS AND POLARIZATION SPECTROMETERS). The principle of a polarimeter is shown in Fig. 16: Linearly polarized, monochromatic, and parallel radiation of a light source *L*, which is in general used in combination with a filter *F* and a condensor *C*, meets the sample *S*. The sample itself may be a piece of matter or a cell filled with a liquid. The analyzer *A* is used to compensate for a possibly introduced rotation of the polarization by the sample; the angle at which the analyzer is rotated is the measurement quantity. Additionally, a Faraday cell FM as an alternating modulator or quartz compensator for precise compensation—for instance by lock-in techniques—may be introduced.

One of the most important applications in polarimetry is sugar content measurement (Emmerich *et al.*, 1991). A great deal of the world's sugar trade relies on the accuracy of

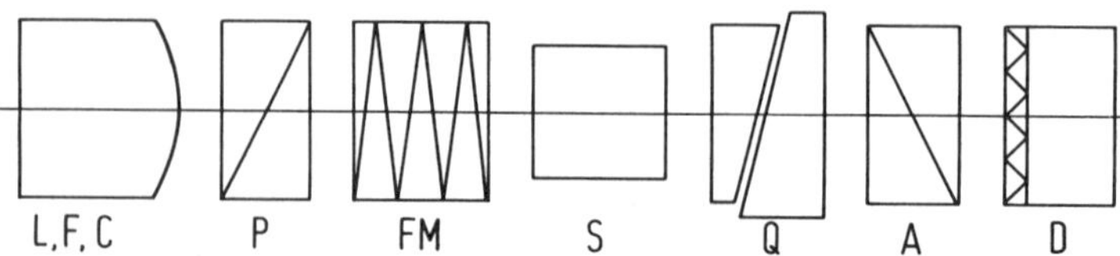

FIG. 16. Scheme of a polarimeter. L, F, C: lamp, filter, condenser; P: polarizer; FM: Faraday modulator; S: sample; Q: quartz-wedge-compensator; A: analyzer; D: detector.

polarimetric measurements. For the measurement, 26.016 g of the substance under investigation is dissolved in distilled water to obtain 100 cm^3 of the solution. The mass of 26.016 g corresponds to a weight of 26.000 g when the sugar is weighed in air at "normal" conditions (temperature 20 °C, air pressure 101 325 Pa, relative humidity 50%). The solution is then filled into a cuvette 200.000 mm in length, and the polarization rotation—because of the optically active substance saccharose in the solution—is measured at a temperature of 20.00 °C or corrected for this temperature. The polarimeters allow the saccharose content to be read directly in percent or in °Z of the International Sugar Scale, the 100 °Z point of which is defined for 26.016 g of purest saccharose in a 100.000 cm^3 solution at normal conditions. The 0 °Z point is defined for pure distilled water. Table 4 contains the values of optical rotation for three different wavelengths for the 100 °Z point, fixed for the International Sugar Scale.

Low-pressure spectral lamps or an He-Ne laser are used as light sources; the wavelength of the radiation has to be known with an uncertainty of $\pm 10^{-4}$ nm if an accuracy of 10^{-3} degrees is desired when using a 200-mm cuvette or an equivalent quartz plate. In commercial polarimeters, the angle is determined with an uncertainty of about $\pm 5 \times 10^{-3}$; the uncertainty in the sugar determination is about ±0.2 °Z.

Polarimeters are calibrated by quartz plates of known rotation, which must be plane within ±0.5 μm; the thickness variation has to be less than 0.3 μm. Quartz plates are usually calibrated by national metrology institutes.

2.5.2 Ellipsometry The study of the change of polarized radiation when interacting with matter, either absorbing or nonabsorbing, is used to obtain information about the value of the complex refractive index $\tilde{n} = n(1 - i\kappa)$ and the local change of that index (see ELLIPSOMETERS). The latter is used to determine the thickness of thin layers on surfaces.

The basis of ellipsometry is the Fresnel formulas given above, where n is replaced by $\tilde{n}$. As a consequence, the components $E_{i\parallel}$ and $E_{i\perp}$ of the incoming electric field E are, when reflected, not only individually damped, but they also have a phase shift δ:

$$E_{r\perp} = E_{i\perp}\rho_\perp e^{i\delta_\perp}, \tag{21}$$

$$E_{r\parallel} = E_{i\parallel}\rho_\parallel e^{i\delta_\parallel}. \tag{22}$$

Here ρ denotes the ratio between the magnitudes of reflected and incident components of E. With the choice of appropriate input conditions—for instance linearly polarized light—an analysis of the reflected light leads to the determination of the phase difference

$$\Delta = \delta_\parallel - \delta_\perp \tag{23}$$

and the azimuth difference Ψ with

$$\tan\Psi = \rho_\parallel/\rho_\perp. \tag{24}$$

The quantities Δ and Ψ, together with the angle of incidence α, are the parameters from which via the Fresnel formulas n, κ, and—with reasonable assumptions and approximations—the thicknesses of thin layers are calculated. The calculations are tedious, and only the use of modern computers has opened the field of ellipsometry for many applications.

The accuracy of the measurement of Δ and Ψ is of the order of 10^{-3} degrees. The spot size may only be a few square micrometers, which allows very small samples to be investigated. Even the thickness of monoatomic layers can be measured, or layers only partially covering a surface can be analyzed.

The analysis of the chemistry of surfaces, the monitoring of the growth of thin layers, the measurement of optical constants and of tension in optical components are the chief fields of application.

Table 4. Values for optical rotation of saccharose solution (200-mm cuvette).

Wavelength (nm)	Value of optical rotation for the 100°Z value
546.2271	(40.777 ± 0.001)°
589.4400	(34.626 ± 0.001)°
632.9914	(29.751 ± 0.001)°

2.6 Discreteness of Photons

"We know that light is made of particles" (Feynman, 1985). We know that three to four of such particles are necessary, to change a silver halide grain of a photographic emulsion in such a manner that the grain may be reduced to silver in a developer. But the photons have to arrive in a certain time interval; otherwise there is the probability that the small speck of silver atoms that, in general, is built up by the photons and starts the development is not stabilized. This is one of numerous examples in nature where the distribution of single photons in time and space can be observed. The flow of photons is measured in the field of radiometry; an application of this flow of single photons is in cooling down the temperature of an atomic beam by radiation.

2.6.1 Radiometry Radiometry is the measurement of flux of the energy of electromagnetic radiation. An appropriate method of performing radiometric measurements of high precision is to absorb the radiation to be measured in some suitable material—called in the following *absorber*—and to measure, for example, the rise of temperature of the absorber. In order to keep corrections for absolute measurements as small as possible, the absorber should in good approximation have the properties of a perfect blackbody, i.e., it should completely absorb the radiation of all wavelengths to be measured. Virtually the best approximation to a blackbody is a cavity with a blackened surface.

A well-known technique for the absolute calibration of a cavity detector is the electrical substitution method (Hengstberger, 1989). Joule heat is generated in a resistor layer attached to the cavity and is substituted for the optical power that gives the same rise in temperature. The electrical power can be measured with high accuracy.

When used at room temperature, this technique may involve considerable problems resulting from the usually rather high heat capacity of the receiver cavity, which leads to slow response times and a low responsitivity of the system, and from the final heat conductivity of the cavity material. The latter gives rise to nonequivalence errors, that is, errors that depend on where the power is applied to the cavity. When, however, the receiver is operated at cryogenic temperatures, the heat capacity is reduced by about three decades and the other effects mentioned above are significantly weakened. Thus, more precise and accurate measurements can be made by so-called cryogenic radiometers (Hoyt and Foukal, 1991; Martin *et al.*, 1985). An alternative method for absolute radiometric calibration of a detector is its exposure to the radiation of a blackbody cavity source. The spectral irradiance of such a source can be calculated by Planck's formula for the blackbody radiation. For practical purposes, suitable photoelectric receivers with high long-term stability that are calibrated against absolute receivers are used as detector transfer standards. Special selected tungsten incandescent lamps may be used for the transfer of the spectral irradiance scale for wavelengths from 250 nm to 2.4 μm.

If the radiant flux is rather low—that means that single photons arrive at the receiver in time intervals that can be resolved by the receiver and the electronics—radiometry may be done by counting the single photons. Special photomultipliers are suitable as receivers, where the operating conditions such as the multiplier voltage are selected to allow photon counting. Counting rates up to 10^8 photons/s are possible, which corresponds (see Table 1) to an irradiation of about 4×10^{-11} W (wavelength 500 nm) at the front side of the receiver.

2.6.2 Laser Cooling According to the kinetic gas theory, the kinetic energy of an atom in a gas per degree of freedom is equal to $\frac{1}{2}kT$, corresponding to $\frac{1}{2}mv^2$, where m is the mass and v the velocity of the atom. The temperature T can therefore be lowered or the gas can be cooled by reducing the velocity and the width of the velocity distribution of the atoms or molecules (Andreev *et al.*, 1981; Phillips and Metcalf, 1982). For the sake of simplicity, let us assume that we have atoms that move in one direction (one degree of freedom only). This assumption can be approximately realized in an experiment in which the atoms leave a closed box through a small hole (see Fig. 17). If a beam of particles hits the atoms from the opposite direction, the velocity of the atoms is diminished by the hits. Strong candidates for such particles are photons, because they do not chemically react with the atoms or pollute the clean gas. Their energy and momentum can easily be controlled by the frequency ν, besides which they may simply be introduced into closed vessels by transparent windows.

If a photon with the momentum $h\nu/c$ just

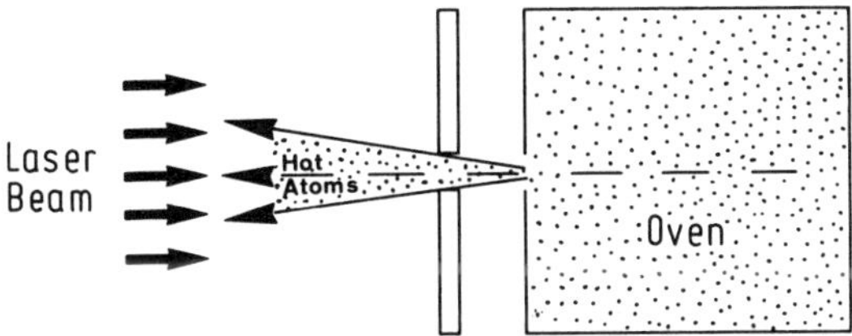

FIG. 17. Cooling of atoms by laser radiation.

in the opposite direction of the movement is absorbed by an atom, its momentum slows down the velocity of the atom. The essential requirement for this process is that the energy of the photon $h\nu$ corresponds to the transition between two states of the atom corrected by the kinetic energy of the atom. If these two states show the same population, the maximum possible retardation rate is reached. Because an atom can only absorb a photon when it has returned to the lower state from the higher one, which may have a mean lifetime τ, the optimum cycle for absorption-emission needs a minimum absorption time of 2τ.

Every hit of a photon, which has a momentum $h\nu/c$, reduces the momentum of the atom by the same amount:

$$-d(mv) = h\nu/c \quad \text{and} \quad -\Delta v = h\nu/mc, \tag{25}$$

when relativity effects are neglected and Δ is written instead of d, since the change of velocity occurs in finite steps. The maximum retardation possible is determined by the time 2τ:

$$-\frac{\Delta v}{2\tau_{max}} = \frac{h\nu}{mc2\tau}. \tag{26}$$

After n hits, the velocity has dropped from the maximum v_{max} to zero:

$$v_{max} = n\Delta v = nh\nu/mc, \tag{27}$$

$$n = v_{max}mc/h\nu. \tag{28}$$

For this process, a minimum time t_{min} with

$$t_{min} = n2\tau = v_{max}mc2\tau/h\nu \tag{29}$$

is necessary. And the minimum distance d at which the atoms may be stopped is given by

$$d = (v_{max}/2)t_{min}. \tag{30}$$

In Table 5, some values are given for the isotope ^{40}Ca, which is a potential candidate for an atomic clock.

It should be mentioned that the velocity of the atoms changes rapidly during the cooling process, and the kinetic energy changes correspondingly. Therefore, either the frequency of the cooling photons or the energy of the transition has to be adjusted accordingly to remain in resonance during the cooling process.

The derivation given above is a rather simple one. A detailed analysis shows that the processes are much more complicated (see ATOMIC COOLING AND TRAPPING). It should be particularly borne in mind that the mean velocity of the atoms may be zero, but that the mean of the squared velocities differs from zero because of the statistical emission and absorption of photons. It can be shown that one particular natural limit of cooling is approximately determined by the energy equivalent to the cooling transition: $2h/2\tau$. The resulting temperatures, depending on the type of atoms and the experimental setup, are around 1 mK.

Cooling is desired for all those applications where the velocity of particles limits the accuracy of measurements, for instance as a consequence of the Doppler effect, or where other techniques are not applicable. These applications include high-resolution spectroscopy, "Coulomb crystallization," the trapping of neutral atoms, and the investigation of collisions of atoms at ultralow energies.

Table 5. Data for cooling an atomic beam of ^{40}Ca atoms.

Temperature (K)	Transition wavelength	Lifetime of $^1P_1$ state	Max. velocity	Number of hits	Cooling time (minimum)	Distance d (minimum)
1000	$^1S_0 \rightarrow {}^1P_1$ 423 nm	4.6 ns	640 m s^{-1}	≈28000	0.26 ms	83 mm

3. CONCLUSIONS

Because of limited space, optical phenomena and their applications could not be discussed in depth in this article; the reader is therefore referred to special articles in the *Encyclopedia* where individual topics are treated in detail. Optical phenomena and particularly their applications are so numerous, as is confirmed by the use of optics in many fields of biology, chemistry, engineering, materials science, medicine, and physics, that a selection was unavoidable. A few examples may illustrate this: The interaction of optical radiation with matter applied in biology, chemistry, photography, etc. is not covered, nor are laser applications in communication, material processing, medicine, and surveying. All the phenomena of squeezing light into ultrashort pulses and the resulting applications in time-resolved spectroscopy, as well as scattering and its application, have also been omitted. Optical phenomena and their application are described in about 100 articles in the *Encyclopedia*, where the reader will find the information he is looking for.

GLOSSARY

Atomic Clock: Accurate clock that depends for its operation on the natural vibration frequencies of an atomic system.

Birefringence: Refraction of light into slightly different directions, depending on polarization, to form two rays.

Blackbody: A body that absorbs all light incident upon it, or when heated radiates ideally according to fundamental physical laws relating energy, frequency, and absolute temperature.

Brewster Angle: Angle at which unpolarized light incident upon the surface of a transparent substance becomes totally polarized when reflected.

Coherence: Relationship of two or more wave trains in phase so as to permit interference.

Color Stimulus: Physical origin of color sensation registered by the eye.

Cryogenic Temperature: Temperature typical of cryogenic fluids, usually below 77 K.

Dichroism: Property possessed by a crystal of showing different colors when viewed in the direction of two different crystal axes.

Diffraction: Modification of light distribution when light passes edges, narrow slits, lines, or holes.

Dispersion: Change of refractive index as a function of wavelength.

Doppler Effect: Change of frequency in dependence of the relative speed between light source and detector.

Focal Plane: Plane through the principal focus of an optical system and perpendicular to its optical axis.

Fourier Analysis: Mathematical technique for analyzing a complex wave form or distribution into a composite of sine and cosine waves whose admixture would reproduce the original distribution.

Fourier Optics: All fields in optics where Fourier transforms and Fourier analysis (see FOURIER AND OTHER MATHEMATICAL TRANSFORMS) are widely used for the treatment of optical problems. Important examples are diffraction phenomena (especially Fraunhofer diffraction), linear transfer of image information, and interference of partially coherent wave fields with particular application in (Fourier) spectroscopy.

Fraunhofer Lines: Dark lines in the solar spectrum characteristic of individual elements in the atmosphere surrounding the incandescent surface of the sun.

Free Spectral Range: The interval of wavelength where no superposition of spectra belonging to different orders occurs.

Goniometer: Instrument for measuring angles.

Ground State: Energy level of an atomic electron system having the least energy of all its possible states.

Hologram: Photographic record of an irradiance pattern produced by interference of the complex wave front emerging from an object with a reference wave front.

Interference: The act that two or more wave trains affect one another in order to augment or diminish the resulting intensity.

Mode Locking: Condition in which all the modes in an optical cavity have the same relative phase.

Paraxial Approximation: The equations of the paraxial approximation are derived by taking the first-order approximations of the equations of a more exact theory. It is assumed that an optical axis can be defined for an optical system and that all light rays as well as all normals to reflecting or refracting surfaces form small angles with the axis.

Photodiode: Two-electrode semiconducting device that converts incident light into a corresponding electrical current.

Polarization: Action of affecting light so that the vibrations of the electric field vector of the light wave are confined to either a single plane or to two mutually perpendicular planes with fixed relationships of phases and amplitudes of the vector components in the planes.

Population Inversion: Condition where the number density of atoms or molecules in a particular excited state exceeds that of atoms or molecules in a lower excited state.

***Q*-Switching:** Switching of the quality of the optical cavity of a laser oscillator, for instance by some special absorber, a rotating mirror, or an electro-optic modulator.

Radiation Collecting Capacity: (in context of spectrometers) Capacity of collecting radiation of an extended incoherent source.

Refraction: Deflection from a straight path undergone by a light ray in passing obliquely from one medium into another in which its velocity is different.

Response Time: Time between an occurrence and a reaction produced by the occurrence.

Speckle Pattern: A random interference pattern of granular appearance that occurs when radiation from a small source is reflected from a rough surface or scattered by a transparent layer, which introduces a spatially random phase variation.

Spectral Resolution: Ratio of a certain wavelength λ to the smallest difference $\delta\lambda$ of two wavelengths λ and $\lambda + \delta\lambda$ that can be resolved by a spectrometer.

Stigmatism: Condition of an optical system in which rays of light from a single point converge into a single point.

Visibility: Modulation of irradiance in an interference pattern.

Zeeman Effect: Splitting of spectral lines into separate polarized lines by the influence of a strong magnetic field.

Works Cited

Andreev, S. V., Balykin, V. I., Letokhov, V. S., Minogin, V. G. (1981), *JETP Lett.* **34**, 442–445.

Born, M., Wolf, E. (1975), *Principles of Optics*, Oxford: Pergamon Press, p. 37.

Burch, J. M. (1963), "The Metrological Application of Diffraction Gratings," in: E. Wolf (Ed.), *Progress in Optics*, Vol. II, Amsterdam: North-Holland, pp. 73–108.

Decker, I. A. (1971), *Appl. Opt.* **10**, 510.

De Marchi, A. (Ed.) (1989), *Frequency Standards and Metrology*, Berlin: Springer.

Demtröder, W. (1981), *Laser Spectroscopy*, Berlin: Springer.

Duffieux, P. M. (1946), *L'Integrade de Fourier et se Applications à l'Optique*, Rennes: Société Anonyme.

Edlén, B. E. (1966), *Metrologia* **2**, 71–80.

Emmerich, A., Zander, K., Seiler, W. (1991), *Zuckerindustrie* **116**, 245–260.

Ennos, A. E. (1970), *Adv. Quantum Electron.* **1**, 199–260.

Fellgett, P. (1958), *J. Phys.* (Paris) **19**, 187, 237.

Feynman, R. P. (1961), *Quantum Electrodynamics*, New York: W. A. Benjamin, Inc.

Feynman, R. P. (1985), *QED—The Strange Theory of Light and Matter*, Princeton: Princeton University Press, p. 14.

Goodman, J. W. (1968), *Introduction to Fourier Optics*, San Francisco: McGraw Hill.

Hengstberger, F. (1989), in: F. Hengstberger (Ed.), *Absolute Radiometry: Electrically Calibrated Thermal Detectors of Optical Radiation*, San Diego, CA: Academic.

Hildebrand, B. P., Haines, K. A. (1967), *J. Opt. Soc. Am.* **57**, 155–162.

Hopkins, H. H. (1951), *Proc. R. Soc. A* **208**, 263.

Hoyt, C. C., Foukal, P. V. (1991), *Metrologia* **28**, 163.

Jennings, D. A., Evenson, K. M., Knight, D. J. E. (1986), *Proc. IEEE* **74**, 168–179.

Klein, M. V. (1986), *Optics*, New York: Wiley.

Labeyrie, A. E. (1976), in: E. Wolf (Ed.), *Progress in Optics*, Vol. XIV, Amsterdam: North-Holland, pp. 47–87.

Leith, E. N., Upatnieks, J. (1967), "Recent Advances in Holography," in: E. Wolf (Ed.), *Progress in Optics*, Vol. VI, Amsterdam: North-Holland, pp. 1–5.

Linnik, W. (1933), *C. R. Acad. Sci. URSS* **1**, 18.

Marathay, A. S. (1982), *Elements of Optical Coherence Theory*, New York: Wiley.

Martin, I. E., Fox, N. P., Key, P. J. (1985), *Metrologia* **21**, 147–155.

Maystre, D. (1984), "Rigorous Vector Theories of Diffraction Gratings," in: E. Wolf (Ed.), *Progress in Optics*, Vol. XXI, Amsterdam: North-Holland, pp. 3–67.

Phillips, W. D., Metcalf, H. (1982), *Phys. Rev. Lett.* **48**, 596–599.

Post, E. J. (1967), *Rev. Mod. Phys.* **39**, 475–493.

Rosenberger, F. (1965), *Die Geschichte der Physik*, Hildesheim: Georg Olms Verlagsbuchhandlung.

Sagnac, G. (1913), *C. R. Acad. Sci.* **157**, 708, 1410.

Stößel, W. (1993), *Fourier-Optik*, Berlin: Springer.

Strutt, J. W. (Lord Rayleigh) (1889), *Philos. Mag.* **28**, 77, 183.

Tamir, T. (Ed.) (1975), *Integrated Optics*, Berlin: Springer.

Telle, H. R., Meschede, D., Hänsch, T. W. (1990), *Opt. Lett.* **15**, 532–534.

Tolansky, S. (1973), *An Introduction to Interferometry*, London: Longman.

Vanasse, G. A., Sakai, H. (1967), "Fourier Spectroscopy," in: E. Wolf (Ed.), *Progress in Optics*, Vol. VI, Amsterdam: North-Holland, pp. 259–330.

Varner, J. R. (1971), *Appl. Opt.* **10**, 212–213.

Weiss, C. O., Ni, Y. C. (1992), "The Speed of Light," in: J. Bortfeldt, B. Kramer (Eds.), *Landolt-Börnstein: Numerical Data and Functional Relationships in Science and Technology*, New Series, *Units and Fundamental Constants in Physics and Chemistry*, Part b, *Fundamental Constants*, Berlin: Springer, pp. 3–37.

Willard, G. W. (1949), *J. Acoust. Soc. Am.* **21**, 10.

Yamaguchi, J. (1985), in: E. Wolf (Ed.), *Progress in Optics*, Vol. XXII, Amsterdam: North-Holland.

Further Reading

Accetta, J. S., Shumaker, D. L. (Exec. Ed.) (1993), *The Infrared and Electro-Optical Systems Handbook (8 Vol.)*, Bellingham, WA: SPIE.

Azzam, R. M. A., Bashara, N. M. (1977), *Ellipsometry and Polarized Light*, Amsterdam: North-Holland.

Berek, M. (1970), *Grundlagen der Praktischen Optik*, Berlin: Walter de Gruyter & Co.

Born, M., Wolf, E. (1986), *Principles of Optics*, Oxford: Pergamon Press.

Bortfeldt, J., Kramer, B. (Eds.) (1992), *Landolt-Börnstein: Numerical Data and Functional Relationships in Science and Technology*, New Series, *Units and Fundamental Constants in Physics and Chemistry*, Parts a and b, Berlin: Springer.

Ditchburn, R. W. (1976), *Light* (2 Vols.), London: Academic Press.

Feynman, R. P., Leighton, R. B., Sands, M. (1977), *Lectures on Physics*, Reading, MA: Addison-Wesley.

Francon, M., Mallick, S. (1971), *Polarization Interferometry*, New York: Wiley.

Hariharan, P. (1986), *Optical Interferometry*, New York: Academic.

Hariharan, P. (1984), *Optical Holography*, Cambridge, U.K.: Cambridge University Press.

Hutley, M. C. (1982), *Diffraction Gratings*, New York: Academic.

Klein, M. V. (1986), *Optics*, New York: Wiley.

Lipson, S. G., Lipson, H. (1969), *Optical Physics*, Cambridge, U.K.: Cambridge University Press.

Naumann, H., Schröder, G. (1992), *Bauelemente der Optik*, München: Carl Hanser Verlag.

Niedrig, H. (Ed.) (1993), *Optik*, Berlin and New York: Walter de Gruyter.

Steel, W. H., (1983), *Interferometry*, Cambridge, U.K.: Cambridge University Press.

Van Heel, A. C. S. (Ed.) (1967), *Advanced Optical Techniques*, Amsterdam: North-Holland.

Zimmer, H. G. (1970), *Geometrical Optics*, Berlin: Springer.

OPTICAL INTERCONNECTIONS

Takao Matsumoto, *NTT Transmission Systems Laboratories, Kanagawa, Japan*

INTRODUCTION

Signal processing in telecommunication systems, computers, and sensors has so far been performed by using electrical circuits. This is because functions such as signal transmission, distribution, connection, conversion, switching, memory, etc., can be done more easily with electronics than with other schemes such as mechanics, optics, bionics, etc. Another reason is that small-size, high-speed, and high-reliability electronic circuits have become available at relatively low cost as a result of the rapid progress of semiconductor technology.

The advancement of semiconductor technology has also made it possible to highly integrate electronic circuits. A very large-scale integrated circuit (VLSI), for example, contains more than 1,000,000 transistors in a single chip. However, it is still very important to reduce circuit size in various types of electronic instruments and systems, especially computers. Thus, it is now believed that, if only electronics continues to be used in circuits, the performance of such instruments and systems will be severely limited. Performance is already being degraded as a result of problems like signal delay and wiring congestion in electrical interconnections.

On the other hand, with the progress of optical-fiber transmission technology in the past 20 years, several kinds of optical devices and components have been developed that have the potential to replace electronic devices and components. Several problems inherent in electrical interconnections can be solved by using optical interconnections, which can be implemented through the use of electronic/optic (E/O) and optic/electronic (O/E) converting devices, and optical-transmission media (Goodman *et al.*, 1984). Future signal processing will greatly depend on the potential of optical interconnections.

1. LIMIT OF ELECTRICAL INTERCONNECTION

The operating speed and logical configuration of digital signal-processing circuits need to be enhanced to improve circuit performance. Figure 1 shows the relationship between clock frequency and number of gates required for various signal-processing systems. State-of-the-art VLSI fabrication technology is also plotted therein. The number of interconnections between the gates can be assumed to be nearly proportional to the number of gates. In high-speed optical-transmission systems, the number of interconnections may be small, but the system hardware is required to operate at as high as over 10

3-527-28134-7/95/$5.00 + .50

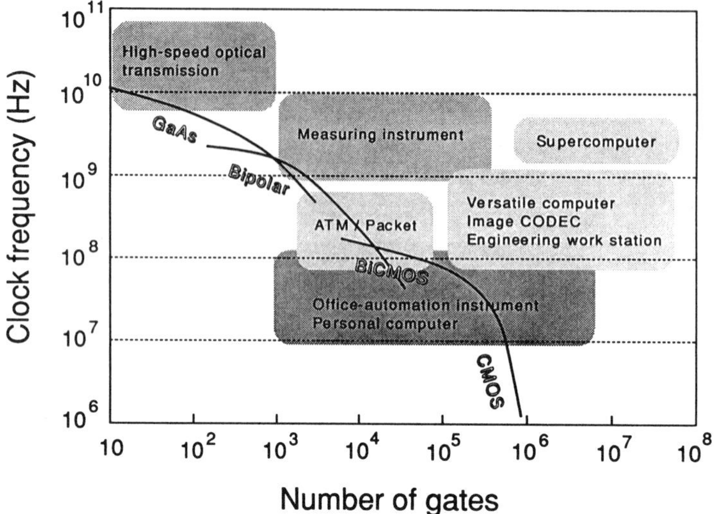

FIG. 1. Relationship between clock frequency and number of gates required for various signal-processing systems.

Gbit/s. Supercomputers may have lower operation speeds than optical-transmission systems, but they require a larger number of interconnections than regular computer systems.

When high performance is required in such systems, the limitations of electrical interconnections become clear. The limitations are evident in the following areas:

1. Signal delay and clock skew. Microstrip lines, which are stripe-shaped conductors implemented on a substrate and which are widely used as electrical interconnections between functional devices, give rise to signal delay as a result of the dielectric constant of the substrate material. In the currently popular alumina substrate, for example, the delay is 100 ps/cm, which is about three times as large as the free-space delay.

   On the other hand, when circuit integration enables the dimensions of a circuit to be reduced, this causes the interconnection delay to become larger than the gate delay. As a simple model, we assume that all the dimensions of a VLSI chip are scaled down by the factor k (Goodman *et al.*, 1984). Since capacitance of a fixed interconnection scales down by k and resistance scales up by k, the resistance-capacitance (RC) time constant and the interconnection delay do not change. On the other hand, gate delays decrease with the scaling because the channel length is decreased by k. Thus, interconnection delay becomes the prime chip-speed limiting factor.

   In digital signal-processing hardware, signals and clock should arrive at a gate without any relative delay, since all circuits operate synchronously according to a common clock. However, the path-length difference between signal and clock causes a relative delay and so breaks the synchronization. This phenomenon, which limits the speed of the processing hardware, is called "clock skew." Technology to decrease interconnection delay is needed to decrease the effect of clock skew.
2. Signal reflection. In electrical interconnections, characteristic impedance (Z_0) is defined for a transmission line of the interconnection. When an electrical device having an impedance $Z \neq Z_0$ is attached to a transmission-line end, the impedance mismatching causes signal reflection. Even if the impedance is matched, signal reflection will occur as a result of stray capacitance and inductance produced by discontinuity in the physical structure of the conductors. Moreover, if an interconnection link is relatively long in comparison with the signal bit length, the effects of signal reflection become very noticeable because of delayed multireflection.

   The bus-interconnection scheme, in which processors and memories are interconnected with a common transmission line, is widely employed in computer systems. The transmission line includes a series of taps, to which the lines from the processors or memories are connected. Impedance mismatching occurs in taps as a matter of course, and this causes signal reflection.

3. Cross talk and noise. As electrical interconnections become more congested, the spacing between interconnection lines decreases and the length of adjacent-line coupling increases. This causes cross talk in the interconnection lines. Some kind of shielding must be devised between coupled lines as well as coupled circuits to avoid such cross talk.

Electromagnetic interference (EMI) from the environment induces noise and so degrades electrical circuit performance. Therefore, anti-EMI countermeasures need to be taken when circuits are to be located near power-supplying units or switching units.

2. POTENTIAL OF OPTICAL INTERCONNECTION

Optical interconnection is based on the propagation of an electromagnetic wave at very high frequency (several THz). Therefore, it is free from the physical restraints due to the electric current or voltage inherent in electrical interconnections. The advantages of optical interconnections are these:

1. Low loss and wide bandwidth. Transmission loss is one of the most important factors that determine the performance of interconnections. Optical interconnections have an extremely low loss characteristic when they are implemented by using optical fibers or free space. Optical-fiber loss around the 1.55-μm wavelength is only 0.1 dB/km. Even if there is a loss problem, present-day optical-fiber amplifier technology can compensate for it. When the effects of diffraction and reflection are ignored, free-space transmission loss is virtually nil, and when optical waveguides are implemented in substrates, the loss is as low as 0.1 dB/cm.

   Optical interconnections make it possible to utilize the extremely wide bandwidths of transmission media. For example, the wavelength region of 1.5–1.6 μm, where optical fibers have low loss characteristics, corresponds to a frequency region of 13 THz. Optical wavelength-division multiplexing technique simultaneously uses multiple channels of different wavelengths in a single transmission medium, and thus enables users to utilize a very wide wavelength region.
2. Small cross talk and small reflection. Since the concept of "current" is not applicable to optical interconnections, the effects of cross talk and noise due to electromagnetic induction can be ignored. Cross talk in optical interconnections can occur if optical signals are leaking from one channel to another, and this problem can be effectively suppressed by carefully designing the optical interconnections in hardware.

   Since the refractive index difference between any two materials normally used in optical interconnections is small, the reflection arising at a joint is very small. Multireflection in a transmission medium is therefore considerably suppressed.
3. Electrical isolation. Since optical interconnections eliminate the need for conductors in signal transmission, transmitters can be electrically isolated from receivers. Furthermore, optical interconnections block the surges of high voltage and/or large current generated by thunder or by malfunctions in the system. As a result, voltage difference can be ignored in designing systems, and system reliability is improved.
4. Nonphysical links. Optical free-space interconnections enable transmitters to be separated from receivers with no need for the physical links that are essential in electrically interconnected systems. This means that we can directly connect them without using a detour circuit, and thus can decrease signal-propagation delay and wiring congestion. Since air flow is not obstructed in the optical free-space interconnection region, devices and components can be easily cooled.
5. Bundle interconnections and distributed interconnections. One of the functions of a lens is to map two-dimensional (2-D) information from one plane to another. This function makes it possible to optically interconnect two groups of 2-D distributed multichannel ports facing each other.

Bundles of multiple channels can be easily guided by serially aligned lenses.

A light wave radiated from a source can illuminate a wide surface. The radiation is suitable for distributing optical signals to multiple detectors. Fanout can be increased with no need for a sophisticated layout design.

6. Crossing of links. Optical beams do not interfere with each other even if they are crossed. Therefore, optical free-space interconnection links can be freely chosen without worrying about whether they cross or not. This is a big advantage in designing interconnection hardware since the hardware configuration can be simple and flexible.

The above-mentioned issues are appropriate in general cases of optical interconnections. In a specific case where a peculiar device is to be implemented, we indicate other possible advantages. Compactness and high-speed operation of optical-interconnection networks were discussed by Kiamilev *et al.* (1991).

As a matter of course, optical interconnections have some disadvantages that should be taken into account in hardware design. The issues are these:

1. Necessity of E/O and O/E converters. Since VLSIs are operated by electrical signals, E/O and O/E converters are necessary to implement optical interconnections. The dimensions and power consumption of E/O and O/E converters should be reduced.
2. Limitation of miniaturization. Miniaturization of interconnection circuits while keeping low-loss throughput is difficult, because of the limitation on the fabrication accuracy and the inherent characteristics of optical waveguides. Bending and scattering losses of optical interconnections are unignorable problems.
3. Power loss due to multiple signal dropping. In electrical interconnections, the high-impedance coupling technique can considerably suppress the signal power loss at a coupling point. In optical interconnections, such a technique is unavailable, and multiple signal dropping at serial taps along an interconnection link causes considerable power loss.

3. OPTICAL-INTERCONNECTION SCHEMES

Optical interconnections can be classified into three schemes in terms of the transmission medium employed, i.e., optical fiber, optical waveguide, and free space. Typical images of these schemes are shown in Fig. 2.

An optical-fiber interconnection scheme can utilize the optical-transmission technology that has already been developed for public telecommunication systems. Since optical interconnection is usually employed in a specific· closed and small region, specifications such as wavelength, fiber parameter, transmission speed, etc., can be chosen more freely than in public telecommunication systems. Today, the mature single-mode fiber technology is useful in a wide variety of applications. Multimode fibers, however, make it possible to reduce interconnection hardware costs because of the ease with which they can be connected. Parallel signal transmission can be easily performed by using a ribbon fiber, which is composed of plural fibers contacting each other side by side. Since an optical ribbon fiber is arranged in a 1-D array structure, optical sources and detectors should

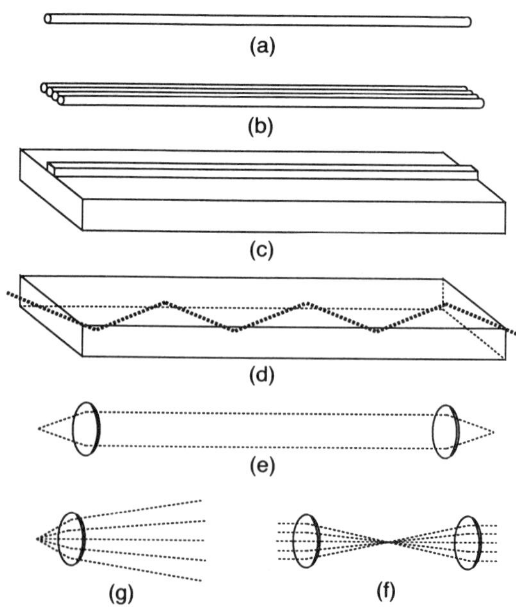

FIG. 2. Images of optical-interconnection schemes: (a) optical fiber, (b) ribbon fiber, (c) stripe waveguide, (d) slab waveguide, (e) point-to-point free-space beam, (f) distributed free-space beam, and (g) bundle free-space beam.

employ the same structure. Optoelectronic integrated-circuit (OEIC) technology, which integrates optical and electrical semiconductor devices in a chip, is suitable for fabricating such array devices.

There are two ways of formulating optical-waveguide interconnection schemes: One way employs stripe waveguides and the other slab waveguides. In the former, it is easy to couple light waves to optical devices such as laser diodes, photodetectors, modulators, couplers, etc., since the light waves are confined to a small section of the waveguide. Therefore, optical-stripe waveguides are indispensable for the realization of optical integrated circuits. With present-day fabrication technology, the substrates used for optical integrated circuits are small, e.g., less than several inches in diameter. Accordingly, the stripe-waveguide interconnection method will be applied to relatively short-length interconnection links.

In the slab-waveguide method, an optical beam is totally reflected at the two surfaces of a slab, i.e., the top and bottom surfaces, and propagates along the waveguide. Since a light wave is easily diffused in a slab, performance degradation factors such as cross talk and signal loss occur. In order to suppress the signal diffusion, some means of focusing light waves should be fabricated on the waveguide surfaces. A Fresnel lens, which can be fabricated in a small size by using ordinary etching or micromachine technology, is suitable for this application.

Optical free-space interconnection schemes include several methods with different beam shapes. Figure 2 shows three examples:

1. a point-to-point interconnection method, which interconnects a pair of optical sources and detectors,
2. a distributed interconnection method, which distributes an optical signal from an optical source to multiple detectors, and
3. a bundle interconnection method, which interconnects multiple optical sources and detectors with one-by-one correspondence in a lump.

Method **1** is used for general interconnections, **2** for broadcasting of signals, and **3** for transmitting multichannel signals or image signals.

Optical free-space interconnection schemes are advantageous in that optical devices can easily be inserted in the middle of an interconnection. Optical beams can be routed and/or switched by inserting mirrors, beam controllers, polarization controllers, polarization prisms, filters, etc. Therefore, a flexible network can be implemented by optical free-space beams.

4. OPTICAL-INTERCONNECTION HIERARCHY

The progress of integrated circuit technology has drastically reduced the size of electrical circuits, which created an interconnection hierarchy in electrical instruments and systems. This is especially true in computers, where interconnections can be classified into several levels because of the sophisticated hardware in which digital signals are processed and transmitted (Fig. 3).

First, in the case of large-scale computers occupying a wide setting space, optical interconnection between racks is required. Since supercomputers and parallel-processing computers handle a large volume of data with a high processing speed, the interconnections between their racks must have a large throughput. Since signals are usually transmitted in byte units (or multiple-byte units) in computers, parallel signal transmission is widely employed. When parallel multiple signals are multiplexed into a serial signal, a higher-speed transmission link is necessary to maintain the system performance. If single-mode fiber transmission is employed, link capacity can be over 10 Gbit/s.

Next, we have board-to-board interconnection, which is used for signal transmission within a single rack. This interconnection corresponds to the backplane interconnection used in conventional systems. When a backplane is used, interconnection-link detours give rise to such problems as signal delay, electromagnetic induction, signal reflection, and wiring congestion. However, when optical fibers or waveguides are used for board-to-board interconnection, the effects of electromagnetic induction and signal reflection are reduced. If optical free-space interconnection is employed for the direct links between adjacent boards, the backplane interconnection can be removed, and so signal delay and wiring congestion can be reduced.

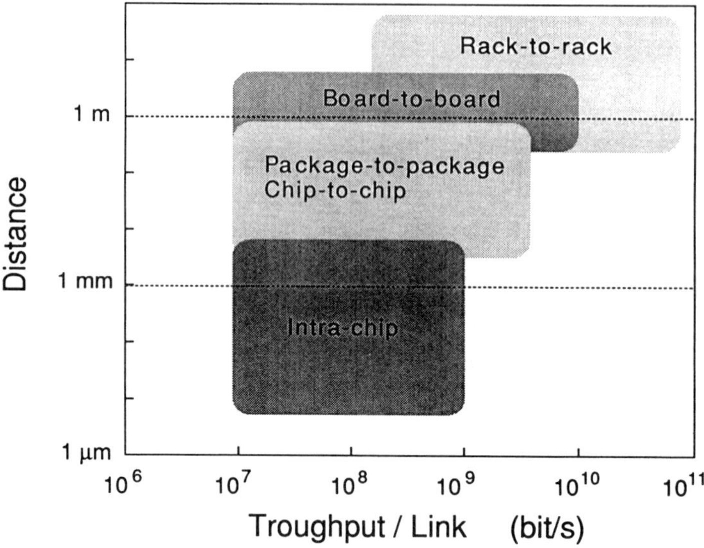

FIG. 3. Relationship between distance and throughput for various hierarchical levels of interconnection.

A third classification is package-to-package or chip-to-chip interconnection. This is used for signal transmission between the VLSI packages or chips accommodated on a board. In the former case, any one of three schemes, i.e., optical fiber, waveguide, and free space, can be employed. However, in the latter case, where link length is quite small, optical-waveguide and free-space schemes are employed.

Finally, intrachip interconnection is used in the smallest regions. When we reduce the size of an electrical circuit in order to improve its high-speed characteristics, the total performance of the circuit is mainly determined by the interconnections. Since the chip area of VLSIs is growing because of the progress of fabrication technology, the link length of intrachip interconnections is increasing. Therefore, optical intrachip interconnection is indispensable for realizing advanced VLSIs.

5. RACK-TO-RACK OPTICAL INTERCONNECTION

Rack-to-rack interconnection technology is the most advanced of the technologies employed in the above-mentioned levels of optical interconnection. This is because the optical-fiber transmission technology, which must be mainly used in rack-to-rack interconnection, is already developed and in use in the commercial field of telecommunications. The keys to making rack-to-rack optical interconnection more useful for practical applications are reducing the cost and size of E/O and O/E modules and increasing the number of channels. To meet these requirements, parallel transmission techniques employing E/O and O/E array modules and ribbon fibers are being widely researched and developed.

There is a rack-to-rack optical-interconnection link using a twelve-channel ribbon fiber (Ota and Swartz, 1991). In this link, the transmitter and receiver modules have an emitter-coupled-logic interface, and the modules contain an InGaAsP/InP light-emitting diode (LED) array and an InGaAs/InP photodiode (PD) array, respectively. A 200-Mbit/s transmission over 1 km has been reported at a signal wavelength of 1.3 μm.

Another example employing a ribbon fiber is a six-channel link comprising a laser diode (LD) array and transmitting a 1-Gbit/s signal per channel (Parker *et al.*, 1992). This link features the use of a silicon "opto-hybrid" submount, which carries all the optoelectronic components, fibers, and active and passive electronic devices. Solder bumps and grooves on the submount are used for mounting LD array chips and ribbon fibers, respectively. This structure effectively reduces the size and cost of the E/O and O/E modules, and power consumption can also be effectively reduced through LD bias-current reduction.

Figure 4 shows the structure of a four-channel transmitter/receiver module employing LD and PD arrays (Jackson *et al.*, 1992). Transmitter and receiver OEICs fabricated in GaAs chips are mounted on a substrate and coupled to silica waveguides fabricated on the substrate by using a bonding technique. Preliminary measurements carried out at a data rate of 1 Gbit/s revealed that good coupling efficiency was obtained at a low coupling loss of between 3.9 and 4.2 dB.

In general, space is more efficiently used in 2-D arrays than in 1-D arrays. Optics is especially suitable for the bundle transmission and processing of 2-D information, as was mentioned in Sec. 3. A 2-D parallel transmission link of 36 channels has been reported, comprising a surface emitting a 2 × 18 LD array, a 2 × 18 PD array, and a multifiber array. 500-Mbit/s transmission at a wavelength of 0.85 μm was confirmed in using it (Hasnain *et al.*, 1992).

6. BOARD-TO-BOARD OPTICAL INTERCONNECTION

There are two methods of carrying out board-to-board optical interconnection, i.e., the adoption of optical-transmission technology to backplane interconnections. One method is simply to replace the electrical interconnections in conventional systems with optical interconnections, and the other is to connect E/O and O/E modules on adjacent boards directly without detouring through the backplane. The former method uses optical fibers, waveguides, and free-space beams, and the latter only free-space beams.

An experimental parallel computer system employing optical backplane interconnection has been reported by a group at Nippon Telegraph and Telephone Corp. (Matsumoto *et al.*, 1990). In this system, electrical signals from every board are converted to optical beams, which are then emitted vertically to the boards along the backplane (Fig. 5). The beams are emitted parallel to each other to prevent them from interfering with each other. A receiver module is used to pick up an optical beam, which is then guided through the optical fiber to an O/E module on a board. By changing the position of the receiver module, the optical beam to be guided to a board is manually selected. A 144-channel × 144-channel optical switch is realized in the backplane section. An experiment on electrically controlled optical-beam switching was performed using a liquid-crystal polarization control device with a 2-D array structure.

Figure 6 shows an example employing a D-fiber backplane (Mackenzie *et al.*, 1992). The D fiber has a core exposed to the external free space. The cross section of the core is one-half of a conventional one and is in the shape of the letter "D." By bringing the flat planes of two fiber cores into contact, it becomes possible to optically couple the fibers. In the rack system, since D fibers are laid on boards as well as on a backplane, the boards can be optically connected and disconnected merely

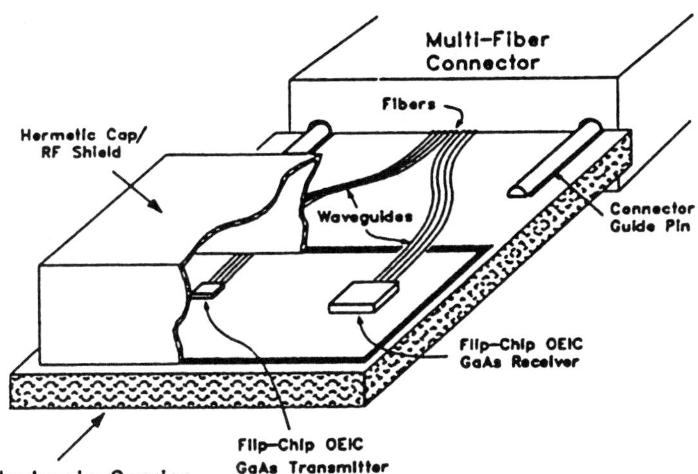

FIG. 4. Structure of the four-channel transmitter/receiver module (Jackson *et al.*, 1992).

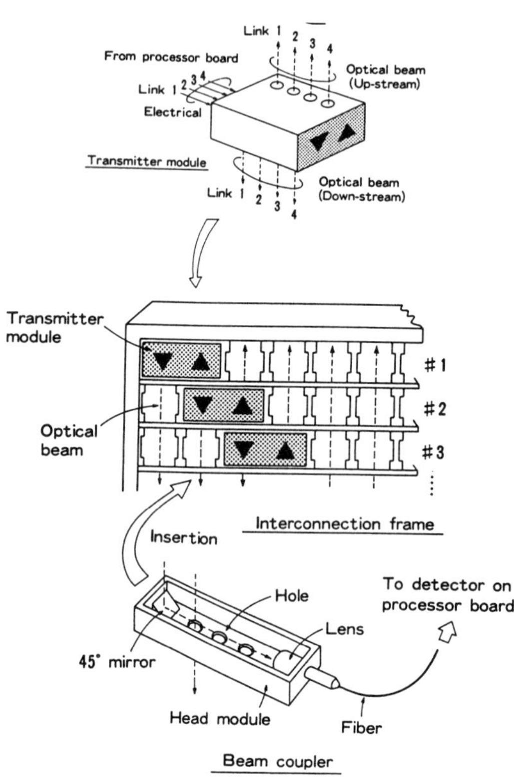

FIG. 5. Optical backplane of a parallel computer system employing fiber and beam interconnection (Matsumoto *et al.*, 1990).

by inserting and removing them. Through an experiment on a rack accommodating boards, it was confirmed that the coupling loss of fibers was −15 dB and that the system ran for over nine months.

Figures 7 and 8 show examples of backplane interconnection using optical waveguides (Sebillotte, 1990; Hamanaka, 1991). In Fig. 7, a glass substrate is used as a backplane and E/O and O/E modules on every board are optically coupled to the glass substrate. A light wave emitted from a laser diode propagates in the waveguide in a zigzag path. Holographic optical elements are fabricated on the glass substrate and used for collimating and routing optical beams. Thus, an optical signal transmitted from one board can be distributed to a number of different boards. An experiment using nine boards was carried out where a 200-Mbit/s optical clock signal was transmitted from each board and distributed to the other eight boards.

In Fig. 8, 1-D arrays of rod lenses are serially aligned so that the arrays function as optical waveguides on a backplane. A LED (or LD) emits an unmodulated optical beam from one end of each rod lens train. Optical transparent modulators and detectors are attached to one of the edges of each board, and the edges are inserted in a slot between the rod lens arrays. By coupling an optical beam with a modulator and a detector, signals can be transmitted between any two boards. It was theoretically shown that an optical bun-

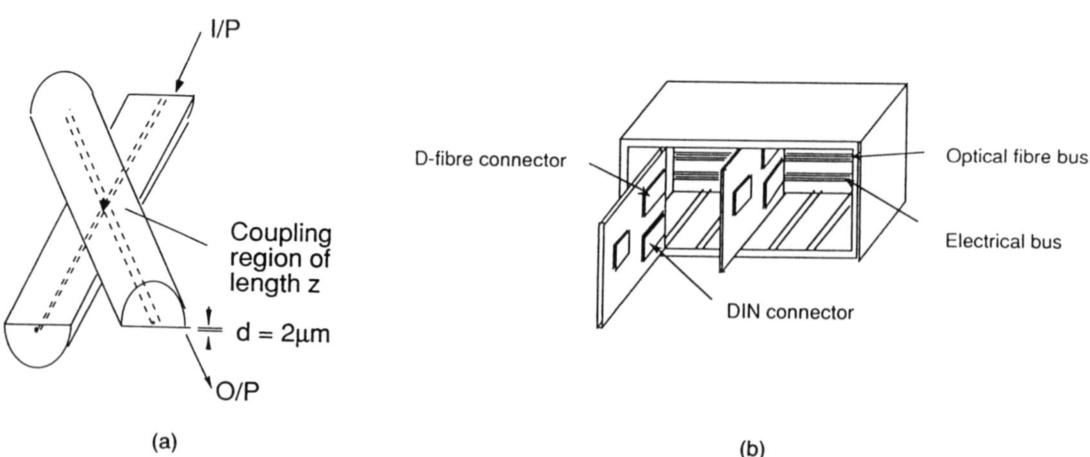

FIG. 6. Backplane interconnection employing D fibers: (a) a D-fiber cross point; (b) boards inserted in a rack (Mackenzie *et al.*, 1992).

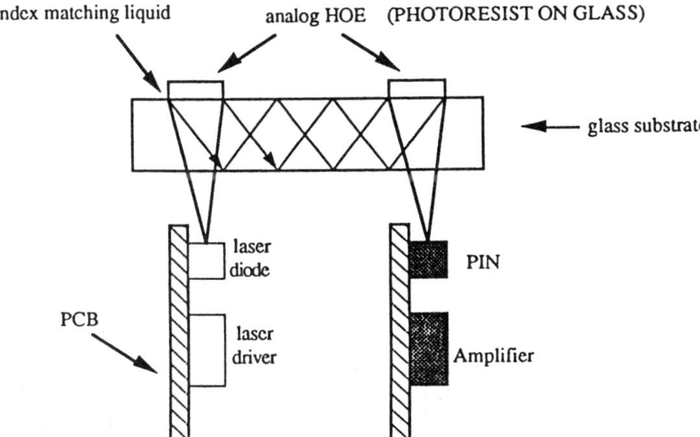

FIG. 7. Schematic representation of a backplane interconnection using a glass substrate (Sebillotte, 1990).

dle beam of 3-mm outer diameter can carry up to 21 channels.

Direct optical interconnection between adjacent boards without detouring through a backplane can reduce both signal delay and wiring congestion. Figure 9 shows a model of a 3-D mesh multiprocessor system (Sakano *et al.*, 1993). It is composed of n stacked processor boards, on each of which n^2 processing units are arranged and interconnected in a two-dimensional square mesh. It is assumed that the backplane is perpendicular to the stacked processor boards. The signal delay between two adjacent processing units for the intraboard interconnection and that for the board-to-board interconnection are assumed to be T and αT, respectively. The minimum delay between an arbitrary pair of processing units using a direct free-space optical interconnection and that between two units using the backplane interconnection were calculated under the assumption that the routing time for relaying signals at each processing unit is ignored. Distribution of the signal delay for $n = 10$ is shown in Fig. 10. The center values and the widths of the curve for the free-space interconnection are smaller than those for the backplane interconnection for both α values. This demonstrates that the

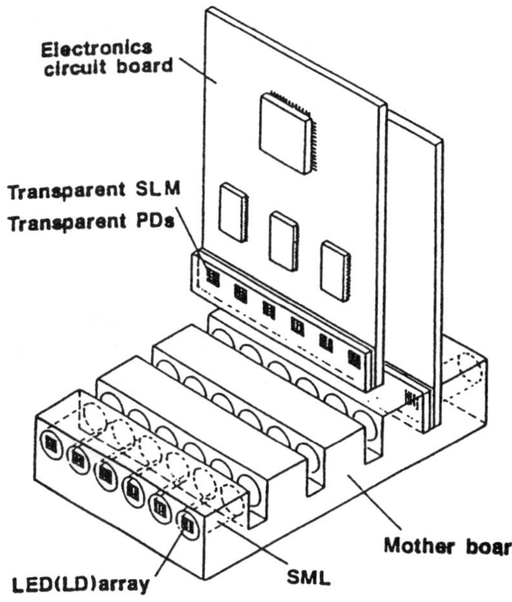

FIG. 8. Schematic representation of a backplane interconnection using rod lens arrays (Hamanaka, 1991).

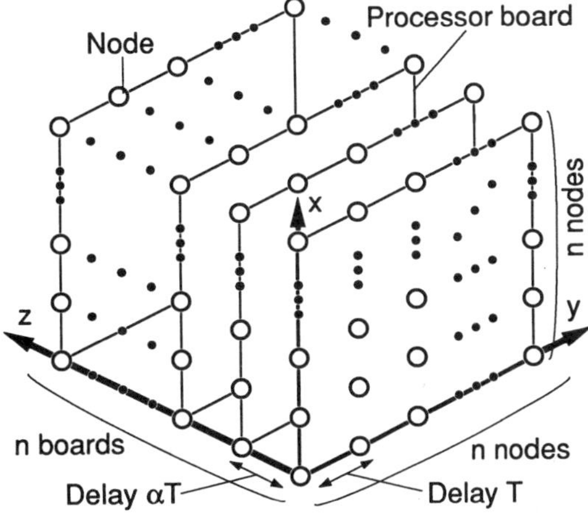

FIG. 9. 3-D mesh multiprocessor system model (Sakano *et al.*, 1993).

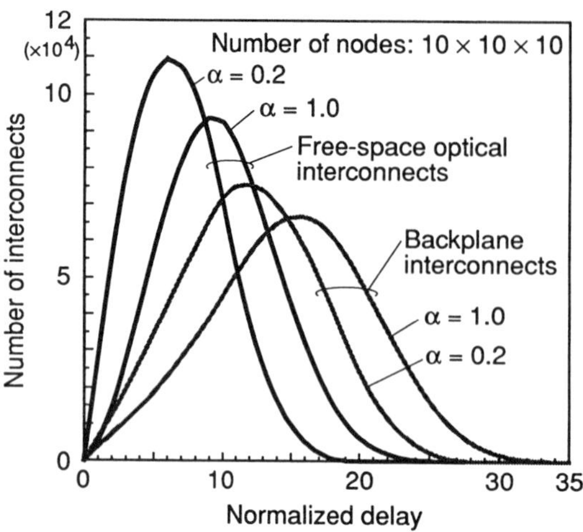

FIG. 10. Distribution of normalized signal delay (Sakano *et al.*, 1993).

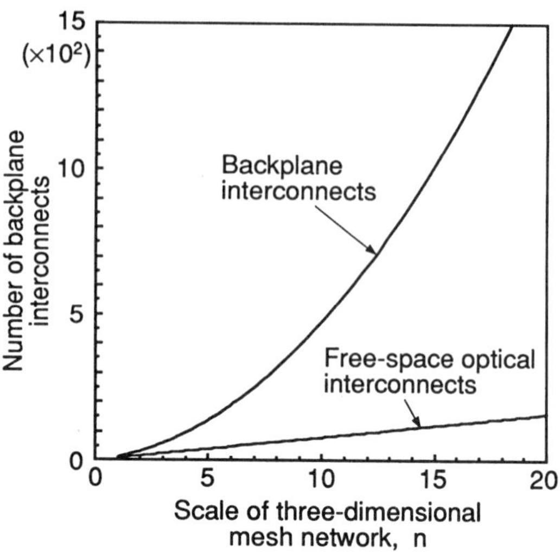

FIG. 11. Relaxation of wiring congestion by employing optical free-space interconnection (Sakano *et al.*, 1993).

direct board-to-board free-space optical-interconnection scheme is effective in reducing signal delay and clock skew.

In the model shown in Fig. 9, each board has $8n$ unused ports, which are allocated on the edges. When the system is extended to a larger system by incorporating with other systems, the unused ports are connected to the added systems through a backplane. Thus, we assume that each board uses $8n$ backplane-interconnection links for system extension, no matter which interconnection scheme is employed. In the backplane-interconnection scheme, board-to-board connection is performed only through the backplane, and each board needs $4n^2$ backplane-interconnection links to form a three-dimensional mesh network. Therefore, the total number of links between the backplane and a board is $4n^2 + 8n$. On the other hand, a free-space optical-interconnection scheme requires only $8n$ links. As shown in Fig. 11, it is apparent that optical free-space interconnection can drastically reduce wiring congestion on a backplane.

Figure 12 shows an example of a multiprocessor system with board-to-board free-space optical interconnections distributed over the processor boards (Sakano *et al.*, 1993). The system, named COSINE-III, consists of 64 processing units interconnected in a 3-D mesh network. Each processing unit has six bidirectional links, four of which are electrically implemented on a board and two of

FIG. 12. Outside view of the 3-D mesh multiprocessor system COSINE-III. One side panel is removed (Sakano *et al.*, 1993).

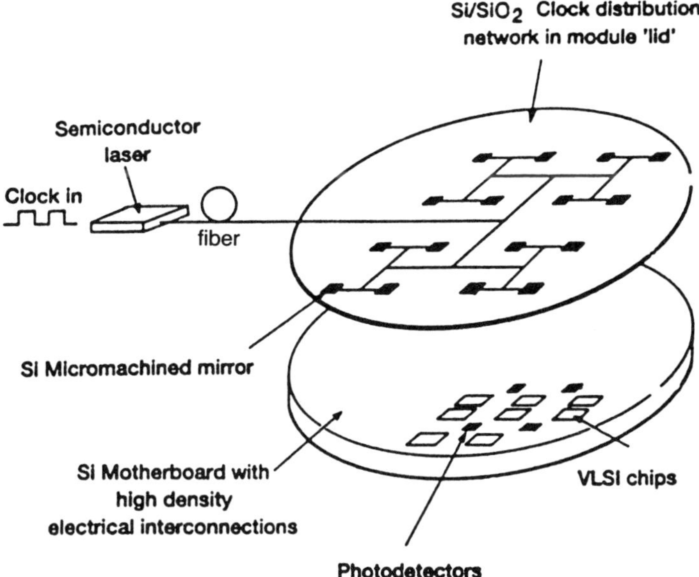

FIG. 13. Optical clock signal distribution scheme using waveguides and beam deflection (Parker, 1990).

which are optically implemented between adjacent boards. A differential signal-transmission method is employed to suppress the effect of received optical fluctuations caused by board displacement and that of environmental light leakage. Optical sources and detectors are attached to the boards with stiffeners to avoid misalignment due to board warp. The optical source is a LED with a wavelength of 0.82 μm. System operation was confirmed by the running of a parallel-processing program.

7. INTRABOARD OPTICAL INTERCONNECTION

The concept of intraboard interconnection includes package-to-package, chip-to-chip, and intrachip interconnections. When electrical interconnection is used, intraboard links are laid in a 2-D arrangement. However, when optical interconnection is employed, a 3-D arrangement becomes possible by adding optical free-space links perpendicular to the 2-D arrangement. Although several types of 2-D and 3-D arrangements for use in optical intraboard interconnections have been proposed, most of them are still in the concept phase, and so their feasibility has not yet been estimated from the viewpoint of practical application.

A layout that distributes a clock signal over a VLSI chip by using optical waveguides and beams is shown in Fig. 13 (Parker, 1990). In this layout, a silica waveguide on a silicon substrate is divided into multiple waveguides in the shape of an H tree. An optical clock signal is coupled to a micromachined mirror allocated at every output port of the waveguide, and deflected onto a receiver on the circuit board. The clock delay is the same in all of the branches because of the H-tree structure. Link fanout can be increased by repeating the division of the branches.

When passive optical waveguides are coupled to active optical devices, the optical-interconnection performance is improved. Figure 14 shows an example of such an optical-waveguide circuit, which is composed of a waveguide and an active tap (Lytel *et al.*, 1991). The active tap picks up a portion of

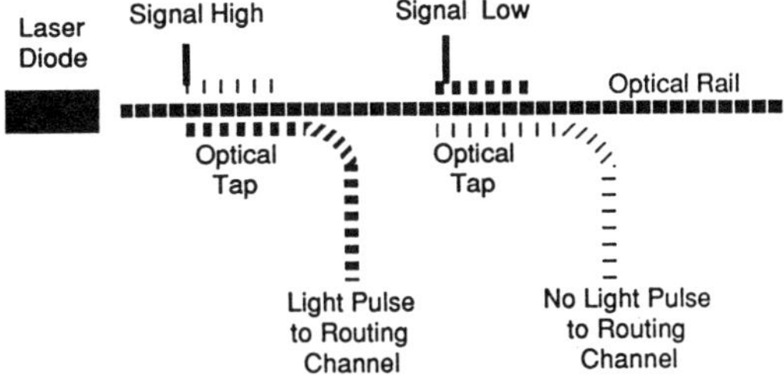

FIG. 14. Schematic representation of optical interconnection composed of waveguides and active taps (Lytel *et al.*, 1991).

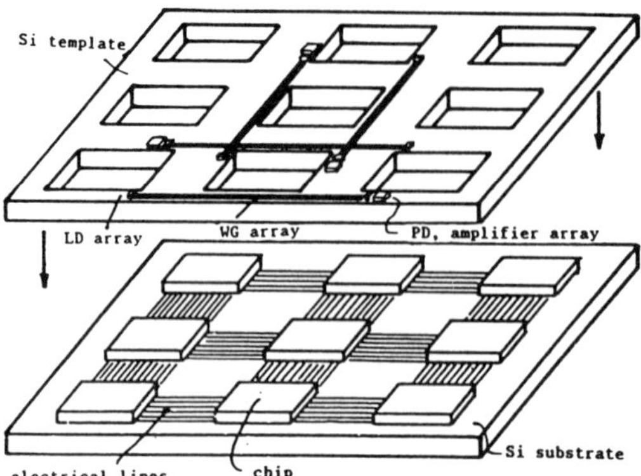

FIG. 15. Multichip module using electrical interconnection on a substrate and optical interconnection on a template (Karstensen *et al.*, 1990).

the light wave transmitted through the waveguide, and modulates the light wave according to a supplied control signal. Because of the thermal effect of optical sources, the number of optical interconnections on a chip is usually limited. When we adopt the arrangement shown in Fig. 14, the number of interconnections can be considerably increased, since a common optical source is used.

Figure 15 shows an example in which optical-waveguide interconnection is applied to a multichip module (Karstensen *et al.*, 1990). The interconnection network of the module is composed of two parts, a silicon substrate on which VLSI chips are mounted and electrical interconnections are implemented, and a silicon template on which optical sources and detectors are mounted and optical-waveguide interconnections are implemented. Since optical interconnections are suitable for long-path interconnections, they constitute an overlay network for long-path signal transmission in a multichip module.

Figure 16 shows an example of optical interconnection based on planar optics (Jahns *et al.*, 1990). A light wave emitted by a microlaser is transmitted from one chip to the next through the repetition of reflection in the substrate. The reflection of the light wave is performed by diffractive optical elements and mirrors.

A clock-distributing method using optical free-space interconnection has been proposed and is shown in Fig. 17 (Zarschizky *et al.*, 1990). In this method, light waves emitted by a LD illuminate a hologram device attached to the ceiling of the package. The hologram device then distributes the light waves and focuses them to the detectors on a VLSI chip. The number and the directions of the distributed light waves can be changed by modifying the design of the hologram device. A preliminary experiment of the method has been carried out.

8. OPTICAL-INTERCONNECTION DEVICES

Optical-interconnection devices can be classified into two categories, one being the devices needed to realize optical-interconnection links and the other being the devices fabricated by employing optical-interconnection technology. Since the devices in the former category have already been discussed in terms of their structure and application, they will not be further described here.

A number of devices can be fabricated by employing optical-interconnection technology; one of them is the artificial neural network chip shown in Fig. 18 (Nitta *et al.*, 1990). This chip is an integrated device in a hybrid

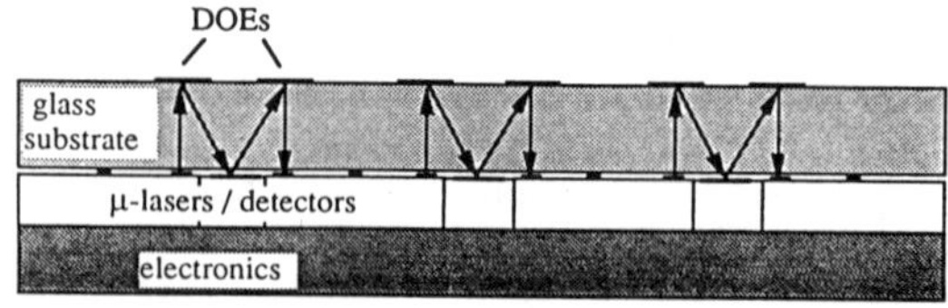

FIG. 16. Planar optics for chip-to-chip interconnection (Jahns *et al.*, 1990).

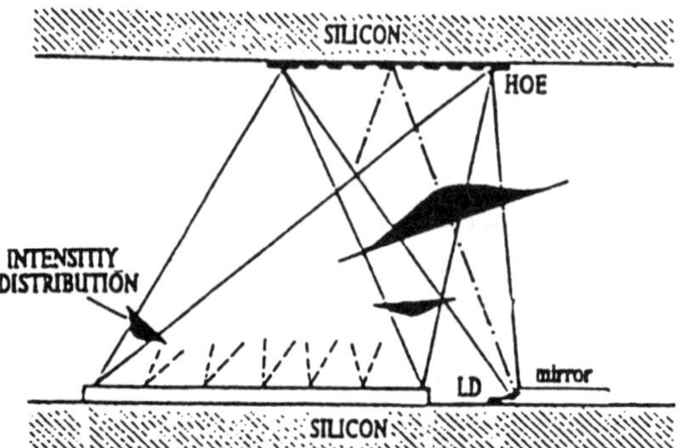

FIG. 17. Optical clock signal distribution through beam reflection on a package ceiling (Zarschizky *et al.*, 1990).

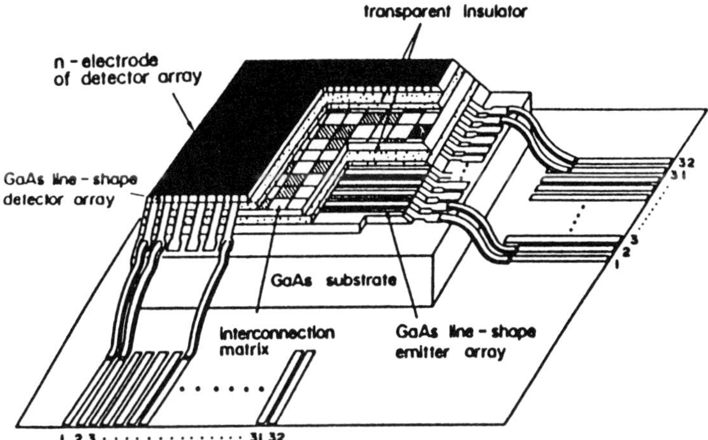

FIG. 18. Schematic view of a neural network chip employing optical interconnection (Nitta *et al.*, 1990).

structure with a 32 line-shaped LED array, a 32 × 32 interconnection matrix mask, and a 32 line-shaped photodetector (PD) array. The stripes of the LED array are arranged to be perpendicular to those of the PD array. This chip has a crossbar-interconnection function, in which the weightings at the cross points can be determined by the transparency of the elements in the matrix mask. An artificial neural network was realized by combining this chip with an electrical circuit having sigmoidal characteristics. An alphabet-recognition experiment was carried out on the network, and the results obtained were verified by computer simulation.

In a parallel-processor system, a common memory significantly increases the overall speed and throughput in computations. Figure 19 shows the cross-sectional structure of a 3-D optically coupled common memory having a multilayered structure (Koyanagi *et al.*, 1990). The figure shows a case with only two layers, each of which is composed of memory cells, LEDs, and photoconductors (PCs). A LED and a PC, which are on different layers and adjacent to each other, are optically coupled in order to transmit signals between the memory cells. LEDs and PCs are produced by using the heteroepitaxial growth of GaAs on silicon after completing the silicon device fabrication process. Computer simulation showed that the device will have a speed of 32 Gbit/s layer.

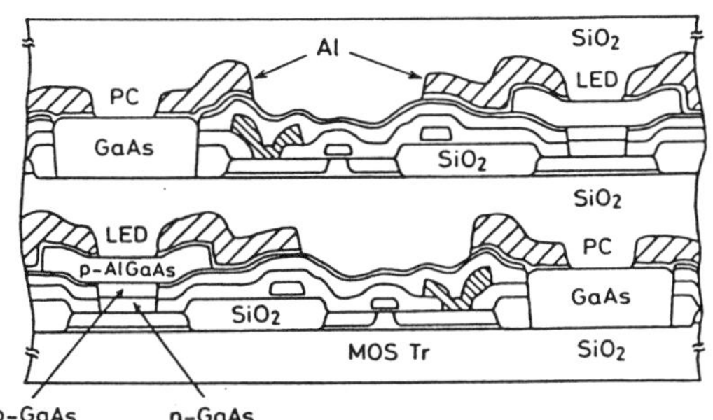

FIG. 19. Cross-sectional view of 3-D optically coupled common memory (Koyanagi *et al.*, 1990).

9. SYSTEMS AND APPLICATIONS

As was described in Sec. 5, rack-to-rack optical interconnection can utilize the technologies developed for telecommunication systems. Although reduction of cost and size of O/E and E/O modules is important and should be continuously striven for, the key technologies required for such interconnection can essentially be regarded as mature technologies.

On the other hand, the optical-interconnection technologies to be applied inside of system frames are still premature, and so the number of systems employing them is very small. Figure 20 shows a system image in which multiple boards in a rack are optically coupled via an array of optical star couplers (Parker *et al.*, 1992). An optical star coupler is a device composed of multiple optical couplers and used to mix the signals in multiple optical fibers. In this system, 8-bit parallel signals transmitted from each board are distributed to other boards through ribbon fibers. A six-channel interconnection at 700 Mbit/s per channel has been reported. The

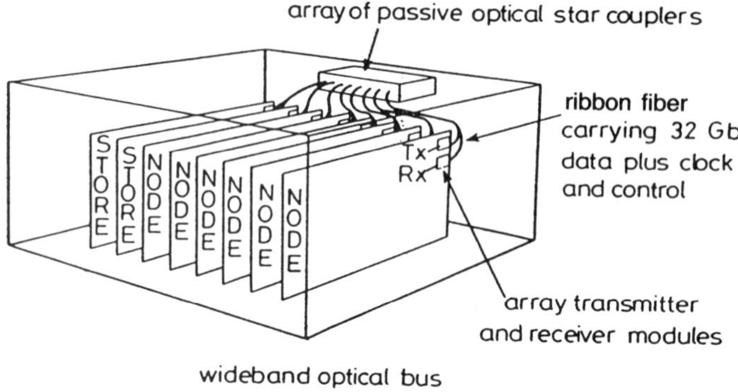

FIG. 20. Multifiber bus using a ribbon-fiber star coupler (Parker *et al.*, 1992).

transmitter and receiver modules comprise a silicon opto-hybrid submount described in Sec. 5.

Figure 21 shows an extended version of a parallel computer system called COSINE-III, which was described in Sec. 6 (Sakano *et al.*, 1993). Since optical beams are transmitted and received on both sides of the system unit, it is easy to combine multiple system units serially. This combination makes it possible to extend the scale of the system. When optical-beam switching sections are inserted in the optical links, the network of the system becomes flexible. Liquid-crystal spatial modulators are suitable for the multichannel optical-beam switching.

Figure 22 shows a multistage optical space-division switching system, in which optical-fiber interconnection is used to link the switching stages (Sawano *et al.*, 1993). Each board accommodates $LiNbO_3$ matrix switches and short fibers terminated by optical connectors. The newly developed component that joins the boards contains a fiber and connectors, and is axially flexible. A test system with a capacity of 128 channels has been developed, and has been confirmed to operate successfully for 600 Mbit/s signal switching at a 1.3-μm wavelength. An orthogonal arrangement of the matrix switch boards can simplify the complicated interconnection network of the switching system. Such an ar-

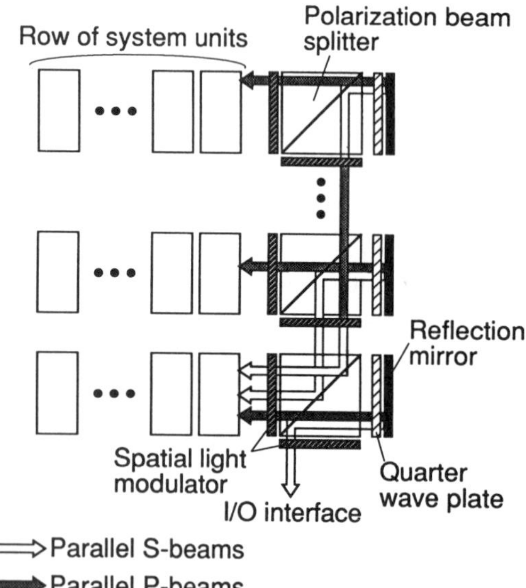

FIG. 21. Schematically represented system extension of the COSINE-III multiprocessor system (Sakano *et al.*, 1993).

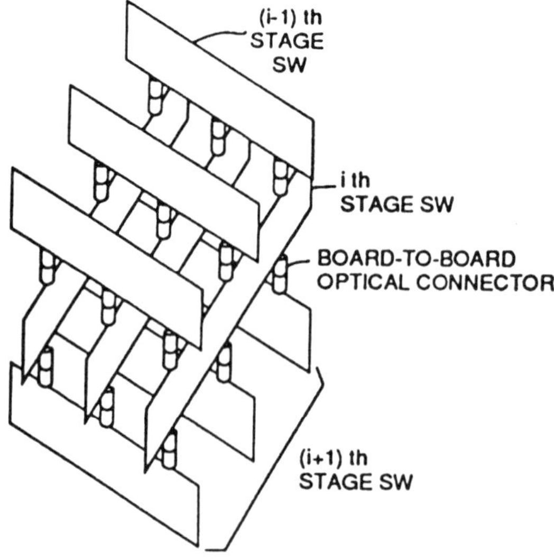

FIG. 22. Optical space-division switching system using board-to-board optical interconnection (Sawano *et al.*, 1993).

Table 1. Technological subjects for realizing optical multidimensional circuits.

Technological subjects \ Circuit type	0-Dimensional (Device unit)	1-Dimensional	2-Dimensional	3-Dimensional
Bandwidth, sensitivity, output power, noise	↕	↕	↕	↕
Cross talk, multichannel, density, uniformity		↕	↕	↕
Surface emission/detection, congestion of input/output ports			↕	↕
Congestion of internal interconnects, thermal radiation				↕

rangement straightens all the links between the boards.

10. FUTURE PROSPECTS

Because of the technical trend toward high-speed and multichannel signal processing in computers as well as communications network nodes, the electrical interconnections used inside and outside of instruments will be gradually replaced by optical interconnections. This will be accelerated by the active research currently being carried out worldwide.

Optical interconnection includes various hierarchical levels as shown in Fig. 3. In the higher levels, since the size of the units to be linked is relatively large, the size of the devices and components is not necessarily minute. Therefore, conventional technologies can be easily applied thereto. However, the development of high-speed, multichannel interconnection links for practical use is strongly needed.

On the other hand, in the lower hierarchical levels of optical interconnection, the scale of circuit should be small, and thus optical-waveguide and integrated-circuit technologies are required to fabricate them. Therefore, novel fabrication technologies and hardware architecture should be aggressively researched. With the progress of these technologies, optical interconnection will gradually penetrate from the higher levels to the lower levels.

The main subjects for future research in the area of optical-interconnection devices are as follows:

1. array implementation of E/O and O/E modules,
2. optical integrated-circuit and waveguide technology,
3. hologram technology,
4. 2-D device technology,
5. OEIC technology, and
6. multiquantum-well laser technology.

These technologies will be advanced by combining research work to be done in the fields of optics and electronics.

As was described in Secs. 6 and 7, optical interconnection is suitable for implementing 3-D circuits. Table 1 shows the problems to be solved for the implementation of optical multidimensional circuits (Matsumoto *et al.*, 1992). As the number of dimensions increases, the number of problems also increases.

GLOSSARY

Artificial Neural Network: A network simulating the structure and the functions of neural networks in a human brain. It is made up of multiple nonlinear elements and interconnections, and implemented by hardware or software.

Characteristic Impedance: The impedance inherent in a distributed-constant line, which is determined by both the structure of the line cross section and the material used. When the line is terminated by the characteristic impedance, signal reflection does not arise at the line end.

Fresnel Lens: A planar lens utilizing the diffraction effect of a circular grating. Miniature Fresnel lenses are easily fabricated by using photolithography and etching techniques.

Multiquantum Well: A structure of semiconductor crystals in which thin layers of two different materials are alternately stacked. The LDs of this structure have low threshold current characteristics compared with conventional types.

***RC* Time Constant:** A number obtained by multiplying the resistance by the capacitance in a circuit. It indicates the response time in an electrical signal transition.

Works Cited

Goodman, J. W., Leonberger, F. J., Kung, S.-Y., Athale, R. A. (1984), *Proc. IEEE* **72**, 850–866.

Hamanaka, K. (1991), *Optical Computing '91 Digest*, pp. 32–35.

Hasnain, G., Novotny, R. A., Wyann, J. D., Leibenguth, R. (1992), *CLEO '92 Digest*, pp. 170–171.

Jackson, K. P., Flint, E. B., Cina, M. F., Lacey, D., Trewhella, J. M. (1992), *Electronic Components and Technology '92 Digest*, pp. 93–97.

Jahns, J., Lee, Y. H., Jewell, J. L. (1990), *Optical Computing '90 Digest*, pp. 164–166.

Karstensen, H., Schneider, H. W., Staudt, A., Zarschizky, H., Gerndt, C., Klement, E., Tischer, H. (1990), in: H. Bartelt (Ed.), *Optical Interconnections and Networks*, SPIE Proceedings Vol. 1281, Bellingham, WA: SPIE, pp. 23–32.

Kiamilev, F. E., Marchand, P., Krishnamoorthy, A. V., Esener, S. C., Lee, S. H. (1991), *IEEE J. Lightwave Technol.* **9**, 1674–1692.

Koyanagi, M., Takata, H., Mori, H., Iba, J. (1990), *IEEE J. Solid-State Circuits* **25**, 109–116.

Lytel, R., Ticknor, A. J., Van Eck, T. E., Lipscomb, G. F. (1991), *Photonics Switching '91 Digest*, pp. 218–221.

Mackenzie, F., Hodgkinson, T. G., Cassidy, S. A., Healey, P. (1992), *Opt. Quantum Electron.* **24**, S491–S504.

Matsumoto, T., Noguchi, K., Koga, M. (1992), *Trans. IEICE Jpn.* **J75-C-I**, 223–234.

Matsumoto, T., Sakano, T., Noguchi, K., Sawabe, T. (1990), *ICCD '90 Digest*, pp. 426–429.

Nitta, Y., Ohta, J., Mitsunaga, K., Kyuma, K. (1990), *Optical Computing '90 Digest*, pp. 319–320.

Ota, Y., Swartz, R. G. (1991), *IEEE Mag. Lightwave Telecommun. Syst.* **2**, 24–32.

Parker, J. W. (1990), *Optical Computing '90 Digest*, pp. 408–413.

Parker, J. W., Ayliffe, P. J., Clapp, T. V., Ceear, M. C., Harrison, P. M., Peall, R. G. (1992), *Electron. Lett.* **28**, 801–803.

Sakano, T., Matsumoto, T., Noguchi, K. (1993), *ICCD '93 Digest*, pp. 278–283.

Sawano, T., Matsuda, K., Suzuki, S., Fujiwara, M. (1993), *Photonics Switching '93 Digest*, pp. 56–59.

Sebillotte, C. (1990), in: G. Arjavalingam, J. Pazaris (Eds.), *Microelectronic Interconnects and Packages: Optical and Electrical Technologies*, SPIE Proceedings Vol. 1389, Bellingham, WA: SPIE, pp. 600–611.

Zarschizky, H., Karstensen, H., Gerndt, Ch., Klement, E., Schneider, H. W. (1990), in: G. Arjavalingam, J. Pazaris (Eds.), *Microelectronic Interconnects and Packages: Optical and Electrical Technologies*, SPIE Proceedings Vol. 1389, Bellingham, WA: SPIE, pp. 484–495.

Further Reading

Feldman, M. R., Esener, S. C., Guest, C. C. Lee, S. H. (1988), *Appl. Opt.* **27**, 1742–1751.

Haugen, P. R., Rychnovsky, S., Husain, A., Hutcheson, L. D. (1986), *Opt. Eng.* **25**, 1076–1085.

Miller, D. A. B. (1989), *Opt. Lett.* **14**, 146–148.

OPTICAL MATERIALS

See TRANSMITTING OPTICAL MATERIALS

OPTICAL MICROSCOPY

THOMAS HELLMUTH, *Carl Zeiss Firma, Oberkochen, Germany*

INTRODUCTION

Optical microscopy is the general term used to describe all processes of image formation with a high magnification, in which electromagnetic waves are utilized as the transfer medium. The range of these wavelengths extends from the region of soft x-ray radiation (nanometers) to the infrared region (micrometers). Longer wavelengths are not employed, as high magnifications are then no longer reasonable because of the limitation in the resolving power caused by the wavelength. Optical microscopy does not encompass electron microscopy and ultrasonic microscopy, although electron optics and ultrasonic optics are also part of the terminology used in these fields—more, however, in an attempt to draw an analogy with light optics than for any other reason.

The first microscope dates back to the Dutch spectacle maker Zacharias Janssen

3-527-28134-7/95/$5.00 + .50

from Middelburg (1588–1632). Other pioneers were Leeuwenhoek (1632–1723) and Musschenbroek (1692–1761). Galileo Galilei presented his first microscope in 1610. It was not until the nineteenth century, however, that Ernst Abbe (1840–1905) paved the way for the systematic development of the light microscope on the basis of the wave theory of light. It was only by collaborating with Carl Zeiss (1816–1888), the founder of the Zeiss Works who had mastered the production of precision lenses, that Abbe was able to turn his theory into powerful microscopes whose resolution reached the limit set by the laws of physics. It was also vital for the properties of the glass materials used to be reproducible and meet the conditions of Abbe's theory. A solution was found to these requirements by the chemist Otto Schott (1851–1935), the founder of the Schott Glaswerke. It was, therefore, the interplay of theory, manufacturing techniques, and material technology that made the production of high-performance microscopes possible in the first place.

Light microscopy was further perfected in the twentieth century. New techniques, such as the phase-contrast microscopy developed by Zernike (1888–1966), made it possible to visualize object structures that had been barely discernible until then. Fluorescence microscopy reveals biochemical functions in cell structures with the aid of selective fluorescent dyes. Laser-scan microscopy and image processing can be used to display the three-dimensional structure of microscopic objects. X-ray microscopy and optical near-field microscopy venture into ranges of resolution that have remained beyond the capabilities of light microscopy until now.

The fields of application for optical microscopy can only be mentioned briefly here. The main areas of use are the materials sciences, where the fine structure of materials is of interest, the semiconductor industry, where microstructures are of special significance, and biology, especially cell biology and medicine. In the future, optical microscopy will play an important role in these fields, some of which are currently witnessing a dramatic pace of development.

1. GEOMETRICAL IMAGE FORMATION

A microscope is an image-forming system that supplies a magnified image of an object. The process of image formation can initially by described by the laws of geometrical optics. These can be used in particular to provide a good explanation not only for such properties as magnification, ray limitation, and field size, but also for aberrations. This shall be the subject of the following sections. However, geometrical optics do not suffice to explain those image-forming properties of a light microscope that are determined by the wave properties of light. The effects of diffraction will be dealt with in Sec. 2.

1.1 Objective and Eyepiece

As shown in Fig. 1, a microscope consists of an objective that produces a magnified, inverted, intermediate image that the observer then views through an eyepiece in much the same way as through a magnifier. Of course, the intermediate image can also be recorded photographically or by a TV-camera target (Inoue, 1986). However, we will not discuss these techniques further.

The intermediate image is located in the focal plane of the eyepiece, with the result that the image lies at infinity for the observer whose eye is then relaxed.

High-performance objectives do not, however, consist of a simple lens as shown in Fig. 1, but of complex lens systems that have been optimized in such a way that the aberrations described in Sec. 1.4 are kept to a minimum or even eliminated. Unlike a thin lens, a microscope objective cannot therefore be characterized by merely specifying its focal length.

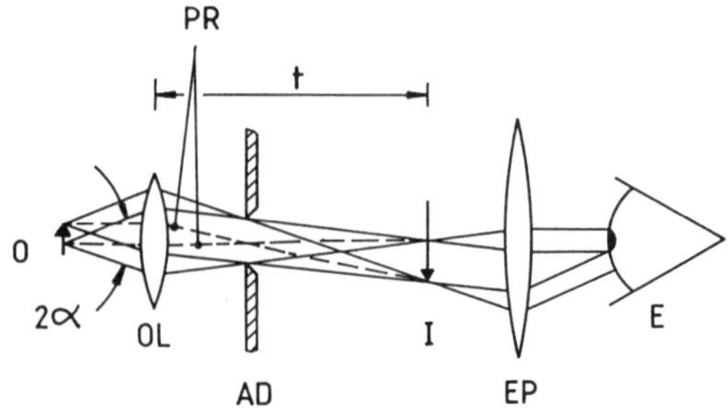

FIG. 1. Image formation in the compound microscope. The object O is imaged to the image I by the objective lens OL. The distance between objective and image plane is called the tube length t. The observer's eye E sees the image I through the eyepiece EP. The aperture angle α is determined by the aperture diaphragm AD of the objective. The principal rays PR run parallel on the object side. This setup is called an object-side telecentric ray path.

An important characterizing parameter is the working distance. A flat object is focused in such a way that its image lies in the focal plane of the eyepiece. The distance from the object plane to the object-side lens vertex of the objective is called the working distance. Objectives with a particularly long working distance are called LD objectives (long distance) and are mainly used in the semiconductor industry for mask and wafer inspection.

An objective also contains stops. These limit, for example, the ray pencil emanating from a point on the optical axis that lies in the object plane (cf. Fig. 1). The angle of this light cone determines the so-called (object-side) numerical aperture

$$NA = \sin\alpha. \quad (1)$$

If a medium—for example, immersion oil—with a refractive index n greater than one is inserted between the objective and object, the numerical aperture increases (Born, 1980):

$$NA = n \sin\alpha. \quad (2)$$

This parameter is generally engraved on the objective mount. In microscope objectives the ray-limiting diaphragm (aperture diaphragm) is usually located in the image-side focal plane of the objective. In this case it is also the exit pupil. Pupils are generally images of diaphragms. The exit pupil is obtained by imaging the limiting diaphragm toward the image plane. As there are no other lenses between the aperture diaphragm and the intermediate-image plane, this diaphragm also constitutes the exit pupil. The entrance pupil is obtained by imaging the diaphragm toward the object. In this case it lies at infinity. The principal rays of the ray pencils emanating from points in the focusing plane are therefore parallel on the object side and meet in the image-side focal point. This is described as an object-side telecentric ray path.

In the objective type discussed until now, the image of a focused object point lies at a finite distance, with the result that the image-side ray pencil has a finite angular aperture (Fig. 1). In many cases, however, as for example in polarization microscopy (cf. Sec. 3.3), it is desirable for the rays to follow a parallel path behind the objective, as shown in Fig. 2. Here, however, an additional lens is required—the so-called tube lens—which focuses the parallel ray bundle in the intermediate-image plane. This is termed infinity optics, as the objectives are corrected in such a way that the image of the focusing plane lies at infinity. If, in addition, the ray-limiting objective diaphragm is located in the focal plane of the tube lens (cf. Fig. 2), the image of the diaphragm, i.e., the exit pupil, lies at infinity. Furthermore, the principal rays on the intermediate-image side run parallel. This setup is then both an image-side and object-side telecentric ray path, with the advantage that the location of a point image does not move when the object is defocused.

1.2 Magnification

The magnification of the microscope is the product of that of the objective and that of the eyepiece. The objective magnification M_{ob} is the ratio between the size of the intermediate image B and that of the object G. We then obtain

$$M_{ob} = B/G = b/g, \quad (3)$$

where g is the object distance and b the image distance. The intermediate image lies in the focal plane of the eyepiece. The distance of the objective from this plane is determined by the length t (tube length; cf. Fig. 1). With high magnifications, the object distance can be roughly equated with the objective focal length f, so that the following

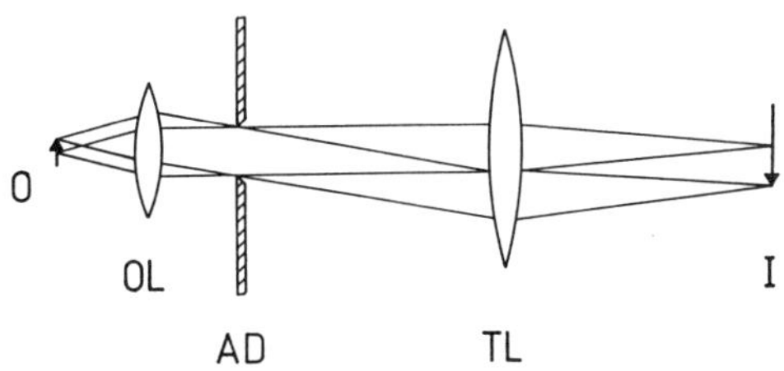

FIG. 2. Infinity optics. The rays from a point of the object *O* run parallel on the image side of the objective lens OL. The ray bundles are limited by the aperture diaphragm AD and are focused to the plane of the image *I* by the tube lens TL. If the aperture diaphragm AD and the image *I* lie in the corresponding focal planes of the tube lens, this setup is called an image-side telecentric ray path.

applies for the objective magnification:

$$M_{ob} = t/f. \tag{4}$$

This figure is generally engraved on the objective mount.

The magnification of the eyepiece is that of a magnifier. If f is the focal length of the eyepiece, the magnification is then given by

$$M_{ep} = s/f_{ep}, \tag{5}$$

where s is the so-called normal viewing distance of 250 mm. The magnification of the eyepiece is also normally engraved on the mount. The total magnification is the product of the objective and eyepiece magnifications:

$$M_{tot} = M_{ob}M_{ep}. \tag{6}$$

In practice, the objective magnification can be anything up to 100×. It is limited by the fact that the focal length of the objective cannot be reduced at will because of the finite lens radii. The typical eyepiece magnification lies between 10× and 20×. A higher magnification is not advisable, as the amount of detail in the intermediate image is limited by the resolving power of the objective as described in Sec. 2. A higher eyepiece magnification would not provide any additional information in the image. This is known as empty magnification (Sec. 2.1.1).

1.3 Image Field and Field Lens

Diaphragms limit not only the aperture, but also the object field. In the microscope diagram shown in Fig. 1, the object field is limited by the edge of the pupil of the observer's eye. Even in simple microscopes, therefore, a lens incorporated close to the intermediate-image plane bends the principal rays in such a way that they intersect in the pupillary plane of the observer's eye. This so-called field lens is dimensioned such that it images the exit pupil of the objective in the pupillary plane of the observer. Figure 3 shows this field lens positioned in the intermediate-image plane. In this case it does not contribute to the magnification of the microscope, but the observer sees the entire image field that is only limited by the lens mount.

In practice, the field lens is not positioned exactly in the intermediate-image plane, as dirt on the lens would otherwise be seen in sharp focus by the observer. The image field is usually limited by the intermediate-image diaphragm in order to prevent stray light at the lens edges. This diaphragm and the field lens are generally mounted on the eyepiece tube.

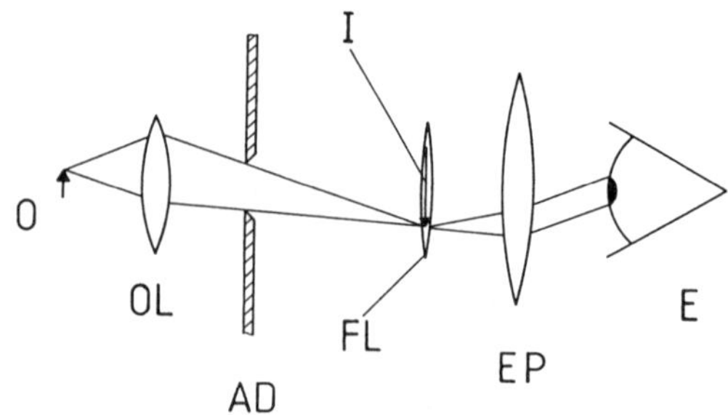

FIG. 3. The field lens FL, which is located in the plane of the image *I*, images, together with the eyepiece EP, the aperture diaphragm AD of the objective lens OL to the pupil of the observer's eye *E*, so that the observer can see the whole image *I*.

1.4 Geometrical Aberrations

Microscope objectives are lens systems that have been optimized such that as many of their geometrical aberrations as possible are eliminated. In rotationally symmetrical systems there are seven types of aberrations, which generally occur in combination. Expressed in simple terms, their effect consists in imaging the points of a point object not as points, as would be expected from ideal geometrical optics, but as extended intensity distributions (disks of confusion). This applies most especially to spherical aberration (also known as aperture aberration), astigmatism, coma, field curvature, axial chromatic aberration, and lateral chromatic aberration. The magnification may also vary over the image field, an aberration that is known as distortion.

In spherical aberration, the marginal rays of a ray pencil emanating from a point object do not intersect at the same location as the rays with a small angular aperture (paraxial rays) [Fig. 4(a)]. Much the same applies for off-axis object points in the case of coma [Fig. 4(b)]. In astigmatism the rays lying in the plane spanned by the object point and the optical axis (meridional) intersect at a dif-

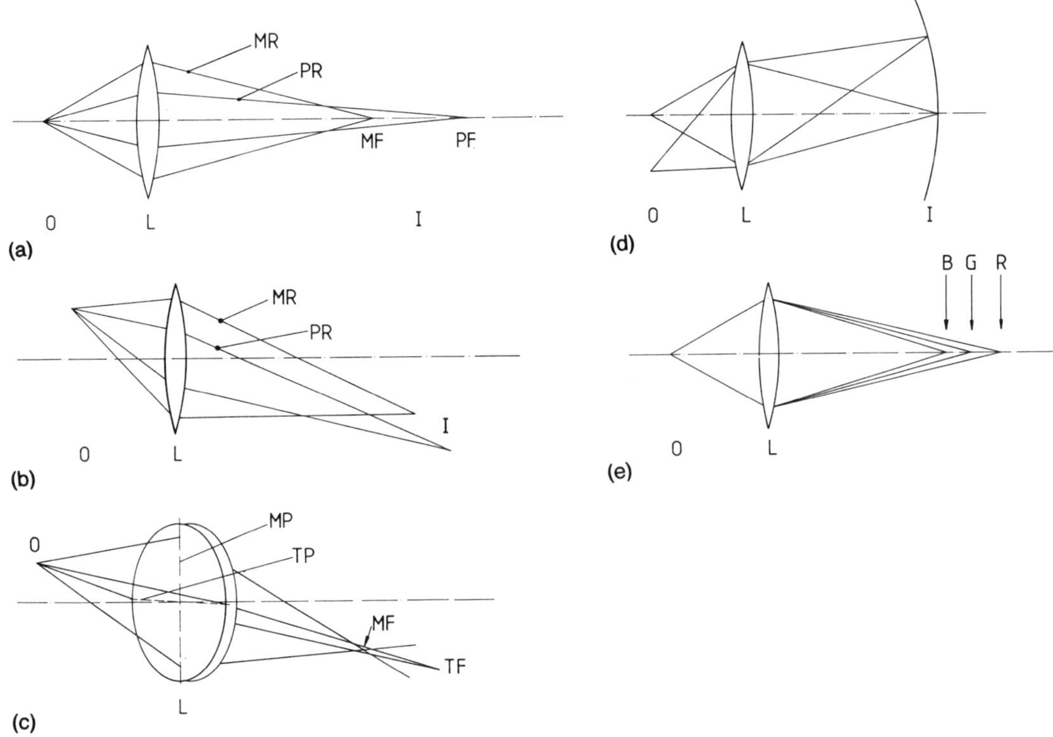

FIG. 4. (a) Spherical aberration. The marginal rays MR from the object *O* are focused to the marginal-ray focus MF by the lens *L*; the paraxial rays PR are focused to the paraxial-ray focus PF. The image *I* of the object point *O* is smeared out. (b) Coma. The marginal rays MR from the off-axis object point *O* are focused to another point on the image side of the lens *L* other than the paraxial rays PR. (c) Astigmatism. The rays from the object point *O* lying in the plane spanned by the object point *O* and the optical axis (meridional plane, MP) intersect at a different point MF than the rays running in the perpendicular plane (tangential plane, TP) focused to the point TF. (d) Field curvature. Points in the object plane *O* are focused to the surface curved around the (paraxial) image plane *L*. (e) Chromatic aberration. The spectral portions *B* (blue), *G* (green), and *R* (red) of the light emanating from the object point *O* are focused in various planes in the image plane.

ferent point than the rays running in the plane perpendicular to this (tangential) [cf. Fig. 4(c)].

In the case of field curvature, object points in the focusing plane are not imaged in the image plane, but on a surface curved around the image plane [Fig. 4(d)]. In axial chromatic aberration the spectral portions of the light emanating from a point object are focused at various points of the intermediate-image space [Fig. 4(e)]. The magnification may, however, also be dependent on the wavelength. This is termed lateral chromatic aberration. Object edges are then surrounded by color fringes.

While the aberrations described above are minimized by optimization of the objective's lens system, the lateral chromatic aberration is frequently also compensated by the eyepiece whose lateral chromatic aberration is selected with the opposite sign during its design. In modern objectives the lateral chromatic aberration is often already completely corrected.

2. DIFFRACTION-LIMITED IMAGE FORMATION

Section 1 described those image-forming properties of the microscope that can be explained using geometrical optics. According to this model, it should be possible to image any object, no matter how small, if the magnification is large enough for the observer's eye to resolve the details—provided, of course, that the objective does not display any aberrations. In reality, however, geometrical optics gives only an approximate description. Light exhibits wave characteristics, which are particularly evident when, for example, the light waves are diffracted at stops. These effects limit the resolving power of the light microscope and reduce the image contrast. This will be the subject of Sec. 2.1.

While, within the context of aberrations, the only property of the illumination to be taken into account by geometrical optics is the wavelength at the very best, it can be shown that the properties of illumination also decisively influence the image contrast and the resolving power if we consider the wave nature of light. This will be discussed in Sec. 2.2.

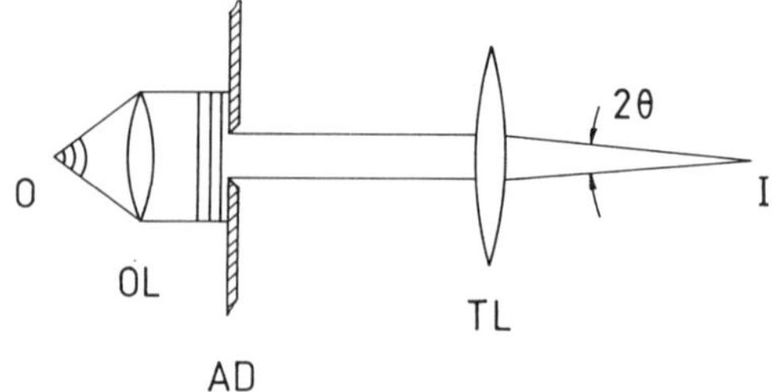

FIG. 5. The spherical wave emanating from the object point *O* is changed into a plane wave by the objective lens OL. It is limited by the aperture diaphragm AD, which is located in the focal plane of the tube lens TL. The intensity distribution of the focus in the image plane *I* is the Airy's pattern.

2.1 Resolution and Contrast

The *resolution* of a microscope is defined as the least distance between two points at which they can still be discerned as separate. This limit is also commonly specified as the grating constant of an optical grating whose lines are just visible in the intermediate image.

The brightness modulation of the intermediate image is a measure of the image *contrast*. Object details lying beyond the limit of resolution do not therefore provide any image contrast. Measures to enhance contrast do not lead to an improvement in resolution, while, however, an increase in resolution generally results in better contrast.

Needless to say, the ability to discern two points as separate or to resolve the image of a grating is dependent on the quality of the observer's vision or on the quality of the film or the camera target. In the following text, ideal observers or detectors will therefore be assumed, and only the limits of resolution and contrast characteristics determined by the microscope will be examined.

2.1.1 Rayleigh's Criterion and Useful Magnification To be able to specify the minimum discernible distance between two object points, we shall first examine the wave-optical formation of the image of a point. This object emits a spherical wave (cf. Fig. 5) that is changed into a plane wave by the objective. We use the infinity ray path described in Sec. 1.1. The plane wave strikes the aperture diaphragm of the objective. The distribution of the electrical field amplitude $A(x,y)$ of the light wave in the plane of the aperture diaphragm is then

$$A(x,y) = A_0, \quad 0 < \sqrt{x^2 + y^2} < a,$$
$$A(x,y) = 0, \quad a < \sqrt{x^2 + y^2}, \tag{7}$$

where A_0 is a constant and a is the radius of the aperture diaphragm. The laws of Fourier optics state that the amplitude distribution of the wave field in the back focal plane of a lens—here the tube lens—equals the Fourier transform of the amplitude distribution of the wave field in the front focal plane, here the aperture diaphragm. We can, therefore, write for the amplitude distribution $A'(x',y')$ in the image plane (Born, 1980)

$$A'(x',y') \sim \frac{J_1((2\pi/\lambda)(a/f_t)r')}{(2\pi/\lambda)(a/f_t)r'},$$
$$r' = \sqrt{x'^2 + y'^2}, \tag{8}$$

where $J_1(r')$ is the first-order Bessel function, λ the wavelength, a the radius of the aperture diaphragm, and f_t the focal length of the tube lens in Fig. 2. The intensity distribution is

$$I(x',y') = |A'(x',y')|^2. \tag{9}$$

This distribution is also known as Airy's disk. The dark rings are called Airy's rings (Fig. 6).

To characterize the resolving power of the microscope, we shall examine a second object point P' that lies at a distance d from the point P. The two objects shall be assumed to be self-luminous (fluorescent point objects), with the result that interference effects between the waves emanating from the points

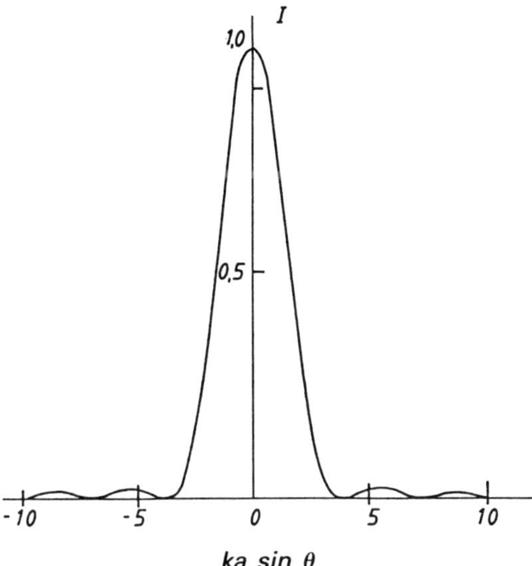

FIG. 6. The normalized radial intensity distribution of the Airy's pattern [cf. Eq. (1)]. The unit of the x axis is given by $ka \sin\theta$, with the wave number $k = 2\pi/\lambda$, the radius of the aperture diaphragm a, and the image-side aperture angle θ (cf. Fig. 5).

can be ignored. The intensity distribution of the images of the two points is then given by the sum of the intensities of the diffraction patterns displayed by each single point.

As the Airy's disks merge more and more into each other, the less the distance d between the two object points becomes. It is a matter of definition as to what distance d constitutes the limit of resolution as the ability to discern the disks as being separate does, of course, vary from one observer to another. According to Lord Rayleigh (1842–1919), the limit of resolution of a microscope is determined by the point distance $d_{\min}$ at which the maximum of the Airy's disk of the image of one object point coincides with the first Airy's ring of the image of the second object point. The following applies:

$$d_{\min} = 0.67\lambda f_t/Ma. \quad (10)$$

The limit of resolution is therefore determined initially by the wavelength of the light λ and by the ratio between the radius a of the aperture diaphragm and the focal length f_t of the tube lens. On the other hand, Abbe's sine condition (Born, 1980) specifies that for every optical system without aberrations, the ratio of the object-side aperture $\mathrm{NA} = n \sin\alpha$ to the image-side aperture $\mathrm{NA}' = n' \sin\theta$ is equal to the lateral magnification M of the optical system:

$$n' \sin\theta/n \sin\alpha = M, \quad (11)$$

where n is the refractive index of the object-side immersion medium, n' that of the image-side immersion medium, α half of the object-side geometrical angular aperture, and θ that on the image side. In the case of the microscope of Fig. 2, the image-side aperture is $\mathrm{NA}' = a/f_t$, as a is small with respect to the focal length of the tube lens f_t. Hence we obtain

$$d_{\min} = 0.67\lambda/\mathrm{NA} \quad (12)$$

for the resolving power, which is therefore only dependent on the numerical aperture of the objective and on the wavelength of the light used.

It thus follows that the resolving power cannot be improved by a higher magnification either, with the use of a high-power eyepiece, for example. On the other hand, the magnification of the eyepiece must be high enough to ensure that the Airy's disk of the point image is larger on the observer's retina than the resolving elements of the human eye (rods and cones). For that reason the magnification of eyepieces is usually not higher than 20×.

2.1.2 Optical Transfer Function Section 2.1.1 describes the formation of an image of a single point. The normalized intensity distribution $i(x',y';x,y)$ of the image of a point with the spatial coordinates (x,y) in the object plane is the so-called point-spread function

$$i(x',y';x,y) = \mathrm{PSF}(x',y';x,y) = [J_1(r)/r]^2 \quad (13)$$

with

$$r = \frac{2\pi}{\lambda}\frac{a}{f_t}\sqrt{\left(x' - \frac{x}{M}\right)^2 + \left(y' - \frac{y}{M}\right)^2};$$

M is the magnification. However, objects generally have a spatial extension. As in the last section, we shall again initially restrict ourselves to self-luminous objects, which we shall image as being composed of individual self-luminous point objects. Hence, we can describe the object by a distribution function $o(x,y)$, which provides the point density distribution. The intensity distribution $i(x',y')$ in the image is then

$$i(x',y') = \int\int_{-\infty}^{\infty} \mathrm{PSF}(x',y';x,y) \times o(x,y)dxdy. \quad (14)$$

If both sides of the equation are subjected to a Fourier transformation, we obtain the simple dependence between the Fourier transform of the object density distribution $\tilde{o}(\omega_x,\omega_y)$ (object spectrum) and the Fourier transform of the image density distribution $\tilde{i}(\omega_x,\omega_y)$ (image spectrum). The "filtering" function OTF is called the optical transfer function:

$$\tilde{i}(\omega_x,\omega_y) = \mathrm{OTF}(\omega_x,\omega_y)\tilde{o}(\omega_x,\omega_y); \quad (15)$$

in other words, the object and the image are expanded into Fourier components (two-dimensionally oriented gratings) with the spatial frequency vector (ω_x,ω_y), $|\boldsymbol{\omega}| = 1/d$ where d is the period of the corresponding spatial Fourier component. The function OTF is the Fourier transform of the point-spread function. The absolute value |OTF| is called modulation transfer function MTF. Optical systems are usually rotationally symmetric. In this case the MTF depends only on the length $\omega = (\omega_x^2 + \omega_y^2)^{1/2}$ of the spatial frequency vector (ω_x,ω_y). The MTF is illustrated in Fig. 7. The MTF describes the modulation of the image of a grating object as a function of its spatial frequency. The modulation disappears when the grating constant is smaller than the resolution limit. This fact can be utilized for an objective definition of the limit of resolution that is independent of the observer. The grating constant corresponding to ω_{lim} is

$$d_{\mathrm{lim}} = 1/\omega_{\mathrm{lim}} = \lambda/2\mathrm{NA} \quad (16)$$

but only on condition that the objects are self-luminous. The influence exerted by the illumination on the contrast and the resolving power will be the subject of the next section.

2.2 Influence of Illumination

Until now, we have based our considerations on self-luminous objects. The light emanating from different object points is not

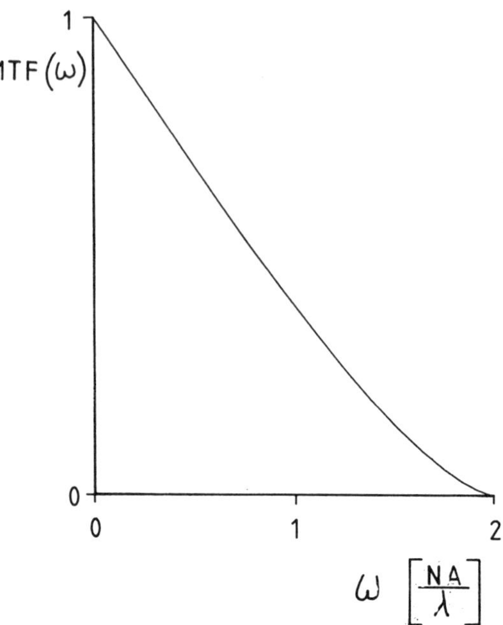

FIG. 7. Incoherent modulation transfer function (MTF). NA is the numerical aperture of the objective and λ is the (mean) wavelength of the light.

capable of interference. In general, however, objects are not self-luminous. Light waves emitted by different object points may interfere in the image plane. The intensity distribution of the image is no longer the sum of the intensity distributions of the point images.

2.2.1 Coherent Illumination In Fig. 8a grating object O with the transmission

$$t(x) = \tfrac{1}{2}(1 + \sin 2\pi x/d) \quad (17)$$

and the grating constant d is illuminated with a plane wave. The latter is formed by substantially decreasing the diameter of the condenser aperture diaphragm (iris diaphragm). The light source is focused in the condenser aperture diaphragm with a transfer optics that is not shown in Fig. 8. The grating diffracts the incident plane wave into the zero order and the two first diffraction orders. The diffraction angle φ is dependent on the wavelength of the light used and on the grating constant d (Born, 1980):

$$a \sin\varphi = \lambda. \quad (18)$$

The plane waves of the various diffraction or-

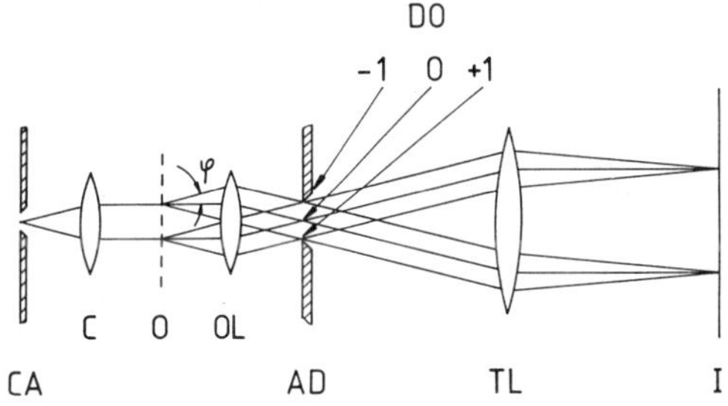

FIG. 8. Image formation with coherent illumination. The light source is focused in the condenser aperture diaphragm CA, which has in this case a small diameter (iris diaphragm). The light wave emanating from there is changed by the condenser *C* into a plane wave illuminating the object *O*, here a grating with a sinusoidal transmission function. The objective lens OL focuses the various diffraction orders DO (−1, 0, +1) into the aperture diaphragm AD. If the diffraction angle φ is not too high, the tube lens TL changes the spherical waves emanating from the aperture diaphragm into three plane waves, which are superimposed in the image plane I.

ders are focused by the objective in the aperture diaphragm plane. The light waves of the various diffraction orders are superposed in the image plane as shown in Fig. 8, resulting in a sinusoidal intensity distribution. If the grating constant d of the object is so small that

$$d < \lambda/\sin\varphi = \lambda/\mathrm{NA}, \tag{19}$$

then the higher diffraction orders do not strike the aperture diaphragm. Only the zero-order plane wave reaches the image plane, with the result that no image contrast is produced.

2.2.2 Coherent Transfer Function With the coherent illuminated object mentioned in Sec. 2.2.1, the image of a sinusoidal grating is, however, only approximately described by a sinusoidal *intensity* distribution if the object is weak, i.e., if the zero order is strong compared with the higher orders, thus allowing the interference among the higher orders to be ignored. It is a different matter with the *amplitude* distribution. Using Huygens's principle, we can write the following for the amplitude distribution in the intermediate image if the object with the amplitude transmission function $T(x,y)$ is illuminated by a plane wave:

$$\begin{aligned} A(x',y') &= \int\int_{-\infty}^{\infty} T(x,y) \\ &\times \frac{J_1((2\pi/\lambda)\sqrt{(x'-x/M)^2+(y'-y/M)^2})}{(2\pi/\lambda)\sqrt{(x'-x/M)^2+(y'-y/M)^2}} \\ &\times dx\,dy, \end{aligned} \tag{20}$$

where $J_1(x)$ is the first-order Bessel function, λ the wavelength, and M the magnification. $A(x',y')$ is a superposition of the amplitude distributions of point images weighted by the amplitude transmission function $T(x,y)$ of the object. If we once again apply the Fourier transformation to both sides of the equation, we obtain

$$\tilde{A}(\omega_x,\omega_y) = \mathrm{CTF}(\omega_x,\omega_y)\tilde{T}(\omega_x,\omega_y), \tag{21}$$

where $\boldsymbol{\omega} = (\omega_x,\omega_y)$ is the spatial frequency vector. A linear dependence therefore connects the image spectrum and the object spectrum via a transfer function. As these are amplitude spectra, however, as opposed to the intensity spectra described in Sec. 2.1.2, the term "coherent transfer function" (CTF) is used to ensure that a distinction is made between the two. The following applies in particular:

$$\begin{aligned} &\mathrm{CTF}(\omega_x,\omega_y) = 1, \quad 0 < \sqrt{\omega_x^2+\omega_y^2} < \omega_{\mathrm{lim}}; \\ &\mathrm{CTF}(\omega_x,\omega_y) = 0, \quad \omega_{\mathrm{lim}} < \sqrt{\omega_x^2+\omega_y^2} \\ &\quad (\omega_{\mathrm{lim}} = \mathrm{NA}/\lambda). \end{aligned} \tag{22}$$

The intensity distribution is

$$I(x,y') = |A(x',y)|^2. \tag{23}$$

As a result of the squaring, it no longer displays a linear dependence on the amplitude transmission distribution of the object.

The coherent transfer function in Fig. 9(a) shows that the limiting resolution is only half as good as in the case of self-luminous objects. However, the CTF remains constant at 1 up to the limit of resolution, as the first diffraction orders in a sinusoidal grating object either strike the aperture diaphragm in full or not at all.

2.2.3 Oblique Coherent Illumination If the aperture of the iris diaphragm is shifted away from the optical axis, the object is illuminated by a plane wave that is incident

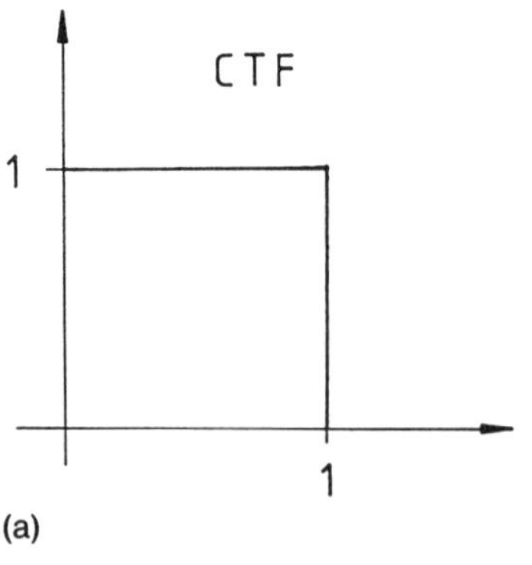

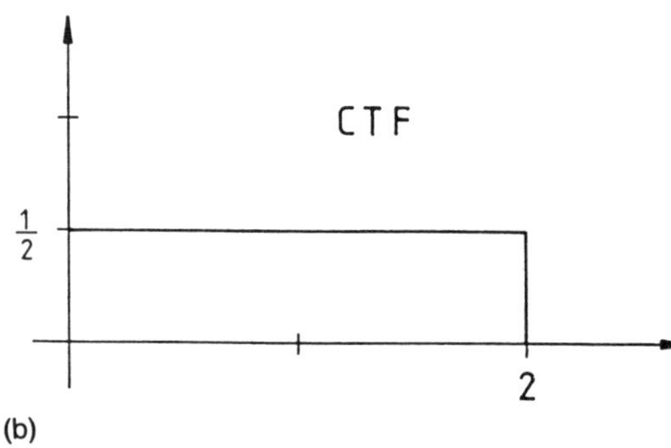

FIG. 9. (a) Coherent transfer function with direct coherent illumination. (b) Coherent transfer function with oblique coherent illumination.

obliquely. If the object is once again an optical grating, the diffraction orders are focused on the aperture diaphragm plane (Fig. 10). As in the straight coherent illumination, the plane waves are superimposed in the intermediate-image plane of the microscope, thus once again producing a sinusoidal intensity distribution (image of the grating). For this to occur, however, not only the zeroth but also the first diffraction order must strike

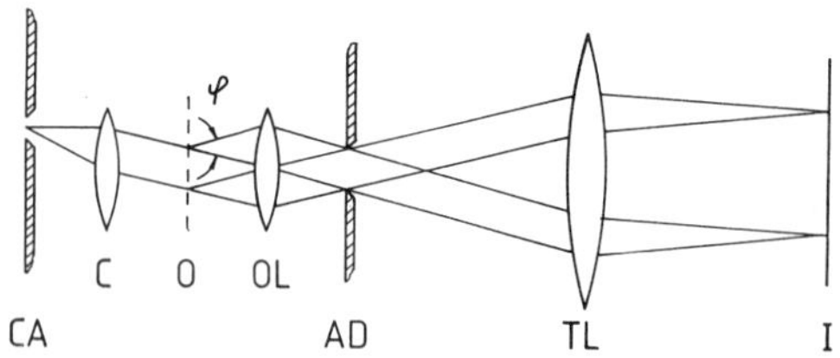

FIG. 10. Oblique coherent illumination. At the resolution limit only the zero order and one first order strike the aperture diaphragm. (Condenser aperture diaphragm CA, condenser lens *C*, grating object *O*, diffraction angle φ, objective lens OL, aperture diaphragm AD, tube lens TL, image *I*.)

the aperture diaphragm. In oblique coherent illumination in which the angle of incidence of the plane wave equals the angular aperture of the objective, as shown in Fig. 10, this condition is still fulfilled for gratings with a spatial frequency that is twice as high as the limit of resolution in straight coherent illumination. The contrast is only half as high, however, as only one first order is superimposed with a zero order. We obtain the CTF shown in Fig. 9(b).

2.2.4 Koehler and Critical Illumination

2.2.4.1 Koehler Illumination. In the coherent illumination already described, only a very small fraction of the light source is utilized. It is, therefore, advisable to image the light source in the condenser-aperture diaphragm in such a way that the image of the light source lies as completely as possible in the diaphragm aperture and fills it as homogeneously as possible. The numerical aperture of the condenser should not be greater than that of the objective, otherwise light will be lost and fail to be picked up by the objective. But it should not be smaller either, as this would result in a loss of resolution.

To show this, we shall break up the image of the light source in the aperture diaphragm of the condenser into point light sources. The latter emit spherical waves, which the condenser lens converts into oblique plane waves that illuminate the object. Each of these waves constitutes an oblique coherent illumination. However, gratings with a high spatial frequency are only relayed to the image if the plane waves of the illumination include some that are incident with sufficient obliqueness to ensure that both zero order and first order strike the aperture diaphragm of the objective. The resolution is thus at its maximum if the condenser aperture is equal to the objective aperture. This configuration is known as Koehler illumination. Although we are not dealing with either coherent illumination or incoherent self-luminous objects here, a transfer function can be specified for weak objects at least. Weak objects are characterized by the first diffraction order of a Fourier component (grating) that is weak compared to the zero order, with the result that interference effects between the higher orders of diffraction can be ignored. The MTF for weak objects in Koehler illumination is therefore made up of the MTFs of the various oblique illumination components. In the end, the same MTF is obtained as for self-luminous objects (cf. Fig. 7) (Born, 1980).

2.2.4.2 Critical Illumination. If the light source is imaged directly in the object plane, the term "critical illumination" is used. While in Koehler illumination the wave field in the object is a mixture of plane waves with different directions of incidence, in critical illumination the points of the light source are imaged in the object plane as Airy's disks. The wave field of critical illumination is hence a mixture of point images. It can be shown that, with the same condenser and objective apertures, the imaging properties in Koehler and critical illumination are equivalent, provided that the illumination of the condenser aperture diaphragm in Koehler illumination and that of the object field in critical illumination are homogeneous (Born, 1980). As light sources are usually not homogeneous enough, Koehler illumination is preferred, although the optics are more complicated.

2.2.5 Image Formation in Scanning Microscopes Optical scanning microscopes will not be dealt with in detail until Sec. 7. However, as image formation is similar in all types, we shall use the example of the laser-scan microscope (LSM) to explain the basic principle. In the LSM, a laser beam is focused on the object (Fig. 11). This beam is periodically deflected by two galvanometer mirrors (one each for the x and y directions). The angular movement is relayed via a telecentric system to the objective pupil that, also in its capacity as a telecentric system, converts the angular movement into a lateral movement of the laser focus in the object plane. The light reflected by the object is directed via a beam splitter to a detector (e.g., a photomultiplier), whose signal is displayed on a monitor. The latter's beam deflection is synchronized with the galvanometer mirrors. Of course, it is also possible to scan the object instead of the laser. However, in practice this technique has the drawback that the object size is limited when fast scanning is necessary.

2.2.5.1 Point-Image Formation. The object point illuminated by the laser focus emits a spherical wave that is converted into a plane wave by the objective and finally strikes the detector. The intensity of the backscattered wave is proportional to the intensity of the laser focus at the location of the object point. Therefore, the detector signal is proportional to the intensity of the laser focus as a function of the location. The point-spread function is the same in the LSM as in conventional microscopy. It hence follows that the MTF is also the same in the case of self-luminous objects. One can also show that the transfer properties are the same as for Koehler illumination.

2.3 Three-Dimensional Image Formation

Until now, only the imaging of focused, plane objects has been described. In practical microscopy, however, objects are usually extended in three dimensions. A fluorescence object can thus be represented by a density function $o(x,y,z)$. From the purely geometrical viewpoint, an object point with the coordinates (x,y,z) is imaged in the image point with the coordinates $(x/M,y/M,z/M)$ (Born,

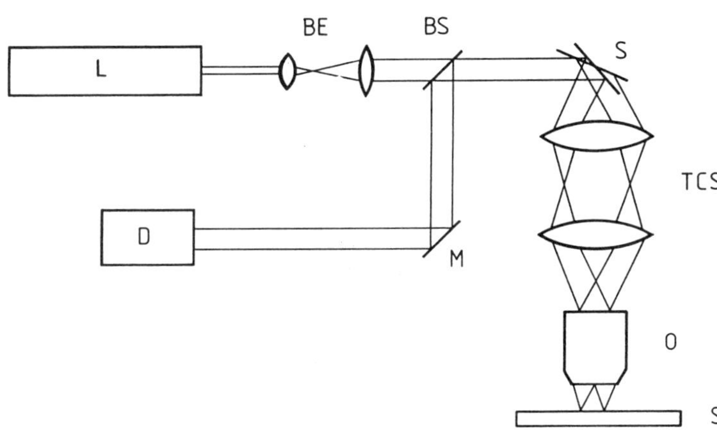

FIG. 11. Laser-scan microscope. The beam of a laser L, expanded by the beam expander BE, is periodically deflected by the scanning mirror S and relayed by the telecentric system TCS to the pupil of the objective O. The light reflected from the sample is directed via the beam splitter BS and the mirror M to the detector D.

1980). Because of diffraction, however, the point image is in reality a three-dimensional intensity distribution that we already know as an Airy's distribution in the image plane. This three-dimensional point-spread function can be represented in an isophote diagram (Fig. 12). The "contours" mark the lines of equal intensity. The intensity distribution of the image $i(x',y',z')$ is related to $o(x,y,z)$:

$$i(x',y',z') = \int\int\int_{-\infty}^{\infty} \mathrm{PSF}(x' - x/M, y' - y/M, z' - z/M)o(x,y,z) \times dxdydz. \quad (24)$$

If we once again apply the Fourier transformation to both sides of the equation, we obtain

$$\tilde{i}(\omega_x,\omega_y,\omega_z) = \mathrm{OTF}(\omega_x,\omega_y,\omega_z)\tilde{o}(\omega_x,\omega_y,\omega_z). \quad (25)$$

Image and object are again expanded into Fourier components (three-dimensional gratings) with the spatial frequency vector $(\omega_x,\omega_y,\omega_z)$. The function OTF $(\omega_x,\omega_y,\omega_z)$ is known as the three-dimensional OTF.

When we only apply the Fourier transformation with respect to the x and y coordinates and let z be the defocusing parameter of a plane object, we get a series of OTFs shown in Fig. 13. As one can expect, the resolution decreases with increasing defocusing,

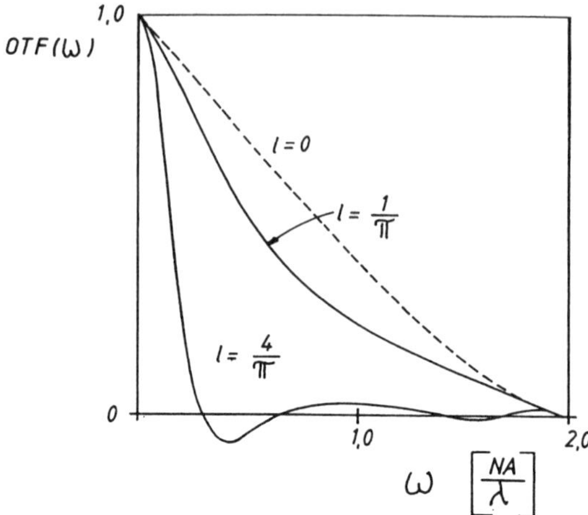

FIG. 13. Optical transfer function for various degrees of defocus z. The unit of the focus error t is $\mathrm{NA}^2z/2\lambda$. NA is the numerical aperture of the objective and λ the wavelength.

i.e., the contrast of the image of a plane grating object disappears with increasing defocusing. At certain values of the defocusing parameter the OTF gets negative. That means that the contrast of the image of the grating object reappears, however, with a phase shift of 180° with respect to the object. This effect is called pseudoresolution.

Strictly speaking, only points of a three-dimensional object are focused that are located in the focusing plane of the objective.

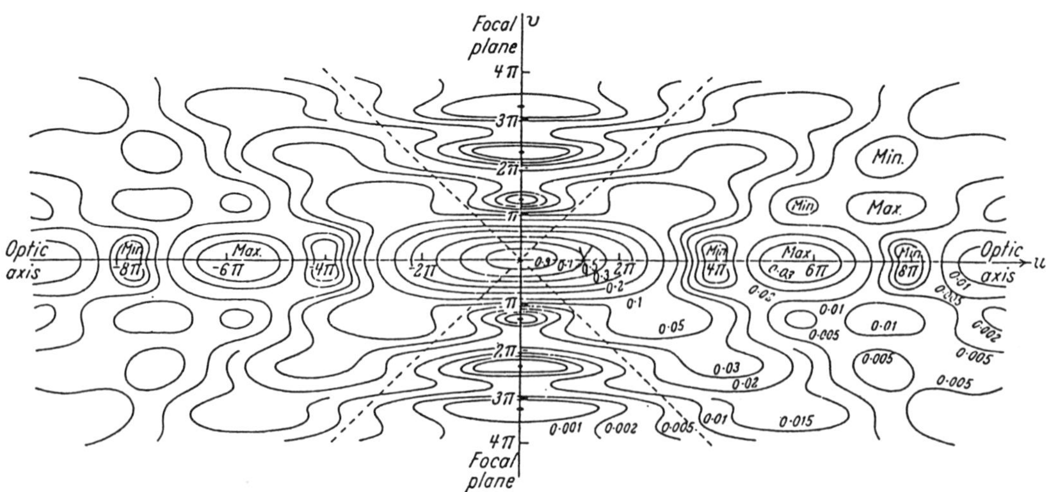

FIG. 12. Isophotes in a meridional plane of the three-dimensional intensity distribution of the image of a point object. The dotted lines represent the marginal rays in the focal region (geometrical optics). [Adapted from Linfoot and Wolf (1956).]

This is, however, a condition that is too rigid. A plane object is defined to be defocused when the defocusing parameter e of the corresponding OTF is equal to 1/4 (c.f. Born, 1980). In other words, object points are defined to be focused when their distance to the focusing plane of the objective is less than

$$\Delta = \lambda/2\mathrm{NA}^2. \tag{26}$$

The range $D = \lambda/\mathrm{NA}^2$ is called the diffraction-limited depth of focus of an objective.

3. METHODS OF LIGHT MICROSCOPY IN THE VISIBLE SPECTRAL REGION

The limit of resolution in light microscopy is determined by both the wavelength and the numerical aperture, and at the very best corresponds to half of the wavelength of the light used. The visible spectral region extends from 390 nm (violet) to 780 nm (red). The eye is at its most sensitive at approximately 550 nm, with the result that the resolution limit of light microscopy in the visible wavelength range lies at around 250 nm.

However, the limit of resolution only specifies what spatial frequencies of the object structure can in principle be imaged. These spatial frequencies do not necessarily contribute to the image contrast. The following sections describe techniques that provide an image containing object information that is either only barely discernible or totally concealed in the image of the basic configuration of the microscope described above.

3.1 Phase-Contrast Microscopy

Until now, we have based our considerations on amplitude objects that can be described by a transmission function. Biological specimens in particular (cells in an aqueous medium) often display practically no absorption, but have a position-dependent refractive index $n(x,y)$ (here we consider only plane objects) that can be expanded using Fourier's theory. We assume that our object is a phase grating with the spatial frequency $\omega = 1/d$ (d is the grating constant) illuminated by a plane wave (coherent illumination). For discussion purposes we assume that the phase grating consists of equidistant zones with index of refraction n and zones with refraction index equal to 1. We can imagine the wave field behind the phase grating (object wave) as being composed of the partial wave emanating from the zones with index n and the partial wave emanating from the zones with index 1. So the object wave is composed of the wave fields of two amplitude gratings A and B displaced by one-half of a grating period with respect to each other. Grating B has the refractive index n in its spaces, and grating A the refractive index 1. In the case of coherent illumination (Fig. 14, below), the incident plane wave is diffracted. The zero and first diffraction orders of each grating are focused on the aperture diaphragm. As gratings A and B are displaced by one-half of a grating period with respect to each other, the

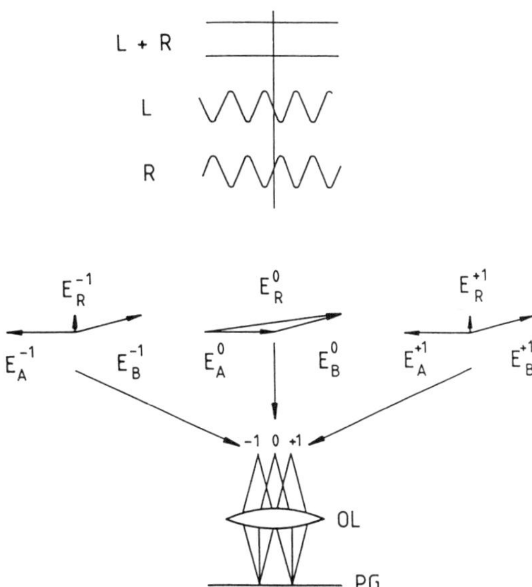

FIG. 14. Image formation with phase objects. The phase grating PG is illuminated by a plane wave (coherent illumination). The objective lens OL focuses the diffraction orders −1, 0, +1 into the back focal plane (below). The diffracted light field is assumed to be composed of the light fields, diffracted by two amplitude gratings A (n = 1) and B (n > 1) that are laterally displaced by one-half of a grating period. Figure 14 (middle) shows the phase diagram of the E field of the focused orders. The field vectors E_B^{+1}, E_B^{-1} are rotated due to the index n of grating B. The resultant field strengths E_R^{+}, E_R^{-1}, E_R^0 determine the strength and the phase relation of three spherical waves emanating from the aperture diaphragm. With the infinity ray path of Fig. 8 the diffraction orders interfere in the image plane. L is the pattern of the interference of order −1 and 0; R is that of order +1 and 0 (Fig. 14, top). The resultant image is $l + R$ with no modulation.

electrical field strength vectors of the first orders of both gratings E_A^{+1}, E_B^{+1} and E_A^{-1}, E_B^{-1} respectively oscillate in a first approximation with a phase shift of 180°. However, as the refractive index in the spaces of grating B is not equal to one, the phase of the field strength vector of this grating's orders are additionally displaced. Figure 14 (middle) shows the phase conditions of the various diffraction orders in a vector diagram. Three spherical waves propagate, therefore, from the aperture diaphragm, with the resultant field strength vectors E_R^{-1} and E_R^{+1} of the two first orders being out of phase by 90° with respect to, and considerably weaker than, the zero order E_R^0. The superposition of the order −1 with the zero order in the intermediate image yields a sinusoidal distribution L that is displaced by one-quarter of a period to the left (Fig. 14, top). The superposition of the order +1 with the zero order yields a sinusoidal intensity distribution R that is displaced by the same amount to the right, with the result that, overall, no grating is discernible in the intermediate image ($L + R$). If a plate (phase plate) with the thickness d and the refractive index n is inserted in the aperture diaphragm such that $nd = \lambda/4$, the phase of the zero order is shifted by 90°. The two sinusoidal gratings R and L in the intermediate image are shifted accordingly with the result that their superposition yields a sinusoidal intensity distribution (Fig. 15). To enhance the contrast, the transmission of the phase plate can be selected such that the amplitude of the zero order is comparable with that of the first orders. This technique is termed the phase-contrast method, which was developed by Zernike in 1932 and for which he received the Nobel prize in physics. In practice, however, the straight, coherent illumination beam path is not used. Instead, an annular diaphragm (partially coherent oblique illumination) is inserted in the condenser pupil, and a layer with the thickness d and the refractive index n—also annular in shape and situated at the image of the annular diaphragm—is deposited in the plane of the aperture diaphragm. This permits almost the entire range of the objective resolution to be utilized.

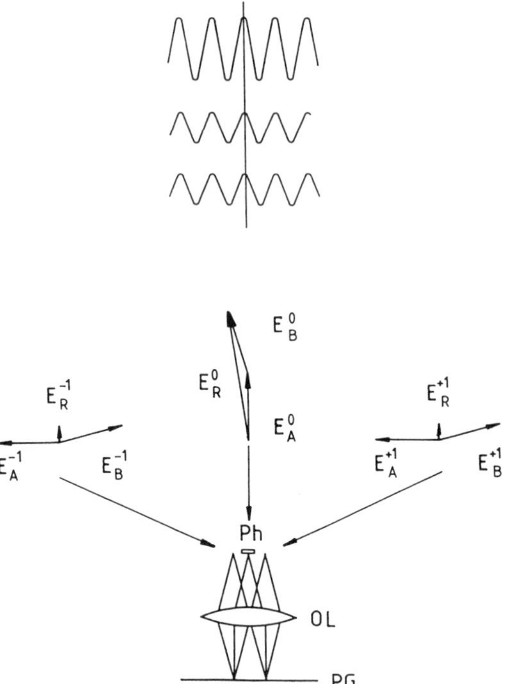

FIG. 15. Image formation of phase-contrast microscopy. The phase plate Ph (bottom) rotates the field vector E_R^0 by 90° (middle). The lowest pattern (top) is due to the interference of order −1 and 0; the other one is due to interference of orders 0 and +1. The resultant pattern (top) has a modulation that is twice as deep as that of the components.

3.2 Interference-Contrast Microscopy

In both phase-contrast microscopy and interference-contrast microscopy, the phase distribution impressed on the illuminating wave field by a phase object is imaged. However, while in phase-contrast microscopy the separation of the zero and first diffraction orders cannot be performed with the same degree of effectiveness for all spatial frequencies because of the finite dimension of the phase plate, this limitation does not exist in interference-contrast microscopy. The latter's technical complexity is, however, considerably greater. A distinction is made between microscope interferometers, which are all based on the Michelson interferometer, and the differential interference-contrast method (DIC). The microscope interferometers include, for example, the Mireau interferometer, the Linnik interferometer, and the Interphako method (cf. Pluta, 1989). The following text describes the interference method using the example of the Mireau interferometer, and the differential interference method using the Nomarsky technique.

3.2.1 Mireau Interferometer Objective We shall assume that critical incident-light illumination is used. As this is (in the ideal case) equivalent to Koehler illumination (cf. Sec. 2.2.4), it does not involve any limitations and makes our discussion of the matter easier. In critical illumination, the light source is imaged in the object plane. As the individual points of the light source are incoherent, the selection of one is sufficient (Fig. 16). The Mireau objective consists of the actual objective, a beam-splitting plate, and a reference mirror. The wave emanating from light source point P is focused on the object plane by the objective. Part of the wave travels via the beam splitter to the reference mirror, which must be the same size as the illuminated object field. The waves reflected from the reference plate and the object are superimposed on the beam splitter and interfere in the intermediate image. Depending on the path difference between the object wave and the reference wave, either constructive or destructive interference results at point P'' in the intermediate image. The path difference can be caused, for example, by a defocusing of the object point P' due to the object topography. This applies to all points of the light source, with the result that an interferogram representing, for example, the object topography is produced in the intermediate image (Fig. 17). Related techniques are so-called coherent correlation techniques (Kino, 1990).

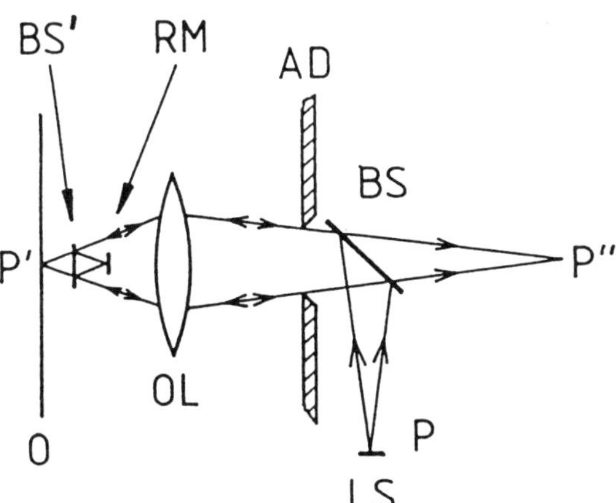

FIG. 16. Mireau interferometer microscope. The point P of the light source LS is focused via a beam splitter BS and the objective OL with the aperture diaphragm AD to the point P' of the object Q, where the light is reflected and imaged to the point P''. A part of the incoming light is reflected by the beam splitter BS′ to the reference mirror RM. This reference beam interferes with the object beam at the image point P''.

3.2.2 Nomarsky DIC We shall once again assume that critical illumination is used, although Koehler illumination is common in practice. The illumination wave (Fig. 18) is linearly polarized by a polarizer and strikes a so-called Wollaston prism. This is a birefringent crystal plate that is cut in such a way that it splits the principal ray of the illumination wave into two rays, which are parallel on the object side of the objective lens. The Wollaston prism is located in the back focal plane of the objective such that the object is illuminated by two foci a and b with the spacing d. If the object point P_a is further from

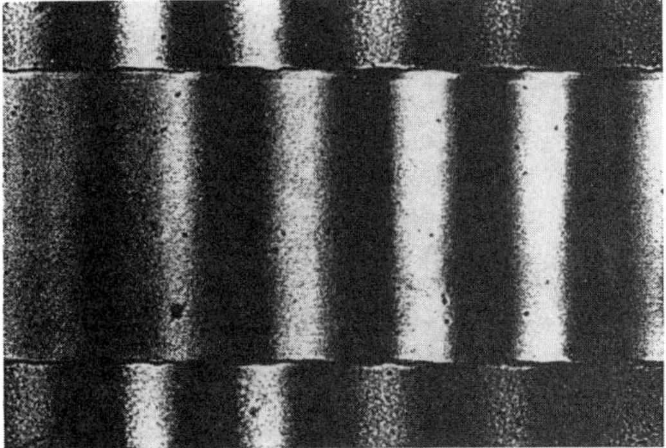

FIG. 17. Plateau on an integrated circuit [incident light; objective, (Carl Zeiss) Epiplan LD 40/0.6 Pol with Mireau interferometer device].

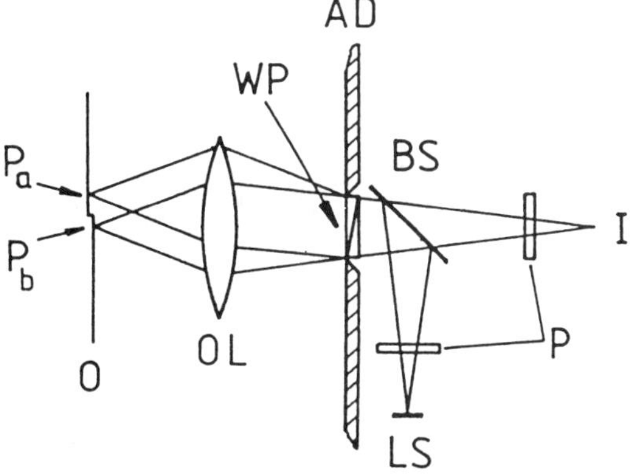

FIG. 18. Nomarsky DIC. The beam from the light source LS is polarized by the polarizer P and split by the Wollaston prism WP. Two foci strike the object at two neighboring points P_a and P_b. The reflected beams interfere at the image.

the objective lens by Δ (fraction of the wavelength) than the point P_b, the reflected wave experiences a phase shift of

$$\delta = 4\pi\Delta/\lambda \qquad (27)$$

with respect to the other wave. Both waves are recombined in the Wollaston prism and superimposed in the intermediate image, where they then interfere. In this method, phase differences between adjacent object points separated by the distance d are represented in the form of brightness contrast in the intermediate image (Fig. 19). This is, therefore, a differential method (DIC). An appropriate setup can also be implemented for transmitted light.

3.3 Polarization Microscopy

A light wave is characterized not only by its phase and amplitude, but also by its polarization. In a normal microscope, polarization is of no significance since unpolarized light is normally used for the illumination. In a polarization microscope the specimen is illuminated with linearly polarized light. The observation ray path contains a further polarization device that polarizes perpendicularly to the polarizer (analyzer). If the specimen does not change the state of polarization, no light reaches the image plane. In the case of anisotropic materials (e.g., anisotropic metals or alloys), the reflectivity is dependent on the direction of polarization of the incident light, with the result that the direction of polarization of the reflected light is rotated with respect to the incident light. In a polycrystalline specimen, the effect may vary strongly from one crystallite to another, so that the crystallite structure appears with high contrast in the intermediate image (Fig. 20). Scattering object details often display a depolarizing effect, which means that—in much the same way as in dark-field microscopy—they appear brighter in the image than the smooth surfaces.

Polarization microscopes require specially selected objectives that exhibit as little strain birefingence as possible. The analyzer and the polarizer should also be arranged in a parallel ray path, since their polarizing efficiency is then at its highest.

3.4 Fluorescence Microscopy

A microscope technique that is of special importance in biology is fluorescence microscopy. Fluorescence is a term used to describe the emission of light emanating from molecules that themselves are excited by light. The mean wavelength of the emission spectrum is generally longer than that of the exciting light. The fluorescence of dyes is particularly strong. This property can be used, for example, to dye biological cells. The dyes are deposited on the different cell structures with varying degrees of intensity. Dye molecules can also be used to mark enzymes or antigens and then to identify in the microscope where and in what concentrations they are dispersed among the various cell organelles (Fig. 21). Many cell substances also fluoresce in their own right, a property known

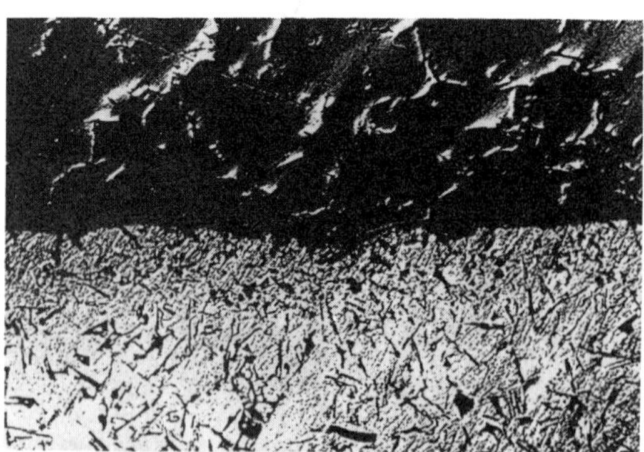

FIG. 19. Piston of a combustion motor (Al-Mg alloy). Objective, (Carl Zeiss) Epiplan-Neofluar 20×/0.50 HD DIC.

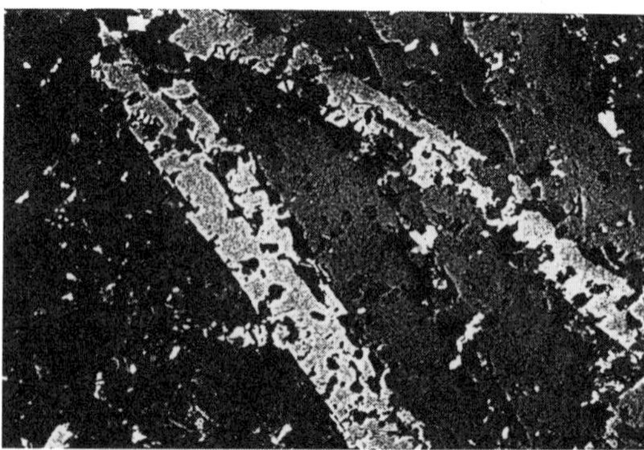

FIG. 20. Mg-Fe-Si alloy. Incident light, polarization contrast. Objective, (Carl Zeiss) Piplan Neofluar 5×/0.15.

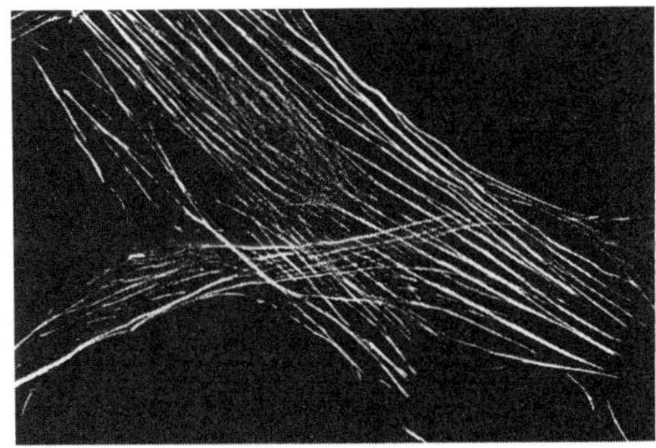

FIG. 21. 3T3 cell, actine cytoskeleton. Incident light fluorescence. Objective, (Carl Zeiss) Plan-Apochromat 63×/1.4 oil.

as autofluorescence. In a fluorescence microscope the illumination light is coupled into the beam path via a dichroic mirror that is in essence a glass plate to which a multicoating system has been applied. This system has been optimized to such an extent that the reflectivity for the spectral region absorbed by the dye is as great as possible. The transmission, on the other hand, is at its maximum for the spectral region of the fluorescence light. This avoids the losses experienced with a normal beam splitter. In fluorescence microscopy, it is generally important that as little fluorescence light as possible is lost between the object and the image plane, since the fluorescence yield is often not particularly high on the one hand, and many dyes decompose if exposed to overly intensive illumination on the other.

3.5 Dark-Field Microscopy

In a dark-field microscope only the light diffracted by the object reaches the image plane. The ray path of a dark-field microscope with coherent illumination is the same as shown in Fig. 8; however, the zero diffraction order is cut off. This makes it possible to see very weak objects that would be swamped by the undiffracted light in normal bright-field illumination. However, while it is precisely weak objects that are imaged linearly in bright-field illumination (cf. Sec. 2.2.2), i.e., a grating object with a sinusoidal transmission appears with sinusoidal contrast in the image, this is no longer the case in dark-field illumination. The first two diffraction orders interfere, with the result that the period of the interference pattern in the intermediate image is half as large as in bright-field illumination. The image in dark-field illumination is, therefore, not always true to the object. In practice, however, it is not coherent illumination that is used because of its limited resolving power, but the configuration shown in Fig. 22, which could be described as indirect oblique illumination.

3.6 Stereomicroscopy

Microscopic objects, with the exception of thin sections, are extended in all three dimensions. Nevertheless, a microscope does not provide a three-dimensional or stereoscopic impression of the specimen, even if a binocular tube is used. In normal vision, the impression of three dimensions is created by the different angles at which the object is seen by the right and left eyes. This can also be achieved in a microscope, as shown in Fig. 23. The left eye, viewing through the left eyepiece of the binocular tube, views the object through the one objective pupil, and the right eye through the other. The principal rays of the respective ray paths strike the object at

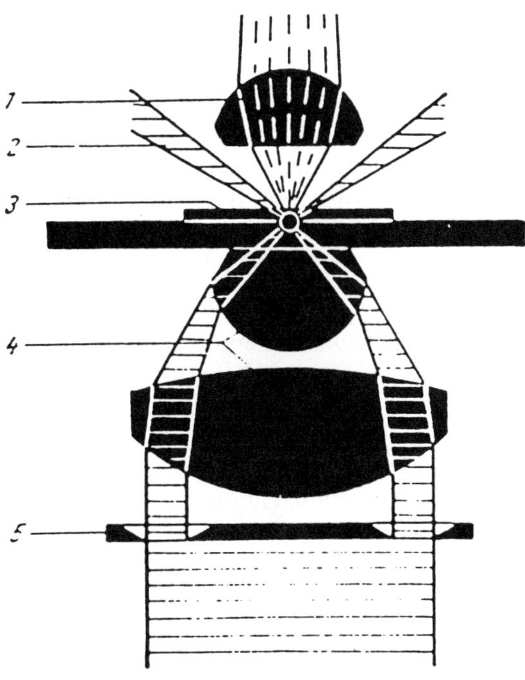

FIG. 22. Dark-field illumination (objective 1, illumination ray bundle 2, object 3, bright-field condenser 4, ring diaphragm 5).

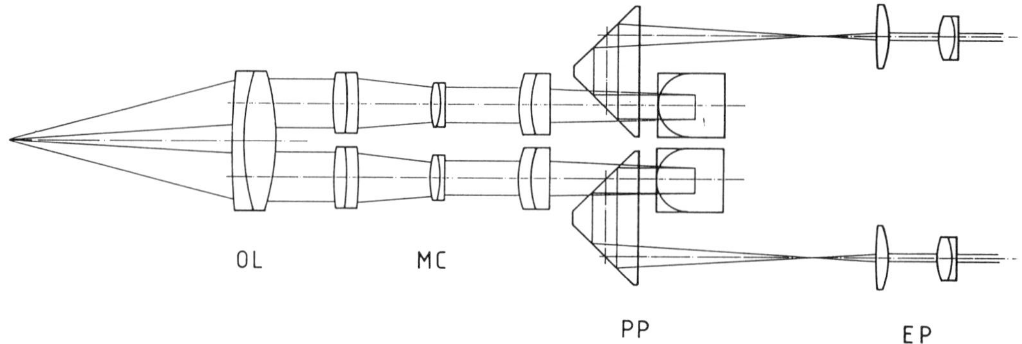

FIG. 23. Stereo microscope (*O* objective, OL objective lens, MC magnification changer, PP path folding prism, EP eyepiece).

different angles. The observer obtains a stereoscopic impression that can, of course, only refer to those parts of the object lying within the depth of focus of the objective. As the depth of field is inversely proportional to the square of the numerical aperture (cf. Sec. 2.3), and since the latter increases with the magnification, stereomicroscopes are normally used with total magnifications of not more than 50×.

4. ULTRAVIOLET MICROSCOPY

If we wish to increase further the resolving power of the light microscope, shorter wavelengths must be used. The visible range of the spectrum extends to approximately 450 mm. Light with a shorter wavelength than this is termed ultraviolet. The light sources used are mainly mercury vapor lamps with spectral lines in the UV. In the case of laser-scan microscopes (cf. Sec. 7.1), HeCd lasers or frequency-doubled argon lasers are also used.

4.1 UV Optics

Only a very limited number of special types of glass (e.g., quartz glass) are suitable for image formation in the UV region. The correction of objectives for UV applications is therefore difficult. Below 200 nm, even the absorption of quartz is too high. UV objectives that can be used down to 250 nm do, however, exist. They can, of course, only be used in conjunction with a special UV microscope whose reflecting prisms and condenser optics are also suitable for UV applications. Reflecting objectives are also used for special applications. It is an inherent property of image-forming reflecting optics that they do not display either chromatic aberrations or considerable absorption losses. In many metallic surfaces the reflecting coefficient is good in the UV region. The correction possibilities offered by reflecting optics are, however, very limited. The most frequently used type of reflecting objective is the Schwarzschild objective (Fig. 24).

4.2 Detectors

Direct, visual observation of the intermediate image with an eyepiece is not possible, of course, in the UV region. Either UV-sensitive films or photoplates are used, or the image is viewed with image-intensifying cameras. In essence, these intensify the photoelectrons triggered by the UV photons and finally image them on a viewing screen.

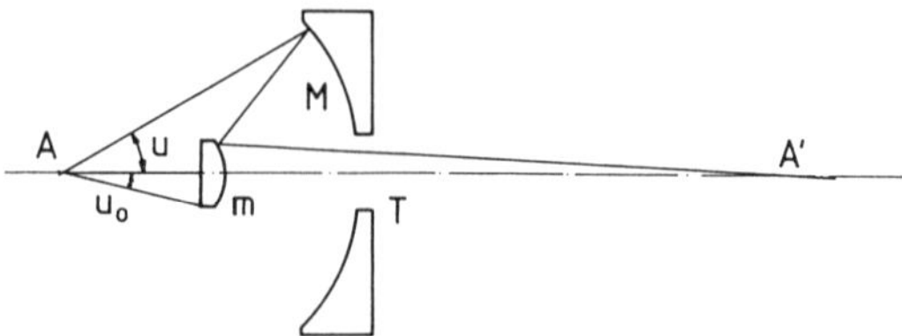

FIG. 24. Schwarzschild objective. Object point *A* is imaged to the image point *A'* via the two mirrors *m* and *M*. The ring-shaped aperture is determined by the angles u_0 and *u*. The image-side numerical aperture is determined by the aperture *T*.

4.3 Applications

The main field of application of UV microscopy is the photometric examination of materials in transmitted light. In crystalline materials, for example, transmission in the UV range is critically dependent on the presence of impurities.

5. INFRARED MICROSCOPY

The infrared region of the spectrum that is relevant in microscopy extends between approximately 0.8 and 20 μm. Although the resolution of a light microscope with a given numerical aperture decreases as the wavelength increases, characteristic properties of infrared radiation can nevertheless be used for three areas: temperature measurement, infrared spectroscopy, and microscopy of semiconductors.

5.1 IR Optics

In the so-called near infrared down to approximately 1.2 μm the glass optics of conventional microscopes can generally still be used without any significant aberrations. The invisibility of this wavelength requires the use of charge-coupled-device (CCD) targets that display a high level of sensitivity precisely in this range. Light with a greater wavelength is no longer sufficiently transmitted by glass, with the result that we have either to resort to refractive optics made of germanium or use reflecting objectives. Both offer only very limited possibilities for the correction of aberrations.

5.2 Detectors

Semiconductor detectors are especially suitable for the detection of infrared radiation. They must be cooled in order to suppress thermal noise, especially when they are used in the long-wave infrared range. In the case of a full-field microscope configuration, the detectors must be arranged in the form of an array (e.g., CCD). However, a single detector is sufficient for a scanning setup (e.g., stage scanner). A comparison of the two setups using an image with 512 × 512 pixels and the same image buildup time shows, however, that the pixel time in a scanning setup is shorter by a factor of 512^2, and the signal per pixel is correspondingly less. Cost reasons and availability, however, frequently favor the selection of a scanning setup.

5.3 Applications

Infrared spectroscopy is a widely used method for the detection of organic compounds whose molecules absorb characteristic line bands in the infrared region. Frequently, an image showing the microscopic distribution of these substances is required. A microscope for infrared spectroscopy therefore consists of a broadband light source with a monochromator for filtering a certain range from its spectrum, a condenser mirror, the object that can be scanned with an *x-y* stage, a reflecting objective, and a fast Fourier-transform spectrometer. The latter is in essence a Michelson interferometer, one mirror of which can be moved rapidly along the interferometer axis. The infrared detector is arranged at the interferometer outlet. The time-resolved signal recorded is subjected to a Fourier transformation by an electronic processor. The result is an infrared spectrum from which a specific substance concentration can be concluded. The object is scanned with the *x-y* stage. An image representing the distribution of a certain substance is obtained. The spatial resolution is, however, limited to a few micrometers.

5.4 Thermography

The technique of thermography used widely in military, building, and materials technology can, of course, also be applied to microscopic objects. According to Planck's law, bodies emit an infrared spectrum that is characteristic of their temperature. It is, therefore, possible to detect the spatial temperature distribution of an object with a resolution totaling mere fractions of a degree, thus allowing, for example, the localization of defective areas on integrated circuits.

5.5 NIR Microscopy

In semiconductors, transmission is relatively good in the near infrared because of the band gap. It is, therefore, possible to image subsurface structures microscopically, although they are not discernible in the visible region of the spectrum. This technique is employed for quality control in the semiconductor industry.

6. X-RAY MICROSCOPY

To increase the resolution of the light microscope further, even shorter wavelengths must be used than that of UV light. The UV region is followed in the spectrum by VUY radiation (vacuum UV, 50 nm $< \lambda <$ 150 nm) which merges into the range of soft x-ray radiation ($\lambda <$ 50 nm). Microscopy in this area is relatively new and is still in a stage of research and development.

Microscopy at wavelengths below the UV region borders on the limits of physics and technology. Since, as already described in Sec. 4, refractive optics can no longer be used for wavelengths smaller than 200 nm, reflecting optics and zone-plate optics remain for shorter wavelengths.

A Fresnel zone plate (Fig. 25) is in principle a rotationally symmetrical grating whose grating constant decreases from the interior to the exterior such that in illumination with parallel light the rays of the first diffraction order intersect in the focal point. The numerical aperture is therefore

$$\mathrm{NA} = \lambda/2dr_n \tag{28}$$

if dr_n is the width of the outermost zone. As in lens optics, the intensity distribution present in the focus in the ideal case is the Airy's pattern with the result that the resolving power of an x-ray microscope is given by the width of the outermost zone. Using electron-beam lithography techniques, zone widths in the order of 10–20 nm are achievable.

The absorption zone plates described only transmit in their gaps, with the result that 50% of the light is lost. The lines of phase zone plates, on the other hand, consist of a material that displays as little absorption as possible, but that also shifts the phase of the transmitted light. The thickness of the lines is selected such that the phase is shifted by 180°. The first diffraction orders of the light transmitted by the lines and that transmitted by the gaps, therefore, interfere constructively in the focus.

As simple optical components, zone plates naturally display greater aberrations than is possible with optimized lens systems. As zone plates are optical gratings, the axial chromatic aberration is especially pronounced. The foci of different wavelengths lie at different points. Monochromatic light must be used. The other aberrations are of no consequence, however, as both the object field and the numerical aperture are small in practice.

Zone plates can, of course, also be used for the x-ray microscope in the form of a condenser. Figure 26 shows the setup of the x-ray microscope on the Berlin electron synchrotron (BESSY). The soft broadband x-ray light emanating from the synchrotron is focused in the object plane by the condenser zone plate. The latter also functions as a monochromator at the same time. As a result of the axial chromatic aberration, the foci of the different wavelengths of the x-ray continuum emitted by the synchrotron lie in different areas of the optical axis. The required wavelength travels through the pinhole stop to the object, a magnified image of which is produced in the image plane by the objective zone plate. Here, a photoplate of a specially prepared CCD target is positioned.

Reflecting objectives are also used in the soft x-ray field whenever high luminous in-

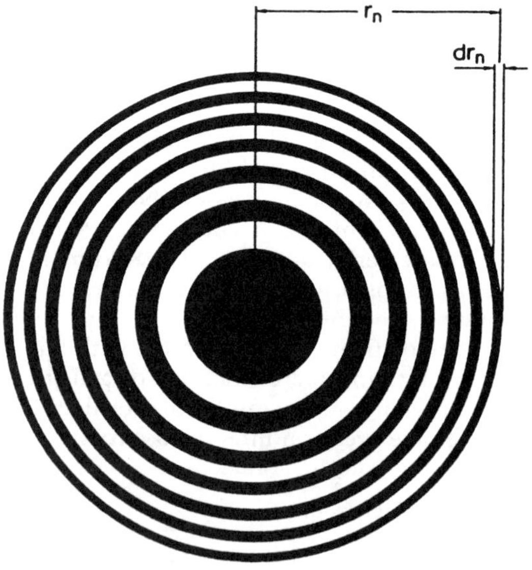

FIG. 25. Zone-plate structure r_n is the radius and dr_n is the width of the outermost zone. (Adapted from Schmahl and Cheng, 1990.)

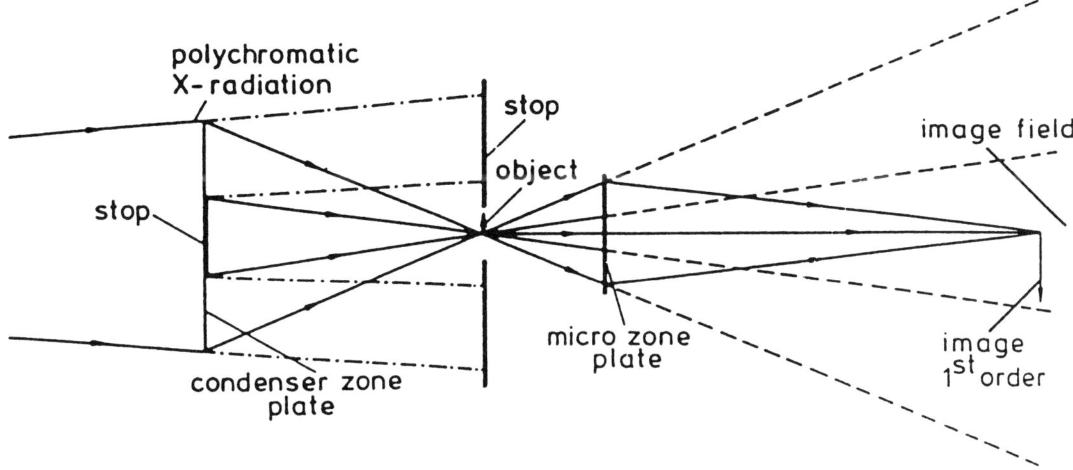

FIG. 26. X-ray optical arrangement of the Göttingen/BESSY x-ray microscope. (Adapted from Schmahl and Cheng, 1990.)

tensity has priority over high resolution. The required reflectivity is obtained by grazing incidence of the x-ray light, e.g., on gold surfaces or by multilayer coating films of, for example, Si and Mo. In the latter case, the thickness of the layers must only lie in the range of $\lambda/4$, with the result that for wavelengths below 10 nm a single film consists of only a few atomic layers, making the production process difficult. The resolution limit of reflecting objectives is given by the producible accuracy of curved mirror surfaces and lies at several hundred nanometers.

Although the resolution attainable in x-ray microscopy (>10 nm) is markedly better than that of light microscopy in the spectral range $\lambda > 200$ nm, it is still not possible to achieve the resolving power of the electron microscope, which lies in the range of angstroms. The advantage offered by x-ray microscopy, however, lies in the possibility of imaging biological specimens in an aqueous medium (Fig. 27). In the wavelength range between 2.3 and 4.5 nm water transmits the x-ray radiation relatively well, whereas carbon and other elements, which form the fixed components of a biological cell, display pronounced absorption and can, therefore, be imaged with a high degree of contrast. Less preparation of the specimen is, therefore, required, and structures that are only stable under the osmotic conditions in the cell can be imaged.

7. SCANNNING MICROSCOPY

If as in UV and IR microscopy, no highly resolving array detectors are available, a scanning technique is an obvious alternative. Roberts and Young used the flying-spot microscope for the first time in UV microscopy, an instrument in which the object is scanned with a light spot. No spatially resolving detector is required. The light transmitted or reflected by the object is picked up by the detector. The signal is displayed as a gray scale value on the monitor synchronously with the position of the light spot on the object. The resolution is, of course, determined by the size of the light spot. It is possible to minimize the size and maximize the brightness of the light spot by focusing a laser.

7.1 Laser-Scanning Microscopy

The principle of a laser-scan microscope (LSM) has already been described in Sec. 2.2.5. UV and IR lasers can be used to obtain a high-resolution image in spectral ranges where, for example, image intensifiers would be necessary in direct-imaging microscopy, resulting in an impairment of image quality. However, laser-scan microscopy is centered on fields of application where it allows the use of contrast-enhancing techniques that are not possible in direct microscopy.

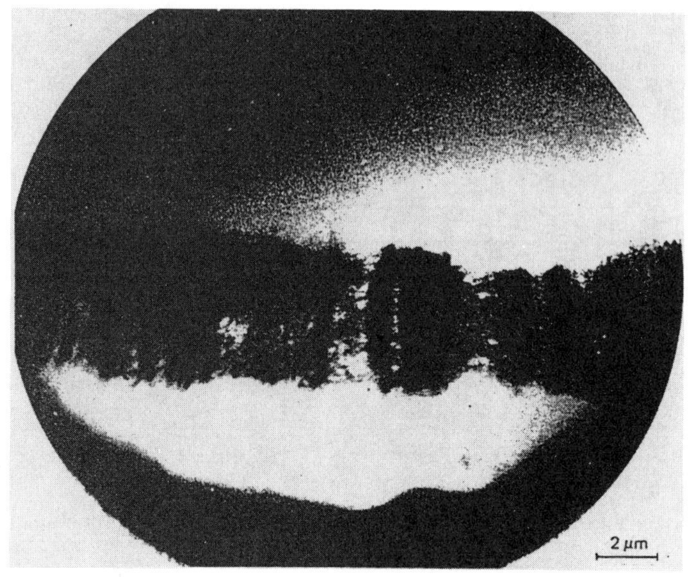

FIG. 27. X-ray image of a part of a giant chromosome in a wet state of the salivary glands of a larva of *Chironomus thummi* taken with the Göttingen/BESSY x-ray microscope at 2.4 nm. (Adapted from Schmahl and Cheng, 1990.)

7.1.1 Confocal Microscopy A confocal microscope setup such as that shown in Fig. 28 makes it possible to image optical sections, i.e., to discriminate light emanating from defocused planes of the object and detect only that light that originates in the focal plane. To achieve this, a lens is arranged in front of the detector to focus the light reflected by the object on a pinhole stop. The pinhole stop is arranged such that it is conjugate to the focal plane of the objective, thus ensuring that only the light coming from this plane can travel past the stop. The object is scanned point by point to obtain an optical section. A measure for the thickness of the optical sections is given by the half-width of the fluorescent point-spread function PSF in the z direction. The PSF is the convolution of the illuminating intensity distribution of the point image near the pinhole. The PSF along the z axis is given by (Wilson and Sheppard, 1984)

$$\mathrm{PSF}(z) = \sin\left(\frac{\sin\left(\frac{\pi}{2\lambda}\,\mathrm{NA}^2 z\right)}{\frac{\pi}{2\lambda}}\,\mathrm{NA}^2 z\right)^4. \qquad (29)$$

The half-width is $1.27\lambda/\mathrm{NA}^2$, where NA is the numerical aperture of the objective and λ is the wavelength of the laser. The image formation for two-dimensional objects is described by the OTF, which is the Fourier

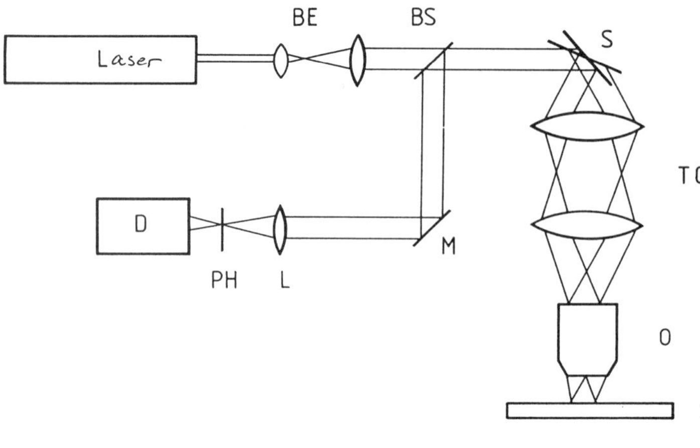

FIG. 28. Confocal laser-scan microscope. The reflected laser beam of the setup of a laser-scan microscope of Fig. 11 is focused on a pinhole PH by a lens *L*. The pinhole is located in a plane that is optically conjugate with the focusing plane of the objective *O*. The light intensity at the detector is at its maximum if the sample *S* is focused (optical sectioning).

transform of the PSF. It is rotationally symmetric and is shown in Fig. 29. The resolution limit is twice the resolution limit of a conventional microscope. However, the value of the OTF is very small for spatial frequencies greater than the conventional limit, so that in practice the effect is usually hidden by photon and detector noise; thus the main feature of confocal microscopy is optical sectioning. By gradual shifting of the object toward the optical axis, it is possible to produce a whole series of optical sections and thus to reconstruct the object in three dimensions (Fig. 30).

7.1.2 Optical Beam Induced Current (OBIC) Another contrast-enhancing technique that is not possible with direct microscopy is the so-called OBIC method. Here, the photoeffect induced by focusing of the laser in the *pn* junctions of a microchip generates a current that, when synchronized with the position of the scanner, can be displayed on a screen. It is, therefore, possible to display not only the topography but also the functional properties of semiconductor chips with microscopic resolution.

7.2 Direct-View Confocal Microscopy

The prerequisite for laser-scan microscopy, the laser, is usually a monochromatic light source. Moreover, lasers in the UV region for fluorescence microscopy are particularly expensive. The flying-spot microscope already mentioned does not require the use of a laser. However, its diffraction-limited light spot that is generated by a conventional light source is very weak, since its radiance is many times lower than that of a laser. It is possible to compensate for this deficiency by simultaneous scanning of the object with many light spots (Petran and Hadravsky, 1968; Kino *et al.*, 1989). In such a so-called direct-view confocal microscope a perforated disk positioned in the intermediate-image plane of a microscope is illuminated by a conventional light source. The pinholes in the perforated disk are imaged on the object. Once again, the reflected light will then only travel through the corresponding pinhole if the reflecting object detail is located in the focal plane of the objective (confocal effect). The intermediate-image plane is imaged by a relay optical system either on a camera target or in the focal plane of an eyepiece, thus allowing the confocal intermediate image to be also viewed directly with the eye. The perforations in the disk are arranged in the form of spirals such that every point of the object is covered at least once in each revolution. The eye or the camera target registers the confocal image of an optical section.

7.3 Scanning Near-Field Optical Microscopy (SNOM)

According to Abbe, the resolving power of an objective is determined by the numerical aperture and the wavelength. In the visible region of the spectrum (λ = 550 nm) it is, therefore, only possible to resolve object de-

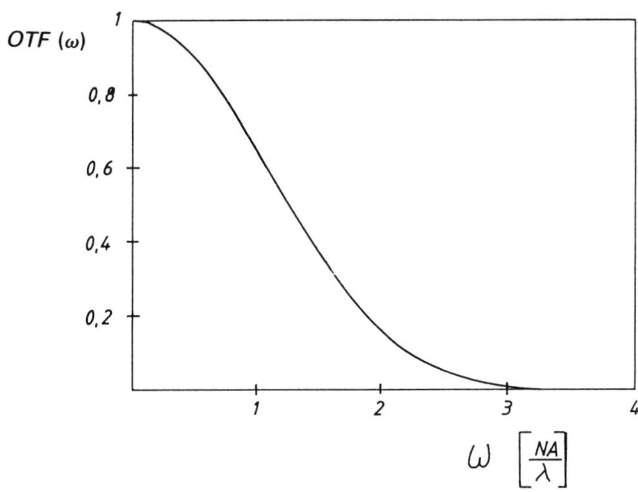

FIG. 29. The OTF of a confocal fluorescence microscope.

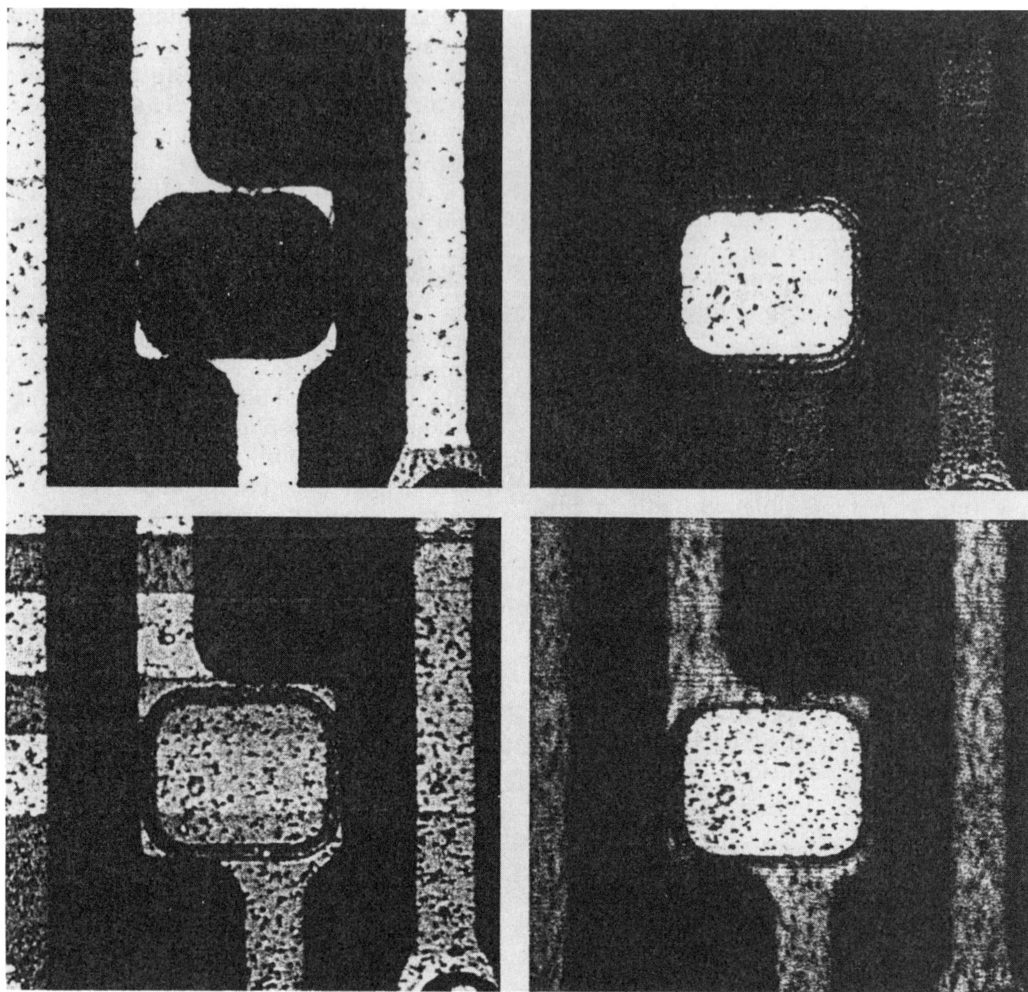

FIG. 30. This figure shows one and the same area of a water surface. In confocal mode "optical sectioning" can be made at levels of focus that are spaced less than 1 μm apart (top left and top right). If several of these images are superimposed, the result is a single, crisp image of extraordinary depth of focus (below left). The nonconfocal laser-scan image shows high resolution contrast but noticeably less depth of focus (below right).

tails down to 200 nm when an immersion objective (NA = 1.4) is used. However, this consideration starts from the assumption that the distance between the objective and the specimen is greater than the wavelength. If, however, a glass fiber with a fine pinhole stop of approximately 50 nm etched into its silvered end is brought to within a few fractions of the light wavelength from the object, and if the object is then illuminated via the glass fiber and the pinhole stop, only light emanating from an area covering approximately 50 nm will enter the glass fiber. The light can be coupled out of the glass fiber using a beam splitter and then detected with a photomultiplier. As in all scanning techniques, the image can be displayed on a monitor. The resolution of this microscope would seem to be better than Abbe's theory permits. This difference only exists, however, as it is the near field of the object wave that is imaged in the scanning microscope described, while Abbe's theory is concerned with the far field.

GLOSSARY

Band Gap: The energy gap between the conducting and the valence band of semiconductors. Photons with energies lower than the band gap are transmitted.

Birefringence: The effect that noncubic crystals split an incoming light wave into two partial rays that are linearly polarized with orthogonal directions of polarization.

CCD (Charge Coupled Device): An array of photosensitive elements that are electronically read out according to a shift register technique.

Diffraction Limited: An optical system is diffraction limited if the resolution is only determined by diffraction and not by aberrations.

Galvanometer Mirror: A small mirror mounted on a torsion suspension periodically driven by a galvanometer coil.

Geometrical Optics: The propagation of light is assumed to be straight (no diffraction).

Image Intensifier: A device that converts an optical image into an intensified electron optical image visualized on a fluorescent screen.

Immersion: The objective is merged into a liquid (oil, water, glycerine, etc.) to enhance resolution.

Interference: The superposition of waves with enhancement (constructive interference) or extinction (destructive interference) or extinction (destructive interference).

Principal Ray: The ray of a ray pencil emanating from an object point striking the entrance pupil of an optical system at its center.

Spatial Frequency: $\omega = 1/x$, x being the spatial period. Analogy to the frequency in the time domain.

Works Cited

Born, M., Wolf, E. (1980), *Principles of Optics*, New York: Pergamon.

Inoue, S. (1986), *Video Microscopy*, New York: Plenum.

Kino, G. S., Chim, S. S. C. (1990), *Appl. Opt.* **29**, 3775–3783.

Kino, G. S., Corle, T. R. (1989), *Phys. Today* **42** (9), 55–62.

Linfoot, E. N., Wolf, E. (1956), *Proc. Phys. Soc. B* **69**, 823–832.

Petran, M., Hadravsky, M. (1968), *J. Opt. Soc. Am.* **58**, 661–664.

Wilson, T., Sheppard, C. (1984), *Theory and Practice of Scanning Microscopy*, New York: Academic.

Schmahl, G., Cheng, P. C. (1990), in: A. G. Michette, G. R. Morrison, C. J. Buckley (Eds.), *X-Ray Microscopy III*, Springer Series in Optical Sciences, Vol. 67, New York: Springer.

Further Reading

Goodman, J. W. (1968), *Introduction to Fourier Optics*, New York: McGraw Hill.

Michette, A. G., Morrison, G. R., Buckley, C. J. (Eds.) (1980), *X-Ray Microscopy III*, Springer Series in Optical Sciences, Vol. 67, New York: Springer.

Pluta, M. (1988), *Advanced Light Microscopy*, Warsaw: PWN-Polish Scientific.

Pohl, D. W. (1991), *Scanning Near-Field Optical Microscopy (SNOM)*, Advances in Optical and Electron Microscopy, Vol. 12, New York: Academic, pp. 243–312.

Sayre, D., Bowels, M., Kitz, J., Rarback, H. (1987) *X-Ray Microscopy II*, Springer Series in Optical Sciences, Vol. 56, New York: Springer.

Schmahl, G. Rudolph, D. (Eds.) (1984), *X-Ray Microscopy: Proceedings of the International Symposium in Göttingen, West Germany, September 14–16, 1983*, Springer Series in Optical Sciences, Vol. 43, New York: Springer.

OPTICAL PROPERTIES OF SOLIDS

ROBERT W. COLLINS AND K. VEDAM, *Materials Research Laboratory and Department of Physics, The Pennsylvania State University, University Park, Pennsylvania, U.S.A.*

INTRODUCTION

A broad discussion of the optical properties of solids from an applied physics standpoint must include the following topics:

1. the macroscopic manifestations of the interaction of light with solids;
2. the microscopic electronic processes in solids that give rise to these manifestations; and
3. the methods by which these processes can be studied and harnessed to control the macroscopic behavior in useful devices.

First, one seeks to understand the spectroscopic reflection behavior of solids, which may be used in simple passive optical components such as mirrors and beamsplitters. For an isotropic solid surface, specular reflection is governed by the index of refraction and extinction coefficient, familiar wavelength-dependent quantities that define the complex dielectric function. Once the complex dielectric function has been determined, one can predict the performance of the solid in a wider variety of advanced applications, beyond simple reflectors. For example, both the

3-527-28134-7/95/$5.00 + .50

spectroscopic reflection and absorption properties of the solid, calculated from the complex dielectric function and thickness, influence the external quantum efficiency of active solid-state devices such as photodetectors and photovoltaic cells.

Because of its importance in describing the reflection and absorption characteristics of solids, the complex dielectric function (or tensor, for an anisotropic solid) will be the central theme of this article from a *macroscopic* viewpoint. In fact, within the condensed-matter physics, applied physics, and materials research communities, when one refers to "the optical properties" of solids the intended meaning is most often "the complex dielectric function versus optical wavelength (or photon energy)." Thus, such a meaning has been adopted for the purposes of this article.

The complex dielectric function (or tensor) is an intrinsic characteristic of a solid, at least when considered on a linear scale. As a result, its real and imaginary parts are often known as "optical constants." (Here they are referred to as "optical functions," however, in order to emphasize their inherent wavelength dependence.) Thus, one of the first steps in research on the optical properties of a new isotropic solid involves a determination of its complex dielectric function (or complex dielectric tensor, for an anisotropic solid). One method of doing this involves measuring, under carefully controlled conditions, the change in polarization that occurs when a polarized monochromatic light wave reflects from the solid at an oblique angle, without a change in frequency. The intrinsic nature of the optical functions of solids implies that optical techniques can be used for convenient, nondestructive characterization of the composition of solids, once a correlation has been established between the optical functions (or parameters deduced from them, e.g., band gaps) and the composition.

Many observed effects that might ordinarily be categorized as optical properties under a broader definition will not be discussed in this article. Such restrictions are necessary in order to avoid sacrificing the phenomenology that relates the macroscopic behavior and the microscopic processes for a variety of solids. Nonlinear optical effects or inelastic processes such as the spontaneous Raman effect or photoluminescence that lead to new frequencies in the emergent wave are relegated to specialized articles in these volumes. For example, photoluminescence in a semiconductor or insulator is a complicated, multistep process that can provide important insights into the electronic characteristics of a material. In contrast to the complex dielectric function, however, the photoluminescence spectrum is often controlled by extrinsic features of the solid (i.e., defects), and its interpretation and relevance to materials performance in a device is highly specific to the material under study. Raman spectroscopy, on the other hand, is more general in its interpretation and is employed to study the atomic and molecular vibrations in the solid.

Because the operating range of optical devices is typically from the near-infrared to the ultraviolet wavelengths of the electromagnetic spectrum, the central theme of this article from a *microscopic* viewpoint will be the intrinsic optical processes involving valence and conduction electrons. This restriction means that the discussion of the complex dielectric function itself will be confined to the range of wavelengths where electronic transitions occur between (or among) valence- and conduction-band states (i.e., states lying within a ~10-eV band centered at the Fermi level). For example, measurements of the complex dielectric function in the infrared range yield information on atomic/molecular vibrations in semiconductors and insulators (complementary to Raman spectroscopy) and on free electrons in metals. Thus, only the latter capabilities will be discussed here. The dielectric function from the deep ultraviolet to the x-ray range provides information on the electronic core levels in the solid and will not be discussed. Usually in the literature, the optical properties in regions outside the near infrared to ultraviolet are given a descriptor, e.g., "the x-ray optical properties," to emphasize their relevance for specialized types of devices, e.g., x-ray mirrors and monochromators, in this case.

In Sec. 1, a relationship is established between the microscopic electric charges/currents induced by electromagnetic fields in a solid and the macroscopic complex dielectric function. Plane-wave solutions to Maxwell's equations then show how the complex dielectric function determines the propagation characteristics of an electromagnetic wave within the solid. Alternative expressions of the optical functions are defined, including the complex index of refraction, the absorption coefficient, and the complex optical conductivity. Because most solids are not optically isotropic, and many optical devices employ natural and induced birefringence and optical activity, part of Sec. 1 is devoted to wave propagation in these materials. In order to understand wave propagation in such materials, the complex dielectric tensor is introduced. Characteristics of the complex dielectric function and tensor dictated by causality are treated at the conclusion of the section.

In Sec. 2, a simple classical mechanical model is presented to describe the interaction of light with an atom in a solid, based on the early work of Drude and Lorentz. This approach yields the induced atomic charge redistribution, or polarization (and hence the dielectric function), from the Newtonian forces of the optical electric field on the electrons of the atom. The model provides a simple intuitive understanding of the origin of absorption and dispersion, or the wavelength dependence of the complex dielectric function. It can also account for the differences in optical properties among insulators, semiconductors, and metals, in simple terms based on the strength of the restoring forces exerted on the electrons by the atomic cores.

One must resort to a full quantum-mechanical description of electronic states in the solid, however, to understand the differences in the dielectric functions observed among different materials within a given group (e.g., between Ag and Au, or between Si and Ge). Such a description is provided in Sec. 3. In the quantum-mechanical model, the one-electron approximation is typically used. Thus, the effect of the field on the many-electron wave function is obtained by determining the response of each separate electron in a self-consistent scalar potential. Then one can consider the optical electric field as inducing transitions of single electrons from their ground Bloch states to higher-lying excited states within the framework of time-dependent perturbation theory. The dielectric function is calculated from the expectation value of the conduction current density taken over a volume that is large compared with atomic dimensions. Three situations will be discussed in detail:

1. direct interband transitions in metals, semiconductors and insulators, i.e., electronic transitions that conserve crystal momentum, from filled bands to partially filled or empty bands;
2. intraband transitions in metals, i.e., transitions of electrons within partially filled bands; and
3. indirect, interband transitions in semiconductors or insulators, i.e., transitions of electrons from filled bands to empty bands, along with simultaneous absorption or emission of a quantum of lattice vibrational energy.

The ability of the complex dielectric function to provide information on the joint density of electronic states in an amorphous or crystalline solid is also discussed.

In Sec. 4, a subject of great importance in applied physics is discussed, namely the effect of external perturbations on the dielectric function. The important external perturbations that reduce the symmetry of the dielectric tensor include the electric field, magnetic field, mechanical stress, or strain. The reduction in symmetry leads to induced birefringence and optical rotation that can be exploited in electro-optic, magneto-optic, piezo-optic, elasto-optic, and acousto-optic devices whose purpose is to control and manipulate the propagation of light. Some of these types of devices are treated in greater detail in specialized articles in these volumes.

This article concludes in Sec. 5 with a discussion of the techniques used to measure the complex dielectric function or tensor of solid materials. The primary focus will be on the purely optical techniques, including reflectance, transmittance, and ellipsometry. Other techniques such as modulation, photoelectron, and electron energy-loss spectroscopies that often bypass complex dielectric function determination in favor of direct information on the electronic densities of states are treated elsewhere in these volumes.

1. PROPAGATION OF LIGHT IN SOLIDS

1.1 Optically Isotropic Solids and the Complex Dielectric Function

In studying the propagation of light waves in solid media, it is convenient to begin with the macroscopic expression of Maxwell's equations (in SI units):

$$\nabla \cdot \mathcal{E} = \rho_{tot}/\epsilon_0, \quad (1a)$$

$$\nabla \times \mathcal{E} = -\frac{\partial \mathbf{B}}{\partial t}, \quad (1b)$$

$$\nabla \cdot \mathbf{B} = 0, \quad (1c)$$

$$\nabla \times \mathbf{B} = \mu_0\left(\epsilon_0 \frac{\partial \mathcal{E}}{\partial t} + \mathbf{J}_{tot}\right), \quad (1d)$$

as presented in numerous texts on electrodynamics and optics (see, for example, Landau and Lifshitz, 1965; Born and Wolf, 1970; Jackson, 1975). In Eqs. (1), ϵ_0 and μ_0 are the permittivity and permeability of free space, given by $\epsilon_0 = 8.8542 \times 10^{-12}\ C^2/N\ m^2$ and $\mu_0 = 4\pi \times 10^{-7}\ N\ s^2/C^2$. In addition, $\mathcal{E}$, $\mathbf{B}$, ρ_{tot}, and $\mathbf{J}_{tot}$ are the macroscopic electric field and magnetic induction associated with the wave and the macroscopic total charge and current densities in the solid, respectively. By "macroscopic," one means that the quantities are obtained by averaging the microscopic fields and charge or current densities over a volume ΔV having linear dimensions of ~10 nm, which is small compared to the wavelength of the light of interest, ~500 nm [see discussion prior to Eq. (11)], but large compared to atomic dimensions, 0.2 nm. For example,

$$\mathcal{E}(\mathbf{r}) = \frac{1}{\Delta V}\int_{\Delta V} \mathcal{E}_{loc}(\mathbf{r} + \mathbf{r}')d^3\mathbf{r}', \quad (2a)$$

$$\rho_{tot}(\mathbf{r}) = \frac{1}{\Delta V}\int_{\Delta V} \rho_{micro}(\mathbf{r} + \mathbf{r}')d^3\mathbf{r}'. \quad (2b)$$

Here $\mathcal{E}_{loc}$ and ρ_{micro} are the microscopic electric field and charge density that vary on the atomic scale (Wooten, 1972). Similar equations also relate $\mathbf{B}_{loc}$ to $\mathbf{B}$ and $\mathbf{J}_{micro}$ to $\mathbf{J}_{tot}$. In a classical model, ρ_{micro} and $\mathbf{J}_{micro}$ are calculated by summing contributions from the individual electrons and nuclei. In a quantum-mechanical model, the electron contribution must be calculated from the electronic wave function $\psi(\mathbf{r})$ of the system, i.e., ρ_{micro}(electrons) $= -e\psi^*(\mathbf{r})\psi(\mathbf{r})$, where $-e$ is the charge of the electron (see, for example, Merzbacher, 1970).

As described in detail by Wooten (1972), Eqs. (1) need to be reformulated in terms that will allow the properties of the solid to be readily incorporated. Contributions to the total charge density arise from bound charges (ρ_{bound}) and external charges (ρ_{ext}) introduced into the system. Similarly, the current density arises from bound charges ($\mathbf{J}_{bound}$) and free charges ($\mathbf{J}_{free}$):

$$\rho_{tot} = \rho_{bound} + \rho_{ext} = -\nabla \cdot \mathbf{P} + \rho_{ext}, \quad (3a)$$

$$\mathbf{J}_{tot} = \mathbf{J}_{bound} + \mathbf{J}_{free} = \frac{\partial \mathbf{P}}{\partial t} + (\nabla \times \mathbf{M}) + (\mathbf{J}_{cond} + \mathbf{J}_{ext}). \quad (3b)$$

Bound macroscopic charge can accumulate in regions of the solid as a result of a spatially nonuniform dipole moment per unit volume, or polarization $\mathbf{P}$. Such polarization can be visualized as a displacement of the electron clouds with respect to the nuclei (see Sec. 2.1). The bound current density in turn includes contributions from a time-dependent polarization and from a spatially nonuniform magnetic dipole moment per unit volume, or magnetization $\mathbf{M}$. In nonmagnetic materials, weak magnetization can result from electron spin and the motion of conduction electrons; however, such effects on the optical properties are small and can be neglected. The effects of the much stronger magnetization in a saturated ferromagnetic solid on its optical properties are significant and will be discussed in Sec. 4.3. In this section, nonmagnetic solids are assumed. The free current in Eq. (3b) consists of the current due to conduction electrons in metals and semiconductors ($\mathbf{J}_{cond}$) as well as the external current ($\mathbf{J}_{ext}$). Note that the former does not generate a net macroscopic charge in Eq. (3a) because of the balance of the positive charge background of the nuclei. By defining two new vector fields, the displacement $\mathbf{D}$ and the magnetic field intensity $\mathbf{H}$, given by

$$\mathbf{D} = \epsilon_0\mathcal{E} + \mathbf{P}, \quad (4a)$$

$$\mathbf{H} = \mu_0^{-1}\mathbf{B} - \mathbf{M}, \quad (4b)$$

then Maxwell's equations can be rewritten as

$$\nabla \cdot \mathbf{D} = \rho_{ext}, \quad (5a)$$

$$\nabla \times \mathcal{E} = -\frac{\partial \mathbf{B}}{\partial t}, \quad (5b)$$

$$\nabla \cdot \mathbf{B} = 0, \quad (5c)$$

$$\nabla \times \mathbf{H} = \frac{\partial \mathbf{D}}{\partial t} + (\mathbf{J}_{cond} + \mathbf{J}_{ext}). \quad (5d)$$

Next, the macroscopic properties of the solid can be introduced through the constitutive relationships

$$\mathbf{P} = \epsilon_0 \chi_e \mathcal{E}, \tag{6a}$$

$$\mathbf{M} = \chi_m \mathbf{H}, \tag{6b}$$

$$\mathbf{J}_{\text{cond}} = 4\pi\epsilon_0 \sigma \mathcal{E}, \tag{6c}$$

where χ_e, χ_m, and σ are the real electric and magnetic susceptibilities and real *optical* conductivity, respectively. Here it is assumed that the response to the fields is linear and the solid is optically isotropic. Nonlinear optical responses are treated in detail elsewhere in these volumes (see OPTICS, NONLINEAR) and an extensive discussion of optically anisotropic solids is provided in Secs. 1.2 and 4. Two parameters alternative to the susceptibilities can be used to describe the medium, defined by

$$\mathbf{D} = \epsilon_0 \epsilon \mathcal{E}, \tag{7a}$$

$$\mathbf{B} = \mu_0 \mu \mathbf{H}. \tag{7b}$$

Here ϵ is the real dielectric function and μ is the magnetic permeability. These dimensionless parameters are related to the susceptibilities by $\epsilon = 1 + \chi_e$ and $\mu = 1 + \chi_m$. Again, the focus here will be on nonmagnetic solids for which $\chi_m = 0$ and $\mu = 1$. In ferromagnetic solids, one can still set $\mu \approx 1$ for the high-frequency optical field; however, one must include the effect of the permanent magnetization on the optical properties as noted earlier (see also Sec. 4.3).

By using Eqs. (6c), (7a), and (7b) to eliminate **D**, **H**, and $\mathbf{J}_{\text{cond}}$ in Eqs. (5), one arrives at a re-expression of Maxwell's equations that explicitly includes the macroscopic properties of the material (i.e., σ, ϵ, μ). However, the new equations are no longer exact since Eqs. (6) and (7) describe only an approximate response of the solid to the fields. Assuming no external charges or currents, and an isotropic, homogeneous (i.e., no spatial variations in σ, ϵ, and μ), and nonmagnetic (i.e., $\mathbf{M} = 0$, $\mu = 1$) material, then Eqs. (5b), (5d), (6c), (7a), and (7b) show that the electric field obeys the following wave equation (Born and Wolf, 1970; Wooten, 1972; Jackson, 1975):

$$\nabla^2 \mathcal{E} = \frac{\epsilon}{c^2}\frac{\partial^2 \mathcal{E}}{\partial t^2} + \frac{4\pi\sigma}{c^2}\frac{\partial \mathcal{E}}{\partial t}. \tag{8}$$

Here c is the speed of light in vacuum. Since $\nabla \cdot \mathcal{E} = 0$, the solutions to Eq. (8) are transverse plane waves that describe the electric field vector of the light wave in the solid. Solutions can be described as functions of an intrinsically complex wave vector $\tilde{\mathbf{q}}$ introduced to account for dissipation of energy via the optical conductivity term at the far right in Eq. (8):

$$\mathcal{E}(\mathbf{r},t) = \mathcal{E}_0 \exp[i(\tilde{\mathbf{q}} \cdot \mathbf{r} - \omega t)]. \tag{9}$$

Here a single frequency component ω appears, based on the assumption that the wave in the medium is monochromatic. The complex wave vector $\tilde{\mathbf{q}} = \tilde{q}\hat{\mathbf{q}}$, perpendicular to $\mathcal{E}_0$, obeys

$$\tilde{\mathbf{q}}^2 = \left(\frac{\omega}{c}\right)^2 \left[\epsilon + i\left(\frac{4\pi\sigma}{\omega}\right)\right] = \left(\frac{\omega}{c}\right)^2 \tilde{N}^2, \tag{10}$$

where $\tilde{N}$ is the frequency-dependent complex index of refraction, given by $\tilde{N} = n + i\kappa$. In this latter expression, n is the (real) index of refraction, and κ is the extinction coefficient. Thus, the electric field of Eq. (9) can be expressed as a wave having phase velocity $v_p = c/n$, wavelength $\lambda_m = 2\pi c/\omega n$, and an exponentially decaying amplitude with a decay length (at $1/e$) of $\omega\kappa/c$:

$$\mathcal{E}(\mathbf{r},t) = \mathcal{E}_0 \exp\left(-\frac{\omega\kappa\hat{\mathbf{q}} \cdot \mathbf{r}}{c}\right) \times \exp\left\{i\left(\frac{\omega n\hat{\mathbf{q}} \cdot \mathbf{r}}{c} - \omega t\right)\right\}. \tag{11}$$

The irradiance I, or the time average of the energy density crossing a unit area normal to $\hat{\mathbf{q}}$, is given by $I(\mathbf{r}) = (\epsilon_0 v_p/2)\epsilon\mathcal{E}^* \cdot \mathcal{E}$ for an isotropic solid. The absorption coefficient α describes the decay of the irradiance due to energy dissipation. Thus,

$$I(x) = I_0 \exp(-\alpha x), \tag{12a}$$

$$\alpha = 2\omega\kappa/c = 4\pi\kappa/\lambda, \tag{12b}$$

where for simplicity $\hat{\mathbf{q}} = \hat{\mathbf{x}}$, and $I_0 = I(x = 0)$. In Eq. (12b), $\lambda = 2\pi c/\omega = n\lambda_m$ is the wavelength for a wave of the same frequency sustained in vacuum. Equations (12) do not apply when $n = 0$ and $\kappa \neq 0$, e.g., for a perfect metal. In this special case, $I(x) = 0$, and since no net energy is transmitted into the material, no energy is dissipated.

In addition to the complex index of refraction, the complex dielectric function $\tilde{\epsilon}$ is commonly used to define the macroscopic optical properties of solids, where

$$\tilde{\epsilon} = \tilde{N}^2 = \epsilon + i(4\pi\sigma/\omega). \tag{13a}$$

The real and imaginary parts of $\tilde{\epsilon}$ are given by

$$\epsilon_1 = \epsilon = n^2 - \kappa^2, \tag{13b}$$

$$\epsilon_2 = 4\pi\sigma/\omega = 2n\kappa. \tag{13c}$$

Thus,

$$n = \{[(\epsilon_1^2 + \epsilon_2^2)^{1/2} + \epsilon_1]/2\}^{1/2}, \tag{13d}$$

$$\kappa = \{[(\epsilon_1^2 + \epsilon_2^2)^{1/2} - \epsilon_1]/2\}^{1/2}. \tag{13e}$$

The complex dielectric function develops naturally in the solution to the wave equation by starting with Maxwell's equations [Eqs. (5)] and redefining a displacement field that is intrinsically complex (denoted by "~") and incorporates both the real displacement [Eq. (7a)] and the current due to conduction electrons [Eq. (6c)] (Wooten, 1972):

$$\tilde{\mathbf{D}} = \epsilon_0\tilde{\epsilon}\mathcal{E} = \mathbf{D} + (i/\omega)\mathbf{J}_{\text{cond}}. \tag{14}$$

This also leads to an intrinsically complex polarization vector and complex electric susceptibility in accordance with Eq. (6a):

$$\tilde{\mathbf{P}} = \epsilon_0\tilde{\chi}_e\mathcal{E} = \epsilon_0(\tilde{\epsilon} - 1)\mathcal{E}. \tag{15}$$

A third definition of the optical properties often used in the study of metals is the complex optical conductivity, given in terms of $\tilde{\epsilon}$ by

$$\tilde{\sigma} = (i\omega/4\pi)(1 - \tilde{\epsilon}) \tag{16a}$$

with the following real and imaginary parts:

$$\sigma_1 = \omega\epsilon_2/4\pi = \sigma, \tag{16b}$$

$$\sigma_2 = \omega(1 - \epsilon_1)/4\pi = \omega(1 - \epsilon)/4\pi = -\omega\chi_e/4\pi. \tag{16c}$$

The complex optical conductivity develops naturally in the solution to the wave equation by defining a new conduction electron current in Maxwell's equations [Eqs. (5)] that is intrinsically complex and incorporates the polarization [Eq. (6a)] along with the real conduction-electron current [Eq. (6c)]:

$$\tilde{\mathbf{J}}_{\text{cond}} = 4\pi\epsilon_0\tilde{\sigma}\mathcal{E} = \mathbf{J}_{\text{cond}} - i\omega\mathbf{P}. \tag{17}$$

Although often impractical, experiments can be devised to measure $\tilde{N}$ for bulk solids on the basis of Eqs. (11) and (12). For example, one can illuminate a disk of material having two optically polished plane-parallel surfaces with a monochromatic plane wave of known vacuum wavelength λ. If the thickness d of the disk is within an order of $1/\alpha$, then n and κ can be obtained by measuring the phase shift (relative to a reference wave, for n) and the irradiance change (for κ) that result when the wave traverses the sample. In practice, the measurements must be made external to the solid; thus, one must account for the reflections at the interfaces between the solid and ambient in the irradiance measurement (see Sec. 5.1.2) (Heavens, 1965). Because the maximum measurable value of α is $\sim 10^2$ cm^{-1} for polished bulk materials with $d \sim 0.1$ cm, reflection rather than transmission measurements must be used to obtain the much larger α values ($>10^5$ cm^{-1}) associated with intraband and direct interband transitions. Furthermore, direct measurements of the large phase shifts relative to vacuum that occur in such a sample geometry are impractical for solids for which n is typically 1.5–4.0. Further details regarding transmission measurements as well as a discussion of reflection and ellipsometry measurements will be presented in Sec. 5.

1.2 Optically Anisotropic Solids and the Dielectric Tensor

In Sec. 1.1, the medium was assumed to be optically homogeneous and isotropic, i.e., uniform with respect to the optical field (**D**, **H**) directions. This latter assumption is not valid for crystalline solids that are neither cubic nor structurally isotropic, and this leads to interesting phenomena including linear and circular birefringence and dichroism. Linear birefringence is a characteristic of anisotropic solids in which two light waves having the same propagation vector, but orthogonal linear polarization modes, experience different indices of refraction and thus travel at different phase speeds in the solid. Circular birefringence is observed in anisotropic solids possessing a helical arrangement of atoms. In this case, orthogonal left and right elliptical polarization modes, in general, experience different indices of refraction. Linear and circular dichroism accompany linear and circular birefringence, respectively, in absorbing media and are characterized by different extinction coefficients for the two polarization modes.

Even if the material is normally optically isotropic, anisotropy and its effects such as linear birefringence and dichroism can be induced by an external perturbation such as an electric field (electro-optics), a magnetic field (magneto-optics), or mechanical stress (piezo-optics) as described in Sec. 4. Hence, a thorough understanding of the optical properties of optically anisotropic solids as well as the propagation of light in such media is necessary if these interesting phenomena are to be exploited in practical applications. For a more extensive discussion than is possible here, one can consult a number of texts on optics, crystal physics, and crystal optics provided under Further Reading.

1.2.1 Nonabsorbing Solids That Are Not Optically Active The scalar dielectric function for an optically isotropic solid, defined by Eq. (7a), becomes a second-rank tensor in an anisotropic solid. For a homogeneous solid that is nonabsorbing, nonmagnetic, and not optically active, one can write (Born and Wolf, 1970)

$$D_x = \epsilon_0(\epsilon_{11}\mathcal{E}_x + \epsilon_{12}\mathcal{E}_y + \epsilon_{13}\mathcal{E}_z), \tag{18a}$$

$$D_y = \epsilon_0(\epsilon_{21}\mathcal{E}_x + \epsilon_{22}\mathcal{E}_y + \epsilon_{23}\mathcal{E}_z), \tag{18b}$$

$$D_z = \epsilon_0(\epsilon_{31}\mathcal{E}_x + \epsilon_{32}\mathcal{E}_y + \epsilon_{33}\mathcal{E}_z), \tag{18c}$$

or for simplicity in tensor notations:

$$D_i = \epsilon_0\epsilon_{ij}\mathcal{E}_j, \tag{19a}$$

$$\mathbf{D} = \epsilon_0[\epsilon]\cdot\boldsymbol{\mathcal{E}}. \tag{19b}$$

Here orthogonal (x,y,z) axes designated by indices $i = 1,2,3$ or $j = 1,2,3$ are fixed in the medium and are chosen for the reference frame. In Eq. (19a), the Einstein convention of summing over repeated indices is employed. When enclosing a single variable as in Eq. (19b), the brackets denote a tensor, in this case of second rank and defined by the elements of a 3×3 matrix. In restricting the discussion of this subsection to nonabsorbing solids, ϵ_{ij} are assumed to be real. In this case, energy conservation requires that $\epsilon_{ij} = \epsilon_{ji}$, i.e., $[\epsilon]$ is symmetric, so that it is always possible to choose the coordinate system such that $[\epsilon]$ is diagonalized. The resulting (x,y,z) directions for this choice are called the principal dielectric axes. For orthorhombic and higher symmetries, the principal dielectric axes lie along the symmetry axes of the crystal and hence are completely specified. Thus, *at most* three independent quantities, the principal dielectric functions, ϵ_{jj} $(j = 1,2,3)$, must be specified in order to define the dielectric tensor and principal axes. In contrast, the absence of symmetry for the triclinic system means that the directions of all three orthogonal principal axes are unspecified. Thus, six independent quantities (ϵ_{jj}, $j = 1,2,3$, plus the cordinate system orientation) must be specified. Table 1 provides the number of independent elements needed to define the diagonalized dielectric tensor and the orientation of the principal axes for different crystal systems.

In nonmagnetic solids, Maxwell's equations exhibit plane-wave solutions in which $\mathbf{D}$, $\mathbf{H}$ (or $\mathbf{B}$), and the propagation vector $\mathbf{q}$ form a right-handed coordinate system. In optically isotropic solids, $\mathbf{D}$ and $\boldsymbol{\mathcal{E}}$ are parallel, so that $\boldsymbol{\mathcal{E}}$ is perpendicular to $\mathbf{q}$. In optically anisotropic solids, Eq. (18) shows that $\mathbf{D}$ and $\boldsymbol{\mathcal{E}}$ are *not* parallel, and since Maxwell's equations require that $\boldsymbol{\mathcal{E}}$ remains perpendicular to $\mathbf{H}$ (or $\mathbf{B}$), then $\boldsymbol{\mathcal{E}}$ cannot remain perpendicular to $\mathbf{q}$. Thus, the direction of *wave* propagation defined by the unit vector $\hat{\mathbf{q}}$, called the wave normal, is not parallel to the direction of *energy* propagation defined by the unit Poynting vector $\hat{\mathbf{S}}$, called the ray direction (which must be perpendicular to $\boldsymbol{\mathcal{E}}$. The angle ξ between the two is given by $\cos\xi = \boldsymbol{\mathcal{E}}\cdot\mathbf{D}/|\boldsymbol{\mathcal{E}}||\mathbf{D}|$. Since the ray and wave front exhibit the same frequency and must remain in phase, the speed of the ray, v_u, must exceed that of the wave front, v_p, by the factor $1/\cos\xi$. Thus, $v_u = v_p/\cos\xi$, where the phase speed is defined as in optically isotropic solids by $v_p = c/n = c(\epsilon_0\boldsymbol{\mathcal{E}}\cdot\mathbf{D})^{1/2}/D$. As a result, the phase speed is the projection of the ray speed on the direction of the wave normal. In the next

Table 1. Optical behavior of transparent solids of different crystal systems. In this table, (a,b,c) designate the principal crystallographic axes, which are aligned with the principal dielectric axes (x,y,z) for crystals of orthorhombic and higher symmetry. The semiaxes of the index ellipsoid define the principal indices of refraction (n_x,n_y,n_z). For biaxial crystals, (x,y,z) are labeled such that $n_x < n_y < n_z$. For uniaxial crystals, the z axis of the index ellipsoid is the optic axis and is parallel to the c crystallographic axis. For these crystals, $n_o = n_x = n_y$ is the ordinary index of refraction, and $n_e = n_z$ is the extraordinary index of refraction.

Crystal systems and point groups	Number of parameters	Nature and orientation of index ellipsoid	Optical behavior	Variation with λ, temperature or hydrostatic pressure	Point groups exhibiting optical activity
Triclinic $(1, \bar{1})$	6	Ellipsoid, principal axes in unspecified directions	Biaxial, optic axes in unspecified directions	Unspecified	1
Monoclinic (b axis unique) $(2, m, 2/m)$	4	Ellipsoid, one principal axis ∥ to b, other two ⊥ to b but in unspecified directions	Biaxial, optic axial plane either ∥ or ⊥ to b	Orientation of one principal axis always along b	$2, m$
Orthorhombic $(2mm, 222, mmm)$	3	Ellipsoid, three principal axes along a, b, and c	Biaxial, optic axial plane ∥ to ab, bc, or ca; acute bisectrix ∥ to one crystal axis	No change in orientation, but only change in length of principal axes, i.e., n_x, n_y, and n_z	$2mm, 222$
Trigonal $(3, \bar{3}, 3m, 32, \bar{3}m)$	2	Spheroid, with unique axis ∥ to c	Uniaxial, optic axis ∥ to c	Optic axis always along c, but n_o and n_e will vary	3, 32
Tetragonal $(4, \bar{4}, 4/m, 4mm, \bar{4}2m, 422, 4/mmm)$	2	Spheroid, with unique axis ∥ to c	Uniaxial, optic axis ∥ to c	Optic axis always along c, but n_o and n_e will vary	$4, 422, \bar{4}, \bar{4}2m$
Hexagonal $(6, \bar{6}, 6/m, \bar{6}m2, 622, 6mm, 6/mmm)$	2	Spheroid, with unique axis ∥ to c	Uniaxial, optic axis ∥ to c	Optic axis always along c, but n_o and n_e will vary	6, 622
Cubic $(23, m3, \bar{4}3m, 432, m3m)$	1	Sphere	Isotropic	Always isotropic, but n will change	23, 432

paragraphs, further consequences of this will be discussed.

The electrical energy density associated with an electromagnetic wave in an optically anisotropic crystal is given by (Born and Wolf, 1970)

$$u_E = \frac{\mathcal{E}\cdot\mathbf{D}}{2} = \frac{1}{2\epsilon_0}\left(\frac{D_x^2}{\epsilon_{11}} + \frac{D_y^2}{\epsilon_{22}} + \frac{D_z^2}{\epsilon_{33}}\right), \tag{20a}$$

expressed in the (x,y,z) coordinate system of the principal axes. Making the correspondence $r \rightarrow \mathbf{D}/(2\epsilon_0 u_E)^{1/2}$ yields

$$1 = \frac{x^2}{n_x^2} + \frac{y^2}{n_y^2} + \frac{z^2}{n_z^2}. \tag{20b}$$

Thus, the surface consisting of the locus of **D** vector end points, corresponding to an energy density of $1/2\epsilon_0$, is called the index ellipsoid (or optical indicatrix), and has semi-axes given by the principal indices of refraction, $n_x = \epsilon_{11}^{1/2}$, $n_y = \epsilon_{22}^{1/2}$, and $n_z = \epsilon_{33}^{1/2}$. The coordinate transformational properties of this ellipsoid are not representative of the dielectric tensor of Eqs. (19), however, but rather its inverse $[a] \equiv [\epsilon]^{-1}$, called the index (or impermeability) tensor given by $\mathcal{E} = \epsilon_0^{-1}[a]\cdot\mathbf{D}$ (Ramachandran and Ramaseshan, 1961).

The construction of Eqs. (20) can be used to solve the practical problem of determining the indices of refraction and directions of **D**, associated with the two independent, linearly polarized waves that can propagate along a given direction $\hat{\mathbf{q}}$ (see Fig. 1). First, consider the plane passing through the origin, normal to $\hat{\mathbf{q}}$. This plane intersects the index ellipsoid to form an ellipse, or central section, that defines the indices of refraction of all possible directions of **D** normal to $\hat{\mathbf{q}}$. The major and minor axes of this ellipse define the minimum and maximum phase speeds, respectively, corresponding to two specific orthogonal directions of **D** (or polarization directions). Thus, for *any* direction of **D** normal to $\hat{\mathbf{q}}$, the displacement can be resolved along the two directions, yielding two orthogonally polarized waves having different phase speeds. When a wave in an optically isotropic medium crosses an interface into an anisotropic medium, the two polarization components will be refracted through different angles, and this is the physical explanation of the unique visual effect of birefringence.

An alternative approach to understanding wave propagation provides a direct connection between the direction of $\hat{\mathbf{q}}$ and the indices of refraction for the two orthogonally polarized waves. For plane-wave solutions to Maxwell's equations in an anisotropic, nonabsorbing solid that is not optically active, the index of refraction (defining the phase speed, c/n) and the wave propagation direction are related through Fresnel's equation (Born and Wolf, 1970):

$$\frac{1}{n^2} = \sum_{j=1}^{3}\left(\frac{q_j}{q}\right)^2\left(\frac{1}{n^2 - n_j^2}\right). \tag{21}$$

As usual, the principal-axis coordinate system is used, and the summation on j is over the (x,y,z) components of $\hat{\mathbf{q}}$, and over the principal indices of refraction. This equation has two positive solutions for the index of refraction n, corresponding to the two orthogonally polarized waves of different phase speeds having wave normals in the direction of $\hat{\mathbf{q}} = (q_x/q, q_y/q, q_z/q)$. Each of the two solutions for $n(\hat{\mathbf{q}})$ can be visualized as a surface in propagation-vector space such that the vector from the origin to a point on the surface in the direction of $\hat{\mathbf{q}}$ has length n. These

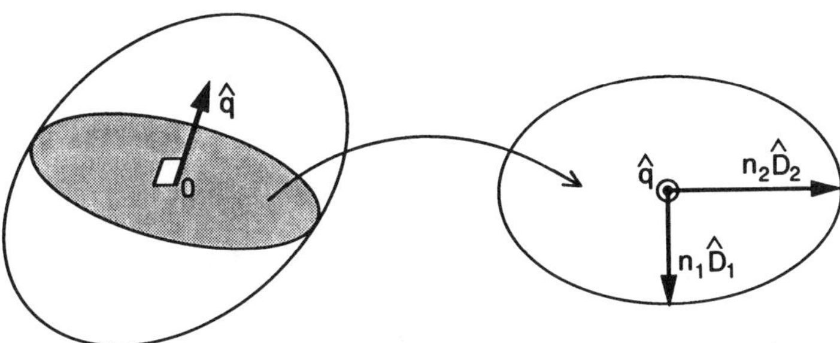

FIG. 1. Index ellipsoid (left) and central section (right) for an anisotropic solid demonstrating the procedure used to determine the two possible values of the index of refraction (n_1 and n_2), as well as the directions of the associated orthogonal linear polarization modes ($\hat{\mathbf{D}}_1$ and $\hat{\mathbf{D}}_2$), given the direction of the wavevector, $\hat{\mathbf{q}}$.

surfaces, along with the corresponding surfaces given by $v_p(\hat{\mathbf{q}}) = c/n(\hat{\mathbf{q}})$, are collectively referred to as normal surfaces. Given the direction of the Poynting vector $\hat{\mathbf{S}}$, similar surfaces called ray surfaces can be constructed. These surfaces provide two possible solutions for the ray or energy index $n_u(\hat{\mathbf{S}})$ and the ray speed, related by $v_u(\hat{\mathbf{S}}) = c/n_u(\hat{\mathbf{S}})$. For a given wave defined by $\hat{\mathbf{q}}$ and $\hat{\mathbf{S}}$, the normal and ray surface solutions for the index are related by $n_u = n \cos\xi$.

In general, two surfaces intersect in a curve; however, the form of the normal surfaces given by Eq. (21) prescribes that they intersect at only four points for crystal systems of orthorhombic and lower symmetry that have three unequal principal indices of refraction (Born and Wolf, 1970). In this case, two optic axes can be defined, each passing through the origin and one pair of intersecting points. For propagation vectors along these two axes, the two orthogonally polarized waves exhibit the same phase speed. Thus, the crystals exhibiting this behavior are called "optically biaxial" as shown in Table 1. In an optically biaxial crystal, the principal coordinate axes are labeled such that the principal indices of refraction obey $n_x < n_y < n_z$. With this convention, the optic axes lie in the x-z plane. The cross section in the x-z plane of a normal surface defined in terms of $n(\hat{\mathbf{q}})$ is shown in Fig. 2(a) for a biaxial crystal.

For the higher-symmetry optically anisotropic crystals (see Table 1), there are two independent principal indices of refraction, the ordinary index, $n_o \equiv n_x = n_y$, and the extraordinary index, $n_e \equiv n_z$. In this case, one normal surface is a sphere that defines the directionally independent ordinary index of refraction, given by $n = n_o$. The other is a spheroid having an axis of symmetry along the z direction, given by (Yariv and Yeh, 1984)

$$\frac{1}{n^2} = \frac{\sin^2\phi}{n_e^2} + \frac{\cos^2\phi}{n_o^2}. \tag{22}$$

where ϕ is the angle of the propagation vector in the x-z plane defined relative to the z axis. Such crystals are called "optically uniaxial," because the normal surfaces now intersect at only two points at $\phi = 0$ in the x-z plane. The line passing through the origin and joining the two intersecting points defines the optic axis, i.e., the z axis in this case. If $n_o < n_e$, then the crystal is said to be positive uniaxial, an example being α-quartz with $n_o = 1.5443$ and $n_e = 1.5534$ at $\lambda = 589.3$ nm. If $n_o > n_e$, it is said to be negative uniaxial, an example being calcite ($CaCO_3$) with $n_o = 1.6584$ and $n_e = 1.4864$ at $\lambda = 589.3$ nm. The intersections of the x-z plane with the normal surfaces of positive and negative uniaxial crystals are shown in Figs. 2(b) and 2(c), respectively.

Finally, consider the problem of plane-wave refraction at an interface between an isotropic ambient (assume air: $n_a = 1$) and an optically uniaxial anisotropic solid. In this specific case, the coordinate system is aligned so that the x-y plane defines the interface. The boundary conditions specify an angle of refraction θ_t given by $\sin\theta_t = \sin\theta/n$, where θ is

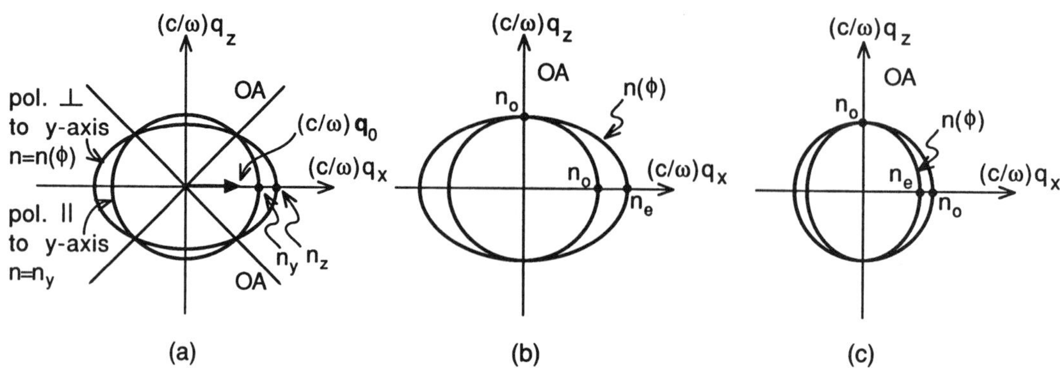

FIG. 2. Intersections of the two normal surfaces with the x-z plane for (a) a biaxial crystal ($n_x < n_y < n_z$); (b) a positive uniaxial crystal ($n_o = n_x = n_y < n_e = n_z$); and (c) a negative uniaxial crystal ($n_e < n_o$). The optic axes, defined by the intersection points of the normal surfaces, are also indicated in each case. For the biaxial crystal in (a), the principal indices of refraction are indicated for $\mathbf{q} = \mathbf{q}_0$ along $\hat{\mathbf{x}}$.

the angle of incidence and n is the index of refraction associated with a given **D** field polarization component. For the component associated with the ordinary wave, the phase speed is independent of direction, and behavior in accordance with Snell's law for an optically isotropic solid with $n = n_o$ is observed. For the orthogonal polarization component associated with the extraordinary wave, the phase speed depends on the propagation direction, and the observed behavior is in accordance with a more complicated expression for $\sin\theta_t$ that includes $n(\phi)$ from Eq. (22).

1.2.2 Absorbing Solids That Are Not Optically Active In optically isotropic solids, energy dissipation due to conduction-electron currents can be accounted for by introducing a complex dielectric function (see Sec. 1.1). Following the procedure used in the development of Eq. (13a), Eq. (19b) for nonabsorbing anisotropic solids can be extended to the case of absorbing solids by introducing a complex dielectric tensor $[\tilde{\epsilon}]$ with components given by

$$\tilde{\epsilon}_{ij} = \epsilon_{ij} + i(4\pi/\omega)\sigma_{ij}. \tag{23}$$

From thermodynamics considerations (i.e., the Onsager relations), it can be shown that the real tensors $[\epsilon]$ and $[\sigma]$ are symmetric (Landau and Lifshitz, 1965). Thus, both tensors can be diagonalized, but the principal axes coincide only for crystal systems of orthorhombic and higher symmetry.

As noted in Sec. 1.2.1, the index ellipsoid given by Eqs. (20) for nonabsorbing solids relates the polarization direction and the phase speed of the wave, and its semiaxes are defined in terms of the components of the diagonalized (real) index tensor $[a]$. The index tensor is complex in absorbing solids, however, and is written as $[\tilde{a}] = [a] + i[b] = [\tilde{\epsilon}]^{-1}$ with components given by $\tilde{a}_{ij} = a_{ij} + ib_{ij}$, where $[b]$ denotes the absorption tensor. The ellipsoids representing $[a]$ and $[b]$ are called the index (as in the case of nonabsorbing solids) and the absorption ellipsoids, respectively (Ramachandran and Ramaseshan, 1961). Again, the principal axes of these ellipsoids coincide only for solids of orthorhombic and higher symmetry. The nature of the waves propagating along any direction $\hat{\mathbf{q}}$ of the crystal depends on the central sections of the index and absorption ellipsoids, normal to $\hat{\mathbf{q}}$. The directions of the major and minor axes of the central sections define the orthogonal polarizations and principal directions of linear birefringence (for the index ellipsoid) and linear dichroism (for the absorption ellipsoid). The magnitudes of the major and minor axes can be interpreted to yield the index of refraction and extinction coefficient for each of the two polarizations. Often, the magnitude of the dichroism is very small compared to the birefringence, and the behavior of the wave is similar to that in the nonabsorbing case, but with an exponential attenuation of the fields [according to Eqs. (12)] that is independent of $\hat{\mathbf{q}}$ and polarization.

1.2.3 Nonabsorbing, Optically Active Solids Optical activity, or optical rotatory power, describes the phenomenon in which the polarization direction of a linearly polarized light wave rotates as the wave traverses the medium (Drude, 1965; Agranovich and Ginzburg, 1966). Typical magnitudes of this effect range from 20° to 2000° per centimeter of path length in optically active solids. The last column of Table 1 lists the point groups that exhibit optical activity. Depending on the crystal symmetry, solids from any of the crystal systems can be optically active along their unique optic-axial directions, including cubic (e.g., $NaClO_3$, $Bi_{12}GeO_{20}$), optically uniaxial (e.g., Se, Te, α-quartz, HgS, and benzil), and biaxial (e.g., cane sugar and Rochelle salt) crystals. Optical activity is generally treated as arising from circular birefringence, i.e., a difference between the indices of refraction n_l and n_r for left- and right-circularly polarized light waves with a typical magnitude of $|n_l - n_r| \sim 10^{-5}$. Optical activity arises from an intrinsic helical arrangement of the atoms in the crystal. Since the medium has the property of a screw axis, the displacement vector **D** at a given point must depend not only on the electric vector $\mathcal{E}$ but also on its spatial derivatives at that point. Hence, for plane-wave propagation in an optically active medium, Eq. (19b) must be replaced by

$$\tilde{\mathbf{D}} = \epsilon_0\{[\epsilon]\cdot\mathcal{E} + i\mathbf{G}\times\mathcal{E}\}, \tag{24a}$$

$$= \epsilon_0([\epsilon] + i[G])\cdot\mathcal{E}, \tag{24b}$$

where the gyration vector **G** is parallel to $\hat{\mathbf{q}}$ and $[\epsilon]$ is the dielectric tensor that would exist in the absence of the optical activity (Yariv and Yeh, 1984). Equation (24b) follows since the vector product $\mathbf{G} \times \mathcal{E}$ can always be represented as the product of an antisymmetric gyration tensor $[G]$ and the field $\mathcal{E}$ such that $G_{23} = -G_{32} = -G_x$; $G_{31} = -G_{13} = -G_y$; $G_{12} = -G_{21} = -G_z$; and $G_{jj} = 0$ ($j = 1,2,3$). Equation (24b) leads to a new complex dielectric tensor for nonabsorbing, optically active solids, given by $[\tilde{\epsilon}^0] = [\epsilon] + i[G]$ and represented by a Hermitian matrix, i.e., $\tilde{\epsilon}^0_{ij} = (\tilde{\epsilon}^0_{ji})^*$. From this result, it can be shown that the divergence of the Poynting vector, $\nabla \cdot \mathbf{S} = \nabla \cdot (\mathcal{E} \times \mathbf{H})$, vanishes so that there is no dissipation.

An equation analogous to Eq. (24a) that employs the index tensor can be written:

$$\mathcal{E} = \epsilon_0^{-1}\{[a] \cdot \tilde{\mathbf{D}} - i\boldsymbol{\Gamma}_o \times \tilde{\mathbf{D}}\}, \tag{25}$$

where $\boldsymbol{\Gamma}_o$ is called the optical activity vector (Ramachandran and Ramaseshan, 1961). Like the gyration vector, $\boldsymbol{\Gamma}_o$ is a function of the wave propagation direction, according to the relation $\boldsymbol{\Gamma}_o = [\gamma_o] \cdot \hat{\mathbf{q}}$, where $[\gamma_o]$ is a second-rank optical activity tensor described in general by a 3×3 matrix.

Because $G_{ij}/\epsilon_{ij} \ll 1$, the normal surfaces of an optically active solid are nearly identical to those without optical activity having $\mathbf{G} \times \mathcal{E} = 0$. However, the interpretation of the resulting wave propagation must be modified from that for solids that are not optically active. As an example, consider the case of a uniaxial crystal such as α-quartz. For $\hat{\mathbf{q}}$ along the optic axis, the two normal surfaces no longer make contact and the resulting two indices of refraction are given by

$$n_l \approx n_o + G/2n_o, \tag{26a}$$

$$n_r \approx n_o - G/2n_o, \tag{26b}$$

where $G = G_z$ and $G_x = G_y = 0$. The polarization modes corresponding to wave propagation along the optic axis at phase speeds of $v_l = c/n_l$ and $v_r = c/n_r$ are no longer linear and orthogonal, but rather left and right circular, respectively. As a result, for a linearly polarized wave propagating along the optic axis, the direction of polarization rotates by $\rho_o = \pi(n_l - n_r)/\lambda = \pi G/\lambda n_o$ per unit path length in the medium. The sign convention most often employed for ρ_o in this application has been adopted from the chemical literature and is opposite that typically chosen by physicists (see also Sec. 4.3), namely that optical rotation is clockwise-positive (looking into the source) for a dextrorotatory crystal with $n_l > n_r$, and counterclockwise-negative for a levorotatory crystal with $n_r > n_l$. The specific rotatory power for α-quartz is 188°/cm for light with $\lambda = 632.8$ nm, which corresponds to $|G|/n_o = |n_l - n_r| = 6.6 \times 10^{-5}$. Thus, one can sensitively determine very small values of the circular birefringence by measuring the rotation that the polarization direction of an incident linearly polarized, monochromatic light wave undergoes when it traverses a known thickness of the solid.

For an optically active biaxial crystal, index splitting similar to that described by Eqs. (26) is observed for wave propagation along the two optic axes, and again the corresponding polarization modes are left and right circular. In cubic crystals the same behavior is observed, irrespective of the direction of wave propagation. Wave propagation in optically active, anisotropic solids in directions other than the optic axes is complicated by the presence of both linear birefringence and optical activity. Under these circumstances, the characteristic polarization modes corresponding to the two possible phase speeds are elliptical. The change in polarization that occurs when a polarized light beam traverses the solid can be evaluated utilizing Jones calculus (Yariv and Yeh, 1984). In this approach, the incident polarization state is described by a two-component, complex column vector (Jones vector) that is operated on by a 2×2 complex matrix, characteristic of the solid (Jones matrix), to obtain the emergent polarization state. The advantage of this approach is that once the Jones matrix is known, the effect of the solid on any incident polarization state can be determined. An alternative approach is to solve the problem geometrically by the method of superposition using the Poincaré sphere (Ramachandran and Ramaseshan, 1961).

1.2.4 Absorbing, Optically Active Solids

When both absorption and optical activity are present, then the relationships between $\mathbf{D}$ and $\mathcal{E}$ take the same form as Eqs. (24) and (25), but the second-rank tensors $[\epsilon]$, $[G]$, $[a]$, and $[\gamma_o]$ defined there, as well as the vectors $\mathbf{G}$ and $\boldsymbol{\Gamma}_o$, are all complex. It is most convenient to express the wave propagation characteristics of the solid by four real tensors $[a]$, $[b]$, $[\gamma_o]$, and $[\beta]$, defined according to $[\tilde{a}] = [a] + i[b]$ and $[\tilde{\gamma}_o] = [\gamma_o] + i[\beta]$. As in the case of absorbing solids that are not optically active, $[a]$ and $[b]$ are the index and absorption tensors that characterize the index and absorption ellipsoids. The term $[\gamma_o]$ is the optical activity tensor, as in Sec. 1.2.3, and $[\beta]$ is called the tensor of circular dichroism. For a given direction of wave propagation, these latter tensors define the complex optical activity vector according to

$$\tilde{\mathbf{\Gamma}}_o = \mathbf{\Gamma}_o + i\mathcal{B} = ([\gamma_o] + i[\beta])\cdot\hat{\mathbf{q}}, \qquad (27)$$

where the real part is the optical activity vector, $\mathbf{\Gamma}_o = [\gamma_o]\cdot\hat{\mathbf{q}}$ [as in Eq. (25)] and the imaginary part is the circular dichroism vector, $\mathcal{B} = [\beta]\cdot\hat{\mathbf{q}}$ (Ramachandran and Ramaseshan, 1961). For crystal systems of orthorhombic and higher symmetries in which wave propagation occurs along the optic axes, only circular birefringence and dichroism need to be considered. In this case, the left and right circular polarization modes experience different indices of refraction and extinction coefficients.

1.3 Dispersion Relationships

The real and imaginary parts of the optical properties of solids, whether given by $\tilde{\chi}_e$, $\tilde{N}$, $\tilde{\epsilon}$, or $\tilde{\sigma}$, are not independent, but connected through dispersion relationships. These relationships arise when considering that the complex electric susceptibility, for example, is a response function that describes the polarization existing at position $\mathbf{r}$ and time t in terms of the electric field acting at all other positions and times (Wooten, 1972). In solids that are not optically active, spatial dispersion can be neglected for the purposes of this discussion because the wavelength is long compared to the atomic scale of the optical processes. As a result, for a given time t one can write

$$\tilde{\mathbf{P}}(t) = \epsilon_0 \int_{-\infty}^{\infty} \tilde{\chi}_e(t - t')\mathcal{E}(t')dt', \qquad (28)$$

which holds for optically isotropic solids. An analogous expression involving the susceptibility tensor components holds for anisotropic solids that are not optically active. (Optically active solids, in which spatial dispersion is important, have been described briefly in Sec. 1.2.) Causality requires that there can be no response $\tilde{\mathbf{P}}$ to the field $\mathcal{E}$ before it is applied; as a result, $\tilde{\chi}_e(t - t') = 0$ in Eq. (28) for $t < t'$. When Eq. (28) is Fourier transformed to complex frequency space, yielding $\tilde{\mathbf{P}}(\tilde{\omega}) = \epsilon_0\tilde{\chi}_e(\tilde{\omega})\mathcal{E}(\tilde{\omega})$, then the condition of causality leads to the dispersion relationships. For the dielectric function, these can be expressed by

$$\epsilon_1(\omega) - 1 = \frac{2}{\pi} \mathrm{P} \int_0^{\infty} \frac{\omega'\epsilon_2(\omega')}{\omega'^2 - \omega^2} d\omega', \qquad (29a)$$

$$\epsilon_2(\omega) = -\frac{2\omega}{\pi} \mathrm{P} \int_0^{\infty} \frac{\epsilon_1(\omega') - 1}{\omega'^2 - \omega^2} d\omega', \qquad (29b)$$

where P denotes the principal value of the integral (Wooten, 1972; Jackson, 1975). In deriving Eqs. (29), it has been assumed that one is interested in response functions that yield real outputs. In the case of $\tilde{\mathbf{P}} = \mathbf{D} - \epsilon_0\mathcal{E} + (i/\omega)\mathbf{J}_{\mathrm{cond}}$, which is intrinsically complex, this implies that $\mathbf{D}$, $\mathcal{E}$, and $\mathbf{J}_{\mathrm{cond}}$ are real. Equations (29), called the Kramers–Kronig relations, have important implications for measurements of optical properties that will be discussed further in Sec. 5.1.1.

2. CLASSICAL DESCRIPTION OF THE OPTICAL PROPERTIES OF SOLIDS

Next, the discussion will focus on simple classical models to explain the photon-energy dependence of the complex dielectric function $\tilde{\epsilon}$ of metals, semiconductors, and insulators as well as the relationship of $\tilde{\epsilon}$ to the fundamental microscopic behavior of electrons in these solids. The reader is also referred to the classical texts on the optical behavior of electrons listed under Further Reading. The classical optical models originate from the work of Lorentz and Drude and are useful starting points from which one can understand the observed optical properties of solids, e.g., the color and reflectivity of metals, from the standpoint of classical mechanics and electromagnetic theory. As will be seen in Sec. 3, the classical models have quantum-mechanical analogs that in fact lead to the same mathematical expressions for the dielectric functions, but with a reinterpretation of some of the terms. The Lorentz oscillator model is applicable to semiconductors and insulators; the Drude model is applicable to free-electron metals; and their quantum-mechanical analogs describe the direct interband transitions of Sec. 3.2 and the intraband transitions of Sec. 3.5, respectively.

2.1 The Lorentz Oscillator Model

For the optical properties of an optically isotropic (nonmagnetic) semiconductor or insulator, the Lorentz model starts with an equation of motion for a spherical charge distribution, having a single electron of charge $-e$ and mass m, bound to a nucleus of charge $+e$ fixed in position in the solid. This system constitutes one "atom." According to Newton's second law,

$$m\frac{d^2\mathbf{r}}{dt^2} = -e\langle\mathcal{E}_{\text{loc}}\rangle - m\omega_0^2\mathbf{r} - m\gamma\frac{d\mathbf{r}}{dt}, \tag{30}$$

where $\mathbf{r}$ is the displacement of the center of charge of the electron cloud from the nucleus located at $\mathbf{r} = 0$ (Wooten, 1972). The first term on the right is a driving force generated by the microscopic electric field, spatially averaged over the electron cloud: $\langle\mathcal{E}_{\text{loc}}\rangle = \langle\mathcal{E}_{0,\text{loc}}\rangle \exp(-i\omega t)$. The factor $\langle\mathcal{E}_{0,\text{loc}}\rangle$ is the spatial average of the amplitude of the field. The second term is a Hooke's-law restoring force with a force constant of $m\omega_0^2$, where ω_0 is the resonance frequency. The third term represents viscous damping of the motion of the electron cloud with a frictional constant γ. This provides an energy dissipation pathway that occurs in solids through interactions with neighboring atoms. In this classical model, two assumptions are made. First, the mass of the nucleus is assumed to be infinite since it forms a rigid lattice with neighboring nuclei. Second, the force $\mathbf{F}_M$ due to the magnetic induction $\mathbf{B}_{\text{loc}}$, $\mathbf{F}_M = -e\langle\mathbf{v} \times \mathbf{B}_{\text{loc}}\rangle$, is negligible since the electron speed v is small compared with the speed of light.

Upon solving Eq. (30) one arrives at a complex charge displacement $\tilde{\mathbf{r}}$, and assuming N_a identical atoms per unit volume, one obtains a complex macroscopic polarization $\tilde{\mathbf{P}}$:

$$\begin{aligned}\tilde{\mathbf{P}} &= N_a\tilde{\mathbf{p}} = -N_ae\tilde{\mathbf{r}}(t) = N_a\tilde{\alpha}_a(\omega)\langle\mathcal{E}_{\text{loc}}\rangle \\ &= \frac{N_ae^2\langle\mathcal{E}_{\text{loc}}\rangle}{m[(\omega_0^2 - \omega^2) - i\omega\gamma]} = \epsilon_0\tilde{\chi}_e(\omega)\mathcal{E}.\end{aligned} \tag{31}$$

Here $\tilde{\mathbf{p}}$ is the complex induced dipole moment for a single atom, and $\tilde{\alpha}_a$ is the complex atomic polarizability. Thus, through this model, the frequency dependence of the optical properties can be understood intuitively. The physical mechanism underlying the dissipation via the conduction electrons in Eqs. (13a), (14), and (15) and the frictional term in Eq. (31) must be the same, since they both lead to a complex $\tilde{\chi}_e(\omega)$. Considering $\omega < \omega_0$ for the moment, if $\gamma \ll (\omega_0^2 - \omega^2)/\omega$, the interaction with neighboring atoms is very weak and the electrons remain "bound." However, if γ is of the order of $(\omega_0^2 - \omega^2)/\omega$, the electrons in the Lorentz model need to be considered as "conduction electrons" that interact strongly with the neighboring atoms. This points out a problem in the separation of Eqs. (3b) and (5d): The same electrons that are bound at one frequency become conduction electrons at another frequency. For this reason, the use of single complex quantity, $\tilde{N}$, $\tilde{\epsilon}$, or $\tilde{\sigma}$, through the application of Eqs. (14) or (17) is preferred to account for this behavior when describing the frequency-dependent optical properties.

It should be pointed out that, in general, the macroscopic field in the solid is not equivalent to the spatial average of the local field over the electron cloud, i.e., $\langle\mathcal{E}_{\text{loc}}\rangle \neq \mathcal{E}$. As a result, $\epsilon_0\tilde{\chi}_e(\omega) \neq N_a\tilde{\alpha}_a(\omega)$. The reason is that the averaging to determine $\langle\mathcal{E}_{\text{loc}}\rangle$ does not include the contribution from the atom at the site of interest (which is "removed" to perform the averaging), nor does it include regions between sites. For conduction electrons, the equality $\langle\mathcal{E}_{\text{loc}}\rangle = \mathcal{E}$ is expected to be reasonably correct. For the opposite extreme where the electrons are strongly localized at the atomic sites so as to be considered point dipoles, Lorentz calculated a field of $\tilde{\mathcal{E}}_{\text{loc}}(\text{site}) = \langle\tilde{\mathcal{E}}_{\text{loc}}\rangle = \mathcal{E} + (\tilde{\mathbf{P}}/3\epsilon_0)$. It is expected that the correct local field is somewhere between these two extremes. The choice $\langle\mathcal{E}_{\text{loc}}\rangle = \mathcal{E}$ gives rise to a complex frequency-dependent dielectric function given by (Wooten, 1972)

$$\tilde{\epsilon} = 1 + \frac{e^2}{\epsilon_0 m}\sum_n \frac{N_{en}}{(\omega_n^2 - \omega^2) - i\omega\gamma_n} \tag{32}$$

with the real and imaginary parts given by

$$\epsilon_1 = 1 + \frac{e^2}{\epsilon_0 m}\sum_n \frac{N_{en}(\omega_n^2 - \omega^2)}{(\omega_n^2 - \omega^2)^2 + \omega^2\gamma_n^2}, \tag{33a}$$

$$\epsilon_2 = \frac{e^2}{\epsilon_0 m}\sum_n \frac{N_{en}\omega\gamma_n}{(\omega_n^2 - \omega^2)^2 + \omega^2\gamma_n^2}. \tag{33b}$$

These equations have been generalized to include the possibility of more than one electron per atom. Here N_{en} is the concentration of electrons having resonance frequency ω_n and frictional constant γ_n. Thus, $N_e = ZN_a = \Sigma_n N_{en}$ is the total electron concentration and Z is the number of electrons per atom. Often in using Eqs. (33) to fit experimental data, the value of unity on the right in Eq. (33a) is replaced by a static (real) dielectric constant, denoted here as ϵ_{0s}, that represents the contribution to ϵ_1 from electronic resonances with ω_n much greater than the measurable frequency range.

The alternative choice for the average local field, $\langle\tilde{\mathcal{E}}_{loc}\rangle = \mathcal{E} + \tilde{\mathbf{P}}/3\epsilon_0$, leads to a dielectric function given by (Guenther, 1990)

$$\frac{\tilde{\epsilon} - 1}{\tilde{\epsilon} + 2} = \frac{e^2}{3\epsilon_0 m} \sum_n \frac{N_{en}}{(\omega_n^2 - \omega^2) - i\omega\gamma_n}. \tag{34}$$

This expression is directly related to the Clausius–Mossotti equation. This can be seen by replacing the atomic polarizability that appears in the Clausius–Mossotti equation with the sum of the electronic polarizabilities, and by including the simple model based on Eqs. (30) and (31) for the frequency dependence of the polarizabilities. For the purposes of this discussion, the simpler expression of Eq. (32), however, suffices to demonstrate the optical behavior of insulators and semiconductors.

Figure 3 shows a schematic of the photon energy ($\hbar\omega$) dependence of the complex dielectric function for a semiconductor, calculated assuming single-electron "atoms" ($N_e = N_a$) for simplicity. In the limit of low frequencies ($\omega \ll \omega_0$) when $\gamma \lesssim \omega_0$, ϵ_1 approaches a constant, $\epsilon_1 = 1 + N_e e^2/\epsilon_0 m\omega_0^2$, and ϵ_2 vanishes. The real part of the dielectric function increases with ω, exhibiting so-called normal dispersion, until $\omega_- = \omega_0[1 - \gamma/\omega_0]^{1/2}$. Above ω_-, ϵ_1 decreases with ω, crossing unity at ω_0. This is the regime of anomalous dispersion, which extends to $\omega_+ = \omega_0[1 + \gamma/\omega_0]^{1/2}$. Above ω_+, normal dispersion resumes and ϵ_1 approaches unity asymptotically as $\epsilon_1 \sim 1 - N_e e^2/\epsilon_0 m\omega^2$ for $\omega \gg \omega_0 \gg \gamma$. When $\omega_0 \gg \gamma$, the regime of anomalous dispersion exhibits a width of γ. Under this condition, ϵ_2 increases with ω for $\omega < \omega_0$, reaches its maximum value of $\epsilon_2 = N_e e^2/\epsilon_0 m\gamma\omega_0$ at $\omega = \omega_0$, and approaches zero asymptotically as $\epsilon_2 \sim N_e e^2\gamma/\epsilon_0 m\omega^3$ for $\omega \gg \omega_0$. For $\omega_0 \gg \gamma$, the full width at half maximum of the band in ϵ_2 is also γ. In Fig. 4, the corresponding frequency dependence of the complex index of refraction is presented, and it shows features similar to those for $\tilde{\epsilon}$ in Fig. 3. Figure 5 includes both the absorption coefficient and normal-incidence reflectance. The former is calculated from the extinction coefficient and wavelength according to Eq. (12b), and the latter is calculated from the index of refraction and extinction coefficient according to the techniques of Sec. 5.1.1, assuming that the index of refraction of the ambient medium is unity.

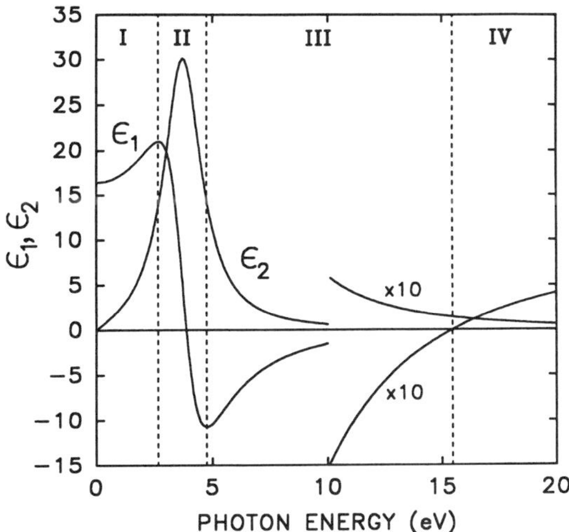

FIG. 3. Photon-energy dependence of the real (ϵ_1) and imaginary (ϵ_2) parts of the complex dielectric function ($\tilde{\epsilon}$) for a single Lorentz oscillator with resonance energy, $\hbar\omega_0 = 3.85$ eV; broadening, $\hbar\gamma = 2.00$ eV; and strength, $e^2\hbar^2 N_e/\epsilon_0 m = (15.1 \text{ eV})^2$. Normal dispersion in ϵ_1 is observed in photon-energy regimes I and III–IV, below and above the resonance energy, respectively. In regime II, anomalous dispersion is observed. The boundary between regimes III and IV is placed at the plasma energy, $\hbar\omega_p$.

In Figs. 3–5, four regimes are delineated as described by Wooten (1972). In regime I corresponding to $\omega < \omega_0$, the electrons are strongly bound, and the solid exhibits relatively weak, but increasing, absorption and low reflectance. The reflectance in this regime is lower for solids with a low electron density and high electronic force constant $m\omega_0^2$. Regime II, with $\omega_0 - \gamma/2 \leq \omega \leq \omega_0 + \gamma/2$, is characterized by weakly bound conduction electrons with a sharply increasing reflectance and absorption coefficient as a function of frequency. In regime III, $\omega > \omega_0$ and the electrons behave as if they were free

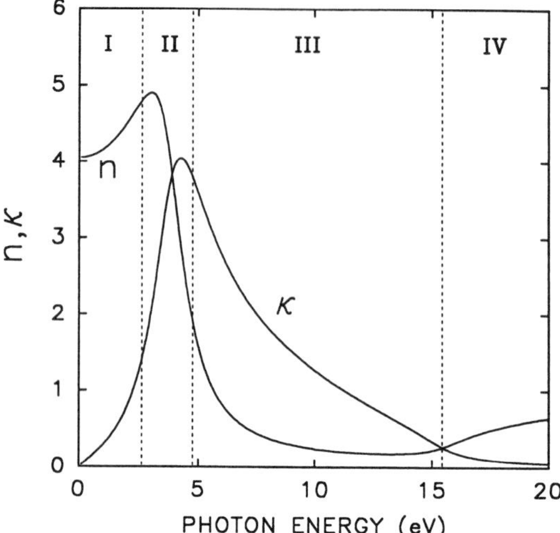

FIG. 4. Photon-energy dependence of the real (n) and imaginary (κ) parts of the complex index of refraction ($\tilde{N}$) for the Lorentz oscillator of Fig. 3. The four regimes are delineated as described in Fig. 3.

electrons in a metal [see Eqs. (35) in Sec. 2.2]. As a result, the solid has a metalliclike reflectance in this regime, but gradually decreasing absorption with increasing ω. The onset of regime IV occurs when $\epsilon_1 = 0$, which defines the plasma frequency ω_p, another characteristic of free electrons to be discussed further in Secs. 2.2 and 2.3. In regime IV, both the reflectance and absorption are relatively low and decrease to zero. If $|\omega_0^2 - \omega^2|/\omega \gg \gamma$, then the plasma frequency is given by $\omega_p^2 = \omega_0^2 + N_e e^2/\epsilon_0 m$, and ϵ_2 varies as $\epsilon_2 = N_e e^2 \gamma\omega/\epsilon_0 m(\omega^2 - \omega_0^2)^2 \rightarrow 0$ in this regime.

Figures 6–8 show the dielectric functions obtained by spectroscopic ellipsometry (see Sec. 5.1.3) for three thin-film materials, hydrogenated amorphous silicon (a-Si:H) and discontinuous elemental aluminum (Al) and silver (Ag), which exhibit single broad resonances similar to that shown in Fig. 3, and bulk crystalline gallium arsenide (GaAs), which appears to exhibit a collection of resonances. Although the dielectric function of the amorphous silicon is qualitatively similar to that of the single oscillator in Fig. 3, a good fit cannot be obtained. The reason is that solids in general are more realistically described in terms of a large collection of Lorentz oscillators enumerated by n, as in Eqs.

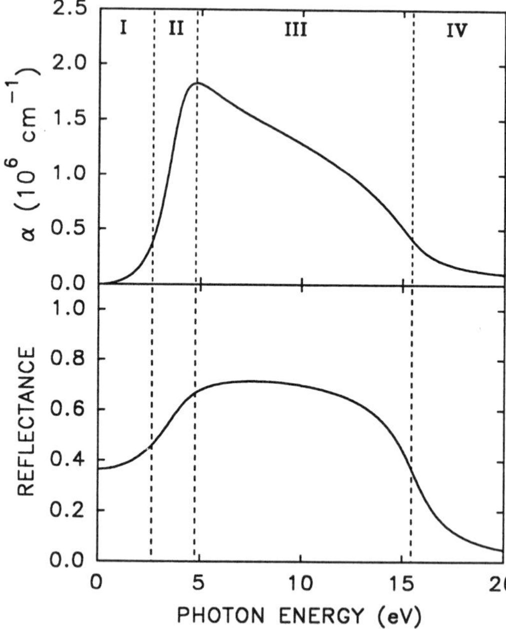

FIG. 5. Photon-energy dependence of the absorption coefficient (α, top) and the normal-incidence reflectance (bottom) for a hypothetical material whose dielectric function is given by the Lorentz oscillator of Fig. 3. The four regimes are delineated as described in Fig. 3.

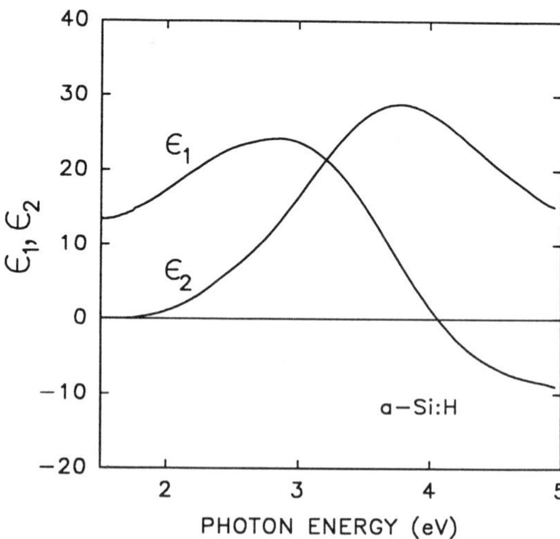

FIG. 6. Real and imaginary parts of the dielectric function of hydrogenated amorphous silicon measured at room temperature by spectroscopic ellipsometry on a thin film prepared by plasma-enhanced chemical vapor deposition. This material was prepared under conditions leading to high electronic quality for photovoltaic applications. Data from Dawson *et al.* (1992).

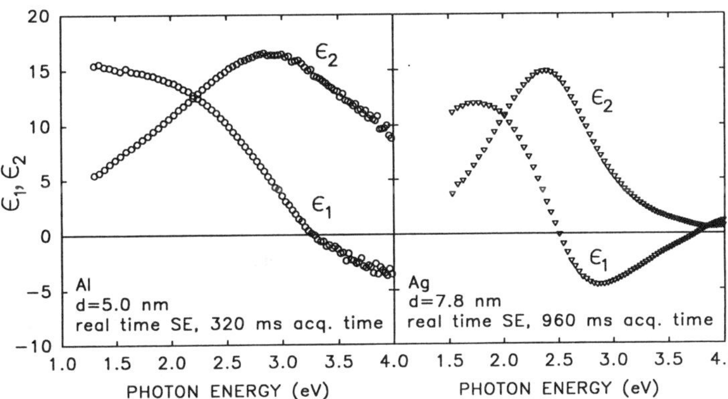

FIG. 7. Real and imaginary parts of the dielectric functions of thin films consisting of isolated aluminum (left) and silver (right) particles, measured by spectroscopic ellipsometry (SE) in real time during preparation at room temperature. The film thicknesses at the measurement times were 5.0 and 7.8 nm for the Al and Ag films, respectively. The acquisition times for the full spectra are provided. Data from Nguyen and Collins (1993).

(33), with closely spaced resonance frequencies ω_n spread over bands, and each of the individual resonances has a width γ_n and electron concentration N_{en} associated with it. In the specific case of *a*-Si:H, there is a single broad band of resonances typical of an amorphous solid, that can be understood in terms of the excitation of the valence electrons (four per Si atom) from σ bonding states across an energy gap (~1.7 eV) to unoccupied antibonding states (Zallen, 1983).

In GaAs the bands of resonances are narrower and tend to be confined to discrete frequency regions typical of a crystalline solid (see Fig. 8). For this material, a very good simulation of the data can be obtained with seven oscillators (broken line in Fig. 8), but without a quantum-mechanical description of the electronic states in the solid, the positions of these resonances cannot be explained. In fact, the resonances correspond to collections of interband transitions, i.e., the excitation of electrons from states in the filled valence band across an energy gap $\hbar\omega_n$ to states in the empty conduction band (see Secs. 3.2–3.3). Such transitions are direct or vertical. Because the magnitude of the photon momentum $\hbar q$ is small compared with the corresponding values $0 \leq \hbar k \leq (2\pi/a_c)\hbar$ (a_c: unit cell length) describing the electron in the crystal, and because only an electron is excited by the photon in a direct transition, the transition does not lead to an appreciable change in the electron wave vector **k**. Indirect transitions can also occur between bands, involving the emission or absorption of quanta of lattice vibrational energy in conjunction with the excitation of electrons by photons, but their probability is much lower (see Sec. 3.4).

The dielectric functions in Fig. 7 for films consisting of an array of discontinuous aluminum and silver particles (5.0 and 7.8 nm, respectively) are of interest because they demonstrate that other physical mechanisms can also give rise to Lorentz oscillator be-

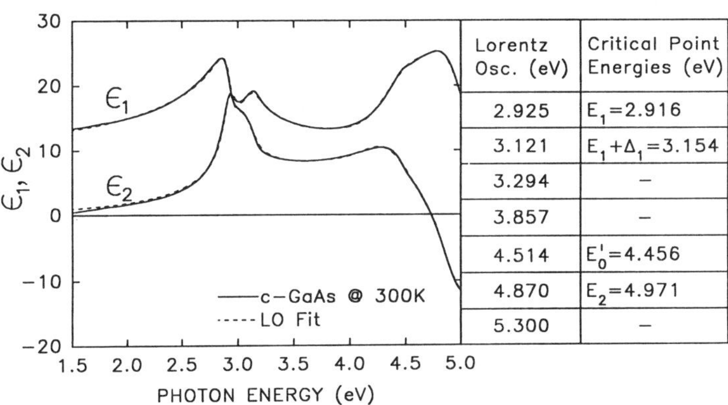

Lorentz Osc. (eV)	Critical Point Energies (eV)
2.925	E_1=2.916
3.121	$E_1+\Delta_1$=3.154
3.294	–
3.857	–
4.514	E_0'=4.456
4.870	E_2=4.971
5.300	–

FIG. 8. Real and imaginary parts of the dielectric function of single-crystal GaAs at room temperature as measured by spectroscopic ellipsometry (solid line). The broken line, nearly coinciding with the data on this scale, is a fit using seven Lorentz oscillators. The table at the right shows the best-fit resonance energies in eV, along with a comparison with the energies obtained in a critical-point analysis of the dielectric function (see Fig. 17). Lorentz oscillator energies that do not correspond to critical points serve to simulate the broad background in $\tilde{\epsilon}$. Data from Aspnes *et al.* (1986).

havior. In these cases, the resonance can be understood in terms of the forced collective oscillation of free electrons within each metal particle with the restoring force provided by the rigid array of positive nuclei. Such free-electron, or plasma, oscillations will be discussed further in Secs. 2.2 and 2.3. The Lorentz oscillator describes other physical situations as well. For example, the vibration of ions in ionic crystals and the orientation of permanent dipoles generate optical property characteristics similar to those of Figs. 3–5 in the infrared and microwave ranges of the electromagnetic spectrum.

Finally, it should be noted that the Lorentz oscillator model can also be extended to anisotropic crystals in order to model wavelength-dependent birefringence and dichroism. For example, consider the case of an orthorhombic crystal that is not optically active, and the direction of linear polarization (defined by the direction of **D**) is parallel to the *a*, *b*, or *c* crystallographic axes. In this case, the approach outlined above applies and leads to one of the three principal complex dielectric functions; however, the resonance frequencies, frictional constants, and resonance electron concentrations associated with the three different polarization directions will be different in general. A mechanical analog of an optically active crystal can also be developed. In this case, the spatial dispersion is simulated by describing the interaction of the electromagnetic wave with two coupled harmonic oscillators that lie in separate planes in the crystal (Guenther, 1990). In Sec. 4.3, the Lorentz oscillator model will be extended to magnetic materials.

2.2 The Drude Model

The Drude model for optically isotropic, nonmagnetic metals can be extracted directly from the Lorentz model by setting the restoring force equal to zero, meaning that the electrons are free to move throughout the collection of atoms (Wooten, 1972). In addition, because the free electrons are uniformly distributed, then $\langle \mathcal{E}_{loc} \rangle \approx \mathcal{E}$. Thus, setting $\omega_0 = \omega_n = 0$ and $\gamma = \gamma_n$ in Eqs. (33) one obtains

$$\epsilon_1 = 1 - \frac{N_e e^2}{\epsilon_0 m(\omega^2 + \gamma^2)}, \tag{35a}$$

$$\epsilon_2 = \frac{N_e e^2 \gamma}{\epsilon_0 m\omega(\omega^2 + \gamma^2)}. \tag{35b}$$

With this approach, however, the physical origin of the damping term in Eq. (30) is not evident.

In fact, Drude derived these results by considering the metal as a free-electron gas electrically compensated by a background of positively charged nuclei (see, for example, Ashcroft and Mermin, 1976). He assumed that the electrons move according to Newton's laws of motion between instantaneous scattering events that occur with a probability per unit time of $1/\tau$ and randomize the velocity direction. The scattering events are the same as those that give rise to Joule heating when current passes through the metal. Although Drude ascribed the scattering to electronic collisions with the positive nuclei, they are in fact due to deviations from perfect periodicity in the solid, e.g., lattice vibrations at room temperature. With such a model, the equation of motion for the average velocity **v** of an electron is

$$\frac{d\mathbf{v}}{dt} = -\frac{\mathbf{v}}{\tau} - \frac{e\mathcal{E}}{m}, \tag{36}$$

assuming no distinction between local and macroscopic fields. However, for this equation to be valid, the macroscopic field must not vary significantly over the mean free path L of the electron, i.e., $L = v\tau \ll \lambda_m$ and $L \ll 2/\alpha$ where λ_m is the wavelength in the material and $2/\alpha$ is the exponential decay length of the field. Assuming $\mathcal{E} = \mathcal{E}_0 \exp(-i\omega t)$, then a steady-state solution of the form $\tilde{\mathbf{v}} = \tilde{\mathbf{v}}_0 \times \exp(-i\omega t)$ yields the complex conduction current $\tilde{\mathbf{J}}_{cond} = -N_e e\tilde{\mathbf{v}} = 4\pi\epsilon_0\tilde{\sigma}\mathcal{E}$, where the complex optical conductivity is given by

$$\tilde{\sigma} = \sigma_1 + i\sigma_2 = \frac{N_e e^2\tau/4\pi\epsilon_0 m}{1 - i\omega\tau} = \frac{\sigma_0/4\pi\epsilon_0}{1 - i\omega\tau}. \tag{37}$$

The real and imaginary parts are

$$\sigma_1 = \frac{\sigma_0/4\pi\epsilon_0}{1 + \omega^2\tau^2}, \tag{38a}$$

$$\sigma_2 = \frac{\sigma_0\omega\tau/4\pi\epsilon_0}{1 + \omega^2\tau^2}, \tag{38b}$$

where $\sigma_0 = 4\pi\epsilon_0\tilde{\sigma}(\omega{=}0) = N_e e^2\tau/m$ denotes the dc electrical conductivity. Equations (35) with $\gamma = 1/\tau$ can be seen to be equivalent to Eqs. (38), using the relationship between the optical conductivity and dielectric function in Eqs. (16).

Figures 9–11 show the complex dielectric function and index of refraction, the absorption coefficient, and the normal-incidence reflectance calculated for two different values of τ derived using Eqs. (12b), (13d), (13e), (16a), and (37), and the techniques of Sec. 5.1.1. Marked changes in the optical characteristics in Figs. 10 and 11 occur at the plasma frequency ω_p given by $\epsilon_1(\omega_p) = 0$, where $\omega_p^2 = N_e e^2/\epsilon_0 m$ when $\omega_p\tau \gg 1$. As ω passes through ω_p, the real part of the index of refraction increases from near zero toward unity, the extinction coefficient drops to zero, and the reflectance decreases abruptly from near unity toward zero. At the plasma frequency, the

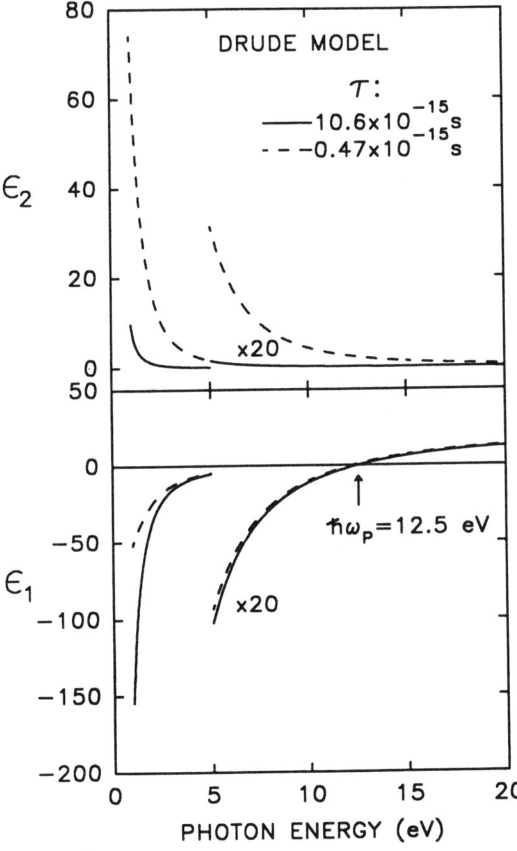

FIG. 9. Photon-energy dependence of the real and imaginary parts of the dielectric functions for hypothetical free-electron metals, according to the Drude model. In this case, the plasma energy is fixed at $\hbar\omega_p$ = 12.5 eV, and two values of the scattering time τ are assumed; 10.6×10^{-15} s (solid line) and 0.47×10^{-15} s (broken line).

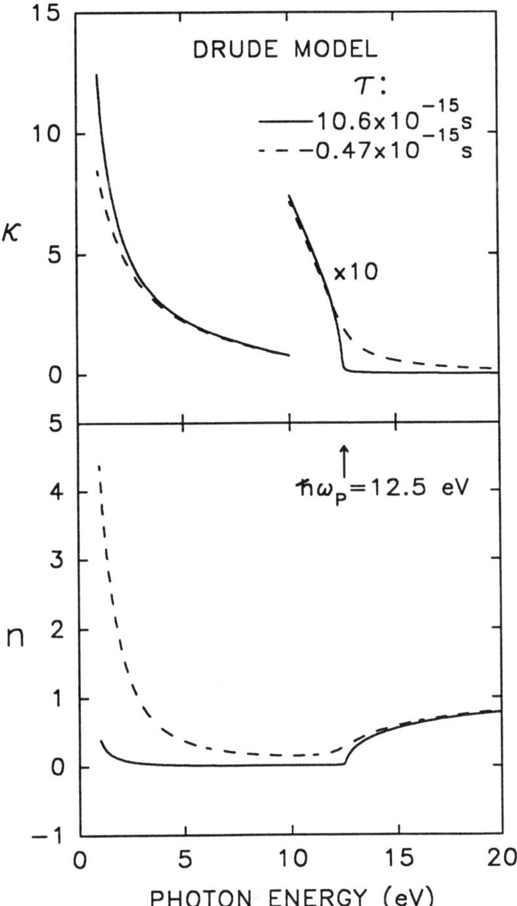

FIG. 10. Photon-energy dependence of the real and imaginary parts of the complex index of refraction, deduced from the two dielectric functions for the free-electron metals given in Fig. 9.

wavelength associated with the electric field in the metal tends to infinity, meaning that the entire electron population oscillates in phase, driven by the electric field. Such oscillations do not result in charge-density fluctuations in the bulk of the material, and thus are not to be confused with plasmons, which are oscillations of bulk charge-density fluctuations. These latter oscillations cannot be excited by transverse electromagnetic fields with $\nabla \cdot \mathcal{E} = 0$, but can be excited by the longitudinal fields associated with incident charged particles. In optically isotropic materials in which the longitudinal and transverse dielectric functions are equal, plasmons are excited at the plasma frequency given by $\omega_p^2 = N_e e^2/\epsilon_0 m$ (when $\omega\tau \gg 1$), as

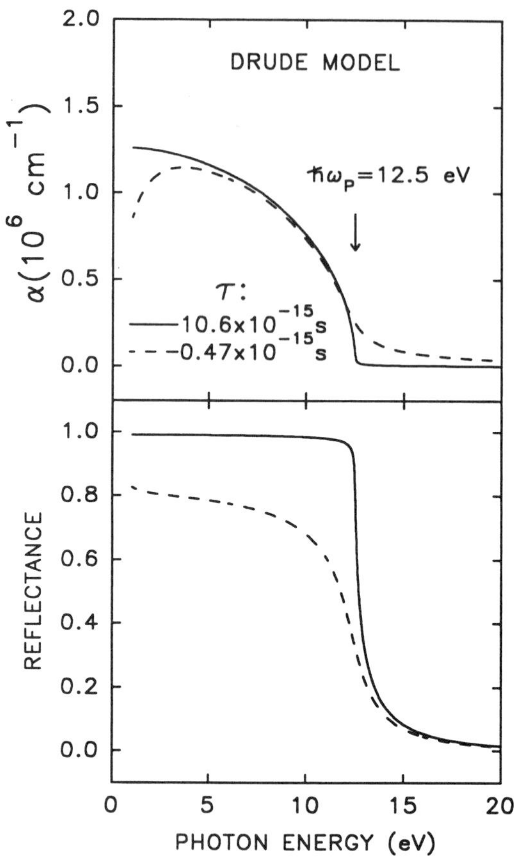

FIG. 11. Photon-energy dependence of the absorption coefficient (α, top) and normal-incidence reflectance (bottom), for the free-electron metals whose dielectric functions are given in Fig. 9.

measured by the transverse optical probe, and they persist even after the longitudinal field is removed. Further discussion of plasmons is provided in Sec. 2.3.

The experimental results for the reflectance and dielectric function of silver are shown in Figs. 12 and 13, and provide an instructive example of the complexity inherent in the optical properties of real metals (Wooten, 1972). The rapid decrease in the reflectance just below 4 eV where $\epsilon_1 = 0$ identifies the plasma energy $\hbar\omega_p$. However, this value is much lower than 9.2 eV, the value expected from the free-electron model. The increase in reflectance above 3.9 eV, along with the features in ϵ_1 and ϵ_2 in this region, suggests that there are also contributions to the optical properties due to a band of bound-electron resonances centered above 4 eV. This can be seen clearly by the broken lines in Fig. 13 that represent the decomposition of the dielectric function into contributions from free (f) and bound (b) electrons, according to $\tilde{\epsilon} = \tilde{\epsilon}_f + \tilde{\epsilon}_b$. In the decomposition, Eqs. (35) are used for the free-electron contribution with parameters obtained by fitting the data for $\tilde{\epsilon}$ below 3.75 eV. In the fit, a static dielectric constant ϵ_{0s} is included in the expression for ϵ_1 [see Eq. (35a)], rather than unity. As noted in Sec. 2.1, this accounts for the higher-energy bound transitions that occur above the fitted spectral range. The best-fit value of $\epsilon_{0s} - 1$ is then included as part of $\tilde{\epsilon}_b$ in the decomposition of Fig. 13. The bound-electron resonances near 4–5 eV arise from direct interband transitions in which electrons are excited from filled d bands to empty states in the partially filled conduction band (see inset, Fig. 12). In contrast, the free-electron contribution can be attributed to the intraband transitions in which electrons are excited from below the Fermi level to above, within the same partially filled conduction band. Such transitions occur only in metals and degenerate semiconductors. On the basis of the decomposition in Fig. 13, it is clear that the contribution to ϵ_1 from the bound electrons leads to a shift to lower energies in the frequency at which $\epsilon_1 = 0$ (i.e., the plasma frequency). This can be understood physically by a screening of the free-electron oscillations by the bound electrons in the d band.

Now, the dielectric functions of the metal particle films in Fig. 7 can be better understood from the standpoint of a simple mechanical model. For thin films consisting of nanometer-size isolated metal particles (i.e., small compared to λ), true oscillations in charge density can in fact be excited by transverse electromagnetic waves. These fluctuations build up and collapse periodically at particle surfaces. A simple model for the optical properties of metal particle films yields a Lorentz oscillator characteristic with a resonance frequency $\omega_0 = \omega_{pp}$, the plasma frequency for the particle. For a spherical particle, an expression for ω_{pp} can be determined from the equation of motion for the displacement $\mathbf{r}$ of the free-electron distribution relative to the fixed, positively charged nuclei:

$$m\frac{d^2\mathbf{r}}{dt^2} = -e\mathbf{E} + \frac{e\mathbf{P}}{3\epsilon_0} = -e\mathbf{E} - \frac{N_e e^2\mathbf{r}}{3\epsilon_0}. \tag{39}$$

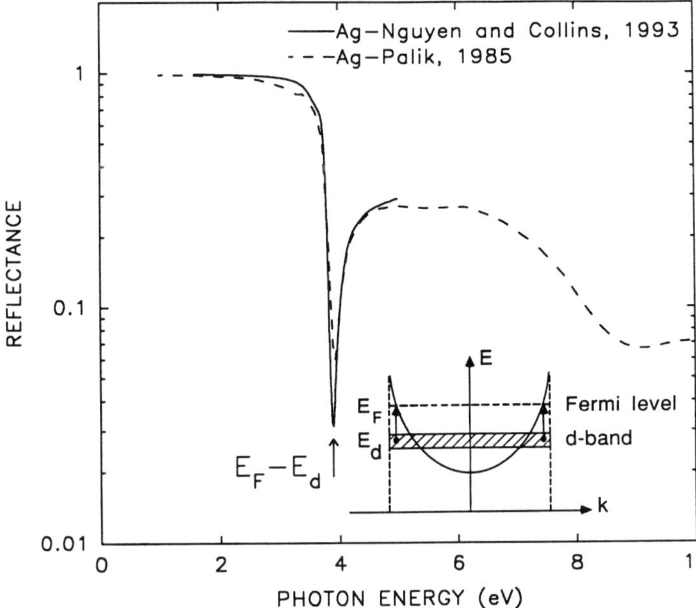

FIG. 12. Normal-incidence reflectance of silver plotted vs photon energy, including handbook data (broken line; from Palik, 1985) and results deduced from spectroscopic ellipsometry measurements on a sample with a higher-quality surface (solid line; from Nguyen and Collins, 1993). Both results reveal a sharp minimum at the (screened) plasma frequency ($\hbar\omega_p$ = 3.9 eV) as well as the onset of interband reflection associated with (*d*-band)-to-(Fermi-level) transitions. The inset shows the direct interband transitions on a simplified electron band-structure diagram.

Here $\mathcal{E} - \mathbf{P}/3\epsilon_0$ is the field within the particle, including the applied optical field and the depolarization field. This equation has the same form as Eq. (30) for an undamped Lorentz oscillator with a resonance frequency of $\omega_{pp} = (N_e e^2/3\epsilon_0 m)^{1/2} = \omega_p/\sqrt{3}$. Certainly more sophisticated optical models for metal particle films have been developed (see, for example, Yamaguchi *et al.*, 1978); however, the simple one described by Eq. (39) will suffice for the purposes of a brief, illustrative discussion.

For the aluminum film, the observed resonance energy in the dielectric function, $\hbar\omega_0 = 2.9$ eV, is much below $\hbar\omega_{pp} = 7.2$ eV, calculated from $\hbar\omega_p$ for bulk aluminum. This reduction in plasma frequency has a similar origin to that observed for bulk silver in Figs. 12 and 13. It results from screening of the fields within the particle by the action of the neighboring particles (which exist at high density on the substrate surface in the case of the aluminum film). Thus, $\hbar\omega_0$ is observed to decrease with increasing volume of aluminum in the discontinuous film. The bound electrons in aluminum associated with interband transitions near 1.5 eV (see Fig. 16) also play a role by providing a pathway for energy dissipation from the free-electron oscillations and increasing γ defined in Eq. (30) and, thus, the width of the resonance compared to that of the Ag particles.

2.3 Plasmons

Volume plasmons are collective oscillations of the electrons in a bulk solid, i.e., oscillations of fluctuations in charge density that are self-sustaining, longitudinal, and quantized with energy $\hbar\omega_p$ (Pines, 1963; Wooten, 1972; Ziman, 1972). Because volume plasmons can be present in the absence of an external field, the oscillations occur when the real and imaginary parts of the (longitudinal) dielectric function satisfy $\epsilon_1 = 0$ and $\epsilon_2 \ll 1$. Because volume plasmons are longitudinal, they can be excited by incident charged particles, but not by light (not even at ω_p). Thus, plasmons can be observed by passing monoenergetic electrons through the sample; peaks occur in the electron energy-loss spectrum at energies corresponding to integral multiples of $\hbar\omega_p$. In fact, plasmons generated by electrons exhibit a range of wave-vector magnitudes q_p up to a characteristic cutoff value, which for a free-electron gas is $q_{pc} \approx \omega_p/v_F$, where v_F is the Fermi velocity (Wooten, 1972). At this point, single-electron excitations occur, and these screen out the long-range Coulomb interaction. Weak dispersion in the plasmon energy $\hbar\omega_{qp}$ occurs as a result of the thermal motions of electrons. In optically isotropic materials in which the longitudinal and transverse dielectric functions are equal, the optical measurement identifies the

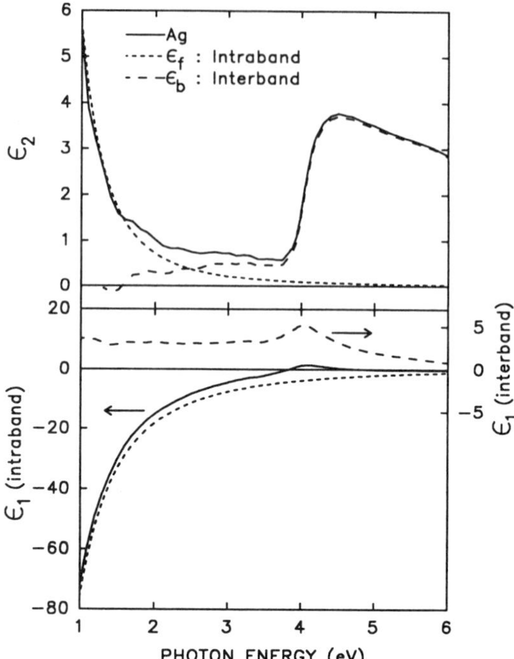

FIG. 13. Real and imaginary parts of the dielectric function of silver (solid line). The short- and long-dashed lines represent the approximate intraband and interband components of the dielectric function, respectively. The appropriate intraband parameters are obtained in a best fit of $\tilde{\epsilon}$ for $\hbar\omega < 3.75$ eV, assuming it follows the Drude relationship of Eqs. (35), but with a photon-energy–independent interband contribution ϵ_{0_s} replacing unity on the right in Eq. (35a). In the diagram, however, part of this contribution ($\epsilon_{0_s} - 1$) is included in the interband component. Data from Palik (1985).

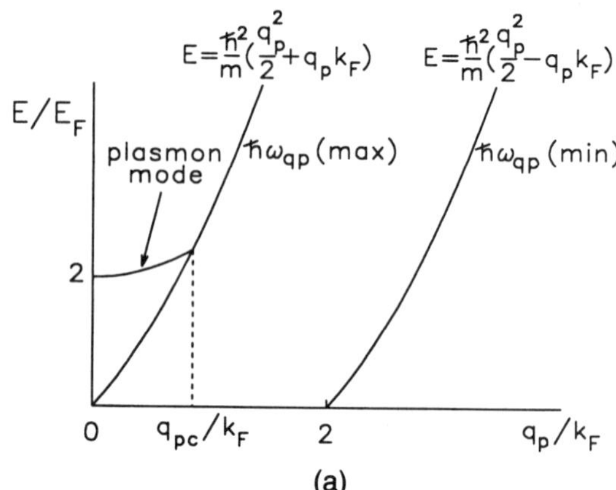

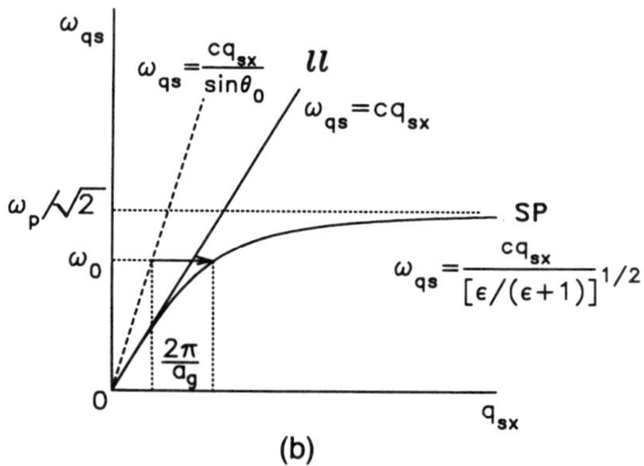

FIG. 14. (a) Dispersion relationships associated with volume excitations of a free-electron gas. The electron energy E and wave vector q_p are normalized to the Fermi energy E_F and wave vector k_F, respectively. The region between the lines designated by $\hbar\omega_{qp}$ (max) and $\hbar\omega_{qp}$ (min) represents possible one-electron excitations. Below a critical electron wave vector designated q_{pc}, a collective electron excitation or volume plasmon mode is observed. In this plot, the plasma energy $\hbar\omega_p$ is chosen to be twice the Fermi energy (after Wooten, 1972; adapted with permission from the author). (b) Dispersion relationship for surface plasmons (SP) in a free-electron gas. The photon line (//) is also shown. The horizontal arrow denotes the excitation of surface plasmons by light of frequency ω_0 using a grating coupler [see Eq. (40)]. Here ϵ is the dielectric function of the grating (real; <-1), θ_0 is the angle of incidence of the light on the grating plane, and a_g is the grating constant.

plasmon energy in the limit $q_p \to 0$. The dispersion characteristics of volume plasmons and one-electron excitations in a free-electron gas are summarized in Fig. 14(a).

Surface plasmons are collective oscillations of electrons in the plane of a solid surface that are self-sustaining, quantized with energy $\hbar\omega_s$, and localized at the surface to within the Thomas–Fermi screening length of ~0.1 nm (Otto, 1974, 1976; Raether, 1977). Assuming a self-sustaining field at the surface, Maxwell's equations yield a dispersion relationship of $q_{sx} = (\omega_{qs}/c)[\epsilon_a\tilde{\epsilon}/(\epsilon_a + \tilde{\epsilon})]^{1/2}$. Here q_{sx} and ω_{qs} are the wave-vector magnitude and frequency of the surface plasmon, assumed to be propagating along the x direction on the surface of a medium. In addition, $\tilde{\epsilon} = \epsilon_1 + i\epsilon_2$ is the dielectric function of the medium, and ϵ_a is the dielectric function of the ambient, assuming that $\epsilon_1 < -\epsilon_a$ and $\epsilon_2 \approx 0$. When $\epsilon_1 \to -\epsilon_a$, q_{sx} becomes large, and the frequency of the surface plasmon approaches $\omega_p(1 + \epsilon_a)^{-1/2}$ for a free-electron metal of volume-plasma frequency ω_p.

Figure 14(b) shows the dispersion relationship for surface plasmons, along with the straight line associated with light in vacuum having the same wave vector and frequency $\omega = \omega_{qs} = cq_{sx}$. From this plot, one can conclude that under usual circumstances light cannot simultaneously satisfy the momentum and energy conditions required to excite surface plasmons. However, one can design various geometrical configurations in which photons can be coupled to surface plasmons. In a grating coupler, for example, light with frequency ω_0 strikes a grating with grating constant a_g and dielectric function ϵ at an angle of incidence θ_0. Coupling can be achieved under the following conditions ($\epsilon_1 < -1$; $\epsilon_2 \approx 0$):

$$\frac{\omega_0 \sin\theta_0}{c} \pm \frac{2\pi\nu}{a_g} = \frac{\omega_0}{c}\left(\frac{\epsilon}{\epsilon+1}\right)^{1/2} = q_{sx}, \tag{40}$$

where ν is an integer. The coupling is shown in Fig. 14(b) by the horizontal arrow (assuming $\nu = 1$). This process is the origin of Wood's anomalies for metallic gratings (Otto, 1974). From analogous reasoning, it is clear that plasmons can be excited on rough surfaces, as well. In fact, weak bands due to surface plasmon absorption are sometimes detected in the reflectance spectra of rough metal films (Wooten, 1972).

2.4 Sum Rules

Sum rules are based on the observation that at sufficiently high frequency, all electrons behave as free electrons with $\epsilon_1(\omega) = 1 - \omega_p^2/\omega^2$ (Ehrenreich, 1966). Using this fact in the Kramers–Kronig relations, one can relate the optical properties of an optically isotropic solid to the concentration of electrons that participate in transitions induced by the electric field of a (transverse) electromagnetic wave:

$$\int_0^\infty \omega\epsilon_2(\omega)d\omega = \frac{\pi}{2}\,\omega_p^2 = \frac{\pi N_e e^2}{2\epsilon_0 m}. \tag{41}$$

It is more practical to employ a finite frequency range, and this leads to

$$\int_0^{\omega_c} \omega\epsilon_2(\omega)d\omega = \frac{\pi N_a e^2}{2\epsilon_0 m} Z_{\text{eff}}(\omega_c), \tag{42}$$

where m is the free-electron mass, N_e and N_a are the concentrations of electrons and atoms, respectively, and $Z_{\text{eff}}(\omega_c)$ is the number of electrons per atom contributing to the optical properties up to frequency ω_c. Similar sum rules have been calculated in which $\text{Im}[-1/\tilde{\epsilon}]$ replaces ϵ_2 in Eqs. (41) and (42). These latter rules are applicable when the transitions are induced by the field of a (longitudinal) electron wave (Wooten, 1972). For optically isotropic materials in the long-wavelength limit, appropriate for plasmons (see Sec. 2.3) as well as for the optical range of electromagnetic waves in Eq. (42), the two approaches are interchangeable. Other sum rules, derived from quantum-mechanical principles and applicable to the oscillator strengths of transitions in atomic and solid-state systems, will be mentioned briefly in Secs. 3.1 and 3.2.

3. QUANTUM-MECHANICAL DESCRIPTION OF THE OPTICAL PROPERTIES OF SOLIDS

In this section, a quantum-mechanical description of the dielectric functions of optically isotropic solids will be presented. First, a quantum-mechanical analog of the classical Lorentz oscillator is presented. In this part, the influence of the optical electric field on the localized electronic wave function of each atom of the solid is determined using first-order time-dependent perturbation theory. Then the dielectric function is calculated from the wave function using quantum-mechanical expressions for the complex polarization or conduction current. In the second part of this section, this expression for the dielectric function is extended to describe the more practical case of delocalized one-electron wave functions, or Bloch waves, associated with electrons in a crystalline solid. The dielectric function in this case provides information on the density of electronic states in the solid, as described in the third part of the section. In the second and third parts, only direct transitions will be considered; in the fourth part, indirect transitions are described in which the excitation of electrons by photons is accompanied by the emission or absorption of phonons. These latter, second-order processes can be observed in indirect semiconductors in the absence of direct transitions, i.e., in the range of photon energy above the indirect gap but below the lowest direct gap. To conclude this section, a quantum-mechanical description of intraband transitions involving electrons in the partially filled bands in metals will be presented. For a more extensive discussion, one can consult the texts on quantum-mechanical principles, the quantum theory of solids, and the modern theories of the optical properties of solids, provided under Further Reading.

As in the case of the Lorentz oscillator model of Sec. 2.1, the treatment of this section can be readily generalized to anisotropic crystals of orthorhombic and higher symmetries that are not optically active. In this case, the results must be expressed in the coordinate system of the principal crystallographic axes that simultaneously diagonalizes both real and imaginary parts of the complex dielectric tensor. In addition, the direction of linear polarization (i.e., the direction of **D**) must lie in one of the principal planes. Then the expression for the principal complex dielectric function is of the same form as that presented below, but the relevant quantum-mechanical matrix element and the joint density of states will depend on the polarization direction. As an example of the approach that must be taken, anisotropy is included explicitly in the expression for the intraband dielectric function, valid for monovalent metals and degenerate semiconductors (see Sec. 3.5).

3.1 Quantum Theory of Absorption and Dispersion

In performing the quantum-mechanical derivation of the dielectric function of a collection of atoms that compose an optically isotropic solid, one begins by considering a single atom with eigenstates ϕ_n given by (Wooten, 1972)

$$\mathcal{H}_0\phi_n = E_n\phi_n, \tag{43}$$

where $\mathcal{H}_0$ is the time-independent Hamiltonian, E_n is the energy of the nth eigenstate, and $n = 0$ denotes the ground state. Application of the perturbation $\mathcal{H}'(t)$ yields the time-dependent Hamiltonian $\mathcal{H} = \mathcal{H}_0 + \mathcal{H}'(t)$, and the resulting wave function ψ for the atom must be obtained from the time-dependent Schrödinger equation:

$$i\hbar\frac{\partial\psi}{\partial t} = [\mathcal{H}_0 + \mathcal{H}'(t)]\psi. \tag{44}$$

The solution can be written as

$$\psi(\mathbf{r},t) = \sum_n a_n(t)\phi_n(\mathbf{r})\exp\left(\frac{-iE_nt}{\hbar}\right), \tag{45}$$

where $a_n(t)$ is the time-dependent amplitude for the atom to be found in eigenstate n. Equation (44) needs to be solved when the perturbation consists of an electric field, $\mathcal{E}_{loc} = \mathcal{E}_{0,loc}\,\hat{\mathbf{x}}$, assumed to be linearly polarized along the x direction, given by

$$\mathcal{H}'(t) = e\mathcal{E}_{loc}\cdot\mathbf{r} = \frac{e\mathcal{E}_{0,loc}x}{2} \times \exp(\gamma_P t)\{\exp(i\omega t) + \exp(-i\omega t)\}. \tag{46}$$

In this case, the perturbation is allowed to build up gradually on the time scale of the optical frequency through the parameter γ_P. Solving Eqs. (44) and (45) for this perturbation yields (Baym, 1973)

$$a_n(t) = -\frac{e\langle\mathcal{E}_{0,loc}\rangle x_{n0}}{2\hbar}\exp(\gamma_p t)\left\{\frac{\exp[i(\omega_{n0}-\omega)t]}{\omega_{n0}-\omega-i\gamma_P} + \frac{\exp[i(\omega_{n0}+\omega)t]}{\omega_{n0}+\omega-i\gamma_P}\right\}. \tag{47}$$

Here it is assumed that the atom is in its ground state $n = 0$ for $t \to -\infty$, and that first-order perturbation theory is applicable, namely that $a_n(t)\mathcal{H}'(t)\phi_n(\mathbf{r})$ is negligible for $n \neq 0$. The $\exp(\gamma_P t)$ term controls the buildup of amplitude and, hence, the lifetime associated with the electron transition; if γ_P is reduced, then the lifetime increases. In addition, $\omega_{n0} = \omega_n - \omega_0$, where $\omega_n = E_n/\hbar$, $\omega_0 = E_0/\hbar$, and x_{n0} is the dipole matrix element given by

$$x_{n0} = \int_{\Delta V_A}\phi_n^*(\mathbf{r})x\phi_0(\mathbf{r})d^3\mathbf{r}. \tag{48}$$

The integral in this expression is taken over the atomic volume ΔV_A. The induced complex polarization is

$$\tilde{\mathbf{P}} = \epsilon_0(\tilde{\epsilon}-1)\mathcal{E}, \tag{49a}$$

$$= N_a\tilde{\mathbf{p}} = -N_a e\hat{\mathbf{x}}\int_{\Delta V_A}\psi^*(\mathbf{r},t)x\psi(\mathbf{r},t)d^3\mathbf{r}, \tag{49b}$$

where $N_a = N_e/Z$ is the concentration of atoms (Wooten, 1972). By applying Eqs. (45) and (47) to Eqs. (49), setting $\mathcal{E} = \langle \mathcal{E}_{\text{loc}} \rangle$, and assuming $\gamma_P t \to 0$, one finds

$$\tilde{\epsilon} = 1 + \frac{N_a e^2}{2\epsilon_0 m} \sum_n \frac{f_{n0}}{\omega_{n0}} \left(\frac{1}{\omega_{n0} + \omega + i\gamma_P} + \frac{1}{\omega_{n0} - \omega - i\gamma_P} \right), \tag{50}$$

where f_{n0} is the oscillator strength defined by

$$f_{n0} = (2m\omega_{n0}/\hbar)|x_{n0}|^2. \tag{51}$$

Equation (50) has the same form as Eq. (32) for the Lorentz oscillator as can be seen by neglecting terms in γ_P^2 in the former, and making the correspondences $N_a f_{n0} \to N_{en}$, $\omega_{n0} \to \omega_n$, and $2\gamma_P \to \gamma_n$. Thus, the concentration of electrons bound with resonance frequency ω_n in Eqs. (32) and (33) simulates the oscillator strength, the resonance energy simulates the energy difference between ground and excited states, and the frictional constant describes the lifetime of the transition.

One can obtain a second description of the dielectric function in terms of the momentum matrix element by starting with a perturbation of the form $\mathcal{H}'(t) = e\mathcal{E}_{\text{loc}} \cdot \mathbf{r} = (e/m)\mathbf{A}_{\text{loc}} \cdot \mathbf{p} = -(i\hbar e/m)\mathbf{A}_{\text{loc}} \cdot \nabla$, where $\mathbf{p}$ is the momentum operator and $\mathbf{A}_{\text{loc}}$ is the local vector potential corresponding to $\mathcal{E}_{\text{loc}}$, in the Coulomb gauge ($\nabla \cdot \mathbf{A} = 0$). (This expression is valid to first order in A_{loc}.) Then the derivation of $\tilde{\epsilon}$ is carried out using the expression (Cardona, 1969; Baym, 1973)

$$\tilde{\mathbf{J}}_{\text{cond}} = N_a \tilde{\mathbf{j}}_{\text{cond}} = i\omega\epsilon_0(1 - \tilde{\epsilon})\mathcal{E}, \tag{52a}$$

$$= -\frac{N_a e}{2m} \int_{\Delta V_A} [\psi^*(\mathbf{r},t)\mathbf{p}\psi(\mathbf{r},t) - \psi(\mathbf{r},t)\mathbf{p}\psi^*(\mathbf{r},t)] d^3\mathbf{r} - \frac{N_a e^2 \mathbf{A}}{m} \int_{\Delta V_A} \psi^*(\mathbf{r},t)\psi(\mathbf{r},t) d^3\mathbf{r}, \tag{52b}$$

where $\tilde{\mathbf{j}}_{\text{cond}}$ is the conduction-current contribution of a single atom. The first and second terms in Eq. (52b) are the paramagnetic and diamagnetic currents, respectively. In the derivation for the response of a system of atoms that interact only weakly (i.e., only via the local field), the diamagnetic current can be ignored; this is not the case for solids, in general. It turns out that the approach described by Eq. (52b) is very useful for determining the dielectric function associated with interband and intraband transitions in solids (see Secs. 3.2 and 3.5).

Alternatively, to obtain the second description of the dielectric function directly, the commutation relation $[p_x, x] = -i\hbar$ can be applied to yield $|p_{x,n0}|^2 = (m\omega_{n0})^2|x_{n0}|^2$ and an equivalent expression for the oscillator strength of Eq. (51):

$$f_{n0} = 2|p_{x,n0}|^2/m\hbar\omega_{n0}, \tag{53}$$

where the polarization direction is again assumed to be along $\hat{\mathbf{x}}$ (Wooten, 1972). Here $p_{x,n0}$ is the momentum matrix element given by the x component of the vector

$$\mathbf{p}_{n0} = -i\hbar \int_{\Delta V_A} \phi_n^*(\mathbf{r}) \nabla \phi_0(\mathbf{r}) d^3\mathbf{r}. \tag{54}$$

For atoms with Z electrons, the general f-sum rule given by $\Sigma_n f_{nm} = Z$ is obeyed, where f_{nm} represents the oscillator strength for electronic transitions from state m to state n.

Another useful quantum-mechanical approach for describing the optical absorption is derived from an expression for the absorption coefficient associated with electronic transitions from state m to n, given by (Wooten, 1972)

$$\alpha_{nm} = \frac{\hbar\omega_{nm} W_{nm}}{I} = \frac{2\hbar\omega_{nm} W_{nm}}{n\epsilon_0 c \mathcal{E}_0^2}. \tag{55}$$

Here $\omega_{nm} = (E_n - E_m)/\hbar$ and $W_{nm} = \Omega^{-1} (d|a_{nm}|^2/dt)$ is the transition rate per excitation volume Ω. Thus, the numerator is the rate of energy absorption per volume, and the denominator is the irradiance in the beam. The quantity $|a_{nm}|^2$ represents the probability of finding an electron in state ϕ_n, having made a transition from state ϕ_m. This can be obtained by generalizing Eqs. (45) and (47). Also in Eq. (55), n and $\mathcal{E}_0$ are the real index of refraction and the amplitude of the optical electric field, respectively.

In real solids, the valence electrons are delocalized so that they cannot all make transitions between the same pair of localized atomiclike states as was assumed for the $N_e = ZN_a$ electrons that determine the fundamental dielectric response of Eq. (50). Thus, in order to extend the above approaches beyond a collection of weakly interacting atoms to a real solid, the one-electron model is usually employed. In this model, the influence of the electric field or vector potential on the many-electron wave function is obtained by determining the response of a single electron in a self-consistent scalar potential. Local field corrections (i.e., $\langle \mathcal{E}_{\mathrm{loc}} \rangle \neq \mathcal{E}$; not considered here) are needed when the self-consistent potential depends on the external field. The fact that the one-electron energy states in solids are spread out over bands means that there will always be a continuum of initial and final states.

In Sec. 3.2, crystalline solids will be considered in which the one-electron wave functions are Bloch functions and one must distinguish between direct and indirect interband transitions. Here the case of an amorphous semiconductor will be considered to demonstrate how the continuum of states enters into the calculation of the optical properties (Cody, 1984). In this case, the lack of periodicity implies that crystal momentum **k** is not a conserved quantity. In such a case, one can write an expression for the absorption coefficient as a function of the photon energy associated with the wave:

$$\alpha(\hbar\omega) = 2\Omega^2 \int \rho_v(E_m) \times \left[\int \alpha(E_n, E_m, \hbar\omega)\rho_c(E_n)dE_n \right] dE_m, \tag{56}$$

where $\alpha(E_n,E_m,\hbar\omega) = \alpha_{nm}$ is defined through Eq. (55). In Eq. (56), $E_n = \hbar\omega_n$, $E_m = \hbar\omega_m$ so that $E_n - E_m = \hbar\omega_{nm}$. In addition, $\rho_v(E_m)dE_m$ and $\rho_c(E_n)dE_n$ are respectively the number of single-spin valence and conduction-band states per volume having energies between E_m and $E_m + dE_m$ and between E_n and $E_n + dE_n$. The factor of 2 accounts for the enhancement in absorption due to the fact that there are two spin subbands, in general. Next, Fermi's "golden rule" for the transition rate per volume can be exploited:

$$\begin{aligned} W_{nm} &= \frac{1}{\Omega}\frac{d|\alpha_{nm}|^2}{dt} \\ &= \frac{2\pi}{\hbar\Omega}|V_{nm}|^2\,\delta(E_n - E_m - \hbar\omega) \\ &= \frac{\pi e^2}{2\hbar\Omega}\mathcal{E}_0^2|x_{nm}|^2\,\delta(E_n - E_m - \hbar\omega), \end{aligned} \tag{57}$$

where V_{nm} is the matrix element of the perturbation (Baym, 1973). Using this expression in Eqs. (55) and (56), one arrives at a useful result for the absorption coefficient:

$$\frac{\alpha(\hbar\omega)n(\hbar\omega)}{\hbar\omega} = \frac{2\pi e^2\Omega}{\epsilon_0 c\hbar}\int_{E_{0c}-\hbar\omega}^{E_{0v}} |x_{nm}|^2 \times \rho_v(E_m)\rho_c(E_m + \hbar\omega)dE_m, \tag{58}$$

where $n(\hbar\omega)$ is the frequency-dependent index of refraction. In Eq. (58), E_{0v} and E_{0c} denote the energies of the highest (occupied) state of the valence band and the lowest (unoccupied) state of the conduction band, respectively. Thus, $E_g = E_{0c} - E_{0v}$ is the band gap of the amorphous semiconductor, which in tetrahedrally bonded solids arises from the energy separation between σ bonding and antibonding levels.

For example, if one assumes that the dipole matrix element is independent of the energy of the initial and final states, and that the densities of states ρ_v and ρ_c increase parabolically into the bands, separated by the gap E_g, then one arrives at the simple expression

$$[\alpha(E)n(E)/E]^{1/2} = C(E - E_g), \tag{59}$$

where $E = \hbar\omega$ and C is a constant. One can derive an alternative expression to Eq. (59) by assuming instead that the momentum matrix element $|p_{x,nm}|^2$ given by Eq. (54) is independent of the initial and final states. The result in this case is

$$[\alpha(E)n(E)E]^{1/2} = C'(E - E_g). \tag{60}$$

This latter expression, called the Tauc law, has been used widely to determine the optical band gap for tetrahedrally bonded amorphous semiconductor thin films such as *a*-Si:H and hydrogenated amorphous germanium (*a*-Ge:H) from reflectance and transmittance measurements over a relatively narrow range of E (Tauc *et al.*, 1966). However, when higher–photon-energy data (for example, from spectroscopic ellipsometry) are employed in the case of *a*-Si:H, it has been found that the relationship of Eq. (59) provides a better overall fit (Cody, 1984). Figure 15 shows this effect for the dielectric function of *a*-Si:H given previously in Fig. 6. Note that the two gap values differ by 0.24

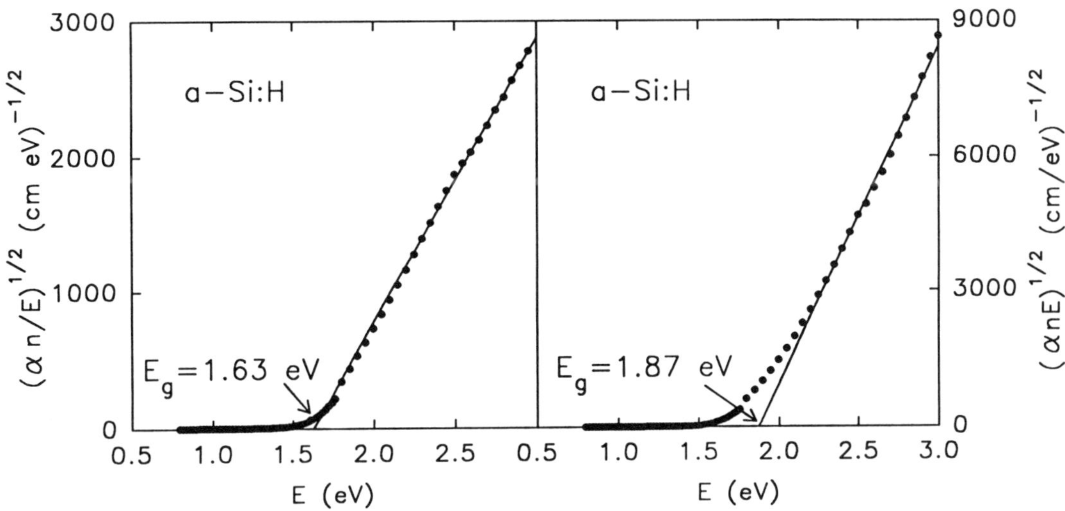

FIG. 15. Optical spectra for hydrogenated amorphous silicon of Fig. 6, plotted as $(\alpha n/\hbar\omega)^{1/2}$ vs photon energy (left) and as $(\alpha n\hbar\omega)^{1/2}$ vs photon energy $E = \hbar\omega$ (right). For parabolic band densities of states and constant dipole (left) or momentum (right) matrix elements, the plot should be linear and extrapolate to the band gap at $\alpha = 0$. Data from Dawson *et al.* (1992).

eV, using the range of photon energy of 2.2–3.0 eV in the linear extrapolation. Because uncertainties exist, not only in the appropriate matrix element but also in the exact densities of states that should be used for different materials, many workers have resorted to an empirical definition of the band gap given by the photon energy at which $\alpha = 10^3$ cm^{-1}, which tends to lie between the two gap values in Fig. 15.

3.2 Direct Interband Transitions

By applying the approach associated with Eq. (52b) in Sec. 3.1, one can derive the following expression for the dielectric function of an optically isotropic crystalline solid:

$$\tilde{\epsilon}(E) = 1 + \frac{e^2\hbar^2}{\epsilon_0 m^2 E^2} \sum_{\mathbf{k},c,v} |\hat{\mathbf{e}} \cdot \mathbf{p}_{cv}(\mathbf{k})|^2 \times \left(\frac{1}{E_{cv}(\mathbf{k}) - E - i\Gamma} + \frac{1}{E_{cv}(\mathbf{k}) + E + i\Gamma} \right), \quad (61)$$

including electronic transitions between valence-band (v) and conduction-band (c) states characterized by the electron wave vector ($\mathbf{k}$), which is conserved in the transition as elucidated below. Here $E = \hbar\omega$, $\Gamma = \hbar\gamma_P$, and $E_{cv}(\mathbf{k}) = E_c(\mathbf{k}) - E_v(\mathbf{k})$ is the energy difference between conduction- and valence-band states (Aspnes, 1980a). The second term in parentheses in Eq. (61) is nonresonant and describes the low-energy behavior of the dielectric function. In fact, the form of this expression is identical to that of Eq. (50) for a collection of weakly interacting atoms. However, the oscillator strength is now expressed in terms of the momentum matrix element $\mathbf{p}_{cv}(\mathbf{k})$ for transitions between valence- and conduction-band states, and the polarization vector $\hat{\mathbf{e}}$ of the electric field is now included explicitly.

What makes Eq. (61) different from its atomic analog is the interpretation of the summation and the matrix element, the latter providing the selection rules for the transition. In deriving Eq. (50) for an atomic system, one assumes that the transitions occur from a single ground state, identical for each atom. In Eq. (61), however, one must view the transitions as occurring from a set of one-electron Bloch states in *completely occupied* valence bands (v). Thus, in deriving Eq. (61), one uses an expression analogous to Eq. (52b) in which

1. the atomic wave function is replaced by a Bloch function; and
2. the factor of N_a is replaced by a summation over separate terms, each one connecting an occupied Bloch state with a state in another band.

In fact, for transitions between pairs of occupied states in different bands (designated ν_1 and ν_2), the (ν_1,ν_2) term will cancel the (ν_2,ν_1) term, so that one need not be concerned with the exclusion principle. For simplicity, however, the sum in Eq. (61) can be interpreted as being over pairs of occupied valence-band (v) and empty conduction-band (c) electronic states. Finally, it should be noted that Eq. (61) arises solely from the paramagnetic conduction current, which is the first term in the expression analogous to Eq. (52b); the diamagnetic current vanishes in summing all possible transitions from completely filled bands (see Sec. 3.5).

The matrix element in Eq. (61) is defined by Eq. (54), but with wave functions $\phi_\upsilon(\mathbf{r},k_\upsilon)$ given by the Bloch functions, i.e., $\phi_\upsilon(\mathbf{r},k_\upsilon) = \Omega^{-1/2}u_{\mathbf{k}\upsilon}(\mathbf{r})\exp(i\mathbf{k}_\upsilon\cdot\mathbf{r})$, where $\upsilon = (c,v)$ (Wooten, 1972):

$$\mathbf{p}_{cv} = -\frac{i\hbar}{\Omega}\int_\Omega [u^*_{\mathbf{k}c}(\mathbf{r})\exp(-i\mathbf{k}_c\cdot\mathbf{r})]\nabla[u_{\mathbf{k}v}(\mathbf{r}) \times \exp(i\mathbf{k}_v\cdot\mathbf{r})]d^3\mathbf{r}, \tag{62a}$$

$$= -\frac{i\hbar}{\Omega}\left\{\sum_l \exp[i(\mathbf{k}_v - \mathbf{k}_c)\cdot\mathbf{R}_l]\right\} \times \int_{\text{cell}} u^*_{\mathbf{k}c}\nabla u_{\mathbf{k}v} d^3\mathbf{r}. \tag{62b}$$

The integrals in Eqs. (62a) and (62b) are over the excited volume Ω of the crystal, and a unit cell of the crystal, respectively. Here $\mathbf{k}_v$ and $\mathbf{k}_c$ are the electron wave vectors for the valence- and conduction-band states, and $u_{\mathbf{k}v}$ and $u_{\mathbf{k}c}$ are the parts of the Bloch functions having the periodicity of the lattice, i.e., $u_{\mathbf{k}\upsilon}(\mathbf{r}) = u_{\mathbf{k}\upsilon}(\mathbf{r} + \mathbf{R}_l)$, where $\mathbf{R}_l$ is a lattice vector. The summation in Eq. (62b) vanishes unless $\mathbf{k}_v - \mathbf{k}_c = \mathbf{K}$, where $\mathbf{K}$ is a reciprocal lattice vector (in which case it equals the number of unit cells in the crystal). In the reduced zone scheme, $\mathbf{K} = 0$ and $\mathbf{k}_v = \mathbf{k}_c$, so that the transition can be described by a single value of $\mathbf{k} \equiv \mathbf{k}_v = \mathbf{k}_c$. Such transitions are called direct, interband transitions. If the integral in Eq. (62b) vanishes, then the direct transitions are forbidden; otherwise they are allowed. For forbidden transitions, it is no longer valid to neglect the spatial dependence $[\exp(i\tilde{\mathbf{q}}\cdot\mathbf{r})]$ of the vector potential, as has been done in the derivations of $\tilde{\epsilon}$ up to this point.

Since $\mathbf{k}$ is conserved in the allowed transitions, then one can convert the summation in Eq. (61) to an integral over k_x, k_y, and k_z values in the first Brillouin zone. These integration variables can be converted to the energy E' and two wave-vector variables k_1 and k_2 that sweep out the points on a constant-energy surface, S, defined by $E_{cv}(\mathbf{k}) = E'$. With this transformation, one obtains the well-known expression (Aspnes, 1980a)

$$\tilde{\epsilon}(E) = 1 + \frac{e^2\hbar^2}{4\pi^3\epsilon_0 m^2E^2} \times \int dE'\int_S dk_1dk_2\frac{|\hat{\mathbf{e}}\cdot\mathbf{p}_{cv}(\mathbf{k})|^2}{|\nabla_\mathbf{k}E_{cv}(\mathbf{k})|} \times \left(\frac{1}{E_{cv}(\mathbf{k}) - E' - i\Gamma} + \frac{1}{E_{cv}(\mathbf{k}) + E' + i\Gamma}\right). \tag{63}$$

If one assumes that the contribution to the first term of the integrand is dominated by $E' = E = \hbar\omega$, then an expression for $\epsilon_2(E)$ can be calculated:

$$\epsilon_2(E) = \frac{e^2\hbar^2}{4\pi^2\epsilon_0 m^2E^2}\int_S dk_1dk_2\frac{|\hat{\mathbf{e}}\cdot\mathbf{p}_{cv}(\mathbf{k})|^2}{|\nabla_\mathbf{k}E_{cv}(\mathbf{k})|}, \tag{64}$$

where the integration is now over the surface with an energy difference given by the specific value of the photon energy $E_{cv}(\mathbf{k}) = E$.

Equation (64) is useful in calculating absorption spectra (Wooten, 1972). For example, in the case of a semiconductor with parabolic valence and conduction bands that are both isotropic in $\mathbf{k}$ space, one has

$$E_{cv}(\mathbf{k}) = E_g + \hbar^2k^2/2\mu_{cv}, \tag{65a}$$

where

$$\mu_{cv}^{-1} = m_e^{-1} + m_h^{-1}. \tag{65b}$$

Here μ_{cv}, m_e, and m_h are the reduced effective mass and the electron and hole effective masses, respectively. If one assumes that the momentum matrix element is nearly constant over the range of $\mathbf{k}$ of interest, and defines a mean square value of $\langle p_{cv}^2\rangle = (1/3)|\hat{\mathbf{e}}\cdot\mathbf{p}_{cv}|^2$, then the absorption spectrum is given by

$$\alpha(E)n(E)E = \frac{3(2\mu_{cv})^{3/2}e^2\langle p_{cv}^2\rangle}{2\pi\epsilon_0\hbar^2m^2c}(E - E_g)^{1/2}. \tag{66}$$

If the oscillator strength [from Eq. (53)] is set

at unity, the reduced effective mass at $m/2$, and the index of refraction n at 4, then an estimate of $\alpha = 2 \times 10^4\ \mathrm{cm}^{-1}$ is obtained for $E - E_g = 0.01$ eV. This is in reasonable agreement with experimental results for Ge, for example. Although Eq. (66) is designed to be used in such applications near the absorption onset for the lowest-energy direct transitions in semiconductors, inclusion of the effects of many-body interactions is required here for a complete understanding of the shape of the experimental absorption spectra. Specifically, a Coulomb interaction can bind the excited electron and the hole left behind in the valence band, forming an exciton (see EXCITONS).

An approach based on Eq. (63) can be undertaken for materials in which sufficient band-structure information is available. The interband transitions in aluminum will be discussed as an example. In the visible-photon energy range, the imaginary part of the dielectric function of aluminum is dominated by the transitions between nearly parallel bands that occur in **k**-space planes parallel to the square (200) Brillouin zone faces (Ashcroft and Mermin, 1976). Assuming that all other interband transitions are negligible, one can deduce a theoretical expression for the interband dielectric function using a nearly-free-electron model in which the first-order perturbation due to the periodic potential accounts for the splitting between the two bands (Ashcroft and Sturm, 1971). Figure 16 shows the measured imaginary part of the dielectric function for aluminum films with thicknesses of 10.6 and 75.0 nm along with fits by a theoretical expression that includes not only the (200) parallel-band contribution, but also the much weaker intraband contribution [see Eqs. (35) and (75)]. Good agreement is obtained with the interband transition energies, E_{PB}, and broadening parameters, $\Gamma_{PB} = \hbar/2\tau_{PB}$, provided in the figure. Here τ_{PB} is the interband relaxation time.

The 70% larger broadening for the thinner film in Fig. 16 is attributed to a smaller grain size. This leads to a reduction in excited-state lifetime due to scattering at grain boundaries. In fact, when grain-boundary scattering dominates, the broadening parameter for direct interband transitions in nanocrystalline and microcrystalline materials can provide information on the grain size L_g through

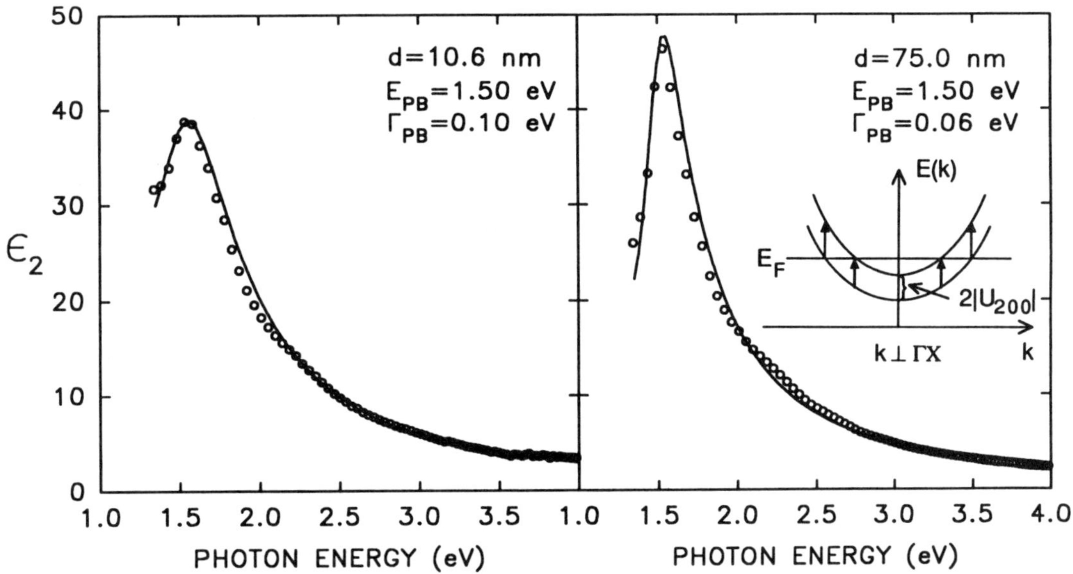

FIG. 16. Imaginary parts of the dielectric function for aluminum films 10.6 nm (left) and 75.0 nm (right) thick along with the best fit by an analytical formula including both interband and intraband contributions. The interband contribution can be described by a nearly-free-electron model and is dominated by transitions between bands that are parallel in planes perpendicular to the ΓX line [and thus parallel to the square (200) Brillouin-zone face (see inset)]. Free parameters in the fit include the interband (or parallel-band) relaxation time $\Gamma_{PB} = \hbar/2\tau_{PB}$ and the interband energy gap $E_{PB} = |2U_{200}|$, where U_{200} is the Fourier component of the crystal potential for the (200) reciprocal lattice vector. Data from Nguyen *et al.* (1993).

a relationship of the form $\Gamma = \Gamma_0 + C_g/L_g$ (Feng and Zallen, 1989). Here Γ_0 is the broadening parameter in the bulk material and $C_g \sim \hbar v$, where v is the electron speed in the excited state. Such an expression is based on the uncertainty principle $\Gamma \sim \Delta E \sim \hbar/\Delta t$, and the assumed additivity of independent broadening processes.

3.3 Critical Points and Band Structure

The dielectric function of Eq. (63) exhibits relatively sharp structure when $\nabla_{\mathbf{k}} E_{cv}(\mathbf{k}) = 0$, and points in **k** space where this condition is met are called critical points or van Hove singularities. At the photon energy corresponding to a critical point, the primary contribution to the dielectric function in Eq. (63) arises from the region of **k** space centered around the critical point, and this leads to the structure in the dielectric function. In the vicinity of a semiconductor critical point, one can replace the normally complicated variation $E_{cv}(\mathbf{k})$ in Eq. (63) with a local parabolic expansion:

$$E_{cv}(\mathbf{k}) = E_g + \frac{\hbar^2}{2}\left(\frac{k_x^2}{\mu_{xx}} + \frac{k_y^2}{\mu_{yy}} + \frac{k_z^2}{\mu_{zz}}\right), \qquad (67)$$

where **k** is expressed in a coordinate system with the origin at the critical point, oriented so as to diagonalize the inverse reduced effective-mass tensor $[\mu]^{-1}$. By substituting this equation into the integral of Eq. (63), neglecting the nonresonant term, and assuming that the matrix element is only a weak function of **k** (and can be brought outside the integral), one finds (Aspnes, 1980a)

$$\tilde{\epsilon}(E) = \frac{Q}{E^2} i^{n_c+1} U_x K_{cy} K_{cz} \times (E - E_g + i\Gamma)^{-1/2} \quad \text{for 1D}, \qquad (68a)$$

$$\tilde{\epsilon}(E) = \frac{Q}{E^2} i^{n_c+2} U_x U_y K_{cz} \times \ln (E - E_g + i\Gamma) \quad \text{for 2D}, \qquad (68b)$$

$$\tilde{\epsilon}(E) = \frac{2\pi Q}{E^2} i^{n_c+1} U_x U_y U_z \times (E - E_g + i\Gamma)^{1/2} \quad \text{for 3D}. \qquad (68c)$$

Here expressions are given for one-, two-, and three-dimensional critical points for which two, one, or none of the effective masses μ_{jj} are infinite, respectively. In addition in Eqs. (68), one has $Q = (e^2\hbar^2/4\pi^2\epsilon_0 m^2)|\hat{\mathbf{e}} \cdot \mathbf{p}_{cv}(\mathbf{k})|^2$; $U_j = (2|\mu_{jj}|\hbar^2)^{1/2}$, $(j = x,y,z)$; and n_c is the critical-point type, designated by the number of effective masses that are negative ($n_c = 0$: minimum; $n_c = 1$: saddle point; $n_c = 2$: saddle point; or $n_c = 3$: maximum). K_{cj} $(j = y,z)$ are the cutoff lengths in the Brillouin zone that are imposed to restrict the limits of integration in Eq. (63) along the z direction for a two-dimensional critical point ($\mu_{zz} \to \infty$) and along the y and z directions for a one-dimensional critical point (μ_{yy}, $\mu_{zz} \to \infty$). In fact, Eqs. (68) are only valid for calculating difference or derivative spectra in which the weaker, smoothly varying background in $\tilde{\epsilon}$ from other regions of **k** space completely disappears. A general form can be written:

$$\tilde{\epsilon}(E) = A\Gamma^{-p} \exp(i\varphi)(E - E_g + i\Gamma)^p, \qquad (69)$$

where A is an amplitude factor, φ is the phase projection factor, and p is the exponent. The latter takes on the values of $-\frac{1}{2}$, 0 (logarithmic), and $\frac{1}{2}$ for 1D, 2D, and 3D critical points, respectively. As an aside, one notes that even the critical-point line shape for discrete excitons can be included, corresponding to $p = -1$, with the phase factor interpreted as a coupling parameter between the exciton and the overlapping interband continuum.

As an example of the use of Eq. (69), Fig. 17 shows the second derivative of the real and imaginary parts of the experimental dielectric function of GaAs at room temperature (points), along with the fits using Eq. (69) (lines). The two prominent lower-energy critical points, E_1 and $E_1 + \Delta_1$, were fitted with excitonic line shapes, while the two prominent higher-energy transitions, E_0' and E_2, were fitted with 2D line shapes. The corresponding transitions are identified in Fig. 18 on the band-structure diagram for GaAs (Cohen and Chelikowsky, 1982). Figure 8 includes the best-fit critical-point energies for comparison with the Lorentz oscillator resonance energies. From this comparison, one can determine which oscillators are actually fitting critical points and which are simulating the broad background in the dielectric function. Because three of the oscillators do not correspond to critical points, one finds that such a model simply provides a good parametrization of the optical properties; however, it lacks a physical basis. An im-

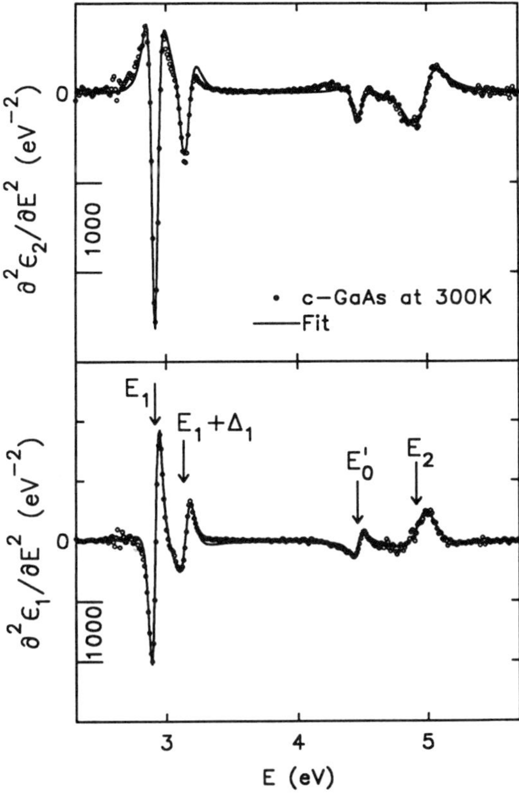

FIG. 17. Second-derivative spectra of the real (bottom) and imaginary (top) parts of the complex dielectric function plotted vs photon energy E for single crystal GaAs, measured at room temperature by spectroscopic ellipsometry (points). Also shown are fits to the experimental results (solid lines) using an excitonic line shape for the two lower-energy transitions (E_1 and $E_1 + \Delta_1$) and a 2D line shape for the two higher-energy transitions (E_0' and E_2) [see Eq. (69)]. The best-fit critical-point energies obtained from these fits are included in Fig. 8. Data from Aspnes *et al.* (1986).

proved model based on Eq. (63) has been developed recently to fit the dielectric function of GaAs and related III-V semiconductors (Kim *et al.*, 1992). Because this model is based on the band structure of GaAs, it is physically correct and, unlike the oscillator model, can simultaneously fit $\tilde{\epsilon}$ and its multiple derivatives.

One can relate the critical points more closely to the band structure by noting that the joint density of states, $J_{cv}(E)$, defined as the number of single-spin electronic states (i.e., not including the factor of 2 for spin) per volume per energy interval, separated by an energy E is given by (Wooten, 1972)

$$J_{cv}(E) = \frac{1}{8\pi^3}\int_S \frac{dk_1dk_2}{|\nabla_{\mathbf{k}}E_{cv}(\mathbf{k})|}. \tag{70}$$

Here the integration is performed over the surface in $\mathbf{k}$ space with $E_{cv} = E$. Thus, if one neglects the $\mathbf{k}$ dependence of the matrix element, one finds by comparing Eqs. (64) and (70) that J_{cv} is directly proportional to ϵ_2. Table 2 includes the functional form of the joint density of states in the neighborhood of one-, two-, and three-dimensional critical points of different types having allowed transitions as calculated from Eq. (70) (Cardona, 1969). For transitions that are forbidden at the critical point, but allowed on moving away from the point, a factor of $|E - E_g|$ is added to the integrand in Eq. (70). This arises from a first-order expansion of the momentum matrix element about the critical point.

3.4 Indirect Interband Transitions

The interactions of an electron with a crystalline array of atoms are taken into account, within the one-electron model, through the introduction of the electron crystal momentum $\hbar\mathbf{k}$, the effective mass, and the one-electron band structure of allowed energy states vs $\mathbf{k}$. Then the interaction of an electromagnetic wave with an electron in a perfect lattice leads to direct transitions between electron states in which $\mathbf{k}$ is conserved [see Eq. (61)]. If one considers interactions of the electron with imperfections of the lattice, then $\mathbf{k}$ conservation in electron transitions is relaxed, and this gives rise to indirect transitions (Pankove, 1971). Quantized lattice vibrations, or phonons, constitute an important source of imperfections and are unavoidable at finite temperature even in a perfect crystal. Furthermore, even at 0 K the electron undergoing a transition can interact with the lattice to generate a phonon. Specifically, transitions between an initial electron state $\phi_v(\mathbf{r},\mathbf{k}_v)$ and a final state $\phi_c(\mathbf{r},\mathbf{k}_c)$ with $\mathbf{k}_v \neq \mathbf{k}_c$ can occur if a phonon of wave vector $\mathbf{k}_p = \pm(\mathbf{k}_c - \mathbf{k}_v) + \mathbf{K}$ is created (−) or destroyed (+). Here $\mathbf{K}$ is a reciprocal-lattice vector.

The fact that such processes are second-order, involving perturbations described by the electron-photon and electron-phonon coupling Hamiltonians, means that the resulting transitions will be observed experi-

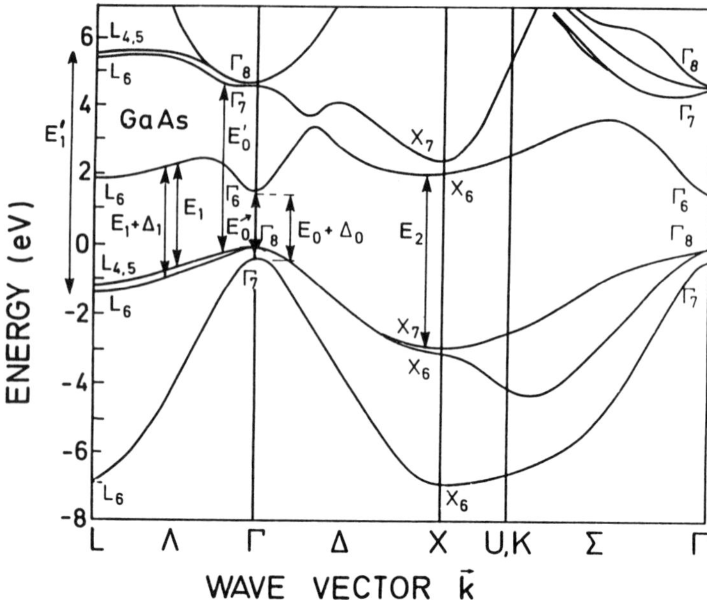

FIG. 18. Band diagram of electron energy vs wave vector for GaAs showing the direct transitions at the critical points that correspond to the features in Figs. 8 and 17. Reprinted with permission from Lautenschlager *et al.* (1987). Copyright 1987 The American Physical Society.

mentally only when no direct processes occur at the same photon energy. Thus, the practical situation of interest is one that occurs in an indirect-gap semiconductor when the photon energy $\hbar\omega$ is below the minimum direct gap (so that no direct transitions can occur), but above the minimum indirect gap. To describe the situation mathematically, one must resort to second-order perturbation theory. It is sufficient here to consider only

1. transitions from a single valence-band maximum at $\mathbf{k} = 0$ to a single conduction-band minimum away from the zone center, and
2. single-phonon absorption or emission.

Because momentum conservation is required in each step of the indirect process, the transition can occur as depicted in Fig. 19. The transition rate per unit volume is given by (Wooten, 1972)

$$W_{cv} = \frac{\pi e^2 A_0^2}{2\Omega m^2 \hbar^2} \times \left(\frac{|V_{ca}^{\text{phon}}|^2 |\hat{\mathbf{e}} \cdot \mathbf{p}_{av}|^2}{(E_{av} - E)^2} + \frac{|V_{bv}^{\text{phon}}|^2 |\hat{\mathbf{e}} \cdot \mathbf{p}_{cb}|^2}{(E - E_{cb})^2} \right) \times \{\delta(E_{cv} - \hbar\omega_{p\mathbf{k}} - \hbar\omega) + \delta(E_{cv} + \hbar\omega_{p\mathbf{k}} - \hbar\omega)\}, \tag{71}$$

where A_0 is the magnitude of the vector potential. In this expression, the matrix-element terms represent the two possible transitions via intermediate conduction-band state a and valence-band state b. In these terms, $|V_{ca}^{\text{phon}}|^2$ and $|V_{bv}^{\text{phon}}|^2$ are the matrix elements of the electron-phonon coupling Hamiltonian, which, unlike electron-photon coupling, must include spatial dispersion. With the initial state of the electron characterized by $\mathbf{k}_v = 0$, then both these matrix elements vanish unless $\mathbf{k}_p = \pm\mathbf{k}_c + \mathbf{K}$. The delta-function terms in Eq. (71) describe energy conservation corresponding to absorption ($-\hbar\omega_{p\mathbf{k}}$) and emission ($+\hbar\omega_{p\mathbf{k}}$) of phonons, where $E_{cv} = E_g$ is the indirect gap. Normally, one of the two matrix-element terms will dominate on account of the squared energy-difference factors in the denominator. These factors arise because the transition rates for the two different pathways increase as the square of the lifetimes in the intermediate states, which according to the uncertainty principle are given by $\hbar/\tau_a = E_{av} - E$ and $\hbar/\tau_b = E_{cb} - E$, for states a and b. Here $E_{av} = E_a - E_v$, and $E_{cb} = E_c - E_b$, where E_a and E_b are the electron energies for the two states.

In practice, transitions according to Eq. (71) occur between two distributions of states near the valence-band maximum and the conduction-band minimum with a joint density of states $J_{cv}(E') = \rho_v(E')\rho_c(E' + \hbar\omega \pm \hbar\omega_{p\mathbf{k}})$, where ρ_v and ρ_c are typically parabolic functions. Then by integrating over all possible

Table 2. The form of the joint density of electronic states J_{cv} versus photon energy E [Eq. (70)] for allowed transitions in the neighborhood of band-structure critical points of different types. The dimensionality of the critical point is determined by the number of principal components of the reduced effective-mass tensor μ_{jj} ($j = x,y,z$) that are infinite. The type of the critical point is determined by the number of components of the diagonalized effective-mass tensor that are negative. Weakly varying background contributions to the joint density of states are assumed in each case. In the expression for $J_{cv}(E)$, c_{mn} and c'_{mn} ($m = 0,1,2,3$; $n = 1,2,3$) are constants independent of photon energy.

						$J_{cv}(E)$	
Dim.	Type		μ_{xx}	μ_{yy}	μ_{zz}	$E < E_g$	$E > E_g$
3D	M_0	minimum	+	+	+	c_{03}	$c_{03} + c'_{03}(E - E_g)^{1/2}$
3D	M_1	saddle	+	+	−	$c_{13} - c'_{13}(E_g - E)^{1/2}$	c_{13}
3D	M_2	saddle	+	−	−	c_{23}	$c_{23} - c'_{23}(E - E_g)^{1/2}$
3D	M_3	maximum	−	−	−	$c_{33} + c'_{33}(E_g - E)^{1/2}$	c_{33}
2D	—	minimum	+	+	∞	c_{02}	$c'_{02} > c_{02}$
2D	—	saddle	+	−	∞	$c_{12} - c'_{12}\ln(E_g - E)$	$c_{12} - c''_{12}\ln(E - E_g)$
2D	—	maximum	−	−	∞	c_{32}	$c'_{32} < c_{32}$
1D	—	minimum	+	∞	∞	c_{01}	$c_{01} + c'_{01}(E - E_g)^{-1/2}$
1D	—	maximum	−	∞	∞	$c_{31} + c'_{31}(E_g - E)^{-1/2}$	c_{31}

transitions in an approach similar to that of Eqs. (56)–(60), one can arrive at the following practical formulas [assuming the first matrix-element term in Eq. (71) dominates]:

$$\alpha(E)n(E)E(E_{av} - E)^2 = \frac{C}{\exp(\beta_{\mathbf{k}}) - 1} \times (E + \hbar\omega_{p\mathbf{k}} - E_g)^2, \quad E > E_{cv} - \hbar\omega_{p\mathbf{k}}, \tag{72a}$$

$$\alpha(E)n(E)E(E_{av} - E)^2 = \frac{C'\exp(\beta_{\mathbf{k}})}{\exp(\beta_{\mathbf{k}}) - 1} \times (E - \hbar\omega_{p\mathbf{k}} - E_g)^2, \quad E > E_{cv} + \hbar\omega_{p\mathbf{k}}, \tag{72b}$$

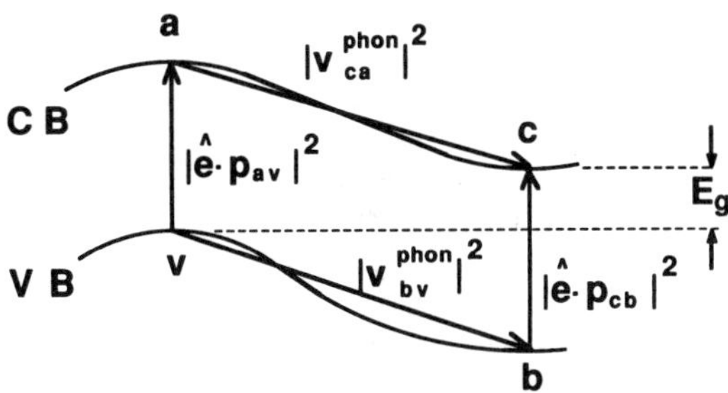

FIG. 19. Semiconductor band diagram showing two possible pathways for the indirect excitation of an electron from the valence-band maximum (*v*) to the conduction-band minimum (*c*). In one case, the electron is excited in a virtual direct transition to *a*, followed by phonon-assisted scattering to *c*. In the other case, an electron is excited in a direct transition from *b* to *c*, and the hole left behind is scattered to *v*. The matrix elements for the transitions from Eq. (71) are included.

where $E = \hbar\omega$ is the photon energy, C and C' are constants, and $\beta_{\mathbf{k}} \equiv \hbar\omega_{p\mathbf{k}}/k_BT$ (Cardona, 1969). For photon energies below the specified ranges, $\alpha = 0$. Equation (72a) is appropriate for phonon absorption and the temperature (T)-dependent term arises from the phonon concentration N_p, which has been extracted from the phonon matrix element. Similarly, Eq. (72b) is appropriate for phonon emission with probability $N_p + 1$. Over a sufficiently narrow range, one can neglect all photon-energy–dependent terms on the left except for α in applying these formulas to identify indirect transitions. Finally, some qualifications are required before accepting Eqs. (72) as a good description of reality. First, in deriving Eqs. (72), one assumes that the momentum matrix element connecting the valence- and conduction-band states near $\mathbf{k} = 0$ is independent of energy. This is not valid when the dominant direct transition here is forbidden. A derivation in this case leads to analogous equations to those of Eqs. (72) but with cubic dependences on $E \pm \hbar\omega_{p\mathbf{k}} - E_g$. Second, when excitons are formed, the results for α near the absorption onset of the indirect gap given by Eqs. (72) are modified, as in the case of the direct transitions.

3.5 Intraband Transitions

In the final part of this section, intraband transitions will be briefly considered from the standpoint of the quantum theory of solids. For the purposes of this discussion, it is assumed that the intraband contribution is large where the interband contribution is nondispersive. This means that the photon energy must be less than the minimum interband transition energy. This situation holds for monovalent metals and degenerate semiconductors. The dielectric function can be extracted by applying the approach associated with Eq. (52b). This leads to the following result:

$$\tilde{\epsilon}_{jj} = (\epsilon_{0s})_{jj} - \frac{e^2/4\pi^3\epsilon_0\hbar^2\omega}{\omega + i(2\Gamma_{mj}/\hbar)} \int \frac{\partial^2 E(\mathbf{k})}{\partial k_j^2} d^3\mathbf{k}, \tag{73}$$

where $j = x,y,z$ (Cardona, 1969). Note first that a more general expression is used here, appropriate for an optically anisotropic solid. Thus, the vector $\mathbf{k}$ is measured in the coordinate system that diagonalizes the real and imaginary parts of the complex dielectric tensor; so orthorhombic or higher symmetries are assumed. The first term on the right in Eq. (73), $(\epsilon_{0s})_{jj}$, originates from the interband term of Eq. (61) when $E = \hbar\omega \ll E_{cv}(\mathbf{k})$, for all $\mathbf{k}$. The energy-dependent, second term originates from the diamagnetic current [see Eq. (52b)], which is nonzero for metals and degenerate semiconductors characterized by partially filled bands. In this term, Γ_{mj} defines the lifetime, given by $\hbar/2\Gamma_{mj}$. The lifetime is introduced by assuming an exponentially increasing magnitude of the vector potential. The integral at the right (performed over all occupied states in $\mathbf{k}$ space) arises from the $\mathbf{k}\cdot\mathbf{p}$ sum rule (Cardona, 1969). This sum rule is derived from a comparison of the second-order terms in a Taylor-series expansion of the electron energy about a point in $\mathbf{k}$ space with the corresponding second-order terms in a perturbation-theory analysis (Wooten, 1972; Ashcroft and Mermin, 1976). The sum rule allows one to eliminate the electron concentration N_e from the diamagnetic current. In fact, since this integral vanishes for completely filled bands, the second term does not appear in Eq. (61).

It is of interest to evaluate Eq. (73) for the practical situation of a cubic crystal in which there is a single set of equivalent, parabolic conduction-band minima, so that in the vicinity of each, one has

$$E(\mathbf{k}) = \frac{\hbar^2}{2}\left(\frac{k_x^2}{m_{e,xx}} + \frac{k_y^2}{m_{e,yy}} + \frac{k_z^2}{m_{e,zz}}\right). \tag{74}$$

With a cubic crystal, the coordinate system for $\mathbf{k}$ has now been fixed in turn at each of the conduction-band minima so as to coincide with the principal directions of the inverse effective-mass tensor. Substituting Eq. (74) into Eq. (73) yields

$$\tilde{\epsilon}(\omega) = \epsilon_{0s} - \frac{N_{ec}e^2/\epsilon_0 m_o\omega}{\omega + i(2\Gamma_m/\hbar)}, \tag{75}$$

where N_{ec} is the electron concentration in the partially filled conduction band and m_o is often called the optical effective mass, given by $m_o^{-1} = (1/3)(m_{e,xx}^{-1} + m_{e,yy}^{-1} + m_{e,zz}^{-1})$. This expression is similar in form to those of Eqs. (35), derived classically. In these classical formulas, if one were to replace the unity term in ϵ_1 with ϵ_{0s}, the free-electron mass m with m_o, the electron concentration N_e with N_{ec}, and the scattering probability τ with $\hbar/2\Gamma_m$, then the quantum-mechanically derived expression would result. If $\tilde{\epsilon}(\omega)$ can be obtained from reflectance or ellipsometry measurements and N_{ec} from Hall-effect measurements, for example, then Eq. (75) can be employed to deduce ϵ_{0s}, m_o, and Γ_m (Pines, 1963).

4. INDUCED OPTICAL EFFECTS

External perturbations that affect the optical properties of any solid can assume the form of a scalar (e.g., temperature or hydrostatic pressure), a vector (e.g., electric field or magnetic field), or a second-rank tensor (e.g., stress or strain). The effect of these external perturbations is to alter the optical parameters that characterize the propagation of light in the solid. The optical parameters employed to quantify these induced effects are the components of the index, absorption, optical activity, and circular dichroism tensors. In order to demonstrate the basic approach, the emphasis in the following will be on the induced changes in the index tensor for nonabsorbing solids that are not optically active. However, the procedures described here are applicable to the other three tensors, as well.

The effect of an external perturbation is to alter each component of the index tensor a_{ij} [see discussion after Eq. (20)] by an amount Δa_{ij}. In general, this leads to changes in both the magnitudes and directions of the principal axes of the index ellipsoid. As a first-order approximation, it is assumed for the purposes of this discussion that the changes induced by the external perturbation are, for a scalar S, proportional to its magnitude and, for a vector $\mathbf{V}$ or tensor $[T]$, a homogeneous, linear function of its components. Using the Einstein notation, i.e., summation over repeated indices, one can write (Ramachandran and Ramaseshan, 1961)

$$\Delta a_{ij} = c_{ij}S \quad (S\text{: scalar}), \tag{76a}$$

$$\Delta a_{ij} = c_{ijk}V_k \quad (V_k\text{: vector}), \tag{76b}$$

$$\Delta a_{ij} = c_{ijkl}T_{kl} \quad (T_{kl}\text{: second-rank tensor}). \tag{76c}$$

From a knowledge of the properties of the tensor $[\Delta a]$, as well as those of S, $\mathbf{V}$, and $[T]$, one can deduce the properties of the tensor $[c]$ in each of Eqs. (76a)–(76c) (Nye, 1976). For example, if $[a]$ is a symmetric tensor of rank two, as in the case of a nonabsorbing crystal that is not optically active, then $[\Delta a]$ will also be a symmetric tensor of rank two. In this situation, the components c_{ij} of the tensor of rank two in Eq. (76a) form a 3×3 symmetric matrix having six independent elements, and the components c_{ijk} in Eq. (76b) denote a tensor of rank three that is symmetric in i and j. Hence, the third-rank tensor $[c]$ can be contracted to a 6×3 matrix representation having 18 independent elements. Finally, for a nonabsorbing crystal that is not optically active, the components c_{ijkl} in Eq. (76c) denote a tensor of rank four that is symmetric in the pair of indices (i,j) as well as in the pair (k,l). It should be noted, however, that this tensor is not symmetric with respect to an interchange of (i,j) and (k,l), since each pair of indices refers to distinctly different aspects of the medium, e.g., optical and elastic properties, respectively. Thus, this tensor can be contracted to a 6×6 matrix representation, having 36 independent elements. In each of these cases, the number of independent components of $[c]$ is reduced if the crystal possesses some symmetry.

If quadratic or higher-order terms in the perturbation are involved, then Eqs. (76) can be modified to include them. Although the quadratic electro-optic effect will be described briefly, such nonlinear effects will not be discussed in detail here since they are treated elsewhere in these volumes. Other nonlinear induced effects such as the photorefractive effect are also of great practical importance. The photorefractive effect is generated by a nonuniform charge redistribution in the solid induced by nonuniform illumination, which causes a semipermanent change in the index of refraction via the linear or quadratic electro-optic effects (see Sec. 4.2). These photoinduced index variations can in fact be erased by flooding the solid with uniform illumination. Further details on this specialized topic and its applications in optical communications, optical computing, and phase conjugation can be found in recent texts (Gunter and Huignard, 1988, 1989). Finally, it should be noted that the changes in the optical properties of solids resulting from phase transitions induced by the external perturbation are also not considered here.

In general, the variation in the optical properties induced by the various perturbations, as described briefly below, can depend on

1. the material and its temperature;
2. the frequency, propagation vector, and amplitude of the optical wave; and
3. the frequency, direction, and amplitude of the applied perturbation.

For example, as the optical photon energy approaches a critical-point energy in the electronic density of states of the solid, the effect of the perturbation on the optical properties increases dramatically. This is the principle underlying the various experimental techniques of modulation spectroscopy, in which the applied perturbation is modulated and the induced modulation of the optical properties is measured vs photon energy (Cardona, 1969). In addition, in ferromagnetic or ferroelectric materials, the effect of the perturbation is strongly enhanced as the Curie or phase-transition temperature is approached. Although operation of the material in a device at such critical points may be impractical in some cases, by a judicious choice of the material, optical wave, and perturbation, it has been possible to develop numerous electro-, magneto-, and acousto-optical devices for a range of industrial applications (Levi, 1980; Yariv and Yeh, 1984; Guenther, 1990). The discovery and perfection of new materials, and the invention of new devices incorporating them, is currently an active area of research and development for physicists, chemists, materials scientists, and optical and electrical engineers.

4.1 Scalar Perturbations: Temperature and Hydrostatic Pressure

As shown in Table 1, the index tensor $[a]$ for all crystals except those of the two lowest-symmetry systems (i.e., monoclinic and triclinic) can be fully described by *at most* three parameters, namely the three principal indices of refraction. Thus, the tensor $[\Delta a]$ describing the effect of the scalar perturbations of temperature or hydrostatic pressure is characterized by a change in the magnitudes of the principal indices of refraction of $[a]$, but without any change in the orientation of the principal axes. For example, a perturbation consisting of a temperature rise ΔT leads to components of tensor $[c]$ of the form

$$c_{jj} = \frac{\Delta a_{jj}}{\Delta T} = -\frac{2}{n_j^3}\frac{dn_j}{dT}, \tag{77}$$

which is diagonal within the principal-axis coordinate system (Ramachandran and Ramaseshan, 1961). Here use has been made of the fact that $\Delta a_{jj} = \Delta(n^{-2})_{jj}$ (Yariv and Yeh, 1984). For similarly perturbed monoclinic crystals, one of the principal axes will continue to coincide with the unique axis, but the other two may change their orientations while remaining in the symmetry plane. For triclinic crystals, on the other hand, a change in temperature (or application of a hydrostatic pressure) can alter the orientations of all three principal axes as well as the magnitudes of the principal indices of refraction. The major contribution to the observed thermo-optic effect in nonabsorbing crystals in general arises from the change in volume due to thermal expansion. Near the absorption onset, however, changes in the electron–phonon interactions dominate the changes in the index and absorption tensors. Further information on the effects of pressure and temperature on the optical properties of solids can be found in various review articles (see, for example, Paul, 1966; Thomas, 1991).

4.2 Electro-optic Effects

When a solid is exposed to an externally applied electric field $\mathcal{E}'$ (the prime is used to distinguish this field from the optical field $\mathcal{E}$), its electron distribution is modified, and this in turn alters its optical properties. For crystals lacking a center of symmetry, the applied electric field produces also a slight deformation of the solid due to the piezoelectric effect, which will be discussed briefly in Sec. 4.4. The net effect of the field $\mathcal{E}'$ is to change the index tensor, and this is generally termed the electro-optic effect. It is important to distinguish between the linear and quadratic electro-optic effects (see Table 3). The linear effect (also called the Pockels effect) occurs only in crystals lacking a center of symmetry, i.e., the same group of crystals that exhibit the piezoelectric effect. The quadratic effect (also called the electro-optic Kerr effect) occurs in all solids, but because it is

Table 3. A summary of electro-optic and magneto-optic effects, along with the approximate order of magnitude of the linear or circular birefringence for typical materials (SI units for the applied electric field and magnetic induction). The Cotton–Mouton effect is observed in solids, but is typically 10^3 times smaller than the Faraday effect. As a result, the former is of limited technological importance, and thus has not been discussed here.

Field/light direction	Order	Description of effect	Name of effect (order-of-magnitude strength Δn in terms of applied fields $\mathcal{E}_0'$ and B_0'): Electro-optic	Magneto-optic
∥	1st	Direction of polarization rotated upon transmission	—	Faraday ($10^{-6}\ B_0'$)
∥	1st	Direction of polarization rotated upon reflection	—	Kerr, magneto-optic
⊥	1st	Birefringence induced	Pockels ($10^{-11}\ \mathcal{E}_0'$)	—
⊥	2nd	Birefringence induced	Kerr, electro-optic $[10^{-19}\ (\mathcal{E}_0')^2]$	Cotton–Mouton $[10^{-9}\ (B_0')^2]$

small, the effect is usually neglected in the noncentrosymmetric crystals because of the dominance of the linear effect. In centrosymmetric solids, the linear effect vanishes and the quadratic effect then becomes dominant. Thus, even in solids that are optically isotropic in their natural state, the quadratic electro-optic effect generates birefringence. Although this discussion is restricted to nonabsorbing crystals, it should be pointed out that the electric field influences the absorption tensor as well. In particular, the shift of the absorption onset toward lower photon energies under the influence of the field is called the Franz–Keldysh effect (Aspnes, 1980a).

The electro-optic coefficients are defined according to

$$\begin{aligned}\Delta a_{ij} &= \Delta(n^{-2})_{ij} = a_{ij}(\mathcal{E}') - a_{ij}(0)\\ &= r_{ijk}\mathcal{E}'_k + R_{ijkl}\mathcal{E}'_k\mathcal{E}'_l, \qquad (78a)\\ &= a_{ij}(\mathbf{P}') - a_{ij}(0)\\ &= f_{ijk}P'_k + F_{ijkl}P'_kP'_l, \qquad (78b)\end{aligned}$$

in the Einstein notation. In Eq. (78), $\mathcal{E}'_k$ and $\mathcal{E}'_l$ are the vector components of the applied electric field $\mathcal{E}'$, and P'_k and P'_l are the components of the polarization $\mathbf{P}'$ (Yariv and Yeh, 1984). The third-rank tensor components r_{ijk} and f_{ijk} are the linear electro-optic (or Pockels) coefficients, and the fourth-rank tensor components R_{ijkl} and F_{ijkl} are the quadratic electro-optic (or Kerr) coefficients. The field and polarization formulations for the coefficients are related according to

$$f_{ijk} = r_{ijk}/\epsilon_0(\epsilon_{kk} - 1), \qquad (78c)$$

$$F_{ijkl} = R_{ijkl}/\epsilon_0^2(\epsilon_{kk} - 1)(\epsilon_{ll} - 1), \qquad (78d)$$

where ϵ_{kk} and ϵ_{ll} are the principal dielectric functions.

A contracted matrix notation is used for these third- and fourth-rank symmetric tensors in order to simplify manipulations. In this notation, for the tensor component r_{ijk}, the two indices (i,j), [or equivalently (j,i)] are replaced by a single index according to the scheme

$$11 \rightarrow 1,\quad 22 \rightarrow 2,\quad 33 \rightarrow 3,\quad 23 \rightarrow 4,\quad 31 \rightarrow 5,\quad 12 \rightarrow 6,$$

while the index k remains the same in the matrix and tensor notations. For example, using these contracted indices, we can write $r_{jjk} = r_{jk}$, for $j = 1,2,3$ and $k = 1,2,3$; $r_{23k} = r_{32k} = r_{4k}$, $r_{31k} = r_{13k} = r_{5k}$, and $r_{12k} = r_{21k} = r_{6k}$, for $k = 1,2,3$. Similarly, the tensor components of R_{ijkl} can also be written in contracted notation. For R_{ijkl}, both (i,j) [or (j,i)] and (k,l) [or (l,k)] are replaced by a single index according to the same scheme. It is important to remember that the contraction of these indices is for convenience only, and that the matrix elements in the contracted notation do not retain the tensor transformational properties.

As noted in the introduction to this section, the number of independent *linear* electro-optic coefficients is 18 for the most general case of the (noncentrosymmetric) triclinic crystal point group 1. For crystals of increasing symmetry, the number of independent coefficients decreases. For cubic crystals of the (noncentrosymmetric) point groups $\bar{4}3m$ and 23, only one coefficient describes the third-rank tensor $[r]$. In the contracted matrix notation, the nonzero elements for such cubic crystals are $r_{41} = r_{52} = r_{63}$. For the quadratic electro-optic effect, the corresponding numbers of independent coefficients are 36 for the triclinic crystals (i.e., the full 6×6 contracted matrix for $[R]$), and 3 for the cubic crystals of point groups $m3m$, $\bar{4}3m$, and 432. For the latter crystals, the nonzero contracted matrix elements are $R_{11} = R_{jj}$, $j = 2, 3$; $R_{44} = R_{jj}$, $j = 5, 6$; and $R_{12} = R_{ij}$, $i, j = 1, 2, 3$ $(i \neq j)$; 12 in all. The form of the matrices in the linear and quadratic electro-optic coefficients, valid for the different point groups, is provided in many standard texts [see, for example, the recent text by Guenther (1990)].

Next, the linear electro-optic effect will be discussed in greater detail, as a typical example of the various induced optical phenomena introduced in this section. Using the coordinate system (x,y,z) of the principal dielectric axes in the unperturbed state, the contracted matrix notation, and the Einstein summation notation, the index ellipsoid given by Eq. (20b) is altered in the presence of the applied electric field $\mathcal{E}'$ to

$$\begin{aligned}&x^2(n_x^{-2} + r_{1j}\mathcal{E}'_j) + y^2(n_y^{-2} + r_{2j}\mathcal{E}'_j)\\ &\quad + z^2(n_z^{-2} + r_{3j}\mathcal{E}'_j) + 2yzr_{4j}\mathcal{E}'_j + 2zxr_{5j}\mathcal{E}'_j\\ &\quad + 2xyr_{6j}\mathcal{E}'_j = 1, \qquad (79)\end{aligned}$$

where n_x, n_y, n_z are the principal indices of refraction (Levi, 1980). Thus, the coefficients r_{ij}, $i = 1, 2, 3$, affect the magnitudes of the principal indices of refraction, but do not affect the orientation of the ellipsoid. Assuming $r_{ij}\mathcal{E}'_j \ll n_i^{-2}$, the changes in the principal indices effected by r_{ij}, $i = 1, 2, 3$, are given by

$$\Delta n_i = -\frac{n_i^3}{2}\sum_{j=1}^{3} r_{ij}\mathcal{E}'_j; \quad i = 1,2,3. \tag{80}$$

The coefficients r_{ij}, $i = 4, 5, 6$, on the other hand, generate a rotation of the ellipsoid axes in addition to a change in their lengths.

As an example of this effect, consider the case of a tetragonal crystal of point group $\bar{4}2m$ [e.g., potassium dihydrogen phosphate (KDP)]. In this case, the z axis in Eq. (79) is the optic axis (which is a fourfold crystallographic axis of symmetry), and there are two mutually orthogonal twofold symmetry axes in the plane normal to z; these are the x-y axes in Eq. (79). There are only two independent nonzero linear electro-optic coefficients, $r_{41} = r_{52}$ and r_{63}; the remainder vanish as a result of symmetry considerations. If the electric field is assumed to lie along the z axis (i.e., $\mathcal{E}' = \mathcal{E}'_0\hat{\mathbf{z}}$, then Eq. (79) becomes

$$\frac{x^2 + y^2}{n_o^2} + \frac{z^2}{n_e^2} + 2xyr_{63}\mathcal{E}'_0 = 1, \tag{81}$$

where n_o and n_e are ordinary and extraordinary indices of refraction for the unperturbed birefringent crystal (Yariv and Yeh, 1984). Thus, because of the applied field $\mathcal{E}'$, the principal axes of the index ellipsoid are no longer parallel to the (x,y,z) crystallographic axes. On choosing a new system of principal axes, (x',y',z') with $z = z'$ and x' and y' rotated by 45° (about z) relative to x and y, then the new principal indices of refraction become

$$n_{x'} = n_o - \tfrac{1}{2}n_o^3 r_{63}\mathcal{E}'_0, \tag{82a}$$

$$n_{y'} = n_o + \tfrac{1}{2}n_o^3 r_{63}\mathcal{E}'_0, \tag{82b}$$

$$n_{z'} = n_e. \tag{82c}$$

Thus, for light propagating along the optic axis ($z = z'$), birefringence of magnitude $n_o^3 r_{63}\mathcal{E}'_0$ is induced by the field. The numerical values of the appropriate parameters for KDP at $\lambda = 632.8$ nm are $n_o = 1.5074$ ($n_e = 1.4669$) and $r_{63} = 11 \times 10^{-12}$ m/V ($r_{41} = r_{52} = 8 \times 10^{-12}$ m/V). Having determined the principal indices of refraction under the influence of the field, as well as the orientation of the principal axes with respect to the crystallographic axes, the optical behavior of the crystal can be evaluated using the procedures of Sec. 1.2. The orientation of the principal axes and the equations needed to evaluate the principal indices of refraction for crystals of the noncentrosymmetric point groups have been tabulated in standard texts. Similar tables have also been generated that allow one to interpret the quadratic electro-optic effect in crystals of various point groups (see, for example, Cook and Jaffe, 1979; Levi, 1980; Narasimhamurty, 1981; Yariv and Yeh, 1984).

Before closing this subsection, it is appropriate to discuss the complication of the linear electro-optic effect by the accompanying piezoelectric effect in noncentrosymmetric crystals. When the frequency of the applied electric field is much lower than the mechanical resonance frequency of the crystal, the crystal is free to deform under the influence of the field, in phase with it, in accordance with the laws of piezoelectricity. As a result, the observed static or low-frequency electro-optic coefficients r^f_{ijk} of the crystal in its freely oscillating state arise from

1. the pure electro-optic effect of the sample (i.e., under constant strain so that the sample is not allowed to deform); and
2. the photo-elastic effect due to the strain induced by the electric field.

In contrast, when the frequency of the applied field is much higher than the mechanical resonance frequency of the crystal, then the response is the same as if the crystal were mechanically clamped. As a result, one measures the electro-optic effect under constant strain, denoted by the coefficients r^c_{ijk}. The sets of coefficients associated with the free and clamped states exhibit the same symmetry and are related according to

$$r^f_{ijk} = r^c_{ijk} + p_{ijlm}d_{lmk}, \tag{83}$$

where p_{ijlm} and d_{lmk} are the elasto-optic and piezoelectric coefficients, respectively (Yariv and Yeh, 1984).

4.3 Magneto-optic Effects

When a linearly polarized light wave traverses a nonabsorbing solid under influence of an applied magnetic field, then the polarization direction is rotated by an angle proportional to

1. the thickness of the solid traversed, and
2. the component of the magnetic field along the direction of propagation.

This phenomenon is known as the Faraday effect, which will be one focus of this subsection. The second focus will be the magneto-optic Kerr effects, which have the same physical origin, but are observed when a light wave is reflected from the surface of the solid under the influence of the applied magnetic field (see Table 3).

Wave propagation along the optic axis in an optically active solid is similar to propagation in an optically isotropic solid under the influence of an applied magnetic field. There is one important difference, however. Natural optical rotation depends on the direction of wave propagation, $\hat{\mathbf{q}}$, whereas the Faraday rotation depends on the magnetic field direction, but not on $\hat{\mathbf{q}}$. In order to describe the Faraday effect, the gyration vector defined by Eq. (24a) is written as $\mathbf{G} = g\mathbf{B}'$, where g is the magnetogyration coefficient of the medium and $\mathbf{B}'$ is the applied magnetic induction. Thus, for plane-wave propagation in a nonabsorbing crystal under the influence of a magnetic field, one can write (Yariv and Yeh, 1984)

$$\tilde{\mathbf{D}} = \epsilon_0\{[\epsilon]\cdot\mathcal{E} - i(g\mathbf{B}' \times \mathcal{E})\}. \tag{84}$$

The Faraday effect arises from the Lorentz force, $\mathbf{F}'_M = -e\mathbf{v} \times \mathbf{B}'$, exerted by the applied field $\mathbf{B}'$ on the electrons in the solid moving (at velocity $\mathbf{v}$) in response to the electric field of the optical wave. The Lorentz force leads to a "lateral" component of the displacement $\mathbf{r}$ of the electrons (i.e., a component perpendicular to the optical electric field). This in turn leads to a contribution to the induced dipole moment proportional to $\mathbf{v} \times \mathbf{B}'$ or, equivalently, to $i\mathbf{B}' \times \mathcal{E}$ as in Eq. (84).

A lateral component of the electron displacement, phase shifted by $\pi/2$ relative to the optical field $\mathcal{E}$, results in left and right circular normal polarization modes (i.e., states that remain unchanged as they traverse the solid). By including the Lorentz force in the oscillator model [Eq. (30)] and neglecting the damping term, it can be shown that the indices of refraction for the two modes are given approximately by

$$n_\pm^2 = 1 + \frac{e^2}{\epsilon_0 m}\sum_n \frac{N_{en}}{(\omega_n \mp \omega_L)^2 - \omega^2}, \tag{85}$$

where the upper and lower sign choices indicate right and left circular polarization, respectively, i.e., $n_+ = n_r$ and $n_- = n_l$ (Freiser, 1968; Guenther, 1990). In Eq. (85), $\omega_{\mathrm{L}} \ll \omega$ is the Larmor procession frequency, given by $\omega_{\mathrm{L}} = eB'/2m$, where B' is the component of the magnetic induction along the propagation direction. Equation (85) shows the features of the normal Zeeman effect. In other words, the magnetic field splits the resonance frequency associated with the left and right circular polarization modes into two new frequencies shifted by ω_{L} above and below the field-free resonance. This splitting gives rise to circular birefringence at frequencies well below the resonance region ($\omega \ll \omega_n$), where the solid is nonabsorbing. When a linearly polarized light beam traverses the solid, the circular birefringence of Eq. (85) predicts a specific rotation of the direction of polarization (rotation per unit path length) given by

$$\rho_o = \frac{\pi(n_r - n_l)}{\lambda} \approx \frac{e^2 N_e \omega_0 \omega_{\mathrm{L}} \omega/\epsilon_0 \bar{n} mc}{(\omega_0^2 + \omega_L^2 - \omega^2)^2 - 4\omega_0^2\omega_{\mathrm{L}}^2}, \tag{86}$$

where λ is the wavelength of the light in vacuum, $\bar{n}$ is the average index of refraction, and a single dominant resonance of frequency ω_0 is assumed. For $\omega_{\mathrm{L}} \ll \omega_0$, Eq. (86) predicts that the rotation is proportional to ω_{L} or B'.

The behavior predicted by this simple model is consistent with the empirical observation that $\rho_o = \mathcal{V}B'$, where $\mathcal{V}$ is the Verdet constant. The sign of ρ_o in Eq. (86) is chosen counterclockwise positive when looking into the beam (opposite to the convention of Secs. 1.2.3–1.2.4 for optical activity). Thus, for positive $\mathcal{V}$, counterclockwise or levorotatory behavior ($n_r > n_l$) is observed in nonmagnetic materials when the vector component of $\mathbf{B}'$ along $\hat{\mathbf{q}}$ is positive. In paramagnetic and diamagnetic materials, the permeability differs little from free space, so that $\rho_o = \mu_0\mathcal{V}H'$, where H' is the applied magnetic field intensity. Some typical values of the Verdet constant are as follows: ZnS, 54° $(T\ \mathrm{cm})^{-1}$ at 546.1 nm; and NaCl, 5.5° $(T\ \mathrm{cm})^{-1}$ at 589.3 nm (Levi, 1980). In a saturated ferromagnetic material, however, the low-frequency (static) relative permeability associated with the applied magnetic field is much larger than unity so that $\mathbf{B}' \approx \mu_0 \mathbf{M}'_s$, where $\mathbf{M}'_s$ is the saturation magnetization. (In spite of this, the relative permeability at the high frequencies of the *optical* magnetic field remains close to unity.) Then the specific rotation is called the specific saturation rotation F_s, given by $F_s = K'M'_s \approx \mu_0 \mathcal{V} M'_s$, where $K' \approx \mu_0 \mathcal{V}$ is Kundt's constant. Some typical values of F_s are as follows: MnBi at 20 °C (low-temperature phase), 90°/μm; EuS at 20 °C, 478°/μm (Chen, 1974). Additional values of F_s and other characteristics of magnetic recording materials can be found in the literature (see Chen, 1974; Levi, 1980; and references therein).

Thus far, the discussion of this section has centered on the modification of the propagation characteristics of a light wave as it traverses a nonabsorbing solid under the influence of an external perturbation. Induced optical effects, of course, can also be observed in reflection, and although the resulting wave modifications are generally smaller, the experimental techniques can be applied to absorbing solids and thin films. The magneto-optic Kerr effect represents a technologically important example. In the general case of the reflection of polarized light from an optically isotropic absorbing solid at a nonnormal angle of incidence, changes in both the irradiance and polarization of a light beam occur (see Sec. 5.1.3). In this case, the normal polarization modes (i.e., those that remain unchanged upon reflection) are the linear p and s states, parallel and perpendicular to the plane of incidence, respectively. In the magneto-optic Kerr effect, the magnitudes of the irradiance and polarization changes depend on the applied field ($\mathbf{B}'$) or magnetization ($\mathbf{M}'$), and the normal polarization modes are elliptical, in general.

Consider here the case in which $\mathbf{M}'$ is normal to the surface of a ferromagnetic solid (defining the x-y plane) and a light wave impinges at normal incidence in the $+z$ direction (the polar Kerr effect). In this high-symmetry situation, the normal polarization modes are left and right circular with different complex indices of refraction $\tilde{N}_l$ and $\tilde{N}_r$. With the inclusion of a damping term, Eq. (85) can provide a model for the circular birefringence and dichroism. In this case, the reflected wave will be elliptically polarized, having an ellipticity e_K and a major axis at an angle θ_K with respect to the incident field, where (Levi, 1980)

$$e_K + i\theta_K = -\left(\frac{\tilde{N}_r - \tilde{N}_l}{\tilde{N}_l\tilde{N}_r - 1}\right). \tag{87}$$

The negative sign is required, since the reflected beam propagates in the $-z$ direction. In the absence of $\mathbf{M}'$, $\tilde{N}_l = \tilde{N}_r$, and Eq. (87) shows that there is no change in polarization upon reflection. Note that the largest values of the Kerr rotation are obtained from ferromagnetic materials at saturation magnetization. Although θ_K in Eq. (87) is rarely observed to exceed 1°, the magnitude of the effect can be enhanced by multiple reflections, achieved by depositing a dielectric film or multilayer structure onto the surface of the ferromagnetic solid. Other configurations in addition to the polar Kerr effect are also important (Freiser, 1968). In the longitudinal and equatorial Kerr effects, the applied $\mathbf{B}'$ (or $\mathbf{M}'$) lies in the plane of the solid surface, either parallel (for longitudinal) or perpendicular (for equatorial) to the plane of incidence. In the equatorial Kerr effect, the normal polarization modes are linear p and s states. Thus, the state of linearly polarized light does not change upon reflection; however, a reversal of $\mathbf{M}'$ in a ferromagnetic material generates a change in the power reflected in the p state.

4.4 Piezo-optic, Elasto-optic, and Acousto-optic Effects

This subsection is devoted to a brief discussion of the changes in the index tensor of nonabsorbing solids subjected to a mechanical stress (piezo-optics), static strain (elasto-optics), or dynamic strain (acousto-optics). The equations describing piezo- and elasto-optic phenomena are usually written in the Einstein notation as

$$\Delta a_{ij} = q_{ijkl}X_{kl} = p_{ijkl}x_{kl}, \tag{88}$$

where q_{ijkl} and p_{ijkl} are the components of the fourth-rank piezo-optic and elasto-optic tensors, respectively; X_{kl} denote the six independent stress components (three normal and three shear); and x_{kl} denote the six independent strain components (Ramachandran and Ramaseshan, 1961). Here tensile stresses and strains are treated as positive. Since all three second-rank tensors $[\Delta a]$, $[X]$, and $[x]$ are symmetric, then the piezo-optic and elasto-optic tensors have a maximum of 36 independent components for the most general crystal system, as noted in the introduction to this section. For all crystals of a given point group, the symmetries of these two tensors (and the number of independent components) are the same, and identical to the symmetries of the quadratic electro-optic tensors $[R]$ and $[F]$ of Eqs. (78).

As in the case of the quadratic electro-optic effect, it is convenient to use the contracted matrix notation:

$$\Delta a_i = q_{ij}X_j = p_{ij}x_j, \tag{89}$$

where the representations for $[q]$ and $[p]$ are 6×6 matrices and the representations for $[\Delta a]$, $[X]$, and $[x]$ are 6-component vectors. The six stress and strain components are interrelated through $X_i = c_{ij}x_j$ and $x_i = s_{ij}X_j$, where c_{ij} are the elastic constants and s_{ij} are the elastic moduli of the solid, also represented by 6×6 matrices. Because these matrices are symmetric, there are 21 independent elastic constants (and moduli) for the most general (triclinic) crystal system. This number reduces to 3 for the cubic crystals. Similarly, for the piezo- and elasto-optic matrices, the 36 independent elements for the triclinic system reduce to 3 for the cubic crystals of point groups 432, $m3m$, and $\bar{4}3m$, the 12 nonzero elements being $p_{11} = p_{jj}$, $j = 2, 3$; $p_{44} = p_{jj}$, $j = 5, 6$; $p_{12} = p_{ij}$, $i, j = 1, 2, 3$ $(i \neq j)$. Finally, if the solid is structurally isotropic such as a glassy solid then the same set of 12 elements is nonzero, but $p_{44} = p_{jj} = \frac{1}{2}(p_{11} - p_{12})$, $j = 5, 6$, and the number of independent elements is reduced to two. Some typical values of the elasto-optic coefficients at 632.8 nm are as follows: vitreous silica, $p_{11} = 0.121$ and $p_{12} = 0.270$; $Y_3Al_5O_{12}$ (YAG), $p_{11} = -0.029$, $p_{12} = 0.0091$, and $p_{44} = -0.0615$ (Yariv and Yeh, 1984).

If the mechanical stress is applied dynamically at high frequency, the stress field creates a strain field in the material and an accompanying periodic modulation of the index of refraction of the material in accordance with the equations of this subsection. The acousto-optic phenomena that result from interaction of acoustic and optical waves in solids are employed in numerous scientific and technological applications. Further details can be found in more specialized articles in these volumes, and in several texts (Sapriel, 1979; Korpel, 1988; Xu and Stroud, 1992).

5. MEASUREMENT TECHNIQUES

In this section, the emphasis will be on the most direct measurement techniques that provide the complex dielectric function (or complex index of refraction) of solids as a function of the photon energy (i.e., the optical functions). This approach is motivated by the fact that the optical functions themselves are of great importance in predicting the performance of solids in a wide range of modern electronic, optical, and magnetic applications, e.g., photodetectors, solid-state lasers, photovoltaic cells, optical and magneto-optical discs for data storage, etc. The spectroscopic measurement approaches to be discussed here include the following:

1. normal-incidence reflectance (i.e., the ratio of the reflected and incident irradiances) for the specular surface of an opaque bulk solid or film (Sec. 5.1.1);
2. combined normal-incidence reflectance and transmittance (the latter being the ratio of the transmitted and incident irradiances) for a semitransparent, plane-parallel platelet having specular surfaces (Sec. 5.1.2) or a substrate-supported thin film (Sec. 5.2); and
3. ellipsometry (*q.v.*), which determines the p-to-s complex-amplitude reflection ratio (i.e., the ratio of the polarization states associated with the reflected and incident beams) for an opaque specular solid or film (Sec. 5.1.3) or a semitransparent film (Sec. 5.2).

In Sec. 5.1.3, application of the ellipsometric techniques to optically anisotropic solids will be discussed briefly. In a number of the simplest cases, the basic optical formulas used to calculate $\tilde{N}$ from the different measurements will be provided. Additional details on the measurement and reduction of the optical data for bulk solids and multilayer films appear in a number of texts given below under Further Reading.

All techniques that involve measuring the reflected or transmitted light from a solid require special attention to surface preparation methods since the presence of oxide layers, surface roughness, and/or polishing damage can severely distort experimental data and limit the accuracy of the methods. As an example, the differences in the reflectance spectra of Fig. 12 for opaque silver films can be explained by the presence of a more extensive surface roughness layer on the sample used in obtaining the handbook data. Chemical and/or chemomechanical treatments (for semiconductors), and electropolishing (for metals), particularly when directed by optical measurements performed in real time, are usually necessary for bulk solids. With the widespread development of epitaxial growth processes for many materials, the best optical functions may yet result from measurements of atomically smooth thin crystalline films *in situ* under ultrahigh vacuum.

Other classes of measurement techniques such as photoemission and electron energy-loss spectroscopy will not be discussed here since the motivation for such experiments is not usually to deduce the full complex dielectric response directly, but rather to elicit information on the electronic states. Modulation spectroscopies (see Sec. 4) also fall into this category. In these latter experiments, one probes the critical points in the electronic density of states by measuring the derivatives of the optical data (e.g., reflectance and transmittance) with respect to a modulated external perturbation, such as an electric field. This approach enhances the optical structures that appear in the reflectance at the energies of the critical points, but suppresses the smoothly varying background. Instead, in Sec. 5.1.3, we briefly discuss spectroscopic ellipsometry (i.e., the *p*-to-*s* reflection ratio measured vs wavelength), which has become very precise and accurate, not only for determination of complex dielectric functions, but also making possible critical-point analysis through multiple differentiation of $\tilde{\epsilon}$ with respect to optical frequency (see Fig. 17).

5.1 Bulk Solids and Opaque Films

5.1.1 Reflectance The reflectance R of an optically isotropic, opaque solid in bulk or thin-film forms (i.e., a solid consisting of a single interface to the ambient) is obtained by reflecting a quasimonochromatic wave from its specular surface at normal incidence and determining the ratio of the reflected irradiance to the incident irradiance of the wave. The reflectance is related to the complex-amplitude reflection coefficient $\tilde{r}$ by $R = |\tilde{r}|^2$. At first glance, one might expect that a measurement of the spectral reflectance R at normal incidence, expressed in terms of the ambient and solid indices of refraction n_a and $\tilde{N} = n + i\kappa$, by (Cardona, 1969)

$$\tilde{r} = R^{1/2}\exp(i\Theta), \qquad (90a)$$

$$= \frac{\tilde{N} - n_a}{\tilde{N} + n_a} = \frac{(n + i\kappa) - n_a}{(n + i\kappa) + n_a}, \qquad (90b)$$

is insufficient to provide both real and imaginary parts of optical functions, since the phase Θ is not measured. However, since both (n,κ) and (R,Θ) are interrelated through dispersion relations [similar to those given in Sec. 1.3 for (ϵ_1,ϵ_2)], measurements of R over a wide spectral range are sufficient to permit calculation of Θ. Specifically, the reflectance and phase are interrelated through

$$\Theta(\omega) = \frac{\omega}{\pi}\mathrm{P}\int_0^{\infty}\frac{\ln R(\omega') - \ln R(\omega)}{\omega^2 - \omega'^2}\,d\omega', \qquad (91)$$

where P denotes the principal value of the integral (Wooten, 1972). Obviously, the limitation in applying Eq. (91) arises from the limited spectral range of any measurement. Typically, one must measure semiconductors and insulators from the near infrared well into the vacuum ultraviolet, and then extrapolate to low photon energies using a constant reflectance (n = constant, $\kappa = 0$) and to high energies using a form for the reflectance based on the asymptotic behavior of Eqs. (33). Thus, one selects $R(\omega) = R_c(\omega_c/\omega)^4$ for $\omega > \omega_c$, where R_c is the reflectance corresponding to the highest-frequency data point at ω_c. For metals, the free-electron behavior can be assumed at both high and low energies.

With such extrapolations, Eq. (91) can be integrated by Simpson's rule since the singularity in the denominator at $\omega' = \omega$ is removed by the $\ln[R(\omega)]$ term in the numerator. The resulting phase spectrum is used in combination with the reflectance spectrum to extract the optical functions through Eqs. (90). The extrapolations at high photon energies are usually the dominant source of error, preventing the deduction of high-accuracy semiconductor optical functions by this method. As a result, ellipsometry measurements that determine both the amplitude and phase of the reflection coefficient are preferred for this application. Another disadvantage of reflectance measurements is their sensitivity to any defects on the solid surface that scatter irradiance from the impinging beam and simulate an absorption process; ellipsometry is less sensitive to scattering defects.

5.1.2 Reflectance and Transmittance of Bulk Solids For optically isotropic solid samples whose thickness d and absorption coefficient α [Eq. (12b)] are such that αd is on the order of unity, the quasimonochromatic, normal-incidence transmittance can be measured in conjunction with the reflectance. With spectra in both parameters, the real and imaginary parts of the optical functions can be deduced without the need for the dispersion relationship of Eq. (91). Because the typical minimum solid sample thickness that can be prepared by polishing is ~1 μm, this approach is only applicable for studying indirect intrinsic or extrinsic (defect) absorption processes in semiconductors or insulators in the limited photon-energy range below the lowest direct gap. Because the absorption coefficient ranges from ~10^5 to >10^6 cm^{-1} for interband transitions above the direct gap in semiconductors and insulators and for intraband transitions in metals, sample preparation to a thickness <100 nm becomes necessary in studying such processes. This is impractical in most situations, and one must resort to measurements on thin films prepared by vacuum deposition (see Sec. 5.2).

In the incoherent limit (typically for $d \gg \lambda$), the measured reflectance $\mathfrak{R}$ and transmittance $\mathfrak{T}$ of a plane-parallel platelet in air or vacuum ($n_a = 1$) are given by

$$\mathfrak{T} = \frac{(1 - R)^2 \exp(-\alpha d)}{1 - R^2 \exp(-2\alpha d)}, \tag{92a}$$

$$\mathfrak{R} = R[1 + \mathfrak{T} \exp(-\alpha d)], \tag{92b}$$

where R is the reflectance of the single ambient/solid interface given by Eqs. (90) with $n_a = 1$ (Cardona, 1969). Equations (92) can be inverted to obtain R and α (or κ), which in turn can be combined to derive n from $R = |\tilde{r}|^2$ using Eqs. (90). Since α and d appear as a product in Eqs. (92), an independent method for determining the platelet thickness must be available. When $\alpha d \gg 1$, the absorbance is near unity; then $\mathfrak{T} = 0$, and no information can be obtained about α from Eqs. (92). Similarly, for $\alpha d \ll 1$, $\mathfrak{T}$ becomes independent of α, and no information is provided in this case, either. The equations corresponding to those of Eqs. (92) for the coherent limit can be derived employing methods described by Heavens (1965):

$$\mathfrak{T} = [T_{af}T_{fs} \exp(-\alpha d)]/D, \tag{93a}$$

$$\mathfrak{R} = \{R_{af} + R_{fs} \exp(-2\alpha d) + 2 \exp(-\alpha d)\mathrm{Re}[(\tilde{r}_{af})^* \tilde{r}_{fs} \exp(i\delta)]\}/D, \tag{93b}$$

where

$$D = 1 + R_{af}R_{fs} \exp(-2\alpha d) + 2 \exp(-\alpha d)\mathrm{Re}[\tilde{r}_{af}\tilde{r}_{fs} \exp(i\delta)], \tag{93c}$$

$$\delta = 4\pi n d/\lambda. \tag{93d}$$

These equations, however, represent the general case in which the incident and transmitted waves are in different transparent media with indices of refraction n_a (ambient) and n_s (substrate). The single-interface (ambient/platelet and platelet/substrate) amplitude reflection and transmission coefficients in Eqs. (93) are $\tilde{r}_{af} = (\tilde{N} - n_a)/(\tilde{N} + n_a)$, $\tilde{r}_{fs} = (n_s - \tilde{N})/(n_s + \tilde{N})$, $\tilde{t}_{af} = 2n_a/(\tilde{N} + n_a)$, and $\tilde{t}_{fs} = 2\tilde{N}/(n_s + \tilde{N})$, and $R_{\mu\nu} = |\tilde{r}_{\mu\nu}|^2$ and $T_{\mu\nu} = (n_\nu/n_\mu)\,|\tilde{t}_{\mu\nu}|^2$ for $(\mu,\nu) = (a,f)$ or (f,s). For the platelet in air, $n_a = n_s = 1$, and one can also neglect the imaginary part of $\tilde{N}$ in the amplitude reflection and transmission coefficients because $n \gg \kappa$ for the maximum accessible κ. In this case, interference fringes are present, and their spacing can be used to determine d independently.

5.1.3 Ellipsometry Reflection ellipsometry is another more powerful approach that can provide both parts of the optical functions of an optically isotropic, opaque solid in bulk or thin-film forms (Azzam and Bashara, 1977). Although the reflection ellipsometry measurement can be done in a number of ways, the rotating-analyzer configuration is the simplest and most popular for spectroscopic measurement, as described elsewhere (see ELLIPSOMETRY). The underlying measurement principle of reflection ellipsometry is the change in polarization state that a polarized, quasimonochromatic light beam undergoes when it reflects from a surface at an oblique angle. In the measurement, one determines the complex-amplitude reflection ratio, $\tilde{\rho} = \tilde{r}_p/\tilde{r}_s = [\tilde{\mathcal{E}}_{rp}/\tilde{\mathcal{E}}_{rs}]/[\tilde{\mathcal{E}}_{ip}/\tilde{\mathcal{E}}_{is}]$, where the complex optical fields $\tilde{\mathcal{E}}$ are specified for the p and s components of the incident i or reflected r beam. If one recognizes that the complex field ratio $[\tilde{\mathcal{E}}_p/\tilde{\mathcal{E}}_s]$ defines the polarization state of the optical beam, then the quantity $\tilde{\rho}$ can be understood as the ratio of the reflected to incident polarization states.

The reflection ellipsometry data are most often expressed in terms of the angles ψ and Δ, defined by $\tilde{\rho} \equiv \tan\psi \exp(i\Delta)$. When the ambient medium has a (real) index of refraction n_a, the complex index of refraction of the reflecting sample $\tilde{N}$ can be calculated from

$$\tilde{N} = n_a \sin\theta \left\{ 1 + \left(\frac{1 - \tilde{\rho}}{1 + \tilde{\rho}} \right)^2 \tan^2\theta \right\}^{1/2}, \qquad (94)$$

where θ is the angle of incidence. A relative uncertainty $|\delta\tilde{\epsilon}/\tilde{\epsilon}|$ of $\sim 10^{-4}$ obtained under optimum circumstances provides the capability of multiple differentiation of $\tilde{\epsilon}$ in order to enhance weak critical-point structures (Aspnes, 1976). Thus, spectroscopic ellipsometry has become preferred over modulation spectroscopy for this purpose in recent years. Obviously, reflection ellipsometry is also preferred over reflectance because of its ability to measure both real and imaginary parts of the optical functions directly. Furthermore, the ellipsometry measurement is not as strongly influenced by surface imperfections that scatter light out of the beam, as is reflectance.

Through the phase angle Δ, however, ellipsometry is very sensitive to native oxide and atomic-scale roughness layers that are often undetected and thus are not included in the data interpretation. In general, it is best to avoid such problems through *in situ* measurements of materials either prepared in ultrahigh vacuum or surface treated in wet-chemical processes. The results for the pseudo-optical functions $\langle \tilde{N} \rangle$ [calculated from $\tilde{\rho}$ using Eq. (94) irrespective of possible overlayers] can often be assessed with the "biggest is best" criterion, i.e., the pseudodielectric function denoted $\langle \tilde{\epsilon} \rangle = (\langle \tilde{N} \rangle)^2$ that exhibits the highest peak $\langle \epsilon_2 \rangle$ magnitude is most representative of the true bulk dielectric function of the solid (Aspnes, 1976). Thus, it is sample surface condition, not instrument accuracy, that limits one's ability to deduce accurate optical functions by ellipsometry. Because of this, one often may not be able to determine accurate κ values much lower than 0.1 in the weak absorption range; then it is best to resort to reflectance and transmittance measurements. In fact, one of the few advantages of supplementing a reflectance measurement by transmittance as in Sec. 5.1.2, rather than performing ellipsometry, is the capability (in the former case) of increasing the sample thickness in order to increase sensitivity in order to detect low κ (or α) values in dielectric and semiconductor materials.

Reflection ellipsometry is also the method of choice to determine the optical functions that characterize strongly absorbing, anisotropic solids that are not optically active. If orthorhombic and higher crystal symmetry is assumed, then the principal axes of the real and imaginary parts of the index of refraction both coincide with the principal crystallographic axes. The general case of reflection from an arbitrary crystal plane is complicated because the normal polarization modes, i.e., those that remain unchanged upon reflection, are not the pure linear p and s states (as for an optically isotropic sample), but are elliptical in general. If one assumes, however, that an orthorhombic crystal is cleaved (or cut and suitably polished) so that the z axis is normal to the surface, and the plane of incidence is parallel to the x-z plane of the crystal, then the normal modes revert to the linear p and s states. For this geometry, the complex-amplitude reflection coefficients are related to the principal complex indices of refraction $\tilde{N}_x$, $\tilde{N}_y$, and $\tilde{N}_z$, according to

$$\tilde{r}_{pp} = \frac{\tilde{N}_x \tilde{N}_z \cos\theta - n_a(\tilde{N}_z^2 - n_a^2 \sin^2\theta)^{1/2}}{\tilde{N}_x \tilde{N}_z \cos\theta + n_a(\tilde{N}_z^2 - n_a^2 \sin^2\theta)^{1/2}}, \qquad (95a)$$

$$\tilde{r}_{ss} = \frac{n_a \cos\theta - (\tilde{N}_y^2 - n_a^2 \sin^2\theta)^{1/2}}{n_a \cos\theta + (\tilde{N}_y^2 - n_a^2 \sin^2\theta)^{1/2}}, \qquad (95b)$$

where $\tan\psi \exp(i\Delta) = \tilde{r}_{pp}/\tilde{r}_{ss}$. In order to obtain accurate results for $\tilde{N}_x$, $\tilde{N}_y$, and $\tilde{N}_x$ using this approach, ellipsometry spectra in (ψ,Δ) are taken on the three different planes of the crystal at multiple angles of incidence θ (Azzam and Bashara, 1977).

If $|\tilde{N}_i^2| \gg 1$ for $i = x, y, z$, then a much simpler approach can be adopted to obtain approximate results for two principal indices of refraction, for example $\tilde{N}_x$ and $\tilde{N}_y$, from two measurements on the face of the orthorhombic crystal normal to z (Aspnes, 1980b; Logothetidis *et al.*, 1985). The two measurements are performed with the plane of incidence parallel to the y-z and z-x planes and are used to obtain first-order corrections $\Delta\tilde{\epsilon}_i$, $i = x, y$, to an isotropic mean dielectric function $\tilde{\epsilon}_m$, according to $\tilde{N}_i^2 = \tilde{\epsilon}_m + \Delta\tilde{\epsilon}_i$, $i = x, y$. Here $\tilde{\epsilon}_m$ is calculated first by converting the (ψ,Δ) measurements to "pseudodielectric functions" using the expression for an isotropic solid [Eq. (94)] and then by averaging the two pseudodielectric functions according to an effective-medium theory based on a composite with 50-50 vol %. This procedure has been found to reproduce the more rigorous approach based on Eqs. (95) with sufficient accuracy for most purposes, e.g., critical-point analysis. The validity of this simple procedure is based on the general fact that, if $|\tilde{N}_i^2| \gg 1$ and if one of the principal planes is normal to the plane of incidence, then the dominant contribution to an ellipsometry measurement arises from the projection of $[\tilde{\epsilon}]$ onto the line of intersection between the surface and the plane of incidence (Aspnes, 1980b).

The characteristics of nonabsorbing, optically anisotropic solids, including linear and circular birefringence and principal-axis orientation, can be studied with a transmission ellipsometer. Such an instrument provides the ratio of the transmitted to incident polarization states, and from this ratio, the characteristics of the solid can be determined using Jones calculus. Although the rotating-analyzer ellipsometer described elsewhere in these volumes can be employed for transmission ellipsometry, it is important to add a compensating element (i.e., a known birefringent crystal as a calibrated phase retarder) either before or after the sample in order to establish the handedness of emergent elliptical polarizations. For example, for a biaxial crystal with plane-parallel faces cut normal to one of the principal axes, a transmission ellipsometry measurement provides the birefringence and the orientation of the other two principal axes in the plane of the surface (measured with respect to the transmission axis of the fixed polarizer of the instrument). The absolute principal indices of refraction are obtained most accurately by the method of minimum deviation using suitably aligned, cut, and polished prisms of the solid.

5.2 Semitransparent Thin Films

In this part, the application of reflectance and transmittance as well as ellipsometry techniques to optically isotropic, semitransparent thin films will be discussed briefly. The case of anisotropic films for a number of high-symmetry situations can be found in the literature (Azzam and Bashara, 1977; Yeh, 1988).

The limitations of reflectance and transmittance measurements of thin films on nonabsorbing substrates are similar to those of bulk platelets; however, the advantage of films is the capability of preparing samples to thicknesses less than 1 μm. Reflectance and transmittance measurements of identically prepared semiconductor films having different thicknesses from 10 nm to 10 μm can provide the index of refraction and values of α from 10^2 to $>10^6$ cm^{-1} vs photon energy along the absorption onset, as long as the film grows uniformly with no thickness dependence in the optical properties. For a single film with a thickness of 0.5 μm, for example, the accessible range of α is one to two orders of magnitude centered near $\alpha \approx 1/d = 2 \times 10^4$ cm^{-1}. For higher absorption coefficients, the film is nearly opaque and ellipsometry can be used (see Sec. 5.1.3). For lower absorption coefficients, photoacoustic, photothermal, photoconductive, or bolometric techniques may be employed. As an example, Fig. 20 shows the absorption spectrum for the *a*-Si:H sample of Figs. 6 and 15, measured over nearly eight orders of magnitude using photoconductivity, reflectance and transmittance, and ellipsometry measurements. The photoconductivity in this case arises from electrons and is defined by $\Delta\sigma_p = G_e e(\mu\tau)_e$, where G_e is the generation rate of photoexcited electrons, which is proportional to α in the weak absorption limit, and $(\mu\tau)_e$ is the mobility–lifetime product for electrons, which is controlled independent of the incident photon energy by applying a white light bias. Rela-

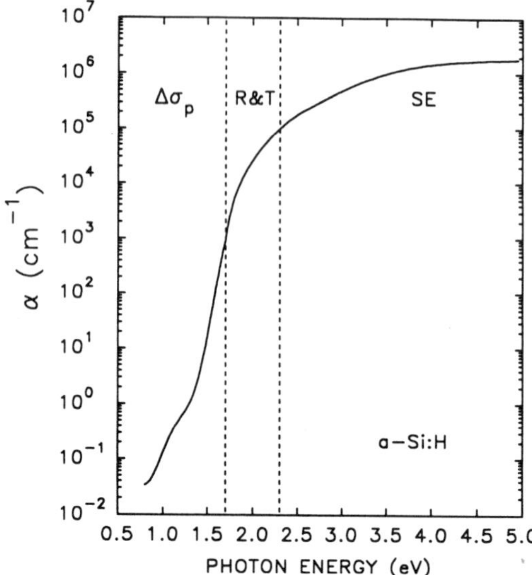

FIG. 20. Optical absorption spectra for the ~1-μm-thick hydrogenated amorphous silicon film of Figs. 6 and 15, obtained using a photoconductive method ($\Delta\sigma_p$) at the lowest photon energies, spectroscopic ellipsometry (SE) at the highest energies, and reflectance and transmittance (R&T) at intermediate energies. Data from Dawson *et al.* (1992).

tive values of α are obtained from this expression and are then normalized to the lowest photon-energy results from reflectance and transmittance.

The analysis of reflectance and transmittance measurements for substrate-supported films is more involved than the platelet case (Heavens, 1965). Thus, numerous variations are possible, most requiring iterative methods of solution. Normally, the substrate is thick enough so that the incoherent limit can be assumed for the wave in the substrate. One noniterative approach designed for semiconductors in the photon-energy range near the band edge involves measuring the transmittance and reflectance with the light impinging on the substrate side, $\mathfrak{T}_s$ and $\mathfrak{R}_s$, respectively. Then these measured spectra can be converted to hypothetical spectra, $\mathfrak{T}'_s$ and $\mathfrak{R}'_s$, for a semi-infinite substrate using

$$\mathfrak{T}_s = \frac{T_{sa}\mathfrak{T}'_s}{1 - R_{sa}\mathfrak{R}'_s}, \tag{96a}$$

$$\mathfrak{R}_s = R_{sa} + \frac{T_{sa}^2\mathfrak{R}'_s}{1 - R_{sa}\mathfrak{R}'_s}, \tag{96b}$$

where R_{sa} and T_{sa} are the reflectance and transmittance of the substrate/ambient interface, defined as in the discussion after Eqs. (93) (Maley, 1992). In fact, $\mathfrak{T}'_s$ and $\mathfrak{R}'_s$ are given by $\mathfrak{T}$ and $\mathfrak{R}$ in Eqs. (93a) and (93b), respectively, but with an interchange of subscripts for the ambient and substrate (a,s). As long as the film is nonabsorbing over part of the spectral range [i.e., $\alpha d \ll 1$, so that $\exp(-\alpha d) \approx 1$], and thick enough so that a number of interference fringes appear over this range, then the real part of the index of refraction (assumed to be weakly dependent on photon energy) and the thickness of the film can be determined from the maximum/minimum values of $\mathfrak{T}'_s$ and the fringe spacing, respectively. The absorption spectrum can be determined from the expression

$$\frac{\mathfrak{T}'_s}{1 - \mathfrak{R}'_s} = \frac{(1 - R_{fa}) \exp(-\alpha d)}{1 - R_{fa} \exp(-2\alpha d)}, \tag{97}$$

where it is assumed that $n \gg \kappa$ for the film so that $\tilde{r}_{fa}$ is real (Maley, 1992).

Spectroscopic ellipsometry has an advantage over reflectance and transmittance measurements because it can be used to determine the optical functions of semitransparent thin films on absorbing substrates (Aspnes, 1976; Azzam and Bashara, 1977). In the ellipsometry data analysis for an ideal film, two wavelength-dependent parameters (n,κ) and one constant d must be determined from two wavelength-dependent experimental measurements, (ψ,Δ). Normally a guess is made for d, and the multilayer optical equations for (ψ,Δ) are inverted to obtain trial (n,κ) spectra. Then (n,κ) are inspected for possible artifacts that may indicate an incorrect choice of d. These may include residual interference fringes or spectroscopic features that derive from the critical points of the substrate. The correct thickness is thus chosen to avoid the artifacts. For a semiconductor film for which it is known that $\kappa = 0$ at the lowest photon energies, d can be chosen to ensure that this condition is met.

Atomic-scale roughness and columnar morphology (i.e., nonuniformity in the density along the direction perpendicular to the surface) are common inhomogeneities in real thin films that limit the accuracy of the methods used to determine optical functions of solids in thin-film form. Ellipsometry spectra collected on nonabsorbing films can be interpreted using simple optical models for the microstructure to yield both microstructural information (void volume fraction vs depth into the film) and accurate optical functions of the thin-film material (Chindaudom and Vedam, 1994). The most advanced ellipsometry instruments employ multichannel detection systems so that full spectra can be collected during thin-film preparation in times as short as 10 ms (Collins, 1990; for examples see Figs. 7 and 16). In this case, the data are not influenced by surface contamination on the film, and sufficient information is provided to extract the evolution of the microstructure and the optical functions.

6. CONCLUDING REMARKS

A broad overview of the optical properties of solids has been presented in this article. Previous extensive treatments of this subject have focused either on the fundamental aspects of electrons in optically isotropic solids (see, for example, Wooten, 1972), or on the applied aspects of electromagnetic wave propagation and induced optical phenomena in anisotropic crystals (see, for example, Yariv and Yeh, 1984). The approach taken here has sought a balance between these diverse aspects of the subject. This approach is based on the recognition that, although the best understood solids are optically isotropic (e.g., Al, Si, diamond), the majority (~80%) of the known crystals are anisotropic and, as a result, have a wider range of novel optical applications. The central theme of this article has been the dielectric function or tensor that characterizes the macroscopic, linear response of a solid to a monochromatic light wave. In Sec. 1, the discussion has progressed in steps, starting with the complex dielectric function that describes wave propagation in an optically isotropic solid, and ending with the complex dielectric and gyration tensors that describe wave propagation in an optically anisotropic, optically active crystal.

A related theme of this article has been the inherent wavelength dependence of the dielectric function or tensor (i.e., the absorption and dispersion) for solids over the wavelength range from the near-infrared to the ultraviolet. Over this range, the interaction of the light wave with the valence and conduction electrons is primarily responsible for the observed wavelength-dependent characteristics. In Secs. 2–3, the discussion has also progressed in steps, starting with a simple classical model for the absorption and dispersion in weakly interacting atoms, and ending with more sophisticated models based on the quantum theory of solids. The former model provides a rudimentary understanding of the differences in the dielectric functions among metals, semiconductors, and insulators; the latter models provide an in-depth understanding of the differences among different materials even within the same group. With the more sophisticated models, the wavelength dependence of the dielectric function yields insights into the underlying electronic processes and the electronic joint density of states in the solid, specifically through comparisons of experimental data for the dielectric function with corresponding results calculated from theoretically derived band structures.

In addition to the fundamental physical information that the dielectric function or tensor can provide on electronic states in solids, it provides indispensable information related to applications, as well. In general, the dielectric function or tensor allows one to predict the performance of solids in a wide range of technologically important optical components and devices. These include passive components such as mirrors, windows, retarders, and thin-film coatings, as well as active devices such as photodetectors, photovoltaic cells, solid-state lasers, liquid crystal displays, and electrochromic thin films. Furthermore, the induced changes in the inverse dielectric tensor can be measured in suitable experimental configurations when well-characterized external perturbations are applied to a solid—for example, an increase in temperature, or an applied electric field, stress, or strain. Such measurements can yield the linear coefficients that relate the perturbation to the observed optical change (e.g., the thermo-optic, electro-optic, piezo-optic, and elasto-optic coefficients of Sec. 4). From these coefficients, tabulated for various solids, one can predict the performance of numerous devices whose purpose is to control and manipulate light waves in technologies such as optical communications, data storage, and computing.

The article has concluded in Sec. 5 with a

brief overview of the techniques used to determine the complex dielectric function and tensor, from near-infrared to ultraviolet wavelengths, for solids in both bulk and thin-film forms. The most powerful single technique for this purpose is spectroscopic ellipsometry. The key to this technique is its ability to obtain both amplitude and phase information, through an analysis of the change in polarization that a fully polarized, monochromatic light beam incurs upon reflection from the specular surface of a solid. As a result, ellipsometry avoids the need to perform Kramers–Kronig analysis on conventional reflectance data, which must be collected well into the vacuum ultraviolet and extrapolated beyond the boundaries of the measurement. In fact, because of the ability of spectroscopic ellipsometry to provide better data than were possible with earlier techniques, renewed interest has been stimulated in the optical properties of solids over the last decade (Boccara *et al.*, 1993; Vedam, 1994). Applications have included, for example, the determination of the electronic structure of new materials such as high-temperature superconductors through *ex situ* spectroscopic ellipsometry studies, and the correlation of microstructural nucleation and coalescence in thin films with their electronic performance through real-time spectroscopic ellipsometry studies. More advanced experimental techniques and applications along these lines are expected to be the driving force for future progress on the optical properties of solids in applied physics fields.

ACKNOWLEDGMENTS

The preparation of this article was supported by the National Science Foundation under Grant No. DMR-8957159 of the PYI Program, and Grant No. DMR-9217169. The authors express their appreciation to Dr. Hien V. Nguyen, Mr. Yiwei Lu, and Prof. Ilsin An for their valuable assistance in the preparation of this article.

GLOSSARY

Absorption Coefficient (α): The reciprocal of the exponential decay length for the irradiance of a light wave traveling within a medium. It is given by $\alpha = 4\pi\kappa/\lambda$, where κ is the extinction coefficient and λ is the wavelength of the wave in vacuum.

Acousto-optic Effect: A periodic modification of the index ellipsoid or other optical characteristics of a solid that results from a periodic strain field (or acoustic wave) created when the solid is subjected to mechanical stress at high frequencies.

Circular Birefringence: Also called optical activity. A characteristic of crystals having a helical atomic structure. Left- and right-circularly polarized waves traveling along an optic axis within the crystal experience different indices of refraction and different phase speeds. This is also a characteristic of molecules with intrinsic handedness, whether in a molecular crystal or in solution.

Circular Dichroism: A characteristic of optically absorbing, optically active, anisotropic crystals, analogous to circular birefringence; however, the two circularly polarized waves experience different extinction coefficients.

Critical Point: Also called a van Hove singularity. For an interband optical transition, a point in electron energy–wave-vector space at which the gradient of the difference between the electron energies in the initial and final states vanishes. This gives rise to structure in the dielectric function.

Dielectric Function ($\tilde{\epsilon}$): A macroscopic complex quantity that provides a complete description of the linear response of an isotropic solid to an electromagnetic wave. It is given by $\tilde{\epsilon} = \tilde{N}^2$, where $\tilde{N}$ is the complex index of refraction.

Dielectric Tensor ($[\tilde{\epsilon}]$): A second-rank, complex tensor that provides a complete description of the linear response to an electromagnetic wave of an anisotropic solid that is not optically active. It relates the optical electric displacement vector $\tilde{\mathbf{D}}$ and field vector $\mathcal{E}$ according to $\tilde{\mathbf{D}} = [\tilde{\epsilon}] \cdot \mathcal{E}$.

Direct Interband Transition: Also called a vertical transition. A transition in which an electron is excited from a Bloch state in one band to a Bloch state in another band such that electron wave vector is conserved (in the reduced Brillouin-zone scheme).

Elasto-optic Effect: A modification of the index ellipsoid or other optical characteristics of a solid when it is subjected to strain under static conditions. In nonabsorbing solids, it is characterized by the fourth-rank elasto-optic tensor that operates on the strain tensor to obtain the change in the index tensor.

Electro-optic Effects: Modifications of the index ellipsoid (or other optical characteristics) induced by an external electric field. The linear electro-optic (or Pockels) effect is observed in noncentrosymmetric crystals, is linear in the applied field, and is characterized by a symmetric third-rank tensor. The quadratic electro-optic (or Kerr) effect is observed in all solids, is quadratic in the applied field, and is described by a symmetric fourth-rank tensor.

Ellipsometry: A procedure designed to determine the optical properties of a specular surface or the thicknesses of film(s) on the surface by measuring the change in the polarization state that occurs when a polarized light beam reflects from the surface.

Exciton: An excitation of a solid-state system, consisting of an excited electron and hole, interacting via their mutual Coulomb attraction. Typically excitons are observed in the optical spectra near the absorption threshold in semiconductors and insulators.

Extinction Coefficient (κ): The imaginary part of the complex index of refraction that describes the exponential decay of the electric field amplitude with distance for a light wave traveling within a medium.

Forbidden Interband Transition: A transition in which the excitation of an electron from one band to another would be possible on the basis of energy and momentum conservation considerations, but does not occur because of selection rules on the momentum matrix element. For example, in a centrosymmetric crystal, a direct transition between valence and conduction-band states of the same parity is forbidden at the center of the Brillouin zone.

Gyration Tensor ($[G]$): A second-rank, antisymmetric tensor, which, together with the real dielectric tensor $[\epsilon]$, provides a complete description of the linear response of a nonabsorbing, optically active solid to an electromagnetic wave.

Index Ellipsoid: Also called the optical indicatrix. A construction consisting of a three-dimensional surface used to characterize wave propagation in an anisotropic solid that is not optically active. For a nonabsorbing solid, the length of the vector drawn in the direction of the optical displacement **D** (or direction of polarization) from the origin to a given point on the surface is equal to the real index of refraction for the wave.

Index of Refraction, Complex ($\tilde{N}$): A complex quantity whose real part is the real index of refraction and whose imaginary part is the extinction coefficient. It describes electromagnetic wave propagation in a medium.

Index of Refraction, Real (n): The real part of the complex index of refraction that defines the phase speed v_p of a light wave traveling in a medium, according to $v_p = c/n$, where c is the speed of light in vacuum.

Index Tensor ($[a]$): Also called the impermeability tensor. A second-rank tensor that is the inverse of the real dielectric tensor in a nonabsorbing solid. It relates the optical electric field vector $\mathcal{E}$ and displacement vector **D** according to $\mathcal{E} = [a] \cdot \mathbf{D}$.

Indirect Interband Transition: A transition in which an electron is excited from a Bloch state in one band to a Bloch state in another band such that the wave vector of the electron is not conserved. The crystal momentum difference between the initial and final states is either emitted or absorbed by a phonon or defect in the crystal.

Intraband Transition: A transition of an electron from a state at or below the Fermi energy to a state above the Fermi energy, but within the same band. Such transitions occur in metals and in degenerately doped semiconductors.

Irradiance: The energy transported by a light beam across a unit area perpendicular to the ray direction, in a unit of time.

Linear Birefringence: A characteristic of anisotropic crystals in which two light waves having the same propagation vector, but orthogonal linear polarizations, experience different real indices of refraction and travel at different phase speeds. One wave has an isotropic propagation velocity (ordinary wave) and the other does not (extraordinary wave).

Linear Dichroism: A characteristic of optically absorbing, anisotropic crystals, analogous to linear birefringence; however, the two orthogonal linearly polarized waves experience different extinction coefficients.

Magneto-optic Effects: Optical anisotropies induced in a solid by the application of an external magnetic field. In solids, the important effects are the Faraday and the magneto-optical Kerr effects, observed in transmission and reflection, respectively. These effects are usually characterized by the rotation of the direction of polarization of an incident linearly polarized wave, such that the rotation is linear in the strength of the applied field.

Normal Surface: Construction alternative to the index ellipsoid that consists of a three-dimensional surface used to characterize wave propagation in an anisotropic solid that is not optically active. For a nonabsorbing solid, the length of the vector drawn in the direction of propagation from the origin to a given point on the surface is equal to either the phase speed or the real index of refraction for the wave, depending on which formalism is used.

Optical Conductivity ($\tilde{\sigma}$): A macroscopic description of the optical properties of an optically isotropic solid, alternative to the complex dielectric function. It is given by $\tilde{\sigma} = (i\omega/4\pi)(1 - \tilde{\epsilon})$, where $\tilde{\epsilon}$ is the complex dielectric function.

Phonon: A quantum of energy associated with the atomic displacement field in a solid.

Photon: A quantum of energy associated with electromagnetic radiation.

Piezo-optic Effect: A modification of the index ellipsoid or other optical characteristics of a solid when it is subjected to mechanical stress under static conditions. In nonabsorbing solids, it is characterized by a fourth-rank piezo-optic tensor that operates on the stress tensor to obtain the change in the index tensor.

Plasmon: A quantum of energy associated with the collective oscillation of electrons (i.e., charge-density fluctuations) in a solid.

Principal Dielectric Axes: Coordinate axes associated with an anisotropic medium (not optically active) for which the real part of the dielectric tensor is diagonalized. For crystalline solids of orthorhombic and higher symmetries, the principal dielectric axes coincide with the crystallographic axes.

Principal Indices of Refraction: The square roots of the diagonal components of the dielectric tensor for a nonabsorbing anisotropic solid that is not optically active, as expressed in the coordinate system of the principal dielectric axes.

Reflectance: A property of single or multiple interfaces that characterizes a specular surface. It is defined as the ratio of the reflected to incident irradiances in a beam that is reflected from the surface.

Transmittance: A property of single or multiple interfaces that characterizes a solid with two plane-parallel, specular surfaces. It is defined as the ratio of the transmitted to incident irradiances in a beam that passes through the solid.

Works Cited

Agranovich, V. M., Ginzburg, V. L. (1966), *Spatial Dispersion in Crystal Optics and the Theory of Excitons*, London: Wiley.

Ashcroft, N. W., Mermin, N. D. (1976), *Solid State Physics*, New York: Holt, Rinehart and Winston.

Ashcroft, N. W., Sturm, K. (1971), *Phys. Rev. B* **3**, 1898–1910.

Aspnes, D. E. (1976), in: B. O. Seraphin (Ed.), *Optical Properties of Solids: New Developments*, Amsterdam: North-Holland, p. 799.

Aspnes, D. E. (1980a), in: M. Balkanski, T. S. Moss (Eds.), *Handbook on Semiconductors*, Vol. 2, Amsterdam: North-Holland.

Aspnes, D. E. (1980b), *J. Opt. Soc. Am.* **70**, 1275–1277.

Aspnes, D. E., Kelso, S. M., Logan, R. A., Bhat, R. (1986), *J. Appl. Phys.* **60**, 754–767.

Azzam, R. M. A., Bashara, N. M. (1977), *Ellipsometry and Polarized Light*, Amsterdam: North-Holland.

Baym, G. (1973), *Lectures on Quantum Mechanics*, Reading: W. A. Benjamin.

Boccara, A. C., Pickering, C., Rivory, J. (Eds.) (1993), *Spectroscopic Ellipsometry*, Elsevier: Amsterdam.

Born, M., Wolf, E. (1970), *Principles of Optics*, Oxford: Pergamon.

Cardona, M. (1969), *Modulation Spectroscopy*, New York: Academic.

Chen, D. (1974), *Appl. Opt.* **13**, 767–778.

Chindaudom, P., Vedam, K. (1994), *Appl. Opt.* **33**, 2664–2671.

Cody, G. D. (1984), in: J. I. Pankove (Ed.), *Semiconductors and Semimetals*, Vol. 21B, New York: Academic, p. 11.

Cohen, M. L., Chelikowsky, J. R. (1982), in: W. Paul (Ed.), *Handbook on Semiconductors*, Vol. 1, Amsterdam: North-Holland.

Collins, R. W. (1990), *Rev. Sci. Instrum*, **61**, 2029–2062.

Cook, W. R., Jaffe, H. (1979), in: K.-H. Hellwege, A. M. Hellwege (Eds.), *Landolt-Börnstein: Numerical Data and Functional Relationships in Science and Technology*, New Series, Group 3, Vol. 11, *Elastic, Piezoelectric, Pyroelectric, Piezooptic, Electrooptic Constants, and Nonlinear Dielectric Susceptibilities of Crystals*, Berlin: Springer, pp. 552–651.

Dawson, R. M., Li, Y. M., Gunes, M., Heller, D., Nag, S., Collins, R. W., Wronski, C. R., Bennett, M.,

Li, Y.-M. (1992), *Mater. Res. Soc. Symp. Proc.* **258**, 595–600.

Drude, P. K. L. (1965), *Theory of Optics*, New York: Dover.

Ehrenreich, H. (1966), in: J. Tauc (Ed.), *The Optical Properties of Solids*, New York: Academic, p. 106.

Feng, G. F., and Zallen, R. (1989), *Phys. Rev. B* **40**, 1064–1073.

Freiser, M. J. (1968), *IEEE Trans. Magn.* **MAG-4**, 152–161.

Guenther, R. (1990), *Modern Optics*, New York: Wiley.

Gunter, P., Huignard, J.-P. (Eds.) (1988; 1989), *Photorefractive Materials and Their Applications*, Vols. 1 and 2, Berlin: Springer.

Heavens, O. S. (1965), *Optical Properties of Thin Solid Films*, New York: Dover.

Jackson, J. D. (1975), *Classical Electrodynamics*, New York: Wiley.

Kim, C. C., Garland, J. W., Raccah, P. M. (1992), *Phys. Rev. B* **45**, 11749–11767.

Korpel, A. (1988), *Acousto-optics*, New York: Marcel Dekker.

Landau, L. D., Lifshitz, E. M. (1965), *Electrodynamics of Continuous Media*, Reading, MA: Addison-Wesley.

Lautenschlager, P., Garriga, M., Logothetidis, S., Cardona, M. (1987), *Phys. Rev. B* **35**, 9174–9189.

Levi, L. (1980), *Applied Optics*, Vol. 2, New York: Wiley.

Logothetidis, S., Viña, L., Cardona, M. (1985), *Phys. Rev. B* **31**, 2180–2189.

Maley, N. (1992), *Jpn. J. Appl. Phys.* **31**, 768–769.

Merzbacher, E. (1970), *Quantum Mechanics*, New York: Wiley.

Narasimhamurty, T. S. (1981), *Photoelastic and Electro-optic Properties of Crystals*, New York: Plenum.

Nguyen, H. V., An, I., Collins, R. W. (1993), *Phys. Rev. B* **47**, 3947–3965.

Nguyen, H. V., Collins, R. W. (1993), The Pennsylvania State University: unpublished data.

Nye, J. F. (1976), *Physical Properties of Crystals*, Oxford: Clarendon.

Otto A. (1974), *Festkoerperprobleme* **14**, 1–37.

Otto, A. (1976), in: B. O. Seraphin (Ed.), *Optical Properties of Solids: New Developments*, Amsterdam: North-Holland, p. 677.

Palik, E. D. (Ed.) (1985), *Handbook of Optical Constants of Solids*, New York: Academic.

Pankove, J. I. (1971), *Optical Processes in Semiconductors*, New York: Dover.

Paul W. (1966), in: J. Tauc (Ed.), *The Optical Properties of Solids*, New York: Academic, p. 257.

Pines, D. (1963), *Elementary Excitations in Solids*, New York: W. A. Benjamin.

Raether, H. (1977), in: G. Hass (Ed.), *Physics of Thin Films*, Vol. 9, New York: Academic, p. 145.

Ramachandran, G. N., Ramaseshan, S. (1961), in: S. Flügge (Ed.), *Handbuch der Physik*, Vol. 25/1, Berlin: Springer, p. 1.

Sapriel, J. (1979), *Acousto-Optics*, New York: Wiley.

Tauc, J., Grigorovici, R., Vancu, A. (1966), *Phys. Status Solidi* **15**, 627–637.

Thomas, R. E. (1991), in: E. D. Palik (Ed.), *Handbook of Optical Constants of Solids II*, New York: Academic, p. 177.

Vedam, K. (Ed.) (1994), *Physics of Thin Films*, Vol. 19, New York: Academic.

Wooten, F. (1972), *Optical Properties of Solids*, New York: Academic.

Yamaguchi, T., Takahashi, H., Sudoh, A. (1978), *J. Opt. Soc. Am.* **68**, 1039–1044.

Yariv, A., Yeh, P. (1984), *Optical Waves in Crystals*, New York: Wiley.

Yeh, P. (1988), *Optical Waves in Layered Media*, New York: Wiley.

Xu, J., Stroud, R. (1992), *Acousto-optic Devices*, New York: Wiley.

Zallen, R. (1983), *The Physics of Amorphous Solids*, New York: Wiley.

Ziman, J. M. (1972), *Principles of the Theory of Solids*, Cambridge, UK: Cambridge University.

Further Reading

Classical Texts on The Optical Behavior of Electrons in Solids

Drude, P. K. L. (1965), *Theory of Optics*, New York: Dover.

Lorentz, H. A. (1952), *The Theory of Electrons*, New York: Dover.

Selected Background Texts on Electromagnetic Waves and Optics

Born, M., Wolf, E. (1970), *Principles of Optics*, Oxford: Pergamon.

Jackson, J. D. (1975), *Classical Electrodynamics*, New York: Wiley.

Landau, L. D., Lifshitz, E. M. (1965), *Electrodynamics of Continuous Media*, Reading, MA: Addison-Wesley.

Selected Background Texts on Quantum Mechanical Principles

Baym, G. (1973), *Lectures on Quantum Mechanics*, Reading, MA: W. A. Benjamin.

Merzbacher, E. (1970), *Quantum Mechanics*, New York: Wiley.

Selected Background Texts on the Theory of Solids

Ashcroft, N. W., Mermin, N. D. (1976), *Solid State Physics*, New York: Holt, Rinehart and Winston.

Phillips, J. C. (1973), *Bonds and Bands in Semiconductors*, New York: Academic.

Ziman, J. M. (1972), *Principles of the Theory of Solids*, Cambridge, UK: Cambridge University.

Selected Background Texts on Crystal Physics

Juretschke, H. J. (1974), *Crystal Physics*, Reading, MA: W. A. Benjamin.

Mason, W. P. (1966), *Crystal Physics of Interaction Processes*, New York: Academic.

Nye, J. F. (1976), *Physical Properties of Crystals*, Oxford: Clarendon.

Readings in Crystal Optics

Ramachandran, G. N., Ramaseshan, S. (1961), in: S. Flügge (Ed.), *Handbuch der Physik*, Vol. 25/1, Berlin: Springer, p. 1.

Szivessy, G. (1928), in: S. Flügge (Ed.), *Handbuch der Physik*, Vol. 20, Berlin: Springer, p. 635.

Yariv, A., Yeh, P. (1984), *Optical Waves in Crystals*, New York: Wiley.

Texts Including Modern Theories of the Optical Properties of Solids

Bassani, F., Pastori Parravicini, G. (1975), *Electronic States and Optical Transitions in Solids*, Oxford: Pergamon.

Greenaway, D. L., Harbeke G. (1968), *Optical Properties and Band Structure of Semiconductors*, Oxford: Pergamon.

Hodgson, J. N. (1970), *Optical Absorption and Dispersion in Solids*, London: Chapman and Hall.

Moss, T. S. (1959), *Optical Properties of Semiconductors*, London: Butterworth.

Sokolov, A. V. (1967), *Optical Properties of Metals*, New York: American Elsevier.

Wooten, F. (1972), *Optical Properties of Solids*, New York: Academic.

Collections of Articles Devoted to Modern Topics in the Optical Properties of Solids

Abeles, F. (Ed.) (1966), *Optical Properties and Electronic Structure of Metals and Alloys*, New York: Academic.

Abeles, F. (Ed.) (1972), *Optical Properties of Solids*, Amsterdam: North-Holland.

Nudelman, S., Mitra, S. S. (Eds.) (1969), *Optical Properties of Solids*, New York: Plenum.

Seraphin, B. O. (Ed.) (1976), *Optical Properties of Solids: New Developments*, Amsterdam: North-Holland.

Tauc, J. (Ed.) (1966), *The Optical Properties of Solids*, New York: Academic.

Sources of Data on the Optical Properties of Solids

Aspnes, D. E., Studna, A. A. (1983), *Phys. Rev. B* **27**, 985–1009.

Palik, E. D. (Ed.) (1985), *Handbook of Optical Constants of Solids*, New York: Academic.

Palik, E. D. (Ed.) (1991), *Handbook of Optical Constants of Solids II*, New York: Academic.

Texts Devoted to Thin-Film Optics

Heavens, O. S. (1965), *Optical Properties of Thin Solid Films*, New York: Dover.

Macleod, H. A. (1986), *Thin Film Optical Filters*, Bristol: Adam Hilger.

Vasicek, A. (1960), *Optics of Thin Films*, Amsterdam: North-Holland.

Yeh, P. (1988), *Optical Waves in Layered Media*, New York: Wiley.

OPTICAL SCANNERS

LEO BEISER, *Leo Beiser Inc., Flushing, New York, U.S.A.*

INTRODUCTION

Optical scanning transforms spatial information $f(x,y,z)$ *to* or *from* a temporal signal $f(t)$, to adapt to electronic processing. A spatial continuum is analyzed or synthesized as an array of picture elements—"pixels," "pels," or "voxels." Scanning in this context is a for-

3-527-28134-7/95/$5.00 + .50

matted convolution of an optical "point" intensity distribution over a field of information, or a sampling by the point when "digitized" (Bouwhuis and Bratt, 1983).

The addressing process may be active or passive. When active, as for "flying spot scanning," a portion of the flux scattered from a spot of light traversing an object is directed to photodetector(s) to form a signal that is representative of the reflectance–scatter–transmittance of the object along the scanned path. When passive (e.g., remote sensing), an optical image of the object is dissected to form a signal—either by focusing the image on an array of photodetector elements, as discussed in the articles SENSORS, OPTICAL and SENSORS, INFRARED, or by interposing an optical scanner that directs sampled portions of the radiant flux from the remote scene to photodetector(s) to form a signal, as considered here.

The reassembly (recording or display) process is almost always active: an intensity-modulated point of light is scanned across an area in a pattern, either for recording on a photosensitive medium or for display.

Table 1 identifies some prominent applications of optical scanning technology—active scanning: inputs (scanners) or outputs (recorders or displays), and passive scanning (remote sensing). Active scanning techniques for interrogating a field of information (inputs) or for reconstructing it (outputs) can be

Table 1. Some applications of optical scanning technology.

Active (laser) scanning	
Inputs	**Outputs**
Image digitizing	Image recording/printing
Bar code reading	Color image reproduction
Optical inspection	Medical image recording
Confocal microscopy	Data marking and engraving
Optical character recognition	Microimage recording
Graphic arts camera	Optical data storage
Color separation	Phototypesetting
Robot vision	Earth resources recording
Laser radar	Data/image display
Passive (remote sensing)	
Night vision	Earth resources sensing
Surveillance	Medical image sensing
Gun sighting	Thermal fault detection
Weather monitoring	Nondestructive testing
Missile tracking	Forest fire monitoring

very similar. Both utilize a "flying spot" of light. While some passive scanners may appear interchangeable with those for active scanning, this option is often precluded by the dispersive characteristics of diffractive devices. Even for (nondispersive) mirrored deflectors, their differently filled apertures impart distinctions in resolution and duty cycle. This is expressed further in Sec. 2. Extensive detail on remote-sensing scanner configurations is provided elsewhere (Beiser and Johnson, 1994; Wolfe, 1989).

The high intensity and purity of the laser has been adapted most effectively since the early 1960s as the preferred light source for scanning at high speed and high resolution. This article concentrates on the field of laser scanning. Continuous-wave (cw) lasers are employed almost exclusively. They allow arbitrary sampling rates and offer a broad spectrum of selectable wavelengths, encompassing the ultraviolet, visible, and infrared.

An interesting application is exemplary. Scanning confocal microscopy (Pawley, 1980) utilizes typically an x-y–scanned laser beam, which is directed through the eyepiece of an optical microscope (see OPTICAL MICROSCOPY). Focused as a fine point and scanned over the specimen, a portion of the scattered light is retrodirected and returned through the same scanning optics ("de-scanned") to form a stabilized and refocused stationary point, optically conjugate to the one that scans the specimen. When filtered spatially with a pinhole, only the flux from a limited focal region is permitted to pass to a photodetector, suppressing scattered noise from neighboring regions of the specimen. This detected electrical signal, after processing, yields a high-resolution and higher contrast x-y image at a selected depth z, allowing tomographic sectioning and analysis of the specimen—composed now of elementary volume elements (x-y-z pixels) termed "voxels." This is classed as an input system in Table 1.

Other input systems operate similarly, but do not necessarily retrodirect the scattered light back through the same optical system for de-scanning and filtering. Instead, the scattered flux may be collected and directed to large-area detectors for conversion to an electrical signal, which forms the input for further processing.

Systems in the Outputs column record or display the processed information. They function more directly, entailing typically no collection nor detection of scattered energy. Instead, the incident energy of the focused scanned beam serves to impart a physical change to a photosensitive medium. By far, the most prominent of this output category is the ubiquitous laser printer. If the incident laser beam energy is high enough, it can provide for data marking or engraving—or even cutting via the controlled absorbed and dissipated thermal energy. In almost all output devices, an additional dimension of intensity modulation of the beam is required while scanning, so that the information can be conveyed to the image medium at the right intensity and at the right place. Although this article is limited in scope to spatial scanning, some perception of a dominant method of light modulation—acousto-optic—will be apparent from the discussion of acousto-optic deflection.

Laser scanners are designated as *preobjective, objective,* and *postobjective* (Beiser 1974, 1992b). Figure 1 indicates the scan regions within a general conjugate imaging system that transfers rays from an object point P_o, referenced to a fixed laser beam, to its image point P_i, which is scanned. The reciprocal nature of conjugate imaging is manifest in remote sensing, where the direction of ray travel is reversed, whereby the radiation from different portions of the source or object (P_i on the right) is accessed point for point by the scanner and imaged to P_o upon a fixed detector on the left. Section 2 discusses further distinctions of the nomenclature and scanner utilization.

Objective scanning appears typically as a *translation* of the objective lens (transversely) with respect to the information medium. This includes translation of the objective lens or the medium, or both. In contrast, *preobjective* and *postobjective* scanning impart an *angular change* to the optical beam. Although lenticular elements can be added to an angular scanner, it is seldom so configured. In holographic scanning, however (Sec. 2.2), the hologram can serve as an objective lens and scanner simultaneously. These principles are expanded in Sec. 1 in discussing scanner architecture.

Section 2 covers the principal scanner devices and techniques, including the dominant rotating mirror, rotating hologram, vibrational mirror, and nonmechanical scanners.

Section 3 presents significant aspects of scanned resolution, including the unique effects of aperture shape and illumination filling upon active (laser) scanning.

Section 4 addresses the need for and techniques of scan error reduction, to achieve high accuracy in beam positioning and freedom from artifacts due to line misplacements.

Finally, Sec. 5 summarizes the characteristics of the principal scanner options.

Though the coherence of the laser beam is relative, this work assumes that the radiation exhibits sufficiently high coherence to develop distinctive diffractive phenomena. High monochromaticity is also implicit. Even the laser diode, which can exhibit significant departure from monochromacity and coherence, unless otherwise identified and qualified, is considered on first analysis as providing coherent radiation. Accommodation for

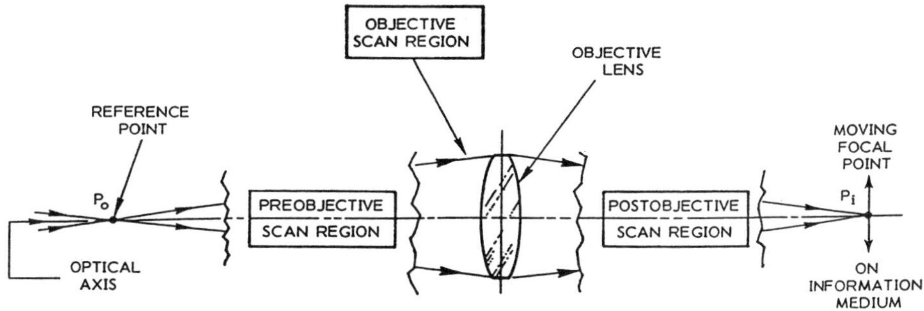

FIG. 1. Conjugate imaging system, with identified scan regions. Ray directions represent active (laser) scanning, where P_o is a point source. Passive (remote sensing) ray directions are reversed, where (scattered) flux from P_i propagates back through the optics to detector at P_o. System and nomenclature distinctions are discussed in Sec. 2.0 (Beiser, 1992b).

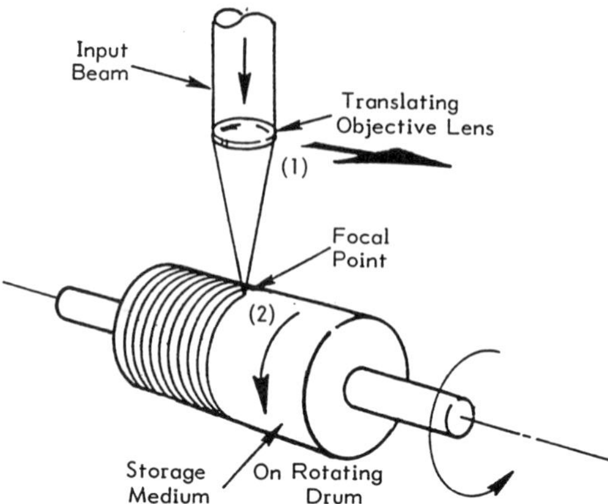

FIG. 2. Drum configuration executing two forms of objective scan: (1) translating objective lens and (2) translating storage medium (on rotating drum) (Beiser, 1992b).

the anticipated departure from coherent perfection and consequential aberration is required, as in any practical design.

1. SCANNER ARCHITECTURE AND OPTICAL TRANSFER

The Introduction presented the scanner classifications, associated with regions shown in Fig. 1, as *preobjective*, *objective*, and *postobjective*—represented within a general conjugate optical transfer. This section expresses their characteristics and typical optical configurations.

1.1 Objective Scan (Translational)

Translation of an objective lens transverse to its axis translates the imaged focal point. (Axial lens translation, which optimizes focus, is not considered scanning in this context.) Translation of the information medium with respect to a fixed objective lens renders the same effect—both termed objective scan. Both forms are illustrated as operating simultaneously in Fig. 2, the configuration of a drum scanner. Objective scan is limited in speed generally by the effort required to move relatively massive components.

1.2 Preobjective Scan (Angular)

Preobjective scan is the principal means for providing a flat image field (notably beyond the small scan angle when $\Theta \approx \tan \Theta$, which may be considered linear and having no dynamic focus to aid forming a flat field). This is exemplified in Fig. 3 by angularly scanning a laser beam into a "flat-field" or "f-Θ" lens—a major related component, discussed further in this section.

1.3 Postobjective Scan (Angular)

Postobjective scan, which is radially symmetric (Sec. 3.3.2) as shown in Fig. 4, generates a perfectly circular scan locus. Departure from radial symmetry (e.g., incident focal point not on the axis) generates a noncircular (e.g., limaçon) scan (Beiser, 1974), except for the following special case.

A postobjective mirror *with its surface on its rotation axis* generates a perfectly circular scan locus, illustrated in Fig. 5. The input beam is focused beyond the axis at point o. Scan magnification (Sec. 3.3.2) $m = 2$.

1.4 Objective Scan (Angular)

An alternative to postobjective scan that generates a perfectly circular locus is that of a beam illuminating a rotary scanner radially symmetric, as illustrated in Fig. 4, with its objective lens located, instead, in the output region of the scanner and coupled rigidly thereto. This is instrumented by moving the

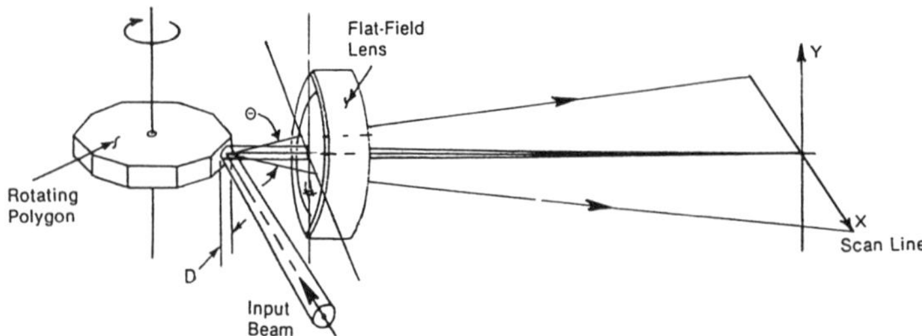

FIG. 3. Preobjective scan, prismatic polygon. Flat-field objective lens transforms angular change to (nominally) linear x displacement. When the mirror surfaces are parallel to the rotating axis, all beams can reside in a plane normal to the axis. Scan magnification $m = 2$ (Beiser, 1992b).

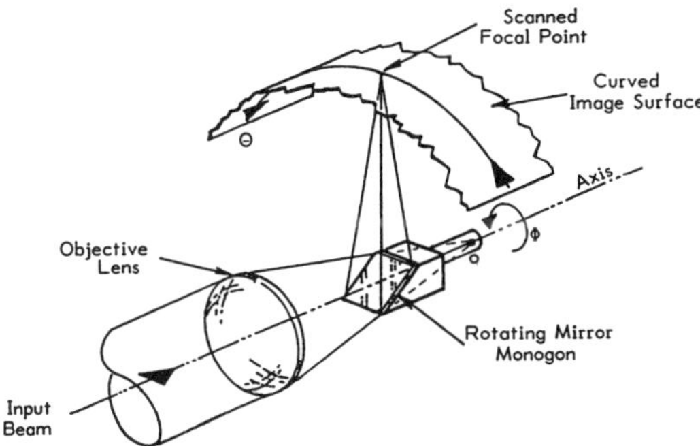

FIG. 4. Postobjective scan, monogon. Generates a curved image. When input beam is focused on the axis (at point *o*), system is radially symmetric, whence image locus forms section of a perfect circle. Scan magnification $m = 1$ (Beiser, 1992b).

objective lens of Fig. 4 (having proper focal length) into the output beam and fastening it effectively to the scanner shaft. In objective scan (angular), the lens is displaced radially from and rotates with the scanner.

1.5 Objective Optics

The objective lens, which converges the scanned beam to a moving focal point, forms an integral part of the architecture. The scanner and its optics are intimately related, as represented throughout many portions of this work. This section expands on some of the characteristics of objective optics.

1.5.1 On-Axis Optics The simplest objective lens is one that appears before the deflector, as in Fig. 4, where it is required only to focus a (typically) monochromatic beam

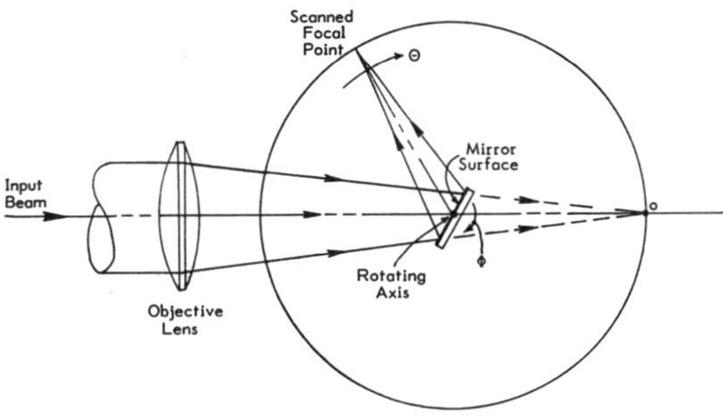

FIG. 5. Postobjective scan with mirror surface on rotating axis. Input beam is intercepted by mirror and reflected to scan a circular locus. Scan magnification $m = 2$ (Beiser, 1992b).

on axis (no field angle). The (postobjective) deflector intercepts the converging beam to scan its focal point—ideally, conducted aberrationlessly. This arrangement, when radially symmetric, can scan a perfectly circular arc—applied, for example, to an "internal drum" scanner, in which the information medium is mounted to the inner cylindrical surface of a fixed drumlike support. The scanner is, in turn, mounted on a linear traverse mechanism that guides it along the axis of the drum as the scanner rotates. The lens, nonrotating, is maintained spaced properly from the scanner.

1.5.2 Flat-Field Optics This important lens configuration forms a flat field by transforming an arced beam scan to a straight-line focal locus. The deflector is preobjective. The most prevalent arrangement is similar to that of Fig. 3, in which the input beam is assumed collimated.

Such lenses adapt to a variety of scanners, including polygonal, galvanometric, acoustooptic, electro-optic, and holographic, all discussed in Sec. 2. The lens must accept the scanned angle Θ from the aperture D and converge the beam throughout the scanned field to a best focus along a straight-line locus (Hopkins and Stephenson, 1991). Depending on the magnitudes of Θ and D, the *f*-number of the converging cone, and the desired perfection of straight-line focus and linearity, the lens assembly can be composed of from one to seven (or more) elements, with an equal number of choices of index of refraction, two to 14 (or more) surfaces, and up to six (or more) lens spacings—all representing the degrees of freedom available to the lens designer to accommodate performance. A typical arrangement of three elements is illustrated in Fig. 6.

1.5.3 Telecentricity Telecentric optics provides for the chief ray of the output beam to be maintained parallel to the optical axis throughout scan. Figure 7 illustrates a high-performance telecentric scan lens, which is represented schematically in Fig. 8. Interposed one focal length f between both the entrance (scanning) aperture D and the flat image surface, the ideal lens transforms the *angular* change at the input to a *translation* of the output cone. The degree of telecentricity is expressed by the angular departure from normal landing of the chief ray.

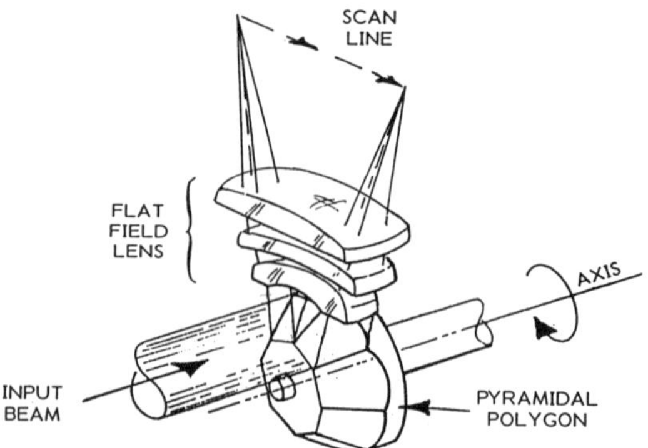

FIG. 6. Pyramidal polygon (overilluminated or overfilled). Paraxial input beam is scanned generally perpendicular to axis. When input beam illuminates two facets, one facet is active at all times, yielding up to 100% duty cycle. Facet is full optical aperture, minimizing polygon size for a given resolution, but wasting illumination beyond the used facet. If operate underilluminated (as in Fig. 11) to conserve illumination efficiency, requires increased polygon size to attain high duty cycle (see Sec. 1.7). Lens elements shown cut for illustration (Beiser, 1992b).

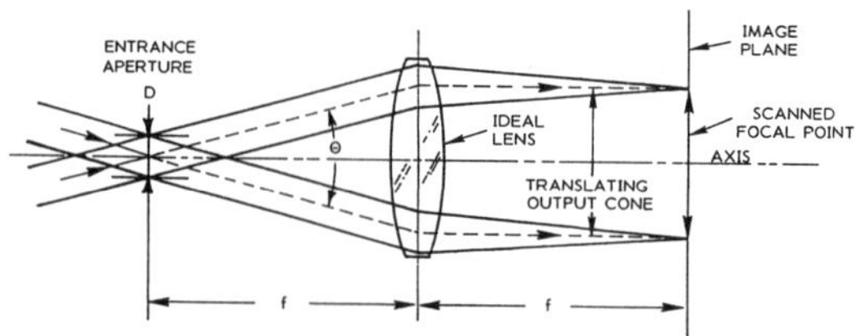

FIG. 8. Telecentric optical schematic. Ideal lens transforms angular scan Θ from aperture D to translational scan, which lands normal to the image plane (Beiser, 1992b).

Telecentricity is applied to restrict the spread or misplacement of the focal spot (due to non-normal projection) and/or to retroreflect the probing beam efficiently for calibration or measurement. This facility comes dearly, however, for the final lens elements (Fig. 7) must be at least as wide as the desired scan format. A further requirement is "f-Θ" correction to the fundamental nonlinearity of the system of Fig. 8, in which the spot displacement is proportional to $f \tan\Theta$ rather than to $f\,\Theta$ directly. This is responsible, in part, for the lens complexity in Fig. 7.

1.5.4 Double-Pass and Beam Expansion

The objective lens may be adapted to double-pass, as depicted in Fig. 9. The lens assembly serves first as the collimating portion of a lenticular beam expander (Beiser 1974; O'Shea, 1985) and, upon reflection of the beam by the scanner, as a conventional flat-field lens. This provides compaction by avoiding allocation of extra space ("pupil relief") for the input beam (Sec. 2.1.3). And, with the illuminating beam approaching normal landing upon the undeflected (neutral position)

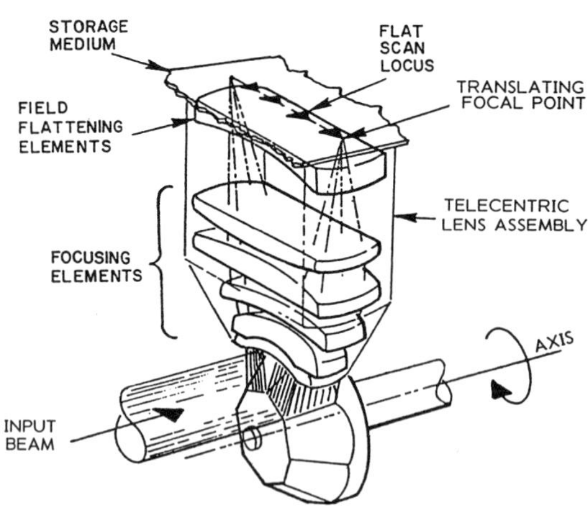

FIG. 7. Telecentric flat-field lens at output of pyramidal polygon scanner operating radially symmetric. Lens elements shown cut for illustration (Beiser, 1992b).

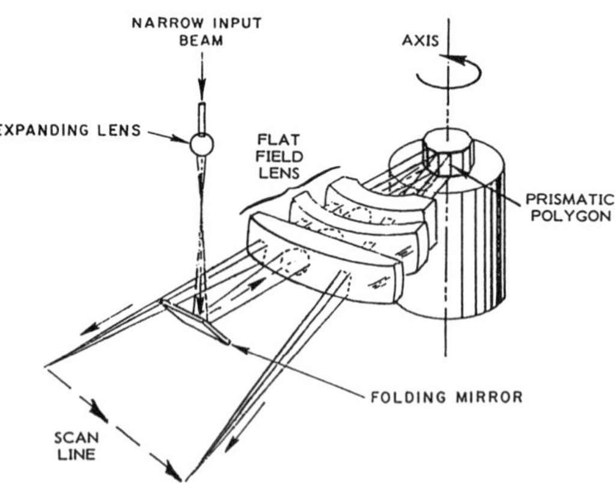

FIG. 9. Prismatic polygon in double-pass configuration. Input beam, focused by positive lens, expands beyond focus and is picked off by folding mirror to be directed through flat-field lens and collimated to illuminate polygon facets. Reflected beam is scanned by polygon and reconverged by flat-field lens to focus on scan line. Input and output beams are slightly skewed above and below lens axis for separation by folding mirror. Lens elements shown cut for illustration (Beiser, 1992b).

facet, the beam and facet undergo minimum enlargement, conserving the size of the deflector. Increased scatter may result, however, because of the traversal of the beam twice through the lens surfaces.

A slight skew of the input and output planes, per Fig. 9, avoids obstruction of the input and scanned beams at the folding mirror. An alternative method is to make the lens aperture wide enough to include injection of the input beam from the side, at an angle sufficiently off axis to avoid obstruction of the reflected scanned beam (Hopkins and Stephenson, 1991). This imposes an off-axis angle, and associated facet enlargement and beam aberration, but allows all beams to remain in the same plane normal to the axis, avoiding a scanned bow that develops in the above skewed method.

The requirement for beam expansion is fundamental for acquisition of the aperture width D that provides a desired scanned resolution. An optical beam that is narrower than required is broadened by propagating it through a (inverted telescope) beam expander—an afocal lens group having the shorter focal length followed by the longer focal length. If reversed, it forms "beam compression." Introduction of the phrase "beam expander" in 1964 by Leo Beiser and its dissemination to generic form is summarized elsewhere (Beiser, 1988, Appendix 1). Beam expansion or compression can also be achieved nonlenticularly with prismatic elements (Lohman and Stork, 1980; Marchant, 1988).

1.6 Power Transfer from a Coherent Source

The determination of power transfer is much simplified with the utilization of a monochromatic (single-line laser) source. Even if it radiates several useful lines, they are each sufficiently narrow, compared to the spectral breadth of most transmission and detection media, that single point evaluations at the radiation wavelengths are usually adequate. Further, in contrast to the typical increase in focused spot size with increased electron beam current of a cathode-ray tube, the size of the focused laser spot remains essentially constant with power variation, allowing for exposure of extremely fine spots with extremely high power densities.

Accordingly, the laser power required to irradiate a photosensitive material or detector is expressed as (Beiser, 1966)

$$P = \frac{sR}{T}\left(\frac{A}{t}\right) \text{ watts,} \tag{1}$$

in which s = material sensitivity, J/cm^2; R = reciprocity failure factor, ≥ 1; T = optical throughput efficiency, ≤ 1; A = exposed area, cm^2; and t = time to scan over area A, s.

The reciprocity failure factor R accommodates an exposure interval that is sufficiently short to elicit a loss in sensitivity of the photosensitive medium (typical of silver halide media). If the A/t value is taken as an entire frame of assumed uniform interval including blanking, then the two-dimensional η (Sec. 1.8) must appear in the denominator to restrict the frame time to that which occupies only the exposed area.

The throughput efficiency T accounts for loss factors including those due to absorption, reflection, scatter, diffraction, polarization, and beam truncation or vignetting—requiring disciplined attention. While the radiation from fundamental-mode laser sources is essentially conserved in propagating through sufficiently large apertures, large scanner apertures burden its size and its required drive power. To evaluate the throughput power consistent with aperture size, Fig. 10 is useful. The data are generalized to an elliptic beam spot, accommodating the irradiance of typical laser diodes. Figure 10(a) shows an irradiance distribution having ellipticity $\epsilon = w_x/w_y$ (w's at $1/e^2$ intensity) apertured by a circle of radius r_0. Figure 10(b) plots the encircled power (%) vs the ellipticity, with the ratio r_0/w_x as a parameter. When $\epsilon = 1$, it represents the circular Gaussian beam. (Nomenclature of Fig. 10 is as published. The w's in the figure correspond to $D/2$ as used here, and the r_0 in the figure corresponds to $W/2$ as used here.)

1.7 Overillumination and Underillumination (Overfilling and Underfilling)

In overillumination, the light flux encompasses the entire useful aperture. This is often implemented by illuminating at least two adjacent apertures (e.g., polygon facets) such

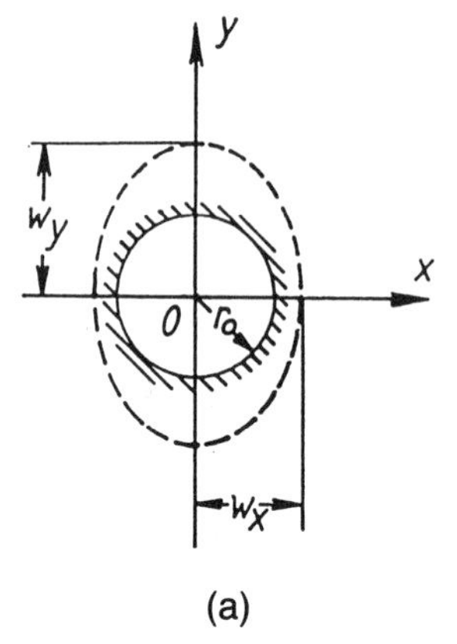

(a)

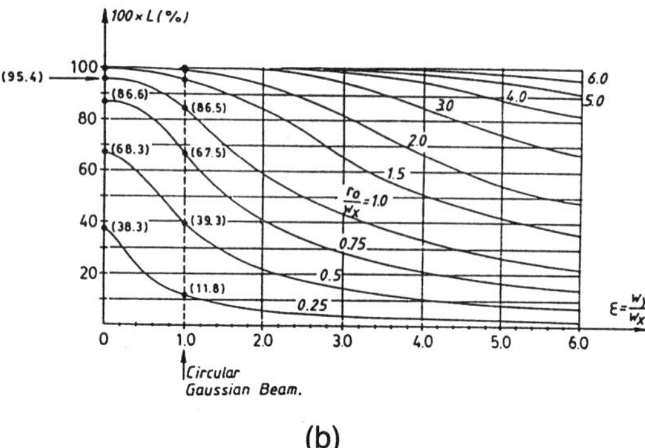

(b)

FIG. 10. Relationships of illuminating Gaussian beam size and its aperature size throughout power transfer. See nomenclature comment closing Sec. 1.6. (a) Irradiance of a single-mode laser beam, generalized to elliptical, centered within a circular aperture of radius r_0. (b) Variations of the encircled energy $100 \times L$ (%) vs ellipticity ϵ and the ratio r_0/w_x as a parameter (Li and Katz, 1991).

that the active one is always filled with light flux. This not only allows unity duty cycle, but provides for resolution to be maximized for two reasons: first, retrace loss may be reduced to zero; and second, the full available aperture width is operative throughout scan. The tradeoff is the loss of illuminating flux beyond the aperture edges and attendant reduction in optical power throughput (refer to Fig. 10). An alternative is to prescan the light flux synchronously with the path of the scanning aperture such that it is filled with illumination during its entire transit (Urbach *et al.*, 1982).

In underillumination, the light flux is incident on a portion of the available aperture. This maximizes the transfer of incident flux to the scanned output. However, if the scanner is a polygon, for example, then a substantive unused (blanking and retrace) interval can result, thereby depleting the duty cycle. Further, the polygon facet width must be made substantially larger than the required beamwidth (to maintain the beam fully on the facet during active scan), widening significantly the polygon facets [Eq. (7)] and its diameter, and imposing thereby constraints on its speed.

1.8 Duty Cycle

The active portion of a scan cycle is almost always accompanied by a blanking or retrace interval. This interval can include short overscan portions (straddling the main format), which may be used for radiometric and timing calibration. The ratio of the active portion to the full scan period is termed the duty cycle η, taken in each (usually quadrature) scan direction, expressed as

$$\eta = 1 - \tau/T, \tag{2}$$

in which τ is the blanking time and T is the scan period. The scan angle is limited to

$$\Theta = \eta\Theta_{\max}, \tag{3}$$

where $\Theta_{\max}$ is the scan angle available over the full scan period. A reduced duty cycle increases instantaneous bandwidth for a given average data rate. It also increases the laser power requirement for a given exposure energy, requiring the determination of the two-dimensional efficiency (product) when scanning an area.

1.9 Image Rotation and Derotation

In discussing the imaged spot (or array of spots), it is useful to utilize an optical descriptor termed the *point spread function* (PSF)—the intensity distribution about the image of an infinitesimal point object. Ideally, this imaged spot size is limited only by diffraction. Its intensity function is determined analytically as the two-dimensional Fourier transform of the intensity distribution of the illuminating beam at the aperture.

The rotation of an isotropic PSF about its axis is typically undetectable. If, however, the PSF is nonisotropic or polarized, or if an array of points is scanned to provide multiplexed beams (Beiser, 1988), certain scanning techniques can cause undesired rotation of the point or the array of points in the image.

Consider a monogon scanner, as in Fig. 4, where the input beam *over*illuminates the rotating mirror. The mirror delimits the beam, establishing a rectangular *x-y* cross section, which sustains its orientation during scan with respect to the image field. Thus, if uniformly illuminated, the focal point [in this case having a PSF of $(\sin^2 x/x^2)(\sin^2 y/y^2)$] maintains the same orientation along its scanned line. If the input beam is polarized, the axis of polarization of the imaged point will, however, rotate directly with mirror rotation within the rectangular PSF. Similarly will rotate any radial asymmetry (e.g., intensity or ellipticity) within the aperture.

If the same scanner is *under*illuminated with, for example, an elliptical beam, the axis of the imaged elliptic spot will rotate directly with the mirror. Similarly, if the scanner is illuminated with multiple beams to generate an in-line array of spots, the imaged array will rotate (generally undesirably) directly with mirror rotation.

This effect is identical in the pyramidal polygon, which is illuminated as shown in Fig. 6. Each mirror imparts the same rotation characteristics as the monogon of Fig. 4 through its scan angle Φ. The mirrors of Fig. 6 are also shown overilluminated, maintaining a stationary geometric PSF during scan (if uniformly illuminated), but subject to rotation of, for example, polarization. Similarly will rotate an elliptic beam or a multiple beam array.

Not so, however, for the mirror mounted as in Fig. 5 or for the prismatic polygon of Fig. 3. When the illuminating beam and the scanned beam are in a plane normal to the axis of mirror rotation, execution of scan does not alter the PSF, except for the possible vignetting of the aperture with variation in incident angle. In the prior cases, the incident angles remained constant while the image is subject to rotation. Here, the incident angles change, while the image develops no rotation.

The distinction is in the symmetry of the illumination/scanning system. The prior examples are radially symmetric (Sec. 3.3.2). The latter cases represent extreme radial asymmetry. While mirrored scanners seldom operate in regions between these two extremes, holographic scanners can (Sec. 2.2.3), creating possible complications with, for example, polarization. This is manifest in the variation in diffraction efficiency of gratings for *p* and *s* polarizations during their rotation (Kramer, 1991).

1.9.1 Image Derotation Methods Interposing another image-rotating component in the optical path can cancel that caused by the scanner. An image rotator inverts the image (Levi, 1968). With continuous rotation, it develops two complete rotations of the image per rotation of the component. Thus, it must be rotated at $\frac{1}{2}$ the angular velocity of the scanner. While the Dove prism* (Levi, 1968; Burke, 1969) is one of the most familiar components used for image rotation, other coaxial image inverters include

1. 3-mirror (K-mirror) inverter—an all-reflective form of the Dove prism;
2. cylindrical/spherical lens optical relay (Johnson, 1974);
3. Pechan prism,** allowing operation in a noncollimated beam (Levi, 1968).

2. SCANNER DEVICES AND TECHNIQUES

As expressed in the Introduction, many of the techniques addressed here for active (laser) scanning apply equally to passive remote sensing. Their reciprocal path characteristics can be realized effectively by reversing the positions of the light source(s) and detector(s) and their ray directions. Preobjective

*Dove prism: Typically, a glass member of square cross section, of length approximately four times its width. Opposite ends are cut at symmetric angles of approximately 45°. The ends and long base are polished flat. A collimated beam enters a (45°) face parallel to the base, is refracted toward the base, and internally reflected thereby toward the opposite (45°) face, where it is again refracted to exit parallel to the input beam.

**Pechan prism: Typically, two glass prisms having a small air space between their parallel faces. Prism angles are such that a beam entering one face is totally internally reflected at the interface, reflected again at another surface to propagate through the interface into the second prism, where it is reflected three times to exit the second prism paraxial to the input beam.

and postobjective laser scanning (Sec. 1) have their counterparts in *image space* and *object space* scanning for remote sensing (Beiser and Johnson, 1994). A notable distinction is in the option of underillumination or overillumination of the deflecting aperture in active imaging, while in remote sensing, the aperture is almost always filled fully with the collected flux from a remote source. This leads to the unique attention devoted to the aperture shape factor (Sec. 3.2) in active input/output imaging, seldom an issue in remote sensing. Typical in remote sensing is the need to accommodate a broader optical spectral range, while active laser operation is most often monochromatic. The use of spectrally broad reflective optics in remote sensing tends to ameliorate this distinction. Thus, spectrally dispersive (e.g., diffractive) devices are seldom utilized in remote sensing. This section forms an overview of the most frequently utilized scanning devices and techniques, with the mirrored reflective ones serving all forms of optical scanning.

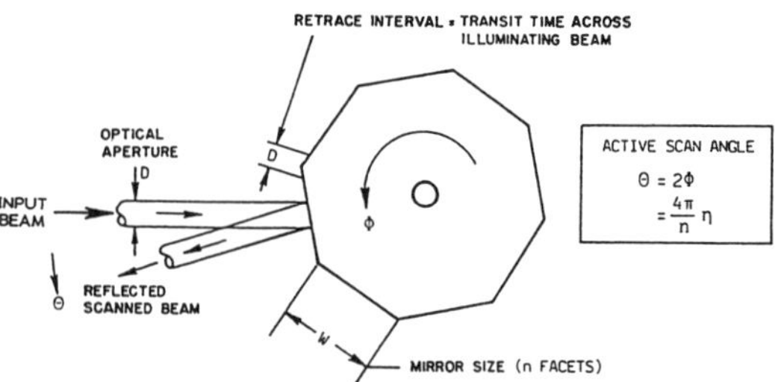

FIG. 11. Prismatic polygon (underilluminated or underfilled). Input beam of width D (perpendicular to axis) illuminates portion of facet width W. Rotation through angle Φ yields scanned angle $\Theta = 2\Phi$ ($m = 2$), till facet corner encounters beam, making scan ineffective for fraction of rotation. This yields duty cycle $\eta = 1 - D_m/W$, per Sec. 1.8 (Beiser, 1992b).

2.1 Rotating-Mirror Scanners

The rotating-mirror polygon is characterized by a multiplicity of reflective facets that are disposed in a regular array about a substrate that is rotatable about an axis. When the number of facets reduces to one, it is identified as a monogon. The modes of operation are typified in Figs. 3–6 and 9 in which an input beam illuminates at least a portion of one facet while its substrate is rotated. The reflected beam propagates through an angularly scanned field and forms a scanned line of light on an image surface. Arrangements of facets are termed *pyramidal* per Fig. 6 or *prismatic* per Fig. 11. The pyramidal form allows the lens to be close to the polygon, while as in Fig. 3, the prismatic configuration requires space for clear passage of the input beam. Design considerations for this important arrangement are provided in Sec. 2.1.3. Figure 4 represents a monogon pyramidal equivalent, and Fig. 5 a prismatic equivalent (single mirror, galvanometer mount).

2.1.1 Monogons The single-facet or monogon scanner exhibits the fundamental advantages of simplicity and avoidance of errors (radiometric and spatial) due to multifacet differences. Errors due to shaft wobble are, however, transmitted to the output beam. Several techniques are available to reduce the effects of these errors, subsequently discussed. The single-facet scanner also may impose a significant loss in duty cycle, unless the image field occupies a major portion of the available 360° scan angle.

2.1.2 Polygons Table 2 provides principal features and distinctions of typical polygon scanners. For example, for a given mechanical rotation angle Φ, the optical scan angle Θ of the prismatic polygon is twice that of the pyramidal one. To obtain equal optical angles Θ of equal beam width D (for equal resolution N) and to realize equal duty cycle η at equal scan rates, the prismatic polygon requires twice the number of facets, is almost twice the diameter, and rotates at half the speed of the pyramidal one. The actual diameter is a function of the aperture shape factor (Sec. 3.2) and the changing projection of the beam cross section on the facet during its rotation, per subsequent discussion.

2.1.3 Design Considerations The prismatic polygon operating in conjunction with a flat-field lens (preobjective) is illustrated in Fig. 3, while the critical region relating the two is detailed in Fig. 12. This prominent configuration merits special attention. The distance P (pupil relief distance) measured from the lens input surface to the scanner facet surface is an important parameter. It is determined primarily by the need for the input beam to clear the edge of the lens (distance Ds). A shallower input beam angle (reduced β) increases P and imposes upon the lens a

Table 2. Features of typical polygonal scanners.

Function	Pyramidal	Prismatic
Input beam direction	Radially symmetric[a] (typically parallel to axis)	Perpendicular to axis[b]
Output beam direction	Arbitrary angle to axis (typically perpendicular)	Perpendicular to axis[b]
Scan magnification $m = d\Theta/d\phi$	1	2[a]
Along-scan error magnification	1	2[a]
Maximum scan angle, Θ_{max}	$2\pi/n$	$4\pi/n$[b]
Output beam rotation about its axis[c]	Yes	No
Aperture shape[d] (overilluminated)	Triangular/keystone	Rectangular
Enlargement of along-scan beamwidth	No	Yes;[e] $D_m = D/\cos\alpha$
Error due to axial shift of polygon[f]	Yes	No
Error due to radial shift of polygon[f]	Yes	Yes

[a]See Fig. 4 and Sec. 3.3.2.
[b]All beams in same plane perpendicular to axis; n = number of facets.
[c]See sec. 2.1.4. Observable when beam is nonisotropic.
[d]See Tables 3 and 4.
[e]α = angular departure from normal landing.
[f]Shift of image focal point in noncollimated light; no error in collimated light.

wider acceptance aperture—and a consequential increase in its size and complexity. While a wider scan angle Θ also increases lens complexity (for correction of aberration and flat-field operation), it relieves some demand on the precision of the scanner by allowing a greater angular error for a given number N of resolution elements. Further consideration of the scanner–lens relationships requires a preliminary estimate of the polygon facet

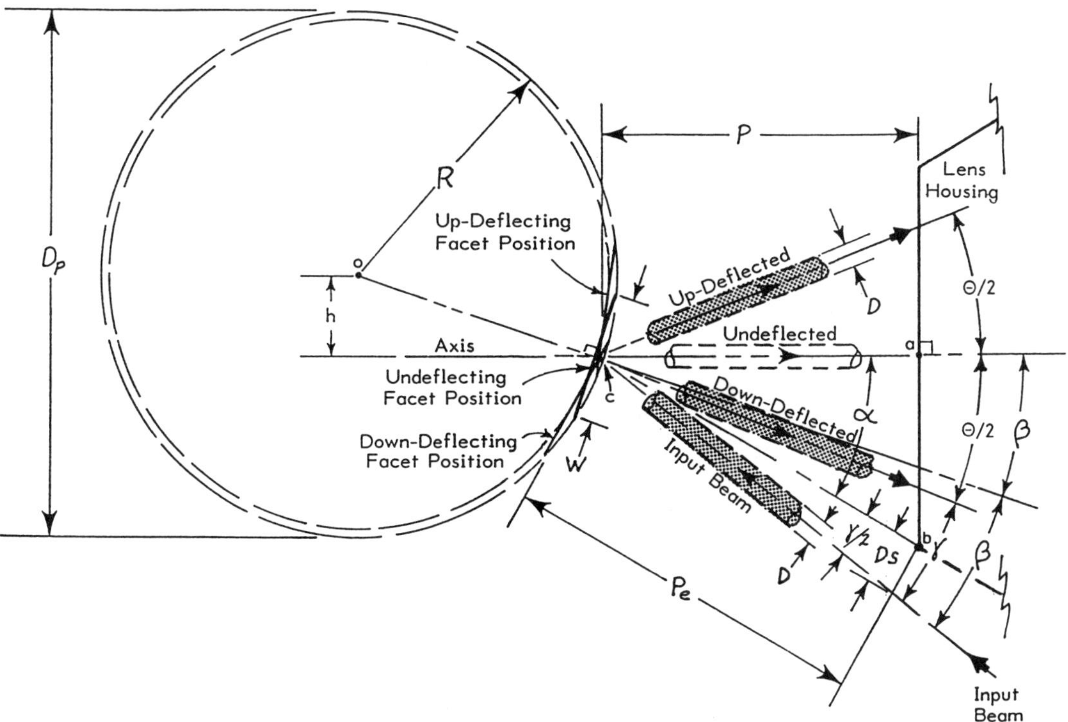

FIG. 12. Rotating-polygon scanning beam and lens relationships, showing undeflected input beam and limit facet and deflected beam positions and angles (Beiser, 1992b).

count, in light of its associated diameter and speed. Its speed is determined by the desired data rates and entails factors that transcend the optogeometric ones developed here (Beiser, 1974). Diffraction-limited relationships are used throughout, requiring adjustment for anticipated aberration in real systems.

Performance characteristics that are usually predisposed are the resolution N, the full optical scan angle Θ, and the duty cycle η. Their interrelationships are presented in Secs. 1.8 and 3.3. The values of N and Θ must be deemed practical for the flat-field lens. Typical high values are $N = 20\,000$ for $\Theta = 1$ rad, achieving reasonably flat and linear scans over format widths greater than 100 mm. Narrower formats generally sacrifice N and Θ. Following these preliminary judgments, the collimated beam width is determined from Eq. (32) as

$$D = Na\lambda/\Theta \tag{4}$$

in which a is the aperture shape factor (Sec. 3.2). The number of facets n is determined from Table 2 and Eq. (3) as

$$n = 4\pi\eta/\Theta \tag{5}$$

whereupon it is adjusted to an integer.

2.1.3.1 Scanner–Lens Relationships. The polygon size and geometry related to the flat-field lens may now be determined. Figure 12 illustrates a typical prismatic polygon with all beams in the same plane. One of n facets of width W is shown in three positions: undeflecting, and in its limit positions. The output beam is shown in corresponding positions. A lens housing edge denotes the input surface of a flat-field lens. Angle γ provides clear separation of the input beam and the down-deflected beam or lens housing. The pupil relief distance P and its slant distance P_e establish angle α as $\cos^{-1}(P/P_e)$—the off-axis illumination on the polygon that broadens the beam width on the facet to D_m. An additional safety factor t ($1 \leq t \leq 1.4$) limits one-sided truncation of the beam by the edge of the facet at the end of scan, yielding

$$D_m = Dt/\cos\alpha. \tag{6}$$

Replacing τ/T in Eq. (2) with D_m/W, the facet width becomes

$$W = D_m/(1 - \eta) \tag{7}$$

from which the outer polygon diameter is expressed by (Beiser, 1991)

$$D_p = \frac{Dt}{\sin(\pi/n)\cos\alpha\,(1 - \eta)}. \tag{8}$$

Solution of Eq. (8) or one of similar form (Kessler *et al.*, 1989) entails determination of α—the angle of off-axis illumination on the facet. This usually requires a detailed layout, similar to that of Fig. 12. Series approximation of $\cos\alpha$ allows its replacement (Beiser, 1991) with more direct dependence on the important pupil relief distance P, yielding

$$D_p = \frac{Dt}{\sin(\pi/n)(1 - \eta)}\,\frac{1 + \Theta Ds/2P}{1 - \Theta^2/8}. \tag{9}$$

Per Fig. 12, $s \approx 2$ is a safety multiplier on D for input-output beam clearance.

Proper orientation also requires the height h—the normal distance from the lens axis to the polygon center, developed as

$$h = R_c \sin(\gamma/2 + \Theta/4), \tag{10}$$

in which R_c is the nominal distance oc during the small shift of reflection point c,

$$R_c = R[1 - \tfrac{1}{4}(\pi/n)^2]. \tag{11}$$

2.2. Holographic Scanners

2.2.1 General Characteristics Holographic scanning (Beiser, 1988) is a method of controlling the direction of an optical beam by moving a hologram (Smith, 1975) in the path of the beam. Holographic scanners utilize many of the concepts representative of polygon scanners: holographic elements are disposed about a rotating substrate to serve as facets, diffracting (rather than reflecting) a fixed incident beam into one that scans. The substrate is typically smooth and radially symmetric. Interferometrically generated holograms serve as diffractive elements, capable of reconstructing wave fronts that are diffraction limited. When properly made and utilized, they are free of the artifacts and background noise prevalent in alternative optical components such as Fresnel lenses and machine-generated diffractors. As with polygons, the number of facets is determined principally by the optical scan angle, while the scanned resolution (number of elements per scan) is determined by the incident beamwidth and the optical scan angle. In a radially symmetric system, the scan can be identical to that of the pyramidal polygon.

The most attractive features of holographic scanners are

1. reduced aerodynamic loading and windage at high-speed rotation, as a result of the elimination of radial dimensional variations on the substrate due to conventional facets;
2. reduced inertial deformation with elimination of above radial variations;
3. reduced optical beam misplacement (due to substrate wobble) when operated in the Bragg regime, subsequently discussed;
4. no physical contact to the facets during holographic exposure. Precision angular indexing of the shaft between exposures provides high accuracy of facet orientation.

Additional favorable factors are these:

1. Operation in transmission allows efficient beam transfer into a closely spaced lens.
2. There is provision for disk-scanner configuration (facets disposed on the periphery of a flat disk), more adaptable to economical fabrication and replication.
3. Filtering in retrocollection (Fig. 13) allows spatial and spectral selection by rediffraction.
4. There is simple adaptability to adjustment of focus, size, and orientation of individual facets during each holographic exposure, providing variable scan capability within a rotation.

Some limiting factors are, however,

1. need for special expertise and facilities in diffractive optics, instrumentation, metrology, and processing chemistry;
2. accommodation of wavelength shift: holographic exposure at one wavelength (of high photosensitivity) and reconstruction at another (for system operation).

Regarding wavelength shift, the equation for first-order diffraction from a transmission grating is given by (Beiser, 1988)

$$\sin\Theta_i + \sin\Theta_o = \lambda/d, \tag{12}$$

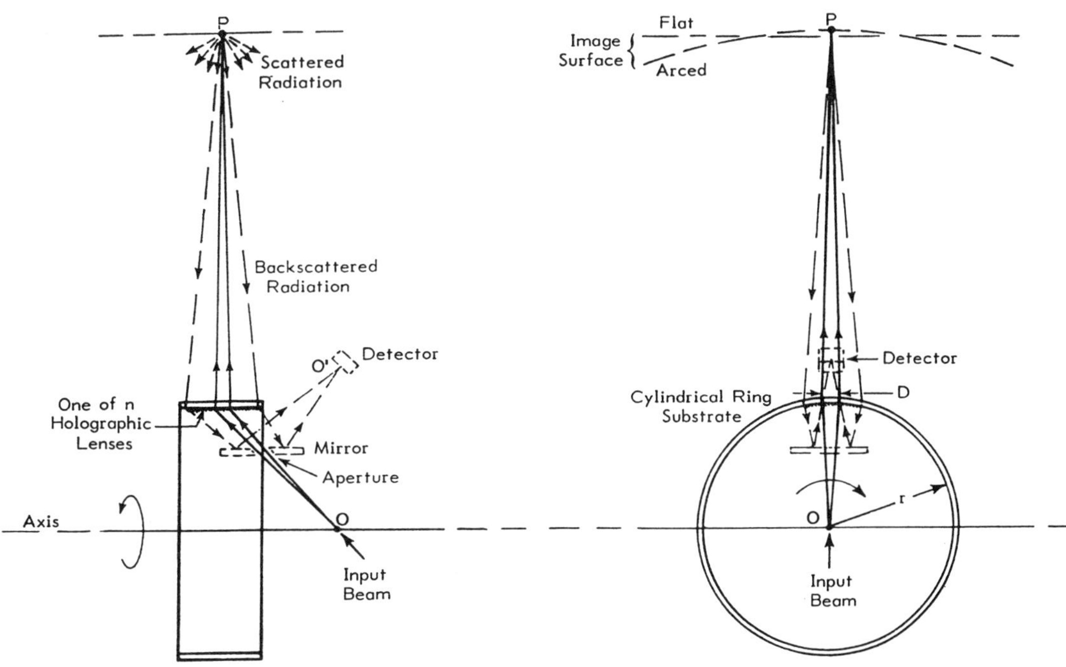

FIG. 13. Transmissive cylindrical holographic scanner, radially symmetric. Input beam underilluminates hololens, which focuses diffracted beam to image surface. Dashed lines designate optional collection of backscattered radiation detection at O′ (Beiser, 1988).

where Θ_i and Θ_o are the input and diffracted output angles with respect to the grating normal and d is the grating spacing. Thus, a plane linear grating illuminated by a collimated beam (of a shifted wavelength) reconstructs a collimated beam at a shifted angle. Collimation is often employed to maintain wave-front purity, though separate (focusing) objective optics is required. When optical power is added to the hologram (to provide self-focusing), its wavelength shift requires compensation for aberration (Beiser, 1988). Further complications arise when the hologram is intended for polychromatic operation, even if a plane linear grating is employed. Also, even small wavelength shifts, as from laser diode thermal changes, can cause beam misplacements, which require corrective action (Beiser, 1988; Kramer, 1991).

2.2.2 Symmetry Factors Departure from radial symmetry develops complex interactions, which require critical balancing to achieve good scan linearity, scan-angle range, wobble correction, radiometric uniformity, and insensitivity to input beam polarization (Beiser, 1988). This is especially demanding in systems having optical power in the holograms. When radial symmetry is retained, however, the configuration may inhibit Bragg-angle compliance and its benefit of wobble reduction, and can require auxiliary (anamorphic) error correction (Sec. 4.2.1).

2.2.3 Scanner Configurations A scanner that demonstrates the characteristics of radial symmetry is represented in Fig. 13 (Beiser, 1988). A cylindrical glass substrate supports an array of equally spaced holographic lens elements (hololenses), which image the input beam at O to the output point at P. Since point O is on the axis, the scanner is radially symmetric, such that point P executes an arc of a circular scan that is concentric with the axis. Scan magnification is maintained at $m = d\Theta/d\Phi = 1$. For convenient signal detection, a portion of the radiation incident at P is backscattered, intercepted by the hololens, rediffracted, and reflected to a detector that is located at the mirror image O' of the point O. The resolution of N scan elements is shown (Beiser, 1988) to be equivalent to that of a radially symmetric pyramidal polygon.

An even closer analogy to the polygon appears in an earlier reflective form, illustrated in Fig. 14, emulating the pyramidal polygon of Fig. 6. Its output is a collimated beam that scans angularly into a flat-field lens to form a scanned focused straight line. This is one of the group of Holofacet scanners (Beiser, 1988), the most prominent of which achieved the highest performance to date in combined resolution and speed: 20 000 elements/scan at 200 Mpixels/s. This 1971 apparatus, the first patented holographic scanner, is now in

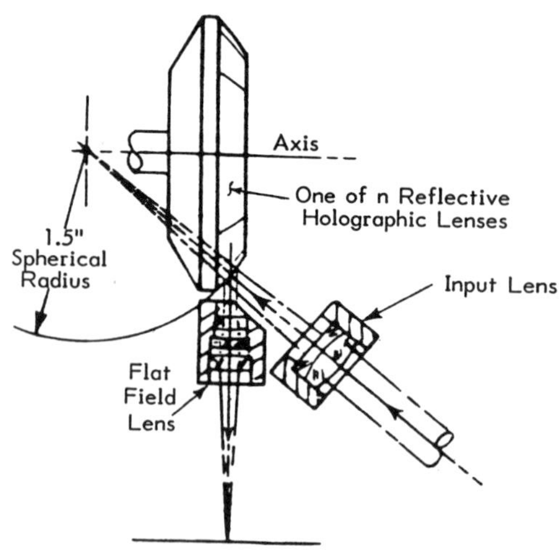

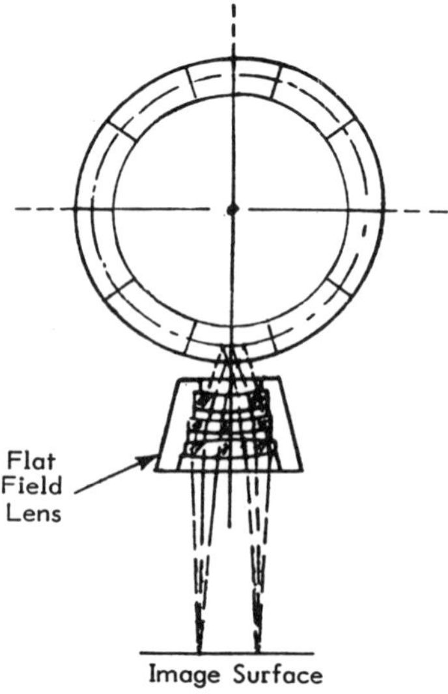

FIG. 14. Reflective Holofacet scanner, radially symmetric, forming flat-field high resolution microimage scanner (Beiser, 1992b).

the permanent collection of the Smithsonian Institution, Division of Electricity and Modern Physics.

2.2.3.1 Operation in the Bragg Regime. The above systems are radially symmetric and utilize substrates that allow the diffracted beam to propagate normal to the rotating axis to generate straight-line scans. It was not until plane linear grating operation in the Bragg regime was introduced (Kramer, 1980) that progress developed in disk configurations operating in *radial asymmetry* (Sec. 3.3.2). Referring to Fig. 15, the input and output beams I and O appear as principal rays directed to and diffracted from the holographic sector HS, forming angles Θ_i and Θ_o with respect to the grating surface normal. In the vicinity of Bragg operation ($\Theta_i \approx \Theta_o$), the differential output angle $d\Theta_o$ for a differential in hologram tilt angle $d\alpha$ within a tilt error $\Delta\alpha$ is given by

$$d\theta_o = \left[1 - \frac{\cos(\theta_i + \Delta\alpha)}{\cos(\theta_o - \Delta\alpha)}\right] d\alpha, \qquad (13)$$

whence, when the Bragg condition is satisfied, a small $\Delta\alpha$ is effectively nulled. While perfect Bragg symmetry may be perturbed during hologram rotation for scan, the reduction in pointing error from scan to scan remains significant.

When $\Theta_i = \Theta_o \approx 45°$, a special case develops in the unbowing of the output scanned beam: its locus resides in an almost perfect straight line that is normal to the plane of the paper of Fig. 15 over a limited but useful range (Beiser, 1988; Kramer, 1991). Further, the incremental angular scan for incremental disk rotation becomes almost uniform: their ratio m at small scan angles is shown to be equal to the ratio λ/d of the grating equation [Eq. (12)]. At 45° Bragg angle, $m = \lambda/d = \sqrt{2}$. This results in the output scan angle being $\sqrt{2}$ larger (in its plane) than the disk rotation angle. It also imposes two restrictions:

1. For high diffraction efficiency from a relief grating (e.g., photoresist), the depth-to-spacing ratio of the grating must be extremely high, while the spacing $d = \lambda/\sqrt{2}$ must be extremely narrow—difficult to maintain and to replicate for high efficiency.
2. Such gratings also exhibit a high polarization sensitivity, imposing a significant variation in diffraction efficiency with grating rotation.

These limitations can be reduced by reducing the Bragg angle and introducing a bow correcting element to straighten the scan line (Kramer, 1991). In Fig. 16, the Bragg angle is reduced to 30°, reducing the magnification to $m = 1 = \lambda/d$. This increases d to equal λ, yielding a more realizable deep-groove grating. It also stabilizes polarization sensitivity

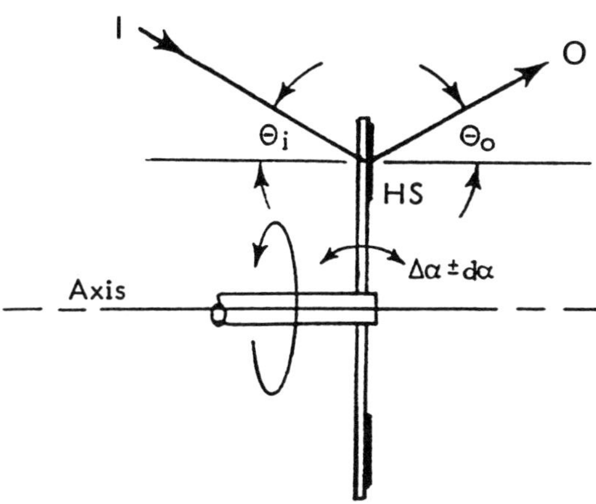

FIG. 15. Holographic scanner operating in Bragg regime ($\Theta_i = \Theta^-$ for input beam *I* and output beam *O*). Output angle Θ^- is stabilized against differential $d\alpha$ and tilt error $\Delta\alpha$ of holographic segment HS (see Sec. 2.2.3.1) (Beiser, 1992b).

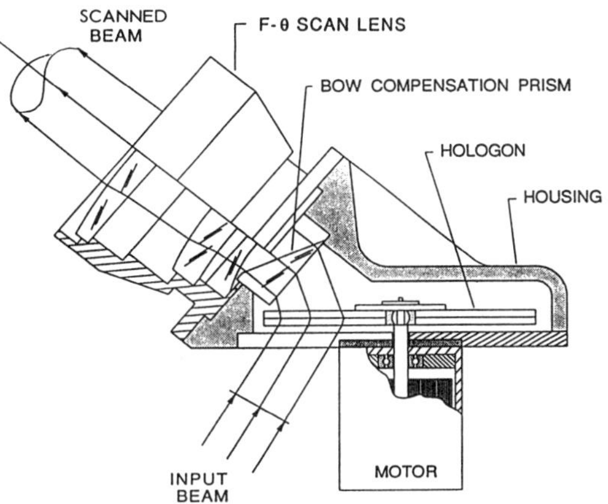

FIG. 16. Plane linear grating (Hologon) holographic disk scanner. Useful portion of scanned beam is in and out of plane of paper. 30° Bragg angle renders fabricatable and polarization-insensitive grating, but requires bow compensation with prism. (Beiser, 1992b.)

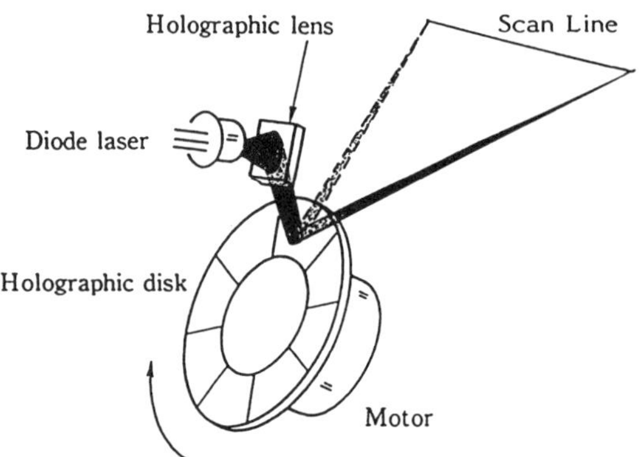

FIG. 17. Holographic disk scanner and corrective holographic lens, operating in Bragg regime, providing complex error balancing (Beiser, 1992b).

during grating rotation. The holograms are plane linear gratings. Thus, a collimated input beam is maintained collimated at the output, requiring a flat-field lens to form the focused spot throughout the scanned field.

The 45° Bragg configuration has been adapted with self-focusing holograms in less demanding tasks, exemplified in Fig. 17 to include a holographic lens to balance the (temperature-dependent) shift of the laser diode wavelength (Ikeda *et al.*, 1987; Kay, 1984; Yamagishi *et al.*, 1986), and to shape the laser output for proper illumination of the scanner. Such multiple-function focusing systems (with optical power) demand, however, critical component centration (Beiser, 1988). Their interrelated characteristics require a carefully balanced design to achieve a specific set of objectives.

2.3 Galvanometer and Resonant Scanners

The scan nonuniformities that can arise from variations among facets are avoided by reducing the number of facets to one. This adapts well, for example, to the internal drum scanner (Fig. 4), achieving a high duty cycle with a very large angular scan within a cylindrical image surface. Flat-field scanning, however, constrains the optical scan angle and the duty cycle. If the mirror is vibrated rather than rotated completely, the wasted scan interval may be reduced. Vibrational scanners include galvanometer and resonant devices, as well as piezoelectrically driven mirror transducers (Beiser, 1974).

2.3.1 The Galvanometer Figure 18(a) shows the cross section of a typical galvanometer driver, in which permanent magnets provide a fixed field that is augmented by the variable field developed from an adjustable current through the stator coils. Seeking a new balanced field, the armature

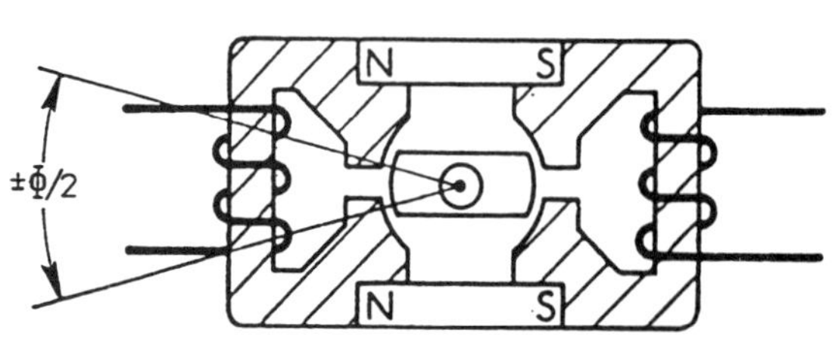

(a) Galvanometer
(Moving Iron or Moving Magnet)

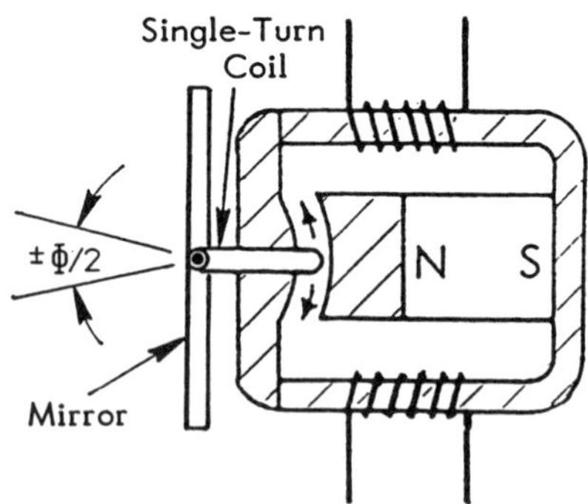

(b) Resonant Scanner
(Moving Coil)

FIG. 18. Examples of galvanometer and resonant scanner transducers. Field of permanent magnet is augmented by variable field from current through stator coils. (a) Galvanometer: mirror surface (not shown) on shaft axis perpendicular to plane of paper. Torque rotates iron or magnetic core. (b) Resonant scanner: torque from field induced into armature coil rotates mirror suspended on axis perpendicular to plane of paper. One stator coil may be nondriven and used for velocity pickoff. (Beiser, 1992b).

executes a limited angular excursion ($\pm\Phi/2$). With the mirror and input beam principal ray per Fig. 5, the reflected beam scans through $\pm\Theta/2$—twice the angle of the rotor. Typical armature types include moving iron or moving magnet as in Fig. 18(a), or moving coil as in Fig. 18(b). High-energy NdFeB rare earth magnetic materials provide the advantage of higher torque at lower inertia (Brosens, 1993) for moving-magnet types.

The galvanometer is a broadband device, damped to scan through a wide range of drive frequencies to an upper value close to its mechanical resonance. Thus, it can provide the "sawtooth" waveform having a longer linearized portion and short retrace time, as represented in Fig. 19 (solid lines). It can also serve for random access and vector scanning—rapid positioning of the mirror to an arbitrary angle within its access time. Thus, the galvanometer was early categorized as a *low inertia* scanner.

2.3.2 The Resonant Scanner When damping is reduced, large vibrations are sustained at or near the resonant frequency, executing nearly perfect sinusoidal oscillations. Figure 19 (dashed lines) shows a sinusoid with the same zero crossings as those of the sawtooth wave form. Contrary to its popular characterization as "low inertia," the resonant scanner provides only rigid time increments. Although the rotary inertia is low, the resonant scanner provides no random access and no scan wave-form shaping, for which the galvanometer was originally identified as low inertia (Beiser, 1974).

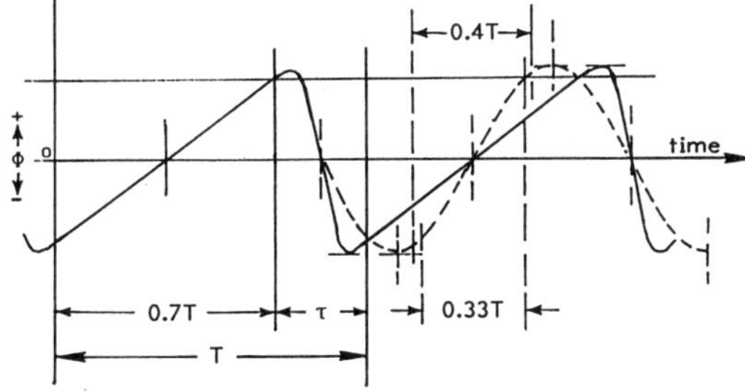

FIG. 19. Wave forms (ϕ vs time) of vibrational scanners having same period and zero crossings. Solid lines: galvanometer with linearized scan providing 70% (typical) duty cycle. Dashed lines: resonant scanner providing 33.3% duty cycle (in one direction) with 2:1 slope change, or 40% duty cycle with $3\frac{1}{4}$:1 slope change. Ratio of maximum slopes at zero crossing = 2.6:1 (Beiser, 1992b).

2.3.3 Suspension Systems The bearings and suspension systems are the principal determinants of scan uniformity. The galvanometer shaft must be sufficiently stiff, though long, to reduce cross-scan wobble due to its own flexure and transverse play. However, to minimize inertia, the armature is restricted in size and mass. Its reciprocating motion does tend to retrace its path (and its perturbations) faithfully over many cycles, making adjacent scans more uniform than if the same shaft rotated completely within the same bearings.

Resonant scanner bearings are often flexure, torsion, or taut-band devices, which exhibit low damping and insert almost no along-scan perturbations (Montagu, 1991). When damped, they can be applied to the galvanometer, suffering a small sacrifice in bandwidth and maximum excursion, but gaining more uniform scan with very low noise and almost unlimited life. Their low radial stiffness imposes possible coupling from torsion (shift of the axis of rotation with scan angle) and possible appearance of spurious modes when lightly damped. These factors require control in commercial instrument designs.

2.3.4 Adaptations and Comparisons Significant correction is required to adapt resonant sinusoidal oscillations to linearized scan. As illustrated in Fig. 19 (dashed lines), one must select from the sine function a central portion that allows equalizing spatial intervals by extracting pixels out of memory at a complementary rate (Tweed, 1985). To limit the pixel rate change to 2:1 (i.e., velocity at zero crossover twice that at the scan limit), then the scan must be restricted to $\frac{2}{3}$ of its peak angle. When scanning with one slope of the sinusoid (to generate uniformly spaced lines), this represents a duty cycle of only 33.3%. To raise the duty cycle to a useful scan of 80% of its full excursion (40% when using one slope), then the velocity variation rises to 3.24 × at crossover compared to that at the scan limit.

The corresponding variation in the dwell time of the pixels result in variation in image exposure or detectivity: 2:1 for 33.3% duty cycle and $3\frac{1}{4}$:1 for 40% duty cycle. This may require compensation using position sensing (Reich, 1985; Tweed, 1985; Blais 1988; Marshall and Gadhok, 1991). In contrast, the broadband galvanometer with feedback can provide highly linearized scans (Montagu, 1991), albeit at generally lower scan rates. When the galvanometer is driven to its high speed limit in sawtooth scan, a useful criterion (Beiser, 1974) is to approach a 70% duty cycle.

2.4 Acoustic-Optic Scanners

As illustrated in Fig. 20, a diffraction grating (of spacing Λ) is synthesized by the wavefront spacing of an acoustic wave traveling through a block of transparent elastic medium (such as glass). The acoustic (piezoelectric) transducer converts an electrical drive signal to a pressure wave that traverses the medium at a velocity v_s. It is absorbed at the far end to suppress standing waves. The pressure wave in the medium forms a corresponding photoelastic variation in its refractive index. An incident light beam of width D is introduced at the Bragg angle (shown exaggerated). An electrical drive signal at the center frequency f_0 develops a variable-index grating of spacing Λ, which diffracts the output beam at Θ_B into position b. When f_0 is increased to $f_s = f_0 + \Delta f$, the grating spacing is decreased, diffracting the output beam through a larger angle, to position c. The small scan angle Θ is effectively proportional to the change in frequency Δf, swept linearly $\pm\Delta f$ for continuous scan.

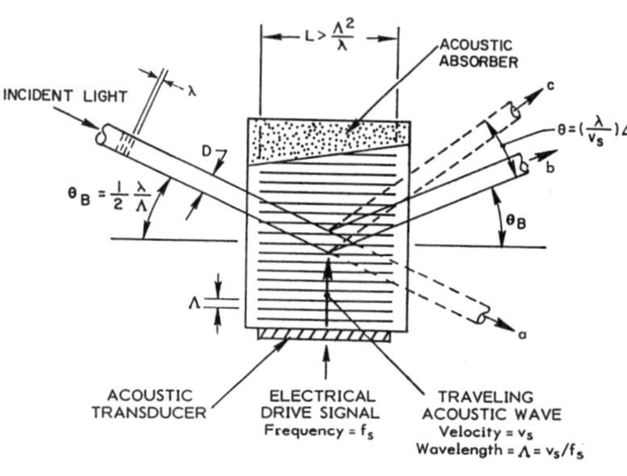

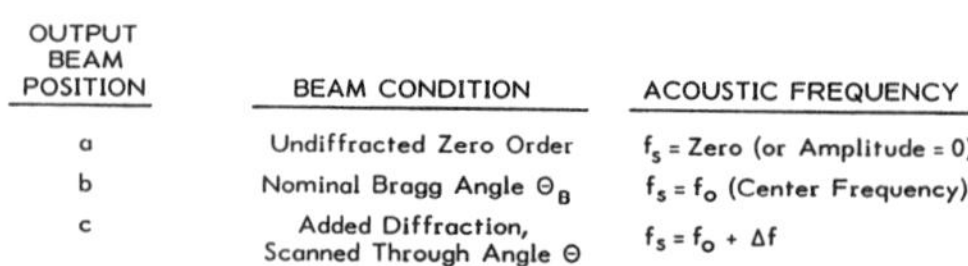

OUTPUT BEAM POSITION	BEAM CONDITION	ACOUSTIC FREQUENCY
a	Undiffracted Zero Order	f_s = Zero (or Amplitude = 0)
b	Nominal Bragg Angle Θ_B	$f_s = f_0$ (Center Frequency)
c	Added Diffraction, Scanned Through Angle Θ	$f_s = f_0 + \Delta f$

FIG. 20. Acousto-optic deflector. Bragg angles exaggerated for illustration. Electrical drive upon acoustic transducer generates traveling acoustic wave in elastic medium. Periodic index change simulates a thick optical grating. Relationship between the output beam positions and the acoustic frequency is tabulated (Beiser, 1992b).

Acousto-optic diffraction can form repetitive linear scans at very high rates or can provide random beam positioning within extremely short access times. It offers, however, relatively low resolution—seldom more than $N = 1000$ elements per scan. Its principles were formulated in 1932 (Debye and Sears, 1932) and applied five years later to the Scophony TV projection system (Okolicsanyi, 1937). Acousto-optic laser scanning was explored intensively in the mid-1960s (Adler, 1967; Gordon, 1966), and remains dominated by operation in the Bragg regime (Beiser, 1974; Korpel, 1980).

2.4.1 Fundamental Characteristics As with the grating equation applied in holographic deflection [Eq. (12)], diffraction from a structure having a periodic spacing Λ (Fig. 20) is expressed as

$$\sin\Theta_i + \sin\Theta_o = n\lambda/\Lambda. \tag{14}$$

in which Θ_i and Θ_o are the input and output beam angles, respectively, n is the diffractive order, and λ is the wavelength. As a "thick" diffractor of length

$$L \geq \Lambda^2/\lambda, \tag{15}$$

all the orders are transferred efficiently to the first. Bragg operation requires that

$$\Theta_i = \Theta_o = \Theta_B, \tag{16}$$

whence for typically small angles, the Bragg angle reduces to

$$\Theta_B = \tfrac{1}{2}\lambda/\Lambda. \tag{17}$$

The small scan angle becomes

$$\Theta = \lambda/\Delta\Lambda = (\lambda/v_s)\Delta f. \tag{18}$$

The beamwidth is traversed during time τ at the acoustic velocity v_s such that

$$D = v_s\tau. \tag{19}$$

Using Eq. (32) and accounting for duty cycle per Eq. (2), the resolution of the acousto-optic scanner, becomes

$$N = \frac{\tau\Delta f}{a}\left(1 - \frac{\tau}{T}\right). \tag{20}$$

The $\tau\Delta f$ component represents the familiar time-bandwidth product—a measure of information-handling capacity.

2.4.2 Deflection Techniques Since the clear aperture of the device is fixed and the beam height (normal to D) can be arbitrarily narrow, anamorphic optics is often used to optimize the illumination of the aperture. A variable aperture width W (in direction of D, which is measured at $1/e^2$ intensity) represents one-dimensional trunction of a Gaussian beam, requiring assignment of an appropriate aperture shape factor a (Sec. 3.2), summarized in Table 4.

Additional topics in acousto-optic deflection are cylindrical lensing due to a linearly swept drive frequency (Korpel, 1980), correction for decollimation in random-access operation (Beiser, 1974), Scophony operation (Johnson, 1979), traveling lens or chirp operation (Bedamian, 1981), correction for color dispersion (Watson and Korpel, 1970), polarization effects and materials selection (Gottlieb, 1991).

2.5 Electro-Optic (Gradient) Scanners

A generalized form of beam scanner is the gradient deflector (Beiser, 1967, 1974) in which the wave front encounters increasing retardation transverse to the beam. The rays, as the orthogonal trajectories of the wave fronts, bend in the direction of the shorter wavelength. Figure 21(a) illustrates a cell having a fixed continuous index gradient. The small bend angle Θ is expressed as

$$\Theta = k(dn/dy)l, \tag{21}$$

where n is the number of wavelengths per unit axial length l, y is the transverse distance, and k is a cell system constant. When the wave front encounters an index change Δn over the beam aperture D and the light rays propagate generally normal to the index change in a cell of length L, then the small deflection angle becomes

$$\Theta = (\Delta n/n_f)L/D, \tag{22}$$

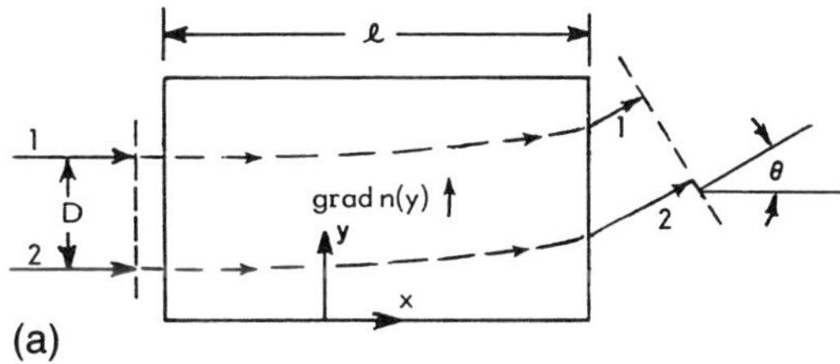

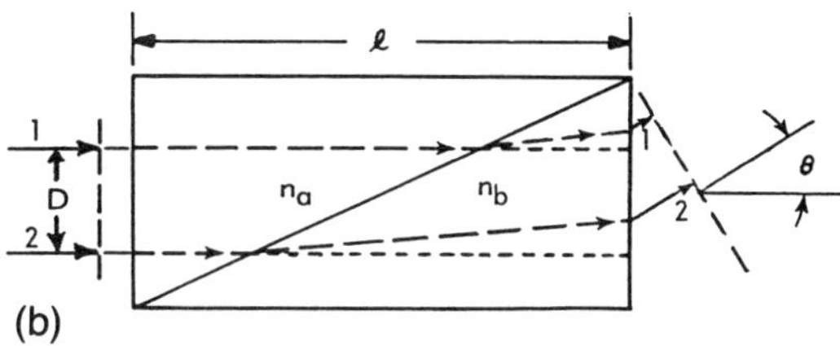

FIG. 21. Equivalent fixed-gradient deflectors. (a) Basic deflector cell having continuous optical index gradient, grad $n(y)$. Ray 1 propagates through a higher index (greater retardation) than ray 2, tipping the wave front through angle θ (including boundary effects). (b) Analogous prismatic cell, in which piecewise $n_a > n_b$ such that ray 1 is retarded more than ray 2, tipping the wave front through angle θ (Beiser, 1992b).

where n_f is the refractive index of the final medium. Following Eq. (32), the corresponding resolution in elements per scan is expressed as

$$N = (\Delta n/n_f)L/a\lambda. \tag{23}$$

A continuous index gradient can be simulated with alternate electro-optic prisms, illustrated in Fig. 21(b) for a single stage. Dynamic control is achieved by utilizing the transverse electro-optic effect and varying the electric field. The Δn is given by

$$\text{(for class I materials)} \quad \Delta n = n_o^3 r_{ij} E_z \tag{24a}$$

$$\text{(for class II materials)} \quad \Delta n = n_e^3 r_{ij} E_z, \tag{24b}$$

where $n_{o,e}$ is the (ordinary, extraordinary) index of refraction, r_{ij} is the electro-optic coefficient, and $E_Z = V/Z$ is the electric field in the z direction.

2.5.1 Implementation Methods A time-dependent index gradient has been proposed (Giarola and Billeter, 1963) utilizing resonant acoustic pressure variations in a cell of a transparent material. Although this appears similar to acousto-optic deflection, it differs fundamentally in that the cell is terminated reflectively to support a standing wave, rather than absorptively. Also, the acoustic wavelength is much longer than the beamwidth for refraction, rather than much shorter for diffraction. A continuous index gradient was formed in bulk electro-optic material by means of shaped bounding electrodes (Fowler *et al.*, 1964; Lospeich, 1968).

Utilizing simple parallel electrodes to develop a transverse electro-optic effect, the iterated prism array of Fig. 22 allows increased length. Successive interfaces impart a cumulative effect. The magnitude and direction of retardation is controlled by the index changes in the material. Significant review, experiment, and test are reported for this electro-optic deflector (Lee and Zook, 1968; Beiser, 1974; Ireland and Ley, 1991).

2.5.2 Drive Power The power dissipated within the electro-optic material is given by

$$P = \tfrac{1}{4}\pi V^2 Cf/Q, \tag{25}$$

where V is the applied (peak-to-peak sinusoidal) voltage in volts, C is the deflector capacitance in farads, f is the frequency in hertz, and Q is the material Q factor [$Q = 1/\text{loss tangent}$]. The capacitance for transverse electroded deflectors of Fig. 22 is approximately that for a rectangular parallel plate capacitor,

$$C = 0.09\kappa LY/Z \quad \text{pF}, \tag{26}$$

where κ is the dielectric constant of the material; length L, width Y, and thickness Z.

The characteristics of some electro-optic materials (Tsai and Kraushaar, 1972; Beiser, 1974; Ireland and Ley, 1991) are often a strong function of frequency beyond 10^5 Hz. A resolution-speed-power figure of merit has been proposed (Zook and Lee, 1970).

2.5.3 Special Considerations Since most electro-optic coefficients are extremely low, high drive voltages are required to achieve even moderate resolutions (to $N \simeq 100$). These devices can, however, scan to very high speeds (to 10^5/s) and suffer effectively no time delay.

3. SCANNED RESOLUTION

3.1 Resolution Criteria

The resolution of an optical scanner is expressed (Beiser, 1974) by the number N of spots of width δ conveyed along a spatial path by the scanner. The path is usually nearly linear and traversed with uniform velocity. While δ is analogous to a pixel or pel (picture element) of spatially digitized scan, active optical scan is typically contiguous. The scanned spot convolves with the object intensity function to form a continuous function that can be divided into elements by intensity modulation or signal digitization. The separation between such spot centers is $w = vt$, where v is the velocity of the scanned spot over the time interval t. The width δ corresponds approximately to w; that is, the useful width of the imaged spot approximates the center spacing from its neighbor.

Spots that exhibit a Gaussian intensity distribution are often considered as overlapping at one of two widths: at their $1/e^2$ in-

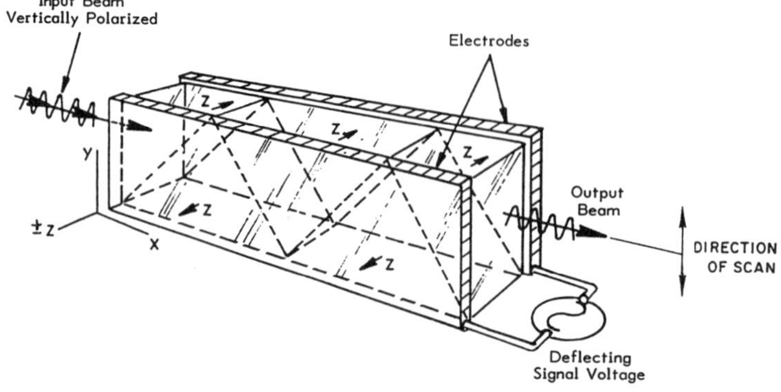

FIG. 22. Iterated electro-optic prism deflector. Alternating crystallographic z axes provide the alternating index changes: form a piecewise-continuous gradient from a uniform deflecting field, which is established by the parallel-plate electrodes (Beiser, 1992b).

tensity points, or at their 50% intensity points (denoted as FWHM; full width at half maximum). The relationship between these two widths is given by

$$\delta_{\mathrm{FWHM}} = 0.589\ \delta_{1/e^2}. \tag{27}$$

The resolution N is a function of its measurement criterion, for the same system will register different apparent N, according to Eq. (27). That is, it will count approximately 1.7× as many spots at FWHM as at $1/e^2$ intensity.

All expressions and numeric values are represented here as diffraction limited. Accommodation is required for systematic imperfection and spot aberration, as in any practical system design. Adjustment is also required for the duty cycle in which a retrace interval may deplete the full scan cycle, causing loss in available resolution, average bandwidth, and sensitivity for a given energy or signal-to-noise requirement.

3.2 Aperture Shape Factor

The aperture shape factor a accommodates differences in measurement criteria, as expressed above, or differences in aperture distributions due typically to vignetting, apodizing, or truncation, discussed below. The value of a for the Gaussian distribution at the $1/e^2$ criterion (Beiser, 1974; O'Shea, 1985) is represented by

$$a_{1/e^2} = 4/\pi = 1.27. \tag{28a}$$

Applying Eq. (27) yields for the FWHM criterion,

$$a_{\mathrm{FWHM}} = 0.75. \tag{28b}$$

When these are substituted into the equation for spot size (Beiser, 1974; O'Shea, 1985)

$$\delta = aF\lambda, \tag{29}$$

in which $F = f/D$ is the f number of the cone converging over the distance f from aperture-width D, and λ is the radiation wavelength, the corresponding Gaussian spot size becomes

$$\delta_{1/e^2} = 1.27\ F\lambda \tag{30a}$$

and

$$\delta_{\mathrm{FWHM}} = 0.75\ F\lambda. \tag{30b}$$

Regarding truncation, a common type appears when the illuminating beam is much larger than the limiting aperture (overillumination) to form the uniformly illuminated aperture. Analytic examples are the rectangular and round (or elliptic) apertures, which develop (for the variable x) the normalized PSF distributions $\sin^2 x/x^2$ and $2J_1^2(x)/x^2$, respectively, where $J_1(x)$ is the first order Bessel function of the first kind. Apodization (Levi, 1968) may be applied to improve the PSF by constricting regions of the aperture that contribute more to aberration. Apodization also applies to central obscuration of an aperture in some reflective optical systems. Vignetting (Levi, 1968) is often a consequence of compromise in wide-angle compound lens design, tolerating gradual obscuration of the edge of the beam as it propagates near its off-axis limit.

Figure 23 illustrates the modulation transfer functions (MTFs) of several uniformly illuminated scanning apertures. The MTF is the ratio of output intensity modulation to that for sinusoidal input modulation, as a function of spatial or temporal frequency. While the illumination and the resulting PSFs are of a coherent wave, scanning forms a sequence of incoherently related measurements of the space-shifting

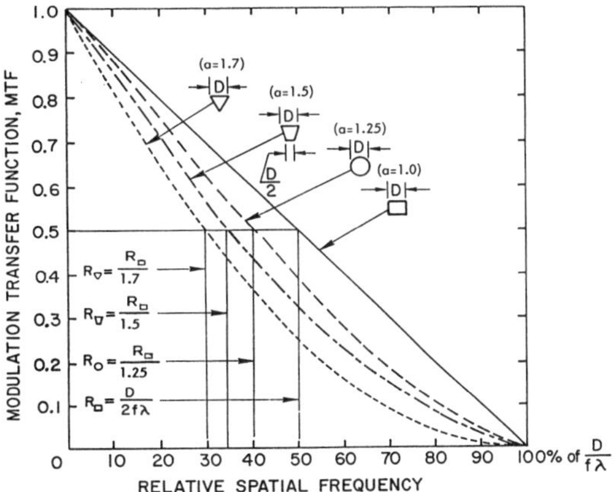

FIG. 23. Modulation transfer function (MTF) vs relative spatial frequency for uniformly illuminated rectangular, round, keystone, and triangular apertures, and their associated *a* values for the 50% MTF criterion (Beiser, 1992b).

function, yielding an incoherent MTF. Intersections with MTF = 0.5 identify the spatial frequencies at which the modulation is 50%. With the rectangular aperture as a reference (a = 1), the intersections of the others with MTF = 0.5 yield relative a values. Since the spatial frequency bandpass is proportional to D/f, the apertures D of others must be widened by their a values to render equivalent response mid-range. It is noteworthy that the a value of 1.25 for the uniformly illuminated round/elliptic aperture corresponds closely to the Rayleigh value of 1.22, while the a value for the uniformly illuminated rectangular aperture corresponds exactly to the Rayleigh criterion of 1.0—the distance from the center of the PSF to its first null.

Table 3 summarizes the aperture shape factors for several useful distributions. Scanning is in the direction of width D. Gaussian beamwidth is measured at $1/e^2$ intensity and centered within its (round/elliptic) limiting aperture of width W. For $W \geq 1.7D$, the Gaussian beam is considered untruncated.

One-dimensional truncation of a Gaussian beam by parallel boundaries is typical for acousto-optic scanning (Sec. 2.4). There, the limiting aperture width W is constant, while the Gaussian width D is variable. Table 4 provides several values of a for this condition (Beiser, 1974). To relate to the data in Table 3, when $\rho = W/D = 0$, illumination is through a narrow slit, corresponding to the uniformly illuminated rectangular aperture, whence a = 1. When ρ = 1, $W = D$, truncating the Gaussian beam at its $1/e^2$ intensity points. The parallel barriers allow more of the Gaussian skirts to fill the aperture, providing a = 1.15 vs 1.38 for round truncation. When ρ = 2, the Gaussian beamwidth reduces to $\frac{1}{2}$ of the boundary width—effectively an untruncated Gaussian beam of half the original width, halving the resolution (a = 1.75 vs 0.85 in Table 3).

Table 3. Aperture shape factor.[a]

Uniformly illuminated[a]		
Shape	→|D|←	a
Rectangular	▭	1.0
Round/elliptic	○	1.25
Keystone	⏢	1.5
Triangular	▽	1.7
Gaussian-illuminated[b]		
δ (Spot overlap)	a (Untruncated) $W \geqq 1.7\ D$	a (Truncated) $W = D$
At $1/e^2$ intensity	1.27	1.83
At $\frac{1}{2}$ intensity	0.75	1.13
For 50% MTF	0.85	1.38

[a]Width D for 50% MTF.
[b]Beam of width D at $1/e^2$ intensity, centered within aperture boundary of width W.

Table 4. Aperture shape factor a for one-dimensional truncation of a Gaussian intensity distribution of width D at $1/e^2$ intensity centered within straight boundary of width W. Scanning is in direction of D and W.

Truncation ratio $\rho = W/D$	Shape factor a for 50% MTF
0	1.0
0.5	1.05
1.0	1.15
1.5	1.35
2.0	1.75

3.3 Scanned Resolution; the Resolution Invariant

The resolution (elements per scan) accessed by an optical beam that is focused to spot size δ and *translated* over path distance S is simply

$$N_s = S/\delta. \tag{31}$$

The resolution developed during *angular* scan of a diffraction-limited beam (Beiser, 1974; O'Shea, 1985) is given by

$$N_\theta = \Theta D_0/a\lambda, \tag{32}$$

in which Θ is the deflected optical angle and D_0 is the aperture width at the deflection nodal center, as represented in Fig. 24. It is independent of (diffraction-limited) spot size δ and dependent only on the scan angle, the aperture distribution, and the wavelength λ. The beam could be collimated, converging, or diverging from the deflector. When collimated, $D_0 = D$, the actual aperture width. When noncollimated, resolution is augmented, as described in subsequent discussion.

The numerator of Eq. (32) represents a resolution invariant—a form of the Lagrange

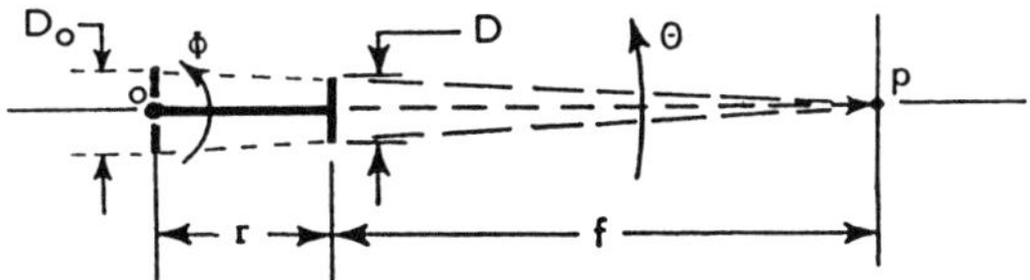

FIG. 24. Deflecting element of width D displaced by (positive) radius r from axis o, propagating a converging beam to point p over (positive) focal distance f. Effective larger aperture D_0 appears at nodal axis (Beiser, 1992b).

invariant for optical transfer (Levi, 1968), expressed in this nomenclature as

$$n\Theta D = n'\Theta' D', \tag{33a}$$

where the primed terms are the refractive index, angular deviation, and aperture width, respectively, in the final image space. For the common condition of $n = n' = 1$, this forms the *resolution invariant*

$$I = \Theta D = \Theta' D'. \tag{33b}$$

Resolution is conserved such that N is invariant with centered optics following the deflector (with $a\lambda$ taken as a relative constant of the system). For example, in preobjective scan, resolution N is determined fundamentally by the scanner and not by the lens. Thus,

$$N = I/a\lambda, \tag{34}$$

bearing no dependence on the optics following the deflector, whose principal function is to distribute and form the N spots—each to size δ along the scanned path.

3.3.1 Augmented Resolution; the Displaced Deflector A scanning system can accumulate resolution by adding the two processes represented by Eqs. (32) and (31)—angular scan and linear translation—to give

$$N = N_\theta + N_s. \tag{35}$$

This occurs, for example, with multiple-element scanners (such as polygons) having deflecting elements (facets) that are displaced from the rotating axis by a distance r, and whose output beam is noncollimated (Beiser, 1983). One active element and its focused output beam are illustrated in Fig. 24, in which the output beam is derived effectively from a larger aperture D_0 located at o. By similar triangles and from Eq. (32),

$$N = (\Theta D/a\lambda)(1 + r/f). \tag{36}$$

This more general equation corresponds to Eq. (35) as the aperture executes a displacement $S \approx r\Theta$ to represent Eq. (31).

It is noteworthy that

1. augmentation goes to zero when $r = 0$ (deflector on nodal axis) or when $f = \infty$ (output beam collimated);
2. augmentation adds when output beam is convergent (f positive) and subtracts when output beam is divergent (f negative);
3. augmentation adds when r is positive and subtracts when r is negative (output derived from opposite side of axis o).

3.3.2 Radial Symmetry and Scan Magnification Some angular scanners exhibit a basic characteristic identified as radial symmetry. *When the illuminating beam converges to or diverges from the rotating axis of an angular scanner* (Fig. 4) *it is said to exhibit radial symmetry* (Beiser, 1988). When an incident beam is collimated and parallel to the axis (Fig. 6), this is a special case of radial symmetry, for the beam derives effectively from a very distant point on the axis. Scanners exhibiting radial symmetry provide unity angular optical change for unity mechanical change. That is, $m = d\Theta/d\phi = 1$, where Θ is the optical scan angle and ϕ is the mechanical angular change. The parameter m is the "scan magnification," discussed in Section 3.3.3. It ranges typically between 1 and 2, depending on the scanner-illumination conditions, as exemplified in Table 2.

The prismatic polygon in its typical operating configuration (Fig. 3) exhibits a variable m, depending on the degree of collimation or focusing of the output beam. When the beam is collimated, $m = 2$. As developed in the next section, when the beam is focused, the value of m shifts from 2 according to Eq. (41).

In holographic scanners that are not radially symmetric, m depends on the angles of incidence and diffraction of the input and output beams (Kramer, 1981; Beiser, 1988),

$$m = \sin\Theta_i + \sin\Theta_o = \lambda/d, \tag{37}$$

where Θ_i and Θ_o are the input and diffracted angles with respect to the grating normal and d is the grating spacing. That is, when $\Theta_i = \Theta_o = 45°$, $m = \sqrt{2}$; when $\Theta_i = \Theta_o = 30°$, $m = 1$.

3.3.3 Augmented Resolution with Scan Magnification Equation (36) was developed for the condition that the optical scan angle Θ is equal to the mechanical angle ϕ. This occurs when the scanner exhibits radial symmetry (per Sec. 3.3.2). When, however, $m = d\Theta/d\phi \neq 1$, as for configurations such as Figs. 11 and 17, account is taken of scan magnification m, yielding the more complete resolution equation (Beiser, 1983)

$$N = (\Theta D/a\lambda)(1 + r/mf), \tag{38}$$

in which, per Fig. 24, Θ = optical scan angle (active), D = scan aperture width, λ = Wavelength (same units as D), a = aperture shape factor, m = scan magnification ($= d\Theta/d\phi$), ϕ = mechanical scan angle about o, r = distance from o to D, f = distance from D to p ($f = \infty$ for collimated; + for convergent; − for divergent). Note: r and f are typically positive, extending from left to right in Fig. 24. Considering $m = \Theta/\phi$ as a constant, another useful form is

$$N = (\phi D/a\lambda)(m + r/f), \tag{39}$$

whose augmenting term shows a composite magnification

$$m' = m + r/f, \tag{40}$$

which, for the typical prismatic polygon, becomes

$$m' = 2 + r/f. \tag{41}$$

4. SCAN ERROR REDUCTION

The high-resolution rotational scanner (supporting deflecting mirrors or holograms) depends on precise shaft orientation about its axis. The angular uniformity of these elements with respect to the shaft axis and of the axis with respect to its frame imposes substantial demand on fabrication procedures and on consequential cost. Figure 25 illustrates the (exaggerated) beam misplacement of a typical polygon scanner due to its instantaneous angular error in the cross-scan direction. Errors in this direction are insidious as compared to the along-scan direction, which is amenable to rectification by controlling the timing of the data stream. Several noteworthy techniques have been developed to alleviate this critical problem.

4.1 Available Methods

A composite of resources for cross-scan error reduction is represented in Fig. 26. *Fabrication accuracy* may serve as the only (and often costly) discipline, or, it may be augmented by any of the alternative methods. The *active* ones utilize high-speed, low-inertia acousto-optic, electro-optic, or piezoelectric deflectors (Beiser, 1974; Donohue, 1979), or lower-speed galvanometers that are programmed to shift the beam to rectify its positional errors. While open-loop methods may

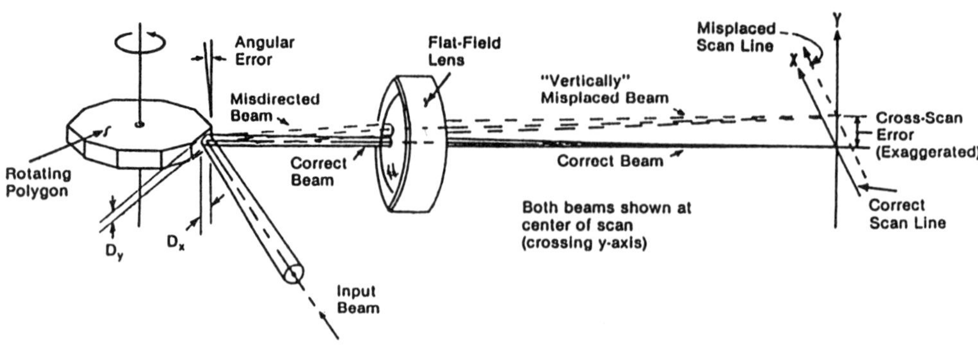

FIG. 25. Typical polygon scanner system having angular error (exaggerated), which misplaces scan lines in the y direction (Beiser, 1992b).

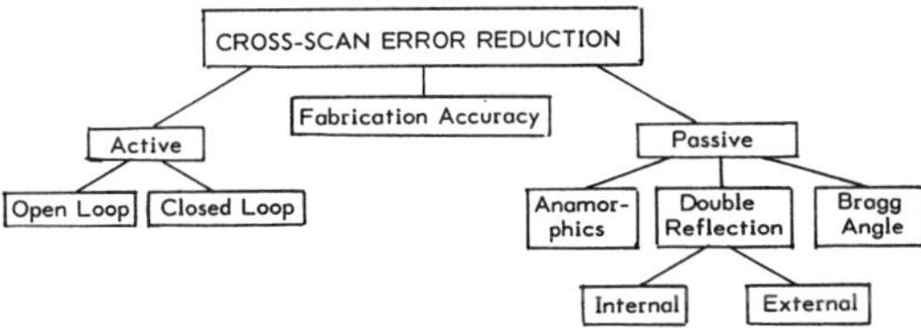

FIG. 26. Techniques for cross-scan error reduction (Beiser, 1992b).

be adequate for repetitive errors, elegant closed-loop methods are often required to correct pseudorandom perturbations. The cost and complexity of this active approach need to be compared to the alternatives of increased fabrication accuracy and of the use of *passive* techniques.

4.2 Passive Methods

Operating continuously and automatically, passive techniques require no programming. Utilizing optical principles in novel configurations, they reduce beam misplacement due to angular error in reflection or diffraction. Error reduction of tilted holographic deflectors operating near the Bragg angle is discussed in Sec. 2.2.3.1. Two additional passive techniques are now described, which employ anamorphic and double-reflection principles.

4.2.1 Anamorphic Error Control The most prominent treatment, anamorphics, may be applied to almost any deflector. The basics and operational characteristics (Beiser, 1988) are summarized here. Anamorphic optics exhibit unequal power along quadrature meridians. A cylindrical lens, for example, exhibits power in one direction only.

Separating the angular resolution equation [Eq. (32)] into quadrature (x,y) components and denoting the critical cross-scan error direction as y, the resolvable error, expressed in terms of the number (usually fractional) of misplaced resolution elements, is

$$N_y = \Theta_y D_y / a\lambda, \tag{42}$$

in which $a\lambda$ is assumed constant, Θ_y is the causative angular error of the output beam in the cross-scan direction, and D_y is the height of the beam illuminating the deflector. The objective is to make $N_y \rightarrow 0$. (N_x would represent the *desired* along-scan resolution.)

The inaccuracies of fabrication and rotational dynamics determine Θ_y (the beam error angle). To reduce N_y, anamorphics are introduced to reduce D_y. This is usually accomplished as illustrated in Fig. 27, with a first cylindrical lens focusing the illuminating beam in the y direction upon the deflector. (D_x remains unaffected.) As D_y is reduced to D_y', so is the y displacement error, per Eq. (42). Following deflection, the y-direction beam distribution is restored with additional anamorphics (a second cylinder)—reestablishing the nominal converging beam angle to form (along with the flat-field lens) the nominal spot size. The x-direction

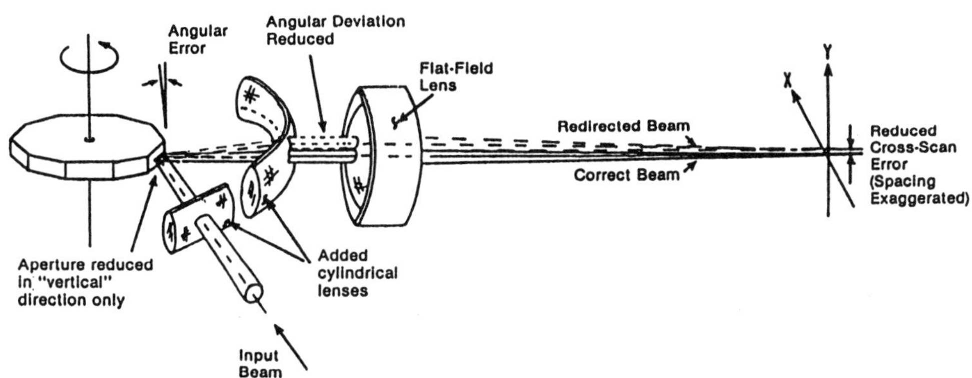

FIG. 27. Input beam (of Fig. 25) is compressed in the y direction with a first cylindrical lens, which reduces D_y and cross-scan error proportionately. A second (toroidal) cylindrical lens restores the beam such that the flat-field lens forms an (ideally) unmodified focal spot. Variations of this basic process entail combinations of 2nd cylinder and flat-field lens (Beiser, 1992b).

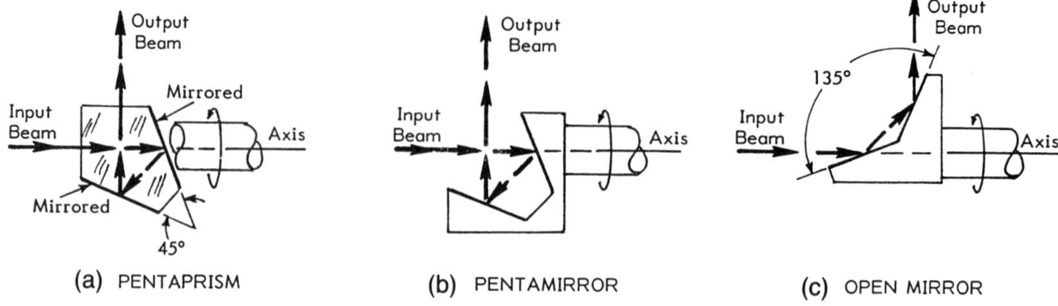

FIG. 28. Monogon scanners employing double reflection to null cross-scan error (Beiser, 1992b).

characteristics remain (ideally) unaffected. Following Eq. (42), the y-error reduction ratio is

$$R = D'_y/D_y, \tag{43}$$

where D'_y is the compressed beam height on the deflector and D_y is the original beam height on the deflector.

A variety of anamorphic configurations have been instituted, with principal variations in concert with the (usually "flat-field") objective lens, to reestablish the nominal converging beam angle and to maintain focused spot quality and uniformity during scan.

4.2.2 Double-Reflection Error Control

In double reflection, the deflector that creates a cross-scan error is reilluminated by the beam in complementary phase to null the error. This is discussed first as applied to monogon configurations, followed by its application in polygons. Double reflection can be conducted in two forms: internal and external.

An internal double-reflection scanner is exemplified by the pentaprism (glass substrate having two of its five surfaces mirrored) monogon of Fig. 28(a) (Starkweather, 1984). Though not shown, the shaft is mechanically coupled to the pentaprism. This is an optically stabilized alternative to the 45° monogon of Fig. 4, operating here preobjective in collimated light. Tipping the pentaprism cross scan (in the plane of the paper) leaves the output beam orientation unaffected. (A resulting minute translation of the collimated output beam is nulled when the beam is focused by a subsequent objective lens.) The pentamirror of Fig. 28(b) (same ray path) requires significant balancing and support of the unbalanced mirrors, for a shift (e.g., inertial) in the nominal 45° include angle doubles the angular error in the output beam.

A stable and simpler double reflector is the open-mirror monogon (Beiser, 1990) of Fig. 28(c). Its nominal 135° angle (which is structurally extremely rigid) serves identically to maintain a constant (90°) output beam angle, independently of cross-scan wobble of the deflector.

Two variations that double the duty cycle, as would a two-faced pyramidal polygon or axe-blade scanner (Beiser and Johnson, 1994), appear in Fig. 29. Figure 29(a) is effectively

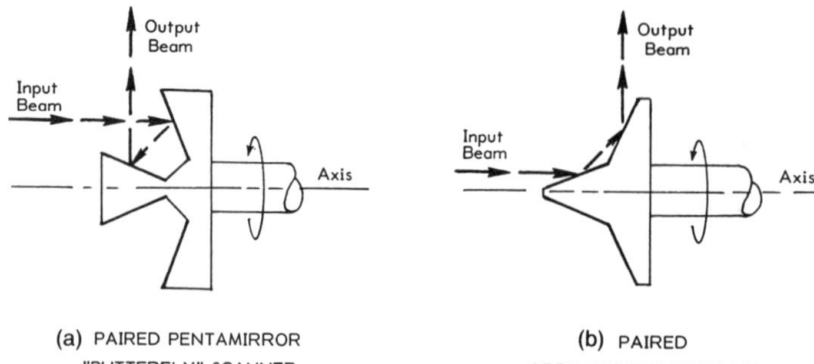

FIG. 29. Paired scanners to double the scan rate and duty cycle, employing double reflection to null cross-scan error (Beiser, 1992b).

two pentamirrors forming a "butterfly" scanner (Marshall *et al.*, 1991) and Fig. 29(b) is effectively a pair of open mirrors (Beiser, 1992a, b). The included angles of each half-section must be made equal to within $\frac{1}{2}$ of the allowed error in the output beam. Also, the center section of the method of Fig. 29(a) must be angularly stable to within $\frac{1}{4}$ of the allowed error, for an increase in included angle on one side forms a corresponding decrease on the other. Dynamic considerations include transverse beam displacements as the device rotates, increasing required mirror widths in proportion to the radial distance of the input beam from the rotating axis. This increases further the bulk of the method in Fig. 29(a) and the need to stabilize the central section against minute angular shifts.

The need for near equality of the included angles of the multiple double reflectors can be eliminated by transferring the accuracy requirement to a fixed external element that redirects the recurrent beam scans (Hanson and Sherman, 1984; Garwin, 1984). One such "external" form of cross-scan error correction is illustrated in Fig. 30, shown in the undeflected position. A prismatic polygon illuminated with a collimated beam of required width D deflects the beam (only principal rays shown) first to a Porro prism (Levi, 1968) or roof mirror, which returns the beam to the same facet for a second deflection toward the (flat-field) objective lens. The roof mirror

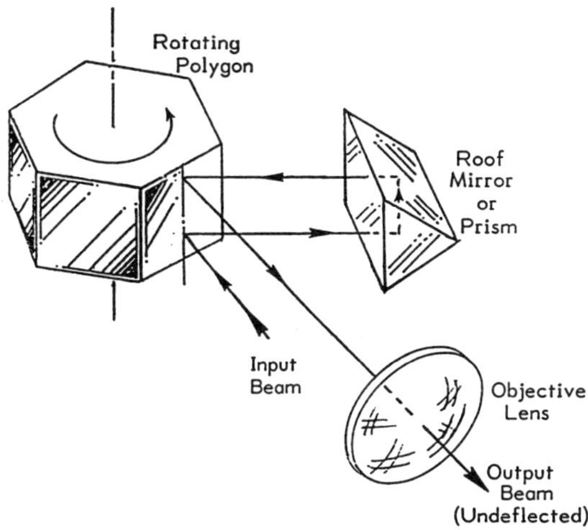

FIG. 30. Method of external double reflection. Principal rays shown in undeflected position. Components and distances not to scale (Beiser, 1992b).

phases the returned beam such as to null the cross-scan error during the second reflection. Several characteristics are noteworthy.

1. The along-scan angle is doubled. Thus, the scan magnification $m = 4$ rather than 2.
2. This requires increasing the number of facets to provide the same scan angle with the same duty cycle.
3. During polygon rotation, the point of second reflection shifts significantly along the facet and sacrifices duty cycle. Since error nulling is based on the second reflection exhibiting the identical perturbation as the first, the mirror surfaces at the two reflections must convey identical errors.
4. The pupil distance from the (flat-field) objective lens is effectively extended by the extra reflections, requiring a larger lens to avoid vignetting.
5. The roof mirror and the objective lens must be positioned to avoid obstruction of the input and scanned beams.

As a consequence of these factors, a detailed layout is essential, combining effectively Figs. 12 and 30, to ensure its viability in the requisite application.

5. SUMMARY

This article presents an overview of the principles and practice of optical scanning. Following introduction of some definitions and useful applications, the subject is divided into four principal sections: system architecture, scanning devices, scanned resolution, and error reduction. While these topics are referenced extensively for more detailed investigation, one that has little published representation is a review of the relative virtues and limitations of the various scanning options. Thus, a summary of the choices for optical scanning is provided here.

The field of remote sensing is dominated by the use of reflective optical devices, because of their relative insensitivity to wavelength variations. There is a trend toward combining refractive and diffractive optics to benefit from their complementary dispersive properties. This "hybrid" technique is now surfacing for use in scanning systems requiring broad or multispectral operation. Lasers, which provide effectively monochromatic radiation, allow for the efficient use of the full gamut of scanning disciplines.

The most prominent scanning device is the rotating polygon, because of its sheer effectiveness in providing a range of utility, such as selectable scan angles, high scan rates, high resolution, and high accuracy and linearity. Almost all laser printers, for example, employ rotating prismatic polygons. The pyramidal polygon is utilized less often, generally limited by its higher fabrication cost. Where, however, optical packaging is advantageous, or where radial symmetry is useful, the pyramidal polygon provides equally fast and highly resolved optical scanning. Both are capable of resolutions to—and beyond—30 000 elements/scan at speeds in the 100-MHz range. The simultaneous combination of such speed and resolution is, however, quite rare, requiring substantial discipline in optical and mechanical design and fabrication skill.

The holographic scanner runs quite parallel to the polygons in utility, flexibility, and systematic performance, offering distinct advantages in Bragg-angle wobble correction and low aerodynamic and inertial loading. Thus, the appropriately designed holographic scanner can outperform the polygon in speed. Its critical technology demands, however, adept expertise for its development and production, constraining its extent of application. This can change when the holographic scanner becomes fabricated in quantity by economical replication techniques on a stable substrate, while retaining the critical detailed requirements of grating accuracy, uniformity, and integrity of its surface relief contour.

Because of its relative functional simplicity, the galvanometer is often selected for experimental and moderate-to-low speed "sawtooth" and random-access scanning. The modern galvanometer with feedback is capable of accuracy comparable to that of the monogon and polygon. At sufficiently low speed and with an appropriately large mirror, it can challenge the polygon in resolution and duty cycle. At higher speeds, the scan angle becomes sufficiently limited that scan rates beyond 1 kHz are rarely useful. The resonant scanner is capable of much higher useful scan rates, albeit generating only sinusoidal oscillations at a fixed frequency. Although the product of speed and resolution can be moderately high, the utility of resonant devices is inhibited by the adaptations that become necessary for "electronic" pixel linearization and exposure energy compensation for nonuniform velocity and a limited (undirectional) duty cycle.

The acousto-optic (AO) scanner is the most popular "nonmechanical" modulator and deflector. It is capable of extremely high scan rates to 20 000 per second. As the scan rate is increased, however, its resolution and duty cycle are reduced, on account of its fixed "retrace" interval. Thus, resolution rarely exceeds 1000 elements per scan. The AO device can serve also as a rapid "random access" device, limited by the time needed to fill the aperture with the new acoustic wave. Unfortunately, for only one Bragg angle (per input angle and drive frequency), a changing drive frequency, which makes the acoustic wave shift the optical Bragg regime, is accompanied by a loss in diffraction efficiency. This is often controlled to within a 10% variation. Throughput efficiencies range typically between 60% and 90%.

The electro-optic (EO) scanner, another nonmechanical device, is capable of the highest speed at, however, the lowest resolution. Requiring only the time to charge and discharge a small value of capacitance, the speed of the EO deflector is limited mainly by the power dissipation (loss tangent) in its electrooptic materials and the attainable voltage gradients at the drive wave forms and frequencies. Although relatively costly, the EO deflector finds application for extremely fast (100 kHz) beam positioning to resolutions within 100 or so elements.

Alternative scanner techniques, not covered explicitly in this article, include piezoelectric beam deflection (movement of mirrors), refractive scanners (as by rotating prisms), phased arrays (optical analog of microwave antenna beam steering), digital deflection (cumulative binary EO switching), and intracavity scanning (mode selection within a laser cavity having transverse degeneracy). Additional topics include beam iteration (beam-angle enhancement and transfer between deflectors) and scan amplification (near-resonant beam iteration for increased resolution). All of the above (except for rotating-prism scanners) are referenced in Beiser (1974). Rotating-prism scanners appear in Wolf (1989). Recent work in liquid crystal phased arrays appears in Dorschner *et al.*,

(1993). Recent work in electron-beam–pumped semiconductor lasers (to select the radiation site) appears in Nasibov *et al.*, (1992).

GLOSSARY

Acousto-Optic Scanner: A deflector that controls the angle of diffraction of a laser beam by changing the grating spacing within an elastic medium, accomplished by varying the drive frequency to an acoustic transducer that launches an acoustic wave through the medium. The acoustic wavelength in the medium determines the spacing of the pressure changes that form refractive index changes (acousto-optic effect). Operation is typically in the Bragg regime.

Along-Scan: The spatial direction executed or developed by an optical deflector, (through the angle Θ, or in the direction x). See **Cross-Scan**.

Aperture, Scanner: The boundary (defined by width D) of the bundle of light incident upon a scanner. See **F-Number**.

Aperture Shape Factor: A factor a ($0.75 \leq a \leq 2$) that adjusts the value of the spot size or resolution as a function of the light intensity distribution at the scanner aperture. See **Apodizing**.

Apodizing: Intentional restriction of the intensity distribution within an aperture, typically in the central or peripheral region. See above two definitions.

Beam Expander: An afocal lens pair having a shorter focal length lens followed by a longer focal length lens; inverted telescope. Serves to widen an illuminating beam. ("Beam compressor" is complement.) Expander or compressor may also be formed in one dimension by, for example, a prism pair.

Bragg Regime: Region of strong first-order diffraction from a thick grating (thickness $L > d^2/\lambda$; d = grating spacing, λ = wavelength) when the input beam angle Θ is equal to the output beam angle, with respect to the grating planes. The "first order" Bragg condition occurs when $2 \sin\Theta = \lambda/d$.

Conjugate Imaging: The property of reciprocal imaging through an optical system of its object and image points.

Cross-Scan: Direction transverse to along-scan (y direction). See **Along-Scan**.

Double-Pass: Forth-and-back traversal of an optical beam through an objective lens. The first pass, toward the scanner, serves as (part of) beam expansion; the second pass, reflected from the scanner, propagates to the image plane.

Double Reflection: The reillumination of a deflector by the scanned beam in such phase that its cross-scan error is nulled.

Duty Cycle: The fraction or percent of useful interval of time with respect to the time for a full cycle of a repetitive process (e.g., recurrent scanning).

Electro-Optic (Gradient) Deflector: Electrically controllable scanner that develops or synthesizes a gradient of index of refraction in electro-optic material. The optical beam is bent thereby toward the direction of the gradient.

f-Θ Lens: Objective lens of focal length f, which provides the displacement s of an imaged spot to be proportional to input beam scan angle Θ, such that $s = f\Theta$.

F-number: $F = f/D$; f = focal length, as measured from beamwidth D.

Field of View; Field Angle: Useful optical angular subtense.

Flat-field Lens: Objective optics that forms a focused spot that may be scanned along a flat surface (which is typically normal to the lens axis). See **f-Θ Lens**.

FWHM: Full width (of spot δ) measured between its half maximum intensity points.

Galvanometer Deflector: Reciprocating mirror scanner having broadband or rapid-access capability. Moving-coil type derived from D'Arsonval electromagnetic meter movement. Resonant type limited to single harmonic frequency.

Holographic Deflector: Diffractive elements (often interferometrically generated) manipulated in a manner analogous to reflective/refractive ones (mirrors/lenses) disposed typically on a rotating substrate.

Magnification, Scan: $m = d\Theta/d\Phi$, where Θ = optical scan angle, Φ = mechanical scan angle.

Modulation Transfer Function (MTF): Ratio of output intensity modulation to input (sinusoidal) intensity modulation, as a function of spatial or temporal frequency.

Monogon: Rotating scanner having single facet (or facet set) whose normals appear in a single plane that includes the rotating axis.

Objective Lens: Optical elements that form the primary or output image of an optical wave. In scanning, the objective lens often images a single, or only a few, object points. See **Scanning**.

Overillumination (Overfilling): Subtending of the full aperture optically, by illuminating it with an oversized beam, to provide relatively uniform intensity distribution within the aperture.

Point Spread Function (PSF): Intensity distribution of a focused spot that is the image of a point object.

Pixel, Pel: Picture element of width x and height y. See **Voxel**.

Polygon: Rotating scanner having a multiplicity of typically reflective facets distributed about its axis. Subdivided into prismatic, having facet surfaces nominally parallel to the rotating axis—and pyramidal, having facet surfaces nominally convergent, forming a pyramid with apex on the rotation axis.

Pupil Relief Distance: In preobjective scanning, the distance between the scanner and the first mechanical surface of the objective lens.

Radial Symmetry: Condition of illuminating beam converging to or diverging from the nodal or rotation axis of an angular scanner. Illumination by a paraxial collimated beam is a special case.

Resolution, Scanned: Number N of diffraction-limited spots subtended by full format width of the image. For most systems, $N = \Theta D/a\lambda$; λ = wavelength; see **Resolution Invariant** and **Aperture Shape Factor** for other symbol definitions.

Resolution Invariant: $I = \Theta D = \Theta' D'$; adaptation of Lagrange invariant. Θ = optical scan angle; D = aperture width. Primed terms are same factors, following an interposed generic coaxial, stigmatic optical system. The invariance of the product ΘD denotes a conservation of scanned resolution and its independence from the traversal of subsequent optics. See **Resolution, Scanned**.

Scanning, Active; Flying Spot: Changing the direction of a beam of light, as from a cathode ray tube (CRT) or a laser.

Scanning, Passive: Collecting and directing portions of scattered flux from a distant object source to a detector (or detectors), as in remote sensing.

Scanning, Objective: Transverse translation of objective lens with respect to information surface, or its converse.

Scanning, Preobjective: Angular change of beam direction before the objective lens.

Scanning, Postobjective: Angular change of beam direction after the objective lens.

Spot Size, δ: Width of the focused spot. If approximately Gaussian, typically measured across $1/e^2$ intensity points or at FWHM. If non-Gaussian, often measured at FWHM.

Telecentric: Describing an optical system that forms an output image having all its principal rays propagating (approximately) parallel to the optical axis.

Truncation: Abrupt obscuration of beam intensity beyond the border of an aperture.

Underillumination (Underfilling): Condition of illuminating flux subtending a portion of the available aperture. See **Overillumination**.

Vignetting: The gradual obstruction of a (typically marginal) portion of a beam in executing its field angle, as it propagates through optical components having limiting boundaries.

Voxel: Volume picture element of width x, height y, and depth z. See **Pixel**.

Works Cited

Adler, R. (1967), *IEEE Spectrum* **4** (5), 42–54.

Bedamian, L. (1981), *Opt. Engr.* **20** (1), 143–149.

Beiser, L. (1966), *Photo. Sci. Engr.* **10** (4), 222–228.

Beiser, L. (1967), *J. Opt. Soc. Am.* **57**, 923–931.

Beiser, L. (1974), in: M. Ross (Ed.), *Laser Applications*, Vol. 2, New York: Academic Press, pp. 53–159.

Beiser, L. (1983), *Appl. Opt.* **22** (20), 3149–315.

Beiser, L. (1988), *Holographic Scanning*, New York: John Wiley & Sons.

Beiser, L. (1990), U.S. Patent, No. 4,963,643.

Beiser, L. (1991), in: G. F. Marshall, L. Beiser (Eds.), *Beam Deflection and Scanning Technologies*, Bellingham, WA: SPIE Proceedings No. 1454, Bellingham, WA: SPIE, pp. 60–66.

Beiser, L. (1992a), U.S. Patent No. 5,114,217.

Beiser, L. (1992b), *Laser Scanning Notebook*, Bellingham, WA: SPIE Optical Engineering Press, Vol. PM13.

Beiser, L., Johnson, R. B. (1994), in: M. Bass (Ed.), *Handbook of Optics*, Vol. 2, Chap. 19, New York: McGraw Hill.

Blais, F. (1988), *Opt. Eng.* **27** (2), 104–110.

Bouwhuis, G., Braat, J. J. M. (1983), in: R. R. Shannon, J. C. Wyant (Eds.) *Applied Optics and Optical Engineering*, Vol. IX, New York: Academic Press, pp. 73–110.

Brosens, P. J. (1993), in: L. Beiser (Ed.), *Recording Systems*, SPIE Proceedings Vol. 1987, 234–241.

Burke, J. R. (1969), U.S. Patent No. 3,428,812.

Debye, P., Sears, F. W. (1932), *Proc. Natl. Acad. Sci.* (U.S.), **18**, 409.

Donohue, J. P. (1979), *Proc. SPIE* **200**, 179–186. Also (1985) in: L. Beiser (Ed.), *Selected Papers on Laser Scanning and Recording*, SPIE Milestone Series Vol. 378, Bellingham, WA: SPIE, pp. 421–428.

Dorschner, T. A., Sharp, R. C., Resler, D. P., Friedman, L. J., Hobbs, D. C. (1993), *Basic Laser Beam Agility Techniques*, Wright Patterson Air Force Base, WL-TR-93-1020 (AD-B-175-883).

Fowler, V. J., Buhrer, C. F., Bloom, L. R. (1964), *Proc. IEEE* **52**, 193.

Garwin, R. J. (1984), U.S. Patent No. 4,429,948.

Gordon, E. I. (1966), *Proc. IEEE*, **54**, 1391–1401.

Giarola, A. J., Billeter, T. R. (1963), *Proc. IEEE* **51**, 1150.

Gottlieb, M. (1991), in: G. F. Marshall (Ed.), *Optical Scanning*, New York: Marcel Dekker, pp. 615–685.

Hanson, D. F., Sherman, R. J. (1984), U.S. Patent No. 4,433,894.

Hopkins, R. E., Stephenson, D. (1991), in: G. F. Marshall (Ed.), *Optical Scanning*, New York: Marcel Dekker, 27–81.

Ikeda, H., *et al.* (1987), *Fujitsu Sci. Tech. J.* **23**, 3.

Ireland, C. L., Ley, J. M. (1991), in: G. F. Marshall (Ed.), *Optical Scanning*, New York: Marcel Dekker, 687–778.

Johnson, R. B. (1979), *Appl. Opt.* **18**, 4030–4038.

Kay, D. B. (1984), U.S. Patent No. 4,428,643.

Kessler, D., DeJager, D., Noethen, M. (1989), in: L. Beiser (Ed.), *Hard Copy Output*, SPIE Proceedings No. 1079, Bellingham, WA: SPIE, pp. 27–35.

Korpel, A. (1980), in: R. Kingslake, B. J. Thompson (Eds.), *Applied Optics and Optical Engineering*, New York: Academic Press, pp. 89–141.

Kramer, C. J. (1980), U.S. Patent No. 4,239,326.

Kramer, C. J. (1981), U.S. Patent No. 4,289,371.

Kramer, C. J. (1991), in: G. F. Marshall (Ed.), *Optical Scanning*, New York: Marcel Dekker, 213–349.

Lee, T. C., Zook, J. D. (1968), *IEEE J. Quant. Electron.* **QE-4**, 442–454.

Levi, L. (1968), *Applied Optics*, Vol. 1, New York: John Wiley & Sons.

Li, Y., Katz, J. (October 1991), *Appl. Opt.* **30**, 4283.

Lohmann, A. W., Stork, W. (1980), *Appl. Opt.* **28**, 1318–1319.

Lospeich, J. F. (1968), *IEEE Spectrum* **5** (2), 45–52.

Marchant, A. B. (1988), U.S. Patent No. 4,759,616.

Marshall, G. F., Gadhok, J. S. (June 1991), *Photonics Spectra* 155–160.

Marshall, G. F., Vettese, T. J., Carosella, J. H. (1991), in: G. F. Marshall, L. Beiser, (Eds.), *Beam Deflection and Scanning Technologies*, SPIE Proceedings No. 1454, Bellingham, WA: SPIE, pp. 37–45.

Montagu, J. (1991), in: G. F. Marshall (Ed.), *Optical Scanning*, New York: Marcel Dekker, 525–613.

Nasibov, A. S., Kozlovsky, V. I., Reznikov, P. V., Skazyrsky, Ya. K., Popov, Yv. M. (1992), *J. Crystal Growth* **117**, 1040–1045.

Okolicsanyi, E. (1937), *Wireless Eng.* **615**, 128–132.

O'Shea, D. C. (1985), *Elements of Modern Optical Design*, New York: John Wiley & Sons.

Pawley, J. B. (Ed.) (1980), *Handbook of Biological Confocal Microscopy*, New York: Plenum.

Reich, S. (1985), in: L. Beiser (Ed.), *Laser Scanning and Recording*, SPIE Milestone Series Vol. 378, Bellingham, WA: SPIE, pp. 229–238.

Smith, H. M. (1975), *Principles of Holography*, New York: John Wiley & Sons.

Starkweather, G. S. (1984), U.S. Patent No. 4,475,787.

Tsai, C. S., Kraushaar, J. M. (1972), in: *Proceedings of the Electro-Optic Systems Design Conference*, New York: Industrial and Scientific Conference Management, Inc., pp. 176–182.

Tweed, D. G. (1985), *Opt. Eng.* **24**, 1018–1022.

Urbach, J. C. *et al.* (1982), *Proc. IEEE* **70**, 597–618.

Watson, W. H., Korpel, A. (1970), *Appl. Opt.* **7**, 1176–1179.

Wolf, W. I. (1989), W. L. Wolf, G. Zissis (Eds.), *The Infrared Handbook*, Chap. 10, MI: ERIM.

Yamagishi, F., *et al.* (1986), *Proc. SPIE* **615**, 128–132.

Zook, J. D., Lee, T. C. (1970), *Proc. SPIE 14th Symposium* (of 1969), 281.

Further Reading

Adler, R. (1967), Interaction between Light and Sound, *IEEE Spectrum* **4** (5), 42–54.

Beiser, L. (1974), "Laser Scanning Systems," in: M. Ross (Ed.), *Laser Applications*, Vol. 2, New York: Academic Press, 53–159.

Beiser, L. (1983), "Generalized Equations for the Resolution of Laser Scanners," *Appl. Opt.* **22**, 3149–3151.

Beiser, L. (Ed.) (1985), *Selected Papers on Laser Scanning and Recording*, SPIE Milestone Series Vol. 378, Bellingham, WA: SPIE.

Beiser, L. (1988). *Holographic Scanning*, New York: John Wiley & Sons.

Beiser, L. (1989), "A Guide and Nomograph for Laser Scanned Resolution," in: L. Beiser (Ed.), *Hardcopy Output*, SPIE Proceedings No. 1079, Bellingham, WA: SPIE, pp. 2–5.

Beiser, L. (1992), *Laser Scanning Notebook*, Bellingham, WA: SPIE Optical Engineering Press.

Dubovic, A. S. (1981), *The Photographic Recording of High Speed Processes*, New York: John Wiley & Sons (translated from Russian, 1964).

Fowler, V. J. (1966), "Survey of Laser Beam Deflection Techniques," *Appl. Opt.* **5**, 1675.

Korpel, A., *et al.* (1966), "A TV Display Having Acoustic Deflection and Modulation of Coherent Light," *Appl. Opt.* **5**, 1667.

Marshall, G. F. (Ed.) (1991), *Optical Scanning*, New York: Marcel Dekker.

Zook, J. D. (1974), "Light Beam Deflector Performance: a Comparative Analysis," *Appl. Opt.* **13**, 875–887.

OPTICAL SENSORS

See SENSORS, OPTICAL

OPTICAL SPECTROMETERS

See SPECTROMETERS, OPTICAL

OPTICAL STORAGE

JAMES J. BURKE, *Optical Data Storage Center, University of Arizona, Tucson, Arizona, U.S.A.*

INTRODUCTION

The purpose of this article is to provide the reader with an overview of optical storage systems and technology as they exist at the end of 1993. Though we will concentrate on digital systems, it is appropriate to acknowledge that the oldest and most universal form of optical recording is photographic film. Equally ubiquitous today is the recording of television (TV) signals on video cassette recorders (VCR). These are analog, as opposed to digital, recordings, wherein the brightness and color distribution of a scene, as framed by a camera, are faithfully recorded for subsequent display by means of a glossy print, slide projector, or TV monitor. Though these forms of optical recording provide excellent means of storing and communicating both entertainment and data to humans, they are incompatible with computers. Our focus in this article is digital storage by optical techniques.

A substantial portion of the research conducted between 1965 and 1975 led to extensive product development efforts in the decade that followed. The commercial success of several digital products introduced in the 1980s rests heavily on lessons learned in the development of less successful laser videodisc products, which had reached maturity by the early 1980s. Ironically, commercial videodiscs employ analog recording techniques—for example, frequency modulation (FM), wherein the length of (and spacing between) recorded marks on the disc varies inversely with the instantaneous amplitude of the video signal. Accordingly, they are not directly compatible with digital computer technology. Though we shall not consider videodisc systems further, we must acknowledge here that the digital systems we will

3-527-28134-7/95/$5.00 + .50

consider, particularly the CD (compact disc) and CD-ROM (CD-read-only memory) systems in wide use today, depend very heavily on technologies developed for videodisc systems. In particular, techniques for writing a master disc for subsequent mass replication, as well as the mass replication technologies themselves, and techniques for keeping the focused laser beam on track and in focus, were all perfected in the development of videodisc systems. The first published book on optical recording (Isailovic, 1985) is devoted almost entirely to videodisc systems. Special issues of the *Philips Technical Review* (Miltenburg, 1982) and the *RCA Review* (Hittinger and Sonnenfeldt, 1978) are devoted to optical and capacitive videodisc technologies, respectively. The book by Bouwhuis *et al.* (1985) is a monograph whose chapters, written by major contributors to the development of optical recording technologies, give excellent, detailed, technical discussions of the optical, mechanical, mastering, coding, and materials technologies used in the development and production of optical disc systems through 1985. We shall refer to these chapters later in this article.

All optical storage systems currently in use or in product development use a focused laser beam to read or write data. The smallest achievable spot size is limited by the wavelength of the laser and the numerical aperture (NA) of the objective lens that focuses the beam on the disc; its full width at half maximum (FWHM) intensity is approximately equal to $0.6\lambda/\mathrm{NA}$. The NAs of systems now in use are about equal to 0.5, while the wavelengths of the laser diodes in commercial systems available today range from 780 to 840 nm. Thus the effective spot size is about 1 μm, yielding a storage density of approximately 10^8 bits/cm^2. In contrast to magnetic disks, all optical disks are removable, and may thus be transported from one drive to another, as with floppy disks. These two attributes, high capacity and removability, are the principle advantages of optical technology. An additional positive attribute, relative to magnetic storage, is the fact that the optical head never touches the optical disk. Rather it is actively held at a distance of several millimeters from the information bearing layer. Thus the *head crashes* that are familiar to users of magnetic disks cannot occur with optical disks.

On most optical disks, including videodiscs and CDs, information is written on spiral tracks, where the track pitch for the 840- and 780-nm laser systems is 1.6 μm. In these digital systems, the data are written and read as a serial bit stream of 1's and 0's, called *channel* bits. These are to be distinguished from *user data* bits, which are the input to (output from) the (de)modulator that drives the laser source (decodes the received signal). User bits, in this context, include bits to make possible error detection and correction (EDAC), plus identifying information, all subsequently processed by the channel's modulation coder, which transforms ECC-encoded data into channel bits (by means of look-up tables, for example; ECC is the acronym for error correction code). Optical recording channels generally employ run-length-limited (RLL) codes, also called (d,k) modulation codes. These prescribe the minimum number, d, and the maximum number, k, of 0's that must follow each 1 in the channel bit stream. (1's represent transitions from marks to background—called *land*—or vice versa.) A commonly used modulation code is the (2,7) code, developed for magnetic disks. We will discuss these matters further in Sec. 3.

Optical disks (discs*), like magnetic disks, come in a variety of sizes. The smallest diameter, at this writing, is 2.5 in. (the Sony Minidisc); larger disks have diameters of 3.5, 5.25, 8, 12, or 14 in. The size of the optical drive is specified by its *form factor,* which tells the computer system designer the size of the standardized *slot* it will occupy in a computer's chassis. For example, a *full-height, 3.5-in.* form factor fits into a slot that is 101.6 mm wide and 83 mm high. Slot depths are not fully standardized; actual drives have lengths that range from 146 to 155 mm (Marchant, 1990, p. 21). As smaller drives emerged, *half-height* and *third-height* form factors came into being. Thus, the form factors of both optical and magnetic drives are continually changing.

In the next section, we will describe the various types of digital optical storage systems available at the end of 1993, and their current applications. In the remaining three

*It is customary, though not universally adopted, to call consumer products discs and players and professional products disks and drives. Recordable CD systems just emerging present a dilemma.

sections, we will describe some of the basic technologies that made possible the production of these systems and of ongoing research and development for their improvement. In the final section, we devote a little space to research in optical *memory*, where the *system*, as well as the applications, are as yet poorly defined. These include studies of holographic recording, photon-echo, and spectral hole-burning techniques. At the present time, it appears that these technologies may play a role in massively parallel computers (perhaps even optical computers) similar to that played by RAM (random access memory) in today's serial computers. The engineering of such memory systems has hardly begun.

1. SYSTEMS AND APPLICATIONS

In this section, we describe attributes of systems currently available commercially and some of their applications. The systems may be divided into three classes: *read only*, *write once*, and *rewritable*. A *read-only* system, such as the CD (also called CD-digital audio disc or CD-DAD) uses a *player* to transform the serial stream of channel bits, represented by depressions (pits) embossed in a polycarbonate disc, into stereo sound (in the case of CD) or computer-compatible digital data (CD-ROM), etc. *Write-once* (WO) systems, also called *WORM* (write once, read many) systems, employ lasers of higher power (10–20 mW) to form marks in media whose optical properties change dramatically when heated locally to high temperatures by the focused laser spot. The written marks are permanent. They may be read in the same drive by employing much lower laser power (1 mW), because modest heating does not mark the medium. (The marking process requires that a threshold power density must be exceeded at the recording layer.) *Erasable* systems, as their name implies, use disks that can be erased and rewritten, in a manner similar to magnetic disk systems. Commercially available erasable systems fall into two classes: *magneto-optic* (MO) systems, where an erasure cycle must precede a rewrite, and *phase-change* systems, wherein new data may be written directly over old. (This is called *direct overwrite*, a capability of all magnetic disk systems, but still a subject of research and development for magneto-optic systems. The Sony Minidisc is the exception. It accomplishes direct overwrite in MO media by modulating the magnetic field of a small magnetic head flying a few micrometers above the MO layer. The laser light focused through the substrate onto the MO layer raises its temperature, permitting a small magnetic field to reverse the magnetization direction.)

1.1 Read-Only Systems—Entertainment and Publishing

These systems are the major beneficiaries of technologies developed for the videodisc. They include the CD audio disc, which has revolutionized the market for music publishing, the CD-ROM, which is now in the process of revolutionizing some forms of text publication, CD-Interactive (CD-I), which makes possible *multimedia* applications, and DVI, digital video interactive, which provides an hour of digital video on a CD. We describe attributes of these systems in the following subsections.

1.1.1 CD—Audio Entertainment In his keynote address at the Joint International Symposium on Optical Memory (ISOM) and Optical Data Storage (ODS), 1993, Senri Miyaoka of Sony Corporation reported that more than 300 million CD discs and 50 million players were sold in 1992 (Miyaoka, 1993). Thus, a product concept developed and standardized by Philips in the Netherlands and Sony in Japan, and introduced in 1982, has become immensely successful. CD discs and CD players are all made to specifications contained in what has become known as the *Red Book*, which Philips provides to all licensees of its technology. [There is also a *Yellow Book* for CD-ROM, a *Green Book* for CD-I, and an *Orange Book* for CD-R. The *Yellow Book* has also been published by the European Computer Manufacturers Association (ECMA) as a standard, known as ECMA-130, entitled *Data Interchange on Read-Only 120 mm Optical Data Discs*, dated July 1988 (Howe, 1993a).]

A CD is a polycarbonate disc whose inner and outer diameters are 15 and 120 mm. All discs are 1.2 mm thick. The discs are made by injection molding; one side of the mold is occupied by a conformable nickel stamper whose surface contains the finely structured pattern [the spiral of bumps that create the depressions (pits) that represent channel data] to be embossed into the polycarbonate. The creation of a single molded disc takes about 10 s. After molding, the patterned side of the disc is coated with an aluminum film, about 100 nm thick. This film is then covered with a plastic film about 10 μm thick to protect the Al film from scratches. The plastic (lacquer) layer is usually an ultraviolet-cured acrylate, applied by spin coating (Marchant, 1990, p. 25). Title information, etc., is printed over the lacquer layer. The (mass) manufacturing costs of this product are substantially less than one U.S. dollar.

The single, tightly wound (1.6-μm pitch), spiral track of embossed pits (0.12 μm in depth and 0.6 μm in width) extends from an inner radius of 25 mm to an outer radius of 58 mm. This is the *physical track* of coded digital channel data. The disc may contain many *logical information tracks*, ranging in number from one to as many as 99. There is also a *logical lead-in* track, which is disc specific in that it tells the player, among other things, how many logical tracks are on the disc and where they are located. For the CD audio discs under discussion, each logical track would contain one piece of stereo music. (As we shall see below, the data in CD-ROM discs are constrained to the same range of logical tracks as CD audio, as is CD-I.)

Each of the two original stereo audio channels is sampled at a rate of 44 100 samples/s and digitized to a precision of 16 bits per sample. Thus, the information rate in the raw digital data is 1.4112 (2 $\times$ 16 $\times$ 44 100) Mb/s (Mb stands for megabit). With the addition of ECC bits and identification information, the raw data rate is about 2 Mb/s. The digitally encoded music data are input to the ECC encoder in *frames* of 24 bytes at a rate of 7350 *frames*/s. (24 bytes equals six samples of the stereo signal.) Each frame goes through two levels of error correction coding, which add a total of eight additional bytes to the input frame. (The code used is called CIRC, the acronym for *cross interleaved Reed-Solomon code*.) An additional byte containing timing and display information (called the *control and display* byte) is then appended; this ends the process of error-correction encoding and control formatting wherein 24 bytes of digital music are expanded into a 33-byte frame of raw data. Before these data are ready for the mastering process, they are further transformed by the modulation encoder in a manner such that each byte of eight bits is converted to 14 channel bits. (This process, accomplished by look-up tables, is called *eight-fourteen modulation* or EFM.) Three additional bits are inserted between adjacent 14-b channel words to ensure that the constraints of the modulation code (a 2, 10 RLL code) are not violated. [Only two bits are required for this purpose. A third bit is added to suppress the low-frequency content of the signal (Bouwhuis *et al.*, 1985, p. 245).] Then a 27-b *sync field* is inserted at the start of the frame, for synchronization and timing purposes. Each frame thus contains 588 channel bits (33 $\times$ 17 + 27), which is referred to as an *EFM frame*. At the original sampling rate of 44 100 samples/s, 7350 such EFM frames are generated each second, yielding a channel bit rate of 4.3218 Mb/s. The binary signal generated in this way is used during the disc mastering process to turn the laser beam on and off.

CDs are mastered and played at a constant linear velocity (CLV) of 1.2 m/s. Thus, the rotation rate of the disc is a function of radius. Though changing the rotation rate when accessing different parts of the disc increases access time, it is still much faster than in tape systems. The maximum capacity of the CD is 72 min (4320 s) of audio, which corresponds to 762 MB (megabytes) of digitized music, or 18.67 $\times$ 10^9 channel bits.

We should also note, because it is important to understanding the other CD-XXX formats, that each successive set of 98 frames (2352 music bytes) forms a *control and display block* (also called a *sector* in CD-ROM). All frames in a given block carry the same block number (0 to 74), which is located in the control and display byte: the block number is thus incremented by 1 every 98 frames. This is part of the identifying information, which also includes minutes (0 to 70) and seconds (0 to 59), that is used by the player to locate a selected piece of music on the disc.

1.1.2 CD-ROM—Document and Data Publication Though the CD system was designed for the publication of music, it was immediately recognized by its designers as a means for publishing digital data of any kind. Large databases, computer software manuals, encyclopedias, etc. were all candidates.

To be appealing to computer users, only modest additions to the CD specification of the *Red Book* were required. These are spelled out in the *Yellow Book*. Because computer data on magnetic disks, both hard and floppy, is stored in *sectors* that contain a number of bytes equal to a power of 2, a CD *block* of 2352 music bytes (one *control and display block*) would become a CD-ROM sector of 2048 user bytes. The remaining 304 bytes in a block would be used for an additional layer of ECC, plus 12 lead-in *sync bytes* used for clocking purposes and four address bytes that identify the sector in terms of minutes, seconds, and block number.

Though Philips and Sony unveiled the specifications for CD-ROM in 1984 (the *Yellow Book*), they did not specify how the data would be organized, indexed, and retrieved from the disc. This led to an initial period of chaos, as publishers each went their own way, developing different software to accomplish these tasks. There was also a problem with the first CD-ROM players to reach the market; the layout of the computer connector had not been standardized. The connection problem was solved with the availability of inexpensive, custom-chip SCSI (small computer system interface) controllers. The software problems were essentially solved when the High Sierra Group submitted its proposal [HSG Proposal (Einberger and Zoellick, 1987a)] for a logical file format for CD-ROM in July of 1987. This proposal, which also accommodated the CD-I specifications announced by Philips and Sony in 1986, is the basis for today's software standards. [The initial members of the High Sierra Group included representatives of Apple Computer, Digital Equipment Corp., Hitachi, LaserData, Microsoft, 3M, Philips, Reference Technology Inc., Sony, TMS, Videotools, and XEBEC. The group takes its name from the location of its first meeting in the fall of 1986 at the High Sierra Casino and Hotel in Lake Tahoe, Nevada (Einberger and Zoellick, 1987b).]

1.1.3 CD-I—Audio, Video, and Text—Training and Entertainment CD-I extends compact-disc technology to multimedia application, including the storage and retrieval of the multimedia application program itself. Its specifications are contained in the *Green Book*. The first CD mixed-media standard appeared somewhat earlier. It prescribed how music and text would be mixed in the CD format. The prescription was announced as an addendum to the *Yellow Book*, called CD-ROM/XA (extended architecture). CD-I is essentially a reworking of CD-ROM/XA to support still and moving pictures and graphics, in addition to sound and text. It does so by specifying a logical file structure that accommodates all these forms of data. Images are stored in compressed form: still images according to the JPEG standard (Joint Photographic Experts Group) and video according to the MPEG standard (Motion Picture Experts Group). CD-I players contain chip sets to decompress the stored data. They can play CD and CD-ROM/(XA) discs as well as CD-I discs; they also play the new writable discs, such as Photo-CD. This is not surprising, because all of these discs must conform to *Red* and *Yellow Book* specifications; in the case of Photo-CD, they must also conform to the specification of the *Orange Book*, which covers writable CD technology.

Though the applications of CD-I technology to education and entertainment appear very exciting, there has not been much commercial activity to the time of this writing. In the age of multimedia, where the hardware and software are priced in a range where millions of people can afford to prepare (premaster) their own productions, the potential is enormous. As with home movies, most will not have the training and experience to create professional productions. But the medium itself may accelerate the acquisition of this learning. That prospect is exciting.

1.1.4 DVI—Video Entertainment DVI (digital video interactive) is a proprietary image-compression technology initiated at the RCA David Sarnoff Laboratories (now Sarnoff Laboratories) in the 1980s (Leenhardt, 1987). At the end of the decade it was licensed to Intel, which is developing chip sets for real-time decompression (already available) and compression. DVI uses standard CD technology to record 60 min of NTSC color television on the 120-mm compact disc. It takes advantage of the substantial redundancy of most video, recording only changes from frame to frame. Compression factors greater than 100 are found to yield aesthetically pleasing video displays. While this technology may be expected to play a substantial role in the future, perhaps even replacing videotape and VCR systems, it has not happened yet. But it is only a matter of time. Because compact discs do not wear out, DVI discs will maintain the original image quality for many years. They should be appearing shortly in your video rental store.

1.2 Write-Once Systems—Document and Image Archiving

Because of the way in which data are written on compact discs, i.e., the data interleaving prescribed by the CIRC coding scheme, the access time to a given user data byte is slow. It is necessary to read and decode 114 CD frames, each containing 24 user bytes, before the user has access to the contiguous set of 24 bytes that was originally sent, during mastering, to the ECC encoder at a particular time. This rather massive interleaving of user bytes was designed to avoid *burst errors* caused by any source of contamination present during disc fabrication, or introduced later by user handling. (Because the data density is so high, any piece of contaminant is likely to be much larger than a CD pit or a single user data byte.) Additionally, the average *access time* to any position on the disc is 600 ms. By comparison, magnetic hard drives in PCs and workstations have access times in the range of 10 to 20 ms. If optical storage were to find a place in the professional market of computer peripherals, much higher performance than that of the CD would be needed. A CD rotates at a rate of about 500 rpm at its inner radius and 200 rpm at its outer radius. The need to change motor speed to accommodate the variable rotation rate also adds to the access time. Typical magnetic drives spin at 3600–5400 rpm. Thus, while the capacity of the CD is excellent, its access time and, to some extent, its data rate, are far from satisfying the needs of most computer applications.

One such application is random access to large databases containing documents and/or images. One or more optical drives in a *jukebox* (also called library or autochanger) containing tens or hundreds of optical disks could make possible fast access (a few seconds) to huge amounts of data—tens or hundreds of gigabytes. The first optical drives developed to serve such applications appeared in 1987. Products were offered by several manufacturers, but, because standards had not then been adopted, they reached only specialized applications for individual, generally large, customers (government agencies, banks, insurance companies), and relatively few of even these potential users. Among the first were 12-in. systems with 306-mm glass substrates, wherein two coated substrates were hermetically sealed together with their active layers facing one another across a thin air gap. Each side of the disk had a capacity of 1 Gbyte. Access time was specified as 150 ms and data rate (for large files) at 262 kB/s. 5.25-in. drives appeared shortly thereafter with 130-mm polycarbonate substrates. Typical capacities were 200 to 300 MB per surface, with data rates from 75 to 120 kB/s. Some had form factors corresponding to those used in the computer industry, such as full height 5.25 in., but most were packaged as stand-alone computer peripherals. Many were marketed as components of desk-size document storage systems, which included, in addition to the WO (write once) drive, a laser scanner for entering documents into the system, a high-resolution monitor, a laser printer, and the software to make such systems productive. Early heavy users of these workstations were in engineering-design environments, often in large, vertically integrated organizations that had, themselves, designed the optical drives, media, and workstations.

Another early system was the 14-in. optical disk drive and jukebox developed by Eastman Kodak company. This used a dye-doped polymer active layer on both sides of an Al substrate. Each side is covered by a 0.1-mm-thick plastic membrane suspended about 0.3 mm above the active layers. The membrane keeps dirt and dust contaminants sufficiently out of focus to prevent them from having any substantial effect on the intensity of the beam. The original disks had a capacity of 3.4 GB/side. Access time was slow (several hundred milliseconds), partly because the disk was partitioned in radial zones, each with a different (though constant within a zone) angular velocity. The comparatively long access time of the drive was not considered a serious deficiency in a jukebox, where the average access time to data on any of 150 disks was 12 s. This system currently has been upgraded to 7 GB/side; disk access time has been reduced to 6 s.

At the time of this writing, standards are in place for WO (write once) drives and media, and both large companies (for example, IBM and Hewlett Packard, Eastman Kodak) and small ones (Pinnacle Micro, Alphatronics, Plasmon Data Systems) offer jukeboxes with tens and hundreds of GB capacity. Typical current systems offer disks with 1-GB capacity in 5.25-in. format, and drives with 90-ms access time and 1.5-MB/s data rate. It should be noted that not all systems available today were designed to meet national and international standards. Potential users should inquire about interchangeability if they need to read their disks on other systems.

1.2.1 Physical Mechanisms All commercially available WO (as well as rewritable) media rely on heating the active layer to a high temperature to effect writing. Many different materials have been studied over the years, and several have found their way into commercial disks. Metals with low melting temperatures, like tellurium and selenium, have proven quite useful. An early paper by Maydan (1971) demonstrated how small holes are written in metal films, describing the optical and thermal characteristics of the process. Simple melting is not sufficient to form a hole (Bouwhuis *et al.*, 1985, p. 216; Kivits *et al.*, 1982; Broer and Vriens, 1983). Ablation (removal of material) must occur to initiate hole formation. A 10-mW laser focused to a 1-μm spot has a power density on the order of 1 MW/cm^2, quite enough energy to ablate tellurium or selenium in less than a microsecond.

Textured films of platinum on a subwavelength-textured *moth eye* substrate (Storey *et al.*, 1988) are substantially less reflective (15%) and more absorptive than flat platinum. When heated by the laser spot, the platinum film flattens locally, giving a reflectivity about equal to that of a perfectly flat film (50%). Plasmon Data Systems manufactures moth eye disks of this kind today, quoting 50-yr life expectancy in their WO systems.

The mechanism for creating holes in dye-doped polymer films is also ablative. As the dye absorbs the laser energy, material is ejected from the surface. Kodak disks using this principle have an additional layer above the dye-doped polymer film, one that deforms into a bubble that maintains its approximately hemispherical shape after the material cools. The extra film prevents the products of the ablation from contaminating other parts of the disk, yielding media with signal-to-noise ratio (SNR) greater than 60 dB.

Phase-change media, wherein an amorphous phase of an alloy changes into a crystalline phase, or vice versa, when abruptly heated by a focused laser beam, are important to both WO and erasable systems. These material systems can be designed to be reversible. They were first employed as WO media, however, and are still used today for this purpose. The mechanism, in the case of media prepared in an amorphous state, is essentially one of annealing, wherein small crystallites form when the material is heated and subsequently adiabatically cooled, yielding higher optical reflectivity than the amorphous phase. This type of WO recording has been demonstrated in Te_3Sb_2 and Se_3Sb_2 (Watanabe *et al.*, 1983). WO phase-change media are also prepared in the crystalline phase. These rapidly absorb enough energy from the laser spot to exceed the melting temperature of the active layer. Rapid quenching leaves an amorphous mark with substantially less reflectivity than the crystalline phase. TeSeSb, in a thin film of about 100 nm thickness, is one such material. In a thinner film version (30 nm) it is ablative material (Bouwhuis *et al.*, 1985, p. 219). We discuss these materials further in connection with erasable systems (Sec. 1.3).

1.2.2 Tape Systems Though a number of optical tape systems were in development in the late 1980s (Marchant, 1990, p. 61), and one was demonstrated (Vogelgesang and Hartmann, 1988), only one is commercially available today. The first commercial systems, developed by CREO Products, Inc., were delivered in 1990. Their specifications were described by Dan Gelbart in March of that year (Gelbart, 1990). Capacity: one TB on a 12-in. reel of 75-μm-thick tape, which can be increased to 4 TB on a 14-in. reel of 25-μm tape. Data rate: 3 MB/s read and write. Access time: 30 s average; 60 s worst case in the 1-TB system. Size: 24 in. wide, 75 in. high, 30 in. deep. Media: ICI WO optical tape or equivalent. (Dow Chemical's hardcoated metal alloy tape is a potential option. We discuss reported developments in this tape medium in Sec. 4.) The CREO OTR (optical tape recorder) is a mainframe peripheral, large and expensive. The Canadian government was its first customer. The drive uses linear, bidirectional scanning across a stationary tape, which is stepped rapidly at the end of each scan. An array of 33 laser diodes provides 32 data tracks plus one servo track with each linear scan. The diodes each provide 30 mW of write power at a wavelength of 830 nm. The written marks are about 1 μm in diameter and are spaced at 1.5 μm in both directions. All of the 33 tracks are read simultaneously using a custom integrated circuit that contain all photodetectors, amplifiers, comparators, and data-recovery and tracking logic. The OTR's tape compartment is maintained as a class-100 clean room using recirculating air. The system contains built-in diagnostic equipment to aid in maintenance, an unusual feature but one that is not surprising in such a new and complex electro-opto-mechanical machine. Efforts to develop new and improved tape systems persist at CREO and other organizations, though there have been no new product announcements.

1.2.3 Optical Card Systems Jerome Drexler of Drexler Technologies developed the *Drexon Laser Card* in 1980 (Time-Life Books, 1987*). It was designed to hold about 2.6 MB on a credit-card–sized substrate, with tracks spaced at approximately 12 μm. SRI was contracted to develop a prototype read/write drive (Pierce, 1988) by Drexler, who then began licensing the card and drive technology to drive manufacturers. One of the first to market was the Nippon Conlux LC-303 Optical Card Reader/Writer (Marchant, 1990, p. 64). Performance specifications were capacity of 2.6 MB; access time of 0.5 s; data rate of 10 kB/s; size 10.6 × 10.6 × 4.3 in. The Drexon Laser Card employs a modified silver-halide photographic process to form the active WO layer. Applications of optical card systems to health care records, software distribution, and financial transactions are under study or in use, particularly in Japan. Other companies, including Omron Corp. (Tsutsui *et al.*, 1990), are developing drives for the Drexon card.

*This is a beautifully illustrated, technically accurate, and very well written description of the history and technologies of magnetic and optical storage.

1.2.4 CD-R and Photo-CD—Recordable CD Technology The specifications of writable compact discs are contained in the *Orange Book*, available from Philips and Sony. There are in fact two Orange Books, *Orange Book* Part I for re-recordable CD (CD-MO—magneto-optic) and *Orange Book* Part II for WO systems (CD-WO) (Howe, 1993b). The latter covers what is commonly known as CD-recordable, or CD-R. At the present time, no magneto-optic media exist that meet the performance specifications of *Orange Book* Part I (in addition to those of *Red* and *Yellow Books*). Most CD-R discs are manufactured by Toiyo Yuden Co. Ltd. and, apparently, by other companies under license to Toiyo Yuden. On top of a polycarbonate substrate with a continuous spiral pregroove, a 150-nm dye layer is spincoated. The active layer is covered with a 70-nm-thick layer of gold, on top of which a UV-curable lacquer is spincoated and cured. In a detailed study reported by Holtslag *et al.* (1991), it was concluded that bumps form at the dye/polycarbonate interface and pits form at the gold/dye interface. These effects are associated with two threshold temperatures. In a room-temperature environment, the first threshold is at about 170 °C and is associated with the softening of the polycarbonate. The second threshold, at about 265 °C, is associated with a permanent change in the dye layer. Studies like these will aid in the continuing search for inexpensive material systems that can meet the specifications of recordable CD systems. A recent report of an interchangability test among eight CD-R disc manufacturers and CD players and CD-R recorders from nine different manufacturers uncovered some operational differences (Inoue *et al.*, 1993); nonetheless, all players were able to play all of the CD-R discs successfully. This success suggests that the high commercial expectations of companies involved with CD-R technology are justified.

Photo-CD, introduced by Kodak in 1992, was designed as a line of hardware and software that would make possible the incorporation of high-quality images, as originally recorded on 35-mm film, into computer, publishing, and home video entertainment applications (Howe, 1993b). Images in the photographic negatives are scanned at a photographic shop to a resolution of 3072 by 2048 pixels. Each pixel (sample) is digitized to 8 bits in each of 3 colors, almost 20 MB of data. Each image is stored as a single file of five contiguous subblocks of data. One subblock corresponds to a lower-resolution rendition of the originally digitized image, suitable for displaying the full scene on a home television monitor. In addition to this *base image*, two smaller versions and two larger versions are stored. The smaller versions (384 × 128 and 192 × 64) are used to get a quick look at all the images recorded in a given *session* (see next paragraph), while the larger images (3 × 2 k and 1.5 × 1 k) are stored in compressed form. Initial versions of Photo-CD players available to consumers do not have the chip sets needed to decompress these higher-resolution images, through they can be used at the photographic shop to make high-resolution prints. In the near future, it is expected that desktop publishing systems capable of decompressing and working with Photo-CD images at all resolutions will be available.

As originally released, the *Orange Book* disc was to conform to all the specification of CD-ROM/XA. This meant that the complete, physical, spiral track had to be recorded in a single, continuous, real-time, nonstop *session*, which begins with the recording of the lead-in logical track, continues with up to 99 logical information tracks, and ends with a lead-out track. This would have been a severe constraint for Photo-CD users as well as most computer users. To address this problem, the *Orange Book* also defines a CD-WO/HYBRID disc that can contain multiple sessions (Howe, 1993b). Though still limited to 99 information tracks, the disk can have up to 45 sessions recorded on it, each with its own lead-in and lead-out track. (If all 45 are used, there is not much space left for information tracks.) The Photo-CD system was designed to accommodate three separate trips to the photographic shop, each involving the recording of one roll of film. Though designed primarily as a consumer system, Photo-CD has found other applications, such as a way of generating computer-compatible telephone bills to large corporate customers of telephone companies. Reams of paper so voluminous that few would take the time to examine them were replaced by a small disc and efficient search software, saving both trees and time.

1.3 Erasable/Rewritable Systems—Computer Peripherals; Network Servers

1.3.1 Magneto-optic Systems: 80 mm, 120 mm, 300 mm Magneto-optic (MO) systems of 5.25 in. diameter, based on rare earth/transition metal (RE/TM) alloys, were the first erasable optical storage systems to reach the market. They were soon followed by 3.5-in. systems. In both systems, national and international standards for media specifications were in place prior to their introduction. Larger, very high-performance, MO systems (12 and 14 in.) were built for special government applications (Levine, 1990). We will focus on commercial systems. Early disks had capacities of 325 MB/side (5.25 in., double sided) and 128 MB/side (3.5 in., single sided). In 1993, several companies introduced 2*X* 5.25-in. systems, with 650 MB/side. The drives have also been incorporated into jukebox (library) applications. The standard for 3*X* media (975 MB/side) is essentially completed, while the 4*X* standard (1300 MB/side) is in negotiation. 3*X* and 4*X* systems will employ laser diodes operating at 680-nm wavelength.

MO media are ferrimagnetic, with their easy axis perpendicular to the surfaces of the active layers. When linearly polarized light is incident on such a magnetic layer, the reflected light is elliptically polarized, with the major axis of the ellipse rotated with respect to the direction of polarization of the incident field. The rotation is clockwise if the material illuminated by the focused beam is magnetized upward, out of the disk. It is counterclockwise for the opposite magnetization. This is the magneto-optic Kerr effect, discovered a century ago. Magneto-optic systems store information by writing small magnetic marks, called domains, of one magnetic orientation in a material that is globally magnetized in the opposite direction. This is done by locally heating the layer in the presence of an external magnet polarized opposite to the direction of the initial magnetization of the film. The marks are read at low power (1 mW) by sensing the rotation of the polarization as the playback spot scans the track of written marks. Signal-to-noise ratios greater than 50 dB are typical in commercial systems. For a further discussion of

MO disk principles, the reader is referred to MAGNETOOPTICAL EFFECTS AND DEVICES.

1.3.2 Sony Minidisc—Direct Overwrite, Magneto-optic Recording The 2.5 in. Sony Minidisc (MD) became commercially available in 1992. It is designed as a consumer product, essentially a CD-like unit that records, as well as plays, stereo music. Because of the small size of the disc, Sony had to make some compromises, relative to the CD, in the quality of the recorded stereo. It is nonetheless a remarkable system. The MD system also includes a read-only disc "for music-software publishing" (Yoshida, 1993), i.e., the player can also read pressed discs. The MD uses modulation and ECC coding very much like that of CD-ROM Mode 2. [The *Yellow Book* specifies three modes for writing discs. The one described in the subsection on CD-ROM in this article is Mode 1, which adds a third layer of error correction to the two layers used in CD. Mode 2 does not add this additional layer, so that 2336 user bytes are contained in a sector (block), along with 12 sync bytes, 3 address bytes and one mode byte, bringing the total to 2352, as in the CD audio system.] Just as in the CD system, the stereo signal is sampled at a 44.1-kHz rate and digitized to 16 bits in each stereo channel. This data must then be compressed by a factor of about 5 in order to record 74 min of stereo on a 2.5-in. disc. The compression is done by a system (chip set) named ATRAC (Adaptive TRansform Acoustic Coding) that uses both the frequency and temporal content of the data to achieve an aesthetically pleasing recording. After coding (CIRC and ATRAC) successive streams of data corresponding to 5632 samples are recorded in two CD blocks (sectors). With respect to the CD, this represents an overall compression of 4.79 (2816 samples in a MD sector vs 588 in a CD sector). On playback the MD uses a memory buffer of 4 Mb between the optical pickup and the ATRAC decoder. This memory can store about 12 s of compressed audio, which can be read from the disc in just a few seconds because of the 4.79/1 compression. In normal playback, the disc is read intermittently to keep the buffer full. Because the MD is designed to be a portable player, the buffer provides adequate time to recover the appropriate track if the player is jostled.

Sony has recently announced that it is developing a nonaudio version of the MD, named MD-DATA. It will be described in an invited paper to be presented at the ODS 1994 meeting at Dana Point, California on 16–18 May 1994 (managed by the Optical Society of America).

1.3.3 Direct Overwrite Phase-Change Systems Matsushita introduced the first rewritable phase-change system in Japan in 1992. The drive and 5.25-in. media were developed in-house. The technology was described in two papers presented in Sapporo in 1991 (Yoshida *et al.*, 1991; Ishibashi *et al.*, 1991). The active medium is a GeTe-Sb_2Te_3-Sb alloy of 20 nm thickness. To construct the disk, a dielectric layer of ZnS-SiO_2 of 150 nm thickness is first deposited on a pregrooved polycarbonate substrate. The active layer is then deposited and is topped by another dielectric layer of ZnS-SiO_2 of 20 nm thickness. This is covered by an Al film 150 nm thick, followed by a plastic protective layer. Two different levels of laser power are used to effect direct overwrite. In the experiments reported by Ishibashi *et al.* (1991), a 0.5 NA objective focused 16 mW of power from a 830-nm laser onto the active layer. The laser was pulsed for 40 or 20 ns to write amorphous marks in the crystalline background (*land*) at a linear velocity of 5 m/s. The pulses were erased by 80- and 60-ns pulses, respectively, at a 7.2-mW power level. The bit period was 200 ns and the bit separation was 800 nm. One million successful overwrites were demonstrated.

Because the optical path (Sec. 2) in phase-change systems is considerably simpler than in MO systems, it appears that the former can compete with the latter as network servers with massive, changing databases, and in emerging multimedia applications. Price/performance is the main issue. Because MO systems have an established manufacturing base (and magnetic systems a much larger one), they have the advantage presently. Nonetheless, a great deal of interest in reversible phase-change systems persists, as evidenced by nine papers presented at the International Symposium on Optical Memory and Optical Data Storage, 1993, by researchers from six companies—Matsushita, Toshiba, NEC, Asahi Chemical, TDK, Ricoh, and Mitsubishi (see also Ito *et al.*, 1993). Some of these involved experiments in rewritable CD formats, which portends an interesting future for these systems.

2. THE OPTICAL PATH

The optical path followed by light emitted from the laser diode in a typical WO or rewritable phase-change drive is illustrated schematically in Fig. 1. The diverging beam from the laser diode is collimated by the lens above it, which is followed, in some systems, by a linear polarizer to ensure that the direction of vibration of the electric field in the collimated beam is parallel to the plane of the figure. (This is called *p* polarization in optics.) The polarizing beam splitter is designed to transmit 100% of the *p*-polarized light incident on the dielectric thin-film multilayer situated along the diagonal of the cube, which is made up of two right-angle prisms cemented together. The collimated beam then passes through a quarter-wave plate whose optical axis is oriented at 45 deg to the plane of the figure. On emerging from this plate, the light is circularly polarized. The objective lens focuses the circularly polarized beam to a spot on the active layer. In almost all current systems, the active layer resides on the upper surface of the substrate, 1.2 mm from the bottom surface of the disk. (In most media today, a two-sided disk is composed of two substrates, plus deposited layers, sealed together with the active layers facing each other across a small air gap.) Thus the light must be focused through a 1.2-mm-thick dielectric plate, the substrate. This introduces spherical aberration in the converging beam.

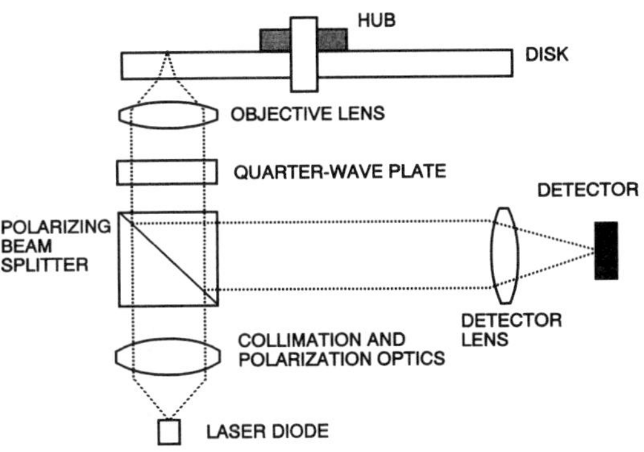

FIG. 1. A schematic diagram of the optical path in write-once systems.

The objective is designed to introduce an equal and opposite amount of spherical aberration, so that the focused spot is diffraction limited, i.e., an Airy disk with FWHM $\approx 0.6\lambda/\mathrm{NA}$. Reflected light from the recording layer is collected by the objective lens, which recollimates it for transmission back through the quarter-wave plate, where its circular polarization is converted to linear polarization, but with the direction of oscillation of the electric field perpendicular to the figure. (This is called *s* polarization.) The polarizing beam splitter reflects 100% of *s*-polarized light. The detector lens then causes the light to converge onto a photodetector. The basic elements of an *optical head* are all those elements in Fig. 1 that reside below the disk, which is rotating on the spindle that passes through the disk and its hub. In high-performance systems, additional optical elements are contained in the head. For example, the intensity distribution of the laser beam after collimation by the first lens above the laser diode is elliptical in shape, being two to three times broader in the direction perpendicular to the figure than in the direction parallel to it. This is corrected by an *anamorphic prism pair* that circularizes the beam (see below). The detector is composed of several independent elements (*split detector*) whose photocurrents are added and subtracted to give not only a signal proportional to the light collected by the objective lens after reflection at the disk, but also focus-error and tracking-error servo signals (FES and TES) used to drive the electromechanical servo system so as to keep the objective lens in focus and on track during both reading and writing. The voice coils of the focus and tracking actuators are also essential parts of the optical head; they support the objective and move it vertically and horizontally to compensate for the *radial* and *axial runout* of the disk. These vertical and horizontal displacements of the disk, relative to a fixed focal spot, during its rotation are each approximately 0.1 mm in peak-to-peak magnitude. They are caused by the modest warping of a polycarbonate disk—it cannot be made optically flat—and by the eccentricity of the tracks. We will discuss these functions further below.

The basic elements of the optical path in a magneto-optic head are illustrated in Fig. 2. We examine it in read (playback) operation. The partially polarizing beam splitter

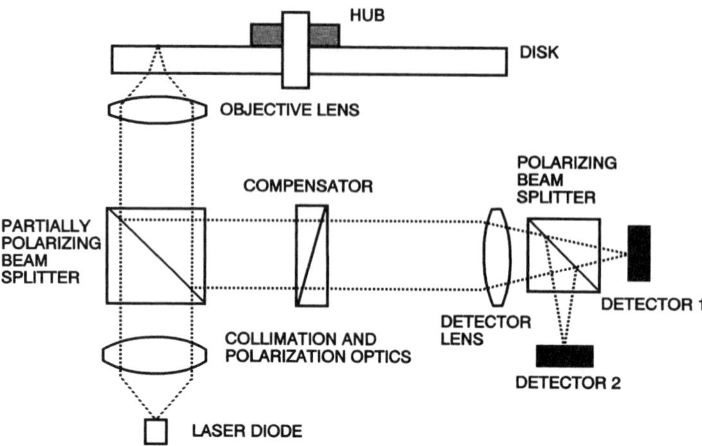

FIG. 2. A schematic diagram of the optical path in a magneto-optic system.

passes about 70% of the *p*-polarized light and reflects all of the *s*-polarized light from the laser [Fig. 3(a)]. This is focused through the substrate to the active MO layer. On reflection the light is elliptically polarized [Fig. 3(b)]. (The elliptical polarization is caused by the interaction of the incident light with the MO medium, which introduces a small amount of *s*-polarized light in the reflected beam, out of phase with the much larger *p*-polarized component.) The major axis of the elliptically polarized beam is rotated about 1 deg above or below the plane of the drawing, depending on whether the local magnetization of the active layer is out of or into the paper. The beam splitter reflects all of the *s*-polarized component of the light returning from the disk and about 30% of the *p*-polarized component. The compensator (often a combination of a quarter-wave plate and a half-wave plate, independently oriented) represents an optical element(s) that changes the elliptical polarization to linear polarization along the direction of the major axis of the elliptically polarized light [Fig. 3(c)] and rotates the *p*-polarized component so that it makes an angle of 45 deg with the plane of the figure [Fig. 3(d)]. In the absence of the *s*-polarized component (no Kerr effect), this would allow half of the light to be transmitted to detector 1 through the polarizing beamsplitter, while the other half would be reflected to detector 2. The difference between the (total) signals received by the two split detectors would be zero. In the presence of the *s*-polarized component, one detector receives more light than the other. The difference between them yields twice the ac signal available from either and its sign indicates whether the magnetization of the active

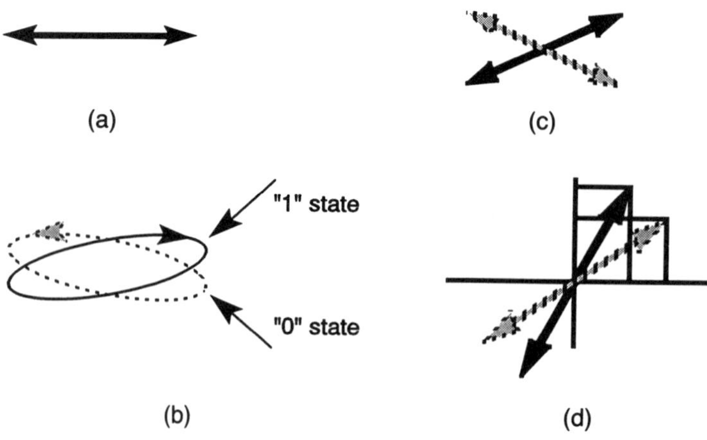

FIG. 3. A schematic representation of the polarization of the beam in the optical path of Fig. 2, as seen looking into the beam. The arrows indicate the path followed, in time, by the tip of the electric field vector at positions: (a) on the way to the disk; (b) after reflection from the disk and recollimation; (c) after bringing the *s* and *p* components into phase; and (d) after rotation to an average angle of 45 deg.

layer is up or down. This is called *differential detection*. An important advantage of this mode of playback is *common-mode rejection*. Laser noise, for example, is identical at the two detectors and is eliminated when their photocurrents are subtracted.

2.1 Laser Diodes for Optical Players and Recorders

The semiconductor laser diode (LD) has made possible not only the optical storage industry (including CD players), but also the optical communication industry. The former consumes orders of magnitude more diodes than the latter. Though some of the earliest CD players used HeNe gas lasers, all of today's systems are fitted with AlGaAs LDs, with operating wavelengths around 830 nm. For WO and erasable drives, 780-nm LDs are most common today. Because the storage industry generally needs low-cost components to reach mass markets, it drove the development of the technologies that enabled mass manufacturing of LDs. Very reliable, long-lived (50 000 h), inexpensive (on the order of 1 dollar per milliwatt of optical power output) devices with excellent beam quality are available from a number of manufacturers—Philips, Sony, Sharp, Toshiba, Hitachi, and Mitsubishi, for example.

A drawing illustrating the most basic structural attributes of a LD is shown in Fig. 4. It is a *double heterostructure* device whose GaAs active layer, where injected electrons and holes combine to produce photons, is bordered by *p*- and *n*-doped layers of GaAlAs that confine the carriers to the active region. The layers are grown epitaxially on an *n*-doped GaAs substrate. A 2–3-μm-wide metal stripe on the top of the chip provides a path for the injected current. The circuit is closed by the metal contact on the bottom of the highly conducting substrate. The dimensions of a typical laser diode chip are length 300 μm, height and width 100 μm. The refractive index of GaAs is higher than GaAlAs, so that the structure also serves as a waveguide confining, in the vertical dimension, the electromagnetic (light) field produced by the coherent stimulated emission of photons. In the lateral direction, the confinement of the field is effected by the presence of the injected electrons, which raises the refractive index. (This is called *gain guiding*.) In more complex and common structures, the lateral confinement of the field is effected by a more complex epitaxial growth and etching process that yields a narrow stripe of GaAs bounded laterally by epitaxially grown material of a lower (effective) refractive index. (This is called *index guiding*.) The diverging wave fronts emanating from the output facet exhibit astigmatism that must be corrected by the optics of the optical head in a storage device. Gain-guided lasers exhibit 10–15 μm of astigmatism; index-guided lasers about 2–3 μm. These numbers are a measure of the axial separation of the two line images (in the geometrical optics limit) that would be formed if the diverging wave front from the laser were turned into a converging one by a perfect lens at unit magnification. Clearly, the index-guided laser is preferable. Nonetheless, the anamorphic prism pair that circularizes the laser beam can be designed to compensate for the rather substantial astigmatism of the gain-guided laser.

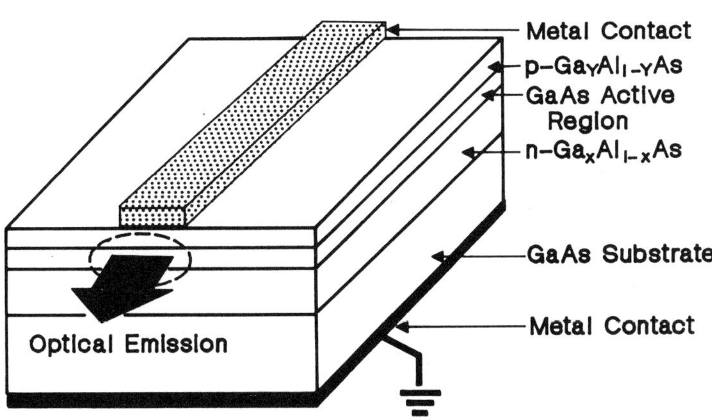

FIG. 4. A schematic diagram of a simple, double-heterostructure laser diode. The stripe contact at the top and the metal contact at the bottom establish the path of injected current. In the active region, holes and electron combine to produce photons, which are coherently emitted from a 0.7 × 2-μm^2 area at both ends of chip, under the metal stripe. The *p*- and *n*-type GaAlAs layers confine the carriers to the active region and also guide the optical field vertically.

Fig. 5 illustrates the asymmetric distribution of the radiance at the laser facet and in the far field (at distances from the facet much greater than d^2/λ, where d is the FWHM of the intensity distribution at the facet). The cross-sectional intensity distributions of typical lasers are modeled in the lower figures, where Gaussian and exponential/Lorentzian models are compared. In read-only optical players, where laser power is not at a premium, this asymmetry can be circumvented by *apodizing* with the collimating lens, which passes only the central portion of the laser beam. In recording systems, however, an anamorphic prism pair must be used to circularize the beam after collimation. This is discussed in the next section.

A typical packaged laser diode is illustrated in Fig. 6. The electric contact to the metal stripe at the top of the diode is indicated, as is a blowup of the diode on its submount. The submount is bonded to a copper heat sink. The package also includes a photodiode situated so as a collect a portion of the light emitted from the bottom (rear) facet of the diode. A glass-windowed cover hermetically seals the package.

The cleaved facets of the laser diode crystal act as mirrors, reflecting about 30% of the light generated in the active stripe back into it. Without this feedback, laser action would not be possible. The transverse dimensions of the active stripe waveguide region are such that the waveguide will support only one spatial mode. That mode gives the intensity distributions illustrated in Fig. 5. The laser cavity, on the other hand, will support several longitudinal modes. These are spaced in frequency by $c/2nL$, where c is the velocity of light in free space, n is the refractive index of GaAs (3.5), and L is the length of the laser chip. Around 780 nm, the corresponding wavelength separation of the longitudinal modes is about 0.3 nm. At a constant temperature, the laser will oscillate in only one longitudinal mode, the one closest to the peak of the gain curve. Well above the threshold current for oscillation, the spectral linewidth of the emission can be as small as 10^{-4} nm. But neither the gain nor the oscillation fre-

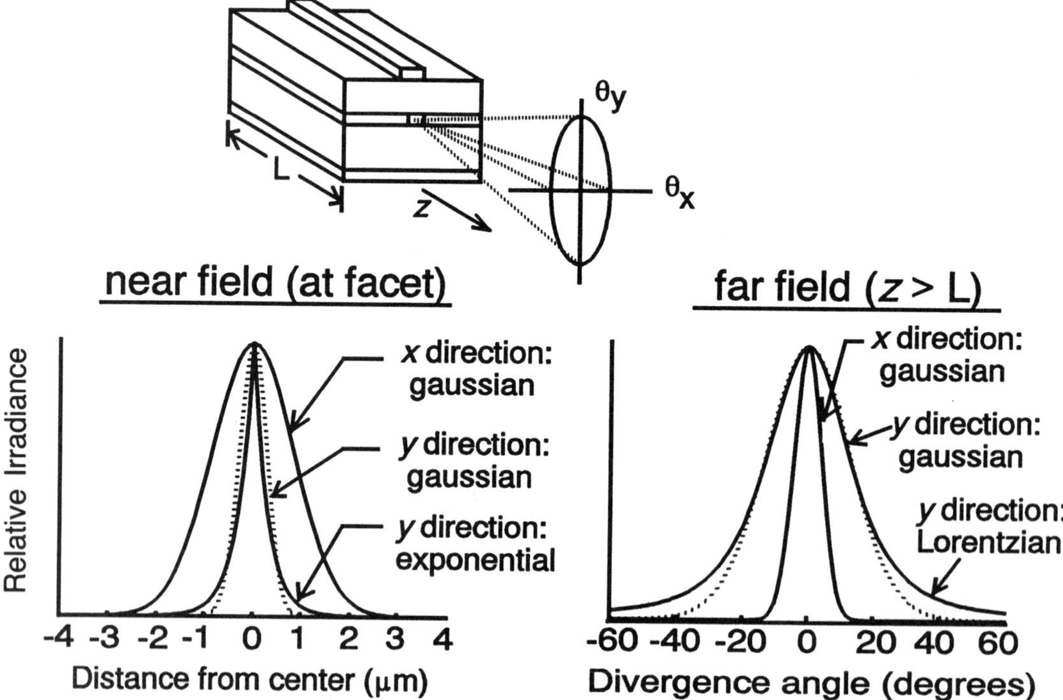

FIG. 5. Upper: Illustrating the elliptical shape of the laser emission in the near field (at the facet) and the far field. Lower: Comparison of two useful analytical descriptions of such intensity distributions.

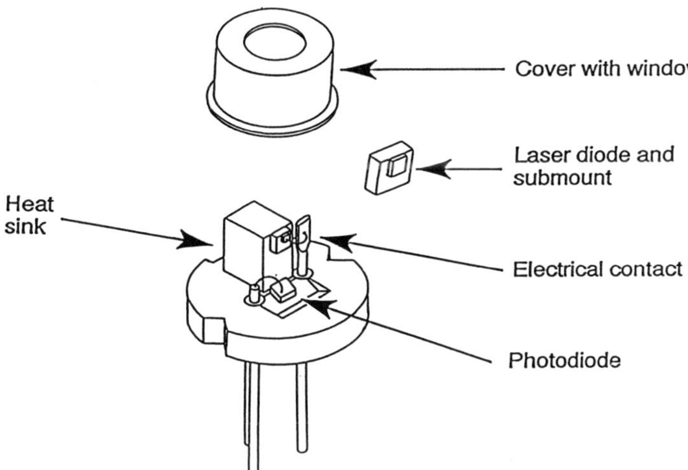

FIG. 6. A typical laser diode package. The tiny laser diode (100 × 100 × 300 μm) is shown on its submount with a wire extending from the metal post to the stripe on the top of the diode.

quency is constant with temperature. The laser can therefore jump from one longitudinal mode to another as the temperature changes, a phenomenon called *mode hopping*. The new mode may produce more or less power than its predecessor, depending on where it lies under the spectral gain curve. This produces a form of laser noise in optical drives. Another form of laser noise is a consequence of feedback of light into the laser cavity from reflections taking place outside the laser diode cavity. In the optical path of an MO drive illustrated in Fig. 2, some of the light reflected from the MO disk is transmitted back through the first beam splitter and into the diode cavity. Thus, the MO film itself forms an external cavity for laser oscillation. The problem is so serious that all MO drives modulate the laser current at a high rf (300–500 MHz) to force the laser into multiple longitudinal-mode oscillation, such that the average output power does not change over time scales characteristic of optical storage (tens of nanoseconds). The rapid modulation of the current never allows the laser to settle down into just one longitudinal mode. The higher the gain in the laser, the more sensitive it is to external feedback.

Much more could be written here about diode lasers (see LASERS, SEMICONDUCTOR). Much more is happening today that will come to the market in the next few years. There have been several demonstrations of linear arrays of diodes on a single chip, reading and writing data on multiple (3–9) tracks at once. Notable recent reports are three from NTT researchers (Murata *et al.*, 1993; Arai *et al.*, 1993; Mizukami *et al.*, 1993). They demonstrated a data transfer rate of 300 Mb/s with an 8-element array. Miniature, diode-pumped, Nd-doped YAG lasers have been frequency-doubled to produce green beams (530 nm) for a video disc recorder (Oka *et al.*, 1989; Oka and Kubota, 1991; Choi *et al.*, 1993; Seo *et al.*, 1993). Nonlinear waveguides that frequency-double the light from laser diodes are being developed for optical storage systems (Risk *et al.*, 1993). These will provide one or more blue (420 nm) light sources, making possible capacities four times as great as that of systems available today. Research into lasers made of II-VI materials has produced demonstrations of green and blue stimulated emission. Serious problems with efficiency and longevity in these new II-VI lasers remain to be solved.

2.2 Optical Head Technology

In describing the optical paths for WO and erasable heads, and in the discussion of the asymmetry of laser diode emission above, we mentioned the need for beam circularizers. Such an element is illustrated in Fig. 7. The laser beam is viewed from a direction exhibiting its narrower divergence—the *x* direction in Fig. 5. The anamorphic prism pair broadens the beam to a diameter equal to that in the perpendicular direction, while maintaining the original direction of the beam. (A single prism is sufficient for circularization, but it would change the direction of the beam.)

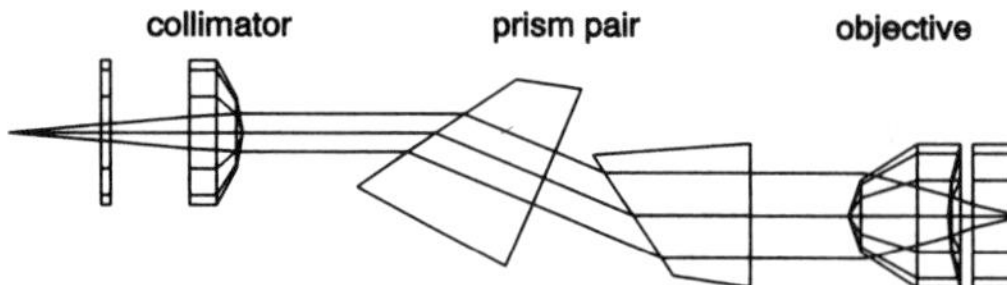

FIG. 7. An anamorphic prism pair used to circularize a beam from a laser diode.

The early CD players employed three-element (*triplet*) lenses, all with spherical surfaces, as objectives. Most of today's systems use molded glass aspheric lenses, as illustrated in Fig. 8. First available only from Eastman Kodak Company, they are now supplied by several vendors. They provide essentially perfect focused spots on the active medium, while presenting a convenient flat annular ring around their outside edge to facilitate mounting.

As noted above, optical storage players and recorders use *split detectors* to generate servo signals for keeping the objective on track and in focus. The depth of focus is about equal to λ/NA^2, about 3.2 μm in systems with 800-nm laser diodes with 0.5 numerical aperture objectives. (This number means that the spot can go inside or outside best focus by about 1.6 μm before serious problems occur.) Systems are typically designed to limit these excursions away from focus to 0.5 μm. Tracking accuracy is more demanding; the objective must stay on track to a tolerance of 0.1 μm. A commonly employed focusing technique, called the *astigmatic technique*, is illustrated in Fig. 9. The detector is made up of four segments, electrically isolated from one another by insulating gaps represented by the solid diagonal lines. Light reflected back to the beam splitter (BS) from the mirror at the right is reflected downward by the beam splitter to an astigmatic lens (often a combination of a conventional, rotationally symmetric lens followed by a weak cylindrical lens, or two cylindrical lenses in tandem). The astigmatic lens has two foci, in that it forms a narrow vertical line in one plane and a narrow horizontal line in another plane. In between these two planes is one wherein the converging light forms a *circle of least confusion*. When the disc is in focus, the detector is illuminated equally in each of its four quadrants. As it moves out of focus, the circular light distribution becomes elliptical, with its major axis either horizontal or vertical, depending on whether the objective is too close to or too far from the disk. The photocurrents in elements I and III are summed, as are those in elements II and IV. The difference between these two sums will be zero if the objective is in focus. When it is nonzero, its sign will tell the servos which way to move the objective to bring it back into focus on the disk.

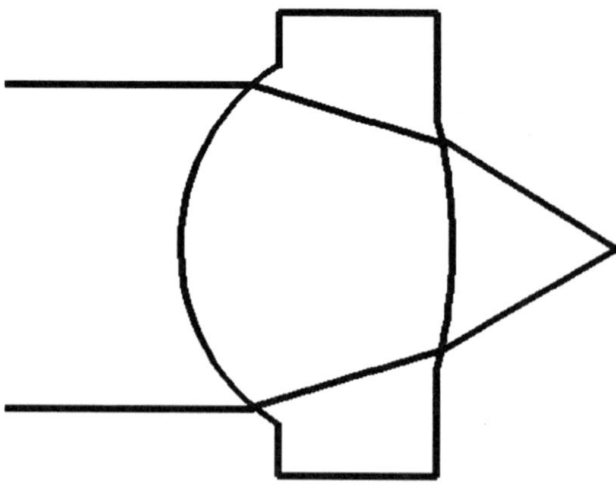

FIG. 8. A molded-glass, aspheric lens commonly used as a single-element objective lens in an optical head.

The arrangement of Fig. 9 can also provide a tracking-error signal. As is apparent, focus control will be effective so long as the axes of the astigmatic lens are oriented properly with respect to the gaps between the detector segments. If we rotate both elements by 22.5 deg, for example, we generate the same focus-error signals. If we orient the detector and astigmatic lens so that the gap separating segments II and III from segments I and IV is parallel to the tracks on the disk, more light will reach the former segments than the latter, or vice versa, as the objective moves off the center of the track. By summing the photocurrents of segments II and III, as well as those of segments I and IV, and then taking their difference, we generate a tracking-error signal for the tracking servo. (This is called *push-pull* tracking. The fact that the light illuminates the detectors in the way described when the objective is moving off track is a consequence of the fact that the light reflected from the disk fills the objective symmetrically only if the focused spot symmetrically illuminates the marks on the disk.)

Several other techniques for generating focus and tracking error signals have been

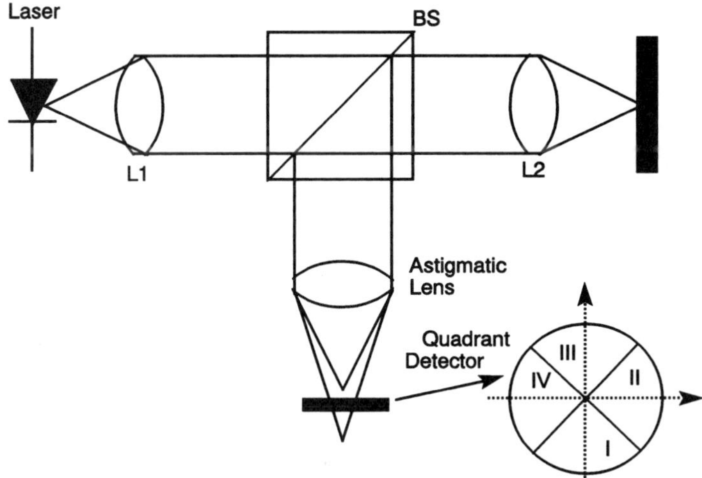

FIG. 9. The astigmatic technique for detecting focus error. When the beam is in focus, all four detectors, each bordered and isolated by solid lines, receive the same amount of light. When in or out of focus, the irradiance on the detector plane is elliptical in shape with the major axis along one or the other of the dotted lines. Photocurrents from detectors I and III are summed, as are those of detectors II and IV. The difference of these sums indicates the amount of defocus, while the sign of the difference tells the drive whether to move the objective closer to or further from the disk.

demonstrated and used in commercial systems. The reader is referred to Chap. 7 of Marchant (1990) and to Chap. 2 of Bouwhuis *et al.* (1985). Chapter 4 of the latter reference also provides a detailed discussion of electromechanical aspects of the focus and tracking servo systems, which we do not discuss in this article. We note only that the bandwidths of servo systems in use today are in the range of 1 to 2 kHz.

The diameters of lenses used in current optical heads are typically in the 3–5-mm range. Thus, the optical head has considerable mass, and substantial force is required to move it rapidly. To improve the access time of optical drives, *split heads* have been introduced. In split heads, only the objective and a 90-deg folding mirror (a mirror making an angle of 45 deg to the optical axis, turning the beam through 90 deg) are moved for tracking and focusing; the remainder of the elements shown in Figs. 1 and 2 are stationary in time. Though more difficult to assemble, splits heads can reduce access time to 40 ms in a 5.25-in. drive. When the entire head must be moved, access time is typically greater than 100 ms.

Efforts to decrease the size, mass, and cost of optical heads are ongoing, particularly in the high-volume markets of CD and CD-ROM players. A particularly noteworthy example of this was recently reported by NEC researches (Nagano *et al.*, 1993). The head is about 25 mm on a side, but only 6.9 mm thick (high), including the distance of the objective lens from the disk. A laser diode and a micromirror are bonded to a silicon chip containing eight independent photodetectors, two of them split detectors, and preamplifiers for the photocurrents generated in the photodetectors. This assembly is hermetically sealed to form an *optical module* 6 mm on a side and 1.64 mm thick. This assembly and the objective lens actuator assembly are mounted and fixed (UV curing resin) to the upper surface of the head, which also contains two folding mirrors with a holographic element between them. All elements are fixed in place after being adjusted to yield appropriate signals at the photodetectors. As is common in most CD heads, only one lens, the objective lens, is used in this design. Light from the laser diode makes three right-angle turns on its way to the objective lens. Light returning from the disc passes through the holographic element, which consists of four regions designed to diffract light to the detectors array so as to yield, in various combination, FES, TES, and readout signals. As noted by the authors, this head is thin enough for use in notebook computers.

A more inventive and ambitious head for CD players was introduced by Suhara, Ura, Nishihara, and Koyama as long ago as 1985. The concept is illustrated schematically in Fig. 10 (Nishihara *et al.*, 1992). It is a guided-wave device grown by techniques and materials well established in the integrated circuit industry. Except for bonding the laser diode to the end of the waveguide on the silicon substrates, the device is fully integrated. The substrate is a silicon chip. An oxidized buffer

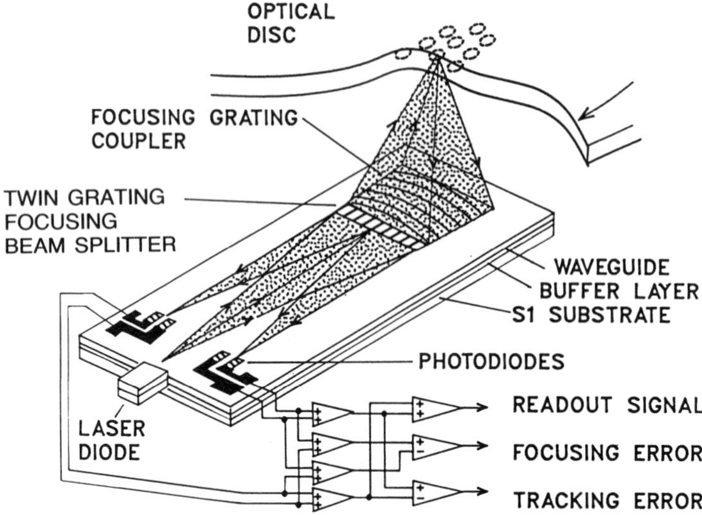

FIG. 10. An integrated-optic head. A laser diode is butt-coupled to a planar waveguide on a Si substrate. Light in the waveguide is diffracted by the focusing-grating coupler so as to form a spot on the disc. Reflected light is coupled back into the waveguide and diffracted by the twin-grating, focusing beam splitter so as to converge on the built-in detectors.

layer of SiO_2 separates the thin-film waveguiding layer from the otherwise highly absorbing silicon. The guiding layer is formed by sputtering Corning 7059 glass, followed by chemical vapor deposition of a thin layer of Si_2N_3 into which two gratings are etched. The *focusing grating coupler* (FGC) has curved grooves whose width and separation increase linearly from beginning to end. Light from the laser diode diverges in the plane of the waveguide in the usual fashion, but is confined in the vertical dimension. Under the FGC, however, it is converted to a converging radiation field, coming to a focus a few millimeters above the device. Light reflected from the disc is coupled back into the waveguide by the FGC. As it passes through the *twin grating focusing beam splitter*, it remains confined in the vertical direction while being diffracted horizontally so as to reach the photodiodes. (Light is also diffracted horizontally by the grating on its way to the disc, an unavoidable loss that is one of the compromises faced by the designer of such devices.) A device like this was built in 1985 and most of its functions were demonstrated. It could not be used in the CD player, however, because its spot size was about three times too large. The FGC was made by an electron beam lithographic system with a 1 mm^2 scanning field. This is too small a field to yield a numerical aperture greater than 0.4, as required in a CD player. There were also problems with the optical power throughput of the device, because of scattering as well as the compromises associated with the twin-grating focusing beam splitter. Work nonetheless continues within several organizations to develop this and other integrated optic heads (Ura *et al.*, 1986). They offer the promise of inexpensive mass manufacturing of devices small enough to make possible multiple-disk, multiple-head storage systems.

3. CODES AND FORMATS

Most recordable disks are rotated at a constant angular velocity (CAV). A few are divided into radial zones, each with a different angular velocity, which is constant within each zone. This is called *zoned* CAV or ZCAV. This is in contrast with the constant linear velocity (CLV) employed by CD systems. The higher performance of recordable systems, in both access time and data rates, requires a substantial sacrifice in capacity; CLV discs have almost twice the capacity of CAV disks.

There are two standardized formats for recordable optical disks, the *continuous/composite servo* (CCS) and the *sampled servo* (SS) format. The disks are divided azimuthally into sectors, each containing typically 1024 user bytes. (512 user byte sectors are an option.) With a 40% overhead for ECC, identification, and synchronization information, this yields about 13 sectors per track on a 3.5-in. disk. In the CCS format, the recording track is defined by a narrow, shallow spiral groove. Recording within each sector is done on the flat region between the turns of the spiral groove. The groove provides, by means of diffraction of the focused spot, continuous tracking-error servo signals, even in areas (sectors) that have not been recorded. The groove is embedded in a thin organic layer on top of the substrate. This layer is prepared by a photopolymerization process (called 2P) where a liquid monomer is injected and flowed between the substrate and a stamper, whose fine-structured upper surface exhibits the inverse of the groove pattern to be embossed in the disk. UV light shining through the transparent substrate polymerizes the thin organic layer, which causes it to adhere to the substrate when it is removed from the stamper. These 2P *substrates* are then sent to sputtering stations, where the active layer(s) and their buffering layers, and any required reflecting layers, are deposited.

At the beginning of each sector, in the region between turns of the spiral groove, is a set of marks that provide synchronization and identification information to the drive. These *preformat marks* are stamped into the thin organic layer during disk manufacture, along with the continuous grooves.

In the SS format, no spiral tracking groove exists. The drive instead periodically measures tracking information on the disk, approximately 1000 times per revolution. This information is manifest by marks offset to either side of the center of the track, preceded by a sequence of centered marks that tell the drive exactly when the offset marks will be encountered, and also synchronize the drive's clock so that all data are recorded in synchronism with the format pattern. The sampled servo marks may be prepared during disk manufacture, as in the CCS format, or they can be recorded in the active layer of each disk. The latter is a relatively expensive practice, requiring recording machinery much more accurate than a commercial drive, and the time to write the marks on each disk. Nonetheless, because it provides flexibility, an organization that makes both disks and drives may choose to use it.

3.1 Error-Correcting Codes

As noted in Sec. 1.2, the CIRC code used in CDs processes contiguous 24-B segments of user data. Each 24-B segment is spread out over 108 33-B EFM frames, and 112 contiguous EFM frames must be read in order to acquire all the data needed to decode the CIRC codeword that contains the 24-B segment of user data. In high-performance systems, this procedure would consume too much time. A Reed-Solomon (RS) code transforms a k-character user word, composed of m bits per character, into an n-character RS code word. The code requires that n must be less than 2^m; k can be any number less than n, with the restriction that $n - k = 2t$, an even number. The $2t$ characters are parity characters, which can be thought of as containing the address and correct value of up to t errors (Marchant, 1990, p. 293). The characters we consider will be 8-b bytes, so $m = 8$ and n must be less than 256. It is natural and efficient to encode by sector, treating each sector independently. Sectors typically contain 512 or 1024 user bytes. Code designers usually choose code words of about 100 bytes that are capable of correcting eight errors. Thus, 16 bytes of each code word are parity bytes. We follow Chap. 11 of Marchant (1990) to describe the encoding of 512-B user-data sectors. First of all, we must note that 18 additional bytes are added to the user data before ECC encoding. Twelve of these are used by the controller to communicate additional information about the sector, such as pointers to directories. The other six bytes represent a *cyclic redundancy check* (CRC) code that provides an additional, quick means for detecting errors. Thus, there are 530 bytes to be ECC encoded for each sector. These bytes are first divided into 106-B words. (Thus, the k of the RS code is 106.) Because the code is designed to recover up to eight errors per word and $n - k = 2t = 16$, the value of n in the RS code is 122. The parity bytes for each code word are calculated one word at a time and stored in a buffer. After the 16 parity bytes are appended to the 106 data bytes of each of the five words, an interleaving process is initiated such that the first bytes of code words one through five are arranged in sequence. These are followed by the second bytes of words one through five, etc. The process continues until the last parity bytes of words one through five have been sequenced. The result is an interleaved stream of 610 bytes that is the output of the ECC encoder. This output serves as input to the channel (modulation) encoder. For raw (before encoding) byte error rates in the range of 10^{-4}–10^{-5}, as are common in data recovered from optical disks, the error rates after decoding the code words read from the disk are substantially smaller than 10^{-12}. Thus, on average, less than one in 10^{12} user bytes returned to the controller by the optical disk drive will be erroneous.

3.2 Modulation Codes

IBM's (2,7) modulation code is commonly used in digital optical and magnetic recording. The entire code is illustrated in the two columns below, where the left-hand column lists the seven unique elemental sequences of user bits that, on concatenation, can represent any arbitrary bit sequence (e.g., the encoded user data output by the ECC encoder) and the right-hand column gives the corresponding elemental bit sequences of the (2,7) modulation code.

10	0100
11	1000
000	000100
010	100100
011	001000
0010	00100100
0011	00001000

The bits that are recorded on or read from the disk are concatinations of the elemental bit sequences on the right in the order that they occurred as elemental sequences on the left.

4. MATERIALS FOR OPTICAL RECORDING

A comprehensive article on optical recording materials, with 264 references to the archival literature, plus abundant references to patent literature, is presented in Johnson *et al.* (1989). For the reader who requires a broad review of this subject, as well as a guide to the important literature, that article is highly recommended. We make no attempt here to be comprehensive. Rather, we highlight a few of the materials and processes that have contributed greatly to today's most successful commercial systems and note some of the near-term and long-term research and development that is expected to lead to the successful systems of the future.

4.1 Substrates

As noted in Sec. 1.1, CD's are usually made by the injection molding of polycarbonate, one side of the mold containing a stamper. The liquid polymer is injected at high pressure into a cold mold; the disc cools inhomogeneously, producing an amount of birefringence (difference in optical path length for normally incident collimated beams polarized in the radial and tangential directions) that varies between the surfaces and the midplane of the disc. CD specifications require that the birefringence be less than 50 nm, with surface roughness less than 15 nm. PMMA (polymethyl methacrylate) has also been used in CDs. The sequence of processes followed in the fabrication of CDs is illustrated in Fig. 11. The first is *mastering*, which is the process whereby the 1's and 0's of the modulation-encoded data are initially written to a disc. This is illustrated at the left of the figure. A glass disc coated with photoresist is mounted on an air-bearing spindle. A modulated Ar laser beam is focused on the resist layer. The radial position of the beam and focusing objective is controlled by a laser interferometer, which ensures that the pitch of the spiral track is maintained at 1.6 μm. The modulation-encoded data drive an acoustic or electro-optic modulator that turns the Ar laser beam on and off at every transition for 1 to 0 or 0 to 1. When writing is complete, the photoresist layer is developed to exhibit the written pits. A layer of silver is then deposited on the resist to complete the *master disc*. A nickel negative, called the *father*, is then electroplated on the master. The father disc is then used to form one or more *mother* discs, each of which is used to form a number of *sons*. These sons are the stampers used in the injection-molding process that produces the CDs. Because they deteriorate in the process of molding CD discs, they must be periodically replaced in the stamper. Large runs require several mother discs and many sons.

The 2P (photopolymerization) process used to form grooves and preformat marks on plastic and glass substates was described in Sec. 3. The active layers of typical WO and erasable disks were discussed in the corresponding subdivisions of Sec. 1. In our discussion of optical tape systems there, we noted that both ICI and Dow Chemical were media suppliers for the CREO optical tape recorder.

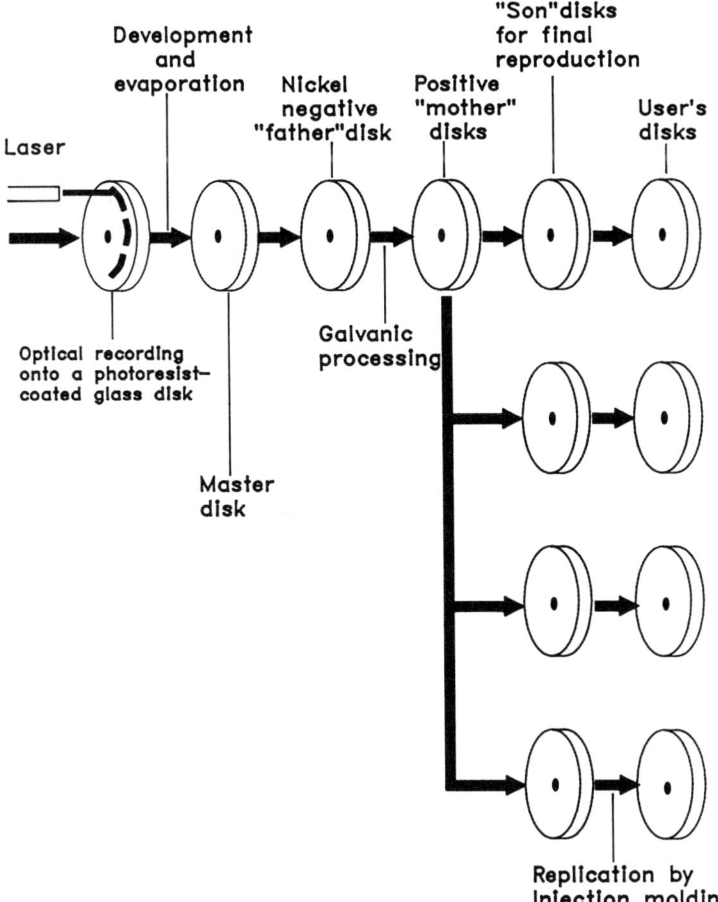

FIG. 11. Schematic of the processes involved in manufacturing read-only discs. At the left, a modulated laser writes the pit pattern into photoresist, which is coated with a silver layer after resist development. A nickel negative *father* disc is grown on the master disc by electroplating. One or more positive *mother* discs are made from the father by the same process. Each mother then produces several *sons*, which are used as stampers in the injection-molding of user discs, which are finally coated with an Al film and a protective layer.

Dow has been working with Southwall Technologies of Palo Alto, California, to develop a very robust tape. The current status of this effort is described by Larsen (1993). The media structure consists of three inorganic layers sequentially sputter-deposited in a web-coating vacuum system onto a PET (polyethylene terephthalate—Mylar) substrate that has been precoated with an organic *subbing* layer on top and an organic backcoat on the bottom of the PET. The subbing layer provides a smooth, hardened surface for the inorganic layers and the backcoat enhances tape handling and wear properties. The active layer is based on Dow's *optical metal* layer (Strandjord *et al.*, 1990), an oxygen-doped Sn/Bi/Cu alloy. After this is sputtered on the subbing layer, it is covered by an *activation* layer of ZrN_x that provides protection and enhances hole formation during laser writing. Finally, an abrasion-resistant coating of $SiCH_x$ serves as the primary protection against scratching and tape wear. In tests at CREO, a raw bit error rate of 5×10^{-5} and an ECC-corrected error rate of 10^{-22} were measured initially on tapes without the backcoat layer under the PET substrate. After 5000 searches at 5 m/s, the ECC-corrected error rate had increased to 10^{-12}. On samples with the backcoat under the PET, it took 50 000 search cycles for the corrected error rate to reach 10^{-12}. Tests are in progress aimed at demonstrating a potential for 100-yr archival life.

4.2 Candidates for New Active Media

Periodic multilayers of Pt/Co and Pd/Co, wherein a few atomic layers of each element are alternately deposited by MBE (molecular beam epitaxy), sputtering, or evaporation to form each period, have been studied extensively as potential candidates for MO media

(Greidanus *et al.*, 1989; Takahashi *et al.*, 1992). They are more responsive to blue light than the RE/TM media now in use in the infrared. They also are very resistant to corrosion, unlike their RE/TM counterparts. Because they are relatively expensive to prepare, however, they have not been commercialized.

A variety of other material candidates for MO media continue to be studied because of their potentially large Kerr effect. MnBi alloys were first studied as long ago as the late 1960s. When vacuum deposited, these films exhibit polycrystalline morphology that introduces excessive *media noise*. Because they also exhibit several degrees of Kerr rotation, they remain interesting. Similarly, Pt-Mn-Sb alloys offer the promise of high Kerr rotation. A cooperative research effort involving several U.S. companies and universities, partially supported by ARPA, is addressing these materials as potential candidates for blue-sensitive media. As noted in the discussion of erasable phase-change systems, however, the accent in Japan seems to be on phase-change materials in the near term.

In the longer term, the prospects of current research are difficult to assess. Holographic storage has been studied since the invention of the laser in the 1960s. Early efforts indicated that the available materials were not sufficiently stable for storage applications. New materials and system concepts have since emerged that may one day be commercialized. Four different experimental holographic memory systems were reported at the ISOM/ODS 1993 conference in Maui. In a large-scale memory experiment, blocks (pages) of 1000 × 1000 bits (binary images) were stored in an iron-doped lithium niobate crystal 1 cm thick (Burr *et al.*, 1993). (The crystal is *photorefractive*, that is, its refractive index changes on exposure to light, particularly shorter wavelength visible light.) These holograms (pages) are stored at an array of 16 × 16 locations on the surface of the crystal. Each location contains a superposition of 4000 pages. Each hologram is written with a unique reference beam, identified by its angle with the normal to the crystal. An acousto-optic beam deflector and a specially designed and fabricated mirror array permit readout of a page in approximately 100 μsec with a measured signal-to-noise ratio of 817. The calculated probability of error was 10^{-28}. No description of the photodetector array used for readout was reported.

In another experiment by the same group, an iron-doped lithium niobate crystal in the shape of a circular disk of 5 mm thickness and 5 cm diameter was used to multiplex 5200 1-Mb binary images in each of about 20 locations around the disk (Li *et al.*, 1993). By spinning the disk and translating the head radially, any location can be quickly accessed. The authors note that their crystal is a dynamic memory device in that its contents must be periodically read out, amplified and re-recorded. A density of 20 b/μm^2 was demonstrated. Theory suggests that 200 b/μm^2 of surface area should be possible in this 5-mm-thick disk. (The volume density is thus much less than that of conventional recorded disks, which have about 1 b/μm^2 in a 100-nm-thick film.) The authors have also demonstrated that photopolymers can support the same density; but they have not yet succeeded in making a photopolymeric disk thicker than 300 μm.

Photorefractive $BaTiO_3$ was used to store 256 filters of 256 × 256 pixels for a holographic image correlation experiment in a system under study in France (Alves *et al.*, 1993). With a reading power of 10 mW, an unknown image can be correlated with all of the filters in 500 μs. This system has also been used in optical interconnect experiments, where holography seems to have a promising future.

In another experimental system (Marchand *et al.*, 1993) that requires no FES or TES, data blocks are stored on the disk as one-dimensional Fourier-transform CGHs (computer generated holograms). Each one of them is calculated to reconstruct one column of a two-dimensional output image. During the sequential recording process, all the CGHs for a given two-dimensional image are laid out side by side and radially shifted from one another until they span the entire radius of the active layer of the disk. A commercial 5.25-in. WO disk was used in the experiments, which involved images of size 16 × 16. These were read out at a rotation rate of 30 rpm. The motionless head is the most interesting aspect of this experiment; it is made possible by the shift-invariant properties of the Fourier transform.

Cerium-doped strontium barium niobate in fiber form has been proposed as a viable candidate for a page-oriented holographic memory (Bashaw *et al.*, 1993; Sugiyama *et al.*, 1991). Though rumors abound, the current status of these systems has not been published. They are said to be very promising.

Interest in the application of spectral hole-burning techniques to optical storage persists around the world in spite of their cryogenic requirements. These systems employ materials that exhibit inhomogeneously broadened absorption spectra associated with dye molecules embedded in amorphous materials. The absorption spectrum of any individual dye molecule depends on the site where it resides in the host material. Its local environment determines where the peak of its absorption spectrum will occur; its temperature sets the width of its absorption band. At a temperature of 50 mK, it is easy to write and read a very well resolved 16-b pattern of 1's and 0's in a 30-GHz band (Ao *et al.*, 1991). At the temperature of 4 K preferred by most researchers, it is possible to burn (1–10 000) spectral holes at a single focused spot. This enormous increase in storage density drives the continuing research into PHB (photochemical hole burning and/or persistent spectral hole burning). The interested reader will find five additional papers on this subject in Ao *et al.* (1991).

Photon-echo techniques also provide a means for storing multiple bits at a single location in a homogeneously broadened system. When illuminated by a sequence of pulses, representing 1's, the atomic or molecular constituents of these materials are set into oscillation over a relatively broad spectrum, each constituent at its own resonant frequency. Because of their frequency differences, they quickly get out of phase, though they remain excited. Applying a read pulse of appropriate width and power will reverse the phases of all the constituents. The medium will then re-emit, at reduced amplitude, the original sequence of pulses at times after the read pulse equal to the difference between the time they were written and the time they were queried by the read pulse. The time between writing and reading is quite limited. It must be less than T_2, the coherence time of the oscillations of the individual constituents. The first to demonstrate this phenomenon were Abella *et al.* (1966), using only one write pulse and one read pulse. They called it *photon echo* in analogy with spin echo in nuclear magnetic resonance discovered by Hahn (1950). Research continues today in several locations. Kachru *et al.* (1992) describe some modern experiments in Pr^{3+}:YAG crystal. These were stimulated by earlier experiments in a Eu^{3+}:$YAlO_2$ crystal that, at cryogenic temperatures, could store data for several hours (Bai *et al.*, 1986). Because of the time limit between writing and reading, these systems are not candidates for optical storage; they may someday find their place as dynamic memory in the computers of the future.

GLOSSARY

Access Time: The time required to move an optical (or magnetic) head to an arbitrary data track on a disk.

Anamorphic Prism Pair: A pair of prisms that change the elliptical cross section of a laser diode beam to a circular cross section, while maintaining the beam direction.

ATRAC: Adaptive TRanform Acoustic Coding; used in MD for data compression.

Block: 98 EFM frames (2352 bytes) of CD audio; blocks are numbered 0 to 74 and hold 1/75 s of an audio program. (A short form of *control and display block.*)

Burst Errors: Errors in all of a group of sequential bytes on a track.

CAV: Constant angular velocity (used in most WO and MO drives).

CCS: Continuous composite servo, a format for recordable disks that has continuous grooves that define its tracks, with short non-grooved sections containing pits to set the tracking servos.

CD: Compact disc for audio (music) publishing.

CD-I: Compact disc–interactive; combines still and moving pictures, graphics, audio, and text in CD format.

CD-R: Recordable CD, only available in WO form (not erasable) at this time.

CD-ROM: Compact disc–read-only memory; computer data in CD format.

CD-ROM/XA: Addendum to *Yellow Book* CD-ROM standard covering mixed audio and data discs.

CIRC: Cross-interleaved Read-Solomon code, the ECC used in CDs.

CLV: Constant linear velocity (along any track of a CD disc).

Differential Detection: A technique employed in MO drives wherein two detectors are used to enhance the detected MO signal.

Direct Overwrite: To write new data over old without an erasure cycle.

DVI: Digital video interactive; a data-compression scheme that produces one hour of digital video on a CD disc.

ECC: Error correcting code.

EDAC: Error detection and correction.

EFM: Eight-fourteen modulation code (used in CDs).

Folding Mirror: A mirror that bends a light beam through 90 deg.

Form Factor: Standardized height and width specifications of computer peripherals.

Frame: An EFM frame that contains 24 bytes of user data plus 8 bytes of ECC plus 1 byte of control information, or equivalently the 588 channel bits into which these are encoded.

Full-Height: Standardized height specification of peripheral: 83 mm.

Green Book: Book containing Philips and Sony specifications for CD-I discs.

Half-Height: Standardized height specification of pherpheral: 41.5 mm.

High Sierra Group: Consortium that formulated the adopted logical file standards for CD-ROM.

ISOM: International Symposium on Optical Memory held biannually in Japan.

Jukebox: A cabinet with a robotic mechanism that can load/unload any one of many optical disks into/out of one or more of its drives.

LD: Laser diode, a semiconductor device that converts injected electrons into coherent light.

Magneto-Optic Media: Media containing one or more active layers with magnetization perpendicular to their surfaces. These change the polarization of the optical field on reflection or transmission.

MD: Minidisc (Sony)—a direct-overwrite MO system for recording and playing music on a 2.5-in. disc.

MD-DATA: A data, or text, standard for the Sony's Minidisc system.

MO: Magneto-optic (media or system).

NA: Numerical aperture, equal to $\sin\theta_{max}$, where θ_{max} is the apex angle of the cone of light converging, in air, to the focal spot.

ODS: Optical Data Storage conference held annually in North America.

Optical Head: The optical, mechanical, and electronic assembly, including diode laser, detectors, optical elements, and mechanical actuators, that reads and writes an optical disk.

Orange Book: Book containing Philips and Sony specifications for recordable CD discs.

Phase-Change Media: Storage media whose microstructure can be changed from crystalline to amorphous, or vice versa, by laser heating.

Photo-CD: Kodak's recordable CD system for storing digitized 35-mm film images.

Polycarbonate: Most commonly used plastic for optical disk substrates, and for eyeglasses.

RAM: Random access memory.

Red Book: Book containing Philips and Sony specifications for CD-Audio discs. (Forms basic specifications for all other CD systems.)

RLL: Run-length limited (a type of modulation code).

Split Detector: A silicon chip containing several independent (electronically isolated) detector elements.

Split Head: An optical head where most of the parts are fixed, while only the objective lens and a folding mirror are moved by the focus and tracking actuators.

SS: Sampled servo, a grooveless format with periodic reference pits around the track to set the tracking servo and the data clocking rate.

Track: An outward spiraling sequence of small marks, representing stored digital data, on the surface of a disc. The pitch of this spiral ranges from 400 to 1 000 turns/mm. Mark lengths range from 0.5 to 5 μm in the direction tangent to the spiral and mark widths range from 0.2 to 1.0 μm in the direction perpendicular to the spiral.

WO: Write-once media or system; data cannot be erased or overwritten.

WORM: Acronym sometime used for *write-once* systems; *write once, read many*.

Yellow Book: Book containing Philips and Sony specifications for CD-ROM discs.

Works Cited

Abella, I. D., Kurnit, N. A., Hartmann, S. R. (1966), *Phys. Rev.* **141**, 391–406.

Alves, C., Aing, P., Pauliat, G., Roosen, G. (1993),

in: IEEE (Ed.), *Conference Digest—Joint International Symposium on Optical Memory and Optical Data Storage*, New York: IEEE, pp. 73–74.

Ao, R., John, S., Kuemmerl, L., Weiner, R., Haarer, D. (1991), in: *JJAP—Series 6, Optical Memory*, Tokyo: Japan Society of Applied Physics.

Arai, R., Mizukami, M., Tanabe, T., Katoh, K., Yoshizawa, T., Yamazaki, H., Murata, S., Tanaka, Y., Sato, I. (1993), *Jpn. J. Appl. Phys.* **32**, Pt. 1, 5411–5416.

Bai, Y. S., Babbit, W. R., Mossberg, T. W. (1986), *Opt. Lett.* **11**, 724–727.

Bashaw, M. C., Aharoni, A., Hesselink, L. (1993), *Opt. Lett.* **18**, 2059–2061.

Bouwhuis, G., Braat, J., Huijser, A., Pasman, J., van Rosmalen, G., Immink, K. S. (1985), *Principles of Optical Disc Systems*, Bristol, UK: Adam Hilger Ltd.

Broer, D., Vriens, L. (1983), *Appl. Phys. A* **32**, 107–123.

Burr, G. W., Mok, F. H., Psaltis, D. (1993), in: IEEE (Ed.), *Conference Digest—Joint International Symposium on Optical Memory and Optical Data Storage*, New York: IEEE, pp. 75–76.

Choi, B. S., Chee, J. K., Hwang, Y. M., Jang, H. Y., Kwon, D., Lee, S. H. (1993), *Jpn. J. Appl. Phys.* **32**, Pt. 1, 5453–5454.

Einberger, J., Zoellick, B. (1987a), in: Suzanne Ropiequet (Ed.), *Optical Publishing—CD ROM 2*, Redmond, CA: Microsoft Press, pp. 195–216.

Einberger, J., Zoellick, B. (1987b), in: Suzanne Ropiequet (Ed.), *Optical Publishing—CD ROM 2*, Redmond, CA: Microsoft Press, p. 41.

Gelbart, D. (1990), in: M. de Hann, Y. Tsunoda (Eds.), *Optical Data Storage*, SPIE Proceedings No. 1316, Bellingham, WA: SPIE.

Greidanus, F. J. A. M., Zeper, W. B., den Broeder, F. F. A., Godlieb, W. F. (1989), *Appl. Phys. Lett.* **54**, 2481–2483.

Hahn, E. L. (1950), *Phys. Rev.* **81**, 580–594.

Hittinger, W. C., Sonnenfeldt, R. W. (1978), Special Issue on Videodisc Technology, *RCA Rev.* **39** (1).

Holtslag, A. H. M., McCord, E. F., Werumeus Buning, G. H. (1991), in: *JJAP—Series 6, Optical Memory*, Tokyo: Japan Society of Applied Physics, pp. 99–108.

Howe, Dennis (1993a), *Opt. Photon. News* **4**(2), 58–59.

Howe, Dennis (1993b), *Opt. Photon. News* **4**(3), 58–59.

Inoue, A., Kablau, J. G. K., Heemskerk, J. P. J., Ogawa, H., Yamauchi, H. (1993), in: IEEE (Ed.), *Conference Digest—Joint International Symposium on Optical Memory and Optical Data Storage*, New York: IEEE, pp. 120–121.

Isailovic, Jordon (1985), *Videodisk and Optical Memory Systems*, Englewood Cliffs, NJ: Prentice-Hall.

Ishibashi, H., Moriya, M., Ohta, T. (1991), in: *JJAP—Series 6, Optical Memory*, Tokyo: Japan Society of Applied Physics, pp. 89–90.

Ito, R., Tsunoda, Y., Ukita, H., Ohta, K., Kubota, S., Ogawa, K., Okuda, M., Irie, M., Mitsumori, S., Nishihara, H. (Eds.) (1993), Special issue on Optical Memory and Optical Data Storage, *Jpn. J. Appl. Phys.* **32** (11B), Pt. 1.

Johnson, G. H., Thomas, G. E., Kloosterboer, H. J. G., Legierse, P. E. J., Gravesteijn, D. J. (1989), in: W. Gerhartz, Y. S. Yamamoto (Eds.), *Ullmann's Encyclopedia of Industrial Chemistry*, 5th ed., Vol. A14, New York: VCH Publishers, pp. 197–239.

Kachru, R., Bai, Y. S., Shen, X. A., Huestis, D. L. (1992), in: A. Jamberdino, W. Niblack (Eds.), *Image Storage and Retrieval Systems*, SPIE Proceedings No. 1662, Bellingham, WA: SPIE, pp. 205–210.

Kivits, P., de Bont, R., Jacobs, B., Zahm, P. (1982), *Thin Solid Films* **87**, 215–231.

Larsen, T. L., Woodard, F. E., Pace, S. J. (1993), *Jpn J. Appl. Phys.* **32**, Pt. 1, 5461–5462.

Leenhardt, C. (1987), *Mem. Opt.* **39**, 20.

Levine, M. L. (1990), in: M. de Haan, Y. Tsunoda (Eds.), *Optical Data Storage*, SPIE Proceedings No. 1316, Bellingham, WA: SPIE, pp. 48–57.

Li, H. Y. S., Curtis, K., Qiao, Y., Psaltis, D. (1993), in: IEEE (Ed.), *Conference Digest—Joint International Conference on Optical Memory and Optical Data Storage*, New York: IEEE, pp. 79–80.

Marchand, P. J., Krishnamoorthy, A. V., Harvey, P. C., Esener, S. C. (1993), in: IEEE (Ed.), *Conference Digest—Joint International Conference on Optical Memory and Optical Data Storage*, New York: IEEE, pp. 77–78.

Marchant, Alan B. (1990), *Optical Recording, A Technical Overview*, Reading, MA: Addison-Wesley.

Maydan, D. (1971), *Bell Syst. Tech. J.* **50**, 1761–1789.

Miltenburg, J. W. (Ed.) (1982), Special Issue on Optical Videodisc Technology, *Philips Tech. Rev.* **40**, 287–309.

Miyaoka, S. (1993), in: IEEE (Ed.), *Conference Digest—International Conference on Optical Memory and Optical Data Storage*, New York: IEEE, pp. 3–4.

Mizukami, M., Yoshizawa, T., Sato, I. (1993), *Jpn. J. Appl. Phys.* **32**, Pt. 1, 5417–5420.

Murata, S., Nakada, H., Abe, T., Tanaka, H., Watabe, A. (1993), *Jpn. J. Appl. Phys.* **32**, Pt. 1, 5284–5291.

Nagano, T., Ueda, E., Onayama, S., Katayama, R., Hamada, H., Ono, Y. (1993), *Jpn. J. Appl. Phys.* **32**, Pt. 1, 5263–5268.

Nishihara, N., Suhara, T., Ura, S. (1992), in: D. Carlin, D. Kay (Eds.), *Optical Data Storage*, SPIE Proceedings No. 1663, Bellingham, WA: SPIE, pp. 26–36.

Oka, M., Kashiwagi, T., Kubota, S. (1989), in: T. Wilson (Ed.), *Optical Storage and Scanning Technology*, SPIE Proceedings No. 1139, Bellingham, WA: SPIE, pp. 149–154.

Oka, M., Kubota, S. (1991), in: *JJAP—Series 6,*

Optical Memory, Tokyo: Japan Society of Applied Physics, pp. 135–140.

Pierce, G. (1988), in: D. B. Carlin, Y. Tsunoda, A. A. Jamberdino (Eds.), *Optical Storage Technology and Applications*, SPIE Proceedings No. 899, Bellingham, WA: SPIE, pp. 31–33.

Risk, W. P., Lau, S. D., Kurdi, B. (1993), *Opt. Lett.* **18**, 1804–1806.

Seo, J. E., Park, I. S., Oh, Y. N., Lee, S. H., Seong, P. Y., Jang, Y. K., Shin, K. H. (1993), *Jpn. J. Appl. Phys.* **32**, Pt. 1, 5455–5456.

Storey, P. A., Longman, R. J., Davies, N. A. (1988), in: D. B. Carlin, Y. Tsunoda, A. A. Jamberdino (Eds.), *Optical Storage Technology and Applications*, SPIE Proceedings No. 899, Bellingham, WA: SPIE, pp. 226–232.

Strandjord, A. J. G., Webb, S. P., Beaman, D. R., Carroll, S. L. B. (1990), in: R. D. Seddon (Ed.), *Optical Thin Films III: New Developments*, SPIE Proceedings No. 1323, Bellingham, WA: SPIE, pp. 127–131.

Sugiyama, Y., Yagi, S., Yokohama, I., Hatakeyama, I. (1991), in: *JJAP—Series 6, Optical Memory*, Tokyo: Japan Society of Applied Physics, pp. 354–358.

Suhara, T., Ura, S., Nishihara, H., Koyama, J. (1985), in: IEEE (Ed.), *Conference Digest—International Conference on Integrated Optics and Optical Fiber Communication, Venice*, New York: IEEE, pp. 117–120.

Takahashi, M., Nakamura, J., Ojima, M., Tatsuno, K. (1992), in: D. Carlin, D. Kay (Eds.), *Optical Data Storage*, SPIE Proceedings No. 1663, Bellingham, WA: SPIE, pp. 250–256.

Time-Life Books (Ed.) (1987), *Memory and Storage*, Alexandria, VA: Time-Life, Inc., pp. 105–106.

Tsutsui, K., Murao, H., Yoda, S., Tsuboi, K., Otsuki, S., Yamashita, T. (1990), in: M. de Haan, Y. Tsunoda (Eds.), *Optical Data Storage*, SPIE Proceedings No. 1316, Bellingham, WA: SPIE, pp. 341–344.

Ura, S., Suhara, T., Nishihara, N., Koyama, J. (1986), *IEEE J. Lightwave Tech.* **LT-4**, 913–918.

Vogelgesang, P., Hartmann, J. (1988), in: D. B. Carlin, Y. Tsunoda, A. A. Jamberdino (Eds.), *Optical Data Storage Technology and Applications*, SPIE Proceedings No. 899, Bellingham, WA: SPIE, pp. 172–177.

Watanabe, K., Oyama, T., Aoki, Y., Sato, N., Wiyaoka, S. (1983), in: D. Chen (Ed.), *Optical Data Storage*, SPIE Proceedings No. 382, Bellingham, WA: SPIE, pp. 191–195.

Yoshida, Y. (1993), in: IEEE (Ed.), *Conference Digest—Joint International Symposium on Optical Memory and Optical Storage*, New York: IEEE, pp. 35–36.

Yoshida, T., Akahira, N., Ohara, S., Nishiuchi, K., Ishida, T. (1991), in: *JJAP—Series 6, Optical Memory*, Tokyo: Japan Society of Applied Physics, pp. 83–88.

Further Reading

Bouwhuis, G., Braat, J., Huijser, A., Pasman, J., van Rosmalen, G., Immink, K. S. (1985), *Principles of Optical Disc Systems*, Bristol, UK: Adam Hilger Ltd.

Mansuripur, M. (1994), *Physical Principles of Magneto-Optical Recording*, Cambridge, UK: Cambridge University Press.

Marchant, Alan B. (1990), *Optical Recording, A Technical Overview*, Reading, MA: Addison-Wesley.

McDaniel, T., Victora, R. (Eds.) (1994), *Magneto-Optical Data Recording—Materials, Subsystems, Techniques*, Park Ridge, NJ: Noyes Publications.

Time-Life Books (1987), *Memory and Storage*, Alexandria, VA: Time-Life, Inc.

OPTICAL SYSTEMS

See OPTICAL COMPONENTS AND SYSTEMS

OPTICAL TELESCOPES

See ASTRONOMICAL TELESCOPES

OPTICAL TOMOGRAPHY

GABOR T. HERMAN AND CONWAY YEE, *University of Pennsylvania, Philadelphia, Pennsylvania, U.S.A.*

INTRODUCTION

The general problem of "tomography" is the following. There is a three-dimensional structure whose internal composition is unknown to us. We subject this structure to some kind of radiation, either by transmitting radiation through the structure or by introducing an emitter of radiation into the structure. We measure the radiation transmitted through or emitted from the structure at a number of points. We desire to reconstruct from these measurements the distribution of the physical parameter(s) inside the structure that have an effect on the measurements. This general area of applied mathematics (or mathematical physics) is commonly identified by the name "inverse problems"; see, e.g., Herman *et al.* (1987). Under this general definition, the title "Optical Tomography" is ambiguous, since it can mean either of the following very different topics.

1. Using optical (rather than digital) devices to achieve reconstruction from tomographic data collected using any of a number of means (x rays, γ rays, sound, etc.). This is in fact not the topic of our article; the interested reader may look, for example, at Edholm (1977) and at Gmitro *et al.* (1990).
2. Sending light into an object, measuring the light that comes out, and then attempting to recover information regarding the internal properties of the object based on these measurements by any digital or analog method that is available. This is the topic of our concern and from now on we use the term "optical tomography" only in this sense. This is an extremely active field of research: most of the February 1993 issue of *Applied Optics: Optical Technology* is devoted to it and articles appear in other relevant journals all the time (e.g., Yamada *et al.*, 1993; Fishkin and Gratton, 1993).

Optical tomography is an imaging technique that uses light in the near-infrared or the infrared range. While the human body is nearly opaque to visible light, it is not too absorbent to near-infrared and infrared; light in this regime that is sent into the body has a reasonable chance to escape. This escaping light contains information about the optical properties of tissue.

There are two essentially different modes of data collection using light. The first takes advantage of the fact that the first photons that reach a detector have undergone the least amount of scattering. Thus, they must have traveled in a relatively straight line between the source and the detector. By using only these photons, it is possible to apply any of the many algorithms that have been previously developed for computerized tomography from x rays and γ rays to reconstruct an image. In this article, we refer to imaging using such a mode of data collection as FAT (short for first-arrival tomography). The disadvantage of this mode of data collection is that the statistics are inherently quite poor.

3-527-28134-7/95/$5.00 + .50

For any given light pulse, only the initial few photons reaching the detector may be used. In fact, for most areas of the body, only a few photons will be able to traverse the body without being scattered significantly—too few to obtain medically useful information. For this reason, the potential medical uses of FAT are restricted to the more translucent parts of the body, such as the female breast.

In the second mode of data collection, we try to make use of all the light photons that are detected over a relatively extended period of time. We refer to imaging using such a mode of data collection as PMI (short for photon-migration imaging). Since this allows us to make use of multiply scattered photons, its potential advantage comes not only from the improved statistics, but also from the ability to place the detector on the same side of the body as the source, making it possible to image organs that cannot be imaged by FAT. For this reason, PMI is the major concern of this article.

1. THE PHYSICS OF OPTICAL TOMOGRAPHY

There are a number of different ways of data collection for the purpose of PMI. One difference is at the detector, which either simply counts up the total number of photons it detects during the data collection time or may have the ability to resolve the data collection time into a number of subintervals and report on the number of photons detected during each one of these. For the latter case, we have also the option to make a choice at the input: we may choose the source to produce a pulse or a continuous modulated light wave. While most of what we say applies to optical tomography in general, much of our discussion is directed toward the special case of PMI, in which the experimental data consist primarily of measurements, at the surface, of the reflected intensity, taken as a function of time and distance from the point of injection (see, e.g., Weiss *et al.*, 1989). Some photons are absorbed within the tissue while others, after being repeatedly scattered, get back to the surface, where they are re-emitted. Typically, laser light is used for the simple reason that it is possible to gate a laser pulse more quickly than any other source of light. Because of scattering, coherence is soon lost—the coherence of laser light is not a significant factor in PMI. (Examples of recent articles relevant to the instrumentation of data collection for optical tomography are Hebden, 1993, and Lilge *et al.*, 1993.)

In x-ray tomography, the relation of the collected data to the object that is imaged is relatively simple since photons travel in straight lines from the source to the detector. The number of photons detected is related to the absorption by the tissues between the source and the detector; the logarithm of the intensity of x rays seen at the detector is related to the line integral of the absorption coefficients (see, e.g., Herman, 1980). A similar situation arises in FAT, but in PMI the situation is much more complicated since photons that are detected are not likely to have traveled on the direct path between the source and the detector; the vast majority of detected photons have taken a much more circuitous route. Thus, a PMI measurement is not describable using a simple line integral.

In order to recover three-dimensional information from the PMI mode of data collection, it appears essential to find a physical model suitable for predicting the distribution and the timing of the photons coming out of the object being imaged. This is usually referred to as the "forward problem." To understand the physics behind photon scattering, multiple-scattering theory must be used. The approaches currently applied to the three-dimensional multiple-scattering problem can be classified as being one of the following three types: analytical theory, Monte Carlo, and transport-theory simulation.

Analytical theory is based on the Maxwell equations, the fundamental equations governing field quantities. As mentioned in Groenhuis *et al.* (1983), although the use of the Maxwell equations is mathematically rigorous, presently available results and tools only allow us to handle situations that are too simple compared to what is likely to be encountered in a medical imaging problem. Thus, unreasonable approximations must be made to obtain results. This limits the practical usefulness of this approach.

In Monte Carlo simulations of photon transport, as described by Carter and Cashwell (1976), photons injected into the media are followed on their paths in the media until detection, escape, or absorption. The pathways are calculated on the basis of the scattering and absorption characteristics of the media and the random distribution of scattering events. Very complex geometries may be handled using Monte Carlo simulations. Unfortunately, such simulations imply a great computational cost, which limits the practical usefulness of Monte Carlo simulations in PMI. (In the Monte Carlo method, packets of photons are followed along their paths until absorption or escape. The number of photons in each packet, w, can be thought of as the weight of the packet. This is preferable to following individual photons, since most of those would disappear prior to contributing to the estimated measurements because of absorption within the media. The problem remains that some of the packet weights may eventually become so small that a considerable fraction of machine time is wasted on packets that do not significantly contribute to the estimate. Because of this problem, a cutoff weight is generally set.)

The application of transport theory to photon scattering has been described by Patterson *et al.* (1989) and Ishimaru (1978a,b). Ishimaru (1978b) has shown that for the limiting case of an infinite number of scatters, the diffusion equation can predict the behavior of light in media. Photon transfer is described simply by the photon density and the net photon flux at every point within the media. Thus, results are obtained by solving the diffusion equation. Analytical solutions to the diffusion equation, however, exist only for very simple geometries such as those found in Crank (1967) and Arridge *et al.* (1992) (e.g., concentric spheres and semi-infinite slabs). Fortunately, solutions to the diffusion equation even for complicated geometries can be obtained using numerical methods.

The objective in PMI is to determine the scattering and absorbing properties of the imaged object. Specifically, the properties measured are the scattering and absorbing cross sections (μ_s and μ_a, respectively). For homogeneous media, the probability that a photon will travel at least a distance r without scattering is $\exp(-\mu_s r)$, while the probability of traveling at least this distance without being absorbed is $\exp(-\mu_a r)$. In this definition, scattering and absorption are assumed to be isotropic. While a comprehensive data set of the optical parameters of all the tissues in the body does not currently exist, a large table of tissues for which the properties are known has been compiled by Cheong *et al.* (1990).

Patterson *et al.* (1989) have analyzed photon migration in turbid media using the diffusion equation, which in this case is

$$\frac{\partial c}{\partial t} = \nabla \cdot \mathfrak{D} \nabla c - \mu_a v c. \tag{1}$$

In this equation, c, $\mathfrak{D}$, and μ_a are functions of the three spatial variables: c is the concentration of light (which is also a function of the temporal variable t) and the other two are the (spatially varying) coefficients of diffusion and absorption, respectively. Here, v is the speed of light in tissue. The coefficient of diffusion $\mathfrak{D}$ is related to the absorption and scattering cross sections by the equation

$$\mathfrak{D} = v/3(\mu_a + \mu_s). \tag{2}$$

The advantage of using transport theory to solve the forward problem is that it can be used to model the complex biologically reasonable geometries of absorption and scattering while avoiding the computation time required for Monte Carlo simulations. It is currently not clear whether using the diffusion equation is the ideal way to describe photon migration; work by Patterson *et al.* (1989) and Madsen *et al.* (1992) indicates that it is adequate for many applications. By simulating photon migration on the basis of the diffusion equation using a cubic grid of small volume elements (voxels), it is possible to simulate the manner in which photons diffuse within the space while retaining the freedom to assign differing tissue properties to individual voxels (see, e.g., Yee and Herman, 1992). This freedom appears to be necessary in any method that aims at solving the forward problem.

In such a simulation, several assumptions are made. It is assumed that the scattering of photons in tissue is isotropic; i.e., it has no directionality. This assumption is certainly incorrect, since scattering is generally biased in the forward direction. However, if photons scatter repeatedly (as is the case in PMI), then they soon lose any directionality they originally had. This condition is easily met if the size of the voxels is large compared to the mean length between scattering events (as is expected to be the case). It can be reasonably argued, from the macroscopic (voxel-to-voxel) point of view, that photon scattering lacks directionality since photons entering a voxel will scatter repeatedly before exiting. Thus, isotropic and anisotropic scattering can be treated identically. Also, it is assumed that photons reaching the surface contribute to the backscattered intensity that is detected. Finally, there are also additional assumptions made in the diffusion equation, namely that diffusion is proportional to the concentration gradient (Fick's law) and that the concentration of light can be treated as a continuum. While not strictly true, all are adequate approximations in the regime of photon scattering of near-infrared light with resolutions in the centimeter range.

For examples of the many papers that report on attempts at solving the forward problem and that also contain some discussion of the success of these attempts, the reader may wish to look at Bonner *et al.* (1987), Patterson *et al.* (1989), Haselgrove *et al.* (1991), and Yee and Herman (1992).

2. TECHNIQUES OF IMAGE PRODUCTION

The analysis of PMI belongs to the general area commonly identified by the name "inverse problems." As in emission and transmission computed tomography, the goal of PMI is to reconstruct the structure of the object that produced the data set. Unlike x rays, the photons used in PMI do not travel in straight lines. Thus, image reconstruction techniques designed for x rays and γ rays as described in Herman (1980) are not applicable. Similarly, techniques developed for ultrasound and impedance tomography are not applicable because of various assumptions made in deriving the algorithms (for details, see Greenleaf, 1983 and Webster, 1990). Consequently, an entirely new approach to reconstruction must be found.

It should be noted that the model developed must be three dimensional in nature. The emerging photons are highly scattered and it is unlikely that many of them would have been scattered entirely within a single plane. Furthermore, it would be impossible to isolate those photons that in fact did scatter only in a single plane. Thus, a two-dimensional model is not appropriate for practical applications of PMI. Unfortunately, the penalty for using a three-dimensional model is increased computational cost.

In recent years, there has been a great deal of interest in the area of photon migration in general and PMI in particular. Thus far, most of the effort has been in trying to solve the forward problem. There have been only a few reports on attempts to solve the inverse problem; examples include Singer *et al.* (1990), Giordana *et al.* (1991), and Hebden and Wong (1993).

2.1 Nonreconstructive Imaging

From the theoretical analyses of Patterson *et al.* (1989) and Bonner *et al.* (1987, 1989), it is known that at times far from the photon injection, the rate at which the detected signal decays in an instrument that detects the photons escaping from the medium is a direct indication of the absorption coefficient of the medium that the photons have passed through. Similar considerations led Chance *et al.* (1988, 1990) to suggest that two-dimensional maps related to absorption coefficients can be obtained directly. Thus far, they have been able to show that placing absorbing objects within an Intralipid™ (a liquid that can be used when a dense random medium is needed in light-diffusion experiments) bath produces a significant change in the resulting data. In addition, they have been able to obtain several preliminary images by injecting a sinusoidally modulated light signal into the medium and collecting phase and amplitude data of the emerging photons at a fixed distance from the light source. By moving the source/detector pair (which are at fixed distance apart) along the surface, phase and amplitude information is obtained for all points on the surface. By plotting these phase and amplitude data, two-dimensional images are obtained. The value assigned to a point in such two-dimensional images depends on the physical characteristics of many points in three-dimensional space and so these images are not "tomographic." Nevertheless, they demonstrate that PMI has the potential of becoming a viable clinical tool.

2.2 Tomography Based on Early Arrivals

In FAT, we can take advantage of the fact that the first bunch of photons that reach a detector have undergone the least amount of scatter and assume that they must have traveled in a relatively straight line between the source and the detector. By using only these photons, it becomes possible to apply any of the many algorithms, such as those described in Herman (1980), that have already been developed for computerized tomography from x rays and γ rays to reconstruct an image. One of the earliest papers on the potential medical use of this approach is Jackson *et al.* (1987); for an informative list of such papers since that time, see the references of Hebden and Wong (1993). Although it has not yet been proven efficacious in medicine, FAT has found a variety of nonmedical applications; e.g., Bennett *et al.* (1984), Snyder and Hesselink (1984), and Puro and Kell (1992).

The typical approach is to reconstruct a single cross section of the body from data collected for lines lying in that cross section and then repeat the process for other cross sections as needed. Following Herman (1980), reconstruction algorithms are characterized either as transform methods or as series expansion methods. While transform methods (which are essentially numerical approximations to a closed-form inversion formula) are more popular in this special application (mainly because of their elegance and their perceived computational efficiency), we discuss here only series expansion methods, since generalizations of them are more likely to be useful in the more difficult problem of PMI using all photons, as opposed to just early arrivals. (Making use of all photons should, in principle, lead to higher-quality reconstructions.)

The series expansion approach assumes that the function f to be reconstructed can be approximated by a linear combination of a finite set of known and fixed basis functions,

$$f(r, \phi) \approx \sum_{j=1}^{J} x_j b_j(r, \phi), \tag{3}$$

and that our task is to estimate the unknowns x_j. If we assume that the measurements depend linearly on the object to be reconstructed (certainly true in the special case of line integrals) and that we know (at least approximately) what the measurements would be if the object to be reconstructed were one of the basis functions (we use $l_{i,j}$ to denote the value of the ith measurement of the jth basis function), then we can conclude (Herman, 1980) that the ith of our measurements of f is approximately

$$\sum_{j=1}^{J} l_{i,j} x_j. \tag{4}$$

Our problem is then to estimate the x_j from the measured approximations to Eq. (4). Substituting these estimated values into Eq. (3) will then provide us with an estimate of the function f.

The simplest way of choosing the basis functions is to subdivide the plane into pixels (or space into voxels) and to choose basis functions, each of whose value is 1 inside a specific pixel (or voxel) and is 0 everywhere else. However, there are other choices which may be preferable; for example, Lewitt (1992) uses spherically symmetric basis functions that are not only spatially limited, but also can be chosen to be smoothly varying. Such smooth basis functions result in smooth reconstructions, while the spherical symmetry allows easy calculation of the $l_{i,j}$.

Advantages of series expansion methods over transform methods are their flexibility (no special relationship needs to be assumed between the object to be reconstructed and the measurements taken, such as that the latter are line integrals of the former) and the ability to control the type of solution we want by specifying the exact sense in which the x_j are to be estimated from the approximate values of Eq. (4) (for example, we can base the estimation on our understanding of the statistical nature of the noise in the data). A recent example of a relevant paper in which such a series expansion approach is investigated for a nonmedical application is Verhoeven (1993).

2.3 Tomography Based on All Arrivals

In the PMI approach that tries to make use of all the photons (not just the early arrivals), the linear approximation of Eq. (4) is no longer valid. Nevertheless, the series expansion method is applicable; what we need to do is to create a discretized version of the diffusion Eq. (1) and then use it to estimate the diffusion and absorption coefficients assigned to individual voxels so that we obtain an optimal, or at least acceptable, match with the observed data.

One way to do this is by using a biased random search, as suggested for example by Singer *et al.* (1990). After a suitable solution to the forward problem is found, it becomes possible to estimate, on the basis of the optical properties of the tissue that is being imaged, the measurements to which that tissue would give rise during our data collection. The aim is then to minimize the "distance" between the measured data set and its estimate based on the conjectured optical properties of the tissue that is being imaged, by randomly changing these properties. To search the entire space of possible images for the correct image is obviously impractical; instead, the optical properties assigned to each voxel are simultaneously changed so that the amount and direction of the change are biased by the effects of recent changes on the "distance" between the data set and the solution of the forward problem based on the conjectured image. By incorporating terms designed to enforce biological reasonableness into the definition of "distance," we can direct the method toward a solution that is medically useful.

Two examples of biased random-search algorithms are ALOPEX (see, e.g., Harth *et al.*, 1988; Herman *et al.*, 1993) and simulated annealing (see, e.g., Aarts and Korst, 1989). While they tend to have a very high computational cost (and therefore should not be used when alternative nonstochastic algorithms exist), biased random-search algorithms have the advantage that they can be applied without modification to a large variety of global optimization problems, including PMI, in which there are expected to be many local minima in the parameter space of scattering and absorption. Nevertheless, because of the expense associated with such an approach, we cannot consider that it provides a practically useful solution to the problem of reconstruction from PMI data.

3. MEDICAL CONSIDERATIONS

The major contributor to absorption of light in the human body is hemoglobin and other heme derivatives. Thus, a distribution of the absorption cross section derived from PMI has the potential to map regions of hemorrhage. In addition, the distribution of the scattering cross section provides a map of anatomic structures. In principle, this will allow registration of regions of hemorrhage with anatomic landmarks.

We know from Robbins *et al.* (1984) and Sabiston (1987) that while in most areas of the body hemorrhage has little clinical significance, within the skull it can be devastating. Since the skull is a relatively closed volume, the effects of the hemorrhage can be fatal. Unlike other parts of the body, the skull cannot accommodate the extra volume. The brainstem is compressed and can herniate, thus causing death. Detection of such hemorrhages is an important clinical task. While existing imaging modalities such as magnetic resonance imaging (MRI) and x-ray computed tomography (CT) can be used, PMI has the potential of accomplishing this task at lower cost and without any exposure to ionizing radiation.

Besides hemorrhages due to trauma, there are other cerebral hemorrhages that are clinically important. One major form is hemorrhagic stroke. Hemorrhagic strokes are generally due to some underlying susceptibility in the blood vessels of the head such as hypertensive vascular disease. From Thorn (1977) and Rowland (1989) we know that the clinical presentation is very similar to that of strokes due to embolism. The treatment, however, is very different. In the latter, anticoagulation therapy with heparin and coumadin is frequently useful. In hemorrhagic strokes, obviously, such therapy is of little use and can in fact be harmful. Thus, the ability to differentiate between the two is quite important; PMI has the potential of providing it.

Recall that hemoglobin and other heme derivatives are a major source of absorption; most other components of biological tissue only scatter light. Thus, PMI has the potential of mapping the distribution of blood in the body. Furthermore, since the absorption spectra for oxy- and deoxy-hemoglobin are different, PMI also has the potential of mapping the oxidative state of tissue by using repeated measurements over several different frequencies. This ability has great relevance in the treatment of tumors since it is known that hypoxic regions of tumors are highly resistant to chemotherapy.

Other possible uses include monitoring in the operating theater (the instrumentation required is likely to be of small size) and the diagnosis of breast tumors (facilitated by secondary effects of microcalcifications) as suggested by Key *et al.* (1991). Since traditional mammography delivers a dose of ionizing radiation which itself can cause tumors, if PMI could be made efficacious for diagnosing breast tumors, then it would be preferable to the current x-ray–based methods.

In the many medical situations in which detailed knowledge of the structure of tissue is required, the applicability of PMI is expected to be limited, as compared to higher-resolution medical imaging modalities such as MRI and CT. On the other hand, PMI offers some advantages. PMI will not subject the patient to the radiation dose that CT requires. Furthermore, the cost of a PMI system should be far less than that of a MRI system. Thus, there will be fewer objections to monitoring a patient with sequential measurements. These advantages make PMI a potentially useful clinical tool in certain medical problems.

4. CONCLUSIONS

The applicability of PMI to medical problems is inherently more limited than MRI or CT; its primary uses involve detection of heme derivatives such as hemoglobin. On the other hand, in clinical situations where there is a need to detect hemorrhages or microcalcifications, PMI offers a potential imaging modality that may become useful as a screening tool that is less expensive than MRI and avoids the radiation dose inherent in CT. Thus, optical tomography has a significant potential as a useful medical tool. At the current state of the art, instrumentation has been developed to collect the data, and quite a large body of knowledge regarding the forward problem has been accumulated, but the inverse problem has not been solved in a way that would be medically useful.

GLOSSARY

Forward Problem: The prediction of the data set obtained from the imaging modality given knowledge of the object that is being imaged.

Inverse Problem: The estimation, based on an acquired data set, of parameters associated with an object that is being investigated.

Photon-Migration Imaging (PMI): A proposed imaging modality that uses infrared light.

Tomography: The process of reconstruction of the distribution of physical parameters at individual points inside a three-dimensional structure from data obtained from radiation transmitted through or emitted from it.

Voxel: Short for volume element. This is analogous to pixel in two-dimensional images.

Works Cited

Aarts, E., Korst, J. (1989), *Simulated Annealing and Boltzmann Machines*, Chichester, UK: Wiley.

Arridge, S., Cope, M., Delpy, D. (1992), "The Theoretical Basis for the Determination of Optical Pathlengths in Tissue: Temporal and Frequency Analysis," *Phys. Med. Biol.* **37**, 1531–1560.

Bennett, K., Faris, G., Byer, R. (1984), "Experimental Optical Fan Beam Tomography," *Appl. Opt.* **23**, 2678–2685.

Bonner, R., Nossal, R., Havlin, S., Weiss, G. (1987), "Model for Photon Migration in Turbid Biological Media," *J. Opt. Soc. Am. A* **4**, 423–432.

Bonner, R., Nossal, R., Weiss, G. (1989), "A Random Walk Theory of Time-Resolved Optical Absorption Spectroscopy in Tissue," in: B. Chance (Ed.), *Photon Migration in Tissues*, New York: Plenum Press, pp. 11–23.

Carter, L., Cashwell, E. (1976), *Particle-Transport Simulation with the Monte Carlo Method*, Los Alamos: U.S. Energy Research and Development Administration.

Chance, B., Maris, M., Sorge, J., Zhang, M. (1990), "A Phase Modulation System for Dual Wavelength Difference Spectroscopy of Hemoglobin Deoxygenation in Tissues," in: *Time-Resolved Laser Spectroscopy in Biochemistry II*, Bellingham, WA: SPIE, pp. 481–491.

Chance, B., Nioka, S., Kent, J., McCully, K., Fountain, M., Greenfeld, R., Holtom, G. (1988), "Time-Resolved Spectroscopy of Hemoglobin and Myoglobin in Resting and Ischemic Muscle," *Anal. Biochem.* **174**, 698–707.

Cheong, W., Prahl, S., Welch, A. (1990), "A Review of the Optical Properties of Biological Tissues," *IEEE Quantum Electron.* **26**, 2166–2185.

Crank, J. (1967), *The Mathematics of Diffusion*, London: Oxford University Press.

Edholm, P. (1977), "Tomogram Reconstruction Using an Opticophotographic Method," *Acta Radiol. Diagnosis* **18**, 126–144.

Fishkin, J., Gratton, E. (1993), "Propagation of Photon-Density Waves in Strongly Scattering Media Containing an Absorbing Semi-Infinite Plane Bounded by a Straight Edge," *J. Opt. Soc. Am. A* **10**, 127–140.

Giordana, C., Mochi, M., Zirilli, F. (1991), "The Numerical Solution of an Inverse Problem for a Class of One-Dimensional Diffusion Equations with Piecewise Constant Coefficients," *SIAM J. Appl. Math.* **52**, 428–441.

Gmitro, A., Tresp, V., Gindi, G. (1990), "Videographic Tomography—Part I: Reconstruction with Parallel-Beam Projection Data," *IEEE Trans. Med. Imag.* **9**, 366–375.

Greenleaf, J. (1983), "Computerized Tomography with Ultrasound," *Proc. IEEE* **71**, 330–337.

Groenhuis, R., Ferwerda, H., Bosch, J. (1983), "Scattering and Absorption of Turbid Materials Determined from Reflection Measurements," *Appl. Opt.* **22**, 2456–2462.

Harth, E., Kalogeropoulos, T., Pandya, A. (1988), "A Universal Optimization Network," in: J. Myklebust and G. Harris (Eds.), *Proc. Spec. Symp. Maturing Technologies and Emerging Horizons in Biomed. Eng.*, New Orleans: IEEE, pp. 97–107.

Haselgrove, J., Leigh, J., Yee, C., Wang, N., Maris, M., Chance, B. (1991), "Monte Carlo and Diffusion Calculations of Photon Migration in Non-Infinite Highly Scattering Media," in: B. Chance (Ed.), *Time-Resolved Spectroscopy and Imaging of Tissue*, SPIE Proceedings, Vol. 1431, Bellingham, WA: SPIE, pp. 30–41.

Hebden, J. (1993), "Line Scan Acquisition for Time-Resolved Imaging through Scattering Media," *Opt. Eng.* **32**, 626–633.

Hebden, J., Wong, K. (1993), "Time-Resolved Optical Tomography," *Appl. Opt.* **32**, 372–380.

Herman, G. (1980), *Image Reconstruction from Projections: The Fundamentals of Computerized Tomography*, New York: Academic Press.

Herman, G., Odhner, D., Yeung, K. (1993), "Optimization for Pattern Classification Using Biased Random Search Techniques," *Ann. Oper. Res.* **43**, 419–427.

Herman, G., Tuy, H., Langenberg, K., Sabatier, P. (1987), *Basic Methods of Tomography and Inverse Problems*, Bristol, UK: Adam Hilger.

Ishimaru, A. (1978a), "Diffusion of a Pulse in Densely Distributed Scatterers," *J. Opt. Soc. Am.* **68**, 1045–1050.

Ishimaru, A. (1978b), *Wave Propagation and Scattering in Random Media*, San Diego: Academic Press.

Jackson, P., Stevens, P., Smith, J., Kear, D., Key, H., Wells, P. (1987), "The Development of a System for Transillumination Computed Tomography," *Brit. J. Radiol.* **60**, 375–380.

Key, H., Davies, E., Jackson, P., Wells, P. (1991), "Optical Attenuation Characteristics of Breast Tissues at Visible and Near-Infrared Wavelengths," *Phys. Med. Biol.* **36**, 579–590.

Lewitt, R. (1992), "Alternatives to Voxels for Image Representation in Iterative Reconstruction Algorithms," *Phys. Med. Biol.* **37**, 705–716.

Lilge, L., Haw, T., Wilson, B. (1993), "Miniature Isotropic Optical Fibre Probes for Quantitative Light Dosimetry in Tissue," *Phys. Med. Biol.* **38**, 215–230.

Madsen, S., Wilson, B., Patterson, M., Park, Y., Jacques, S., Hefetz, Y. (1992), "Experimental Tests of a Simple Diffusion Model for the Estimation of Scattering and Absorption Coefficients of Turbid Media from Time-Resolved Diffuse Reflectance Measurements," *Appl. Opt.* **31**, 3509–3517.

Patterson, M., Chance, B., Wilson, B. (1989), "Time Resolved Reflectance and Transmittance for the Non-Invasive Measurement of Tissue Optical Properties," *Appl. Opt.* **28**, 2331–2336.

Puro, A., Kell, K. (1992), "Complete Determination of Stress in Fiber Preforms of Arbitrary Cross Section," *J. Lightwave Tech.* **10**, 1010–1014.

Robbins, S., Cotran, R., Kumar, V. (1984), *Pathologic Basis of Disease*, 3rd ed., Philadelphia: W.B. Saunders Co.

Rowland, L. (Ed.) (1989), *Merritt's Textbook of Neurology*, 8th ed., Philadelphia: Lea & Febiger.

Sabiston, D., Jr. (1987), *Sabiston's Essentials of Surgery*, Philadelphia: W.B. Saunders Co.

Singer, J., Grünbaum, F., Kohn, P., Zubelli, J. (1990), "Image reconstruction of the interior of bodies that diffuse radiation," *Science* **248**, 990–993.

Snyder, R., Hesselink, L. (1984), "Optical tomography for flow visualization of the density field around a revolving helicopter motor blade," *Appl. Opt.* **23**, 3650–3656.

Thorn, G. (1977), *Harrison's Principles of Internal Medicine*, 8th ed., New York: McGraw-Hill.

Verhoeven, D. (1993), "Multiplicative Algebraic Computed Tomographic Algorithms for the Reconstruction of Multidimensional Interferometric Data," *Opt. Eng.* **32**, 410–419.

Webster, J. G. (1990), *Electrical Impedance Tomography*, Bristol, UK: Adam Hilger.

Weiss, G., Nossal, R., Bonner, R. (1989), "Statistics of Penetration Depth of Photons Re-Emitted from Irradiated Tissue," *J. Modern Opt.* **36**, 349–359.

Yamada, Y., Hasegawa, Y., Maki, H. (1993), "Simulation of Time-Resolved Optical Computer Tomography Imaging," *Opt. Eng.* **32**, 634–641.

Yee, C., Herman, G. (1992), "A Convergent Algorithm Modelling Photon Migration in Tissue," in: *Proc. 14th Ann. Int. Conf. IEEE Eng. Med. Biol. Soc.*, Piscataway, NJ: IEEE, pp. 2043–2045.

Further Reading

Chance, B., (Ed.) (1989), *Photon Migration in Tissues*, New York: Plenum Press.

Cheong, W., Prahl, S., Welch, A. (1990), "A Review of the Optical Properties of Biological Tissues," *IEEE Quantum Electron.* **26**, 2166–2185.

Groenhuis, R., Ferwerda, H., Bosch, J. (1983), "Scattering and Absorption of Turbid Materials Determined from Reflection Measurements," *Appl. Opt.* **22**, 2456–2462.

Herman, G. (1980), *Image Reconstruction from Projections: The Fundamentals of Computerized Tomography*, New York: Academic Press.

Singer, J., Grünbaum, F., Kohn, P., Zubelli, J. (1990), "Image Reconstruction of the Interior of Bodies that Diffuse Radiation," *Science* **248**, 990–993.

OPTICS, ATMOSPHERIC

CRAIG F. BOHREN, *Department of Meteorology, Pennsylvania State University, University Park, Pennsylvania, U.S.A.*

INTRODUCTION

Atmospheric optics is nearly synonymous with light scattering, the only restrictions being that the scatterers inhabit the atmosphere and the primary source of their illumination is the sun. Essentially all light we see is scattered light, even that directly from the sun. When we say that such light is unscattered we really mean that it is scattered in the forward direction; hence it is *as if* it were unscattered. Scattered light is radiation from matter excited by an external source. When the source vanishes, so does the scattered light, as distinguished from light emitted by matter, which persists in the absence of external sources.

Atmospheric scatterers are either molecules or particles. A particle is an aggregation of sufficiently many molecules that it can be ascribed macroscopic properties such as temperature and refractive index. There is no

3-527-28134-7/95/$5.00 + .50

canonical number of molecules that must unite to form a *bona fide* particle. Two molecules clearly do not a quorum make, but what about 10, 100, 1000? The particle size corresponding to the largest of these numbers is about 10^{-3} μm. Particles this small of water substance would evaporate so rapidly that they could not exist long under conditions normally found in the atmosphere. As a practical matter, therefore, we need not worry unduly about scatterers in the shadow region between molecule and particle.

A property of great relevance to scattering problems is *coherence*, both of the array of scatterers and of the incident light. At visible wavelengths, air is an array of incoherent scatterers: the radiant power scattered by N molecules is N times that scattered by one (except in the forward direction). But when water vapor in air condenses, an incoherent array is transformed into a coherent array: uncorrelated water molecules become part of a single entity. Although a single droplet is a coherent array, a cloud of droplets taken together is incoherent.

Sunlight is incoherent but not in an absolute sense. Its lateral coherence length is tens of micrometers, which is why we can observe what are essentially interference patterns (e.g., coronas and glories) resulting from illumination of cloud droplets by sunlight.

This article begins with the color and brightness of a purely molecular atmosphere, including their variation across the vault of the sky. This naturally leads to the state of polarization of skylight. Because the atmosphere is rarely, if ever, entirely free of particles, the general characteristics of scattering by particles follow, setting the stage for a discussion of atmospheric visibility.

Atmospheric refraction usually sits by itself, unjustly isolated from all those atmospheric phenomena embraced by the term scattering. Yet refraction is another manifestation of scattering, coherent scattering in the sense that phase differences cannot be ignored.

Scattering by single water droplets and ice crystals, each discussed in turn, yields feasts for the eye as well as the mind. The curtain closes on the optical properties of clouds.

1. COLOR AND BRIGHTNESS OF MOLECULAR ATMOSPHERE

1.1 A Brief History

Edward Nichols began his 1908 presidential address to the New York meeting of the American Physical Society as follows: "In asking your attention to-day, even briefly, to the consideration of the present state of our knowledge concerning the color of the sky it may be truly said that I am inviting you to leave the thronged thoroughfares of our science for some quiet side street where little is going on and you may even suspect that I am coaxing you into some blind alley, the inhabitants of which belong to the dead past."

Despite this depreciatory statement, hoary with age, correct and complete explanations of the color of the sky still are hard to find. Indeed, all the faulty explanations lead active lives: the blue sky is the reflection of the blue sea; it is caused by water, either vapor or droplets or both; it is caused by dust. The true cause of the blue sky is not difficult to understand, requiring only a bit of critical thought stimulated by belief in the inherent fascination of all natural phenomena, even those made familiar by everyday occurrence.

Our contemplative prehistoric ancestors no doubt speculated on the origin of the blue sky, their musings having vanished into it. Yet it is curious that Aristotle, the most prolific speculator of early recorded history, makes no mention of it in his *Meteorologica* even though he delivered pronouncements on rainbows, halos, and mock suns and realized that "the sun looks red when seen through mist or smoke." Historical discussions of the blue sky sometimes cite Leonardo as the first to comment intelligently on the blue of the sky, although this reflects a European bias. If history were to be written by a supremely disinterested observer, Arab philosophers would likely be given more credit for having had profound insights into the workings of nature many centuries before their European counterparts descended from the trees. Indeed, Möller (1972) begins his brief history of the blue sky with Jakub Ibn Ishak Al Kindi (800–870), who explained it as "a mixture of the darkness of the night with the light of the dust and haze particles in the air illuminated by the sun."

Leonardo was a keen observer of light in nature even if his explanations sometimes fell short of the mark. Yet his hypothesis that "the blueness we see in the atmosphere is not intrinsic colour, but is caused by warm vapor evaporated in minute and insensible atoms on which the solar rays fall, rendering them luminous against the infinite darkness of the fiery sphere which lies beyond and includes it" would, with minor changes, stand critical scrutiny today. If we set aside Leonardo as *sui generis*, scientific attempts to unravel the origins of the blue sky may be said to have begun with Newton, that towering pioneer of optics, who, in time-honored fashion, reduced it to what he already had considered: interference colors in thin films. Almost two centuries elapsed before more pieces in the puzzle were contributed by the experimental investigations of von Brücke and Tyndall on light scattering by suspensions of particles. Around the same time Clausius added his bit in the form of a theory that scattering by minute bubbles causes the blueness of the sky. A better theory was not long in coming. It is associated with a man known to the world as Lord Rayleigh even though he was born John William Strutt.

Rayleigh's paper of 1871 marks the beginning of a satisfactory explanation of the blue sky. His scattering law, the key to the blue sky, is perhaps the most famous result ever obtained by dimensional analysis. Rayleigh argued that the field E_s scattered by a particle small compared with the light illuminating it is proportional to its volume V and to the incident field E_i. Radiant energy conservation requires that the scattered field diminish inversely as the distance r from the particle so that the scattered power diminishes as the square of r. To make this proportionality dimensionally homogeneous requires the inverse square of a quantity with the dimensions of length. The only plausible physical variable at hand is the wavelength of the incident light, which leads to

$$E_s \propto E_i V/r\lambda^2. \qquad (1)$$

When the field is squared to obtain the scattered power, the result is Rayleigh's inverse fourth-power law. This law is really only an often—but not always—very good approximation. Missing from it are dimensionless properties of the particle such as its refractive index, which itself depends on wavelength. Because of this *dispersion*, therefore, nothing scatters exactly as the inverse fourth power.

Rayleigh's 1871 paper did not give the complete explanation of the color and polarization of skylight. What he did that was not done by his predecessors was to give a law of scattering, which could be used to test quantitatively the hypothesis that selective scattering by atmospheric particles could transform white sunlight into blue skylight. But as far as giving the agent responsible for the blue sky is concerned, Rayleigh did not go essentially beyond Newton and Tyndall, who invoked particles. Rayleigh was circumspect about the nature of these particles, settling on salt as the most likely candidate. It was not until 1899 that he published the capstone to his work on skylight, arguing that air molecules themselves were the source of the blue sky. Tyndall cannot be given the credit for this because he considered air to be *optically empty*: when purged of all particles it scatters no light. This erroneous conclusion was a result of the small scale of his laboratory experiments. On the scale of the atmosphere, sufficient light is scattered by air molecules to be readily observable.

1.2 Molecular Scattering and the Blue of the Sky

Our illustrious predecessors all gave explanations of the blue sky requiring the presence of water in the atmosphere: Leonardo's "evaporated warm vapor," Newton's "Globules of water," Clausius's bubbles. Small wonder, then, that water still is invoked as the cause of the blue sky. Yet a cause of something is that without which it would not occur, and the sky would be no less blue if the atmosphere were free of water.

A possible physical reason for attributing the blue sky to water vapor is that, because of selective *absorption*, liquid water (and ice) is blue upon transmission of white light over distances of order meters. Yet if all the water in the atmosphere at any instant were to be compressed into a liquid, the result would be a layer about 1 cm thick, which is not sufficient to transform white light into blue by selective absorption.

Water vapor does not compensate for its hundredfold lower abundance than nitrogen and oxygen by greater scattering per molecule. Indeed, scattering of visible light by a water molecule is slightly *less* than that by either nitrogen or oxygen.

Scattering by atmospheric molecules does not obey Rayleigh's inverse fourth-power law exactly. A least-squares fit over the visible spectrum from 400 to 700 nm of the *molecular scattering coefficient* of sea-level air tabulated by Penndorf (1957) yields an inverse 4.089th-power scattering law.

The molecular scattering coefficient β, which plays important roles in following sections, may be written

$$\beta = N\sigma_s \,, \tag{2}$$

where N is the number of molecules per unit volume and σ_s, the scattering cross section (an average because air is a mixture) per molecule, approximately obeys Rayleigh's law. The form of this expression betrays the incoherence of scattering by atmospheric molecules. The inverse of β is interpreted as the scattering *mean free path*, the average distance a photon must travel before being scattered.

To say that the sky is blue because of Rayleigh scattering, as is sometimes done, is to confuse an agent with a law. Moreover, as Young (1982) pointed out, the term Rayleigh scattering has many meanings. Particles small compared with the wavelength scatter according to the same law as do molecules. Both can be said to be Rayleigh scatterers, but only molecules are necessary for the blue sky. Particles, even small ones, generally diminish the vividness of the blue sky.

Fluctuations are sometimes trumpeted as the "real" cause of the blue sky. Presumably, this stems from the fluctuation theory of light scattering by media in which the scatterers are separated by distances small compared with the wavelength. In this theory, which is associated with Einstein and Smoluchowski, matter is taken to be continuous but characterized by a refractive index that is a random function of position. Einstein (1910) stated that "it is remarkable that our theory does not make *direct* use of the assumption of a discrete distribution of matter." That is, he circumvented a difficulty but realized it could have been met head on, as Zimm (1945) did years later.

The blue sky is really caused by scattering by molecules—to be more precise, scattering by bound electrons: free electrons do not scatter selectively. Because air molecules are separated by distances small compared with the wavelengths of visible light, it is not obvious that the power scattered by such molecules can be added. Yet if they are completely uncorrelated, as in an ideal gas (to good approximation the atmosphere is an ideal gas), scattering by N molecules is N times scattering by one. This is the only sense in which the blue sky can be attributed to scattering by fluctuations. Perfectly homogeneous matter does not exist. As stated pithily by Planck, "a chemically pure substance may be spoken of as a vacuum made turbid by the presence of molecules."

1.3 Spectrum and Color of Skylight

What is the spectrum of skylight? What is its color? These are two different questions. Answering the first answers the second but not the reverse. Knowing the color of skylight we cannot uniquely determine its spectrum because of *metamerism*: A given perceived color can in general be obtained in an indefinite number of ways.

Skylight is not blue (itself an imprecise term) in an absolute sense. When the visible spectrum of sunlight outside the earth's atmosphere is modulated by Rayleigh's scattering law, the result is a spectrum of scattered light that is neither solely blue nor even peaked in the blue (Fig. 1). Although blue does not predominate spectrally, it does predominate perceptually. We perceive the sky to be blue even though skylight contains light of all wavelengths.

Any source of light may be looked upon as a mixture of white light and light of a single wavelength called the *dominant wavelength*. The *purity* of the source is the relative amount of the monochromatic component in the mixture. The dominant wavelength of sunlight scattered according to Rayleigh's law is about 475 nm, which lies solidly in the blue if we take this to mean light with wavelengths between 450 and 490 nm. The purity of this scattered light, about 42%, is the upper limit for skylight. Blues of real skies are less pure.

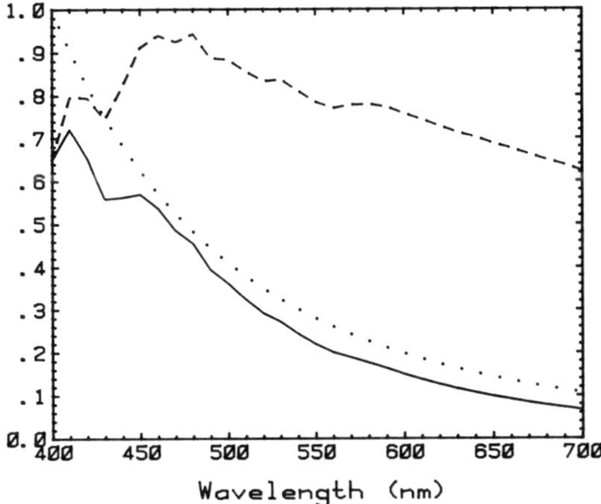

FIG. 1. Rayleigh's scattering law (dots), the spectrum of sunlight outside the Earth's atmosphere (dashes), and the product of the two (solid curve). The solar spectrum is taken from Thekaekara and Drummond (1971).

Another way of conveying the color of a source of light is by its *color temperature*, the temperature of a blackbody having the same perceived color as the source. Since blackbodies do not span the entire gamut of colors, not all sources of light can be assigned color temperatures. But many natural sources of light can. The color temperature of light scattered according to Rayleigh's law is infinite. This follows from Planck's spectral emission function $e_{b\lambda}$ in the limit of high temperature,

$$e_{b\lambda} \approx 2\pi ckT/\lambda^4, \quad hc/\lambda \ll kT, \tag{3}$$

where h is Planck's constant, k is Boltzmann's constant, c is the speed of light *in vacuo*, and T is absolute temperature. Thus, the emission spectrum of a blackbody with an infinite temperature has the same functional form as Rayleigh's scattering law.

1.4 Variation of Sky Color and Brightness

Not only is skylight not pure blue, but its color and brightness vary across the vault of the sky, with the best blues at zenith. Near the astronomical horizon the sky is brighter than overhead but of considerably lower purity. That this variation can be observed from an airplane flying at 10 km, well above most particles, suggests that the sky is inherently nonuniform in color and brightness (Fig. 2). To understand why requires invoking multiple scattering.

Multiple scattering gives rise to observable phenomena that cannot be explained solely by single-scattering arguments. This is easily demonstrated. Fill a blackened pan with clean water, then add a few drops of milk. The resulting dilute suspension illuminated by sunlight has a bluish cast. But when more milk is added, the suspension turns white. Yet the properties of the scatterers (fat globules) have not changed, only their *optical thickness*: the blue suspension being optically thin, the white being optically thick.

Optical thickness is physical thickness in units of scattering mean free path, and hence is dimensionless. The optical thickness τ between any two points connected by an arbitrary path in a medium populated by (incoherent) scatterers is an integral over the path:

$$\tau = \int_1^2 \beta ds. \tag{4}$$

FIG. 2. Even at an altitude of 10 km, well above most particles, the sky brightness increases markedly from the zenith to the astronomical horizon.

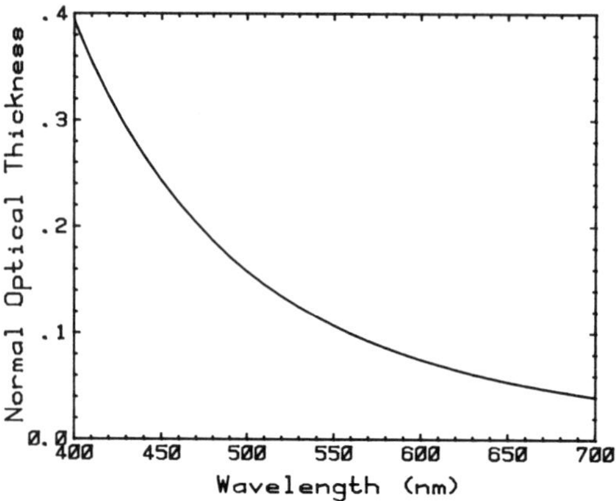

FIG. 3. Normal optical thickness of a pure molecular atmosphere.

The *normal optical thickness* τ_n of the atmosphere is that along a radial path extending from the surface of the Earth to infinity. Figure 3 shows τ_n over the visible spectrum for a purely molecular atmosphere. Because τ_n is generally small compared with unity, a photon from the sun traversing a radial path in the atmosphere is unlikely to be scattered more than once. But along a tangential path, the optical thickness is about 35 times greater (Fig. 4), which leads to several observable phenomena.

Even an intrinsically black object is luminous to an observer because of *airlight*, light

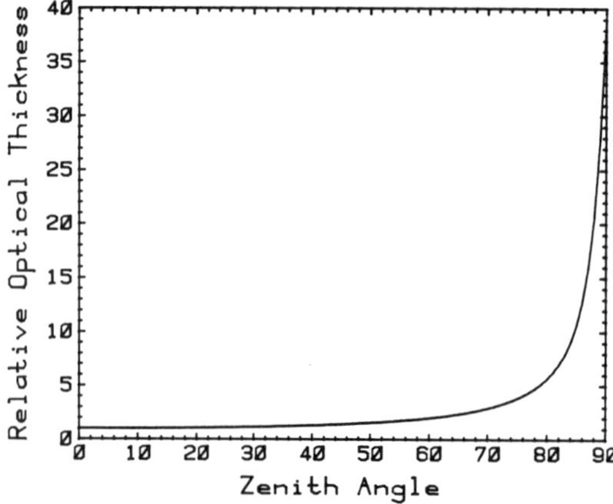

FIG. 4. Optical thickness (relative to the normal optical thickness) of a molecular atmosphere along various paths with zenith angles between 0° (normal) and 90° (tangential).

scattered by all the molecules and particles along the line of sight from observer to object. Provided that this is uniformly illuminated by sunlight and that ground reflection is negligible, the airlight radiance L is approximately

$$L = GL_0 (1 - e^{-\tau}), \tag{5}$$

where L_0 is the radiance of incident sunlight along the line of sight with optical thickness τ. The term G accounts for geometric reduction of radiance because of scattering of nearly monodirectional sunlight in all directions. If the line of sight is uniform in composition, $\tau = \beta d$, where β is the scattering coefficient and d is the physical distance to the black object.

If τ is small ($\ll 1$), $L \approx GL_0\tau$. In a purely molecular atmosphere, τ varies with wavelength according to Rayleigh's law; hence the distant black object in such an atmosphere is perceived to be bluish. As τ increases so does L but not proportionally. Its limit is GL_0: The airlight radiance spectrum is that of the source of illumination. Only in the limit $d = 0$ is $L = 0$ and the black object truly black.

Variation of the brightness and color of dark objects with distance was called *aerial perspective* by Leonardo. By means of it we estimate distances to objects of unknown size such as mountains.

Aerial perspective belongs to the same family as the variation of color and brightness of the sky with zenith angle. Although the optical thickness along a path tangent to the Earth is not infinite, it is sufficiently large (Figs. 3 and 4) that GL_0 is a good approximation for the radiance of the horizon sky. For isotropic scattering (a condition almost satisfied by molecules), G is around 10^{-5}, the ratio of the solid angle subtended by the sun to the solid angle of all directions (4π). Thus, the horizon sky is not nearly so bright as direct sunlight.

Unlike in the milk experiment, what is observed when looking at the horizon sky is not multiply scattered light. Both have their origins in multiple scattering but manifested in different ways. Milk is white because it is weakly absorbing and optically thick, and hence all components of incident white light are multiply scattered to the observer even though the blue component traverses a shorter average path in the suspension than the red component. White horizon light has escaped being multiply scattered, although multiple scattering is why this light is white (strictly, has the spectrum of the source). More light at the short-wavelength end of the spectrum is scattered *toward* the observer than at the long-wavelength end. But long-wavelength light has the greater likelihood of being transmitted to the observer without being scattered *out of* the line of sight. For a long optical path, these two processes compensate, resulting in a horizon radiance spectrum which is that of the source.

Selective scattering by molecules is not sufficient for a blue sky. The atmosphere also must be optically thin, at least for most zenith angles (Fig. 4) (the blackness of space as a backdrop is taken for granted but also is necessary, as Leonardo recognized). A corollary of this is that the blue sky is not inevitable: an atmosphere composed entirely of nonabsorbing, selectively scattering molecules overlying a nonselectively reflecting earth need not be blue. Figure 5 shows calculated spectra of the zenith sky over black ground for a molecular atmosphere with the present normal optical thickness as well as for hypothetical atmospheres 10 and 40 times thicker. What we take to be inevitable is accidental: If our atmosphere were much thicker, but identical in composition, the color of the sky would be quite different from what it is now.

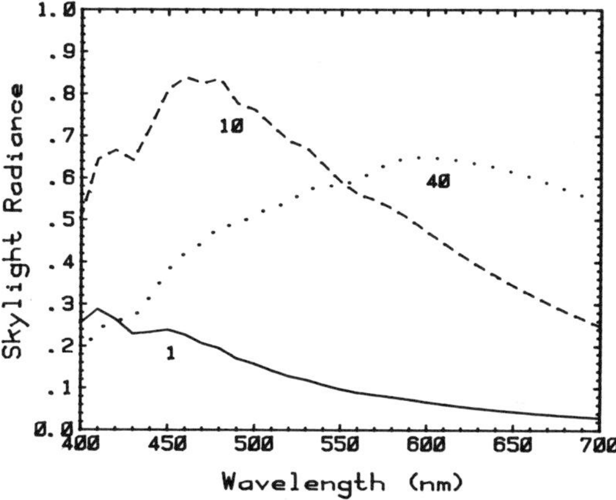

FIG. 5. Spectrum of overhead skylight for the present molecular atmosphere (solid curve), as well as for hypothetical atmospheres 10 (dashes) and 40 (dots) times thicker.

1.5 Sunrise and Sunset

If short-wavelength light is preferentially scattered out of direct sunlight, long-wavelength light is preferentially transmitted in the direction of sunlight. Transmission is described by an exponential law (if light multiply scattered back into the direction of the sunlight is negligible):

$$L = L_0 e^{-\tau}, \tag{6}$$

where L is the radiance at the observer in the direction of the sun, L_0 is the radiance of sunlight outside the atmosphere, and τ is the optical thickness along this path.

If the wavelength dependence of τ is given by Rayleigh's law, sunlight is *reddened* upon transmission: The spectrum of the transmitted light is comparatively richer than the incident spectrum in light at the long-wavelength end of the visible spectrum. But to say that transmitted sunlight is reddened is not the same as saying it is red. The perceived color can be yellow, orange, or red, depending on the magnitude of the optical thickness. In a molecular atmosphere, the optical thickness along a path from the sun, even on or below the horizon, is not sufficient to give red light upon transmission. Although selective scattering by molecules yields a blue sky, reds are not possible in a molecular atmosphere, only yellows and oranges. This can be observed on clear days, when the horizon sky at sunset becomes successively tinged with yellow, then orange, but not red.

Equation (6) applies to the radiance only in the direction of the sun. Oranges and reds can be seen in other directions because reddened sunlight illuminates scatterers not lying along the line of sight to the sun. A striking example of this is a horizon sky tinged with oranges and pinks in the direction *opposite* the sun.

The color and brightness of the sun changes as it arcs across the sky because the optical thickness along the line of sight changes with solar zenith angle Θ. If the Earth were flat (as some still aver), the transmitted solar radiance would be

$$L = L_0 e^{\tau_n / \cos\Theta}. \tag{7}$$

This equation is a good approximation except near the horizon. On a flat earth, the optical thickness is infinite for horizon paths. On a spherical earth, optical thicknesses are finite although much larger for horizon than for radial paths.

The normal optical thickness of an atmosphere in which the number density of scatterers decreases exponentially with height z above the surface, $\exp(-z/H)$, is the same as that for a uniform atmosphere of finite thickness:

$$\tau_n = \int_0^\infty \beta dz = \beta_0 H, \tag{8}$$

where H is the *scale height* and β_0 is the scattering coefficient at sea level. This equivalence yields a good approximation even for the tangential optical thickness. For any zenith angle, the optical thickness is given approximately by

$$\frac{\tau}{\tau_n} = \sqrt{\frac{R_e^2}{H^2}\cos^2\Theta + \frac{2R_e}{H} + 1} - \frac{R_e}{H}\cos\Theta, \tag{9}$$

where R_e is the radius of the Earth. A flat earth is one for which R_e is infinite, in which instance Eq. (9) yields the expected relation

$$\lim_{R_e \to \infty} \frac{\tau}{\tau_n} = \frac{1}{\cos\Theta}. \tag{10}$$

For Earth's atmosphere, the molecular scale height is about 8 km. According to the approximate relation Eq. (9), therefore, the horizon optical thickness is about 39 times greater than the normal optical thickness. Taking the exponential decrease of molecular number density into account yields a value about 10% lower.

Variations on the theme of reds and oranges at sunrise and sunset can be seen even when the sun is overhead. The radiance at an observer an optical distance τ from a (horizon) cloud is the sum of cloudlight transmitted to the observer and airlight:

$$L = L_0 G(1 - e^{-\tau}) + L_0 G_c e^{-\tau}, \tag{11}$$

where G_c is a geometrical factor that accounts for scattering of nearly monodirectional sunlight into a hemisphere of directions by the cloud. If the cloud is approximated as an isotropic reflector with reflectance R and illuminated at an angle Φ, the geometrical factor G_c is $\Omega_s R \cos\Phi/\pi$, where Ω_s is the solid angle subtended by the sun at the Earth. If $G_c > G$, the observed radiance is redder (i.e., enriched in light of longer wavelengths) than the incident radiance. If $G_c < G$, the observed radiance is bluer than the incident radiance. Thus, distant horizon clouds can be reddish if they are bright or bluish if they are dark.

Underlying Eq. (11) is the implicit assumption that the line of sight is uniformly illuminated by sunlight. The first term in this equation is airlight; the second is transmitted cloudlight. Suppose, however, that the line of sight is shadowed from direct sunlight by clouds (that do not, of course, occlude the distant cloud of interest). This may reduce the first term in Eq. (11) so that the second term dominates. Thus, under a partly overcast sky, distant horizon clouds may be reddish even when the sun is high in the sky.

The zenith sky at sunset and twilight is the exception to the general rule that molecular scattering is sufficient to account for the color of the sky. In the absence of molecular absorption, the spectrum of the zenith sky would be essentially that of the zenith sun (although greatly reduced in radiance), hence would not be the blue that is observed. This was pointed out by Hulburt (1953), who showed that absorption by ozone profoundly affects the color of the zenith sky when the sun is near the horizon. The Chappuis band of ozone extends from about 450 to 700 nm and peaks at around 600 nm. Preferential absorption of sunlight by ozone over long horizon paths gives the zenith sky its blueness when the sun is near the horizon. With the sun more than about 10° above the horizon, however, ozone has little effect on the color of the sky.

2. POLARIZATION OF LIGHT IN A MOLECULAR ATMOSPHERE

2.1 The Nature of Polarized Light

Unlike sound, light is a vector wave, an electromagnetic field lying in a plane normal to the propagation direction. The polarization state of such a wave is determined by the degree of correlation of any two orthogonal components into which its electric (or magnetic) field is resolved. Completely polarized light corresponds to complete correlation; completely unpolarized light corresponds to no correlation; partially polarized light corresponds to partial correlation.

If an electromagnetic wave is completely polarized, the tip of its oscillating electric field traces out a definite elliptical curve, the *vibration ellipse*. Lines and circles are special ellipses, the light being said to be linearly or circularly polarized, respectively. The general state of polarization is elliptical.

Any beam of light can be considered an incoherent superposition of two collinear beams, one unpolarized, the other completely polarized. The radiance of the polarized component relative to the total is defined as the *degree of polarization* (often multiplied by 100 and expressed as a percentage). This can be measured for a source of light (e.g., light from different sky directions) by rotating a (linear) polarizing filter and noting the minimum and maximum radiances transmitted by it. The degree of (linear) polarization is defined as the difference between these two radiances divided by their sum.

2.2 Polarization by Molecular Scattering

Unpolarized light can be transformed into partially polarized light upon interaction with matter because of different changes in amplitude of the two orthogonal field components. An example of this is the partial polarization of sunlight upon scattering by atmospheric molecules, which can be detected by looking at the sky through a polarizing filter (e.g., polarizing sunglasses) while rotating it. Waxing and waning of the observed brightness indicates some degree of partial polarization.

In the analysis of any scattering problem, a plane of reference is required. This is usually the *scattering plane*, determined by the directions of the incident and scattered waves, the angle between them being the *scattering angle*. Light polarized perpendicular (parallel) to the scattering plane is sometimes said to be vertically (horizontally) polarized. Vertical and horizontal in this context, however, are arbitrary terms indicating orthogonality and bear no relation, except by accident, to the direction of gravity.

The degree of polarization P of light scattered by a tiny sphere illuminated by unpolarized light is (Fig. 6)

$$P = \frac{1 - \cos^2\theta}{1 + \cos^2\theta}, \tag{12}$$

where the scattering angle θ ranges from 0° (forward direction) to 180° (backward direction); the scattered light is partially linearly polarized perpendicular to the scattering plane. Although this equation is a first step toward understanding polarization of skylight, more often than not it also has been a false step, having led countless authors to assert that skylight is completely polarized at 90° from the sun. Although $P = 1$ at $\theta = 90°$ according to Eq. (12), skylight is never 100% polarized at this or any other angle, and for several reasons.

Although air molecules are very small compared with the wavelengths of visible light, a requirement underlying Eq. (12), the dominant constituents of air are not spherically symmetric.

The simplest model of an asymmetric molecule is a small spheroid. Although it is indeed possible to find a direction in which the light scattered by such a spheroid is 100% polarized, this direction depends on the spheroid's orientation. In an ensemble of randomly oriented spheroids, each contributes its mite to the total radiance in a given direction, but each contribution is partially polarized to varying degrees between 0 and 100%. It is impossible for beams of light to

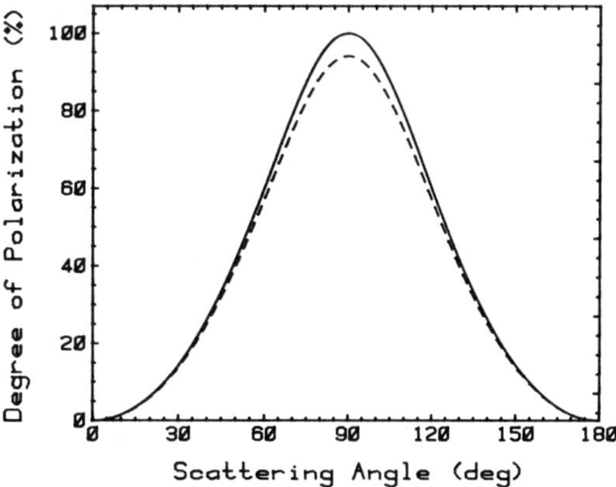

FIG. 6. Degree of polarization of the light scattered by a small (compared with the wavelength) sphere for incident unpolarized light (solid curve). The dashed curve is for a small spheroid chosen such that the degree of polarization at 90° is that for air.

be incoherently superposed in such a way that the degree of polarization of the resultant is greater than the degree of polarization of the most highly polarized beam.

Because air is an ensemble of randomly oriented asymmetric molecules, sunlight scattered by air never is 100% polarized. The intrinsic departure from perfection is about 6%. Figure 6 also includes a curve for light scattered by randomly oriented spheroids chosen to yield 94% polarization at 90°. This angle is so often singled out that it may deflect attention from nearby scattering angles. Yet, the degree of polarization is greater than 50% for a range of scattering angles 70° wide centered about 90°.

Equation (12) applies to air, not to the atmosphere, the distinction being that in the atmosphere, as opposed to the laboratory, multiple scattering is not negligible. Also, atmospheric air is almost never free of particles and is illuminated by light reflected by the ground. We must take the atmosphere as it is, whereas in the laboratory we often can eliminate everything we consider extraneous.

Because of both multiple scattering and ground reflection, light from any direction in the sky is not, in general, made up solely of light scattered in a single direction relative to the incident sunlight but is a superposition of beams with different scattering histories, hence different degrees of polarization. As a consequence, even if air molecules were perfect spheres and the atmosphere were completely free of particles, skylight would not be 100% polarized at 90° to the sun or at any other angle.

Reduction of the maximum degree of polarization is not the only consequence of multiple scattering. According to Fig. 6, there should be two *neutral points* in the sky, directions in which skylight is unpolarized: directly toward and away from the sun. Because of multiple scattering, however, there are three such points. When the sun is higher than about 20° above the horizon there are neutral points within 20° of the sun, the *Babinet point* above it, the *Brewster point* below. They coincide when the sun is directly overhead and move apart as the sun descends. When the sun is lower than 20°, the *Arago point* is about 20° above the antisolar point, the direction opposite the sun.

One consequence of the partial polarization of skylight is that the colors of distant objects may change when viewed through a rotated polarizing filter. If the sun is high in the sky, horizontal airlight will have a fairly high degree of polarization. According to the previous section, airlight is bluish. But if it also is partially polarized, its radiance can be diminished with a polarizing filter. Transmitted cloudlight, however, is unpolarized. Because the radiance of airlight can be reduced more than that of cloudlight, distant clouds may change from white to yellow to orange when viewed through a rotated polarizing filter.

3. SCATTERING BY PARTICLES

Up to this point we have considered only an atmosphere free of particles, an idealized state rarely achieved in nature. Particles still would inhabit the atmosphere even if the human race were to vanish from the Earth. They are not simply by-products of the "dark satanic mills" of civilization.

All molecules of the same substance are essentially identical. This is not true of particles: They vary in shape and size, and may be composed of one or more homogeneous regions.

3.1 The Salient Differences between Particles and Molecules: Magnitude of Scattering

The distinction between scattering by molecules when widely separated and when packed together into a droplet is that between scattering by incoherent and coherent arrays. Isolated molecules are excited primarily by incident (external) light, whereas the same molecules forming a droplet are excited by incident light and by each other's scattered fields. The total power scattered by an incoherent array of molecules is the sum of their scattered powers. The total power scattered by a coherent array is the square of the total scattered field, which in turn is the sum of all the fields scattered by the individual molecules. For an incoherent array we *may* ignore the wave nature of light, whereas for a coherent array we *must* take it into account.

Water vapor is a good example to ponder because it is a constituent of air and can condense to form cloud droplets. The difference between a sky containing water vapor and the same sky with the same amount of water but in the form of a cloud of droplets is dramatic.

According to Rayleigh's law, scattering by a particle small compared with the wavelength increases as the sixth power of its size (volume squared). A droplet of diameter 0.03 μm, for example, scatters about 10^{12} times more light than does one of its constituent molecules. Such a droplet contains about 10^7 molecules. Thus, scattering per molecule as a consequence of condensation of water vapor into a coherent water droplet increases by about 10^5.

Cloud droplets are much larger than 0.03 μm, a typical diameter being about 10 μm. Scattering per molecule in such a droplet is much greater than scattering by an isolated molecule, but not to the extent given by Rayleigh's law. Scattering increases as the sixth power of droplet diameter only when the molecules scatter coherently in phase. If a droplet is sufficiently small compared with the wavelength, each of its molecules is excited by essentially the same field and all the waves scattered by them interfere constructively. But when a droplet is comparable to or larger than the wavelength, interference can be constructive, destructive, and everything in between, and hence scattering does not increase as rapidly with droplet size as predicted by Rayleigh's law.

The figure of merit for comparing scatterers of different size is their scattering cross section per unit volume, which, except for a multiplicative factor, is the scattering cross section per molecule. A scattering cross section may be looked upon as an effective area for removing radiant energy from a beam: the scattering cross section times the beam irradiance is the radiant power scattered in all directions.

The scattering cross section per unit volume for water droplets illuminated by visible light and varying in size from molecules (10^{-4} μm) to raindrops (10^3 μm) is shown in Fig. 7. Scattering by a molecule that belongs to a cloud droplet is about 10^9 times greater than scattering by an isolated molecule, a striking example of the virtue of cooperation. Yet in molecular as in human societies there are limits beyond which cooperation becomes dysfunctional: Scattering by a molecule that belongs to a raindrop is about 100 times less than scattering by a molecule that belongs to a cloud droplet. This tremendous variation of scattering by water molecules depending on their state of aggregation has profound observational consequences. A cloud is optically so much different from the water vapor out of which it was born that the offspring bears no resemblance to its parents. We can see through tens of kilometers of air laden with water vapor, whereas a cloud a few tens of meters thick is enough to occult the sun. Yet a rainshaft born out of a cloud is considerably more translucent than its parent.

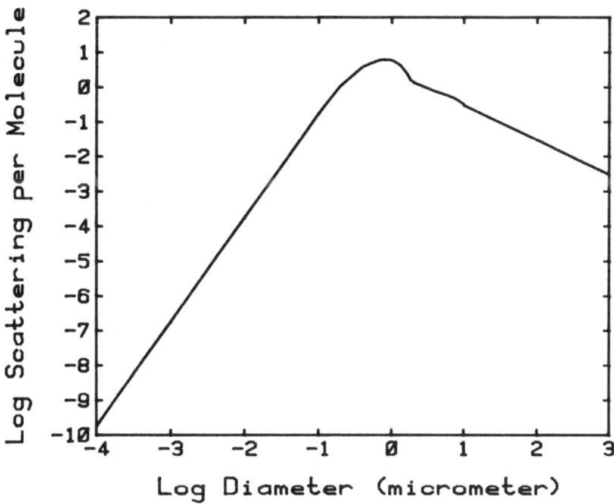

FIG. 7. Scattering (per molecule) of visible light (arbitrary units) by water droplets varying in size from a single molecule to a raindrop.

3.2 The Salient Differences between Particles and Molecules: Wavelength Dependence of Scattering

Regardless of their size and composition, particles scatter approximately as the inverse fourth power of wavelength if they are small compared with the wavelength and absorption is negligible, two important caveats. Failure to recognize them has led to errors, such as that yellow light penetrates fog better because it is not scattered as much as light of shorter wavelengths. Although there may be perfectly sound reasons for choosing yellow instead of blue or green as the color of fog lights, greater transmission through fog is not one of them: Scattering by fog droplets is essentially independent of wavelength over the visible spectrum.

Small particles are selective scatterers;

large particles are not. Particles neither small nor large give the reverse of what we have come to expect as normal. Figure 8 shows scattering of visible light by oil droplets with diameters 0.1, 0.8, and 10 μm. The smaller droplets scatter according to Rayleigh's law; the larger droplets (typical cloud droplet size) are nonselective. Between these two extremes are droplets (0.8 μm) that scatter long-wavelength light more than short-wavelength. Sunlight or moonlight seen through a thin cloud of these intermediate droplets would be bluish or greenish. This requires droplets of just the right size, and hence it is a rare event, so rare that it occurs once in a blue moon. Astronomers, for unfathomable reasons, refer to the second full moon in a month as a blue moon, but if such a moon were blue it would be only by coincidence. The last reliably reported outbreak of blue and green suns and moons occurred in 1950 and was attributed to an oily smoke produced in Canadian forest fires.

3.3 The Salient Differences between Particles and Molecules: Angular Dependence of Scattering

The angular distribution of scattered light changes dramatically with the size of the scatterer. Molecules and particles that are small compared with the wavelength are nearly isotropic scatterers of unpolarized light, the ratio of maximum (at 0° and 180°) to minimum (at 90°) scattered radiance being only 2 for spheres, and slightly less for other spheroids. Although small particles scatter the same in the forward and backward hemispheres, scattering becomes markedly asymmetric for particles comparable to or larger than the wavelength. For example, forward scattering by a water droplet as small as 0.5 μm is about 100 times greater than backward scattering, and the ratio of forward to backward scattering increases more or less monotonically with size (Fig. 9).

The reason for this asymmetry is found in the singularity of the forward direction. In this direction, waves scattered by two or more scatterers excited solely by incident light (ignoring mutual excitation) are always in phase regardless of the wavelength and the separation of the scatterers. If we imagine a particle to be made up of N small subunits, scattering in the forward direction increases as N^2, the only direction for which this is always true. For other directions, the wavelets scattered by the subunits will not necessarily all be in phase. As a consequence, scattering in the forward direction increases with size (i.e., N) more rapidly than in any other direction.

Many common observable phenomena depend on this forward-backward asymmetry. Viewed toward the illuminating sun, glistening fog droplets on a spider's web warn us of its presence. But when we view the web with our backs to the sun, the web mysteriously disappears. A pattern of dew illuminated by

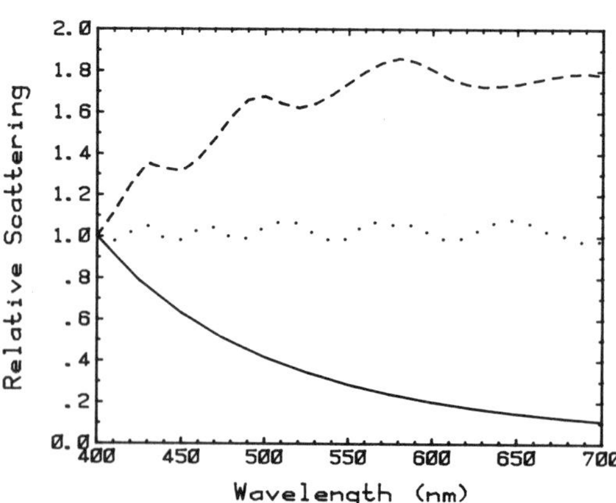

FIG. 8. Scattering of visible light by oil droplets of diameter 0.1 μm (solid curve), 0.8 μm (dashes), and 10 μm (dots).

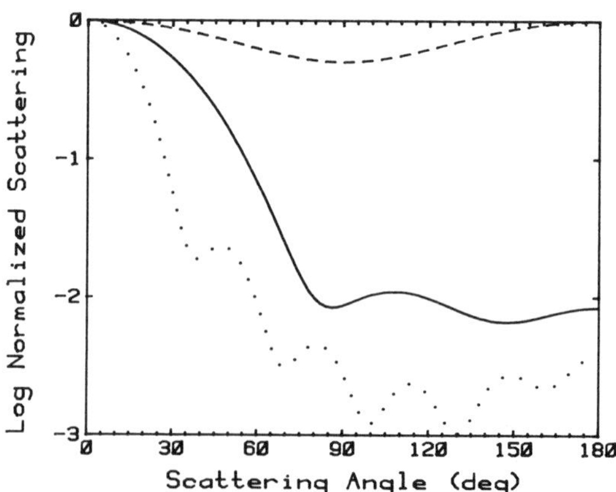

FIG. 9. Angular dependence of scattering of visible light (0.55 μm) by water droplets small compared with the wavelength (dashes), diameter 0.5 μm (solid curve), and diameter 10 μm (dots).

the rising sun on a cold morning seems etched on a window pane. But if we go outside to look at the window, the pattern vanishes. Thin clouds sometimes hover over warm, moist heaps of dung, but may go unnoticed unless they lie between us and the source of illumination. These are but a few examples of the consequences of strongly asymmetric scattering by single particles comparable to or larger than the wavelength.

3.4 The Salient Differences between Particles and Molecules: Degree of Polarization of Scattered Light

All the simple rules about polarization upon scattering are broken when we turn from molecules and small particles to particles comparable to the wavelength. For example, the degree of polarization of light scattered by small particles is a simple function of scattering angle. But simplicity gives way to complexity as particles grow (Fig. 10), the scattered light being partially polarized parallel to the scattering plane for some scattering angles, perpendicular for others.

The degree of polarization of light scattered by molecules or by small particles is essentially independent of wavelength. But this is not true for particles comparable to or larger than the wavelength. Scattering by such particles exhibits *dispersion of polarization*: The degree of polarization at, say, 90° may vary considerably over the visible spectrum (Fig. 11).

In general, particles can act as polarizers or retarders or both. A polarizer transforms unpolarized light into partially polarized light. A retarder transforms polarized light of one form into that of another (e.g., linear into elliptical). Molecules and small particles, however, are restricted to roles as polarizers. If the atmosphere were inhabited solely by such scatterers, skylight could never be other than partially linearly polarized. Yet particles comparable to or larger than the wavelength often are present; hence skylight can acquire a degree of ellipticity upon multiple scattering: Incident unpolarized light is partially linearly polarized in the first scattering event, then transformed into partially elliptically polarized light in subsequent events.

Bees can navigate by polarized skylight. This statement, intended to evoke great awe for the photopolimetric powers of bees, is rarely accompanied by an important caveat: The sky must be clear. Figures 10 and 11 show two reasons—there are others—why bees, remarkable though they may be, cannot do the impossible. The simple wavelength-independent relation between the position of the sun and the direction in which skylight is most highly polarized, an underlying necessity for navigating by means of polarized skylight, is obliterated when clouds cover the sky. This was recognized by the decoder of

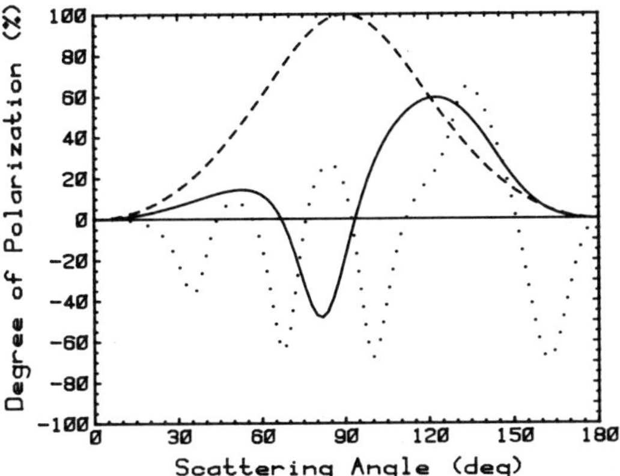

FIG. 10. Degree of polarization of light scattered by water droplets illuminated by unpolarized visible light (0.55 μm). The dashed curve is for a droplet small compared with the wavelength; the solid curve is for a droplet of diameter 0.5 μm; the dotted curve is for a droplet of diameter 1.0 μm. Negative degrees of polarization indicate that the scattered light is partially polarized parallel to the scattering plane.

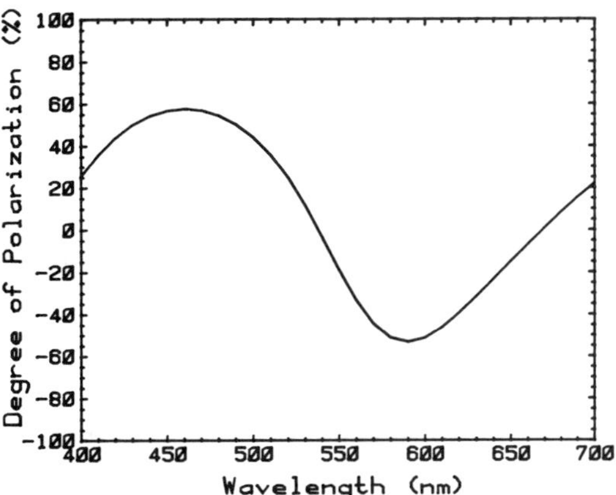

FIG. 11. Degree of polarization at a scattering angle of 90° of light scattered by a water droplet of diameter 0.5 μm illuminated by unpolarized light.

bee dances himself (von Frisch, 1971): "Sometimes a cloud would pass across the area of sky visible through the tube; when this happened the dances became disoriented, and the bees were unable to indicate the direction to the feeding place. Whatever phenomenon in the blue sky served to orient the dances, this experiment showed that it was seriously disturbed if the blue sky was covered by a cloud." But von Frisch's words often have been forgotten by disciples eager to spread the story about bee magic to those just as eager to believe what is charming even though untrue.

3.5 The Salient Differences between Particles and Molecules: Vertical Distributions

Not only are the scattering properties of particles quite different, in general, from those of molecules; the different vertical distributions of particles and molecules by themselves affect what is observed. The number density of molecules decreases more or less exponentially with height z above the surface: $\exp(-z/H_m)$, where the molecular scale height H_m is around 8 km. Although the decrease in number density of particles with height is also approximately exponential, the scale height for particles H_p is about 1–2 km. As a consequence, particles contribute disproportionately to optical thicknesses along near-horizon paths. Subject to the approximations underlying Eq. (9), the ratio of the tangential (horizon) optical thickness for particles τ_{tp} to that for molecules τ_{tm} is

$$\frac{\tau_{tp}}{\tau_{tm}} = \frac{\tau_{np}}{\tau_{nm}} \sqrt{\frac{H_m}{H_p}}, \tag{13}$$

where the subscript t indicates a tangential path and n indicates a normal (radial) path. Because of the incoherence of scattering by atmospheric molecules and particles, scattering coefficients are additive, and hence so are optical thicknesses. For equal normal optical thicknesses, the tangential optical thickness for particles is at least twice that for molecules. Molecules by themselves cannot give red sunrises and sunsets; molecules need the help of particles. For a fixed τ_{np}, the tangential optical thickness for particles is greater the more they are concentrated near the ground.

At the horizon the relative rate of change of transmission T of sunlight with zenith angle is

$$\frac{1}{T}\frac{dT}{d\Theta} = \tau_n \frac{R_e}{H}, \tag{14}$$

where the scale height and normal optical thickness may be those for molecules or particles. Not only do particles, being more concentrated near the surface, give disproportionate attenuation of sunlight on the horizon, but they magnify the angular gradient of attenuation there. A perceptible change in color across the sun's disk (which subtends about 0.5°) on the horizon also requires the help of particles.

4. ATMOSPHERIC VISIBILITY

On a clear day can we really see forever? If not, how far can we see? To answer this question requires qualifying it by restricting viewing to more or less horizontal paths during daylight. Stars at staggering distances can be seen at night, partly because there is no skylight to reduce contrast, partly because stars overhead are seen in directions for which attenuation by the atmosphere is least.

The radiance in the direction of a black object is not zero, because of light scattered along the line of sight (see Sec. 1.4). At sufficiently large distances, this airlight is indistinguishable from the horizon sky. An example is a phalanx of parallel dark ridges, each ridge less distinct than those in front of it (Fig. 12). The farthest ridges blend into the horizon sky. Beyond some distance we cannot see ridges because of insufficient contrast.

Equation (5) gives the airlight radiance, a radiometric quantity that describes radiant power without taking into account the portion of it that stimulates the human eye or by what relative amount it does so at each wavelength. Luminance (also sometimes called brightness) is the corresponding photometric quantity. Luminance and radiance are related by an integral over the visible spectrum:

$$B = \int K(\lambda)L(\lambda)d\lambda, \tag{15}$$

FIG. 12. Because of scattering by molecules and particles along the line of sight, each successive ridge is brighter than the ones in front of it even though all of them are covered with the same dark vegetation.

where the luminous efficiency of the human eye K peaks at about 550 nm and vanishes outside the range 385–760 nm.

The *contrast* C between any object and the horizon sky is

$$C = (B - B_\infty)/B_\infty, \tag{16}$$

where B_∞ is the luminance for an infinite horizon optical thickness. For a uniformly illuminated line of sight of length d, uniform in its scattering properties, and with a black backdrop, the contrast is

$$C = -\frac{\int KGL_0 \exp(-\beta d)\, d\lambda}{\int KGL_0 d\lambda}. \tag{17}$$

The ratio of integrals in this equation defines an average optical thickness:

$$C = -\exp(-\langle\tau\rangle). \tag{18}$$

This expression for contrast reduction with (optical) distance is mathematically, but not physically, identical to Eq. (6), which perhaps has engendered the misconception that atmospheric visibility is reduced because of attenuation. Yet as there is no light from a black object to be attenuated, its finite visual range cannot be a consequence of attenuation.

The distance beyond which a dark object cannot be distinguished from the horizon sky is determined by the *contrast threshold*: the smallest contrast detectable by the human observer. Although this depends on the particular observer, the angular size of the object observed, the presence of nearby objects, and the absolute luminance, a contrast threshold of 0.02 is often taken as an average. This value in Eq. (18) gives

$$-\ln|C| = 3.9 = \langle\tau\rangle = \langle\beta d\rangle. \tag{19}$$

To convert an optical distance into a physical distance requires the scattering coefficient. Because K is peaked at around 550 nm, we can obtain an approximate value of d from the scattering coefficient at this wavelength in Eq. (19). At sea level, the molecular scattering coefficient in the middle of the visible spectrum corresponds to about 330 km for "forever": the greatest distance at which a black object can be seen against the horizon sky assuming a contrast threshold of 0.02 and ignoring the curvature of the earth.

We also observe contrast between elements of the same scene, a hillside mottled with stands of trees and forest clearings, for example. The extent to which we can resolve details in such a scene depends on sun angle as well as distance.

The airlight radiance for a nonreflecting object is Eq. (5) with $G = p(\Theta)\Omega_s$, where $p(\Theta)$ is the probability (per unit solid angle) that light is scattered in a direction making an angle Θ with the incident sunlight and Ω_s is the solid angle subtended by the sun. When the sun is overhead, $\Theta = 90°$; with the sun at the observer's back, $\Theta = 180°$; for an observer looking directly into the sun, $\Theta = 0°$.

The radiance of an object with a finite reflectance R and illuminated at an angle Φ is given by Eq. (11). Equations (5) and (11) can be combined to obtain the contrast between reflecting and nonreflecting objects:

$$C = \frac{Fe^{-\tau}}{1 + (F - 1)e^{-\tau}},$$
$$F = (R \cos\Phi)/\pi p(\Theta). \tag{20}$$

All else being equal, therefore, contrast decreases as $p(\Theta)$ increases. As shown in Fig. 9, $p(\Theta)$ is more sharply peaked in the forward direction the larger the scatterer. Thus, we expect the details of a distant scene to be less distinct when looking toward the sun than away from it if the optical thickness of the line of sight has an appreciable component contributed by particles comparable to or larger than the wavelength.

On humid, hazy days, visibility is often depressingly poor. Haze, however, is not water vapor but rather water that has ceased to be vapor. At high relative humidities, but still well below 100%, small soluble particles in the atmosphere accrete liquid water to become solution droplets (haze). Although these droplets are much smaller than cloud droplets, they markedly diminish visual range because of the sharp increase in scattering with particle size (Fig. 7). The same number of water molecules when aggregated in haze scatter vastly more than when apart.

5. ATMOSPHERIC REFRACTION

5.1 Physical Origins of Refraction

Atmospheric refraction is a consequence of molecular scattering, which is rarely stated given the historical accident that before light and matter were well understood refraction and scattering were locked in separate compartments and subsequently have been sequestered more rigidly than monks and nuns in neighboring cloisters.

Consider a beam of light propagating in an optically homogeneous medium. Light is scattered (weakly but observably) laterally to this beam as well as in the direction of the beam (the forward direction). The observed beam is a coherent superposition of incident light and forward-scattered light, which was excited by the incident light. Although refractive indices are often defined by ratios of phase velocities, we may also look upon a refractive index as a parameter that specifies the phase shift between an incident beam and the forward-scattered beam that the incident beam excites. The connection between (incoherent) scattering and refraction (coherent scattering) can be divined from the expressions for the refractive index n of a gas and the scattering cross section σ_s of a gas molecule:

$$n = 1 + \tfrac{1}{2}\alpha N, \tag{21}$$

$$\sigma_s = (k^4/6\pi)|\alpha|^2, \tag{22}$$

where N is the number density (not mass density) of gas molecules, $k = 2\pi/\lambda$ is the wavenumber of the incident light, and α is the polarizability of a molecule (induced dipole moment per unit inducing electric field). The appearance of the polarizability in Eq. (21) but its square in Eq. (22) is the clue that refraction is associated with electric fields whereas lateral scattering is associated with electric fields squared (powers). Scattering, without qualification, often means incoherent scattering in all directions. Refraction, in a nutshell, is coherent scattering in a particular direction.

Readers whose appetites have been whetted by the preceding brief discussion of the physical origins of refraction are directed to a beautiful paper by Doyle (1985) in which he shows how the Fresnel equations can be dissected to reveal the scattering origins of (specular) reflection and refraction.

5.2 Terrestrial Mirages

Mirages are not illusions, any more so than are reflections in a pond. Reflections of plants growing at its edge are not interpreted as plants growing into the water. If the water is ruffled by wind, the reflected images may be so distorted that they are no longer recognizable as those of plants. Yet we still would not call such distorted images illusions. And so is it with mirages. They are images noticeably different from what they would be in the absence of atmospheric refraction, creations of the atmosphere, not of the mind.

Mirages are vastly more common than is realized. Look and you shall see them. Contrary to popular opinion, they are not unique to deserts. Mirages can be seen frequently even over ice-covered landscapes and highways flanked by deep snowbanks. Temperature *per se* is not what gives mirages but rather temperature gradients.

Because air is a mixture of gases, the polarizability for air in Eq. (21) is an average over all its molecular constituents, although their individual polarizabilities are about the same (at visible wavelengths). The vertical refractive index gradient can be written so as to show its dependence on pressure p and (absolute) temperature T:

$$\frac{d}{dz}\ln(n-1)=\frac{1}{p}\frac{dp}{dz}-\frac{1}{T}\frac{dT}{dz}. \tag{23}$$

Pressure decreases approximately exponentially with height, where the scale height is around 8 km. Thus, the first term on the right-hand side of Eq. (23) is around 0.1 km^{-1}. Temperature usually decreases with height in the atmosphere. An average lapse rate of temperature (i.e., its decrease with height) is around 6 °C/km. The average temperature in the troposphere (within about 15 km of the surface) is around 280 K. Thus, the magnitude of the second term in Eq. (23) is around 0.02 km^{-1}. On average, therefore, the refractive-index gradient is dominated by the vertical pressure gradient. But within a few meters of the surface, conditions are far from average. On a sun-baked highway your feet may be touching asphalt at 50 °C while your nose is breathing air at 35 °C, which corresponds to a lapse rate a thousand times the average. Moreover, near the surface, temperature can increase with height. In shallow surface layers, in which the pressure is nearly constant, the temperature gradient determines the refractive index gradient. It is in such shallow layers that mirages, which are caused by refractive-index gradients, are seen.

Cartoonists by their fertile imaginations unfettered by science, and textbook writers by their carelessness, have engendered the notion that atmospheric refraction can work wonders, lifting images of ships, for example, from the sea high into the sky. A back-of-the-envelope calculation dispels such notions. The refractive index of air at sea level is about 1.0003 (Fig. 13). Light from empty space incident at glancing incidence onto a uniform slab with this refractive index is displaced in angular position from where it would have been in the absence of refraction by

$$\delta=\sqrt{2(n-1)}. \tag{24}$$

This yields an angular displacement of about 1.4°, which as we shall see is a rough upper limit.

Trajectories of light rays in nonuniform

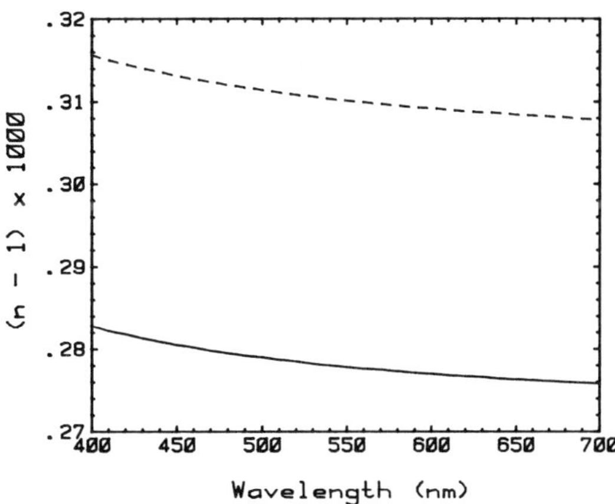

FIG. 13. Sea-level refractive index vs wavelength at −15 °C (dashes) and 15 °C (solid curve). Data from Penndorf (1957).

media can be expressed in different ways. According to Fermat's principle of least time (which ought to be extreme time), the actual path taken by a ray between two points is such that the path integral

$$\int_1^2 n\,ds \tag{25}$$

is an extremum over all possible paths. This principle has inspired piffle about the alleged efficiency of nature, which directs light over routes that minimize travel time, presumably freeing it to tend to important business at its destination.

The scale of mirages is such that in analyzing them we may pretend that the Earth is flat. On such an earth, with an atmosphere in which the refractive index varies only in the vertical, Fermat's principle yields a generalization

$$n\sin\theta=\text{constant} \tag{26}$$

of Snel's law, where θ is the angle between the ray and the vertical direction. We could, of course, have bypassed Fermat's principle to obtain this result.

Under the assumption that θ is small compared with one, Eq. (26) yields the following differential equation satisfied by a ray:

$$\frac{d^2z}{dy^2}=\frac{dn}{dz}, \tag{27}$$

where y and z are its horizontal and vertical coordinates, respectively. For a constant refractive-index gradient, which to good approximation occurs for a constant temperature gradient, Eq. (27) yields parabolas for ray trajectories. One such parabola for a constant temperature gradient about 100 times the average is shown in Fig. 14. Note the vastly different horizontal and vertical scales. The image is displaced downward from what it would be in the absence of atmospheric refraction; hence the designation *inferior* mirage. This is the familiar highway mirage, seen over highways warmer than the air above them. The downward angular displacement is

$$\delta = \frac{1}{2} s \frac{dn}{dz}, \tag{28}$$

where s is the horizontal distance between object and observer (image). Even for a temperature gradient 1000 times the tropospheric average, displacements of mirages are less than a degree at distances of a few kilometers.

If temperature increases with height, as it does, for example, in air over a cold sea, the resulting mirage is called a *superior mirage*. Inferior and superior are not designations of lower and higher caste but rather of displacements downward and upward.

For a constant temperature gradient, one and only one parabolic ray trajectory connects an object point to an image point. Multiple images therefore are not possible. But temperature gradients close to the ground are rarely linear. The upward transport of energy from a hot surface occurs by molecular conduction through a stagnant boundary layer of air. Somewhat above the surface, however, energy is transported by air in motion. As a consequence, the temperature gradient steepens toward the ground if the energy flux is constant. This variable gradient can lead to two observable consequences: magnification and multiple images.

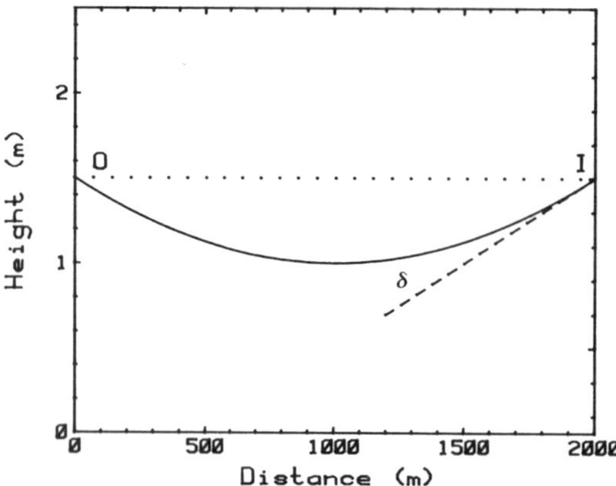

FIG. 14. Parabolic ray paths in an atmosphere with a constant refractive-index gradient (inferior mirage). Note the vastly different horizontal and vertical scales.

According to Eq. (28), all image points at a given horizontal distance are displaced downward by an amount proportional to the (constant) refractive index gradient. A corollary is that the closer an object point is to a surface, where the temperature gradient is greatest, the greater the downward displacement of the corresponding image point. Thus, nonlinear vertical temperature profiles may magnify images.

Multiple images are seen frequently on highways. What often appears to be water on the highway ahead but evaporates before it is reached is the inverted secondary image of either the horizon sky or horizon objects lighter than dark asphalt.

5.3 Extraterrestrial Mirages

When we turn from mirages of terrestrial objects to those of extraterrestrial bodies, most notably the sun and moon, we can no longer pretend that the Earth is flat. But we can pretend that the atmosphere is uniform and bounded. The total phase shift of a vertical ray from the surface to infinity is the same in an atmosphere with an exponentially decreasing molecular number density as in a hypothetical atmosphere with a uniform number density equal to the surface value up to height H.

A ray refracted along a horizon path by this hypothetical atmosphere and originating from outside it had to have been incident on it from an angle δ below the horizon:

$$\delta = \sqrt{\frac{2H}{R}} - \sqrt{\frac{2H}{R} - 2(n-1)}, \tag{29}$$

where R is the radius of the Earth. Thus, when the sun (or moon) is seen to be on the horizon it is actually more than halfway below it, δ being about 0.36°, whereas the angular width of the sun (or moon) is about 0.5°.

Extraterrestrial bodies seen near the horizon also are vertically compressed. The simplest way to estimate the amount of compression is from the rate of change of angle of refraction θ_r with angle of incidence θ_i for a uniform slab

$$\frac{d\theta_r}{d\theta_i} = \frac{\cos\theta_i}{\sqrt{n^2 - \sin^2\theta_i}}, \tag{30}$$

where the angle of incidence is that for a curved but uniform atmosphere such that the refracted ray is horizontal. The result is

$$\frac{d\theta_r}{d\theta_i} = \sqrt{1 - \frac{R}{H}(n-1)}, \tag{31}$$

according to which the sun near the horizon is distorted into an ellipse with aspect ratio about 0.87. We are unlikely to notice this distortion, however, because we expect the sun and moon to be circular, and hence we see them that way.

The previous conclusions about the downward displacement and distortion of the sun were based on a refractive-index profile determined mostly by the vertical pressure gradient. Near the ground, however, the temperature gradient is the prime determinant of the refractive-index gradient, as a consequence of which the sun on the horizon can take on shapes more striking than a mere ellipse. For example, Fig. 15 shows a nearly triangular sun with serrated edges. Assigning a cause to these serrations provides a lesson in the perils of jumping to conclusions. Obviously, the serrations are the result of sharp changes in the temperature gradient—or so one might think. Setting aside how such changes could be produced and maintained in a real atmosphere, a theorem of Fraser (1975) gives pause for thought. According to this theorem, "In a horizontally (spherically) homogeneous atmosphere it is impossible for more than one image of an extraterrestrial object (sun) to be seen above the astronomical horizon." The serrations on the sun in Fig. 15 are multiple images. But if the refractive index varies only vertically (i.e., along a radius), no matter how sharply, multiple images are not possible. Thus, the serrations must owe their existence to horizontal variations of the refractive index, a consequence of gravity waves propagating along a temperature inversion.

FIG. 15. A nearly triangular sun on the horizon. The serrations are a consequence of horizontal variations in refractive index.

5.4 The Green Flash

Compared to the rainbow, the green flash is not a rare phenomenon. Before you dismiss this assertion as the ravings of a lunatic, consider that rainbows require raindrops as well as sunlight to illuminate them, whereas rainclouds often completely obscure the sun. Moreover, the sun must be below about 42°. As a consequence of these conditions, rainbows are not seen often, but often enough that they are taken as the paragon of color variation. Yet tinges of green on the upper rim of the sun can be seen every day at sunrise and sunset given a sufficiently low horizon and a cloudless sky. Thus, the conditions for seeing a green flash are more easily met than those for seeing a rainbow. Why then is the green flash considered to be so rare? The distinction here is that between a rarely observed phenomenon (the green flash) and a rarely observable one (the rainbow).

The sun may be considered to be a collection of disks, one for each visible wavelength. When the sun is overhead, all the disks coincide and we see the sun as white. But as it descends in the sky, atmospheric refraction displaces the disks by slightly different amounts, the red less than the violet (see Fig. 13). Most of each disk overlaps all the others except for the disks at the extremes of the visible spectrum. As a consequence, the upper rim of the sun is violet or blue, its lower rim red, whereas its interior, the region in which all disks overlap, is still white.

This is what would happen in the absence of lateral scattering of sunlight. But refraction and lateral scattering go hand in hand, even in an atmosphere free of particles. Selective scattering by atmospheric molecules and particles causes the color of the sun to change. In particular, the violet-bluish upper rim of the low sun can be transformed to green.

According to Eq. (29) and Fig. 13, the angular width of the green upper rim of the low sun is about 0.01°, too narrow to be resolved with the naked eye or even to be seen against its bright backdrop. But depending on the temperature profile, the atmosphere itself can magnify the upper rim and yield a second image of it, thereby enabling it to be seen without the aid of a telescope or binoculars. Green rims, which require artificial magnification, can be seen more frequently than green flashes, which require natural magnification. Yet both can be seen often by those who know what to look for and are willing to look.

6. SCATTERING BY SINGLE WATER DROPLETS

All the colored atmospheric displays that result when water droplets (or ice crystals) are illuminated by sunlight have the same underlying cause: light is scattered in different amounts in different directions by particles larger than the wavelength, and the directions in which scattering is greatest depends on wavelength. Thus, when particles are illuminated by white light, the result can be angular separation of colors even if scattering integrated over all directions is independent of wavelength (as it essentially is for cloud droplets and ice crystals). This description, although correct, is too general to be completely satisfying. We need something more specific, more quantitative, which requires theories of scattering.

Because superficially different theories have been used to describe different optical phenomena, the notion has become widespread that they are caused by these theories. For example, coronas are said to be caused by diffraction and rainbows by refraction. Yet both the corona and the rainbow can be described quantitatively to high accuracy with a theory (the Mie theory for scattering by a sphere) in which diffraction and refraction do not explicitly appear. No fundamentally impenetrable barrier separates scattering from (specular) reflection, refraction, and diffraction. Because these terms came into general use and were entombed in textbooks before the nature of light and matter was well understood, we are stuck with them. But if we insist that diffraction, for example, is somehow different from scattering, we do so at the expense of shattering the unity of the seemingly disparate observable phenomena that result when light interacts with matter. What is observed depends on the composition and disposition of the matter, not on which approximate theory in a hierarchy is used for quantitative description.

Atmospheric optical phenomena are best classified by the direction in which they are seen and by the agents responsible for them. Accordingly, the following sections are arranged in order of scattering direction, from forward to backward.

When a single water droplet is illuminated by white light and the scattered light projected onto a screen, the result is a set of colored rings. But in the atmosphere we see a mosaic to which individual droplets contribute. The scattering pattern of a single droplet is the same as the mosaic provided that multiple scattering is negligible.

6.1 Coronas and Iridescent Clouds

A cloud of droplets narrowly distributed in size and thinly veiling the sun (or moon) can yield a spectacular series of colored concentric rings around it. This corona is most easily described quantitatively by the Fraunhofer diffraction theory, a simple approximation valid for particles large compared with the wavelength and for scattering angles near the forward direction. According to this approximation, the differential scattering cross section (cross section for scattering into a unit solid angle) of a spherical droplet of radius a illuminated by light of wavenumber k is

$$|S|^2/k^2, \tag{32}$$

where the *scattering amplitude* is

$$S = x^2 \frac{1 + \cos\theta}{2} \frac{J_1(x \sin\theta)}{x \sin\theta}. \tag{33}$$

The term J_1 is the Bessel function of first order and the size parameter $x = ka$. The quantity $(1 + \cos\theta)/2$ is usually approximated by 1 since only near-forward scattering angles θ are of interest.

The differential scattering cross section, which determines the angular distribution of the scattered light, has maxima for $x \sin\theta = 5.137, 8.417, 11.62, \ldots$ Thus, the dispersion in the position of the first maximum is

$$\frac{d\theta}{d\lambda} \approx \frac{0.817}{a} \tag{34}$$

and is greater for higher-order maxima. This dispersion determines the upper limit on drop size such that a corona can be observed. For the total angular dispersion over the visible spectrum to be greater than the angular width of the sun (0.5°), the droplets cannot be larger than about 60 μm in diameter. Drops in rain, even in drizzle, are appreciably larger than this, which is why coronas are not seen through rainshafts. Scattering by a droplet of diameter 10 μm (Fig. 16), a typical cloud droplet size, gives sufficient dispersion to yield colored coronas.

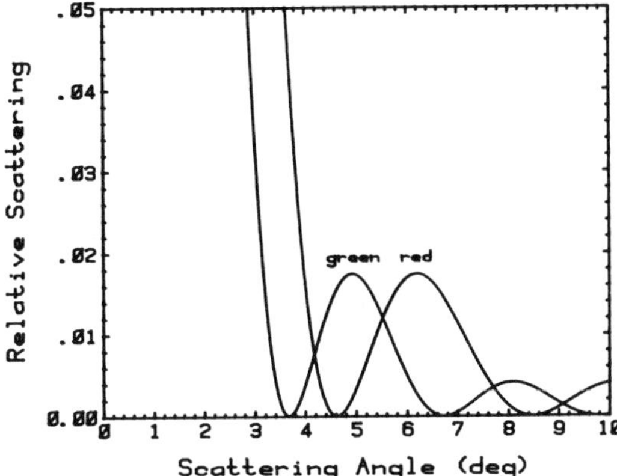

FIG. 16. Scattering of light near the forward direction (according to Fraunhofer theory) by a sphere of diameter 10 μm illuminated by red and green light.

Suppose that the first angular maximum for blue light (0.47 μm) occurs for a droplet of radius a. For red light (0.66 μm) a maximum is obtained at the same angle for a droplet of radius $a + \Delta a$. That is, the two maxima, one for each wavelength, coincide. From this we conclude that coronas require narrow size distributions: if cloud droplets are distributed in radius with a relative variance $\Delta a/a$ greater than about 0.4, color separation is not possible.

Because of the stringent requirements for the occurrence of coronas, they are not observed often. Of greater occurrence are the corona's cousins, iridescent clouds, which display colors but usually not arranged in any obviously regular geometrical pattern. Iridescent patches in clouds can be seen even at the edges of thick clouds that occult the sun.

Coronas are not the unique signatures of spherical scatterers. Randomly oriented ice columns and plates give similar patterns according to Fraunhofer theory (Takano and Asano, 1983). As a practical matter, however, most coronas probably are caused by droplets. Many clouds at temperatures well below freezing contain subcooled water droplets. Only if a corona were seen in a cloud at a temperature lower than −40 °C could one assert with confidence that it must be an ice-crystal corona.

6.2 Rainbows

In contrast with coronas, which are seen looking toward the sun, rainbows are seen looking away from it, and are caused by water drops much larger than those that give coronas. To treat the rainbow quantitatively we may pretend that light incident on a transparent sphere is composed of individual rays, each of which suffers a different fate determined only by the laws of specular reflection and refraction. Theoretical justification for this is provided by van de Hulst's (1957, p. 208) *localization principle*, according to which terms in the exact solution for scattering by a transparent sphere correspond to more or less localized rays.

Each incident ray splinters into an infinite number of scattered rays: externally reflected, transmitted without internal reflection, transmitted after one, two, and so on internal reflections. At any scattering angle θ, each splinter contributes to the scattered light. Accordingly, the differential scattering cross section is an infinite series with terms of the form

$$\frac{b(\theta)}{\sin\theta}\frac{db}{d\theta}. \tag{35}$$

The *impact parameter* b is $a \sin\Theta_i$, where Θ_i is the angle between an incident ray and the normal to the sphere. Each term in the series corresponds to one of the splinters of an incident ray. A *rainbow angle* is a singularity (or *caustic*) of the differential scattering cross section at which the conditions

$$\frac{d\theta}{db} = 0, \qquad \frac{b}{\sin\theta} \neq 0 \tag{36}$$

are satisfied. Missing from Eq. (35) are various reflection and transmission coefficients (Fresnel coefficients), which display no singularities and hence do not determine rainbow angles.

A rainbow is not associated with rays externally reflected or transmitted without internal reflection. The succession of rainbow angles associated with one, two, three . . . internal reflections are called primary, secondary, tertiary . . . rainbows. Aristotle recognized that "Three or more rainbows are never seen, because even the second is dimmer than the first, and so the third reflection is altogether too feeble to reach the sun (Aristotle's view was that light streams outward from the eye)". Although he intuitively grasped that each successive ray is associated with ever-diminishing energy, his statement about the nonexistence of tertiary rainbows in nature is not quite true. Although reliable reports of such rainbows are rare (unreliable reports are as common as dirt), at least one observer who can be believed has seen one (Pledgley, 1986).

An incident ray undergoes a total angular deviation as a consequence of transmission into the drop, one or more internal reflections, and transmission out of the drop. Rainbow angles are angles of minimum deviation.

For a rainbow of any order to exist,

$$\cos\Theta_i = \sqrt{\frac{n^2 - 1}{p(p+1)}} \tag{37}$$

must lie between 0 and 1, where Θ_i is the angle of incidence of a ray that gives a rainbow after p internal reflections and n is the refractive index of the drop. A primary bow therefore requires drops with refractive index less than 2; a secondary bow requires drops with refractive index less than 3. If raindrops were composed of titanium dioxide ($n \approx 3$), a commonly used opacifier for paints, primary rainbows would be absent from the sky and we would have to be content with only secondary bows.

If we take the refractive index of water to be 1.33, the scattering angle for the primary rainbow is about 138°. This is measured from the forward direction (solar point). Measured from the antisolar point (the direction toward which one must look in order to see rainbows in nature), this scattering angle corresponds to 42°, the basis for a previous assertion that rainbows (strictly, primary rainbows) cannot be seen when the sun is above 42°. The secondary rainbow is seen at about 51° from the antisolar point. Between these two rainbows is *Alexander's dark band*, a region into which no light is scattered according to geometrical optics.

The colors of rainbows are a consequence of sufficient dispersion of the refractive index over the visible spectrum to give a spread of rainbow angles that appreciably exceeds the width of the sun. The width of the primary bow from violet to red is about 1.7°; that of the secondary bow is about 3.1°.

Because of its band of colors arcing across the sky, the rainbow has become the paragon of color, the standard against which all other colors are compared. Lee and Fraser (1990) (see also Lee, 1991), however, challenged this status of the rainbow, pointing out that even the most vivid rainbows are colorimetrically far from pure.

Rainbows are almost invariably discussed as if they occurred literally in a vacuum. But real rainbows, as opposed to the pencil-and-paper variety, are necessarily observed in an atmosphere, the molecules and particles of which scatter sunlight that adds to the light from the rainbow but subtracts from its purity of color.

Although geometrical optics yields the positions, widths, and color separation of rainbows, it yields little else. For example, geometrical optics is blind to *supernumerary bows*, a series of narrow bands sometimes seen below the primary bow. These bows are a consequence of interference, and hence fall outside the province of geometrical optics. Since supernumerary bows are an interference phenomenon, they, unlike primary and secondary bows (according to geometrical optics), depend on drop size. This poses the question of how supernumerary bows can be seen in rain showers, the drops in which are widely distributed in size. In a nice piece of detective work, Fraser (1983) answered this question.

Raindrops falling in a vacuum are spherical. Those falling in air are distorted by aerodynamic forces, not, despite the depictions of countless artists, into teardrops but rather into nearly oblate spheroids with their axes more or less vertical. Fraser argued that supernumerary bows are caused by drops with a diameter of about 0.5 mm, at which diameter the angular position of the first (and second) supernumerary bow has a minimum: interference causes the position of the supernumerary bow to increase with decreasing size whereas drop distortion causes it to increase with increasing size. Supernumerary patterns contributed by drops on either side of the minimum cancel, leaving only the contribution from drops at the minimum. This cancellation occurs only near the tops of rainbows, where supernumerary bows are seen. In the vertical parts of a rainbow, a horizontal slice through a distorted drop is more or less circular, and hence these drops do not exhibit a minimum supernumerary angle.

According to geometrical optics, all spherical drops, regardless of size, yield the same rainbow. But it is not necessary for a drop to be spherical for it to yield rainbows independent of its size. This merely requires that the plane defined by the incident and scattered rays intersect the drop in a circle. Even distorted drops satisfy this condition in the vertical part of a bow. As a consequence, the absence of supernumerary bows there is compensated for by more vivid colors of the primary and secondary bows (Fraser, 1972). Smaller drops are more likely to be spherical, but the smaller a drop, the less light it scatters. Thus, the dominant contribution to the luminance of rainbows is from the larger drops. At the top of a bow, the plane defined by the incident and scattered rays intersects the large, distorted drops in an ellipse, yielding a range of rainbow angles varying with the amount of distortion, and hence a pastel rainbow. To the knowledgeable observer, rainbows are no more uniform in color and brightness than is the sky.

Although geometrical optics predicts that all rainbows are equal (neglecting background light), real rainbows do not slavishly follow the dictates of this approximate theory. Rainbows in nature range from nearly colorless fog bows (or cloud bows) to the vividly colorful vertical portions of rainbows likely to have inspired myths about pots of gold.

6.3 The Glory

Continuing our sweep of scattering directions, from forward to backward, we arrive at the end of our journey: *the glory*. Because it is most easily seen from airplanes it sometimes is called the *pilot's bow*. Another name is *anticorona*, which signals that it is a corona around the antisolar point. Although glories and coronas share some common characteristics, there are differences between them other than direction of observation. Unlike coronas, which may be caused by nonspherical ice crystals, glories require spherical cloud droplets. And a greater number of colored rings may be seen in glories than in coronas because the decrease in luminance away from the backward direction is not as steep as that away from the forward direction. To see a glory from an airplane, look for colored rings around its shadow cast on clouds below. This shadow is not an essential part of the glory, it merely directs you to the antisolar point.

Like the rainbow, the glory may be looked upon as a singularity in the differential scattering cross section Eq. (35). Equation (36) gives one set of conditions for a singularity; the second set is

$$\sin\theta = 0, \qquad b(\theta) \neq 0. \tag{38}$$

That is, the differential scattering cross section is infinite for nonzero impact parameters (corresponding to incident rays that do not intersect the center of the sphere) that give forward (0°) or backward (180°) scattering. The forward direction is excluded because this is the direction of intense scattering accounted for by the Fraunhofer theory.

For one internal reflection, Eq. (38) leads to the condition

$$\sin\Theta_i = \frac{n}{2}\sqrt{4 - n^2}, \tag{39}$$

which is satisfied only for refractive indices between 1.414 and 2, the lower refractive index corresponding to a grazing-incidence ray. The refractive index of water lies outside this range. Although a condition similar to Eq. (39) is satisfied for rays undergoing four or more internal reflections, insufficient energy is associated with such rays. Thus, it seems that we have reached an impasse: the theoretical condition for a glory cannot be met by water droplets. Not so, says van de Hulst (1947) in a seminal paper. He argues that 1.414 is close enough to 1.33 given that geometrical optics is, after all, an approximation. Cloud droplets are large compared with the wavelength, but not so large that geometrical optics is an infallible guide to their optical behavior. Support for the van de Hulstian interpretation of glories was provided by Bryant and Cox (1966), who showed that the dominant contribution to the glory is from the last terms in the exact series for scattering by a sphere. Each successive term in this series is associated with ever larger impact parameters. Thus, the terms that give the glory are indeed those corresponding to grazing rays. Further unraveling of the glory and vindication of van de Hulst's conjectures about the glory were provided by Nussenzveig (1979).

It sometimes is asserted that geometrical optics is incapable of treating the glory. Yet the same can be said for the rainbow. Geometrical optics explains rainbows only in the sense that it predicts singularities for scattering in certain directions (rainbow angles). But it can predict only the angles of intense scattering, not the amount. Indeed, the error is infinite. Geometrical optics also predicts a singularity in the backward direction. Again, this simple theory is powerless to predict more. Results from geometrical optics for both rainbows and glories are not the end but rather the beginning, an invitation to take a closer look with more powerful magnifying glasses.

7. SCATTERING BY SINGLE ICE CRYSTALS

Scattering by spherical water drops in the atmosphere gives rise to three distinct displays in the sky: coronas, rainbows, and glories. Ice particles (crystals) also can inhabit the atmosphere, and they introduce two new variables in addition to size: shape and orientation, the second a consequence of the first. Given this increase in the number of degrees of freedom, it is hardly cause for wonder that ice crystals are the source of a greater variety of displays than are water drops. As with rainbows, the gross features of ice-crystal phenomena can be described simply with geometrical optics, various phenomena arising from the various fates of rays incident on crystals. Colorless displays (e.g., sun pillars) are generally associated with reflected rays, colored displays (e.g., sun dogs and halos) with refracted rays. Because of the wealth of ice-crystal displays, it is not possible to treat all of them here, but one example should point the way toward understanding many of them.

7.1 Sun Dogs and Halos

Because of the hexagonal crystalline structure of ice it can form as hexagonal plates in the atmosphere. The stable position of a plate falling in air is with the normal to its face more or less vertical, which is easy to demonstrate with an ordinary business card. When the card is dropped with its edge facing downward (the supposedly aerodynamic position that many people instinctively choose), the card somersaults in a helter-skelter path to the ground. But when the card is dropped with its face parallel to the ground, it rocks back and forth gently in descent.

A hexagonal ice plate falling through air and illuminated by a low sun is like a 60° prism illuminated normally to its sides (Fig. 17). Because there is no mechanism for orienting a plate within the horizontal plane, all plate orientations in this plane are equally probable. Stated another way, all angles of incidence for a fixed plate are equally probable. Yet all scattering angles (deviation angles) of rays refracted into and out of the plate are not equally probable.

Figure 18 shows the range of scattering angles corresponding to a range of rays incident on a 60° ice prism that is part of a hex-

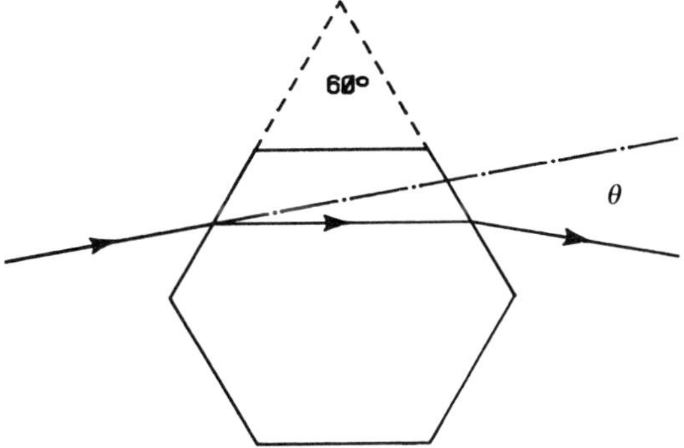

FIG. 17. Scattering by a hexagonal ice plate illuminated by light parallel to its basal plane. The particular scattering angle θ shown is an angle of minimum deviation. The scattered light is that associated with two refractions by the plate.

agonal plate. For angles of incidence less than about 13°, the transmitted ray is totally internally reflected in the prism. For angles of incidence greater than about 70°, the transmittance plunges. Thus, the only rays of consequence are those incident between about 13° and 70°.

All scattering angles are not equally probable. The (uniform) probability distribution $p(\theta_i)$ of incidence angles θ_i is related to the probability distribution $P(\theta)$ of scattering angles θ by

$$P(\theta) = \frac{p(\theta_i)}{d\theta/d\theta_i}. \qquad (40)$$

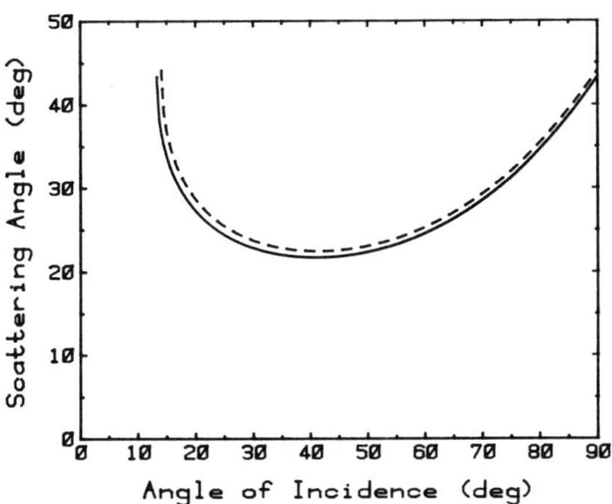

FIG. 18. Scattering by a hexagonal ice plate (see Fig. 17) in various orientations (angles of incidence). The solid curve is for red light, the dashed for blue light.

At the incidence angle for which $d\theta/d\theta_i = 0$, $P(\theta)$ is infinite and scattered rays are intensely concentrated near the corresponding angle of minimum deviation.

The physical manifestation of this singularity (or caustic) at the angle of minimum deviation for a 60° hexagonal ice plate is a bright spot about 22° from either or both sides of a sun low in the sky. These bright spots are called *sun dogs* (because they accompany the sun) or *parhelia* or *mock suns*.

The angle of minimum deviation θ_m, hence the angular position of sun dogs, depends on the prism angle Δ (60° for the plates considered) and refractive index:

$$\theta_m = 2\sin^{-1}\left(n \sin\frac{\Delta}{2}\right) - \Delta. \qquad (41)$$

Because ice is dispersive, the separation between the angles of minimum deviation for red and blue light is about 0.7° (Fig. 18), somewhat greater than the angular width of the sun. As a consequence, sun dogs may be tinged with color, most noticeably toward the sun. Because the refractive index of ice is least at the red end of the spectrum, the red component of a sun dog is closest to the sun. Moreover, light of any two wavelengths has the same scattering angle for different angles of incidence if one of the wavelengths does not correspond to red. Thus, red is the purest color seen in a sun dog. Away from its red inner edge a sun dog fades into whiteness.

With increasing solar elevation, sun dogs move away from the sun. A falling ice plate is roughly equivalent to a prism, the prism angle of which increases with solar elevation. From Eq. (41) it follows that the angle of minimum deviation, hence the sun dog position, also increases.

At this point you may be wondering why only the 60° prism portion of a hexagonal plate was singled out for attention. As evident from Fig. 17, a hexagonal plate could be considered to be made up of 120° prisms. For a ray to be refracted twice, its angle of incidence at the second interface must be less than the critical angle. This imposes limitations on the prism angle. For a refractive index 1.31, all incident rays are totally internally reflected by prisms with angles greater than about 99.5°.

A close relative of the sun dog is the 22° halo, a ring of light approximately 22° from the sun (Fig. 19). Lunar halos are also possible and are observed frequently (although less frequently than solar halos); even moon dogs are possible. Until Fraser (1979) analyzed halos in detail, the conventional wisdom had been that they obviously were the result of randomly oriented crystals, yet another example of jumping to conclusions. By combining optics and aerodynamics, Fraser showed that if ice crystals are small enough to be randomly oriented by Brownian motion, they are too small to yield sharp scattering patterns.

But completely randomly oriented plates are not necessary to give halos, especially ones of nonuniform brightness. Each part of a halo is contributed to by plates with a different tip angle (angle between the normal to the plate and the vertical). The transition from oriented plates (zero tip angle) to randomly oriented plates occurs over a narrow range of sizes. In the transition region, plates can be small enough to be partially oriented yet large enough to give a distinct contribution to the halo. Moreover, the mapping between tip angles and azimuthal angles on the halo depends on solar elevation. When the sun is near the horizon, plates can give a distinct halo over much of its azimuth.

FIG. 19. A 22° solar halo. The hand is not for artistic effect but rather to occlude the bright sun.

When the sun is high in the sky, hexagonal plates cannot give a sharp halo but hexagonal columns—another possible form of atmospheric ice particles—can. The stable position of a falling column is with its long axis horizontal. When the sun is directly overhead, such columns can give a uniform halo even if they all lie in the horizontal plane. When the sun is not overhead but well above the horizon, columns also can give halos.

A corollary of Fraser's analysis is that halos are caused by crystals with a range of sizes between about 12 and 40 μm. Larger crystals are oriented; smaller particles are too small to yield distinct scattering patterns.

More or less uniformly bright halos with the sun neither high nor low in the sky could be caused by mixtures of hexagonal plates and columns or by clusters of bullets (rosettes). Fraser opines that the latter is more likely.

One of the by-products of his analysis is an understanding of the relative rarity of the 46° halo. As we have seen, the angle of minimum deviation depends on the prism angle. Light can be incident on a hexagonal column such that the prism angle is 60° for rays incident on its side or 90° for rays incident on its end. For $n = 1.31$, Eq. (41) yields a minimum deviation angle of about 46° for $\Delta = 90°$. Yet, although 46° halos are possible, they are seen much less frequently than 22° halos. Plates cannot give distinct 46° halos although columns can. Yet they must be solid and most columns have hollow ends. Moreover, the range of sun elevations is restricted.

Like the green flash, ice-crystal phenomena are not intrinsically rare. Halos and sun dogs can be seen frequently—once you know what to look for. Neuberger (1951) reports that halos were observed in State College, Pennsylvania, an average of 74 days a year over a 16-yr period, with extremes of 29 and 152 halos a year. Although the 22° halo was by far the most frequently seen display, ice-crystal displays of all kinds were seen, on average, more often than once every four days at a location not especially blessed with clear skies. Although thin clouds are necessary for ice-crystal displays, clouds thick enough to obscure the sun are their bane.

8. CLOUDS

Although scattering by isolated particles can be studied in the laboratory, particles in the atmosphere occur in crowds (sometimes called clouds). Implicit in the previous two sections is the assumption that each particle is illuminated solely by incident sunlight; the particles do not illuminate each other to an appreciable degree. That is, clouds of water droplets or ice grains were assumed to be optically thin, and hence multiple scattering was negligible. Yet the term cloud evokes fluffy white objects in the sky, or perhaps an overcast sky on a gloomy day. For such clouds, multiple scattering is not negligible, it is the major determinant of their appearance. And the quantity that determines the degree of multiple scattering is optical thickness (see Sec. 1.4).

8.1 Cloud Optical Thickness

Despite their sometimes solid appearance, clouds are so flimsy as to be almost nonexistent—except optically. The fraction of the total cloud volume occupied by water substance (liquid or solid) is about 10^{-6} or less. Yet although the mass density of clouds is that of air to within a small fraction of a percent, their optical thickness (per unit physical thickness) is much greater. The number density of air molecules is vastly greater than that of water droplets in clouds, but scattering per molecule of a cloud droplet is also much greater than scattering per air molecule (see Fig. 7).

Because a typical cloud droplet is much larger than the wavelengths of visible light, its scattering cross section is to good approximation proportional to the square of its diameter. As a consequence, the scattering coefficient [see Eq. (2)] of a cloud having a volume fraction f of droplets is approximately

$$\beta = 3f\frac{\langle d^2\rangle}{\langle d^3\rangle}, \tag{42}$$

where the brackets indicate an average over the distribution of droplet diameters d. Unlike molecules, cloud droplets are distributed in size. Although cloud particles can be ice particles as well as water droplets, none of the results in this and the following section hinge on the assumption of spherical particles.

The optical thickness along a cloud path of physical thickness h is βh for a cloud with uniform properties. The ratio $\langle d^3\rangle/\langle d^2\rangle$ defines a mean droplet diameter, a typical value for which is 10 μm. For this diameter and $f = 10^{-6}$, the optical thickness per unit meter of physical thickness is about the same as the normal optical thickness of the atmosphere in the middle of the visible spectrum (see Fig. 3). Thus, a cloud only 1 m thick is equivalent optically to the entire gaseous atmosphere.

A cloud with (normal) optical thickness about 10 (i.e., a physical thickness of about 100 m) is sufficient to obscure the disk of the sun. But even the thickest cloud does not transform day into night. Clouds are usually translucent, not transparent, yet not completely opaque.

The scattering coefficient of cloud droplets, in contrast with that of air molecules, is more or less independent of wavelength. This is often invoked as the cause of the colorlessness of clouds. Yet wavelength independence of scattering by a single particle is only sufficient, not necessary, for wavelength independence of scattering by a cloud of particles (see Sec. 1.4). Any cloud that is optically thick and composed of particles for which absorption is negligible is white upon illumination by white light. Although absorption by water (liquid and solid) is not identically zero at visible wavelengths, and selective absorption by water can lead to observable consequences (e.g., colors of the sea and glaciers), the appearance of all but the thickest clouds is not determined by this selective absorption.

Equation (42) is the key to the vastly different optical characteristics of clouds and of the rain for which they are the progenitors. For a fixed amount of water (as specified by the quantity fh), optical thickness is inversely proportional to mean diameter. Rain drops are about 100 times larger on average than cloud droplets, and hence optical thicknesses of rain shafts are correspondingly smaller. We often can see through many kilometers of intense rain whereas a small patch of fog on a well-traveled highway can result in carnage.

8.2 Givers and Takers of Light

Scattering of visible light by a single water droplet is vastly greater in the forward ($\theta <$ 90°) hemisphere than in the backward ($\theta >$

90°) hemisphere (Fig. 9). But water droplets in a thick cloud illuminated by sunlight collectively scatter much more in the backward hemisphere (reflected light) than in the forward hemisphere (transmitted light). In each scattering event, incident photons are deviated, on average, only slightly, but in many scattering events most photons are deviated enough to escape from the upper boundary of the cloud. Here is an example in which the properties of an ensemble are different from those of its individual members.

Clouds seen by passengers in an airplane can be dazzling, but if the airplane were to descend through the cloud these same passengers might describe the cloudy sky overhead as gloomy. Clouds are both givers and takers of light. This dual role is exemplified in Fig. 20, which shows the calculated diffuse downward irradiance below clouds of varying optical thickness. On an airless planet the sky would be black in all directions (except directly toward the sun). But if the sky were to be filled from horizon to horizon with a thin cloud, the brightness overhead would markedly increase. This can be observed in a partly overcast sky, where gaps between clouds (blue sky) often are noticeably darker than their surroundings. As so often happens, more is not always better. Beyond a certain cloud optical thickness, the diffuse irradiance decreases. For a sufficiently thick cloud, the sky overhead can be darker than the clear sky.

Why are clouds bright? Why are they dark?

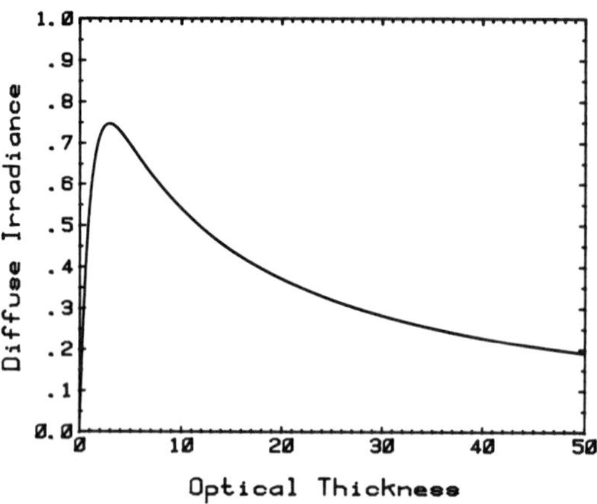

FIG. 20. Computed diffuse downward irradiance below a cloud relative to the incident solar irradiance as a function of cloud optical thickness.

No inclusive one-line answers can be given to these questions. Better to ask, Why is that particular cloud bright? Why is that particular cloud dark? Each observation must be treated individually; generalizations are risky. Moreover, we must keep in mind the difference between brightness and radiance when addressing the queries of human observers. Brightness is a sensation that is a property not only of the object observed but of its surroundings as well. If the luminance of an object is appreciably greater than that of its surroundings, we call the object bright. If the luminance is appreciably less, we call the object dark. But these are relative rather than absolute terms.

Two clouds, identical in all respects, including illumination, may still appear different because they are seen against different backgrounds, a cloud against the horizon sky appearing darker than when seen against the zenith sky.

Of two clouds under identical illumination, the smaller (optically) will be less bright. If an even larger cloud were to appear, the cloud that formerly had been described as white might be demoted to gray.

With the sun below the horizon, two identical clouds at markedly different elevations might appear quite different in brightness, the lower cloud being shadowed from direct illumination by sunlight.

A striking example of dark clouds can sometimes be seen well after the sun has set. Low-lying clouds that are not illuminated by direct sunlight but are seen against the faint twilight sky may be relatively so dark as to seem like ink blotches.

Because dark objects of our everyday lives usually owe their darkness to absorption, nonsense about dark clouds is rife: they are caused by pollution or soot. Yet of all the reasons that clouds are sometimes seen to be dark or even black, absorption is not among them.

GLOSSARY

Airlight: Light resulting from scattering by all atmospheric molecules and particles along a line of sight.

Antisolar Point: Direction opposite the sun.

Astronomical Horizon: Horizontal direction determined by a bubble level.

Brightness: The attribute of sensation by which an observer is aware of differences of luminance (definition recommended by the 1922 Optical Society of America Committee on Colorimetry).

Contrast Threshold: The minimum relative luminance difference that can be perceived by the human observer.

Inferior Mirage: A mirage in which images are displaced downward.

Irradiance: Radiant power crossing unit area in a hemisphere of directions.

Lapse Rate: The rate at which a physical property of the atmosphere (usually temperature) decreases with height.

Luminance: Radiance integrated over the visible spectrum and weighted by the spectral response of the human observer. Also sometimes called **photometric brightness**.

Mirage: An image appreciably different from what it would be in the absence of atmospheric refraction.

Neutral Point: A direction in the sky for which the light is unpolarized.

Normal Optical Thickness: Optical thickness along a radial path from the surface of the earth to infinity.

Optical Thickness: The thickness of a scattering medium measured in units of photon mean free paths. Optical thicknesses are dimensionless.

Radiance: Radiant power crossing a unit area and confined to a unit solid angle about a particular direction.

Scale Height: The vertical distance over which a physical property of the atmosphere is reduced to $1/e$ of its value.

Scattering Angle: Angle between incident and scattered waves.

Scattering Coefficient: The product of scattering cross section and number density of scatterers.

Scattering Cross Section: Effective area of a scatterer for removal of light from a beam by scattering.

Scattering Plane: Plane determined by incident and scattered waves.

Solar Point: The direction toward the sun.

Superior Mirage: A mirage in which images are displaced upward.

Tangential Optical Thickness: Optical thickness through the atmosphere along a horizon path.

Works Cited

Many of the seminal papers in atmospheric optics, including those by Lord Rayleigh, are bound together in Bohren, C. F. (Ed.) (1989), *Selected Papers on Scattering in the Atmosphere*, Bellingham, WA: SPIE Optical Engineering Press. Papers marked with an asterisk are in this collection.

*Bryant. H. C., Cox, A. J. (1966), *J. Opt. Soc. Am.* **56**, 1529–1532.

Doyle, W. T. (1985), *Am. J. Phys.* **53**, 463–468.

Einstein, A. (1910), *Ann. Phys (Leipzig)*. **33**, 175. English translation in: J. Alexander (Ed.) (1926), *Colloid Chemistry*, Vol. I. New York: The Chemical Catalog Company, p. 323.

Fraser, A. B. (1972), *J. Atmos. Sci.* **29**, 211–212.

*Fraser, A. B. (1975), *Atmosphere* **13**, 1–10.

*Fraser, A. B. (1979), *J. Opt. Soc. Am.* **69**, 1112–1118.

*Fraser, A. B. (1983), *J. Opt. Soc. Am.* **73**, 1626–1628.

*Hulburt, E. O. (1953), *J. Opt. Soc. Am.* **43**, 113–118.

Lee, R., Fraser, A. (1990), *New Scientist* **127** (1 September), 40–42.

Lee, R., (1991), *Appl. Opt.* **30**, 3401–3407.

Möller, F. (1972), "Radiation in the Atmosphere," in: D. P. McIntyre (Ed), *Meteorological Challenges: A History*, Ottawa: Information Canada, p. 43.

Neuberger, H. (1951), *Introduction to Physical Meteorology*, University Park, PA: College of Mineral Industries, Pennsylvania State University.

*Nussenzveig, H. M. (1979), *J. Opt. Soc. Am.* **69**, 1068–1079.

*Penndorf, R. (1957), *J. Opt. Soc. Am.* **47**, 176–182.

Pledgley, E. (1986), *Weather* **41**, 401.

Takano, Y., Asano, S. (1983), *J. Meteor. Soc. Jpn.* **61**, 289–300.

Thekaekara, M. P., Drummond, A. J. (1971), *Nature Phys. Sci.* **229**, 6–9.

*van de Hulst, H. C. (1947), *J. Opt. Soc. Am.* **37**, 16–22.

van de Hulst, H. C. (1957), *Light Scattering by Small Particles*, New York: Wiley-Interscience.

von Frisch, K. (1971), *Bees: Their Vision, Chemical Senses, and Language*, rev. ed., Ithaca, NY: Cornell University Press, p. 116.

*Young, A. T. (1982), *Phys. Today* **35** (1), 2–8.

Zimm, B. H. (1945), *J. Chem. Phys.* **13**, 141–145.

Further Reading

Minnaert, M. (1954), *The Nature of Light and Colour in the Open Air*, New York: Dover, is the bible for those interested in atmospheric optics. Like accounts of natural phenomena in the *Bible*, those in Minnaert's book are not always correct, despite which, again like the *Bible*, it has been and will continue to be a source of inspiration.

A history of light scattering, "From Leonardo to the Graser: Light Scattering in Historical Perspective," was published serially by Hey, J. D. (1983), *S. Afr. J. Sci.* **79**, 11–27, 310–324; (1985), *op. cit.* **81**, 77–91, 601–613; (1986), *op. cit.* **82**, 356–360. The history of the rainbow is recounted by Boyer, C. B. (1987), *The Rainbow*, Princeton, NJ: Princeton University Press.

Special issues of *Journal of the Optical Society of America* (August 1979 and December 1983) and *Applied Optics* (20 August 1991 and 20 July 1994) are devoted to atmospheric optics.

Several monographs on light scattering by particles are relevant to and contain examples drawn from atmospheric optics: van de Hulst, H. C. (1957), *Light Scattering by Small Particles*, New York: Wiley-Interscience [reprint (1981), New York: Dover]; Deirmendjian, D. (1969), *Electromagnetic Scattering on Polydispersions*, New York: Elsevier; Kerker, M. (1969), *The Scattering of Light and Other Electromagnetic Radiation*, New York: Academic; Bohren, C. F., Huffman, D. R. (1983), *Light Scattering by Small Particles*, New York: Wiley-Interscience; Nussenzveig, H. M. (1992), *Diffraction Effects in Semiclassical Scattering*, Cambridge, U.K.: Cambridge University Press.

The following books are devoted to a wide range of topics in atmospheric optics: Tricker, R. A. R. (1970), *Introduction to Meteorological Optics*, New York: Elsevier; McCartney, E. J. (1976), *Optics of the Atmosphere*, New York: Wiley; Greenler, R. (1980), *Rainbows, Halos, and Glories*, Cambridge, U.K.: Cambridge University Press. Monographs of more limited scope are those by Middleton, W. E. K. (1952), *Vision Through the Atmosphere*, Toronto: University of Toronto Press; O'Connell, D. J. K. (1958), *The Green Flash and Other Low Sun Phenomena*, Amsterdam: North Holland; Rozenberg, G. V. (1966), *Twilight: A Study in Atmospheric Optics*, New York: Plenum; Henderson, S. T. (1977), *Daylight and its Spectrum*, 2nd ed., New York: Wiley; Tricker, R. A. R. (1979), *Ice Crystal Haloes*, Washington, D.C.: Optical Society of America; Können, G. P. (1985), *Polarized Light in Nature*, Cambridge, U.K.: Cambridge University Press; Tape, W. (1994), *Atmosphere Halos*, Washington, DC: American Geophysical Union.

Although not devoted exclusively to atmospheric optics, Humphreys, W. J. (1964), *Physics of the Air*, New York: Dover, contains a few relevant chapters. Two popular science books on simple experiments in atmospheric physics are heavily weighted toward atmospheric optics: Bohren, C. F. (1987), *Clouds in a Glass of Beer*, New York: Wiley; Bohren, C. F. (1991), *What Light Through Yonder Window Breaks?*, New York: Wiley.

For an expository article on colors of the sky see Bohren, C. F., Fraser, A. B. (1985), *Phys. Teacher* **23**, 267–272.

An elementary treatment of the coherence properties of light waves was given by Forrester, A. T. (1956), *Am. J. Phys.* **24**, 192–196. This journal also published an expository article on the observable consequences of multiple scattering of light: Bohren, C. F. (1987), *Am. J. Phys.* **55**, 524–533.

Although a book devoted exclusively to atmospheric refraction has yet to be published, an elementary yet thorough treatment of mirages was given by Fraser, A. B., Mach, W. H. (1976), *Sci. Am.* **234** (1), 102–111.

Colorimetry, the often (and unjustly) neglected component of atmospheric optics, is treated in, for example, Optical Society of America Committee on Colorimetry (1963), *The Science of Color*, Washington, D. C.: Optical Society of America; Billmeyer, F. W., Saltzman, M. (1981), *Principles of Color Technology*, 2nd ed., New York: Wiley-Interscience; MacAdam, D. L. (1985), *Color Measurement*, 2nd ed., Berlin: Springer.

Understanding atmospheric optical phenomena is not possible without acquiring at least some knowledge of the properties of the particles responsible for them. To this end, the following are recommended: Pruppacher, H. R., Klett, J. D. (1980), *Microphysics of Clouds and Precipitation*, Dordrecht, Holland: Reidel; Twomey, S. A. (1977), *Atmospheric Aerosols*, New York: Elsevier.

OPTICS, GEOMETRICAL

ROLAND SHACK, *Optical Sciences Center, University of Arizona, Tucson, Arizona, U.S.A.*

INTRODUCTION

Optics is blessed with several levels of theory. At the top are quantum optics and electromagnetic theory, which account respectively for the microscopic and the macroscopic interaction of light and matter. This is the domain of optical physics, and all optical phenomena are encompassed by them.

However, in applications such as electromagnetic diffraction theory, exact solutions to problems are extremely difficult. To deal

3-527-28134-7/95/$5.00 + .50

practically with such problems there is a simpler approximate theory, scalar wave theory, which provides an adequately accurate result in most applications involving interference and diffraction. It does not, however, account for polarization nor does it predict the quantitative interactions with matter. This is the domain of physical optics.

At a still lower level of approximation, where not even interference and diffraction are accounted for, is geometrical optics. Geometrical optics is the lowest level of optical theory available today. It does not compete with physical optics in explaining optical phenomena, and can only provide a crude approximation to the detailed behavior of light. Why is it still in use?

Its overwhelming virtue is in its simplicity. It can provide useful, if somewhat inexact, answers to many practical problems, where the complexity of a more accurate theory would be prohibitive. It has been applied successfully to problems where one would not expect useful results, e.g., in the analysis of waveguides.

Its primary area of application, however, is in optical design. It has proven itself to be indispensable for the efficient development of optical systems. No useful alternative has yet been developed.

Until the development of the computer and its application to optical design, virtually no consideration was given to the physical nature of light. However, in recent years the application of the wave theory of geometrical optics has made it possible to convert from geometrical optics to physical optics in the final image space of an optical system, and to calculate the structure of the image taking diffraction into account. The hard work of tracing the wave front through the optical system is done geometrically.

In this article, we deal with the basic concepts of both the corpuscular theory and the wave theory of geometrical optics. Applications as an approximation to physical optics and to radiometry are touched on, but the major part of this article deals with the geometrical concepts relevant to an understanding of image-forming optics.

1. BASIC CONCEPTS

Historically, there have been two different approaches to the theory of geometrical optics. The oldest and most prevalent one is the corpuscular theory in which it is assumed that light consists of particles that transport energy from a source to a receiver. The other is a wave theory, which goes as far back as Christian Huygens, and which is somewhat subtly different from physical wave optics.

1.1 Corpuscular Theory

1.1.1 Rays as Trajectories of Energetic Particles In the corpuscular theory, light is assumed to consist of energetic particles. Particles of different colors have different energies, the blue being more energetic than the red. In this respect, they can be thought of as simplified versions of physical photons.

These particles are emitted from a source and travel along trajectories to a receiver. The trajectories are identified as ray paths, or simply rays. In classical geometrical optics, the study of rays and their properties is paramount.

One important characteristic of geometrical particles of light is that they never collide or interfere with each other regardless of the intensity of the light beam. Intense beams can intersect each other with no interaction.

1.1.2 Refractive Index, Color, and Dispersion In a vacuum, all light particles travel at the well-known speed of light. However, in transparent media they travel more slowly, and particles of different colors travel at different speeds. The ratio of the speed of light in a vacuum to the speed of light in a given medium is known as the refractive index of the medium. Blue light travels more slowly in the medium than red light, and so its refractive index is higher. The variation of refractive index with color is what is known as dispersion.

The designation of color by name results in a rather imprecise determination of dispersion, as indeed it was before Fraunhofer's discovery of spectrum lines. Today the wavelengths of the light producing the spectrum lines are used for a precise designation of color.

1.1.3 Fermat's Principle and Optical Paths A fundamental property of ray paths is described by Fermat's principle. Given two points along a ray, the time it takes for light to go from one point to the other is given by

$$t_A - t_B = \int_{t_A}^{t_B} \mathrm{d}t = \frac{1}{c} \int_A^B n ds, \qquad (1)$$

which is stationary, usually a minimum, along all possible neighboring paths. If the refractive index of the medium is constant along the path, the path will be a straight line. If the refractive index is not constant, the path will vary in such a way that Fermat's principle is satisfied.

The integral of the product of refractive index and geometrical path length is called the optical path length, and is the distance that light would travel in a vacuum in the same time interval. Because the speed of light in a vacuum is constant, Fermat's principle can be expressed in terms of optical path as well as transit time.

1.1.4 Snell's Law Most optical systems consist of reasonably homogeneous media with constant refractive indices separated by abrupt discontinuities, e.g., lenses in air. Given a boundary separating two dissimilar media, a ray will change its direction in going through the boundary unless it is normal to the boundary. This change in direction is governed by Snell's law,

$$n' \sin i' = n \sin i, \qquad (2)$$

where the primed quantities are in the emergent space, i is the angle between the incident ray and the normal to the boundary at the intersection point, and i' is the angle between the emergent ray and the normal.

Snell's law is the basic tool in designing and analyzing optical systems. It is simple and powerful, and can be derived from Fermat's principle.

1.2 Wave Theory

1.2.1 The Point Source The basic elementary source in geometrical optics is the point source. It is an infinitesimal region of space that emits light. Associated with every point source is a family of rays that trace the trajectories of the light particles emitted. Different point sources and their ray families behave independently of each other.

1.2.2 Geometrical Wave Fronts Associated with each point source and its family of rays is a family of surfaces. These surfaces are surfaces of constant optical path from the source measured along the rays, and they are called geometrical wave fronts. Although there is no characteristic wavelength associated with these surfaces, they are called wave fronts because they are generally good approximations to physical wave fronts except in the neighborhood of geometrical shadow boundaries. Christian Huygens was the first to use the idea of geometrical wave fronts in discussing the properties of light.

1.2.3 Connection with Physical Optics As stated above, the precise specification of a color in geometrical optics is in terms of the wavelength of spectral lines. This wavelength is of course a physical quantity that in itself has no other significance in geometrical optics. However, it is a unit of length, and geometrical optical path lengths can be measured in wavelength units. This combined with the geometrical wave front gives us a geometrical model of a physical wave field.

1.2.4 Significance in Imaging Systems Traditionally, optical image-forming systems have been designed and evaluated using ray optics almost exclusively. Before the advent of computers, the stupendous amount of work required to do it any other way was out of the question. However, today it is possible to trace geometrical wave fronts through an optical system (admittedly using rays to do it) to the final image space where one can convert to physical optics in order to account for diffraction in the image-forming process.

2. GENERAL APPLICATIONS

There are a number of areas where geometrical optics is commonly applied. What they have in common is that the geometrical optical model is much easier to deal with than a more accurate higher-level model, and that, for these applications, the assumptions and approximations are adequate for the purpose. Two application areas are discussed briefly in this section, and the remainder of the article deals with the most important area, that of image-forming optics.

2.1 Physical Optics

2.1.1 Wave-Front Propagation The principal general area of physical optics where geometrical optical tools are commonly employed is in the analysis and evaluation of instruments in which wave-front transmission is the principal phenomenon, such as spectrometers and interferometers. Geometrical optics has even been applied successfully to waveguide analysis.

2.1.2 The Use of *Ad Hoc* Properties A number of physical wave-field properties do not directly occur in the geometrical optical model. Only by assuming them as *ad hoc* properties can they be accounted for. One such property already mentioned is the use of wavelength as the unit for measuring optical paths. Other properties are wave amplitude, reflectance, transmittance, and polarization. When used with care, *ad hoc* properties are a significant enhancement to the geometrical optical model.

2.1.3 Diffracted Rays One interesting special application is in extending the geometrical model to include diffraction, at least by sharp-edged apertures. It is based on the Rabinowicz model of diffraction in which he decomposes a diffracted field into the coherent superposition of a standard geometrical field truncated by the aperture and an induced field that appears as that produced by a coherent line source on the edge of the aperture. In the geometrical model, additional "diffracted" rays are created on the boundary.

2.2 Radiometry

In most cases, a geometrical optical model is adequate for dealing with the radiometry of optical systems. Although point sources are certainly included in radiometric phenomena, extended sources generally play a greater role. In the geometrical model, an extended source is simply a dense array of independent point sources. The independence of the point sources means that in radiometry, extended sources are assumed to be noncoherent.

3. IMAGING SYSTEMS: FIRST ORDER

The principal application of geometrical optics is in the area of image-forming systems. Principles and concepts specific to imaging systems are fairly extensive, and the rest of this article deals with these.

No optical system forms images perfectly, and departures from ideal behavior are called aberrations. However, to deal with aberrations we must first define what constitutes ideal behavior. Although there are subtle differences between the terms, ideal behavior is variously described as first order, collinear, paraxial, and Gaussian. Each of these will be dealt with in further detail in the following conceptual development.

3.1 Objects and Images: Optical Spaces

In every optical image-forming system, there is at least one object space and at least one image space. In compound imaging systems, each element takes the image space of the preceding element as its object space and reimages that into its own image space, which in turn is the object space of the next element. At its most basic level, the elements of a compound system are the optical surfaces. A single lens is a compound of two surfaces and has a total of three spaces. A three-lens system has a total of seven spaces.

These spaces are called optical spaces and are of infinite extent. They are not bounded by the optical surfaces. That part of the optical space that lies between adjacent surfaces is called the real part. The rest of the optical space is its virtual part. In a conventional drawing of an optical system showing ray paths from surface to surface, only the real parts of each space are shown.

3.2 Ideal Behavior: Collinear Mapping

An object consists of a three-dimensional array of point sources. Its image consists ideally of a corresponding array of point images. The ideal image is related to the object through a mapping process.

The most popular mapping to represent ideal behavior is a collinear mapping. The basic rule is that, in addition to there being a one-to-one correspondence between object and image points, there is also a one-to-one correspondence between straight lines in the two spaces. As a consequence, there is in addition a one-to-one correspondence between object and image planes. Corresponding elements are called conjugate elements.

3.3 Axially Symmetric Systems

In the vast majority of optical systems, every surface and aperture in the system is intended to be rotationally symmetric about a single common mechanical axis. The basic concepts in the geometric theory of image-forming systems were developed in the context of such systems, and most of the following discussion assumes axial symmetry.

3.3.1 Focal Systems and Gaussian Imagery Each space in an axially symmetric system has its own axis of symmetry. They all coincide with the mechanical axis of symmetry, but each space is distinct in all its properties.

Each point on the object axis has one and only one point on the image axis corresponding to it. If an object point at infinity on the axis has a conjugate at a finite location on the image axis, the conjugate image point is the rear focal point of the system. There is a corresponding front focal point that has as its conjugate a point at infinity on the image axis. Such a system is called a focal system.

3.3.1.1 Cardinal Points. The two focal points of a focal system are two of the six cardinal points defined by Gauss. The other four are defined as follows.

A point off axis in the object space will have a conjugate image point off axis. The distance of such a point from the axis is called the height of the point, and the ratio of the image height to the object height is called the magnification of the pair.

An object plane perpendicular to the axis and containing the given point has as its conjugate an image plane, also perpendicular to the image axis, that contains the given image point. All conjugate pairs of points in the two planes have the same magnification. Every pair of conjugate planes has a unique magnification connecting them. No two pairs have the same magnification. The pair of planes having a unit positive magnification are called the principal planes of the system. Their axial points are called the principal points. These are also two of the cardinal points.

The distance from the principal point to the focal point in image space is called the rear focal length of the system, and the corresponding distance in object space is called the front focal length.

The last pair of cardinal points are the nodal points. These are conjugate axial points located so that a line passing through the front nodal point making some angle with the axis will have as its conjugate a line passing through the rear nodal point making the same angle with the axis. These make it possible to determine the image size in the rear focal plane of an object at infinity but with a finite angular extent.

The cardinal points of a system are used to establish the constants used in the collinear mapping of object space into image space. The term Gaussian imagery is used to describe this application of the cardinal points, but the underlying model is collinear.

3.3.1.2 Gaussian Imagery. It is important in developing the mapping equations to be consistent in the assumed sign convention. We assume a standard Cartesian coordinate system in each space with the standard Cartesian sign conventions. We also identify the axis with the optical axis in each of the spaces.

The usual mapping equations take the origin of the coordinate system in each space to be located at the principal point. The distances to a pair of conjugate planes are given by

$$n'/z' = n/z + \phi, \tag{3}$$

where

$$\phi = n'/f' = -n/f = 1/f_e \tag{4}$$

is the power of the system, f and f' are the front and rear focal lengths respectively, and f_e, the reciprocal of the power, is the effective focal length. The magnification is given by

$$m = \frac{z'/n'}{z/n}, \tag{5}$$

and the transverse coordinates by

$$x' = mx, \quad y' = my. \tag{6}$$

3.3.1.3 Gaussian Reduction. It is a relatively simple thing to determine the Gaussian properties (the power and the location of the principal points) of a compound system if the Gaussian properties of the components are known. The reduction is carried out two components at a time. The most elementary component is a single refracting surface, and the Gaussian properties of a single lens can be obtained by this method. Once one has the Gaussian properties of the individual lenses in a lens system, the properties of the system as a whole can be obtained by continuing the process.

If we have two components, where the power of the first is ϕ_1 and the power of the second is ϕ_2, there are three spaces involved. The object space for the first component is the object space for the system, and the image space of the second component is the image space for the system. Let the refractive index of the object space be n_o and the image space n'_o.

The front principal point P_1 of the first component is in the object space, as is the front principal point P of the system. The rear principal point P'_2 of the second component is in the image space, as is the rear principal point P' of the system.

The space between the two components is common to both. It is the image space of the first component and the object space of the second component, and contains respectively the rear principal point P'_1 of the first component and the front principal point P_2 of the second component. Let the distance from P'_1 to P_2 be t and the refractive index of the space be n. Then the power of the system is given by

$$\phi = \phi_1 + \phi_2 - \phi_1\phi_2 t/n, \tag{7}$$

and the principal points of the system are located by

$$\frac{\overline{P_1P}}{n_o} = \left(\frac{\phi_2}{\phi}\right)\frac{t}{n}, \quad \frac{\overline{P'_2P'}}{n'_o} = -\left(\frac{\phi_1}{\phi}\right)\frac{t}{n}, \tag{8}$$

where $\overline{P_1P}$ is the distance in object space from the front principal point of the first component to the front principal point of the system, and $\overline{P'_2P'}$ is the distance in image space from the rear principal point of the second component to the rear principal point of the system. The resultant system can in turn be a component in a larger system.

3.3.2 Afocal Systems It is also possible for a system to have an image point at infinity corresponding to an object point at infinity. Such a system is called an afocal system.

Afocal systems have no cardinal points. Instead, they have a constant characteristic magnification; that is, all pairs of conjugate planes have the same magnification. An arbitrarily selected pair of conjugate planes can be used to establish the origins of the coordinate systems. The mapping equations are then

$$z'/n = m^2z/n, \quad x' = mx, \quad y' = my. \tag{9}$$

3.3.3 Ray Tracing and Paraxial Optics The general procedure for tracing a ray through an optical system is sequential. An incident ray at an optical surface is defined by its location on the surface and its direction of propagation. By applying Snell's law the ray is refracted, determining the direction of the emergent ray. The ray is transferred to the next surface, for which it is the incident ray, and the process is repeated. The operations are fairly complex, and very tedious if done by hand.

However, in a region sufficiently close to the axis that angles, sines, and tangents cannot be distinguished from each other, the ray tracing is considerably simplified. Both the refraction and the transfer operations reduce to simple linear equations. This paraxial behavior of the optical system is congruent with the paraxial behavior of the collinear model of the optical system, and because the ray-tracing equations are linear, they can be applied to the extended collinear model of the optical system without restriction to its paraxial region.

If the real ray-tracing equations had been represented by a power-series expansion, the first-order terms would have to be congruent with the paraxial behavior. This is where the idea of first-order optics comes from.

In the collinear model that represents the ideal optical system, every refracting surface is represented by its principal planes, which are coincident at the surface. In the ray tracing, refraction occurs at the principal planes and transfer occurs between adjacent principal planes. The rays should properly be called collinear rays, but they are commonly called paraxial rays, even though they are not restricted to the paraxial region.

In the refraction process, a ray incident on a surface is determined by its height at the surface and the angle that it makes with the optical axis. The emergent ray is determined by the refraction equation

$$n'u' = nu - y\phi, \tag{10}$$

where $\phi = (n'n)c$ is the power of the surface and c is the curvature of the surface. The transfer to the next surface is governed by

$$y' = y + n'u'(t'/n'), \tag{11}$$

where t' is the axial distance from the first surface to the second. This ray is now incident on the second surface, and the process can be repeated until the ray trace is complete.

The cardinal points of the complete system can be determined by tracing just two rays without concern for the cardinal points of the components (see Fig. 1). In general, Gaussian reduction is more efficient if there are no more than three or four surfaces. For larger numbers, the ray trace is much more efficient.

FIG. 1. Collinear (first-order) modeling. The top part represents a real system with real ray paths. The middle part shows the collinear model with collinear (paraxial) rays. The bottom part compares the real ray paths with the collinear ray paths.

3.3.4 Fields and Pupils It is customary in describing an optical image-forming system to identify a unique object plane perpendicular to the axis as the object, and its conjugate as the image. The useful areas of these surfaces are also usually limited in extent, either by a physical aperture known as a field stop, or by a virtual limit imposed by the application. The resulting limited surfaces are called the object and image fields, and paraxial (collinear) ray tracing is used to find the location and size of the image field if the object field is given.

For an object point on axis, only a limited beam of light can actually get through the optical system to form the image point on axis. The aperture in the system that limits the size of this beam is called the aperture stop of the system, and it often is located internally. The object-space conjugate to the aperture stop is called the entrance pupil and the image-space conjugate the exit pupil. The beam in object space appears to be limited by the entrance pupil and in image space by the exit pupil.

3.3.5 Marginal and Chief Rays Because of the axial symmetry assumed for the optical system, the axial beam of light is completely characterized by a single ray, one that goes from the axial object point to the edge of the aperture stop and on to the axial image point. This ray is called the marginal ray.

For an object point at the edge of the field, the beam connecting it with its image point is also limited by the aperture stop, and to a first approximation has the same cross section everywhere as the axial beam except for being displaced from the axis. The displacement is entirely characterized by a ray going from the point at the edge of the field through the center of the aperture stop and on to the conjugate image point. This ray is called the chief ray.

Only these two rays are necessary to specify the fields and pupils. Wherever the marginal ray crosses the axis we have an image, and its size is given by the chief ray height at that location. Wherever the chief ray crosses the axis we have a pupil, and its size is given by the marginal ray.

The chief ray properties are distinguished from the marginal ray properties by the variables being barred, whereas the marginal ray variables remain unbarred.

3.3.6 The Optical (Lagrange) Invariant At any plane in the optical system, the marginal ray is characterized by its height on the plane and its angle with respect to the axis. The chief ray is similarly characterized. A very interesting and useful relationship between the two rays is obtained by forming the combination

$$L = n\bar{u}y - nu\bar{y}. \tag{12}$$

This quantity is constant throughout the system and is called the optical, or Lagrange, invariant. It is particularly useful at the field and pupil planes, having the same value at all of them.

One simple application involves the Lagrange invariant at the object and image planes. At the object and image planes respectively,

$$L = -nu\bar{y} = -n'y'\bar{y}', \tag{13}$$

and so we can express the magnification connecting the image with the object in terms of the marginal ray angles:

$$m = y'/y = nu/n'u'. \tag{14}$$

3.3.7 Numerical Aperture and the *f* Number The magnitude of the quantity nu is a paraxial approximation to a quantity known as the numerical aperture, which is strictly given by $n \sin u$ and is usually represented by the symbol NA. The numerical aperture is an important quantity for several reasons. In physical optics, it determines the resolution of the system, but even in geometrical optics the numerical aperture in the image space determines the irradiance on the image plane; the irradiance varies as the square of the numerical aperture.

In the early days of photography, it was recognized that some simple yet effective way of controlling the exposure of the photographic negative was needed. The amount of light getting to the film clearly was proportional to the area of the entrance pupil. Also, the irradiance on the film plane for a given pupil area was inversely proportional to the focal length of the lens. A long–focal-length lens with the same pupil area put less light on the film plane in the same exposure time. Thus, the ratio of the focal length of the lens to its entrance pupil diameter, or f number, was established as a quantity that could be used to control the exposure. Lenses of the same f number give the same exposure in the same exposure time, regardless of the focal length.

When the object is at infinity, the numerical aperture is equal to the reciprocal of twice the f number.

Complications arose when the object was so close to the lens that the image distance was considerably larger than the focal length. The exposure of course went down, and because the definition of the f number involved the focal length, not the image distance, some modification of the exposure factor involving the magnification was in order. Various fixes, some more ungainly than others, have persisted to this day. The fundamental problem is that the numerical aperture is the quantity that actually controls the exposure, and the f number was an empirical creation, adequate for the conditions it was designed for. Today, the numerical aperture is as easy to measure as the f number, and because it is completely general, one might, for the convenience of those who persist in using the f number, define an effective f number that is the reciprocal of twice the numerical aperture.

3.4 Systems with No Axial Symmetry

Most of the above properties seem to have depended on the presence of an axis of rotational symmetry. If you do not have an axis, how can you have paraxial behavior?

The fact is, we do have a kind of axis. There is some point on the object defined as the center of the object field, and there is some element in the system that limits the beam of light coming from that point. The latter is the aperture stop, and its image in object space is the entrance pupil. A real ray traced from the central object point through the center of the entrance pupil will eventually pass through the center of the aperture stop and on through the system until it reaches the image surface at what is therefore defined as the center of the image field. This real ray is called the optical-axis ray, and is the nearest thing we have to an optical axis.

Unfortunately, rays that are paraxial to the optical-axis ray can in general have large angles of incidence at the refracting or reflecting surfaces, and the ray-tracing equations do not simplify the way they do for an axially symmetric system. On the other hand, the collinear mapping process does not require any imposed symmetry conditions. For a focal system, unique front and rear focal planes exist, and object planes parallel to the front focal plane have as conjugates planes parallel to the rear focal plane. Moreover, there is a line normal to the front focal plane that has as its conjugate a line normal to the rear focal plane. These lines can be identified as the axes in their respective spaces.

The mapping between any conjugate pair of the planes parallel to the focal planes is affine, that is, the image is at most anamorphic. This means that there are a pair of orthogonal directions in the image plane where the magnification is constant, although it may be different in the two directions. We can select the orientation of the x and y axes to correspond with these directions. The result is a set of mapping equations that are very nearly as simple as those for an axially symmetric system. The only added feature is the anamorphism between object and image planes.

The tough part of the problem is how to obtain the collinear mapping parameters from the system properties. A general solution to this problem has not yet been obtained.

4. IMAGING SYSTEMS: ABERRATIONS

4.1 Aberrations as Departures from Ideal Behavior

The collinear model represents ideal behavior in an optical image-forming system. Real systems depart from this behavior, and these departures are called aberrations.

From a ray point of view, all real rays from any object point should pass through the ideal collinear image point. If a real ray misses the ideal image point, we say that the ray is aberrated. The transverse displacement of the ray intersection with image plane from the ideal image point is the usual measure of a ray aberration.

From a wave-front point of view, the wave front emerging from the exit pupil should be spherical and centered on the ideal image point. If it is not spherical, or if it is not centered at the ideal image point, we say that the wave front is aberrated. The ideal wave front is usually called the reference sphere from which the wave aberration is measured. At any given point on the reference sphere, the wave aberration is the optical path distance from the reference sphere to the aberrated wave front along the ray passing through that point. The wave aberration function is the wave aberration as a function of the position on the reference sphere. The latter is the proper surface of reference for the pupil.

4.2 Wave and Ray Aberrations

Rays are everywhere normal to geometrical wave fronts, and so ray aberrations and wave aberrations must be connected. As a function of pupil coordinates, the wave aberration function is a scalar function, whereas the transverse ray aberrations are vector quantities. If the pupil coordinates are chosen to be the transverse coordinates on the reference sphere, it can be shown that, to an excellent approximation, the transverse ray aberrations are proportional to the gradient of the wave aberration function.

Between the two types of aberration, the wave aberration is a simpler function of the pupil coordinates, and the ray aberrations are simply derived from it. The reverse is not true.

Another significant advantage of the wave aberration is that the geometrical wave front is a good approximation to a physical wave front, especially in the exit pupil where Fresnel diffraction effects are minimal. The wave aberration function is easily and simply converted into a phase variation over the reference sphere, the latter being the proper surface of integration for a diffraction integral.

4.3 Monochromatic Aberrations

In general, different wavelengths of light result in different aberration functions, and these variations with wavelength are identified as chromatic variations. Before dealing with them, we first must understand the aberrations at a given wavelength, the monochromatic aberrations.

4.3.1 Expansions and Classification The wave aberration function is a function of pupil coordinates and is also parametrically a function of field coordinates. That is, for every point in the field there is a corresponding wave aberration function that may be different for different field points. Taking into account the axial symmetry of the optical system, it is possible to represent the wave aberration function by a two-dimensional power-series expansion in the pupil coordinates, the terms of which are the elementary wave aberrations.

The principal classification of the terms is by order determined dimensionally (see Fig. 2).

4.3.2 Elementary (Third-Order) Aberrations The lowest nontrivial order for monochromatic aberrations is called third order, although the terms are dimensionally fourth degree. The labeling was historically determined by a ray classification, and the ray aberrations are determined by the gradient of the wave aberration function. There are five significant third-order aberrations.

4.3.2.1 Spherical Aberration. Spherical aberration is the only aberration that is independent of the field. It is therefore the only monochromatic aberration to affect the axial image. Although its effect is felt over the entire field, it is simpler to describe as an axial aberration.

The wave aberration function is a rotationally symmetric, fourth-degree departure from the reference sphere. All the rays in a given pupil zone come to a common focus on the axis, but the axial location of these zonal foci varies with the radius of the zone. In addition to this zonal focal shift, the rays produce a trumpet-shaped external caustic, seen as the envelope of the rays, that is characteristic of spherical aberration (see Fig. 3).

The geometrically predicted image is represented by tracing a uniformly dense mesh of rays in the pupil and observing the array of intersection points with an observation plane in the neighborhood of the ideal image. This array of points is called a spot diagram. In the case of spherical aberration, if we move the observation plane through focus (see Fig. 4), we find that there is no symmetry with respect to focus.

4.3.2.2 Coma. Coma is an aberration that varies linearly with field height and is therefore absent on axis. Its wave aberration function has a cubic cross section, and the ray aberrations are all on one side of the ideal image, either toward the center of the field or away from it, depending on the sign of the aberration coefficient. Moreover, at best focus the rays are contained within a 60° angle with the ideal image at its vertex.

Coma is symmetrical through focus, as shown in Fig. 5.

4.3.2.3 Astigmatism. Astigmatism is an aberration that varies as the square of the field height and is absent on axis. The wave aberration function is a quadratic cylinder, constant in a direction perpendicular to the meridional plane containing the ideal field point. (A meridional plane is a plane that contains the axis of rotational symmetry.)

The distinguishing feature of astigmatism

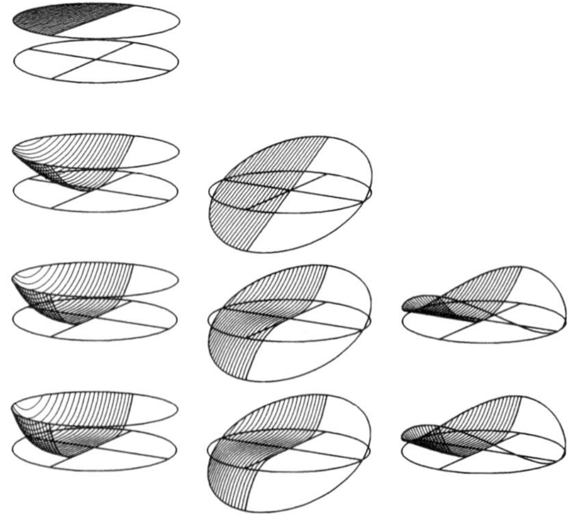

FIG. 2. Wave aberration functions. Each row represents a different order of aberration. The third row down shows the third-order aberration functions for spherical aberration, coma, and astigmatism.

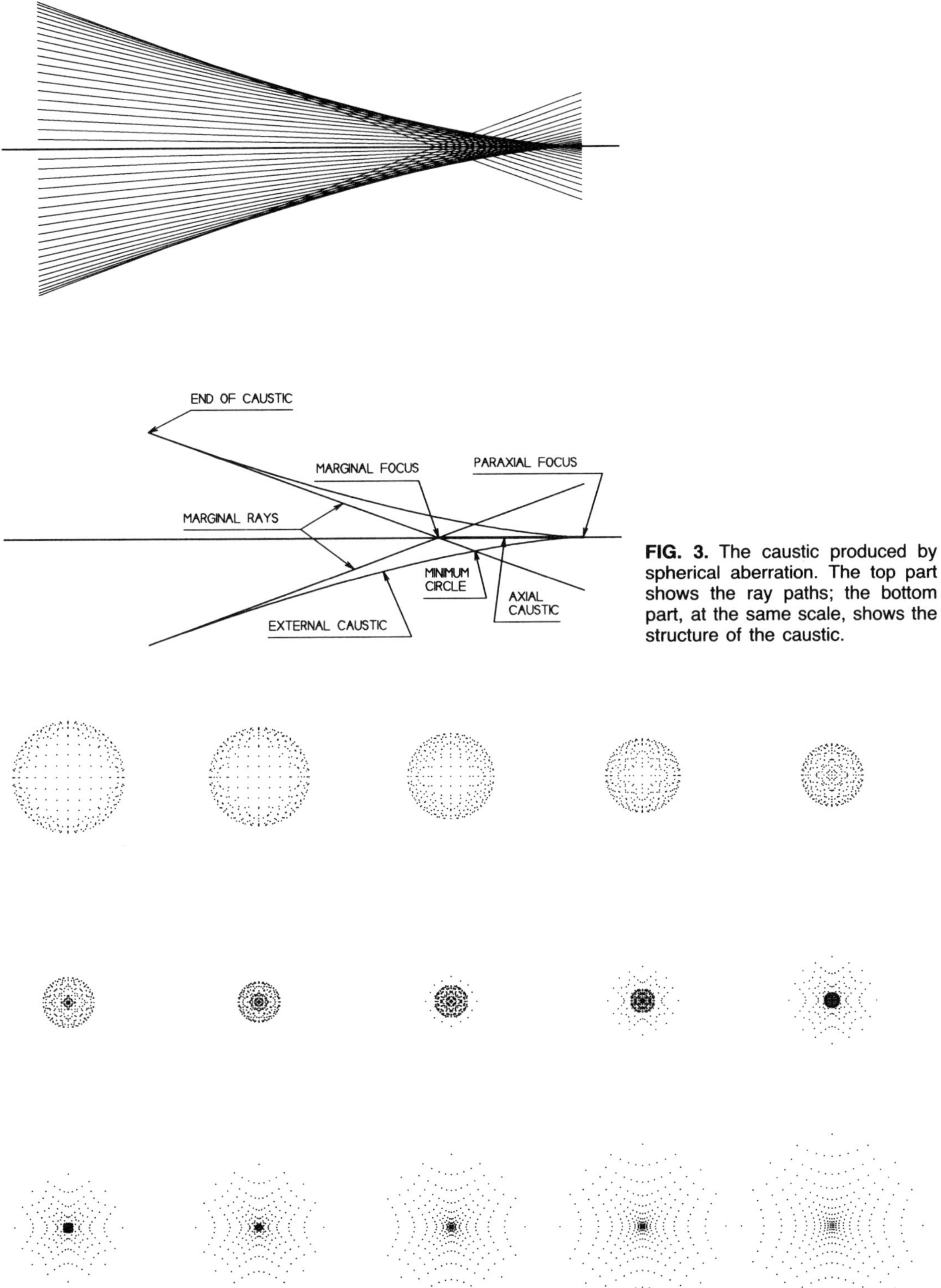

FIG. 3. The caustic produced by spherical aberration. The top part shows the ray paths; the bottom part, at the same scale, shows the structure of the caustic.

FIG. 4. A through-focus sequence of spot diagrams for spherical aberration. In this figure, and in Figs. 5 and 6, the sequence of fifteen focal positions is from left to right along each row and from top row to bottom row. Note the asymmetry through focus.

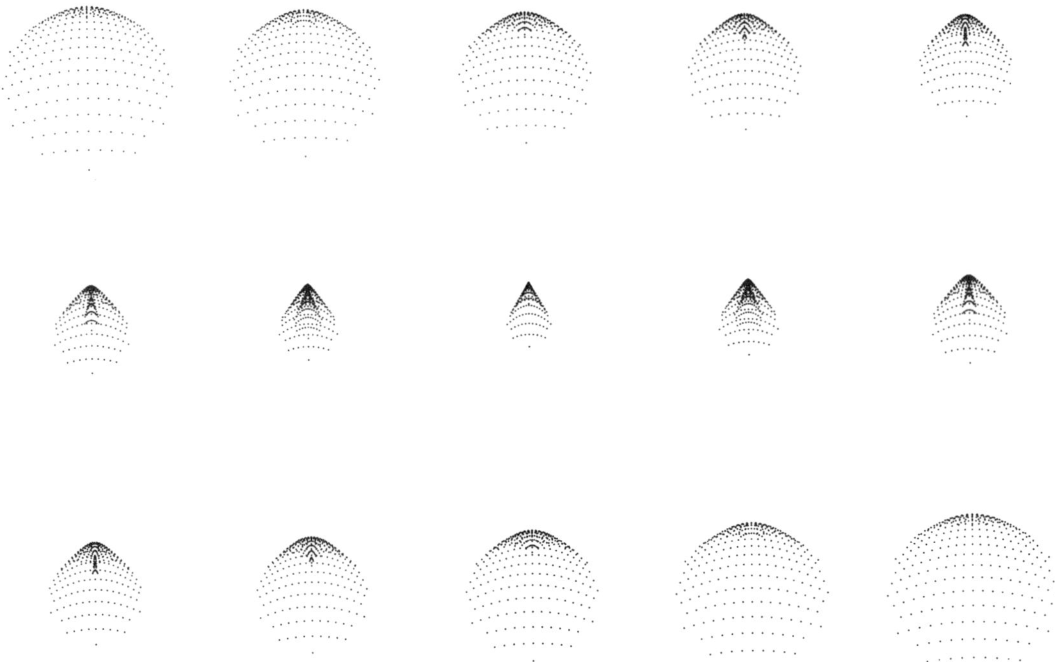

FIG. 5. A through-focus sequence of spot diagrams for coma. The images are symmetrical through focus, but they are asymmetrical in the vertical direction.

is the presence of two linear foci, one in the meridional plane and the other perpendicular to it (see Fig. 6). The first is called the sagittal focus and the second the tangential focus. The two foci are separated along the chief ray by a distance that is proportional to the amount of astigmatism. In the absence of other aberrations the sagittal field is flat, but because the focus shift to the tangential image varies as the square of the field height, the surface containing the tangential images is curved.

4.3.2.4 Field Curvature. Field curvature is an aberration in which the wave front is spherical and all the rays for a given image pass through a single point, but this point is shifted along the chief ray by an amount that increases quadratically with field height. As a consequence, in the absence of other aberrations, the images are locally very good, but they lie on a curved surface.

4.3.2.5 Distortion. Distortion is also an aberration in which the wave front is spherical and all the rays pass through a single point, but in this case the points remain in the image plane. However, they are displaced toward or away from the axis by an amount that varies as the cube of the field height. The consequence of this is that images are radially distorted. A square centered on the axis is imaged with curved sides. If the sides are concave the result is called pincushion distortion, whereas if the sides are convex it is called barrel distortion (see Fig. 7).

4.4 Chromatic Aberrations

Chromatic aberrations arise because the optical properties of a lens system depend on the refractive indices of the components, and the refractive indices vary with wavelength.

4.4.1 Elementary (Primary) Chromatic Aberrations Even the first-order properties of a system can change with wavelength. The resulting aberrations are the primary chromatic aberrations. There are only two, longitudinal chromatic aberration and transverse chromatic aberration. The first is a change in focal position with wavelength and the second is a change in image size, or magnification, with wavelength.

Mirror systems do not suffer from chromatic aberrations because they do not contain dispersive elements.

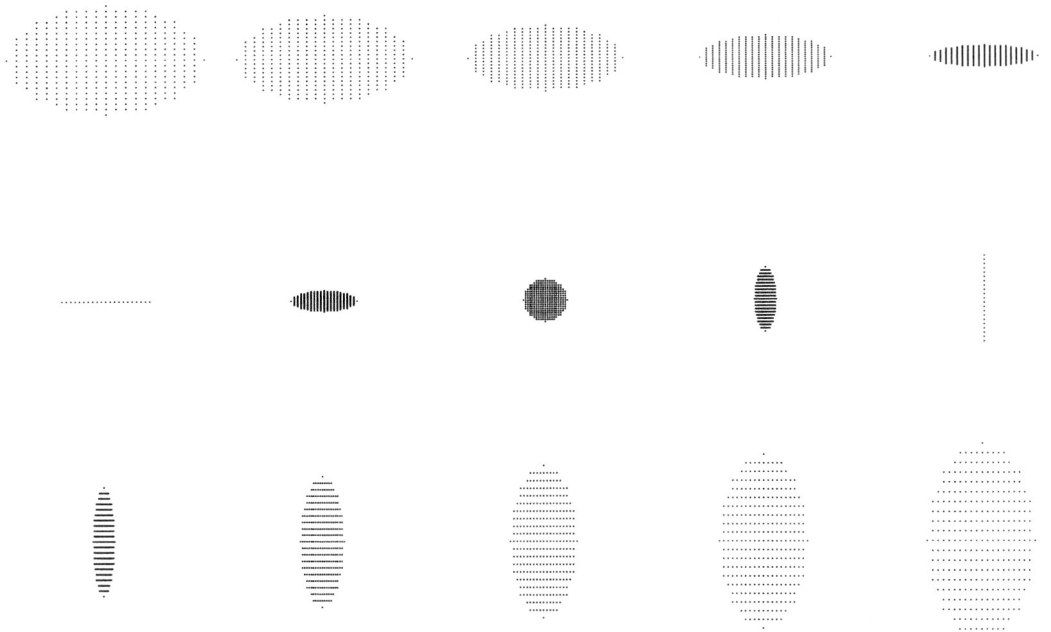

FIG. 6. A through-focus sequence of spot diagrams for astigmatism. The images are symmetrical about the vertical and horizontal axes, but there is a twist symmetry through focus.

4.4.2 Chromatic Variation of Monochromatic Aberrations In addition to the primary chromatic aberrations, the aberration types that have been identified as monochromatic can also vary with wavelength. In some cases, the chromatic variation of these aberrations is common enough for them to be given their own names; for example, the chromatic variation of spherical aberration is called spherochromatism.

4.5 The Control of Aberrations in Optical Design

Some of the aberrations of a single lens can be modified by bending the lens (changing the shape of the lens without changing its power) and by locating the aperture stop away from the lens. Others—for example, longitudinal chromatic aberration—are insensitive to these variables. At least two lenses of different

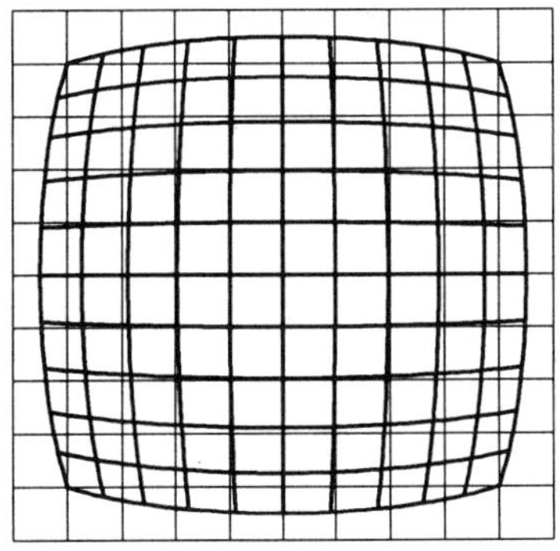

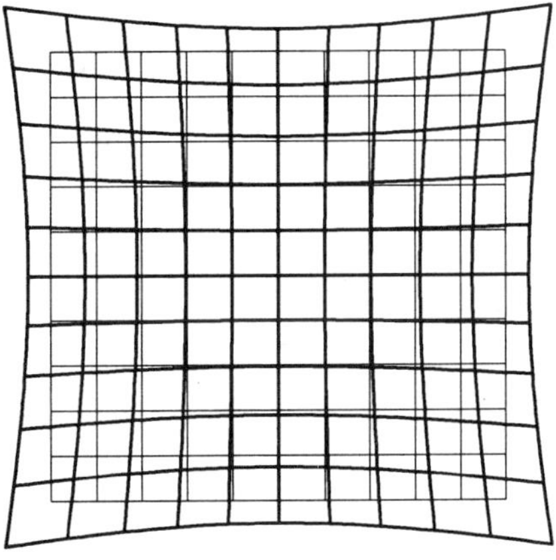

FIG. 7. Distortion. The part on the left shows barrel distortion; the part on the right shows pincushion distortion.

glasses are required to correct chromatic aberrations.

All of the third-order aberrations and the primary chromatic aberrations are correctable with only three lenses, but the performance of this triplet, although fairly good, is limited by higher-order aberrations. The complexity of a high-speed camera lens is due entirely to the need for a high degree of aberration correction.

Even in a well-corrected lens, the state of correction is the result of balancing sometimes surprisingly large aberrations from one or more elements with correspondingly large aberrations of opposite sign from other elements. As a result, manufacturing tolerances are usually very tight.

In the case of mirror systems, there are no chromatic aberrations to be corrected. However, a mirror cannot be "bent" like a lens to modify its performance. Aberrations are primarily controlled by making the mirror surfaces aspheric.

Lens systems can also use aspheric surfaces, but until recently this has been avoided because of the expense involved in producing them. Aspheric mirrors are expensive too, but there is no other recourse.

GLOSSARY

Aberration: A departure from ideal behavior.

Afocal Systems: Systems that do not contain focal planes. An object plane at infinity has its conjugate image plane at infinity. Rays parallel to the axis in object space have conjugate rays parallel to the axis in image space.

Aperture Stop: That aperture that limits the size of a beam going from the axial object point to the axial image point.

Cardinal Points: The **Focal Points, Principal Points**, and **Nodal Points** of a system taken as a set.

Caustic: The envelope of all rays emanating from a point and reflected or refracted by a curved surface.

Chief Ray: A ray that goes from the edge of the object field to the edge of the image field, passing through the center of the aperture stop on the way.

Collinear Mapping: A mapping of an object space into an image space in which there is a one-to-one correspondence between points, between straight lines, and between planes in the two spaces.

Conjugate Elements: Object and image elements (points, lines, or planes) that are in one-to-one correspondence with each other.

Dispersion: The variation in refractive index of a medium.

Effective *f* Number: The reciprocal of twice the numerical aperture. It is a property of the space (object or image) of the system.

Entrance Pupil: The image of the aperture stop in the object space of the system.

Exit Pupil: The image of the aperture stop in the image space of the system.

***f* Number:** The ratio of the effective focal length of a system to its entrance pupil diameter. It is a property of the system.

Focal Length (Effective): The reciprocal of the power of a focal system.

Focal Lengths (Front and Rear): The distances from the principal points to the focal points.

Focal Systems: Systems that contain focal planes. An object plane at infinity has a conjugate plane at a finite distance from the optical system in the image space.

Front Focal Plane: The object plane that is conjugate to an image plane at infinity.

Front Focal Point: The axial point in the front focal plane.

Height: The distance from the axis of an object or image point.

Ideal Behavior: A Collinear Mapping.

Image Space: The optical space that contains the image of an optical element.

Magnification: The ratio of the image height to the object height for an off-axis object point.

Marginal Ray: A ray that goes from the axial object point to the axial image point, touching the edge of the aperture stop on its way.

Meridional Plane: A plane in an optical system that contains the optical axis.

Nodal Points (Front and Rear): Conjugate axial points where any object ray passing through the front nodal point and its conjugate ray passing through the rear nodal point make equal angles with the optical axis.

Numerical Aperture: The sine of the angle between the real marginal ray in the object or image space of a system multiplied by the refractive index of the space. It is a property of the space, and not of the system as a whole.

Object Space: The optical space that contains the object of an optical element.

Optical Angles: Ray angles multiplied by the refractive index of the space that contains them.

Optical Path Length: The product of the refractive index and the geometrical path length along a ray.

Optical Space: The object or image space of an optical element.

Paraxial: Close to the axis.

Power: The refractive strength of a system. It is equal to the reciprocal equivalent length of a focal system. It is zero for afocal systems.

Principal Planes (Front and Rear): The conjugate object and image planes that have a unit positive magnification connecting them.

Principal Points (Front and Rear): The axial points of the principal planes.

Ray: (1) The trajectory along which light particles travel from a source to a receiver. (2) The normal to a wave front.

Ray Aberration (Transverse): The departure of the ray intersection with the image plane from its ideal position.

Real Image: An image formed in the real part of an image space.

Real Part of a Space: That part of an optical space that is normally seen in a conventional drawing of an optical system.

Rear Focal Plane: The image plane that is conjugate to an object plane at infinity.

Rear Focal Point: The axial point in the rear focal plane.

Reduced Distances: Axial distances divided by the refractive index of the space in which they exist.

Refractive Index: The ratio of the speed of light in vacuum to the speed of light in a medium.

Virtual Part of a Space: That part of an optical space that is outside the real part.

Wave Aberration: The departure of an aberrated wave front in the exit pupil from the ideal wave front, measured along the ray.

Wave Front: A surface of constant optical path from the source.

Further Reading

Optical Design

Kingslake, R. (1978), *Lens Design Fundamentals*, New York: Academic.

Kingslake, R. (1983), *Optical System Design*, New York: Academic.

O'Shea, D. C. (1985), *Elements of Modern Optical Design*, New York: Wiley-Interscience.

Smith, W. J. (1990), *Modern Optical Engineering*, New York: McGraw-Hill.

Theoretical

Buchdahl, H. A. (1993), *An Introduction to Hamiltonian Optics*, New York: Dover.

Hansen, R. C. (Ed.) (1981), *Geometric Theory of Diffraction*, New York: IEEE Press.

Luneburg, R. K. (1966), *Mathematical Theory of Optics*, Berkeley, CA: Univ. of California Press.

Sommerfeld, A. (1964), *Optics*, New York: Academic.

Stavroudis, O. N. (1972), *The Optics of Rays, Wavefronts, and Caustics*, New York: Academic.

Welford, W. T. (1986), *Aberrations of Optical Systems*, London: Adam Hilger.

OPTICS, LINEAR

A. Dorsel, *Abteilung PGM, Carl Zeiss Jena GmbH, Jena, Germany*

3-527-28134-7/95/$5.00 + .50

INTRODUCTION

When optical pumping and especially the invention of the laser opened the vast and fascinating field of *non*linear optics in 1960, it became meaningful if not necessary to introduce the distinction of linear optics for all optics that had been dealt with until then. Linear optics can be defined as that field of science that covers all phenomena of light propagation in media (including vacuum) in which the response of the medium (i.e., the polarization of that medium) is directly proportional to the field strength of the electromagnetic wave used to describe the light. What then *is* light? Originally restricted to the visible part of the spectrum ranging from 380 to 780 nm, the notion of light is commonly used to describe an extremely wide range of wavelengths of electromagnetic waves reaching from tens and hundreds of micrometers in the far-infrared (FIR) region down to the extreme ultraviolet (XUV) or even to x rays with nanometer wavelengths. The theoretical descriptions of light follow a dual approach in trying to describe its nature: the particle model and the wave model. In addition, geometrical optics may be regarded as a third less fundamental but most successful way of solving optical problems.

A most important feature of linear optics is the principle of linear superposition, which states that the electric field at a given point in space and time is the vector sum of the fields from all sources contributing:

$$E = E_1 + E_1 + E_3 + \cdots . \tag{1}$$

Historically, optics may be traced back to ancient Greece where Euclid noticed that "light travels along straight lines in homogeneous media." Hero of Alexandria was able to derive the law of reflection in the first century B.C. starting from the additional assumption that light travels the shortest possible path between two points, but it took more than 15 centuries until Willebrord Snell (1580–1626) discovered the correct law of refraction as the third principle of geometrical optics.

Christiaan Huygens (1629–1695) introduced a first description of light using waves that was able to explain phenomena like birefringence. Thomas Young (1773–1829) and Augustin Fresnel (1788–1827) made most important contributions to the development of the wave model, introducing the concepts of interference and diffraction. The model of geometrical or ray optics was included in the wave theories as a special case with wavelengths approaching zero. James Clerk Maxwell (1831–1879) set the foundations for the description of light as an electromagnetic wave, which is still valid today.

In the 20th century, the emission and absorption or detection of radiation were included in the theoretical treatment of quantum optics. Here the particle model employing the photon to describe light and dating back to René du Perron Descartes (1596–1650), Pierre de Fermat (1601–1665), and Isaac Newton (1642–1727) has found renewed interest among physicists. Linear optics can be generalized to include effects of special and general relativity, although these are often not observable with present equipment.

1. FUNDAMENTAL PROPERTIES OF LIGHT

1.1 The Velocity of Light

The determination of the velocity of light has been the object of many ingenious experiments. Interest in measurements of the velocity of light was caused by Einstein's theory of relativity, which declared it to be a fundamental constant, an unattainable limiting speed for all objects with finite rest mass, and, in its general form, defined the energy of any object as the product of its mass and the square of the velocity of light in vacuum.

As early as 1675, the Danish astronomer Ole Rømer found the velocity of light to be finite as he determined it from the apparent times of revolution of the moons of Jupiter; these are apparently changing as Jupiter approaches or flees a terrestrial observer, because of the relative orbital motion of the two planets. Many experiments were performed to get a more and more accurate value of the velocity of light. Only about 50 years later, Bradley used stellar aberration and obtained a much more precise result. Fizeau, Foucault, and Michelson each made valuable contributions in this field until this noble competition was ended in 1983 by the *Conférence Générale des Poids et Mésures*, which defined the velocity of light in vacuum at its then-known value of $c = 299\,792\,458$ m/s.

This permits one to measure distance (as a multiple of the wavelength of an optical radiation source) by measuring the frequency of that source; this, however, can be traced back to radiofrequency counting and thus to one of the most advanced precision measurement techniques available today [see also Quinn (1992) and CONSTANTS, FUNDAMENTAL]. These measurements yielded many other interesting insights. When Foucault extended the measurement of the velocity of light to other media than air and vacuum—if we allow ourselves for simplicity to call vacuum a medium—he found that the velocity in water, an optically denser medium than air, was lower than in air. This was in agreement with the predictions of the wave model and contradicted the particle model of Newton, which explained refraction by an *increased* velocity of light in denser media.

These remarks refer to the phase velocity of a signal only; it is the velocity of a point of constant phase. For a modulated (i.e., information-carrying) wave, the group velocity describes the speed at which a point of constant amplitude is advancing; only in nondispersive media (i.e., strictly speaking only in vacuum) are these two velocities equal.

The ratio of the velocity of light in vacuum, c_0, as compared to its velocity in a medium, c, is defined as the refractive index n of that medium:

$$n = c_0/c. \tag{2}$$

1.2 Dispersion

It was soon found out that this index varied with wavelength and thus (angular) frequency for all media except vacuum: $n = n(\omega)$. This effect is called dispersion, an increase of refractive index with frequency corresponding to normal dispersion and a decrease to anomalous dispersion. Today, quantum mechanics can explain quantitatively this characteristic behavior of the index of refraction from the behavior of atoms and molecules acting as forced oscillators under the influence of the radiation field.

In addition, a distinction had to be made between phase velocity and signal velocity as these are only identical in the absence of dispersion, i.e., in vacuum. Also, these effects should not be confused with modal dispersion in optical wave guides, which is caused by bulk effects (i.e., material dispersion as just described) *and* spatial boundary conditions.

1.3 The Doppler Effect

When a source of light moves relative to an observer, its spectrum is shifted for that observer: When it approaches the observer, all frequencies are increased in his frame of reference; when it moves away from him, they are decreased. This is due to the Doppler effect, the latter describing similar phenomena for all wave processes. For a monochromatic source with frequency f and vacuum wavelength $\lambda_0 = c_0/f$ decreasing its distance from the observer with a velocity v, the observed wavelength is

$$\lambda_0' = \lambda_0 \sqrt{\frac{c_0 - v}{c_0 + v}} \tag{3}$$

and the corresponding frequency is

$$f' = f \sqrt{\frac{c_0 + v}{c_0 - v}}. \tag{4}$$

The physical explanation, e.g., for a receding source is that consecutive crests or hollows of the wave have to travel an increasing distance to the observer, so that the time between the arrival of two crests is greater than for a source at rest; thus, the measured oscillation period is longer for that observer, the frequency lower, and the wavelength longer. For an approaching source, all these effects are simply reversed.

For values of v small compared to c_0, these expressions may be approximated by the simpler forms

$$\lambda_0' = \lambda_0(1 - v/c_0) \tag{5}$$

and

$$f' = f(1 + v/c_0), \tag{6}$$

whence this effect is sometimes also called the linear or first-order Doppler effect. This attribute became necessary as the theory of special relativity yields an additional effect, the quadratic or second-order Doppler effect, sometimes also called the transverse Doppler effect as it is more easily observed when the source moves transverse to the observer's line of sight, thus suppressing any first-order effect. In this case, the formulas are

$$f' = f\sqrt{1 - v^2/c_0^2} \tag{7}$$

and

$$\lambda_0' = \lambda_0/\sqrt{1 - v^2/c_0^2}, \tag{8}$$

showing that only a redshift is possible. For a detailed description of these effects, the reader should consult the article RELATIVITY, SPECIAL.

1.4 Polarization

As light waves are transverse, they may have specific states of polarization, i.e., of the direction of oscillation of the electric field vector:

If the field vector oscillates (almost) exclusively in one linear direction, thus defining a plane together with the propagation direction, the light wave is called linearly or plane polarized. For a given direction of propagation, any state of linear polarization can be described as the linear superposition of two orthogonal components of plane-polarized light that are in phase.

If two orthogonal components are superposed at a phase difference different from zero, elliptically polarized light will result. In case the phase difference is ±90° and the magnitudes of the two components are identical, we have circularly polarized light: The tip of the electric field vector moves around the origin on a circle either clockwise or counterclockwise.

When the polarization characteristics of light were first described, the electromagnetic wave model was yet to come, and unfortunately, the plane of polarization was at that time defined in such a way that it is normal to the electric field vector. Naturally, ignoring the distinction between these two definitions quite often has led to confusion. In this article (see, e.g., Sec. 2.3), this historical convention is followed simply because it seems to be most commonly used today.

Originally, reflection phenomena or crystals were employed to polarize light. Today, light is often polarized using polarizing films (Polaroid) that absorb light of one linear polarization while transmitting the orthogonal polarization. Thin-film beam splitters can be designed to show polarizing behavior, too.

1.5 Linear and Angular Momentum of Light Waves

1.5.1 Linear Momentum of Electromagnetic Waves Electromagnetic waves carry linear momentum, which is changed when a light wave is changed in its direction—e.g., when reflected by a plane mirror. As conservation of momentum must be satisfied, an ideally reflecting mirror will experience a resulting force directed along the normal of the mirror. This effect, often referred to as radiation pressure or photon pressure, can be explained by classical electrodynamics, but is more easily explained in the photon model: A light beam with power P and of frequency ν has a photon rate n of

$$n = P/h\nu, \tag{9}$$

where h is Planck's constant, $h = 6.626 \times 10^{-34}$ J·s. Each of these photons has a linear momentum of $p = E/c_0 = h\nu/c_0$. Thus, for normal incidence on a mirror (where $\mathbf{p}$ changes to $-\mathbf{p}$) we have a momentum transfer rate or force F of

$$F = 2np = 2P/c_0. \tag{10}$$

For a perfect absorber, the force would be half this value. Although these forces are usually very small, they were detected near the beginning of the 20th century, and the invention of the laser was followed by particle levitation and handling using laser beams; applications in spacecraft propulsion based on photon recoil are still in the research stage.

1.5.2 Angular Momentum of Electromagnetic Waves The photons in a circularly polarized light beam each carry an angular momentum $l = h/2\pi$. Transition through a half-wave plate (see Sec. 5.1.2.2) changes light of left circular polarization to right circular polarization and vice versa. The conservation of angular momentum then requires a torque to be exerted on the half-wave plate, which has been detected experimentally (Beth, 1936).

2. OPTICS OF PLANE INTERFACES AND ISOTROPIC MEDIA

As wave properties of light are more evident in phenomena like interference (see Sec. 4.1) and diffraction (see Sec. 4.2), we will restrict ourselves to ray optics in this section. In general, it is quite justified to identify a ray with the wave normal of an (almost) infinite plane wave.

2.1 Reflection

When a light ray is incident on an interface, i.e., on the (locally) plane boundary surface between two adjacent media, at least some portion of it may be reflected. The plane containing the normal to the interface and the incident ray is called the plane of incidence. The fundamental law of reflection in geometrical optics states that the reflected beam will lie in the plane of incidence and will form the same angle with the normal to the interface as the incident ray (Fig. 1).

Let us assume a vector representation that uses the normal of the interface as one of its basis vectors. All components of the reflected wave's unit vector are then identical to those of the incident wave's unit vector, except that the normal component changes sign.

2.2 Refraction

The law of refraction or Snell's law is somewhat more complicated. At a plane interface separating regions with different indices of refraction n_1 and n_2, a transmitted ray will be changed in its direction (see Fig. 2). It will still lie in the plane of incidence, and for the values of the angles between the incident (in n_1) and transmitted (in n_2) beams and the interface normal the relation

$$n_1 \sin\alpha_1 = n_2 \sin\alpha_2 \tag{11}$$

is valid. If no solution of this equation for real values of α_1 or α_2 exists, total internal reflection occurs (see Sec. 2.4).

2.3 Transmission

The transmission and reflection of light waves at interfaces are dependent on the state of polarization. If the plane of incidence is coincident with the plane of oscillation of the electric field vector, the light is called π polarized. If, instead, the plane of polarization

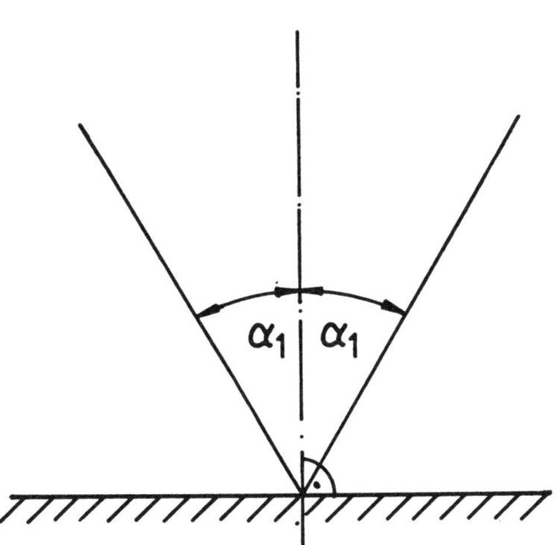

FIG. 1. The law of reflection.

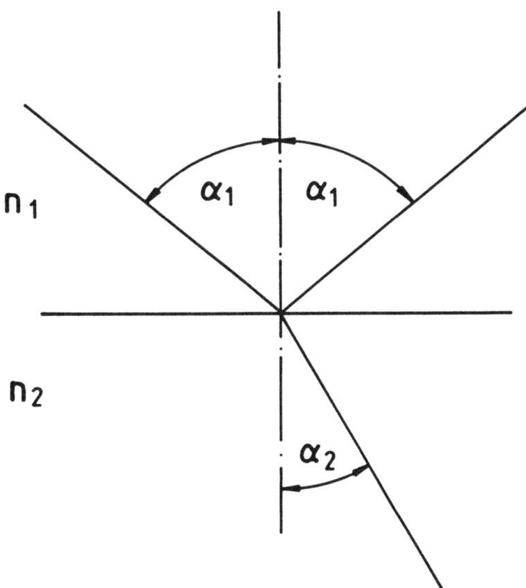

FIG. 2. The law of refraction.

coincides with the plane of incidence, the light is called σ polarized (from the German *senkrecht*). The reflection and transmission of energy at an interface separating regions of refractive indices n_1 and n_2 are described by the Fresnel formulas; for light incident from the side with refractive index n_1 with an angle of incidence α_1 and an angle α_2 between the refracted ray and the interface normal, these are (compare Fig. 3)

$$R_\pi = \left(\frac{n_1 \cos\alpha_2 - n_2 \cos\alpha_1}{n_1 \cos\alpha_2 + n_2 \cos\alpha_1}\right)^2, \qquad (12)$$

$$R_\sigma = \left(\frac{n_1 \cos\alpha_1 - n_2 \cos\alpha_2}{n_1 \cos\alpha_1 + n_2 \cos\alpha_2}\right)^2, \qquad (13)$$

$$T_\pi = \frac{n_2 \cos \alpha_2}{n_1 \cos \alpha_1}\left(\frac{2n_1 \cos\alpha_1}{n_1 \cos\alpha_2 + n_2 \cos\alpha_1}\right)^2, \qquad (14)$$

$$T_\sigma = \frac{n_2 \cos \alpha_2}{n_1 \cos \alpha_1}\left(\frac{2n_1 \cos\alpha_1}{n_1 \cos\alpha_1 + n_2 \cos\alpha_2}\right)^2. \qquad (15)$$

One should note that these formulas are describing *energy* transfer. When light with mixed polarization is incident, the fields have to be described as superpositions of σ-polarized and π-polarized waves.

For nonabsorbing media, n_1 and n_2 are real and R_π vanishes for $\tan\alpha_1 = n_2/n_1$. This angle is called Brewster's angle or polarizing angle. The physical model for this situation is that the dipoles in the refracted ray cannot radiate in their direction of oscillation, i.e., orthogonally to the direction of propagation of that ray; thus, Brewster's angle is equivalent to the refracted and reflected rays being orthogonal to each other. If a plane wave is incident on an interface at Brewster's angle, the reflected wave is completely polarized. Stacks of plates or thin films at or near Brewster's angle can be used as polarizers.

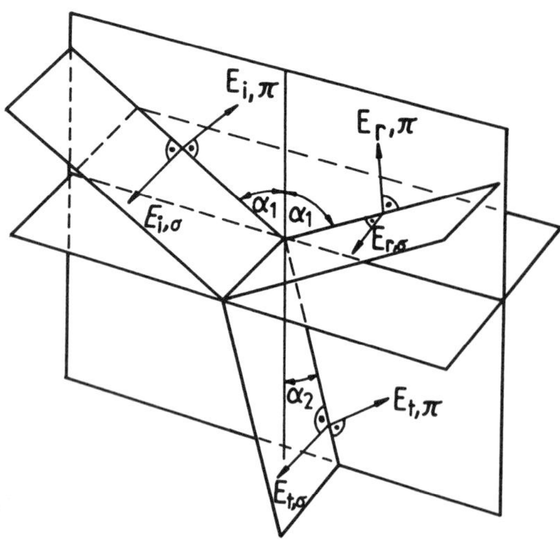

FIG. 3. Transmission and reflection (after Alonso and Finn, 1976).

2.4 Total Internal Reflection

The relation between angles in the law of refraction (Snell's law) shows that for $n_1 > n_2$ a choice of $\sin\alpha_1 > n_2/n_1$ will lead to $\sin\alpha_2 > 1$, which is not defined for a real value of α_2. For these values, total internal reflection occurs: Only a sharply decaying evanescent field probes into the region with index n_2, and the entire energy of the wave is reflected at the interface. The reflected beam may be attenuated through absorption or scattering, however. In the wave picture, the evanescent wave advances so much faster than the incident wave that the wave path (ray) is bent back into the medium from which it is incident.

Frustrated total internal reflection occurs when a second interface to a medium with sufficiently high index of refraction (e.g., n_1) is brought within the region where the evanescent field is still significant. In this case, the Fresnel equations have to be used with the complex angle that satisfies the law of refraction, and interference from the two closely spaced interfaces has to be taken into account.

2.5 Absorption

Materials absorb light waves in different regions of the electromagnetic spectrum. The absorbed light may be reemitted at the same frequency or at one or more different frequencies, or its energy may be used for completely different processes—e.g., simply heating the medium. When these processes are due to electronic, vibrational, or rotational excitation of crystals, molecules, or atoms, they are called absorption.

In a homogeneous absorbing medium with absorption coefficient α, the irradiance I, i.e., the energy per unit area per unit time, of a plane wave traveling in the positive x direction is described by Beer's law,

$$I = I_0 \exp(-\alpha x), \tag{16}$$

where I_0 is the irradiance at $x = 0$. It is derived from the assumption that the decrease dI caused by passing through an infinitesimal length dx of absorbing material is the product of the local irradiance I and of the absorption coefficient: $dI = -\alpha I dx$.

2.6 Extinction

The attenuation of light by absorption *and* scattering is called extinction. The processes most relevant in scattering normally are Rayleigh and Mie scattering. These two are generally characterized by the relative sizes of the scatterers and the wavelength of the light involved. For Rayleigh scattering to occur, the scattering particles must be small compared to the wavelength of the light scattered. In this case, the strength of the scattering is proportional to the fourth power of the light frequency. Although originally developed to cover particles of arbitrary size, the expression Mie scattering is mostly reserved for scatterers of a size similar to the wavelength of the light under consideration.

The scattered light is normally changed in frequency by quite small amounts by the Doppler effect from the moving scatterers. As the angular distribution of the scattered light is normally not isotropic and may have a pronounced peak in the forward direction, care has to be taken in experiments to discern the unscattered light from light scattered through small angles.

Other scattering processes such as Brillouin scattering and Raman scattering may be accompanied by significant amounts of frequency shifts; these imply energy transfer from solids or molecules.

3. GEOMETRICAL OPTICS

Optics is closely linked to vision and thus to images. Accordingly, imaging optics play a major role in the applications of optics. Later on, we shall see how diffraction limits the performance of different optical devices. In the limit of the wavelength λ approaching zero, the wave-optical description converges toward the ray-optical image, i.e., geometrical optics, as diffraction effects become negligible. For the time being, we will make the approximation that the wavelength of light is negligibly small compared to system dimensions so that a ray-optical treatment characteristic of geometrical optics is sufficient (see OPTICS, GEOMETRICAL).

Most of the treatment presented here is limited to the so-called paraxial case, where the light rays considered deviate so little from the axis of the optical system that the sine of the angle they form with the optical axis is sufficiently approximated by its argument: $\sin(x) \approx x$. For a more complete approach, the article OPTICS, GEOMETRICAL and the literature, e.g., Born and Wolf (1980), should be consulted.

3.1 Plane Mirrors

The most simple kind of imaging is performed by a plane mirror: The image position can be derived by geometrically mirroring the object at the plane of the mirror, which is then the plane of symmetry between the object and its image. The image thus created is a virtual image: If a screen is brought to its apparent position, the image will not be apparent on it.

3.2 Spherical Mirrors

Spherical surfaces for mirrors and lenses are most common as they are comparatively easy to produce. Spherical mirrors can be convex or concave (this distinction naturally is only justified for a nontransmitting mirror): A convex mirror has a protruding center while a concave mirror has a receding center.

3.2.1 Concave Mirrors When a bundle of parallel rays of light is incident on a concave mirror parallel to its axis, these will be deflected so that they intersect at a common point F. This point is called the focus of the mirror. Its distance from the center of the mirror is the focal length (cf. Fig. 4). A concave mirror of radius R has a focal length $f = R/2$.

Vice versa, rays impinging on the mirror surface after going through the focus will afterward be parallel to the axis. In addition, a ray going through the center of curvature of the mirror will always be reflected into it-

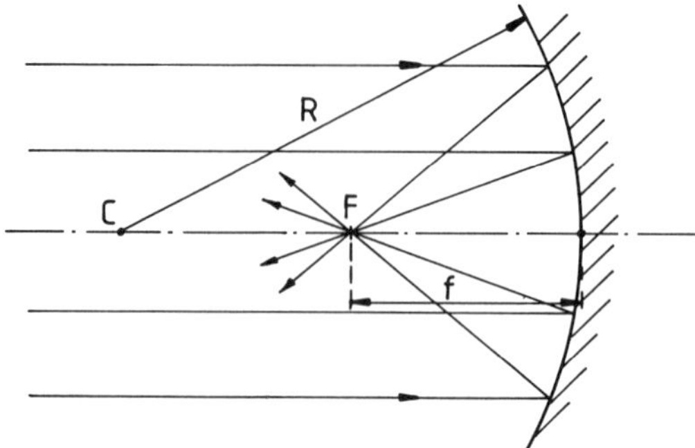

FIG. 4. Focusing of parallel light rays by a concave mirror (after Hecht and Zajac, 1974).

self, as it is always perpendicular to the mirror surface. With these rules, it is easy to construct the image point for any given object point, as follows.

Rays are drawn toward the mirror from the object point in directions parallel to the mirror axis, toward the focus, and toward the center of curvature. At their point of incidence on the mirror, they are then prolonged according to the above rules. The intersection of the reflected rays is the image point. If these do not intersect, either they are parallel, which means that the image is at infinity, or their prolongations on the opposite side of the mirror will intersect to form the virtual image point.

The mathematical treatment of this simple imaging situation yields the relation between the object and image distances (measured along the mirror axis) s_1 and s_2 and the focal length:

$$\frac{1}{f}=\frac{1}{s_1}+\frac{1}{s_2}. \tag{17}$$

Positive values for s_1 and s_2 correspond to an image or object position in front of the mirror; negative values indicate that the (virtual) object or image seems to lie behind the mirror. With $s_i' = s_i - f$ for $i = 1, 2$, this imaging equation can be written in a simpler form as

$$s_1's_2' = f^2. \tag{18}$$

The magnification (or demagnification) can be derived from Fig. 5 directly:

$$M = h_i/h_o = -s_2/s_1. \tag{19}$$

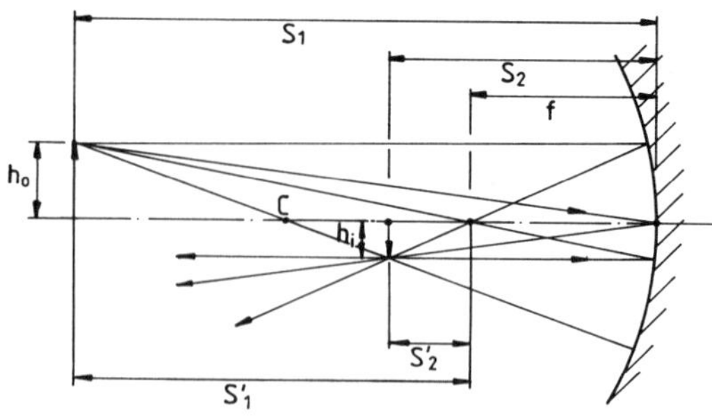

FIG. 5. Image formation for a concave mirror (after Hecht and Zajac, 1974).

If $s_1 < f$, we get a virtual upright magnified image.

3.2.2 Convex Mirrors If a bundle of parallel rays is incident on a convex mirror parallel to the mirror axis, it will be reflected in such a way that the emerging rays seem to have a common point of origin; this point is called the focus of that mirror. For a mirror with a radius of curvature R, this point is situated at a distance $-f = R/2$ behind the mirror measured along the ray that defines the optical axis. For reasons of symmetry, the optical axis of a spherical mirror alone is not defined but has to be derived from other components in a system or from the illumination conditions under which the mirror is used. For a mirror alone and an incident bundle of parallel rays, it is given by the ray that impinges on the mirror at normal incidence.

Figure 6 shows the situation just described. As all light paths can be reversed, it is immediately obvious that a bundle of rays converging toward the focus of such a mirror

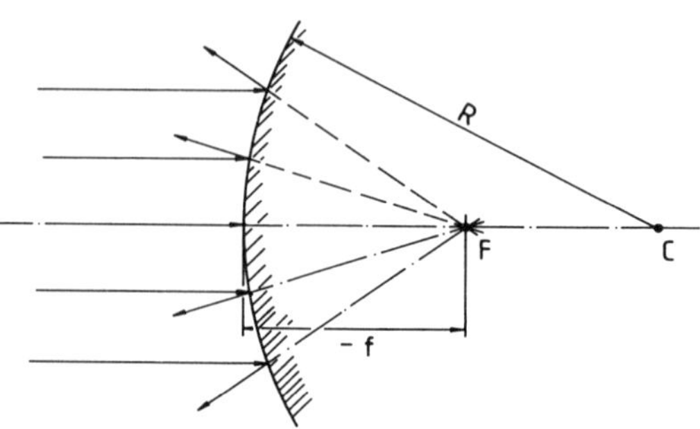

FIG. 6. Convex mirror with parallel incident bundle (after Hecht and Zajac, 1974).

will be collimated to a parallel bundle after reflection.

Imaging of a real object is represented in Fig. 7. A convex mirror will always yield a virtual image for a real object. Geometrical construction of the image point for a given object point is simple when referring to the situation in Fig. 6: A ray parallel to the mirror axis and a ray toward the focus are drawn from those points to the mirror surface, and from there they are drawn further according to the rules given above. They then seem to emerge from a common origin behind the mirror, the image point. This procedure can be performed for all object points so that a complete image can be constructed.

Sometimes it may be helpful to use a third kind of ray in constructing the image of a given object: A ray incident toward the mirror's center of curvature C will always simply be reversed in direction upon (normal-incidence) reflection. The mathematical description of the image position is quite simple:

$$\frac{1}{f} = \frac{1}{s_1} + \frac{1}{s_2}. \tag{20}$$

Here s_1 and s_2 are the object and image separations from the mirror (measured along the mirror axis in paraxial optics). The term f is negative for a convex mirror. A negative value for s means that an image (or a virtual object) is situated behind the mirror. Note that Eqs. (17) and (20) for concave and convex mirrors are identical and that only the focal length changes its sign.

A virtual object may be created using a concave mirror. If the position of such a virtual object lies between the focus and the

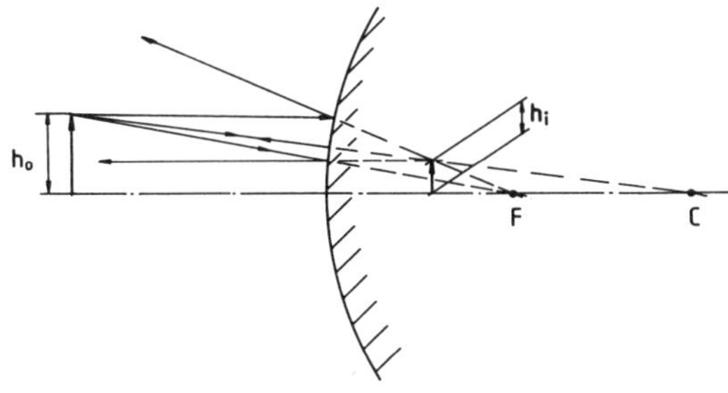

FIG. 7. Real object imaged with convex mirror (after Hecht and Zajac, 1974).

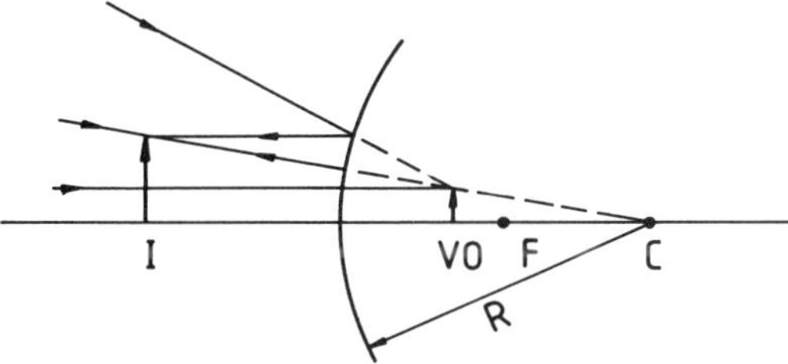

FIG. 8. Virtual object imaged with convex mirror. *VO* denotes the virtual object and *I* the image obtained therefrom.

mirror surface, even a convex mirror will yield a real image (Fig. 8).

3.3 Spherical Lenses

When refraction occurs at a spherical interface, (de)focusing effects will occur. Here we deal with spherical surfaces only, but advanced optical systems often employ aspherical surfaces also. However, the deviations from a sphere are often small for imaging systems with f numbers corresponding to paraxial optics. (The f number is the ratio of the focal length to the diameter of the entrance pupil. In paraxial optics, we typically encounter f numbers of about 10 or more.)

3.3.1 Single-Interface Lens Although most technical lenses have two surfaces, there are two reasons why the single-interface lens is a worthwhile object for study: First, any given lens can be treated mathematically as an arrangement of consecutive single-surface lenses, and second, the single-interface lens is a first approximation to the imaging system of the human eye, where object space and image space are not filled with the same medium.

Figure 9 shows the imaging situation for a lens consisting of a spherical surface with radius R separating the object space with re-

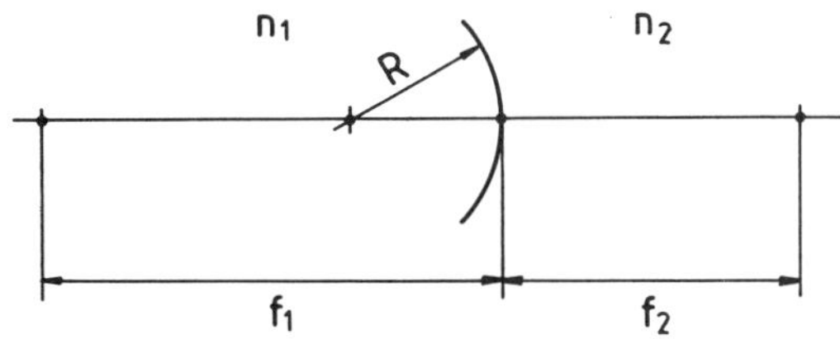

FIG. 9. Single-interface lens.

fractive index n_1 from the image space with refractive index n_2 where $n_1 > n_2$ is assumed (the situation may, e.g., be given for a spherical fish bowl). In this case, two focal lengths appear, one for each medium:

$$f_1 = Rn_1/(n_1 - n_2) \tag{21}$$

and

$$f_2 = Rn_2/(n_1 - n_2). \tag{22}$$

The imaging equation still has a form similar to that of mirrors though some changes become necessary:

$$\frac{n_1}{s_1} + \frac{n_2}{s_2} = \frac{n_1 - n_2}{R}. \tag{23}$$

If one medium is air, its refractive index can be replaced by 1 for most practical purposes, and we can omit the suffix for the refractive index of the other medium; Eq. (23) then simplifies to

$$\frac{n}{s_1} + \frac{1}{s_2} = \frac{n - 1}{R}. \tag{24}$$

The lateral magnification accordingly has to be changed to

$$M = -\frac{n_1}{n_2}\frac{s_2}{s_1}. \tag{25}$$

This quantity gives the ratio of image size to object size; it is positive for an erect image and negative for an inverted image.

3.3.2 Double-Interface Lens; Thin Lens A double-interface lens is usually called a thin lens if its center thickness D is small compared to its focal length and the object and image distances. For such a thin lens of index n in air as shown in Fig. 10, the focal length is given by

$$1/f = (n - 1)(1/R_1 + 1/R_2), \tag{26}$$

where the ambient medium is assumed to be vacuum. Comparing this to Eq. (24) shows that this is simply the sum of the refractive powers of the two interfaces. The quantity $1/f$ is often called the power of a lens. It should be noted that R_1 and R_2 are positive for a biconvex lens. The magnification is then simply given by $M = -s_2/s_1$.

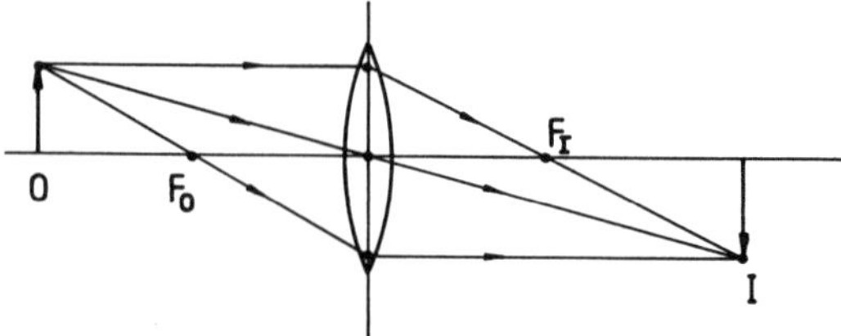

FIG. 10. Double-interface lens. F_O and F_I denote the focal points on the object side and on the image side. (After Klein and Furtak, 1986.)

3.4 Thick Lenses

Whereas for thin lenses in paraxial treatment it is normally sufficient to assume that a ray is deflected at the central plane of the lens, for thick lenses refraction at the entrance and exit faces of the lens must be treated separately as well as the propagation within the lens. The situation is depicted in Fig. 11 where the (paraxial) tracing is shown for the three rays normally used in geometrical constructions: the ray through the focus on the object side, the ray through the focus on the opposite side, and the ray through the lens center. The central plane used for thin lenses is split into two principal planes; between these, all rays are treated as if they traveled parallel to the optical axis (which they usually do not), and the figure indicates the directional changes for the different rays at the principal planes.

For a lens with center thickness t and refractive index n and surfaces with radii of curvature R_1 and R_2 of the surfaces facing the adjacent media with refractive indices n_1 and n_2, the lens power is then

$$P = P_1 + P_2 - P_1P_2t/n. \tag{27}$$

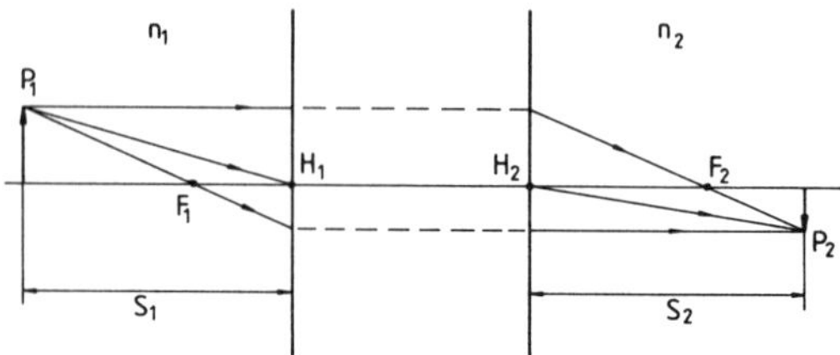

FIG. 11. Fundamental rays through a thick lens.

Here P_1 is the refractive power of the surface with radius R_1 and P_2 is defined analogously. Here we have used $P_1 = (n_1 - n)/R_1$ and $P_2 = (n - n_2)/R_2$ with R_i negative for a surface convex to the left. Then P is used in the relation $P = n_1/f_1 = n_2/f_2$.

The separations between the vertices and the principal planes are given by $D_1 = -n_1P_2t/P_n$ and $D_2 = -n_2P_1t/P_n$ where these values are positive when the separation points out of the lens as seen from the vertex.

For the common simple case of $n_1 = n_2 = 1$, the magnification is the same as that for the thin lens described in Sec. 3.3.2, but the focal length changes to

$$\frac{1}{f} = (n - 1)\left(\frac{1}{R_1} + \frac{1}{R_2} - \frac{(n - 1)D}{nR_1R_2}\right). \tag{28}$$

For a more detailed treatment, the reader should refer to OPTICS, GEOMETRICAL.

3.5 Matrix Optics

High-performance optics are usually characterized by diffraction-limited performance for large numerical apertures (see Sec. 4.2.3 for a definition of numerical aperture). For the design of these, it is always necessary to trace a ray through the system without approximations, i.e., by determining the ray's intersections with different interfaces and applying the law of refraction (or reflection) for each change of ray direction caused at such an interface. For many applications, however, it is sufficient (at least for a first sketch) to use paraxial optics with principal planes and a linear approximation of the law of refraction.

The resulting approximate equations describe the change of two variables characterizing a ray passing through an element: the distance of that ray from the optical axis and the angle between that ray and the optical axis times the refractive index. The equations are linear in these two variables, and, therefore, 2 × 2 matrices may be used to calculate free-space propagation and directional changes due to optical elements. This greatly simplifies paraxial ray tracing, as it permits a modular treatment of optical systems: The matrix representing the entire system is nothing but the (matrix) product of the matrices representing the optical elements constituting that system. Thus, after determining that system matrix, a single matrix multiplication suffices for tracing a ray through the entire system (see OPTICS, GEOMETRICAL).

3.6 Simple Systems

Vision is man's "optical" sense. Thus, we have chosen as the most prominent examples of (simple) optical systems the eye and the instruments devised to extend its reach into the microscopic and macroscopic world. For a much more detailed presentation, the reader can consult OPTICAL COMPONENTS AND SYSTEMS.

3.6.1 The Eye The eye consists of a single air-solid interface lens, namely the cornea, and an embedded variable-power lens. The lens power is changed in a unique way: The elastic lens is deformed by muscles and thus changes its radii of curvature. Image space is filled by the aqueous body, and in the image plane—which actually is a spherical surface in this case; cf. Fig. 12—we find the retina, which acts as a detector. A variable stop contributes a little to light regulation, whereas the main task here is performed by adaption (see OPTICS, PHYSIOLOGICAL.) Some typical data of the eye are given in Table 1.

The properties of the eye vary considerably with age and from one individual to another. Thus, small children may be able to focus at distances smaller than 100 mm. (See also OPTICS, PHYSIOLOGICAL.)

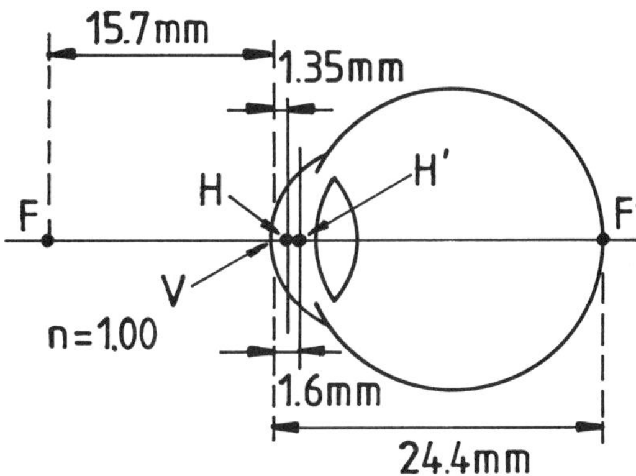

FIG. 12. The human eye. H and H' are the two principal planes and F and F' are the focal points. (After Klein and Furtak, 1986.)

Table 1. Some typical data of the human eye.

Property	Value	Unit
Back focal length	20	mm
Pupil diameter	1–7	mm
Refractive index of		
cornea	1.376	
lens	1.386	
aqueous body	1.336	
Length	25	mm
Lens radii of curvature, far accommodation		
front	10	mm
rear	6	mm
center thickness	3.6	mm
Near accommodation		
front	5.33	mm
rear	5.33	mm
center thickness	4	mm
Near object distance	250	mm

An optical camera is quite similar to the eye: The main differences are the absence of an aqueous body (which would lead to unacceptable weight), focusing by changing the separation between lens and film or detector plane, and the replacement of the curved retina by other, usually planar, detection or registration means (film, CCD array, etc.)

3.6.2 The Magnifying Glass or Simple Microscope The magnifying glass is one of the simplest optical systems: It consists of a single (possibly compound) lens of positive power with a focal length that is smaller than the nearest object distance for the eye. If the object is in the focal plane of the lens, the image lies at infinity and a magnification factor

$$M = s_0/f \tag{29}$$

results. Alternatively, the object distance is chosen so that a virtual image at about 250 mm behind the lens results. Thus, the image can just be focused when the eye is brought close to the lens (cf. Fig. 13). In this case, we have the magnification

$$M = 1 + s_0/f. \tag{30}$$

For weak magnifiers, this contribution can be useful, whereas it is normally neglected for strong magnifiers with small focal length, e.g., microscope or telescope eyepieces.

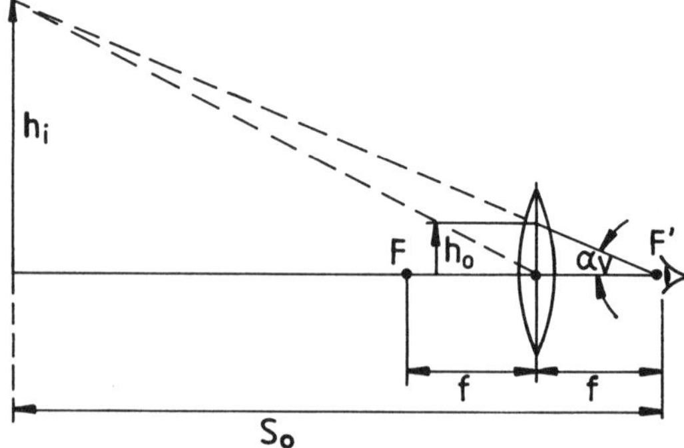

FIG. 13. The magnifying glass (after Klein and Furtak, 1986).

3.6.3 The Compound Microscope The microscope (see also OPTICAL MICROSCOPY) normally has two lenses or rather lens systems (cf. Fig. 14): the objective and the eyepiece. The objective is used to create a real magnified image of the object. This intermediate image is then viewed with the eyepiece serving as a magnifying glass. It is common usage to call the distance t from the back focal plane of the objective to the intermediate-image plane the tube length. This leads to a simple expression for the magnification factor of the objective:

$$M_{obj} = t/f_{obj}. \tag{31}$$

The total magnification of an optical system is the product of the magnifications of its subsystems. Thus, the total magnification of the microscope is

$$M_{mic} = \frac{t}{f_{obj}}\frac{s_0}{f_{ep}}, \tag{32}$$

where f_{ep} is the focal length of the eyepiece.

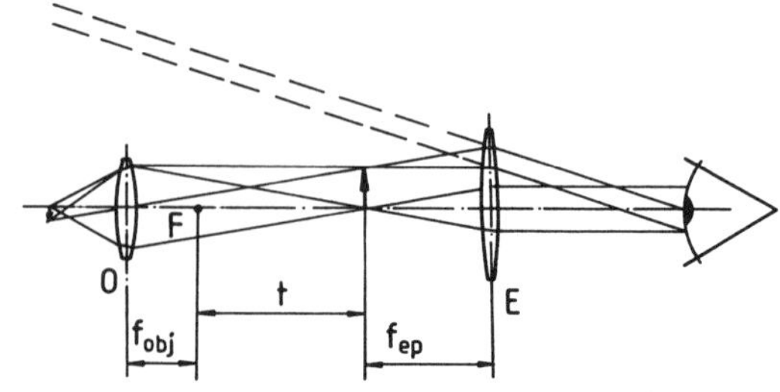

FIG. 14. The compound microscope. *O* is the objective, *E* is the eyepiece, and *F* denotes the back focal point of the objective. (After Hecht and Zajac, 1974.)

3.6.4 The Telescope A telescope usually consists of two lenses with a common focus. Thus, a parallel bundle of rays will remain a parallel bundle when passing through a telescope and will not be focused outside of it; therefore, telescopes and telescopical subsystems are sometimes called afocal systems.

3.6.4.1 The Astronomical Telescope. The astronomical telescope consists of an objective lens of large focal length and diameter and an eyepiece of small focal length and diameter (Fig. 15). Both the object and the image then lie at infinity, and the angular magnification is given by

$$M = -f_{\mathrm{obj}}/f_{\mathrm{ep}}. \tag{33}$$

For astronomical telescopes, both lenses often have positive power so that a negative magnification (i.e., an inverted image) results. As the amount of light that enters the telescope from a given star depends only on the apparent luminance of that star and the objective diameter, the number of stars visible will not depend on magnification but on that diameter (and optical transmission) only.

Larger telescopes are normally built using mirrors that can be supported from the back; larger lenses would experience sag under their own weight accompanied by severe image-quality deterioration. (See also ASTRONOMICAL TELESCOPES; SPACE TELESCOPES.)

3.6.4.2 The Terrestrial Telescope. The terrestrial telescope uses an eyepiece with negative focal length and thus positive magnification, i.e., upright images (Fig. 16). The magnification is given by the same formula as for the astronomical telescope. However, this construction yields an upright image and a shorter system, which besides is often folded using prisms for further compactness. These prisms may also assure an upright image for an eyepiece of positive power. For more comfortable observation, two telescopes are often aligned in a parallel pair, called binoculars. Two figures are normally used to characterize terrestrial telescopes. The first gives the magnification and the second the objective diameter in millimeters.

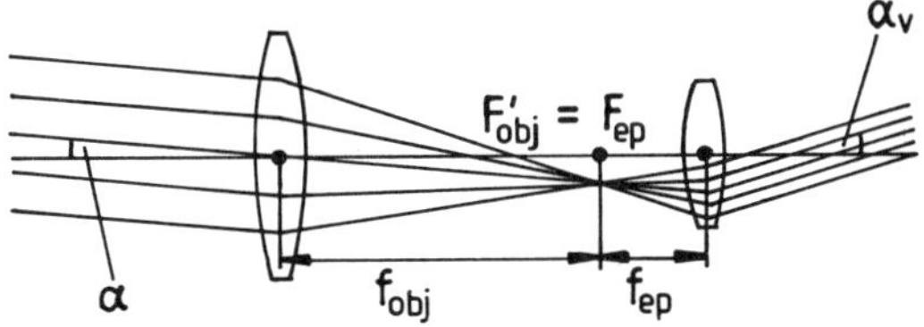

FIG. 15. The astronomical telescope. α denotes the angular size of the object without and α_v with the use of the telescope. (After Klein and Furtak, 1986.)

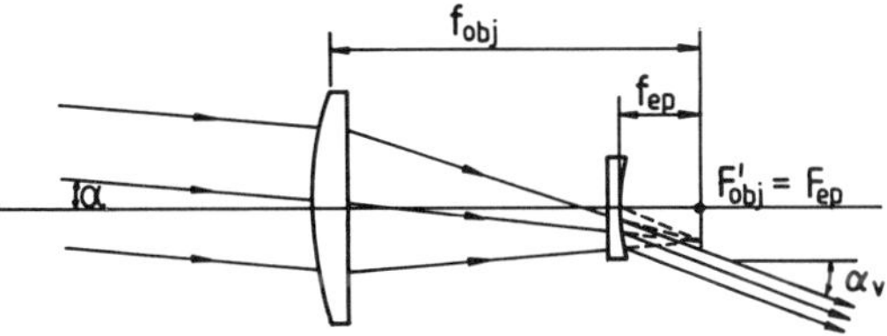

FIG. 16. The terrestrial telescope (after Klein and Furtak, 1986).

The useful diameter of the input pupil is the maximum diameter of the eye's pupil times the magnification. Light entering larger input pupils will be blocked by the iris. However, larger exit pupils may be useful for a telescope used in a vibrating environment, as they permit relative motion of the eye and the telescope.

It is not even possible to give an appropriate overview of the many designs that have been invented for optical systems. For further information, we have to refer the reader to the articles on OPTICAL INSTRUMENTATION; OPTICS, GEOMETRICAL; and OPTICAL COMPONENTS AND SYSTEMS of this Encyclopedia and to the literature, e.g., Kingslake (1978, 1983).

4. INTERFERENCE PHENOMENA

Interference phenomena are very widespread in their applications: From reflective (or antireflective) coatings to gratings and zone plates, they demonstrate the wave properties of light. One of the most important consequences of these is the limits set to the resolution of optical instruments like microscopes, telescopes, or spectrographs as well as to the propagation of diffraction-limited beams.

One important characteristic of interference phenomena is that the wavelength dependence here is often much stronger than for refractive systems. For example, whereas the focal length of a refractive lens may change by a few percent over the visible spectrum range, the focal length of a diffractive lens will vary almost by a factor of 2 for this case.

4.1 Interference

It is extremely difficult to observe interference of the waves from two different sources, as explained in Sec. 4.1.1. Therefore, in most practical cases the different sources are derived from the same source using beam splitters, etc. Unless otherwise stated, the treatment in this section restricts itself to monochromatic plane waves.

In order for interference to occur, several conditions must be fulfilled, as follows.

The interfering waves must have nonorthogonal polarizations. For identical polarization, fringe contrast is maximum, and for orthogonal states of polarization the interference pattern vanishes.

For static interference patterns to occur for finite path differences, the interfering waves usually must have the same constant frequency, i.e., the light should be monochromatic. (Note, however, that for zero path difference interference may be even observed using white light.) Otherwise, the interference pattern will change with time. If, in this case, the oscillation period associated with the difference frequency becomes shorter than the integration time of the device used for the detection of the interference pattern (eye, CCD camera, film, etc.), the interference pattern will be washed out. This is also the reason why it is quite difficult to observe interference between waves from different sources. Normally, their frequencies will differ considerably so that the interference pattern changes so rapidly that the interference fringes are washed out for time-integrating detectors like the eye.

Finally, the interfering waves have to be coherent. As ideal monochromaticity is never really obtained, the phase of a light source varies constantly. As the different waves for an interference setup are normally derived from the same source using beam splitters, the phase is controlled by choosing different optical path lengths to the point of observation for different beams. If the phase of the light source changes considerably during the time it takes the light to pass through the difference of optical path lengths, the interference pattern will also change in phase and will again be washed out for a time-integrating detector. As a quantitative measure for this property, the coherence length defines that path difference at which the fringe contrast drops to half of its maximum value. Coherence lengths vary from a couple of microns for white light to several thousand kilometers for highly stabilized lasers.

It should be noted that these last two conditions—monochromaticity and coherence—are closely linked as the frequency (difference) is directly proportional to the time derivative of the phase (difference) of the waves.

4.1.1 Two-Beam Interference In many cases, just two beams interfere or it is sufficient to take two of the interfering beams into account. The irradiance for a given point is proportional to the time-averaged square of the electric field at that point,

$$I = \epsilon\langle \mathbf{E}\cdot\mathbf{E}\rangle. \tag{34}$$

For $\mathbf{E} = \mathbf{E}_1 + \mathbf{E}_2$, this yields

$$\begin{aligned} I &= \epsilon\langle(\mathbf{E}_1^2 + \mathbf{E}_2^2 + 2\mathbf{E}_1\cdot\mathbf{E}_2)\rangle \\ &= I_1 + I_2 + 2\epsilon\langle\mathbf{E}_1\cdot\mathbf{E}_2\rangle, \end{aligned} \tag{35}$$

where we have used the abbreviation $\mathbf{E}_i^2 = \mathbf{E}_i\cdot\mathbf{E}_i$ for the scalar product of a vector with itself.

Often complex numbers are used instead of vectors to represent the fields. The physical field is then regarded as the real part of these quantities. For two waves with amplitudes A_1 and A_2, the fields may be represented as

$$E_1 = A_1\exp(i\phi_1),\quad E_2 = A_2\exp(i\phi_2) \tag{36}$$

with the phases $\phi_1 = \omega t$ and $\phi_2 = \omega t + \delta$. In this case, the resultant field is given by

$$E = E_1 + E_2 = A\exp(i\Phi) \tag{37}$$

with

$$A = [A_1^2 + A_2^2 + 2A_1A_2\cos\delta]^{1/2} \tag{38}$$

and

$$\Phi = \omega t + \Phi'. \tag{39}$$

The value of Φ' can be determined from the relations

$$\cos\Phi' = (A_1 + A_2\cos\delta)/A \tag{40}$$

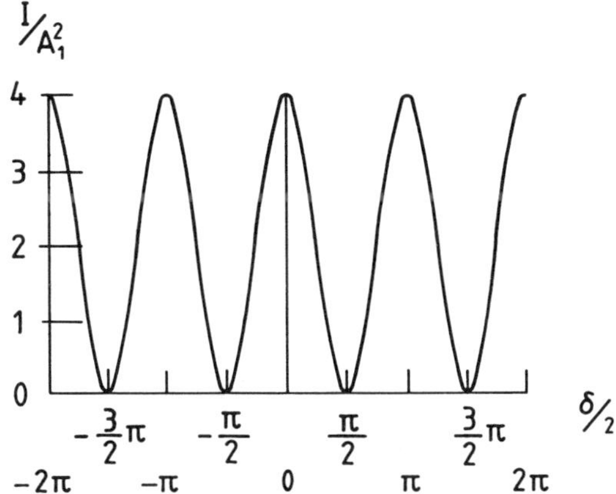

FIG. 17. Interference of two waves. Normalized irradiance as a function of phase difference. (After Klein and Furtak, 1986.)

and

$$\sin\Phi' = A_2 \sin\delta/A. \tag{41}$$

For $A_1 = A_2$, we obtain for the irradiance

$$I = |E^2| = 2A_1^2(1 + \cos\delta) = 4A_1^2 \cos^2(\delta/2), \tag{42}$$

which is represented in Fig. 17.

Figure 18 shows two setups for two-source interference. In both cases, the phase difference is given by

$$\delta \approx \frac{2\pi}{\lambda}\frac{ax}{D}, \tag{43}$$

and a pattern of $\cos^2$ shape on the screen can be detected. Here we have neglected possible phase changes upon reflection at the mirror. The term a is the separation of the two sources, D is the distance between the planes of the sources and that of the screen, and x is the position in the screen plane measured from the axis of symmetry.

4.1.1.1 Newton's Rings and Fizeau Fringes. When a (plano-)convex lens is placed on a plane glass surface, rings around the point of contact will occur, which are especially well pronounced in reflection and for monochromatic light (Fig. 19). The radius of the mth dark ring can be shown to be approximately given by $x_m = \sqrt{m\lambda R}$ where R is the radius of curvature of the lens side facing the plane. This is valid for near-normal incidence.

Figure 20 shows another situation that is normally described with sufficient precision using two-beam interference. Again near-normal incidence lighting and viewing with monochromatic light are supposed. In this case, the relation between wedge angle α and fringe separation s is given by

$$\alpha s = \lambda/2. \tag{44}$$

In both cases, the waves reflected from the

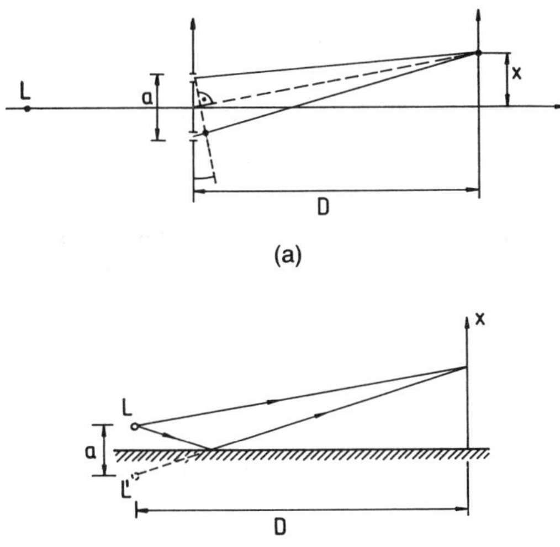

FIG. 18. (a) Young's double-slit experiment and (b) Lloyd's mirror. For details see text. (After Klein and Furtak, 1986.)

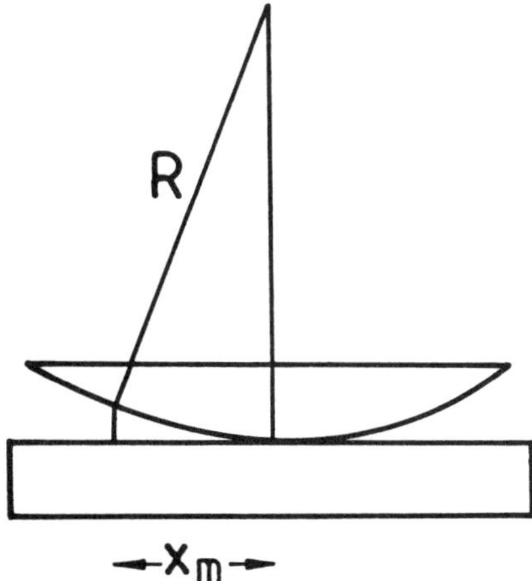

FIG. 19. Setup for the observation of Newton's rings (after Klein and Furtak, 1986).

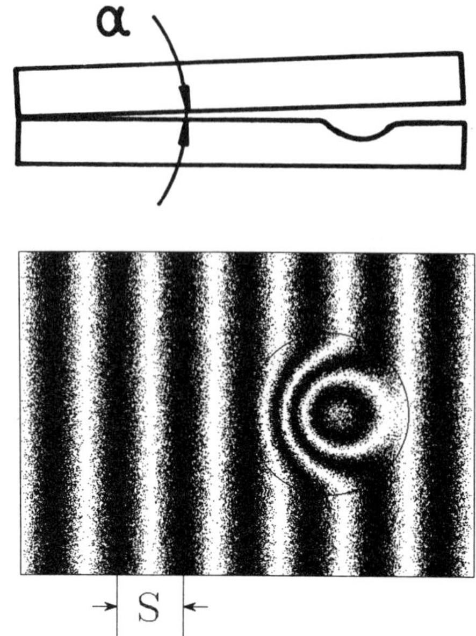

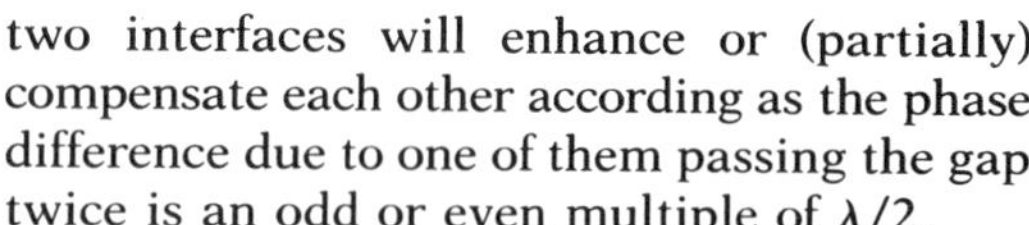

FIG. 20. Calculated Fizeau fringes caused by an air wedge with a circular depression. The depth of the depression is strongly exaggerated for clarity. The actual depth used for computing the fringe pattern is approximately one wavelength. (After Klein and Furtak, 1986.)

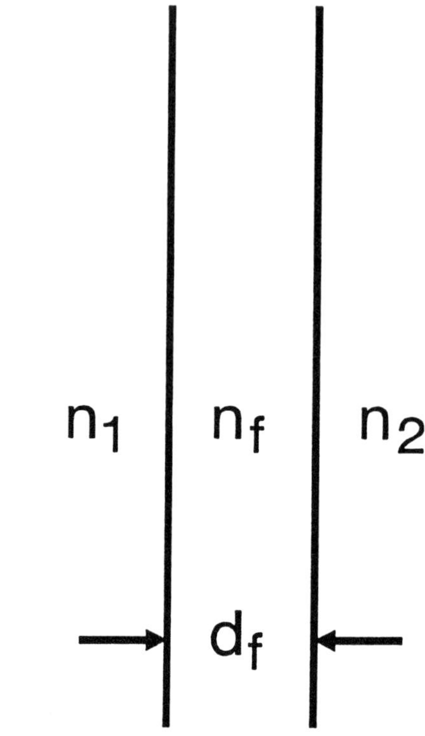

FIG. 21. Single-layer antireflection coating.

two interfaces will enhance or (partially) compensate each other according as the phase difference due to one of them passing the gap twice is an odd or even multiple of $\lambda/2$.

4.1.1.2 Antireflection Coating. Figure 21 shows the most simple antireflection coating. To reduce Fresnel reflection losses at (air-glass) interfaces, a thin film of transparent material can be deposited on the surface using, e.g., vacuum deposition. For normal incidence and a given wavelength, the reflection vanishes if the film thickness d_f satisfies the condition

$$d_f = \lambda/4n_f, \tag{45}$$

where n_f is the refractive index of the film material and is chosen to be

$$n_f = \sqrt{n_1 n_2}, \tag{46}$$

where n_1 and n_2 are the refractive indices of the adjacent media. The reflections from the two interfaces are then approximately equal in amplitude but out of phase by half a wavelength and thus interfere destructively, canceling almost completely. In practical systems, more elaborate coatings consisting of a stack of several thin-film layers are used to achieve reflex suppression over an extended wavelength range. At the same time, the permissible range of angles of incidence normally increases.

For other thin-film coatings such as high-reflection coatings or dichroic mirrors, the reader is referred to the appropriate articles of this Encyclopedia.

4.1.1.3 The Michelson Interferometer. Interferometers have found widespread applications in precision measurements of lengths, positions, and shapes. Microscope interferometers are used for surface analysis or process quality control. As these are handled in detail in a dedicated entry (see INTERFEROMETERS), only one type of interferometer is presented here.

The Michelson interferometer is depicted in Fig. 22. Light waves coming from the source are split by a semireflecting beam splitter and propagate through the two arms of the interferometer; here the reflected wave passes through a compensator plate, is reflected from

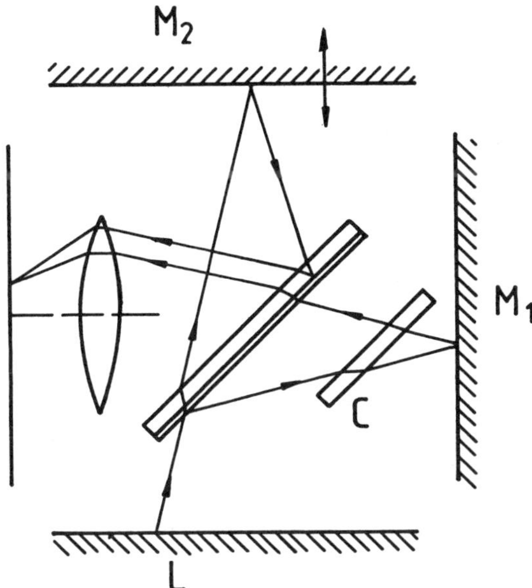

FIG. 22. The Michelson interferometer. M_1 is the fixed mirror and M_2 is the moving mirror. C denotes the compensator plate. (After Klein and Furtak, 1986.)

a fixed mirror M_1, and traverses the compensator plate once again, while the transmitted wave is reflected from a movable mirror M_2 forming the second arm of the interferometer.

The compensator plate, made of the same material as the beam splitter substrate, assures symmetry of optical path lengths. Otherwise, dispersion effects would complicate considerably the theoretical description of this device under broadband illumination. Besides, this plate symmetrizes the interferometer paths. Although this can be done by simply moving one of the mirrors for plane waves, this is not possible for divergent (or convergent) waves without such a plate.

After passing through the arms of the interferometer, the two waves are recombined by the same partially transmitting mirror, thus creating interference. Tiny differences in the surface profiles of the two mirrors appear as interference fringes if the system is adjusted at a mean path difference of zero between the two arms. Minute tilts of the mirrors will result in parallel fringes in the output.

Alternatively, with a collimated source, length measurements are possible. For simplicity, normal incidence on the plane mirrors is assumed. Moving one mirror will then result in a modulation of output light intensity following a $\cos^2$ law. The zeros of this signal being separated by mirror movements of $\lambda/2$ can be counted or even interpolated for length measurements. For extreme precision, it is necessary to account for the refractive index of air, if the measurements are not performed in vacuum.

For a plane wave incident normally on the mirrors, the irradiance on the output screen will be given by

$$I = I_0 \cos^2(\delta/2) \tag{47}$$

with $\delta = 2\pi\Delta x/\lambda$ and Δx the difference in path length in the two arms.

Michelson interferometers are often used to measure the wavelengths of lasers by comparing them to known lengths (e.g., gauges) or to the known wavelength of another laser. These devices are often called wave meters.

4.1.2 Multiple-Beam Interference The case of multiple-beam interference is treated similarly. With the extension of the above notation to N beams, the total field is

$$E = A_1 \exp(i\phi_1) + A_2 \exp(i\phi_2) + \cdots + A_N \exp(i\phi_N). \tag{48}$$

For equal amplitudes and with $\phi_n = \omega t + (n - 1)\delta$, this yields

$$E = A_1 \exp\left[i\left(\omega t + \frac{(N-1)\delta}{2}\right)\right] \frac{\sin(N\delta/2)}{\sin(\delta/2)}. \tag{49}$$

For the irradiance for a given value of δ there follows hence

$$I(\delta) = I(0)\left[\frac{\sin(N\delta/2)}{N\sin(\delta/2)}\right]^2. \tag{50}$$

This distribution has maxima for $\delta = 2m\pi$ where m is an integer. Between these principal maxima, $N - 1$ minima occur, separated by $N - 2$ minor maxima. For large values of N, the distribution $I(\delta)$ approaches limiting values in the (minor and major) maxima that depend only on δ.

4.1.2.1 Diffraction Gratings. A diffraction grating is a periodic array of transmitting or reflecting structures. With these spaced at equal distances [Fig. 23(a)], the light that is detected at an angle β for a plane wave incident at an angle α is given by the equations presented in Sec. 4.1.2 with $\delta = g(\sin\alpha + \sin\beta)$. The separation of the repetitive structures g is called the grating period. The irradiance distribution according to Eq. (49) for $N = 10$ is shown in Fig. 23(b). Usually, N is much larger, however. For higher values of δ, additional maxima in this curve occur; the diffraction order is the corresponding value of δ divided by 2π (e.g., first order). For negative δ, negative diffraction orders occur (e.g., −second order).

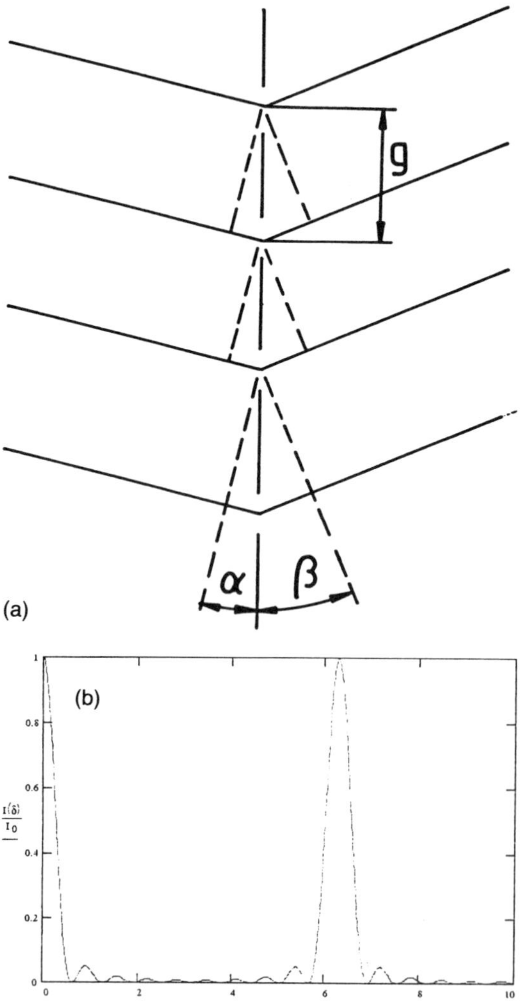

FIG. 23. (a) Schematic view of a diffraction grating (after Klein and Furtak, 1986). (b) Irradiance distribution for a grating with $N = 10$ grooves. $N = 10$, $I_0 = 1$, $I(\delta) = I_0[\sin(N\delta/2)/N\sin(\delta/2)]^2$, $\delta = 0.00001$, 0.00501 ... 10.000.

Gratings can be amplitude or (for higher efficiency) phase gratings. An amplitude-transmission grating has transmitting and opaque stripes, whereas for a phase-transmission grating grooves are produced as the surface profile of a transparent carrier. An amplitude-reflection grating has reflecting and nonreflecting stripes, whereas a reflection phase grating is a mirror with a periodic surface profile. The spectral resolution of gratings is discussed in Sec. 4.2.3.

Phase gratings with planar groove surfaces can be *blazed* for a given wavelength by making the angle of the triangular groove profile satisfy the law of refraction for a transmission grating or the law of reflection for a reflection grating. In this case, a reflection grating, e.g., consists of many microscopic strip mirrors reflecting the incident light in the direction of the desired order. Blazing implies that the angle of incidence also is known at least approximately. Blazing concentrates the diffracted light into one desired diffraction order.

4.1.2.2 The Fabry-Pérot Interferometer. A Fabry-Pérot interferometer consists, e.g., of two plane high-quality mirrors adjusted parallel to each other with a separation d (Fig. 24). Again, multiple-beam interference occurs, but in this case each consecutive beam is weaker as a result of the reflection coefficient, which will always be smaller than unity. The mathematical treatment is a bit different as an infinite number of beams has to be taken into account, and for (identical) transmission T_m and reflection R_m of the two mirrors the transmission T of the arrangement

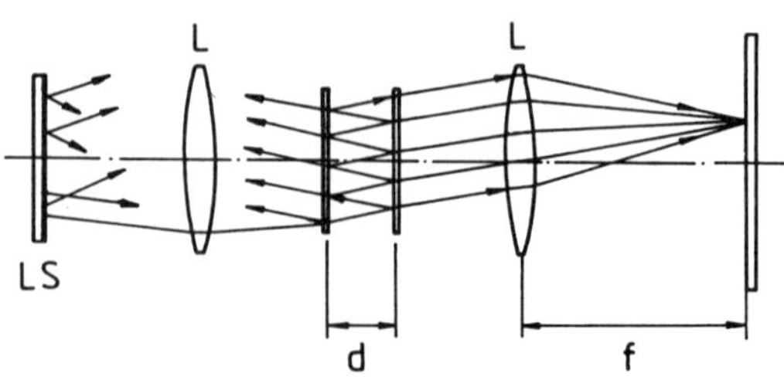

FIG. 24. A plane Fabry-Pérot interferometer. *LS* is the light source, *L* is used to denote lenses, and *f* is the focal length of the second lens, which is used for imaging the interference fringes onto a screen. (After Hecht and Zajac, 1974.)

is

$$T = \frac{T_m^2}{1 + R_m^2 - 2R_m \cos\delta}. \tag{51}$$

The total phase difference δ between successive beams is the sum of the propagation difference $\delta_1 = (2\pi/\lambda)nd \cos\theta$ and twice the phase change δ_2 experienced upon reflection. The maximum transmission is then given by

$$T_{\max} = T_m^2/(1 - R_m)^2. \tag{52}$$

Using $T_{\max}$ and $F = 4R_m/(1 - R_m)^2$, the expression for T can be rewritten as

$$T = \frac{T_{\max}}{1 + F \sin^2(\delta/2)}. \tag{53}$$

The ratio of the separation of two adjacent transmission maxima to the width of these maxima for monochromatic light (both measured in units of delta) is called the finesse $\mathcal{F}$:

$$\mathcal{F} = \frac{\pi\sqrt{F}}{2} \approx \frac{\pi}{1 - R_m}. \tag{54}$$

4.2 Diffraction

Diffraction can be regarded as the interference of infinitely many infinitely weak sources. So most of the sums in the previous section turn into integrals. However, diffraction is more than that. The diffraction integral is one of the cornerstones of linear optics. It should be noted that diffraction also limits the validity of the fundamental statement of geometrical optics that light travels in straight lines. Light may actually "bend" around obstacles and thus illuminate the geometrical shadow area.

A rigorous theory based on Maxwell's equations and appropriate boundary conditions is available for a vectorial treatment. However, this rigorous treatment is often dispensable and various simplified scalar treatments have been developed, e.g., by Young, Fresnel, and Kirchhoff. The necessity of using a vectorial (i.e., polarization-sensitive) treatment is self evident in the case of deflection angles where the electric field vectors from different elementary waves are not sufficiently parallel anymore. Other examples are pinholes with openings comparable to the wavelength of the diffracted light and gratings with a transmission function varying considerably over dimensions of about one wavelength. For these demanding applications of diffraction theory, the reader is directed to the Further Reading.

4.2.1 Fresnel Diffraction For most practical purposes, the Fresnel–Kirchhoff integral of diffraction is sufficient. It states the scalar field relation

$$E(r) = \frac{i}{\lambda} \int E_{\text{in}}(\tilde{r}) \frac{e^{-ikR'}}{R'} \frac{\cos\theta_{\text{in}} + \cos\theta_{\text{out}}}{2} \, d\sigma. \tag{55}$$

For monochromatic light of wavelength λ, it yields the field at a given point r in space, if the field of the incident wave E_{in} is known for all points $\tilde{r}$ on the surface σ of the diffracting aperture(s), these points being separated by distances R' from the point r. The wave number k is defined as $k = 2\pi/\lambda$. The terms θ_{in} and θ_{out} are the angles between the directions of propagation of the incident or diffracted wave relative to the normal to σ at $\tilde{r}$. The only deviation from a straightforward extension of interference (summation) to diffraction (integration) seems to lie in the directional cosines that describe the component of the incoming and outgoing waves and the normal to the surface of integration.

The factor $e^{-ikR'}/R'$ in the diffraction integral represents a spherical wave emerging from the aperture plane toward the detection plane. Several approximations to the actual phase differences for different coordinate pairs in these two planes can be made depending on the particular geometric limits under consideration.

A first approximation neglects all terms contributing to R', and thus to the phase, that are of higher than quadratic order in the aperture coordinates. This parabolic or Fresnel approximation for the phase is often used when the exact solution is too complicated and a linear or Fraunhofer approximation would result in insufficient accuracy. For the arrangement depicted in Fig. 25, the Fresnel integral then takes the simplified form

$$E(P') = \frac{i}{2} E_0(P') \iint \tau(\eta_x, \eta_y) \times \exp\left[-\frac{i\pi}{2}(\eta_x^2 + \eta_y^2)\right] d\eta_x d\eta_y \tag{56}$$

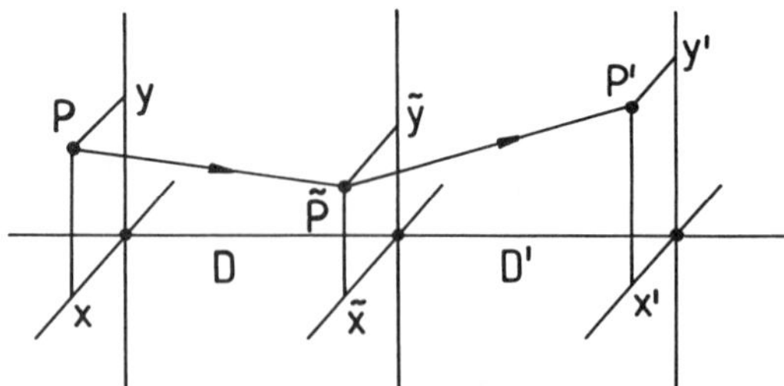

FIG. 25. Configuration for Fresnel diffraction for a point source at P (after Klein and Furtak, 1986).

with the normalized aperture-plane coordinates

$$\eta_x = \sqrt{\frac{2}{\lambda d}}(\tilde{x} - \tilde{x}_u), \tag{57}$$

$$\eta_y = \sqrt{\frac{2}{\lambda d}}(\tilde{y} - \tilde{y}_u), \tag{58}$$

and

$$1/d = 1/D + 1/D'. \tag{59}$$

Here τ is the transmission function of the aperture, whose value determines the amplitude-transmission coefficient, which may be complex to include phase shifts. The tilde denotes aperture-plane coordinates, and the suffix u corresponds to that point in the aperture plane that lies on the straight line connecting the source point and the observation point. The terms D and D' are the distances of the source plane and of the plane of observation from the aperture plane.

Here we limit our presentation to three well-known examples of Fresnel diffraction.

4.2.1.1 Rectangular Aperture. In this case, the transmission function can be factored:

$$\tau(\eta_x, \eta_y) = \tau_x(\eta_x)\tau_y(\eta_y), \tag{60}$$

and thus the Fresnel integral can be solved separately for the two coordinates x and y:

$$E(P) = \frac{i}{2} E_0(P) I_x I_y, \tag{61}$$

with

$$I_x = \int_{-\infty}^{+\infty} \tau_x(\eta_x) \exp\left(-\frac{i\pi}{2}\eta_x^2\right) d\eta_x, \tag{62}$$

$$I_y = \int_{-\infty}^{+\infty} \tau_y(\eta_y) \exp\left(-\frac{i\pi}{2}\eta_y^2\right) d\eta_y. \tag{63}$$

The complex Fresnel integral

$$I(\eta) = \int_0^{\eta} e^{-(i\pi/2)u^2} du = C(\eta) + iS(\eta) \tag{64}$$

with

$$C(\eta) = \int_0^{\eta} \cos\left(\frac{\pi}{2}u^2\right) du \tag{65}$$

and

$$S(\eta) = \int_0^{\eta} \sin\left(\frac{\pi}{2}u^2\right) du \tag{66}$$

may be used to express these integrals as

$$I_x = I(\eta_{x2}) - I(\eta_{x1}), \tag{67}$$

$$I_y = I(\eta_{y2}) - I(\eta_{y1}), \tag{68}$$

where $x1$, $x2$, $y1$, $y2$ denote the upper and lower limits of the interval wherein the transmission function is unity, whereas its value is zero outside. The plot of the Fresnel integral in the complex plane yields the well-known Cornu spiral. The values of the integral can be found in tabulated form. The Fresnel integral simply gives the relative amplitudes of the diffracted wave's field as compared to the amplitude that would result with the incident wave reaching the point of observation without a diffracting obstacle.

4.2.1.2 Half-Plane Obstruction. An opaque screen ranging from $\tilde{x} = -\infty$ to $\tilde{x} = \tilde{x}_1$ and from $\tilde{y} = -\infty$ to $\tilde{y} = +\infty$ can be treated as a special case of a rectangular aperture by using Eqs. (56), (57), (66), and (67) to calculate I_x and I_y. We then have $I_y = 1 - i$ and $\eta_{x2} = \infty$. Changing to the representation in observation-plane coordinates using

$$\eta_{x1} = (x_1' - x')/F \tag{69}$$

with

$$F = [\lambda D'(D + D')/2D]^{1/2}, \tag{70}$$

we can then calculate I_x for any point in the observation plane. The resulting field is obtained simply as the vector joining the point corresponding to η_{x1} with the center of the $+\infty$ lobe of the Cornu spiral (Fig. 26). When x' has large negative values, η_{x1} is quite large and the resulting points on the curve lie close to those for $\eta_{x1} = +\infty$; the vector representing I_x is small and increases monotonically with increasing x'. For $x'_1 = x'$, the value of the integral is $\frac{1}{2}(1 - i)$ with an absolute value half as big as that obtained in the absence of the obstacle (as is to be expected from symmetry arguments); for further increase of x', the absolute value of the integral converges with decreasing oscillations toward the limiting value for $x' = +\infty$. At the point of the geometrical shadow edge, the field has half its maximum value resulting in an irradiance of a quarter of the maximum value. This result is important when determining the position of an edge from its measured shadow.

4.2.1.3 Arago's Bright Spot. An interesting phenomenon occurs for circular obstacles: At a distance behind the obstacle large enough to assure that the *inclination factor*, i.e., the term $\frac{1}{2}(\cos\theta_{\text{in}} + \cos\theta_{\text{out}})$ in Eq. (54), is still approximately unity, Fresnel diffraction theory correctly predicts that the intensity on the axis be practically equal to the value without an obstacle, a phenomenon called Arago's bright spot, a reference to the French researcher who first demonstrated this somewhat surprising phenomenon.

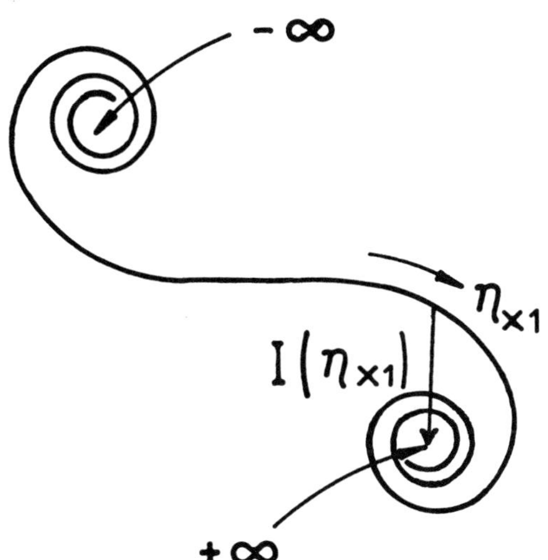

FIG. 26. Discussion of half-plane diffraction using the Cornu spiral (after Klein and Furtak, 1986). Actually, $I(\eta_{x1})$ stands for I_x here.

4.2.2 Fraunhofer Diffraction When it is sufficient to take into account the phase changes that are linear in the aperture coordinates, the quadratic terms are discarded and Fraunhofer diffraction is obtained. This is true, for example, in a situation where both the source and the detection planes are at very large distances from the diffracting aperture so that the waves impinging on the aperture are virtually plane waves (these correspond to bundles of parallel rays; see Sec. 4.2.2.1 and especially Fig. 27) and so that waves converging toward a given point in the detection plane are essentially plane over the area of the aperture. This situation is quite similar to the one treated above in discussing the diffraction grating.

For simplicity, we assume a plane wave impinging normally on the aperture plane with coordinates ξ and η. In this case, the amplitude of a plane wave diffracted with directional cosines p and q in the directions defined by the aperture-plane coordinates is given by the integral

$$U(p, q) = C \iint G(\xi, \eta) e^{-(2\pi i/\lambda)(p\xi + q\eta)} d\xi d\eta, \tag{71}$$

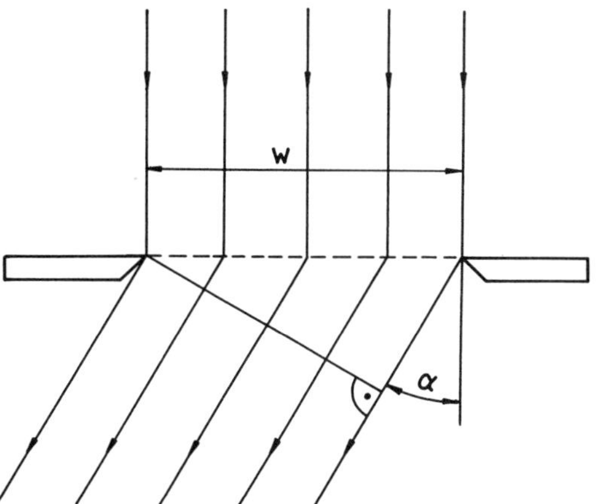

FIG. 27. Fraunhofer diffraction. The incident wave is normal to the aperture plane for reasons of simplicity only. (After Alonso and Finn, 1976.)

where C is a constant and G is the pupil function describing the phase and amplitude transmission of the aperture plane. The physical meaning of this equation is very important: The far-field diffraction pattern is the Fourier transform of the diffracting aperture. This has led to important applications of Fraunhofer diffraction in Fourier optics and other fields.

Again examples are given for the simplest cases only, i.e., for a linear slit and for a circular and an annular aperture.

4.2.2.1 Linear Slit, Rectangular Aperture. The diffraction integral is a function deriving the field in an arbitrary point from the known field distribution on a given surface, usually a plane or sphere. It is useful for calculating the resulting field distribution on another surface. For an infinitely long straight slit (see Fig. 27), the diffraction integral in linear approximation is

$$I = I_0\left(\frac{\sin u}{u}\right)^2 \tag{72}$$

with $u = \pi w \sin\alpha/\lambda$ where w is the slit width, α is the diffraction angle shown in Fig. 27, and λ is the wavelength. This is evidently proportional to the square of the Fourier transform of the transmission function, so that a rectangular transmission profile yields the well-known sinc function.

A rectangular aperture can be factorized to be described by the product of two orthogonal slits. The resultant field distribution is then given by the product of two functions of the above form, each with a value of w_1 or w_2 that corresponds to the width of the opening in that direction and with its value of α_1 or α_2 representing the deviation in that coordinate.

4.2.2.2 Circular Aperture. For a circular aperture such as the common stops in optical instruments like telescopes and microscopes, the far-field diffraction pattern is given by

$$I = I_0\left[\frac{2J_1(ka\sin\alpha)}{ka\sin\alpha}\right]^2, \tag{73}$$

where the common abbreviation $k = 2\pi/\lambda$ has been employed. Here a is the radius of the diffracting aperture supposedly illuminated by a plane wave at normal incidence, J_1 is the first-order Bessel function, and α is the diffraction angle. This formula describes diffraction for a plane wave incident normally on the aperture plane. It is valid if the angular deviation is not too large so that the inclination factor $Q = \frac{1}{2}(\cos\theta_i + \cos\theta_d)$ is almost equal to 1. Here θ_i and θ_d are the angles that the wave normals of the incident and diffracted wave form with the surface normal of the aperture plane. This well-known diffraction pattern is characterized by a central peak encircled by a dark ring and containing some 84% of the total energy. It is usually called the Airy disk when observed in the image plane of the source.

4.2.2.3 Annular Aperture. An annular aperture can be handled with this formula quite simply as the weighted difference of the fields for two circular apertures. The irradiance is then obtained from the square of that difference:

$$I = I_0\left[\frac{a^2}{a^2-b^2}\frac{2J_1(ka\sin\alpha)}{ka\sin\alpha} - \frac{b^2}{a^2-b^2}\frac{2J_1(kb\sin\alpha)}{kb\sin\alpha}\right]^2, \tag{74}$$

where a and b are the outer and inner radii of the aperture.

4.2.3 Fundamental Consequences of Diffraction When the lens edge is the diffracting aperture, as for the objective lens of an astronomical telescope with focal length f and diameter d, the diameter of the Airy disk is given by $D = 2f\alpha$ where $\alpha = 1.22\lambda/d$ is the angle at which the first minimum of the diffraction pattern occurs. According to a widespread resolution criterion (or convention), two equally strong point sources are resolved if their mutual separation is equal to this angle. This is the angular resolution of a telescope with a circular aperture.

Similarly, for a microscope, the spatial resolution D is given by the same formula as above, where the object distance is assumed to be about equal to the focal length of the objective lens. Then D is given by $D = 2.44\lambda f/d$. The product of the refractive index n of the medium filling the object space and of the sine of the angle α subtended by the system axis and the most oblique rays passing the objective to contribute to the image is called the numerical aperture NA of the system. For small values of α, we have $\sin\alpha \approx d/2f$ so that

NA is roughly given by NA $= d/2f = r/f$ and the resolution is $D = 1.22\lambda/\text{NA}$.

The resolution $R = \lambda/\Delta\lambda$ of spectrographs is also fundamentally based on the limiting effect of grating or prism boundaries; the results are

$$R = N \times m \tag{75}$$

for the resolution of a grating spectrometer where N is the number of grooves illuminated and m is the diffraction order used. For a prism spectrometer, we obtain

$$R \approx b\frac{dn}{d\lambda} \tag{76}$$

for the resolution, where b is the maximum geometrical path differences for the limiting rays in the prism.

Obviously, these theoretical limits are only attained for high-quality equipment with wave fronts deviating by $\lambda/10$ or even less from their theoretical shape, which is therefore often characterized as "diffraction limited" in its performance.

For completeness, we include the resolution of the Fabry-Pérot interferometer also:

$$R = \frac{\pi\sqrt{F}}{2}m \tag{77}$$

with $\pi\sqrt{F}/2 \approx \pi/(1 - r)$ for mirror reflectivities r close to unity and m the interference order.

4.3 Thin-Film Optics

Dielectric antireflection coatings of a single layer are a very simple example of thin-film optics. With stacks of thin layers, almost arbitrary spectral reflection/transmission curves can be designed. With a matrix formalism, the calculation of these curves for thin-film systems with known parameters (film thicknesses and refractive indices including dispersion) is possible. For any interface, the relation between the (complex) amplitudes of waves impinging from and leaving toward both sides is described by a 2×2 matrix. In the same way, the passage of waves through a homogeneous (film) slab of material can be described. Simple matrix multiplication then permits calculation of the performance of more complex coatings; for a more detailed treatment, the reader should refer to the literature.

In practice, however, thin-film properties may differ from bulk parameters because of voids, film tension, etc. For coatings used at non-normal incidence, polarization-dependent designs are also possible. Computer programs for coating design are commercially available.

High-pass, low-pass, and band-pass filters, dichroic mirrors, polarizing beam splitters, and dielectric high reflectors for lasers are only some examples of useful specialized coating designs. However, until now no simple algorithm allows even the approximate inverse determination of the coating design from a desired or measured spectral reflectivity/transmission.

4.4 Holography

As a result of interference, it is possible to record not only the amplitude or intensity information of an electric field but also its phase distribution so that the information stored, e.g., in a photographic emulsion, may be used to restore the wave front completely. Such wave-front recordings are called holograms. Figure 28(a) shows a setup that may be used to record such a hologram. A monochromatic source is split into two waves. One serves as a reference wave, and the other is used to illuminate the object of which a hologram is to be taken. The scattered wave from the object is the wave to be recorded, with the other wave serving as a phase reference.

In Fig. 28(b), the inverse process is shown. The processed emulsion is illuminated by the reference wave, and the interference pattern recorded on the film diffracts the incident wave in such a way that a wave with the recorded phase front (and approximately the recorded amplitude) is generated. Thus, an apparently three-dimensional image of the object used in recording the hologram is obtained. These reconstructed wave fronts permit depth perception in the image, unlike normal photographs. For a photograph, the source points for (spherical) elementary waves will lie in the plane of the emulsion, whereas for a hologram wavefront reconstruction permits them to lie behind or in front of that plane as well.

As a film permits sequential recording of

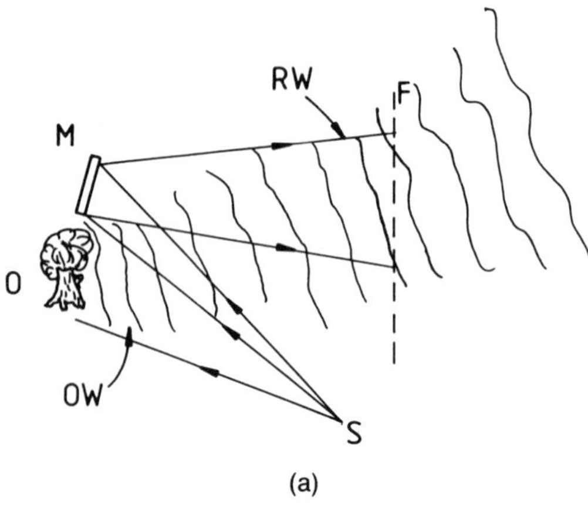

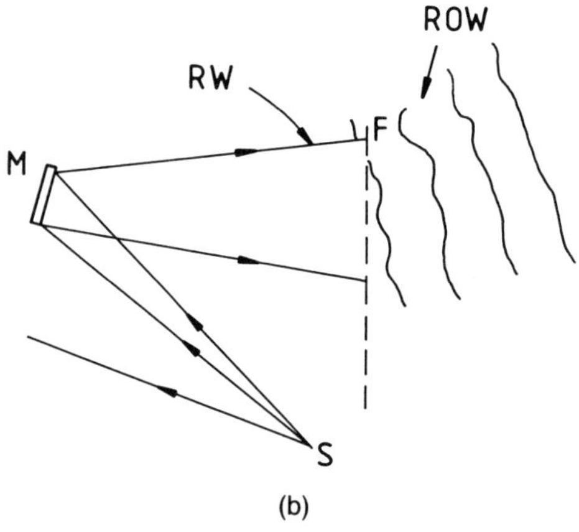

FIG. 28. Holography: (a) recording setup and (b) reconstruction setup. *S* denotes the source, *O* the object, *M* a mirror, *F* the recording film, *OW* the object wave, *RW* the reference wave, and *ROW* the reconstructed object wave.

several wave fronts, these may be superimposed using holography. This effect is used in interferometric holography where an object wave is recorded before and after deformation. Thus, deformations become directly visible as interference fringes in the reconstructed holographic recording.

4.5 Fourier Optics

As mentioned above, the far field of a given structure is proportional to its Fourier transform assuming plane-wave illumination. (The diffraction pattern is therefore proportional to the square of this transform.) If a lens is inserted, the far field is imaged from infinity to the focal plane of that lens. This plane is therefore sometimes called the Fourier plane. Insertion of stops in the Fourier plane and retransformation with a second lens thus permit changing the distribution of spatial frequencies for a given image. Two applications shall be considered here, as follows.

If a pinhole is placed in the Fourier plane, all spatial noise, i.e., high spatial frequency deviations, e.g., from a Gaussian profile, is suppressed whereas the TEM_{00} mode of a laser (see LASER PHYSICS), having an essentially Gaussian field distribution, would pass without attenuation. Such a device is therefore called a spatial filter or mode filter.

If a picture has a periodic structure, e.g., from rastering or from successive recording (like many space photographs), these periodic disturbances will be manifest in the Fourier plane as pronounced peaks spaced regularly in the corresponding direction(s). If these are suppressed by an appropriate stop, the periodic disturbance in the picture will vanish almost without loss of information but with a tremendous increase in subjective image quality.

5. ANISOTROPIC MEDIA

Anisotropic media have opened up many new fields of applications for optics, e.g., in polarization optics and electro-optic modulation and switching. The most important anisotropic media are crystals of various symmetry classes. However, fluids may become anisotropic under external fields, too. Recently, liquid crystals have found an increasing number of applications for which modulating speed requirements are less stringent.

5.1 Linear Birefringence

Linear birefringence or double refraction, as it is sometimes called, is observed in media that are anisotropic by nature (crystals) or by external effects (e.g., electric fields). Birefringent crystals have an optical axis; light traveling along the optical axis has a refractive index, and thus a velocity, independent of its polarization. Any plane containing the optical axis is called a principal plane. For light waves traveling in such a plane, two states of polarization can be identified. The ordinary ray is characterized by its electric field vector being normal to the optical axis and thus also perpendicular to the principal plane. The electric field vector of the extraordinary ray, however, lies in the principal plane. For the ordinary ray, the propagation velocity is independent of direction, the elementary waves and the index surfaces being spheres. For the extraordinary beam, the elementary waves and the index surfaces are ellipsoids tangent to this sphere at the points determined by the optical axis through the sphere's center (Fig. 29). If the extraordinary ray has a larger wave velocity than the ordinary ray, the material is said to exhibit positive birefringence (e.g., crystalline quartz), whereas the opposite situation is called negative birefringence (e.g., calcite).

When a ray of unpolarized light is incident normally on a birefringent plate, the ordinary beam is not deflected upon transmission through each surface, just as one would expect from Snell's law. The extraordinary ray behaves quite differently: The ray is deflected inside and displaced outside the crystal. Only when the optical axis lies in the face plane of the crystal or is orthogonal to it will the transmitted ray be normal to the surface inside the crystal.

This behavior of the extraordinary ray is illustrated in Fig. 30. While the light travels from A to B outside the crystal (wave fronts normal to ray vector), its ordinary and extraordinary components travel from B to C and from B to D inside the crystal (wave fronts not normal to ray vector in the latter case). So for this situation, the behavior of the extraordinary ray is really most extraordinary. Much more complicated phenomena are observed for biaxial crystals, which we leave to be studied in the literature.

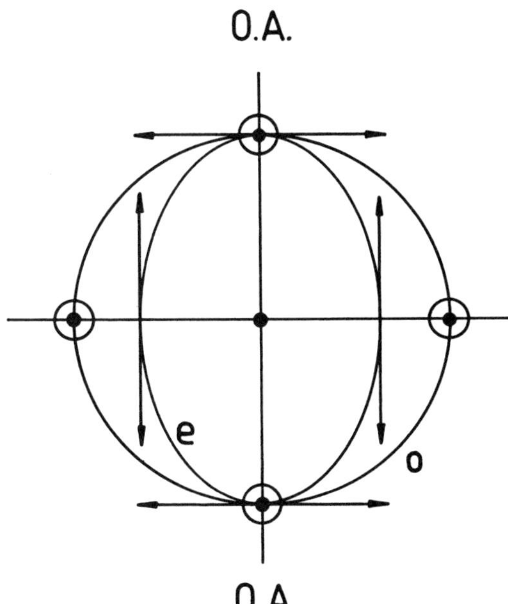

FIG. 29. Elementary wave surfaces for ordinary (*o*) and extraordinary (*e*) rays in a uniaxial birefringent crystal. *O.A.* denotes the optical axis. (After Klein and Furtak, 1986.)

5.1.1 Crystal Polarizers As stated above (see Sec. 2.3), Fresnel reflection can be used to polarize a light wave. For more demanding applications, crystal polarizers are used. These can be divided into two groups: those that use total internal reflection and those that use birefringence to transmit both polarization states at different angles through an internal interface.

For brevity, only one example of each group will be shown. The Nicol prism is one of the oldest polarizers, though less common nowadays. Two prisms are cut from a birefringent calcite crystal as shown in Fig. 31. A beam entering the prism is split into the ordinary and extraordinary components as shown. At the interface between the first crystal and the optical-cement layer, the ordinary beam is totally internally reflected whereas the extraordinary beam passes on through the second prism and exits parallel to the input beam though with a lateral offset.

The Rochon prism depicted in Fig. 32 passes the ordinary beam undeflected and without an offset whereas the extraordinary beam is deviated in this case. Most polarization optical elements are made from quartz or calcite. For polarization prisms, calcite is

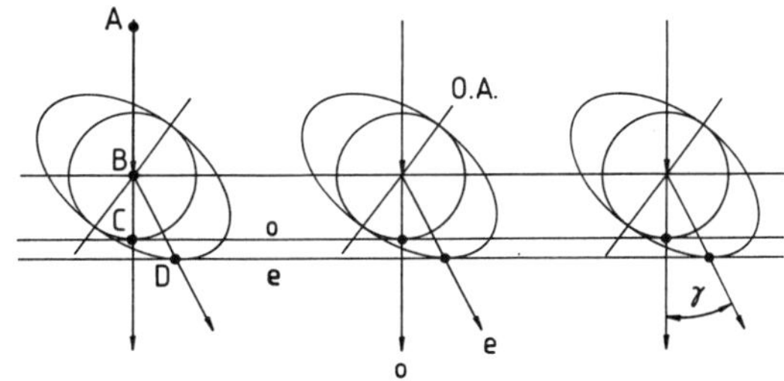

FIG. 30. Refraction of an extraordinary ray entering a birefringent crystal. γ is the deviation from straight transition. The crystal surface is parallel to the drawn wave fronts. (After Klein and Furtak, 1986.)

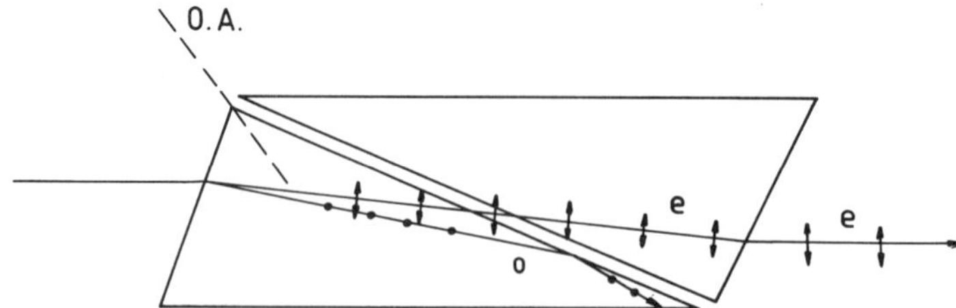

FIG. 31. The Nicol prism (after Klein and Furtak, 1986).

usually advantageous because of its larger difference of refractive indices for the ordinary and extraordinary beams.

5.1.2 Retardation Plates Retardation plates are plates cut from birefringent crystals with the optical axis usually lying in the plane of the parallel faces. The phases of the ordinary and extraordinary rays change by different amounts when passing through such a plate:

$$\Delta\phi_o = (2\pi/\lambda)n_o d \tag{78}$$

and

$$\Delta\phi_e = (2\pi/\lambda)n_e d. \tag{79}$$

Two cases for the value of the phase difference $\Delta\phi = \Delta\phi_e - \Delta\phi_o$ are of special interest, namely $\Delta\phi = \pm\pi/2$ and $\Delta\phi = \pm\pi$. A retardation plate satisfying the first condition is called a quarter-wave plate: a plate satisfying the second is called a half-wave plate.

5.1.2.1 Quarter-Wave Plate. For a quarter-wave plate, the equation

$$|n_e - n_o|d = (4n + 1)\,\lambda/4 \tag{80}$$

holds. If the integer n is zero, the quarter-wave plate is said to be a "zero-order" plate; this indicates a relatively small chromatic dependency of the phase difference. [It should be noted, however, that the chromaticity is still much more pronounced than for a Fresnel rhomb (Fig. 33). There, total internal reflection is used to create the phase difference between the polarization component with the electric field vector in the plane of incidence and the component with the electric field vector orthogonal to that plane. Such a rhomb, however, has the disadvantage of introducing a lateral offset between input and output rays.] In practice, the thickness of retardation plates may be much larger than the value corresponding to the above equation. Two plates for which the phase differences differ by $\pi/2$ are cemented together with their optical axes orthogonal to each other, thus yielding a net phase difference of $\pi/2$ and greater mechanical stability.

Linearly polarized light incident on such a plate will maintain its state of polarization when the electric field is parallel or orthogonal to the optical axis. For all other orientations of the plate, the light is split into two components, which will be retarded by $\pm\pi/2$ relative to each other after passing through

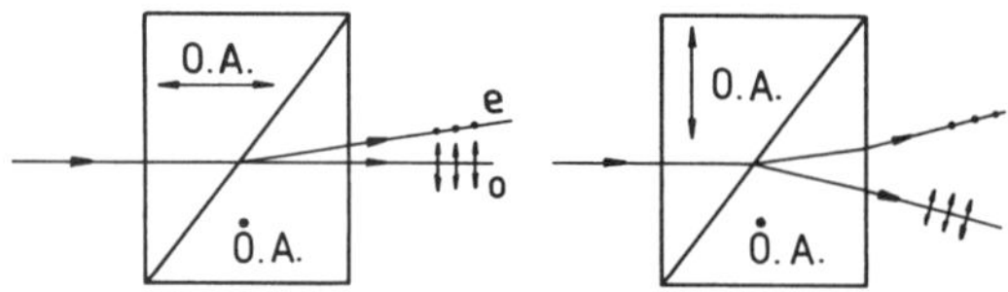

FIG. 32. The Rochon prism. A bold dot indicates that the optical axis *O.A.* is normal to the plane of the drawing. (After Klein and Furtak, 1986.)

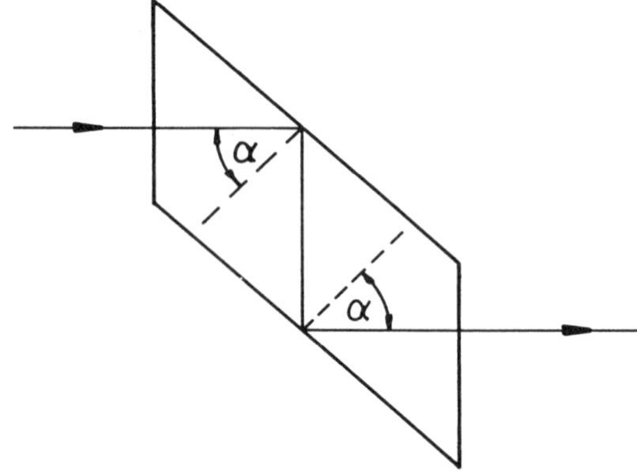

FIG. 33. A Fresnel rhomb. The phase shift upon reflection is obtained by inserting the complex angles that correspond to total internal reflection into the equations for the coefficients of reflection. (After Klein and Furtak, 1986.)

the plate; thus, generally elliptically polarized light will result. When the angle between the electric field vector and the optical axis is 45°, the two components will be equal (neglecting slightly different Fresnel losses), and circularly polarized light results. The transformation of linear into circular polarization states and vice versa is one of the most important applications of quarter-wave plates.

5.1.2.2 Half-Wave Plate. A half-wave plate has a thickness satisfying the equation

$$|n_e - n_o|d = (2n + 1)\,\lambda/2. \tag{81}$$

Naturally, two Fresnel rhombs in series will serve the same purpose and in this case may compensate each other's offset. For linearly polarized light passing through such a plate, the plane of polarization will be rotated by 2α when the plate is rotated by α. Thus, when the plate's axis is oriented at 45° to the electric field vector, the outgoing ray will be polarized orthogonal to the incident ray.

For circularly polarized light, the same exchange between the two orthogonal polarization states is obtained. Left circular light will be transformed into right circular light and vice versa.

5.1.3 Compensators Compensators are retardation plates with adjustable phase difference. Their main application is the measurement of unknown birefringences. A sample of the material under inspection and the compensator are placed between crossed polarizers, and the compensator is adjusted for minimum transmission of the analyzer. For this adjustment, the (known) retardation introduced by the compensator compensates the retardation due to, e.g., the crystal under study modulo an integer multiple of 2π; a measurement with a second plate of suitable thickness is then used to exclude this modulo uncertainty.

A widespread example of a compensator is the Soleil–Babinet compensator (Fig. 34). Two wedges with equal wedge angles and parallel optical axes form a plate of variable thickness. A plane-parallel plate with its optical axis orthogonal to the wedges allows adjustment to both positive and negative retardation values.

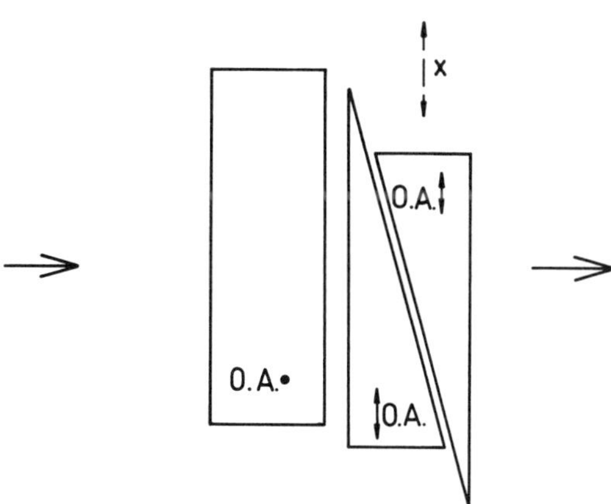

FIG. 34. A Soleil–Babinet compensator. *x* is the lateral displacement between the two wedges.

5.2 Circular Birefringence

For chiral systems (i.e., systems that are not invariant under inversion), circular birefringence occurs. The refractive indices for left and right circular polarized light are different. This effect is isotropic. Instead of circular birefringence, the expression optical activity is often used for this phenomenon.

For linearly polarized light passing through a medium exhibiting optical activity, the plane of polarization is rotated by an amount proportional to the path traveled. According to the sign of this rotation, right- and left-rotating substances are distinguished. Optical activity is usually caused by a substance in a solution. In this case, the rotation angle is also proportional to the concentration of that substance. Thus, polarimetry, i.e., the measurement of the rotation angle, is an important means of determining the concentration of such a substance in a solution (see POLARIMETERS AND POLARIZING SPECTROMETERS).

Some materials have different modifications that have the same chemical constitution but differ in molecular structure and thus may be left, right, or even nonrotating. In this case, the different effects add up to form a net rotation. Other substances, e.g., quartz, exhibit both linear birefringence and optical activity. In this case, an exact treatment is more complicated as the energy of a light beam is continuously redistributed among different polarization states.

5.3 Electro-optical Effects

Crystals and liquids can be polarized or oriented by electric fields that are constant or change with frequencies much lower than those of light oscillations. These phenomena have many important applications for modulation of the phase and the amplitude of light waves.

5.3.1 The Kerr Effect The Kerr effect was the first electro-optic effect to be discovered. When isotropic substances are placed in a homogeneous electric field, they behave like a uniaxial crystal with its optical axis parallel to the applied field. The difference between the two refractive indices—which can be seen in analogy to n_o and n_e—is given by

$$\Delta_n = n_e - n_o = \lambda K E^2, \tag{82}$$

where λ is the wavelength, E is the electric field, and K is the Kerr constant, which is usually but not necessarily positive. Because of the quadratic field dependence, the Kerr effect is sometimes called the quadratic electro-optic effect.

Today, crystals are commonly employed in Kerr modulators. The medium exhibiting the Kerr effect is placed in a parallel-plate capacitor so that an applied voltage U results in a phase shift

$$\Delta\phi = 2\pi K L E^2 = 2\pi K L (U/d)^2 \tag{83}$$

between the two polarization components (cf. Fig. 35). Here L is the length of the Kerr cell and d is the electrode separation. If such a cell is placed between crossed polarizers oriented at 45° relative to the electric field, the transmitted light power P as a function of input power P_0 is given by

$$P = P_0 \sin^2(\Delta\phi/2). \tag{84}$$

Unfortunately, typical values for a change from complete blocking to complete transmission, called half-wave voltage as the cell then behaves like a half-wave plate, are often in the high voltage range. Possibly the diminishing of the required modulation voltages is an important advantage of integrated optics (see below).

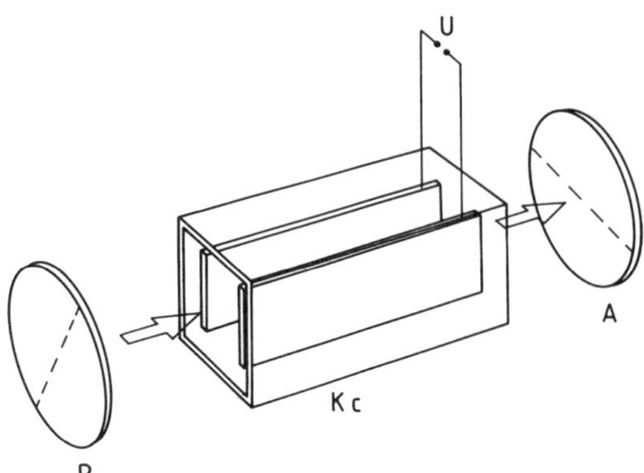

FIG. 35. A Kerr cell *Kc* between polarizer *P* and analyzer *A* (after Hecht and Zajac, 1974).

5.3.2 The Pockels Effect The Pockels effect or linear electro-optic effect is restricted to crystals lacking a center of symmetry. For these, the change in birefringence is linearly dependent on the applied fields and typically requires much lower voltages for device operation than the Kerr effect.

Figure 36 gives a sketch of Pockels cells. In (a) a longitudinal device is shown for which the electric field causing the retardation change is parallel to the propagation direction of the modulated light. In (b) the light travels orthogonal to the applied modulation field, thus forming a transverse configuration. These devices are advantageous as the voltage required for a phase change of π decreases with increasing cell length. Transit-time limitations to the modulation frequency can be overcome by using a traveling-wave electrode arrangement. Enormous progress has been achieved in integrated optical modulators requiring only modulation voltages of a few volts, which can be provided easily (see below).

5.4 Magneto-optical Effect, Faraday Effect

The magneto-optical effect having presumably the largest number of applications is the Faraday effect. Light traveling (anti-)parallel to a magnetic field experiences polarization rotation in media that exhibit this effect. For a homogeneous magnetic field, the rotation angle is given by

$$\beta = VBL, \tag{85}$$

where V is the Verdet constant characteristic for a given material, B is the value of the

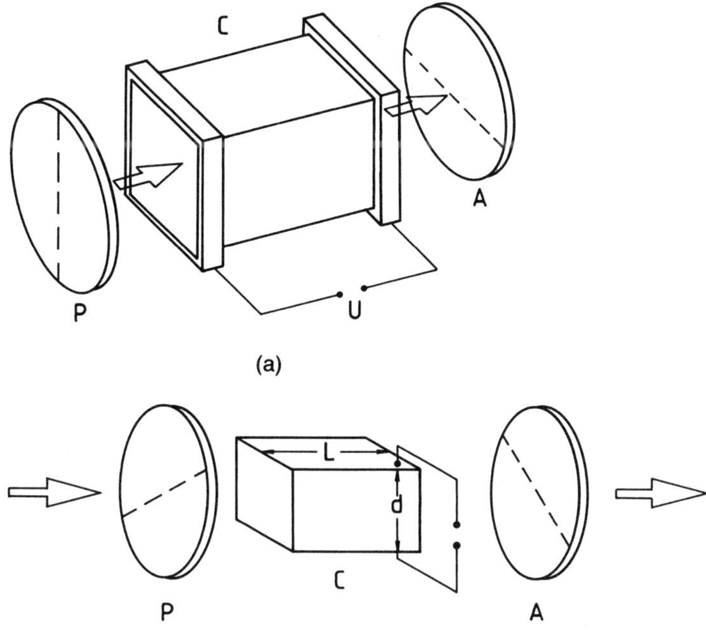

FIG. 36. Pockels cells: (a) longitudinal cell (after Hecht and Zajac, 1974), (b) transverse cell (after Yariv, 1991).

magnetic field, and L is the path length in the material.

A positive Verdet constant corresponds to a substance for which a magnetic field parallel to the direction of propagation seemingly turns the direction of the electric field counterclockwise when one looks toward the source.

A very important feature of the Faraday effect is its nonreciprocity. When linearly polarized light passing through a half-wave plate is reflected off a mirror to pass back through the same retardation plate, the net rotation of the plane of polarization is zero. When the retardation plate is replaced by a Faraday rotator, i.e., a medium exhibiting the Faraday effect placed in a magnetic field, however, the rotation is not zero but twice the value for a single pass.

This effect is used in optical isolators. Light with linear polarization passing first through a polarizer adjusted for maximum transmission, then through a Faraday rotator, and finally through a half-wave plate compensating the rotation caused by the Faraday rotator will pass through an analyzer oriented parallel to the first polarizer. If the light passes through the arrangement in the opposite direction, however, the effects of half-wave plate and Faraday rotator add up; if both elements rotate the plane of polarization by 45°, the light is then blocked by the polarizer. Therefore, these devices are also called optical diodes.

5.5 Acousto-optical Effects

Acoustic waves in gases, liquids, and (preferably) solids are accompanied by density variations, and usually a monochromatic acoustic plane wave will create a phase grating due to the pressure (or density) dependence of the refractive index. Such a grating can be used for the diffraction of light waves when the grating constant and angle of incidence are chosen appropriately. Acousto-optical modulators based on this effect are used to modulate the amplitude of (laser) light beams or for shifting the frequency of a monochromatic light ray (see MODULATORS AND DEMODULATORS, OPTICAL). They may also be employed as variable deflectors as the diffraction angle can be controlled by changing the frequency of the acoustic frequency and thus the grating constant.

6. RECENT DEVELOPMENTS

In this section, we cannot even attempt to be exhaustive in our treatment of the subject; thus, it is rather intended as an arbitrary choice of some examples of developments that played a major role in recent years. The foundations of many of the fields treated date back much further, though. The reader is encouraged to take this section as a kind of appetizer to interest him in further studies in this field of physics which—not only because of its close link to one of man's most important senses, namely vision—is still full of new possibilities for applications.

6.1 Fiber Optics and Light Guides

The principle of total internal reflection can be used to guide light in conduits made of glass, plastics, and other materials transparent for the radiation under consideration. Development of new fibers has to a large extent been promoted by the aim to use optical transmission through fibers for telecommunications. Today, optical fibers can be divided into three basic types: step-index multimode fibers and gradient-index multimode and monomode fibers (see Fig. 37), the latter permitting polarization-maintaining designs (see FIBER OPTICS).

Optical fibers already play a major role in communication and data transfer. They are employed in many new sensor types. Endoscopes are another important field of application for fiber optics. They may be used as mode scramblers or simply as light guides. Monomode fibers are implemented in interferometers, e.g., of Sagnac or Mach–Zehnder type (see INTERFEROMETERS AND INTERFEROMETRY). Directional couplers can be produced for optical fibers in close analogy to microwave techniques.

6.2 Integrated Optics

Integrated optics has the aim to transfer the success of planar mass production processes known from semiconductor electronics to the field of optics (see QUANTUM OPTOELECTRONIC DEVICES). Using photolithography, planar or linear waveguides can be created, e.g., on a glass substrate using ion-exchange processes (Fig. 38).

Components produced in this way usually look and act either like projections of optical systems into the substrate plane (e.g., grating spectrometers) or like analogies to microwave components (e.g., directional couplers). Combining such elements can lead to novel devices for communication systems or sensors (see, e.g., Fig. 39).

One of the fundamental advantages of integrated optics is the ability to confine optical waves to a narrow cross section (i.e., a single-mode waveguide) over an extended length. This makes possible the construction of devices unequaled in classical optics.

A single-mode waveguide in a suitable material should yield much higher conversion efficiencies for frequency doubling (see OPTICS, NONLINEAR) than bulk systems. It may be necessary to change substrate properties periodically to achieve phase matching here. The effect has been demonstrated though so far technical problems seem to have blocked superior results.

In a material exhibiting the electro-optic effect, single-mode waveguides permit a very close electrode spacing for an electro-optic modulator (see MODULATORS AND DEMODULATORS, OPTICAL). Using this advantage has led to modulators using voltages of a few volts

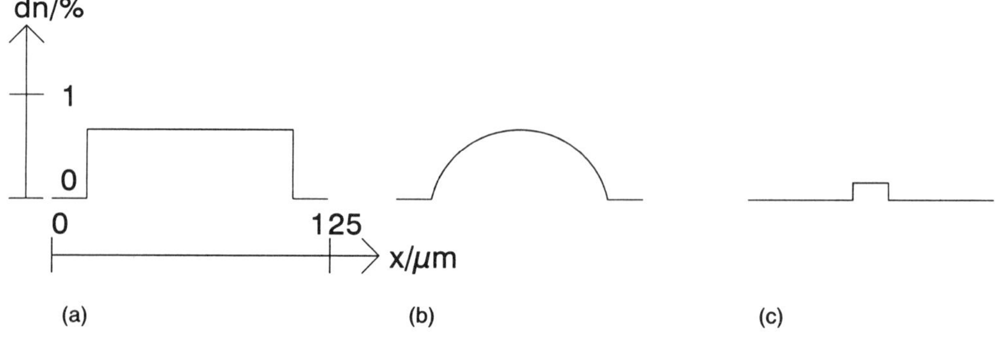

FIG. 37. Optical fibers: (a) step-index multimode, (b) gradient-index multimode, and (c) monomode fiber (refractive index profiles).

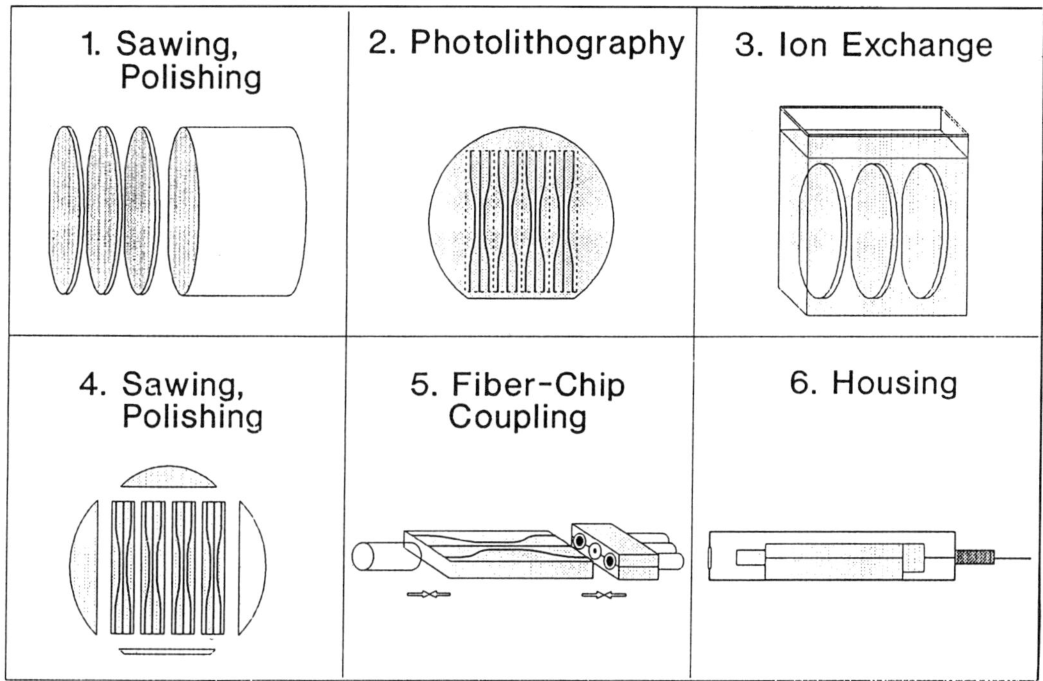

FIG. 38. Integrated-optics production process (picture courtesy of IOT Integrierte Optik GmbH, Jena).

for a phase change of π and multimegahertz modulation frequencies.

6.3 Diffractive Optics

Diffractive optics extends the applications of diffraction from gratings and holography to two different areas: the correction of optical imaging systems using diffractive structures and the use of diffractive structures to reconfigure light distributions in many ways.

6.3.1 Optical-Imaging System Correction

Refractive lenses and diffractive elements (e.g., blazed Fresnel-zone plates) both exhibit chromatic aberration. However, normally the two aberrations differ considerably in their

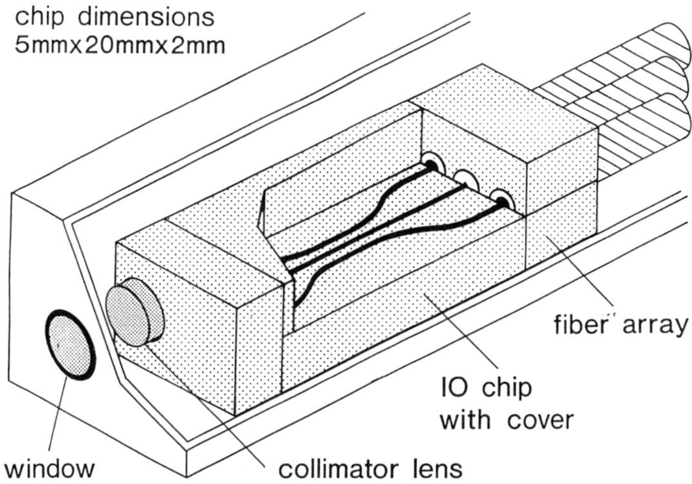

FIG. 39. An integrated optical interferometer system (picture courtesy of IOT Integrierte Optik GmbH, Jena).

value and are of opposite sign. It is therefore obvious to try to combine a strong refractive lens with a weak diffractive lens in order to obtain a system in which these chromatic aberrations compensate each other.

This approach has been proven useful, and diffractive structures are not only used to compensate chromatic aberration, but play an important role as an additional (set of) free parameter(s) for minimizing the various errors present in optical systems. However, practical applications are predominantly in the infrared region of the spectrum, as production of diffractive structures with the required precision using, e.g., diamond turning is much easier at the longer wavelengths. As diamond turning is expensive and lithographic production of diffracting elements is in practice mostly limited to plane surfaces, applications are presently mainly in areas where high cost is acceptable because the required performance cannot be achieved otherwise.

6.3.2 Unconventional Applications Possibly, purely diffractive elements are even more promising. Using photolithography from computer-generated mask patterns, they permit virtually any change of wave front one can think of—provided that the light used is monochromatic. The borderline between diffractive optics and (computer-generated) holography is somewhat arbitrary but will not be considered here.

Laser-beam shaping is a prominent example of such application. In material processing, the beam profile of the laser employed is often inappropriate for a specific application. Laser beams are mostly of three different types: TEM_{00} (i.e., with a transverse field distribution described by a Gaussian), doughnut modes, and multimode beams. For beams with a well-defined mode (and to a lesser extent for multimode beams), diffractive optical elements (DOEs) permit the generation of virtually any desired intensity distribution. Multiple-spot illumination for simultaneous welding of all leads of an integrated circuit, specific intensity profiles advantageous for cutting and hardening, or even character generation for marking can be obtained. For a multimode beam, these possibilities may be somewhat limited, but still beam homogenization is possible.

6.3.3 Binary Optics Binary optics is a special type of diffractive optics. It uses lithography to create blazed diffractive optical elements. Binary masks, i.e., either transmitting or opaque locally, are used to create binary, i.e., two-level, profile structures in resist layers, and from these the profile may be transferred to the substrate using etching techniques. By stacking consecutive two-level profiles, while reducing the profile step by a factor of 2 for consecutive masks, multilevel profiles with higher efficiency can be obtained in a way very similar to the generation of binary numbers. When a light wave passes through such a structure, the surface profile results in a phase profile of the wave front.

Figure 40 shows different gratings having the same grating constant. In (a) an amplitude grating is shown, and about 10% of the power incident on it is diffracted into the first order. With a blazed phase grating as shown in (b), almost 100% can be concentrated into this diffraction order. Binary optics uses consecutively refined mask structures to approximate this blaze by a staircaselike structure. In (c) with two phase levels an efficiency of about 40% is achieved; this value is roughly doubled when two masks are used to generate four different depth levels. With four levels as in (d), a value of about 80% is reached, but the finest features on the required mask are also two times smaller than on the mask used for the two-level structure.

6.4 Adaptive Optics

Adaptive optics are used to change the imaging behavior, e.g., of an optical telescope in order to compensate for disturbances whose origins lie in the density variations and turbulence of the atmosphere (seeing). Because of these disturbances, earthbound telescopes normally achieve a much lower effective resolution than the diffraction limit due to finite aperture diameter.

Two subsystems are necessary for the implementation of an adaptive optical system: a wave-front sensor for the detection of deviations of the incoming wave front from its ideal shape, and a servo element such as a deformable mirror for the correction of these.

As this technique is easier to implement for longer wavelengths, it was first successfully tested in near-infrared astronomy, but

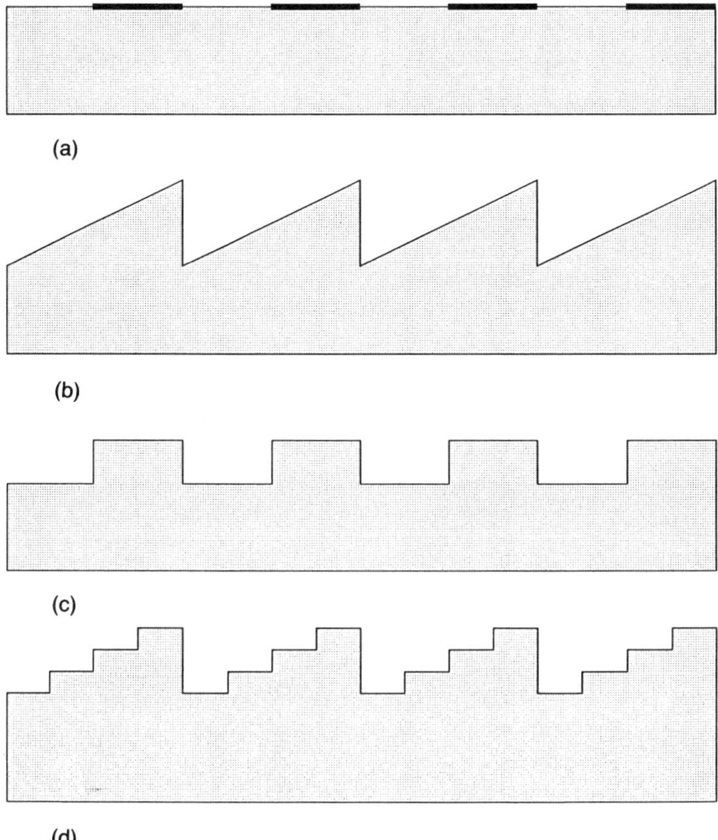

FIG. 40. Binary optics. For details see text.

it is being transferred to the visual region as well with impressive results. However, this does not mean that space telescopes will be unnecessary in the future. Absorption by the earth's atmosphere will always limit the wavelength range accessible to earthbound telescopes.

It is not possible to give a complete impression of the wide variety of optical systems and applications. Interesting fields like x-ray optics, stress birefringence, or intensity correlation experiments have not even been sketched in this article. However, this outlook may motivate readers to engage themselves more deeply in the study of this fascinating field of physics.

GLOSSARY

Aperture: A stop that limits the object area that is imaged by an optical system.

Chromatic Aberration: The dependence of the focal length on wavelength due to the dispersion of refractive materials.

Coherence: The ability of light (waves) to interfere, i.e., their phase relation and its spatial and temporal behavior.

Diffraction: The deviation of light waves from straight propagation that is caused by obstacles.

***f* Number:** The ratio r of focal length and diameter of an objective. Often this value is written as f/r, e.g., $f/4$ for a 200-mm lens with 50-mm diameter.

Interference: The superposition of coherent fields yielding phase-sensitive phenomena.

Kerr Cell: An electro-optic modulator based on the Kerr or quadratic electro-optic effect.

Magnification: Two similar quantities are called magnification: The ratio of the sizes of image and object is usually called magnification. The same expression is used for the ratio between the apparent sizes (corresponding to the viewing angles) of an object with and without an optical system like a microscope or a telescope. For clarity, it is useful to speak of linear magnification in the first case and of angular magnification in the second.

Numerical Aperture: This expression can have two different significations depending on the field of optics in which it is employed:

1. The sine of the half-angle of the cone of light entering a microscope objective from an object point. Its value sets a limit to the resolution that can be obtained.
2. The maximum value for the sine of the angle of incidence for light rays incident upon the front face of an optical fiber at which these rays will still be guided by total internal reflection.

Optical Activity: Circular birefringence.

Optical Axis: Axis of a uniaxial crystal along which the refractive index for the extraordinary ray has the same value as for the ordinary ray.

Phase Velocity: The velocity with which a point of constant phase travels within a wave.

Plane of Incidence: Plane containing the propagation vector of a plane wave and the normal to an interface upon which this wave is incident.

Pockels Cell: An electro-optic modulator based on the Pockels or linear electro-optic effect.

Polarizer: A device selectively transmitting one linearly polarized component of light, or both (orthogonally polarized) components into different directions thus separating them.

Pupil: A stop limiting the solid angle of rays contributing to the image of a point for a given imaging situation.

Refractive Index: The ratio of the velocity of light in vacuum and in a medium. A complex value indicates extinction (or amplification).

Resolution: The ability of an optical system to distinguish two neighboring points in the image of an object or of a spectrum. According to a criterion given by Lord Rayleigh, these are considered to be resolved if the intensity decreases to about 80% of the maximum between two maxima of equal strength.

Scattering: The generation of secondary waves by light traveling in a medium and the propagation of these. The secondary waves do not necessarily have the same frequency as the exciting wave.

Signal Velocity: The velocity with which a point of constant relative amplitude travels in a modulated wave.

Spectrograph: A device (usually employing prisms or gratings) for the spectral analysis of light, and a film or optoelectronic detector for the recording of the spectral distribution.

Stellar Aberration: The apparent movement of the stars due to observation from the Earth moving in its orbit. A relativistic vector addition of the Earth's velocity and the velocity vector of the impinging photons yields the actual telescope pointing direction.

Verdet Constant: A coefficient characterizing the strength of the Faraday effect for a given material and for a given wavelength.

Works Cited

Alonso, M., Finn, E. J. (1976), *Fundamental University Physics*, Vol. 2, Reading, MA: Addison-Wesley, pp. 789 ff.

Beth, R. A. (1936), *Phys. Rev.* **50**, 115–125.

Born, M., Wolf, E. (1980), *Principles of Optics*, 6th ed., Oxford: Pergamon Press.

Hecht, E., Zajac, A. (1974), *Optics*, Reading, MA: Addison-Wesley.

Kingslake, R. (1978), *Lens Design Fundamentals*, New York: Academic Press.

Kingslake, R. (1983), *Optical System Design*, Orlando, FL: Academic Press.

Klein, M. V., Furtak, T. E. (1986), *Optics*, New York: Wiley.

Quinn, T. J. (1992), "*Mise en Pratique* of the Definition of the Metre," *Metrologia 1993/1994* **30**, 523–541.

Further Reading

General

Born, M., Wolf, E. (1980), *Principles of Optics*, 6th ed., Oxford: Pergamon Press.

Hecht, E. (1975), *Optics*, New York: McGraw-Hill.

Jenkins, F. A., White, H. E. (1981), *Fundamentals of Optics*, Singapore: McGraw-Hill.

Klein, M. V., Furtak, T. E. (1986), *Optics*, New York: Wiley.

Selected Topics

Anan'ev, Y. A. (1991), *Laser Resonators and the Beam Divergence Problem*, Bristol: Adam Hilger.

Arnaud, J. A. (1976), *Beam and Fiber Optics*, New York: Academic Press.

Beckmann, P., Spizzichino, A. (1987), *The Scattering of Electromagnetic Waves from Rough Surfaces*, Norwood, MA: Artech House.

Caulfield, H. J., Lu, S. (1970), *The Applications of Holography*, New York: Wiley.

Collett, E. (1993), *Polarized Light*, New York: Marcel Dekker.

Cox, A. (1964), *A System of Optical Design*, London: Focal Press.

Duffieux, P. M. (1983), *The Fourier Transform and Its Applications to Optics*, New York: Wiley.

Felsen, L. B. (Ed.) (1984), *Hybrid Formulation of Wave Propagation and Scattering*, Dordrecht, The Netherlands: Martinus Nijhoff.

Flügge, S. (Ed.) (1967), *Encyclopedia of Physics*, Vol XXIX, *Optical Instruments*, Berlin: Springer-Verlag.

Frieden, B. R. (Ed.) (1980), *The Computer in Optical Research*, Berlin: Springer-Verlag.

Gaskill, J. D. (1978), *Linear Systems, Fourier Transforms, and Optics*, New York: Wiley.

Gerrard, A., Burch, J. M. (1975), *Introduction to Matrix Methods in Optics*, London: Wiley.

Ghatak, A. K., Thyagarajan, K. (1989), *Optical Electronics*, Cambridge, UK: Cambridge University Press.

Goodman, J. W. (1984), *Statistical Optics*, New York: Wiley.

Guenther, R. D. (1990), *Modern Optics*, New York: Wiley.

Haferkorn, H. (1986), *Bewertung Optischer Systeme*, Berlin: VEB Deutscher Verlag der Wissenschaften.

Hall, D. R., Jackson, P. E. (1989), *The Physics and Technology of Laser Resonators*, Bristol: Adam Hilger.

Haus, H. A. (1984), *Waves and Fields in Optoelectronics*, Englewood Cliffs, NJ: Prentice-Hall.

Hecht, E., Zajac, A. (1974), *Optics*, Reading, MA: Addison-Wesley.

Hecht, E. (1975), *Optics*, New York: McGraw-Hill.

Horne, D. F. (1975), *Lens Mechanism Technology*, London: Adam Hilger.

Hutley, M. C. (1982), *Diffraction Gratings*, London: Academic Press.

Iizuka, K. (1983), *Engineering Optics*, New York: Springer-Verlag.

Kersten, R. Th. (1983), *Einführung in die optische Nachrichtentechnik*, Berlin: Springer-Verlag.

Kingslake, R. (Founding Ed.) (1965–), *Applied Optics and Optical Engineering*, Vol. 1- , New York: Academic Press.

Kingslake, R. (1978), *Lens Design Fundamentals*, New York: Academic Press.

Kingslake, R. (1983), *Optical System Design*, Orlando, FL: Academic Press.

Kliger, D. S., Lewis, J. W., Randall, C. E. (1990), *Polarized Light in Optics and Spectroscopy*, Boston: Academic Press.

Korsch, D. (1991), *Reflective Optics*, Boston: Academic Press.

Laikin, M. (1990), *Lens Design*, New York: Marcel Dekker.

Lauterborn, W., Kurz, T., Wiesenfeldt, M. (1993), *Kohärente Optik*, Berlin: Springer-Verlag.

Mahajan, V. N. (1991), *Aberration Theory Made Simple*, Bellingham, WA: SPIE.

Malacara, D. (1978), *Optical Shop Testing*, New York: Wiley.

Marathay, A. S. (1982), *Elements of Optical Coherence Theory*, New York: Wiley.

Meyers, R. A. (Ed.) (1991), *Encyclopedia of Lasers and Optical Technology*, San Diego, CA: Academic Press.

Neumann, E.-G. (1988), *Single-Mode Fibers*, Berlin: Springer-Verlag.

Nieto-Vesperinas, M. (1991), *Scattering and Diffraction in Physical Optics*, New York: Wiley.

Ogilvy, J. A. (1991), *Theory of Wave Scattering from Random Rough Surfaces*, Bristol, UK: Adam Hilger.

O'Neill, E. L. (1963), *Introduction to Statistical Optics*, Reading, MA: Addison-Wesley.

Parker, S. P. (Ed.) (1988), *Optics Source Book*, New York: McGraw-Hill.

Petit, R. (Ed.) (1980), *Electromagnetic Theory of Gratings*, Berlin: Springer-Verlag.

Petykiewicz, J. (1990), *Wave Optics*, Dordrecht: Kluwer Academic Publishers.

Reynolds, G. O., DeVelis, J. B., Parrent, G. B., Thompson, B. J. (1989), *The New Physical Optics Notebook*, Bellingham, WA: SPIE.

Saleh, B. E. A., Teich, M. C. (1991), *Fundamentals of Photonics*, New York: Wiley.

Schroeder, D. J. (1987), *Astronomical Optics*, San Diego: Academic Press.

Slyusarev, G. G. (1984), *Aberration and Optical Design Theory*, Bristol: Adam Hilger.

Snyder, A. W., Love, J. D. (1983), *Optical Waveguide Theory*, London: Chapman and Hall.

Solimeno, S., Crosignani, B., Di Porto, P. (1986), *Guiding, Diffraction, and Confinement of Optical Radiation*, Orlando, FL: Academic Press.

Stamnes, J. J. (1986), *Waves in Focal Regions*, Bristol: Adam Hilger.

Stover, J. C. (1990), *Optical Scattering*, New York: McGraw-Hill.

Tamir, T. (Ed.) (1979), *Integrated Optics*, Berlin: Springer-Verlag.

Tamir, Th. (Ed.) (1988), *Guided-Wave Optoelectronics*, Berlin: Springer-Verlag.

Tyson, R. K. (1991), *Principles of Adaptive Optics*, Boston: Academic Press.

van Etten, W., van der Plaats, J. (1990), *Fundamentals of Optical Fiber Communications*, New York: Prentice-Hall.

Voronovich, A. G. (1994), *Wave Scattering from Rough Surfaces*, Berlin: Springer-Verlag.

Weichel, H. (1990), *Laser Beam Propagation in the Atmosphere*, Bellingham, WA: SPIE.

Welford, W. T. (1986), *Aberrations of Optical Systems*, Bristol: Adam Hilger.

Welford, W. T., Winston, R. (1989), *High Collection Nonimaging Optics*, San Diego: Academic Press.

Williams, C. S. (1989), *Introduction to the Optical Transfer Function*, New York: Wiley.

Yariv, A., Yeh, P. (1983), *Optical Waves in Crystals*, New York: Wiley.

Yariv, A. (1991), *Optical Electronics*, Fort Worth: Saunders College Publishing.

Yu, F. T. S. (1985), *White-Light Optical Signal Processing*, New York: Wiley.

Appetizers

Falk, D., Brill, D., Stork, D. (1986), *Seeing the Light*, New York: Harper & Row.

Léna, P., Blanchard, A. (1990), *Lumières*, Paris: InterEditions.

Minnaert, M. (1954), *Light & Colour*, Dover Publications.

Schulz, G. (1974), *Paradoxa aus der Optik*, Leipzig: Johann Ambrosius Barth.

OPTICS, NONLINEAR

CHRISTOS FLYTZANIS, *Laboratoire d'Optique Quantique, Ecole Polytechnique, 91128 Palaiseau, cédex, France*

3-527-28134-7/95/$5.00 + .50

1. INTRODUCTION

Nonlinear optics is the area of optics comprising phenomena that occur when the response of a material system to an applied electromagnetic field is nonlinear in the amplitude of the field. Such phenomena can only be observed with intense coherent light sources, for instance lasers, and it has become a tradition now to place the beginning of nonlinear optics with the first observation of the second harmonic of the ruby laser frequency in quartz (Franken *et al.*, 1961). Up to that time, the optical phenomena seemed to obey the linear regime and the amplitude of the field did not matter. However, some nonlinear effects had already crept into the area of optics and were extensively studied (e.g., Born and Huang, 1954) and used long before the advent of the laser although in a different spirit and context from the present ones. They were namely treated as parametric effects and to some extent are still treated as such; here we have in mind the electro-optic effects, like the Pockels and Kerr ones, or the light-scattering (e.g., Fabelinskii, 1968) ones like the Raman, Brillouin, and Rayleigh effects, to name a few. All these effects can be actually related (Bloembergen, 1964) to the nonlinear interaction of the charges with the electromagnetic fields and constitute special cases of the area of nonlinear optics.

That the optical properties of matter cannot remain insensitive to the field amplitude when the latter increases can be anticipated on very simple fundamental grounds. Indeed, when this amplitude becomes comparable to the cohesive fields E_c that hold the charges tied together in the matter, for instance the electron to the proton in the hydrogen atom, the displacements of the charges from their equilibrium positions cannot be treated any longer within the harmonic approximation that constituted (e.g., Born and Wolf, 1975) the basis of optics with incoherent light sources; anharmonic effects must now be included, which invariably lead to a nonlinear dependence of the charge displacements on the field amplitude and concomitantly of the induced dipole moments and polarizations. To fix the ideas, with the light sources before the invention of the laser one could only achieve electric field amplitudes of the order of 10–10^3 V/m at the optical-frequency range while E_c for the hydrogen is 10^{11} V/m, a field easily attainable nowadays with lasers. Anharmonicity in the mutual interactions of the charges, however, is not the sole origin of nonlinearities. Indeed, even in the absence of such interactions the response will be nonlinear because of the intrinsic nonlocality built into the interaction between a charge and the electromagnetic field as can be inferred by the expression of the Lorentz force (e.g., Landau and Lifshitz, 1960). This is a propagation-retardation effect, an intrinsic feature of the Maxwell equations, and because of its cumulative character can lead to nonlinearities comparable to the ones due to the anharmonicity. In the last analysis (Bloembergen, 1964; Flytzanis, 1975) anharmonicity in the charge motion and nonlocality in its interaction with an electromagnetic field constitute the two major mechanisms that at the microscopic level are responsible for the nonlinear response of the matter to intense electromagnetic fields.

The scope of nonlinear optics, however, is not limited to the observation of the nonlinear terms in the induced polarization; it also concerns the impact that these may have on the propagation of intense electromagnetic fields in the matter (Armstrong *et al.*, 1962). As a matter of fact, the spectacular growth that the field witnessed (e.g., Bloembergen, 1964; Robin and Tang 1975; Shen, 1984) within the span of the past three decades heavily relies on the anticipation of dramatic technological developments and breakthroughs that are expected in return—in particular, in future telecommunications and computer technologies. It is expected that within the foreseeable future the linear technology will be replaced with a nonlinear one in both transmission and information treatment (e.g., Hasegawa, 1989; Agrawal, 1989).

The theoretical framework (Armstrong *et al.*, 1962; Bloembergen, 1964) for the description and study of the nonlinear optical phenomena remains the same as for the other optical phenomena in general, namely, the macroscopic Maxwell equations

$$\nabla \times \mathbf{E} = -\frac{1}{c}\frac{\partial \mathbf{B}}{\partial t}, \quad \nabla \cdot \mathbf{E} = 4\pi\rho,$$
$$\nabla \times \mathbf{B} = \frac{1}{c}\frac{\partial \mathbf{D}}{\partial t}, \quad \nabla \cdot \mathbf{B} = 0, \tag{1}$$

with

$$\mathbf{D} = \mathbf{E} + 4\pi \int_{-\infty}^{t} \mathbf{J}(t')dt', \tag{2}$$

where for simplicity we assumed that there are no extraneous charges and currents and **J** is the induced current density, which can be related to the fields **E** and **B** through the solution of the equation of motion of the charges in the presence of the electromagnetic field; in principle, this is the Schrödinger equation, but classical equations of motions can also be used for a phenomenological description, for instance the forced damped anharmonic oscillator equation

$$\ddot{\mathbf{r}} + \Gamma\dot{\mathbf{r}} + \omega_0^2\mathbf{r} + \beta\mathbf{r}^2 + \gamma\mathbf{r}^3 = (e/m)\mathbf{E}(t) \tag{3}$$

for bound charges in dielectrics, or the equation for a free charge in a Lorentz force field

$$\dot{\mathbf{v}} + \Gamma\mathbf{v} = (e/m)[\mathbf{E}(t) + \mathbf{v} \times \mathbf{B}(t)] \tag{4}$$

for unbound electrons in metals, where Γ takes into account the unavoidable random interactions of the charge with its environment (bath). The two Eqs. (3) and (4) at the elementary level reflect the two basic mechanisms that induce a nonlinear behavior in matter, namely anharmonicity and nonlocality.

The current density **J** can be expanded in multipole series

$$\mathbf{J}(t) = \frac{\partial}{\partial t}(\mathbf{P} - \nabla \cdot \mathbf{Q}) + c(\nabla \times \mathbf{M}), \tag{5}$$

where **P**, **Q**, and **M** are the electric dipole, electric quadrupole, and magnetic dipole polarization densities respectively with $\mathbf{H} = \mathbf{B} - 4\pi\mathbf{M}$. When perturbation theory applies, these can be cast in the form of power-series expansions in the field amplitudes; in certain cases, analytical expressions in closed form can be derived. In general, one may write $\mathbf{P} = \mathbf{P}_L + \mathbf{P}_{NLS}$ and similarly for **Q** and **M**, where $\mathbf{P}_L$ is the part of the dipole polarization linear in the field amplitude **E** and $\mathbf{P}_{NLS}$ is the nonlinear part. These nonlinear terms act as polarization sources generating new fields with drastically different spatiotemporal characteristics from those expected in the linear regime. Actually, the energy transfer involved,

$$\frac{dW_{NL}}{dt} = -\left\langle \mathbf{E} \cdot \frac{\partial}{\partial t}\{[\mathbf{P}_{NLS} - \nabla \cdot \mathbf{Q}_{NLS}] + c\nabla \times \mathbf{M}_{NLS}\}\right\rangle, \tag{6}$$

where the brackets indicate time averaging, can be substantial, and the new fields can attain large amplitudes with specific spatiotemporal characteristics.

Nonlinear optics is concerned with all these aspects and in addition with their implementation in devices and other applications (Arecchi and Schulz-Dubois, 1975; Miller *et al.*, 1994; Ostrowsky and Reinisch, 1992). In this respect, the enhancement of the efficiency of the nonlinear processes by appropriate choice of nonlinear optical materials and interaction configurations constitutes a central concern in all these studies (Digiorgio and Flytzanis, 1994). The future development of nonlinear optics is intimately connected with the progress in nonlinear optical materials. Nonlinear spectroscopy is another byproduct of these efforts, and much ingenuity has been developed to use the nonlinear effects for diagnostic purposes (e.g., Letokhov and Chebotayev, 1977; Levenson and Kano, 1988).

Nonlinear phenomena in general and the related nonlinear equations are exceedingly complex and very quickly become intractable without the introduction of simplifications and approximations (Whitham, 1974; Newell and Moloney, 1992). We shall highlight the main issues of nonlinear optics with a sample of effects that are simple and representative of the whole field. Most of them are chosen among those related to the second- and third-order nonlinear terms in the series expansion of the dipole polarization in powers of the electric field amplitude $\mathbf{P} = \mathbf{P}^{(1)} + \mathbf{P}^{(2)} + \mathbf{P}^{(3)} + \cdots + \mathbf{P}^{(n)} + \cdots$ or

$$\mathbf{P} = \chi^{(1)} \cdot \mathbf{E} + \chi^{(2)}:\mathbf{EE} + \chi^{(3)}:\mathbf{EEE} + \cdots, \tag{7}$$

where the coefficient $\chi^{(n)}$ is the dipolar susceptibility of order n, a tensor of rank $n + 1$, and $\mathbf{P}^{(1)} = \mathbf{P}_L$; it is customary to introduce a Fourier analysis of these polarization terms and the fields:

$$\mathbf{E} = \mathbf{E}(\omega)e^{i(\mathbf{k}_i\mathbf{r}-\omega t)} + \text{c.c.},$$

$$\mathbf{P} = \mathbf{P}(\omega)e^{i(\mathbf{k}_i\mathbf{r}-\omega t)} + \text{c.c.},$$

and write (Bloembergen, 1964; Flytzanis, 1975; Shen, 1984)

$$P_i^{(1)}(\omega_1) = \chi_{ij}^{(1)}(\omega_1)E_j(\omega_1), \tag{8a}$$

$$P_i^{(2)}(\omega_1 + \omega_2) = D_2\chi_{ijk}^{(2)}(\omega_1,\omega_2)E_j(\omega_1)E_k(\omega_2), \tag{8b}$$

$$P_i^{(3)}(\omega_1 + \omega_2 + \omega_3) = D_3\chi_{ijkl}^{(3)}(\omega_1,\omega_2,\omega_3) \times E_j(\omega_1)E_k(\omega_2)E_l(\omega_3) \tag{8c}$$

for the linear, second-, and third-order terms and similarly for the higher-order ones. The D_n are degeneracy factors equal to the number of distinct permutations of the applied field Fourier components; with this definition, the value of $\gamma^{(n)}$ for $\omega_i \rightarrow 0$ is independent of the chosen path. The precise behavior of the susceptibilities as a function of the frequencies can be obtained only from the equation of motion of the charges, but some general statements can be made by referring to the general properties that all response functions must obey: invariance with respect to translation of the time origin, causality, reality, and symmetrization. The quasitotality of material systems of interest in nonlinear optics, crystalline or not at the microscopic level, is formed from a repeat unit of volume v, and one may also define the corresponding macroscopic polarizabilities through the relations

$$\chi^{(1)} = \tilde{\alpha}/v,\ \chi^{(2)} = \tilde{\beta}/v,\ \chi^{(3)} = \tilde{\gamma}/v, \tag{9}$$

where the linear and second- and third-order polarizabilities $\tilde{\alpha}$, $\tilde{\beta}$, and $\tilde{\gamma}$, respectively, are understood as being those of an effective polarizable unit that fills the volume v that is the inverse of their number density N, or $v = 1/N$. A rough order-of-magnitude estimation of the dipolar nonlinear susceptibilities can be obtained through

$$\chi^{(n+1)} \approx 1/E_c^n, \tag{10}$$

where E_c is the effective cohesive field that keeps attached the polarizable charges or units to each other. The mechanisms that contribute to the nonlinear polarization are numerous (Flytzanis, 1975) and can be related to the different displacements and deformations that the charge distributions undergo within a polarizable unit under the action of the strong light field and also because of translational or rotational motion of these polarizable units under the action of these same light fields or other externally mediated forces. These two types of motion, the intramolecular and intermolecular, differ in several qualitative and quantitative aspects because of the vastly different masses and cohesive forces involved. Their study is an essential aspect in nonlinear material research, and nonlinear optical spectroscopy.

Broadly speaking, nonlinear optics are used either to shift the optical carrier frequency of the available fields or to remodel their spatiotemporal characteristics. Accordingly, one can separate the nonlinear effects into two classes:

1. the frequency-shifting effects, which serve to generate fields at new frequencies in which case the coherence or equivalently the field amplitude matters and one is invariably faced with the problem of phase matching, the classic example here being the second-harmonic generation;
2. the frequency-preserving effects, which modify the spatiotemporal characteristics of a light beam, in particular its spatiotemporal profile and polarization state, but preserve its carrier frequency, in which case the phase-matching problem can be automatically solved. Here one has either the light self-action effects or photoinduced ones where only the spatiotemporal profile of the beam intensity matters and not the field amplitude *per se*, the classic examples being the optical Kerr-nonlinearity mediated effects; or the parametric effects where an external parameter is used to modulate the optical characteristics of the medium and by the same token those of the light beam, the classic examples here being the electro-optic (Pockels) and acousto-optic effects. In Table 1, we summarize the most commonly used second- and third-order processes.

Table 1. Nonlinear optical effects.

Nonlinear effect	Incident frequency	Created frequencies	Nonlinear	Some applications
Second harmonic[a]	ω, ω	2ω	$\chi^{(2)}(\omega,\omega)E_\omega E_\omega$	Near-uv generation
Optical rectification	$\omega, -\omega$	0	$2\chi^{(2)}(\omega,-\omega)E_\omega E_\omega$	Ultrashort electrical pulses
Electro-optic effect (Pockels)	$\omega, 0$	ω	$2\chi^{(2)}(\omega,0)E_\omega E_0$	Electro-optic modulation
Frequency summation[a]	ω_1, ω_2	$\omega_1 + \omega_2$	$2\chi^{(2)}(\omega_1,\omega_2)E_{\omega 1}E_{\omega 2}$	Near-uv generation; up-conversion
Frequency difference[a]	$\omega_1, -\omega_2$	$\omega_1 - \omega_2$	$2\chi^{(2)}(\omega_1,-\omega_2)E_{\omega 1}E_{\omega 2}$	IR radiation, parametric amplification
Third harmonic[a]	ω, ω, ω	3ω	$\chi^{(3)}(\omega,\omega,\omega)E_\omega E_\omega E_\omega$	
Frequency summation[a]	$\omega_1, \omega_2, \omega_3$	$\omega_1 + \omega_2 + \omega_3$	$6\chi^{(3)}(\omega_1,\omega_2,\omega_3)E_{\omega 1}E_{\omega 2}E_{\omega 3}$	Near- and far-uv generation
Frequency mixing[a]	$\omega_1, \omega_2, -\omega_2$	$2\omega_1 - \omega_2$	$3\chi^{(3)}(\omega_1,\omega_1,-\omega_2)E_{\omega 1}E_{\omega 1}E^*_{\omega 2}$	CARS if $\omega_1 - \omega_2 \approx \omega_R$[b]
Optical Kerr effect	$\omega, -\omega, \omega$	ω	$3\chi^{(3)}(\omega,-\omega,\omega)\|E_\omega\|^2E_\omega$	Optical modulation, optical phase, optical gratings, two-photon absorption if $2\omega = \omega_c$[b]
Two-photon absorption or Stimulated Raman	$\omega_1, -\omega_1, \omega_2$	ω_2	$6\chi^{(3)}(\omega_1,-\omega_1,\omega_2)\|E_{\omega 1}\|^2E_{\omega 2}$	if $\omega_1 + \omega_2 = \omega_c$[b]; if $\omega_1 - \omega_2 = \omega_R$[b]
Static Kerr effect	$\omega, 0, 0$	ω	$3\chi^{(3)}(\omega,0,0)E_\omega E_0^2$	Electro-optic modulation
Induced second harmonic[a] and	$\omega, \omega, 0$	2ω	$3\chi^{(3)}(\omega,\omega,0)E_\omega^2E_0$	
Optical rectification	$\omega, -\omega, 0$	0	$6\chi^{(3)}(\omega,-\omega,0)\|E_\omega\|^2E_0$	

[a]Phase matching required.
[b]ω_c, ω_R: material frequencies.

2. NONLINEAR PROPAGATION

2.1 Basic Nonlinear Propagation Equation

As previously stated, the nonlinear polarization sources induced in a medium generate new fields, and the basic problem (Armstrong *et al.*, 1962; Bloembergen, 1964) here is to derive the propagation equation for the amplitude of a monochromatic electric field of frequency Ω in a nonlinear medium where a nonlinear polarization has been induced at the same frequency Ω. Restricting ourselves to transverse plane waves, or $\nabla \cdot \mathbf{E} = 0$, in a nonmagnetic dielectric we obtain from Maxwell equations the propagation equation

$$\Delta \mathbf{E} - \frac{4\pi}{c^2}\frac{\partial^2 \mathbf{D}_L}{\partial t^2} = \frac{4\pi}{c^2}\frac{\partial^2 \mathbf{P}_{NL}}{\partial t^2}, \tag{11}$$

where $\mathbf{D}_L = \mathbf{E} + 4\pi\mathbf{P}_L$ is the linear induction. In reality, the intense coherent light sources we are dealing with do not deliver plane waves but pulses, namely wave packets whose envelope has a finite spatiotemporal extension; their temporal extension (pulse duration) can range from a few nanoseconds (1 ns $= 10^{-9}$ s) to a few picoseconds (1 ps $= 10^{-12}$ s) down to a few femtoseconds (1 fs $= 10^{-15}$ s); we remind the reader that in 1 ps, light travels 300 μm in vacuum. Accordingly, the induced nonlinear polarization may have comparable spatiotemporal extensions, and the problem then is to derive the equation that governs the spatiotemporal evolution of the envelope of a coherent light pulse generated by such a pulsed polarization source. As can be inferred from Eq. (7), P_{NL} is a sum of nonlinear polarization sources. For most purposes, we may single out one such polarization source as being the only one relevant.

Quite generally, the latter can be written in the form

$$\mathbf{P}_{NL}(\mathbf{r},t) = Re\{\mathcal{P}_{NL}(\mathbf{r},t)e^{i(\mathbf{K}_\Sigma \cdot \mathbf{r} - \Omega t)}\}, \tag{12a}$$

where $\mathcal{P}_{NL}(\mathbf{r},t)$ is its envelope that varies slowly over the carrier period $T = 2\pi/\Omega$ and $\mathbf{K}_\Sigma$ is the vector sum of the wave vectors of all the fields that interact to set up the nonlinear polarization source [Eq. (12a)]. The electric field generated in the same frequency Ω can be also written in the form

$$\mathbf{E}(\mathbf{r},t) = Re\{\hat{\mathbf{e}}A(\mathbf{r},t)e^{i(\mathbf{K}\cdot\mathbf{r}-\Omega t)}\}, \tag{12b}$$

where K satisfies the dispersion relation

$$K = \Omega\sqrt{\mu_0\epsilon(\Omega)} \tag{13}$$

and in general $\mathbf{K} \neq \mathbf{K}_\Sigma$; $A(\mathbf{r},t)$ is the envelope of the electric field amplitude and varies slowly over a wavelength $\Lambda = 2\pi/K$ and over a period $T = 2\pi/\Omega$. Actually in Eq. (13), although not explicitly indicated, $\epsilon(\Omega) = n_\omega^2$ is an appropriately contracted direction-dependent form of the dielectric constant $\boldsymbol{\epsilon}(\Omega) = 1 + 4\pi\boldsymbol{\chi}^{(1)}(\Omega)$, which in general is a second-rank tensor, and n_ω is the refractive index for a given direction.

Replacing Eqs. (12a) and (12b) in Eq. (11), introducing the wave variable transformation $\xi = x$, $\eta = y$, $\zeta = z$, $\tau = t - z/v_g$ where the group velocity is defined as $v_g = (\partial K/\partial\Omega)^{-1}$ but is actually a vector relation, and inserting the slowly varying envelope approximation (Armstrong *et al.*, 1962) by setting $|K\partial A/\partial\zeta| \gg |\partial^2 A/\partial\zeta^2|$, $v_g^{-1}|\partial^2 A/\partial\zeta\partial\tau|$ and in addition $|\Omega^2\mathcal{P}_{NL}| \gg |\partial^2\mathcal{P}_{NL}/\partial t^2|$, one obtains

$$\Delta_\perp A + \frac{\partial A}{\partial\zeta} - \frac{K''}{2i}\frac{\partial^2 A}{\partial\tau^2} = \left(\frac{iK}{2\epsilon(\Omega)}\mathcal{P}_{NL} - \frac{1}{c}\frac{\partial\mathcal{P}_{NL}}{\partial\tau}\right)e^{i\Delta K\zeta} \tag{14}$$

after projecting over the unit vector $\hat{\mathbf{e}}$, which we assume is unaffected by the nonlinear interaction. In Eq. (14), $\Delta K = K_\Sigma - K$ is the wave vector mismatch along direction ζ, $\Delta_\perp = \partial^2/\partial x^2 + \partial^2/\partial y^2$, and $K'' = \partial^2 K/\partial\Omega^2 = -(\partial v_g/\partial\Omega)/v_g^2$ is the group velocity dispersion. The slowly varying envelope approximation is essentially equivalent to retaining only the forward-propagating field component generated by $\mathcal{P}_{NL}$ in the solution of Eq. (11) and neglecting the backward-propagating one. Equation (14) is the basic nonlinear propagation equation valid for coherent light pulses down to a few femtoseconds and finite transverse extension. The term $\Delta_\perp A$ takes care of the diffraction and is important whenever the beam profile varies strongly. Note that the amplitude A is dephased by $\pi/2$ with respect to the polarization term $\mathcal{P}_{NL}$ at $z = 0$.

If $\Delta_\perp A$ is negligible in Eq. (14), one has

$$\frac{\partial A}{\partial\zeta} - \frac{K''}{2i}\frac{\partial^2 A}{\partial\tau^2} = \left(i\frac{2\pi K}{\epsilon(\Omega)}\mathcal{P}_{NL} - \frac{1}{c}\frac{\partial\mathcal{P}_{NL}}{\partial\tau}\right)e^{i\Delta K\zeta}, \tag{15}$$

which for $|\Omega\mathcal{P}_{NL}| \gg |\partial\mathcal{P}_{NL}/\partial\tau|$ reduces to

$$\frac{\partial A}{\partial\zeta} - \frac{K''}{2i}\frac{\partial^2 A}{\partial\tau^2} = i\frac{2\pi K}{\epsilon(\Omega)}\mathcal{P}_{NL}e^{i\Delta K\zeta} \tag{16}$$

and for a Kerr-nonlinearity medium becomes the nonlinear Schrödinger equation that admits as solutions the temporal solitons (Zakharov and Shabat, 1972). For $K'' \approx 0$ and reverting to the initial variables, one has

$$\frac{\partial A}{\partial z} + \frac{1}{v_g}\frac{\partial A}{\partial t} = i\frac{2\pi K}{\epsilon(\Omega)}\mathcal{P}_{NL}e^{i\Delta Kz}, \tag{17}$$

which in the stationary regime reduces to

$$\frac{\partial A}{\partial z} = i\frac{2\pi K}{\epsilon(\Omega)}\mathcal{P}_{NL}e^{i\Delta Kz}, \tag{18}$$

but this equation can actually be used even for pulses in the range of few nanoseconds. If in the stationary regime diffraction becomes important, one has

$$-\frac{1}{2iK}\Delta_\perp A + \frac{\partial A}{\partial z} = i\frac{2\pi K}{\epsilon(\Omega)}\mathcal{P}_{NL}e^{i\Delta Kz}, \tag{19}$$

which with appropriate normalization and renaming of variables can be cast in the form (16), and indeed Eq. (19) for a Kerr-nonlinearity medium admits as solutions the spatial solitons (Hasegawa, 1989; Agrawal, 1989). For simplicity, in the previous equations we tacitly assumed an isotropic medium, but provisions for anisotropic ones can be easily introduced [see Eq. (20) below].

For the solutions of the nonlinear propagation equation (14) and the subsequently derived ones, Eqs. (15)–(19), one clearly needs the expression of $\mathcal{P}_{NL}$ in terms of a susceptibility tensor and the appropriate product of field amplitudes or a sum of such terms; each of the field amplitudes involved in the interaction in principle is affected by corresponding nonlinear polarization sources and obeys a similar nonlinear equation coupled to those of all fields involved in the interaction, and one must actually solve a coupled-amplitude system of equations, a rather formidable and more often intractable problem unless some drastic simplifications are made. We have chosen some few key cases to illustrate these problems and highlight some major features of nonlinear optical processes (Armstrong *et al.*, 1962; Shen, 1984).

2.2 Second-Order Processes

The second-order processes are strictly coherent processes involving three waves (ω_i, $\mathbf{k}_i$, $\hat{\mathbf{e}}_i$), $i = 1, 2, 3$, with $\omega_3 = \omega_1 + \omega_2$, interacting in a medium that lacks inversion symmetry (Fig. 1). No absorption loss to the matter is associated with such process, and this is true for all even-order processes. Assuming a lossless medium and no distortion due to dispersion, the mutual interaction and energy transfer between these waves can be described by a set of three coupled-amplitude equations of the form

$$-\tan\alpha_3 \frac{\partial A_3}{\partial x} + \frac{\partial A_3}{\partial z} + \frac{1}{v_{g_3}}\frac{\partial A_3}{\partial t} = i\frac{4\pi^2}{n_3\lambda_3}\frac{\chi_e}{\cos\alpha_3}A_1A_2e^{-i\Delta Kz} \tag{20}$$

and similar ones for A_1 and A_2 obtained by permuting the indices 1, 2, and 3; here χ_e is

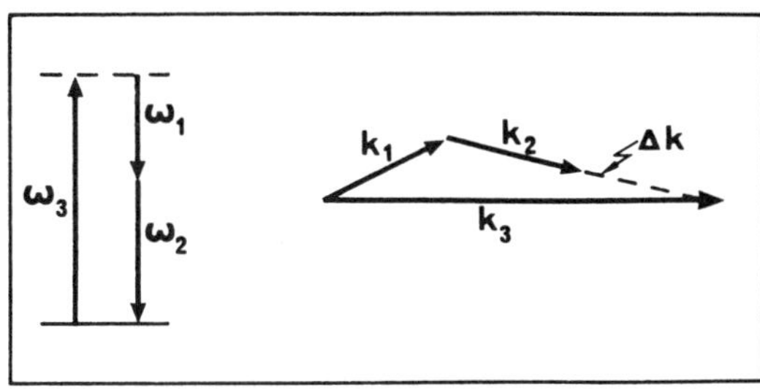

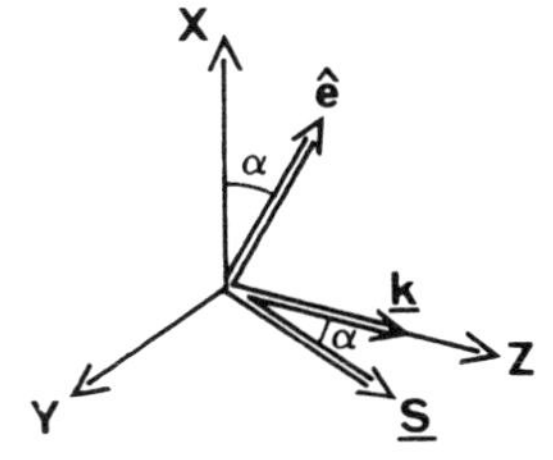

FIG. 1. Frequencies, wave vectors, and energy flow configurations in three-wave interaction.

an effective second-order susceptibility with all frequencies in the same transparency region so that Kleinman's relation applies or $\chi_e \equiv \hat{\mathbf{e}}_3 \cdot \boldsymbol{\chi}^{(2)}(\omega_1,\omega_2):\hat{\mathbf{e}}_1\hat{\mathbf{e}}_2 = \hat{\mathbf{e}}_2 \cdot \boldsymbol{\chi}^{(2)}(\omega_3,-\omega_1):\hat{\mathbf{e}}_3\hat{\mathbf{e}}_1 = \hat{\mathbf{e}}_1 \cdot \boldsymbol{\chi}^{(2)}(\omega_3,-\omega_1):\hat{\mathbf{e}}_3\hat{\mathbf{e}}_1$; α_1 is the direction angle of $\hat{\mathbf{e}}_i$, and n_i and λ_i are the refractive index and wavelength respectively at frequency ω_i.

2.2.1 Conservation Laws Introducing the intensities, which are the magnitudes of the time-averaged Poynting vector $\mathbf{S} = c\mathbf{E}\chi\mathbf{H}/4\pi$,

$$I_i = \frac{c}{2\pi} n_i \cos^2\alpha_i A_iA_i^* \tag{21}$$

and the total number of photons crossing a given plane normal to the z axis,

$$N_i(z) = \int_{-\infty}^{\infty} dxdydtS_i(x,y,z,t)/\hbar\omega_i,$$

one gets after some simple algebraic operations in Eqs. (20)

$$\frac{dN_1}{dz} = \frac{dN_2}{dz} = -\frac{dN_3}{dz} \tag{22}$$

and $d(N_1\hbar\omega_1 + N_2\hbar\omega_2 + N_3\hbar\omega_3)/dz = 0$, which expresses energy conservation between the waves only; the medium takes no part in the energy transfer for infinite transverse extent. One also has

$$\frac{d}{dz}\left(\frac{S_3}{\hbar\omega_3} + \frac{S_2}{\hbar\omega_2}\right) = \frac{d}{dz}\left(\frac{S_3}{\hbar\omega_3} + \frac{S_1}{\hbar\omega_1}\right) = \frac{d}{dz}\left(\frac{S_2}{\hbar\omega_2} - \frac{S_1}{\hbar\omega_1}\right) = 0. \tag{23}$$

These are the Manley–Rowe relations (Manley and Rowe, 1959; Armstrong *et al.*, 1962) and express simply the fact that when the wave at ω_3 gains or loses one photon the ones at ω_1 and ω_2 simultaneously lose or gain one, a process that symbolically can be represented by $\hbar\omega_3 \rightleftarrows \hbar\omega_1 + \hbar\omega_2$.

2.2.2 Second-Harmonic Generation In this case, we have only two frequencies, ω and 2ω, and two coupled-amplitude equations

$$\frac{dA_1}{dz} = i\frac{8\pi^2}{\lambda_1 n_1}\chi_e A_2A_1^*e^{-i\Delta kz}, \tag{24a}$$

$$\frac{dA_2}{dz} = i\frac{4\pi^2}{\lambda_2 n_2}\chi_e A_1^2e^{i\Delta kz}, \tag{24b}$$

with $\Delta k = 2k_1 - k_2$, where we have simplified by assuming $\cos\alpha_i \approx 1$ and the relation $\hat{\mathbf{e}}_1 \cdot \boldsymbol{\chi}^{(2)}(2\omega,-\omega):\hat{\mathbf{e}}_2\hat{\mathbf{e}}_1 = 2\hat{\mathbf{e}}_2 \cdot \boldsymbol{\chi}(\omega,\omega):\hat{\mathbf{e}}_1\hat{\mathbf{e}}_1 = \chi_e$ was used. This set of equations can be solved quite generally in closed form (Armstrong *et al.*, 1962).

As long as $\Delta k \neq 0$, the solutions will be oscillating in space and will never grow sufficiently to deplete the pump wave at ω_1. If this is the case, we may assume $|A_1|^2$ constant, and for a crystal of length L along the beam propagation one obtains the solution

$$A_2(L) = i\frac{8\pi^2}{\lambda_1}\frac{1}{n_2}\chi_e A_1^2 L\left\{\frac{\sin(\Delta kL/2)}{\Delta kL/2}\right\}e^{i\Delta kL/2}$$

for the amplitude of the second harmonic at the exit crystal face if we assume $A_2(0) \approx 0$; or, in terms of the intensity (21),

$$I_2 = \frac{128\pi^5}{c\lambda_1^2}\frac{\chi_e^2}{n_1^2 n_2}I_1^2\left\{\frac{\sin(\Delta kL/2)}{\Delta kL/2}\right\}^2 L^2. \qquad (25)$$

As anticipated, I_2 oscillates in space with period $L_c = \pi/\Delta k = \pi c/|n_1 - n_2|$, the coherence length, and the efficiency of the process decreases as Δk increases as was experimentally demonstrated with the Maker fringes (Fig. 2) (Maker *et al.*, 1962). Actually, this behavior provides a very precise method to determine the dispersion of the refractive index.

In the general case, where the depletion of the pump at ω_1 cannot be neglected, one must solve the set of the two coupled equations (24a) and (24b). Setting $A_j = (2\pi I/n_j c)^{1/2} a_j \times \exp(i\varphi_j)$ there, with $I = I_1 + I_2$ and a_j real, introducing the interaction length $L_i = (n_1^2 n_2 c^3/8\pi^3\omega^2 I\chi_e^2)^{1/2}$ and taking real and imaginary parts, these reduce to

$$\frac{da_1}{d\zeta} = -a_1 a_2 \sin\theta, \quad \frac{da_2}{d\zeta} = -a_1^2 \sin\theta,$$

$$\frac{d\theta}{d\zeta} = \Delta s + \tan\theta\frac{d}{d\zeta}(\ln a_1^2 a_2),$$

where $\Delta s = \Delta kL_i$, $\zeta = z/L_i$, and $\theta = 2\varphi_1 - \varphi_2 + \Delta kz$, which can be solved in terms of elliptic integrals (Armstrong *et al.*, 1962). In the case of phase matching, $\Delta k = 0$, one can easily show that there are two conserved quantities: $a_1^2 + a_2^2 = 1$, which expresses the energy flux conservation, and $a_1^2 a_2 \cos\theta = \Gamma$, where Γ only depends on the values at the boundary; by inserting these relations in the previous equations, one obtains $da_2^2/d\zeta = \pm 2[(1 - a_2^2)^2 a_2^2 - \Gamma^2]^{1/2}$, which can be readily solved and expressed in terms of the Jacobi elliptic functions when $\Gamma = 0$, namely when either one of the amplitudes is zero at $z = 0$ or $\cos\theta = 0$; then one gets $a_2(z) = \tanh\zeta$, $a_1(z) = \mathrm{sech}\zeta$, and $\varphi_2(z) = 2\varphi_1(z) + \pi/2$. The solution shows that whenever $\Gamma = 0$ the second harmonic will continuously grow at the expense of the fundamental, which after a distance L_i, the interaction length, will be reduced by about 75% (Fig. 3).

Without elaborating further, we point out here that because of birefringence, which is required to achieve phase matching (see Sec. 2.2.3 below), the fundamental and the har-

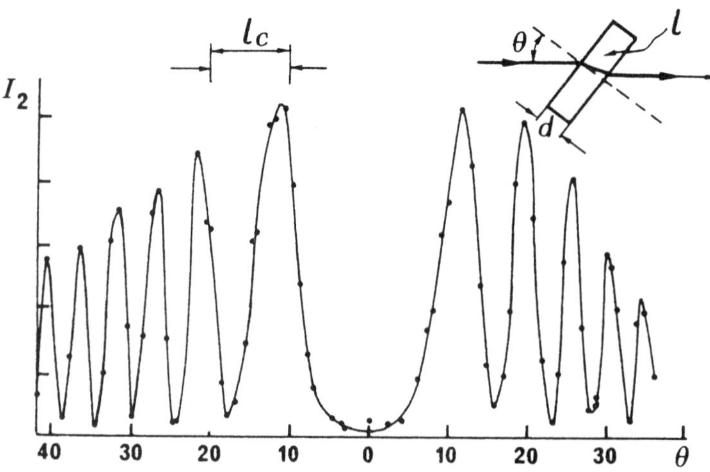

FIG. 2. Maker fringes in second-harmonic generation.

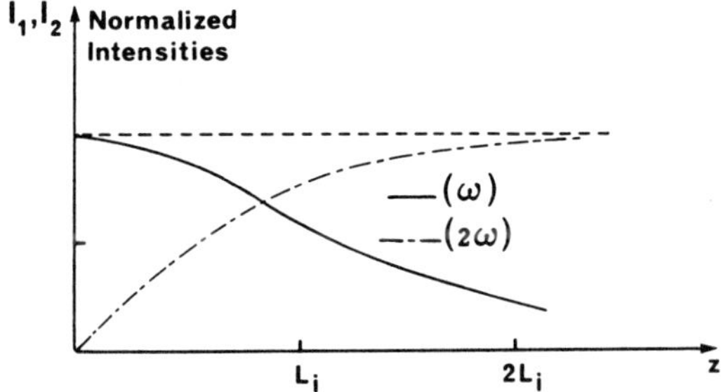

FIG. 3. Fundamental (continuous) and harmonic (dashed) intensities in phase-matched second-harmonic generation.

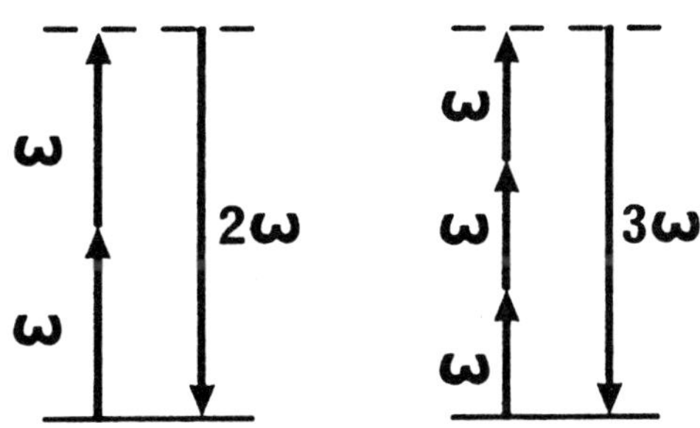

FIG. 4. Second- and third-harmonic frequency-generation schemes.

monic will not propagate in the same direction. Hence, beams of finite cross sections will eventually cease to overlap after they propagate a certain distance (walk off) and the conversion efficiency will be reduced. This can be accounted for by including transverse derivatives of A as indicated in Eq. (20); this problem and the finite-beam profile problem have been extensively studied (e.g., Shen, 1984) for Gaussian beams, as it is crucial for several applications involving nonlinear processes with frequency conversion. A similar effect also occurs in the time domain when short light pulses are used for second-harmonic generation or other frequency-converting nonlinear process; because the group velocities, say at ω and 2ω, are different, the two pulses eventually will not overlap in time and the efficiency will be reduced here too. This problem too has been extensively studied for Gaussian shape pulses.

2.2.3 Phase Matching It is evident from Eq. (25) that one must achieve the condition $\Delta k = 0$, namely to phase match the two waves, in order to obtain appreciable energy transfer. This is a major problem in nonlinear optics and dramatically conditions the efficiency of a given nonlinear process, in particular when these involve a frequency change (Fig. 4).

In the case of three-wave interaction with $\omega_3 = \omega_1 + \omega_2$, the phase-matching condition is $\mathbf{k}_3 = \mathbf{k}_1 + \mathbf{k}_2$, which for collinear beams becomes $n_1\omega_1 + n_2\omega_2 = n_3\omega_3$. For an optically isotropic medium, this condition cannot be satisfied when all three frequencies ω_1, ω_2, and ω_3 fall within the same transparency region of the medium because of the normal dispersion. This is clearly evident for the special case of second-harmonic generation where the previous condition reduces to $n(\omega) = n(2\omega)$ which cannot be satisfied in the normal dispersion region. In principle, it can be achieved if one of the frequencies falls in the region of anomalous dispersion, which is a rather awkward and impractical situation.

The most convenient and common method for achieving phase matching is to exploit the linear (Giordmaine, 1962; Maker *et al.*, 1962) or circular birefringence that many crystals exhibit. The linear birefringence is a consequence of their optical anisotropy. If such is the case for a given propagation direction, there are two indices for two orthogonal polarizations (Fig. 5), namely the ordinary and extraordinary with indices n_o and n_e, respec-

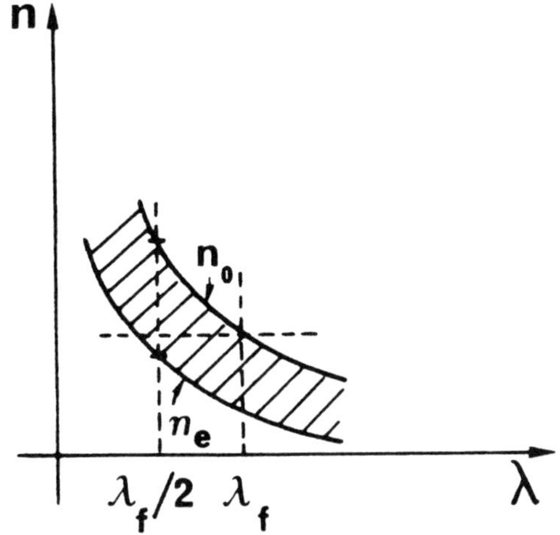

FIG. 5. Construction for phase matching in uniaxial crystal with $n_e > n_0$. The curve of n_e sweeps the hatched region as the direction of the wave vector is varied; the curve for n_0 is independent of the wave-vector direction.

tively, which can be either $n_e > n_o$ (positive uniaxial) or $n_o > n_e$ (negative uniaxial). By appropriate choice of the polarizations of the beams, one can offset the phase mismatch because of the normal dispersion with the birefringence (Giordmaine, 1962); one distinguishes type I and type II phase matching as follows:

	$n_e > n_o$	$n_e < n_o$
Type I	$e,e \to o$	$o,o \to e$
Type II	$o,e \to o$	$e,o \to e$

Clearly, for each set of frequencies different values of the refractive indices are required to achieve phase matching. The change of the refractive indices in birefringent crystals can be obtained most conveniently either with angle or temperature tuning, but modification of the birefringence by a static electric field or pressure can also be used.

Phase matching can also be achieved by exploiting the circular anisotropy of gyrotropic media (natural rotatory power) where the phase velocities for right- and left-circular polarization states are different; these can be solids or even liquids or other isotropic media. Circular birefringence can also be achieved artificially with a magnetic field (Faraday effect) and used for phase matching.

The phase matching sets very stringent restrictions on the nonlinear processes and constitutes the most efficient way for selectively enhancing or suppressing a given process with respect to others.

2.2.4 Second-Harmonic Generation in Reflection The reflection of the second harmonic from an interface or a surface has been given much attention because it constitutes a powerful technique for surface diagnostics (Shen, 1986, 1989). The nonlinear reflection laws can be obtained by imposing the continuity conditions on the fields and inductions at the interface between two media (Bloembergen, 1964; Shen, 1984). We only point out here that this leads to $n_2 \sin\theta_2 = n_1 \sin\theta_1$ for the case of the second harmonic where θ_1 and θ_2 are the reflection angles for the fundamental and the harmonic; thus, the second harmonic is reflected at a different angle from the fundamental. One can also estimate the reflected harmonic intensity with respect to the transmitted one for normal incidence as

$$\frac{I_{2R}}{I_{2T}} = \left(\frac{2k_1 - k_2}{2k_1 + k_2}\right)^2 = \left(\frac{\Delta n}{2n}\right)^2, \tag{26}$$

which is several orders of magnitude smaller. This was implicitly assumed to be the case in the previous discussion of the second-harmonic generation where we assumed that $A_2 = 0$ at the entrance of the crystal, which, in view of Eq. (26), is an approximation, albeit a good one.

2.2.5 Sum- and Difference-Frequency Generation and Parametric Amplification One can proceed to treat other cases along similar lines (e.g., Armstrong *et al.*, 1962; Shen, 1984; Boyd, 1992); we shall briefly consider two cases that are important for device applications, namely, the sum- and difference-frequency generation with one beam, the pump, remaining undepleted. The case of sum-frequency generation $\omega_1 + \omega_2 = \omega_3$ where one of the two input beams, say ω_2, is strong and remains undepleted while the other, ω_1, is weak can be treated analytically in the stationary regime. Since $|A_2|$ is unaffected by the nonlinear process, Eqs. (20) reduce to a set of two coupled-amplitude equations for A_1 and A_3 only:

$$\frac{dA_1}{dz} = i\frac{4\pi^2}{n_1\lambda_1}\chi^{(2)}A_3A_2^*e^{-i\Delta kz} \equiv \tfrac{1}{2}\kappa_1A_3e^{-i\Delta kz}, \tag{27a}$$

$$\frac{dA_3}{dz} = i\frac{4\pi^2}{n_3\lambda_3}\chi^{(2)}A_1A_2e^{i\Delta kz} \equiv \tfrac{1}{2}\kappa_3A_1e^{i\Delta kz}, \tag{27b}$$

where as previously $\cos\alpha_i \approx 1$, $\Delta k = k_1 + k_2 - k_3$. Introducing $\kappa^2 = -\kappa_1\kappa_3^*$ and $g = (\kappa^2 + \Delta k^2)^{1/2}/2$, one obtains

$$A_3(z) = (\kappa_3/g)A_1(0)\sin gze^{i\Delta kz/2}$$

when $A_3(0) = 0$, which shows that the efficiency of the up-conversion process decreases with increasing phase mismatch Δk (Fig. 6). For perfect phase match, $\Delta k = 0$, and $A_3(0) = 0$ the solutions of Eqs. (27a) and (27b) become

$$A_1(z) = A_1(0)\cos\kappa z, \tag{28a}$$

$$A_3(z) = -A_1(0)(\kappa/\kappa_1)\sin\kappa z, \tag{28b}$$

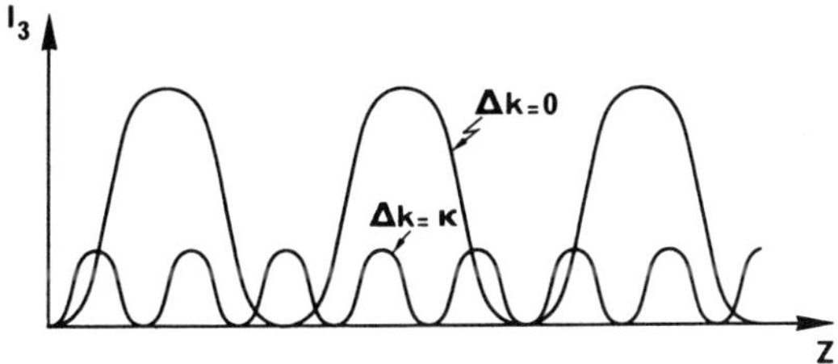

FIG. 6. Sum-frequency generation for two values of the phase mismatch Δk ($\Delta k = 0$ and $\Delta k = \kappa$).

which exhibit an oscillatory behavior with period $L = 2\pi/\kappa$. The previous discussion is of particular relevance in the case of frequency up-conversion where a weak infrared signal can be amplified and upconverted in frequency—for instance, in the visible—by letting it interact with an intense laser beam.

Another important problem is the difference-frequency generation $\omega_2 = \omega_3 - \omega_1$, where the beam at ω_3 will be assumed strong and undepleted by the nonlinear process (Akhmanov and Khokhlov, 1962). The set of coupled-amplitude equations in the stationary regime is now

$$\frac{dA_1}{dz} = i\frac{4\pi^2}{n_1\lambda_1}\chi^{(2)}A_3A_2^*e^{i\Delta kz} = \tfrac{1}{2}\,\kappa_1A_2^*e^{i\Delta kz}, \qquad (29a)$$

$$\frac{dA_2}{dz} = i\frac{4\pi^2}{n_2\lambda_2}\chi^{(2)}A_3A_1^*e^{i\Delta kz} = \tfrac{1}{2}\,\kappa_2A_1^*e^{i\Delta kz}, \qquad (29b)$$

which can be exactly solved in terms of sinhγz and coshγz; in Eqs. (29a) and (29b), we may neglect the dispersion of $\chi^{(2)}$. For the important case of $A_1(0) \neq 0$ but $A_2(0) = 0$, one has

$$A_1(z) = A_1(0)\left\{\cosh\gamma z + i\frac{i\Delta k}{2\gamma}\sinh\gamma z\right\}$$

$$A_2(z) = iA_1(0)\sinh\gamma z$$

for the signal and idler respectively with $\gamma^2 = \gamma_0^2 - (\Delta k)^2$ and $\gamma_0^2 \sim I_3$. Thus, monotonic growth of an initially weak signal wave (ω_1) and concomitantly of the idler (ω_2), which initially is zero, will occur at the expense of the pump only if γ is a real number—that is, if the pump intensity I_3 satisfies the condition $\gamma_0^2 > (\Delta k)^2$ or

$$I_3 > I_c = \frac{c^3(\Delta k)^2 n_1 n_2 n_3}{32\pi^3|\chi^{(2)}|\omega_1\omega_2}. \qquad (30)$$

Thus, if phase matching is not achieved there will be a threshold for the onset of amplification. Above this threshold, $I_1(z) = I_1(0)[(\gamma_0/\gamma)^2\sin^2\gamma z + 1]$. Clearly, for perfect phase matching and no absorption losses there is no threshold condition.

If in the perfect phase-matching case an appropriate feedback is provided, for instance by inserting the system between two mirrors to form an optical resonator, oscillation can occur (Akhmanov and Khokhlov, 1962) as a result of the gain of the parametric amplification process without having beams at frequencies ω_1 and ω_2 initially. The process is initiated by the quantum noise provided by spontaneous breakup of pump photons ω_3 into signal and idler photons, ω_1 and ω_2, respectively, and is called spontaneous parametric scattering; its treatment needs the quantification of the electromagnetic field modes. The optical parametric oscillation process does not involve any resonances of the material and is exempt from any limitations regarding time response and relaxation of material excitations; it has found several applications in devices—for instance, optical parametric oscillators down to the femtosecond regime.

2.3 Third-Order Processes

There is a fundamental difference between second- and third-order processes—and, for that matter, between even- and odd-order processes in general—which has deep implications in several respects. This difference can be mostly easily perceived by inspection of the energy transfer due to the nonlinear polarization in Eq. (6). Inserting there expression (7) and separating the terms of different orders, $dW^{(n+1)}/dt = \langle \mathbf{E}\cdot d\mathbf{P}^{(n)}/dt\rangle$, one sees that the energy-transfer term related to second-order processes cannot be fixed independently of the phases of the involved electric field amplitudes since $dW^{(3)}/dt$ is an odd power in the latter; and similar considerations apply for all even-order processes. In contrast, the energy-transfer term $dW^{(4)}/dt$ related to third-order processes, being an even power in the field amplitudes, for certain frequency and wave-vector configurations, namely whenever the condition $\omega_i + \omega_j = 0$ is satisfied for any pair, can be fixed independently of the phases of the electric fields, and similarly for all higher odd-order processes. Otherwise stated, if such is the case the imaginary part of $\chi^{(3)}$ can be related to a real energy exchange between the field and the matter through a nonlinear mechanism involving two photons and a material excitation while the real part of $\chi^{(3)}$ is related to an intensity-dependent dispersion.

Whenever the condition $\omega_i + \omega_j = 0$ is not satisfied for any frequency pair, one cannot make such a separation, and then one has only coherent processes with energy transfer between the fields without the matter being involved, similar to the second-order processes although more complex because of the higher-order dependence on the fields. This is the case of third-harmonic generation, frequency mixing, and parametric amplification and oscillation, which can be discussed along the same lines as outlined in the previous section. The phase-matching conditions here can be more easily satisfied than in the corresponding second-order processes as they involve the closing of a quadrilateral and not a triangle. In particular, one can achieve perfect phase matching even in the case of isotropic media like gases or liquids by appropriate mixing of two components to introduce adjustable dispersion. These processes can be enhanced with one-photon resonances, but this to some extent is compensated by the concomitant important linear losses.

Here we shall only consider effects that can occur whenever the condition $\omega_i + \omega_j = 0$ is satisfied for any pair of frequencies; it is automatically satisfied for the other frequency pair as well. We consider the case of two waves of frequencies ω and ω' and the degenerate one with $\omega = \omega'$. As hinted above, in these cases the real and imaginary parts of $\chi^{(3)}$ can be related to intensity-dependent refractive index and absorption, respectively. Indeed, from

$$\begin{aligned} P(\omega) &= P^{(1)}(\omega) + P^{(3)}(\omega) \\ &= [\chi^{(1)}(\omega) + 3\chi^{(3)}(\omega,-\omega',\omega')|E(\omega')|^2]E(\omega) \\ &\equiv \tilde{\chi}(\omega,I_{\omega'})E(\omega) \end{aligned}$$

one may define (Shen, 1984) an effective refraction index

$$\tilde{n} = n_0 + n_2 I, \tag{31a}$$

where $n_2 = 12\pi^2\chi^{(3)}/cn_0^2$, and similarly for the absorption

$$\tilde{\alpha} = \alpha_0 + \alpha_2 I. \tag{31b}$$

We point out here that Eqs. (31a) and (31b) cannot be related to each other in a straightforward manner by a Kramers–Kronig relation as is the case between the linear coefficients n_0 and α_0, respectively. We shall refer to the cases (31a) and (31b) as optical Kerr nonlinearities, dispersive and absorptive, respectively. In connection with the effective refractive index (31a), there is an additional complication compared to the linear case; in general, the nonlinear one does not follow instantly the temporal intensity variations. To a good approximation, its temporal evolution obeys (Akhmanov *et al.*, 1975) a Debye-type equation

$$\tau\dot{\tilde{n}}(t) + \tilde{n}(t) = n_2 I(t) \tag{32}$$

or in integral form

$$\tilde{n}(t) = n_0 + \frac{n_2}{\tau}\int_{-\infty}^{\infty} I(t)e^{-(t-s)/\tau}ds,$$

where τ is a phenomenological response time for the Kerr nonlinearity.

Another crucial feature of the effective indices (31a), (31b), and (32) is that these may now become space and time dependent through their dependence on the beam intensity $I(\mathbf{r},t)$, and one has spatial and temporal lensing and grating effects that can lead to dramatic modifications of the spatiotemporal characteristics of a beam; the optical lensing effect or photoinduced effects result from the light self-action and include self-focusing and -defocusing, self-phase modulation, spatial or temporal soliton formation, modulational instabilities, and other similar effects. The optical grating effects result from the coupling of two beams and the interference pattern they set up, which gives rise to effects like optical phase conjugation, optical bistability, instabilities, and chaotic behavior. We shall briefly describe these effects as they pertain to the dispersive Kerr nonlinearity (31a), although the absorptive case can also be treated along similar lines but several of these effects cannot occur there because of the inherent energy loss.

2.3.1 Degenerate Four-Wave Interactions; Nonlinear Grating Effects Here we review two cases of optical gratings that proceed through a nonlinear interaction: optical phase conjugation and optical bistability. Permanent gratings have been extensively studied and used in optics. They result from permanent spatial modulation of the refractive index or the coefficient of absorption for instance by the interference of two light waves; they constitute the basic elements in several important applications like interference filters and holography and are used to make important operations on optical beams. Similar gratings can be set up (e.g., Eichler, 1980) by nonlinear processes and in particular by exploiting the coupling of two intense beams provided by the Kerr-type nonlinearities (31a) and (31b), which affect both the refractive index and the coefficient of absorption. In contrast to the permanent gratings, which are static, however, the ones that result from such nonlinear processes are transient or dynamic and decay more or less fast after the optical beams have been switched off. Accordingly, their uses and applications are similar to those of the permanent gratings but transposed in real time. As a matter of fact, the temporal evolution of these gratings and in particular their rise and decay times are essential features and condition all their applications, which most often concern the self-diffraction or the diffraction of a third beam from the interference pattern of the two beams.

The cause of the formation of these gratings is the real or virtual excitation of some material mode of the medium that couples to either the refractive index or the coefficient of absorption. This excitation actually proceeds through an intermediate two-photon process previously mentioned and constitutes an essential characteristic of third-order processes and of Kerr nonlinearities in particular. One can have stationary optical gratings or propagating ones depending on whether the two interfering beams are degenerate in frequency or not (e.g., Eichler, 1980).

The degenerate case is particularly simple to present and can serve as a basis for several effects that will be succinctly discussed below. Let us consider two plane waves of frequency ω, wave vectors $\mathbf{k}_F$ and $\mathbf{k}_B$, and amplitudes $\mathbf{A}_F$ and $\mathbf{A}_B$, respectively; we chose the Oxz plane to coincide with the one defined by $(\mathbf{k}_F,\mathbf{k}_B)$ with the z axis along $\mathbf{q} = \mathbf{k}_F - \mathbf{k}_B$ (Fig. 7). Then one can easily see that the intensity distribution in an optically isotropic medium is

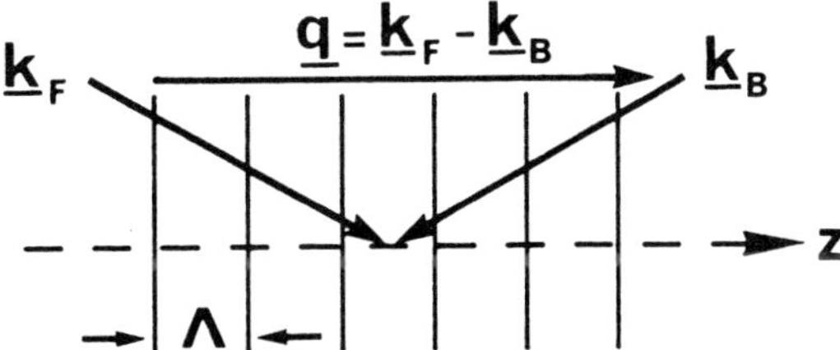

FIG. 7. Optical gratings (stationary).

$$I = I_F + I_B + 2\Delta I \cos 2q_z z, \tag{33}$$

where $\Delta I = n\mathbf{A}_B \cdot \mathbf{A}_F^* 2\pi$; for anisotropic media, this is a tensorial relation. Thus, the compound intensity (33) is periodically modulated with a spatial period $\Lambda = 2\pi/q = 2\lambda \times \sin(\theta/2)$ where θ is the angle between $\mathbf{k}_B$ and $\mathbf{k}_F$. If a Kerr medium is placed in the region of interference of the two beams, this modulation is "printed" on the refractive index or the absorption coefficient by means of the corresponding Kerr nonlinearity (31a) or (31b), respectively; one has the primary optical gratings that in addition can give rise to secondary gratings by the ensuing relaxation processes in the medium.

These gratings can be probed by diffracting a third optical beam of different frequency ω or eventually of same frequency in which case self-diffraction can also be used. The diffracted beams actually appear only in certain directions with respect to $\mathbf{q} = \mathbf{k}_F - \mathbf{k}_B$ and the probe beam direction. The amplitudes of the different orders $m = \pm 1, \pm 2, \cdots$ give a measure of the modulation; they are strongly modified by the thickness of the sample, and one distinguishes thin gratings when the diffraction contains contributions due to the finite thickness of the sample and thick gratings when these contributions are negligible. Certain complications also arise from the finite transverse extension and profile of the beams. If the diffraction intensity is very weak, heterodyne-type techniques must be used to extract the scattered light from stray light. The case of propagating gratings is slightly more complicated than the stationary one.

The formation and detection of optical gratings, either stationary or propagating, constitute the basis of numerous laser-diagnostic techniques (Eichler, 1980) in particular for the study of the dynamics of the material excitations where the spatiotemporal evolution of the gratings gives direct information about their relaxation processes and propagation or diffusion characteristics. These are also implemented in devices like the acousto-optic modulation and deflection to be discussed below.

2.3.1.1 Optical Phase Conjugation. Phase conjugation in optics is used to correct aberrations and consists in the operation of replacing a wave $\mathbf{E}_s(\mathbf{r},t) = \mathbf{E}_s e^{-i\omega t}$ + c.c. by its phase conjugate in space, $\mathbf{E}_c(\mathbf{r},t) = \rho \mathbf{E}_s^* e^{-i\omega t}$ + c.c. where $\mathbf{E}_s = \hat{\mathbf{e}}_s A_s e^{i\mathbf{k}_s \cdot \mathbf{r}}$, or equivalently $\mathbf{E}_c(\mathbf{r},t) = \rho \mathbf{E}_s(\mathbf{r},-t)$, which can also be termed time-reversed wave front; ρ is a scalar that characterizes the phase-conjugation operation. Thus, the polarization state and the amplitude are replaced by their complex conjugates and the wave-vector direction is reversed. These operations result in aberration correction as can be easily deduced by the propagation equation or diffraction theory. It should be noted that phase conjugation is also relevant in holography.

The four-wave mixing processes provide very efficient interaction schemes for real-time optical phase conjugation (Hellwarth, 1977). The classic configuration concerns the degenerate four-wave interaction scheme depicted in Fig. 8. Two counterpropagating intense beams of frequency ω and wave vectors $\mathbf{k}_F$ and $\mathbf{k}_B = -\mathbf{k}_F$, the forward and backward pumps respectively, interact in a nonlinear medium with a third weak signal beam of the same frequency but arbitrary propagation direction $\mathbf{k}_s$ to form a beam of frequency ω counterpropagating and phase conjugated to the signal beam; this operation is achieved through the induced third-order polarization at frequency ω and amplitude $P_{NL} = 6\chi^{(3)} A_F A_B A_s^* \exp(-i\mathbf{k}_s \cdot \mathbf{r})$ since $\mathbf{k}_F + \mathbf{k}_B = 0$ and $\mathbf{k}_s + \mathbf{k}_c = 0$ so that the phase-matching condition is automatically satisfied. Actually, the total induced third-order polarization contains several other terms, but the corresponding fields in general are suppressed because of phase mismatch.

If we assume that the pumps remain undepleted, the amplitudes of the signal and its phase-conjugated beams in the stationary regime satisfy (e.g., Fisher, 1983; Zel'dovich *et al.*, 1985) the coupled equations

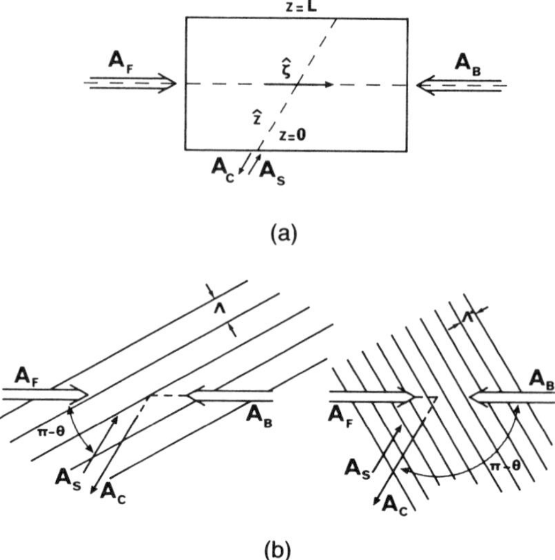

FIG. 8. Optical phase conjugation: (a) interaction scheme and (b) interpretation in terms of spatial gratings.

$$\frac{dA_s}{dz} = i\kappa A_c^*, \tag{34a}$$

$$\frac{dA_c}{dz} = -i\kappa A_s^*, \tag{34b}$$

where $\kappa = (12\pi\omega/nc)\chi^{(3)} A_F A_B$, which can be readily solved. For $A_c(L) = 0$, one obtains

$$A_s(L) = A_s(0)/\cos(|\kappa|L),$$

$$A_c(0) = (i\kappa^*/|\kappa|)A_s^*(0)\tan(|\kappa|L).$$

Thus, the transmitted signal wave is always amplified while its phase-conjugated wave can have a value ranging from zero to infinity depending on the value of $|\kappa|L$, and the whole system acts as a phase-conjugated mirror with reflectivity $R = |r|^2 = \tan^2|\kappa|L$, which can exceed 100% evidently at the expense of the pumps. Note that in analogy with the optical parametric oscillator for $|\kappa|/L = \pi/2$ one can achieve oscillation in any direction by introducing a feedback for the photon noise that results from two twin counterpropagating pump photons being scattered to two twin counterpropagating photons in another direction.

Actually, the whole process can be also visualized in terms of optical gratings as previously discussed (e.g., Fisher, 1983; Zel'dovich *et al.*, 1985). Indeed, the induced nonlinear polarization can be written quite generally

$$\mathbf{P}_{NL} = a(\mathbf{A}_F \cdot \mathbf{A}_s^*)\mathbf{A}_B + b(\mathbf{A}_B \cdot \mathbf{A}_s^*)\mathbf{A}_F + c(\mathbf{A}_F \cdot \mathbf{A}_B)\mathbf{A}_s^*, \tag{35}$$

and the three terms in Eq. (35) can be interpreted as the diffraction of each beam in turn by the optical gratings set up by the other two; actually, the first two terms correspond to spatial gratings (Fig. 9), the third one to a temporal grating oscillating at frequency 2ω. As previously stated, the optical phase conjugation by degenerate four-wave mixing can be used to correct beam aberrations in real time in a wide range of applications (Fig. 10). In particular, all operations and effects that are envisaged in holography can be transposed here in real time. Optical phase conjugation by nonlinear processes has been extensively studied (e.g., Fisher, 1983; Zel'dovich *et al.*, 1985) for several configurations and in particular in media with periodic refractive index and inside a nonlinear Fabry–Pérot cavity where bistable behavior, different types of instabilities, and chaotic behavior can occur.

2.3.1.2 Optical Bistability. Optical bistability or multistability is related to the situation when the output intensity resulting from a given operation on a beam can have two or more values and accordingly must "decide" on which one the system will operate. This can be achieved by abruptly rendering the system opaque or transparent by slight changes of some external parameter along different paths. Several third-order processes and in particular the ones related to the optical Kerr nonlinearity, dispersive or absorptive, provide a simple scheme to realize such a situation in real time if an appropriate feedback is introduced—for instance, by inserting the Kerr medium in a Fabry–Pérot cavity; other configurations can also be devised for the same purpose.

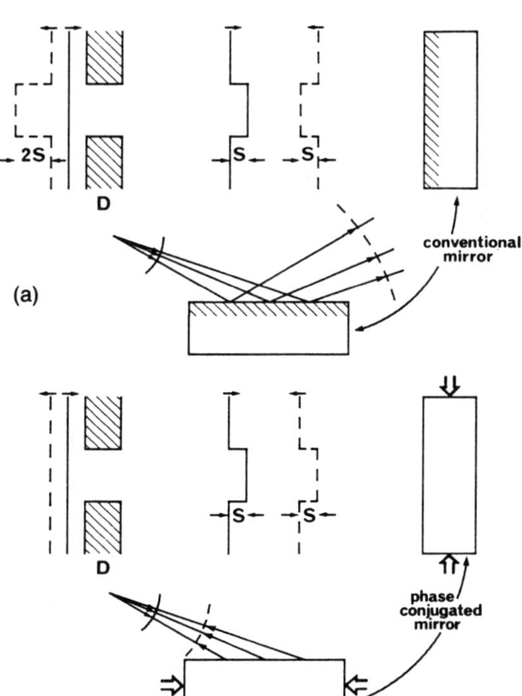

FIG. 9. Comparison of reflectivity (a) from a conventional mirror and (b) from a phase-conjugating mirror.

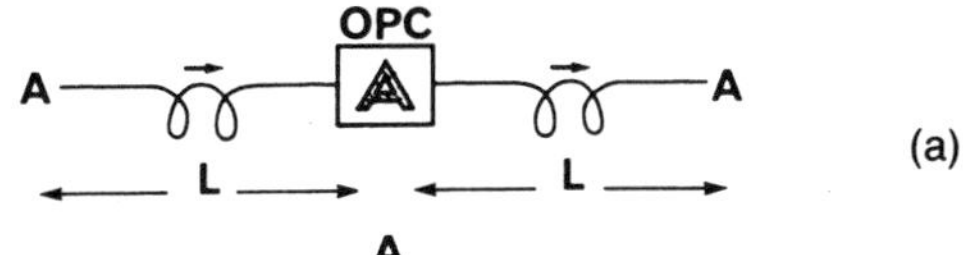

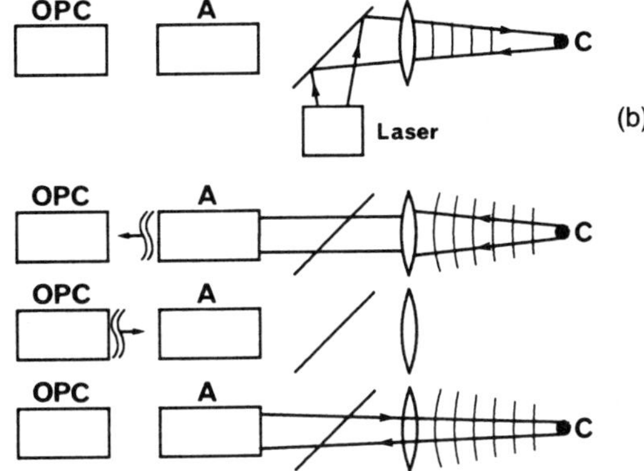

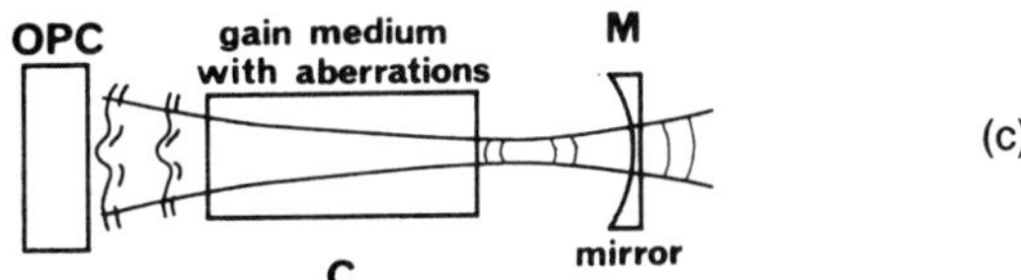

FIG. 10. Applications of optical phase conjugation: (a) wavefront correction in a fiber, (b) wavefront correction in a laser ablation design system, and (c) aberration correction in a laser cavity.

We succinctly discuss the main features and ingredients of the optical bistability in the case of the dispersive (Felber and Marburger, 1976, 1978) nonlinear Fabry–Pérot cavity assuming plane monochromatic waves. A monochromatic wave of frequency ω normally incident on a Fabry–Pérot cavity containing a dispersive Kerr isotropic medium sandwiched between two parallel-plane mirrors sets up an electric field

$$E(z,t) = Re[(A_+e^{ikz} + A_-e^{-ikz})e^{-i\omega t}]$$

inside the cavity whose forward- and backward-propagating components, A_+ and A_- respectively, satisfy the two coupled-amplitude equations

$$\frac{\partial A_+}{\partial \zeta} + \frac{\partial A_+}{\partial \tau} = i\kappa[|A_+|^2 + 2|A_-|^2]\, A_+, \qquad (36a)$$

$$\frac{\partial A_-}{\partial \zeta} + \frac{\partial A_-}{\partial \tau} = -i\kappa[|A_-|^2 + 2|A_+|^2]\, A_-, \qquad (36b)$$

where $\zeta = kz$, $\tau = \omega t$, and $\kappa = 12\pi\chi^{(3)}/n_2c$. In the stationary regime, these can be easily solved and give $A_\pm = A_\pm^0 \exp(\pm\kappa[|A_\pm|^2 + 2|A_\mp|^2])\zeta$ where $A_\pm^0$ are determined by the boundary conditions. Since we assume no losses, linear or nonlinear, $|A_\pm|^2$ remain constant but their relative phase ϕ changes with z or $\phi = \phi_+ - \phi_- = 2\zeta + 2\kappa\{|A_+|^2 + |A_-|^2\}\zeta$.

The continuity conditions for the Poynting vector (flux conservation) across the boundaries give for the transmissivity of the nonlinear Fabry–Pérot cavity

$$\mathcal{T} \equiv \frac{I_t}{I_0} = \frac{1}{1 + F^2\sin^2\phi/2}, \qquad (37a)$$

where $\phi = \phi_0 + \phi' I_t$; the latter can also be rewritten as

$$\mathcal{T} \equiv I_t/I_0 = (\phi - \phi_0)/\phi' I_0, \qquad (37b)$$

where $\phi_0 = 4\pi n_0L/\lambda$, $\phi' = 4\pi n_2(1 + R)L/(1 - R)$, R is the mirror intensity reflectivity, I_0 and I_t are the incident and transmitted beam intensities respectively, and $F = 2\sqrt{R}/(1 - R)$ is the finesse of the cavity.

The operation point of the system is defined as the intersection of the two transmissivity curves (37a) and (37b) and can be obtained with the graphical procedure depicted in Fig. 11; one has bistable operation and hysteresis. As I_0 increases, one obtains additional hysteresis loops and multistability. The hysteresis is obtained because the intersection of the straight line (37b) and the dashed part of the Airy function (37a) is an unstable operation point where even the smallest deviation will grow monotonically with time. To prove this, one must use the nonstationary equations for the amplitudes $A_\pm$ and the phase $\phi = 4\pi\tilde{n}L/\lambda$ where $\tilde{n}$ satisfies the Debye equation (42). One finds that there is critical slowing down similar to that occurring in first-order phase transitions. Actually, a stability analysis of this system of equations shows that the nonstationary solutions are unstable in a certain region of values of the external parameters and one also has transition to chaos. It is difficult however to characterize this transition for realistic systems where transverse effects become important and lead to insurmountable computational difficulties.

Optical bistability has been extensively

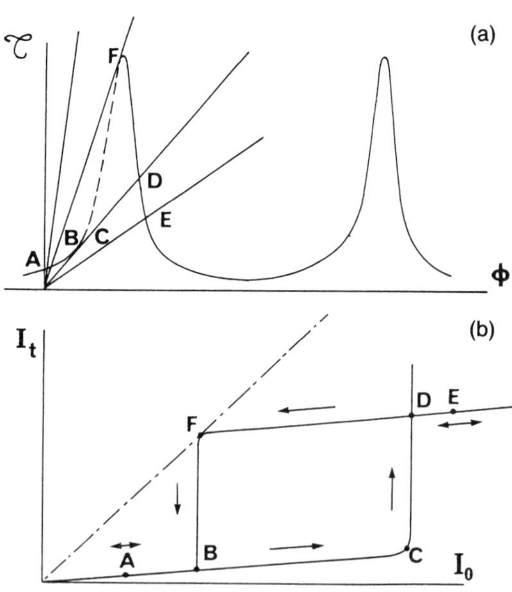

FIG. 11. In (a) the oscillatory curve represents the Airy function of Eq. (37a) and the straight lines represent Eq. (37b) for increasing values of the input intensity *I*. The points where the two curves cross are the operation points of the nonlinear Fabry–Pérot cavity. At the points *F* and *C*, the straight lines are tangent to the Airy curve; on the dashed-line portion, between *F* and *C*, the operation is unstable. In (b) with I_t vs I_0, the operation points are depicted where also the hysteresis loop is shown with arrows.

studied (e.g., Gibbs, 1985). There is a whole class of beam-interaction configurations that exhibit bistable behavior; essentially, all optical systems that can be switched from opaque to transparent and vice versa exhibit bistable operation if the switching operation can be achieved by a nonlinear process. Thus, in the absorptive regime this is achieved by the saturation of absorption. Bistable operation has also been observed in the total reflection effect, in periodic structures, in intracavity optical phase conjugation, in hybrid systems, etc. Evidently, the interest in the optical bistable operation relies on the possibility of having optical memories in real time by exploiting the hysteresis.

2.3.2 Light Self-Action; Nonlinear Lensing Effects In contrast to the previous effects that involve the coupling of two fields, there is a rich class of nonlinear propagation effects that result from the self-action (e.g., Akhmanov *et al.*, 1975; Shen, 1984) of a light beam on itself, the coupling being provided by the optical Kerr nonlinearity. These self-action effects are of considerable fundamental and technological interest and are relevant in a medium with dispersive Kerr nonlinearity. We shall briefly outline their principles and main features.

2.3.2.1 Self-Focusing, -Defocusing. A light beam of finite transverse extension can be focused, defocused, or trapped by appropriately reshaping its equiphase surfaces. In conventional linear optics, this is achieved by lens systems or spatial modulation of the optical density of the medium. The optical Kerr nonlinearity induced in the medium by the beam itself can provoke similar effects since the effective refractive index has different values in different parts of the beam profile (Akhmanov *et al.*, 1975). This self-action can be readily seen when we consider a Gaussian beam $\mathbf{E}(\rho,z,t) = Re\{\hat{\mathbf{e}}A(\rho)\exp(ikz - i\omega t)\}$ with $A(\rho) = A_0 \exp\{-\rho^2/2a_0^2\}$ where a_0 measures the transverse beam extension and ρ is the radial coordinate in an Oxy plane measured from the propagation axis Oz. From Eq. (31a), the effective refractive index is now $\tilde{n} = n_0 + n_2 I(\rho)$ where $I(\rho) = |A_0|^2 \exp\{-\rho^2/a_0^2\}$ with $I_0 = nc|A_0|^2/2\pi$. This dependence of $\tilde{n}$ on ρ effectively introduces a spatially modulated optical density; accordingly, the optical rays are bent inward or outward from the Oz axis depending on whether $n_2 < 0$ or $n_2 > 0$ leading to self-focusing and defocusing respectively (Fig. 12); both effects have been extensively studied.

The self-focusing effect has a threshold because it enters into competition with the diffraction, which increases as the beam diameter decreases through the aperture effect (e.g., Akhmanov *et al.*, 1975; Shen, 1984). The compound effect on the propagation of the positive Kerr nonlinearity and the diffraction evidently can be treated with Eq. (19) for $\Delta K = 0$ where the term $\Delta_\perp A$ now becomes important as the beam shrinks. The threshold condition however can be derived to a good approximation without making appeal to the differential equation but using the simple geometric construction of Fig. 13. The condition that the central ray and the one at the borderline of the beam reach the focal point z_f with the same phase, assuming rectilinear propagation and negligable nonlinearity for the borderline ray, gives $n_0 l = (n_0 + n_2 I_0)z_f$; furthermore, the condition for self-focusing to occur reads $\theta_f > \theta_d$, where $\theta_d = \lambda/2a_0$ is the diffraction angle through an aperture of radius a_0. Hence, the threshold condition for self-focusing to occur reads $I_0 > (n_0/8n_2)(\lambda/\alpha_0)^2 \equiv I_c$. The analytical treatment with a Gaussian profile beam leads to a similar expression multiplied by a factor of 1.22.

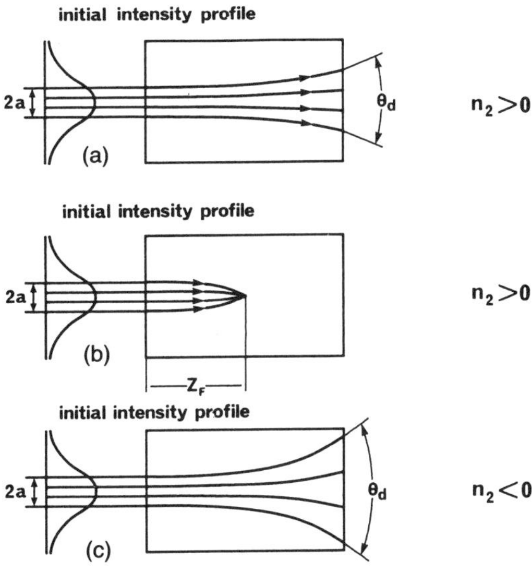

FIG. 12. Schemes for self-focusing and self-defocusing: (a) beam propagation when diffraction dominates with $n_2 > 0$ and $I < I_c$; (b) self-focusing $n_2 > 0$ and $I > I_c$; and (c) self-defocusing when $n_2 < 0$.

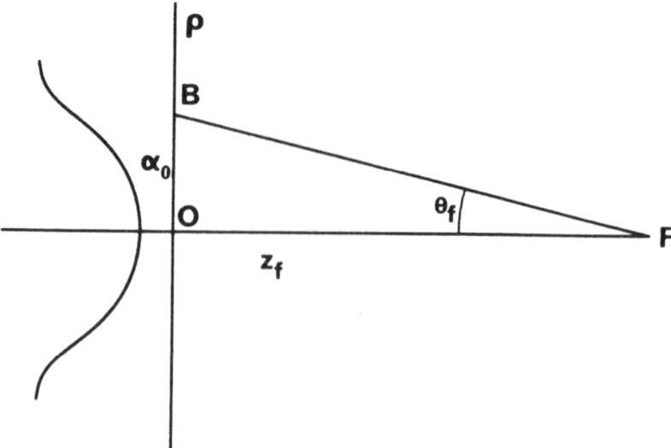

FIG. 13. Geometrical ray configuration for the derivation of the threshold condition for self-focusing. $\alpha_0 = OB$ is a characteristic radius of the Gaussian beam profile (waist) at the entrance of the Kerr medium, and θ_d is the diffraction angle from a circular aperture of radius α_0. The curvature of the ray *BF* is neglected.

2.3.2.2 Light Self-Trapping. Spatial Solitons. If for a certain beam profile the condition $\theta_f = \theta_d$ can be achieved throughout the propagation, this particular beam profile can be maintained unchanged; this is the light self-trapping effect (Akhmanov *et al.*, 1975). Because of the initially irregular transverse profile, this condition may be satisfied simultaneously for different parts of the beam profile and lead to the beam splitting into several filaments. Although the analytical treatment of the problem is very complex, one can gain insight with the following simple model. Assuming $\partial^2 A/\partial y^2 = 0$ and setting $P_{NL} = 3\chi^{(3)}|A|^2 A$ with $\Delta K = 0$ in Eq. (19), the equation reads

$$i\frac{\partial\psi}{\partial\zeta} = \frac{1}{2}\frac{\partial^2\psi}{\partial\xi^2} - |\psi|^2\psi, \tag{38}$$

with $\psi = A/(6\pi\omega\chi^{(3)}/n_0c)^{1/2} \equiv A\sqrt{\kappa}$, $\zeta = z$, and $\zeta = x\sqrt{k}$. Equation (38) is the nonlinear Schrödinger or Landau–Ginzburg equation (Zakharov and Shabat, 1972; Whitham, 1974; Newell and Moloney, 1992; Hasegawa, 1989) that admits a class of stable solutions (eigenstates) termed spatial solitons ψ_N with N an integer, $N = 0, 1, 2 \ldots$; for $N = 1$, $\psi_1 = 1/\cosh\xi \equiv \operatorname{sech}\xi$ while the higher-order solitons, $N > 1$, are periodic functions satisfying the boundary condition $\psi_N = N/\cosh\xi \equiv N \operatorname{sech}\xi$ for $\xi = 0$ and N integer. These spatial solitons play a very important role in guided nonlinear optics (e.g., Hasegawa, 1989; Agrawal, 1989; Ostrowsky and Reinisch, 1992; Miller *et al.*, 1994) and have recently been observed in an increasing number of nonlinear systems.

2.3.2.3 White Self-Transparency. In the previous cases, the medium is assumed homogeneous with an index of refraction n_0 that has the same value throughout the medium. If n_0 fluctuates randomly in space but not in time, the spatial fluctuations of n may be compensated with the optical Kerr nonlinearity and the losses due to elastic scattering are then suppressed, leading to a new class of self-action effects and in particular to self-transparency. To simplify, let us assume that the inhomogeneous medium is a composite consisting of small spherical dielectric particles of refractive index n_B and number density N randomly dispersed in a homogeneous dielectric of refractive index n_A, both without absorption losses. A beam propagating in the z direction experiences losses because of elastic scattering, which in the Rayleigh and Rayleigh–Debye–Gans regimes, neglecting multiple scattering, is expressed with an elastic diffusion constant $\alpha_s \sim (\Delta n_0)^2$ where $\Delta n_0 = n_A - n_B$. If the incident beam intensity is high, then each refractive index is modified by the optical Kerr nonlinearity or $\tilde{n}_A = n_A + n_{2A}I$ and $\tilde{n}_B = n_B + n_{2B}I$. If we assume that the elastic diffusion constant can be written

$$\tilde{\alpha}_s \sim (\Delta\tilde{n})^2 \sim (\Delta n_0 + \Delta n_2 I)^2, \tag{39}$$

where $\Delta n_2 = n_{2A} - n_{2B}$, one easily sees that for $\operatorname{sgn}(\Delta n_0 \Delta n_2) = -1$ there is a light intensity $I_c = |\Delta n_0/\Delta n_2|$ for which $\tilde{\alpha}_s$ in Eq. (39) vanishes; below this intensity, the beam is totally attenuated and above it is transmitted but with intensity I_c; in a certain way, the photons in the scattered modes are recycled into the incident one. In particular, that part of the spatiotemporal intensity profile $I(\mathbf{r},t)$ of a beam that has intensity below I_c is attenuated while the part that has higher intensities propagates without further loss once its intensity has been attenuated to I_c; one can obtain square-form spatiotemporal profiles. One also can achieve bistable operation with a single mirror and several other interesting effects.

2.3.2.4 Self Phase Modulation. The temporal counterparts of the previous effects within the class of third-order processes lead to quite striking effects of fundamental and technological interest. These occur whenever different parts of a light pulse experience different effective refractive indices because of the time dependence of the refractive index imposed by the light pulse itself through the optical Kerr nonlinearity; it results in a rearrangement of its spectral density that may have striking repercussions on the pulse form (Gustafsson *et al.*, 1969).

From Eq. (31a) we see that a short light pulse of intensity $I(t)$ experiences an effective refractive index $\tilde{n}(t) = n_0 + n_2 I(t)$ where for simplicity we assume that the Kerr nonlinearity is instantaneous and exactly follows the pulse form. After propagation over a distance l, the beam experiences a phase modulation $\Delta\phi(t) = \omega n_2 I(t) l/c$ and a frequency modulation $\Delta\omega(t) = -\partial(\Delta\phi)/\partial t$ that leads to a spectral broadening with a spectral density

$$S(\omega) = \left| \int_0^\infty I(t) \exp[i\omega_0 t + i\Delta\phi(t)] dt \right|^2 .$$

If $I(t)$ is symmetric, $\Delta\phi$ will also be symmetric while $\Delta\omega(t)$ will be antisymmetric (Fig. 14) with respect to the carrier frequency ω_0 and with maximum extension $|\Delta\omega|_{\max} = |\partial(\Delta\phi)/\partial t|_{\max}$ on either side of ω_0, the Stokes and anti-Stokes regions; this is the self-phase modulation (Shen, 1984; Hasegawa and Brinkman, 1980). One can easily see that the spectrum will show an approximately periodic structure because there are always on $\Delta\phi(t)$ two points with the same slope that correspond to two waves with the same frequency but different phases; this can be easily inferred from the curve of $\Delta\omega = -\partial(\Delta\phi)/\partial t$ in Fig. 14 since any horizontal line, $\Delta\omega =$ const, with $|\Delta\omega| < \Delta\omega_{\max}$ crosses the curve at two points. Hence, a constructive or destructive interference takes place depending on the phase difference of each such pair, which introduces a succession of valleys and crests, their total number being approximately

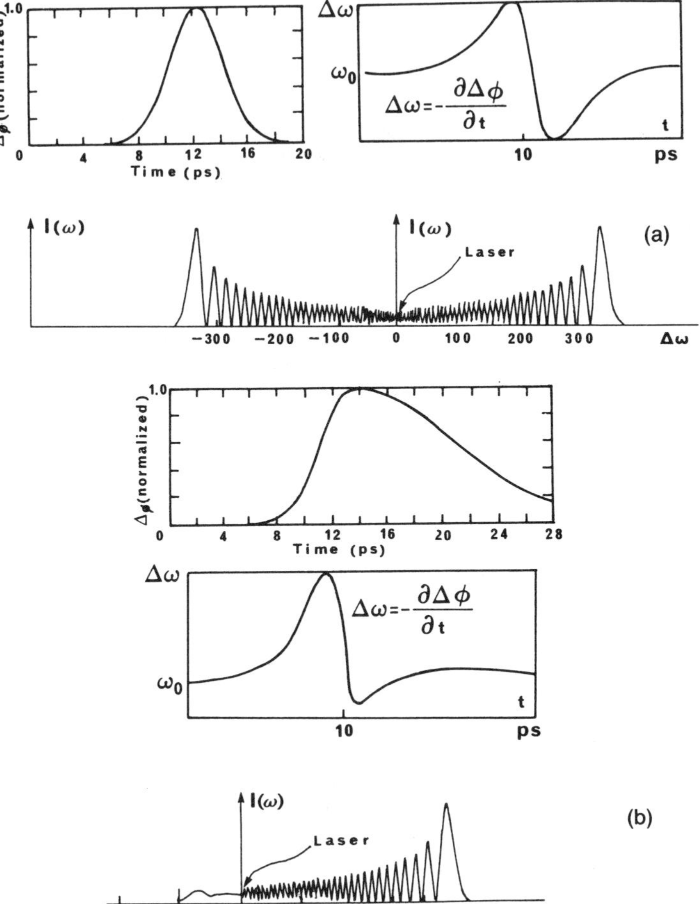

FIG. 14. Simplified description of self phase modulation: (a) without damping, (b) with damping.

$|\Delta\omega_{max}|/2\pi$, with the crest heights increasing as one moves toward the two extremes. Actually, one does not see a symmetric periodic structure with respect to ω_0 but an asymmetric one, the anti-Stokes side being weaker and almost becoming a tail, because in reality the Kerr nonlinearity does not respond instantaneously to the pulse form but obeys Eq. (32) with a finite relaxation time. Self-phase modulation plays a very important role in pulse shaping and also in achieving broadband light pulses (e.g., Shen, 1984; Arecchi and Schulz-Dubois, 1975).

2.3.2.5 Temporal Solitons. Modulational Instability. As previously mentioned, spatial solitons result from the compensation of self-focusing with diffraction, the former being related to the optical Kerr nonlinearity and the latter to the second-order transverse spatial derivatives of the amplitude in Eq. (19); they are special solutions of the nonlinear Schrödinger equation (38). Their temporal analogs, namely the temporal solitons (Zakharov and Shabat, 1972; Hasegawa, 1989), exist and result from a compensation of the self-phase modulation and the group-velocity dispersion, the latter being related to a second-order partial time derivative of the amplitude. Assuming plane waves, neglecting diffraction, setting $\Delta K = 0$, and inserting $P_{NL} = \chi^{(3)}|A|^2 A$ in Eq. (16), one gets

$$i\frac{\partial\psi}{\partial\zeta} = -\tfrac{1}{2}\frac{\partial^2\psi}{\partial\tau^2} + |\psi|^2\psi, \tag{40}$$

with $\tau = (t - z/v_g)/\tau_0$, $\zeta = 2|\kappa|z$; $\kappa = -k''/2\tau_0^2$ where k'' is the group velocity dispersion and $\psi = A/(3\pi\omega\kappa\chi^{(3)}/n_0c)$; Eq. (40), previously derived by renaming the variables, is identical to Eq. (38), the nonlinear Schrödinger for the planar spatial solitons. Thus, the fundamental soliton solution is $\psi = 1/\cosh\tau = \text{sech}\tau$ or explicitly $A = A_0\,\text{sech}(\tau/\tau_0)\exp(ikz)$ with $|A_0|^2 = -k''c/2n_2\omega\tau_0^2$ as imposed by the pulse area condition. Clearly, k'' and n_2 must have opposite signs, as one intuitively expects for the compensation between the group-velocity dispersion and the self-phase modulation to occur. There are higher-order soliton solutions ψ_N, which have the property $\psi_N(\tau) = \text{sech}\tau$ at the input boundary and satisfy the pulse area condition $A_N = NA_0$ (e.g., Hasegawa, 1989; Agrawal, 1989). These are periodic functions of ζ with period $\zeta_0 = \pi/2$ (Fig. 15). Because of their remarkable stability, they are being extensively studied in connection with short-pulse reshaping and shortening and long-distance optical signal transmission (Mollenauer *et al.*, 1980); their implications in the latter technology are still difficult to grasp, but they may fundamentally revolutionize the field of telecommunications. Higher-order solitons may scatter to lower-order solitons. One also distinguishes between bright and dark solitons, the previous ones being the bright solitons and their complementary ones in time being the dark solitons.

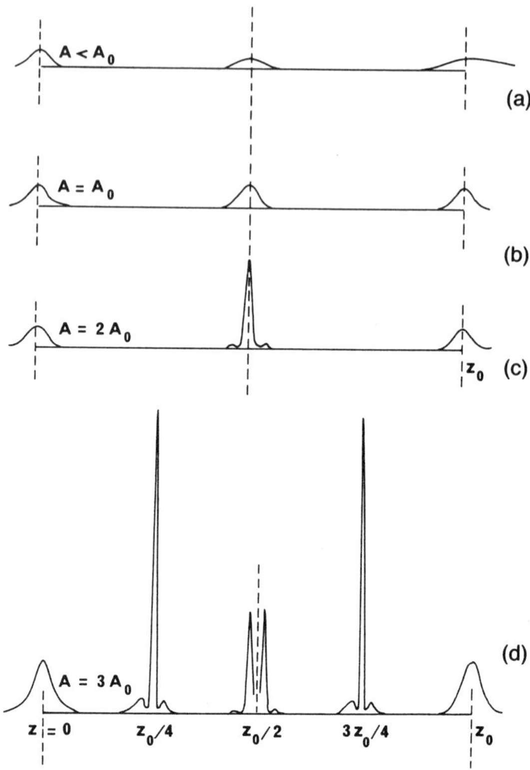

FIG. 15. Soliton propagation: (a) ordinary no soliton propagation, (b) fundamental, (c) second-order, and (d) third-order soliton propagation over the spatial period z_0.

The nonlinear Schrödinger equation exhibits a very interesting instability, namely the modulational instability (Hasegawa and Brinkman, 1980) where energy is transferred from the primary pump wave to a set of sidebands symmetrically disposed about the fundamental frequency; it can be used for high-rate modulation of a beam and other interesting applications.

2.4 Parametric Processes

In the nonlinear optical effects previously discussed, the frequencies of the interacting fields were tacitly assumed to be comparable: All frequencies fall within the same transparency region of the medium so that all fields couple to the same degrees of freedom of the medium. For the transparency region that extends above the vibrational frequencies and below the onset of electronic transitions, for instance, only the valence electrons contribute to the optical nonlinearities and their response can be assumed practically instantaneous.

If any of the frequencies, however, is lowered into a different transparency region, then the corresponding field couples with additional or even different degrees of freedom. For instance, for frequencies below the vibrational frequencies the ionic motion and its coupling with the electrons sets up additional mechanisms for nonlinear polarization as also does the molecular reorientation; these additional mechanisms, besides changing the magnitude and even the sign of the nonlinear coefficients, may modify some of the symmetry properties of these coefficients—for instance, the Kleinman symmetry relations (Kleinman, 1962) will break down—and also introduce different time constants, which can be quite crucial when considering very short light-pulse interactions.

Similarly, when any of the frequencies lies much higher than the electronic transitions and falls in the x-ray region the electrons essentially respond as free and the dipole approximation breaks down and, by the same token, the symmetry properties related to this approximation, the most conspicuous one being $\chi^{(2n)} \equiv 0$ for media with inversion symmetry. Actually, the multipolar expansion must be abandoned, and one must proceed from the microscopic Maxwell–Lorentz equations (e.g., Landau and Lifshitz, 1960) without introducing the space and time averaging that is implicitly assumed in the macroscopic equations (1); the relevant quantity now is the microscopic electron density and current distribution.

For such cases where fields of widely different frequencies are interacting, it is more convenient to treat high- and low-frequency fields differently and regard the latter as external parameters that modify the optical properties of the medium felt by the former. This is the case in particular when one of the fields is static or equivalently when its frequency is zero, but other situations can be envisaged that involve static magnetic fields or other low-frequency disturbances like acoustic waves. The fields of electro-optics, magneto-optics, and acousto-optics are related to some very important applications that have great impact on device technology.

2.4.1 Electro-optics This is related to changes of the refractive index of a material brought in by a dc or low-frequency electric field $\mathbf{E}_0$. The change linear in the static field gives rise to the linear electro-optic or Pockels effect while the change quadratic in the static field gives rise to the quadratic electro-optic effect or static Kerr effect. The former is related to a second-order nonlinear polarization

$$P_i^{(2)}(\omega) = 2\chi_{ijk}^{(2)}(\omega,0)E_j(\omega)E_{0k} \tag{41a}$$

and can occur only for materials without inversion symmetry while the latter is related to a third-order polarization

$$P_i^{(3)}(\omega) = 3\chi_{ijkl}^{(3)}(\omega,0,0)E_j(\omega)E_{0k}E_{0l}, \tag{41b}$$

which can occur in any material. As previously stated, one usually proceeds by first deriving the changes of the refractive index induced by the static or low-frequency field and then analyzing their impact on the propagation properties of an optical field in the linear regime for the latter. One important reason for adopting such a procedure is that phase matching is irrelevant for the processes described by Eqs. (41a) and (41b).

More precisely, one actually evaluates the modifications of the ellipsoid of the optical indices (optical indicatrix) of the material induced by the static electric field (e.g., Lines and Glass, 1979). For the case of the Pockels effect, one thus writes

$$\Delta\left(\frac{1}{n^2}\right)_i = \sum_j r_{ij}E_j,$$

where $(1/n^2)_i$ are the optical constants that describe the optical indicatrix in an arbitrary coordinate system and can be expressed in terms of the principal-axis optical constants; r_{ij} are the electro-optic coefficients and can be related to $\chi^{(2)}_{ijk}(\omega,0)$. An important application of the Pockels effect is the electro-optic modulation (e.g., Lines and Glass, 1979; Yariv, 1971) of an optical beam where the x and y components of a light beam are forced to experience different phase shifts when propagating through the crystal and its polarization state will be modified in a way that depends on the applied static field along the z axis. The refractive indices in the x and y directions are $n_x = n_0 - n_0^3 r_{63}E_z/2$ and $n_y = n_0 + n_0^3 r_{63}E_z/2$, respectively, and the phase difference is $\Gamma = (n_y - n_x)\omega L/c = n_0^3 r_{63}E_z\omega L/c = n_0^3 r_{63}\omega V/c$, where $V = E_zL$ is the voltage applied in the z direction, the propagation direction of the optical beam, and r_{63} is the relevant electro-optic coefficient. This effect and its variants have found several applications in optoelectronic devices.

Actually, not only the refractive index of a material is modified by a static electric field; its absorption too can be modified, and many applications of the electroabsorption for device operation have been proposed, but here the absorption losses can be substantial.

2.4.2 Magneto-optics The oscillating magnetic field that accompanies the oscillating electric field in a light beam gives rise to very weak effects in comparison with those induced by the electric field, but a static magnetic field can be very intense and induce substantial changes in the optical properties as a consequence of the Zeeman splittings that the degenerate magnetic levels undergo and the concomitant changes of the energy spacings and oscillator strengths.

The lowest-order magneto-optic effect is related to the polarization

$$P^{(2)}_M(\omega) = \chi^{(2)}_M(\omega,0)E(\omega)H_0 \tag{42a}$$

and gives rise to the Faraday effect when the direction of H and that of the light-beam propagation coincide. As for the Pockels effect previously discussed, it is more convenient first to analyze the modifications that the refractive index undergoes in the presence of the static magnetic field. One finds that the refractive indices for right- and left-circular polarization become different and these two components experience different phase shifts after propagating a distance L inside the medium; the polarization state undergoes a rotation by an angle $\theta_F = \omega L(n_- - n_+)/nc$, the Faraday rotation angle.

In terms of susceptibilities, we also have

$$n^2_\mp = 1 + 4\pi(\chi^{(1)} + \Delta\chi^{(1)}_\mp) = n^2 + 4\pi\Delta\chi^{(1)}_\mp,$$

where $\Delta\chi^{(1)}_\mp$ is the static magnetic-field–induced change of the linear susceptibility $\chi^{(1)}$ and can be related to $\chi^{(2)}_M(\omega,0)$.

Very interesting is the photoinduced Faraday rotation effect that is related to the nonlinear polarization

$$P^{(4)}_M(\omega) = \chi^{(4)}_M|E(\omega)|^2E(\omega)H_0 \tag{42b}$$

and leads to a photoinduced change of the Faraday rotation angle (Frey *et al.*, 1992); its implications can best be assessed as follows. If the light intensity $I \sim E^2$ is high, the two refractive indices $n_\mp$ undergo a photoinduced modification and can be written in the form

$$\tilde{n}^2_\mp = 1 + 4\pi(\chi^{(1)} + \Delta\chi^{(1)}_\mp + \chi^{(3)}I + \Delta\chi^{(3)}_\mp I), \tag{43}$$

where $\chi^{(3)}$ is the optical Kerr susceptibility and $\Delta\chi^{(3)}_\mp$ is its magnetic–field–induced modification or equivalently the photoinduced modification of $\Delta\chi^{(1)}_\mp$. Accordingly, the Faraday rotation angle becomes now

$$\begin{aligned}\tilde{\theta}_F &= \frac{\omega L}{nc}(\tilde{n}_- - \tilde{n}_+)\\ &= \frac{\pi\omega L}{nc}\{(\Delta\chi^{(1)}_- - \Delta\chi^{(1)}_+) + (\Delta\chi^{(3)}_- - \Delta\chi^{(3)}_+)I\}\\ &= \theta_F + \theta_{NL};\end{aligned} \tag{44}$$

the second term θ_{NL} is the light-induced modification of the linear Faraday rotation angle. The photoinduced Faraday effect can have important applications when fast modulation of the polarization state of a beam is required for instance in nonreciprocal devices.

2.4.3 Acousto-optics A sound wave in a medium is related to density and pressure fluctuations that modulate (e.g., Fabelinskii, 1968; Yariv, 1971) the refractive index of the material and concomitantly the propagation characteristics of an incident light beam. For an anisotropic solid, the changes in the refractive index are characterized by

$$\Delta\left(\frac{1}{\epsilon}\right)_{ij} = \sum_{kl} p_{ijkl}S_{lk},$$

where S_{kl} is the strain tensor and p_{ijkl} the photoelastic tensor; for an isotropic medium, liquid or gas, one can use a scalar relation. In the case of an acoustic wave of wave vector **q** and frequency Ω, this effect induces a variation $\Delta\epsilon = \Delta\epsilon_0 \exp(i\mathbf{q}\cdot r - i\Omega t) + \text{c.c.}$ in the dielectric constant. An incident beam $\mathbf{E}_{i^-} = Re[\hat{\mathbf{e}}A_i \exp(i\mathbf{k}_i\cdot\mathbf{r} - i\omega_i t)]$ is scattered off this variation and produces a beam $\mathbf{E}_{s^-} = Re[\hat{\mathbf{e}}_s A_s \exp(i\mathbf{k}_s\cdot\mathbf{r} - i\omega_s t)]$ whose characteristics depend on the interaction length. If the interaction length is sufficiently long, then phase matching is important and imposes a single diffracted beam; this is the Bragg regime. If the interaction length is very short, then the phase matching is irrelevant and several diffracted beams are obtained; this is the Raman–Nath regime (Fig. 16).

In the Bragg regime, the single diffracted beam is such that $\omega_s = \Omega + \omega_i$ and $\mathbf{k}_s \approx \mathbf{k}_i + \mathbf{q}$, nearly phase matched. The total optical field $\mathbf{E} = \mathbf{E}_s + \mathbf{E}_i$ obeys the wave equation

$$\nabla^2 E - \frac{\epsilon + \Delta\epsilon}{c^2}\frac{\partial^2 E}{\partial t^2} = 0,$$

which in the slowly varying envelope approximation leads to the two coupled-amplitude equations

$$\frac{dA_i}{dz} = i\kappa A_s e^{-i\Delta kz}, \quad (45a)$$

$$\frac{dA_s}{dz} = i\kappa^* A_i e^{i\Delta kz}, \quad (45b)$$

where $\kappa = \omega^2\Delta\epsilon/2k_z c^2$ and $\Delta\mathbf{k} = \mathbf{k}_s - \mathbf{k}_i - \mathbf{q}$ has only a z component. For perfect phase matching, namely when E_i is incident at the Bragg angle, or $\Delta k = 0$, the two equations essentially reduce to the ones encountered in the treatment of the optical phase conjugation (34a) and (34b), and the solutions given there can be used here too. For $\Delta k \neq 0$, the solution is slightly more complicated but still analytically tractable. One can show that for the Bragg scattering to occur the condition $\lambda L/\Lambda^2 \gg 1$ must be satisfied where λ and Λ are the light and acoustic wave wavelengths respectively and L is thickness of the medium. The main applications of the Bragg scattering are the production of amplitude-modulated waves and the deflection of a beam (e.g., Yariv, 1971).

For the Raman–Nath regime, one instead has $\lambda L/\Lambda^2 < 1$, and one finds that the diffracted light is a superposition of waves with frequencies $\omega + m\Omega$ and wave vectors $\mathbf{k}_i + m\mathbf{q}$. The intensities of the diffracted light components can be estimated (e.g., Yariv, 1971; Boyd, 1992).

2.4.4 Nonlinear X-Ray Optics This field is still in a speculative state but with the progress made in the brightness of incoherent x-ray sources and in achieving coherent ones either by laser operation or nonlinear processes, nonlinear x-ray optics will move into the forefront. Nonlinear effects with x rays that seem most accessible to observation at the present stage of x-ray sources are the parametric ones, where x-ray photons are diffracted off a crystal in the presence of an intense coherent field in the optical frequency region.

When x rays are involved, the macroscopic Maxwell equations (1) cannot be used as such. Equations (1), which are used in the optical frequency range, are actually derived from the microscopic Maxwell–Lorentz

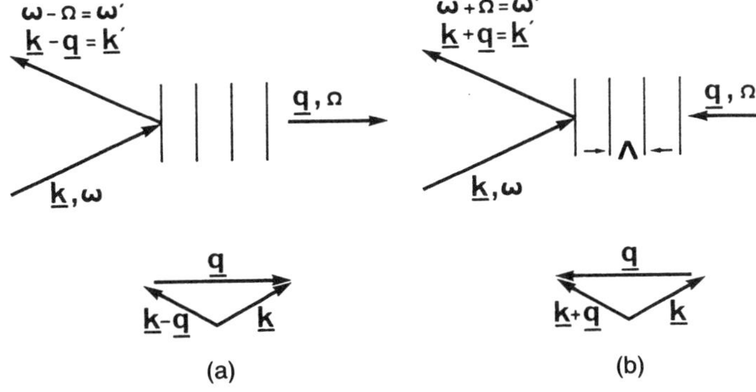

FIG. 16. Diffraction by acoustic waves in forward and backward directions. Λ is the grating spacing.

equations

$$\nabla \times \mathbf{e} = -\frac{1}{c}\frac{\partial \mathbf{b}}{\partial t}, \quad \nabla \times \mathbf{e} = 4\pi\rho,$$
$$\nabla \times \mathbf{b} = \frac{1}{c}\frac{\partial \mathbf{e}}{\partial t} + \frac{4\pi}{c}\mathbf{j}, \quad \nabla \times \mathbf{b} = 0,$$
$$\dot{\mathbf{v}} = q(\mathbf{e} + \mathbf{v} \times \mathbf{b}) \tag{46}$$

by smoothing out fluctuations in the microscopic electric and magnetic fields, **e** and **b** respectively, and charge- and current-density distributions, ρ and **i** respectively, that occur within time and space scales much shorter than the optical period and wavelength; **v** is the charge velocity and $\mathbf{j} = \Sigma q\mathbf{v}$. In the x-ray region, this averaging cannot be performed since the relevant wavelengths become comparable or smaller than the charge extension; one must use the microscopic charge and current densities instead of the polarization densities. In a crystal, these are space-periodic quantities with period the unit cell, and one can use Fourier series expansion in the inverse lattice space to simplify the treatment.

Let us very briefly discuss the case of parametric nonlinear x-ray diffraction in a crystal in the presence of an intense coherent field in the optical frequency region. Because of this intense optical field of frequency ω, the charge-density distribution is modified, and we may write $\rho = \rho_0 + \rho_\omega$ where ρ_0 is the static periodic charge distribution and ρ_ω is the photoinduced charge distribution at frequency ω and has the same crystalline periodicity as ρ_0 since the optical field is essentially uniform over a very large number of unit cells. Incident x-rays at frequency Ω will be diffracted off this periodic charge distribution, and besides the usual diffracted components at the incident frequency Ω (related to ρ_0) and in directions given by the Bragg condition, there will also be superimposed components in the same directions at the shifted frequencies $\Omega \pm \omega$ due to ρ_ω; if the optical laser frequency ω is larger than the x-ray width, one can detect them using the very sensitive photodetection techniques available for x-ray photons. These additional components can be related to the Fourier components of the photoinduced charge-density distribution ρ_ω; one can also show that these are related to the local field corrections for optical frequencies. This is the only way to determine the local field corrections experimentally (Flytzanis, 1975). Because $\Omega \gg \omega$, the cross sections for such scattering can be substantial and detectable despite the fact that the x-ray nonlinearities are weak, much weaker than in the optical frequency range.

3. NONLINEAR POLARIZATION

In the previous treatment of the nonlinear optical effects, the attention was concentrated on the propagation effects and the polarization sources were treated phenomenologically. However rich the class of processes and effects that can be studied with this one-legged approach, one cannot proceed too far without a consistent description of the optical nonlinearities at the macroscopic and microscopic levels. Actually, the nonlinear propagation problem cannot be adequately treated without a proper knowledge of the origin of the optical nonlinearities and the dynamics of the underlying mechanisms. In fact, the nonlinear propagation and the behavior of the optical nonlinearities are intimately connected when the frequencies of the interacting waves or combinations thereof are close to the eigenfrequencies of the material excitations. This is particularly true in the case of the effective odd-order nonlinearities in general and the third-order ones in particular where material excitations can be coherently driven by a two-photon combination frequency. Knowledge of the behavior of the nonlinear optical response close to such resonances is an indispensable condition and a prerequisite for the description of such important effects as saturation, self-induced transparency, stimulated light-scattering processes (like Raman, Brillouin, or Rayleigh), and two-photon absorption, to name a few (e.g., Levenson and Kano, 1988; Letokhov and Chebotayev, 1977). In a different context, the occurrence of instabilities and chaotic behavior is quite often conditioned by the dynamics of the nonlinear optical response, for instance by the Debye-type evolution (32) of the optical Kerr nonlinearity in the case of two counterpropagating beam interaction, as was pointed out in connection with the optical bistability and elsewhere in the previous section. More importantly, for the ongoing intensive research for adequate nonlinear optical materials to be implemented in nonlinear optical devices and in other applications of nonlinear optics, one urgently needs methods to predict the nonlinear optical susceptibilities close to or away from resonances; this too cannot be achieved without a precise description of the nonlinear mechanisms. The future progress in nonlinear optics is in fact subject to progress in nonlinear optical materials with precisely designed microscopic features.

3.1 Macroscopic Aspects. Susceptibilities

The nonlinear dipolar polarization terms $P^{(n)}$ result from the response of the medium to the time-dependent electric fields. Consequently, their interrelation must satisfy certain conditions such as invariance with respect to translation of the time origin, causality, reality, and symmetrization and must also reflect the symmetry of the material system (e.g., Flytzanis, 1975; Shen, 1984; Butcher and Cotter, 1991). We will assume that the interaction is local or, plainly stated, that the polarization at a given space point is only related to the electric field amplitude at the same point and not to its variation in the neighborhood of this point.

The most general relation between the nth-order polarization term and the fields then is

$$\mathbf{P}^{(n)}(t) = \int_{-\infty}^{\infty} d\tau_1 \cdots \int_{-\infty}^{\infty} d\tau_n \mathbf{R}^{(n)}(\tau_1 \cdots \tau_n) \cdot \mathbf{E}(t-\tau_1) \cdot \cdots \cdot \mathbf{E}(t-\tau_n), \tag{47}$$

where $\mathbf{R}^{(n)}$ is the nth order response function. It is a real $(n+1)$th rank tensor, which vanishes for negative values of any of its time arguments τ_n. It can always be cast into a symmetrized form so that $R^{(n)}_{ij_i \cdots j_n}(\tau_1 \cdots \tau_n)$ is invariant under any permutation of the pairs $(j_1,\tau_1) \cdots (j_n,\tau_n)$, and this is the intrinsic permutation symmetry. The number of the independent tensor components, of which there are at most 3^{n+1}, can be substantially reduced if the material system possesses a symmetry; in the dipole approximation assumed here, the even-order polarization response functions $R^{(2n)}$ identically vanish for systems with inversion symmetry. The temporal profile of $E(\tau)$ can be arbitrary, but in the impulsive regime, with deltalike pulses, it is easily seen from Eq. (47) that the polarization terms coincide with the corresponding polarization functions.

By introducing in Eq. (47) the Fourier representation in frequency space for the polarization and fields,

$$\mathbf{P}^{(n)}(t) = \int_{-\infty}^{\infty} d\omega \mathbf{P}^{(n)}(\omega) \exp(-i\omega t)$$

and

$$\mathbf{E}(t) = \int_{-\infty}^{\infty} d\omega \mathbf{E}(\omega) \exp(-i\omega t)$$

respectively one gets

$$\mathbf{P}^{(n)}(\omega) = \int_{-\infty}^{\infty} d\omega_1 \cdots \int_{-\infty}^{\infty} d\omega_n \boldsymbol{\chi}^{(n)}(\omega_1 \cdots \omega_n) \cdot \mathbf{E}(\omega_1) \cdot \cdots \cdot \mathbf{E}(\omega_n)\delta(\omega-\omega_0), \tag{48}$$

$$\boldsymbol{\chi}^{(n)}(\omega_1, \ldots, \omega_n) = \int_{-\infty}^{\infty} d\tau_1 \cdots \times \int_{-\infty}^{\infty} d\tau_n \mathbf{R}^{(n)}(\tau_1 \cdots \tau_n) \times \exp\left(i \sum_{i=1}^{n} \omega_i \tau_i\right) \tag{49}$$

with $\omega_0 = \Sigma_{i=1}^{n} \omega_i$. Relation (49) formally defines the nth-order susceptibility, an $(n+1)$th rank tensor with complex components, and implies certain properties for $\boldsymbol{\chi}^{(n)}$. Thus, because of causality $\boldsymbol{\chi}^{(n)}$ can be analytically continued in the upper half-plane for all frequencies, and $\boldsymbol{\chi}^{(n)}(\omega_1 \cdots \omega_n)^* = \boldsymbol{\chi}^{(n)}(-\omega_1^*, \ldots, -\omega_n^*)$ because of the reality of $R^{(n)}$. Furthermore, because of the intrinsic permutation symmetry $\chi^{(n)}_{jj_1 \cdots j_n}(\omega_1 \cdots \omega_n)$ is invariant under all $n!$ permutations of the pairs (j_i,ω_i), and the symmetry of the material substantially reduces the number of the independent tensor components. Expression (48) is greatly simplified when the fields are delta functions in frequency space, namely monochromatic waves; one then gets

$$\mathbf{P}^{(n)}_{\omega_0} = D_n \boldsymbol{\chi}^{(n)}(\omega_1, \ldots, \omega_n) \cdot \mathbf{E}_{\omega_1} \cdot \cdots \cdot \mathbf{E}_{\omega_n}, \tag{50}$$

where D_n is a numerical factor equal to the number of distinct permutations of $(\omega_1, \ldots, \omega_n)$; with this convention, the limiting value

$\chi^{(n)}(0, \ldots, 0)$ is independent of the path chosen for the frequencies to reach the values 0.

The analyticity property of the nonlinear susceptibilities that follows from causality implies that one may derive Kramers–Kronig type relations analogous to the ones that exist between the real and imaginary parts of the linear susceptibility $\chi^{(1)}(\omega)$ (e.g., Born and Wolf, 1975; Landau and Lifshitz, 1960). Up to the present date, attempts to derive such relations have not produced any useful or physically transparent expressions (e.g., Bloembergen, 1964; Flytzanis, 1975). In some isolated cases, K-K relations between the dispersive and absorptive Kerr nonlinearities (31a) and (31b) have been used with scarce attention as far as their applicability is concerned.

The difficulty in setting up K-K relations for nonlinear susceptibilities and also relating them to physically simple processes is connected with the fact that nonlinear optical processes can lead both to real nonlinear absorption losses to the material system as well as to energy transfer from one coherent field component to a large number of other field components through several interaction paths. At a more fundamental level, these difficulties can be traced to the lack of a proper formulation of stationarity and equilibrium in the nonlinear regime.

3.2 Microscopic Description. Polarizabilities

In the previous discussion, the optical susceptibilities were introduced in *ad hoc* manner as phenomenological parameters. We will now succinctly outline how their expressions can be derived from a microscopic description starting with the equation of charge motion in the presence of intense coherent electromagnetic fields. The steps that lead from this equation to the expressions for the experimentally accessible macroscopic nonlinear susceptibilities previously introduced are several, and all involve certain simplifications and approximations in order to cope with the inherent complexity of the many-body interactions and obtain meaningful and useful information.

3.2.1 Local Approximation One assumes that the material system in the absence of external fields is described by a Hamiltonian $H_0(\mathbf{p}_i,\mathbf{r}_i)$ with known eigenvalue and eigenstate spectrum, where i runs over all the charges. Under the influence of an intense electromagnetic field represented by a vector potential $\mathbf{A}(\mathbf{r},t)$, the state of the charges making up the medium will be perturbed and oscillating currents will be induced in the system. The Hamiltonian describing the system is now $H(t) = H_0(\boldsymbol{\pi}_i,\mathbf{r}_i)$ where $\boldsymbol{\pi}_i = \mathbf{p}_i - e\mathbf{A}_i/c$ with $\mathbf{A}_i = \mathbf{A}(\mathbf{r}_i,t)$; and, assuming that H_0 is quadratic in the momenta $\mathbf{p}$, one has

$$H(t) = H_0(\mathbf{p}_i,\mathbf{r}_i) - \sum_i \frac{e_i}{2m_ic}\{\mathbf{A}_i\cdot\mathbf{p}_i + \mathbf{p}_i\cdot\mathbf{A}_i\} + \sum_i \frac{e_i^2}{2m_ic^2}A_i^2. \tag{51}$$

A Coulomb gauge with zero scalar potential is used, and spin effects are neglected here. Beyond this stage, one must resort to simplifying considerations in order to obtain tractable and useful expressions for the response of the system, and must take into account the actual constitution of the medium and in particular the diversity of the charges and polarizable units involved as well as their mutual interactions. At the outset, we have electrons, ions, and polarizable units of molecular size, the latter in general being nonspherical. Because of their vastly different masses, their motion is characterized by eigenfrequencies that fall into well-separated spectral regions.

Let us for simplicity consider monochromatic waves characterized by a wavelength λ and period T; the electromagnetic field described by $\mathbf{A}(\mathbf{r},t)$ actually samples the space and time over extensions less than λ and T and averages out fluctuations in the charge motion that vary faster in space and time. Otherwise stated, the response of the charges to the electromagnetic field is insensitive to all details of their motion that occur within λ and T. This qualitative argument can also be cast into a rigorous form and sets a scaling procedure.

For fast-oscillating electromagnetic fields with frequencies much higher than all eigenfrequencies, the charges behave as free and the response is directly related to the electronic charge- and current-density distribution as in the case of x-ray scattering where the wavelength is much less than the extension of electronic wave functions.

As the frequencies of the electromagnetic field are lowered into the optical spectrum and the wavelength becomes much longer than the charge extension in a dielectric, the dipole approximation can be introduced; the dipolar response to the electromagnetic field is then described in terms of the electronic polarizabilities of polarizable units. As a consequence of the adiabatic Born–Oppenheimer approximation, these coefficients are functions of ion coordinates and the orientation of the polarizable unit; accordingly, they may be modulated by the corresponding vibrational and orientational motion. In addition, the polarizable units may change their relative positions, and this motion introduces fluctuations in the dielectric constant. The low-frequency fluctuations due to intramolecular and intermolecular motions result in the different light-scattering effects, but they can also be directly or indirectly coupled to the electromagnetic fields and strongly modulate the induced polarization.

It is evident from the above qualitative discussion that for most purposes in nonlinear optics one may start (e.g., Bloembergen, 1964; Flytzanis, 1975) by setting up the expression for the induced polarization in the dipole approximation. We will restrict ourselves to the most commonly recurring case where the material is composed of identical repeat polarizable entities—atoms, molecules, or clusters—of dimensions much smaller than the optical wavelength; these will be either randomly but homogeneously distributed, like for instance in a gas, a liquid, or an amorphous solid, or set in periodic array, for instance like in a crystalline solid. To the extent that the charges are more or less localized in each such entity, the total induced polarization density **P** can be set as

$$\mathbf{P} = \sum \mathbf{p} = N\mathbf{p}, \qquad (52)$$

namely as the sum of the effective dipoles **p** induced on each entity in a volume unit by the electromagnetic field that actually acts on each entity.

The expression for the dipole moment **p** in power series of this field will be obtained using perturbation theory in the dipole approximation, namely,

$$\mathbf{p} = \boldsymbol{\alpha}^{(1)}\cdot\mathbf{E} + \boldsymbol{\alpha}^{(2)}:\mathbf{EE} + \boldsymbol{\alpha}^{(3)}:\mathbf{EEE}, \qquad (53)$$

and the coefficients $\boldsymbol{\alpha}^{(n)}$ of the different powers are the microscopic polarizabilities. Their relation to the previously defined macroscopic polarizabilities and susceptibilities requires (e.g., Flytzanis, 1975) the knowledge of the relation between the effective field locally acting on the charges within each entity and the macroscopic field that enters the Maxwell equations (1). The latter in the optical frequency domain is actually the average of the microscopic field over volumes of dimensions much less than the optical wavelength but still containing several polarizable entities. The two differ because of the mutual interactions that arise between the induced dipoles and other induced interactions. Because of the complexity of these interactions, some simple models have been devised to account for their effect. The outlined procedure actually pertains to dielectrics with charges more or less well localized within each entity—for instance gases, liquids, ionic, or amorphous dielectrics. For semiconductors or metals where electrons are very delocalized, one must use a different procedure based on band theory, and in addition it is more appropriate to calculate there the induced current density $\mathbf{J}(t)$, which is related to the polarization density $\mathbf{P}(t)$ through $\mathbf{J} = \partial\mathbf{P}/\partial t$; the relevant expressions for the susceptibilities in this case have been extensively analyzed in the literature (e.g., Flytzanis, 1975; Butcher and Cotter, 1991).

3.2.2 Microscopic Polarizabilities From Eq. (51), one can show that the Hamiltonian of each entity in the presence of an electric field in the dipole approximation can be written as $H = H_0 + H_c + H_R$ where H_0 represents the Hamiltonian of the entity in the absence of the field, and H_c is the interaction term with the electric field, which in the dipole approximation reads

$$H_c = -\boldsymbol{\mu}\cdot\mathbf{E}_e \qquad (54)$$

where $\mathbf{E}_e$ is the effective field in the entity and $\boldsymbol{\mu}$ is the total dipole-moment operator of all charges of the entity (e.g., Bloembergen, 1964; Flytzanis, 1975). The term H_R stands for all random perturbations of the charges in the entity by the environment, which is represented as a bath with an infinite number of degrees of freedom; they are usually assumed to be stationary Markoffian processes with

$\langle H_R(t)\rangle = 0$, $\langle H_R(t)H_R(t')\rangle = D^2 \exp(-|t - t'|/\tau_c)$, and uncorrelated to the coherent interaction H_c; D is a measure of their strength and τ_c a correlation time that for most purposes will be very short, much shorter than any relevant time of the problem. These random perturbations provoke relaxation processes (Bloembergen, 1964) that under certain simplifying approximations can be incorporated into the density-matrix operator equation, which then reads

$$\frac{d\rho}{dt} = \frac{1}{i\hbar}[H_0 + H_c,\rho] + \left.\frac{\partial\rho}{\partial t}\right|_R, \tag{55}$$

$$\left.\frac{\partial\rho_{mn}}{\partial t}\right|_R = -\Gamma_{mn}\rho_{mn} = -\frac{1}{T_{mn}}\rho_{mn}, \tag{56a}$$

$$\left.\frac{\partial\rho_{mm}}{\partial t}\right|_R = \sum_{n\neq m}[W_{mn}\rho_{nn} - W_{nm}\rho_{mm}], \tag{56b}$$

where Γ_{mn} and W_{mn} are positive quantities that are related to the coherence and population relaxation rates of the transition between states m and n or equivalently to the transverse and longitudinal relaxation times T_2 and T_1 respectively; we remind the reader that the nondiagonal elements ρ_{nm} relate to the coherence between states n and m and the diagonal ones ρ_{mm} to the population in state m. The induced dipole moment is obtained from $\mathbf{p} = \mathrm{Tr}(\rho\boldsymbol{\mu})$. Assuming H_c a perturbation and expanding in powers of its strength, one can set $\rho = \rho^{(0)} + \rho^{(1)} + \rho^{(2)} + \cdots$ in Eq. (55). Making a Fourier analysis of the field and the induced dipole and solving the resulting hierarchy of equations by an iterative procedure, one finds $\mathbf{p} = \mathbf{p}^{(0)} + \mathbf{p}^{(1)} + \mathbf{p}^{(2)} + \cdots$, where

$$\mathbf{p}^{(n)} = \mathrm{Tr}(\rho^{(n)}\boldsymbol{\mu}) = \boldsymbol{\alpha}^{(n)}\cdot\mathbf{E}\cdot\mathbf{E}\cdot\ \cdots\ \cdot\mathbf{E}, \tag{57}$$

which defines the nth-order microscopic polarizability. The relation (57) is tensorial and requires explicit reference to the n Fourier components of the field that are involved and the nth-order dipole they induce together. For the three lowest-order terms, the linear and second and third orders, they read

$$p_i^{(1)}(\omega) = \alpha_{ij}(\omega)E_j(\omega), \tag{58a}$$

$$p_i^{(2)}(\omega_1 + \omega_2) = D_2\beta_{ijk}(\omega_1,\omega_2)E_j(\omega_1)E_k(\omega_2), \tag{58b}$$

$$p_i^{(3)}(\omega_1 + \omega_2 + \omega_3) = D_3\gamma_{ijkl}(\omega_1,\omega_2,\omega_3) \times E_j(\omega_1)E_k(\omega_2)E_l(\omega_3), \tag{58c}$$

where

$$\alpha_{ij}^{(1)}(\omega) \equiv \alpha(\omega) = \frac{1}{\hbar}\sum_{g,r} f_g\left\{\frac{\langle\mu_i\rangle_{gr}\langle\mu_j\rangle_{rg}}{(\omega_{rg} - \omega - i\Gamma_{rg})} + \frac{\langle\mu_i\rangle_{gr}\langle\mu_j\rangle_{rg}}{(\omega_{rg} + \omega + i\Gamma_{rg})}\right\}, \tag{59a}$$

$$\alpha_{ijk}^{(2)}(\omega_1,\omega_2) \equiv \beta_{ijk}(\omega_1,\omega_2) = \frac{1}{\hbar^2}\sum_{g,r,s} f_g\left\{\frac{\langle\mu_i\rangle_{gr}\langle\mu_j\rangle_{rs}\langle\mu_k\rangle_{sg}}{(\omega_{rg} - \omega_1 - \omega_2 - i\Gamma_{rg})(\omega_{sg} - \omega_2 - i\Gamma_{sg})} + 7\ \text{terms}\right\}, \tag{59b}$$

$$\begin{aligned}\alpha_{ijkl}^{(3)}(\omega_1,\omega_2,\omega_3) &\equiv \gamma_{ijkl}(\omega_1,\omega_2,\omega_3)\\ &= \frac{1}{\hbar^3}\sum_{g,r,s,t} f_g\left\{\frac{\langle\mu_i\rangle_{gr}\langle\mu_j\rangle_{rs}\langle\mu_k\rangle_{st}\langle\mu_l\rangle_{tg}}{(\omega_{rg} - i\Gamma_{rg} - \omega_1 - \omega_2 - \omega_3)(\omega_{sg} - i\Gamma_{sg} - \omega_2 - \omega_3)(\omega_{tg} - i\Gamma_{tg} - \omega_3)}\right.\\ &\qquad \left. + 47\ \text{terms}\right\},\end{aligned} \tag{59c}$$

where $f_s = \rho_{ss}^{(0)}$ are the equilibrium populations and $\omega_{mn} = (E_m - E_n)/\hbar$ is the transition frequency between states m and of the material system. For the linear and second- and third-order microscopic polarizabilities (59a)–(59c), we used their more conventional notations α, β, and γ instead of $\alpha^{(1)}$, $\alpha^{(2)}$, and $\alpha^{(3)}$ respectively; these coefficients and the higher ones are tensors of second, third, fourth, and higher order respectively.

It is clear that in contrast to the linear polarizabilities the nonlinear ones possess a very rich and complicated multiresonant behavior. Furthermore, in the third-order polarizability (59c), and this can actually be generalized to the case for all odd-order polarizabilities, a new type of intermediate resonance appears whenever a sum of any two frequencies is close to a material transition frequency, say $\omega_2 + \omega_3 = \omega_{sg} \equiv \omega_c$, which is quite distinct from the primary ones that occur whenever any of the incident and created frequencies are close to a dipole-allowed transition of the medium. The behavior of the nonlinear polarizability close to this intermediate resonance can be related to the stimulated counterpart of the different spontaneous light-scattering effects that will be discussed in Sec. 3.3.

3.2.3 Local Field Corrections As previously stated, the connection between macroscopic susceptibilities and the microscopic polarizabilities requires the relation between the macroscopic and effective fields (e.g., Flytzanis, 1975). This is a central problem in the whole theory of dielectrics and touches its very fundamental aspects. To gain an insight into the problem, let us assume that the charge clouds of the different entities do not overlap, and we subdivide the space into identical cells, each containing one polarizable entity, and filling all the space. The field E_e that acts on a given entity is the one that results from all sources outside its cell; it is the microscopic field E_{mic}, which is the same in every cell with the exclusion of the field from the entity itself E_s, or $E_e = E_{\text{mic}} - E_s$. The macroscopic field that enters Maxwell equations is the space average of E_{mic} over many cells, which in our case reduces to its average within any cell or $\mathbf{E} = \langle E_{\text{mic}} \rangle_c$. The field E_e is actually not constant throughout its cell, and the effective field will be defined as a weighted average of E_e over the cell, the weight being roughly the charge-density distribution, and similarly for E_s so that $\langle E_e \rangle = \langle E_{\text{mic}} \rangle - \langle E_s \rangle$. We define the average $\langle E_e \rangle$ as the effective field E_f or $E_f \equiv \langle E_e \rangle$.

Without going into lengthier analysis, it is quite evident from the above considerations that the effective field correction is a measure of the inhomogeneity of the charge-density distribution within a cell; the smoother this distribution is, the less important are the local field corrections. In the extreme case of a metal where the charge density is uniform throughout each cell, the local field corrections disappear, while in the other extreme case of pointlike induced dipoles, the Lorentz–Lorenz model may apply and for an isotropic dielectric one finds $\mathbf{E}_f = \mathbf{E} + 4\pi\mathbf{P}/3$. For the cases intermediate between these two extremes, the calculation of the local field corrections becomes exceedingly difficult because of the overlap of charge distributions in different cells. Let us assume for simplicity that

$$\mathbf{E}_f = \mathbf{E} + \mathbf{L} \cdot \mathbf{P}, \tag{60}$$

where $\mathbf{L}$ is a frequency-independent second-rank tensor whose components vanish for a metal and reduce to the Lorentz–Lorenz expression ($L = 4\pi/3$) in a dielectric with pointlike nonoverlapping induced dipoles. Thus, the effective field contains a contribution from the induced polarization whose dynamics is described by Eqs. (56a) and (56b). A more rigorous theory in crystals relates the local field corrections to the Fourier series expansion of the microscopic field in the inverse lattice space; they can be determined through parametric nonlinear x-ray diffraction in a crystal as discussed in Sec. 2.4.4 (Flytzanis, 1975).

3.2.4 Macroscopic Polarizabilities. Susceptibilities Assuming an isotropic medium and using expression (60) with L a scalar, inserting in Eq. (57) and substituting then in Eq. (52) one obtains

$$\begin{aligned}\mathbf{P} = {} & N\alpha(\mathbf{E} + L\mathbf{P}) + N\beta(\mathbf{E} + L\mathbf{P})(\mathbf{E} + L\mathbf{P}) \\ & + N\gamma(\mathbf{E} + L\mathbf{P})(\mathbf{E} + L\mathbf{P})(\mathbf{E} + L\mathbf{P}),\end{aligned} \tag{61}$$

where for simplicity no explicit reference to the frequencies is made. Setting $f(\omega) = 1/[1 - N\alpha(\omega)L]$, rearranging, and keeping track of terms of the same order in the electric field in Eq. (61), one can cast this expression into the macroscopic form

$$\begin{aligned}\mathbf{P} &= N\tilde{\boldsymbol{\alpha}} \cdot \mathbf{E} + N\tilde{\boldsymbol{\beta}} : \mathbf{EE} + N\tilde{\boldsymbol{\gamma}} \vdots \mathbf{EEE} \\ &= \boldsymbol{\chi}^{(1)} \cdot \mathbf{E} + \boldsymbol{\chi}^{(2)} : \mathbf{EE} + \boldsymbol{\chi}^{(3)} \vdots \mathbf{EEE},\end{aligned} \tag{62}$$

$$\tilde{\boldsymbol{\alpha}}(\omega) = f(\omega)\boldsymbol{\alpha}, \tag{63a}$$

$$\tilde{\boldsymbol{\beta}}(\omega_1,\omega_2) = f(\omega_1 + \omega_2)f(\omega_1)f(\omega_2)\boldsymbol{\beta}(\omega_1,\omega_2). \tag{63b}$$

The expression for $\tilde{\gamma}$ is somewhat involved and will not be reproduced here (Flytzanis, 1975); it suffices to state that it contains additional terms to γ that correspond to microscopic cascading processes. Relation (62) also defines the macroscopic polarizabilities, which are connected to the susceptibilities through relations (9). For optically anisotropic media, the local field factor f is a second-rank tensor, the inverse of $\mathbf{1} - N\mathbf{L}\cdot\boldsymbol{\alpha}$, and the previous relations must be accordingly transcribed.

Although local field corrections may be substantial, in the following we shall assume that they most often lead to a renormalization of the oscillator strengths and will be lumped together with them in the expressions of the susceptibilities. These oscillator strengths in the following will be understood to be the renormalized ones.

3.2.5 Cascading Processes At this level, we wish to point out that for third- or higher-order polarizations additional contributions must be included that result from the so-called cascading processes (Flytzanis, 1975). These processes result from the interaction of the primary beams with intermediate ones that are generated by lower-order induced nonlinear polarizations compatible with the symmetry of the medium. At the macroscopic level, these cascading processes can be lumped together with the direct contribution since they have the same total wave-vector, frequency, and electric field dependence; the resulting effective susceptibility has the same symmetry properties as the direct one but also depends on the phase matching at the intermediate interactions. The cascading processes introduce a nonlocality in the nonlinear interactions that can be related to the retardation effect of electromagnetic fields.

3.3 Nonresonant Regime. Electronic Nonlinearities

The nonlinear susceptibilities when all primary and intermediate frequencies are within the transparency region of the medium, the one that extends above the vibrational frequencies and below the onset of electronic transitions, are only due to the polarization of the valence electrons. Accordingly, the nonlinear susceptibilities there can be related to some important structural characteristics of the valence-electron charge distribution. Besides its intrinsic fundamental interest, this relationship plays a key role in the ongoing nonlinear optical material research, and we devote a few remarks to it here.

The values of nonresonant electronic nonlinear polarizabilities of a molecule or a unit cell in a dielectric can be in principle calculated by letting all frequencies vanish in Eqs. (59a)–(59c) and interpreting μ as the dipole-moment operator of the electrons with the ions fixed in the equilibrium position. Several computational techniques have been devised that in addition rely on some simplifications concerning the electron charge distribution (e.g., Flytzanis, 1975). These studies clearly showed that the second-order polarizability is very sensitive to the charge asymmetry over a polarizable unit and actually vanishes for a molecule or unit cell with inversion symmetry. On the other hand, the third-order polarizability is sensitive to the electron delocalization over several identical polarizable units. This clearly indicates that β is very sensitive to the local arrangement of asymmetric bonds within a unit cell or a molecule while γ is more sensitive to the electron delocalization in periodic systems.

Along with the detailed calculations that are being developed to account for the magnitude of these coefficients, some simple models allow one to grasp the main characteristics of the charge density that determine β and γ and by the same token $\chi^{(2)}$ and $\chi^{(3)}$ with appropriate provisions for local field corrections. We very briefly discuss two such models that illustrate the statements previously made.

3.3.1 Second-Order Polarizabilities Here one divides the molecule or unit cell into bonds or other primary polarizable units and assumes additivity and transferability for their second-order polarizability. We concentrate our attention only on asymmetric units, the simplest one being a heteropolar bond.

Let us represent the electronic distribution in a heteropolar bond AB, of length R, with two pointlike electronic charges $(1 - \epsilon)e$ and $(1 + \epsilon)e$ around the atoms A and B respectively. When an electric field E is applied in the direction from A to B, a charge amount $e\Delta\epsilon$ flows from B to A such that $\alpha E = e\Delta\epsilon R$ where α is the bond polarizability, which within the Unsöld approximation can be written $\alpha = (1 - \epsilon^2)^2 R^4/4a_0$. The intrabond charge flow results in a modification of this polarizability by replacing ϵ with $\epsilon - \Delta\epsilon$. Developing this electric field–dependent polarizability into powers of the electric field, one can identify (Tang and Flytzanis, 1971)

$$\beta = (1 - \epsilon^2)^3 \epsilon R^7 / 4ea_0^2, \tag{64}$$

which clearly shows that β vanishes for $\epsilon = 0$ (homopolar bond) and $\epsilon = 1$ (ionic bond), the ions being spherical; in Eq. (64) $a_0 = \hbar^2/me^2$ is the Bohr radius. It acquires a maximum value for $\epsilon = 1/\sqrt{7}$, a result that can be used as a guide for the optimal choice of heteropolar bonds with large β; ϵ can be related to the electronegativity difference of atoms A and B. From Eq. (64) one in addition sees that β increases with the polarizability of the bond and its length.

For polyatomic molecules, one can use similar models (e.g., Chemla and Zyss, 1987). Thus, for the monosubstituted benzenes X-⬡, by assuming that the charge distortion is induced by an equivalent electric field E_R such that $\mu = \alpha E_R$, where μ is the electron dipole moment of X-⬡ and α the polarizability of the benzene ring, one can relate β to other molecular parameters and account for all its trends. Alternatively, one can also simplify the spectrum of such molecules by simulating the system as a quantum two-level system with a ground (g) and an excited (e) state. For a two-level system, one has from Eq. (59b)

$$\beta = 3\mu_{ge}^2(\mu_{gg} - \mu_{ee})/E_{eg}^2. \tag{65}$$

The important point to notice in Eq. (65) is that the β coefficient may be large even for a molecule with small or vanishing permanent dipole moment μ_{gg}. These and other aspects play a very important role in guiding the research for new nonlinear optical materials.

3.3.2 Third-Order Polarizabilities An analysis of Eq. (59c) shows that large values of the third-order polarizability in the transparency region are sensitive to the electron delocalization over several identical repeat units, as is the case in crystalline covalent semiconductors or conjugated chains. In this case, it is more suitable to describe the electrons with Bloch band states; within the two-band approximation, the expressions for $\chi^{(1)}$ and $\chi^{(3)}$ are (Agrawal *et al.*, 1978)

$$\chi^{(1)} = \frac{4e^2}{\hbar V} \int_{BZ} \Omega_{cv} S_{cv} dk, \tag{66a}$$

$$\chi^{(3)} = \frac{8e^4}{\hbar V} \int_{BZ} \left[\frac{1}{\omega_{cv}} \left(\frac{\partial S_{cv}}{\partial k} \right) \left(\frac{\partial S_{vc}}{\partial k} \right) - \Omega_{vc} S_{cv} S_{vc} S_{cv} \right] dk, \tag{66b}$$

where $\hbar\omega_{cv} = \epsilon_c - \epsilon_v$. Here Ω_{vc} is the transition dipole-moment matrix element between the two bands, a valence (v) and a conduction (c) band of energies ϵ_v and ϵ_c respectively, and $S_{vc} = \Omega_{vc}/\omega_{cv}$. An important point to notice here is that $\chi^{(3)}$ is determined by the competition of an intraband term and an interband term; in contrast, the linear susceptibility (66a) only involves interband terms. For highly delocalized systems (strong overlap between wave functions of unit cells), the quantities ω_{cv} and Ω_{cv} vary strongly over the Brillouin zone and therefore the intraband term in $\chi^{(3)}$ becomes the dominant one. Another important point that reflects Eq. (66b) is that the main contribution to $\chi^{(3)}$ comes from a few nonoverlapping critical points in the joint density of states, the ones where $\nabla_k \omega_{cv} = 0$, which greatly simplifies the behavior of $\chi^{(3)}$ and allows one to establish (Agrawal *et al.*, 1978; Flytzanis, 1987) scaling laws and study the impact of the dimensionality. So in contrast to $\chi^{(2)}$ where bond additivity—a property of real space—can be assumed and is used to account for its trends, in $\chi^{(3)}$ one can assume critical-point contribution additivity, a k-space property. Thus, for a one-dimensional semiconductor or an infinite conjugated chain one finds

$$\chi^{(3)} \sim (E_F/E_g)^6, \tag{67}$$

where E_F is the valence-electron Fermi level, essentially the bandwidth, and E_g is the optical gap; this behavior of $\chi^{(3)}$ is characteristic of one-dimensional semiconductors. The ratio $N_d = E_F/E_g$ is actually a measure of the electron delocalization, and $L_d = N_d a$, where a is the length of the unit cell, is an electron delocalization length. For chains of length $L < L_d$, the behavior (67) breaks down and instead one finds $\gamma \sim L^5$, which is the behavior predicted (Rustagi and Ducuing, 1974) for free electrons in a box of length L. This change of behavior is a manifestation of quantum confinement, a feature that is increasingly used in artificial materials to enhance the cubic nonlinearities; this occurs when very delocalized electrons are confined to regions smaller than their natural delocalization length. This is particularly relevant for third-order nonlinearities, which are very sensitive to electron delocalization. This is the case of several artificial semiconductor structures like quantum wells, wires, and dots. Here, however, the resonant behavior of the nonlinearities close to the quantum confined transitions is more interesting and more directly related to the nonlinear optical material research that will be briefly discussed in Sec. 4. For large dye molecules where the electron can be described by the free-electron-in-a-box model, one finds that $\gamma \sim L^5$ where L is an effective molecular dimension.

It is quite plausible from these simple considerations that β and γ are related to important features of the electron charge-density distribution in molecules and solids and justifies the effort that is being concentrated to calculate and predict their trends. These studies provide a wealth of information concerning the structure of molecules and solids as also did previously the studies of the linear polarizability.

3.4 Resonant Regime. Stimulated Nonlinear Processes

The resonant behavior of the nonlinear susceptibilities is related to several very important nonlinear optical phenomena. As can be inferred from their expressions, the nonlinear susceptibilities can be resonantly enhanced if one or more of the complex frequency/energy denominators have their real part close to or equal to zero. A simple inspection of Eqs. (59b) and (59c) shows that these resonances fall into two categories. The first one concerns the "direct" or "primary" resonances where one or more frequencies of the fields involved in the interaction are close to electric-dipole–allowed transition frequencies of the medium; either these resonances lead to a trivial resonant enhancement of the nonlinearity, which however is counterbalanced with the concomitant linear absorption losses at the same frequency, or in the case of very intense beams they lead to a class of nonlinear phenomena that can be accounted for only by a nonperturbative analysis of an equivalent two-level system resonantly interacting with a monochromatic field as is discussed in Sec. 3.5. The other category concerns the "intermediate" resonances where an algebraic sum of primary frequencies is close to an appropriately allowed transition frequency of the medium; under certain conditions then energy can be transferred between the field modes through the intermediary of the material mode (e.g., Bloembergen, 1964; Shen, 1984). These indirect or secondary resonances and related energy-exchange processes between the field and material modes are distinct features of the odd-order nonlinearities; as a matter of fact, one can show that the related processes can even grow from photon noise and give rise to stimulated effects.

In this section, we shall give a brief account (e.g., Shen, 1984; Boyd, 1992) of these intermediate resonances in the third-order nonlinearities and the effects they give rise to; these occur whenever $\omega_i + \omega_j = \omega_{mn} \equiv \omega_e$ and the relevant susceptibility is $\chi^{(3)}(\omega_1, \omega_2, -\omega_2)$ where ω_e is any two-photon–allowed excitation of the medium (Fig. 17). Following the remark of Sec. 2.4.2, one can define a nonlinear energy transfer between the two electromagnetic modes ω_1 and ω_2, the difference in energy being stored in the material excitation. These effects can grow even from photon noise and give rise to stimulated effects: two-photon stimulated emission, and stimulated Raman, Brillouin, Rayleigh, and other light-scattering effects. We first derive the expression for the two-photon transition matrix element and relate it to the light-scattering cross section.

3.4.1 Two-Photon Resonances From Eq. (59c) one can easily show (e.g., Flytzanis, 1975)

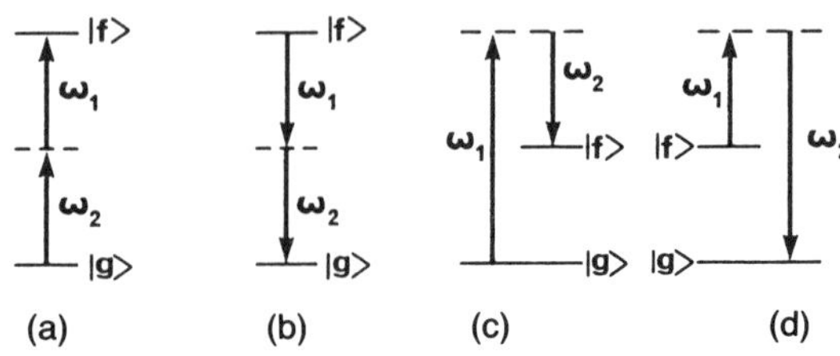

FIG. 17. Two-photon transitions: (a) two-photon absorption, (b) two-photon emission, (c) Raman Stokes, and (d) Raman anti-Stokes.

that for $\omega_i + \omega_j = \omega_{sg} \equiv \omega_e$ and keeping only resonant terms

$$\chi^{(3)}_{ijkl}(\omega_1,\omega_2,-\omega_2) = \frac{\Delta\rho\alpha_{ij}(\omega_1,\omega_2)\alpha^*_{kl}(\omega_1,\omega_2)}{\hbar[\omega_e - (\omega_1 - \omega_2 - i\Gamma_{fg})]}, \tag{68}$$

where $\Delta\rho = f_g - f_s$ and

$$\alpha_{ij}(\omega_1,\omega_2) = \sum_n \left\{ \frac{\langle g|\mu_i|n\rangle\langle n|\mu_j|s\rangle}{\hbar(\omega_{ng} - \omega_1)} + \frac{\langle g|\mu_j|n\rangle\langle n|\mu_i|s\rangle}{\hbar(\omega_{ng} + \omega_2)} \right\} \tag{69}$$

is a generalized polarizability which for $s \equiv g$ and $\omega_1 \equiv \omega_2$ reduces to the linear polarizability (59a); it can also be regarded as the matrix element between states s and g of an effective two-photon operator h_2 whose diagonal elements for $\omega_1 \equiv \omega_2$ are the linear polarizabilities. The two-photon resonant susceptibility (68) is in general complex because of the combined effect of the energy denominator and the generalized polarizability (69). If $i = k$ and $j = l$, the product $\alpha_{ij}(\omega_1,\omega_2)\alpha^*_{kl}(\omega_1,\omega_2)$ is a real number; then the separation of Eq. (68) into real and imaginary parts (Fig. 18) is related only to that of the energy denominator and the associated energy transfer $dW^{(4)}_i/dt = \langle \mathbf{E}_i \cdot d\mathbf{P}^{(3)}_i/dt\rangle$ is proportional to the energy-density flows $I_i \sim E_iE^*_i$ only, $i = 1, 2$, and unaffected by the phases of the electric fields. Hence, one can unambiguously define a nonlinear energy gain or loss in the involved modes 1 or 2 of the electromagnetic field, the difference being taken or given up by the material mode ω_e so that energy conservation is satisfied.

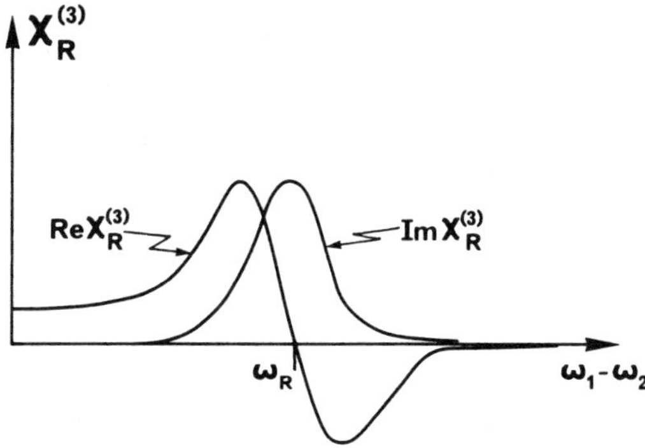

FIG. 18. Schematic representation of real and imaginary parts of the Raman or two-photon resonant susceptibility.

Since the energy-density flow can be expressed as a photon-density flow or $I_i = N_i\hbar\omega_i c/n_i$, where N_i is the photon number density in mode i and n_i is the refractive index at frequency ω_i, if one introduces the compound light-matter states $|N_1N_2a\rangle$ of energy $N_1\hbar\omega_1 + N_2\hbar\omega_2 + E_a$ the whole energy-transfer process can be interpreted as transitions between such isoenergetic states. The main two-photon processes we shall briefly consider here are the ones depicted in Fig. 17, namely the two-photon absorption and emission processes and the stimulated light-scattering processes, which include the Raman (Brillouin), Stokes, and anti-Stokes processes and the Rayleigh-type processes. As can be inferred from the previous comments and Fig. 17, one can describe and analyze these effects within a unified frame. This, however, is quite cumbersome to set up, and we only present a brief analysis of each effect separately with particular emphasis on those related to the light-scattering processes—Raman, Brillouin, and Rayleigh.

The relation between the spontaneous and stimulated light-scattering processes can be easily established starting (e.g., Bloembergen, 1964; Shen, 1984; Boyd, 1992) with the expression $dP/dt = BN_1(N_2 + 1)$ for the probability per unit time for mode 1 losing one photon, mode 2 gaining one, and the material system transiting from state g to state s such that $\omega_{sg} \equiv \omega_e = \omega_1 - \omega_2$ or equivalently the transition from state $|N_1N_2g\rangle$ to state $|N_1 - 1, N_2 + 1, s\rangle$; the coefficient D can be derived from the Fermi "golden rule" using the appropriate two-photon transition operator h_2 (e.g., Heitler, 1954). If the propagation direction of mode i coincides with the z axis and we assume a stationary regime, then the above expression can be cast into the form

$$\frac{dN_2}{dz} = B'N_1(N_2 + 1). \tag{70}$$

For $N_2 \ll 1$, the spontaneous scattering is dominant and $N_2(L) \sim B'N_1L$ where L is the interaction length while for $N_2 \gg 1$ the stimulated effect becomes dominant and grows as

$$N_2(L) = N_2(0)e^{GL}, \tag{71}$$

where $G = B'N_1$. This growth can be initiated by injected photons at frequency ω_2 or by the photon noise provided by the spontaneous process.

Before proceeding to analyze the most important stimulated processes (e.g., Bloembergen, 1964; Shen, 1984; Arecchi and Schulz-Dubois, 1975), we wish to make here more explicit their connection with the light-scattering processes (e.g., Fabelinskii, 1968). The latter are due to the temporal and spatial inhomogeneities of the dielectric constant resulting from the intrinsic fluctuations of the different degrees of freedom of the polarizable units. The fluctuations introduce additional terms in the induced polarization that oscillate at frequencies shifted below (Stokes side) and above (anti-Stokes side) the incident frequency by amounts equal to the eigenfrequencies of the degrees of freedom. Because the dielectric properties of a medium at a given frequency ω corresponding to a wavelength λ are essentially obtained by averaging the charge motion over space and time domains of extensions less than λ and $T = 2\pi/\omega$ respectively, only fluctuations with eigenfrequencies much less than ω are relevant in light scattering. With this in mind, we may write $\epsilon = 1 + 4\pi N\alpha$ for the dielectric constant of a dielectric medium formed by identical polarizable units of effective polarizability α and number density N; both α and N are fluctuating quantities because they are modulated by internal and external degrees of freedom respectively. The internal degrees of freedom, for instance vibration or rotation, being confined in space, are local and essentially possess a discrete spectrum; in contrast, the external ones, for instance the translation of the polarizable unit, involve extended motion of a large number of units and can be related to thermodynamic fluctuations like density, pressure, and temperature whose evolution is described by the hydrodynamic equations of continuity, momentum transfer, and heat transport.

The incident light then is scattered off the fluctuations of either the polarizability or the density, $\Delta\alpha$ and ΔN respectively, to produce Stokes and anti-Stokes components in addition to the incident one. For an intense incident light, these additional components can become intense because each added photon in the scattered-light modes enhances the probability for getting an additional one in the same mode, provoking an avalanche of photons in a single mode that leads to the buildup of a large-amplitude coherent electric field. The generation of this additional field on the Stokes side is accompanied by the buildup of a large-amplitude fluctuation at the relevant material eigenfrequency, which, because of its intrinsic resonant coupling, enhances the energy transfer. This coupling can be best understood by recalling that the energy stored in a polarizable unit interacting with an electric field is $W = -\alpha E^2(z,t)/2$ where $\mathbf{E} = \mathbf{E}_i + \mathbf{E}_s$, $\mathbf{E}_i$ and $\mathbf{E}_s$ being the electric fields in the incident and scattered modes. Since this energy is a function of the internal and external degrees of freedom of the polarizable unit, it can be regarded as potential energy and one can define a force conjugate to each degree of freedom q by

$$F_q = \nabla_q(\alpha E^2/2) \tag{72}$$

or $F_q = (\partial\alpha/\partial q)E^2/2$ and $\alpha\nabla E^2/2$ for an internal and external (position) degree of freedom, respectively. This force can resonantly drive the corresponding conjugate degree of freedom to a large amplitude. This is the origin of the stimulated process.

3.4.2 Stimulated Raman Scattering We consider the case of a dielectric in the presence of an electromagnetic field

$$\mathbf{E}(z,t) = \hat{\mathbf{e}}_1 A_1 \exp(ik_1z - i\omega_1 t) + \hat{\mathbf{e}}_2 A_2 \exp(ik_2z - i\omega_2 t) + \text{c.c.}$$

We assume for simplicity that the dielectric is isotropic and possesses inversion symmetry and that its scalar dielectric constant is $\epsilon = 1 + 4\pi\alpha$ while the induced dipole is $p = \alpha E$ and the polarization density $P = N\alpha E$. Let us now assume that the polarizability α can be modulated by a single internal degree of freedom of coordinate q and can be written $\alpha = \alpha_0 + q\partial\alpha/\partial q$ to first order in this coordinate; this can be a vibration or rotation coordinate. The conjugate force (72) drives the coordinate q according to the equation (Garmire *et al.*, 1963)

$$\ddot{q} + \Gamma_q\dot{q} + \omega_q^2 q = \frac{F_q}{m_q} = \frac{1}{2m_q}\frac{\partial\alpha}{\partial q}E^2, \tag{73}$$

where we also introduce a phenomenological

damping for the material mode coordinate q and ω_q and m_q are the eigenfrequency and effective mass associated to q.

The driving force, being $\sim E^2$, has components at several frequencies and in particular at frequency $\omega_1 - \omega_2$; if $\omega_1 - \omega_2 \approx \omega_q$, only this component is relevant for driving the coordinate q to appreciable amplitude, which can be written as

$$q(z,t) = \frac{1}{2m_q}\frac{\partial\alpha}{\partial q} \times \mathrm{Re}\left\{\frac{A_1A_2^*}{\omega_q^2 - (\omega_1 - \omega_2)^2 - i(\omega_1 - \omega_2)\Gamma_q} \times \exp[i(k_1 - k_2)z - i(\omega_1 - \omega_2 t)]\right\}.$$

The modulation of the polarization $P = N\alpha E$ then gives rise to a nonlinear term $P_{NL}^{(3)} = N(\partial\alpha/\partial q)qE$ cubic in the field intensities. With a collinear geometry of the wavevectors $\mathbf{k}_1$ and $\mathbf{k}_2$, because of the phase-matching condition, only the Stokes component in this polarization

$$P_2^{(3)} = \mathrm{Re}\{6\chi_R^{(3)}(\omega_2)A_1A_1^*A_2 \exp(ik_2z - i\omega_2t)\},$$

with

$$\chi_R^{(3)}(\omega_2) = -iN\left(\frac{\partial\alpha}{\partial q}\right)^2(12m_q\omega_q\Gamma_q)^{-1},$$

for the exact resonant case $\omega_1 - \omega_2 = \omega_q$, is relevant and the others can be disregarded in $P_{NL}^{(3)}$; from Eq. (18) then one has $dA_2/dz = gA_2$ with

$$g = \frac{2\pi^2\omega_2}{n_1n_2c^2}\frac{N}{m_q\omega_q\Gamma_q}\left(\frac{\partial\alpha}{\partial q}\right)^2$$

for the spatial evolution of the field in the stationary regime. Assuming undepleted pump, one gets $A_2(z) = A_2(0)\exp(gz)$, which implies exponential growth the same as that derived by Eq. (71) if we notice that $G = 2g$. It is more convenient to relate the gain coefficient to the Raman cross section related to mode q; one has

$$G = \frac{4\pi^3Nc^2}{\hbar\omega_1^2\omega_2n_2^l}\left(\frac{\partial^2\sigma}{\partial\omega\partial\Omega}\right),$$

where $\partial^2\sigma/\partial\omega\partial\Omega$ is the differential spectral cross section.

Actually, the anti-Stokes component in $P_{NL}^{(3)}$ too is phase matched, but the corresponding gain coefficient is negative, leading to attenuation in the collinear wave-vector geometry. If the collinear geometry is relaxed however, the previous approach is incomplete since it disregards a term in $P_{NL}^{(3)}$ that couples the laser Stokes and anti-Stokes field components and leads, under certain conditions, to a gain transfer from the Stokes to the anti-Stokes component; this can be easily derived (Bloembergen, 1964; Shen, 1984; Arecchi and Schulz-Dubois, 1975) from the previous model if we notice that the total third-order polarizations at the Stokes and anti-Stokes frequencies $\omega_s = \omega_1 - \omega_q$ and $\omega_a = \omega_1 + \omega_q$ respectively are

$$P_a^{(3)} = 6\chi_R^{(3)}(\omega_a)|A_1|^2A_ae^{ik_az} + 6\chi_R^{(3)}(\omega_a)A_1^2A_s^*e^{i\Delta kz},$$

$$P_s^{(3)} = 6\chi_R^{(3)}(\omega_s)|A_1|^2A_se^{ik_sz} + 6\chi_R^{(3)}(\omega_s)A_1^2A_a^*e^{i\Delta kz},$$

and the spatial evolutions of the corresponding field amplitudes are

$$\frac{dA_s}{dz} = -\alpha_sA_s + \kappa_sA_a^*e^{i\Delta kz},$$

$$\frac{dA_a}{dz} = -\alpha_aA_a + \kappa_aA_s^*e^{i\Delta kz}$$

with $\alpha_i = -12\pi i\omega_i\chi_R^{(3)}(\omega_i)|A_i|^2/n_ic$ and $\kappa_i = 12\pi i\omega_i\chi_R^{(3)}(\omega_j)A_i^2/n_jc$; they can be easily solved by assuming undepleted pump and setting $A_i = (A_i^+e^{g_+z} + A_i^*e^{g_-z})e^{i\Delta kz/2}$. This gives

$$g_\pm = [(\alpha_1 + \alpha_2^*) \pm \tfrac{1}{2}[(\alpha_1 - \alpha_2^* + i\Delta k) + 4\kappa_1^*\kappa_2]]/2,$$

which implies a gain for the positive root (Bloembergen, 1964; Shen, 1984). Thus, gain at the anti-Stokes component can be achieved at the expense of the Stokes component.

The above approach is valid whenever the coordinate q is related to a motion with well-defined frequency ω_q, vibrational or rotational, of the polarizable unit; in quantum mechanics, this implies discrete level spacing, and their dynamics in the presence of the driving fields can still be described by an equation in the form of Eq. (73). If the thermal energy kT however becomes comparable to or larger than the level spacing, as is commonly the case for rotation where $kT \geq \hbar\omega_q$, then the motion becomes diffusive and appropriate provisions must be made in the equation of motion (73). In the case of rotation, this leads to the classical reorientational motion of the polarizable units that can be described by the Debye model (e.g., Shen, 1984; Boyd, 1992). Writing $\langle\alpha\rangle = \alpha_0 + \alpha_{NL}$ for the average polarizability where $\alpha_0 = (\alpha_\parallel + 2\alpha_\perp)/3$, and $\tau\dot{\alpha}_{NL} + \alpha_{NL} = \alpha_2\overline{E^2}$ with $\alpha_2 = 8\pi(\alpha_\parallel - \alpha_\perp)^2/45kT$, one obtains the solution

$$\alpha_{NL}(z,t) = 2\alpha_2(|A_i|^2 + A_j|^2) + \frac{2\alpha_2 A_i A_j^* e^{iqz-i\Omega t}}{1 - i\Omega\tau}$$

for the modulated part of $P_{NL}^{(3)}$ in the stationary regime. One can then proceed as previously to solve the coupled-envelope equations.

3.4.3 Stimulated Brillouin and Rayleigh Scattering The dielectric constant can also fluctuate as the result of fluctuations in the external degrees of freedom of the polarizable units, namely changes of their position, which can be related to density fluctuations $\Delta\rho$ by $\Delta\epsilon = (\partial\epsilon/\partial\rho)\Delta\rho$ (e.g., Fabelinskii, 1968). For density variations, one has from thermodynamics $\Delta\rho = (\partial\rho/\partial p)_s\Delta p + (\partial\rho/\partial s)_p\Delta s$ where p and s are pressure and entropy variations; the first term is related to the Brillouin scattering and the second to the Rayleigh scattering. Since s and p are independent thermodynamic variables to a good approximation, one may treat the two effects independently (e.g., Shen, 1984; Boyd, 1992).

If we focus on pressure fluctuations, then the equation for the corresponding density fluctuations is (e.g., Fabelinskii, 1968)

$$\frac{\partial^2\Delta\rho}{\partial t^2} - \Gamma\nabla^2\frac{\partial\Delta\rho}{\partial t} - v^2\nabla^2\Delta\rho = \nabla\cdot\mathbf{f}, \tag{74}$$

where Γ is a damping parameter, v is the sound velocity, and $\mathbf{f}$ is the force density related to Eq. (72). The coupling being strongest between the incident and Stokes components in the collinear geometry, let us assume

$$E(z,t) = A_1(z,t)e^{ik_1z-i\omega_1t} + A_2(z,t)e^{-ik_2z-i\omega_2t} + \text{c.c.}$$

for the electric field and $\rho(z,t) = \rho(z,t)\exp(iqz - i\Omega t) + \text{c.c.}$ with $\Omega = \omega_1 - \omega_2$ and $\mathbf{q} = \mathbf{k}_1 + \mathbf{k}_2 \approx 2\mathbf{k}_1$ for the density fluctuation associated to the acoustic wave whose amplitude is resonantly forced by the nonlinear interaction. One finds

$$\nabla\cdot\mathbf{f} = \gamma_e q^2 A_1 A_2^* \exp(iqz - i\Omega t)/4\pi + \text{c.c.}$$

where $\gamma_e = \rho(\partial\epsilon/\partial\rho)$. The acoustic waves being heavily damped and propagating very slowly, one may use the slowly varying envelope approximation to get

$$-2i\Omega\frac{\partial\rho}{\partial t} + (\Omega_B^2 - \Omega^2 - i\Omega\Gamma_B)\rho - 2iqv^2\frac{\partial\rho}{\partial z} = \frac{\gamma_e q^2}{4\pi}A_1A_2^* \tag{75a}$$

for the density-amplitude equation where $\Gamma_B = q^2\Gamma$ is the reciprocal of the acoustic phonon lifetime, while for the field amplitudes

$$\frac{\partial A_1}{\partial z} + \frac{n}{c}\frac{\partial A_1}{\partial t} = \frac{i\omega\gamma_e}{2nc\rho_0}\rho A_2,$$

$$-\frac{\partial A_2}{\partial z} + \frac{n}{c}\frac{\partial A_2}{\partial t} = \frac{i\omega\gamma_e}{2nc\rho_0}\rho^* A_1,$$

where $\omega = \omega_1 \approx \omega_2$ since $\Omega \ll \omega_1$. In the stationary regime, introducing the beam intensities $I_i = (nc/2\pi)A_iA_i^*$ and eliminating ρ one gets $dI_1/dz = -dI_2/dz = gI_1I_2$ where

$$g \approx \frac{\gamma_e^2\omega^2}{nvc^3\rho_0\Gamma_B}\frac{(\Gamma_B/2)^2}{(\Omega_B - \Omega)^2 + (\Gamma_B/2)^2}$$

is the gain for stimulated Brillouin scattering.

We note that the Stokes component in the stimulated Brillouin scattering is emitted in the backward direction as in the case of optical phase conjugation with degenerate four-wave interaction discussed in Sec. 2.3.1.1. As a matter of fact, the first detailed analysis of optical phase conjugation by nonlinear process was provided by stimulated Brillouin scattering, and a large amount of work was devoted to this aspect, which has important applications (e.g., Fisher, 1983; Zeldovich *et al.*, 1985).

The Rayleigh scattering is related to the second term in the density fluctuations, namely, the one resulting from entropy fluctuations, which obey a diffusion equation

$$\rho c_p \frac{\partial s}{\partial t} - \kappa \nabla^2 s = 0 \tag{75b}$$

the same as the temperature fluctuations, in contrast to the pressure fluctuations, which obey a wave equation (e.g., Fabelinskii, 1968; Shen, 1984; Boyd, 1992): Hence, entropy fluctuations do not propagate, and as a result the scattered light is not shifted in frequency. This is the major difference between Brillouin and Rayleigh scattering. The stimulated Rayleigh scattering is obtained by adjoining in the right-hand member of Eq. (75b) the electrostriction and proceeding as for the stimulated Brillouin scattering.

Actually, a rigorous theory requires (e.g., Shen, 1984; Boyd, 1992) the two to be treated simultaneously, and instead of Eqs. (75a) and (75b) one gets

$$-\left(\Omega^2 + i\Omega\Gamma_B - \frac{v^2 q^2}{\gamma}\right)\rho + \frac{v^2 \beta_p \rho_0 q^2}{\gamma} T = \frac{\gamma_e q^2}{4\pi} A_1 A_2^*, \tag{76a}$$

$$-(i\Omega - \tfrac{1}{2}\gamma\Gamma_R)T + \frac{i(\gamma - 1)\Omega}{\beta_p \rho_0}\rho = \frac{nc\alpha}{2\pi c_v \rho_0} A_1 A_2^*, \tag{76b}$$

where Γ_B is the Rayleigh linewidth and $\gamma = c_T/c_s$, which in the stationary regime can be solved together with the equations for the field amplitudes in the slowly varying envelope approximation to derive the corresponding gains. One can treat along similar lines other stimulated light-scattering effects and in particular the ones related to concentration fluctuations in liquid mixtures (e.g., Shen, 1984).

3.4.4 Optical Balance Very frequently, in multiphoton excitation processes one sees suppression of a process instead of its enhancement when the exciting frequencies are tuned to intermediate resonances. Such effects have been reported in numerous cases. This suppression occurs because another concurrent process that shares the same intermediate resonance gets enhanced and drains the energy available to the compound process (Wynne, 1984). The order of the competing processes may be very different as are also the resulting spectral and spatial characteristics. Thus, the system may switch from one regime to the other and vice versa on changing of certain parameters. This optical balance behavior has been reported in numerous cases—for instance, in the competition of high harmonic generation and multiphoton ionization or fluorescence or in the Stokes–anti-Stokes competition discussed above.

3.5 Nonlinear Effects in Two-Level Systems

A very important class of phenomena in nonlinear optics is related to the behavior of the matter when the frequency of an intense optical field is in close resonance with an electric-dipole–allowed transition frequency of the medium. The nonlinear susceptibilities then are multiresonantly enhanced, increasingly so as their order increases, and the first few terms of the perturbation expansion in powers of the electric field may not adequately describe the nonlinear response of the system even for moderately intense optical fields. However, because all other transitions of the medium contribute very little in comparison to the resonant one, one may altogether ignore them and model the behavior with that of an assembly of two-level systems resonantly interacting with the optical field (Sargent *et al.*, 1974). The latter to a large extent can be treated analytically without recourse to perturbation theory. Because this approach constitutes the basis of numerous investigations, we quickly review this model and present a sample of effects that one may expect if such is the case.

3.5.1 Bloch Equations Let N be the number density of the two-level systems; the two levels are labeled a and b with energies E_a and E_b respectively ($E_b > E_a$) and $h\omega_0 = E_b - E_a$. The density-matrix operator is a 2 × 2 matrix, and its four elements satisfy the equations

$$\dot{\rho}_{ba} = -\left(i\omega_{ba} + \frac{1}{T_2}\right)\rho_{ba} + \frac{i}{\hbar} H_{ba}(t)\Delta,$$

$$\dot{\Delta} = -\frac{\Delta - \Delta^{(0)}}{T_1} - \frac{2i}{\hbar}(H_{ba}\rho_{ab} - \rho_{ba}H_{ab}),$$

where $\Delta = \rho_{bb} - \rho_{aa}$, $\mathrm{Tr}\rho = \text{const}$, T_1 and T_2

are the population and coherence relaxation times, and $H_{ba} = -\mu_{ba}(Ae^{-i\omega t} + A^*e^{i\omega t})$ is the dipole interaction term in the Hamiltonian with $\mu_{ba} \neq 0$ and $\mu_{aa} = \mu_{bb} = 0$. Since $\omega \approx \omega_0$, we can use the rotating-wave approximation by suppressing antiresonant terms (Sargent *et al.*, 1974). With a slight rearrangement of the terms and introducing the expectation value of the induced transition dipole moment $p = \mathrm{Tr}(\rho\mu) = \mu_{ba}(\rho_{ab} + \rho_{ba})$, one gets

$$\ddot{p} + \frac{2}{T_2}\dot{p} + \left(\omega_0^2 + \frac{1}{T_2^2}\right)p = -\frac{2\omega_0}{\hbar}|\mu_{ba}|^2 A(t)\Delta, \tag{77a}$$

$$\dot{\Delta} + \frac{\Delta - \Delta^{(0)}}{T_1} = -\frac{2}{\hbar\omega_0}E\left(\dot{p} + \frac{p}{T_2}\right) \cong -\frac{2}{\hbar\omega_0}E\dot{p}. \tag{77b}$$

Equations (77a) and (77b) are the essential starting point for deriving the behavior of the quantal two-level system interacting with an electric field of frequency $\omega \approx \omega_0$. In particular, Eqs. (77a) and (77b) show that the problem is reduced to that of a harmonic oscillator forced by a quantal force that is connected with an energy-rate equation. With the adjunction of the field-envelope equation (21), one can also treat their impact on the propagation of the field.

Actually, Bloch equations can be transcribed in vector form (Feynman *et al.*, 1957) by use of the fact that any 2×2 matrix M can be cast into the form $M = M_0\sigma_0 + \mathbf{M}\cdot\boldsymbol{\sigma}$ where $\sigma = \{\sigma_x,\sigma_y,\sigma_z\}$ and $M = \{M_x,M_y,M_z\}$ with

$$\sigma_0 = \begin{vmatrix} 1 & 0 \\ 0 & 1 \end{vmatrix}, \quad \sigma_x = \begin{vmatrix} 0 & 1 \\ 1 & 0 \end{vmatrix},$$

$$\sigma_y = \begin{vmatrix} 0 & -i \\ i & 0 \end{vmatrix}, \quad \sigma_z = \begin{vmatrix} 1 & 0 \\ 0 & -1 \end{vmatrix},$$

and $M_0 = (M_{aa} + M_{bb})/2$, $M_x = (M_{ab} + M_{ba})/2$, $M_y = (M_{ba} - Ma_b)/2i$, $M_z = (M_{aa} - M_{bb})/2$. Using this property and setting $H = \hbar(\Omega_0\sigma_0 + \mathbf{\Omega}\cdot\sigma)$, $\rho = (R_0\sigma_0 + \mathbf{R}\cdot\sigma)$, one obtains the equation

$$\frac{d\mathbf{R}}{dt} = (\mathbf{\Omega} \times \mathbf{R}) + \left.\frac{d\mathbf{R}}{dt}\right|_d \tag{78}$$

in the vector space of the Pauli matrices σ_i; the last term takes into account the damping. Equation (78) is the same as for the precession of a magnetic moment in a magnetic field in real space; in deriving Eq. (78), the rotating-wave approximation was used, which now has a very plausible geometrical justification. The vector model has been also extended to the two-photon resonant case (Grischkowsky, *et al.*, 1975).

In terms of this equation, several effects can be given a simple geometrical interpretation by noticing that $\mathbf{\Omega} = \{\Omega_R(t),0,\Delta\omega\}$ where $\Omega_R(t) = 2|\mu_{ba}|A(t)/\hbar$ is the on-resonance Rabi frequency and $\Delta\omega = \omega_0 - \omega$ is the frequency detuning so that $\Omega = \sqrt{\Omega_R^2(t) + \Delta\omega^2}$ is the precession frequency of the state. Effects like superradiance, photon echos, Rabi precession, self-induced transparency, and others can be given a simple geometrical interpretation in terms of the vector model (Sargent, 1974); we shall give a brief account of the self-induced transparency because it strikingly shows the impact of resonant interaction on the nonlinear propagation (McCall and Hahn, 1967, 1969).

3.5.2 Nonlinear Response The induced polarization in the steady-state regime is $P = \tilde{\chi}A$ with

$$\tilde{\chi} = \frac{N|\mu_{ba}|^2\Delta}{\hbar(\omega - \omega_0 + i/T_2)} = -\frac{N|\mu_{ba}|^2\Delta_0 T_2(\Delta\omega T_2 + i)}{\hbar\{[1 + (\Delta\omega T_2)^2] + \Omega_R^2 T_1 T_2\}} \tag{79}$$

and $\Omega_R = 2|\mu_{ba}|A/\hbar$. This expression clearly exhibits (e.g., Levenson and Kano, 1988; Letokhov and Chebotayev, 1977; Sargent, 1974) saturation behavior with a saturation intensity $|A_s|^2 = \hbar^2/2|\mu_{ba}|^2 T_1 T_2$ at zero detuning. Besides obtaining the saturation behavior, which is widely used as a nonlinear spectroscopy technique, with the adjunction of the field-envelope equation (18), one can analyze several other effects like absorptive optical bistability (e.g., Gibbs, 1985), optical phase conjugation (e.g., Fisher, 1983), etc.

Simple results can also be obtained in the limit of the adiabatic following approximation where the relaxation is assumed to be negligible and the dipole exactly follows the optical pulse. Setting now $\Omega_R(t) = 2|\mu_{ba}|A(t)/\hbar$ one gets

$$P = -\frac{1}{2}\frac{N\mu_{ab}\Omega_R(t)}{\sqrt{\Delta\omega^2 + \Omega_R^2(t)}}\,\mathrm{sgn}\,\Delta\omega. \tag{80}$$

It is easy to verify that for low field intensities, by expanding in powers of the intensity one recovers the odd-order susceptibilities of a two-level system in the appropriate limits.

Quite instructive for the understanding of the phenomena that can take place in an assembly of two-level systems resonantly driven by a monochromatic optical field is the expression of the induced dipole moment when T_1 and T_2 are infinite, in which case the equations have the simple solutions $\rho_{aa} = \cos^2(\Omega t/2) + (\Delta\omega/\Omega)^2 \sin^2(\Omega t/2)$, $\rho_{bb} = (\Omega_R/\Omega)^2 \times \sin^2(\Omega t/2)$ for the initial condition $\rho_{aa}(0) = 1$ and $\rho_{bb}(0) = 0$ and

$$\rho_{ab} = \mu_{ab}\frac{\Omega_R}{\Omega}\left[-\frac{\Delta\omega}{2\Omega}e^{-i\omega t} + \tfrac{1}{4}\left(\frac{\Delta\omega}{\Omega} - 1\right)e^{-i(\omega+\Omega)t} + \tfrac{1}{4}\left(\frac{\Delta\omega}{\Omega} + 1\right)e^{-i(\omega-\Omega)t}\right],$$

which clearly shows that the induced transition dipole radiates not only at frequency ω but also at the sidebands $\omega \pm \Omega$, and accordingly we expect enhanced interactions with signal beams at these frequencies.

The problem of coupling of two beams, a pump and a signal, of frequencies ω and $\omega \pm \delta$, respectively, has been extensively studied as it has relevance for saturation spectroscopy but also for optical phase conjugation by quasidegenerate wave mixing and other parametric processes (Boyd, 1992). Assuming a stationary regime, setting $A = A_p + A_s e^{-i\delta t}$ with $|A_p| \gg |A_s|$, $p = p_p + p_s e^{-i\delta t} + p_c e^{i\delta t}$, and $\Delta = \Delta_p + \Delta_s e^{-i\delta t} + \Delta_c e^{i\delta t}$ in Eq. (80) and expanding up to linear terms in A_s around A_p, the expressions of p_s and p_c are obtained as well as the corresponding effective nonlinear susceptibilities $\tilde{\chi}(\omega \pm \delta)$ that depend on the pump intensity. The amplitudes of the fields generated at these sideband frequencies A_s and A_c are then obtained from Eq. (18) in the slowly varying amplitude approximation after inserting the expressions of the polarizations p_s and p_c with a phase mismatch $\Delta k = 2k_p - k_s - k_c$ where the wave vectors k_p, k_s, and k_c are obtained from the linear refractive index. One obtains two coupled-amplitude equations,

$$\frac{dA_s}{dz} = -\alpha_s A_s + \kappa_c A_c^* e^{i\Delta kz},$$

$$\frac{dA_c}{dz} = -\alpha_c A_c + \kappa_s A_s^* e^{i\Delta kz},$$

which can readily be solved for arbitrary pump intensities. One can have gain, which gets enhanced when $\delta \approx \Omega$, the Rabi precession frequency.

3.5.3 Self-Induced Transparency In terms of the vector model, one can anticipate that when $T_1 \gg \tau_\pi$, where τ_π is the optical pulse duration, and when in addition $\int_{-\infty}^{\infty} \Omega_R(t)dt = 2\pi$ is satisfied for $\tau_\pi < T_2$ then the state vector coherently undergoes a complete Rabi precession where in the first half it gains "energy" from the field, which it restores back to it in the second half of the Rabi precession period; this implies a threshold condition $|A| \geq \pi\hbar/T_2\mu_{ba}$, and the optical pulse propagates through the medium without substantial loss but suffers a delay as a consequence of the Rabi precession that is needed to recover the energy. This is the self-induced transparency effect (McCall and Hahn, 1967, 1969; Slusher and Gibbs, 1972) that only occurs close to a resonance, in contrast to the white self-induced transparency in a composite dielectric that can occur at any frequency of the transparency region (see Sec. 2.3.4).

The problem can be treated analytically and constitutes another manifestation of soliton behavior. Separating resonant and nonresonant contributions in the nonlinear response and using the slowly varying envelope approximation, one gets

$$\frac{\partial A}{\partial z} + \frac{n}{c}\frac{\partial A}{\partial t} = \frac{2\pi i\omega N}{c}p,$$

where $dp/dt = -i|\mu_{ba}|^2 A\Delta/\hbar$ and $d\Delta/dt = -4iAp/\hbar$. This set of equations can be solved, and one finds for the field amplitude

$$A(z,t) = \frac{\hbar\sqrt{D}}{\mu_{ba}}\operatorname{sech}\left[\sqrt{D}\left(t - \frac{z}{v_g}\right)\right], \qquad (81a)$$

where

$$D = \frac{2\pi N\omega|\mu_{ab}|^2}{n\hbar c(1/v - n/c)},$$

propagating with velocity

$$\frac{1}{v} = \frac{n}{c} + \frac{N\pi\omega}{2n\hbar c}|\mu_{ab}|^2\tau^2, \qquad (81b)$$

where τ is the extent of the pulse. We see from

Eqs. (81a) and (81b) that the solution is a soliton that propagates with a lower velocity than the one pertinent to a weak field.

By extending the above approaches, one can study several other effects and in particular the stability conditions, the occurrence of instabilities, and transition to chaotic behavior.

4. NONLINEAR OPTICAL MATERIALS AND DEVICES

Notwithstanding the fundamental interest for pursuing the study of the nonlinear effects, the main driving force behind the growth that the field has witnessed since the first demonstration of an all-optical nonlinear effect in 1961 and the justification for the continuing effort are the potential uses and applications in future technologies and in particular the ones related to signal processing and transmission, computing, and optoelectronics (Arecchi and Schulz-Dubois, 1975; Miller *et al.*, 1994; Ostrowski and Reinisch, 1992). Implementation of nonlinear effects in devices that will serve the goals of these future technologies and also surpass the performances of electronic devices or open new possibilities there that are inaccessible or inconceivable with present-day electronics technology are key issues in the present research effort in nonlinear optics. In this respect, the nonlinear optical materials constitute the cornerstone and progress here will condition all future developments (e.g., Digiorgio and Flytzanis, 1994).

4.1 Nonlinear Optical Materials

In contrast to the nonlinearities in acoustics or electronics, the ones in optics are very weak, in particular in the optical transparency region of the materials where one wishes to operate in order to reduce the absorption losses and achieve fast operation. Indeed, the relative photoinduced changes of the refractive index $\Delta n_{NL}/n_0$ in general are exceedingly weak even for intense light beams, several orders of magnitude smaller than the corresponding figure of merit in acoustics or electronics. The situation hardly improves even when resonances are involved. Thus, while the nonlinear interactions in acoustics and electronics can be considered local, in nonlinear optics one needs interactions over several wavelengths for appreciable energy transfer between the beams, and one is faced with the phase-matching problem, which introduces a nonlocal behavior in the nonlinear optical interactions; the nonlocality in several nonlinear optical interactions is a real nuisance but at the same time can be exploited to suppress unwelcome processes and selectively enhance others.

The potential applications of nonlinear optics in devices suffer very much from this confrontation with electronics; the latter in the meantime is continuously progressing, both in miniaturization and in speed and other performances and postpones the introduction of all-optical nonlinear devices.

For these reasons, progress in the performance of nonlinear optical materials (Chemla and Zyss, 1987; Digiorgio and Flytzanis, 1994) well above the presently known ones is an important step toward the desired breakthrough of nonlinear optical devices in the coming technologies. In improving the performance of the nonlinear materials or conceiving new ones, one cannot restrict oneself to the traditional physicochemical crystal-growth techniques but must resort as well to the new growth techniques that have been devised for the fabrication of artificial microstructures. Despite the progress that has been made with homogeneous materials, inorganics or organics, their performance regarding optical nonlinearities (see Table 2) is still below the breakthrough point and attention is now also directed toward artificial ones obtained by interfacing materials of very different chemical constitutions. These new fabrication techniques, besides allowing control of the form and dimension on the atomic scale, also introduce a synergy of materials with different chemical constitutions, which may considerably improve their nonlinear properties and simultaneously meet other requirements for reliable miniaturized devices. At this level, one must differentiate the requirements imposed on materials for second-order processes from those for third-order processes.

4.1.1 Materials for Second-Order Effects

Because $\chi^{(2)}$ vanishes for centrosymmetric or random media, these materials must be formed from asymmetric polarizable units arranged so that the material lacks inversion

Table 2. Values of nonlinear susceptibilities (transparent region).

Second order	
Material	$\chi^{(2)}$ (10^9 esu)
α-SiO_2	1
KDP	1
$LiNbO_3$	10
$BaTiO_3$	50
GaAs	10^3
CdSe	10^3
Urea	1
MAP	10
MOP	10
Poled polymers	~10
Third order	
Material	$\chi^{(3)}$ (10^{12} esu)
Glass	10^{-4}
NaCl	10^{-2}
CS_2	1
Si	1
GaAs	10
Ge	10^2
PTS (polydiacetylene)	10^2

symmetry. This can only be achieved in crystals without inversion symmetry or materials formed from helical molecules, the ones that also possess rotatory power; the latter has never been seriously considered and the effort has been exclusively concentrated on asymmetric crystalline solids, inorganics or organics, and poled polymers. We wish to point out here that exploitation of resonances in second-order processes is not compatible with the coherent character of these processes and in addition introduces substantial linear absorption that may limit the nonlinear process within a very short optical penetration depth.

For the selection of materials for second-order processes, it was initially proposed to use the conjecture that the Miller coefficient $\delta_{ijk} = \chi^{(2)}_{ijk}/\chi^{(1)}_{ii}\chi^{(1)}_{jj}\chi^{(1)}_{kk}$ is roughly constant for a large class of crystalline solids. Although δ varies (e.g., Lines and Glass, 1979) over a narrower range of orders of magnitude than $\chi^{(2)}$, its use as a figure of merit was quickly abandoned in favor of the coefficient $f = \chi^{(2)}/n_0$, which allows a better assessment of the materials for potential use in devices; for applications related with the electro-optic effect, the figure of merit $f = \chi^{(2)}/n_0^3$ is more appropriate (Lines and Glass, 1979; Yariv, 1971).

Whatever the precise definition of the figure of merit, it is quite evident that many other factors as well, not related to the magnitude of $\chi^{(2)}$, determine the choice of a material for use in devices and actually set the ultimate criteria. We may mention here high optical quality and stability over large dimensions and very long times for operation with high-repetition-rate pulse trains, typically several terahertz; wide transparency region; resistance to mechanical, thermal, electrical, and other strains; processability and adaptability, etc. We also wish to point out here that in the long run the electro-optic effects, in particular the Pockels effect, will be the most used ones among the second-order ones for miniaturized optoelectronic devices, and this will condition the future material research here; in particular, the photoconductivity and electrical conductivity will become serious issues, but the phase matching will not be one.

From the very beginning of nonlinear optics, the effort has been directed (Lines and Glass, 1979; Bordui and Fejer, 1993; Chen and Liu, 1986) toward the ferroelectrics or related crystals like the ABO_3-isomorphs [lithium niobate ($LiNbO_3$)], the ADP-isomorphs [ammonium diphosphate ($NH_4H_2PO_4$ or ADP), potassium diphosphate (KH_2PO_4 or KDP) and their deuterated versions], and other related crystals like $Ba_2NaNb_5O_{15}$ and more recently materials like potassium niobate ($KNbO_3$), potassium titanyl phosphate ($KTiOPO_4$ or KTP), and their derivatives such as $KTiOAsO_4$, and also the β-barium borate (β-BaB_2O_4 or BBO); all these crystals are polarizable, anisotropic for phase matching, and transparent over a wide spectral range, up to ~2500 Å, roughly the fourth harmonic of the YAG laser. Attention was also directed toward anisotropic polar semiconductors, the ones with wurtzite structure or similar ones; unfortunately, the ones with highest nonlinear coefficients, like GaAs or InSb, and well-controlled growth capability are isotropic and hence not phase matchable or possess a very narrow transparency region, which leads to substantial two-photon losses. As previously stated, the second-order effect being intrinsically coherent the overall efficiency is not improved by the use of resonances or the creation of photocarriers. For these reasons, the polar semiconductors have been only envisaged for far-infrared coherent light generation by optical frequency mixing. Recently, much progress was made in stacking semiconductor layers with alternating polar direction, which greatly improves the nonlinear interaction efficiency.

Because of the limitations encountered with the inorganic crystals, another line of material research was developed around the molecular crystals, mainly organics (Chemla and Zyss, 1987). To the extent that the values of $\chi^{(2)}$ are directly related to the value of β of the constituent molecules in the unit cell, the line of approach here has been to concentrate the effort first in selecting molecules with large β eventually by introducing chemical modifications on existing ones and then proceeding to grow crystals without inversion symmetry. Particular attention is paid to molecules with helicity (they always give rise to noncentrosymmetric crystals) or with weak permanent dipole moment (the impact of dipole–dipole interactions is reduced) or with hydrogen bonding (to counterbalance forces that tend to lead to centrosymmetric structures), etc. Molecular engineering has been quite helpful (Chemla and Zyss, 1987), and several organic molecular crystals have been obtained like methyl-(2,4-dinitrophenyl)-aminopropanate (MPA), 3-methyl-4-nitropyridine-1-oxide (POM), *N*-(4-nitrophenyl)-(L)-prolinol (NPP) and urea [$CO(NH_2)_2$], which certainly have high values of $\chi^{(2)}$, comparable to or higher than those of the commonly used ferrolectrics, are phase matchable, and possess a transparency region that extends up to 4000 Å, roughly the third harmonic of the YAG. They have however severe drawbacks in other respects as regards photochemical stability, processability, optical quality of large-size crystals, and many others. The only nonlinear molecular crystal that shows some promise and favorably compares in most respects with the ferroelectrics is the urea crystal.

As a general rule, the more polarizable one desires to render a crystal the more difficult it is to obtain high-optical-quality perfect crystals. This is because polarizable units with large polarizabilities are associated with soft charge distributions held by low cohesive fields [compare Eq. (10)] and hence very sensitive to external perturbances as well as internal ones like stresses, impurities, compositional disorder, and inhomogeneity. These introduce defects and nonuniformity of the optical properties and in addition render the crystal very susceptible to photochemical degradation or introduce electrical conductivity and photorefractive effects that limit their use for electro-optical devices.

In view of these and other problems with crystalline solids, the poled polymers (e.g., Digiorgio and Flytzanis, 1994) constitute a real breakthrough since in most respects they match the performance of such a crystal as $LiNbO_3$, which at present seems to be the best candidate for device applications using second-order effects. These materials are obtained by introducing and attaching asymmetric chromophores as pendant side groups in polymeric chains, aligning them, and cross-linking the polymer chains concurrently. By attaching each chromophore at several points, their orientation ability is substantially frozen, and one obtains a medium without inversion symmetry that can be easily processed, implemented, and adapted in devices. Eventually, the chromophore alignment somewhat relaxes with time, and the value of $\chi^{(2)}$ is accordingly reduced but stabilizes within a few hours to values comparable to those of $LiNbO_3$ for years and temperatures up to 80 °C.

Of some interest also is the recently observed second-harmonic generation in glass fibers (Osterberg and Margulis, 1986); the underlying mechanism however has not been elucidated yet.

4.1.2 Materials for Third-Order Effects

Here the effort has been mainly directed toward materials with large photoinduced changes of either the refractive index or the absorption for applications related to the optical Kerr nonlinearities. In contrast to the second-order effects, which will be exploited only in small-size devices, the third-order effects, the ones related to the optical Kerr nonlinearity, will be exploited both over very long interaction lengths, for instance in soliton transmission in fibers over several thousand kilometers, and in small-size devices, for optical switches, windows, bistable devices, etc. The phase-matching problem not being an issue here, isotropic materials can be considered and as matter of fact are preferred; but otherwise all material considerations relevant for second-order processes apply here too.

For the optical Kerr nonlinearities depending on the operation regime, dispersive or absorptive, one uses either of the two parameters $f_d = \omega\chi^{(3)}/n_0^2$ or $f_a = \chi^{(3)}/\alpha\tau$ to assess the merit of the material for nonlinear operation in the transparency and absorption regions respectively where τ is a relaxation time, essentially the population (energy) relaxation time T_1 for the transition. The choice of either regime certainly depends on the magnitude of the nonlinearity and other optical parameters but also on the speed that is required in the operation; in this respect, the speed in the absorptive regime is limited by the relaxation rates of the photoinduced absorption rates, essentially the population decay rates of the photocarriers.

To the extent that $\chi^{(3)}$ depends on the charge delocalization over many repeat units, the periodicity or crystallinity plays a role but not necessarily over regions larger than the natural delocalization length of the electrons. For these reasons, the attention here was mainly concentrated on semiconductors and conjugated polymers (Sauteret *et al.*, 1976) that are one-dimensional organic semiconductors; among the latter, the polydiacetylenes show certain promise. As a general rule however, the electronic Kerr nonlinearities in the transparency region are rather low, with the polydiacetylene PTS showing the highest value for f_d, but its optical quality and other features still leave much to be desired. In the absorption region, the photoinduced absorption changes can be large in particular in semiconductors, but the absorption losses become high and in addition the photoinduced population decay times are usually in the nanosecond range, which is rather slow.

The behavior of the optical Kerr nonlinearity in semiconductors is determined to a large extent by the state and band-filling mechanisms. Here the electrons photoexcited into the conduction band by an optical pulse are quickly thermalized and fill the band states up to a level E_f fixed by their density, the Pauli principle, and their recombination time (or pulse duration, whichever is shorter) and bar these states from further occupation; the apparent gap is blueshifted. Similar mechanisms apply in polydiacetylenes as well when the frequencies are close to or above the onset of optical absorption. Unfortunately, these mechanisms lead to figures of merit that are quite insensitive to frequency, and in addition their values are not sufficiently large for the bulk materials.

For these reasons, one has introduced artificially grown materials where the confinement and dimensionality have definite impact on the optical properties, linear and nonlinear. Here one employs wave or charge confinement to enhance the efficiency of the optical processes. The purpose here is to increase the interaction length or the optical nonlinearity respectively. The two confinements however cannot be implemented simultaneously in the same material as they address aspects incompatible with each other.

The first is achieved by confining the interacting optical beams in guides or resonators whose minimal dimension must be larger than the optical wavelength (Hasegawa, 1989; Agrawal, 1989; Miller, *et al.*, 1994; Ostrowsky and Reinisch, 1992); it concerns the class of materials for nonlinear guided optics in their transparency region, where the nonlinearities, even for the most favorable ones, are weak but the absorption losses can be kept minimal. This is in practice essential for long-distance soliton transmission and here glasses are exclusively used.

The second consists in enhancing the nonlinearity of the materials with very delocalized valence electrons, like metals, semiconductors, or conjugated polymers, by artificially confining (Schmitt-Rink *et al.*, 1989; Flytzanis *et al.*, 1991) the valence electrons in regions much shorter than their natural delocalization length in the bulk, which extends over many unit cells or even to infinity; these are the dielectric and quantum confinements. The dielectric confinement results in an enhancement of the polarizing field close to the so-called surface plasma resonances, which leads to substantial enhancement of the Kerr nonlinearity since it enters its expression in the fourth power; the quantum confinement gives rise to discrete optical resonances whose position, oscillator strength, and dynamics depend on the extension of the artificial confinement and hence can be modified to meet certain requirements. Both these cases can be studied in materials formed by uniformly dispersing semiconductor or metal nanocrystals in glass in small concentrations; the dielectric confinement is important for metal crystallites while the quantum confinement is in the semiconductor crystallites. These classes of composite materials and other artificial ones are called to play an important role in future generations of nonlinear materials for third-order processes because of their large and tunable nonlinearities and their stability, processability, and adaptability for miniaturized device fabrication.

Regarding materials for third-order processes, we wish to mention here that crystals without inversion symmetry may possess large and useful third-order nonlinearities through the cascading processes that can take place there (Stegeman *et al.*, 1993). These can contribute to the optical Kerr effect. In this respect, we also mention the photorefractive crystals that have been extensively used in real-time holography and optical phase conjugation.

4.2 Nonlinear Optical Devices

Most nonlinear optical effects we reviewed in the previous sections and many others can be used and implemented in devices if appropriate nonlinear materials are available, and we shall not even try to enumerate them here (Miller *et al.*, 1994; Ostrowsky and Reinisch, 1992; Yariv, 1971). At present, one cannot pretend that this is yet the case, and much effort is still left in order to meet the stringent requirements that even the most simple device operation puts on the materials. In addition, as was earlier stated the performance of electronic devices has steadily improved although with slower pace than previously, and in addition they start reaching some of their conceptual limits. Whatever the point of view may be, discussion of the principles and performance of the nonlinear optical devices, existing or potential ones, is meaningless without a comparison with the electronic ones or other types of devices, and this falls outside the scope of this review.

At present, it seems that nonlinear optical devices based on some second-order processes can be developed for large-scale manufacturing and implementation in high technology to perform several operations. In this respect, electro-optic devices, modulators, deflectors, and others are promising future developments. Frequency mixers in the far infrared too if adaptable to miniaturized lasers may also find applications. With the continuous progress in lasers, however, many nonlinear effects to produce sum or difference frequencies will become obsolete in several cases except for large power laser systems. Optical parametric amplifiers and oscillators too show promise for future developments, and their adoption in modern technology may constitute the turning point for all-optical technology.

The devices based on optical Kerr nonlinearities however still encounter severe material limitations for their development and implementation in large-scale integrated systems for signal processing and computing or other operations. For instance, neither optical bistable devices nor optical phase conjugators seem to live up to the expectations that arose in their beginning although no effort was spared here to reach optimal conditions with the materials that heavily rely on modern fabrication techniques. Here we wish to mention that the situation is similar with all other proposals for all-optical devices like optical memories based on photochemical hole burning and similar ones. On the other hand, here too, like in the case of second-order processes, devices that exploit parametric processes like acousto-optic modulation and deflection or the static Kerr modulation and similar ones have already found many applications.

From the previous succinct comments, it is clear that only parametric processes have presently led to device development that can be used, incorporated, and adapted to the already established technologies, which up until now essentially relied on microelectronics and micromechanics. Accordingly, the first generation of devices that will be used to make operations on optical beams and signals will be hybrid ones. The situation concerning the nonlinear all-optical devices, however, may change as it depends on several factors whose synergy is not mastered yet in order to reach the threshold that makes their need urgent and necessary.

5. NONLINEAR OPTICAL SPECTROSCOPY AND DIAGNOSTICS

There is a major body (Levenson and Kano, 1988; Letokhov and Chebotayev, 1977) of spectroscopic techniques that has emerged from nonlinear optics that exploits several features of the nonlinear susceptibilities and propagation to obtain crucial information concerning the spectral and dynamical characteristics of the excitations that can take place in the different states of matter; in addition, they constitute the basis for some very powerful nondestructive diagnostic techniques to study most diverse processes that can take place on microscopic or macroscopic scales, at close or far distances, and in other extreme conditions in a laboratory, factory, the environment, or outer space. These techniques have several advantages over the linear ones and can address problems inaccessible by the latter; furthermore, they best exploit the characteristics of the lasers.

Despite their apparent diversity, these techniques share in common several features and follow roughly similar patterns. Most frequently, they rely on the spectral and temporal behavior of the nonlinear susceptibilities close to the intermediate resonances or multiphoton transitions; they also exploit several features of the nonlinear propagation in particular regarding the phase matching, which sets very severe restrictions. Other techniques exploit the tensor character of the nonlinear susceptibilities, and nonlinear polarization-state spectroscopy is being extensively used; finally, the light-scattering technique, modulation spectroscopy, and other parametric ones in essence are nonlinear techniques.

Here we shall briefly illustrate some specific aspects (Levenson and Kano, 1988; Letokhov and Chebotayev, 1977) with the techniques based on third-order nonlinear processes and to a lesser degree with those related to the second-order ones; actually, most frequently the latter are used in cascade and effectively form third-order processes. Some remarks concerning their extensions to higher-order processes will also be included. One may roughly distinguish three classes of nonlinear spectroscopic techniques, namely in the frequency domain, in the time domain, and polarization-state sensitive although these aspects are more or less simultaneously present in most techniques.

5.1 Nonlinear Optical Spectroscopy in the Frequency Domain

Linear spectroscopy in the frequency domain is related to the one-photon resonances in the linear susceptibility as they show up in dispersion or absorption; they are related to electric-dipole–allowed transitions. Nonlinear spectroscopy is essentially related (Shen, 1984; Levenson and Kano, 1988; Letokhov and Chebotayev, 1977) to the intermediate multiphoton resonances in the nonlinear susceptibilities—for instance, the two-photon resonances in the case of the third-order susceptibility—as these show up in dispersion or absorption; the selection rules for the involved transitions can be quite different from the electric-dipole–allowed ones, and this can be tested by appropriate selection of the polarization states of the beams.

5.1.1 Two-Photon (Sum) Spectroscopy This is related to transitions between states g and f of energies E_g and E_f respectively such that $\omega_1 + \omega_2 = \omega_{fg} \equiv (E_f - E_g)/\hbar$ where ω_1 and ω_2 are positive frequencies; the expression for the resonant third-order susceptibility close to such resonance was derived in Sec. 3.3. The quantity $\alpha_{ij}(\omega_1,\omega_2)$ defined there plays a role analogous to the transition dipole moment μ_i in one-photon resonances; it determines the strength of the two-photon transition, and the selection rules of the latter roughly coincide with those for quadrupole-allowed transitions. We only retain here that for a system with inversion symmetry where parity is a good quantum number and the states are either odd or even with respect to space inversion, only states of the same parity, odd or even, are connected by two-photon transitions while the one-photon transitions connect opposite-parity states. Besides this feature, two-photon spectroscopy possesses others quite distinct from those of one-photon spectroscopy; $\boldsymbol{\alpha}(\omega_1,\omega_2)$ is actually a second-rank tensor while $\boldsymbol{\mu}$ is a vector, and in addition the former can be enhanced by one-photon resonant transitions. All these features are currently exploited in two-photon spectroscopy. Two-photon spectroscopic techniques mainly rely on the measurement of the two-photon absorption losses related to $\mathrm{Im}\chi_R^{(3)}(\omega_1,-\omega_1,\omega_2)$ when $\omega_1 + \omega_2 = \omega_{fg}$ and to a lesser degree on the determination of the dispersion of $\chi_R^{(3)}$ around the resonance. Two-photon absorption spectroscopy has been extensively used in gases, atomic or molecular, liquids and solids, molecules in solutions, and atoms or ions in solid matrices. Actually, two-photon absorption spectroscopy has some definite advantages over one-photon spectroscopy even when the parity is not an issue. Thus, by choosing ω_1 and ω_2 away from any resonances one can study transitions in the bulk and avoid the surface absorption layer that creates severe problems in one-photon absorption; or by tuning one of these frequencies close to a one-photon resonance, one obtains information about excited-state configurations and oscillator strengths between two excited states. In another respect, two-photon absorption spectroscopy has a definite advantage since it allows the suppression of the Doppler broadening in gases. This can only be achieved in the degenerate two-photon counterpropagating beam configuration, namely, when $\omega_1 = \omega_2 = \omega$ and $\mathbf{k}_1 = -\mathbf{k}_2 = \mathbf{k}$. Here each atom with velocity vector $\mathbf{v}$ sees the two twin counterpropagating photons with frequencies $|\omega \pm \mathbf{k}\cdot\mathbf{v}|$ so that the condition $2\omega = \omega_{fg}$ can be satisfied for all atoms irrespective of their velocity component along the beam-propagation direction and one sees only the natural linewidth. The same degenerate counterpropagating beam configuration in solids allowed for the first time the direct observation of the upper exciton-polariton branch in polar semiconductors. Since two-photon absorption losses are the main operation losses in many nonlinear optical systems, two-photon spectroscopy is very useful in identifying them. One can also use such information to devise pumping schemes for laser action; there have been attempts to obtain stimulated two-photon emission for laser action. We wish to point out that the determination of the absolute two-photon absorption cross sections is quite tedious and time consuming and one must resort to relative measurements.

Two-photon dispersion spectroscopy although less used can have some interesting applications too since it is sensitive to the interference of the resonant and nonresonant contributions in $\chi^{(3)}(\omega_1,\omega_2,-\omega_2)$ or $\chi^{(3)}(\omega_1,-\omega_1,\omega_2)$ close to a two-photon resonance $\omega_1 + \omega_2 = \omega_{fg}$, or $\boldsymbol{\chi}^{(3)} = \boldsymbol{\chi}_R^{(3)} + \boldsymbol{\chi}_{NR}^{(3)}$ where $\chi_R^{(3)}$ is given by Eq. (68); $\chi^{(3)}(\omega_1,\omega_2,-\omega_2)$ is actually related to the optical Kerr nonlinearity in either the degenerate or the nondegenerate frequency configuration. Thus, the optical Kerr coefficient can be enhanced by two-photon resonances, and this can have an impact on the efficiency of several effects discussed in Sec. 2.3 like optical gratings, optical phase conjugation and bistability, self-action effects, etc. Furthermore, by tuning the sum frequency $\omega_1 + \omega_2$ on either side of the two-photon resonance the coefficient $\mathrm{Re}\chi^{(3)}$ vanishes for a value of $\omega_1 + \omega_2$ either above or below ω_{fg} depending on the absolute sign, + or − respectively, of the value of $\chi_{NR}^{(3)}$, that of $\mathrm{Re}\chi_R^{(3)}$ being known since it only depends on the sign of the denominator $\omega_i + \omega_j - \omega_{fg}$. This constitutes a technique to determine precisely the two-photon absorption cross sections.

5.1.2 Raman Spectroscopy The third-order susceptibility $\chi^{(3)}(\omega_1,-\omega_2,\omega_3)$ where the ω_i are positive exhibits a resonant behavior also for $\omega_1 - \omega_2 = \omega_R$ whenever ω_R is the frequency of a Raman-allowed transition, and this allows one to incorporate into the nonlinear spectroscopy all the classic light-scattering effects like Raman, but also Brillouin, Rayleigh, and others where ω_R are their characteristic frequencies. In Sec. 3.4, it was discussed how the two-beam nonlinear coupling with the related spontaneous fluctuations of the dielectric constant leads to stimulated effects that can be used for the generation of coherent light sources with characteristics different from those of the incident pump beams. Here we turn our attention to the impact that the light-scattering effects have in nonlinear spectroscopy.

The expression for the resonant contribution of $\chi^{(3)}(\omega_1,-\omega_2,\omega_3)$ when $\omega_1 - \omega_2 \approx \omega_R$ was given in Eq. (68), and the discussion there made plausible the connection between the Raman tensor and the coefficient $\boldsymbol{\alpha}(\omega_i;\, \omega_j)$. The Raman tensor and the Stokes or anti-Stokes components in a typical light-scattering experiment are determined by collecting the light scattered off an incident beam from the fluctuations of the polarizability. Typically, it is observed in a different direction from that of the incident beam and quite often over a wide angle; one is usually faced with unfavorable signal-to-noise ratio and other inconveniences related to experimental configuration. The observation of the coherent light generated at the frequency $\omega_i - \omega_j + \omega_k$ on the other hand in a direction specified by the phase-matching condition allows one to circumvent some of these problems. This is in particular the case with the coherent anti-Stokes Raman spectroscopy, where $\omega_1 - \omega_2 \approx \omega_R$ and one measures the intensity at frequency $2\omega_i - \omega_j \approx \omega_i + \omega_R$. Clearly, one or both sources must be tunable in frequency. In the usual light-scattering experiments, one actually measures losses, and the scattering cross section is related to $\mathrm{Im}\chi^{(3)}(\omega_i,-\omega_i,\omega_j)$ with $\omega_i - \omega_j = \omega_R$, while the coherent anti-Stokes Raman scattering (CARS) technique (Levenson and Kano, 1988; Letokhov and Chebotayev, 1977) is in essence a nonlinear dispersion technique. More precisely, one determines the coefficient

$$\chi^{(3)} = \frac{\alpha(\omega_1,\omega_2)\alpha^*(\omega_1,\omega_2)}{\hbar(\omega_1 - \omega_2 - \omega_{fg} - i\Gamma_{fg})} + \chi_{NR}^{(3)}$$

whose real part vanishes for a value of $\omega_1 - \omega_2$ either above or below ω_{fg} depending on the absolute sign, + or − respectively, of $\chi_{NR}^{(3)}$. This feature is exploited in CARS and provides a very precise method for determining the magnitude and sign of χ_{NR} if the Raman cross section is known (Flytzanis, 1975). Note that one can have simultaneously $\omega_1 + \omega_2 = \omega_e$ and $\omega_1 - \omega_2 = \omega_R$ where ω_e and ω_R are electronic and vibrational (or rotational) frequencies respectively, and this leads to interferences between Raman and two-photon dispersions.

The CARS technique and its variants have found extensive applications as a diagnostic in gases in high-temperature conditions, combustion or flames, and molecules in solutions or solid matrices. Clearly, the technique can be used with $\omega_1 - \omega_2$ close to an electronic transition or to study spectral features related to the Brillouin and Rayleigh light-scattering effects.

5.1.3 Saturation Spectroscopy This exploits the saturation effect in Eq. (79) by an intense resonant beam on a homogeneously broadened transition and is extensively used in hole-burning spectroscopy to study inhomogeneously broadened transitions (Shen, 1984; Levenson and Kano, 1988; Letokhov and Chebotayev, 1977; Sargent *et al.*, 1974); it concerns dipole-allowed transitions. This powerful technique has found a wide range of applications and favorably competes with most other nonlinear techniques that address similar problems—for instance, in Doppler-free spectroscopy. Actually, saturation spectroscopy can be used with two-photon transitions as well. The two-photon saturation effect inside a Fabry–Pérot cavity also leads to an interactive two-beam bistable behavior (e.g., Gibbs, 1985).

5.2 Nonlinear Spectroscopy in the Time Domain

The previous techniques pertain to the frequency domain. For the direct study of the dynamics and relaxation of excitations on the molecular or atomic level, the time-resolved nonlinear techniques have definite advantages over the previous ones. These are connected with the important developments that took place in the art of producing short and very intense light pulses of any duration down to a few femtoseconds, possessing very clean and steep fronts, well-defined forms, and tunability over wide spectral ranges.

Several time-resolved nonlinear techniques (De Martini and Ducuing, 1966; Laubereau and Kaiser, 1978; Flytzanis, 1993) have been devised for the study of relaxation processes in a wide range of excitations. Despite their apparent complexity and diversity, these usually follow the following scenario.

1. Excitation stage: The system is perturbed by an "instantaneous" external force $F_e(t)$ that resonantly couples to a "coordinate" q of the system and drives it off equilibrium at time "t" = 0; as a consequence, the coordinate acquires a phase and an amplitude.
2. Free precession: The system may be left to "precess" freely for a time interval t_d after the excitation stage till it settles in the relaxation regime we wish to study.
3. Interrogation stage: The system is probed at time t_d with an external weak force $F_p(t)$ that couples to a coordinate of the system q' that acquired phase and amplitude following the excitation, but this coordinate may not necessarily be the same as the one directly driven off equilibrium in the excitation stage; by measuring the signal as a function of the delay t_d, one obtains the relaxation time appropriate to the chosen relaxation regime or channel.

The exciting and probing forces, $F_e(t)$ and $F_p(t)$ respectively, are powers of the electric field fixed by the degree of the nonlinear interaction involved in the excitation and probing stages respectively. Because the off-equilibrium amplitude of the coordinate q' necessarily depends on the external force F_e

applied at the excitation stage and in addition the signal also depends on F_p, time-resolved spectroscopy is by essence nonlinear. Furthermore, because the excitation and probing stages must be of short duration one must use short light-pulse techniques. We anticipate that the time resolution is not fixed by the pulse duration but by the decay of the appropriate correlation function of F_e and F_p; this is actually determined by the decay time of F_e and rise time of F_p.

The excitation and the interrogation stages may or may not coincide in space, and accordingly we distinguish between local and nonlocal time-resolved techniques respectively; the latter are used whenever relaxation and propagation effects must be separated as in the case for fast-propagating excitations like the polaritons.

5.2.1 Local Time-Resolved Techniques

These concern the study of relaxation processes in localized excitations; the interrogation stage coincides in space with the excitation stage. The most developed and used time-resolved nonlinear technique for the study of vibrational relaxation processes is time-resolved CARS (Laubereau and Kaiser, 1978). Here two short, intense light pulses of frequencies ω_1 and ω_2 interact simultaneously inside the medium and coherently drive a Raman-active transition of frequency ω_R in the region of their spatial overlap; the frequencies ω_1 and ω_2 are such that $\omega_1 - \omega_2 \approx \omega_R$. The evolution of the coherence and population difference between the two states involved in the transition is probed with a delayed weak short pulse of frequency ω_3 by measuring the anti-Stokes intensity at frequency $\omega_p = \omega_3 + \omega_R$. In the coherent regime, the detection is made in the phase-matching condition and one measures the phase relaxation time T_2, while in the incoherent regime the detection is made at a right angle to the phase-matching condition and one measures population (energy) relaxation time T_1. At present, this constitutes the most powerful technique for measuring T_1. There are several parameters that can be varied to improve the selectivity and other performances of this technique. Note in particular that the excitation can be made at any point inside the bulk of the medium if the latter is transparent to the frequencies involved. In addition, in molecules with several coupled modes, one may wish to study intermode energy and coherence transfer. Such information can be obtained by detecting the anti-Stokes intensity from another mode than the one that has been coherently driven in the excitation stage; one can also study overtones and other aspects of anharmonic interactions.

Clearly, the time-resolved CARS technique can only be used for Raman-active modes. For infrared-active (dipole allowed) modes, one may replace the coherent excitation stage by one-photon (infrared) excitation resonant with the material transition, followed by a delayed excitation to a fluorescent state. The measure of the fluorescence as a function of the delay between the two excitation stages gives information about the relaxation of the infrared-active mode. Because of the lack of tunable infrared sources with short pulses, this technique or similar ones have not been used to the same extent as the CARS technique. Another limitation is that the two excitation stages are restricted inside the optical penetration depth of the sample near the surface whose properties can be different from those in the bulk. We also mention the time-resolved luminescence technique that has been successfully used to study fast photocarrier relaxation processes in semiconductors but can also be used in other instances. Here the fast luminescence signal subsequent to a first ultrashort pulse is sampled at a given frequency by up-converting it with a second delayed ultrashort pulse inside a crystal without inversion center; the intensity of the up-converted signal as function of the delay between the two pulses gives important information about the decay.

The CARS technique can also be used for the study of relaxation processes in inhomogeneously broadened Raman-active transitions. For electric dipole transitions, however, techniques like photon echos or time-resolved hole burning provide (Levenson and Kano, 1988; Letokhov and Chebotayev, 1977) a more direct insight into this question. The photon-echo technique has been extensively developed to study coherent relaxation processes in the visible involving electronic transitions; in the infrared where vibrational transitions occur, the results have been scarce. There are several variants of the original photon-echo technique like the stimulated-echo and accumulated-echo techniques. Some attempts to observe Raman echoes have been made, but they seem inconclusive. The time-resolved hole-burning technique is straightforward. One measures the transmission characteristics of a weak short pulse tuned in the spectral range of a inhomogeneously broadened transition that has been previously excited by a very intense short pulse. The decay of the hole burned and its width give information about population and coherence relaxation times.

A very powerful time-resolved technique to study relaxation processes is the transient optical gratings (Eichler, 1980) technique, which can also be designated as real-time holography and was briefly discussed in Sec. 2.4. In its simplest configuration, two pulsed beams of the same frequency ω close to a transition of the medium interact in the medium via a refractive index or photoinduced absorption change and set up a phase or population grating respectively; a third pulsed beam at the same or a different frequency ω' delayed with respect to the first two diffracts off this pattern, and the evolution of the diffracted intensity as a function of the time gives information about the amplitude and phase relaxation of the grating and by the same token about energy migration and loss of coherence. There are several variants and extensions of the original transient grating technique described above. One such variant is the transient degenerate four-wave mixing or the transient optical phase conjugation, which can be described as the superposition of three gratings [compare Eq. (35)]; by appropriate delays between the pulsed beams and judicious choice of beam polarizations, one can study different relaxation and diffusion processes of the material excitation at frequency ω.

Another variant of the transient optical grating technique is the impulsive stimulated Raman technique with ultrashort pulses of duration τ_p in the few-femtosecond range. Such pulses have a frequency width $\Delta\omega_p \approx 1/\tau_p$ that can extend up to a few hundreds of inverse centimeters, and under certain conditions one can, with a single ultrashort pulse, coherently drive vibrational or librational or other low-lying states of frequencies that fall within $\Delta\omega_p$; diffraction of a single ultrashort pulse off the interference pattern set by the first pulse gives information about the relaxation processes of these low-lying states. There have been several investigations with this technique or similar ones. One has observed in particular Raman quantum beats and studied relaxation processes of low-lying vibrational modes.

All the previous techniques employ coherent pulses. There has been recently proposed and demonstrated the use of incoherent light to study relaxation processes and in particular extract information concerning the relaxation times T_1 and T_2. Here one exploits the fact that a temporally incoherent light of wide spectral width possesses a very short correlation time τ_c, much smaller than the light-pulse duration τ_p. In an autocorrelation measurement, such a light appears like a single pulse of duration τ_c. Accordingly, this technique consists in spatially spliting a pulsed light beam of frequency ω into two beams and temporally delaying the one relative to the other by τ before mixing them in a resonant medium. The energy of the output beams in the phase-matching direction is measured as a function of τ to obtain a kind of correlation profile associated with both the incident light and the resonant material response. The time resolution of this technique is set up by the light correlation time τ_c and not by the duration of the light pulse itself. This technique has been applied in some simple cases, but its exploitation on a larger scale and in more complicated cases is not that straightforward.

5.2.2 Nonlocal Time-Resolved Techniques The previous techniques are essentially designed to study relaxation processes of localized excitations in a molecular entity embedded in a condensed matrix. Here the energy and coherence relaxation occur within the immediate environment of the molecule. For delocalized excitations in more or less periodic molecular systems, the problem can be fundamentally different and also becomes more complex because the relaxation process within the immediate molecular environment can take place concurrently with the escape of the excitation from the excited molecule and its resonant transfer to other identical molecules. At present, there are two techniques that allow one to address this problem and to disentangle the relaxation from the propagation: the electro-optic Cherenkov effect (Auston *et al.*, 1984) and nonlocal (Gale *et al.*, 1986) (space resolved) time-resolved CARS. In addition, some of the previous techniques, for instance the transient optical grating techniques, can be used to study relaxation and diffusion processes of delocalized excitations.

The electro-optic Cherenkov technique with femtosecond pulses in the visible exploits the optical rectification or inverse electro-optic effect in a noncentrosymmetric crystal to generate an extremely fast far-infrared electromagnetic transient, a few femtoseconds in duration, and excite propagating infrared-active excitations in the same crystal. This technique already has been demonstrated for polaritons in ferroelectric crystals; its selectivity is, however, limited, and the separation of temporal and spatial features is not straightforward. Furthermore, this method is restricted to low-frequency polaritons, with an upper frequency limit given by the infrared femtosecond-pulse bandwidth.

The nonlocal time-resolved CARS technique does not suffer from these limitations. The technique has been recently proposed and demonstrated for the study of polariton pulses in a crystal without inversion center. In this technique, the coherent excitation and the probe stages of the CARS technique take place at different time and space points. Inside the crystal, it allows separation of relaxation and propagation effects on the spatiotemporal evolution of propagating polariton modes and can be used both for phonon and exciton polaritons. We wish to point out here that the storage, propagation, and redistribution of electromagnetic energy close to resonance in crystals are intimately connected with the features of polariton propagation and in particular the polariton coherence.

Time-resolved nonlinear optical techniques—local or nonlocal ones—are called upon to play an important role in the study of irreversible processes at the microscopic level and are presently extended for the study of non-Markoffian processes where memory effects are important. Such effects are expected to become relevant whenever the pulse duration is comparable to or shorter than the correlation time τ_c of the random processes.

5.3 Nonlinear Polarization-State Spectroscopy

Several nonlinear optical techniques in the frequency or time domain have been developed that exploit the tensor character of the nonlinear susceptibilities and the high selectivity that results by the control of the polarization state of the interacting beams. For instance, the polarization state of the signal in a four-beam interaction can be modified by intermediate resonances due to either two-photon or Raman transitions.

The space symmetries of the nonlinear susceptibilities are also extensively used for surface-specific diagnostic techniques (e.g., Shen, 1986, 1989). In this respect, extremely sensitive techniques based on second-order processes have been developed for the study of the constitution of the surface layer, the phase transitions that it may undergo, the bond configuration in reconstructed layers, the orientation of molecules adsorbed on surfaces, and other aspects.

6. CONCLUSIONS AND TRENDS

The field of nonlinear optics still undergoes rapid changes and cannot be cast in a form where conclusions can be easily drawn concerning its future evolution. We take the risk however to formulate certain trends for future research.

Although the field of nonlinear optics at first sight seems to have gained in coherence and grown to a discipline with conceptual and structural tools, it is important to stress that this has been done up till now by exploiting only a part of these initial assets. In the course of its rapid evolution, which up till now was partially dictated by a massive transfer and adoption of techniques and concepts from related areas and also by an effort to reach results quickly for applications and devices, a large part of its intrinsic aspects were never adequately addressed and others were ignored. This is to be blamed also on the lack of availability of laser sources with appropriate characteristics.

Thus, the previous presentation essentially relies on the dipolar approximation and the assumption that the interactions are local, which also implies that the nonlinearities are provoked by the anharmonicity. Thus, only the temporal variations of the fields are taken into account and not their spatial variations. This of course suppresses the nonlocality both in the interactions, as reflected in the Lorentz force, and in the intrinsic structure of Maxwell equations. The essential implication of nonlocality is that the optical properties become sensitive to the wave vector and new symmetry aspects appear that profoundly modify the interactions, both quantitatively and qualitatively, as can be inferred by the spatial dispersion and the Onsager relations in the linear regime; their generalization to the nonlinear regime has not been undertaken yet. Other manifestations of nonlocality are already present in nonlinear optics and attract growing attention. They result from the retardation that is inherently built into the propagation equation; the phase mismatch is such a manifestation but also the cascading processes that were briefly mentioned in the text. This class of retardation effects can certainly be accounted for in an *ad hoc* manner within the present framework; however, a more appropriate approach to include these effects and grasp their implications would be the introduction, from the beginning, of the true eigenstates of the matter-field system.

With the improvement of laser performance regarding intensity and pulse duration, another aspect that will receive a certain attention is the breakdown of the perturbative treatment of the response of the matter to such fields. There the condition that the average value of the dipole interaction $H_e = \boldsymbol{\mu} \cdot \mathbf{E}$ in Eq. (54) is smaller than the level spacing is not satisfied, and a nonperturbative approach is necessary; this is in particular the case when a continuum of states is involved. It does not suffice here to keep higher- and higher-order terms in a development of Eq. (53) or to use the Bloch equations (55), (56a), and (56b) as such, since the assumptions used to derive them are not valid any longer when very intense fields are acting within very short time scales. In particular, noninteger power dependence on the field amplitude or photon number is a common feature in many of the observed effects (for instance, in multiphoton ionization together with restructure of the continuum spectrum); for very short times, the Bloch equations must be reconsidered too for several reasons (correlation at initial time, memory effects, etc.). The whole problem of optical breakdown, a highly nonlinear problem, will be reconsidered in the regime of ultrashort light pulses. Note that in 1 fs the light travels less than 3000 Å in vacuum, much less than its wavelength in the optical range.

In addition, with very intense and ultrashort pulses the slowly varying envelope approximation must be reconsidered in particular in a medium with rapid spatial variations where in addition nonlocality may lead to quite complicated interaction configurations. In this respect, in the near future the field of nonlinear optics in random dielectric media will receive great attention in particular in connection with the photon trapping and localization problem. This will also affect the conditions for occurrence of instabilities and chaotic behavior that up to the present day have been studied in the framework of the Bloch equations and the slowly varying envelope approximation.

Along with these fundamental considerations, interest will be concentrated on the development of nonlinear optical materials and their implementation in nonlinear optical devices. It is a fact that the choice of such materials cannot reside only on the figures of merit for the nonlinear operation but must also take into account many other criteria crucial for the design, production, and maintenance of the nonlinear devices. Actually, in view of the rather limited nonlinear optical performances of the available materials, these other criteria will be the most crucial ones in making a choice. And this choice must be made with less ambitious goals than one would wish or had wished as our understanding of nonlinear integrated optics is progressing. There are now certain materials that can be used not only to conceive prototype nonlinear optical devices but also to proceed to their massive development and production and their insertion to large-scale technology.

At first hand, these concern materials and devices that exploit quadratic (second order) optical processes. $LiNbO_3$ and poled polymers here seem to provide a good starting point for such development. It is doubtful whether organic crystals at this stage can be used to initiate this process. There are still severe problems associated with their stability, long-term optical quality, etc.

The situation is less bright for materials and devices that will exploit cubic (third order) optical processes. Here either the nonlinearities are weak or they can be made large but with a high price paid in absorption losses and reduced operation speed. At present, only the composites offer some limited possibilities, but their fabrication technique must be substantially improved. The advantage of these materials is that one can artificially modify several of their properties by external control. Another possibility is to exploit cascading processes, but here we need better control of the different factors that affect these processes. The advantage with these processes is that one can use the same class of materials in devices that exploit quadratic and cubic nonlinear optical effects.

But one can still ask the question whether there is room for improvement of the figures of merit by an amount that will allow one to use much lower beam intensities and thus avoid some of the problems related to material stability under high fluences. This seems a difficult task as nonlinearities are associated with very loosely bound electrons and hence the material is apt to undergo irreversible photochemical changes. The relation between the nonlinearity and the photochemical stability is still not understood, and much work remains to be done there.

The nonlinear optical spectroscopic and diagnostic techniques will also continue to improve and be used for several investigations in gases, liquids, or solids but also in surfaces, plasmas, and other extreme conditions. With development of coherent x-ray sources, the field of nonlinear x-ray optics will become accessible, and here a whole class of fundamental investigations can be envisaged.

GLOSSARY

Dipole Approximation: The assumption that the wavelength is much larger than the charge-density extension.

Local Field Corrections: The difference between unweighted space-averaged macroscopic field in the Maxwell equations and the weighted microscopic field averaged over the extension of the polarizable charge density; they are related with the inhomogeneities of the latter.

Manley–Rowe Relations: Relations between the intensities and photon fluxes in three-wave interactions when all three frequencies fall within the same transparency region and the absorption losses are negligible.

Nonlinear Optics: The area of optics that is concerned with generation, modification, and propagation of electromagnetic waves in the optical range when the response of the material is nonlinear in the amplitude of the field.

Nonlinear Polarizabilities: The coefficients of the series expansion of the induced nonlinear dipole in integer powers of the amplitude of the effective field.

Nonlinear Spectroscopy: All spectroscopy techniques that exploit the several spatiotemporal features of nonlinear polarization and propagation and in particular the specific resonant and symmetry features of the nonlinear susceptibilities related to intermediate resonances.

Nonlinear Susceptibilities: The coefficients of the series expansion of the nonlinear polarization in integer powers of the amplitude of the electromagnetic field.

Optical Bistability, Instabilities, and Chaos: The nonlinear optical operation that leads to a double-valued or multivalued beam output for a single beam input when the nonlinear coupling is varied beyond certain values; eventually, this may lead to instabilities or transition to chaos.

Optical Kerr Effect: The ability of a medium to modify its refraction and absorption at a given frequency by the intensity of the electromagnetic field of same or different frequency; the misnomer photoinduced effect is also used.

Optical Phase Conjugation: The operation that replaces the polarization state and amplitude of a beam by its complete conjugate and its vector direction reversed.

Parametric Processes: Processes in which the nonlinear interaction involves a constant or very slowly varying amplitude field; the latter is assumed only to modulate the optical properties of the medium.

Phase Matching: This is related to the difference of the wave vectors of the polarization source and the electric field that the medium can sustain in the same frequency.

Rabi Frequency: The frequency at which precesses the transition dipole amplitude of a two-level system resonantly driven by an intense optical field.

Second- and Third-Order Processes: The ones that result from direct or indirect polarization sources second- and third-order respectively in the involved field amplitudes; it is implied that these are dipolar polarization terms.

Self-Action Effects: The class of effects that result from the ability of a beam to modify its own spatiotemporal characteristics through the optical Kerr nonlinearity that induces inside a medium.

Self-Induced Transparency: The ability of a beam to exploit the nonlinearity of a medium to redistribute and compensate its losses there and be transmitted in a stable form.

Slowly Varying Envelope Approximation: When used in nonlinear optics implies that the spatiotemporal modifications of the envelope of the field amplitude due to the nonlinearity are very weak over a carrier wavelength and period.

Solitons: Stable spatiotemporal beam forms that result from a balanced competition between a linear effect—for instance the dispersion—and a nonlinear one, most frequently the optical Kerr nonlinearity.

Stimulated Processes: In third-order processes, these result from resonant coupling and energy transfer between two different modes of the electromagnetic field and a material excitation; they can grow from the spontaneous (quantal) fluctuations of the material excitation. All and only odd-order processes can show such effects.

Transient (Dynamical) Optical Gratings: The periodic spatiotemporal modulation of the refraction or absorption by the nonlinear coupling of two waves through the optical Kerr nonlinearity.

Two-photon or Multiphoton Absorption: When by adding the energies of two or more photons one induces a real transition between two states in a medium.

Works Cited

Agrawal, G. P. (1989), *Nonlinear Fiber Optics*, Boston: Academic Press.

Agrawal, G. P., Cojan, C., Flytzanis, C. (1978), *Phys. Rev. B* **17**, 776–789.

Akhmanov, S. A., Khokhlov, R. V. (1962), *Sov. Phys. JETP* **16**, 252–254.

Akhmanov, S. A., Khokhlov, R. V., Sukhurov, I. (1975), in: T. Arecchi, E. O. Schultz-Dubois (Eds.), *Laser Handbook*, Amsterdam: North-Holland.

Arecchi, T., Schulz-Dubois, E. O. (Eds.) (1975), *Laser Handbook*, Amsterdam: North-Holland.

Armstrong, J. A., Bloembergen, N., Ducuing, J., Pershan, P. (1962), *Phys. Rev.* **127**, 1918–1939.

Auston, D. H., Cheung, K. P., Valdmanis, J. A., Kleiman, D. A. (1984), *Phys. Rev. Lett.* **53**, 1555–1558.

Bloembergen, N. (1964), *Nonlinear Optics*, New York: Benjamin.

Bordui, F., Fejer, M. M. (1993), *Annu. Rev. Mater. Sci.* **23**, 321–379.

Born, M., Huang, K. (1954), *Dynamical Theory of Lattices*, Oxford, UK: Oxford University Press.

Born, M., Wolf, E. (1975), *Principles of Optics*, Oxford, UK: Pergamon Press.

Boyd, R. W. (1992), *Nonlinear Optics*, New York: Academic Press.

Butcher, P. N., Cotter, D. (1991), *The Elements of Nonlinear Optics*, Cambridge, UK: Cambridge University Press.

Chemla, D. S., Zyss, J. (Eds.) (1987), *Nonlinear Optical Properties of Organic Molecules and Crystals*, Vols. 1 and 2, New York: Academic Press.

Chen, G. I., Liu, G. Z. (1986), *Annu. Rev. Mater. Sci.* **16**, 203–243.

De Martini, F., Ducuing, J. (1966), *Phys. Rev. Lett.* **17**, 117–119.

Digiorgio, V., Flytzanis, C. (Eds.) (1994), *Nonlinear Optical Materials. Principles and Applications*, Amsterdam: North-Holland Co., to appear.

Eichler, H. (1980), *Optical Gratings*, Berlin: Springer-Verlag.

Fabelinskii, I. L. (1968), *Molecular Scattering of Light*, New York: Plenum.

Felber, F. S., Marburger, J. H. (1976), *Appl. Phys. Lett.* **28**, 231; (1978), *Phys. Rev. A* **17**, 335–342.

Feynman, R. P., Vernon, F. L., Hellwarth, R. W. (1957), *J. Appl. Phys.* **28**, 49–53.

Fisher, R. A. (Ed.) (1983), *Optical Phase Conjugation*, Orlando, FL: Academic Press.

Flytzanis, C. (1975), in: H. Rabin, C. L. Tang (Eds.), *Quantum Electronics: A Treatise*, Vol. 1A, New York: Academic Press.

Flytzanis, C. (1987), in: D. S. Chemla, J. Zyss (Eds.), *Nonlinear Optical Properties of Organic Molecules and Crystals*, Vols. 1 and 2, New York: Academic Press.

Flytzanis, C., Hache, F., Klein, M. C., Ricard, D., Roussignol, Ph. (1991), in: E. Wolf (Ed.), *Progress in Optics*, Vol. 29, Amsterdam: Elsevier, pp. 321–411.

Flytzanis, C. (1993), in: W. Bron (Ed.), *Ultrashort Processes in Condensed Matter*, NATO Advanced Study Institutes, Series B, Physics, Vol. 314, Amsterdam: Kluwer Publ. Co.

Franken, P. A., Hill, A. E., Peters, C. W., Weinreich, G. (1961), *Phys. Rev. Lett.* **7**, 118–119.

Frey, J., Frey, R., Flytzanis, C. (1992), *Phys. Rev. B* **45**, 4056–4064.

Gale, G., Vallée, F., Flytzanis, C. (1986), *Phys. Rev. Lett.* **57**, 1867–1870.

Garmire, E., Pandarese, E., Townes, C. H. (1963), *Phys. Rev. Lett.* **11**, 160–163.

Gibbs, H. M. (1985), *Optical Bistability*, Orlando, FL: Academic Press.

Giordmaine, J. (1962), *Phys. Rev. Lett.* **8**, 19–20.

Grischkowsky, D., Loy, M. T., Liao, P. F. (1975), *Phys. Rev.* **12**, 2514–2533.

Gustafson, T. K., Taran, J. P. E., Haus, H. A., Lisfitz, J. R., Kelley, P. L. (1969), *Phys. Rev.* **177**, 306–313.

Hasegawa, A., Brinkman, J. (1980), *IEEE Trans. Quantum Electron.* **QE-16**, 694–697.

Hasegawa, A. (1989), *Optical Solitons in Fibers*, Berlin: Springer-Verlag.

Heitler, W. (1954), *The Quantum Theory of Radiology*, London: Oxford Univ. Press.

Hellwarth, R. W. (1977), *J. Opt. Soc. Am.* **67**, 1–3.

Kleinman, D. A. (1962), *Phys. Rev.* **126**, 1977–1979.

Landau, L. D., Lifshitz, E. M. (1960), *Electrodynamics of Continuous Media*, New York: Pergamon Press.

Laubereau, A., Kaiser, W. (1978), *Rev. Mod. Phys.* **50**, 607–665.

Letokhov, V. S., Chebotayev, V. P. (1977), *Nonlinear Laser Spectroscopy*, Berlin: Springer-Verlag.

Levenson, M. D., Kano, S. S. (1988), *Introduction to Nonlinear Laser Spectroscopy*, New York: Academic Press.

Lines, M. E., Glass, A. M. (1979), *Principles and Applications of Ferroelectrics and Related Materials*. Oxford: Clarendon Press.

Maker, P. D., Terhune, R. W., Nisenoff, M., Savage, C. M. (1962), *Phys. Rev. Lett.* **8**, 21–22.

Manley, J. M., Rowe, H. E. (1959), *Proc. IRE* **47**, 2115–2116.

McCall, S. L., Hahn, E. L. (1967), *Phys. Rev. Lett.* **18**, 908–911; (1969), *Phys. Rev.* **183**, 457–471.

Miller, A., Daino, B., Welford, K. (Eds.) (1995), *Nonlinear Optical Materials and Devices and Their Applications in Information Technology*, NATO ASI Series, Dordrecht, The Netherlands: Kluwer Publ.

Mollenauer, L. F., Stolen, R. H., Gordon, J. P. (1980), *Phys. Rev. Lett.* **45**, 1095–1098.

Newell, A. C., Moloney, J. V. (1992), *Nonlinear Optics*, New York: Addison-Wesley.

Osterberg, V., Margulis, W. (1986), *Opt. Lett.* **11**, 516–518.

Ostrowsky, D. B., Reinisch, R. (Eds.) (1992), *Guided Wave Nonlinear Optics*, NATO Advanced Study Institutes, Series E, Applied Science, Vol. 214, Dordrecht, The Netherlands: Kluwer Publ.

Rabin, H., Tang, C. L. (Eds.) (1975), *Quantum Electronics: A Treatise*, New York: Academic Press.

Rustagi, K., Ducuing, J. (1974), *Opt. Commun.* **10**, 258–261.

Sargent, M., Scully, M. O., Lamb, W. E., Jr (1974), *Laser Physics*, Reading, MA: Addison-Wesley.

Sauteret, C., Hermann, J. P., Frey, R., Pradère, F., Ducuing, J., Chance, R. R., Baughman, R. H. (1976), *Phys. Rev. Lett.* **36**, 956–959.

Schmitt-Rink, S., Chemla, D. S., Miller, D. A. B. (1989), *Adv. Phys.* **38**, 89–188.

Shen, Y. R. (1984), *The Principles of Nonlinear Optics*, New York: Wiley.

Shen, Y. R. (1986), *Annu. Rev. Mater. Sci.* **16**, 69–86.

Shen, Y. R. (1989), *Annu. Rev. Phys. Chem.* **40**, 327–350.

Slusher, R. E., Gibbs, H. M. (1972), *Phys. Rev. A* **5**, 1634–1659.

Stegeman, G. I., Sheik-Bahae, M., Van Stryland, E., Assanto, G. (1993), *Opt. Lett.* **18**, 13–15.

Tang, C. L., Flytzanis, C. (1971), *Phys. Rev. B* **4**, 2520–2524.

Whitham, G. B. (1974), *Linear and Nonlinear Waves*, New York: Wiley.

Wynne, J. J. (1984), *Phys. Rev. Lett.* **52**, 751–754.

Yariv, A. (1971), *Introduction to Optical Electronics*, New York: Holt, Rinehart and Winston.

Zakharov, V. E., Shabat, A. B. (1972), *Sov. Phys. JETP* **34**, 62–69.

Zel'dovich, B. Ya., Pilipetsky, N. F., Shknov, V. V. (1985), *Principles of Phase Conjugation*, Berlin: Springer-Verlag.

Further Reading

Agrawal, G. P. (1989), *Nonlinear Fiber Optics*, Boston: Academic Press.

Bloembergen, N. (1964), *Nonlinear Optics*, New York: Benjamin.

Boyd, R. W. (1992), *Nonlinear Optics*, New York: Academic Press.

Digiorgio, V., Flytzanis, C. (Eds.) (1994), *Nonlinear Optical Materials. Principles and Applications*, Amsterdam: North-Holland Co., to appear.

Eichler, H. (1980), *Optical Gratings*, Berlin: Springer-Verlag.

Fisher, R. A. (Ed.) (1983), *Optical Phase Conjugation*, Orlando, FL: Orlando Academic Press.

Flytzanis, C. (1975), in: H. Rabin, C. L. Tang (Eds.), *Quantum Electronics: A Treatise*, Vol. 1A, New York: Academic Press.

Gibbs, H. M. (1985), *Optical Bistability*, Orlando, FL: Academic Press.

Letokhov, V. S., Chebotayev, V. P. (1977), *Nonlinear Laser Spectroscopy*, Berlin: Springer-Verlag.

Levenson, M. D., Kano, S. S. (1988), *Introduction to Nonlinear Laser Spectroscopy*, New York: Academic Press.

Miller, A., Daino, B., Welford, K. (Eds.) (1995), *Nonlinear Optical Materials and Devices and Their Applications in Information Technology*, NATO ASI Series, Dordrecht, the Netherlands: Kluwer Publ.

Ostrowsky, D. B., Reinisch, R. (Eds.) (1992), *Guided Wave Nonlinear Optics*, Advanced Study Institutes, Series E, Applied Science, Vol. 214, Dordrecht, the Netherlands: Kluwer Publ.

Rabin, H., Tang, C. L. (Eds.) (1975), *Quantum Electronics: A Treatise*, New York: Academic Press.

Shen, Y. R. (1984), *The Principles of Nonlinear Optics*, New York: Wiley.

Whitham, G. B. (1974), *Linear and Nonlinear Waves*, New York: Wiley.

OPTICS, PHYSIOLOGICAL

RAINER RÖHLER, *University of Munich, Munich, Germany*

INTRODUCTION

Physiological optics is concerned with a scientific understanding and explanation of the visual process in the eye and in the neuronal pathways in the brain. Its ultimate goal is the understanding of visual pattern recognition and discrimination, but at present we are far from having reached this.

The main scientific methods in exploring the visual process are biophysics of the eye and anatomy of the neuronal pathway, neurophysiology, psychophysics of the visual system, and development of theoretical (mainly mathematical) models for the signal processing at particular levels or in the entire visual system.

In this article, the discussion is confined to the human visual system. In neurophysiology, direct access to this system obviously is confined to the surgery of brain injuries. Therefore, extrapolation from animal experiments is necessary.

The understanding of visual pattern recognition not only would provide a major insight into the function of the brain, probably it could also induce a breakthrough in improving our concepts for robot vision. Such an improvement would in turn improve the possibilities of automatic handling and quality control of worked articles in manufacturing lines. Even at a lower level of understanding, the knowledge of the properties of the human visual system is important for the compensation of visual deficiencies, for solving problems of illumination of working places, streets, vehicles, etc., for designing signal systems and color combinations for interior decoration, clothing, etc., and for numerous other applications.

3-527-28134-7/95/$5.00 + .50

1. BIOPHYSICS OF THE EYE

1.1 Optical System and Retinal Image

The optical imaging process performed by the optical system of the vertebrate eye, in particular the human eye, was first described correctly by J. Kepler in 1604. In the time before Kepler, philosophers and scholars considered the fact that simple optical imaging would generate an inverted image on the retina as an insurmountable obstacle, and they built very complicated models for an explanation of the fact that we see our environment upright. Chr. Scheiner demonstrated in 1619 that, with an enucleated eye, the retinal image in fact is inverted. Considering the modern knowledge of the neuronal image processing in the visual system, it is of course no problem to understand that this inversion can easily be compensated by a suitable wiring (similar to a TV setup). The first graphical illustration of the imaging process was given by R. Descartes in 1637.

A modern, somewhat simplified illustration of the optical system of the eye is given in Fig. 1. The components that perform the optical imaging are the *cornea* and the *crystalline lens*, in conjunction with the adjacent optical media: the *anterior chamber*, between cornea and crystalline lens, and the *vitreous chamber* between lens and retina. The cornea exerts the main part of the refractive power of the system that generates a sharp retinal image of the external world. In its optically relevant region, the cornea is nearly spherical.

The crystalline lens is the element that performs the *accommodation* of the eye. This is the capability of focusing on the retina objects that are located at different distances. The adequate variation of the shape and hence the refractive power of the lens is performed by the *ciliary muscle*. The lens capsule is elastic—at least in juvenile age—and tends to form a sphere in the relaxed state. It is hindered from that by fibers that connect the rim of the lens with the ciliary muscle, so that the lens is flattened. With an accommodation effort, the muscle slackens the fibers, so that curvature and refractive power of the lens increase. The total refractive power of the eye has an average value of 58 diopters with a variance of about 10 diopters.

In front of the lens is located the *iris*, which forms the aperture stop of the eye with a central opening, the pupil. The iris contains two antagonistic muscles, one of which encircles the pupil, while the other has radial fibers. The two muscles can contract or dilate the pupil, covering a diameter range from 2 to 8 mm. The actual size of the pupil depends on an equilibrium of these muscles and is mainly determined by the intensity of the illumination of the cornea. The pupil serves as a relatively quick mechanism for the regulation of the illumination of the retina (a few tenths of a second).

The retina forms the image surface of the eye. It covers a little more than the rear half of the eye capsule. In its center is the *fovea*, the site of the best *visual acuity*. The line connecting the center of the fovea with the center of the pupil is the *fixation line* or the *line of sight*. The complex structure of the retina is discussed in the next section.

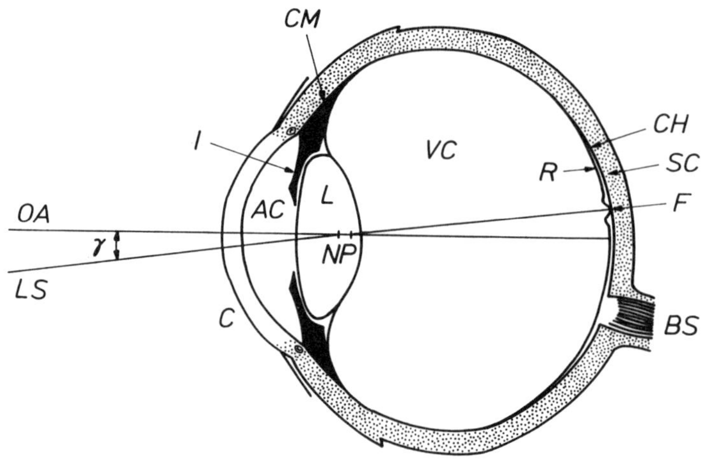

FIG. 1. Optical system of the eye. AC: anterior chamber; BS: blind spot; C: cornea; CH: choroid; CM: ciliar muscle; F: fovea; I: iris; L: crystalline lens; LS: line of sight; NP: nodal points; OA: optical axis; R: retina; SC: sclera; VC: vitreous chamber; γ: angle between LS and OA.

The eye is surrounded by a very tough tissue, the *sclera*, which gives the eyeball its mechanical stability. Another thin membrane, the *choroid*, is situated between sclera and retina. It consists mainly of blood vessels, which supply the outer layers of the retina. In direct contact with the choroid is the *pigment epithelium*, a heavily pigmented layer of the retina. It serves to absorb light that has penetrated the retina, so that backscatter is reduced. The sclera has an opening through which the optical nerve leaves the eyeball. This area is called the *blind spot*, because the retina has no photoreceptors in this region. Under normal visual conditions, the blind spot is not noticed subjectively as a result of a neuronal *fill-in mechanism*.

The eye is not a correctly centered optical system. The optical axis of the eye, i.e., the normal on the anterior surface of the cornea through the center of the pupil, includes an angle γ of about 5° of visual angle with the fixation line.

The central part of the cornea, with about 4 mm diameter, has a fairly good spherical shape. When the pupil is dilated to large diameters, other parts of the cornea contribute to the imaging process and introduce large spherical aberrations. Under usual visual conditions, large pupil diameters occur only under dim illumination conditions, when visual acuity is reduced. Therefore, large aberrations are tolerable.

Frequently, the eye is considered in some way as a photographic or television camera. But the eye is not a stationary, fixed-focus optical instrument. Under stationary conditions, only a visual field of about 4°–5° of visual angle is imaged on the fovea with an adequate visual acuity. To get an overview of larger areas of the visual environment, or to fix on moving objects, we have to perform eye movements, which are effected by six pairs of eye muscles. Several control processes are necessary to combine the mosaic of different foveal image elements into a coherent picture. The underlying problem is easily recognized when we try to get an overview over a room by looking through a keyhole: In this experiment, our control processes for combining image elements are not active.

Moreover, the importance of eye movements for vision is easily demonstrated by producing a stabilized retinal image. This is accomplished by suppressing involuntary eye movements that would cause shifts of the retinal image against the retina. Under this condition, every brightness or color structure fades away in several seconds.

An elegant method for the stabilization of the retinal image is to detect eye movements by a quick image-processing system, and the optical stimulus that is generated on a monitor is moved to compensate the motion of the retinal image caused by the eye movement.

Several types of eye movements can be distinguished (see Alpern, 1972): smooth movements, saccades (quick sudden jumps), and tremor. They serve several important functions of vision.

1. *Slow drift movements* during voluntary fixation: These are involuntary smooth movements with an average amplitude of ca. 2.5 min of arc, uncorrelated between the two eyes, and they seem to be responsible for preventing the fading of the retinal image just mentioned.
2. *Microsaccades*: These are rapid, irregular movements with amplitudes usually less than 20′. Probably they are necessary for correcting the drifting away of the fixation position.
3. *Vergence movements* (fusional movements): Since we have two eyes, which provide us with stereoscopic vision, the optical axes of our eyes have to be adapted to the distance of the fixated object. This is an involuntary process that is related to accommodation and needs about 0.8 s to change convergence between objects of different distances.
4. *Version movements*: These are parallel motions of both eyes. Voluntary *saccadic movements* occur (partly in combinations with vergences) when the view is directed to another area of interest or when the visual field is explored over larger areas. Involuntary smooth versions occur during tracking of a moving object.
5. The *vestibular-oculomotoric reflex*: When movements of head or body are made, dramatic shifts of the retinal image are induced. Under certain conditions, reflectory eye movements are made to compensate the shift of the image, at least partially.
6. *Tremor*: This occurs under steady fixation. It is a high-frequency movement (70–90 Hz) of low amplitude (ca. 30 s of arc). It

is not clear how such movement is related to vision or visual acuity.

The neuronal mechanisms involved in the control processes of eye movements are very complicated and by no means completely understood. They will not be discussed in this article (Lennerstrand and Bach-y-Rita, 1975).

Visual deficiencies that result from defects of the optical system of the eye can easily be explained. With increasing age, the crystalline lens gradually loses its elasticity. Therefore, an accommodation effort does not result in a corresponding increase of the refractive power. This is called *presbyopia*. As a consequence, older people have difficulty in accommodating near objects. They need, e.g., convex lenses for reading.

The main reason for other deficiencies is a discrepancy between the length of the eyeball and the optical refractive power (see Fig. 2). The eye may be too short, so that under normal accommodation the image plane of the system lies behind the retina. This is the case of *hyperopia*. Such persons have to make an accommodation effort to bring distant objects into sharp focus on the retina. Near objects then may lie out of the accommodation range. Naturally, this difficulty increases with increasing presbyopia. Convex lenses are needed for the correction of this deficiency.

When the eyeball is too long, a sharp optical image of distant objects lies in front of the retina. Near objects are focused on the retina with no or little accommodation effort. This defect is called *myopia* and may be corrected with concave lenses.

Moreover, the cornea may be deformed to an aspherical shape. In simple cases, the deformation causes astigmatism, which can be corrected by compensating astigmatic lenses. In more complicated cases, the astigmatic correction serves as a first approximation. Contact lenses may serve better.

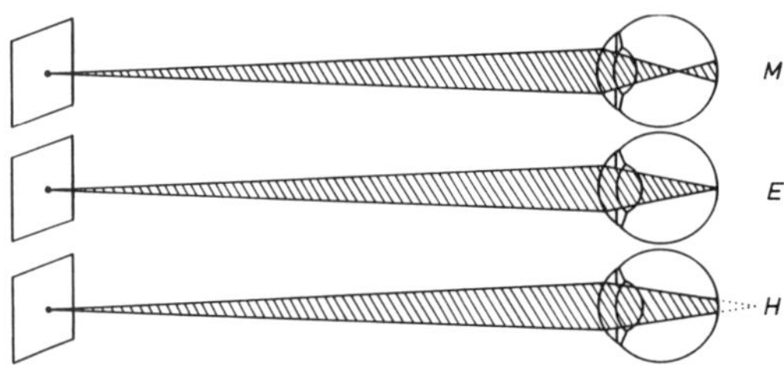

FIG. 2. Visual deficiencies. M: myopia; H: hyperopia; E: emmetropia (for comparison).

Several methods for the diagnosis of the refractive state of the eye are available. The simplest is the use of probing glasses: The subject has to decide what glasses provide him with optimal visual acuity. Of course, this is the ultimate criterion, but it depends on the strategy of the experimenter whether he arrives at the best combination of glasses in a reasonable time.

An objective method is to analyze light that is reflected from the rear of the eye. The *ophthalmoscope*, invented by H. v. Helmholtz, is perhaps the simplest of these instruments, at least in its original form. Detailed descriptions can be found in any textbook on ophthalmology or optometry. Several refinements of this principle have been realized, including automatic analysis of the reflected light.

1.2 Structure and Function of the Retina

The retina has a rather complicated structure, which has been modified significantly during vertebrate evolution. In the following, only mammalian and, more specifically, primatian retinae will be considered (Bornschein and Hanitzsch, 1978). The retina is the site of the photoreceptors, which transform the energy of absorbed light into electrical neuronal excitation. In vertebrate retinae, roughly two types of photoreceptors can be distinguished: *rods* and *cones*. The rods are very sensitive to light and are active mainly under dim illumination conditions. They do not provide color vision. Cones are less sensitive than rods; they are active in daylight. Primates have three types of cones, which differ in their absorption spectra. From the signals of these different cones, the information for *color vision* is extracted. The cones are concentrated mainly in the fovea. In the central region of the fovea, called the *foveola*, which covers an area of ca. 1.3° of visual angle, only cones are present. In outer parts of the fovea, which extend to about 2.5° of visual angle, rods are also present. In more peripheral regions of the retina—the *parafovea*, extending to about 25°—the fraction of rods increases gradually. In the outer periphery, cones are very rare, and no color vision is possible in this region. Because of the absence of rods in the foveola, the fixation area is blind in dim light. This can be probed by trying to fix on a faint star on the nocturnal sky.

The structure of mammalian photoreceptors is demonstrated in Fig. 3. Two main parts can be discriminated, the *inner segment*, which contains all the usual organelles of a cell such as nucleus and mitochondria, and the *outer segment*, which contains the photopigment. These two segments are connected by a thin *cilium*. The photopigment in the outer segment is embedded in a membrane system, consisting of stacks of flattened sacks. The photopigment is a protein with a chromophore, *retinal*. The protein has slightly different structures in the several types of receptors; this is the basis for the differences in the absorption spectra.

The spectral sensitivities of human rods and cones are shown in Fig. 4. The sensitivity of rods can be measured easily by a psychophysical method to be explained in Sec. 4.1. The photopigment can be extracted from the retinae of enucleated eyes and the absorption spectrum measured spectrophotometrically. The combined sensitivity of all types of cones also can be measured psychophysically. It may

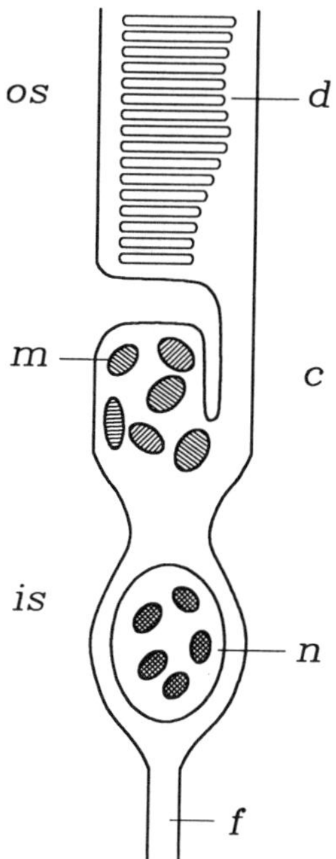

FIG. 3. Mammalian photoreceptor. c: cilium; d: discs; f: nerve fiber; is: inner segment; m: mitochondria; n: nucleus with chromatin; os: outer segment.

be seen from Fig. 4 that this overall sensitivity is shifted to longer wavelengths as compared with the sensitivity of rods. With rod vision, pure red light is not detected. The brightness ratio for light of short or long wavelengths is larger for rods than for cones (the so-called *Purkinje shift*).

It is more difficult to measure the spectral sensitivities of the different types of cones. One useful method is *fundus reflectometry* (Rushton, 1964). By this procedure, the spectral composition of light that is reflected from the central fovea is analyzed. Then the photopigment is bleached by strong illumination, and another analysis of the reflected light is made. The resulting observed difference is due to a decrease of photopigment concentration. Individuals with deficient color vision lack one of the three cone pigments. With this information the individual absorption spectra of different cone types can be isolated.

Another more straightforward method has been applied more recently: namely, the photometric measurement of the spectral absorption of single cones (Marks *et al.*, 1964). These measurements are rather difficult, since the light that is needed for the measurement bleaches the receptors and thereby changes the absorption spectra. Sophisticated computer-assisted methods have been used for correcting such errors.

The detailed molecular processes during photodetection are best known for rods with their photopigment *rhodopsin* (Dartnall, 1962). Absorption of a light quantum by a rhodopsin molecule initiates a series of conformation changes of the molecule, which on their part induce further complicated enzymatic reactions in the receptor cell (Kühn, 1980). The first step in the conformation changes of the molecule is a steric isomerization of the retinal. Its normal structure in the molecule is the 11-cis configuration (see Fig. 5). The effect of the absorption of a light quantum is to change this structure into an all-*trans* configuration (light isomerization). Following this change, the molecule undergoes a series of intermediate states until finally the all-*trans* retinal separates from the molecule, leaves the receptor cell, and is regenerated to its original 11-*cis* configuration in the pigment epithelium. The ultimate consequence of the light absorption is a change of the electrical potential of the cell membrane, i.e., the generation of a receptor potential. In contrast to

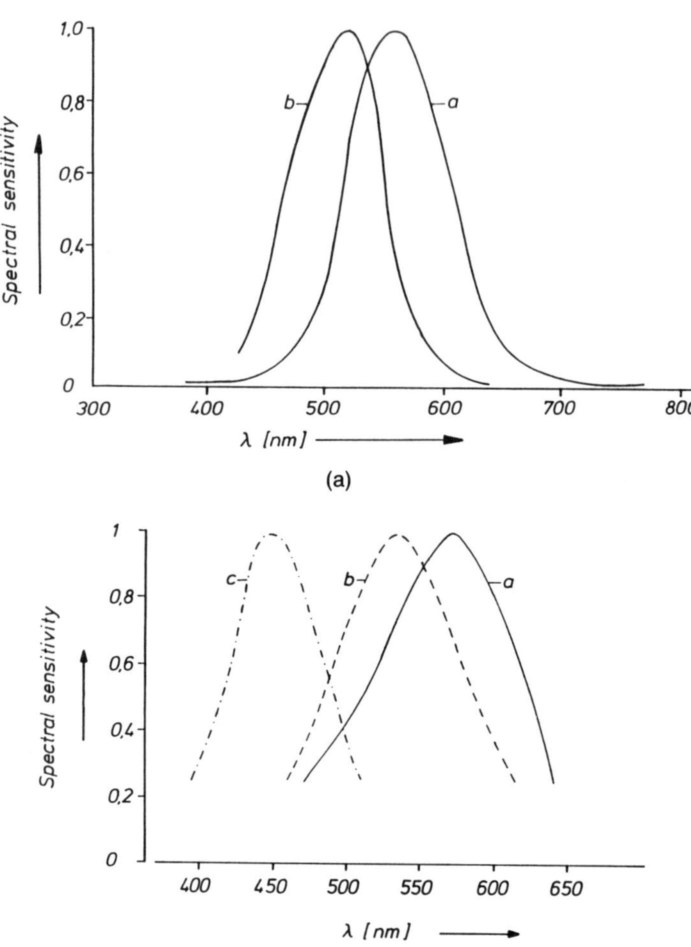

FIG. 4. Spectral sensitivities of human photoreceptors. (a) Curve a: overall sensitivity of cones (V_λ, maximum at 550 nm), curve b: sensitivity of rods (V'_λ, maximum at 507 nm); (b) sensitivities of three types of cones. Maxima are normalized to unity.

most receptors, an excitation causes a hyperpolarization, not a depolarization. The hyperpolarization means that, in darkness, many membrane pores of the cell are opened for ionic currents. During light excitation, these pores are closed. Therefore, considerable metabolic energy is expended to maintain an ionic dark current. A plausible theory to explain the physiological advantage of this effort is that, similar to electronic amplifiers, a rather high dark current ensures a quick and steep detector characteristic.

With a quantum efficiency of about 50%, the absorption of one light quantum in a rhodopsin molecule generates a detectable receptor potential. Since 10^3 to 10^4 ions have to pass the cell membrane for the generation of a local receptor potential, a nearly noise-free amplification of this order of magnitude has to be performed by the cell. The mechanism of this process has been clarified only recently (1980 and later). Several enzymatic reactions have to cooperate in a cascade. The most important participants are a guanosine-nucleotide binding protein (transducin), guanosine-triphosphate, phosphodiesterase, and cyclic guanosine-monophosphate (see, e.g., Kühn, 1980). The detailed process cannot be explained here (but compare the article BIOPHYSICS).

Not only is the retina the site of photoreceptors, but also the first steps of neuronal signal processing are executed here. Therefore, the retina is generally considered as a part of the brain. This is very important, since the retina is accessible to visual inspection with an ophthalmoscope, so that beginning defects of blood vessels or neurons in the brain cortex may be detected.

Each receptor delivers its signals to a *bi-*

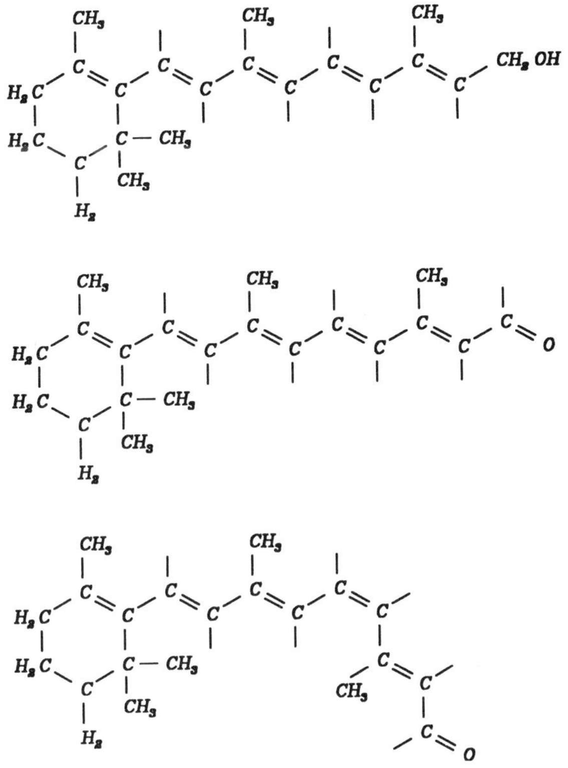

(a)

(b)

(c)

FIG. 5. Chemical structure of retinal. (a) Structure of vitamin A (for comparison); (b) all-*trans* retinal (excited state); (c) 11-*cis* retinal (unexcited state).

polar cell, but usually several receptor signals are led to the same bipolar cell, separate for rods and cones. Rod bipolars and cone bipolars send their signals to *ganglion cells,* again with a convergence of signals from different bipolars as sources to the same ganglion cell. Besides this convergence of signals, also a divergence takes place, since one receptor may send its signals to different bipolars and ganglion cells. Apart from these "vertical" neuronal connections, "horizontal" connections exist between different receptors and receptor types. They are mediated by *horizontal cells.* Another type of horizontal connection is mediated by *amacrine* cells on the ganglion cell level.

Neuronal signals from receptors, bipolars, horizontals, and amacrines are slow, graded potentials. Only the ganglion cells generate spike discharges. They send their axons to higher centers in the brain, as will be discussed in the next section. The retina contains approximately 1.2×10^8 receptors, 6×10^6 of which are cones. On the other hand, the retina has 10^6 ganglion cells, and the same number of axons leaves the eyeball, forming the *optical nerve.* Hence, on the average, an information compression to a fraction 10^{-2} takes place already in the retina. However, the ratio of compression depends on the retinal locus. According to recent results, in the very center of the fovea an extension rather than a compression of information takes place. The underlying strategy will be discussed in following sections.

Roughly two types of ganglion cells can be discerned, some with relatively small cell bodies and small receptive fields (parvo cells) and some with larger cell bodies and larger receptive fields (magno cells). They mediate different modalities of visual information, as is explained in more detail in later sections (Zrenner *et al.*, 1990).

2. NEURONAL SIGNAL PROCESSING

2.1 Methods

2.1.1 The Electroretinogram (ERG) This is the simplest method to detect a neuronal activity of the visual system. In darkness or under moderate constant illumination, the eye is an electrical dipole, the cornea being about 10 mV positive relative to the rear pole of the eye. This electrical activity is attributed to neuronal activities in the retina or in the pigment epithelium or both. Under illumination by a light flash of a few seconds duration, this potential shows temporal variations of less than 1 mV in both directions. These variations are classified into several components (Bornschein and Hanitzsch, 1978).

The ERG can be recorded by attaching a corneal lens with an electrode onto the cornea, the negative electrode being fixed to a conducting tissue in mouth or nose. A local ERG can be recorded when the light stimulus is focused on selected small areas of the retina. In order to discriminate the very small signals generated by local illumination from noise, elaborate correlation methods of signal processing have to be used. The easy accessibility of the ERG makes it a useful medical method for the early diagnosis of degenerative retinal diseases (e.g., retinitis pigmentosa).

2.1.2 Visually Evoked Cortical Potentials (VECPs) A visually evoked neuronal activity in the visual cortex may be recorded with electrodes that are fixed at the skin of the skull just above the visual cortex, i.e., at the back of the head. To distinguish these potentials from other neuronal activities of the cortex, the optical stimuli (e.g., checkerboards, steps in luminance, color, or structure) have to be flickered, and a lock-in summing and averaging technique has to be employed (Riggs and Wooten, 1972). Reference to this method will be made in later sections.

2.1.3 Recordings from Microelectrodes Microelectrodes are glass capillaries, the tips of which have a diameter of less than 1 μm. They are filled with an electrolytic solution, mostly with KCl, and can penetrate cell membranes without destroying the cell. Thus, intracellular electrical potentials can be recorded with them. In exploring the visual system with this method, only animals can be used. The electrode is moved through the opened skull into the neuronal region under investigation (e.g., the retina, higher parts of the visual pathway, or the visual cortex). The gaze of the immobilized animal is directed toward a screen, which is illuminated by optical stimuli, either steady or flickering. The usual technique is to immobilize the animal with curare and to maintain respiration and heart beat artificially. When the electrode has penetrated a cell that reacts to optical stimuli, its receptive field, i.e., the retinal area to which the cell is sensitive, may be explored by moving a local stimulus over the screen.

The electrical reaction of single cells can be recorded with this method. By directing the electrode with the help of a micromanipulator along a predescribed path, the properties of cells on this path can be tested in succession, thereby exploring signal processing structures in cellular assemblages.

Microelectrode recording can be considered as the most powerful electrophysiological method for studying signal processing in the visual pathway. Frequent reference will be made to this method in the following text.

2.2 The Structure of the Neuronal Visual System

A schematic drawing of the structure of the visual system is shown in Fig. 6 (see also Baumgartner, 1978). The axons of retinal ganglion cells are grouped into the optical nerve, which leaves the eyeball and leads the signals to the lateral geniculate body or corpus geniculatum laterale (CGL), the next higher neuronal structure. Before reaching the CGL, the axons have to pass the *chiasma*, a structure so called because of its similarity to the Greek letter chi. Here the fibers from both eyes are sorted so that the fibers from the right halves of both eyes are sent to the

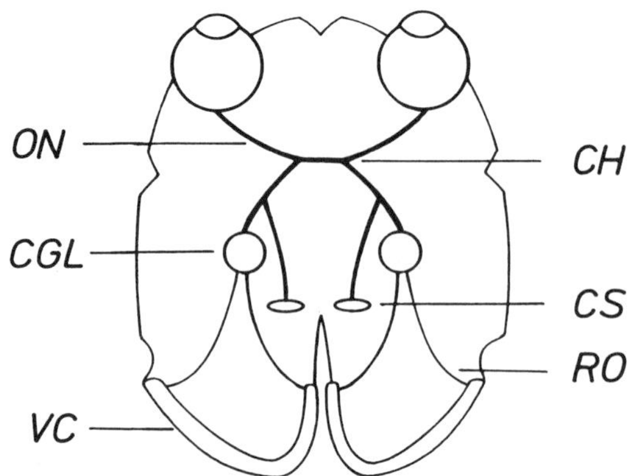

FIG. 6. Structure of the visual system (schematic). ON: optical nerve; CH: chiasma; CGL: corpus geniculatum laterale; CS: colliculus superior; RO: radiatio optica; VC: visual cortex.

right CGL and correspondingly for the left halves. So (because of the inversion of the optical image in the eye) the information from the left half of the visual field is processed in the right half of the brain, while on the other hand information from the right half of the visual field is processed in the left half of the brain. This seems to be important, since in binocular vision the information from both eyes has to be integrated into a stereoscopical image. Naturally, this is facilitated when corresponding neurons from both eyes are close together.

The CGL is the first neuronal "switching station" after the retina. Here the fibers of the optical nerve are connected with fibers of the *radiatio optica*, which runs to the optical cortex. This connection is mediated by interneurons and relay cells, which probably receive signals from the visual cortex by *corticofugal* neuronal fibers, so that the cortex can influence these connections by changing the amplification and setting up spatial filters.

The CGL can be divided into a parvo and a magno cellular subsystem. The parvo system gets its input from parvo retinal ganglion cells. It is a color-sensitive system with high spatial but low temporal resolution. The magno system gets its input from magno retinal ganglion cells; it has a high temporal and a low spatial resolution and is not color sensitive.

The visual cortex is the site of neuronal structures mainly responsible for a signal processing that finally generates visual perception. Several subregions, so-called areas, may be discriminated: They are classified as areas 17, 18, 19. Another more detailed classification is V1, V2, V3, V4, V4a, V5. The fibers of the radiatio optica first reach area 17 or V1 from where the signals are distributed into the different areas. Multiple interconnections exist between different areas. The imaging process from the retina over the CGL into area 17 is retinotopic, i.e., neighboring retinal regions are represented in neighboring area-17 regions. However, the process does not preserve the metric. According to the dominating importance of the fovea, a foveal region with an extension of 1° of visual angle is represented in area 17 in a region of 6-mm extension. A similar retinal region situated 10^0 extrafoveally occupies only 1 mm of area 17. V1 has a substructure, where different visual modalities are processed in parallel: form, color, and motion.

The signal channel from the retina over the CGL into area 17 is called the primary visual channel. Moreover, a secondary channel exists, which leads optical fibers to the *colliculi superiores*, a region in the midbrain. From here, connections lead to the visual cortex as well as to motor centers that are responsible for eye and body movements and to centers that control body balance and spatial orientation. This channel also mediates spatial information on visual targets. Roughly speaking, the primary channel supplies information on "What do I see?" and the secondary channel on "Where do I see something?". Even when the primary channel is interrupted between CGL and cortex, some residual information on "Where?" may be available.

2.3 Receptive Fields and Spatial Frequency Channels

As is already evident from the structure of the retina, usually several or many receptors have connections to the same retinal ganglion cell. This means that the set of these receptors forms the receptive field of their ganglion cell. Because of horizontal connections in the retina by horizontal and amacrine cells, the receptors of the receptive field do not contribute proportionally to their own excitation to the excitation of the ganglion cell. This is demonstrated in Fig. 7, where the rate of neuronal spikes of a retinal ganglion cell is plotted against the position of a small light spot, which is moved across the receptive field. When the light spot is outside of the receptive field, only the pulse rate of the resting state is observed. When the spot reaches the area of the receptive field, the rate drops below the resting level. This phenomenon is very common in neuronal signal processing and is called lateral inhibition. By lateral connections in the retina, the excitation of a receptor in the periphery of a receptive field inhibits the excitation (or the resting pulse rate) generated from the receptors in the field center (see also Fiorentini *et al.*, 1990).

By averaging and smoothing these records, one arrives at a profile of the receptive field sensitivity, which is shown in Fig. 8. Since the receptive fields of retinal ganglion

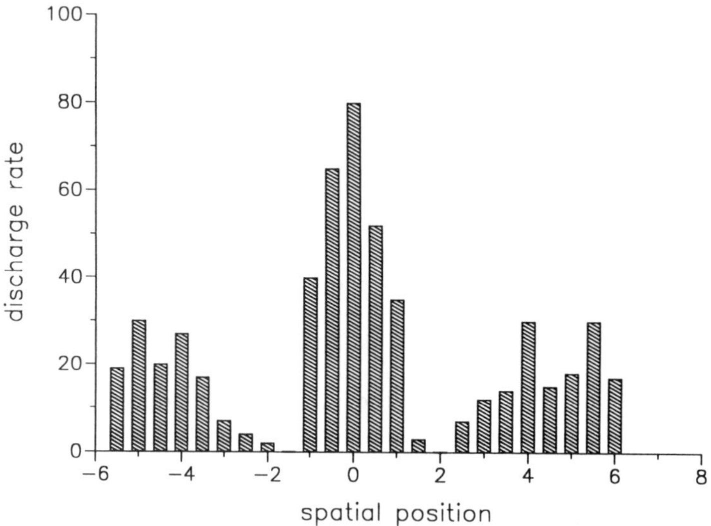

FIG. 7. Typical receptive field of a retinal ganglion cell. The neuronal discharge rate is plotted as a function of the spatial position of the light stimulus on the retina in relation to the field center. Both coordinates are in relative units, since they depend on the retinal locus and the stimulus strength.

cells have circular symmetry, the profile of Fig. 8 has been named the "Mexican hat". For mathematical modeling, an approximation of this structure by two Gaussian functions of different signs seems to be adequate:

$$f(r) = a_1 e^{-r^2/2\sigma_1^2} - a_2 e^{-r^2/2\sigma_2^2}, \tag{1}$$

where r is the distance from the field center, and a_1, a_2, σ_1, σ_2 are constants to be adapted to the experimental data. The first term on the right-hand side describes the excitatory field center, the second term the inhibitory field periphery. For an example, see Fig. 8.

Besides the convergence of neuronal connections from receptors to ganglion cells, it is also evident that one receptor usually makes connections to several bipolars and in turn to several ganglion cells. Therefore, also a divergence of information takes place, implying that receptive fields of neighboring ganglion cells overlap.

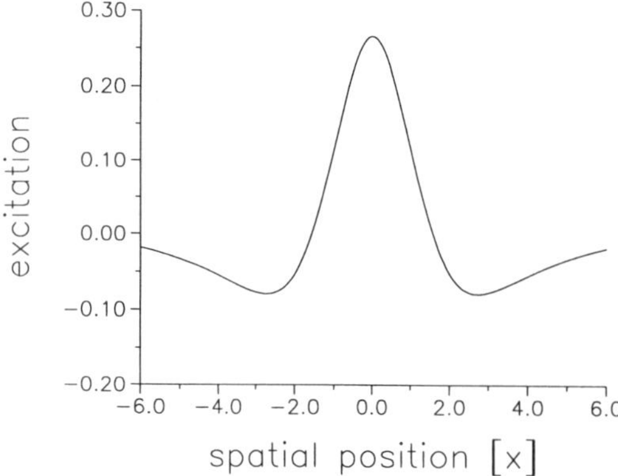

FIG. 8. Analytical approximation of an on-center receptive field profile. Coordinates as in Fig. 7. The zero level represents the resting discharge rate.

Another type of retinal ganglion cells shows properties that are *antagonistic* to the behavior just described. A light stimulus in the field center inhibits the cell, a stimulus in the periphery causes excitation. These reactions are best observed when the test light spot is superposed on a constant background of illumination. The first cell type is called "on-center cell," the second "off-center cell."

It can be seen that the existence of such pairs of antagonistic cells is essential for effective image processing. Because of adaptation, the resting discharge rate of a neuron is rather low, even under illumination with a bright stationary background. Therefore, an on-center cell has only a very limited dynamic range for dark signals. With two types of antagonistic cells available, bright signals can be managed by the on system, dark signals by the off system. At a higher level, these signals have to be set off against each other. Even a small number of on-off-center cells exists. They show a short, quickly adapting response to on as well as to off signals.

Apart from this differentiation between on and off cells, further differences among retinal ganglion cells have been detected, but here variations between different species have to be taken into account. In cats, the organization is relatively simple. Cells that have been referred to so far belong to the X-cell type. These cells have a receptive field of circular symmetry and linear summation of excitatory and inhibitory signals. The Y cells also have circular symmetry but no linear summation. Furthermore, the X cells have good spatial and rather poor temporal resolution, these properties being reversed in the Y cells. In addition, a W type exists, the receptive field of which has no circular symmetry. In primates, a similar antagonistic organization into P (parvo) and M (magno) cells exists, as has been mentioned already (Lennie *et al.*, 1990). They also have different functions in color vision and temporal signal processing, which will be discussed in more detail in the corresponding sections.

Since in physical optics during recent decades a description of imaging processes with the help of Fourier transformation has been proven helpful, this concept has also been applied to imaging processes within the visual system, and even here successfully. Considering the resting discharge rate as the zero level, Eq. (1) is easily transformed according to Fourier. For receptive fields of circular symmetry, it is sufficient to consider a one-dimensional transformation. Substituting a linear variable x for the circular variable r, the transform of $f(x)$ in Eq. (1) is a function

$$F(u) = \int_{\infty}^{-\infty} f(x)e^{-2\pi ixu}\,dx \qquad (2)$$

or

$$F(u) = \sqrt{2\pi}\,[a_1\sigma_1 e^{-2\pi^2\sigma_1^2u^2} - a_2\sigma_2 e^{-2\pi^2\sigma_2^2u^2}]. \qquad (3)$$

The transform of the example given in Fig. 8 is shown in Fig. 9. As is well known from physical optics or signal theory, the transform of a line-spread function with negative side lobes is a frequency characteristic with bandpass structure, details depending heavily on the constants. For neurons, "negative" means inhibitory responses.

To test the idea that the visual system may have neurons that react specifically on targets with selective spatial frequency structures, grids with various spatial frequencies have been presented as optical targets to immobilized cats and the reactions of retinal,

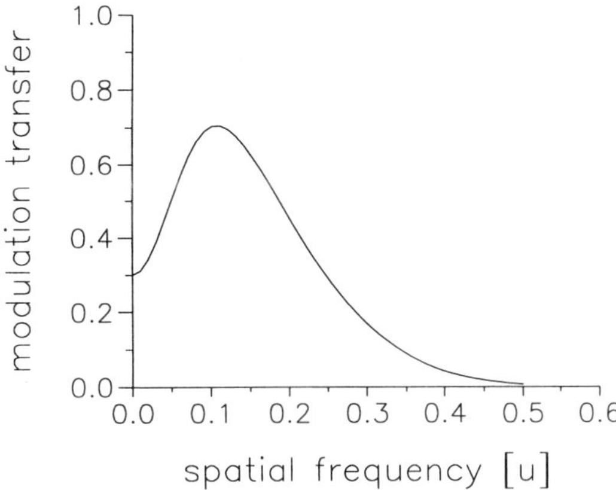

FIG. 9. Spatial frequency profile of the on-center receptive field according to Fig. 8. The modulation transfer function (see Glossary and Sec. 4.3) is plotted against the spatial frequency u (relative units). Since this function is symmetrical for $\pm u$, only the part for positive u is shown.

CGL, and cortical ganglion cells have been recorded by microelectrodes inserted into the cells (Maffei and Fiorentini, 1973). The results show that retinal on-center cells of the X type show a weak bandpass structure, while in CGL and cortical cells the bandpass structure is much more pronounced. The sensitivity of such cells obviously is tuned to an individual spatial frequency that is the center of a very limited sensitive spatial frequency band. Such cells or their underlying mechanisms are called "spatial frequency channels." A detailed discussion of these results will be given in the following section in connection with their psychophysical counterparts.

Receptive fields of cortical neurons have a more complicated structure than retinal or CGL cells. On the basis of microelectrode recordings, they have been classified by Hubel and Wiesel (1970) into *simple, complex*, and *hypercomplex* cells. Most neurons of area 17 are of the simple type. They are no longer circularly symmetric but have a pronounced preference for a specific orientation of the stimulus. Optimal stimuli are bars or grid structures or bright-dark borderlines with that specific orientation. The stimuli have to be moved over the receptive field in order to generate a detectable excitation. Stimuli with other orientations generate lower or negligible excitations.

The cortex is organized into a three-dimensional array of columns of neurons. In those columns that are perpendicular to the surface of the cortex, the neurons are sensitive to the same angular stimulus orientation. In columns that are parallel to the surface, the specific orientation varies, while the retinal locus remains constant. When moving the electrode again parallel to the surface but perpendicular to these orientational columns, a systematic variation of the corresponding retinal loci is observed. Moreover, the paths in this direction are subdivided into regions with different dominances from the two eyes.

This organization of area 17 is clearly observed only in cats. In higher animals, the distribution of simple cells in area 17 and complex cells in areas 18 and 19 is not so clear. Complex cells show optimal reactions to stimuli with definite size and angular structures. Not only the orientation of a stimulus but also the direction of movement may be of importance. Hypercomplex cells have an even more complicated receptive field structure (von Seelen *et al.*, 1987).

Some further comments on this subject will be given in the next section in connection with local spatial frequency analysis.

3. PSYCHOPHYSICAL METHODS

3.1 Threshold Definition and Measurement

Psychophysics is confronted with two problems: the physical definition of the input signal to the visual system and the measurement of the output signal, which is the visual perception. Clearly, the latter is the central problem of this method, but also the input problem has to be analyzed carefully and will be discussed first.

Physical optics usually deals with the transmission and detection of radiation energy; physiological optics deals with the transmission and perception of light. In this sense, light is radiation energy in the visible part of the spectrum, weighted according to the spectral sensitivities of the eye V_λ or V'_λ (see Fig. 4) under high or low illumination, respectively. V_λ and V'_λ are spectral weighting factors, normalized to unity at their maximum [see Eq. (5)]. Therefore, the physical units for the measurement of radiation have to be completed by a set of analogous psychophysical units for the measurement of light.

Let Φ_e be the radiant flux (the radiant power that is emitted by a light source):

$$\Phi_e = \int_0^\infty \Phi_e(\lambda) d\lambda. \tag{4}$$

The corresponding psychophysical unit is Φ_v, the *lumen*, i.e., the luminous flux emitted by a standard light source. The connection between radiant flux and lumen is given by

$$\Phi_v = K \int_0^\infty V_\lambda \Phi_e(\lambda) d\lambda. \tag{5}$$

Here K is the conversion factor between watt and lumen:

$$K = 680 \text{ l/W}. \tag{6}$$

Other important psychophysical units are as follows:

- *illuminance* E_v [lux] = [lm/m^2], corresponding to *irradiance* E_e = [W/m^2], m^2 being the projected area of the illuminated (irradiated) area.
- *luminous intensity* I_v [candle, cd] = [lm/sr], corresponding to *radiant intensity* I_e = [W/sr], sr being the unit solid angle in steradian.
- *luminance* L_v = [lm/(sr · m^2)] = [cd/m^2] corresponding to *radiance* L_e = [W/(sr · m^2)], m^2 being the projected area of the source.

Viewed from the input side, in psychophysical optics the most important unit is luminance, since usually extended test fields are used. From the luminance of the test field, the illuminance of the cornea can be calculated or measured without difficulty. However, from a physiological point of view, the illuminance of the retina is the dominating unit that determines adaptation and visual performance. Corneal and retinal illuminances are not proportional, since the pupil diameter, and therefore the aperture of the eye, depends on the retinal illuminance. Since the functional relation between pupil diameter and retinal illuminance is very well known (including the dependence on the age of the observer), the influence of pupil size can be eliminated easily. This is accomplished by the introduction of a unit *troland*. A troland is the retinal illuminance that is caused in a normal eye with a pupil area of 1 mm^2 by an extended source of 1 cd/m^2 luminance.

The output signal of the visual system is the visual perception. In order to get quantitative results, the visual perception has to be quantified. One important method to do this is the measurement of thresholds. A threshold is the just discriminable difference between two stimuli. An example is, viewing a bipartite photometer field, to discriminate between a homogeneous field and two halves different in luminance or color values. Evidently, such judgments are purely subjective, so that such tests have to be made many times, with many subjects, and under controlled conditions.

The basic "law" that has been established in psychophysical threshold measurement is *Weber's law*, which was found by E. H.Weber in 1834:

$$\Delta G/G = \text{const.} \tag{7}$$

Here ΔG is the difference that has to be added or subtracted from a permanent stimulus to be just noticeable. For large values of G, Eq. (7) fails because of physiological saturation effects.

The specification of these relations as "laws" may well be questioned from a physical standpoint, since they cannot be deduced from a very general "principle." Nevertheless, Weber's law is one of the psychophysical relations that are confirmed experimentally in several sensory modalities, especially in vision, and is parallelled by the sensitivity dependence of physical analog measuring devices.

3.2 Suprathreshold Vision

Measurement of thresholds is the oldest and the most reliable psychophysical method. But visual tasks in the threshold region are somewhat unusual in normal life; generally, they are limited to artificial laboratory conditions. Normal visual tasks are related to the detection and interpretation of luminance or color differences far above threshold.

First, in 1860 G. T. Fechner attacked this problem by integrating Weber's law. To do this, he assumed that thresholds can be added linearly to give the size of a suprathreshold sensation. So he replaced the constant in Weber's equation by $\Delta\Psi$, Ψ being the strength of sensory excitation. Then Eq. (7) is easily integrated to give

$$\log G = k\Psi, \tag{8}$$

k being a factor of proportionality. Equation (8) is called the law of Weber and Fechner. Fechner's assumption is questionable, and this law has a much more limited field of validity than Weber's law.

Weber's law can also be integrated with the assumption that the increment of a sensory excitation at threshold depends on the excitation level in a similar manner as the stimulus increment on stimulus level. In that case, Eq. (7) can be written

$$\Delta G/G = k\,\Delta\Psi/\Psi,$$

which easily is integrated to give

$$G = c\Psi^k, \tag{9}$$

c being an integration constant. This relation is called *Stevens's law*.

The exponent k has to be determined experimentally (Krantz, 1972). It varies for different sensory modalities and is usually smaller than one. In such cases, the power law [Eq. (9)] and the logarithmic law [Eq. (8)] are not very different from each other in a limited range, so that an experimental decision between the two laws is uncertain.

An experimental possibility for the estimation of the constants in Stevens' law is *psychophysical scaling* (Krantz, 1972). Two different suprathreshold stimuli, e.g., two fields of different luminance, are presented to a subject. Thereafter, the subject has the task to adjust a variable stimulus so that the sensation strength is precisely in the middle between those of the comparison stimuli. By repeated tests of this kind, a scale of equally distant sensations and their corresponding stimuli can be obtained.

3.3 Data Acquisition and Evaluation

Besides the usual sources of noise that occur in physical measurements, additional sources have to be considered in psychophysics, as has been mentioned before. Statistical methods have to be used to estimate mean values, variances, and other parameters of the underlying probability distributions. But since in this case humans serve as detectors, specific problems occur. To accumulate sufficient data, many tests have to be performed, which usually take a long time during which the human observer may become tired and his or her detector performance decrease. So the duration of test sessions has to be limited. On the other hand, because of parameters that are mostly unregulated, the day-to-day performance variations usually are much larger than the variations in one session. Usually, results from sessions on different days can be related to each other only by correlation methods. Therefore, it is essential to limit the daily sessions and at the same time to extract as much information as possible from these sessions.

The simplest but also the most inaccurate method for the measurement of thresholds is to let the subject adjust a test field in luminance or color against a fixed control field according to a just noticeable difference. More accurate, but also more time consuming and more expensive, are statistical methods of presentation (Sixtl, 1967). Subjects are confronted in statistical order with stimuli of different strengths. The only task they have to perform is to answer the presentation with "seen" or "not seen." The reliability of the subject is tested by blank trials.

Here signal detection theory comes into consideration (Nachmias, 1972). The subject may adjust his detection criterion at a balance between two possible errors: missed signals or false positives. The balance may be influenced by costs of false and awards of correct answers. By systematic variation of the setting of the criterion, the false positives rate may be varied and the *receiver operating characteristics* can be measured. By this method, results from detectors with different criterion settings may be compared. Applications of this theory to threshold measurements have shown that the probability of false positives can be neglected in this type of experiment, and that it is not possible to influence the criterion settings of human observers in a systematic way. Therefore, signal detection theory is not very useful in this field.

These findings suggest that human subjects usually set their threshold criterion too high, so that they miss some low contrast signals. This is usually the case, as can be verified by the *forced choice* method. Here several spatial or temporal fields are presented, and the subject has to decide in which field the signal did appear. The subject knows that the signal is actually presented in one of these fields. The answer "not recognized" is not accepted, which usually is annoying for inexperienced observers, since they believe that this is pure guessing. That this is not the case can be demonstrated easily by the fact that usually the ratio of correct answers is significantly larger than what could have been expected from pure guessing.

In the following discussion, it is assumed that the experimental procedure is the presentation of single stimuli. The subject has to answer a presentation with "seen" or "not seen" or simply with "yes" or "no." The relative frequency P of correct answers is plotted against stimulus strength X. Such a procedure is called *quantal response measurement*, and the plot $P(X)$ (see Fig. 10) is the basis of all further considerations. It is assumed that, with increasing number of presentations, $P(X)$ converges to the underlying probability distribution $\hat{P}(X)$, which is called the *psychometric function*. The threshold is defined as the value X_s where $\hat{P}(X_s)$ has a predestined value, usually 0.5. Since the correct psychometric function is unknown at least in the beginning of an experiment, the usual practice is to adopt one of a few standard models for this function, e.g., the cumulative normal function or the logistic function. The parameters of the adopted model are estimated from the experimental data (Malinvaud, 1978; Draper and Smith, 1981).

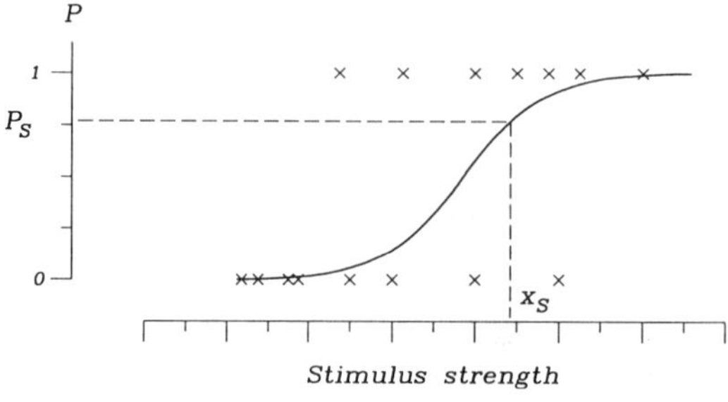

FIG. 10. Quantal response curve. The relative frequency P of a correct recognition of a stimulus is plotted against stimulus strength X. P_s is the relative frequency, which is adopted as threshold criterion; X_s is the corresponding stimulus threshold.

The classical strategy of data collection with a sound statistical foundation is the method of constant stimuli. The interesting range of stimulus sizes is defined and divided into a significant number of intervals, and for each interval a representative value of stimulus size is selected. These stimuli are presented to the subjects at random until the desired statistical significance has been accumulated. Obviously, it is not very efficient to make many presentations in the regions where $P(X)$ is either nearly zero or nearly one, since the answers of the subject are nearly certain in advance. On the other hand, the experimenter does not know in advance where these regions are. So in order to get maximal information from as few as possible presentations, it is useful to adopt an adaptive strategy, beginning with rough checks, and to proceed to more accurate tests according to the results of previous presentations. Very efficient in this sense are modern computer programs, which drive an electro-optical stimulus generator. Relatively simple is the *staircase method* (Fig. 11) (Cornsweet, 1962). The session begins, e.g., with a relatively strong stimulus. When the subject has recognized the signal in one or two successive presentations, the strength is reduced by one step. Similarly, when the subject has missed the signal in one or some successive presentations, the strength is again increased. The step size may be reduced during the session. Since in most cases the subjects know the strategy of this procedure, they may use this knowledge, consciously or not, to improve their recognition rate. Therefore, two or more

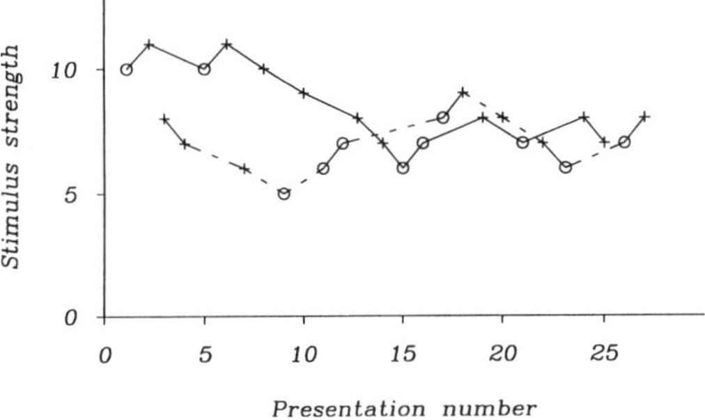

FIG. 11. Example of a staircase measurement. The stimulus strength of successive presentations is plotted against the presentation number for two interlaced branches (answer correct: cross; answer incorrect: circle).

independent staircase branches are interlaced, so that the strength of the next presentation cannot be predicted.

By evaluating the collected data, estimations of the threshold and of the inherent statistical uncertainty may be extracted. Especially with programs that use a variable step size, this is no simple statistical problem, and the discussion of this point is not yet settled.

4. BRIGHTNESS SENSATION

4.1 Absolute Threshold

The absolute threshold, i.e., the highest sensitivity for the detection of light, is attained by the visual system after an adaptation time of at least 45 min in absolute darkness. This threshold depends on several parameters, such as size, form, wavelength, presentation time, and retinal site of the stimulus. For small retinal stimuli, the whole receptor excitation in the stimulus area is completely summed spatially, so that, e.g., with a doubled stimulus area the threshold is reached with half of the retinal illuminance (*Riccò's law*). The maximal diameter of stimuli for the validity of this law depends on the retinal site, being about 10 min of arc in a parafoveal region (see Sec. 1.2) with 5°–15° of eccentricity. This region has maximal sensitivity under such illumination conditions, while the fovea is rather insensitive, because there are no or only few rods. For stimuli with sizes that exceed the range of Riccò's law, the excitation is summed only partially, being proportional to the square root of the stimulus area (*Piper's law*). For even larger areas, no further summation is found.

Besides the spatial, also a temporal summation is observed within a range of ca. 0.1-s duration (*Bloch's law*). For stimuli the duration of which exceeds this limit, also a partial summation proportional to the square root of the presentation time is valid (*Pieron's law*). For presentation times of more than 15 s, no further summation takes place.

Reaching threshold is determined by the number of absorbed light quanta, not by the absorbed energy. The quanta have to be absorbed by the photopigment rhodopsin. A light quantum that is absorbed by a rhodopsin molecule triggers a conformation change in the molecule and an enzymatic amplification, as has been mentioned in Sec. 1. Already the absorption of one quantum can cause an excitation of the whole receptor. Whether this excitation actually occurs depends on statistical effects (quantum efficiency, fluctuations in the amplification process). Only the excitation of several receptors within the range of spatial and temporal summation evokes a visual sensation, probably in order to exclude the possibility that the sensation threshold is surpassed by pure internal noise. The interesting question of how many receptors have to be excited in order to evoke a sensation is more difficult to answer because, as a result of the complicated retinal structure, many optical effects including waveguide effects have to be considered in estimating the probability of a physiologically effective quantum absorption. It is established that, under optimal conditions (wavelength 507 nm—the maximum of V'_λ—and the maximal sensitive parafoveal region), about 80 quanta have to hit the cornea in the pupil area in order to evoke a sensation. This corresponds to an energy of 3×10^{-17} J. The best estimations of the necessary number of effectively *absorbed retinal quanta* are in the range of 5–15.

Further information on this number may be extracted from the shape of the quantal response curve, in this case the probability of detecting a light flash as a function of its average number of corneal (pupil) quanta. The probability distribution of the number of absorbed quanta in a detected light flash should be a Poisson distribution (under some simplifying assumptions). The steepness of this distribution for the transition from "not seen" to "seen" increases with the number of quanta involved in the detection process. With 5–7 quanta, the best correlation between the Poisson distribution and the quantal response curve is reached (Hecht *et al.*, 1942; Pirenne and Marriott, 1959).

4.2 Luminance Difference Threshold

As is well known from daily life, the visual system can adapt to a huge range of luminances, from the nocturnal sky (without moon) with a luminance of 10^{-4} cd/m^2 and even targets in the region of the absolute threshold with approximately 10^{-6} cd/m^2 up to very bright scenes, e.g., white paper or snow, illuminated by bright sunshine, with a luminance of ca. 3×10^4 cd/m^2. This is mainly accomplished by the two types of photoreceptors in our eyes. The scotopic range, mediated by the rods, covers the range from 10^{-6} to 10^{-3} cd/m^2. The photopic range, mediated by the cones, covers the range from 10 to ca. 3×10^4 cd/m^2. The intermediate range is called *mesopic*, a region where both rods and cones are active. An interaction of these two types of receptors may be possible even in the photopic range, especially in the periphery.

The threshold for the detection of luminance differences depends heavily on the adaptation level. It is measured with a typical photometric arrangement, e.g., a circular bipartite field or two concentric circular fields, the larger with the reference luminance L, the smaller with the incremented (or decremented) test luminance $L + \Delta L$. It is essential that the photometer field be surrounded by a large field with an adapting luminance. The results usually are plotted as $\Delta L/L$ against L (see Fig. 12). The term $\Delta L/L$ is called the *Weber fraction*. It can be seen that this unit decreases significantly with increasing luminance, having a sharp bend at about $10^{-2.3}$ cd/m^2. The reason for this bend is the transition from scotopic to photopic vision. For large luminances, the Weber fraction tends to become constant, as is predicted by Weber's law.

The shape of this curve is rather insensitive to the wavelength composition of the stimulus. Only for dark red light (700–780 nm) is no bend observed, in accordance with the insensitivity of rods for such light. The stimulus should cover retinal regions with rods and cones. If only the central fovea is illu-

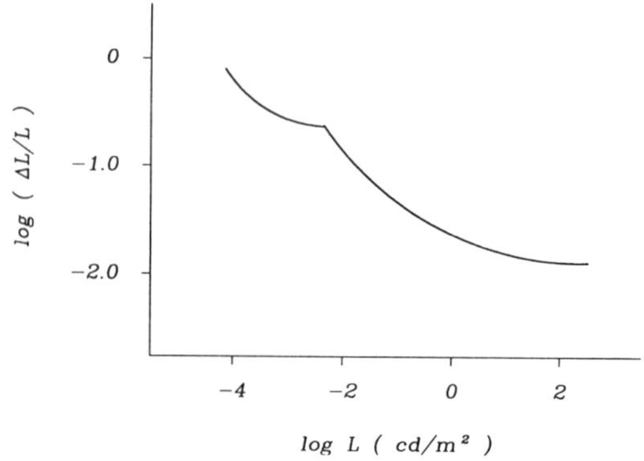

FIG. 12. The Weber fraction $\Delta L/L$ as a function of L (logarithmic scales) (after Wyszecki and Stiles, 1967).

minated, no rods are involved and no bend is observed. Practical use of the insensitivity of rods for red light is made by red illumination of instrument panels and red eye glasses of radiologists to preserve rod dark adaptation.

Two theories to predict (or explain) luminance difference threshold—especially in the range of Weber's law—have been set up. The first has been formulated by Hecht (1937). It is based on two assumptions: first, that the concentration of rhodopsin in the receptors is in balance between bleaching and regeneration, and second, that deviations from equilibrium are measured psychophysically according to Fechner's integration model. While the assumptions on the bleaching of receptor pigments partly have been confirmed by experiments on fundus reflectometry by Rushton (1964), the assumptions on the metric of thresholds are still open for discussion.

The second theory on luminance thresholds has been set up by de Vries (1943), Pirenne (1967), and Rose (1973). It is based on the consideration of statistical fluctuations of absorbed light quanta. When a detector array (e.g., the retina) is illuminated homogeneously by a flash, each unit area is hit on an average by, say, n_0 quanta. The root mean square of the number of hits in an area of f units is then $\sqrt{fn_0}$ (according to the Poisson distribution). For an ideal detector, the threshold luminance increment ΔL for the detection of such an area has to exceed the variance by a significant amount. The retina does not reach this ideal performance, but a tendency in this direction can be observed (Rose, 1973).

4.3 Detection of Spatial Luminance Structures

One of the basic properties of an imaging system is its spatial resolution. In physiological optics, it is usually expressed as the reciprocal value of the limit of angular resolution of the visual system and termed *visual acuity*. As in optical instruments, the resolution is limited by blur, caused by diffraction at the entrance pupil and geometrical optical aberrations of the imaging system, and by detector properties. According to the first two influence factors, a critical pupil size can be defined for minimal blur, when diffraction and aberrations cause approximately the same amount of blur. For the eye, the critical pupil has a diameter of ca. 2 mm.

The Landolt C (a dark ring the width of which is one-fifth of its outer diameter and that has a square gap in the ring whose position must be detected) is one of the more reliable tests of visual acuity. Grids with light and dark bars are also very important.

Roughly, the human eye has a resolution of the order of 1 min of arc, or a visual acuity of 1 under steady fixation, and this corresponds roughly to the separation of the retinal cones in the center of the fovea. In detail, the acuity depends on adaptation level of the subject and contrast level of the test structure. The results with various tests usually differ from one another and may well be better than 1. They cannot be compared very easily, because the distribution of the retinal illuminance in the test image has to be calculated, and this can be done reliably only when the point- or line-spread function of the imaging system is known. Equivalent to this function is its Fourier transform, the modulation transfer function (MTF) of the eye, since one function can be calculated from the other by Fourier transform. The MTF concept is widely used in geometrical and physical optics (see Optics, Linear). The MTF of the eye can be measured

1. by analyzing the light that leaves the eye after reflection from the retina, or
2. by comparing the contrast thresholds for grids that are produced on the retina either by normal imaging or by an interferometric method that bypasses the optical imaging system of the eye (Campbell and Green 1965).

Besides the illuminance difference in the retinal image, also receptor statistics has to be taken into account. For example, the resolution limits in the image of two dots and two parallel lines are about equal. But the illuminance difference between the two maxima and the minimum between them is much larger for points than for parallel lines. This may be explained by the fact that, in averaging along the line images, the signal-to-noise ratio is much better than in the dot images.

Another type of visual acuity is the *vernier acuity*, the minimal lateral displacement of two half lines relative to each other so that they can be discriminated from an uninterrupted line. The resolution limit for this task has a value of a few seconds of arc. Clearly, this task cannot be compared with those which have been discussed before. Here the criterion is not illuminance difference but difference in the localization of border lines. This type of task will be mentioned again in the section on binocular vision.

Similar to the eye, also the whole visual system can be treated by Fourier and MTF methods. However, the visual process is nonlinear so that, strictly speaking, linear Fourier methods are not valid. Only with small luminance modulation may an approximately linear response be assumed. A further assumption is that, at the receptor level, independent of spatial frequency the same contrast is needed for the detection of a modulation. Then the MTF of the visual system may be defined as the reciprocal value of the threshold modulation of a sine wave grid target as a function of spatial frequency. This value is also called contrast sensitivity. Measurements of this function (Campbell and Robson, 1968) are shown in Fig. 13 for sine-wave and square-wave targets. Obviously, this function has bandpass structure. From the discussion in Sec. 2.3, it follows that this structure is equivalent to receptive fields with a Mexican-hat structure (see also Fiorentini *et al.*, 1990).

A correlate to this receptive field structure are Mach bands, which are easily observed at the border between a bright and a dark homogeneous field or, better, at a staircase target consisting of a row of successive gray steps. The effect is that, at the border, the dark field has a still darker and the bright field a still brighter rim. This is exactly what is obtained by integrating Eq. (1) for one dimension along x:

$$\int_{-\infty}^{x} f(x')dx' = \int_{-\infty}^{x} [a_1 e^{-x'^2/2\sigma_1^2} - a_2 e^{-x'^2/2\sigma_2^2}]dx', \tag{10}$$

giving the response of a continuous array of receptive fields to a step function.

Figure 13 has still another interesting aspect. From the Fourier decomposition of a square wave rect(x') (peak-to-peak amplitude

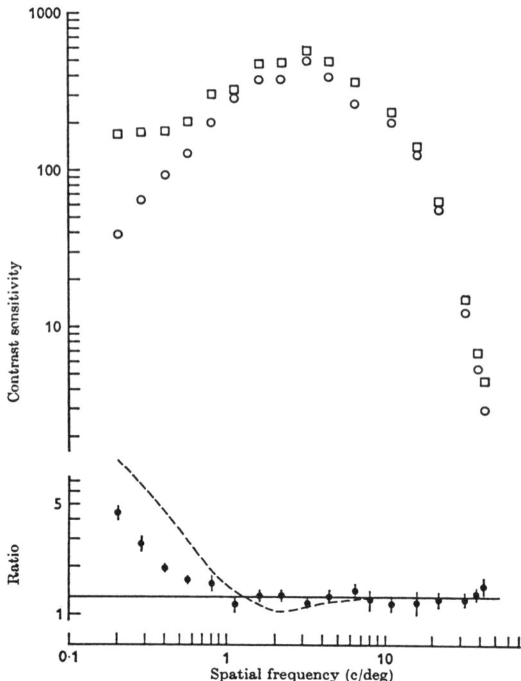

FIG. 13. The modulation transfer function of the eye. Contrast sensitivities for sine-wave gratings (open circles) and square-wave gratings (open squares) are plotted against spatial frequency (cycles/degree of visual angle, luminance 500 cd/m^2). The ratio of the contrast sensitivities at each spatial frequency is plotted at the bottom curve of the figure (with standard error of mean indicated) (from Campbell and Robson, 1968).

= 2, period X) we have

$$\text{rect}(x') = \frac{4}{\pi}\left\{\sin\frac{2\pi x'}{X} + \frac{1}{3}\sin 3\frac{2\pi x'}{X} + \frac{1}{5}\sin 5\frac{2\pi x'}{X} + \cdots\right\},$$

with

$$\text{rect}(x') = \begin{cases} 1 & \text{if } nX < x' \leq (2n+1)X/2, \\ -1 & \text{if } (2n+1)X/2 < x' \leq (n+1)X, \end{cases}$$

and $-\infty < n < \infty$. It is evident that the amplitude of the fundamental harmonic of a square wave is larger by a factor of $4/\pi$ than that of a sine wave with the same modulation as the square wave. This is consistent with the results of Fig. 13 for medium and high spatial frequencies. In this frequency range, the third harmonic is already attenuated below threshold. In the low–spatial-frequency range, the third harmonic of the square-wave grid contributes to perception; therefore, the difference of contrast sensitivity between square-wave and sine-wave increases. This is also the region where the square-wave grating can be discriminated from the sine wave grating. This result confirms convincingly the assumptions on the linearity of the visual system in the threshold region.

But Fourier analysis is more than a calculus for the visual system. This has been shown by adapting the system to high-contrast gratings (Blakemore and Campbell, 1969). During adaptation, the retinal image has to be shifted across the retina, so that the adaptation is to the grating and not to the individual bright and dark bars. After adaptation to a grating of a medium spatial frequency, the contrast sensitivity has decreased in a spatial frequency band of about two octaves, the maximal relative decrease of a factor 2 occurring at the adapting frequency. This adaptation effect is confined to gratings with the same orientation as the adapting grating. This behavior is attributed to a neuronal mechanism in the visual cortex called *spatial frequency channel.* In the range from 3.5 to 14 cycles per degree of visual angle (cpd), many independent spatial frequency channels were found. These results agree very well with the neurophysiological measurements discussed in Sec. 2.3.

It can be concluded from these results that the MTF shown in Fig. 13 is the result of a superposition of numerous spatial frequency channels of relatively small bandwidth. On this basis, the hypothesis has been discussed whether the visual system performs a Fourier decomposition of an image, a selective filtering in the individual channels, and a final synthesis. But a global Fourier analysis and synthesis seems to be unlikely, because of the lateral inhomogeneity of the retina. Moreover, it has been established that with locally limited adapting gratings, the adapted region is limited to the same area. Therefore, at present the hypothesis is favored that signal processing in the visual system is best described on the basis of a *local spatial frequency decomposition.* This concept was created by Gabor (1946) for temporal signals and easily can be applied to spatial signals (Daugman, 1985). The elements of this decomposition are sine waves under a Gaussian envelope, and a suitable set of such functions with discrete values of space and spatial frequency forms a complete set for the decomposition of spatial luminance distributions. In these functions, spatial and spatial frequency coordinates have the same weight. Therefore, the discussion in literature—which has lasted for decades—on the question of whether the visual system is described adequately by receptive fields or by spatial frequency channels seems to have no conceptual or experimental basis.

This is not so for suprathreshold signals, which, of course, are the most frequent and most important ones. Here the visual system is definitely nonlinear, so that a linear decomposition into a set of elementary functions is not possible. Several detector mechanisms (Fukushima, 1969), such as line, border, and bar detectors and local and global mechanisms (see, e.g., Burr *et al.*, 1986), have been identified by psychophysical methods. Some theoretical attempts to understand visual pattern recognition will be mentioned in the last section of this article.

5. BINOCULAR VISION

Binocular vision is the basis of depth perception. This notion is ascribed to Johannes Kepler, and in principle his conception is still valid. When we fixate a target P with our two eyes, their axes usually have to converge, so that the target is imaged into the central foveae of both eyes (see Fig. 14). It appears that both retinae have implanted more or less equal coordinate systems. *Corresponding sites* of the two retinae are sites with the same coordinates. When a second target A is presented so that the angles α_r and α_l are equal, it is imaged on corresponding sites of both retinae and the target is seen as a single target. When one is viewing straight ahead in an erect position, all sites in the so defined horizontal plane the images of which fall on corresponding retinal sites can be shown to lie on a circle, the Vieth–Mueller circle (VMC) (Vieth, 1818). They appear as single targets at the same subjective distance as target P. When another target B is presented that does not lie on the VMC, the angles β_r, β_l are not equal and the target may produce two separate perceptions from the two eyes.

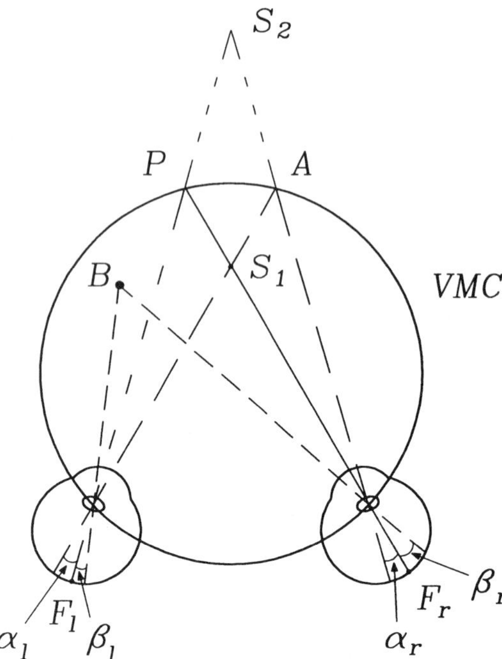

FIG. 14. Geometry of binocular depth perception. For explanation, see text.

The difference $\beta_r - \beta_l$ is called disparity. When the disparity for target B is not too large, it may be perceived as a single target and usually at a distance smaller or larger than that of P, according to the geometric situation and the sign of the disparity. In that way, stereoscopical depth perception is created. It has to be emphasized, however, that the fusion of two retinal images with limited disparity to a single perception does not implicate stereoscopical depth perception. Many people with good fusion lack stereoscopical perception, often without knowing it. Lack of the ability of fusion, however, may cause binocular twin images, which is a serious defect. One reason for that can be anisometropia, a difference in ocular optical magnification, which causes a mismatch between retinal coordinates and corresponding retinal image elements. When targets with a nonzero disparity are fused, their retinal sites are not strictly corresponding. The areas around corresponding sites from which fusion is possible are called *Panum areals*.

The target sites with zero disparity are not confined to the VMC but fill a surface in three-dimensional space, the *horopter*. Its shape depends on the fixation distance, and, moreover, it cannot be derived purely with geometrical considerations but has to be determined experimentally, because of some individual deviations of the internal retinal coordinates from pure geometry. In detail, this is a rather complicated matter (Ogle, 1950).

In monocular vision, the perceived direction at which a target appears is identical with the line of sight. When viewed binocularly, the lines of sight of the two eyes are different, and the target appears in a direction intermediate between the two lines of sight. To cope with this situation, H. v. Helmholtz and E. Hering introduced the concept of a single imaginary cyclopean eye somewhere between the two natural eyes, which determines the binocularly perceived direction. This cyclopean eye is not necessarily in the middle of the forehead as a result of dominance of one eye over the other.

Considerable effort has been devoted to the *correspondence* problem that arises with Kepler's theory. Referring to Fig. 14, each eye sees two targets. When the targets are identical or similar, how does the brain know which ones have to be fused? Instead of fusing P with P and A with A, it could also fuse P with A or vice versa, arriving at S_1 or S_2. Under normal viewing conditions, usually the correct correspondence is found. An obvious hypothesis was that retinal images are fused so that maximal correspondence between image elements is reached, implying that monocular form perception precedes binocular fusion.

That such a conclusion is not or not always correct has been demonstrated by Julesz with random dot stereograms (e.g., Julesz, 1971). He generated computer random dot twin images that were identical except that, in a special area, an additional disparity (defined before in this section) between the dots was introduced. Of course, this area could not have been identified monocularly, but only after fusion. So binocular fusion may well precede form discrimination, and the correspondence problem is left to more sophisticated theories (Regan, 1990).

Stereoacuity is defined as the smallest disparity that can be discriminated from zero, i.e., objects with such a disparity just can be perceived to have another distance than the fixation point. The displacement of such objects against corresponding sites amounts to 5–10 s of arc. This is of the same order of magnitude as the vernier acuity (see Sec. 4.4), and obviously the same type of visual task is involved.

Stereoacuity is improved considerably by enlarging the effective separation of the two eyes. This is accomplished by steroscopes or by stereotelescopes. Stereoscopical vision is also improved by a rapid movement of the observer, e.g., in a car, or perhaps by skiing in a terrain with small luminance contrasts. Even persons lacking binocular vision can develop stereoscopic vision under such conditions.

When an object is viewed binocularly from different or varying distances, within rather wide limits its perceived size remains constant while the size of the retinal image undergoes large changes. The main mechanism for this *perceived size constancy* is vergence (see Sec. 1.1). Vergence and accommodation are linked together in natural binocular vision, since near objects require strong accommodation and strong convergence, while distant objects require small accommodation and small convergence. However, the linking is not very strong, and especially for aged people this linking does not work because of presbyopia.

Neurophysiological mechanisms for binocular vision have been explored by numerous authors. Two types of cells have been found in areas V1 and V2 of monkeys. One type is tuned to individual disparities. In their entirety, these cells cover a large disparity range. Another type is excited generally by disparities of one sign and is inhibited by disparities of the opposite sign (Poggio, 1984).

6. TEMPORAL SIGNAL PROCESSING AND MOTION PERCEPTION

6.1 Temporal Signal Processing

Analogous to the limited spatial resolution, also the temporal resolution of the visual system is limited. Normally, the light sensation that is evoked by a short light flash has a longer duration than the stimulus. The lower limit for temporal discrimination is given by Bloch's law (see Sec. 4.1) on temporal summation and usually is in the order of 0.1 s, but under high adaptation levels and for small stimulus areas, this summation time may decrease by one or two orders of magnitude. The critical interstimulus interval (cISI) between two stimuli depends on the stimulus length. Under otherwise constant conditions, a small stimulus length of 1 ms leads to a cISI of 59 ms, while a stimulus length of 5000 ms results in a cISI of 1 ms. Moreover, the cISI depends on the retinal site, being shorter in the fovea than in the periphery.

The cISI also depends on the number of light pulses presented in succession; it decreases with increasing pulse number. When the pulse number is very large, the cISI is usually expressed as the flicker fusion frequency or critical flicker frequency (CFF). The CFF varies with the stimulus amplitude approximately in a logarithmic relation (*Ferry-Porter law*). Above flicker fusion, the brightness sensation of a modulated light stimulus is matched by a steady light stimulus of the same time-average luminance (*Talbot-Plateau law*).

Similarly to spatial resolution, temporal resolution can be described by a *temporal modulation transfer function* (*tMTF*). A test field with a sinusoidally modulated luminance

$$L(t) = L_0(1 + m \sin \omega t) \qquad (11)$$

is presented to the observer, and the modulation m is adjusted to flicker threshold. The reciprocal value of the threshold is the *modulation sensitivity*, which, plotted as a function of the modulation frequency $f = \omega/2\pi$, is the tMTF. The first extensive measurements of this function were performed by De Lange (De Lange, 1958) (see Fig. 15). Because of the significant attenuation of higher frequencies, the threshold mainly is determined by the fundamental Fourier component of some more complicated wave forms, e.g., rectangular or saw-tooth grids.

The abscissa where the high-frequency branch of a de Lange curve reaches a modulation sensitivity of unity is the ordinary CFF. De Lange and successors established by Fourier methods that temporal signal transmission in the visual system is linear under threshold conditions (Kelly, 1972).

In reaction to a sudden jump in illumination L (say in the positive direction), the threshold ΔL changes according to the constancy of the Weber fraction. When investigating the time course of ΔL in reaction to

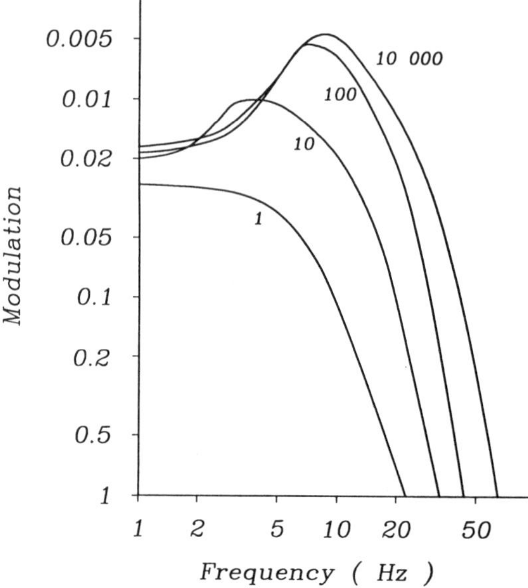

FIG. 15. Flicker fusion. The modulation sensitivity of the visual system against temporally modulated light is plotted against modulation frequency for a large range of retinal illuminances. Parameter is troland of retinal illumination (after de Lange, 1958).

the jump in L, one arrives at curious results. The threshold increases some 100 ms *before* the jump in L is applied. This seems to be a violation of causality. The explanation is that the strong signal of the change in L overrides the weak ΔL signal. This effect is called *masking*. It is a powerful psychophysical method for the investigation of temporal effects in neuronal signal processing (Breitmeyer *et al.*, 1981).

Temporal transmission properties of the visual systems of man and of various animals also have been measured electrophysiologically by recording the ERG and the synchronized VECP (see Sec. 2.1). The latter method is especially suited for the recording of the phase transmission function by using cross correlation techniques.

6.2 Motion Perception

Motion is a spatiotemporal event. Nevertheless, motion perception is an independent, specific sensory quality. For an animal, motion in its natural environment usually means danger or prey, and the perception and analysis of moving objects have preceded the perception of stationary objects in phylogenetic evolution. In the cortex of cats and monkeys, direction- and speed-sensitive neurons have been found, which may contribute to motion perception. Also, motion-segregation cells have been identified.

Motion can be recognized in two different modes:

1. the motion of a retinal illuminance pattern with respect to the stationary retina (when a moving object crosses the otherwise stationary visual field), or
2. when a moving object is kept under fixation by eye, head, or body movements. In the first mode, the subjectively perceived velocity of the moving object is about twice as large as in the second mode.

In the first mode and for small object velocities, up to 10° of visual angle per second, visual acuity is increased compared with stationary objects. This may be due to local adaptation for stationary objects, which decreases the contrast. Furthermore, the acuity decreases to about one-tenth of the stationary value for velocities as large as 100° s^{-1}. This is very remarkable since one would expect a much higher degradation due to blurring of the retinal image. Therefore, a neuronal mechanism has to be assumed that isolates and compares successive, unblurred retinal images or the corresponding neuronal excitation patterns. In the second mode, a steady decrease of the acuity has been observed that is proportional to the third power of the angular velocity.

A moving retinal illuminance structure causes a motion sensation when this structure is generated by a moving object. In contrast to that, no motion sensation is generated when the motion of the retinal image is generated by a movement of the eye, the head, or the body of the observer. When we move ourselves in a stationary environment, no motion sensation is generated despite dramatically moving retinal images. A hypothesis has been advanced (von Holst and Mittelstaedt, 1950) to resolve this paradox: the *reafference principle*. According to this hypothesis, a copy of the efferent neuronal signal that triggers the body movement is compared with the afferent signal from the retina. When these signals correspond to one another, no motion sensation is generated. Unfortunately, such an efference copy has not been found so far by neurophysiological research. A more sophisticated model has been proposed by MacKay (1973).

The perception of motion can also be generated by two or more stationary, but slightly different, targets that are presented in temporal succession, as is well known from cinema or television. This apparent motion has been used extensively during the last decade for the investigation of the mechanisms of motion detection, because such presentations can be generated easily with a computer on a monitor. For example, random-dot structures are very interesting targets, because several artifacts can be avoided that occur with simpler targets, and they may lead to better understanding of this subject, which has important practical applications (van Doorn and Koenderink, 1982).

7. COLOR VISION

The manifold of different human color sensations can be imbedded in a three-dimensional space. This has been established mainly by additive color mixture experiments. The term *additive color mixture* has generated many misunderstandings between different groups of people who are interested in or working with color, such as physicists, physiologists, artists, designers, and architects. Additive color mixture in the physical sense means addition of lights of specified wavelength distributions. This can be accomplished by projecting lights of different spectral composition from different sources one upon another onto a screen. A second method is to present different colors in rapid succession above flicker fusion. This can be done, e.g., with Maxwell's top, a rotating disk with sectors of different colors. In a certain sense, all colors can be generated by mixing three basic colors in varying proportions or intensities. This is explained in more detail in the article OPTICAL INSTRUMENTATION. Additive color mixing in this sense is *not* mixing of color pigments (Wyszecki and Stiles, 1967).

The appearances of two colors may differ in hue, saturation, and luminance. This is one of several possibilities to construct a three-dimensional color space. On the other hand, physical light stimuli may differ in their spectral composition. In order to construct a space for all possible different spectra, one dimension for each wavelength is needed, so that this space would have infinitely many dimensions. Therefore, infinitely many spectral distributions can generate the same color sensation. Such colors of equal appearance but of different spectral composition are called *metameric* colors. They are said to have the same *color valence.*

In a mathematical sense, color spaces that are based on additive color mixing are three-dimensional affine vector spaces. For practical applications, it would be desirable to introduce a metric, i.e., to construct a line element, which ensures that colors with the same perceptional difference are represented by points with the same geometrical distance. Much work has been devoted to this problem without arriving at a generally accepted solution. However, it seems to be clear that only a non-Euclidean metric can solve the problem (Wyszecki and Stiles, 1967).

That human color vision has three dimensions is easily understood by modern research on the three types of cones that are found in the human retina and that have different absorption spectra. These different receptor types were already postulated by Thomas Young and in more detail by H. v. Helmholtz, who founded a color vision theory based on such three hypothetical receptor types on the basis of color mixture experiments.

On the other hand, already at the time of Helmholtz many effects in color vision were known that could be explained only with difficulty by the three-receptor theory. Some of these effects are the following:

1. Existence of two independent pairs of *complementary colors,* e.g., red-green and blue-yellow. A mixture of such complementary colors in adequate portions results in a neutral white. Independent pairs are formed when a green hue is selected that is neither bluish nor yellowish and a yellow hue that is neither reddish nor greenish.
2. *Color contrast*: When, e.g., a red field is presented to an observer for about half a minute and then it is exchanged for a white field, this white appears in an unsaturated greenish, very near to the complementary color of red, and analogously for other colors. This effect is called *successive* color contrast. When a small field of neutral

white is surrounded by a large red field, the white looks greenish. This is called *simultaneous* contrast. Very saturated colors that are not found in the spectrum can be generated by color contrast. Also, the sensation *black* is not perceived in a dark room but can only be generated by contrast, but in contrast to black and white borders, color borders (without brightness differences) do not induce Mach bands. Therefore, different mechanisms seem to be involved in brightness or color contrast.

3. *Afterimages*: When an observer is exposed for about half a minute to a strong colored light with high saturation, he afterward perceives an afterimage of the light source, the color of which is either similar to the light source itself or to the complementary color, depending on whether the background for this observation is bright or dark. Also, temporal changes between the colors of these afterimages occur (Wasserman, 1978).

These and other similar effects motivated E. Hering, a contemporary of Helmholtz, to advance a color vision theory that he placed in comparison with the three-receptor theory. His theory postulated three antagonistic processes: red-green, blue-yellow, and white-black, similar to the neuronal antagonism of excitation and inhibition.

While the receptor theory is confirmed by current knowledge on the cone absorption spectra, Hering's theory is supported by neurophysiological findings. In primate retinae, special color ganglion cells have been found. Some have receptive fields with an on center, which is excited by *L cones* (long-wavelength sensitivity). The periphery of these receptive fields is inhibitory and is connected to *M cones*, which are sensitive to the median range of the spectrum. Such ganglion cells are called $L^+ - M^-$ cells. In addition, $M^+ - L^-$, $-L^- + M^+$, and $-M^- + L^+$ cells are found (Zrenner, 1990). In some cells, the short-wavelength region is opposed to the sum of L- and M- cone signals: $S^+ - (L + M)^-$; others have a nonopponent field structure with a sensitivity according to the V_λ curve. Color sensitive ganglion cells project mainly to the parvo system of the CGL. It is evident that many features of the Hering theory find striking analogies in these ganglion structures.

These results support a *zone* color vision theory with a receptor mechanism according to the Young–Helmholtz model and an information processing according to Hering's model (Bornschein and Hanitzsch, 1978).

Similarly to brightness sensation, also color sensation may adapt. When an observer has been exposed for some time to a strong colored illumination, his spectral color sensitivity is shifted in the direction of the complementary color. This effect has to be accounted for when performing color matches.

Sky light is subjected to considerable changes in its spectral composition, varying from direct sunshine to blue or clouded sky, from dawn to dusk. Even more pronounced is the change from daylight to artificial illumination. Such changes of illumination induce changes in the spectral composition of the light that is then reflected from illuminated structures. Nevertheless, the perceived color of such structures is remarkably stable for an observer who is exposed to such changes of illumination. This stability is due to a special form of color adaptation to a general color shift in the surrounding. Such color constancy of known objects against changes of illumination certainly facilitates pattern recognition. Comprehensive information on color vision and color measurement is found in Wyszecki and Stiles (1967).

8. OPTICAL ILLUSIONS

The visual system is not simply analogous to a television system, because its task is primarily not to produce an undistorted image of the environment but to perform pattern analysis and recognition. Therefore, complicated nonlinear image processing occurs, which leads to specific distortions. Usually, these distortions are not noticed in daily life, because we are accustomed to them from early childhood, but in special situations and under laboratory conditions they can become evident (Schober and Rentschler, 1972). They are known as *optical illusions*. Some have been mentioned already (Mach bands in Sec. 4.3, color and brightness contrast in Sec. 7).

Adaptation to brightness and color: Without special precautions, no reliable estimation of luminance and color can be made by human observers. Even in absolute darkness, one can have brightness sensations (e.g., the so-called *Eigengrau*, a brightness sensation in long-lasting absolute darkness).

Mach bands emphasize luminance borders, which are important features for pattern discrimination.

Optical illusions have been studied widely in order to analyze special procedures performed by the visual system in the course of pattern recognition, but their knowledge may be useful also in daily life in order to avoid misinterpretations of visual information.

8.1 Geometric Illusions

Since we look at the surrounding world with two eyes, the geometrical conditions are rather complicated. Luneburg (1950) has shown that this results in a non-Euclidean geometry of our visual space. A help is the knowledge that, e.g., a dog running through our visual field remains the same dog and retains its overall structure, no matter the distortions that the lower neuronal centers may detect.

In cases when the geometry is not clear in advance, several illusions may be induced. A very popular one is the *Poggendorff* illusion (Fig. 16). The oblique line that is covered partially by a bar seems to consist of two parts that are displaced laterally relative to one another. In fact, this is not true. Also, the subjective judgment of the vertical may be influenced by oblique line structures in the visual field.

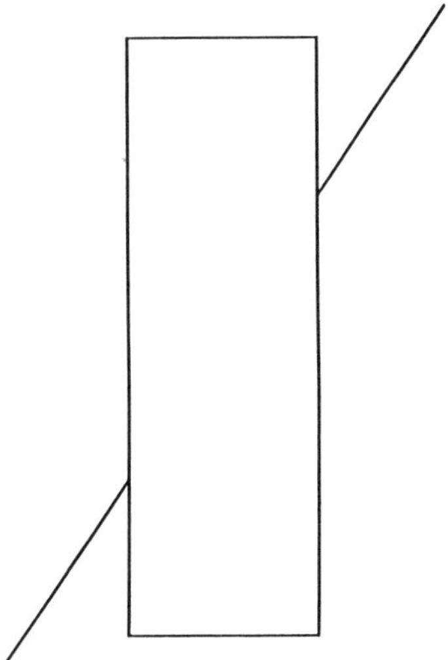

FIG. 16. Poggendorff illusion. The oblique line that is covered partially by a bar seems to be interrupted so that the two parts are displaced laterally relative to each other. That this in fact is not the case is easily verified with a straightedge.

A vertical line compared with a horizontal line of equal length always appears longer. Much work has been done to explain this type of illusion.

8.2 3D Illusions

As is very well known, three-dimensional illusions can be generated by two-dimensional structures, mainly by use of perspective. Another possibility is shadowing (Ramachandran, 1988). Usually, the visual system does select the simplest interpretation, even when this interpretation is contradictory to common sense or the laws of physics, as has been demonstrated by numerous drawings by Escher (Locher, 1986). In many cases, more realistic, but more complicated, interpretations of such "impossible" structures can be found.

A very interesting 3D illusion, which is based on the chromatic aberration of the eye, has been known already to medieval painters. When an observer accommodates (see Sec. 1.1) on a painting, because of the chromatic aberration of the eye red structures will be focused behind the retina, blue structures in front of it. But also, near or far visual objects would produce such differences in focus. Therefore, chromatic aberration can generate 3D illusions. In religious, medieval paintings, persons with a high rank in the hierarchy (e.g., the Madonna) usually are displayed in blue clothes, while more earthbound persons are in red.

8.3 Aftereffects

Afterimages have been mentioned before. When they are imposed on moving structures, curious effects may be generated. The strength of an aftereffect depends on the foregoing stimulus strength and its duration. The reaction of transient neuronal systems may be driven into saturation. Therefore, when one is looking at moving objects that carry pronounced brightness structures, the visual perception may be distorted considerably (Walker and Saunders, 1985; Cavanagh and Anstis 1986).

As has been mentioned before, the afterimage of a colored light flash depends on whether the afterimage is observed against a bright or a dark background. This is easily demonstrated with a rotating disk that is divided into a white and a black half. In the periphery of this disk, a small notch is positioned just at the white/black border, such that a stationary red lamp is uncovered for the observer during the rotation and a red flash is generated. Depending on the sense of rotation, the red flash is followed by a bright field, in which case a somewhat unsaturated green afterimage is perceived, or by a dark field, in which case a clear red afterimage is perceived. This illusion was described already by Goethe as *flashing poppy*. Years ago, it was the reason for a misinterpretation of a red railway signal in the subway of Hamburg—with disastrous consequences.

Movement aftereffects can easily be observed. After one has viewed, e.g., a waterfall for some time, stationary objects seem to move in the opposite direction. Similarly, a rotating structure seems to move in the opposite direction when it is stopped suddenly.

Very interesting is the *contiguous* aftereffect, described by McCollough in 1965 (Shute, 1979). A green/black vertical rectangular grid and a red/black horizontal rectangular grid are presented in periodic succession for several minutes to an observer. Afterward, white/black vertical grids will appear reddish while white/black horizontal grids appear greenish. Depending on the adaptation time, this aftereffect can last for hours or days. From this effect, it is evident that some mutual influence exists between form and color.

9. THEORETICAL MODELS OF VISUAL PERCEPTION

Considerable work has been devoted to the understanding and interpretation of neuronal signal processing in the visual system. In fact, this system is far from being understood in detail. On the other hand, it is accessible relatively easily with neurological and psychophysical methods, and therefore it may be a good entry for the understanding of other brain structures. Since the system is too complex for a purely experimental analysis, theoretical concepts have to be developed.

Fundamental to visual signal processing are receptive fields. The structural organization of such fields into an excitatory center and an inhibitory surrounding (or vice versa) opens up several possibilities of sophisticated signal processing, as, e.g., the transmission of image information with high spatial resolution by neuronal channels of moderate capacity. Application of sampling theory shows how receptive fields should be distributed over the visual field in order that no spatial information is lost.

Hierarchical neuronal networks have been designed to explain visual pattern recognition. In successive levels of information processing, features of increasing complexity are extracted from original input signals, beginning with lines and contours and ending up with complicated patterns such as faces. Associative memory and selective attention can be incorporated into such models (Fukushima, 1986). Advanced signal processing theory has been used to develop networks that can explain the mapping of spatial information from the retina into the cortex, and adaptive filtering. They are able to adapt to special classes of structures such as compound Gabor signals (Caelli *et al.*, 1987).

While in the threshold region linear system theory may be a suitable approximation, superthreshold signal processing is nonlinear and has to be modeled accordingly. A large class of nonlinearities can be described with a method devised by Volterra for temporal relationships. The simple convolution integral

$$y(\tau) = \int_0^{\tau} h_1(t)x(\tau - t)dt, \tag{12}$$

which relates the output signal y of a linear system to the input signal x by a convolution with an impulse reaction h_1, is supplemented by a series of nonlinear terms,

$$\int_0^{\tau}\int_0^{\tau} h_2(t_1, t_2)x(\tau - t_1)x(\tau - t_2)dt_1dt_2 + \cdots \tag{13}$$

with functions $h_2 \cdots$ depending on two or more separate impulses. The individual kernels h_1, $h_2 \cdots$ of this series are not independent of one another, but this difficulty can be eliminated by a method devised by N. Wiener (Mishkin and Braun, 1961). The relations (12) and (13) can be generalized to include spatial variables (Reichardt and Poggio, 1981).

Neurophysiological measurements of the extension of receptive fields and psychophysical measurements of localized spatial frequency channels indicate that a continuum of receptive fields of different lateral extensions is associated with each retinal position. A general magnification of such fields proportional to the retinal eccentricity is implicit. However, by modeling the subjective visual performance for detecting selected test structures, it can be shown that four independent receptive field mechanisms are sufficient to account for the experimental results (Wilson and Bergen 1979). These receptive fields differ in their size and their mode (sustained or transient channels). Moreover, two different detection mechanisms can be discriminated: local and global processing. Global visual processing occurs when a structure is viewed in its entirety. This is demonstrated with a checkerboard pattern from which selected spatial frequencies are removed. Under global vision, other structures are dominant as under local inspection (Burr *et al.*, 1986). Another example is texture, which is perceived globally, and special local deviations from the texture pattern are not recognized (Julesz, 1981).

As has been mentioned, the mapping of retinal into cortical geometry is topological and continuous. That means that neighboring structures in the retinal image are mapped into neighboring loci in the visual cortex. Nevertheless, this mapping implies a profound change in geometry, since a retinal system of polar coordinates is mapped into Cartesian coordinates in the cortex. Interesting conclusions on the neuronal signal processing can be drawn from the inherent geometrical deformations. As has been mentioned already in Sec. 5, the extraction of three-dimensional information from the retinal images of two eyes is not understood without difficulties. It has been argued that the visual system needs items from local geometry for the generation of 3D images (Koenderink and van Doorn, 1987). Topological methods are used to investigate the inherent problems.

Line singularities are more important than specular points, since contours are made up by line elements (Koenderink and van Doorn, 1976). The adequate method to treat line elements is Lie algebra. Numerous problems of visual geometry can be formulated with a Lie transformation group (Hoffman, 1966; Dodwell, 1983).

Very important for the discussion of retinal–cortical mapping is a consistent method to describe image geometry in terms of internal parameters that are invariant against neuronal mapping. Generalized autocorrelation methods have been used for this purpose (Glünder, 1986). Also, concepts of differential geometry have been used to describe neuronal mapping and extraction of shape information (Zetsche and Barth, 1990; Barth *et al.*, 1991).

As may be seen from this short account, the theory of vision is by no means complete. This subject is attacked along several lines and with sophisticated methods.

GLOSSARY

Afferent Fibers: Neuronal fibers that transmit information from receptors to the brain.

Bipolar Cells: A specialized type of neuron in the retina. Their input and output endings are not so clearly differentiated into dendrites and axons as in normal neurons. They transmit only slow, graded potentials, not spikes.

Chromophore: A special molecular configuration that is part of a larger protein molecule and is responsible for the absorption of radiation within a certain spectral region. Such absorption can activate the molecule.

Cortifugal Fibers: Neuronal fibers that lead from the cortex to neurons at lower levels and can influence their sensitivity or activity.

Efferent Fibers: Neuronal fibers that lead signals from the brain to effectors, i.e., structures to be influenced by the brain to perform some action (muscles, glands, etc.).

Fill-In Mechanism: The process by which retinal areas that are insensitive to light (the blind spot or scotomae) are not easily recognized by an observer, because they are not perceived as "black" but are "filled in" with an average sensation from the surroundings. Areas that are marked only by appropriate border line structures are also filled in with brightness and color.

Hue: The main feature of color, determined by the spectral composition of light and characterized by an equivalent wavelength.

This wavelength is determined so that the corresponding spectral light, when mixed with an appropriate amount of white, matches the color in question.

Mitochondrium: A cell organelle that serves the supply of the cell with metabolic energy, gained from nutriment.

MTF: Modulation transfer function, a function of spatial or temporal frequency, generally complex valued, that describes the changes in amplitude and phase that a signal undergoes while passing a linear transmission system.

Photopic Vision: Vision under bright illumination conditions with sensitive cones, color perception, and good visual acuity in the central fovea.

Saturation: The fraction of white light that has to be added to the spectral light of the hue-equivalent wavelength in order to match the color stimulus, varying from "one" for the spectral light to zero for neutral white.

Scotopic Vision: Vision under dim illumination conditions with only rods active, no color vision, insensitive central fovea, and reduced visual acuity.

Works Cited

Alpern, M. (1972), "Eye Movements," in: D. Jameson, L. M. Hurvich (Eds.), *Handbook of Sensory Physiology*, Vol. 7, Pt. 4, Berlin: Springer, pp. 303–330.

Barth, E., Caelli, T., Zetsche, C. (1991), "Efficient Visual Representation and Reconstruction from Generalized Curvature Measures," in: B. C. Vemuri (Ed.), *Geometric Methods in Computer Vision*, SPIE Proceedings No. 1570, Bellingham, WA: SPIE, pp. 86–95.

Baumgartner, G. (1978), "Physiologie des Zentralen Sehsystems," in: O. H. Gauer, K. Kramer, R. Jung (Eds.), *Physiologie des Menschen*, Vol. 13, *Sehen*, München: Urban & Schwarzenberg, pp. 263–356.

Blakemore, C., Campbell, F. W. (1969), "On the Existence of Neurones in the Human Visual System Selectively Sensitive to the Orientation and Size of Retinal Images," *J. Physiol.* **203**, 237–260.

Bornschein, I., Hanitzsch, R. (1978), "Die Netzhaut," in: O. H. Gauer, K. Kramer, R. Jung (Eds.), *Physiologie des Menschen*, Vol. 13, *Sehen*, München: Urban & Schwarzenberg, pp. 205–262.

Breitmeyer, B., Levi, D. M., Harwerth, R. S. (1981), "Flicker Masking in Spatial Vision," *Vision Res.* **21**, 1377–1385.

Burr, D. C., Morrone, M. C., Ross, J. (1986), "Local and Global Visual Processing," *Vision Res.* **26**, 759–757.

Caelli, T., Rentschler, I., Scheidler, W. (1987), "Visual Pattern Recognition in Humans," *Biol. Cybern.* **57**, 233–240.

Campbell, F. W., Robson, J. G. (1968), "Application of Fourier Analysis to the Visibility of Gratings," *J. Physiol.* **197**, 551–566.

Campbell, F. W., Green, D. G. (1965), "Optical and Retinal Factors Affecting Visual Resolution," *J. Physiol.* **181**, 576–593.

Cavanagh, P., Anstis, S. M. (1986), "Brightness Shift in Drifting Ramp Gratings Isolates a Transient Mechanism," *Vision Res.* **26**, 899.

Cornsweet, T. (1962), "The Staircase-Method in Psychophysics," *Am. J. Psychol.* **75**, 485–491.

Dartnall, H. J. A. (1962), in: H. Davson (Ed.), *The Eye*, New York: Acadmic Press.

Daugman, J. G. (1985), "Uncertainty Relation for Resolution in Space, Spatial Frequency, and Orientation Optimized by Two-Dimensional Visual Cortical Filters," *J. Opt. Soc. Am. A* **2**, 1160–1169.

de Lange, H. (1958), "Research into the Dynamic Nature of the Human Fovea-Cortex System with Intermittent and Modulated Light," *J. Opt. Soc. Am.* **44**, 777–789.

de Vries, H. (1943), "The Quantum Character of Light and Its Bearing upon the Threshold of Vision, the Differential Sensitivity and Visual Acuity of the Eye," *Physica* **10**, 553–564.

Dodwell, P. C. (1983), "The Lie Transformation Group Model of Visual Perception," *Percept. Psychophys.* **34**, 1–16.

Draper, N. R., Smith, H. (1981), *Applied Regression Analysis*, New York: Wiley.

Fiorentini, A., *et al.* (1990), "The Perception of Brightness and Darkness: Relations to Neuronal Receptive Fields," in: L. Spillmann, J. S. Werner (Eds.), *Visual Perception/The Neurophysiological Foundations*, San Diego: Academic, pp. 129–162.

Fukushima, K. (1969), "Visual Feature Extraction by a Multilayered Network of Analog Threshold Elements," *IEEE Trans. Syst. Sci. Cybern.* **SSC-5**, 322–333.

Fukushima, K. (1986), "A Neuronal Network Model for Selective Attention in Visual Pattern Recognition," *Biol. Cybern.* **55**, 5–15.

Gabor, D. (1946), "Theory of Communication," *J. Inst. Electr. Eng.* **93**, 429–457.

Glünder, H. (1986), "Neural Computation of Inner Geometric Pattern Relations," *Biol. Cybern.* **55**, 239–251.

Hecht, S. (1937), "Rods, Cones, and the Chemical Basis of Vision," *Physiol. Rev.* **17**, 239–290.

Hecht, S., Shlaer, S., Pirenne, M. H. (1942), "Energy, Quanta, and Vision," *J. Gen. Physiol.* **25**, 819–840.

Hoffman, W. C. (1966), "The Lie Algebra of Visual Perception," *J. Math. Psychol.* **3**, 65–98.

Hubel, D. H., Wiesel, T. N. (1970), "Cells Sensitive to Binocular Depth in Area 18 of the Macaque Monkey Cortex," *Nature (London)* **325**, 41–42.

Julesz, B. (1971), *Foundations of Cyclopean Perception*, Chicago: University of Chicago Press.

Julesz, B. (1981), "Nonlinear and Cooperative Processes in Texture Perception," in: W. E. Reichardt, T. Poggio (Eds.), *Theoretical Approaches in Neurobiology*, Cambridge, MA: MIT Press, pp. 93–108.

Kelly, D. H. (1972), "Flicker," in: D. Jameson, L. M. Hurvich (Eds.), *Handbook of Sensory Physiology*, Vol. 7, Pt. 4, *Visual Psychophysics*, Berlin: Springer, pp. 273–302.

Koenderink, J. J., van Doorn, A. J. (1976), "The Singularities of the Visual Mapping," *Biol. Cybern.* **24**, 51–59.

Koenderink, J. J., van Doorn, A. J. (1987), "Representation of Local Geometry in the Visual System," *Biol. Cybern.* **55**, 367–375.

Krantz, D. H. (1972), "Visual Scaling," in: D. Jameson, L. H. Hurvich (Eds.), *Handbook of Sensory Physiology*, Vol. 7, Pt. 4, Berlin: Springer, pp. 660–731.

Kühn, H. (1980), "Light- and GTP-Regulated Interaction of GTPase and Other Proteins with Bovine Photoreceptor Membranes," *Nature (London)*, **283**, 587–589.

Lennerstrand, G., Bach-y-Rita, P. (Eds.) (1975), *Basic Mechanisms of Ocular Motility and Their Clinical Implications*, Proceedings of International Symposium, Stockholm, 1974, Oxford: Pergamon Press.

Lennie, P., Prevarten, C., van Essen, D., Wässle, H. (1990), "Parallel Processing of Visual Information," in: L. Spillmann, J. S. Werner (Eds.), *Visual Perception/The Neurophysiological Foundations*, San Diego: Academic, pp. 103–128.

Locher, J. L. (Ed.) (1986), *Leben und Werk M. C. Escher*, Eltville (Rhein): Rheingauer.

Luneburg, R. K. (1950), "The Metric of Binocular Visual Space," *J. Opt. Soc. Am.* **40**, 627–642.

MacKay, D. M. (1973), "Visual Stability and Voluntary Eye Movements," in: R. Jung (Ed.), *Handbook of Sensory Physiology*, Vol. 7, Pt. 3, Berlin: Springer, pp. 307–331.

Maffei, L., Fiorentini, A. (1973), "The Visual Cortex as a Spatial Frequency Analyser," *Vision Res.* **13**, 1255–1267.

Malinvaud, E. (1978), *Statistical Methods of Econometrics*, Amsterdam: North-Holland.

Marks, W. B., Dobelle, W. H., MacNichol, E. F., Jr. (1964), "Visual Pigments of Single Primate Cones," *Science* **143**, 1181–1183.

Mishkin, E., Braun, L. (1961), *Adaptive Control Systems*, New York: McGraw-Hill.

Nachmias, J. (1972), "Signal Detection Theory and its Application to Problems in Vision," in: D. Jameson, L. H. Hurvich (Eds.), *Handbook of Sensory Physiology*, Vol. 7, Pt. 4, Berlin: Springer, pp. 56–77.

Ogle, K. N. (1950), *Researches in Binocular Vision*, Philadelphia: W.B. Saunders.

Pirenne, M. H. (1967), *Vision and the Eye*, London: Science Paperbacks.

Pirenne, M. H., Marriott, F. H. C. (1959), "The Quantum Theory of Light and the Psycho-Physiology of Vision," in: S. Koch (Ed.), *Psychology: A Study of a Science*, Vol. 1, New York: McGraw-Hill, pp. 288–361.

Poggio, G. F. (1984), "Processing of Stereoscopic Information in Monkey Visual Cortex," in: W. M. Cowan, W. E. Gall (Eds.), *Dynamic Aspects of Neocortical Functions*, New York: Wiley, pp. 613–635.

Ramachandran, V. S. (1988), "Perceiving Shape from Shading," *Sci. Am.* **259** (2), 76–83.

Regan, D., Frisby, J. P., Poggio, G. F., Schorr, C. M., Tyler, C. W., (1990), "The Perception of Stereodepth and Steromotion," in: L. Spillmann, J. S. Werner (Eds.), *Visual Perception/The Neurophysiological Foundations*, San Diego: Academic.

Reichardt, W. E., Poggio, T. (1981), *Theoretical Approches in Neurobiology*, Cambridge, MA: MIT Press, pp. 193–196.

Riggs, L. A., Wooten, B. R. (1972), "Electrical Measures and Psychophysical Data on Human Vision," in: D. Jameson, L. M. Hurvich (Eds.), *Handbook of Sensory Physiology*, Vol. 7, Pt. 4, Berlin: Springer, pp. 690–731.

Rose, A. (1973), *Vision: Human and Electronic*, New York: Plenum Press.

Rushton, W. A. H. (1964), "Colour Blindness and Cone Pigments," *Am. J. Optom. Am. Acad. Optom.* **41**, 265–282.

Schober, H., Rentschler, I. (1972), *Das Bild als Schein der Wirklichkeit*, München: H. Moos.

Shute, C. C. D. (1979) *The Mac Collough Effect*, Cambridge, U.K.: Cambridge Univ. Press.

Sixtl, F. (1967), *Meßmethoden der Psychologie*, Weinheim: Verlag Chemie.

van Doorn, A. J., Koenderink, J. J. (1982), "Temporal Properties of the Visual Detectability of Moving Spatial White Noise," *Exp. Brain Res.* **45**, 179–188.

Vieth, G. U. A. (1818), "Über die Richtung der Augen," *Ann. Physik (Leipzig)* **58**, 233–253.

von Holst, E., Mittelstaedt, H. (1950), "Das Reafferenzprinzip (Wechselwirkung zwischen Zentralnervensystem und Peripherie", *Naturwissenschaften* **37**, 464–476.

von Seelen, W., Mallot, H. A., Giannakopoulos, F. (1987), "Characteristics of Neuronal Systems in the Visual Cortex," *Biol. Cybern.* **56**, 37–49.

Walker, J. T., Saunders, R. D. (1985), "Form Perception: Some Effects of Brightness and Motion," *Percept. Psychophys.* **38**, 471–475.

Wasserman, G. S. (1978), *Color Vision: An Historical Introduction*, New York: Wiley.

Wilson, H. R., Bergen, J. R. (1979), "A Four Mechanism Model for Threshold Spatial Vision," *Vision Res.* **19**, 19–32.

Wyszecki, G., Stiles, W. S. (1967), *Color Science*, New York: Wiley.

Zetsche, C., Barth, E. (1990), "Image Surface Predicates and the Neuronal Encoding of Two-Dimensional Signal Variations," in: J. P. Allebach, B. E. Rogowitz (Eds.), *Human Vision and Electronic Imaging: Models, Methods, and Applications*, SPIE Proceedings No. 1249, Bellingham, WA: SPIE.

Zrenner, E., Abramov, J., Akita, M., Cowey, A., Livingstone, M., Valberg, A. (1990), "Color Perception," in: L. Spillmann, J. S. Werner (Eds.), *Visual Perception/The Neurological Foundations*, San Diego: Academic, pp. 163–204.

Further Reading

General

Davson, H. (Ed.) (1962), *The Eye*, New York: Academic, 3 volumes. Very comprehensive.

Jameson, D., Hurvich, L. M. (Eds.), (1972), *Handbook of Sensory Physiology*, Vol. 7, Pt. 4, Berlin: Springer. Not all themes covered, but very profound articles. Other volumes and parts are also useful.

Leibovic, K. N. (Ed.), (1990), *Science of Vision*, New York: Springer.

The Eyeball

Snyder, A. W., Menzel, R. (Eds.) (1975), *Photoreceptor Optics*, Berlin: Springer. Optical properties of photreceptors; also insect eyes. Very interesting physical effects including waveguides.

Stenström, S. (1964), *Optics and the Eye*, London: Butterworth. Optical system of the eye, optical and ophthalmological instruments.

Vinnikov, Ya. A. (1974), *Sensory Reception*, Berlin: Springer. Cytology, molecular mechanisms, and evolution, vision included.

Neuronal System

Pöppel, E., Held, R., Dowling, J. E. (1977), "Neuronal Mechanisms in Visual Perception," *Neurosci. Res. Prog. Bull.* **15** (3).

Spillmann, L., Werner, J. S. (1990), *Visual Perception/The Neurological Foundations*, San Diego: Academic.

Optical Illusions

Gombrich, E. H. (1960), *Art and Illusion*, New York: Pantheon. A brilliant report on the profound connections between sight and insight; many illustrations.

Kienle, G. (1968), *Die Optischen Wahrnehmungsstörungen und die Nichteuklidische Struktur des Sehraumes*, Stuttgart: G. Thieme. Interesting experiments with the perception of unusual optical configurations.

Pirenne, M. H. (1970), *Optics, Painting & Photography*, Cambridge, U.K.: Cambridge Univ. Press. Geometry of optical imaging in the eye, perspective, visual interpretation of projected or painted images, curvatures in visual geometry.

OPTICS, QUANTUM

See QUANTUM OPTICS

OPTICS, UNDERWATER

LEV S. DOLIN, *Institute of Applied Physics, Nizhnij Novgorod, Russia*
IOSIF M. LEVIN, *P.P. Shirshov Institute of Oceanology, St. Petersburg, Russia*

INTRODUCTION

Underwater optics is the science of the interaction between light and the natural waters (oceans, seas, rivers, and lakes). Solar radiation supplies most of the energy input to the sea and supports biological productivity. It determines water temperature and protective coloration of native underwater inhabitants, and makes possible the viewing of underwater scenes. Light backscattered by the sea and reflected from its surface gives information about suspended and dissolved substances, turbulent processes, near-surface wind velocity, and oil pollution. Artificial light sources make possible exploration of the sea bottom at large depths by means of imaging systems. Thus, underwater optics helps us to understand many vital processes in the ocean and gives the means of studying its ecological state and its natural resources.

From the theoretical point of view, hydrooptics is related to the optics of the atmosphere, astrophysics, and the theory of elementary particle scattering in matter, since the mathematical background of all these sciences is radiative transfer theory, which describes, on the phenomenological level, the transfer of radiant energy through a medium

3-527-28134-7/95/$5.00 + .50

absorbing, scattering, or emitting energy. Radiative transfer theory itself is a part of electromagnetic theory. The main subjects of experimental hydro-optics are, traditionally, the optical properties of natural waters and light propagation in the sea. It has been developed in close connection with geophysics, biology (especially the photosynthesis problem), ecology, and underwater engineering. Thus, Spinrad (1989) had every reason to say, "The beauty and detriment of hydrologic optics is that it has and always will be an eclectic science."

Hydro-optics is a relatively young science. The first measurements of the underwater radiant energy were performed in 1885; but till the middle of the 20th century experimental hydro-optics developed slowly because of the poor level of measuring instruments, although some important theoretical results had appeared already in 1939 (Gershun, Le Grand). In the 1960s, rapid development of both theoretical and experimental research began, as the interest in studies of the World Ocean had strongly increased and new technical means—in particular, lasers—had appeared. Since then, a great number of papers and a few reviews and monographs on different problems of hydro-optics have been published (Duntley, 1963; Jerlov, 1976; Preisendorfer, 1976; Monin, 1983; Gordon and Morel, 1983; Kirk, 1983; Shifrin, 1988; Zege *et al.*, 1991; Dolin and Levin, 1991; Dera, 1992; Mobley, 1994). Today, hydro-optics is an autonomous part of applied physics.

Within the wide range of problems and aspects of underwater optics, four main subjects seem worth distinguishing as the most important, both for fundamental studies of natural water properties and interaction between light and water and for various technical applications. These are

1. inherent optical properties of natural waters and their correlations with other physical, chemical, and biological water properties;
2. propagation of natural and artificial light in the sea;
3. optical methods for investigation of natural waters; and
4. underwater imaging.

These four subjects form the main contents of this article and determine the subtitles of its four sections.

The task of the article is to consider briefly all the main aspects of hydro-optics, giving preference to the theoretical methods providing approximate and simple final results. Special attention is paid to the aspects that are important from both industrial and social points of view, in particular, to estimates of bioproductivity of natural waters and to underwater imaging.

1. INHERENT OPTICAL PROPERTIES OF WATER (IOP)

Two kinds of hydro-optical characteristics are used, inherent (IOP) and apparent (AOP) optical properties. The IOP are inherent in the sense that their magnitudes for each wavelength depend only on the substances comprising the hydrosol. The AOP are functions of radiance and irradiance, in particular, downward $[E_d(z)]$ and upward $[E_u(z)]$ irradiances at a depth z. They depend not only on properties of the water itself, but also on illumination conditions. The AOP are generally easier to measure than the IOP, which is why some of the AOP, depending on the illumination condition only slightly, are often called quasi-inherent and included in the set of IOP. This section deals with inherent and quasi-inherent properties.

1.1 Definitions

The fundamental IOP are the attenuation coefficient and the volume scattering function (VSF). They describe the influence of an elementary volume of the medium on the elementary (narrow, monodirectional, monochromatic) light beam. The elementary volume is meant to be small enough to affect the light beam intensity only slightly, but, at the same time, much bigger than the average distance between local irregularities. This allows us to define the averaged characteristics of the substance.

For the elementary beam passing through the elementary volume ΔV with a length Δl, the radiant power lost through absorption and scattering, $-\Delta P$, is proportional to the initial radiant power P_0 and Δl:

$$\Delta P = -cP_0\Delta l, \tag{1}$$

where c is called the attenuation coefficient. The VSF $\beta(\Theta)$ describes the power ΔP_Θ of radiation scattered by the elementary volume within the limits of solid angle $\Delta\Omega$ at an an-

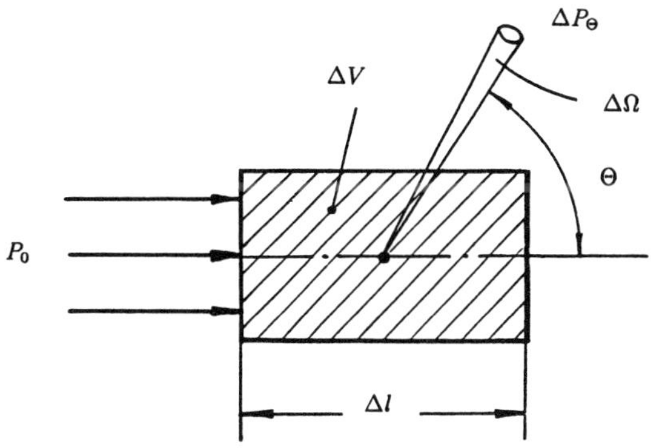

FIG. 1. An elementary light beam passing through the elementary water volume ΔV. P_0 is the initial radiant power, ΔP_θ the power scattered within the solid angle $\Delta\Omega$.

gle Θ with respect to the direction of incident light beam (Fig. 1).

The notation and definitions of the IOP are presented in Table 1. The IOP 3–9 in Table 1 are derived from the fundamental IOP. The absorption coefficient a and scattering coefficient b describe the absorption and integral scattering (that is, scattering over all directions) of radiation by the elementary volume. The scattering phase function $\tilde{\beta}(\Theta)$ is the angular distribution of scattered radiation. The definitions of the parameters b and $\tilde{\beta}(\Theta)$ imply the following normalization condition:

$$2\pi \int_0^{\pi} \tilde{\beta}(\Theta) \sin\Theta d\Theta = 1. \tag{2}$$

In order to take into account the state of polarization of incident and scattered beams, the scattering matrix is used instead of the VSF. For the description of transfer through the air-water surface another IOP is used, namely, the refractive index of water, n_w (in the visible spectral range $n_w \simeq 1.33$).

Quasi-inherent properties, namely, the vertical attenuation coefficient K and irradiance reflectance R, depend not only on the fundamental IOP, but also on the solar zenith angle ϑ_0, the state of the atmosphere, and the waves on the surface and thus are typical AOP. However, if one restricts the conditions of observation to the upper homogeneous layer of the sea, a cloudless sky, the sun not far from zenith, wind velocity $v_w \leq 17$ m/s, and wavelength $\lambda = 500$–550 nm, then K and R depend only on the IOP with an accuracy about 10%, as is shown in the right-hand sides of formulas 10 and 11 of Table 1.

The Secchi depth z_D (that is, the depth where a white disk of 300 mm in diameter becomes invisible) has been used as an esti-

Table 1. Inherent and quasi-inherent optical properties of water.

	Name	Formula	Dimension
		Fundamental IOP	
1	Attenuation coefficient	$c = -\dfrac{\Delta P}{P_0 \Delta l}$	m^{-1}
2	Volume scattering function (VSF)	$\beta(\Theta) = -\dfrac{\Delta P_\theta}{P_0 \Delta\Omega \Delta l}$	m^{-1} sr^{-1}
		Additional IOP	
3	Scattering coefficient	$b = 2\pi \int_0^{\pi} \beta(\Theta) \sin\Theta d\Theta$	m^{-1}
4	Absorption coefficient	$a = c - b$	m^{-1}
5	Scattering phase function	$\tilde{\beta}(\Theta) = \beta(\Theta)/b$	sr^{-1}
6	Scattering albedo	$\omega_0 = b/c$	dimensionless
7	Backscattering coefficient	$b_b - 2\pi \int_{\pi/2}^{\pi} \beta(\Theta) \sin\Theta d\Theta$	m^{-1}
8	Backscattering probability	$\tilde{b}_b = b_b/b$	dimensionless
9	Average cosine of $\tilde{\beta}(\Theta)$	$\langle\cos\Theta\rangle = 2\pi \int_0^{\pi} \cos(\Theta)\tilde{\beta}(\Theta) \sin\Theta d\Theta$	dimensionless
		Quasi-inherent properties	
10	Vertical attenuation coefficient	$K = -\dfrac{d \ln E_d(z)}{dz} \simeq 1.1(a + b_b)$	m^{-1}
11	Irradiance reflectance	$R = \dfrac{E_u(z)}{E_d(z)} \simeq 0.33 \dfrac{b_b}{a + b_b}$	dimensionless
12	Secchi depth	$z_D \simeq 7.5/c$	m

mate of transparency of the water since the middle of the 19th century. Because of the simplicity of measurement, this is the most well-examined optical parameter of the water. It also depends on the conditions of illumination; however, if the same restrictions (a high sun and a flat sea) are included, it may be considered as a function of c only, with accuracy 20% (in the green spectral range) (Gordon and Wouters, 1978).

1.2 Methods of Measuring the IOP

The apparent and quasi-inherent optical properties are measured by devices similar to the usual instruments intended for measuring irradiance and radiance. A typical modern marine spectral radiometer (Doss and Wells, 1992; Moore *et al.*, 1993) consists of collectors sensitive to upward and downward irradiance or radiance. Light from the collectors is usually relayed locally by fiber-optic cables leading to the slit position of a sealed spectrometer. The small size of these collectors minimizes the effect of self-shading, which reduces the accuracy of underwater AOP measurements (Gordon and Ding, 1992). For measurements of angular light distribution, underwater electro-optic camera systems have been used successfully (Voss, 1989b; Maffione *et al.*, 1991).

To measure the IOP in a laboratory or *in situ*, special methods and devices are used. There are also remote sensing methods, which have the advantage of speed (especially from a satellite), but allow measurement of only the parameters of the upper sea layer.

The attenuation coefficient c is determined by the change in radiant power of a light beam that has traversed a water layer of thickness l, on the basis of the Bouguer law:

$$P(l) = P_0 \exp(-cl), \tag{3}$$

which is the consequence of Eq. (1) and is valid for $l \ll c^{-1}$. The remote methods for determination of c are based on measuring the radiance of backscattered light of a continuous or pulsed collimated beam, so that radiance of two cuts of the illuminated column of water (in the former case) and at two different moments (in the latter) are compared. The VSF is usually measured by means of rotating a receiver with a narrow viewing angle around a small scattering volume. The value of the scattering coefficient is determined either by integrating the scattered light using an integrating sphere or by measuring the amount of forward-angle scattering (Fournier *et al.*, 1992; Dolin *et al.*, 1994). A few methods and devices are available for determining the absorption coefficient. In particular, there are the reflective tube absorption meter (Zaneveld *et al.*, 1990, 1994), the integrating cavity absorption meter (Pope *et al.*, 1990; Fry *et al.*, 1992), and also photoacoustic as well as photothermal techniques (Trees and Voss, 1990). When it is determined *in situ*, parameters of the natural light field and the light field from a point source are generally used. In the former case, the method is based on the Gershun equation,

$$a = -\frac{1}{E_0}\frac{d}{dz}(E_d - E_u), \tag{4}$$

where E_0 is a scalar irradiance. The irradiances E_d, E_u, and E_0 are expressed by simple integrals of radiance $L(z,\vartheta,\phi)$ of a light field at a depth z in the direction determined by polar and azimuth angles ϑ, ϕ. Thus, having measured $L(z,\vartheta,\phi)$ or directly E_d, E_u, and E_0, it is possible to determine a (Voss, 1989a; Doss and Wells, 1992). However, measurements of E_0 and $L(\vartheta,\phi)$ require rather complicated experimental devices, and so approximate formulas are often used to determine a on the basis of measurements of $E_d(z)$ and $E_u(z)$ only (Kirk, 1994, and the formulas for K and R in Table 1). The method of determining the absorption coefficient from the point isotropic source is based on Gershun's law in spherical coordinates and requires the measurements of radiance and irradiance from an underwater point source (Maffione *et al.*, 1993).

1.3 Physical Factors of IOP

Light absorption and scattering in the sea are due to pure water, dissolved organic and nonorganic substances, and suspended particles of both mineral and organic origin. As the set of salts (NaCl, KCl, MgCl, $MgSO_4$, $CaSO_4$, and some others) dissolved in the water is relatively constant, hydro-optics usually deals with the properties of "pure seawater," meaning the water with dissolved salts, but without organic and suspended substances. Detailed tables for absorption and scattering coefficients a_w and b_w for pure seawater are given by Morel and Prieur (1977), Smith and Baker (1981), and Shifrin (1988). Along the whole spectral range, a_w (see Fig. 2) is much greater than b_w. The angular distribution of scattering light by pure water and pure seawater is $\beta_w(\Theta) \simeq \beta_w(90°)(1 + 0.84 \times \cos^2\Theta)$, that is, the forward and backward scattering are equal.

Among the organic matters dissolved in water and affecting light, the main one is the so-called "yellow substance," which consists of phenol-humic acids, carbohydrate-humic acids (or melanoidines), and some other compounds. The spectral distribution of absorption by yellow substance (Fig. 2) can be approximated by

$$a_y(\lambda) = a_0(\lambda_0) \exp[-\mu_y(\lambda - \lambda_0)], \qquad (5)$$

where λ_0 is the given wavelength (nm), usually 390 or 440 nm, and μ_y is the index, varying, according to various data, from 0.011 to 0.017 nm^{-1}. The amount of yellow substance in water is rather unstable; the value of a_y(390 nm) in the open ocean ranges from 0.001 to 0.1 m^{-1}, and increases up to 0.6–0.8 m^{-1} in coastal waters.

The particles suspended in natural waters are mineral (terrestrial) particles, phytoplankton cells, bacteria, and detritus (residua of phyto- and zooplankton). Their scattering effect is determined mainly by concentration, sizes, and refractive index n_p. In surface waters, the concentration of suspended particles is 0.05–0.5 g m^{-3}; in deep waters, 0.001–0.25 g m^{-3}. In coastal waters and in rivers, it can be 100 or even 1000 g m^{-3}.

The particle size distribution $N(r_p)$ is usually approximated by the Junge distribution $N(r_p) \sim r_p^{-\mu_p}$, where r_p is particle radius, and $\mu_p = 3$–5. For the mineral particles in the ocean, $r_p = 0.01$–2 μm; for the organic ones, $r_p = 1$–20 μm. In coastal waters and in rivers, it can be much larger (up to 100 μm).

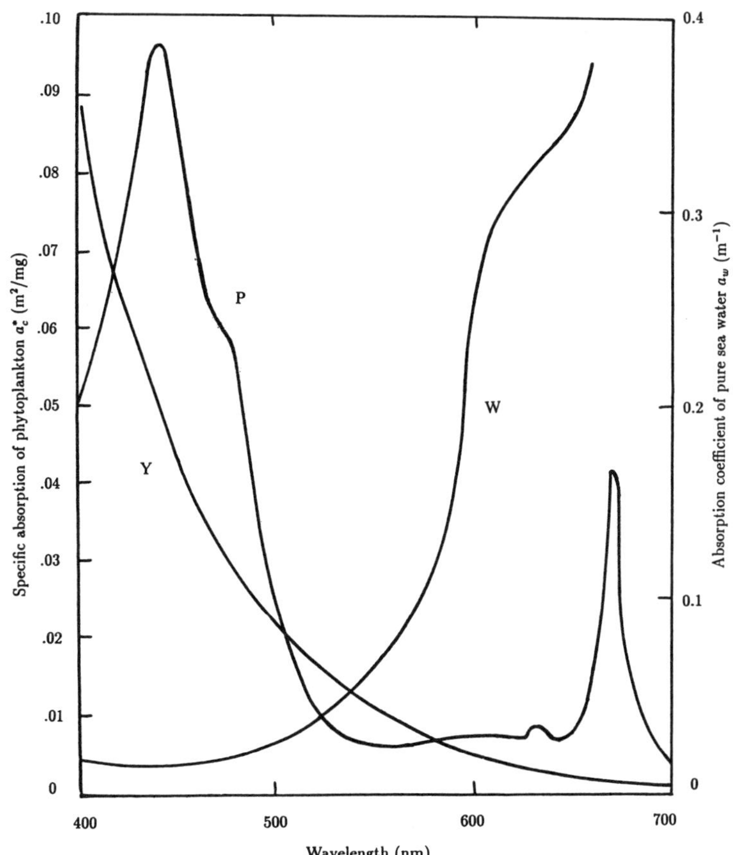

FIG. 2. Spectral absorption of sea water ingredients: pure sea water (W), phytoplankton (P), and yellow substance (relative units) (Y).

The relative refractive index $n = n_p/n_w$ is generally accepted to be $n = 1.15$ for mineral particles and $n = 1.02$ for the organic ones.

The VSF can be expressed in terms of the scattering function of an individual particle β':

$$\beta(\lambda,\Theta,n) = N_p \int_0^\infty N(r_p)\beta'(\lambda,\Theta,n,r_p)dr_p, \quad (6)$$

where N_p is the number of particles per unit volume. The values of β' are calculated according to the Mie theory (Van de Hulst, 1957). The VSF strongly depends on particle size: The larger the particles are, the more light is scattered within small angles.

Absorption in mineral particles is not large and thus is generally neglected. As for organic particles, in particular, phytoplankton cells, they have reasonable influence on the value of a. The specific absorption by phytoplankton pigments (mainly by chlorophyll a) $a_c^*(\mathrm{m^2/mg}) = a_c(\mathrm{m^{-1}})/N_c(\mathrm{mg/m^3})$ (a_c and N_c are absorption coefficient and concentration of phytoplankton) is presented in Fig. 2. According to the value of N_c, there are three types of ocean water: oligotrophic (poor, $N_c \leq 0.1$ mg/m^3), mesotrophic (average), and eutrophic (rich, $N_c > 1$ mg/m^3).

1.4 Data on Spatial and Spectral Distribution of IOP in Natural Waters

Among all the optical properties of water, the most well-studied are the easily measurable quasi-inherent ones, that is, the Secchi depth z_D, the vertical attenuation coefficient K in the upper sea layer for $\lambda = 500$–550 nm, and the attenuation coefficient c for the same spectral region. For these parameters, there exist detailed maps (Monin, 1983; Arnone *et al.*, 1984; Simonot and Le Treut, 1986). The value of the Secchi depth is generally $z_D = 15$–35 m in the open ocean and $z_D = 3$–7 m in coastal waters. The maximum z_D of 62 m has been registered in the Sargasso Sea.

The scattering coefficient is generally $b = 0.1$–0.2 m^{-1} for the surface layer of the open ocean, and $b = 0.05$–0.1 m^{-1} in the deep waters. However, in the purest waters, at large depth lesser values of b are also possible, while in coastal waters b can even be 2–3 m^{-1}. The scattering coefficient slowly grows with decrease of wavelength. The depth distribution varies from uniform to strongly heterogeneous, with layers of especial turbidity where the value of b significantly increases.

Table 2. Wavelength and value of the minimum adsorption.

	λ_{min} (nm)	a_{min} (m^{-1})
Pure waters at large depths	470–490	0.003–0.01
Surface waters in the open ocean	510	0.01–0.04
Turbid coastal waters	550	0.04–0.07
Very turbid waters (Baltic Sea)	570	0.15–0.25

In contrast with b, the absorption coefficient a depends less on the depth, and more on the wavelength. The latter is due to the spectral selectivity of the matter involved (pure water, phytoplankton, and yellow substance); see Fig. 2. It varies the most in the short-wave spectral range. For $\lambda > 550$ nm, a usually does not differ much from the value of a_w for pure seawater. The spectral minimum of absorption λ_{min} depends on a_{min}, so that the stronger the absorption, the greater λ_{min}; see Table 2.

The VSF $\beta(\Theta)$ is highly anisotropic. More than 95% of light is scattered forward (at $\Theta = 0°$–$90°$), while $\beta(1°)/\beta(90°) = 10^4$–$10^5$. The backscattering part of β ($\Theta = 90°$–$180°$) is more or less uniform. Table 3 gives typical values of some parameters of the VSF ($\lambda = 550$ nm) in various regions of the World Ocean, according to Kopelevich's data. The wavelength does not have much influence on VSF shape. That is, for a decrease of λ from 700 to 400 nm the backscattering probability $\tilde{b}_b$ increases by only 25%.

The scattering albedo $\omega_0 = b/(a + b)$ within the spectral range 500–550 nm generally changes from 0.6 through 0.8 (the purer the

Table 3. Typical values of VSF parameters.

Region	$\langle\cos\Theta\rangle$	$\tilde{b}_b$	$\beta(1°)$ (m^{-1} sr^{-1})	$\beta(45°)$ (m^{-1} sr^{-1})
Southern Ocean	0.96	0.013	19	0.0020
Atlantic Ocean (north)	0.90	0.030	5.5	0.0029
Indian Ocean (south)	0.92	0.026	9.3	0.0025
Pacific Ocean (north)	0.94	0.021	11	0.0017

water, the smaller ω_0); e.g., in the Indian Ocean its average value is $\omega_0(550) = 0.75$. The value of ω_0 can increase to 0.9 in coastal waters, because of large values of b, and in pure waters because of small values of a. It can also decrease to 0.3 in the ultraviolet and red spectral ranges, because of the increase of a.

The sea scattering matrices are not studied sufficiently. Most often, the degree of polarization of the light scattered at an angle of 90° is measured. This value ranges from 40% to 80%, decreasing by increase of $\beta(90°)$. The more the scattering angle deviates from 90°, the less is the degree of polarization (Ivanoff, 1974; Voss and Fry, 1984).

Besides scattering and absorption, there also occur in water Raman scattering (Marshall and Smith, 1990; Stavn, 1993); light-induced fluorescence emitted by dissolved organic matter, chlorophyll, and pollution; and bioluminescence emitted by bacteria and zooplankton (Karabashev, 1987; Preisendorfer and Mobley, 1988). Chlorophyll *a* fluoresces in the red spectral range. The bioluminescence spectrum has a maximum at 450–550 nm. The duration of pulses ranges from 0.1 to 10 s. A single organism is capable of providing irradiance 10^{-2}–10^{-1} μW/cm^2 at a distance of 1 cm.

The IOP do not remain constant but uninterruptedly vary with time (Burenkov, 1983; Cullen *et al.*, 1994). Small-scale and synoptical variabilities of IOP can usually be distinguished. The first are mainly due to turbulent and biological processes in the ocean and to internal waves. The representative data on synoptical variability were obtained during the joint Soviet-American expedition POLIMODE in the Sargasso Sea, where the IOP were measured during almost two years in a rather small test area. A strong effect of eddies and current meanders on the optical structure of water was found.

1.5 Few-Parametric Models of IOP

One of the main tasks of underwater optics is the development of simple models for relatively accurate calculations of the spatial, angular, and spectral distributions of IOP, by means of a few measured parameters. There are two types of such models: physical, that is, based on account of the main physical factors of IOP, and empirical, resulting from statistical processing of the measurements.

1.5.1 Physical Models The model presented by Kopelevich (1983, 1991) is based on the assumption that the scattering is determined by pure water and the large (more than 1 μm) organic and small (less than 1 μm) mineral suspended particles, while the absorption is determined by pure seawater, phytoplankton, and yellow substance. Correspondingly the VSF and the absorption coefficient are

$$\beta(\lambda,\Theta) = \beta_w(550,\Theta)\left(\frac{550}{\lambda}\right)^{4.2} + \beta_l^*(550,\Theta)N_l\left(\frac{550}{\lambda}\right)^{0.3} + \beta_s^*(550,\Theta)N_s\left(\frac{550}{\lambda}\right)^{1.7}, \tag{7}$$

$$a(\lambda) = a_w(\lambda) + N_c a_c^*(\lambda) + a_y(\lambda), \tag{8}$$

where β_l^* and β_s^* are the specific VSF of large and small particles (m^2 cm^{-3} sr^{-1}), and N_l and N_s are the corresponding volume concentrations (cm^3 m^{-3}). The specific absorption of phytoplankton $a_c^*(\lambda) = a_c(\lambda)/N_c$ is shown in Fig. 2, and absorption of yellow substance $a_y(\lambda)$ is determined by Eq. (5) (where the recommended values are $\lambda_0 = 390$ nm, $\mu_y = 0.017$ nm^{-1} within the range $\lambda = 280$–500 nm, and $\lambda_0 = 500$ nm, $\mu_y = 0.011$ nm^{-1} within the range $\lambda = 500$–700 nm). The values of the parameters β_w, β_l^*, and β_s^* involved in Eq. (7) for $\lambda = 550$ nm and some angles Θ are presented in Table 4. By using the VSF for $\lambda = 550$ nm measured in two directions [e.g., $\beta(550,1°)$ and $\beta(550,45°)$] and the data from Table 4, two unknowns (N_l and N_s) can be found from two

Table 4. Components of the physical model: water (β_w), small (β_s^*) and large (β_l^*) particles.

Θ (deg)	β_w (m^{-1} sr^{-1})	β_l^* (m^2 cm^{-3} sr^{-1})	β_s^* (m^2 cm^{-3} sr^{-1})
0	2.4×10^{-4}	140	5.3
1	2.4×10^{-4}	46	5.2
6	2.3×10^{-4}	1.1	3.9
15	2.3×10^{-4}	0.05	1.3
45	1.8×10^{-4}	6.2×10^{-4}	9.8×10^{-2}
90	1.3×10^{-4}	6.3×10^{-5}	1.2×10^{-2}
135	1.8×10^{-4}	2.0×10^{-5}	7.4×10^{-3}
180	2.0×10^{-4}	7.0×10^{-5}	8.1×10^{-3}

equations (7). Then $\beta(\lambda,\Theta)$ can be calculated from Eq. (7) for any values of λ and Θ.

In order to calculate $a(\lambda)$, only N_c and $a(\lambda_0)$ are to be measured. Then $a_y(\lambda_0) = a(\lambda_0) - a_w(\lambda_0) - N_c a_c^*(\lambda_0)$, and the value of a for any λ is to be calculated by means of Eqs. (8) and (5).

The scattering coefficient $b(\lambda)$ is found by changing in Eq. (7) $\beta(\lambda,\Theta)$ to $b(\lambda)$ and $\beta_w(550,\Theta)$, $\beta_l^*(550,\Theta)$, and $\beta_s^*(550,\Theta)$ to $b_w(550) = 0.17 \times 10^{-2}\ \mathrm{m}^{-1}$, $b_l^*(550) = 0.312\ \mathrm{m}^2\ \mathrm{cm}^{-3}$, and $b_s^*(550) = 1.34\ \mathrm{m}^2\ \mathrm{cm}^{-3}$, respectively.

Another model, which is used in a number of papers (e.g., Prieur and Sathyendranath, 1981; Sathyendranath *et al.*, 1989), assumes that pure water, phytoplankton, and nonchlorophyllous particles determine scattering and absorption, the latter being also affected by yellow substance:

$$b(\lambda) = b_w(\lambda) + b_c(\lambda) + b_x(\lambda), \tag{9}$$

$$b_b(\lambda) = \tilde{b}_{bw} b_w(\lambda) + \tilde{b}_{bc} b_c(\lambda) + \tilde{b}_{bx} b_x(\lambda), \tag{10}$$

$$a(\lambda) = a_w(\lambda) + a_c(440)a_c'(\lambda) + a_x(440)a_x'(\lambda) + a_y(440)a_y'(\lambda), \tag{11}$$

where $a_c'(\lambda) = a_c(\lambda)/a_c(440)$, $a_x'(\lambda) = a_x(\lambda)/a_x(440)$, and $a_y'(\lambda) = a_y(\lambda)/a_y(440)$ are dimensionless specific absorption coefficients (relative to $\lambda = 440$ nm) for phytoplankton, nonchlorophyllous particles, and yellow substance, respectively. Detailed tables of $a_c'(\lambda)$, $a_x'(\lambda)$, and $a_y'(\lambda)$ are given by Prieur and Sathyendranath (1981). The backscattering probability for pure water is $\tilde{b}_{bw} = 0.5$, for phytoplankton $\tilde{b}_{bc} = 0.005$, and for particles $\tilde{b}_{bx} = 0.015$. The values of $b_c(\lambda)$, $b_x(\lambda)$, $a_c(440)$, and $a_x(440)$ involved in Eqs. (9)–(11) are expressed through the phytoplankton concentration N_c and scattering coefficient $b(550)$ by simple empirical equations, e.g., $a_c(440) = 0.06N_c^{0.602}$ (N_c in mg/m^3). This model provides the full spectral distribution of $b(\lambda)$, $a(\lambda)$, and $b_b(\lambda)$, granted that N_c, $a(440)$, and $b(550)$ are measured. Other variants of the same model do not distinguish between scattering due to phytoplankton and nonchlorophyllous particles (as Kopelevich's model), while the absorption by the nonchlorophyllous particles is neglected (Sathyendranath and Platt, 1989; Morel, 1991).

The models considered satisfactorily predict average values of IOP. The variability in measured IOP values for a given chlorophyll concentration in pure ocean waters can be explained in terms of variability in the detailed microbial composition of the water (Mobley and Stramski, 1994).

1.5.2 Empirical Models The first empirical model was the waters classification presented by Jerlov (1976), who distinguished ten types of natural waters (I, IA, IB, II, III for ocean, 1, 3, 5, 7, 9 for coastal waters) and found the typical spectral distribution of vertical attenuation coefficient K in the upper homogeneous layer for each. For $K(\lambda_0)$ measured for one wavelength λ_0, this classification gives the full spectral distribution of $K(\lambda)$. Similar models for $K(\lambda)$ constructed according to more recent measurements are presented by Pelevin (1983) and Smith and Baker (1978). However, data on $K(\lambda)$ make it possible to find an approximate spectrum of absorption only for $b_b \ll a$ (see Row 10 in Table 1).

Another model, which was suggested by Kopelevich (1983,1991), is based on statistical analysis of a large amount of measured absorption and scattering spectra by means of principal components analysis, that is, with the help of expansion of functions $\log\beta(\Theta)$ and $a(\lambda)$ along the eigenvectors of the covariance matrix:

$$\log\beta(\Theta) = \langle \log\beta(\Theta) \rangle + \sum_{i=1}^{3} k_i \psi_i(\Theta),$$
$$a(\lambda) = \langle a(\lambda) \rangle + \sum_{i=1}^{2} k_i' \psi_i'(\lambda). \tag{12}$$

The average values $\langle \log\beta(\Theta) \rangle$ and $\langle a(\lambda) \rangle$ and the values of the eigenvectors ψ_i and ψ_i' are presented by Kopelevich (1983). Given measured values of $\beta(\Theta)$ for three angles (the recommended values are 1°, 6°, and 45°), $a(\lambda)$ for two wavelengths (e.g., 390 and 550 nm), and coefficients k_i and k_i' found from Eq. (12), the model makes it possible to find the angular distribution of $\beta(\Theta)$ as well as the spectral distribution of $a(\lambda)$.

1.5.3 Vertical Structure of IOP The models discussed deal with the upper layer (0–50 m) of the ocean. In order to find IOP in deeper waters, a three-layer structure is included in a model (Kopelevich, 1991): depths (1) 0–50 m, (2) 50–100 m, (3) 100–200 m. Scattering coefficients in various layers are found according to Eq. (7). The concentrations of large and small suspended particles in the deep layers ($N_{l2},N_{l3},N_{s2},N_{s3}$) are calculated through those in the upper layer (N_{l1},N_{s1}) as follows: $N_{l2} = 0.1(4 - 0.36/N_{l1})$ cm^3 m^{-3}, $N_{l3} = 0.05(4 - 0.36/N_{l1})$ cm^3 m^{-3}, $N_{s2} = 0.03$ cm^3 m^{-3}, $N_{s3} = 0.02$ cm^3 m^{-3}. Absorption coefficients are found according to Eq. (12), where the coefficients k_i' are taken to be half their values for the upper layer. Different models that describe the vertical structure of chlorophyll distribution together with absorption and scattering in pure ocean waters are also used (Morel and Berthon, 1989; Kitchen and Zaneveld, 1990).

1.5.4 Simplified Semiempirical Models The models discussed require a few (but no more than three) measurements of IOP. Sometimes it is necessary to estimate the set of IOP without any measurements at all, using only the maps of z_D and K(500–550 nm) distributions in the World Ocean (Monin, 1983; Arnone *et al.*, 1984; Simonot and Le Treut, 1986). Simplified models that are used for this purpose give only approximate values of IOP. In order to find the distribution of $a(\lambda)$, one can find $K(\lambda)$ from K(500–550 nm) (see Sec. 1.5.2), and $a(\lambda)$ from $K(\lambda)$ for $b_b = 0$ (see Row 10 in Table 1). For evaluating b, there is a semiempirical model (Dolin and Levin, 1991) where, on the basis of an accurate theory of Secchi depth and measured data on c and ω_0, approximate dependences $cz_D(\omega_0)$ and $z_D(\omega_0)$ are built for the spectral region 500–550 nm (Fig. 3). Those dependences can be used to find the value of $b = c\omega_0$, which can be taken to be approximately constant within the whole spectral range. For the parameters $\tilde{b}_b$ and $\langle\cos\Theta\rangle$ one can use an approximate semiempirical formula (Man'kovsky, 1984) $(1 - \tilde{b}_b)/\tilde{b}_b = 350b + 10$ and the regression

$$\langle\cos\Theta\rangle = 1 - 2.9[1 + (1 - \tilde{b}_b)/\tilde{b}_b]. \tag{13}$$

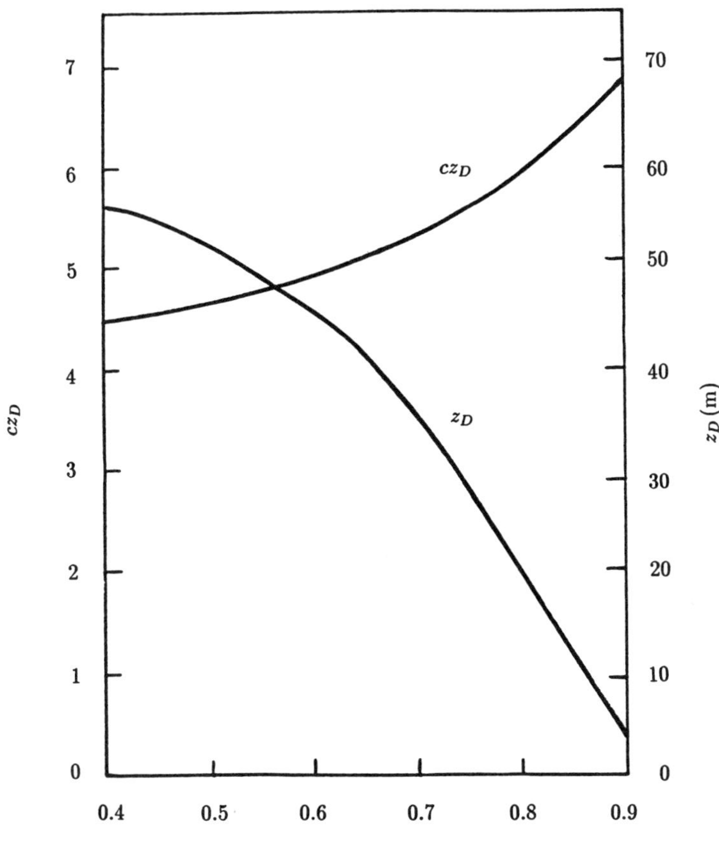

FIG. 3. Semiempirical dependences of Secchi depth z_D and dimensionless optical Secchi depth (cz_D) on the scattering albedo in the spectral region 500–550 nm.

2. THE UNDERWATER LIGHT FIELDS (ULF)

Light propagation in the sea can be expressed quantitatively in terms of the spectral, spatial, and angular distributions of radiance from various light sources, both natural and artificial. Such distributions are termed the underwater light fields (ULF). They are affected by a number of factors shown in Fig. 4. The structure of the ULF depends mainly on the properties and the position of the light source as well as on the IOP. The natural (solar) ULF radiance is also influenced by the atmosphere and the sea surface. The same may be said about laser radiation directed into the sea from a satellite or an airplane. The sea bottom takes part in forming the natural ULF if the sea depth is small enough. Besides the factors that have a direct influence on the structure of ULF, Fig. 4 presents the geophysical fields (wind, sea currents, etc.) that influence it through changing the optical properties of the water or the surface.

This section is aimed at presenting the main results (both theoretical and experimental) on the ULF, which gives information as to how their properties depend on the parameters of the light source, the IOP, and the state of the sea surface. The problems related to the geophysical fields, the underwater targets, and the sea bottom are considered in Secs. 3 and 4.

2.1 Radiative Transfer Equation (RTE) and the Methods of Solving It

The central problem of the ULF theory is the correct account of the multiple scattering of light in water. The theory deals with photometric quantities and is based on the mathematical techniques of linear transport theory. The ULF are described by the spectral radiance $L_\lambda(t,\mathbf{r},\mathbf{\Omega})$, which is, in the general case, a function of time t, point of space $\mathbf{r}$ (with coordinates x,y,z), and direction specified by the unit vector $\mathbf{\Omega}$ (with polar and azimuth angles ϑ,ϕ). The parameters of the light polarization are also used, as well as a number of quantities derived from L_λ, such as radiance in a given spectral region $L = \int_{\lambda_1}^{\lambda_2} L_\lambda d\lambda$ and irradiances E_0 and E. The scalar irradiance is $E_0(t,\mathbf{r}) = \int L d\Omega$, where $d\Omega = \sin\vartheta d\vartheta d\phi$ is a differential element of solid angle, the integration to be taken over the entire solid angle 4π sr. The irradiance of a

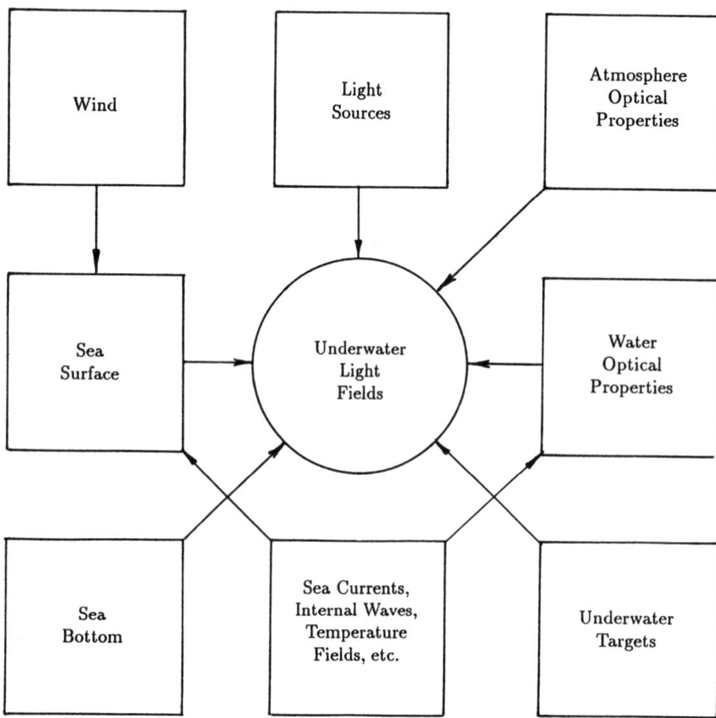

FIG. 4. The factors directly or indirectly affecting the ULF parameters.

plane element with a given orientation is $E(t,\mathbf{r}_s,\mathbf{\Omega}_s) = \int |\eta| L d\Omega$, where $\eta = (\mathbf{\Omega} \cdot \mathbf{\Omega}_s)$ and $\mathbf{\Omega}_s$ is a unit normal to the plane element at the point $\mathbf{r}_s$, the integration to be taken over a solid angle 2π for $\eta \geq 0$ or $\eta \leq 0$, depending on the side illuminated. For the natural ULF, irradiances of the upper (E_d) and the lower (E_u) sides of a horizontal plane are generally considered.

The calculation of the ULF from a given source amounts to solving the radiative transfer equation (RTE) for the spectral radiance,

$$\left(v^{-1}\frac{\partial}{\partial t} + \mathbf{\Omega}\cdot\nabla + c\right)L_\lambda = L_* + Q_\lambda, \qquad (14)$$

with necessary boundary conditions. The left-hand side of the RTE contains derivatives with respect to time and direction; v is the light speed in water, $Q_\lambda(t,\mathbf{r},\mathbf{\Omega})$ the spectral intensity density (radiant power per unit volume per unit solid angle) of sources. Scattering and absorption are taken into account by means of the term cL_λ and the path function $L_*(t,\mathbf{r},\mathbf{\Omega}) = \int L_\lambda(t,\mathbf{r},\mathbf{\Omega}')\beta(\tilde{\theta})d\Omega'$, where $\tilde{\theta}$ is the angle between $\mathbf{\Omega}$ and $\mathbf{\Omega}'$, $d\Omega' = \sin\vartheta' d\vartheta' d\phi'$ is a differential element of solid angle around the direction $\mathbf{\Omega}'$, and the integration is to be taken over the entire solid angle 4π [for c and $\beta(\Theta)$ see Table 1]. The radiance L also satisfies Eq. (14) with sources $Q = \int_{\lambda_1}^{\lambda_2} Q_\lambda d\lambda$ for $\Delta\lambda = \lambda_2 - \lambda_1$ small enough [$\Delta\lambda \ll c/(dc/d\lambda)$]. With this in mind, the terms and indices that indicate the spectral properties are omitted below where possible. Generally speaking, a theory of ULF should be based on a system of equations for Stokes parameters, which would take into account the polarization. However, the results of calculating L provided by these equations and by a more simple Eq. (14) differ slightly.

The main subjects of ULF theory are as follows: How do the natural irradiance and its spectral composition change with depth and how much natural radiation emerges from the sea? How do the radiant power of a narrow beam and the irradiance distribution in its cross section attenuate as the distance from the light source is increased? How does water affect the angular radiance distribution of a point source? How is the shape of the laser pulse transformed as it propagates in the water? The answers to these questions are necessary for solving the problems of photosynthesis, ecological monitoring, underwater imaging, optical communications, etc.

The important problem of the ULF theory is the development of mathematical models for the ULF of a unidirectional point source (UPS), since the ULF of an arbitrary source may be expressed in terms of it (Case and Zweifel, 1967).

The theory of ULF is, to a considerable extent, based on approximate analytical solutions of the RTE, which are found with account taken of the peculiarities of the IOP and ULF—that is, the small value of the backscattering probability $\tilde{b}_b$, the appreciable difference of scattering albedo ω_0 from 1, and, as a consequence, the strong peakedness of the angular radiance distribution in the direction from source to receiver. These are, first of all, various small-angle approximations, and formulas for the diffuse component of ULF in the approximation of single scattering at large angles.

The small-angle approximation for the light field of a UPS is found from the RTE, simplified on the basis of the small value of the variance of ϑ (ϑ is the angle of photon deviation from the beam axis $\mathbf{z}$). This approximation successfully describes the irradiance distribution for a narrow beam if the distance from the source is not too great. The expression for UPS radiance has the simplest form in the "small-angle diffusion" approximation. It requires the photons to be scattered many times, but the angles of their deviation from the $\mathbf{z}$ direction to be small. This approximation, in contrast with the small-angle one, badly describes the detailed structure of the ULF, but is good for evaluation of the integral ULF parameters (Ishimaru, 1978).

A refined version of the small-angle diffusion approximation (the so-called self-similar approximation) takes into account the distribution of photon multipaths. As a result, the irradiance and the shape of the light pulse can be successfully evaluated at any depth or distance from the source (Dolin, 1983a). To calculate the diffusion component of ULF, which determines backscattered solar radiation in the sea and backscattered lidar signals, the approximation of single scattering at large angles (the so-called quasisingle approximation) is used (Dolin and Savel'ev, 1971; Gordon, 1973; Golubitskiy *et al.*, 1974). The most widespread numerical method of studying ULF is to use Monte Carlo simulation, which, as distinct from field experiments, easily provides a wide range of IOP, source parameters, and experiment geometry. The Monte Carlo technique (*q.v.*) is very convenient for corroborating the approximate analytical solutions of the RTE, and it is used for detailed investigations of the ULF, both natural (Plass and Kattawar, 1972; Gordon *et al.*, 1975; Gordon, 1989) and artificial (Funk, 1973). Effective numerical models that combine Monte Carlo simulations with algebraic and analytical technique have been applied for computing the underwater radiance and irradiance with regard to a rough sea surface and sea bottom (Mobley, 1989; Mobley *et al.*, 1993).

In the next subsections, several solutions of the RTE in the above-listed approximations are considered.

2.2 Solar Radiation in the Sea

The modern concept of the daylight distribution in the sea resulted mainly from experimental investigations (e.g., Tyler, 1960; Jerlov, 1976). According to experimental data, most of the daylight energy is absorbed in the upper sea layer of about 20 m. A relatively small part of it returns to the atmosphere by reflection from the sea surface and by backscattering within the water (a few percent). The spectral density of downward and upward radiation decreases with increase of depth (z) according to a law close to the exponential, the spectral composition growing poorer and the angular distribution being reorganized.

As is shown in Fig. 5, the radiance angle distribution immediately under the sea surface in cloudless weather in the plane of the sun (curve 1) has a narrow peak in the direction of the refracted image of the sun, that is for $\vartheta = \vartheta_{0w} = \arcsin(\sin\vartheta_0/n_w)$, where ϑ_0 is the solar zenith angle. For downward radiation, the refraction angle cannot be more than the critical angle $\vartheta_c = \arcsin(1/n_w) \simeq 48.6°$ (equal to the angle of total internal reflection). Thus, in cloudy weather (under diffuse illumination) the radiance angle distribution under the surface is stepped (curve 1′). If the day is clear, the radiance decreases abruptly at $\vartheta = \vartheta_c$ (Snell's circle) as well. The light gets into the interval $\vartheta_c < \vartheta < 90°$ as a result of backscattering in the water and of reflection from the surface. Almost all the light scattered at an angle $180° - \vartheta_c < \vartheta < 180°$ goes out from the sea into the atmosphere.

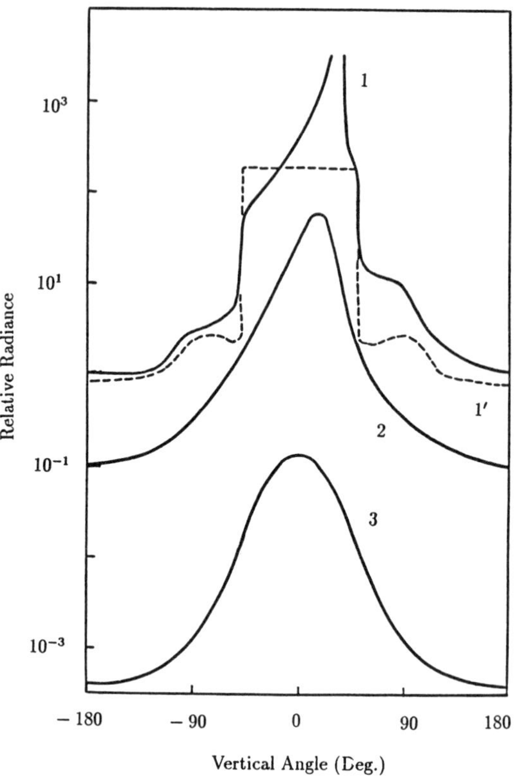

FIG. 5. Radiance distributions in the vertical plane of the sun: just beneath the sea surface (curves 1, 1′), at optical depths $\tau = cz \sim 5$ (curve 2) and $\tau \sim 20$ (curve 3) for a clear (1,2,3) and a cloudy (1′,3) day. The vertical angle ϑ is measured from the nadir direction.

As the optical depth $\tau = cz$ increases, the radiance distribution loses details (the sun disk image, Snell's circle) and becomes less sharply peaked, while the direction of its maximum shifts toward the nadir (curve 2). At last, at the depth $z \sim 20/c$, the distribution becomes asymptotic (curve 3), independent of the surface illumination conditions and determined only by the IOP.

The downward spectral irradiance attenuates with increase of depth according to the law

$$E_d(z) = E_d(0) \exp\left[-\int_0^z K(z)dz\right],$$

K being slightly greater than $a(z)$, but still less than $c(z)$. The changes in the spectral distribution of the downward irradiance are due to the dependence on $a(\lambda)$. Beginning with z of about 20 m, there appears a clear maximum at a wavelength of minimum absorption λ_{min} (see Table 2). While the depth increases, the spectrum becomes more and more narrow; for $z \sim$ 50–100 m, it has a width of about 100 nm.

The upward spectral irradiance E_u generally amounts to a few percent of E_d, and the ratio $R = E_u/E_d$ slightly depends on depth. The spectral distributions of E_u and E_d for z > 5–10 m do not differ considerably.

The quasisingle approximation of the transfer theory sets simple dependences (Gordon, 1973, 1989; Gordon *et al.*, 1975; Golubitskiy and Levin, 1980):

$$K = A_K(a + b_b), \tag{15}$$

$$R = A_R X, \quad \rho = A_\rho X, \quad X = b_b/(a + b_b), \tag{16}$$

where $\rho = \pi L_u/E_d$ is the radiance coefficient, L_u is the upward radiance, and A_K, A_R, and A_ρ take into account the influence on K, R, and ρ by the condition of the illumination. For direct sunlight, $A_K \simeq 1.04/\cos\vartheta_{0w}$, for diffuse illumination $A_K \simeq 1.25$; A_R ranges from 0.31 (the sun in zenith) to 0.37 (total cloudiness); $A_\rho = 0.5/(1 + \cos\vartheta_{0w})$ for direct sunlight, $A_\rho = 0.27$ under diffuse illumination.

More accurate and complete data on how the downward and upward irradiances as well as the water-leaving radiance just above the sea surface depend on the conditions of irradiation and on the IOP can be found from dependences that generalize the results of Monte Carlo simulations (Kirk, 1984, 1994; Gordon, 1989; Greysukh and Levin, 1990; Morel and Gentili, 1991, 1993). These results also show that self-similar solutions of the RTE can be successfully used for calculating $E_d(z)$ and some other integral characteristics of ULF, namely, the angle between directions of radiance angular distribution maximum and nadir ϑ_m, the variation of this distribution $D(z)$, and its asymptotic ($z \to \infty$) value D_∞:

$$E_d(z) = \frac{E_d(0)}{\cosh\zeta + u_1 \sinh\zeta} \times \exp\left[-(1 - \omega_{sa})\tau - \frac{u_2 \tanh\zeta}{1 + u_1 \tanh\zeta}\right], \tag{17}$$

$$\vartheta_m(z) = \vartheta_m(0)/(\cosh\zeta + u_1 \sinh\zeta), \tag{18}$$

$$D(z) = D_\infty \frac{u_1 + \tanh\zeta}{1 + u_1 \tanh\zeta}, \quad D_\infty = \left[\frac{2\omega_{sa}\langle\Theta_{45}^2\rangle}{1 - \omega_{sa}}\right]^{1/2}, \tag{19}$$

where $\zeta = 0.5D_\infty(1 - \omega_{sa})\tau$ is the "effective" depth; $\tau = cz$, the optical depth; $\omega_{sa} = \omega_0(1 - \tilde{b}_{45})$ is the small-angle scattering albedo; $u_1 = D(0)/D_\infty$, $u_2 = [\vartheta_m(0)]^2/D_\infty$; $\tilde{b}_{45}$ and $\langle\Theta_{45}^2\rangle$ are the scattering probability for $\Theta > 45°$ and variance of the scattering angle for $\Theta < 45°$:

$$\tilde{b}_{45} = 2\pi \int_{\pi/4}^{\pi} \tilde{\beta}(\Theta) \sin\Theta d\Theta, \quad \langle\Theta_{45}^2\rangle = \frac{2\pi}{1 - \tilde{b}_{45}} \int_0^{\pi/4} \Theta^2 \tilde{\beta}(\Theta) \sin\Theta d\Theta. \tag{20}$$

These parameters are related to the average cosine of the phase function by regressions, $\tilde{b}_{45} = 0.880 - 0.885\langle\cos\Theta\rangle$ and $\langle\Theta_{45}^2\rangle = 0.285 - 0.264\langle\cos\Theta\rangle$, and thus can be expressed through $\tilde{b}_b$ by means of Eq. (13).

In the case of direct sun illumination, it is possible to accept $D(0) = 0$, $\vartheta_m(0) = \vartheta_{0w}$, and for diffuse illumination $D(0) = 0.35$, $\vartheta_m(0) = 0$.

Although Eqs. (17)–(19) have been written for the case ω_{sa} = const, within the stepped approximation of $\omega_{sa}(z)$ they can be used for consecutive calculations inside all the homogeneous layers, beginning with the upper.

Curves 1 and 2 in Fig. 6 show how D and ϑ_m depend on ζ for $D(0) = 0$, $\vartheta_m(0) = \vartheta_{0w}$. They demonstrate that expansion (D) and shift (ϑ_m) of the radiance distribution stop, practically, at $\zeta \simeq 2$. Thus, it may be assumed that the structure of ULF under direct sun illumination becomes asymptotic (see curve 3 in Fig. 5) for $\zeta > 2$. For "average" seawater, this corresponds to an optical depth cz = 20–30. Under diffuse illumination [when $D(0) \approx D_\infty$], the asymptotic structure sets in at significantly smaller depths.

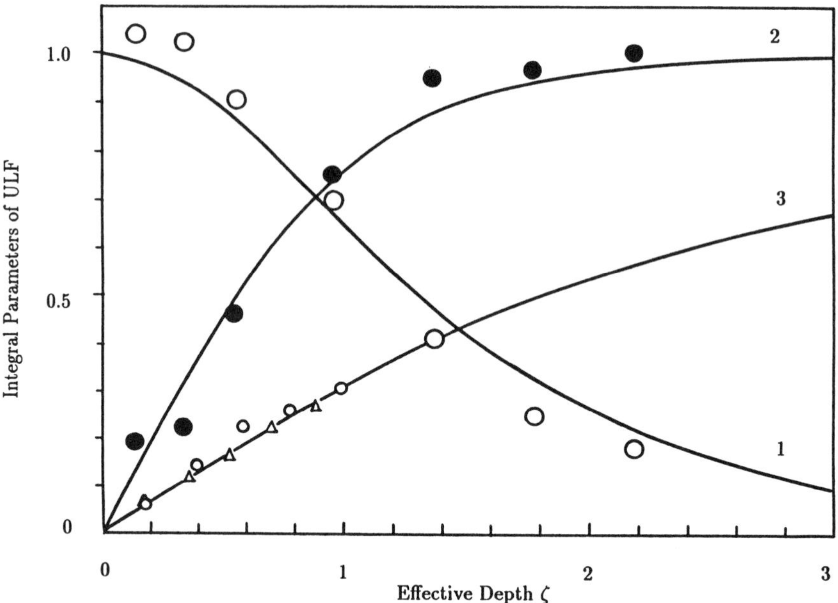

FIG. 6. Direction of maximal radiance with respect to the nadir $\vartheta_m(\zeta)/\vartheta_m(0)$ (curve 1) and the variance of radiance angular distribution $D(\zeta)/D_\infty$ (curves 2,3) as functions of effective depth ζ for daylight (curves 1,2) and omnidirectional point source (curve 3). The curves are calculated in self-similar approximations from Eqs. (18), (19), and (26). Points: ○ and ● are the results of processing of Tyler's (1960) measurements (D_∞ was found for depths 53.7 and 66.1 m); △ and ∘ are the Duntley measurements and Monte Carlo simulation data, respectively (Funk, 1973).

The results of computations of the asymptotic and near-asymptotic ULF parameters are presented also by Zaneveld (1989) and Gordon *et al.* (1993).

Numerous empirical data on relationships between downward irradiance spectra, the vertical distribution of chlorophyll, the IOP, and *in situ* primary production were obtained during nine large Polish-Russian expeditions to various oceanic regions in 1978–1991. On the basis of these data, models and algorithms for assessing primary production in the sea were developed (Wozniak *et al.*, 1992).

2.3 Artificial ULF

2.3.1 Narrow Light Beam Experimental investigations of narrow-beam propagation in water have shown that measurements match well the small-angle theory (Bravo-Zhivotovskiy *et al.*, 1969a). In this theory, for description of the beam structure, the function $F(\vartheta_\Sigma,z) = P_\Sigma/P_\infty$ is used. Here P_Σ is the light flux through a round cross section Σ with center at the beam axis placed at a distance z from the UPS, the angular radius of Σ being ϑ_Σ (see the diagram in Fig. 7); $P_\infty = P_0 \exp(-az)$ is the entire light flux through a plane z = const; P_0 is the source light flux. The function F is given by

$$F = \frac{\vartheta_\Sigma}{\Theta_*} \exp(-bz) \int_0^\infty (x + \sqrt{1 + x^2})^{bz/x} \times J_1(x\vartheta_\Sigma/\Theta_*)dx, \tag{21}$$

where J_1 is the Bessel function, and Θ_* a parameter of the scattering phase function, which has been represented in the form $\tilde{\beta}(\Theta) = (2\pi\Theta_*\Theta)^{-1} \exp(-\Theta/\Theta_*)$. The parameter Θ_* can be expressed through $\langle\cos\Theta\rangle$, and, by means of Eq. (13), through $\tilde{b}_b$, using the regression $\Theta_*^2 = 0.142 - 0.132 \langle\cos\Theta\rangle$. For seawater, the typical values are $\langle\cos\Theta\rangle = 0.9$–$0.96$ (see Table 3) and $\Theta_* = 0.12$–0.15.

Figure 7 provides the computation of F as a function of bz. The bottom curve is the exponential law for unscattered light: $P_\Sigma/P_\infty = P_0 \exp(-cz)/P_0 \exp(-az) = \exp(-bz)$.

The irradiance distribution in a cross section of the beam, $E(z,\vartheta)$ [where ϑ is the polar

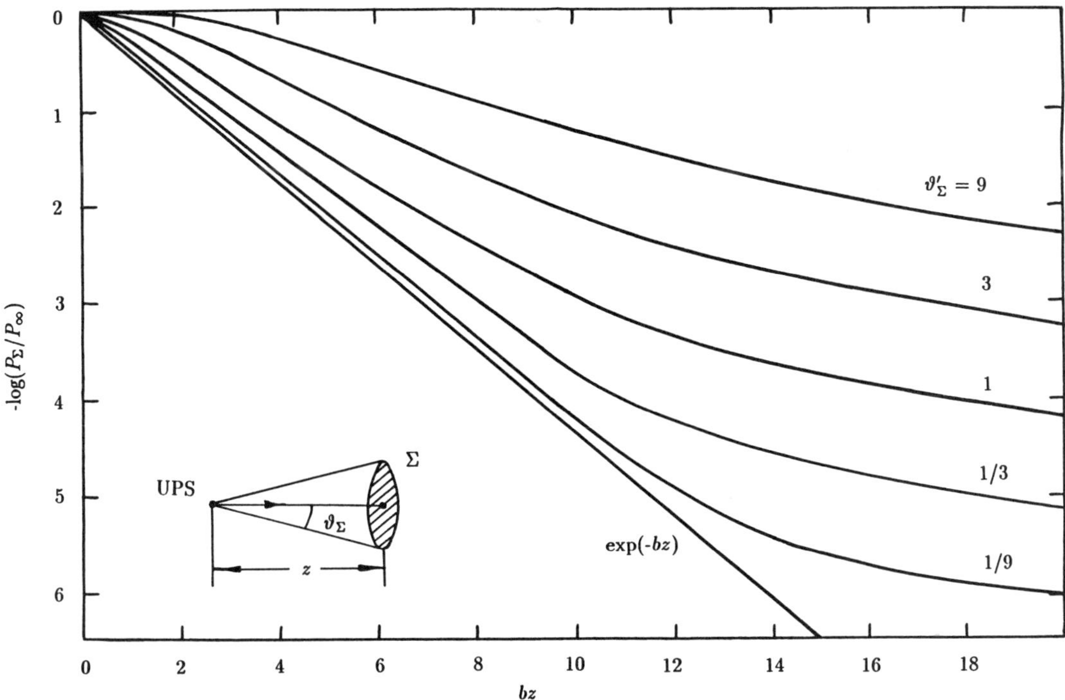

FIG. 7. Logarithm of the radiant flux through a round area Σ related to the entire flux through a plane z = const ($\Sigma = \infty$) as a function of dimensionless distance from the UPS, bz. The reduced angular radii of the area $\vartheta'_\Sigma = 2\vartheta_\Sigma b/\Theta_*$ are 1/9, 1/3, 1, 3, and 9.

angle of the point (see diagram in Fig. 8)], can be found by differentiation of Eq. (21):

$$E(z,\vartheta) = \left[\frac{P_0 \exp(-az)}{2\pi\vartheta z^2}\right]^{-1} \left[\frac{dF}{d\vartheta_\Sigma}\right]_{\vartheta_\Sigma=\vartheta}. \tag{22}$$

Curves in Figs. 8 and 9 provide the computation of E. The relative irradiance distribution in the cross section of a narrow beam $E(z,\vartheta)/E(z,0)$ is called the beam spread function (BSF). The computation has shown that in the initial part of the optical distance $\tau = cz$ the irradiance on the beam axis decreases according to the Bouguer law $E \propto \exp(-\tau)$, and then, at $\tau \simeq 10$, the decrease becomes slower (Fig. 9). At the same distances, the irradiance cross distribution loses its clear peak formed by the unscattered light (see Fig. 8).

Attenuation of a beam with a finite width (l) depends on the value of $\alpha = bl/\sqrt{\langle\Theta_{45}^2\rangle}$. The Bouguer law describes the process well if $\alpha < 0.1$ in an interval $z_{\min} < z < z_{\max}$, the limits of which are determined by the conditions $bz_{\min} = \alpha$, $bz_{\max} = 3 \ln bz_{\max} - 2 \ln\alpha$ (Dolin, 1966). If $z > z_{\max}$, the main contribution to E is formed by the scattered light, which makes the attenuation slower [according to the small-angle theory, $E \propto \exp(-az)/z^3$ for $z > z_{\max}$].

The theoretical models for the narrow-beam radiance distribution have been developed by Romanova (1968), Arnush (1972), and Dolin (1983a).

2.3.2 ULF from an Isotropic Source The ULF of an omnidirectional point source (OPS) is one of the simplest, as it depends only on two variables, the distance from the source (r) and the angle of deviation of the ray from the radial direction (ϑ'). According to measurements (Duntley, 1963; Voss and Chapin, 1990), the shape of the radiance angular distribution $L^{\text{OPS}}(\vartheta')$ changes, as the distance from the source increases, under the influence of two factors with opposite effects. The increase of the optical length of the water layer between the source and the receiver makes the distribution wider. The decrease of the angle size of the glow resulting from the scattering near the source makes it more narrow. As a result, the width of the angular

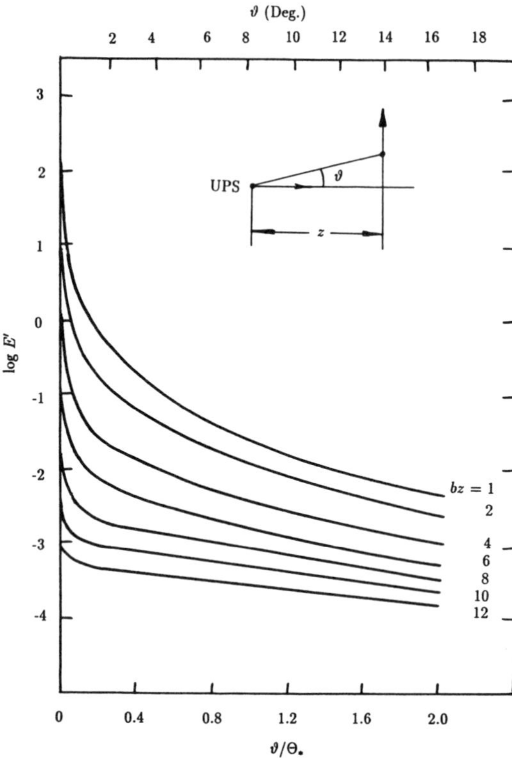

FIG. 8. Irradiance distributions in the cross section of a narrow beam at various distances (bz) from a UPS. $E' = E\Theta_*^2/b^2P_\infty$ is the dimensionless irradiance. On the upper scale, the angle ϑ is given for a typical value of $\Theta_* = 0.14$.

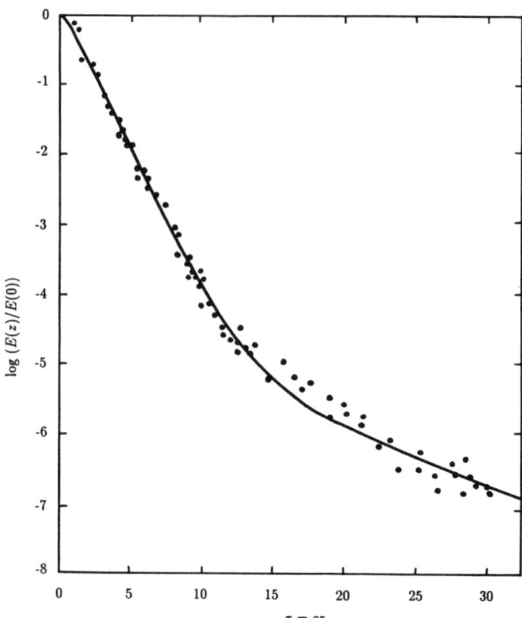

FIG. 9. Logarithm of the relative irradiance on the axis of a narrow light beam as a function of the optical distance from a source. The curve is given by the small-angle theory; the points present the results of experiments (in the Black Sea; the diameter of the beam is 8 cm, $c = 0.23\ \text{m}^{-1}$, $\omega_0 = 0.9$).

radiance distribution increases significantly more slowly than within the infinitely wide beam. As for the irradiance of a surface r = const, the scattering affects it only slightly: The deviations from the law $E^{\text{OPS}} \propto \exp(-ar)/r^2$ (which describes the case without scattering) are difficult to catch by measurements.

With the approximation $\tilde{\beta}(\Theta) = \Theta_0/[2\pi(\Theta_0^2 + \Theta^2)^{3/2}]$ for the phase function, the small-angle scattering theory gives (Wells, 1973)

$$L^{\text{OPS}}(r,\vartheta') = 2\pi \int_0^\infty \exp[-cr + brf(x)] \times J_0(2\pi\vartheta' x)x\,dx, \qquad (23)$$

$$f(x) = [1 - \exp(-2\pi\Theta_0 x)](2\pi\Theta_0 x)^{-1} \qquad (24)$$

where the parameter Θ_0 is about $\langle\Theta_{45}^2\rangle/(\pi/4)$. The results of calculations according to Eq. (23) match well the experimental data for $br > 2$, $\vartheta' < 0.1$ rad, despite the inaccuracy of the approximation $\tilde{\beta}$ (McLean and Voss, 1991).

According to the self-similarity theory, the irradiance E^{OPS} and the variance D^{OPS} of the angle radiance distribution for the case of OPS are presented in the form

$$E^{\text{OPS}} = \frac{P_0}{4\pi r^2}\left(\frac{\zeta}{\sinh\zeta}\right)\exp[-(1 - \omega_{sa})cr], \qquad (25)$$

$$D^{\text{OPS}} = D_\infty\left(\frac{1}{\tanh\zeta} - \frac{1}{\zeta}\right), \qquad (26)$$

where $\zeta = 0.5\, D_\infty(1 - \omega_{sa})cr$; for D_∞, see Eq. (19).

Equation (25) shows that the small-angle scattering makes a considerable contribution to the irradiance attenuation only while $\zeta > 1$. The value of D^{OPS} starts to grow closer to its asymptotic value D_∞, beginning with $\zeta > 3$ (Fig. 6), which corresponds to $r \simeq 0.5$ to 1 km in pure waters. Thus, the asymptotic distribution of radiance L^{OPS} is impossible to observe in such waters.

From the optical reciprocity theorem, it

follows (Levin, 1969; Gordon, 1994) that in the small-angle approximation, $E^{UPS}(z,\vartheta)/E^{UPS}(z,0) = L^{OPS}(r = z,\vartheta' = \vartheta)/L^{OPS}(r = z,0)$, i.e., the relative irradiance distribution (BSF) in the beam cross section at the distance z from a UPS is equal to the relative radiance angular distribution at the same distance from an OPS.

2.3.3 Propagation of a Light Pulse While a laser pulse is passing through a water layer, its duration increases because of random changes of photon multipaths and the time taken for their propagation from the source to the receiver. In sea experiments (Gol'din *et al.*, 1983) with the source and receiver placed coaxially and the initial pulse duration $t_0 = 6$ ns, the durations of the received pulses t_r are 7, 8.5, and 10 ns after paths $r = 50$, 75, and 100 m, respectively (Indian Ocean, $c = 0.23\ \text{m}^{-1}$). With noncoaxial source and receiver, the registered pulse is less in amplitude and greater in duration.

Figure 10 presents the theoretical curves that allow estimation of the duration t_r of an initially δ-pulsed signal ($t_0 = 0$) along its path r for unidirectional light beams with different initial cross-sectional areas S_0 (Dolin, 1983b). One can see that the wider the beam (greater S_0), the faster its duration grows. For a narrow beam, there is a region of small distances ζ where the speed of pulse extension decreases. This region coincides with the region of Bouguer attenuation (up to $\tau \simeq 10$) in a continuous narrow beam; see Fig. 9. The value of t_r limits the maximum frequency $(1/t_r)$ of the transmitted signal in underwater optical communication systems as well as the accuracy of laser depth sounding.

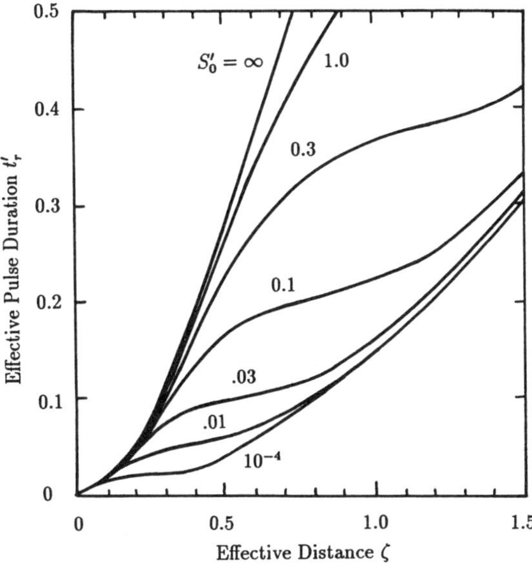

FIG. 10. Dimensionless effective pulse duration $t_r' = 2vat_r$ as a function of the effective distance ζ from a source for an initially δ-pulsed light beam with different initial beam cross sections S_0. The values of dimensionless beam cross section $S_0' = 0.25a^2D_\infty S_0$ are indicated for each curve; a receiver with wide field angle is placed on the beam axis.

The existing theoretical models also allow study of the distortions of the pulse shape resulting from small-angle scattering (Ishimaru, 1978; Remizovich *et al.*, 1983).

2.3.4 Backscattering When a light beam is passing through water, its energy, partly scattered at large angles, constitutes the diffuse component of ULF. It forms an obstacle for underwater imaging systems, as it hides the useful signal, while for remote sensing systems it itself constitutes the useful signal. Experimental data and theoretical models (Mertens and Replogle, 1977; Dolin and Levin, 1991; Zege *et al.*, 1991) give information on how the pulsed and continuous backscattered signals depend on the parameters of the source and the receiver, their mutual placement, and IOP.

For example, the power of backscattered light at a receiver separated from a continuous UPS by the distance l_{sr} (the optical axes of the source and receiver being parallel and normal to l_{sr}) is given by

$$P_b = \pi^{-1}P_0S_rb_b[k_l^{-1}\exp(-k_wk_l) + k_w\text{Ei}(-k_wk_l)]. \tag{27}$$

Here P_0 is the initial power of the UPS, S_r is the area of the receiver lens, $k_w = a + 2b_b$, $k_l = l_{sr}\tan^{-1}(0.25\vartheta_r)$, ϑ_r is the receiver field angle, and Ei is the exponential integral function. For a pulsed UPS, the power P_b is a function of time. For a large enough time interval t between pulse emitting and receiving (corresponding to the typical viewing distances $r = tv/2$ in the underwater imaging problems),

$$P_b(t) = \frac{P_{0p}S_rb_bt_0}{\pi vt^2}\exp(-avt), \tag{28}$$

where P_{0p} and t_0 are the initial power and duration of pulse.

2.4 Influence of the Sea Surface on Light Fields

Because of the difference between refraction indexes of water ($n_w = 1.33$) and air, the sea surface refracts and partly reflects the incident radiation. In the case of a flat sea surface, the ratio between the radiances of incident ($L^{\downarrow}$) and specularly reflected ($L^{\uparrow}$) beams is determined by Fresnel's reflectance $R_F^{\downarrow}(\sin\vartheta_i)$, which is not larger than 0.06 within the incident angle range $\vartheta_i < 60°$, and converges to 1 when ϑ_i approaches 90°. For the light beam incident from below, at an angle ϑ_i to the normal, the reflectance is $R_F^{\uparrow} = R_F^{\downarrow}(n_w \sin\vartheta_i)$ for $\vartheta_i < 48.6°$ and becomes 1 at $\vartheta_i > 48.6°$ (the effect of total internal reflection). Penetrating into the water, the light changes its direction according to Snell's refraction law. The radiance at the direction of the refracted beam is $L_r = n_w^2(1 - R_F^{\downarrow})L^{\downarrow}$. At $\vartheta_i < 40°$, the radiance increases as a result of decrease in the divergence angle of the light beam, $L_r/L^{\downarrow} \simeq 1.7$. At $\vartheta > 83°$, this ratio becomes less than 1 because of the abrupt growth of $R_F^{\downarrow}$. While the light beam is going out from the water into the atmosphere, its radiance changes by a factor $n_w^{-2}(1 - R_F^{\uparrow})$, that is, always decreases.

The wind waves make the sea surface rough, leading to fluctuations in radiance of the light reflected from and penetrating into the water. Under direct illumination, the statistically average angular distributions of reflected and refracted radiances $\langle L^{\uparrow}\rangle$ and $\langle L_r\rangle$ repeat in a relevant scale the surface slope probability density function, which can be approximated well by the normal distribution with variance σ^2 linearly depending on the wind velocity (Cox and Munk, 1956). The widths of the angle distributions $\langle L^{\uparrow}\rangle$ and $\langle L_r\rangle$ at the level $1/e$ are approximately 4σ and $\sigma/2$. The directions of their maxima are close to those of reflection and refraction of the incident light by a flat surface. A statistically average ULF satisfies the RTE. Thus, its calculation amounts to solving Eq. (14) for $\langle L\rangle$ with the boundary condition $\langle L\rangle = \langle L_r\rangle$ for $z = 0$, $(\boldsymbol{\Omega}\cdot\boldsymbol{\Omega}_s) \geq 0$. On this basis, an analytical model for narrow-beam propagation through the marine boundary layer was suggested (Lutomirski, 1978).

The statistically average irradiance $\langle E^{\uparrow}\rangle$ created by the reflected light right above the sea surface is generally represented as $\langle E^{\uparrow}\rangle = R_s E^{\downarrow}$, where R_s is the reflectance of the sea surface and $E^{\downarrow}$ its irradiance. Under direct sunlight and wind velocity $0 < v_w < 16$ m/s, $R_s \leq 0.025$ if the zenith solar angle $\vartheta_0 < 30°$. While ϑ_0 increases from 60° to 90°, R_s increases from 0.06 to 1.00 without waves, and from 0.07 to 0.30 if the roughness is strong ($v_w = 16$ m/s). Under cloudy illumination, R_s slightly depends on v_w and ranges within 0.04 to 0.05 (Austin, 1974). The average downward irradiance above the sea surface is $\langle E_d\rangle \simeq (1 - R_s)E^{\downarrow}$. The results of numerical computations of the time-averaged radiance and polarization of the underwater light field under a rough sea surface are presented by Kattawar and Adams (1989), Mobley (1989), and Mobley *et al.* (1993).

The fluctuations of ULF are found at depths less than 100 m (Monin, 1983; Dera *et al.*, 1993). The variation coefficient σ_E of the downward irradiance depends on z nonmonotonically, with a maximum of 0.1–0.4 at depth $z \simeq 10$ m, where the effect of light focusing by the wind waves appears the strongest. The width of spatial and temporal spectra of these fluctuations does not exceed 2 m^{-1} and 5 Hz, respectively. While the depth increases, σ_E decreases approximately as $1/z$, the spectra of fluctuations growing more narrow. The theoretical models of fluctuations (Luchinin and Sergiyevskaya, 1982) match well the experimental data.

3. OPTICAL METHODS FOR INVESTIGATION OF THE OCEAN AND OTHER NATURAL WATERS

Optical methods have a number of advantages with respect to other methods used for investigation of natural waters: speed of measurements, high resolution, getting information without changes in the physical properties of the water. Remote sensing is highly developed. It makes possible monitoring of large areas in real time, by means of shipboard, airborne, or satellite measurements. In this section, we deal with the main methods of underwater optics, namely, detection of phytoplankton and suspended and dissolved matter, retrieval of the parameters of sea waves, laser remote sensing for the detection of pollutants on the sea surface, observation of internal waves, and depth sounding.

3.1 Detection of Phytoplankton and of Suspended and Dissolved Matter

In various domains, in particular, in ocean geology, laboratory methods for the determination of suspended-matter composition by the properties of the scattered light are generally used. These methods make use of the solution to Eq. (6), the particle size distribution $N(r_p)$ being calculated through measured values of $\beta(\lambda,\Theta,n)$ (the inverse problem of light scattering). Such problems generally turn out incorrect, that is, their solutions are highly sensitive to small errors in measurements. The inverse problems are solved by means of regularization, that is, by using additional information about the solution. The most widespread method of solving Eq. (6) is the method of small angles, which is based on representation of the kernel of Eq. (6), $\beta'(\lambda,\Theta,n,r_p)$, by a simple analytical function for large particles ($2\pi r_p \gg \lambda$) and small angles, $\Theta \ll 1$. Besides that, there are methods of "spectral transparency," "complete indicatrix," and "fluctuations." All of them are based on the measurements of the VSF or its fluctuations in the laboratory, by means of rather complicated instruments, followed by computer calculations of the parameters of the suspended particles (Shifrin, 1988). If the particle size is known, their concentration N_p can be determined (in the laboratory or *in situ*) by measurement of the scattering coefficient b according to a calibration curve $b(N_p)$, which has to be built by means of standard samples with known N_p. For the calculation of phytoplankton (N_c) and yellow substance (N_y) concentrations, one may use the physical models of absorption (see Sec. 1).

The remote sensing of near-surface concentrations is based on measurements of the water-leaving radiance spectra $L_w(\lambda)$. The idea in most algorithms used for extracting N_c from the radiance data is to find a regression between N_c and the color index (green-blue ratio) $I = L_w(445\ \text{nm})/L_w(550\ \text{nm})$. According to Eq. (16), the ratio of the water-leaving radiance coefficients at two wavelengths is approximately inversely proportional to the ratio of the associated values of absorption coefficients (usually $b_b \ll a$). Since the phytoplankton absorbs radiation at 445 nm but not at 550 nm, the correlation between I and N_c proves to be high enough. According to Gordon et al. (1983), the linear regression is $\log N_c = -0.116 - 1.33 \log I$ (N_c is in mg m^{-3}). For the case where N_c is much greater than the concentration of inorganic suspended matter N_x [Case 1 according to Morel and Prieur (1977) waters classification], $\log N_c = 0.053 - 1.71 \log I$. In the satellite measurements, for large N_c, $L_w(445)$ is generally too small relative to the radiance resulting from the atmospheric scattering to be retrieved accurately enough. In this case, the ratio $I_1 = L_w(520)/L_w(550)$ is recommended instead of I; then $\log N_c = 0.622 - 2.44 \log I_1$.

Sathyendranath *et al.* (1989) developed a detailed algorithm for retrieving the concentrations N_c, N_x, and N_y from the measured spectra $L_w(\lambda)$ on the basis of principal component analysis. They presented linear regressions between the concentrations and $L_w(\lambda)$ for five wavelengths: $\lambda = 440, 445, 520, 560$, and 640 nm. The closest correlation was found between L_w and N_x (the wavelength recommended for measurement is 640 nm), and so the sediment concentration is most easily detectable. The algorithm for the retrieval of N_c and N_y from the radiance spectra can be improved by using a preliminary identification of the water type. The influence of a nonuniform vertical chlorophyll profile on the remote sensing of the ocean color spectrum is discussed by Platt *et al.* (1991), Gordon (1992), and Arnone *et al.* (1994).

Besides the ocean color method, the chlorophyll fluorescence ($\lambda = 685$ nm) method is used for passive remote sensing of the oceanic phytoplankton. A criterion may be the value of $L_w(685) - L_w(665)$, which demonstrates high correlations with N_c (Fisher *et al.*, 1986). The two methods do not yield the same information. The ocean color is influenced by the physical characteristics of the plankton population (such as concentration and composition of intracellular pigment), whereas the fluorescence is affected by the physiological state of the cells (such factors as light inhibition and circadian rhythm). Thus, these methods complement each other.

The accuracy of concentration retrieval from the measured radiance spectra can be maximized by the optimum choice of spectral channels for measurements on the basis of the theory of the experimental design (Kozlov *et al.*, 1992; Levin and Zolotuchin, 1994).

3.2 Optical Methods for Investigation of the Sea Surface

Optical images of the sea surface provide extensive information on sea waves, which are, in their turn, a sensitive indicator of dynamic processes in the atmosphere and in the upper layer of the ocean (wind, inhomogeneous currents, internal waves). The images also show the signs of oil pollution and natural organic films (a result of the plankton life activity). The effectiveness of optical methods is clearly demonstrated by the measurements of the wave slope distribution through the image of the sun glitter (Cox and Munk, 1956). Within the domain of sun glitter, it is also possible to observe from an aircraft the ripples and spatial variations of the wave slopes due to the inhomogeneity of the properties of small waves (which cannot be observed themselves). The spectral analysis of high-resolution aerial photographs makes it possible to investigate the spatial wave spectrum (Stilwell and Pilon, 1974), if there are necessary conditions for observations, namely, that the apparent radiance of a surface element is proportional to its slope. Thus, to measure the wave spectra, it is necessary to observe the part of the surface that reflects the light of that part of the sky with linear radiance distribution. The development of the sea-surface imaging theory (Titov, 1982) and of methods of quick Fourier analysis of the images without preliminary registration has led to the creation of optical on-board waveplotters, which provide the possibility to measure two-dimensional wave spectra in the range from centimeters to a few tens of meters in real time. With their help, as well as by digital sea-surface TV-image processing, the influence of the internal waves and oil films on the sea waves was investigated in detail (Gotwols *et al.*, 1988).

The oil films become visible not only as a result of slicks (which results from the damping of waves by the film), but also because of the significant difference in reflectance between pure and polluted water (Osadchy *et al.*, 1994). The highest contrast of the oil films on the water background can be obtained by means of CO_2 lidar (wavelength $\lambda = 10.6\ \mu m$).

3.3 Laser Remote Sensing of the Ocean

The lidars of blue-green spectral range are used for remote sensing of some physical parameters of the upper ocean layer and for depth sounding (which was the first oceanological application of lasers). During the first experiment (Hickman and Hogg, 1969), the reflection of a light pulse from the bottom of a lake 8 m deep was obtained from the height of 180 m. Modern lasers provide the opportunity to measure the bottom profile up to 100 m deep with resolution about 0.3–1.0 m. Depth is determined by measuring the time interval between pulses reflected from the sea surface and the sea bottom. The laser sounding increases the productivity of shoals mapping by using high-speed air vehicles (Lilycrop and Banic, 1993).

The methods of laser remote sensing of water properties are based on the analysis of temporal and spectral parameters of the backscattering signal. Using the decrease in signal intensity, it is possible to determine the attenuation coefficient c or the absorption coefficients a by a narrow or wide receiver, respectively. IOP in the upper ocean layer generally depend on depth (z). Internal waves deform the boundaries between the layers with constant values of a and b, just as surface waves deform the water-air boundary. As a consequence, it is possible, by measurements of $a(z)$ and $b(z)$ from an aircraft or a ship, to register the internal waves, as well as the temperature fronts and boundaries of sea currents, where changes of a and b also occur (Hoge *et al.*, 1988; Feigels and Kopilevich, 1993).

The most significant oceanological information is provided by the analysis of the backscattering signal spectrum (Fadeev, 1992). There exists a method of remote sensing of water temperature, based on analysis of the shape of a spectral band 3000–3700 cm^{-1} of Raman scattering by water molecules (Leonard *et al.*, 1979). For measurements *in situ*, the accuracy of this method is rather low. Lasers are also used for remote sensing of organic matter concentration in the water by intensity of its fluorescence, the Raman-scattering signal being used as the control signal (Klyshko and Fadeev, 1978; Hoge and Swift, 1981, 1986). The phytoplankton concentration is successfully detected by using a laser with wavelength $\lambda = 532$ nm, as the spectral band of its fluorescence is near that of Raman scattering and can be easily distinguished on the background of fluorescence of other admixtures. The fluorescence bands of dissolved organic matter and of oil products intersect, so that the problem of their distinct detection requires rather complicated methods.

There are also some difficulties due to the effects of double passing of light through the sea surface, their result being abnormally large fluctuations in the signal received. The theory of these effects is developed by Luchinin (1979) and Veber and Sergievskaya (1992).

Some radically new laser methods for studying ocean slicks of natural and artificial origin are described by Frysinger *et al.* (1992) and Choquet *et al.* (1993).

3.4 Atmospheric Corrections of Satellite-Measured Data of Ocean Remote Sensing

Let us consider the remote sensing of the ocean from space. The total measured radiance L consists of the "useful" water-leaving radiance L_w and radiance backscattered in the atmosphere:

$$L(\lambda) = L_R(\lambda) + L_a(\lambda) + T_R(\lambda)T_a(\lambda)T_{oz}(\lambda)L_w(\lambda), \quad (29)$$

where $L_R(\lambda)$ and $L_a(\lambda)$ correspond to the contributions to $L(\lambda)$ arising from Rayleigh and aerosol scattering in the atmosphere, and $T_R(\lambda)$, $T_a(\lambda)$, and $T_{oz}(\lambda)$ are diffuse transmittances of the atmosphere due to Rayleigh and aerosol scattering and absorption in the ozone layer, respectively. The atmosphere radiance is typically an order of magnitude larger than L_w. Therefore, satellite-measured radiance data should be corrected. The basis for the atmospheric correction was developed by Gordon (1978). A detailed algorithm of retrieving $L_w(\lambda)$ from a measured $L(\lambda)$ in passive remote sensing for the Coastal Zone Color Scanner (CZCS) on Nimbus-7, which views the ocean in five spectral bands (443, 520, 550, 670, and 750 nm), is described by Gordon *et al.* (1983) (see also Andre and Morel, 1991; Gordon and Wang, 1992). The main ideas are as follows. It is assumed that

1. $T_a(\lambda) = 1$ (observations are possible only when the atmosphere is pure enough);
2. L_a may be calculated in single-scattering approximation, so that $L_a(\lambda) \sim E_0'(\lambda) \times \omega_a(\lambda)\tilde{\beta}_a(\lambda,\pi - \theta)\tau_a(\lambda)$ [here $E_0'(\lambda)$ is extraterrestrial solar irradiance, ω_a, $\tilde{\beta}_a$, and τ_a are aerosol scattering albedo, scattering phase function, and optical thickness, and θ is the angle between directions of sun rays and the receiver axis]; this is also valid for a pure enough atmosphere;
3. ω_a and $\tilde{\beta}_a$ are independent of wavelength; and
4. $L_w = 0$ at $\lambda = 670$ nm (all the radiation is absorbed by the water).

Under these assumptions, Eq. (29) implies

$$L_w(\lambda) = T_R^{-1}(\lambda)T_{oz}^{-1}(\lambda) \times \left[L(\lambda) - L_R(\lambda) - [L(670) - L_r(670)] \times \frac{E_0'(\lambda)\tau_a(\lambda)}{E_0'(670)\tau_a(670)}\right]. \quad (30)$$

Since $E_0'(\lambda)$ is known, $T_R(\lambda)$, $T_{oz}(\lambda)$, and $L_R(\lambda)$ can be calculated exactly, so that to retrieve $L_w(\lambda)$ from a measured $L(\lambda)$ only the ratio $\epsilon(\lambda) = \tau_a(\lambda)/\tau_a(670)$ should be determined. Here the concept of clear water is used. According to this concept, in pure waters ($N_c < 0.25$ mg m^{-3}) $L_w(\lambda)$ is constant and exactly known. Thus, if a region with such N_c can be located, Eq. (30) can be used to determine $\epsilon(\lambda)$.

The next generation of ocean-color sensors such as the Sea Viewing Wide-Field-of-View Sensor (SeaWiFS) will have a radiometric sensitivity that is superior to that of the CZCS. They will also be equipped with additional spectral bands, e.g., bands centered on 765 and 865 nm, to aid atmospheric correction. Gordon and Wang (1994) have proposed a preliminary algorithm that utilizes the near-infrared bands for obtaining SeaWiFS atmospheric corrections. This algorithm, unlike the CZCS correction algorithm, takes into account multiple scattering by aerosol and is supposed to recover water-leaving radiance with an error of no more than 5%.

In active remote sensing by a pulsed laser placed on a satellite at a height H, the radiation scattered in the atmosphere can be cut out by locking the receiver and unlocking it at time $t = 2H/v$ after pulse emitting, i.e., at the moment when the signal from the water comes to the receiver. Thus, the atmospheric signals L_R and L_a that come to the receiver earlier would be removed. In this case, there remains only one problem: Since H is rather large, the satellite-board laser must have a lot of light power and, consequently, a lot of electrical power.

4. UNDERWATER IMAGING

The average sighting range (SR) in water is a thousand times shorter than the SR in clear air. The peculiarities of the underwater imaging process are as follows. In daylight, or if the light source is placed adjacent to the observer, the radiation backscattered throughout the illuminated path of sight forms a detrimental light veil (haze). The latter virtually does not depend on the presence of a target. The image-forming light signal from the target, on the other hand, decreases rapidly through absorption and scattering when the distance r between the target and the observer grows. Correspondingly, the ratio of the image-forming light to haze decreases, and at some distance (generally, $cr = 3$–6) it becomes less than the contrast threshold of the eye or the camera. So the target becomes hidden by the haze. Haze may be almost entirely eliminated by placing a lamp close to the target or by using a pulsed light source. When observing a self-luminous object, the haze is absent entirely. However, the target becomes invisible beyond a certain distance ($cr = 15$–20) even without haze. It disappears against a background of glow formed by the light that propagates from the target in various directions and is scattered toward the observer. If the target details are small, the image-forming signal derives from light that is transferred from the target to the image plane without having been scattered or absorbed. It means that the image-forming radiance L_r and inherent target radiance L_0 are related by $L_r = L_0 \exp(-cr)$. For large target details, the part of the light scattered forward at small angles is also image-forming. Therefore, the apparent contrast and SR of the large details are greater than of the small ones. The quantitative relationship between apparent contrast and target element size is described by the modulation transfer function (MTF); see below.

The underwater imaging theory aims to find the relations between the parameters of the imaging system, viewed target, light source, and water. Overall, this theory is to recommend the parameters of the imaging systems so that the SR in water would be maximal.

4.1 "Classic" Duntley–Preisendorfer Visibility Theory

For a long time, the only means of underwater viewing were the human eye and the photo camera. With regard to such viewing, Duntley and Preisendorfer had developed a visibility theory in a series of papers dated 1949–1957 (Duntley, 1963), which after that was included in all the major monographs on hydro-optics. This theory is intended to calculate the SR of a small Lambertian target against the water background under daylight. If the target is at depth z_t, and the observer is at the depth z, the distance between them being r and the zenith sighting angle ϑ ($z_t - z = r \cos\vartheta$) (see Fig. 11), the apparent radiance of the target is

$$L_r = L_0 \exp(-cr) + L_w(z,\vartheta) - L_w(z_t,\vartheta) \exp(-cr). \quad (31)$$

The first term on the right-hand side of this equation represents the residual image-forming light from the target, the second term is the radiance due to scattering of light in the water along the line of sight, the third is the radiance of the water column "hidden" by the target. If the inherent and apparent contrasts are denoted by

$$C_0 = [L_0 - L_w(z_t,\vartheta)]/L_w(z_t,\vartheta), \quad (32)$$

$$C_r = [L_r - L_w(z,\vartheta)]/L_w(z,\vartheta), \quad (33)$$

and it is supposed, by analogy with irradiance, that $L_w(z_t,\vartheta) = L_w(z,\vartheta) \exp[-(z_t - z)K]$, which is exactly the case for large depths (see

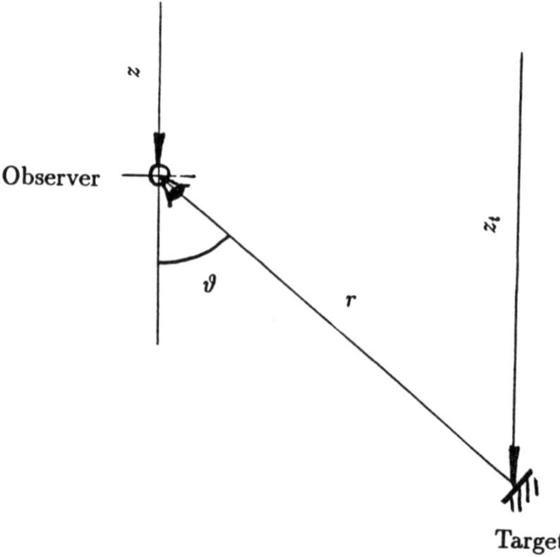

FIG. 11. Diagram of viewing for Duntley–Preisendorfer visibility model.

Sec. 2), then from Eq. (31) it follows that

$$C_r = C_0 \exp[-(K\cos\vartheta + c)r]. \tag{34}$$

If K and C_0 are known, changing C_r to the contrast threshold C^* (for the eye $C^* \simeq 2\%$ for high irradiance) and C_0 to $|C_0|$ and solving Eq. (34) with respect to r, we find SR ($r_{\max}$) in the given direction. For downward viewing of a Lambertian target ($\cos\vartheta = 1$), $C_0 = [R_t - R(z_t)]/R(z_t)$, that is, the radiances of the target and the water are replaced by their reflectances (see formulas for R in Sec. 2). For upward viewing ($\cos\vartheta = -1$), the background is always much brighter than the target, and $C_0 \simeq -1$. For horizontal viewing ($\cos\vartheta = 0$), C_0 depends on the azimuth viewing angle with respect to the sun. If $\cos\vartheta = 0$ and the target is absolutely black ($|C_0| = 1$), Eq. (34) gives (for $C^* = 0.02$) the well-known formula for "meteorological sighting range" $cr_{\max} = \ln 50 = 3.9$, which shows that a black target is distinguishable by the eye at a distance approximately 4 times greater than the "attenuation length" $1/c$. For $\cos\vartheta = 1$, Eq. (34) gives the approximate formula for the Secchi depth,

$$r_{\max} = z_D \simeq \frac{1}{c + K} \ln \frac{R_t - R}{C^* R}. \tag{35}$$

A more accurate, but more complicated, theory of Secchi depth, which takes into account scattering of the radiation reflected from the disk, is developed by Levin (1980).

New underwater imaging systems and expansion of the domain of their application made it necessary to develop a more universal, "instrumental" underwater imaging theory, which is dealt with in the following subsections.

4.2 Image Transfer Theory

In the general case, the underwater imaging process involves a light source (transmitter), an optical receiver, a target, and, of course, water. The light source is either natural (the sun and the sky) or artificial. The receiver consists of an objective lens and a detector. The detector may be the retina in the eye, the film in a photo camera, or the image tube or photomultiplier in a TV camera. The configuration of the imaging system may be described by the simplified diagram of Fig. 12. In all cases, the image is a set of elements, but the way the image is formed depends on the type of detector.

If it is a photomultiplier, the image results from scanning the target plane by a narrow light beam, usually a laser. The size of a resolution element in the target plane is determined by the transmitting angle (beam divergence) of the light source ϑ_s. The light flux forming one image resolution element is distributed uniformly over the entire area of the detector photocathode. The receiving angle ϑ_r (the angle of the cone within which the light forming one resolution element is obtained by the receiver) is determined by the photocathode size. This angle can be either wide or narrow. In the latter case, the transmitter and receiver beams are scanned synchronously across the target plane. The systems with a narrow light beam and a photomultiplier as the detector are referred to as scanning or flying-spot systems.

In all other types of systems (eye, photo, and TV cameras), the receiving angle ϑ_r is determined by the size of the detector resolution element. All elements of the detector

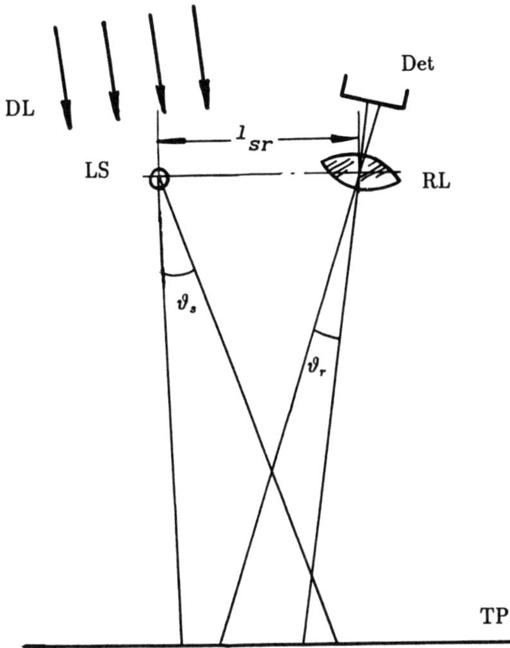

FIG. 12. Schematic diagram of imaging system configuration. LS, light source; RL, receiver objective lens; Det, detector; TP, target plane; DL, daylight. All elements are assumed to be submerged in water.

simultaneously receive the light reflected from the target plane, which is illuminated uniformly by a broad-beam source or by the daylight.

It is usually assumed that the target is a Lambertian plane with reflectance $R_t(x_t,y_t)$, dependent on the coordinates (x_t,y_t) on this plane. The image is defined as the radiant power forming a resolution element $P(x_0,y_0)$ as a function of coordinates x_0, y_0 in the image plane (for scanning systems, x_0, y_0 are the coordinates of the point of intersection between the scanning axis and the target plane). Correspondingly, the image transfer equation is an equation describing the relationship between $P(x_0,y_0)$ and $R_t(x_t,y_t)$. On the assumption that the isoplanatism hypothesis holds, that is, that the structure of the image of a point is independent of the position of the point on the target plane, the image transfer equation has the form (Levin, 1969; Bravo-Zhivotovskiy *et al.*, 1969b; Mertens and Replogle, 1977; Jaffe, 1992)

$$P(x_0,y_0) = P_b + P_t(x_0,y_0), \tag{36}$$

$$P_t(x_0,y_0) = P_{t\infty} \iint R_t(x_t,y_t) \times \mathrm{PSF}(x_t - x_0, y_t - y_0)dx_t dy_t, \tag{37}$$

where P_t, $P_{t\infty}$, P_b are radiant powers forming one image element, P_t being related to the target observed, $P_{t\infty}$ to the hypothetical infinite and ideally white target ($R_t \equiv 1$), P_b to the light backscattered from the water without interaction with a target (haze); PSF is the point spread function of the imaging system, which characterizes the contribution to the image of each point of the target (x_t,y_t); the integration is to be taken over the entire target plane. The values of $P_{t\infty}$, P_b, and PSF depend on the parameters of the system, the IOP, and the distance r.

Equation (37) is a convolution integral. According to the convolution theorem,

$$\tilde{P}_t(\nu_x,\nu_y) = P_{t\infty}\tilde{R}_t(\nu_x,\nu_y)\,\widetilde{\mathrm{PSF}}(\nu_x,\nu_y), \tag{38}$$

where the tilde means the two-dimensional Fourier transform of the function. Equation (38) shows that the imaging system in water acts as a linear filter of spatial frequencies ν_x, ν_y. The gain-frequency characteristic of this filter, $\widetilde{\mathrm{PSF}}(\nu_x,\nu_y)$, is termed the modulation transfer function (MTF), by analogy with the corresponding characteristic of conventional optical systems, although it takes into account the influence of the scattering medium (water) on the image. Insofar as the PSF in water is rotationally symmetric, its Fourier transform is replaced by the Hankel transform, and the argument of the MTF is $\nu = \sqrt{\nu_x^2 + \nu_y^2}$. Let us consider an infinite target with a sinusoidal distribution of reflectance (Fig. 13):

$$R_t(l_t) = \langle R_t\rangle[1 + C_0 \cos(\nu l_t)], \tag{39}$$

where $l_t = \sqrt{x_t^2 + y_t^2}$, $\langle R_t\rangle = 0.5(R_{t1} + R_{t2})$, R_{t1}, R_{t2}, and $\langle R_t\rangle$ are maximum, minimum, and average values of R_t, and $C_0 = (R_{t1} - R_{t2})/(R_{t1} + R_{t2})$. The latter is the inherent contrast, or percentage modulation of R_t. Substituting Eq. (39) into Eq. (37), we get

$$P_t(l_0) = \langle R_t\rangle P_{t\infty}[1 + C_r' \cos(\nu l_0)]. \tag{40}$$

where

$$C_r' = (P_{t1} - P_{t2})/(P_{t1} + P_{t2}) = C_0\mathrm{MTF}(\nu) \tag{41}$$

is the contrast of the image without haze; P_{t1} and P_{t2} are the maximum and minimum of P_t; $\langle P_t\rangle = \langle R_t\rangle P_{t\infty}$; and $l_0 = \sqrt{x_0^2 + y_0^2}$. Thus, the sinusoidal structure Eq. (39) is transferred in water without distortions, with only amplitude changes (see Fig. 13). The MTF is the ratio of apparent and inherent contrasts as a function of spatial frequency ν.

If a scanning system with narrow transmitter and receiver beams is not considered, and the receiver lens is supposed to be "ideal" (that is, the image distortions by the lens may be disregarded with respect to those in water), the PSF is equal to the angular distribution of radiance from the omnidirectional point light source $L^{\mathrm{OPS}}(r,\vartheta')$. The latter coincides with the beam spread function (BSF), which is the irradiance distribution in the cross section of a narrow beam at the distance r from the source (see Sec. 2). That is, PSF and MTF depend only on r and IOP. For example (Bravo-Zhivotovskiy *et al.*, 1969b),

$$\mathrm{MTF}(r,\psi) = \exp[-brA(\psi)], \quad A(\psi) = 1 - \psi^{-1}\ln[\psi_* + \sqrt{1 + \psi_*^2}], \tag{42}$$

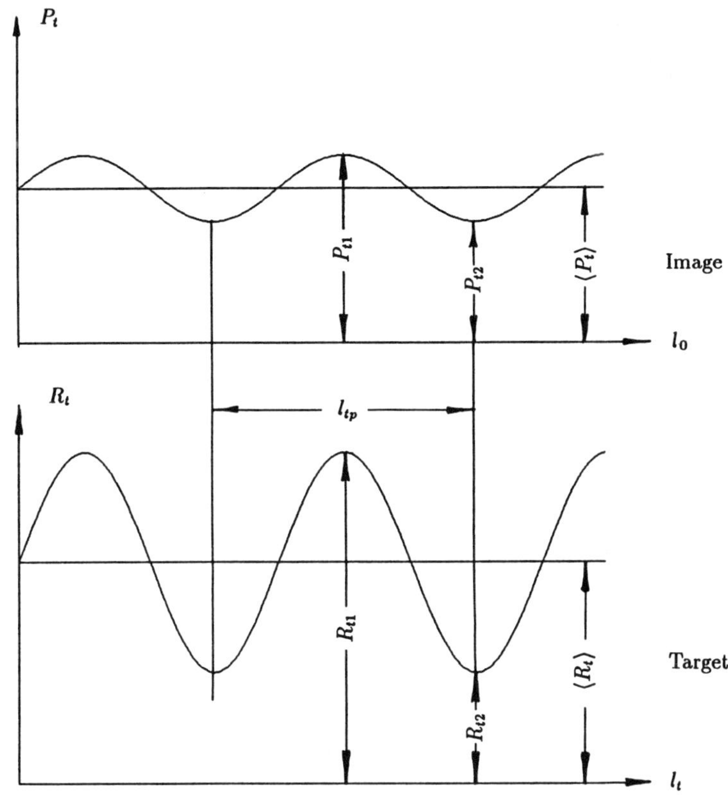

FIG. 13. Image of an underwater sinusoidal target.

where $\psi_* = 2\pi\Theta_*\psi$, $\psi = \nu r/2\pi = r/l_{tp}$ is the angular frequency, l_{tp}, the period of strips on the target (the length of a "white-black" pair: see Fig. 13), Θ_* is a parameter of the phase function (see Sec. 2). A slightly different formula for $A(\psi)$ is given by Wells (1969); the results of direct measurements of the PSF in the sea are presented by Voss (1990, 1991). Equation (42) shows that MTF $\rightarrow$ 1 as $\psi \rightarrow 0$ and MTF $\rightarrow \exp(-br)$ as $\psi \rightarrow \infty$.

The most typical problem of the underwater imaging theory is distinguishing small details on a target of limited size, generally treated as a disk with diameter d_t and a periodic distribution of reflectance [Eq. (39)], with period $l_{tp} \ll d_t$. To estimate the visibility of such a target, besides the characteristics of the imaging system, it is enough to know the useful ($\langle P_t \rangle$) and detrimental (P_b) components of the average radiant power in the image, and the real image contrast C_r. The formulas relating $\langle P_t \rangle$ and C_r to the IOP are as follows (the target is in the center of the field of view):

$$\langle P_t \rangle = P_{t\infty} \langle R_t \rangle F_t, \tag{43}$$

$$C_r = C_0 F_t^{-1} \,\mathrm{MTF}(r,\psi) \langle P_t \rangle / (\langle P_t \rangle + P_b), \tag{44}$$

$$\begin{aligned} P_{t\infty}^{(d)} &= \pi^{-1} S_r \Omega_r E_d(z_t) \exp(-ar), \\ P_{t\infty}^{(1)} &= \pi^{-1} r^{-2} P_0 S_r \exp(-2ar), \end{aligned} \tag{45}$$

$$P_{t\infty}^{(1)} / P_{t\infty}^{(2)} = P_b^{(1)} / P_b^{(2)} = N_f. \tag{46}$$

Here F_t is the integral of PSF over the target plane; as PSF = BSF, the values of $F_t(\vartheta_t, br)$ are equal to those of $F(\vartheta_\Sigma, bz)$ (see Sec. 2.3.1), $\vartheta_t = d_t/2r$ being in this case the angular radius of the target; S_r is the area of the receiver lens, $E_d(z_t)$ the irradiance under the daylight at the depth z_t, P_0 the average transmitted power, $\Omega_r = 2\pi[1 - \cos(\vartheta_r/2)]$ the solid receiving angle, and N_f the number of elements in the image frame. The upper indexes (d), (1), and (2) indicate whether the characteristic belongs to the case of daylight or to that of an artificial light source with narrow (1) or wide (2) beam (system with narrow transmitter and receiver beams is not considered here). Equation (46) is derived on the assumption that the source-receiver separations in systems (1) and (2) are the same: $l_{sr}^{(1)} = l_{sr}^{(2)}$. The transmitting and receiving angles are supposed to be "reversible", i.e., $\vartheta_s^{(1)} = \vartheta_r^{(2)}$, $\vartheta_r^{(1)} = \vartheta_s^{(2)}$. Formulas for backscattered light power P_b are given in Sec. 2.3.4.

In the most general case, the transmitting or receiving angle, the shape of the underwater target, and its position in the viewing field are arbitrary. The target may be viewed from air or space through the sea surface. For this case, the formulas and tables for computation of the image parameters $\langle P_t \rangle$, P_b, and C_r are given by Levin and Levin (1989) and Dolin and Levin (1991).

4.3 Sighting Range and Spatial Resolution in Water

The sighting range $r_{\max}(\psi)$ and the spatial resolution $\psi_{\max}(r)$ are the maximal distance where elements of a given spatial frequency are distinguishable on a given target, and the maximal spatial frequency distinguishable at a given distance. For visual and photo systems, and also in all the cases when the contrast threshold of the system C^* is known (for example, vidicon TV systems working with high irradiances), $r_{\max}$ and $\psi_{\max}$ can be determined by solving the equation $C_r(r,\psi) = C^*$ with respect to r or ψ (having fixed the other variable), using Eq. (44). The value of $C^* = 2\%$ for an eye corresponds to high irradiance and $\psi < 100\ \mathrm{rad}^{-1}$. For small irradiances and large ψ, the value of C^* increases (Rose, 1973). For the photo and vidicon systems, the value of C^* is usually also several percent. For highly sensitive TV systems (with photomultiplier, supervidicon, or image converter tube), SR is determined mainly by photon (shot) image noise (Hodara and Marquedant, 1968). In this case, $r_{\max}(\psi)$ and $\psi_{\max}(r)$ should be found from the equation $\delta(r,\psi) = \delta^*$ where δ is the signal/noise ratio and δ^* is its threshold. The value of δ due to shot noise is

$$\delta(r,\psi) = C_r \sqrt{2t_e P \eta_{ph}/e}, \tag{47}$$

where t_e is the duration of forming one element, η_{ph} the spectral sensitivity of the photocathode (A/W), and e the elementary change. The threshold δ^* depends on a given probability of the target detection and is usually accepted to be $\delta^* = 2$–5. While observing through the rough sea surface, the noise due to waves must be taken into account; when the water is very pure, the fluctuations of the IOP are to be considered as well (Duntley, 1974).

Underwater imaging theory provides a way to compare the capabilities of various imaging systems. Visual, photo, and vidicon systems ensure approximately the same SR since all of them have similar angular parameters (wide light beam and narrow receiving angle) and similar contrast threshold. Let us compare two highly sensitive noise-limited systems with the same frame period t_f:

1. the scanning system with narrow light beam and wide receiving angle ($\vartheta_r \gg \vartheta_s$); and
2. the conventional TV system ($\vartheta_s \ll \vartheta_r$).

In the latter, radiant energy on the image element is accumulated by the detector (image tube or image converter tube) during the entire frame period, and $t_e = t_f$. In the scanning system, the elements are formed one by one, and $t_e = t_f/N_f$. On the other hand, from Eqs. (36) and (43)–(46) it follows that in the scanning system **1** the image radiant power (P) is N_f times greater than in the conventional TV system **2** while the contrasts in both systems are equal (granted that these systems have the same values of P_0 and S_r and the same source mode, continuous or pulsed). Thus, the product $t_e P$ in both systems turns out to be equal. Therefore, as follows from Eq. (47), the compared systems ensure the same signal/noise ratio and thus the same SR.

Imaging systems with a pulsed nanosecond source and gated imaging reception increase the contrast and SR to a reasonable degree, through reduction of the haze P_b. For this purpose, the receiver is opened exactly at the moment when the pulse reflected from the target reaches the detector. The haze, which mainly comes earlier, is gated out.

Other methods of increasing the SR are also used under different conditions. The haze decreases with the increase of the transmitter-receiver separation ("base") l_{sr}. The optimum base in water of average transparency is 2–3 m; in pure ocean waters, it goes up to 10 m (if the base is longer, the contrast begins to decrease because of decrease of the signal from the target). A large horizontal base is difficult to provide; thus, in pure water the light source is sometimes put behind the camera (Patterson, 1975). The haze can also be decreased by the use of polarized light. If cross polarizers are put on the source and the receiver, the contrast grows 3–10 times—the shorter the base, the greater the increase. However, usually it is worthier to make the base longer; the polarizers are to be used when the base for technical reasons cannot be made long. One more way to increase the SR is synchronous scanning of the target by narrow transmitter and receiver beams. In this case, PSF $\propto$ $(\mathrm{BSF})^2$; thus, its Fourier transform MTF increases, compared with the MTF of a system with one narrow beam where PSF = BSF. On the other hand, the useful signal P_t in the dual scanning system is lesser. So this method is adequate only if there is a reserve of the source power. Besides, it is possible to increase the SR by means of a larger entrance pupil S_r and frame period t_f. In the conventional TV system, S_r is limited by the size of the photocathode of the image tube. In the scanning system, it can be made large enough, for example, by means of mosaic consists of photomultipliers. In addition, the frame period may be increased in such a system through slow scanning. Finally, the laser systems have one more advantage: When using lasers with variable wavelength, the latter can be chosen so as to correspond to the minimum absorption at the point of observation.

The SR depends not only on the system type and water transparency, but also on the size of details viewed: The SR increases with decrease of the spatial frequency ψ. For example, if a target with reflectance $\langle R_t \rangle = 0.1$ is viewed in the spectral region $\lambda = 500$–550 nm in daylight, and Secchi depth $z_D = 10$–50 m, then SR = 10–40 m for $\psi = 100$ rad^{-1} and SR = 20–55 m for $\psi = 3$ rad^{-1}. For a pulsed scanning system with large receiver input area $S_r = 100$ cm^2 and large frame period $t_f = 6$ s, the SR of the same target in the same range of λ and z_D are SR = 20–85 m for $\psi = 100$ rad^{-1} and SR = 40–100 m for $\psi = 3$ rad^{-1}. It can be seen that the pulsed illumination increases SR not more than twice. Note that the elements of a target with $\psi = r/l_{tp} = 3$ can be seen only from the air at large height $H \gg r$. The SR in the ocean can be much larger than 100 m if the observation is performed in the spectral range corresponding to the minimum absorption. Estimates show that in this case very large targets can be seen from space in the purest water and under the most favorable observation conditions at SR of about 700 m (Dolin and Levin, 1994).

MTF is almost equal to its asymptotical value $\exp(-br)$ for the spatial frequency $\psi >$ 100 rad^{-1}. Therefore, the resolution $\psi_{\max} =$ 100–1000 rad^{-1} is achieved without difficulty for relatively small distances. However, if $\psi > 10^3$–10^4 rad^{-1} the image would be distorted, even in very pure water and for small distances, by the turbulent fluctuations of the water refractive index (Duntley, 1974; Bogucki, *et al.*, 1994). Thus, $\psi_{\max} = 1000$ rad^{-1} may be considered as the limiting resolution in water.

GLOSSARY

Absorption Coefficient: Reciprocal of the length of water column that attenuates the light beam by a factor of *e* due to absorption.

Apparent Contrast: Contrast of elements in the image of an underwater target.

Apparent Optical Properties (AOP): Functions of radiance and irradiance in the sea that depend on the IOP and the illumination conditions.

Attenuation Coefficient: Sum of absorption and scattering coefficients.

Beam Spread Function (BSF): Irradiance distribution in the cross section of a narrow light beam at some distance from a UPS.

Contrast: Ratio of the difference in radiances of two adjacent elements to the sum of these radiances (or to one of them).

Dissolved Matter: Salts, humic acids, melanoidines, and some other substances dissolved in water and absorbing light.

Few-Parametric Models of IOP: Set of equations that make possible the retrieval of the full spectral or angular distribution of IOP using measurements at a few (not more than three) wavelengths or angles.

Inherent Contrast: Contrast of elements of an underwater target.

Inherent Optical Properties (IOP): Absorption and scattering coefficients, VSF, and other water parameters that determine the light absorption and scattering by the water elementary volume and depend on the composition and concentration of water ingredients, but do not depend on the illumination conditions.

Lambertian Target: Object with rough surface that reflects the same radiance in all directions.

Modulation Transfer Function (MTF): Ratio of the apparent and inherent contrasts for an underwater self-luminous target with sinusoidally varying reflectance as a function of target spatial frequency.

Omnidirectional Point Source (OPS): Source with a small spatial size and isotropic angular distribution of radiant intensity.

Optical Distance/Depth: Dimensionless product of distance/depth by attenuation coefficient.

Optical Methods for Investigation of the Ocean: Retrieval of data on ocean physical properties (phytoplankton and sediment concentrations, depth, surface and internal waves, surface pollution) by measuring various light-field parameters.

Point Spread Function (PSF): Distribution of the irradiance in the image of a self-luminous point observed through a water layer.

Radiative Transfer Equation (RTE): Principal equation that connects the radiance with the IOP and the light-source parameters and determines the underwater light field structure.

Scattering Coefficient: Reciprocal of the length of water column that attenuates a light beam by a factor of e due to scattering.

Secchi Depth: Depth where a white disk of 300 mm diameter becomes invisible from the surface.

Sighting Range (SR): Maximal distance where elements of a given spatial frequency are distinguishable on an underwater target.

Spatial Resolution: Maximal spatial frequency distinguishable on an underwater target at a given distance.

Suspended Matter: Mineral and organic particles (sediments, phytoplankton cells, bacteria, detritus, and others) suspended in water and scattering and absorbing light.

Underwater Light Fields (ULF): Spatial and angular distributions of spectral radiance in the sea.

Unidirectional Point Source (UPS): Source of narrow light beam with a small spatial size, e.g., laser.

Vertical Attenuation Coefficient: Reciprocal of the thickness of water layer that attenuates natural underwater irradiance by a factor of e.

Volume Scattering Function (VSF): Angular distribution of intensity of light scattered by an elementary water volume.

Works Cited

Andre, J.-M., Morel, A. (1991), *Oceanol. Acta* **14**, 3–22.

Arnone, R. A., Tucker, S. P., Hilder, F. A. (1984), in: M. A. Blizard (Ed.), *Ocean Optics VII*, SPIE Proceedings No. 489, Bellingham, WA: SPIE, pp. 195–201.

Arnone, R. A., Gould, R. W., Oriol, R. A., Terrie, G. (1994), in: J. Jaffe (Ed.), *Ocean Optics XII*, SPIE Proceedings No. 2258, Bellingham, WA: SPIE, pp. 322–331.

Arnush, D. (1972), *J. Opt. Soc. Am.* **62**, 1109–1111.

Austin, R. W. (1974), in: N. G. Jerlov, E. S. Nielsen (Eds.), *Optical Aspects of Oceanography*, London: Academic, p. 317.

Bogucki, S., Dickey, T. D., Domaradzki, A., Zaneveld, J. R. (1994), in: J. Jaffe, (Ed.), *Ocean Optics XII*, SPIE Proceedings No. 2258, Bellingham, WA: SPIE, pp. 247–255.

Bravo-Zhivotovskiy, D. M., Dolin, L. S., Luchinin, A. G., Savel'yev, V. A. (1969a), *Izv. Atmos. Ocean. Phys.* **5**, 83–87.

Bravo-Zhivotovskiy, D. M., Dolin, L. S., Luchinin, A. G., Savel'yev, V. A. (1969b), *Izv. Atmos. Ocean. Phys.* **5**, 388–393.

Burenkov, V. I. (1983), in: A. Monin (Ed.), *Optika Okeana*, Vol. 2, Moscow: Nauka, p. 42.

Case, K. M., Zweifel, P. F. (1967), *Linear Transport Theory*, London: Addison-Wesley.

Choquet, M., Heon, R., Vaudreuil, G., Monchalin, G.-P., Pabioleau, C., Goodman, R. H. (1993), in: *Proceedings of the International Oil Spill Conference (Prevention, Preparedness, Response), March 29–April 1, 1993, Tampa, Florida*, American Petroleum Institute Publication No. 4580, Washington, DC: American Petroleum Institute, pp. 531–536.

Cox, C., Munk, W. (1956), *Bull. Scripps Inst. Oceanogr.* **6**, 401–488.

Cullen, J. J., Ciotti, A. M., Lewis, M. R. (1994), in: J. Jaffe (Ed.), *Ocean Optics XII*, SPIE Proceedings No. 2258, Bellingham, WA: SPIE, pp. 105–115.

Dera, J. (1992), *Marine Physics*, Amsterdam: Elsevier.

Dera, J., Sagan, S., Stramski, D. (1993), *Oceanologia* **34**, 13–25.

Dolin, L. S. (1966), *Izv. Vuz. Radiofiz.* **9**, 61–71.

Dolin, L. S. (1983a), *Izv. Atmos. Ocean. Phys.* **19**, 296–299.

Dolin, L. S. (1983b), *Izv. Vuz. Radiofiz.* **26**, 300–309.

Dolin, L. S., Levin, I. M. (1991), *Spravochnik po Teorii Podvodnogo Videnia*, Leningrad: Gidrometroizdat (in Russian).

Dolin, L. S., Levin, I. M. (1994), in: J. Jaffe (Ed.), *Ocean Optics XII*, SPIE Proceedings No. 2258, Bellingham, WA: SPIE, pp. 588–596.

Dolin, L. S., Savel'yev, V. A. (1971), *Izv. Atmos. Ocean. Phys.* **7**, 328–331.

Dolin, L. S., Levin, I. M., Radomysl'skaya, T. M. (1994), in: J. Jaffe (Ed.), *Ocean Optics XII*, SPIE Proceedings No. 2258, Bellingham, WA: SPIE, pp. 522–528.

Doss, W., Wells, W. (1992), *Appl. Opt.* **31**, 4268–4274.

Duntley, S. Q. (1963), *J. Opt. Soc. Am.* **53**, 214–233.

Duntley, S. Q. (1974), in: N. G. Jerlov, E. S. Nielsen (Eds.), *Optical Aspects of Oceanography*, London: Academic, p. 135.

Fadeev, V. V. (1992), in: J. E. I. Korppi-Tommola (Ed.), *Laser Study of Macroscopic Biosystems*, SPIE Proceedings No. 1992, Bellingham, WA: SPIE.

Feigels, V. I., Kopilevich, Y. I. (1993), in: H. C. Eilertsen (Ed.), *Underwater Light Measurements*, SPIE Proceedings No. 2048, Bellingham, WA: SPIE, pp. 352–353.

Fisher, J., Doerffer, R., Grassl, H. (1986), *Appl. Opt.* **25**, 448–456.

Fournier, G. R., Forand, L., Pelletier, G., Pace, P. (1992), in: G. D. Gilbert (Ed.), *Ocean Optics XI*, SPIE Proceedings No. 1750, Bellingham, WA: SPIE, pp. 114–125.

Fry, E. S., Kattawar, G. W., Pope, R. M. (1992), *Appl. Opt.* **31**, 2055–2065.

Frysinger, G. S., Asher, W. E., Korenowski, G. M., Barger, W. R., Klusty, M. A., Frew, N. M., Nelson, R. K. (1992), *J. Geophys. Res. C* **97**, 52253–52269.

Funk, C. J. (1973), *Appl. Opt.* **12**, 301–313.

Gershun, A. (1939), *J. Math. Phys.* **18**, 51–151.

Gol'din, Yu. A., Dolin, L. S., Pelevin, V. N. (1983), in: A. Monin (Ed.), *Optika Okeana*, Vol. 1, Moscow: Nauka, p. 307 (in Russian).

Golubitskiy, B. M., Levin, I. M. (1980), *Izv. Atmos. Ocean. Phys.* **16**, 775–780.

Golubitskiy, B. M., Levin, I. M., Tantashev, M. V. (1974), *Izv. Atmos. Ocean. Phys.* **10**, 766–768.

Gordon, H. R. (1973), *Appl. Opt.* **12**, 2803–2804.

Gordon, H. R. (1978), *Appl. Opt.* **17**, 1631–1636.

Gordon, H. R. (1989), *Limnol. Oceanogr.* **34**, 1389–1409, 1484–1489.

Gordon, H. R. (1992), *Appl. Opt.* **31**, 2116–2129.

Gordon, H. R. (1994), *Appl. Opt.* **33**, 1120–1122.

Gordon, H. R., Ding, K. (1992), *Limnol. Oceanogr.* **37**, 491–500.

Gordon, H. R., Morel, A. Y. (1983), *Remote Assessment of Ocean Color for Interpretation of Satellite Visible Imagery: A Review*, New York: Springer.

Gordon, H. R., Wang, M. (1992), *Appl. Opt.* **31**, 4247–4267.

Gordon, H. R., Wang, M. (1994), *Appl. Opt.* **33**, 443–452.

Gordon, H. R., Wouters, A. W. (1978), *Appl. Opt.* **17**, 3341–3343.

Gordon, H. R., Brown, O. B., Jacobs, M. M. (1975), *Appl. Opt.* **14**, 417–427.

Gordon, H. R., Clark, D. K., Brown, O. B., Evans, R. H., Broenkow, W. W. (1983), *Appl. Opt.* **22**, 20–36.

Gordon, H. R., Ding, K., Gong, W. (1993), *Appl. Opt.* **32**, 1606–1620.

Gotwols, P. L., Sterner, R. E., Thompson, D. R. (1988), *J. Geophys. Res. C* **93**, 12256–12281.

Greysukh, V. M., Levin, I. M. (1990), *Izv. Atmos. Ocean. Phys.* **26**, 62–66.

Hickman, G. D., Hogg, J. E. (1969), *Remote Sens. Envir.* **1**, 47–58.

Hodara, H., Marquedant, R. E. (1968), *Appl. Opt.* **7**, 527–534.

Hoge, F. E., Swift, R. N. (1981), *Appl. Opt.* **20**, 3197–3205.

Hoge, F. E., Swift, R. N. (1986), *Appl. Opt.* **25**, 2571–2583.

Hoge, F. E., Wright, C. W., Krabill, W. B., Buntzen, R. R., Gilbert, G. D., Swift, R. N., Yungel, J. K., Berry, R. E. (1988), *Appl. Opt.* **27**, 3969–3977.

Ishimaru, A. (1978), *Wave Propagation and Scattering in Random Media*, New York: Academic.

Ivanoff, A. (1974), in: N. G. Jerlov, E. S. Nielsen (Eds.), *Optical Aspects of Oceanography*, London: Academic.

Jaffe, J. (1992), in: G. D. Gilbert (Ed.), *Ocean Optics XI*, SPIE Proceedings No. 1750, Bellingham, WA: SPIE.

Jerlov, N. G. (1976), *Marine Optics*, New York: Elsevier.

Karabashev, G. S. (1987), *Fluorescencia v Okeane*, Leningrad: Gidrometroizdat (in Russian).

Kattawar, G. W., Adams, C. N. (1989), *Limnol. Oceanogr.* **34**, 1453–1472.

Kirk, J. T. O. (1983), *Light and Photosynthesis in Aquatic Ecosystems*, New York: Cambridge Univ. Press.

Kirk, J. T. O. (1984), *Limnol. Oceanogr.* **29**, 350–356.

Kirk, J. T. O. (1994), *Appl. Opt.* **33**, 3276–3278.

Kitchen, J. C., Zaneveld, J. R. V. (1990), *J. Geophys. Res. C* **95**, 20237–20246.

Klyshko, D. N., Fadeev, V. V. (1978), *Dokl. Akad. Nauk SSSR* **238**, 320–323.

Kopelevich, O. V. (1983), in: A. Monin (Ed.), *Optika Okeana*, Vol. 1, Moscow: Nauka, p. 150 (in Russian).

Kopelevich, O. V. (1991), in: A. P. Ivanov (Ed.), *Rasseyanie i Pogloshchenie Sveta v Estestvennyk i Iskustvennyk Rasseivayushchik Sredak*, Minsk: Institute of Physics, p. 289 (in Russian).

Kozlov, V. P., Levin, I. M., Zolotuchin, I. V. (1992), in: *Proceedings of the PORSEC-'92*, Shizuoka, Japan: Pacific Ocean Remote Sensing Conference Secretariat, p. 1073.

Le Grand, Y. (1939), *Ann. Inst. Oceanogr.* **19**, 393–436.

Leonard, D. A., Capito, B., Hoge, F. E. (1979), *Appl. Opt.* **18**, 1732–1745.

Levin, E. I., Levin, I. M. (1989), *Izv. Atmos. Ocean. Phys.* **25**, 719–725.

Levin, I. M. (1969), *Izv. Atmos. Ocean. Phys.* **5**, 32–39.

Levin, I. M. (1980), *Izv. Atmos. Ocean. Phys.* **16**, 678–682.

Levin, I. M., Zolotuchin, I. V. (1994), in: J. Jaffe (Ed.), *Ocean Optics XII*, SPIE Proceedings No. 2258, Bellingham, WA: SPIE, pp. 861–869.

Lilycrop, W. J., Banic, J. R. (1993), *Marine Geod.* **15**, 177–185.

Luchinin, A. G. (1979), *Izv. Atmos. Ocean. Phys.* **15**, 531–534.

Luchinin, A. G., Sergiyevskaya, I. A. (1982), *Izv. Atmos. Ocean. Phys.* **18**, 656–661.

Lutomirski, R. F. (1978), in: M. B. White (Ed.), *Ocean Optics V*, SPIE Proceedings No. 160, Bellingham, WA: SPIE, pp. 110–121.

Maffione, R. A., Honey, R. C., Brown, R. A. (1991), in: R. W. Spinrad (Ed.), *Underwater Imaging, Photography, and Visibility*, SPIE Proceedings No. 1537, Bellingham, WA: SPIE, pp. 115–126.

Maffione, R. A., Voss, K. J. Honey, R. C. (1993), *Appl. Opt.* **32**, 3273–3279.

Man'kovsky, V. I. (1984), *Okeanologia* **24**, 63–69.

McLean, J. W., Voss, K. J. (1991), *Appl. Opt.* **30**, 2027–2030.

Marshall, B. R., Smith, R. C. (1990), *Appl. Opt.* **29**, 71–84.

Mertens, L. A., Replogle, F. S. (1977), *J. Opt. Soc. Am.* **67**, 1105–1117.

Mobley, C. D. (1989), *Limnol. Oceanogr.* **34**, 1473–1483.

Mobley, C. D. (1994), *Light and Water: Radiative Transfer in Natural Waters*, San Diego: Academic.

Mobley, C. D., Stramski, D. (1994), in: J. Jaffe (Ed.), *Ocean Optics XII*, SPIE Proceedings No. 2258, Bellingham, WA: SPIE, pp. 184–193.

Mobley, C. D., Gentili, B., Gordon, H. R., Jin, Z., Kattawar, G. W., Morel, A., Reinersman, P., Stamnes, K., Stavn, R. H. (1993), *Appl. Opt.* **32**, 7484–7504.

Monin, A. S. (Ed.) (1983), *Optika Okeana*, Vols. 1 and 2, Moscow: Nauka (in Russian).

Moore, K. D., O'Mongain, E., Plunkett, S., Doerffer, R., Bree, M. (1993), in: H. C. Eilertsen (Ed.), *Underwater Light Measurements*, SPIE Proceedings No. 2048, Bellingham, WA: SPIE, pp. 153–167.

Morel, A. (1991), *Prog. Oceanogr.* **26**, 263–306.

Morel, A., Berthon, J.-F. (1989), *Limnol. Oceanogr.* **34**, 1545–1562.

Morel, A., Gentili, B. (1991), *Appl. Opt.* **30**, 4427–4438.

Morel, A., Gentili, B. (1993), *Appl. Opt.* **32**, 6864–6879.

Morel, A., Prieur, L. (1977), *Limnol. Oceanog.* **22**, 709–722.

Osadchy, V. Y., Shifrin, K. S., Gurevich, I. Y. (1994), in: J. Jaffe (Ed.), *Ocean Optics XII*, SPIE Proceedings No. 2258, Bellingham, WA: SPIE, pp. 747–758.

Patterson, R. B. (1975), *Opt. Eng.* **14**, 357–365.

Pelevin, V. N. (1983), in: A. Monin (Ed.), *Optika Okeana*, Vol. 1, Moscow: Nauka, p. 249 (in Russian).

Plass, G. N., Kattawar, G. W. (1972), *J. Phys. Oceanogr.* **2**, 139–145.

Platt, T. C., Caverhill, S., Sathyendranath, S. (1991). *J. Geophys. Res. D*, **93**, 10831–10855.

Pope, R. M., Fry, E. S., Montgomery, R. L., Sogandares, F. (1990), in: R. Spinrad (Ed.), *Ocean Optics X*, SPIE Proceedings No. 1302, Bellingham, WA: SPIE, pp. 165–175.

Preisendorfer, R. W. (1976), *Hydrologic Optics*, Honolulu: NOAA.

Preisendorfer, R. W., Mobley, C. D. (1988), *J. Geophys. Res. D* **93**, 10831–10855.

Prieur, L., Sathyendranath, S. (1981), *Limnol. Oceanogr.* **26**, 671–689.

Remizovich, V. S., Rogozkin, D. B., Ryazanov, M. I. (1983), *Izv. Atmos. Ocean. Phys.* **19**, 796–801.

Romanova, L. M. (1968), *Izv. Atmos. Ocean. Phys.* **4**, 175–179.

Rose, A. (1973), *Vision Human and Electronic*, New York: Plenum.

Sathyendranath, S., Platt, T. (1989), *Appl. Opt.* **28**, 490–495.

Sathyendranath, S., Prieur, L., Morel, A. (1989), *Int. J. Remote Sens.* **10**, 1373–1394.

Shifrin, K. S. (1988), *Physical Optics of Oceanic Water*, New York: American Institute of Physics.

Simonot, J.-Y., Le Treut, H. (1986), *J. Geophys. Res.* **91**, 6642–6646.

Smith, R. C., Baker, K. S. (1978), *Limnol. Oceanogr.* **23**, 260–267.

Smith, R. C., Baker, K. S. (1981), *Appl. Opt.* **20**, 177–184.

Spinrad, R. W. (Ed.) (1989), Special Issue on Hydrologic Optics, *Limnol. Oceanogr.* **34** (8).

Stavn, R. H. (1993), *Appl. Opt.* **32**, 6853–6863.

Stilwell, D., Pilon, R. O. (1974), *J. Geophys. Res.* **79**, 1277–1284.

Titov, V. I. (1982), *Izv. Atmos. Ocean. Phys.* **18**, 168–169.

Trees, C. C., Voss, K. J. (1990), in: R. Spinrad (Ed.), *Ocean Optics X*, SPIE Proceedings No. 1302, Bellingham, WA: SPIE, pp. 149–156.

Tyler, J. E. (1960), *Bull. Scripps Inst. Oceanogr.* **7**, 363–412.

Van de Hulst, H. C. (1957), *Light Scattering by Small Particles*, New York: Wiley.

Veber, V. L., Sergievskaya, I. A. (1992), *Izv. Atmos. Ocean. Phys.* **28**, 244–249.

Voss, K. J. (1989a), *Limnol. Oceanogr.* **34**, 1614–1622.

Voss, K. J. (1989b), *Opt. Eng.* **28**, 241–247.

Voss, K. J. (1990), in: R. Spinrad (Ed.), *Ocean Optics X*, SPIE Proceedings No. 1302, Bellingham, WA: SPIE, pp. 355–362.

Voss, K. J. (1991), *Appl. Opt.* **30**, 2647–2651.

Voss, K. J., Chapin, A. L. (1990), *Appl. Opt.* **29**, 3638–3642.

Voss, K. J., Fry, E. S. (1984), *Appl. Opt.* **23**, 4427–4439.

Wells, W. H. (1969), *J. Opt. Soc. Am.* **59**, 686–691.

Wells, W. H. (1973), in: *NATO AGARD Lecture Series No. 61*, Paris: NATO, p. 3.3-1.

Wozniak, B., Dera, J., Koblentz-Mishke, O. J. (1992), *Oceanologia* **33**, 5–38.

Zaneveld, J. R. V. (1989), *Limnol. Oceanogr.* **34**, 1442–2452.

Zaneveld, J. R. V., Bartz, R., Kitchen, J. C. (1990), in: R. Spinrad (Ed.), *Ocean Optics X*, SPIE Proceedings No. 1302, Bellingham, WA: SPIE, pp. 124–236.

Zaneveld, J. R. V., Kitchen, J. C., Moore, C. C. (1994), in: J. Jaffe (Ed.), *Ocean Optics XII*, SPIE Proceedings No. 2258, Bellingham WA: SPIE, pp. 44–55.

Zege, E. P., Ivanov, A. P., Katsev, I. L. (1991), *Image Transfer Through a Scattering Medium*, Heidelberg: Springer.

Further Reading

Allan, T. D. (1992), *Int. J. Remote Sens.* **13**, 1261–1276.

Dera, J. (1992), *Marine Physics*, Amsterdam: Elsevier.

Gordon, H. R., Morel, A. Y. (1983), *Remote Assessment of Ocean Color for Interpretation of Satellite Visible Imagery: A Review*, New York: Springer.

Ivanoff, A. (1975), *Introduction à L'océanographie*, Vol. 2, Paris: Librairie Vuibert.

Jerlov, N. G. (1976), *Marine Optics*, New York: Elsevier.

Jerlov, N. G., Nielsen, E. S. (Eds.) (1974), *Optical Aspects of Oceanography*, London: Academic.

Mobley, C. D. (1994), *Light and Water: Radiative Transfer in Natural Waters*, San Diego: Academic.

Preisendorfer, R. W. (1976), *Hydrologic Optics*, Honolulu: NOAA.

Shifrin, K. S. (1988), *Physical Optics of Oceanic Water*, New York: AIP.

Spinrad, R. W. (Ed.) (1989), Special Issue on Hydrologic Optics, *Limnol. Oceanogr.* **34** (8).

Tyler, J. E. (Ed.) (1977), *Light in the Sea*, Benchmark Papers in Optics, Vol. 3, Stroudsburg, PA: Dowden, Hutchinson & Ross, Inc.

CONTENTS OF PREVIOUS VOLUMES

Volume 1

Volume 2

Volume 3

Volume 4

Volume 5

Volume 6

Volume 7

Volume 8

Volume 9

Volume 10

Volume 11

LIST OF RECOMMENDED UNITS AND SYMBOLS

RECOMMENDED UNITS AND CONVERSION FACTORS

The SI system provides six basic units: meter m, kilogram kg, second s, ampere A, kelvin K, candela cd, and mole mol.

Some important derived units are also allowed and bear special names, e.g.:

1 N (newton) = 1 kg m s^{-2}
1 J (joule) = 1 N m = 1 kg m^2 s^{-2}
1 W (watt) = 1 J s^{-1} = 1 kg m^2 s^{-3}
1 Pa (pascal) = 1 N m^{-2} = 1 kg m^{-1} s^{-2}

For mass, the gram g, or the metric ton t which equals 1000 kg, may be used instead of kilogram kg.

The so-called "long ton" (UK) and "short ton" (US) have been abandoned and will not be used in the *Encyclopedia of Applied Physics*.

For pressure, the bar (name and symbol alike) may be used, and for temperature, the degree celsius °C.

From all of these, decimal multiples or fractions can be derived:

Power of ten	Prefix	Symbol	Power of ten	Prefix	Symbol
10	deca	da	10^{-1}	deci	d
10^2	hecto	h	10^{-2}	centi	c
10^3	kilo	k	10^{-3}	milli	m
10^6	mega	M	10^{-6}	micro	μ
10^9	giga	G	10^{-9}	nano	n
10^{12}	tera	T	10^{-12}	pico	p
10^{15}	peta	P	10^{-15}	femto	f
10^{18}	exa	E	10^{-18}	atto	a

For mass, multiples or fractions are derived from g, not from kg (since the latter already contains a prefix), e.g., mg.

Units with the prefixes are considered as one entity and can, therefore, be raised to any power, e.g., cm^3.

The liter is now considered as synonymous with dm^3 (which does not hold in the older literature!). Please use the capital letter L as unit symbol (following a recent IUPAC recommendation). The use of the Ångström unit is discouraged; it should be replaced by fractional meters (1 Å = 100 pm = 0.1 nm).

SELECTED QUANTITIES, UNITS, AND SYMBOLS

Name of quantity[a]	Symbol[b]	SI unit[c]	Name of unit	Other units
Space and time				
length*	l	m	meter[d]	
breadth, width	b	m		
height	h	m		
radius	r	m		
thickness	d	m		
area	A, S	m^2		a (are) h (hectare)
volume	V	m^3		L, l(liter)[e]
plane angle	$\alpha, \beta, \gamma, \vartheta, \varphi$	1, rad	radian	° (degree) ′ (minute) ″ (second)
solid angle	ω, Ω	1, sr	steradian	
wavelength	λ	m		
wave number	σ, ν	m^{-1}		
time*	t	s	second	min (minute) h (hour) d (day)
frequency	ν, f	s^{-1}		Hz (hertz)[f]
relaxation time	τ	s		
velocity	u, v	$m\ s^{-1}$		km/h
acceleration	a	$m\ s^{-2}$		
Mechanics				
mass*	m	kg	kilogram	g (gram) t (tonne)[g]
(mass) density	ρ	$kg\ m^{-3}$		g/cm^3
momentum	$\mathbf{p}$	$kg\ m\ s^{-1}$		
angular momentum	$\mathbf{L}$	$kg\ m^2\ s^{-1}$		
force	$\mathbf{F}$	N	newton[f]	
moment of force	$\mathbf{M}$	N m		
weight	G, W	N		
pressure	p	Pa	pascal	bar (bar)[h]
energy	E, W	J	joule	W h (watt hour)[i] eV (electron volt)[j]
work	W, A	J		
power	P	W	watt	J/s, V A[k]
Molecular Physics and Thermodynamics				
thermodynamic temperature*	T	K	kelvin	°C (degrees Celsius)
Celsius temperature	ϑ, t			°C
number of entities	N			
Avogadro constant[l]	N_A, L	mol^{-1}	(particles) per mole	
Boltzmann constant[m]	k, k_B	$J\ K^{-1}$		
Planck constant[n]	h	J s		
(molar) gas constant[o]	R	$J\ mol^{-1}\ K^{-1}$		
(quantity of) heat	Q	J		
entropy[p]	S	$J\ K^{-1}$		
internal energy[p]	U	J		
Helmholtz function,[p] (Helmholtz) free energy, Helmholtz energy	F, A	J		
enthalpy[p]	H	J		
Gibbs function,[p] (Gibbs) free energy, Gibbs energy	G	J		
heat capacity[p]	C_p, C_V	$J\ K^{-1}$		

Name of quantity[a]	Symbol[b]	SI unit[c]	Name of unit	Other units
Chemical Physics				
amount of substance*	n	mol	mole	
relative atomic mass	A_r	1		
relative molecular mass	M_r	1		
atomic mass constant	m_u	kg		u (atomic mass unit)[q]
mass of a portion (of substance B)	m_B,m(B)	kg		g (gram)
molar mass (of substance B)	M_B,M(B)	kg mol^{-1}		
concentration (of substance B)	c_B,c(B)	mol m^{-3}		mol/L
mole fraction[r] (of substance B)	κ_B,κ(B)	1		
mass fraction (of substance B)	ω_B,ω(B)	1		%,‰,ppm,ppb
volume fraction (of substance B)	φ_B,φ(B) ϕ_B,ϕ(B)	1		%,‰,ppm,ppb
mass concentration	ρ	kg m^{-3}		g/L
molality		mol kg^{-1}		mmol/kg
volume concentration[s]	σ	1		
molar volume	V_m	m^3 mol^{-1}		L/mol
molar heat capacity	C_m	J mol^{-1} K^{-1}		
molar conductivity	Λ_m	S m^2 mol^{-1}	(S: siemens)	
Faraday constant[t]	F	C mol^{-1}		
Electricity and Magnetism				
quantity of electricity[u]	Q	C	coulomb	
charge density	ρ	C m^{-3}		
electric potential	ϕ,V	V	volt	
electric potential difference, voltage	U,$\Delta\phi$,ΔV	V		
electric dipole moment	$\mathbf{p}$, $\mathbf{p}_c$	C m		
electric current*	I	A	ampere	
electric current density	j	A m^{-2}		
electric field strength	$\mathbf{E}$	V m^{-1}		
electric displacement	$\mathbf{D}$	C m^{-2}		
capacitance	C	F	farad	
permittivity	ε	F m^{-1}		
relative permittivity	ε	1		
dielectric polarization	$\mathbf{P}$	C m^{-2}		
electric susceptibility	χ_c	1		
polarization (of a particle)	α	m^2 C V^{-1}		
magnetic flux	Φ	Wb	weber	
magnetic flux density	$\mathbf{B}$	T	tesla	
magnetic field strength	$\mathbf{H}$	A m^{-1}		
permeability	μ	H m^{-1}, N A^{-2}	(H: henry)	
relative permeability	μ_r	1		
magnetization	M	A m^{-1}		
magnetic susceptibility	χ	1		
molar magnetic susceptibility	χ_m	m^3 mol^{-1}		
(electrical) resistance	R	Ω	ohm	
(electrical) conductance	G	S	siemens	
(electrical) resistivity	ρ	Ω m		
(electrical) conductivity	κ,o	S m^{-1}		
self-inductance	L	H	henry	
Radiation				
radiant energy	Q, W,Q_c	J		
luminous intensity*	I	cd	candela	
radiant intensity	I_c	W sr^{-1}, W		
emissivity, emittance	ε	1		

Name of quantity[a]	Symbol[b]	SI unit[c]	Name of unit	Other units
absorptance	α	1		
reflectance	ρ, R	1		
transmittance	τ	1		
absorption coefficient:				
linear (decadic)	a	m^{-1}		
molar (decadic)	ε	$m^2\ mol^{-1}$		
refractive index	n	1		
molar refraction	R_m	$m^3\ mol^{-1}$		
angle of optical rotation	α	1, rad		
Transport Properties				
flux of quantity X	J_X, J	(varies)		
mass flow rate	$q_m, \dot{m}$	$kg\ s^{-1}$		
volume flow rate	$q_V, \dot{V}$	$m^3\ s^{-1}$		
heat flow rate	Φ	W		
thermal conductivity	κ, k, λ	$W\ m^{-1}\ K^{-1}$		
coefficient of heat transfer	h	$W\ m^{-2}\ K^{-1}$		
thermal diffusivity	a	$m^2\ s^{-1}$		
diffusion coefficient	D	$m^2\ s^{-1}$		
thermal diffusion coefficient	D_T	$m^2\ s^{-1}$		
viscosity	η, μ	Pa s		
kinematic viscosity	ν	$m^2\ s^{-1}$		

[a] SI base quantities are marked by asterisks (*)
[b] Recommended by IUPAC
[c] SI base units as well as derived and supplementary units are listed; all are to be used with prefixes as needed
[d] do not use "metre"
[e] do not use "litre"; $1\ L = 10^{-3}\ m^3$
[f] $1\ Hz = 1\ s^{-1}$; $1\ N = 1\ kg\ m\ s^{-2}$
[g] Formerly metric ton; $1\ t = 10^3\ kg$
[h] $1\ bar = 10^5\ Pa$
[i] $1\ W\ h = 3.6 \times 10^3\ J$
[j] $1\ eV = 1.602\ 189 \times 10^{-19}\ J$
[k] $1\ W = 1\ J/s = 1\ VA$
[l] $N_A = 6.022\ 136\ 7 \times 10^{23}\ mol^{-1}$
[m] $k = 1.380\ 658 \times 10^{-23}\ J\ K^{-1}$
[n] $h = 6.626\ 075\ 5 \times 10^{-34}\ J\ s$
[o] $R = 8.314\ 510\ J\ mol^{-1}\ K^{-1}$
[p] Molar quantities can be distinguished from the quantity of a system by adding the subscript m; e.g., molar internal energy U_m, in $J\ mol^{-1}$
[q] $1\ u = 1.660\ 565\ 5 \times 10^{-27}\ kg$
[r] A more accurate, but rather uncommon, name is "amount-of-substance fraction"
[s] σ refers to the total volume of a mixture, whereas the volume fraction φ relates the volume of a substance to the volume of several components before mixing
[t] $F = 9.648\ 530\ 9 \times 10^4\ C\ mol^{-1}$
[u] Also called electric charge